TRIGONOMETRY

Definition of the Six Trigonometric Functions

Right triangle definitions, where $0 < \theta < \pi/2$.

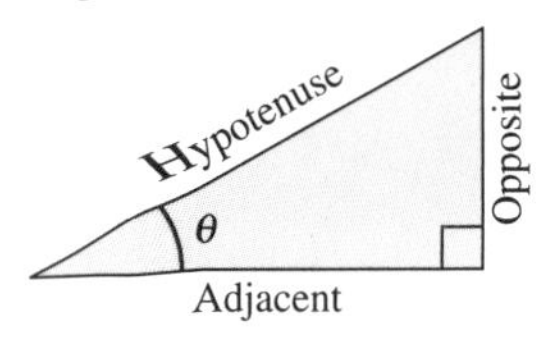

$$\sin\theta = \frac{\text{opp}}{\text{hyp}} \qquad \csc\theta = \frac{\text{hyp}}{\text{opp}}$$

$$\cos\theta = \frac{\text{adj}}{\text{hyp}} \qquad \sec\theta = \frac{\text{hyp}}{\text{adj}}$$

$$\tan\theta = \frac{\text{opp}}{\text{adj}} \qquad \cot\theta = \frac{\text{adj}}{\text{opp}}$$

Circular function definitions, where θ *is any angle.*

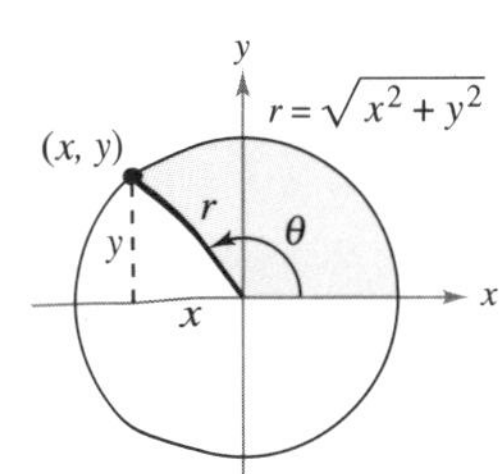

$$\sin\theta = \frac{y}{r} \qquad \csc\theta = \frac{r}{y}$$

$$\cos\theta = \frac{x}{r} \qquad \sec\theta = \frac{r}{x}$$

$$\tan\theta = \frac{y}{x} \qquad \cot\theta = \frac{x}{y}$$

$(0, 1)$ $\frac{\pi}{2}$ 90°
$\left(\frac{1}{2}, \frac{\sqrt{3}}{2}\right)$ $\frac{\pi}{3}$ 60°
$\left(\frac{\sqrt{2}}{2}, \frac{\sqrt{2}}{2}\right)$ $\frac{\pi}{4}$ 45°
$\left(\frac{\sqrt{3}}{2}, \frac{1}{2}\right)$ $\frac{\pi}{6}$ 30°
$(1, 0)$ 0° 0, 360° 2π
$\left(-\frac{1}{2}, \frac{\sqrt{3}}{2}\right)$ $\frac{2\pi}{3}$ 120°
$\left(-\frac{\sqrt{2}}{2}, \frac{\sqrt{2}}{2}\right)$ $\frac{3\pi}{4}$ 135°
$\left(-\frac{\sqrt{3}}{2}, \frac{1}{2}\right)$ $\frac{5\pi}{6}$ 150°
$(-1, 0)$ π 180°
$\left(-\frac{\sqrt{3}}{2}, -\frac{1}{2}\right)$ $\frac{7\pi}{6}$ 210°
$\left(-\frac{\sqrt{2}}{2}, -\frac{\sqrt{2}}{2}\right)$ $\frac{5\pi}{4}$ 225°
$\left(-\frac{1}{2}, -\frac{\sqrt{3}}{2}\right)$ $\frac{4\pi}{3}$ 240°
$(0, -1)$ $\frac{3\pi}{2}$ 270°
$\left(\frac{1}{2}, -\frac{\sqrt{3}}{2}\right)$ $\frac{5\pi}{3}$ 300°
$\left(\frac{\sqrt{2}}{2}, -\frac{\sqrt{2}}{2}\right)$ $\frac{7\pi}{4}$ 315°
$\left(\frac{\sqrt{3}}{2}, -\frac{1}{2}\right)$ $\frac{11\pi}{6}$ 330°

Reciprocal Identities

$$\sin x = \frac{1}{\csc x} \qquad \sec x = \frac{1}{\cos x} \qquad \tan x = \frac{1}{\cot x}$$

$$\csc x = \frac{1}{\sin x} \qquad \cos x = \frac{1}{\sec x} \qquad \cot x = \frac{1}{\tan x}$$

Tangent and Cotangent Identities

$$\tan x = \frac{\sin x}{\cos x} \qquad \cot x = \frac{\cos x}{\sin x}$$

Pythagorean Identities

$$\sin^2 x + \cos^2 x = 1$$

$$1 + \tan^2 x = \sec^2 x \qquad 1 + \cot^2 x = \csc^2 x$$

Cofunction Identities

$$\sin\left(\frac{\pi}{2} - x\right) = \cos x \qquad \cos\left(\frac{\pi}{2} - x\right) = \sin x$$

$$\csc\left(\frac{\pi}{2} - x\right) = \sec x \qquad \tan\left(\frac{\pi}{2} - x\right) = \cot x$$

$$\sec\left(\frac{\pi}{2} - x\right) = \csc x \qquad \cot\left(\frac{\pi}{2} - x\right) = \tan x$$

Reduction Formulas

$$\sin(-x) = -\sin x \qquad \cos(-x) = \cos x$$

$$\csc(-x) = -\csc x \qquad \tan(-x) = -\tan x$$

$$\sec(-x) = \sec x \qquad \cot(-x) = -\cot x$$

Sum and Difference Formulas

$$\sin(u \pm v) = \sin u \cos v \pm \cos u \sin v$$

$$\cos(u \pm v) = \cos u \cos v \mp \sin u \sin v$$

$$\tan(u \pm v) = \frac{\tan u \pm \tan v}{1 \mp \tan u \tan v}$$

Double-Angle Formulas

$$\sin 2u = 2 \sin u \cos u$$

$$\cos 2u = \cos^2 u - \sin^2 u = 2\cos^2 u - 1 = 1 - 2\sin^2 u$$

$$\tan 2u = \frac{2\tan u}{1 - \tan^2 u}$$

Power-Reducing Formulas

$$\sin^2 u = \frac{1 - \cos 2u}{2}$$

$$\cos^2 u = \frac{1 + \cos 2u}{2}$$

$$\tan^2 u = \frac{1 - \cos 2u}{1 + \cos 2u}$$

Sum-to-Product Formulas

$$\sin u + \sin v = 2\sin\left(\frac{u+v}{2}\right)\cos\left(\frac{u-v}{2}\right)$$

$$\sin u - \sin v = 2\cos\left(\frac{u+v}{2}\right)\sin\left(\frac{u-v}{2}\right)$$

$$\cos u + \cos v = 2\cos\left(\frac{u+v}{2}\right)\cos\left(\frac{u-v}{2}\right)$$

$$\cos u - \cos v = -2\sin\left(\frac{u+v}{2}\right)\sin\left(\frac{u-v}{2}\right)$$

Product-to-Sum Formulas

$$\sin u \sin v = \frac{1}{2}[\cos(u - v) - \cos(u + v)]$$

$$\cos u \cos v = \frac{1}{2}[\cos(u - v) + \cos(u + v)]$$

$$\sin u \cos v = \frac{1}{2}[\sin(u + v) + \sin(u - v)]$$

$$\cos u \sin v = \frac{1}{2}[\sin(u + v) - \sin(u - v)]$$

Calculus I

Early Transcendental Functions

Fourth Edition

Calculus I

Early Transcendental Functions

Fourth Edition

Ron Larson
The Pennsylvania State University
The Behrend College

Robert Hostetler
The Pennsylvania State University
The Behrend College

Bruce H. Edwards
University of Florida

Houghton Mifflin Company Boston New York

Publisher: Richard Stratton
Sponsoring Editor: Cathy Cantin
Development Manager: Maureen Ross
Associate Editor: Yen Tieu
Editorial Associate: Elizabeth Kassab
Supervising Editor: Karen Carter
Senior Project Editor: Patty Bergin
Editorial Assistant: Julia Keller
Art and Design Manager: Gary Crespo
Executive Marketing Manager: Brenda Bravener-Greville
Senior Marketing Manager: Danielle Curran
Director of Manufacturing: Priscilla Manchester
Cover Design Manager: Tony Saizon

We have included examples and exercises that use real-life data as well as technology output from a variety of software. This would not have been possible without the help of many people and organizations. Our wholehearted thanks goes to all for their time and effort.

Cover photograph: "Music of the Spheres" by English sculptor John Robinson is a three-foot-tall sculpture in bronze that has one continuous edge. You can trace its edge three times around before returning to the starting point. To learn more about this and other works by John Robinson, see the Centre for the Popularisation of Mathematics, University of Wales, at *http://www.popmath.org.uk/sculpture/gallery2.html.*

Printed in the U.S.A.

Library of Congress Control Number: 2005933920

ISBN 13: 978-0-618-60626-9
ISBN 10: 0-618-60626-2

2 3 4 5 6 7 8 9 DOW 10 09 08 07

Contents

Additional Appendices The following appendices are available at the textbook website at *college.hmco.com/pic/larsoncalculusletf4e*, on the HM mathSpace® Student CD-ROM, and the HM ClassPrep™ with HM Testing CD-ROM.

A Word from the Authors

Welcome to *Calculus I: Early Transcendental Functions*, Fourth Edition. With each edition, we have listened to you, our users, and incorporated many of your suggestions for improvement.

1st 2nd 3rd

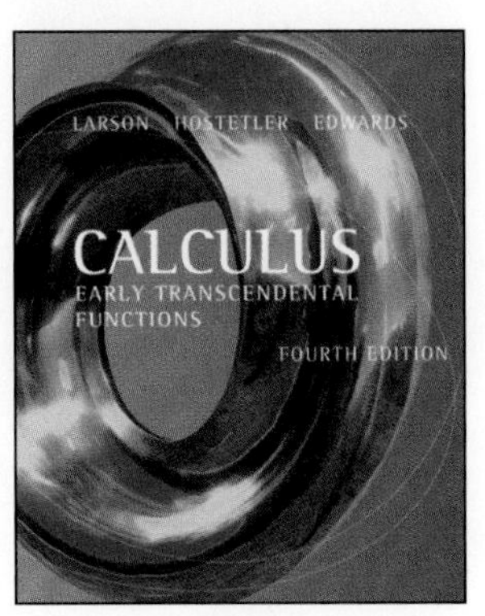

4th

A Text Formed by Its Users

Through your support and suggestions, the text has evolved over four editions to include these extensive enhancements:

- Comprehensive exercise sets containing a wide variety of problems such as skill-building exercises, applications, explorations, writing exercises, critical thinking exercises, and theoretical problems
- Abundant real-life applications that accurately represent the diverse uses of calculus
- Many open-ended activities and investigations
- Clear, uncluttered text presentation with full annotations and labels and a carefully planned page layout
- Comprehensive, four-color art program
- Comprehensive and mathematically rigorous text
- Technology used throughout as both a problem-solving tool and an investigative tool
- A comprehensive program of additional resources available in print, on CD-ROM, and online
- With 5 different volumes of the text available, you can choose the sequence, amount of content, and teaching approach that is best for you and your students (see pages xiv–xv)
- References to the history of calculus and to the mathematicians who developed it, including over 50 biographical sketches available on the HM mathSpace® Student CD-ROM
- References to over 50 articles from mathematical journals are available at *www.MathArticles.com*

What's New and Different in the Fourth Edition

In the Fourth Edition, we continue to offer instructors and students a text that is pedagogically sound, mathematically precise, and still comprehensible. There are many changes in the mathematics, prose, art, and design; the more significant changes are noted here.

- ***New Chapter Openers*** Each Chapter Opener has two parts: a description of the concepts that are covered in the chapter and a thought-provoking question about a real-life application from the chapter.
- ***New Introduction to Differential Equations*** The topic of differential equations is now introduced in Chapter 6 in the first semester of calculus, to better prepare students for their courses in disciplines such as engineering, physics, and chemistry. The chapter contains six sections: 6.1 *Slope Fields and Euler's Method*, 6.2 *Differential Equations: Growth and Decay*, 6.3 *Differential Equations: Separation of Variables,* 6.4 *The Logistic Equation,* 6.5 *First-Order Linear Differential Equations*, and 6.6 *Predator-Prey Differential Equations.*
- **Revised Exercise Sets** The exercise sets have been carefully and extensively examined to ensure they are rigorous and cover all topics suggested by our users. Many new skill-building and challenging exercises have been added.
- **Updated Data** All data in the examples and exercise sets have been updated.
- Eduspace Eduspace® combines numerous dynamic resources with online homework and testing materials to create a comprehensive online learning system. Students benefit from having immediate access to algorithmic tutorial practice, videos, and resources such as a color graphing calculator. Instructors benefit from time-saving grading resources, as well as dynamic instructional tools such as animations, explorations, and Computer Algebra System Labs.
- ***Study and Solutions Guides*** The worked-out solutions to the odd-numbered text exercises are now provided on a CD-ROM, in Eduspace®, and at *www.CalcChat.com.*

Although we carefully and thoroughly revised the text by enhancing the usefulness of some features and topics and by adding others, we did not change many of the things that our colleagues and the over two million students who have used this book have told us work for them. The *Calculus: Early Transcendental Functions*, Fourth Edition, program offers comprehensive coverage of the material required by students in calculus courses, including carefully stated theories and proofs.

We hope you will enjoy the Fourth Edition. We welcome any comments, as well as suggestions for continued improvement.

Ron Larson Robert Hostetler Bruce H. Edwards

Acknowledgments

We would like to thank the many people who have helped us at various stages of this project over the years. Their encouragement, criticisms, and suggestions have been invaluable to us.

For the Fourth Edition

Andre Adler
Illinois Institute of Technology

Evelyn Bailey
Oxford College of Emory University

Katherine Barringer
Central Virginia Community College

Robert Bass
Gardner-Webb University

Joy Becker
University of Wisconsin Stout

Michael Bezusko
Pima Community College

Bob Bradshaw
Ohlone College

Robert Brown
The Community College of Baltimore County (Essex Campus)

Joanne Brunner
DePaul University

Minh Bui
Fullerton College

Fang Chen
Oxford College of Emory University

Alex Clark
University of North Texas

Jeff Dodd
Jacksonville State University

Daniel Drucker
Wayne State University

Pablo Echeverria
Camden County College

Angela Hare
Messiah College

Karl Havlak
Angelo State University

James Herman
Cecil Community College

Xuezhang Hou
Towson University

Gene Majors
Fullerton College

Suzanne Molnar
College of St. Catherine

Karen Murany
Oakland Community College

Keith Nabb
Moraine Valley Community College

Stephen Nicoloff
Paradise Valley Community College

James Pommersheim
Reed College

James Ralston
Hawkeye Community College

Chip Rupnow
Martin Luther College

Mark Snavely
Carthage College

Ben Zandy
Fullerton College

For the Fourth Edition Technology Program

Jim Ball
Indiana State University

Marcelle Bessman
Jacksonville University

Tim Chappell
Penn Valley Community College

Oiyin Pauline Chow
Harrisburg Area Community College

Julie M. Clark
Hollins University

Jim Dotzler
Nassau Community College

Murray Eisenberg
University of Massachusetts at Amherst

Arek Goetz
San Francisco State University

John Gosselin
University of Georgia

Shahryar Heydari
Piedmont College

Douglas B. Meade
University of South Carolina

Teri Murphy
University of Oklahoma

Howard Speier
Chandler-Gilbert Community College

Reviewers of Previous Editions

Raymond Badalian
Los Angeles City College

Norman A. Beirnes
University of Regina

Christopher Butler
Case Western Reserve University

Dane R. Camp
New Trier High School, IL

Jon Chollet
Towson State University

Barbara Cortzen
DePaul University

Patricia Dalton
Montgomery College

Luz M. DeAlba
Drake University

Dewey Furness
Ricks College

Javier Garza
Tarleton State University

Claire Gates
Vanier College

Lionel Geller
Dawson College

Carollyne Guidera
University College of Fraser Valley

Irvin Roy Hentzel
Iowa State University

Kathy Hoke
University of Richmond

Howard E. Holcomb
Monroe Community College

Gus Huige
University of New Brunswick

E. Sharon Jones
Towson State University

Robert Kowalczyk
University of Massachusetts–Dartmouth

Anne F. Landry
Dutchess Community College

Robert F. Lax
Louisiana State University

Beth Long
Pellissippi State Technical College

Gordon Melrose
Old Dominion University

Bryan Moran
Radford University

David C. Morency
University of Vermont

Guntram Mueller
University of Massachusetts–Lowell

Donna E. Nordstrom
Pasadena City College

Larry Norris
North Carolina State University

Mikhail Ostrovskii
Catholic University of America

Jim Paige
Wayne State College

Eleanor Palais
Belmont High School, MA

James V. Rauff
Millikin University

Lila Roberts
Georgia Southern University

David Salusbury
John Abbott College

John Santomas
Villanova University

Lynn Smith
Gloucester County College

Linda Sundbye
Metropolitan State College of Denver

Anthony Thomas
University of Wisconsin–Platteville

Robert J. Vojack
Ridgewood High School, NJ

Michael B. Ward
Bucknell University

Charles Wheeler
Montgomery College

During the past four years, several users of the Third Edition wrote to us with suggestions. We considered each and every one of them when preparing the manuscript for the Fourth Edition. A special note of thanks goes to the instructors and to the students who have used earlier editions of the text.

We would like to thank the staff at Larson Texts, Inc., who assisted with proofreading the manuscript, preparing and proofreading the art package, and checking and typesetting the supplements.

On a personal level, we are grateful to our wives, Deanna Gilbert Larson, Eloise Hostetler, and Consuelo Edwards, for their love, patience, and support. Also, a special note of thanks goes to R. Scott O'Neil.

If you have suggestions for improving this text, please feel free to write to us. Over the years we have received many useful comments from both instructors and students, and we value these very much.

Ron Larson   Robert Hostetler   Bruce H. Edwards

Integrated Learning System for Calculus

Over 25 Years of Success, Leadership, and Innovation

The bestselling authors Larson, Hostetler, and Edwards continue to offer instructors and students more flexible teaching and learning options for the calculus course.

Calculus Textbook Options

CALCULUS: Early Transcendental Functions

The early transcendental functions calculus course is available in a variety of textbook configurations to address the different ways instructors teach—and students take—their classes.

Designed for the three-semester course

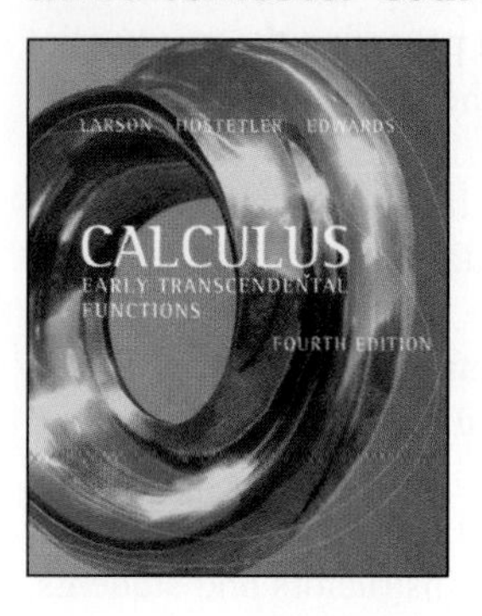

Designed for the two-semester course

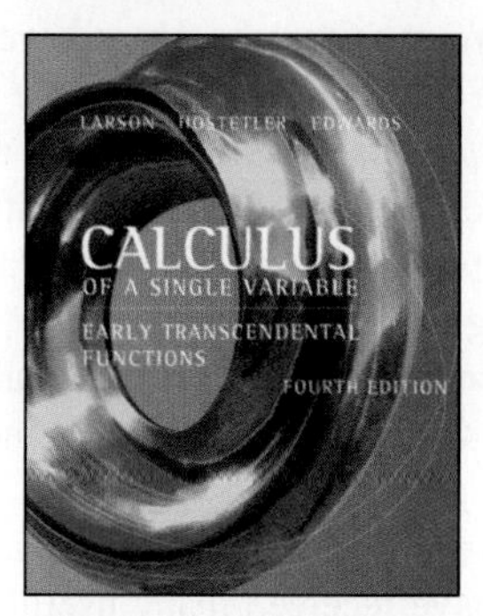

Designed for third semester of Calculus

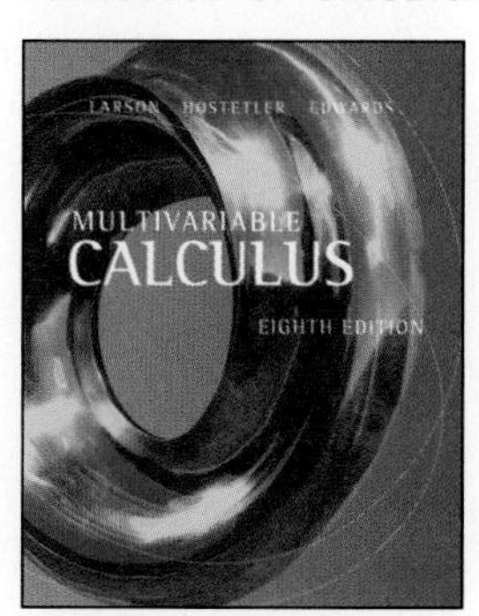

Designed for single-semester course

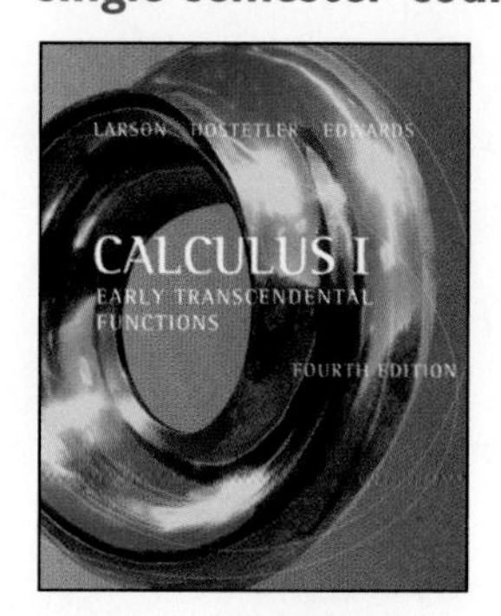

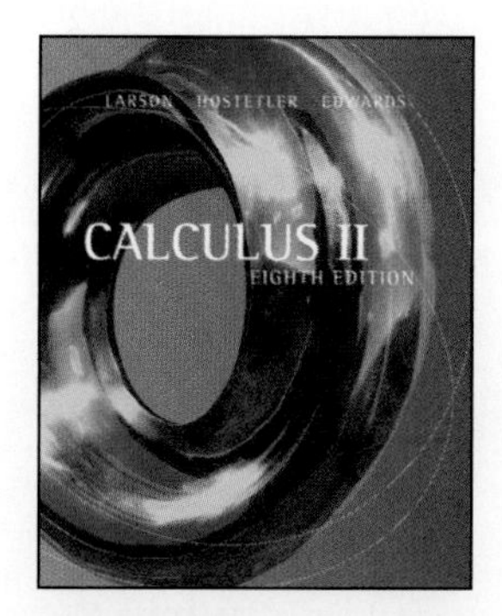

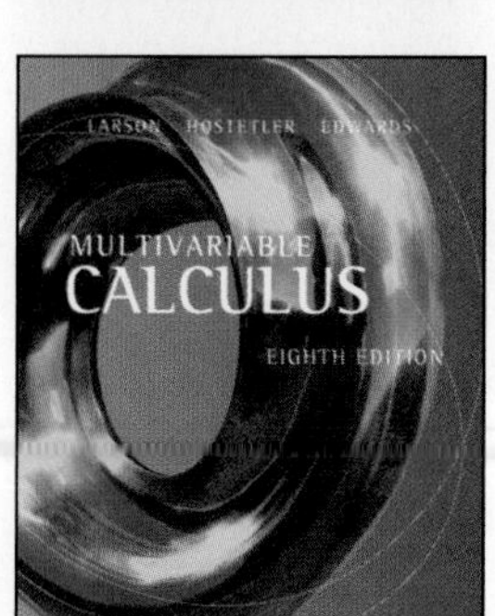

Also available for the *Calculus: Early Transcendental Functions*, Fourth Edition, program by Larson, Hostetler, and Edwards

- Eduspace® online learning system
- HM mathSpace® Student CD-ROMs
- Instructional DVDs and videos

For more information on these—and more—electronic course materials, please turn to pages xvii-xix.

CALCULUS

For instructors who prefer the traditional calculus course sequence, the following textbook sequences are available.

- Calculus I, II, and III
- Calculus I and II and Calculus III
- Calculus I, Calculus II, and Calculus III

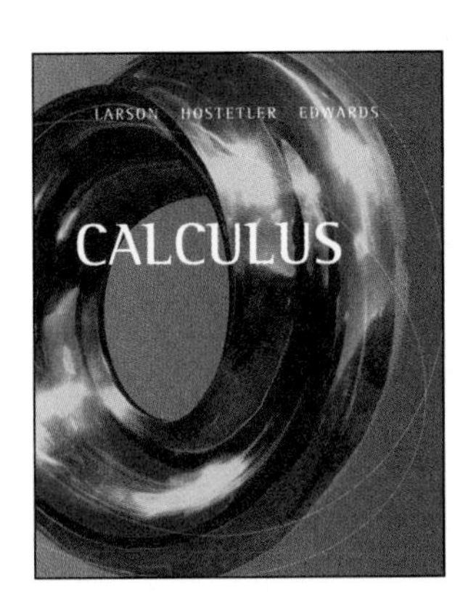

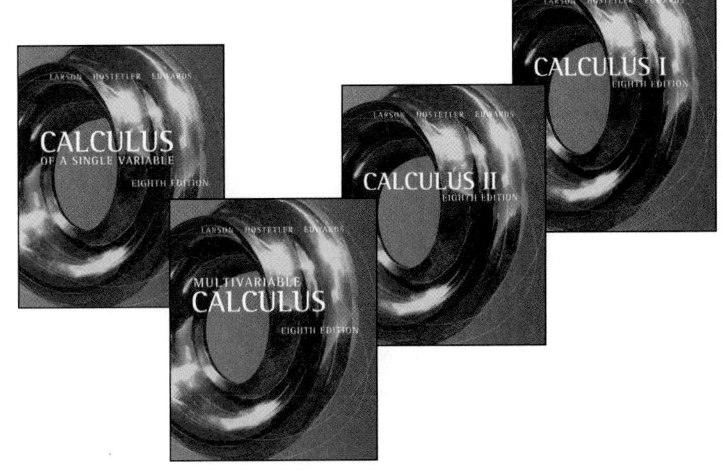

CALCULUS WITH PRECALCULUS

To give more students access to calculus by easing the transition from precalculus, the following textbook sequence is available.

- Precalculus and Calculus I, Calculus II, and Calculus III

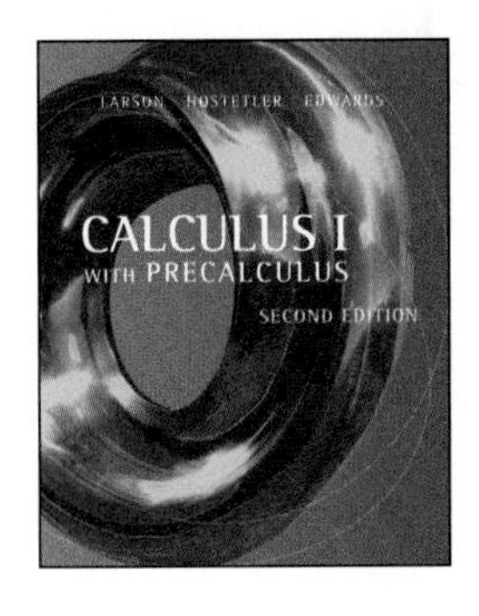

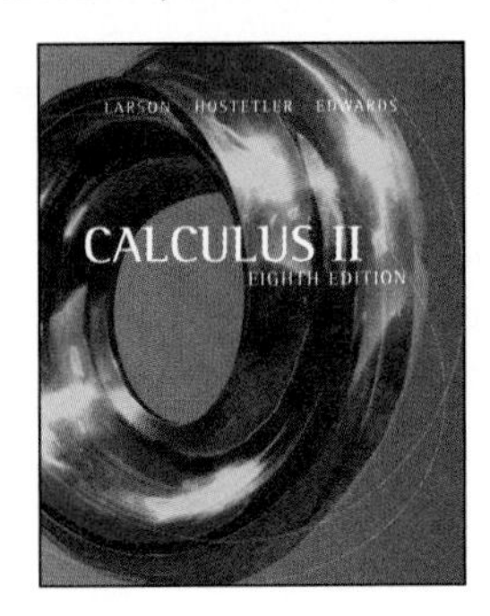

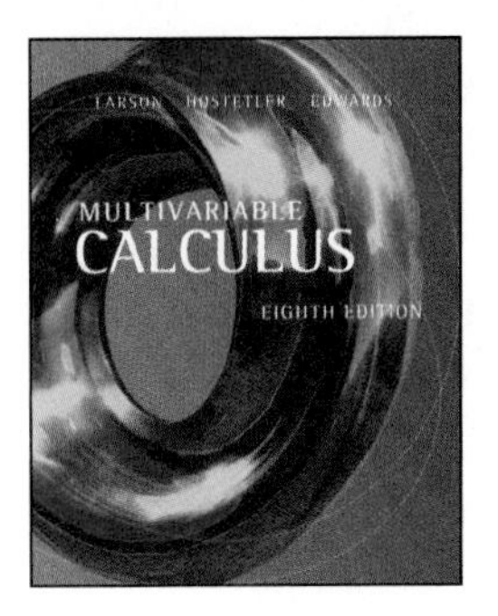

CALCULUS WITH LATE TRIGONOMETRY

For instructors who introduce the trigonometric functions in the second semester, the following textbook is available.

- Calculus I, II, and III

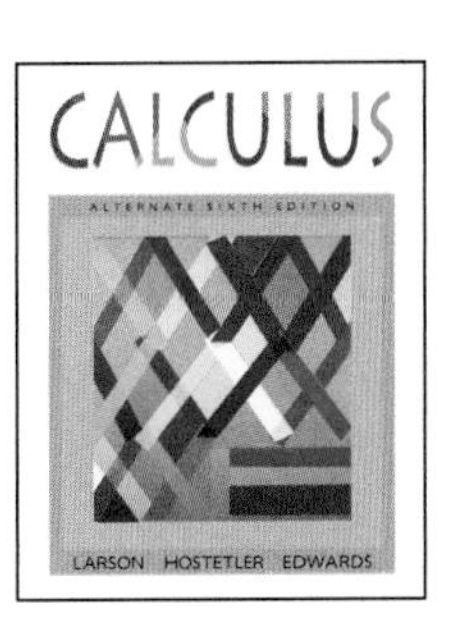

Integrated Learning System for Calculus

Comprehensive Calculus Resources

The Integrated Learning System for the *Calculus: Early Transcendental Functions*, Fourth Edition, program addresses the changing needs of today's instructors and students. Recognizing that the calculus course is presented in a variety of teaching and learning environments, we offer extensive resources that support the textbook program in print, CD-ROM, and online formats.

- Online homework practice
- Testing
- Tutoring
- Graded homework
- Classroom management
- Online course
- Interactive resources

ONE
INTEGRATED LEARNING SYSTEM

The teaching and learning resources you need in the format you prefer

The Integrated Learning System for the *Calculus: Early Transcendental Functions*, Fourth Edition, program offers dynamic teaching tools for instructors and interactive learning resources for students in the following flexible course delivery formats.

- Eduspace® online learning system
- HM mathSpace® Student CD-ROM
- Instructional DVDs and videos
- HM ClassPrep™ with HM Testing CD-ROM
- Companion Textbook Websites
- *Study and Solutions Guide* in two volumes available in print and electronically
- *Complete Solutions Guide* in three volumes (for instructors only) available only electronically

Enhanced! Eduspace® Online Calculus

Eduspace®, powered by Blackboard®, is ready to use and easy to integrate into the calculus course. It provides comprehensive homework exercises, tutorials, and testing keyed to the textbook by section.

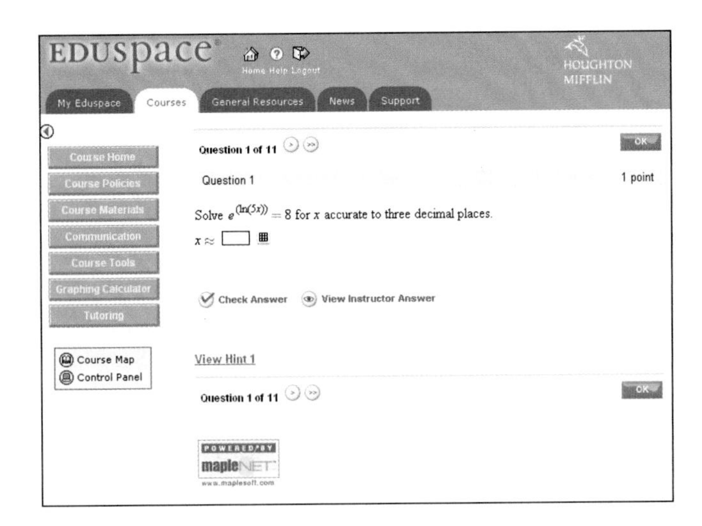

Features

- Algorithmically generated tutorial exercises for unlimited practice
- Comprehensive problem sets for graded homework
- Interactive (multimedia) textbook pages with video lectures, animations, and much more.
- SMARTHINKING® live, online tutoring for students
- Color graphing calculator
- Ample prerequisite skills review with customized student self-study plan
- Chapter tests
- Link to CalcChat
- Electronic version of all textbook exercises
- Links to detailed, stepped-out solutions to odd-numbered textbook exercises

Enhanced! HM mathSpace® Student CD-ROM

For the student, HM mathSpace® CD-ROM offers a wealth of learning resources keyed to the textbook by section.

Features

- Algorithmically generated tutorial questions for unlimited practice of prerequisite skills
- Point-of-use links to additional tools, animations, and simulations
- Link to CalcChat
- Color graphing calculator
- Chapter tests

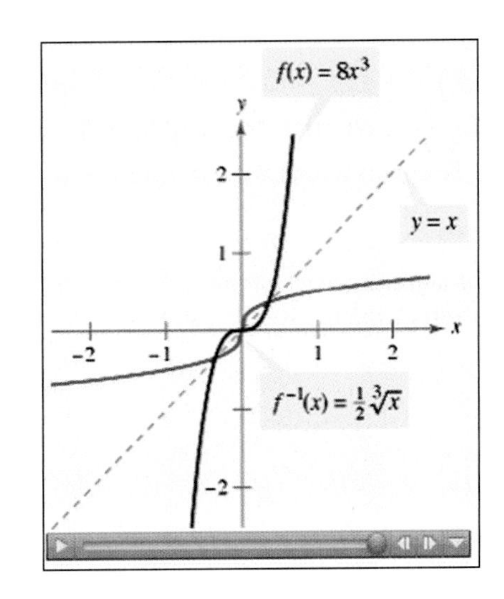

For additional information about the Larson, Hostetler, and Edwards Calculus program, go to *college.hmco.com/info/larsoncalculus*.

Integrated Learning System for Calculus

New! HM ClassPrep™ with HM Testing Instructor CD-ROM

This valuable CD-ROM contains an array of useful **instructor resources** keyed to the textbook.

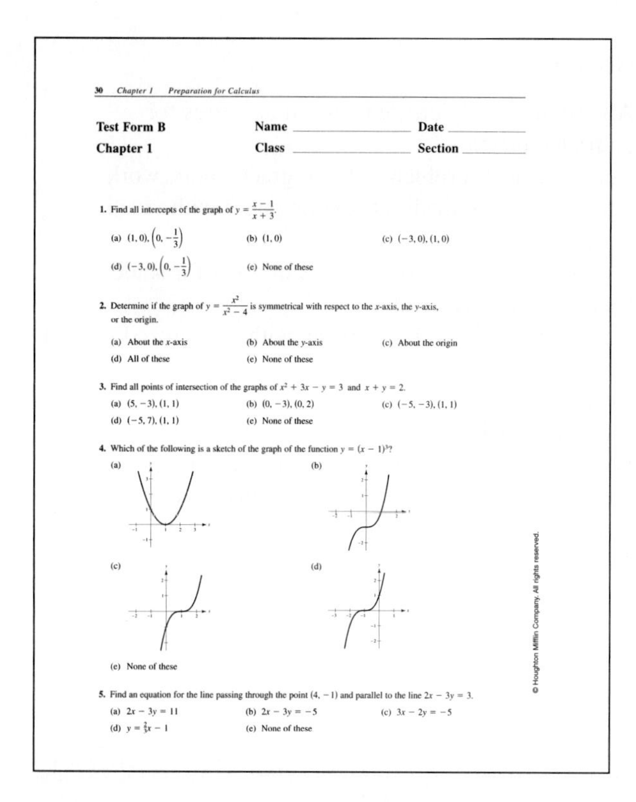

30 *Chapter 1 Preparation for Calculus*

Test Form B **Name** ________ **Date** ________
Chapter 1 **Class** ________ **Section** ________

1. Find all intercepts of the graph of $y = \frac{x-1}{x+3}$.

(a) $(1, 0), \left(0, -\frac{1}{3}\right)$ (b) $(1, 0)$ (c) $(-3, 0), (1, 0)$
(d) $(-3, 0), \left(0, -\frac{1}{3}\right)$ (e) None of these

2. Determine if the graph of $y = \frac{x^2}{x^2-4}$ is symmetrical with respect to the x-axis, the y-axis, or the origin.

(a) About the x-axis (b) About the y-axis (c) About the origin
(d) All of these (e) None of these

3. Find all points of intersection of the graphs of $x^2 + 3x - y = 3$ and $x + y = 2$.

(a) $(5, -3), (1, 1)$ (b) $(0, -3), (0, 2)$ (c) $(-5, -3), (1, 1)$
(d) $(-5, 7), (1, 1)$ (e) None of these

4. Which of the following is a sketch of the graph of the function $y = (x-1)^3$?

(a) (b)

(c) (d)

(e) None of these

5. Find an equation for the line passing through the point $(4, -1)$ and parallel to the line $2x - 3y = 3$.

(a) $2x - 3y = 11$ (b) $2x - 3y = -5$ (c) $3x - 2y = -5$
(d) $y = \frac{2}{3}x - 1$ (e) None of these

Features

- **Complete Solutions Guide** by Bruce Edwards
  This resource contains worked-out solutions to all textbook exercises in electronic format. It is available in three volumes: Volume I covers Chapters 1–6, Volume II covers Chapters 7–11, and Volume III covers Chapters 11–15.
- **Instructor's Resource Guide** by Ann Rutledge Kraus
  This resource contains an abundance of resources keyed to the textbook by chapter and section, including chapter summaries, teaching strategies, multiple versions of chapter tests, final exams, and gateway tests, and suggested solutions to the Chapter Openers, Explorations, Section Projects, and Technology features in the text in electronic format.
- **Test Item File** The *Test Item File* contains a sample question for every algorithm in HM Testing in electronic format.
- HM Testing test generator
- Digital textbook art
- Textbook Appendices D–F, containing additional presentations with exercises covering precalculus review, rotation and the general second degree equation, and complex numbers.
- Downloadable graphing calculator programs

New! HM Testing (powered by Diploma™)

For the instructor, HM Testing is a robust test-generating system.

Features

- Comprehensive set of algorithmic test items
- Can produce chapter tests, cumulative tests, and final exams
- Online testing
- Gradebook function

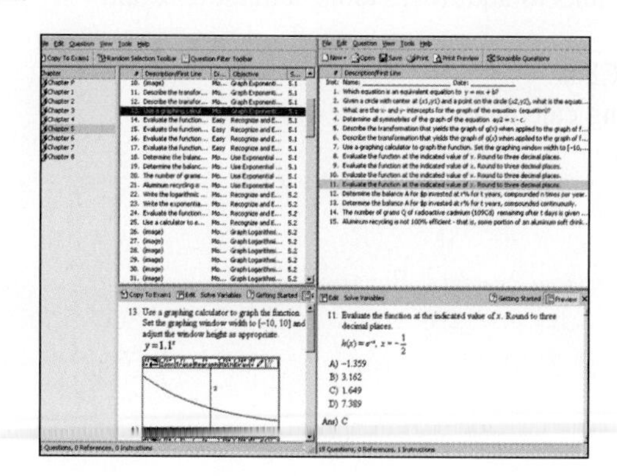

Enhanced! Instructional DVDs and Videos

These comprehensive DVD and video presentations complement the textbook topic coverage and have a variety of uses, including supplementing an online or hybrid course, giving students the opportunity to catch up if they miss a class, and providing substantial course material for self-study and review.

Features

- Comprehensive topic coverage from Calculus I, II, and III
- Additional explanations of calculus concepts, sample problems, and applications

Enhanced! Companion Textbook Website

The free Houghton Mifflin website at *college.hmco.com/pic/larsoncalculusIetf4e* contains an abundance of instructor and student resources.

Features

- Downloadable graphing calculator programs
- Textbook Appendices D–F, containing additional presentations with exercises covering precalculus review, rotation and the general second-degree equation, and complex numbers
- Algebra Review Summary
- Calculus Labs
- 3-D rotatable graphs

Printed Resources

For the convenience of students, the Study and Solutions Guides are available as printed supplements, but are also available in electronic format.

Study and Solutions Guide by Bruce Edwards

This student resource contains detailed, worked-out solutions to all odd-numbered textbook exercises. It is available in two volumes: Volume I covers Chapters 1–10 and Volume II covers Chapters 11–15.

For additional information about the Larson, Hostetler, and Edwards Calculus program, go to *college.hmco.com/info/larsoncalculus*.

Features

Chapter Openers

Each chapter opens with a real-life application of the concepts presented in the chapter, illustrated by a photograph. Open-ended and thought-provoking questions about the application encourage the student to consider how calculus concepts relate to real-life situations. A brief summary with a graphical component highlights the primary mathematical concepts presented in the chapter, and explains why they are important.

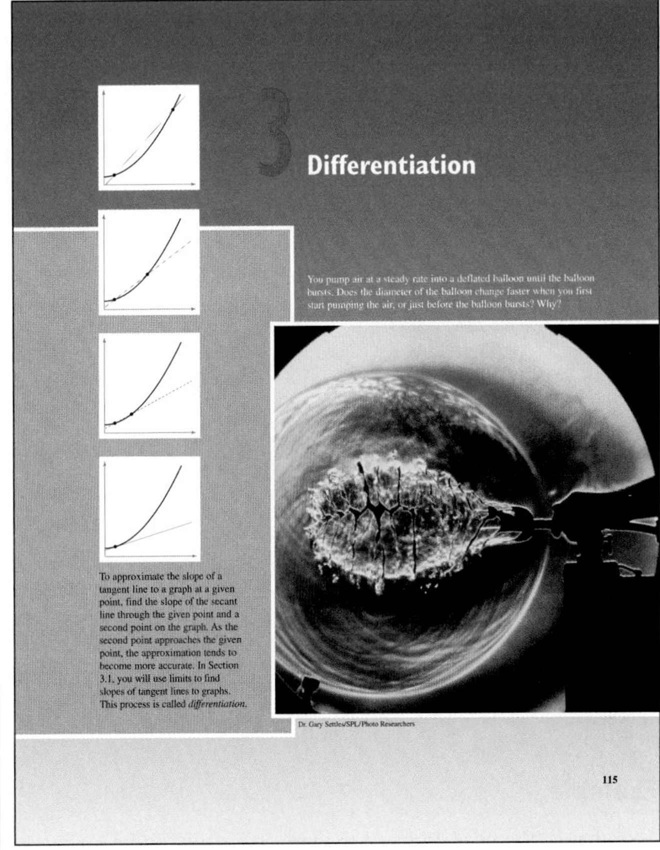

3 Differentiation

You pump air at a steady rate into a deflated balloon until the balloon bursts. Does the diameter of the balloon change faster when you first start pumping the air, or just before the balloon bursts? Why?

To approximate the slope of a tangent line to a graph at a given point, find the slope of the secant line through the given point and a second point on the graph. As the second point approaches the given point, the approximation tends to become more accurate. In Section 3.1, you will use limits to find slopes of tangent lines to graphs. This process is called *differentiation*.

Dr. Gary Settles/SPL/Photo Researchers

115

Section Openers

Every section begins with an outline of the key concepts covered in the section. This serves as a class planning resource for the instructor and a study and review guide for the student.

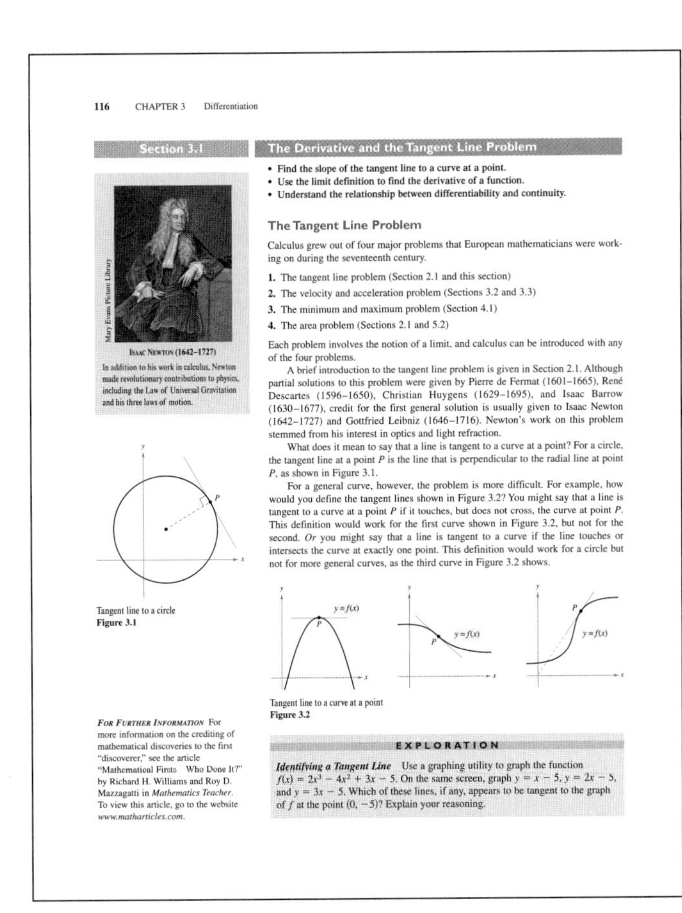

116 CHAPTER 3 Differentiation

Section 3.1 The Derivative and the Tangent Line Problem

- **Find the slope of the tangent line to a curve at a point.**
- **Use the limit definition to find the derivative of a function.**
- **Understand the relationship between differentiability and continuity.**

Mary Evans Picture Library

ISAAC NEWTON (1642–1727)

In addition to his work in calculus, Newton made revolutionary contributions to physics, including the Law of Universal Gravitation and his three laws of motion.

The Tangent Line Problem

Calculus grew out of four major problems that European mathematicians were working on during the seventeenth century.

1. The tangent line problem (Section 2.1 and this section)
2. The velocity and acceleration problem (Sections 3.2 and 3.3)
3. The minimum and maximum problem (Section 4.1)
4. The area problem (Sections 2.1 and 5.2)

Each problem involves the notion of a limit, and calculus can be introduced with any of the four problems.

A brief introduction to the tangent line problem is given in Section 2.1. Although partial solutions to this problem were given by Pierre de Fermat (1601–1665), René Descartes (1596–1650), Christian Huygens (1629–1695), and Isaac Barrow (1630–1677), credit for the first general solution is usually given to Isaac Newton (1642–1727) and Gottfried Leibniz (1646–1716). Newton's work on this problem stemmed from his interest in optics and light refraction.

What does it mean to say that a line is tangent to a curve at a point? For a circle, the tangent line at a point P is the line that is perpendicular to the radial line at point P, as shown in Figure 3.1.

For a general curve, however, the problem is more difficult. For example, how would you define the tangent lines shown in Figure 3.2? You might say that a line is tangent to a curve at a point P if it touches, but does not cross, the curve at point P. This definition would work for the first curve shown in Figure 3.2, but not for the second. *Or* you might say that a line is tangent to a curve if the line touches or intersects the curve at exactly one point. This definition would work for a circle but not for more general curves, as the third curve in Figure 3.2 shows.

Tangent line to a circle
Figure 3.1

Tangent line to a curve at a point
Figure 3.2

FOR FURTHER INFORMATION For more information on the crediting of mathematical discoveries to the first "discoverer," see the article "Mathematical Firsts—Who Done It?" by Richard H. Williams and Roy D. Mazzagatti in *Mathematics Teacher*. To view this article, go to the website *www.matharticles.com*.

EXPLORATION

Identifying a Tangent Line Use a graphing utility to graph the function $f(x) = 2x^3 - 4x^2 + 3x - 5$. On the same screen, graph $y = x - 5$, $y = 2x - 5$, and $y = 3x - 5$. Which of these lines, if any, appears to be tangent to the graph of f at the point $(0, -5)$? Explain your reasoning.

Explorations

For selected topics, Explorations offer the opportunity to discover calculus concepts before they are formally introduced in the text, thus enhancing student understanding. This optional feature can be omitted at the discretion of the instructor with no loss of continuity in the coverage of the material.

Historical Notes

Integrated throughout the text, Historical Notes help students grasp the basic mathematical foundations of calculus.

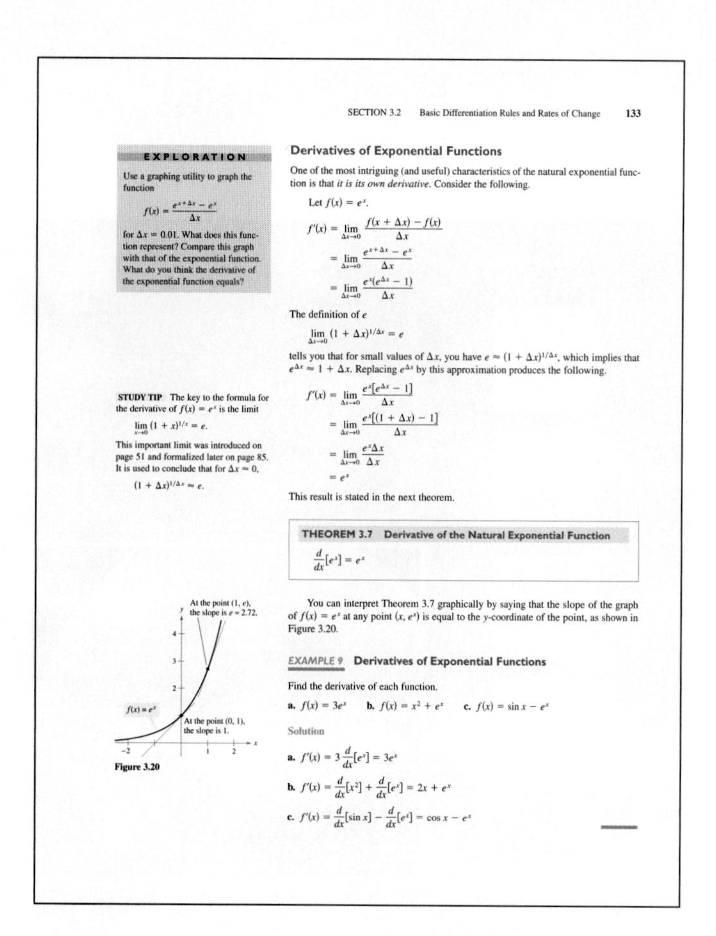

SECTION 3.2 Basic Differentiation Rules and Rates of Change 133

EXPLORATION

Use a graphing utility to graph the function

$$f(x) = \frac{e^{x+\Delta x} - e^x}{\Delta x}$$

for $\Delta x = 0.01$. What does this function represent? Compare this graph with that of the exponential function. What do you think the derivative of the exponential function equals?

Derivatives of Exponential Functions

One of the most intriguing (and useful) characteristics of the natural exponential function is that *it is its own derivative*. Consider the following.

Let $f(x) = e^x$.

$$f'(x) = \lim_{\Delta x \to 0} \frac{f(x+\Delta x) - f(x)}{\Delta x} = \lim_{\Delta x \to 0} \frac{e^{x+\Delta x} - e^x}{\Delta x} = \lim_{\Delta x \to 0} \frac{e^x(e^{\Delta x} - 1)}{\Delta x}$$

The definition of e

$$\lim_{\Delta x \to 0} (1 + \Delta x)^{1/\Delta x} = e$$

tells you that for small values of Δx, you have $e \approx (1 + \Delta x)^{1/\Delta x}$, which implies that $e^{\Delta x} \approx 1 + \Delta x$. Replacing $e^{\Delta x}$ by this approximation produces the following.

$$f'(x) = \lim_{\Delta x \to 0} \frac{e^x[e^{\Delta x} - 1]}{\Delta x} = \lim_{\Delta x \to 0} \frac{e^x[(1+\Delta x) - 1]}{\Delta x} = \lim_{\Delta x \to 0} \frac{e^x \Delta x}{\Delta x} = e^x$$

STUDY TIP The key to the formula for the derivative of $f(x) = e^x$ is the limit

$$\lim_{x \to 0} (1 + x)^{1/x} = e.$$

This important limit was introduced on page 51 and formalized later on page 85. It is used to conclude that for $\Delta x \approx 0$,

$$(1 + \Delta x)^{1/\Delta x} \approx e.$$

This result is stated in the next theorem.

THEOREM 3.7 Derivative of the Natural Exponential Function

$$\frac{d}{dx}[e^x] = e^x$$

You can interpret Theorem 3.7 graphically by saying that the slope of the graph of $f(x) = e^x$ at any point (x, e^x) is equal to the y-coordinate of the point, as shown in Figure 3.20.

Figure 3.20

EXAMPLE 9 **Derivatives of Exponential Functions**

Find the derivative of each function.

a. $f(x) = 3e^x$ **b.** $f(x) = x^2 + e^x$ **c.** $f(x) = \sin x - e^x$

Solution

a. $f'(x) = 3\frac{d}{dx}[e^x] = 3e^x$

b. $f'(x) = \frac{d}{dx}[x^2] + \frac{d}{dx}[e^x] = 2x + e^x$

c. $f'(x) = \frac{d}{dx}[\sin x] - \frac{d}{dx}[e^x] = \cos x - e^x$

Theorems

All Theorems and Definitions are highlighted for emphasis and easy reference. Proofs are shown for selected theorems to enhance student understanding.

Study Tip

Located at point of use throughout the text, Study Tips advise students on how to avoid common errors, address special cases, and expand upon theoretical concepts.

Graphics

Numerous graphics throughout the text enhance student understanding of complex calculus concepts (especially in three-dimensional representations), as well as real-life applications.

Example

To enhance the usefulness of the text as a study and learning tool, the Fourth Edition contains numerous Examples. The detailed, worked-out Solutions (many with side comments to clarify the steps or the method) are presented graphically, analytically, and/or numerically to provide students with opportunities for practice and further insight into calculus concepts. Many Examples incorporate real-data analysis.

Open Exploration

Eduspace® contains Open Explorations, which investigate selected Examples using computer algebra systems (*Maple*, *Mathematica*, *Derive*, and *Mathcad*). The icon (www) identifies these Examples.

Notes

Instructional Notes accompany many of the Theorems, Definitions, and Examples to offer additional insights or describe generalizations.

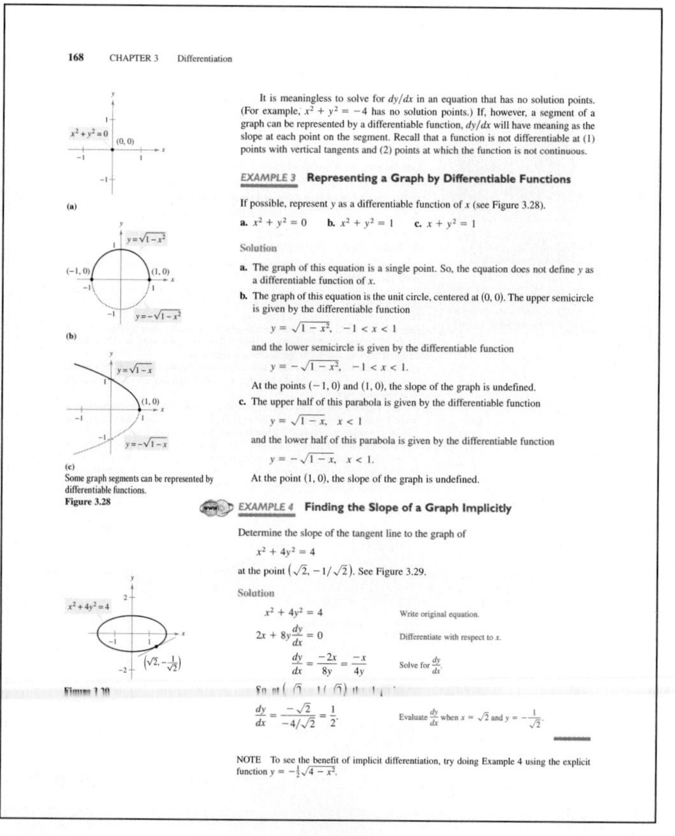

168 CHAPTER 3 Differentiation

It is meaningless to solve for dy/dx in an equation that has no solution points. (For example, $x^2 + y^2 = -4$ has no solution points.) If, however, a segment of a graph can be represented by a differentiable function, dy/dx will have meaning as the slope at each point on the segment. Recall that a function is not differentiable at (1) points with vertical tangents and (2) points at which the function is not continuous.

EXAMPLE 3 **Representing a Graph by Differentiable Functions**

If possible, represent y as a differentiable function of x (see Figure 3.28).

a. $x^2 + y^2 = 0$ **b.** $x^2 + y^2 = 1$ **c.** $x + y^2 = 1$

Solution

a. The graph of this equation is a single point. So, the equation does not define y as a differentiable function of x.

b. The graph of this equation is the unit circle, centered at $(0, 0)$. The upper semicircle is given by the differentiable function

$$y = \sqrt{1 - x^2}, \quad -1 < x < 1$$

and the lower semicircle is given by the differentiable function

$$y = -\sqrt{1 - x^2}, \quad -1 < x < 1.$$

At the points $(-1, 0)$ and $(1, 0)$, the slope of the graph is undefined.

c. The upper half of this parabola is given by the differentiable function

$$y = \sqrt{1 - x}, \quad x < 1$$

and the lower half of this parabola is given by the differentiable function

$$y = -\sqrt{1 - x}, \quad x < 1.$$

At the point $(1, 0)$, the slope of the graph is undefined.

Some graph segments can be represented by differentiable functions.
Figure 3.28

EXAMPLE 4 **Finding the Slope of a Graph Implicitly**

Determine the slope of the tangent line to the graph of

$$x^2 + 4y^2 = 4$$

at the point $(\sqrt{2}, -1/\sqrt{2})$. See Figure 3.29.

Solution

$$x^2 + 4y^2 = 4 \qquad \text{Write original equation.}$$

$$2x + 8y\frac{dy}{dx} = 0 \qquad \text{Differentiate with respect to } x.$$

$$\frac{dy}{dx} = \frac{-2x}{8y} = \frac{-x}{4y} \qquad \text{Solve for } \frac{dy}{dx}.$$

[illegible]

$$\frac{dy}{dx} = \frac{-\sqrt{2}}{-4/\sqrt{2}} = \frac{1}{2}. \qquad \text{Evaluate } \frac{dy}{dx} \text{ when } x = \sqrt{2} \text{ and } y = -\frac{1}{\sqrt{2}}.$$

[illegible]

NOTE To see the benefit of implicit differentiation, try doing Example 4 using the explicit function $y = -\frac{1}{2}\sqrt{4 - x^2}$.

Exercises

The core of every calculus text, Exercises provide opportunities for exploration, practice, and comprehension. The Fourth Edition contains over 6000 Section and Chapter Review Exercises, carefully graded in each set from skill-building to challenging. The extensive range of problem types includes true/false, writing, conceptual, real-data modeling, and graphical analysis.

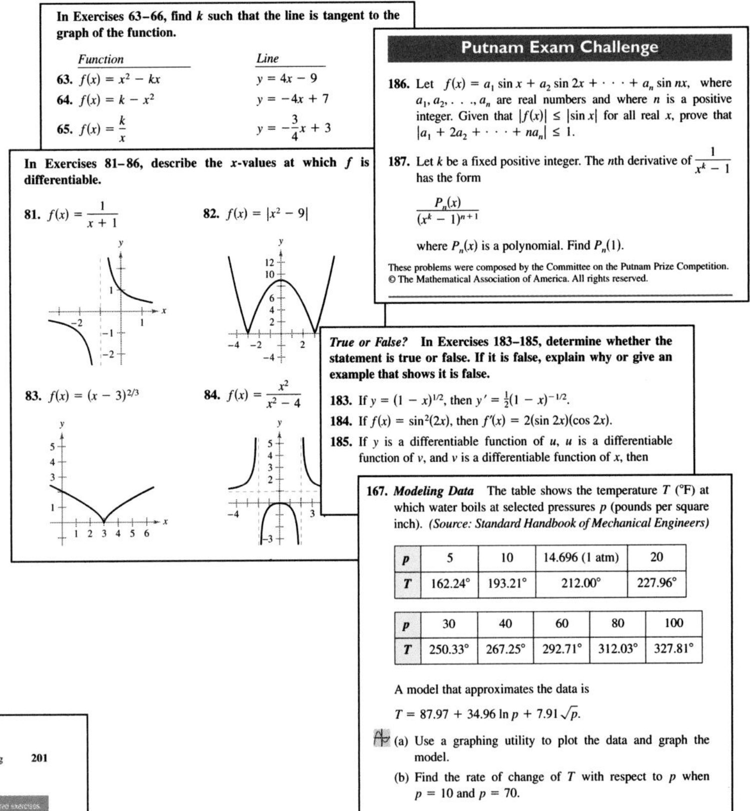

In Exercises 63–66, find k such that the line is tangent to the graph of the function.

	Function	*Line*
63.	$f(x) = x^2 - kx$	$y = 4x - 9$
64.	$f(x) = k - x^2$	$y = -4x + 7$
65.	$f(x) = \frac{k}{x}$	$y = -\frac{3}{4}x + 3$

In Exercises 81–86, describe the x-values at which f is differentiable.

81. $f(x) = \dfrac{1}{x+1}$

82. $f(x) = |x^2 - 9|$

83. $f(x) = (x - 3)^{2/3}$

84. $f(x) = \dfrac{x^2}{x^2 - 4}$

Putnam Exam Challenge

186. Let $f(x) = a_1 \sin x + a_2 \sin 2x + \cdots + a_n \sin nx$, where $a_1, a_2, \ldots, a_n$ are real numbers and where n is a positive integer. Given that $|f(x)| \le |\sin x|$ for all real x, prove that $|a_1 + 2a_2 + \cdots + na_n| \le 1$.

187. Let k be a fixed positive integer. The nth derivative of $\dfrac{1}{x^k - 1}$ has the form

$$\frac{P_n(x)}{(x^k - 1)^{n+1}}$$

where $P_n(x)$ is a polynomial. Find $P_n(1)$.

These problems were composed by the Committee on the Putnam Prize Competition. © The Mathematical Association of America. All rights reserved.

True or False? **In Exercises 183–185, determine whether the statement is true or false. If it is false, explain why or give an example that shows it is false.**

183. If $y = (1 - x)^{1/2}$, then $y' = \frac{1}{2}(1 - x)^{-1/2}$.

184. If $f(x) = \sin^2(2x)$, then $f'(x) = 2(\sin 2x)(\cos 2x)$.

185. If y is a differentiable function of u, u is a differentiable function of v, and v is a differentiable function of x, then

167. ***Modeling Data*** The table shows the temperature T (°F) at which water boils at selected pressures p (pounds per square inch). *(Source: Standard Handbook of Mechanical Engineers)*

p	5	10	14.696 (1 atm)	20
T	162.24°	193.21°	212.00°	227.96°

p	30	40	60	80	100
T	250.33°	267.25°	292.71°	312.03°	327.81°

A model that approximates the data is

$T = 87.97 + 34.96 \ln p + 7.91\sqrt{p}$.

(a) Use a graphing utility to plot the data and graph the model.

(b) Find the rate of change of T with respect to p when $p = 10$ and $p = 70$.

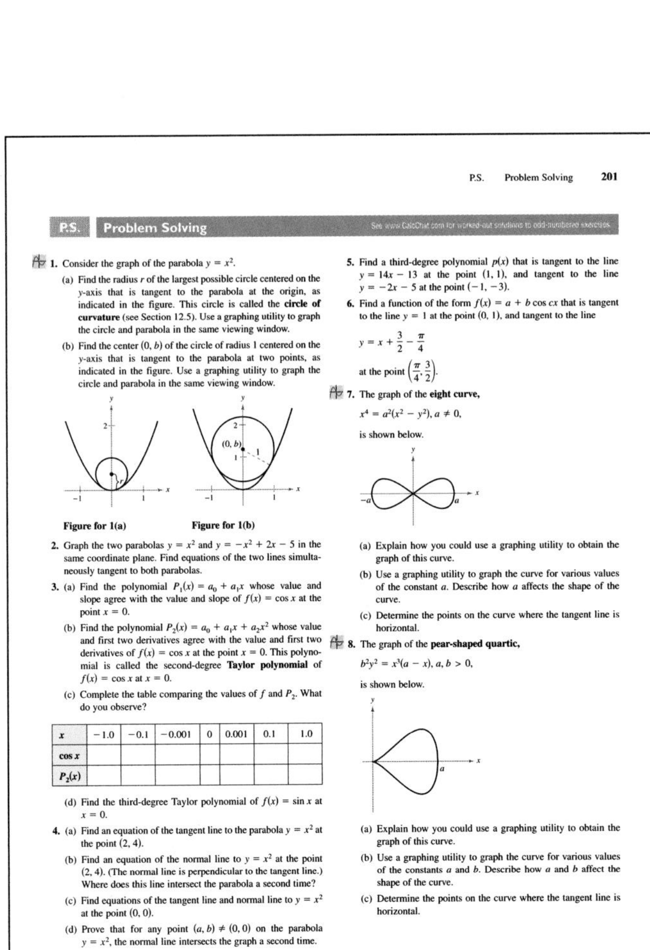

P.S. Problem Solving 201

P.S. Problem Solving

See www.CalcChat.com for worked-out solutions to odd-numbered exercises.

1. Consider the graph of the parabola $y = x^2$.

(a) Find the radius r of the largest possible circle centered on the y-axis that is tangent to the parabola at the origin, as indicated in the figure. This circle is called the **circle of curvature** (see Section 12.5). Use a graphing utility to graph the circle and parabola in the same viewing window.

(b) Find the center $(0, b)$ of the circle of radius 1 centered on the y-axis that is tangent to the parabola at two points, as indicated in the figure. Use a graphing utility to graph the circle and parabola in the same viewing window.

Figure for 1(a) Figure for 1(b)

2. Graph the two parabolas $y = x^2$ and $y = -x^2 + 2x - 5$ in the same coordinate plane. Find equations of the two lines simultaneously tangent to both parabolas.

3. (a) Find the polynomial $P_1(x) = a_0 + a_1x$ whose value and slope agree with the value and slope of $f(x) = \cos x$ at the point $x = 0$.

(b) Find the polynomial $P_2(x) = a_0 + a_1x + a_2x^2$ whose value and first two derivatives agree with the value and first two derivatives of $f(x) = \cos x$ at the point $x = 0$. This polynomial is called the second-degree **Taylor polynomial** of $f(x) = \cos x$ at $x = 0$.

(c) Complete the table comparing the values of f and P_2. What do you observe?

x	-1.0	-0.1	-0.001	0	0.001	0.1	1.0
$\cos x$							
$P_2(x)$							

(d) Find the third-degree Taylor polynomial of $f(x) = \sin x$ at $x = 0$.

4. (a) Find an equation of the tangent line to the parabola $y = x^2$ at the point $(2, 4)$.

(b) Find an equation of the normal line to $y = x^2$ at the point $(2, 4)$. (The normal line is perpendicular to the tangent line.) Where does this line intersect the parabola a second time?

(c) Find equations of the tangent line and normal line to $y = x^2$ at the point $(0, 0)$.

(d) Prove that for any point $(a, b) \ne (0, 0)$ on the parabola $y = x^2$, the normal line intersects the graph a second time.

5. Find a third-degree polynomial $p(x)$ that is tangent to the line $y = 14x - 13$ at the point $(1, 1)$, and tangent to the line $y = -2x - 5$ at the point $(-1, -3)$.

6. Find a function of the form $f(x) = a + b \cos cx$ that is tangent to the line $y = 1$ at the point $(0, 1)$, and tangent to the line

$$y = x + \frac{3}{2} - \frac{\pi}{4}$$

at the point $\left(\frac{\pi}{4}, \frac{3}{2}\right)$.

7. The graph of the **eight curve,**

$x^4 = a^2(x^2 - y^2), a \ne 0,$

is shown below.

(a) Explain how you could use a graphing utility to obtain the graph of this curve.

(b) Use a graphing utility to graph the curve for various values of the constant a. Describe how a affects the shape of the curve.

(c) Determine the points on the curve where the tangent line is horizontal.

8. The graph of the **pear-shaped quartic,**

$b^2y^2 = x^3(a - x), a, b > 0,$

is shown below.

(a) Explain how you could use a graphing utility to obtain the graph of this curve.

(b) Use a graphing utility to graph the curve for various values of the constants a and b. Describe how a and b affect the shape of the curve.

(c) Determine the points on the curve where the tangent line is horizontal.

P.S. Problem Solving

Each chapter concludes with a set of thought-provoking and challenging exercises that provide opportunities for the student to explore the concepts in the chapter further.

Technology

Throughout the text, the use of a graphing utility or computer algebra system is suggested as appropriate for problem-solving as well as exploration and discovery. For example, students may choose to use a graphing utility to execute complicated computations, to visualize theoretical concepts, to discover alternative approaches, or to verify the results of other solution methods. However, students are not required to have access to a graphing utility to use this text effectively. In addition to describing the benefits of using technology to learn calculus, the text also addresses its possible misuse or misinterpretation.

Additional Features

Additional teaching and learning resources are integrated throughout the textbook, including Section Projects, journal references, and Writing About Concepts Exercises.

Preparation for Calculus

Two types of racecars designed and built by NASCAR teams are short track cars and super-speedway (long track) cars. Super-speedway racecars are subjected to extensive testing in wind tunnels like the one shown in the photo. Short track racecars and super-speedway racecars are designed either to allow for as much downforce as possible or to reduce the amount of drag on the racecar. Which design do you think is used for each type of racecar? Why?

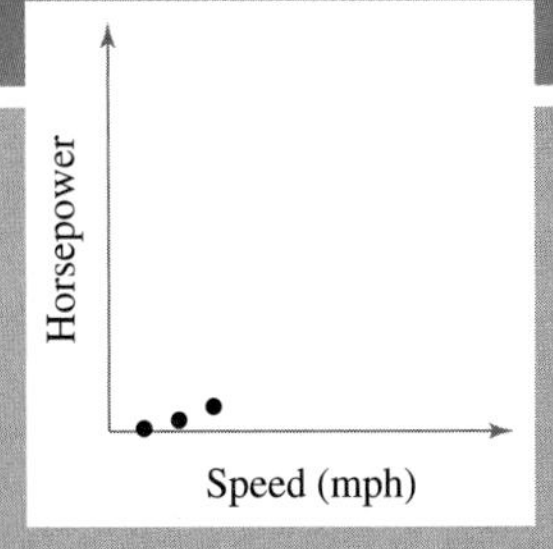

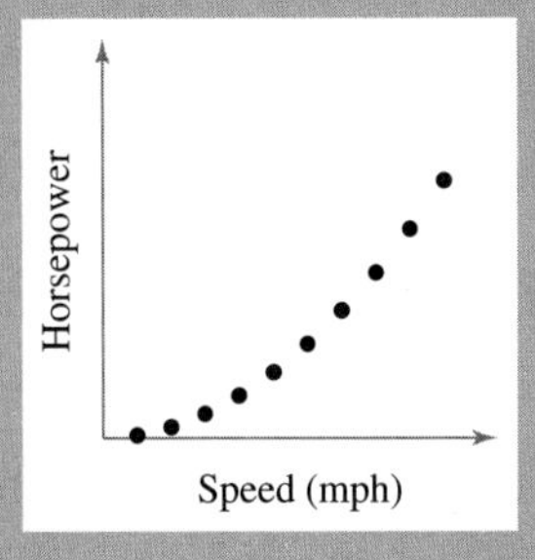

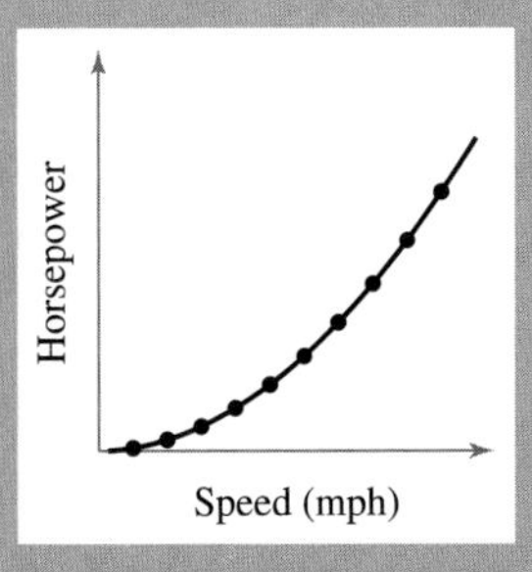

Mathematical models are commonly used to describe data sets. These models can be represented by many different types of functions such as linear, quadratic, cubic, rational, and trigonometric functions. In Chapter 1, you will review how to find, graph, and compare mathematical models for different data sets.

Section 1.1 Graphs and Models

- **Sketch the graph of an equation.**
- **Find the intercepts of a graph.**
- **Test a graph for symmetry with respect to an axis and the origin.**
- **Find the points of intersection of two graphs.**
- **Interpret mathematical models for real-life data.**

RENÉ DESCARTES (1596–1650)

Descartes made many contributions to philosophy, science, and mathematics. The idea of representing points in the plane by pairs of real numbers and representing curves in the plane by equations was described by Descartes in his book *La Géométrie*, published in 1637.

The Graph of an Equation

In 1637, the French mathematician René Descartes revolutionized the study of mathematics by joining its two major fields—algebra and geometry. With Descartes's coordinate plane, geometric concepts could be formulated analytically and algebraic concepts could be viewed graphically. The power of this approach is such that within a century, much of calculus had been developed.

The same approach can be followed in your study of calculus. That is, by viewing calculus from multiple perspectives—*graphically*, *analytically*, and *numerically*—you will increase your understanding of core concepts.

Consider the equation $3x + y = 7$. The point $(2, 1)$ is a **solution point** of the equation because the equation is satisfied (is true) when 2 is substituted for x and 1 is substituted for y. This equation has many other solutions, such as $(1, 4)$ and $(0, 7)$. To systematically find other solutions, solve the original equation for y.

$$y = 7 - 3x \qquad \text{Analytic approach}$$

Then construct a table of values by substituting several values of x.

x	0	1	2	3	4
y	7	4	1	-2	-5

Numerical approach

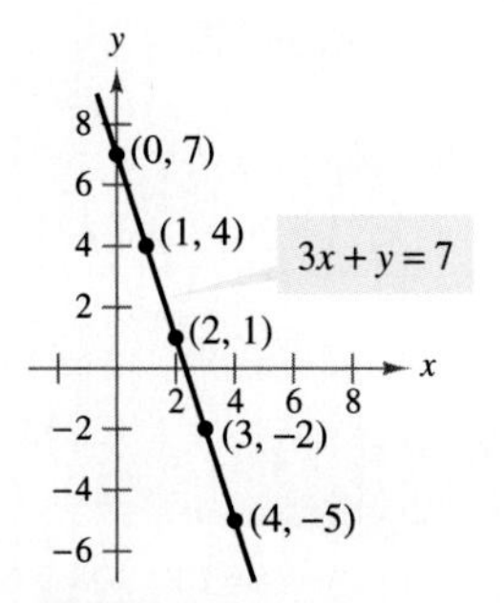

Graphical approach: $3x + y = 7$
Figure 1.1

From the table, you can see that $(0, 7)$, $(1, 4)$, $(2, 1)$, $(3, -2)$, and $(4, -5)$ are solutions of the original equation $3x + y = 7$. Like many equations, this equation has an infinite number of solutions. The set of all solution points is the **graph** of the equation, as shown in Figure 1.1.

NOTE Even though we refer to the sketch shown in Figure 1.1 as the graph of $3x + y = 7$, it really represents only a *portion* of the graph. The entire graph would extend beyond the page.

In this course, you will study many sketching techniques. The simplest is point plotting—that is, you plot points until the basic shape of the graph seems apparent.

EXAMPLE 1 Sketching a Graph by Point Plotting

Sketch the graph of $y = x^2 - 2$.

Solution First construct a table of values. Then plot the points shown in the table.

x	-2	-1	0	1	2	3
y	2	-1	-2	-1	2	7

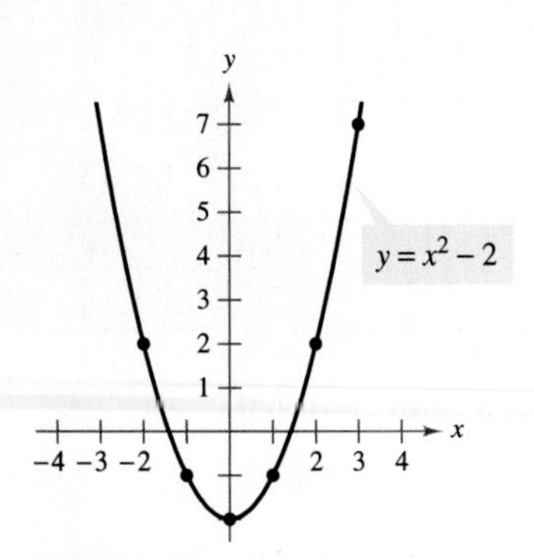

The parabola $y = x^2 - 2$
Figure 1.2

Finally, connect the points with a *smooth curve*, as shown in Figure 1.2. This graph is a **parabola.** It is one of the conics you will study in Chapter 10.

One disadvantage of point plotting is that to get a good idea about the shape of a graph, you may need to plot many points. With only a few points, you could misrepresent the graph. For instance, suppose that to sketch the graph of

$$y = \frac{1}{30}x(39 - 10x^2 + x^4)$$

you plotted only five points:

$$(-3, -3), (-1, -1), (0, 0), (1, 1), \text{ and } (3, 3)$$

as shown in Figure 1.3(a). From these five points, you might conclude that the graph is a line. This, however, is not correct. By plotting several more points, you can see that the graph is more complicated, as shown in Figure 1.3(b).

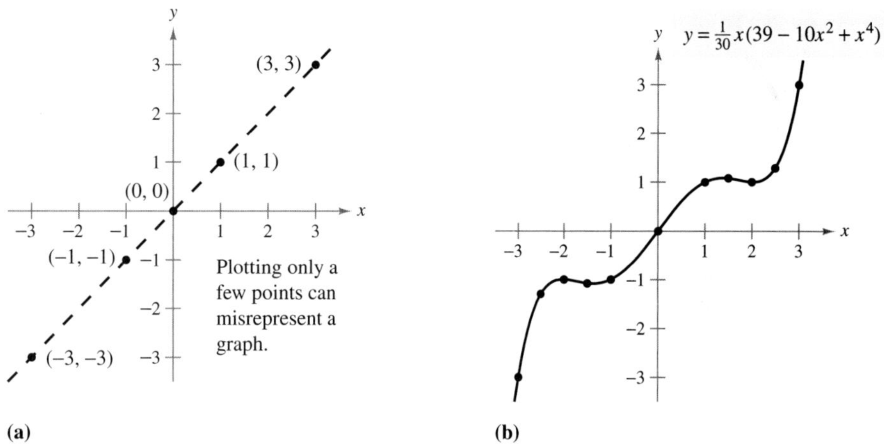

Figure 1.3

EXPLORATION

Comparing Graphical and Analytic Approaches Use a graphing utility to graph each of the following. In each case, find a viewing window that shows the important characteristics of the graph.

a. $y = x^3 - 3x^2 + 2x + 5$

b. $y = x^3 - 3x^2 + 2x + 25$

c. $y = -x^3 - 3x^2 + 20x + 5$

d. $y = 3x^3 - 40x^2 + 50x - 45$

e. $y = -(x + 12)^3$

f. $y = (x - 2)(x - 4)(x - 6)$

A purely graphical approach to this problem would involve a simple "guess, check, and revise" strategy. What types of things do you think an analytic approach might involve? For instance, does the graph have symmetry? Does the graph have turns? If so, where are they?

As you proceed through Chapters 2, 3, and 4 of this text, you will study many new analytic tools that will help you analyze graphs of equations such as these.

TECHNOLOGY Technology has made sketching of graphs easier. Even with technology, however, it is possible to misrepresent a graph badly. For instance, each of the graphing utility screens in Figure 1.4 shows a portion of the graph of

$$y = x^3 - x^2 - 25.$$

From the screen on the left, you might assume that the graph is a line. From the screen on the right, however, you can see that the graph is not a line. Thus, whether you are sketching a graph by hand or using a graphing utility, you must realize that different "viewing windows" can produce very different views of a graph. In choosing a viewing window, your goal is to show a view of the graph that fits well in the context of the problem.

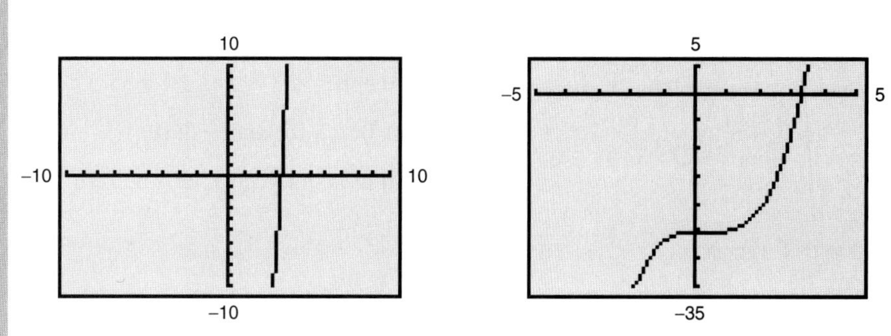

Graphing utility screens of $y = x^3 - x^2 - 25$
Figure 1.4

NOTE In this text, we use the term *graphing utility* to mean either a graphing calculator or computer graphing software such as *Maple*, *Mathematica*, *Derive*, *Mathcad*, or the *TI-89*.

Intercepts of a Graph

Two types of solution points that are especially useful when graphing an equation are those having zero as their x- or y-coordinate. Such points are called **intercepts** because they are the points at which the graph intersects the x- or y-axis. The point $(a, 0)$ is an ***x*-intercept** of the graph of an equation if it is a solution point of the equation. To find the x-intercepts of a graph, let y be zero and solve the equation for x. The point $(0, b)$ is a ***y*-intercept** of the graph of an equation if it is a solution point of the equation. To find the y-intercepts of a graph, let x be zero and solve the equation for y.

NOTE Some texts denote the x-intercept as the x-coordinate of the point $(a, 0)$ rather than the point itself. Unless it is necessary to make a distinction, we will use the term *intercept* to mean either the point or the coordinate.

It is possible for a graph to have no intercepts, or it might have several. For instance, consider the four graphs shown in Figure 1.5.

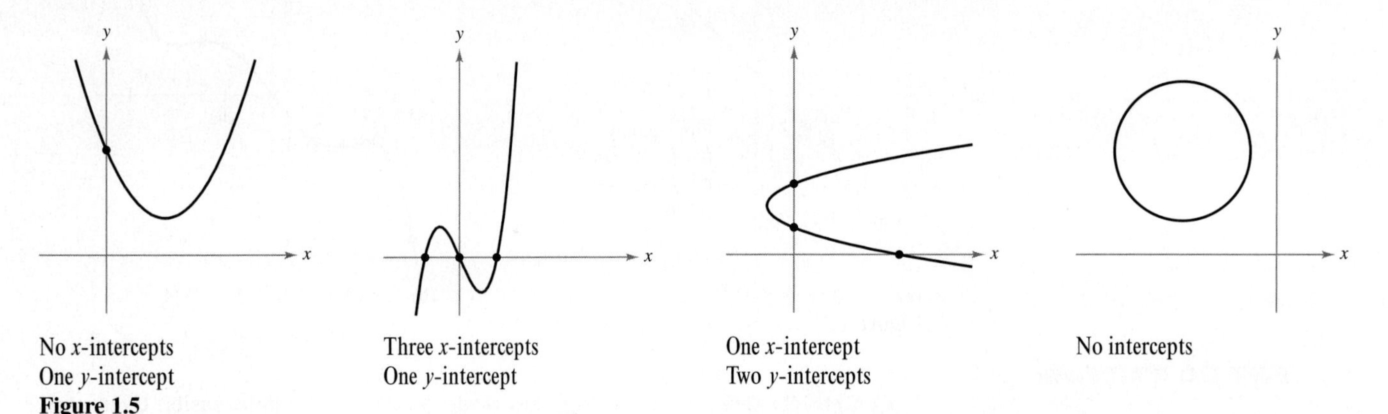

Figure 1.5

EXAMPLE 2 Finding x- and y-intercepts

Find the x- and y-intercepts of the graph of $y = x^3 - 4x$.

Solution To find the x-intercepts, let y be zero and solve for x.

$$x^3 - 4x = 0 \qquad \text{Let } y \text{ be zero.}$$

$$x(x - 2)(x + 2) = 0 \qquad \text{Factor.}$$

$$x = 0, 2, \text{ or } -2 \qquad \text{Solve for } x.$$

Because this equation has three solutions, you can conclude that the graph has three x-intercepts:

$$(0, 0), (2, 0), \text{ and } (-2, 0). \qquad x\text{-intercepts}$$

To find the y-intercepts, let x be zero. Doing so produces $y = 0$. So, the y-intercept is

$$(0, 0). \qquad y\text{-intercept}$$

(See Figure 1.6.)

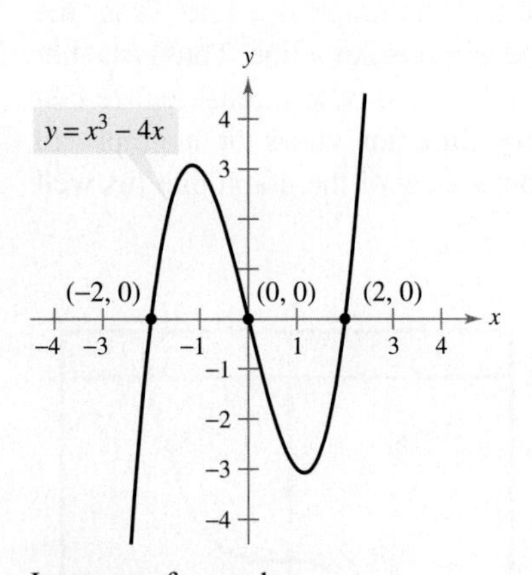

Intercepts of a graph
Figure 1.6

TECHNOLOGY Example 2 uses an analytic approach to finding intercepts. When an analytic approach is not possible, you can use a graphical approach by finding the points where the graph intersects the axes. Use a graphing utility to approximate the intercepts.

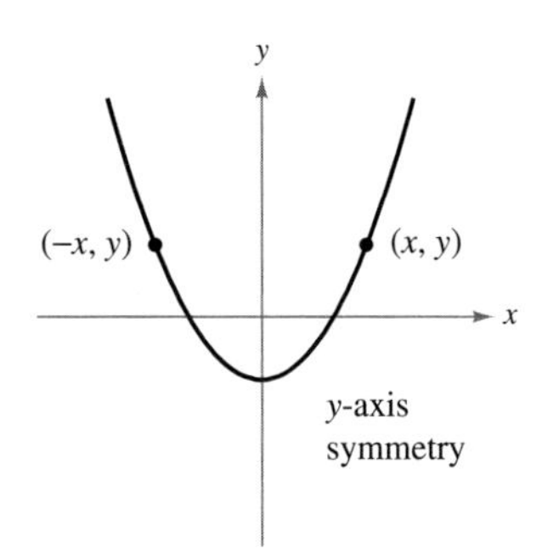

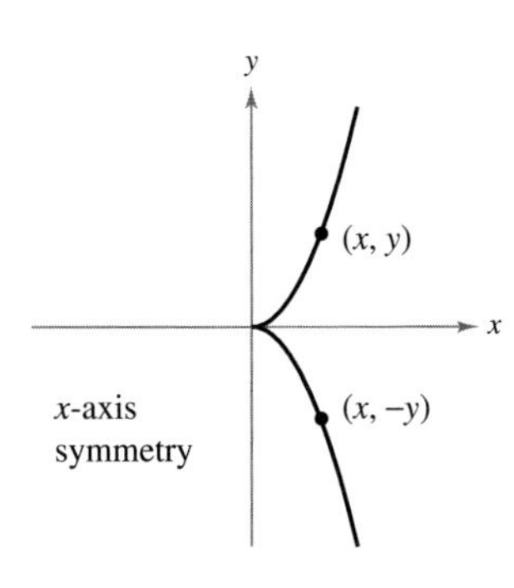

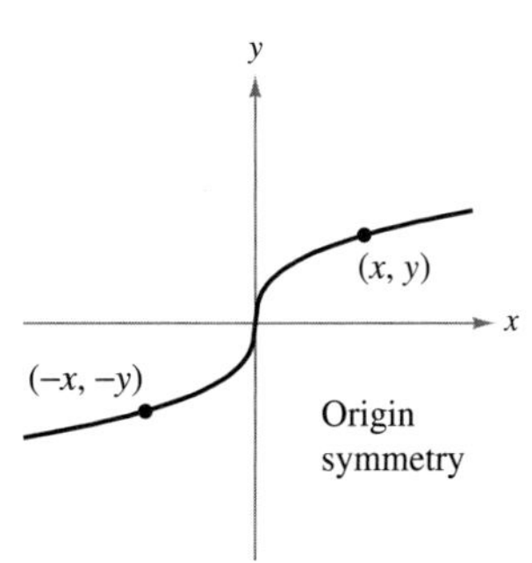

Figure 1.7

Symmetry of a Graph

Knowing the symmetry of a graph *before* attempting to sketch it is useful because you need only half as many points to sketch the graph. The following three types of symmetry can be used to help sketch the graph of an equation (see Figure 1.7).

1. A graph is **symmetric with respect to the y-axis** if, whenever (x, y) is a point on the graph, $(-x, y)$ is also a point on the graph. This means that the portion of the graph to the left of the y-axis is a mirror image of the portion to the right of the y-axis.
2. A graph is **symmetric with respect to the x-axis** if, whenever (x, y) is a point on the graph, $(x, -y)$ is also a point on the graph. This means that the portion of the graph above the x-axis is a mirror image of the portion below the x-axis.
3. A graph is **symmetric with respect to the origin** if, whenever (x, y) is a point on the graph, $(-x, -y)$ is also a point on the graph. This means that the graph is unchanged by a rotation of 180° about the origin.

Tests for Symmetry

1. The graph of an equation in x and y is symmetric with respect to the y-axis if replacing x by $-x$ yields an equivalent equation.
2. The graph of an equation in x and y is symmetric with respect to the x-axis if replacing y by $-y$ yields an equivalent equation.
3. The graph of an equation in x and y is symmetric with respect to the origin if replacing x by $-x$ and y by $-y$ yields an equivalent equation.

The graph of a polynomial has symmetry with respect to the y-axis if each term has an even exponent (or is a constant). For instance, the graph of

$$y = 2x^4 - x^2 + 2 \qquad y\text{-axis symmetry}$$

has symmetry with respect to the y-axis. Similarly, the graph of a polynomial has symmetry with respect to the origin if each term has an odd exponent, as illustrated in Example 3.

EXAMPLE 3 Testing for Origin Symmetry

Show that the graph of

$$y = 2x^3 - x$$

is symmetric with respect to the origin.

Solution

$$\begin{aligned} y &= 2x^3 - x && \text{Write original equation.} \\ -y &= 2(-x)^3 - (-x) && \text{Replace } x \text{ by } -x \text{ and } y \text{ by } -y. \\ -y &= -2x^3 + x && \text{Simplify.} \\ y &= 2x^3 - x && \text{Equivalent equation} \end{aligned}$$

Because the replacement produces an equivalent equation, you can conclude that the graph of $y = 2x^3 - x$ is symmetric with respect to the origin, as shown in Figure 1.8.

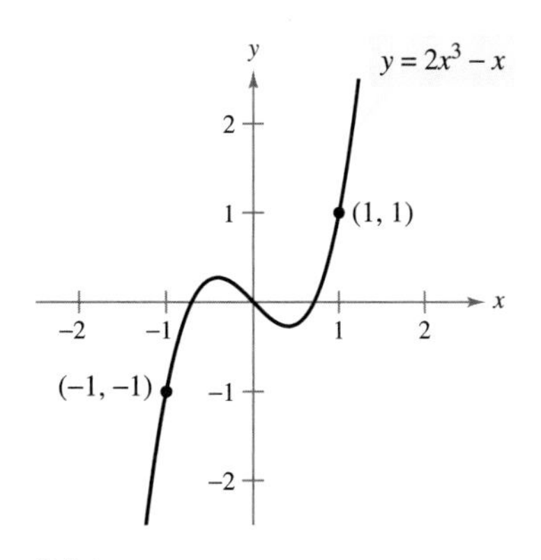

Origin symmetry
Figure 1.8

EXAMPLE 4 Using Intercepts and Symmetry to Sketch a Graph

Sketch the graph of $x - y^2 = 1$.

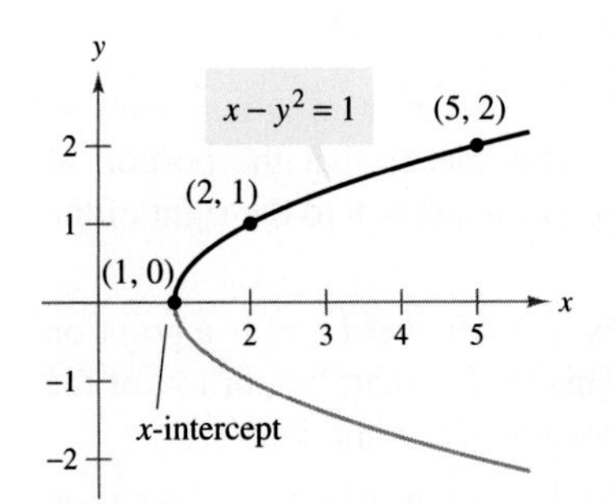

Figure 1.9

Solution The graph is symmetric with respect to the x-axis because replacing y by $-y$ yields an equivalent equation.

$$x - y^2 = 1 \quad \text{Write original equation.}$$
$$x - (-y)^2 = 1 \quad \text{Replace } y \text{ by } -y.$$
$$x - y^2 = 1 \quad \text{Equivalent equation}$$

This means that the portion of the graph below the x-axis is a mirror image of the portion above the x-axis. To sketch the graph, first sketch the portion above the x-axis. Then reflect in the x-axis to obtain the entire graph, as shown in Figure 1.9.

TECHNOLOGY Graphing utilities are designed so that they most easily graph equations in which y is a function of x (see Section 1.3 for a definition of **function**). To graph other types of equations, you need to split the graph into two or more parts *or* you need to use a different graphing mode. For instance, to graph the equation in Example 4, you can split it into two parts.

$$y_1 = \sqrt{x - 1} \quad \text{Top portion of graph}$$
$$y_2 = -\sqrt{x - 1} \quad \text{Bottom portion of graph}$$

Points of Intersection

A **point of intersection** of the graphs of two equations is a point that satisfies both equations. You can find the points of intersection of two graphs by solving their equations simultaneously.

EXAMPLE 5 Finding Points of Intersection

Find all points of intersection of the graphs of $x^2 - y = 3$ and $x - y = 1$.

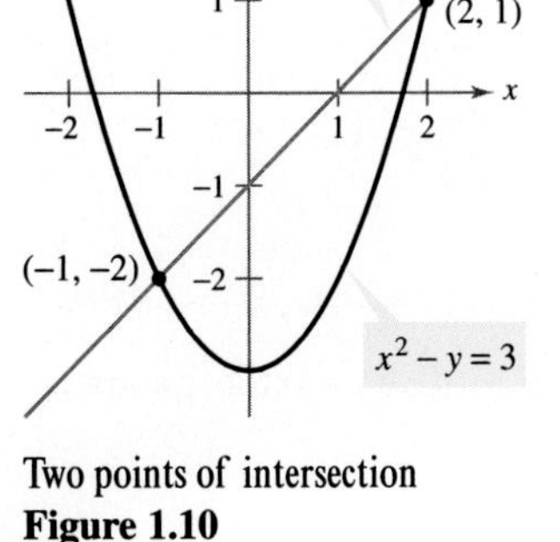

Two points of intersection
Figure 1.10

STUDY TIP You can check the points of intersection from Example 5 by substituting into *both* of the original equations or by using the *intersect* feature of a graphing utility.

Solution Begin by sketching the graphs of both equations on the *same* rectangular coordinate system, as shown in Figure 1.10. Having done this, it appears that the graphs have two points of intersection. To find these two points, you can use the following steps.

$$y = x^2 - 3 \quad \text{Solve first equation for } y.$$
$$y = x - 1 \quad \text{Solve second equation for } y.$$
$$x^2 - 3 = x - 1 \quad \text{Equate } y\text{-values.}$$
$$x^2 - x - 2 = 0 \quad \text{Write in general form.}$$
$$(x - 2)(x + 1) = 0 \quad \text{Factor.}$$
$$x = 2 \text{ or } -1 \quad \text{Solve for } x.$$

The corresponding values of y are obtained by substituting $x = 2$ and $x = -1$ into either of the original equations. Doing this produces two points of intersection:

$$(2, 1) \quad \text{and} \quad (-1, -2). \quad \text{Points of intersection}$$

indicates that in the HM mathSpace® *CD-ROM and the online* Eduspace® *system for this text, you will find an Open Exploration, which further explores this example using the computer algebra systems* Maple, Mathcad, Mathematica, *and* Derive.

Mathematical Models

Real-life applications of mathematics often use equations as **mathematical models.** In developing a mathematical model to represent actual data, you should strive for two (often conflicting) goals—accuracy and simplicity. That is, you want the model to be simple enough to be workable, yet accurate enough to produce meaningful results. Section 1.4 explores these goals more completely.

EXAMPLE 6 Comparing Two Mathematical Models

Gavriel Jecan/Corbis

The Mauna Loa Observatory in Hawaii has been measuring the increasing concentration of carbon dioxide in Earth's atmosphere since 1958.

The Mauna Loa Observatory in Hawaii records the carbon dioxide concentration y (in parts per million) in Earth's atmosphere. The January readings for various years are shown in Figure 1.11. In the July 1990 issue of *Scientific American*, these data were used to predict the carbon dioxide level in Earth's atmosphere in the year 2035. The article used the quadratic model

$$y = 316.2 + 0.70t + 0.018t^2 \qquad \text{Quadratic model for 1960–1990 data}$$

where $t = 0$ represents 1960, as shown in Figure 1.11(a).

The data shown in Figure 1.11(b) represent the years 1980 through 2002 and can be modeled by

$$y = 306.3 + 1.56t \qquad \text{Linear model for 1980–2002 data}$$

where $t = 0$ represents 1960. What was the prediction given in the *Scientific American* article in 1990? Given the new data for 1990 through 2002, does this prediction for the year 2035 seem accurate?

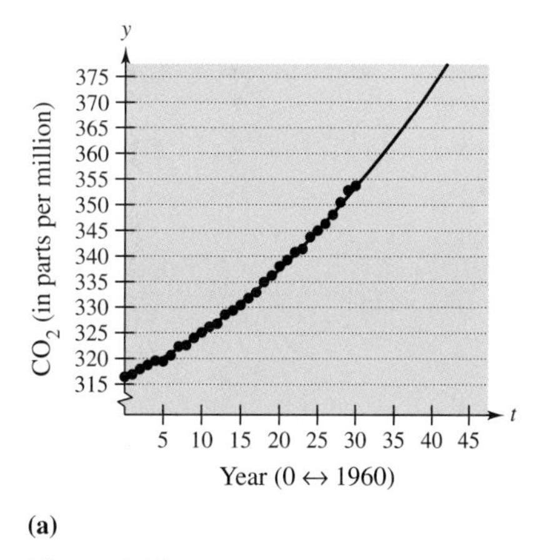

(a)

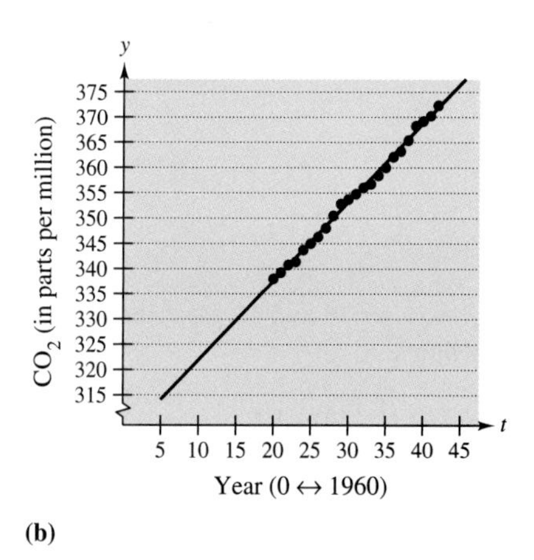

(b)

Figure 1.11

Solution To answer the first question, substitute $t = 75$ (for 2035) into the quadratic model.

$$y = 316.2 + 0.70(75) + 0.018(75)^2 = 469.95 \qquad \text{Quadratic model}$$

So, the prediction in the *Scientific American* article was that the carbon dioxide concentration in Earth's atmosphere would reach about 470 parts per million in the year 2035. Using the linear model for the 1980–2002 data, the prediction for the year 2035 is

$$y = 306.3 + 1.56(75) = 423.3. \qquad \text{Linear model}$$

So, based on the linear model for 1980–2002, it appears that the 1990 prediction was too high.

NOTE The models in Example 6 were developed using a procedure called *least squares regression* (see Section 13.9). The quadratic and linear models have correlations given by $r^2 = 0.997$ and $r^2 = 0.996$, respectively. The closer r^2 is to 1, the "better" the model.

Exercises for Section 1.1

See www.CalcChat.com for worked-out solutions to odd-numbered exercises.

In Exercises 1–4, match the equation with its graph. [Graphs are labeled (a), (b), (c), and (d).]

(a)

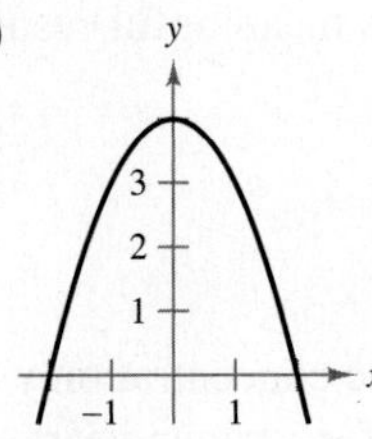

(b)

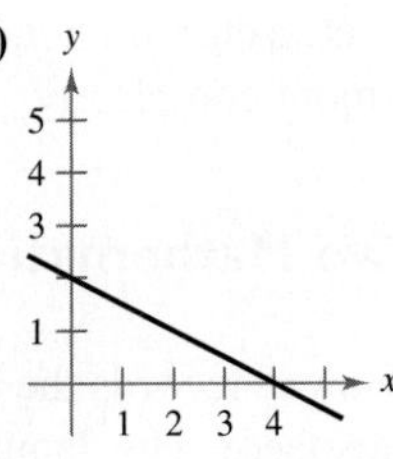

(c)

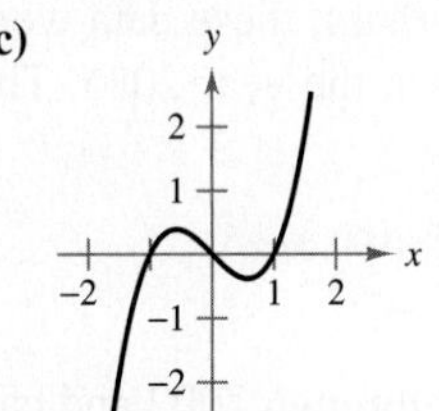

(d)

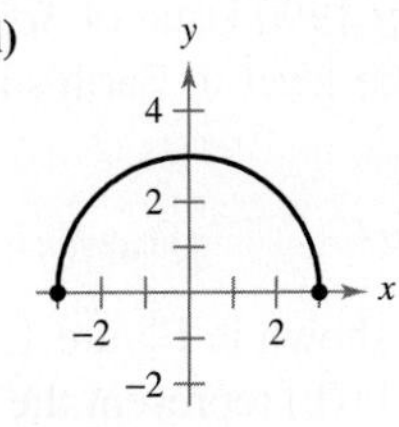

1. $y = -\frac{1}{2}x + 2$
2. $y = \sqrt{9 - x^2}$
3. $y = 4 - x^2$
4. $y = x^3 - x$

In Exercises 5–14, sketch the graph of the equation by point plotting.

5. $y = \frac{3}{2}x + 1$
6. $y = 6 - 2x$
7. $y = 4 - x^2$
8. $y = (x - 3)^2$
9. $y = |x + 2|$
10. $y = |x| - 1$
11. $y = \sqrt{x} - 4$
12. $y = \sqrt{x + 2}$
13. $y = \dfrac{2}{x}$
14. $y = \dfrac{1}{x - 1}$

 **In Exercises 15 and 16, describe the viewing window that yields the figure.**

15. $y = x^3 - 3x^2 + 4$

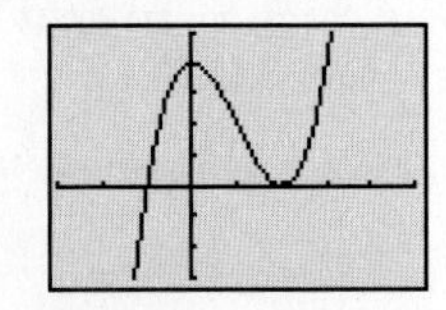

16. $y = |x| + |x - 10|$

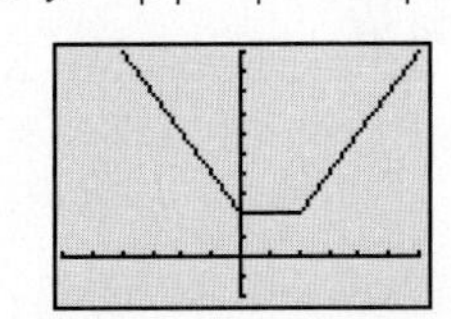

In Exercises 17 and 18, use a graphing utility to graph the equation. Move the cursor along the curve to approximate the unknown coordinate of each solution point accurate to two decimal places.

17. $y = \sqrt{5 - x}$ (a) $(2, y)$ (b) $(x, 3)$
18. $y = x^5 - 5x$ (a) $(-0.5, y)$ (b) $(x, -4)$

In Exercises 19–26, find any intercepts.

19. $y = x^2 + x - 2$
20. $y^2 = x^3 - 4x$
21. $y = x^2\sqrt{25 - x^2}$
22. $y = (x - 1)\sqrt{x^2 + 1}$
23. $y = \dfrac{3(2 - \sqrt{x})}{x}$
24. $y = \dfrac{x^2 + 3x}{(3x + 1)^2}$
25. $x^2y - x^2 + 4y = 0$
26. $y = 2x - \sqrt{x^2 + 1}$

In Exercises 27–38, test for symmetry with respect to each axis and to the origin.

27. $y = x^2 - 2$
28. $y = x^2 - x$
29. $y^2 = x^3 - 4x$
30. $y = x^3 + x$
31. $xy = 4$
32. $xy^2 = -10$
33. $y = 4 - \sqrt{x + 3}$
34. $xy - \sqrt{4 - x^2} = 0$
35. $y = \dfrac{x}{x^2 + 1}$
36. $y = \dfrac{x^2}{x^2 + 1}$
37. $y = |x^3 + x|$
38. $|y| - x = 3$

In Exercises 39–56, sketch the graph of the equation. Identify any intercepts and test for symmetry.

39. $y = -3x + 2$
40. $y = -\frac{1}{2}x + 2$
41. $y = \frac{1}{2}x - 4$
42. $y = \frac{2}{3}x + 1$
43. $y = 1 - x^2$
44. $y = x^2 + 3$
45. $y = (x + 3)^2$
46. $y = 2x^2 + x$
47. $y = x^3 + 2$
48. $y = x^3 - 4x$
49. $y = x\sqrt{x + 2}$
50. $y = \sqrt{9 - x^2}$
51. $x = y^3$
52. $x = y^2 - 4$
53. $y = \dfrac{1}{x}$
54. $y = \dfrac{10}{x^2 + 1}$
55. $y = 6 - |x|$
56. $y = |6 - x|$

In Exercises 57–60, use a graphing utility to graph the equation. Identify any intercepts and test for symmetry.

57. $y^2 - x = 9$
58. $x^2 + 4y^2 = 4$
59. $x + 3y^2 = 6$
60. $3x - 4y^2 = 8$

In Exercises 61–68, find the point(s) of intersection of the graphs of the equations.

61. $x + y = 2$
    $2x - y = 1$
62. $2x - 3y = 13$
    $5x + 3y = 1$
63. $x^2 + y = 6$
    $x + y = 4$
64. $x = 3 - y^2$
    $y = x - 1$

The symbol indicates an exercise in which you are instructed to use graphing technology or a symbolic computer algebra system. The solutions of other exercises may also be facilitated by use of appropriate technology.

65. $x^2 + y^2 = 5$
$x - y = 1$

66. $x^2 + y^2 = 25$
$2x + y = 10$

67. $y = x^3$
$y = x$

68. $y = x^3 - 4x$
$y = -(x + 2)$

 **In Exercises 69–72, use a graphing utility to find the point(s) of intersection of the graphs. Check your results analytically.**

69. $y = x^3 - 2x^2 + x - 1$
$y = -x^2 + 3x - 1$

70. $y = x^4 - 2x^2 + 1$
$y = 1 - x^2$

71. $y = \sqrt{x + 6}$
$y = \sqrt{-x^2 - 4x}$

72. $y = -|2x - 3| + 6$
$y = 6 - x$

 **73.** ***Modeling Data*** The table shows the Consumer Price Index (CPI) for selected years. *(Source: Bureau of Labor Statistics)*

Year	1970	1975	1980	1985	1990	1995	2000
CPI	38.8	53.8	82.4	107.6	130.7	152.4	172.2

(a) Use the regression capabilities of a graphing utility to find a mathematical model of the form $y = at^2 + bt + c$ for the data. In the model, y represents the CPI and t represents the year, with $t = 0$ corresponding to 1970.

(b) Use a graphing utility to plot the data and graph the model. Compare the data with the model.

(c) Use the model to predict the CPI for the year 2010.

 **74.** ***Modeling Data*** The table shows the average number of acres per farm in the United States for selected years. *(Source: U.S. Department of Agriculture)*

Year	1950	1960	1970	1980	1990	2000
Acreage	213	297	374	426	460	434

(a) Use the regression capabilities of a graphing utility to find a mathematical model of the form $y = at^2 + bt + c$ for the data. In the model, y represents the average acreage and t represents the year, with $t = 0$ corresponding to 1950.

(b) Use a graphing utility to plot the data and graph the model. Compare the data with the model.

(c) Use the model to predict the average number of acres per farm in the United States in the year 2010.

75. ***Break-Even Point*** Find the sales necessary to break even ($R = C$) if the cost C of producing x units is

$C = 5.5\sqrt{x} + 10{,}000$ Cost equation

and the revenue R for selling x units is

$R = 3.29x.$ Revenue equation

 **76.** ***Copper Wire*** The resistance y in ohms of 1000 feet of solid copper wire at 77°F can be approximated by the model

$$y = \frac{10{,}770}{x^2} - 0.37, \quad 5 \le x \le 100$$

where x is the diameter of the wire in mils (0.001 inch). Use a graphing utility to graph the model. If the diameter of the wire is doubled, the resistance is changed by about what factor?

Writing About Concepts

In Exercises 77 and 78, write an equation whose graph has the indicated property. (There may be more than one correct answer.)

77. The graph has intercepts at $x = -2$, $x = 4$, and $x = 6$.

78. The graph has intercepts at $x = -\frac{5}{2}$, $x = 2$, and $x = \frac{3}{2}$.

79. Each table shows solution points for one of the following equations.

(i) $y = kx + 5$ (ii) $y = x^2 + k$

(iii) $y = kx^{3/2}$ (iv) $xy = k$

Match each equation with the correct table and find k. Explain your reasoning.

(a)

x	1	4	9
y	3	24	81

(b)

x	1	4	9
y	7	13	23

(c)

x	1	4	9
y	36	9	4

(d)

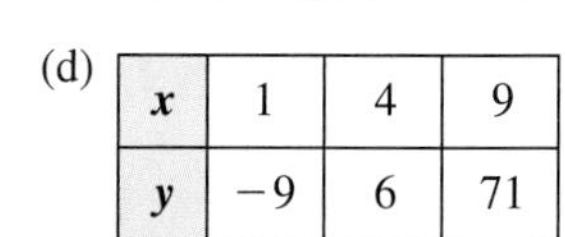

x	1	4	9
y	−9	6	71

80. (a) Prove that if a graph is symmetric with respect to the x-axis and to the y-axis, then it is symmetric with respect to the origin. Give an example to show that the converse is not true.

(b) Prove that if a graph is symmetric with respect to one axis and to the origin, then it is symmetric with respect to the other axis.

True or False? **In Exercises 81–84, determine whether the statement is true or false. If it is false, explain why or give an example that shows it is false.**

81. If $(1, -2)$ is a point on a graph that is symmetric with respect to the x-axis, then $(-1, -2)$ is also a point on the graph.

82. If $(1, -2)$ is a point on a graph that is symmetric with respect to the y-axis, then $(-1, -2)$ is also a point on the graph.

83. If $b^2 - 4ac > 0$ and $a \ne 0$, then the graph of $y = ax^2 + bx + c$ has two x-intercepts.

84. If $b^2 - 4ac = 0$ and $a \ne 0$, then the graph of $y = ax^2 + bx + c$ has only one x-intercept.

In Exercises 85 and 86, find an equation of the graph that consists of all points (x, y) having the given distance from the origin. (For a review of the Distance Formula, see Appendix D.)

85. The distance from the origin is twice the distance from $(0, 3)$.

86. The distance from the origin is $K (K \ne 1)$ times the distance from $(2, 0)$.

Section 1.2 Linear Models and Rates of Change

- **Find the slope of a line passing through two points.**
- **Write the equation of a line given a point and the slope.**
- **Interpret slope as a ratio or as a rate in a real-life application.**
- **Sketch the graph of a linear equation in slope-intercept form.**
- **Write equations of lines that are parallel or perpendicular to a given line.**

The Slope of a Line

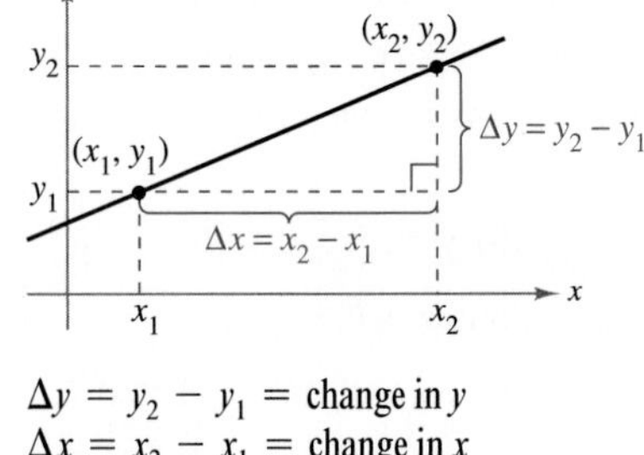

$\Delta y = y_2 - y_1 =$ change in y
$\Delta x = x_2 - x_1 =$ change in x
Figure 1.12

The **slope** of a nonvertical line is a measure of the number of units the line rises (or falls) vertically for each unit of horizontal change from left to right. Consider the two points (x_1, y_1) and (x_2, y_2) on the line in Figure 1.12. As you move from left to right along this line, a vertical change of

$$\Delta y = y_2 - y_1 \qquad \text{Change in } y$$

units corresponds to a horizontal change of

$$\Delta x = x_2 - x_1 \qquad \text{Change in } x$$

units. (Δ is the Greek uppercase letter *delta*, and the symbols Δy and Δx are read "delta y" and "delta x.")

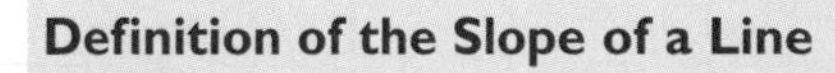

Definition of the Slope of a Line

The **slope** m of the nonvertical line passing through (x_1, y_1) and (x_2, y_2) is

$$m = \frac{\Delta y}{\Delta x} = \frac{y_2 - y_1}{x_2 - x_1}, \qquad x_1 \neq x_2.$$

Slope is not defined for vertical lines.

NOTE When using the formula for slope, note that

$$\frac{y_2 - y_1}{x_2 - x_1} = \frac{-(y_1 - y_2)}{-(x_1 - x_2)} = \frac{y_1 - y_2}{x_1 - x_2}.$$

So, it does not matter in which order you subtract *as long as* you are consistent and both "subtracted coordinates" come from the same point.

Figure 1.13 shows four lines: one has a positive slope, one has a slope of zero, one has a negative slope, and one has an "undefined" slope. In general, the greater the absolute value of the slope of a line, the steeper the line is. For instance, in Figure 1.13, the line with a slope of -5 is steeper than the line with a slope of $\frac{1}{5}$.

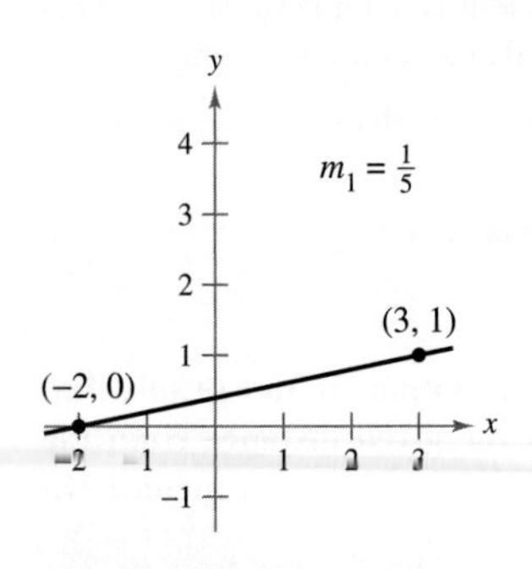

If m is positive, then the line rises from left to right.
Figure 1.13

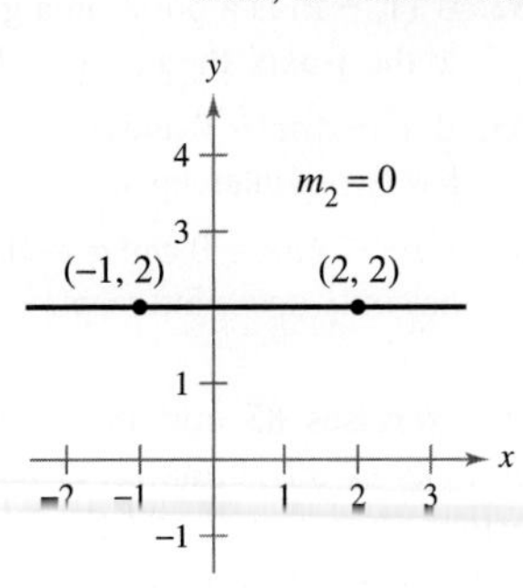

If m is zero, then the line is horizontal.

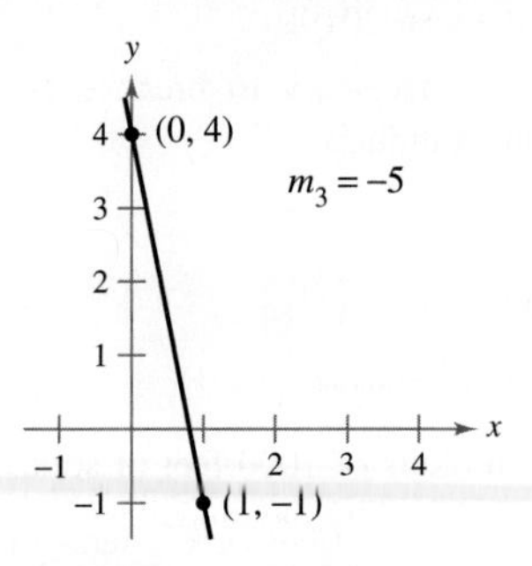

If m is negative, then the line falls from left to right.

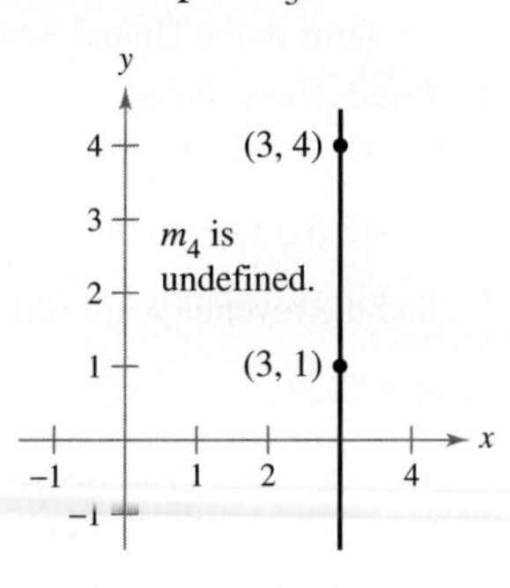

If m is undefined, then the line is vertical.

Equations of Lines

Any two points on a nonvertical line can be used to calculate its slope. This can be verified from the similar triangles shown in Figure 1.14. (Recall that the ratios of corresponding sides of similar triangles are equal.)

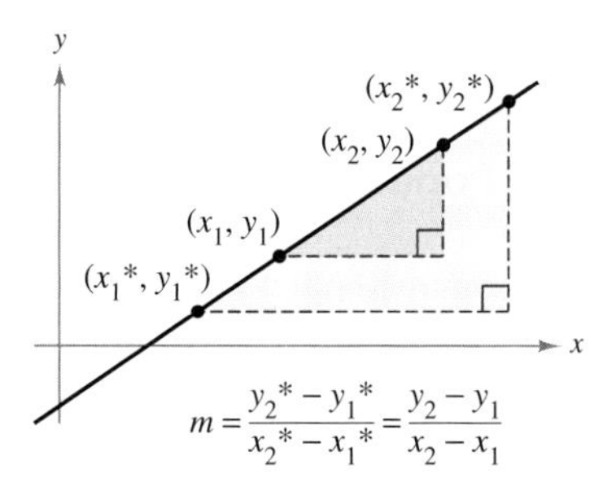

Any two points on a nonvertical line can be used to determine its slope.
Figure 1.14

EXPLORATION

Investigating Equations of Lines Use a graphing utility to graph each of the linear equations. Which point is common to all seven lines? Which value in the equation determines the slope of each line?

a. $y - 4 = -2(x + 1)$
b. $y - 4 = -1(x + 1)$
c. $y - 4 = -\frac{1}{2}(x + 1)$
d. $y - 4 = 0(x + 1)$
e. $y - 4 = \frac{1}{2}(x + 1)$
f. $y - 4 = 1(x + 1)$
g. $y - 4 = 2(x + 1)$

Use your results to write an equation of the line passing through $(-1, 4)$ with a slope of m.

You can write an equation of a nonvertical line if you know the slope of the line and the coordinates of one point on the line. Suppose the slope is m and the point is (x_1, y_1). If (x, y) is any other point on the line, then

$$\frac{y - y_1}{x - x_1} = m.$$

This equation, involving the two variables x and y, can be rewritten in the form

$$y - y_1 = m(x - x_1)$$

which is called the **point-slope equation of a line.**

> **Point-Slope Equation of a Line**
>
> An equation of the line with slope m passing through the point (x_1, y_1) is given by $y - y_1 = m(x - x_1)$.

EXAMPLE 1 Finding an Equation of a Line

Find an equation of the line that has a slope of 3 and passes through the point $(1, -2)$.

Solution

$y - y_1 = m(x - x_1)$	Point-slope form
$y - (-2) = 3(x - 1)$	Substitute -2 for y_1, 1 for x_1, and 3 for m.
$y + 2 = 3x - 3$	Simplify.
$y = 3x - 5$	Solve for y.

(See Figure 1.15.)

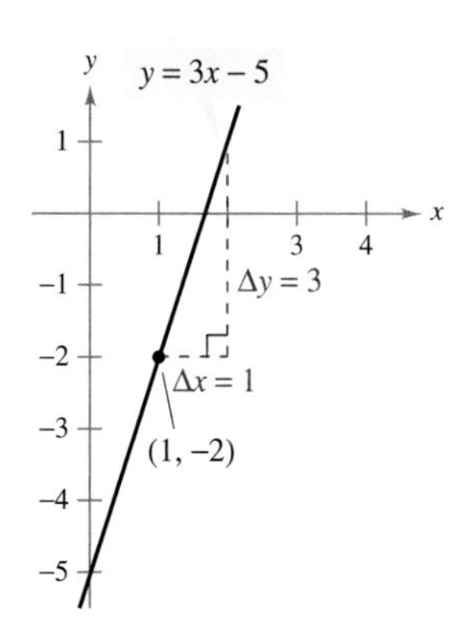

The line with a slope of 3 passing through the point $(1, -2)$
Figure 1.15

NOTE Remember that only nonvertical lines have a slope. Vertical lines, on the other hand, cannot be written in point-slope form. For instance, the equation of the vertical line passing through the point $(1, -2)$ is $x = 1$.

Ratios and Rates of Change

The slope of a line can be interpreted as either a *ratio* or a *rate*. If the x- and y-axes have the same unit of measure, the slope has no units and is a **ratio.** If the x- and y-axes have different units of measure, the slope is a rate or **rate of change.** In your study of calculus, you will encounter applications involving both interpretations of slope.

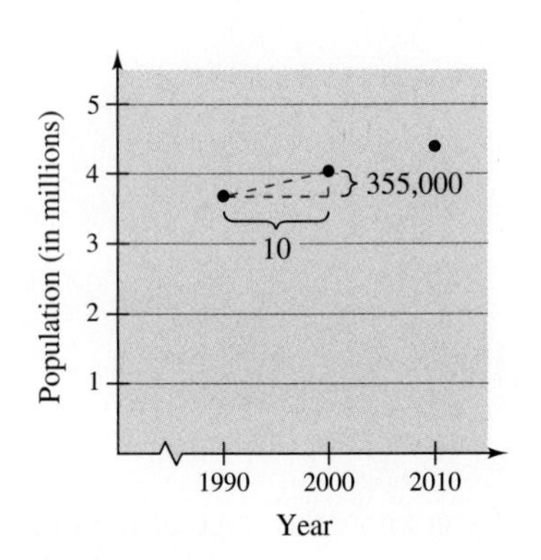

Population of Kentucky in census years
Figure 1.16

EXAMPLE 2 Population Growth and Engineering Design

a. The population of Kentucky was 3,687,000 in 1990 and 4,042,000 in 2000. Over this 10-year period, the average rate of change of the population was

$$\begin{aligned}\text{Rate of change} &= \frac{\text{change in population}}{\text{change in years}}\\ &= \frac{4{,}042{,}000 - 3{,}687{,}000}{2000 - 1990}\\ &= 35{,}500 \text{ people per year.}\end{aligned}$$

If Kentucky's population continues to increase at this rate for the next 10 years, it will have a population of 4,397,000 in 2010 (see Figure 1.16). *(Source: U.S. Census Bureau)*

b. In tournament water-ski jumping, the ramp rises to a height of 6 feet on a raft that is 21 feet long, as shown in Figure 1.17. The slope of the ski ramp is the ratio of its height (the rise) to the length of its base (the run).

$$\begin{aligned}\text{Slope of ramp} &= \frac{\text{rise}}{\text{run}} && \text{Rise is vertical change, run is horizontal change.}\\ &= \frac{6 \text{ feet}}{21 \text{ feet}}\\ &= \frac{2}{7}\end{aligned}$$

In this case, note that the slope is a ratio and has no units.

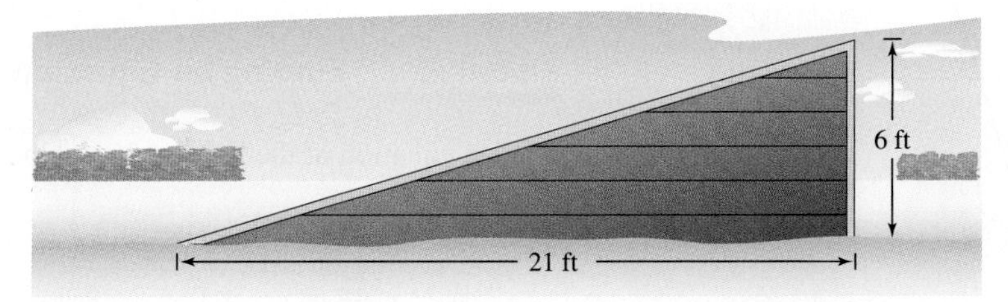

Dimensions of a water-ski ramp
Figure 1.17

The rate of change found in Example 2(a) is an **average rate of change.** An average rate of change is always calculated over an interval. In this case, the interval is [1990, 2000]. In Chapter 3 you will study another type of rate of change called an *instantaneous rate of change.*

Graphing Linear Models

Many problems in analytic geometry can be classified into two basic categories: (1) Given a graph, what is its equation? and (2) Given an equation, what is its graph? The point-slope equation of a line can be used to solve problems in the first category. However, this form is not especially useful for solving problems in the second category. The form that is better suited to sketching the graph of a line is the **slope-intercept** form of the equation of a line.

> **The Slope-Intercept Equation of a Line**
>
> The graph of the linear equation
>
> $$y = mx + b$$
>
> is a line having a *slope* of m and a *y-intercept* at $(0, b)$.

EXAMPLE 3 Sketching Lines in the Plane

Sketch the graph of each equation.

a. $y = 2x + 1$ **b.** $y = 2$ **c.** $3y + x - 6 = 0$

Solution

a. Because $b = 1$, the y-intercept is $(0, 1)$. Because the slope is $m = 2$, you know that the line rises two units for each unit it moves to the right, as shown in Figure 1.18(a).

b. Because $b = 2$, the y-intercept is $(0, 2)$. Because the slope is $m = 0$, you know that the line is horizontal, as shown in Figure 1.18(b).

c. Begin by writing the equation in slope-intercept form.

$$3y + x - 6 = 0 \quad \text{Write original equation.}$$

$$3y = -x + 6 \quad \text{Isolate } y\text{-term on the left.}$$

$$y = -\frac{1}{3}x + 2 \quad \text{Slope-intercept form}$$

In this form, you can see that the y-intercept is $(0, 2)$ and the slope is $m = -\frac{1}{3}$. This means that the line falls one unit for every three units it moves to the right, as shown in Figure 1.18(c).

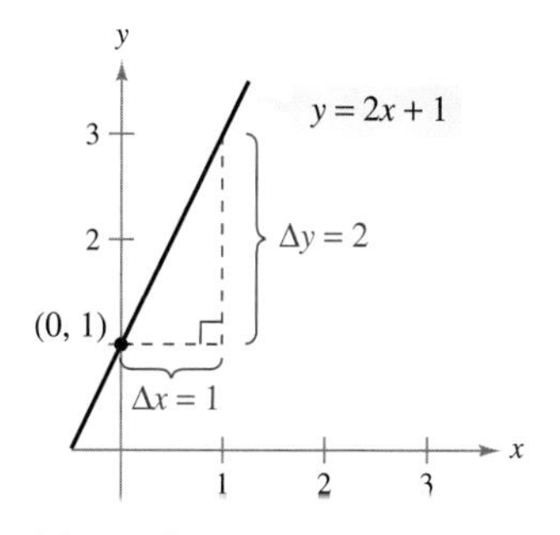

(a) $m = 2$; line rises

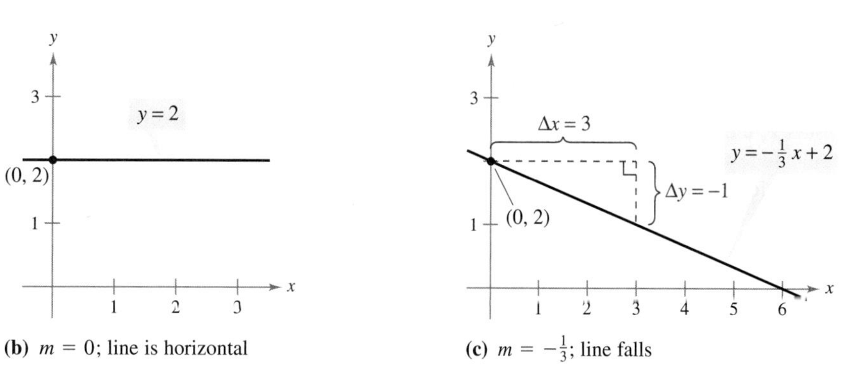

(b) $m = 0$; line is horizontal

(c) $m = -\frac{1}{3}$; line falls

Figure 1.18

Because the slope of a vertical line is not defined, its equation cannot be written in the slope-intercept form. However, the equation of *any* line can be written in the **general form**

$$Ax + By + C = 0$$ General form of the equation of a line

where A and B are not *both* zero. For instance, the vertical line given by $x = a$ can be represented by the general form $x - a = 0$.

Summary of Equations of Lines

1. General form: $Ax + By + C = 0$
2. Vertical line: $x = a$
3. Horizontal line: $y = b$
4. Point-slope form: $y - y_1 = m(x - x_1)$
5. Slope-intercept form: $y = mx + b$

Parallel and Perpendicular Lines

The slope of a line is a convenient tool for determining whether two lines are parallel or perpendicular, as shown in Figure 1.19. Specifically, nonvertical lines with the same slope are parallel and nonvertical lines whose slopes are negative reciprocals are perpendicular.

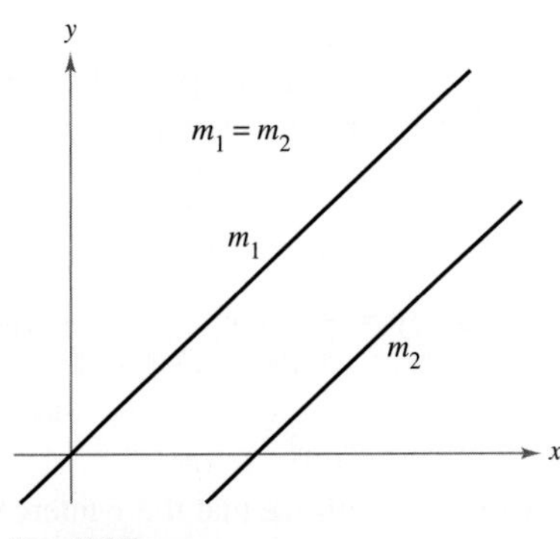

Parallel lines

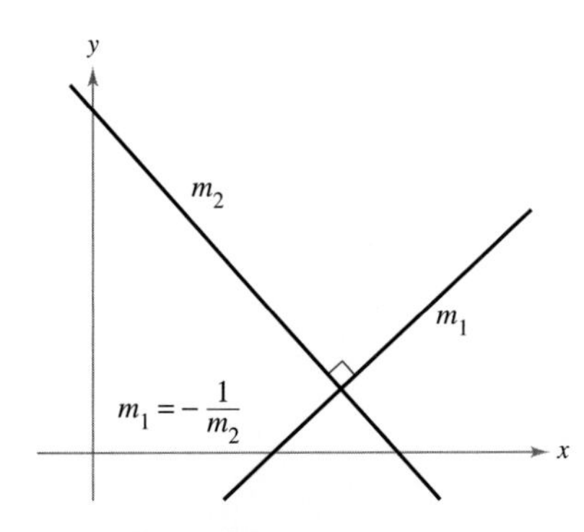

Perpendicular lines

Figure 1.19

STUDY TIP In mathematics, the phrase "if and only if" is a way of stating two implications in one statement. For instance, the first statement at the right could be rewritten as the following two implications.

a. If two distinct nonvertical lines are parallel, then their slopes are equal.

b. If two distinct nonvertical lines have equal slopes, then they are parallel.

Parallel and Perpendicular Lines

1. Two distinct nonvertical lines are **parallel** if and only if their slopes are equal—that is, if and only if

$$m_1 = m_2.$$

2. Two nonvertical lines are **perpendicular** if and only if their slopes are negative reciprocals of each other—that is, if and only if

$$m_1 = -\frac{1}{m_2}.$$

EXAMPLE 4 Finding Parallel and Perpendicular Lines

Find the general form of the equation of the line that passes through the point $(2, -1)$ and is

a. parallel to the line $2x - 3y = 5$. **b.** perpendicular to the line $2x - 3y = 5$.

(See Figure 1.20.)

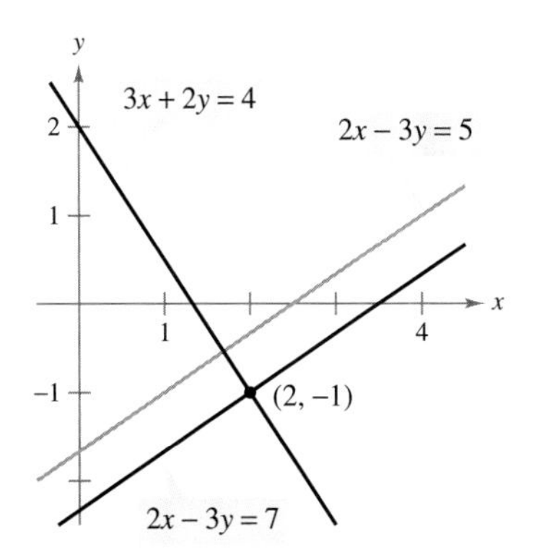

Lines parallel and perpendicular to $2x - 3y = 5$
Figure 1.20

Solution By writing the linear equation $2x - 3y = 5$ in slope-intercept form, $y = \frac{2}{3}x - \frac{5}{3}$, you can see that the given line has a slope of $m = \frac{2}{3}$.

a. The line through $(2, -1)$ that is parallel to the given line also has a slope of $\frac{2}{3}$.

$$y - y_1 = m(x - x_1) \qquad \text{Point-slope form}$$
$$y - (-1) = \tfrac{2}{3}(x - 2) \qquad \text{Substitute.}$$
$$3(y + 1) = 2(x - 2) \qquad \text{Simplify.}$$
$$2x - 3y - 7 = 0 \qquad \text{General form}$$

Note the similarity to the original equation.

b. Using the negative reciprocal of the slope of the given line, you can determine that the slope of a line perpendicular to the given line is $-\frac{3}{2}$. So, the line through the point $(2, -1)$ that is perpendicular to the given line has the following equation.

$$y - y_1 = m(x - x_1) \qquad \text{Point-slope form}$$
$$y - (-1) = -\tfrac{3}{2}(x - 2) \qquad \text{Substitute.}$$
$$2(y + 1) = -3(x - 2) \qquad \text{Simplify.}$$
$$3x + 2y - 4 = 0 \qquad \text{General form}$$

TECHNOLOGY PITFALL The slope of a line will appear distorted if you use different tick-mark spacing on the x- and y-axes. For instance, the graphing calculator screens in Figures 1.21(a) and 1.21(b) both show the lines given by

$$y = 2x \qquad \text{and} \qquad y = -\tfrac{1}{2}x + 3.$$

Because these lines have slopes that are negative reciprocals, they must be perpendicular. In Figure 1.21(a), however, the lines don't appear to be perpendicular because the tick-mark spacing on the x-axis is not the same as that on the y-axis. In Figure 1.21(b), the lines appear perpendicular because the tick-mark spacing on the x-axis is the same as that on the y-axis. This type of viewing window is said to have a *square setting*.

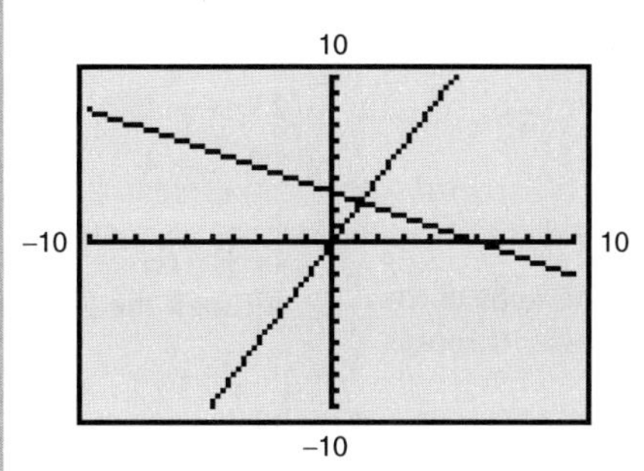

(a) Tick-mark spacing on the x-axis is not the same as tick-mark spacing on the y-axis.

6
−9
9
−6

(b) Tick-mark spacing on the x-axis is the same as tick-mark spacing on the y-axis.

Figure 1.21

Exercises for Section 1.2

See www.CalcChat.com for worked-out solutions to odd-numbered exercises.

In Exercises 1–6, estimate the slope of the line from its graph. To print an enlarged copy of the graph, go to the website *www.mathgraphs.com*.

1.

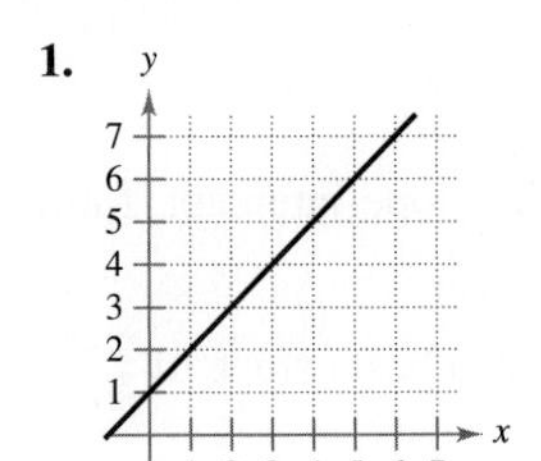

2.

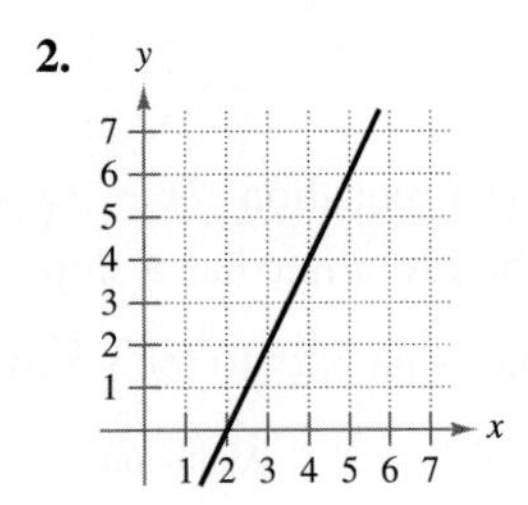

3.

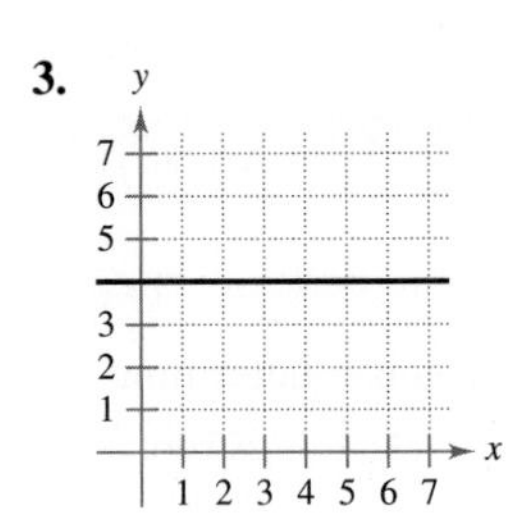

4.

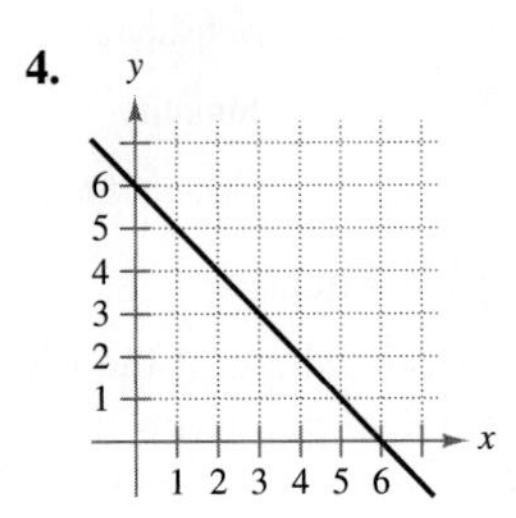

5.

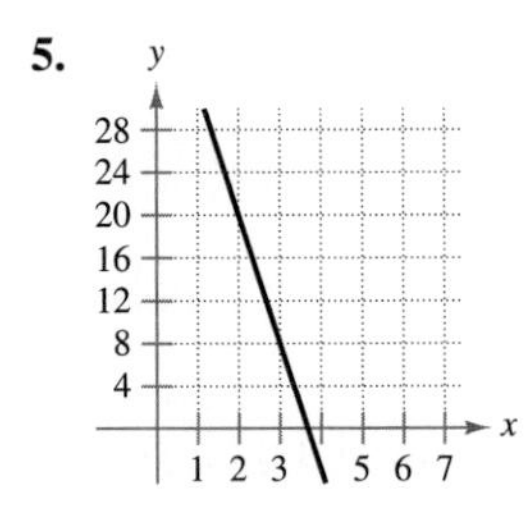

6.

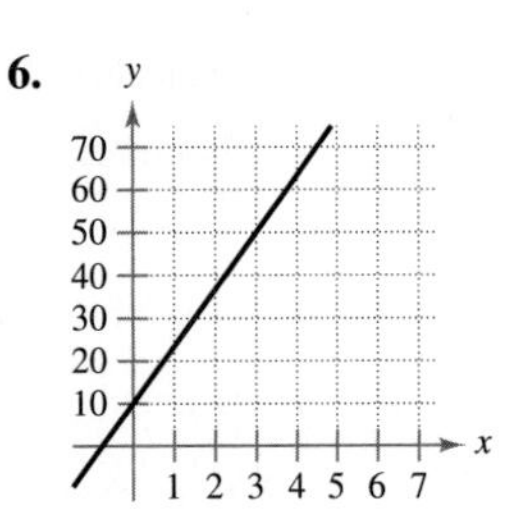

In Exercises 7 and 8, sketch the lines through the given point with the indicated slopes. Make the sketches on the same set of coordinate axes.

	Point	*Slopes*			
7.	$(2, 3)$	(a) 1	(b) -2	(c) $-\frac{3}{2}$	(d) Undefined
8.	$(-4, 1)$	(a) 3	(b) -3	(c) $\frac{1}{3}$	(d) 0

In Exercises 9–14, plot the points and find the slope of the line passing through them.

9. $(3, -4), (5, 2)$

10. $(1, 2), (-2, 4)$

11. $(2, 1), (2, 5)$

12. $(3, -2), (4, -2)$

13. $\left(-\frac{1}{2}, \frac{2}{3}\right), \left(-\frac{3}{4}, \frac{1}{6}\right)$

14. $\left(\frac{7}{8}, \frac{3}{4}\right), \left(\frac{5}{4}, -\frac{1}{4}\right)$

In Exercises 15–18, use the point on the line and the slope of the line to find three additional points that the line passes through. (There is more than one correct answer.)

	Point	*Slope*		*Point*	*Slope*
15.	$(2, 1)$	$m = 0$	**16.**	$(-3, 4)$	m undefined
17.	$(1, 7)$	$m = -3$	**18.**	$(-2, -2)$	$m = 2$

19. ***Conveyor Design*** A moving conveyor is built to rise 1 meter for each 3 meters of horizontal change.

(a) Find the slope of the conveyor.

(b) Suppose the conveyor runs between two floors in a factory. Find the length of the conveyor if the vertical distance between floors is 10 feet.

20. ***Rate of Change*** Each of the following is the slope of a line representing daily revenue y in terms of time x in days. Use the slope to interpret any change in daily revenue for a one-day increase in time.

(a) $m = 400$ (b) $m = 100$ (c) $m = 0$

21. ***Modeling Data*** The table shows the population y (in millions) of the United States for 1996–2001. The variable t represents the time in years, with $t = 6$ corresponding to 1996. *(Source: U.S. Bureau of the Census)*

t	6	7	8	9	10	11
y	269.7	272.9	276.1	279.3	282.3	285.0

(a) Plot the data by hand and connect adjacent points with a line segment.

(b) Use the slope of each line segment to determine the year when the population increased least rapidly.

22. ***Modeling Data*** The table shows the rate r (in miles per hour) that a vehicle is traveling after t seconds.

t	5	10	15	20	25	30
r	57	74	85	84	61	43

(a) Plot the data by hand and connect adjacent points with a line segment.

(b) Use the slope of each line segment to determine the interval when the vehicle's rate changed most rapidly. How did the rate change?

In Exercises 23–26, find the slope and the y-intercept (if possible) of the line.

23. $x + 5y = 20$

24. $6x - 5y = 15$

25. $x = 4$

26. $y = -1$

In Exercises 27–32, find an equation of the line that passes through the point and has the indicated slope. Sketch the line.

	Point	*Slope*		*Point*	*Slope*
27.	$(0, 3)$	$m = \frac{3}{4}$	**28.**	$(-1, 2)$	m undefined
29.	$(0, 0)$	$m = \frac{2}{3}$	**30.**	$(0, 4)$	$m = 0$
31.	$(3, -2)$	$m = 3$	**32.**	$(-2, 4)$	$m = -\frac{3}{5}$

In Exercises 33–42, find an equation of the line that passes through the points, and sketch the line.

33. $(0, 0), (2, 6)$
34. $(0, 0), (-1, 3)$
35. $(2, 1), (0, -3)$
36. $(-3, -4), (1, 4)$
37. $(2, 8), (5, 0)$
38. $(-3, 6), (1, 2)$
39. $(5, 1), (5, 8)$
40. $(1, -2), (3, -2)$
41. $\left(\frac{1}{2}, \frac{7}{2}\right), \left(0, \frac{3}{4}\right)$
42. $\left(\frac{7}{8}, \frac{3}{4}\right), \left(\frac{5}{4}, -\frac{1}{4}\right)$

43. Find an equation of the vertical line with x-intercept 3.

44. Show that the line with intercepts $(a, 0)$ and $(0, b)$ has the following equation.

$$\frac{x}{a} + \frac{y}{b} = 1, \qquad a \neq 0, b \neq 0$$

In Exercises 45–48, use the result of Exercise 44 to write an equation of the line.

45. x-intercept: $(2, 0)$
y-intercept: $(0, 3)$

46. x-intercept: $\left(-\frac{2}{3}, 0\right)$
y-intercept: $(0, -2)$

47. Point on line: $(1, 2)$
x-intercept: $(a, 0)$
y-intercept: $(0, a)$
$(a \neq 0)$

48. Point on line: $(-3, 4)$
x-intercept: $(a, 0)$
y-intercept: $(0, a)$
$(a \neq 0)$

In Exercises 49–56, sketch a graph of the equation.

49. $y = -3$
50. $x = 4$
51. $y = -2x + 1$
52. $y = \frac{1}{3}x - 1$
53. $y - 2 = \frac{3}{2}(x - 1)$
54. $y - 1 = 3(x + 4)$
55. $2x - y - 3 = 0$
56. $x + 2y + 6 = 0$

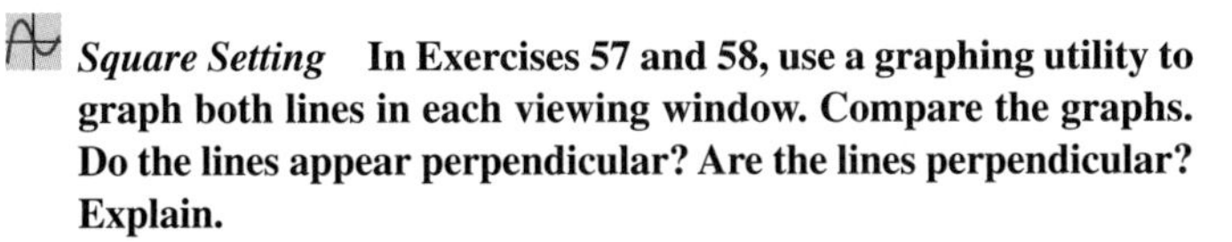

Square Setting **In Exercises 57 and 58, use a graphing utility to graph both lines in each viewing window. Compare the graphs. Do the lines appear perpendicular? Are the lines perpendicular? Explain.**

57. $y = x + 6,\ y = -x + 2$

(a)
Xmin = -10
Xmax = 10
Xscl = 1
Ymin = -10
Ymax = 10
Yscl = 1

(b)
Xmin = -15
Xmax = 15
Xscl = 1
Ymin = -10
Ymax = 10
Yscl = 1

58. $y = 2x - 3,\ y = -\frac{1}{2}x + 1$

(a)
Xmin = -5
Xmax = 5
Xscl = 1
Ymin = -5
Ymax = 5
Yscl = 1

(b)
Xmin = -6
Xmax = 6
Xscl = 1
Ymin = -4
Ymax = 4
Yscl = 1

In Exercises 59–64, write an equation of the line through the point (a) parallel to the given line and (b) perpendicular to the given line.

	Point	*Line*		*Point*	*Line*
59.	$(2, 1)$	$4x - 2y = 3$	**60.**	$(-3, 2)$	$x + y = 7$
61.	$\left(\frac{3}{4}, \frac{7}{8}\right)$	$5x - 3y = 0$	**62.**	$(-6, 4)$	$3x + 4y = 7$
63.	$(2, 5)$	$x = 4$	**64.**	$(-1, 0)$	$y = -3$

Rate of Change **In Exercises 65–68, you are given the dollar value of a product in 2004 *and* the rate at which the value of the product is expected to change during the next 5 years. Write a linear equation that gives the dollar value V of the product in terms of the year t. (Let $t = 0$ represent 2000.)**

	2004 Value	*Rate*
65.	\$2540	\$125 increase per year
66.	\$156	\$4.50 increase per year
67.	\$20,400	\$2000 decrease per year
68.	\$245,000	\$5600 decrease per year

In Exercises 69 and 70, use a graphing utility to graph the parabolas and find their points of intersection. Find an equation of the line through the points of intersection and graph the line in the same viewing window.

69. $y = x^2$
$y = 4x - x^2$

70. $y = x^2 - 4x + 3$
$y = -x^2 + 2x + 3$

In Exercises 71 and 72, determine whether the points are collinear. (Three points are *collinear* if they lie on the same line.)

71. $(-2, 1), (-1, 0), (2, -2)$
72. $(0, 4), (7, -6), (-5, 11)$

Writing About Concepts

In Exercises 73–75, find the coordinates of the point of intersection of the given segments. Explain your reasoning.

73.

(b, c)
$(-a, 0)$ $(a, 0)$

Perpendicular bisectors

74.

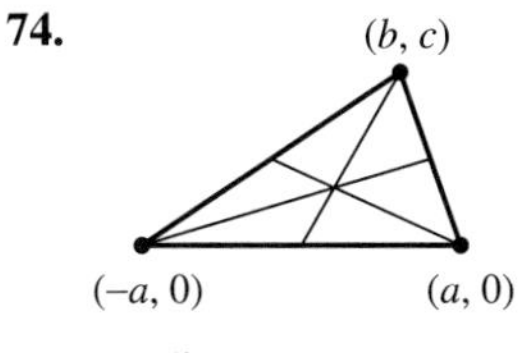

Medians

75.

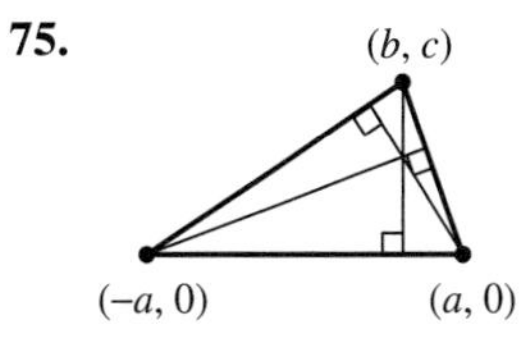

Altitudes

76. Show that the points of intersection in Exercises 73, 74, and 75 are collinear.

77. ***Temperature Conversion*** Find a linear equation that expresses the relationship between the temperature in degrees Celsius C and the temperature in degrees Fahrenheit F. Use the fact that water freezes at 0°C (32°F) and boils at 100°C (212°F). Use the equation to convert 72°F to degrees Celsius.

78. ***Reimbursed Expenses*** A company reimburses its sales representatives \$150 per day for lodging and meals plus 34¢ per mile driven. Write a linear equation giving the daily cost C to the company in terms of x, the number of miles driven. How much does it cost the company if a sales representative drives 137 miles on a given day?

79. ***Career Choice*** An employee has two options for positions in a large corporation. One position pays \$12.50 per hour *plus* an additional unit rate of \$0.75 per unit produced. The other pays \$9.20 per hour *plus* a unit rate of \$1.30.

(a) Find linear equations for the hourly wages W in terms of x, the number of units produced per hour, for each option.

(b) Use a graphing utility to graph the linear equations and find the point of intersection.

(c) Interpret the meaning of the point of intersection of the graphs in part (b). How would you use this information to select the correct option if the goal were to obtain the highest hourly wage?

80. ***Straight-Line Depreciation*** A small business purchases a piece of equipment for \$875. After 5 years, the equipment will be outdated, having no value.

(a) Write a linear equation giving the value y of the equipment in terms of the time x in years, $0 \le x \le 5$.

(b) Find the value of the equipment when $x = 2$.

(c) Estimate (to two-decimal-place accuracy) the time when the value of the equipment is \$200.

81. ***Apartment Rental*** A real estate office handles an apartment complex with 50 units. When the rent is \$580 per month, all 50 units are occupied. However, when the rent is \$625, the average number of occupied units drops to 47. Assume that the relationship between the monthly rent p and the demand x is linear. (*Note:* The term *demand* refers to the number of occupied units.)

(a) Write a linear equation giving the demand x in terms of the rent p.

(b) *Linear extrapolation* Use a graphing utility to graph the demand equation and use the *trace* feature to predict the number of units occupied if the rent is raised to \$655.

(c) *Linear interpolation* Predict the number of units occupied if the rent is lowered to \$595. Verify graphically.

82. ***Modeling Data*** An instructor gives regular 20-point quizzes and 100-point exams in a mathematics course. Average scores for six students, given as ordered pairs (x, y) where x is the average quiz score and y is the average test score, are (18, 87), (10, 55), (19, 96), (16, 79), (13, 76), and (15, 82).

(a) Use the regression capabilities of a graphing utility to find the least-squares regression line for the data.

(b) Use a graphing utility to plot the points and graph the regression line in the same viewing window.

(c) Use the regression line to predict the average exam score for a student with an average quiz score of 17.

(d) Interpret the meaning of the slope of the regression line.

(e) The instructor adds 4 points to the average test score of everyone in the class. Describe the changes in the positions of the plotted points and the change in the equation of the line.

83. ***Tangent Line*** Find an equation of the line tangent to the circle $x^2 + y^2 = 169$ at the point (5, 12).

84. ***Tangent Line*** Find an equation of the line tangent to the circle $(x - 1)^2 + (y - 1)^2 = 25$ at the point $(4, -3)$.

Distance **In Exercises 85–90, find the distance between the point and the line, or between the lines, using the formula for the distance between the point (x_1, y_1) and the line $Ax + By + C = 0$:**

$$\textbf{Distance} = \frac{|Ax_1 + By_1 + C|}{\sqrt{A^2 + B^2}}.$$

85. Point: (0, 0)
Line: $4x + 3y = 10$

86. Point: (2, 3)
Line: $4x + 3y = 10$

87. Point: $(-2, 1)$
Line: $x - y - 2 = 0$

88. Point: (6, 2)
Line: $x = -1$

89. Line: $x + y = 1$
Line: $x + y = 5$

90. Line: $3x - 4y = 1$
Line: $3x - 4y = 10$

91. Show that the distance between the point (x_1, y_1) and the line $Ax + By + C = 0$ is

$$\text{Distance} = \frac{|Ax_1 + By_1 + C|}{\sqrt{A^2 + B^2}}.$$

92. Write the distance d between the point (3, 1) and the line $y = mx + 4$ in terms of m. Use a graphing utility to graph the equation. When is the distance 0? Explain the result geometrically.

93. Prove that the diagonals of a rhombus intersect at right angles. (A rhombus is a quadrilateral with sides of equal lengths.)

94. Prove that the figure formed by connecting consecutive midpoints of the sides of any quadrilateral is a parallelogram.

95. Prove that if the points (x_1, y_1) and (x_2, y_2) lie on the same line as (x_1^*, y_1^*) and (x_2^*, y_2^*), then

$$\frac{y_2^* - y_1^*}{x_2^* - x_1^*} = \frac{y_2 - y_1}{x_2 - x_1}.$$

Assume $x_1 \ne x_2$ and $x_1^* \ne x_2^*$.

96. Prove that if the slopes of two nonvertical lines are negative reciprocals of each other, then the lines are perpendicular.

True or False? **In Exercises 97 and 98, determine whether the statement is true or false. If it is false, explain why or give an example that shows it is false.**

97. The lines represented by $ax + by = c_1$ and $bx - ay = c_2$ are perpendicular. Assume $a \ne 0$ and $b \ne 0$.

98. It is possible for two lines with positive slopes to be perpendicular to each other.

Section 1.3 Functions and Their Graphs

- **Use function notation to represent and evaluate a function.**
- **Find the domain and range of a function.**
- **Sketch the graph of a function.**
- **Identify different types of transformations of functions.**
- **Classify functions and recognize combinations of functions.**

Functions and Function Notation

A **relation** between two sets X and Y is a set of ordered pairs, each of the form (x, y), where x is a member of X and y is a member of Y. A **function** from X to Y is a relation between X and Y having the property that any two ordered pairs with the same x-value also have the same y-value. The variable x is the **independent variable,** and the variable y is the **dependent variable.**

Many real-life situations can be modeled by functions. For instance, the area A of a circle is a function of the circle's radius r.

$A = \pi r^2$ A is a function of r.

In this case r is the independent variable and A is the dependent variable.

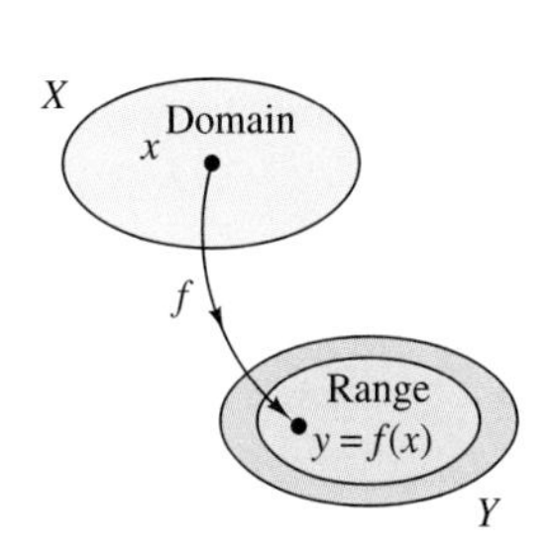

A real-valued function f of a real variable x
Figure 1.22

Definition of a Real-Valued Function of a Real Variable

Let X and Y be sets of real numbers. A **real-valued function f of a real variable x** from X to Y is a correspondence that assigns to each number x in X exactly one number y in Y.

The **domain** of f is the set X. The number y is the **image** of x under f and is denoted by $f(x)$, which is called the **value of f at x.** The **range** of f is a subset of Y and consists of all images of numbers in X (see Figure 1.22).

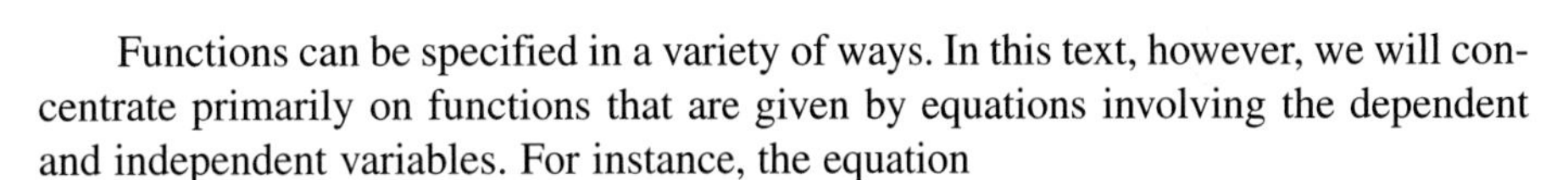

Functions can be specified in a variety of ways. In this text, however, we will concentrate primarily on functions that are given by equations involving the dependent and independent variables. For instance, the equation

$x^2 + 2y = 1$ Equation in implicit form

defines y, the dependent variable, as a function of x, the independent variable. To **evaluate** this function (that is, to find the y-value that corresponds to a given x-value), it is convenient to isolate y on the left side of the equation.

$y = \frac{1}{2}(1 - x^2)$ Equation in explicit form

Using f as the name of the function, you can write this equation as

$f(x) = \frac{1}{2}(1 - x^2).$ Function notation

The original equation, $x^2 + 2y = 1$, **implicitly** defines y as a function of x. When you solve the equation for y, you are writing the equation in **explicit** form.

Function notation has the advantage of clearly identifying the dependent variable as $f(x)$ while at the same time telling you that x is the independent variable and that the function itself is "f." The symbol $f(x)$ is read "f of x." Function notation allows you to be less wordy. Instead of asking "What is the value of y that corresponds to $x = 3$?" you can ask, "What is $f(3)$?"

Function Notation

The word *function* was first used by Gottfried Wilhelm Leibniz in 1694 as a term to denote any quantity connected with a curve, such as the coordinates of a point on a curve or the slope of a curve. Forty years later, Leonhard Euler used the word *function* to describe any expression made up of a variable and some constants. He introduced the notation $y = f(x)$.

In an equation that defines a function, the role of the variable x is simply that of a placeholder. For instance, the function given by

$$f(x) = 2x^2 - 4x + 1$$

can be described by the form

$$f(\quad) = 2(\quad)^2 - 4(\quad) + 1$$

where parentheses are used instead of x. To evaluate $f(-2)$, simply place -2 in each set of parentheses.

$$f(-2) = 2(-2)^2 - 4(-2) + 1 \qquad \text{Substitute } -2 \text{ for } x.$$
$$= 2(4) + 8 + 1 \qquad \text{Simplify.}$$
$$= 17 \qquad \text{Simplify.}$$

NOTE Although f is often used as a convenient function name and x as the independent variable, you can use other symbols. For instance, the following equations all define the same function.

$$f(x) = x^2 - 4x + 7 \qquad \text{Function name is } f \text{, independent variable is } x.$$
$$f(t) = t^2 - 4t + 7 \qquad \text{Function name is } f \text{, independent variable is } t.$$
$$g(s) = s^2 - 4s + 7 \qquad \text{Function name is } g \text{, independent variable is } s.$$

EXAMPLE 1 Evaluating a Function

For the function f defined by

$$f(x) = x^2 + 7,$$

evaluate each of the following.

a. $f(3a)$ **b.** $f(b - 1)$ **c.** $\dfrac{f(x + \Delta x) - f(x)}{\Delta x}, \quad \Delta x \neq 0$

Solution

a. $f(3a) = (3a)^2 + 7$ Substitute $3a$ for x.

$= 9a^2 + 7$ Simplify.

b. $f(b - 1) = (b - 1)^2 + 7$ Substitute $b - 1$ for x.

$= b^2 - 2b + 1 + 7$ Expand binomial.

$= b^2 - 2b + 8$ Simplify.

c.
$$\frac{f(x + \Delta x) - f(x)}{\Delta x} = \frac{[(x + \Delta x)^2 + 7] - (x^2 + 7)}{\Delta x}$$
$$= \frac{x^2 + 2x\Delta x + (\Delta x)^2 + 7 - x^2 - 7}{\Delta x}$$
$$= \frac{2x\Delta x + (\Delta x)^2}{\Delta x}$$
$$= \frac{\cancel{\Delta x}(2x + \Delta x)}{\cancel{\Delta x}}$$
$$= 2x + \Delta x, \quad \Delta x \neq 0$$

STUDY TIP In calculus, it is important to clearly communicate the domain of a function or expression. For instance, in Example 1(c), the two expressions

$$\frac{f(x + \Delta x) - f(x)}{\Delta x} \quad \text{and} \quad 2x + \Delta x,$$
$\Delta x \neq 0$

are equivalent because $\Delta x = 0$ is excluded from the domain of each expression. Without a stated domain restriction, the two expressions would not be equivalent.

NOTE The expression in Example 1(c) is called a *difference quotient* and has a special significance in calculus. You will learn more about this in Chapter 3.

The Domain and Range of a Function

The domain of a function may be described explicitly, or it may be described *implicitly* by an equation used to define the function. The implied domain is the set of all real numbers for which the equation is defined, whereas an explicitly defined domain is one that is given along with the function. For example, the function given by

$$f(x) = \frac{1}{x^2 - 4}, \qquad 4 \le x \le 5$$

has an explicitly-defined domain given by $\{x\colon\ 4 \le x \le 5\}$. On the other hand, the function given by

$$g(x) = \frac{1}{x^2 - 4}$$

has an implied domain that is the set $\{x\colon\ x \ne \pm 2\}$.

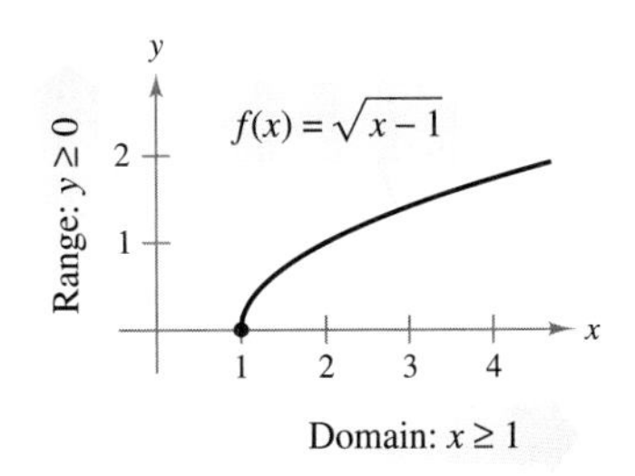

(a) The domain of f is $[1, \infty)$ and the range is $[0, \infty)$.

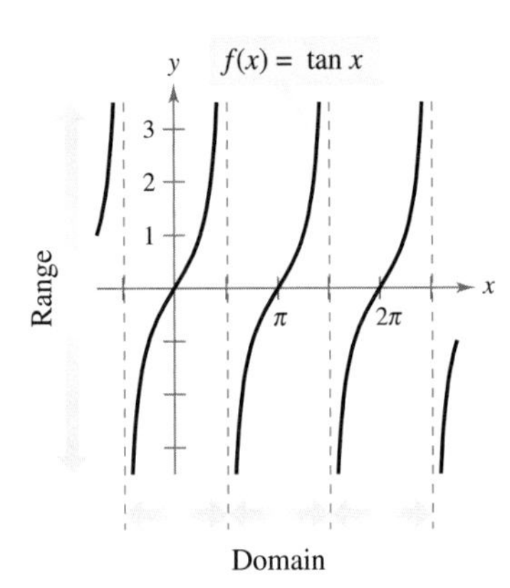

(b) The domain of f is all x-values such that $x \ne \frac{\pi}{2} + n\pi$ and the range is $(-\infty, \infty)$.

Figure 1.23

EXAMPLE 2 Finding the Domain and Range of a Function

a. The domain of the function

$$f(x) = \sqrt{x - 1}$$

is the set of all x-values for which $x - 1 \ge 0$, which is the interval $[1, \infty)$. To find the range, observe that $f(x) = \sqrt{x - 1}$ is never negative. So, the range is the interval $[0, \infty)$, as indicated in Figure 1.23(a).

b. The domain of the tangent function, shown in Figure 1.23(b),

$$f(x) = \tan x$$

is the set of all x-values such that

$$x \ne \frac{\pi}{2} + n\pi, \qquad n \text{ is an integer.} \qquad \text{Domain of tangent function}$$

The range of this function is the set of all real numbers. For a review of the characteristics of this and other trigonometric functions, see Appendix D.

EXAMPLE 3 A Function Defined by More than One Equation

Determine the domain and range of the function.

$$f(x) = \begin{cases} 1 - x, & \text{if } x < 1 \\ \sqrt{x - 1}, & \text{if } x \ge 1 \end{cases}$$

Solution Because f is defined for $x < 1$ and $x \ge 1$, the domain is the entire set of real numbers. On the portion of the domain for which $x \ge 1$, the function behaves as in Example 2(a). For $x < 1$, the values of $1 - x$ are positive. So, the range of the function is the interval $[0, \infty)$. (See Figure 1.24.)

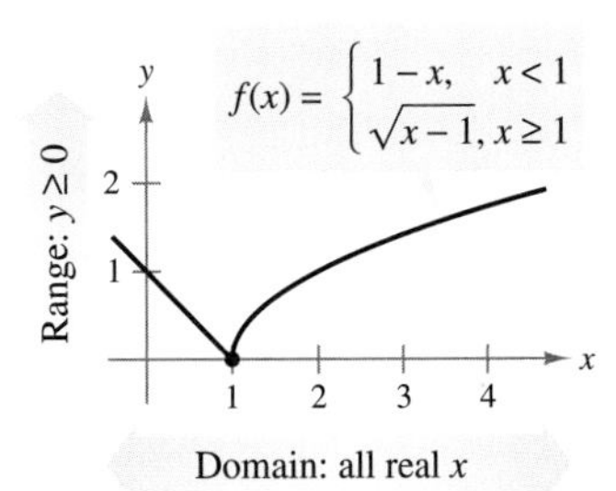

The domain of f is $(-\infty, \infty)$ and the range is $[0, \infty)$.

Figure 1.24

A function from X to Y is **one-to-one** if to each y-value in the range there corresponds exactly one x-value in the domain. For instance, the function given in Example 2(a) is one-to-one, whereas the functions given in Examples 2(b) and 3 are not one-to-one. A function from X to Y is **onto** if its range consists of all of Y.

The Graph of a Function

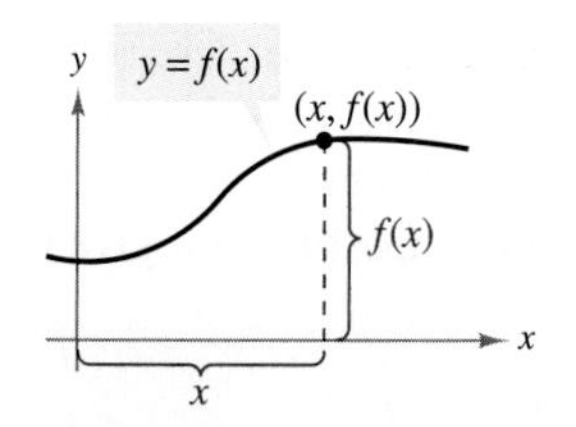

The graph of a function
Figure 1.25

The graph of the function $y = f(x)$ consists of all points $(x, f(x))$, where x is in the domain of f. In Figure 1.25, note that

$$x = \text{the directed distance from the } y\text{-axis}$$
$$f(x) = \text{the directed distance from the } x\text{-axis.}$$

A vertical line can intersect the graph of a function of x at most *once*. This observation provides a convenient visual test, called the **Vertical Line Test,** for functions of x. That is, a graph in the coordinate plane is the graph of a function of x if and only if no vertical line intersects the graph at more than one point. For example, in Figure 1.26(a), you can see that the graph does not define y as a function of x because a vertical line intersects the graph twice. In Figures 1.26(b) and (c), the graphs do define y as a function of x.

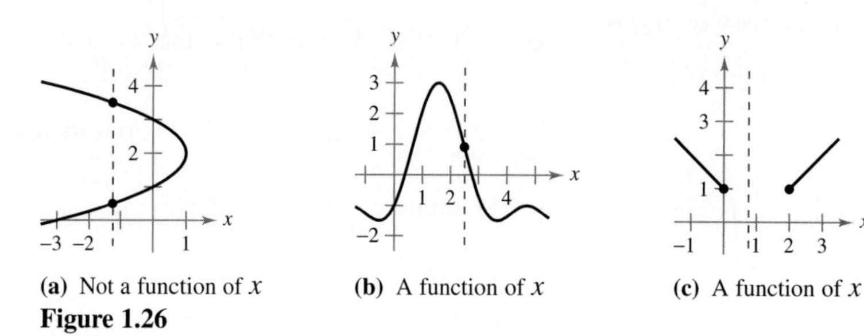

(a) Not a function of x **(b)** A function of x **(c)** A function of x
Figure 1.26

Figure 1.27 shows the graphs of eight basic functions. You should be able to recognize these graphs. (Graphs of the other four basic trigonometric functions are shown in Appendix D.)

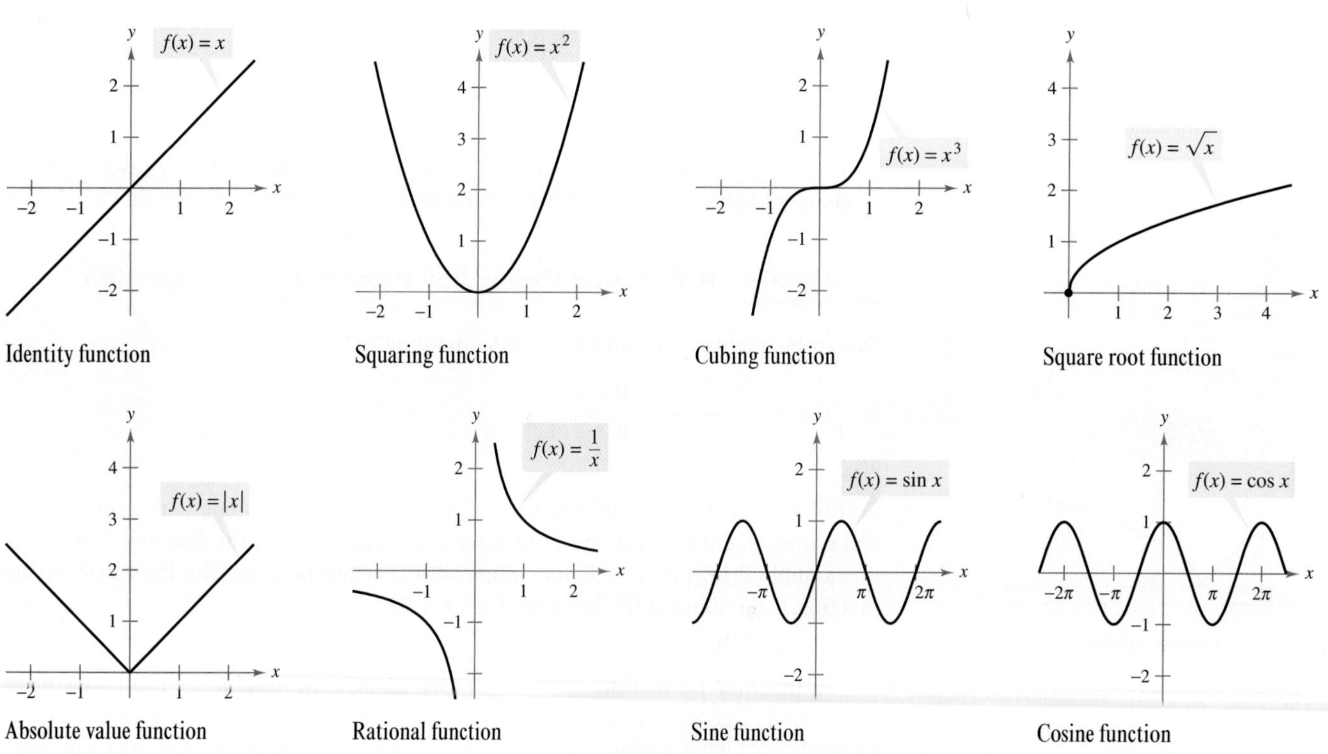

Identity function, Squaring function, Cubing function, Square root function, Absolute value function, Rational function, Sine function, Cosine function

The graphs of eight basic functions
Figure 1.27

EXPLORATION

Writing Equations for Functions
Each of the graphing utility screens below shows the graph of one of the eight basic functions shown on page 22. Each screen also shows a transformation of the graph. Describe the transformation. Then use your description to write an equation for the transformation.

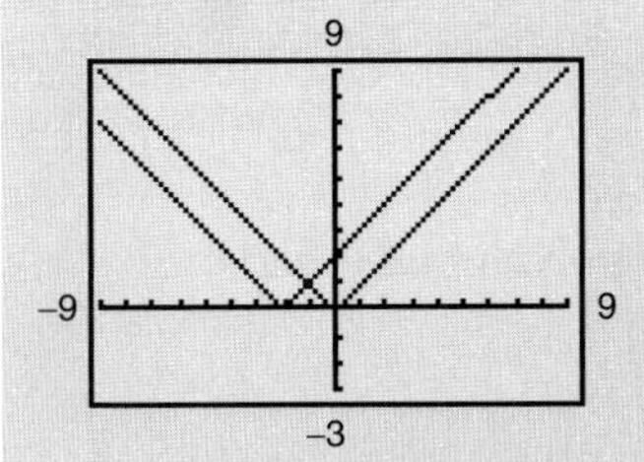

a.

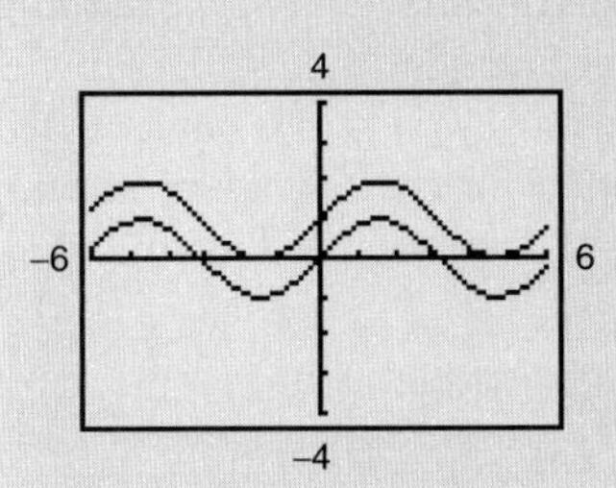

b.

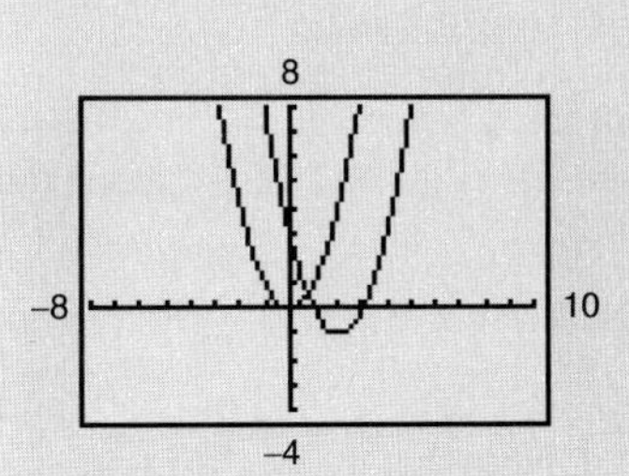

c.

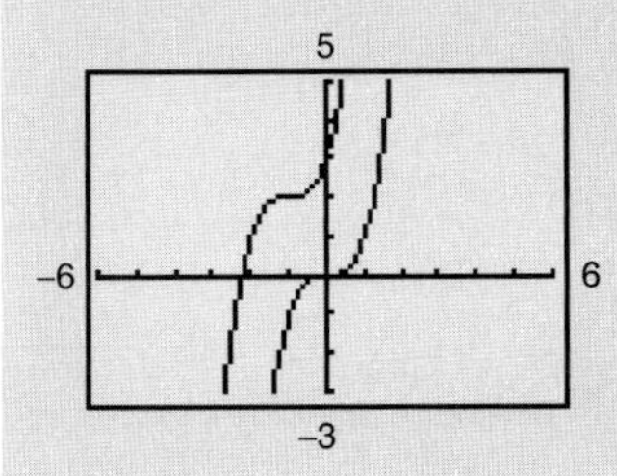

d.

Transformations of Functions

Some families of graphs have the same basic shape. For example, compare the graph of $y = x^2$ with the graphs of the four other quadratic functions shown in Figure 1.28.

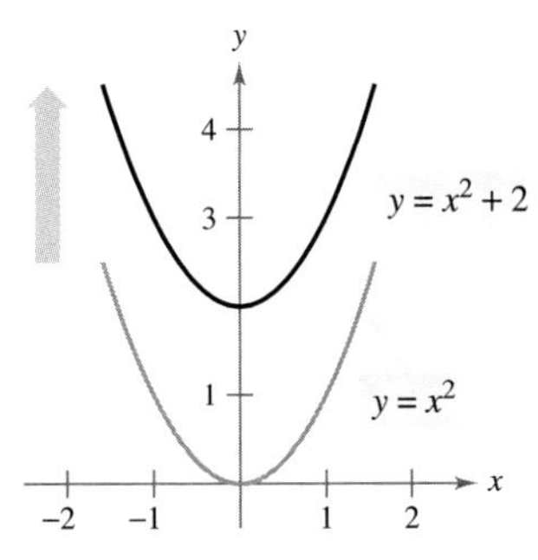

(a) Vertical shift upward

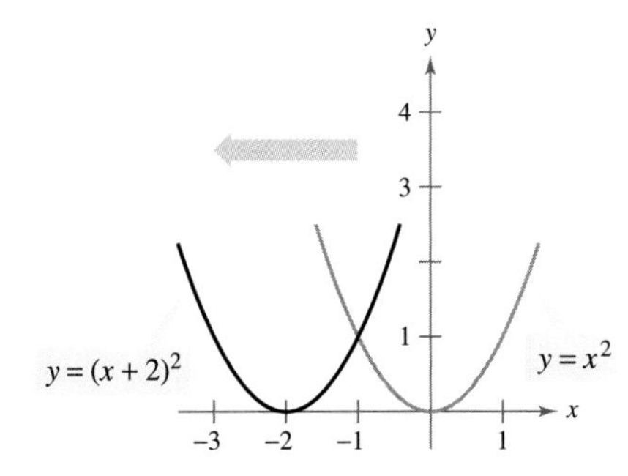

(b) Horizontal shift to the left

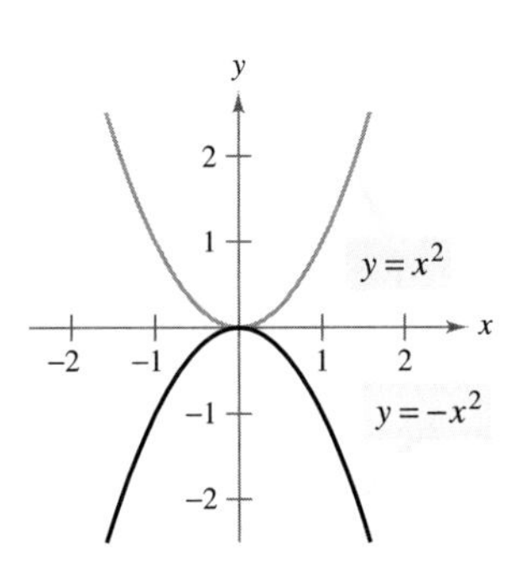

(c) Reflection

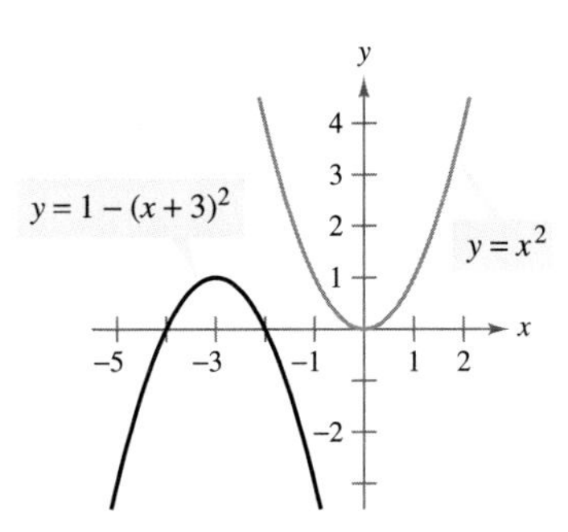

(d) Shift left, reflect, and shift upward

Figure 1.28

Each of the graphs in Figure 1.28 is a **transformation** of the graph of $y = x^2$. The three basic types of transformations illustrated by these graphs are vertical shifts, horizontal shifts, and reflections. Function notation lends itself well to describing transformations of graphs in the plane. For instance, if $f(x) = x^2$ is considered to be the original function in Figure 1.28, the transformations shown can be represented by the following equations.

$y = f(x) + 2$ Vertical shift up two units

$y = f(x + 2)$ Horizontal shift to the left two units

$y = -f(x)$ Reflection about the x-axis

$y = -f(x + 3) + 1$ Shift left three units, reflect about the x-axis, and shift up one unit

Basic Types of Transformations ($c > 0$)

Original graph:	$y = f(x)$
Horizontal shift c units to the **right**:	$y = f(x - c)$
Horizontal shift c units to the **left**:	$y = f(x + c)$
Vertical shift c units **downward**:	$y = f(x) - c$
Vertical shift c units **upward**:	$y = f(x) + c$
Reflection (about the x-axis):	$y = -f(x)$
Reflection (about the y-axis):	$y = f(-x)$
Reflection (about the origin):	$y = -f(-x)$

LEONHARD EULER (1707–1783)

In addition to making major contributions to almost every branch of mathematics, Euler was one of the first to apply calculus to real-life problems in physics. His extensive published writings include such topics as shipbuilding, acoustics, optics, astronomy, mechanics, and magnetism.

Classifications and Combinations of Functions

The modern notion of a function is derived from the efforts of many seventeenth- and eighteenth-century mathematicians. Of particular note was Leonhard Euler, to whom we are indebted for the function notation $y = f(x)$. By the end of the eighteenth century, mathematicians and scientists had concluded that many real-world phenomena could be represented by mathematical models taken from a collection of functions called **elementary functions.** Elementary functions fall into three categories.

1. Algebraic functions (polynomial, radical, rational)
2. Trigonometric functions (sine, cosine, tangent, and so on)
3. Exponential and logarithmic functions

You can review the trigonometric functions in Appendix D. The other nonalgebraic functions, such as the inverse trigonometric functions and the exponential and logarithmic functions, are introduced in Sections 1.5 and 1.6.

The most common type of algebraic function is a **polynomial function**

$$f(x) = a_n x^n + a_{n-1} x^{n-1} + \cdot \cdot \cdot + a_2 x^2 + a_1 x + a_0$$

where n is a nonnegative integer. The numbers a_i are **coefficients,** with a_n the **leading coefficient** and a_0 the **constant term** of the polynomial function. If $a_n \neq 0$, then n is the degree of the polynominal function. The zero polynomial $f(x) = 0$ is not assigned a degree. It is common practice to use subscript notation for coefficients of general polynomial functions, but for polynomial functions of low degree, the following simpler forms are often used. (Note that $a \neq 0$.)

Zeroth degree:	$f(x) = a$	Constant function
First degree:	$f(x) = ax + b$	Linear function
Second degree:	$f(x) = ax^2 + bx + c$	Quadratic function
Third degree:	$f(x) = ax^3 + bx^2 + cx + d$	Cubic function

FOR FURTHER INFORMATION For more on the history of the concept of a function, see the article "Evolution of the Function Concept: A Brief Survey" by Israel Kleiner in *The College Mathematics Journal.* To view this article, go to the website *www.matharticles.com.*

Although the graph of a polynomial function can have several turns, eventually the graph will rise or fall without bound as x moves to the right or left. Whether the graph of

$$f(x) = a_n x^n + a_{n-1} x^{n-1} + \cdot \cdot \cdot + a_2 x^2 + a_1 x + a_0$$

eventually rises or falls can be determined by the function's degree (odd or even) and by the leading coefficient a_n, as indicated in Figure 1.29. Note that the dashed portions of the graphs indicate that the **Leading Coefficient Test** determines *only* the right and left behavior of the graph.

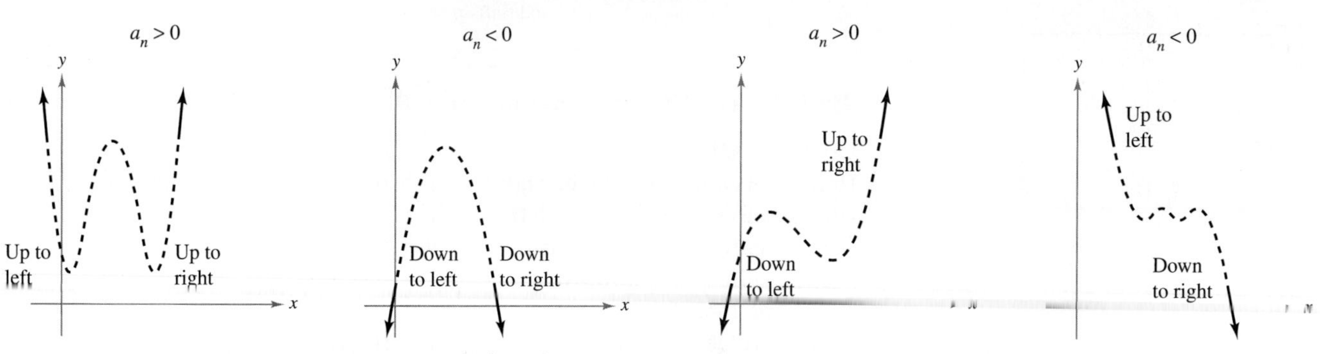

The Leading Coefficient Test for polynomial functions
Figure 1.29

Just as a rational number can be written as the quotient of two integers, a **rational function** can be written as the quotient of two polynomials. Specifically, a function f is rational if it has the form

$$f(x) = \frac{p(x)}{q(x)}, \quad q(x) \neq 0$$

where $p(x)$ and $q(x)$ are polynomials.

Polynomial functions and rational functions are examples of **algebraic functions.** An algebraic function of x is one that can be expressed as a finite number of sums, differences, multiples, quotients, and radicals involving x^n. For example, $f(x) = \sqrt{x+1}$ is algebraic. Functions that are not algebraic are **transcendental.** For instance, the trigonometric functions are transcendental.

Two functions can be combined in various ways to create new functions. For example, given $f(x) = 2x - 3$ and $g(x) = x^2 + 1$, you can form the following functions.

$$(f + g)(x) = f(x) + g(x) = (2x - 3) + (x^2 + 1) \qquad \text{Sum}$$
$$(f - g)(x) = f(x) - g(x) = (2x - 3) - (x^2 + 1) \qquad \text{Difference}$$
$$(fg)(x) = f(x)g(x) = (2x - 3)(x^2 + 1) \qquad \text{Product}$$
$$(f/g)(x) = \frac{f(x)}{g(x)} = \frac{2x - 3}{x^2 + 1} \qquad \text{Quotient}$$

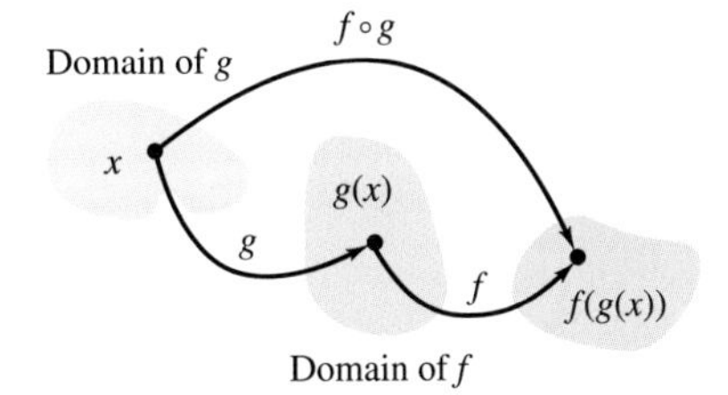

The domain of the composite function $f \circ g$
Figure 1.30

You can combine two functions in yet another way, called **composition.** The resulting function is called a **composite function.**

Definition of Composite Function

Let f and g be functions. The function given by $(f \circ g)(x) = f(g(x))$ is called the **composite** of f with g. The domain of $f \circ g$ is the set of all x in the domain of g such that $g(x)$ is in the domain of f (see Figure 1.30).

The composite of f with g is not generally equal to the composite of g with f.

EXAMPLE 4 Finding Composites of Functions

Given $f(x) = 2x - 3$ and $g(x) = \cos x$, find the following.

a. $f \circ g$ **b.** $g \circ f$

Solution

a.
$$\begin{aligned}(f \circ g)(x) &= f(g(x)) && \text{Definition of } f \circ g\\ &= f(\cos x) && \text{Substitute } \cos x \text{ for } g(x).\\ &= 2(\cos x) - 3 && \text{Definition of } f(x)\\ &= 2\cos x - 3 && \text{Simplify.}\end{aligned}$$

b.
$$\begin{aligned}(g \circ f)(x) &= g(f(x)) && \text{Definition of } g \circ f\\ &= g(2x - 3) && \text{Substitute } 2x - 3 \text{ for } f(x).\\ &= \cos(2x - 3) && \text{Definition of } g(x)\end{aligned}$$

Note that $(f \circ g)(x) \neq (g \circ f)(x)$.

EXPLORATION

Graph each of the following functions with a graphing utility. Determine whether the function is *even, odd,* or *neither.*

$$f(x) = x^2 - x^4$$
$$g(x) = 2x^3 + 1$$
$$h(x) = x^5 - 2x^3 + x$$
$$j(x) = 2 - x^6 - x^8$$
$$k(x) = x^5 - 2x^4 + x - 2$$
$$p(x) = x^9 + 3x^5 - x^3 + x$$

Describe a way to identify a function as odd or even by inspecting its equation.

In Section 1.1, an x-intercept of a graph was defined to be a point $(a, 0)$ at which the graph crosses the x-axis. If the graph represents a function f, the number a is a **zero** of f. In other words, *the zeros of a function f are the solutions of the equation $f(x) = 0$.* For example, the function $f(x) = x - 4$ has a zero at $x = 4$ because $f(4) = 0$.

In Section 1.1 you also studied different types of symmetry. In the terminology of functions, a function is **even** if its graph is symmetric with respect to the y-axis, and is **odd** if its graph is symmetric with respect to the origin. The symmetry tests in Section 1.1 yield the following test for even and odd functions.

Test for Even and Odd Functions

The function $y = f(x)$ is **even** if $f(-x) = f(x)$.

The function $y = f(x)$ is **odd** if $f(-x) = -f(x)$.

NOTE Except for the constant function $f(x) = 0$, the graph of a function of x cannot have symmetry with respect to the x-axis because it then would fail the Vertical Line Test for the graph of the function.

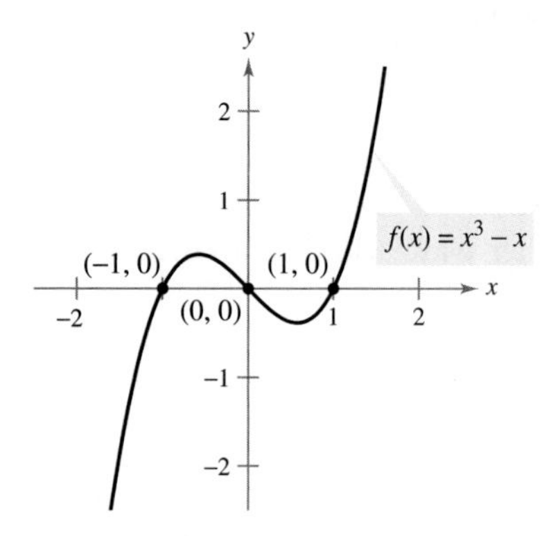

(a) Odd function

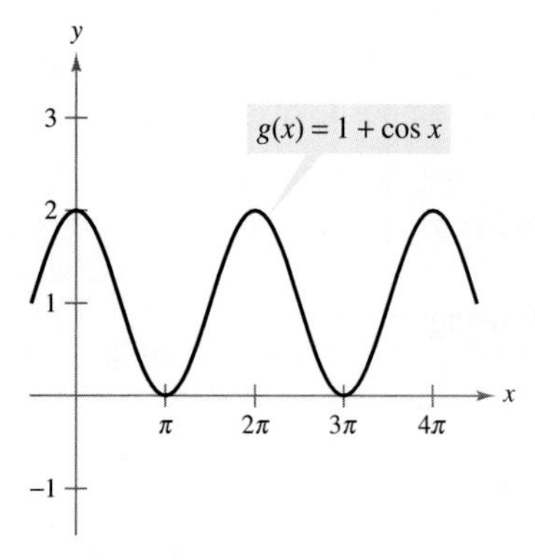

(b) Even function

Figure 1.31

EXAMPLE 5 Even and Odd Functions and Zeros of Functions

Determine whether each function is even, odd, or neither. Then find the zeros of the function.

a. $f(x) = x^3 - x$ **b.** $g(x) = 1 + \cos x$

Solution

a. This function is odd because

$$f(-x) = (-x)^3 - (-x) = -x^3 + x = -(x^3 - x) = -f(x).$$

The zeros of f are found as shown.

$x^3 - x = 0$	Let $f(x) = 0$.
$x(x^2 - 1) = 0$	Factor.
$x(x - 1)(x + 1) = 0$	Factor.
$x = 0, 1, -1$	

See Figure 1.31(a).

b. This function is even because

$$g(-x) = 1 + \cos(-x) = 1 + \cos x = g(x). \quad \cos(-x) = \cos(x)$$

The zeros of g are found as shown.

$1 + \cos x = 0$	Let $g(x) = 0$.
$\cos x = -1$	Subtract 1 from each side.
$x = (2n + 1)\pi$, n is an integer	Zeros of g

See Figure 1.31(b).

NOTE Each of the functions in Example 5 is either even or odd. However, some functions, such as $f(x) = x^2 + x + 1$, are neither even nor odd.

Exercises for Section 1.3

See www.CalcChat.com for worked-out solutions to odd-numbered exercises.

In Exercises 1 and 2, use the graphs of f and g to answer the following.

(a) Identify the domains and ranges of f and g.

(b) Identify $f(-2)$ and $g(3)$.

(c) For what value(s) of x is $f(x) = g(x)$?

(d) Estimate the solution(s) of $f(x) = 2$.

(e) Estimate the solutions of $g(x) = 0$.

1.

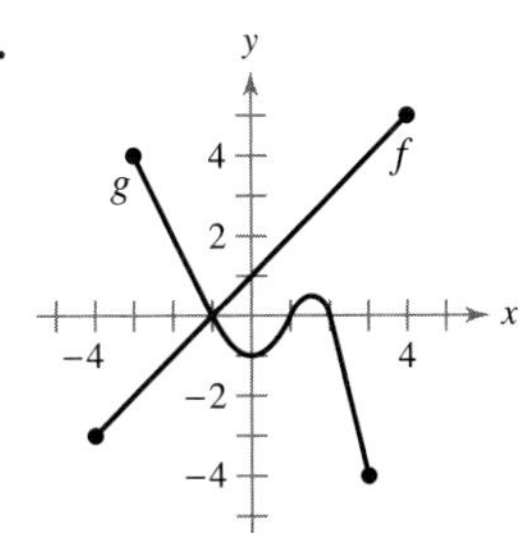

2.

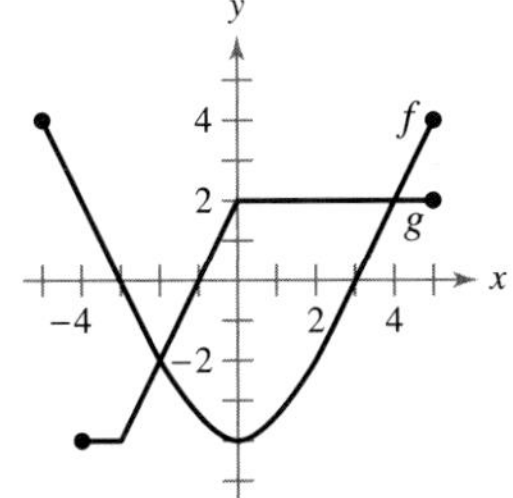

In Exercises 3–12, evaluate (if possible) the function at the given value(s) of the independent variable. Simplify the results.

3. $f(x) = 2x - 3$

(a) $f(0)$
(b) $f(-3)$
(c) $f(b)$
(d) $f(x - 1)$

4. $f(x) = \sqrt{x + 3}$

(a) $f(-2)$
(b) $f(6)$
(c) $f(-5)$
(d) $f(x + \Delta x)$

5. $g(x) = 3 - x^2$

(a) $g(0)$
(b) $g(\sqrt{3})$
(c) $g(-2)$
(d) $g(t - 1)$

6. $g(x) = x^2(x - 4)$

(a) $g(4)$
(b) $g(\frac{3}{2})$
(c) $g(c)$
(d) $g(t + 4)$

7. $f(x) = \cos 2x$

(a) $f(0)$
(b) $f(-\pi/4)$
(c) $f(\pi/3)$

8. $f(x) = \sin x$

(a) $f(\pi)$
(b) $f(5\pi/4)$
(c) $f(2\pi/3)$

9. $f(x) = x^3$

$$\frac{f(x + \Delta x) - f(x)}{\Delta x}$$

10. $f(x) = 3x - 1$

$$\frac{f(x) - f(1)}{x - 1}$$

11. $f(x) = \dfrac{1}{\sqrt{x - 1}}$

$$\frac{f(x) - f(2)}{x - 2}$$

12. $f(x) = x^3 - x$

$$\frac{f(x) - f(1)}{x - 1}$$

In Exercises 13–18, find the domain and range of the function.

13. $h(x) = -\sqrt{x + 3}$

14. $g(x) = x^2 - 5$

15. $f(t) = \sec \dfrac{\pi t}{4}$

16. $h(t) = \cot t$

17. $f(x) = \dfrac{1}{x}$

18. $g(x) = \dfrac{2}{x - 1}$

In Exercises 19-24, find the domain of the function.

19. $f(x) = \sqrt{x} + \sqrt{1 - x}$

20. $f(x) = \sqrt{x^2 - 3x + 2}$

21. $g(x) = \dfrac{2}{1 - \cos x}$

22. $h(x) = \dfrac{1}{\sin x - \frac{1}{2}}$

23. $f(x) = \dfrac{1}{|x + 3|}$

24. $g(x) = \dfrac{1}{|x^2 - 4|}$

In Exercises 25–28, evaluate the function as indicated. Determine its domain and range.

25. $f(x) = \begin{cases} 2x + 1, & x < 0 \\ 2x + 2, & x \geq 0 \end{cases}$

(a) $f(-1)$ (b) $f(0)$ (c) $f(2)$ (d) $f(t^2 + 1)$

26. $f(x) = \begin{cases} x^2 + 2, & x \leq 1 \\ 2x^2 + 2, & x > 1 \end{cases}$

(a) $f(-2)$ (b) $f(0)$ (c) $f(1)$ (d) $f(s^2 + 2)$

27. $f(x) = \begin{cases} |x| + 1, & x < 1 \\ -x + 1, & x \geq 1 \end{cases}$

(a) $f(-3)$ (b) $f(1)$ (c) $f(3)$ (d) $f(b^2 + 1)$

28. $f(x) = \begin{cases} \sqrt{x + 4}, & x \leq 5 \\ (x - 5)^2, & x > 5 \end{cases}$

(a) $f(-3)$ (b) $f(0)$ (c) $f(5)$ (d) $f(10)$

In Exercises 29 and 30, write the function whose graph is given in the figure.

29.

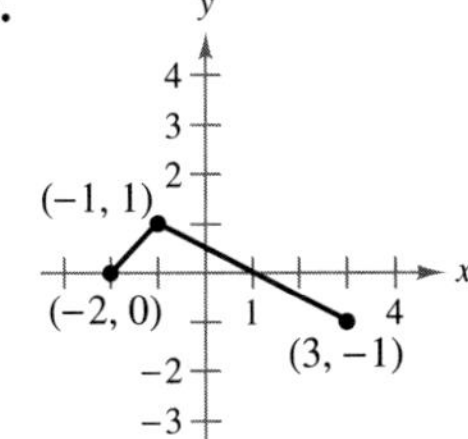

30.

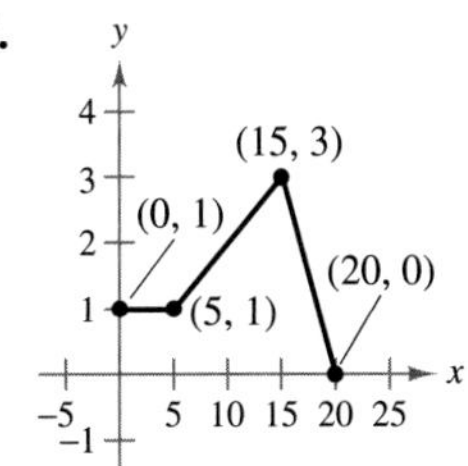

In Exercises 31–38, sketch a graph of the function and find its domain and range. Use a graphing utility to verify your graph.

31. $f(x) = 4 - x$

32. $g(x) = \dfrac{4}{x}$

33. $h(x) = \sqrt{x - 1}$

34. $f(x) = \frac{1}{2}x^3 + 2$

35. $f(x) = \sqrt{9 - x^2}$

36. $f(x) = x + \sqrt{4 - x^2}$

37. $g(t) = 2 \sin \pi t$

38. $h(\theta) = -5 \cos \dfrac{\theta}{2}$

Writing About Concepts

39. The graph of the distance that a student drives in a 10-minute trip to school is shown in the figure. Give a verbal description of characteristics of the student's drive to school.

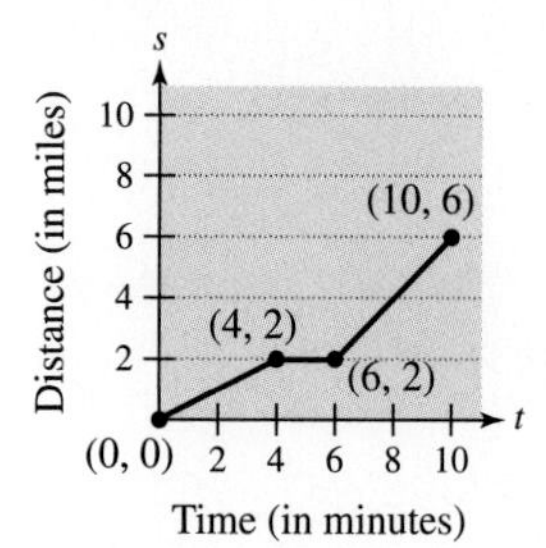

40. A student who commutes 27 miles to attend college remembers, after driving for a few minutes, that a term paper that is due has been forgotten. Driving faster than usual, the student returns home, picks up the paper, and once again starts toward school. Sketch a possible graph of the student's distance from home as a function of time.

In Exercises 41–44, use the Vertical Line Test to determine whether y is a function of x. To print an enlarged copy of the graph, go to the website *www.mathgraphs.com*.

41. $x - y^2 = 0$

42. $\sqrt{x^2 - 4} - y = 0$

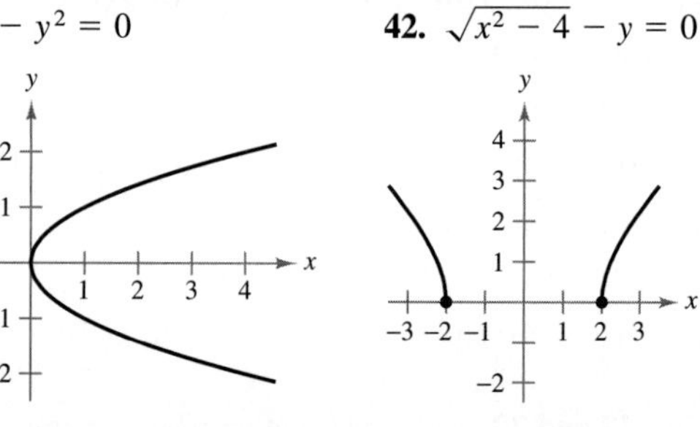

43. $y = \begin{cases} x + 1, & x \le 0 \\ -x + 2, & x > 0 \end{cases}$

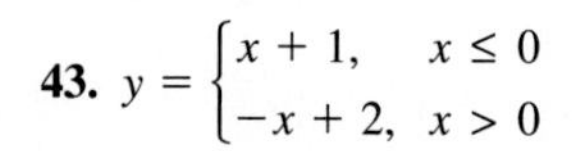

44. $x^2 + y^2 = 4$

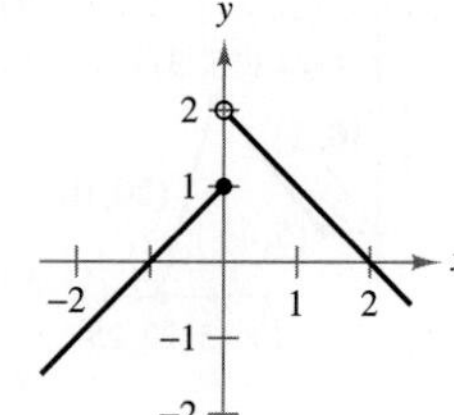

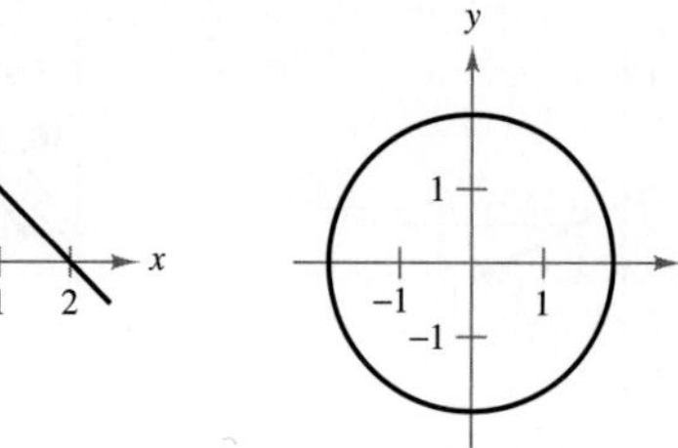

In Exercises 45–48, determine whether y is a function of x.

45. $x^2 + y^2 = 4$

46. $x^2 + y = 4$

47. $y^2 = x^2 - 1$

48. $x^2y - x^2 + 4y = 0$

In Exercises 49–54, use the graph of $y = f(x)$ to match the function with its graph.

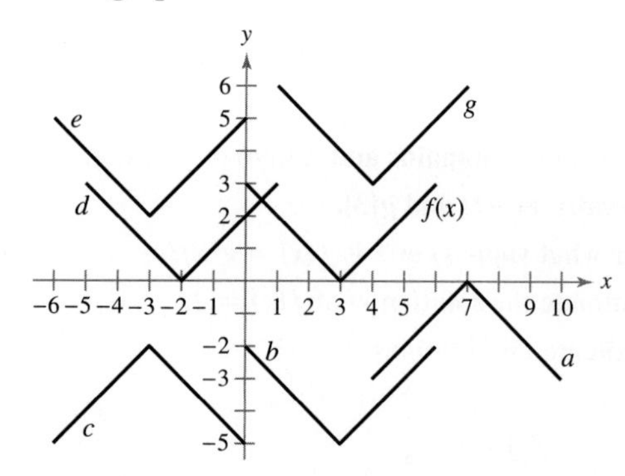

49. $y = f(x + 5)$

50. $y = f(x) - 5$

51. $y = -f(-x) - 2$

52. $y = -f(x - 4)$

53. $y = f(x + 6) + 2$

54. $y = f(x - 1) + 3$

55. Use the graph of f shown in the figure to sketch the graph of each function. To print an enlarged copy of the graph, go to the website *www.mathgraphs.com*.

(a) $f(x + 3)$ (b) $f(x - 1)$
(c) $f(x) + 2$ (d) $f(x) - 4$
(e) $3f(x)$ (f) $\frac{1}{4}f(x)$

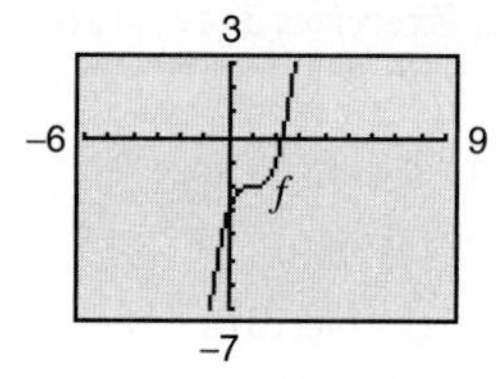

56. Use the graph of f shown in the figure to sketch the graph of each function. To print an enlarged copy of the graph, go to the website *www.mathgraphs.com*.

(a) $f(x - 4)$ (b) $f(x + 2)$
(c) $f(x) + 4$ (d) $f(x) - 1$
(e) $2f(x)$ (f) $\frac{1}{2}f(x)$

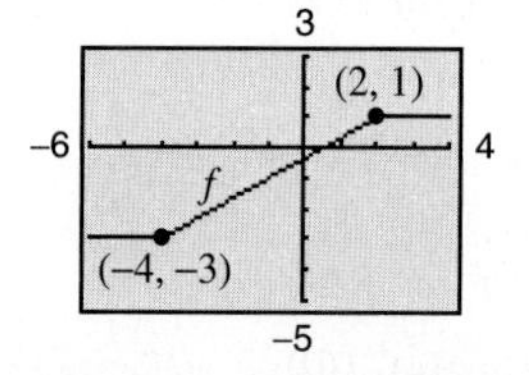

57. Use the graph of $f(x) = \sqrt{x}$ to sketch the graph of each function. In each case, describe the transformation.

(a) $y = \sqrt{x} + 2$ (b) $y = -\sqrt{x}$ (c) $y = \sqrt{x - 2}$

58. Specify a sequence of transformations that will yield each graph of h from the graph of the function $f(x) = \sin x$.

(a) $h(x) = \sin\left(x + \dfrac{\pi}{2}\right) + 1$ (b) $h(x) = -\sin(x - 1)$

59. Given $f(x) = \sqrt{x}$ and $g(x) = x^2 - 1$, evaluate each expression.

(a) $f(g(1))$ (b) $g(f(1))$ (c) $g(f(0))$
(d) $f(g(-4))$ (e) $f(g(x))$ (f) $g(f(x))$

60. Given $f(x) = \sin x$ and $g(x) = \pi x$, evaluate each expression.

(a) $f(g(2))$ (b) $f\left(g\left(\dfrac{1}{2}\right)\right)$ (c) $g(f(0))$

(d) $g\left(f\left(\dfrac{\pi}{4}\right)\right)$ (e) $f(g(x))$ (f) $g(f(x))$

In Exercises 61–64, find the composite functions $(f \circ g)$ and $(g \circ f)$. What is the domain of each composite function? Are the two composite functions equal?

61. $f(x) = x^2$
$g(x) = \sqrt{x}$

62. $f(x) = x^2 - 1$
$g(x) = \cos x$

63. $f(x) = \dfrac{3}{x}$
$g(x) = x^2 - 1$

64. $f(x) = \dfrac{1}{x}$
$g(x) = \sqrt{x + 2}$

65. Use the graphs of f and g to evaluate each expression. If the result is undefined, explain why.

(a) $(f \circ g)(3)$ (b) $g(f(2))$
(c) $g(f(5))$ (d) $(f \circ g)(-3)$
(e) $(g \circ f)(-1)$ (f) $f(g(-1))$

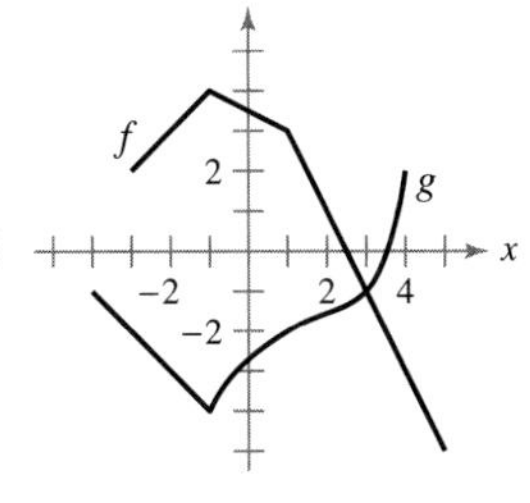

66. ***Ripples*** A pebble is dropped into a calm pond, causing ripples in the form of concentric circles. The radius (in feet) of the outer ripple is given by $r(t) = 0.6t$, where t is the time in seconds after the pebble strikes the water. The area of the circle is given by the function $A(r) = \pi r^2$. Find and interpret $(A \circ r)(t)$.

Think About It **In Exercises 67 and 68, $F(x) = f \circ g \circ h$. Identify functions for f, g, and h. (There are many correct answers.)**

67. $F(x) = \sqrt{2x - 2}$

68. $F(x) = -4 \sin(1 - x)$

In Exercises 69–72, determine whether the function is even, odd, or neither. Use a graphing utility to verify your result.

69. $f(x) = x^2(4 - x^2)$

70. $f(x) = \sqrt[3]{x}$

71. $f(x) = x \cos x$

72. $f(x) = \sin^2 x$

Think About It **In Exercises 73 and 74, find the coordinates of a second point on the graph of the function f if the given point is on the graph and the function is (a) even and (b) odd.**

73. $\left(-\frac{3}{2}, 4\right)$

74. $(4, 9)$

75. The graphs of f, g, and h are shown in the figure. Decide whether each function is even, odd, or neither.

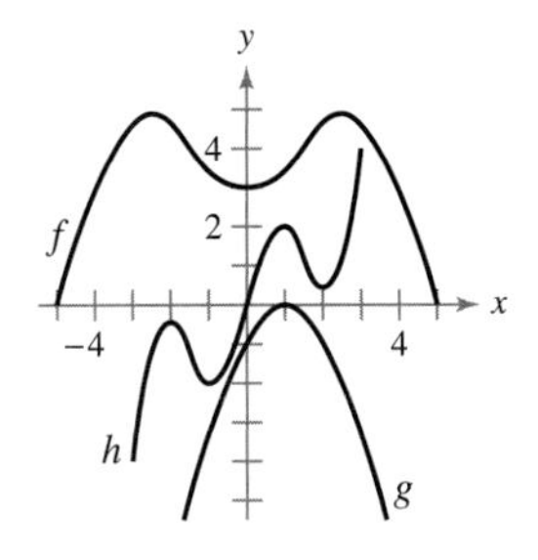

Figure for 75

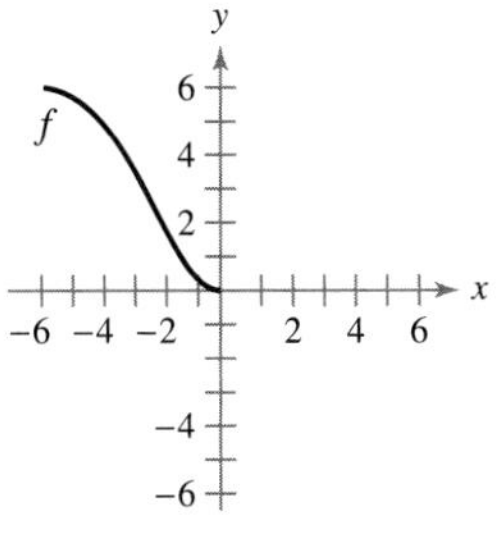

Figure for 76

76. The domain of the function f shown in the figure is $-6 \le x \le 6$.

(a) Complete the graph of f given that f is even.

(b) Complete the graph of f given that f is odd.

Writing Functions **In Exercises 77–80, write an equation for a function that has the given graph.**

77. Line segment connecting $(-4, 3)$ and $(0, -5)$

78. Line segment connecting $(1, 2)$ and $(5, 5)$

79. The bottom half of the parabola $x + y^2 = 0$

80. The bottom half of the circle $x^2 + y^2 = 4$

Modeling Data **In Exercises 81–84, match the data with a function from the following list.**

(i) $f(x) = cx$ **(ii)** $g(x) = cx^2$

(iii) $h(x) = c\sqrt{|x|}$ **(iv)** $r(x) = c/x$

Determine the value of the constant c for each function such that the function fits the data shown in the table.

81.

x	-4	-1	0	1	4
y	-32	-2	0	-2	-32

82.

x	-4	-1	0	1	4
y	-1	$-\frac{1}{4}$	0	$\frac{1}{4}$	1

83.

x	-4	-1	0	1	4
y	-8	-32	Undef.	32	8

84.

x	-4	-1	0	1	4
y	6	3	0	3	6

85. ***Graphical Reasoning*** A thermostat is programmed to lower the temperature during the night automatically (see figure). The temperature T in degrees Celsius is given in terms of t, the time in hours on a 24-hour clock.

(a) Approximate $T(4)$ and $T(15)$.

(b) The thermostat is reprogrammed to produce a temperature $H(t) = T(t - 1)$. How does this change the temperature? Explain.

(c) The thermostat is reprogrammed to produce a temperature $H(t) = T(t) - 1$. How does this change the temperature? Explain.

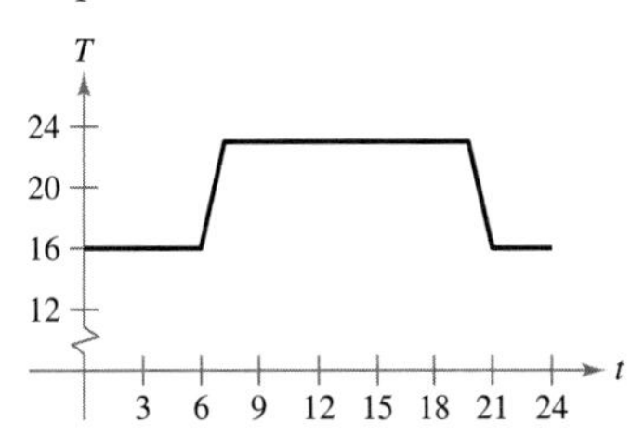

86. Water runs into a vase of height 30 centimeters at a constant rate. The vase is full after 5 seconds. Use this information and the shape of the vase shown to answer the questions if d is the depth of the water in centimeters and t is the time in seconds (see figure).

(a) Explain why d is a function of t.

(b) Determine the domain and range of the function.

(c) Sketch a possible graph of the function.

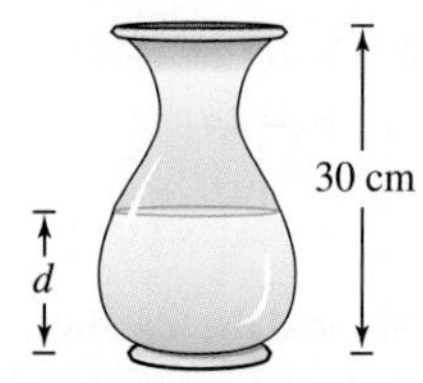

87. ***Modeling Data*** The table shows the average number of acres per farm in the United States for selected years. *(Source: U.S. Department of Agriculture)*

Year	1950	1960	1970	1980	1990	2000
Acreage	213	297	374	426	460	434

(a) Plot the data, where A is the acreage and t is the time in years, with $t = 0$ corresponding to 1950. Sketch a freehand curve that approximates the data.

(b) Use the curve in part (a) to approximate $A(15)$.

88. ***Automobile Aerodynamics*** The horsepower H required to overcome wind drag on a certain automobile is approximated by

$$H(x) = 0.002x^2 + 0.005x - 0.029, \qquad 10 \le x \le 100$$

where x is the speed of the car in miles per hour.

(a) Use a graphing utility to graph H.

(b) Rewrite the power function so that x represents the speed in kilometers per hour. [Find $H(x/1.6)$.]

89. ***Think About It*** Write the function

$$f(x) = |x| + |x - 2|$$

without using absolute value signs. (For a review of absolute value, see Appendix D.)

90. ***Writing*** Use a graphing utility to graph the polynomial functions $p_1(x) = x^3 - x + 1$ and $p_2(x) = x^3 - x$. How many zeros does each function have? Is there a cubic polynomial that has no zeros? Explain.

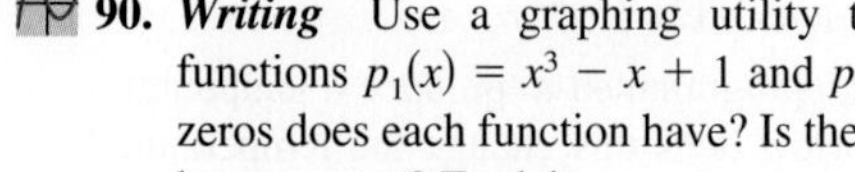

91. Prove that the function is odd.

$$f(x) = a_{2n+1}x^{2n+1} + \cdots + a_3x^3 + a_1x$$

92. Prove that the function is even.

$$f(x) = a_{2n}x^{2n} + a_{2n-2}x^{2n-2} + \cdots + a_2x^2 + a_0$$

93. Prove that the product of two even (or two odd) functions is even.

94. Prove that the product of an odd function and an even function is odd.

95. ***Volume*** An open box of maximum volume is to be made from a square piece of material 24 centimeters on a side by cutting equal squares from the corners and turning up the sides (see figure).

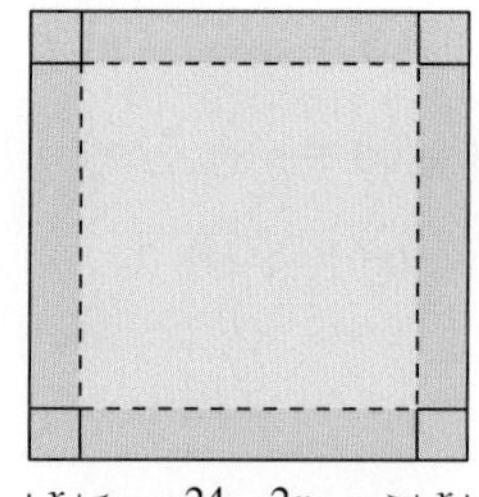

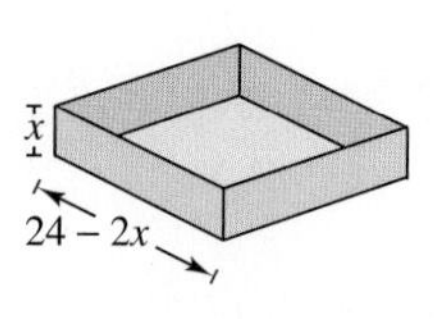

(a) Write the volume V as a function of x, the length of the corner squares. What is the domain of the function?

(b) Use a graphing utility to graph the volume function and approximate the dimensions of the box that yield a maximum volume.

(c) Use the *table* feature of a graphing utility to verify your answer in part (b). (The first two rows of the table are shown.)

Height, x	Length and Width	Volume, V
1	$24 - 2(1)$	$1[24 - 2(1)]^2 = 484$
2	$24 - 2(2)$	$2[24 - 2(2)]^2 = 800$

96. ***Length*** A right triangle is formed in the first quadrant by the x- and y-axes and a line through the point $(3, 2)$. Write the length L of the hypotenuse as a function of x.

True or False? **In Exercises 97–100, determine whether the statement is true or false. If it is false, explain why or give an example that shows it is false.**

97. If $f(a) = f(b)$, then $a = b$.

98. A vertical line can intersect the graph of a function at most once.

99. If $f(x) = f(-x)$ for all x in the domain of f, then the graph of f is symmetric with respect to the y-axis.

100. If f is a function, then $f(ax) = af(x)$.

Putnam Exam Challenge

101. Let R be the region consisting of the points (x, y) of the Cartesian plane satisfying both $|x| - |y| \le 1$ and $|y| \le 1$. Sketch the region R and find its area.

102. Consider a polynomial $f(x)$ with real coefficients having the property $f(g(x)) = g(f(x))$ for every polynomial $g(x)$ with real coefficients. Determine and prove the nature of $f(x)$.

These problems were composed by the Committee on the Putnam Prize Competition.

Section 1.4 Fitting Models to Data

- **Fit a linear model to a real-life data set.**
- **Fit a quadratic model to a real-life data set.**
- **Fit a trigonometric model to a real-life data set.**

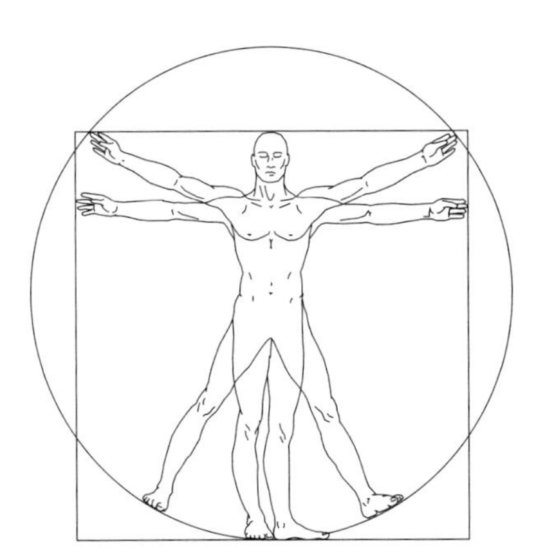

A computer graphics drawing based on the pen and ink drawing of Leonardo da Vinci's famous study of human proportions, called *Vitruvian Man*

Fitting a Linear Model to Data

A basic premise of science is that much of the physical world can be described mathematically and that many physical phenomena are predictable. This scientific outlook was part of the scientific revolution that took place in Europe during the late 1500s. Two early publications connected with this revolution were *On the Revolutions of the Heavenly Spheres* by the Polish astronomer Nicolaus Copernicus, and *On the Structure of the Human Body* by the Belgian anatomist Andreas Vesalius. Each of these books was published in 1543 and each broke with prior tradition by suggesting the use of a scientific method rather than unquestioned reliance on authority.

One method of modern science is gathering data and then describing the data with a mathematical model. For instance, the data given in Example 1 are inspired by Leonardo da Vinci's famous drawing that indicates that a person's height and arm span are equal.

EXAMPLE 1 Fitting a Linear Model to Data

A class of 28 people collected the following data, which represent their heights x and arm spans y (rounded to the nearest inch).

$$(60, 61), (65, 65), (68, 67), (72, 73), (61, 62), (63, 63), (70, 71),$$
$$(75, 74), (71, 72), (62, 60), (65, 65), (66, 68), (62, 62), (72, 73),$$
$$(70, 70), (69, 68), (69, 70), (60, 61), (63, 63), (64, 64), (71, 71),$$
$$(68, 67), (69, 70), (70, 72), (65, 65), (64, 63), (71, 70), (67, 67)$$

Find a linear model to represent these data.

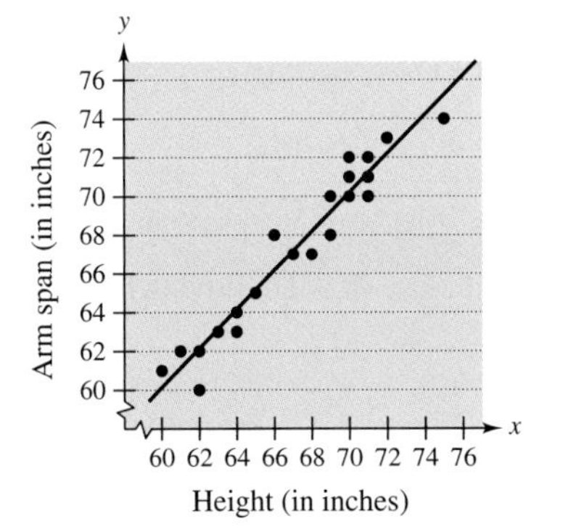

Linear model and data
Figure 1.32

Solution There are different ways to model these data with an equation. The simplest would be to observe that x and y are about the same and list the model as simply $y = x$. A more careful analysis would be to use a procedure from statistics called linear regression. (You will study this procedure in Section 13.9.) The least squares regression line for these data is

$$y = 1.006x - 0.23. \qquad \text{Least squares regression line}$$

The graph of the model and the data are shown in Figure 1.32. From this model, you can see that a person's arm span tends to be about the same as his or her height.

TECHNOLOGY Many scientific and graphing calculators have built-in least squares regression programs. Typically, you enter the data into the calculator and then run the linear regression program. The program usually displays the slope and y-intercept of the best-fitting line and the *correlation coefficient* r. The correlation coefficient gives a measure of how well the model fits the data. The closer $|r|$ is to 1, the better the model fits the data. For instance, the correlation coefficient for the model in Example 1 is $r \approx 0.97$, which indicates that the model is a good fit for the data. If the r-value is positive, the variables have a positive correlation, as in Example 1. If the r-value is negative, the variables have a negative correlation.

Fitting a Quadratic Model to Data

A function that gives the height s of a falling object in terms of the time t is called a position function. If air resistance is not considered, the position of a falling object can be modeled by

$$s(t) = \frac{1}{2}gt^2 + v_0t + s_0$$

where g is the acceleration due to gravity, v_0 is the initial velocity, and s_0 is the initial height. The value of g depends on where the object is dropped. On Earth, g is approximately -32 feet per second per second, or -9.8 meters per second per second.

To discover the value of g experimentally, you could record the heights of a falling object at several increments, as shown in Example 2.

EXAMPLE 2 Fitting a Quadratic Model to Data

A basketball is dropped from a height of about $5\frac{1}{4}$ feet. The height of the basketball is recorded 23 times at intervals of about 0.02 second.* The results are shown in the table.

Time	0.0	0.02	0.04	0.06	0.08	0.099996
Height	5.23594	5.20353	5.16031	5.0991	5.02707	4.95146

Time	0.119996	0.139992	0.159988	0.179988	0.199984	0.219984
Height	4.85062	4.74979	4.63096	4.50132	4.35728	4.19523

Time	0.23998	0.25993	0.27998	0.299976	0.319972	0.339961
Height	4.02958	3.84593	3.65507	3.44981	3.23375	3.01048

Time	0.359961	0.379951	0.399941	0.419941	0.439941
Height	2.76921	2.52074	2.25786	1.98058	1.63488

Find a model to fit these data. Then use the model to predict the time when the basketball will hit the ground.

Solution Draw a scatter plot of the data, as shown in Figure 1.33. From the scatter plot, you can see that the data do not appear to be linear. It does appear, however, that they might be quadratic. To find a quadratic model, enter the data into a calculator or computer that has a quadratic regression program. You should obtain the model

$$s = -15.45t^2 - 1.302t + 5.2340. \qquad \text{Least squares regression quadratic}$$

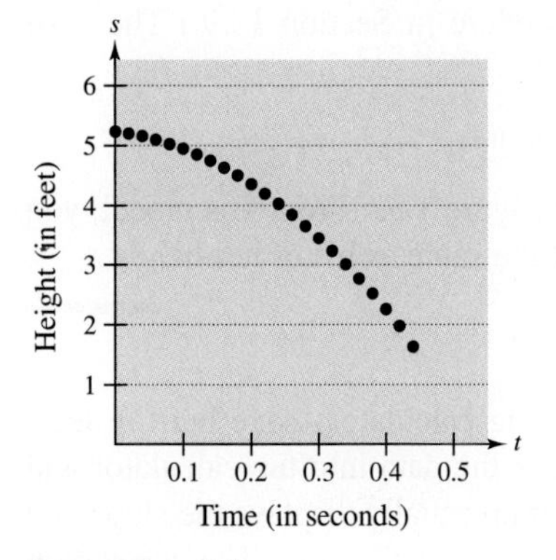

Scatter plot of data
Figure 1.33

Using this model, you can predict the time when the basketball hits the ground by substituting 0 for s and solving the resulting equation for t.

$$0 = -15.45t^2 - 1.302t + 5.2340 \qquad \text{Let } s = 0.$$

$$t = \frac{1.302 \pm \sqrt{(-1.302)^2 - 4(-15.45)(5.2340)}}{2(-15.45)} \qquad \text{Quadratic Formula}$$

$$t \approx 0.54 \qquad \text{Choose positive solution.}$$

The solution is about 0.54 second. In other words, the basketball will continue to fall for about 0.1 second more before hitting the ground.

* *Data were collected with a Texas Instruments CBL (Calculator-Based Laboratory) System.*

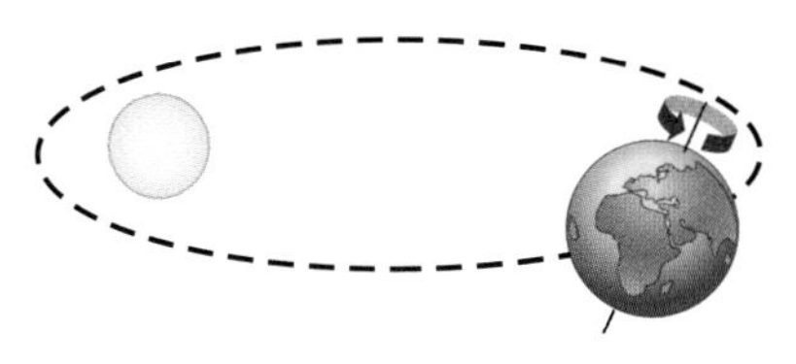

The plane of Earth's orbit about the sun and its axis of rotation are not perpendicular. Instead, Earth's axis is tilted with respect to its orbit. The result is that the amount of daylight received by locations on Earth varies with the time of year. That is, it varies with the position of Earth in its orbit.

Fitting a Trigonometric Model to Data

What is mathematical modeling? This is one of the questions that is asked in the book *Guide to Mathematical Modeling*. Here is part of the answer.*

1. Mathematical modeling consists of applying your mathematical skills to obtain useful answers to real problems.
2. Learning to apply mathematical skills is very different from learning mathematics itself.
3. Models are used in a very wide range of applications, some of which do not appear initially to be mathematical in nature.
4. Models often allow quick and cheap evaluation of alternatives, leading to optimal solutions that are not otherwise obvious.
5. There are no precise rules in mathematical modeling and no "correct" answers.
6. Modeling can be learned only by *doing*.

EXAMPLE 3 Fitting a Trigonometric Model to Data

The number of hours of daylight on Earth depends on the latitude and the time of year. Here are the numbers of minutes of daylight at a location of 20° N latitude on the longest and shortest days of the year: June 21, 801 minutes; December 22, 655 minutes. Use these data to write a model for the amount of daylight d (in minutes) on each day of the year at a location of 20° N latitude. How could you check the accuracy of your model?

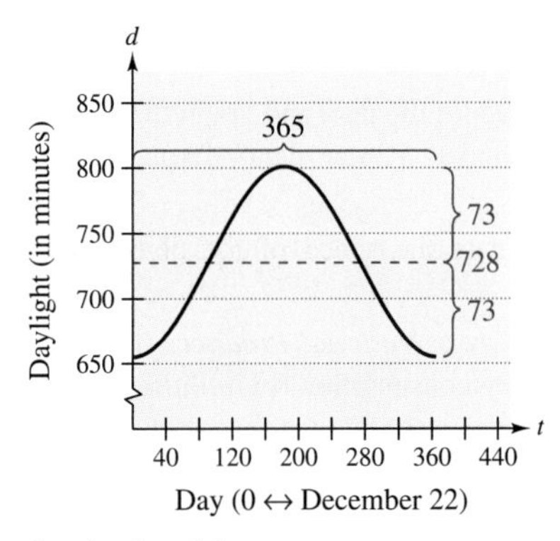

Graph of model
Figure 1.34

Solution Here is one way to create a model. You can hypothesize that the model is a sine function whose period is 365 days. Using the given data, you can conclude that the amplitude of the graph is $(801 - 655)/2$, or 73. So, one possible model is

$$d = 728 - 73\sin\left(\frac{2\pi t}{365} + \frac{\pi}{2}\right).$$

In this model, t represents the number of the day of the year, with December 22 represented by $t = 0$. A graph of this model is shown in Figure 1.34. To check the accuracy of this model, we used a weather almanac to find the numbers of minutes of daylight on different days of the year at the location of 20° N latitude.

Date	*Value of t*	*Actual Daylight*	*Daylight Given by Model*
Dec 22	0	655 min	655 min
Jan 1	10	657 min	656 min
Feb 1	41	676 min	672 min
Mar 1	69	705 min	701 min
Apr 1	100	740 min	739 min
May 1	130	772 min	773 min
Jun 1	161	796 min	796 min
Jun 21	181	801 min	801 min
Jul 1	191	799 min	800 min
Aug 1	222	782 min	785 min
Sep 1	253	752 min	754 min
Oct 1	283	718 min	716 min
Nov 1	314	685 min	681 min
Dec 1	344	661 min	660 min

You can see that the model is fairly accurate.

---

* *Text from Dilwyn Edwards and Mike Hamson,* Guide to Mathematical Modelling *(Boca Raton: CRC Press, 1990). Used by permission of the authors.*

Exercises for Section 1.4

See www.CalcChat.com for worked-out solutions to odd-numbered exercises.

In Exercises 1–4, a scatter plot of data is given. Determine whether the data can be modeled by a linear function, a quadratic function, or a trigonometric function, or that there appears to be no relationship between *x* and *y*. To print an enlarged copy of the graph, go to the website *www.mathgraphs.com*.

1.

2.

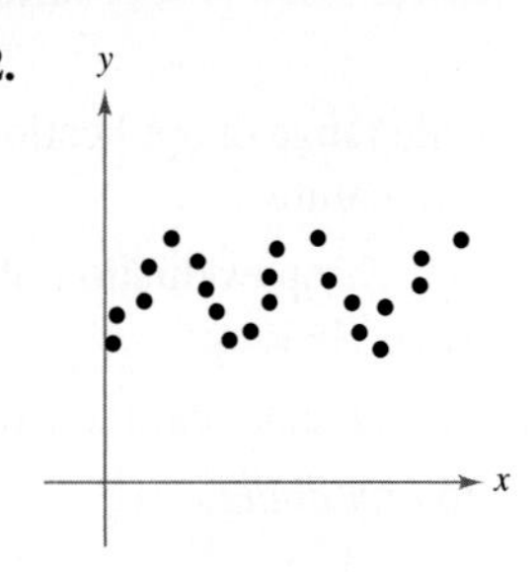

3.

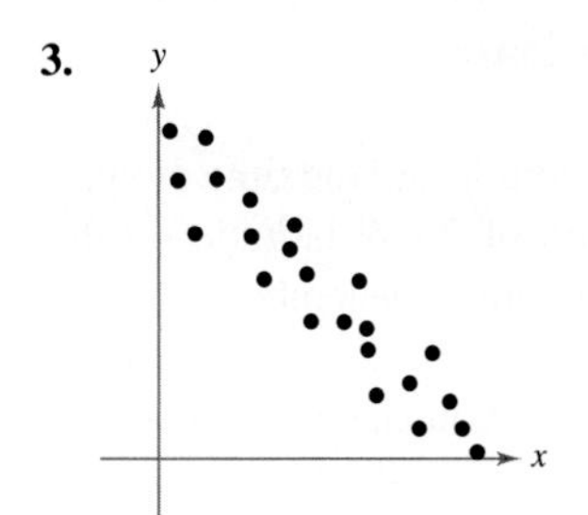

4.

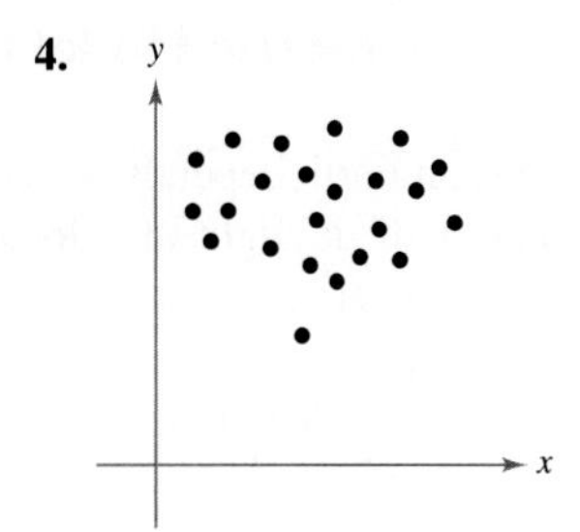

5. ***Carcinogens*** Each ordered pair gives the exposure index x of a carcinogenic substance and the cancer mortality y per 100,000 people in the population.

(3.50, 150.1), (3.58, 133.1), (4.42, 132.9),
(2.26, 116.7), (2.63, 140.7), (4.85, 165.5),
(12.65, 210.7), (7.42, 181.0), (9.35, 213.4)

(a) Plot the data. From the graph, do the data appear to be approximately linear?

(b) Visually find a linear model for the data. Graph the model.

(c) Use the model to approximate y if $x = 3$.

6. ***Quiz Scores*** The ordered pairs represent the scores on two consecutive 15-point quizzes for a class of 18 students.

(7, 13), (9, 7), (14, 14), (15, 15), (10, 15), (9, 7),
(14, 11), (14, 15), (8, 10), (15, 9), (10, 11), (9, 10),
(11, 14), (7, 14), (11, 10), (14, 11), (10, 15), (9, 6)

(a) Plot the data. From the graph, does the relationship between consecutive scores appear to be approximately linear?

(b) If the data appear to be approximately linear, find a linear model for the data. If not, give some possible explanations.

7. ***Hooke's Law*** Hooke's Law states that the force F required to compress or stretch a spring (within its elastic limits) is proportional to the distance d that the spring is compressed or stretched from its original length. That is, $F = kd$, where k is a measure of the stiffness of the spring and is called the *spring constant*. The table shows the elongation d in centimeters of a spring when a force of F newtons is applied.

F	20	40	60	80	100
d	1.4	2.5	4.0	5.3	6.6

Table for 7

(a) Use the regression capabilities of a graphing utility to find a linear model for the data.

(b) Use a graphing utility to plot the data and graph the model. How well does the model fit the data? Explain.

(c) Use the model to estimate the elongation of the spring when a force of 55 newtons is applied.

8. ***Falling Object*** In an experiment, students measured the speed s (in meters per second) of a falling object t seconds after it was released. The results are shown in the table.

t	0	1	2	3	4
s	0	11.0	19.4	29.2	39.4

(a) Use the regression capabilities of a graphing utility to find a linear model for the data.

(b) Use a graphing utility to plot the data and graph the model. How well does the model fit the data? Explain your reasoning.

(c) Use the model to estimate the speed of the object after 2.5 seconds.

9. ***Energy Consumption and Gross National Product*** The data show the per capita electricity consumption (in millions of Btu) and the per capita gross national product (in thousands of U.S. dollars) for several countries in 2000. *(Source: U.S. Census Bureau)*

Argentina	(73, 12.05)	Bangladesh	(4, 1.59)
Chile	(68, 9.1)	Egypt	(32, 3.67)
Greece	(126, 16.86)	Hong Kong	(118, 25.59)
Hungary	(105, 11.99)	India	(13, 2.34)
Mexico	(63, 8.79)	Poland	(95, 9)
Portugal	(108, 16.99)	South Korea	(167, 17.3)
Spain	(137, 19.26)	Turkey	(47, 7.03)
United Kingdom	(166, 23.55)	Venezuela	(113, 5.74)

(a) Use the regression capabilities of a graphing utility to find a linear model for the data. What is the correlation coefficient?

(b) Use a graphing utility to plot the data and graph the model.

(c) Interpret the graph in part (b). Use the graph to identify the three countries whose data points differ most from the linear model.

(d) Delete the data for the three countries identified in part (c). Fit a linear model to the remaining data and give the correlation coefficient.

10. *Brinell Hardness* The data in the table show the Brinell hardness H of 0.35 carbon steel when hardened and tempered at various temperatures t (degrees Fahrenheit). *(Source: Standard Handbook for Mechanical Engineers)*

t	200	400	600	800	1000	1200
H	534	495	415	352	269	217

(a) Use the regression capabilities of a graphing utility to find a linear model for the data.

(b) Use a graphing utility to plot the data and graph the model. How well does the model fit the data? Explain your reasoning.

(c) Use the model to estimate the hardness when t is 500°F.

11. *Automobile Costs* The data in the table show the variable costs for operating an automobile in the United States for several recent years. The functions y_1, y_2, and y_3 represent the costs in cents per mile for gas and oil, maintenance, and tires, respectively. *(Source: American Automobile Manufacturers Association)*

Year	y_1	y_2	y_3
0	5.40	2.10	0.90
1	6.70	2.20	0.90
2	6.00	2.20	0.90
3	6.00	2.40	0.90
4	5.60	2.50	1.10
5	6.00	2.60	1.40
6	5.90	2.80	1.40
7	6.60	2.80	1.40

(a) Use the regression capabilities of a graphing utility to find a cubic model for y_1 and linear models for y_2 and y_3.

(b) Use a graphing utility to graph y_1, y_2, y_3, and $y_1 + y_2 + y_3$ in the same viewing window. Use the model to estimate the total variable cost per mile in year 12.

12. *Beam Strength* Students in a lab measured the breaking strength S (in pounds) of wood 2 inches thick, x inches high, and 12 inches long. The results are shown in the table.

x	4	6	8	10	12
S	2370	5460	10,310	16,250	23,860

(a) Use the regression capabilities of a graphing utility to fit a quadratic model to the data.

(b) Use a graphing utility to plot the data and graph the model.

(c) Use the model to approximate the breaking strength when $x = 2$.

13. *Health Maintenance Organizations* The bar graph shows the number of people N (in millions) receiving care in HMOs for the years 1990 through 2002. *(Source: Centers for Disease Control)*

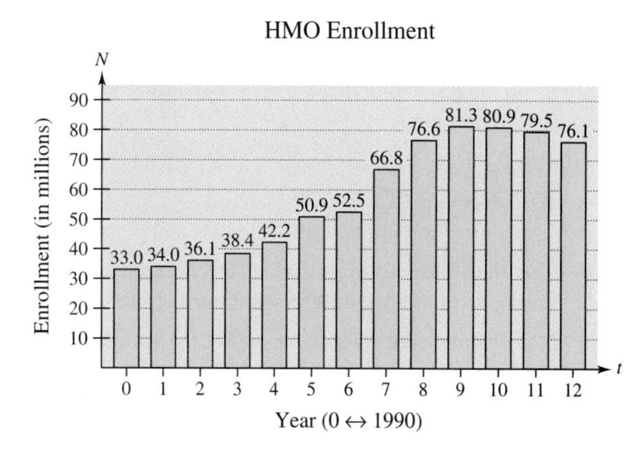

(a) Let t be the time in years, with $t = 0$ corresponding to 1990. Use the regression capabilities of a graphing utility to find linear and cubic models for the data.

(b) Use a graphing utility to graph the data and the linear and cubic models.

(c) Use the graphs in part (b) to determine which is the better model.

(d) Use a graphing utility to find and graph a quadratic model for the data.

(e) Use the linear and cubic models to estimate the number of people receiving care in HMOs in the year 2004.

(f) Use a graphing utility to find other models for the data. Which models do you think best represent the data? Explain.

14. *Car Performance* The time t (in seconds) required to attain a speed of s miles per hour from a standing start for a Dodge Avenger is shown in the table. *(Source: Road & Track)*

s	30	40	50	60	70	80	90
t	3.4	5.0	7.0	9.3	12.0	15.8	20.0

(a) Use the regression capabilities of a graphing utility to find a quadratic model for the data.

(b) Use a graphing utility to plot the data and graph the model.

(c) Use the graph in part (b) to state why the model is not appropriate for determining the times required to attain speeds less than 20 miles per hour.

(d) Because the test began from a standing start, add the point (0, 0) to the data. Fit a quadratic model to the revised data and graph the new model.

(e) Does the model from part (d) more accurately model the behavior of the car for low speeds? Explain.

15. ***Car Performance*** A V8 car engine is coupled to a dynamometer and the horsepower y is measured at different engine speeds x (in thousands of revolutions per minute). The results are shown in the table.

x	1	2	3	4	5	6
y	40	85	140	200	225	245

(a) Use the regression capabilities of a graphing utility to find a cubic model for the data.

(b) Use a graphing utility to plot the data and graph the model.

(c) Use the model to approximate the horsepower when the engine is running at 4500 revolutions per minute.

16. ***Boiling Temperature*** The table shows the temperatures T (in degrees Fahrenheit) at which water boils at selected pressures p (pounds per square inch). *(Source: Standard Handbook for Mechanical Engineers)*

p	5	10	14.696 (1 atmosphere)	20
T	162.24°	193.21°	212.00°	227.96°

p	30	40	60	80	100
T	250.33°	267.25°	292.71°	312.03°	327.81°

(a) Use the regression capabilities of a graphing utility to find a cubic model for the data.

(b) Use a graphing utility to plot the data and graph the model.

(c) Use the graph to estimate the pressure required for the boiling point of water to exceed 300°F.

(d) Explain why the model would not be correct for pressures exceeding 100 pounds per square inch.

17. ***Harmonic Motion*** The motion of an oscillating weight suspended by a spring was measured by a motion detector. The data collected and the approximate maximum (positive and negative) displacements from equilibrium are shown in the figure. The displacement y is measured in centimeters and the time t is measured in seconds.

(a) Is y a function of t? Explain.

(b) Approximate the amplitude and period of the oscillations.

(c) Find a model for the data.

(d) Use a graphing utility to graph the model in part (c). Compare the result with the data in the figure.

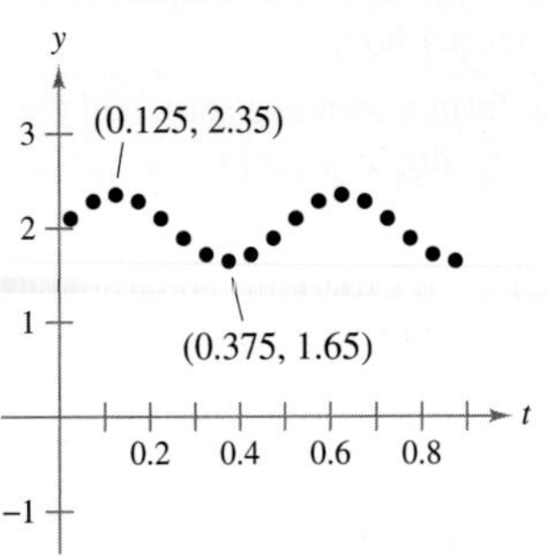

18. ***Temperature*** The table shows the normal daily high temperatures for Honolulu H and Chicago C (in degrees Fahrenheit) for month t, with $t = 1$ corresponding to January. *(Source: NOAA)*

t	1	2	3	4	5	6
H	80.1	80.5	81.6	82.8	84.7	86.5
C	29.0	33.5	45.8	58.6	70.1	79.6

t	7	8	9	10	11	12
H	87.5	88.7	88.5	86.9	84.1	81.2
C	83.7	81.8	74.8	63.3	48.4	34.0

(a) A model for Honolulu is

$$H(t) = 84.40 + 4.28 \sin\left(\frac{\pi t}{6} + 3.86\right).$$

Find a model for Chicago.

(b) Use a graphing utility to graph the data and the model for the temperatures in Honolulu. How well does the model fit the data?

(c) Use a graphing utility to graph the data and the model for the temperatures in Chicago. How well does the model fit the data?

(d) Use the models to estimate the average annual temperature in each city. What term of the model did you use? Explain.

(e) What is the period of each model? Is it what you expected? Explain.

(f) Which city has a greater variability of temperatures throughout the year? Which factor of the models determines this variability? Explain.

Writing About Concepts

19. Search for real-life data in a newspaper or magazine. Fit the data to a model. What does your model imply about the data?

20. Describe a real-life situation for each data set. Then describe how a model could be used in the real-life setting.

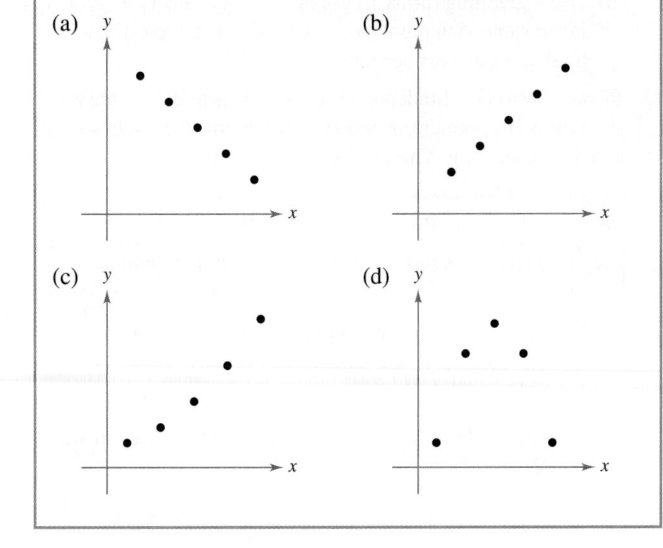

Section 1.5

Inverse Functions

- **Verify that one function is the inverse function of another function.**
- **Determine whether a function has an inverse function.**
- **Develop properties of the six inverse trigonometric functions.**

Inverse Functions

Recall from Section 1.3 that a function can be represented by a set of ordered pairs. For instance, the function $f(x) = x + 3$ from $A = \{1, 2, 3, 4\}$ to $B = \{4, 5, 6, 7\}$ can be written as

$$f: \{(1, 4), (2, 5), (3, 6), (4, 7)\}.$$

By interchanging the first and second coordinates of each ordered pair, you can form the **inverse function** of f. This function is denoted by f^{-1}. It is a function from B to A, and can be written as

$$f^{-1}: \{(4, 1), (5, 2), (6, 3), (7, 4)\}.$$

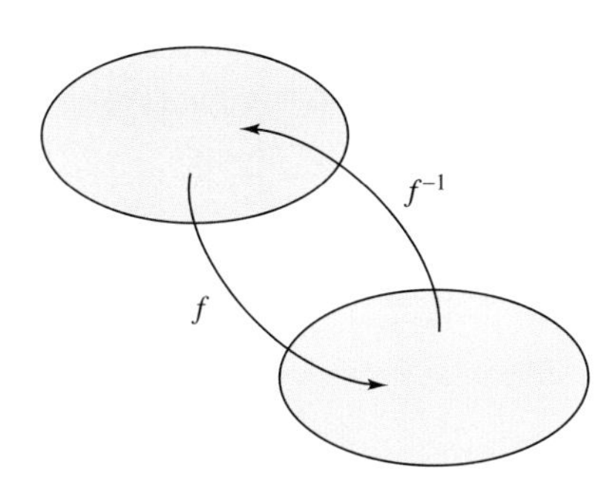

Domain of f = range of f^{-1}
Domain of f^{-1} = range of f
Figure 1.35

Note that the domain of f is equal to the range of f^{-1} and vice versa, as shown in Figure 1.35. The functions f and f^{-1} have the effect of "undoing" each other. That is, when you form the composition of f with f^{-1} or the composition of f^{-1} with f, you obtain the identity function.

$$f(f^{-1}(x)) = x \quad \text{and} \quad f^{-1}(f(x)) = x$$

EXPLORATION

Finding Inverse Functions Explain how to "undo" each of the following functions. Then use your explanation to write the inverse function of f.

a. $f(x) = x - 5$
b. $f(x) = 6x$
c. $f(x) = \dfrac{x}{2}$
d. $f(x) = 3x + 2$
e. $f(x) = x^3$
f. $f(x) = 4(x - 2)$

Use a graphing utility to graph each function and its inverse function in the same "square" viewing window. What observation can you make about each pair of graphs?

Definition of Inverse Function

A function g is the **inverse function** of the function f if

$$f(g(x)) = x \text{ for each } x \text{ in the domain of } g$$

and

$$g(f(x)) = x \text{ for each } x \text{ in the domain of } f.$$

The function g is denoted by f^{-1} (read "f inverse").

NOTE Although the notation used to denote an inverse function resembles *exponential notation*, it is a different use of -1 as a superscript. That is, in general, $f^{-1}(x) \neq 1/f(x)$.

Here are some important observations about inverse functions.

1. If g is the inverse function of f, then f is the inverse function of g.
2. The domain of f^{-1} is equal to the range of f, and the range of f^{-1} is equal to the domain of f.
3. A function need not have an inverse function, but if it does, the inverse function is unique (see Exercise 143).

You can think of f^{-1} as undoing what has been done by f. For example, subtraction can be used to undo addition, and division can be used to undo multiplication. Use the definition of an inverse function to check the following.

$f(x) = x + c$ and $f^{-1}(x) = x - c$ are inverse functions of each other.

$f(x) = cx$ and $f^{-1}(x) = \dfrac{x}{c}$, $c \neq 0$, are inverse functions of each other.

EXAMPLE 1 Verifying Inverse Functions

Show that the functions are inverse functions of each other.

$$f(x) = 2x^3 - 1 \qquad \text{and} \qquad g(x) = \sqrt[3]{\frac{x+1}{2}}$$

Solution Because the domains and ranges of both f and g consist of all real numbers, you can conclude that both composite functions exist for all x. The composite of f with g is given by

$$\begin{aligned} f(g(x)) &= 2\left(\sqrt[3]{\frac{x+1}{2}}\right)^3 - 1 \\ &= 2\left(\frac{x+1}{2}\right) - 1 \\ &= x + 1 - 1 \\ &= x. \end{aligned}$$

The composite of g with f is given by

$$\begin{aligned} g(f(x)) &= \sqrt[3]{\frac{(2x^3 - 1) + 1}{2}} \\ &= \sqrt[3]{\frac{2x^3}{2}} \\ &= \sqrt[3]{x^3} \\ &= x. \end{aligned}$$

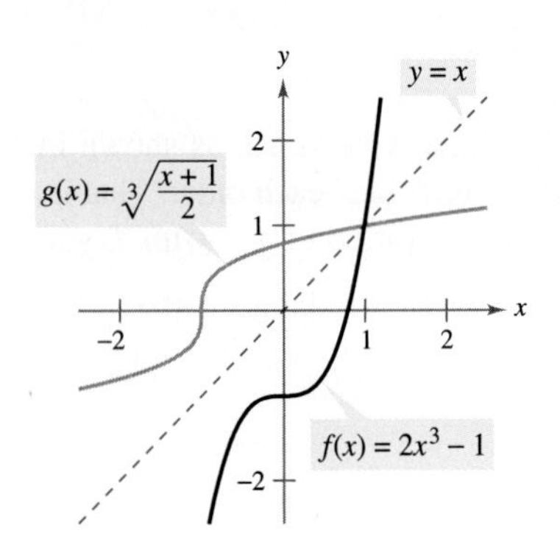

f and g are inverse functions of each other.
Figure 1.36

Because $f(g(x)) = x$ and $g(f(x)) = x$, you can conclude that f and g are inverse functions of each other (see Figure 1.36).

STUDY TIP In Example 1, try comparing the functions f and g verbally.

For f: First cube x, then multiply by 2, then subtract 1.

For g: First add 1, then divide by 2, then take the cube root.

Do you see the "undoing pattern"?

In Figure 1.36, the graphs of f and $g = f^{-1}$ appear to be mirror images of each other with respect to the line $y = x$. The graph of f^{-1} is a **reflection** of the graph of f in the line $y = x$. This idea is generalized as follows.

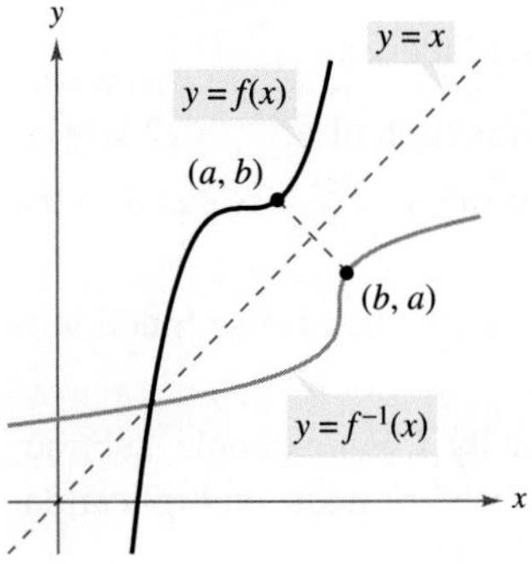

Reflective Property of Inverse Functions

The graph of f contains the point (a, b) if and only if the graph of f^{-1} contains the point (b, a).

The graph of f^{-1} is a reflection of the graph of f in the line $y = x$.
Figure 1.37

To see this, suppose (a, b) is on the graph of f. Then $f(a) = b$ and you can write

$$f^{-1}(b) = f^{-1}(f(a)) = a.$$

So, (b, a) is on the graph of f^{-1}, as shown in Figure 1.37. A similar argument will verify this result in the other direction.

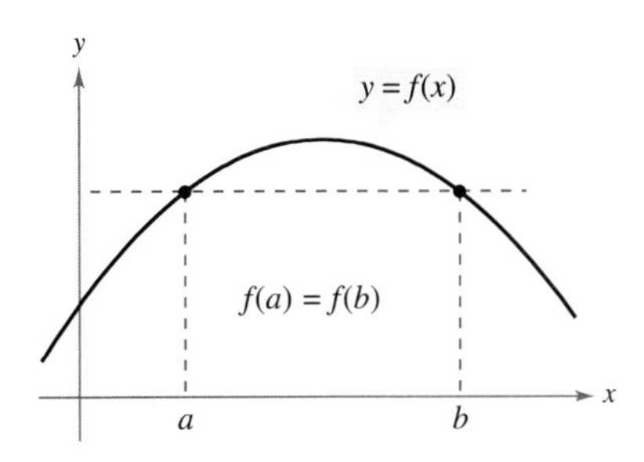

If a horizontal line intersects the graph of f twice, then f is not one-to-one.
Figure 1.38

Existence of an Inverse Function

Not every function has an inverse, and the Reflective Property of Inverse Functions suggests a graphical test for those that do—the **Horizontal Line Test** for an inverse function. This test states that a function f has an inverse function if and only if every horizontal line intersects the graph of f at most once (see Figure 1.38). The following formally states why the Horizontal Line Test is valid.

> **The Existence of an Inverse Function**
>
> A function has an inverse function if and only if it is one-to-one.

EXAMPLE 2 The Existence of an Inverse Function

Which of the functions has an inverse function?

a. $f(x) = x^3 - 1$ **b.** $f(x) = x^3 - x + 1$

Solution

a. From the graph of f given in Figure 1.39(a), it appears that f is one-to-one over its entire domain. To verify this, suppose that there exist x_1 and x_2 such that $f(x_1) = f(x_2)$. By showing that $x_1 = x_2$, it follows that f is one-to-one.

$$f(x_1) = f(x_2)$$
$$x_1^3 - 1 = x_2^3 - 1$$
$$x_1^3 = x_2^3$$
$$\sqrt[3]{x_1^3} = \sqrt[3]{x_2^3}$$
$$x_1 = x_2$$

Because f is one-to-one, you can conclude that f must have an inverse function.

b. From the graph in Figure 1.39(b), you can see that the function does not pass the Horizontal Line Test. In other words, it is not one-to-one. For instance, f has the same value when $x = -1, 0$, and 1.

$$f(-1) = f(1) = f(0) = 1 \quad \text{Not one-to-one}$$

Therefore, f does not have an inverse function.

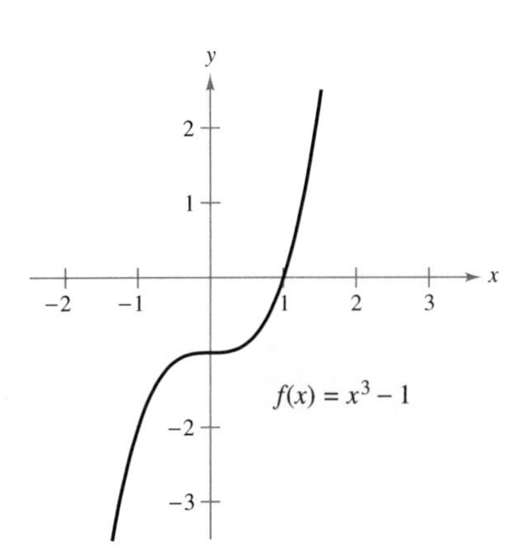

(a) Because f is one-to-one over its entire domain, it has an inverse function.

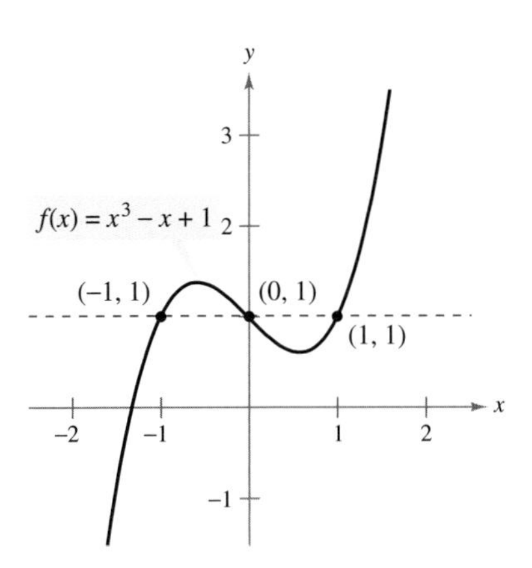

(b) Because f is not one-to-one, it does not have an inverse function.
Figure 1.39

NOTE Often it is easier to prove that a function has an inverse function than to find the inverse function. For instance, by sketching the graph of $f(x) = x^3 + x - 1$, you can see that it is one-to-one. Yet it would be difficult to determine the inverse of this function algebraically.

> **Guidelines for Finding an Inverse of a Function**
>
> 1. Determine whether the function given by $y = f(x)$ has an inverse function.
> 2. Solve for x as a function of y: $x = g(y) = f^{-1}(y)$.
> 3. Interchange x and y. The resulting equation is $y = f^{-1}(x)$.
> 4. Define the domain of f^{-1} to be the range of f.
> 5. Verify that $f(f^{-1}(x)) = x$ and $f^{-1}(f(x)) = x$.

EXAMPLE 3 Finding an Inverse Function

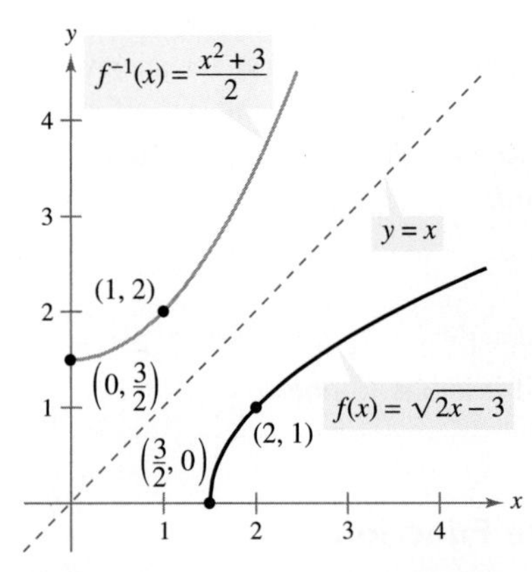

The domain of f^{-1}, $[0, \infty)$, is the range of f.
Figure 1.40

Find the inverse function of

$$f(x) = \sqrt{2x - 3}.$$

Solution The function has an inverse function because it is one-to-one on its entire domain (see Figure 1.40). To find an equation for the inverse function, let $y = f(x)$ and solve for x in terms of y.

$$\begin{aligned}
\sqrt{2x - 3} &= y && \text{Let } y = f(x). \\
2x - 3 &= y^2 && \text{Square both sides.} \\
x &= \frac{y^2 + 3}{2} && \text{Solve for } x. \\
y &= \frac{x^2 + 3}{2} && \text{Interchange } x \text{ and } y. \\
f^{-1}(x) &= \frac{x^2 + 3}{2} && \text{Replace } y \text{ by } f^{-1}(x).
\end{aligned}$$

The domain of f^{-1} is the range of f, which is $[0, \infty)$. You can verify this result as follows.

$$f(f^{-1}(x)) = \sqrt{2\left(\frac{x^2 + 3}{2}\right) - 3} = \sqrt{x^2} = x, \quad x \geq 0$$

$$f^{-1}(f(x)) = \frac{\left(\sqrt{2x - 3}\right)^2 + 3}{2} = \frac{2x - 3 + 3}{2} = x, \quad x \geq \frac{3}{2}$$

NOTE Remember that any letter can be used to represent the independent variable. So,

$$f^{-1}(y) = \frac{y^2 + 3}{2}, \qquad f^{-1}(x) = \frac{x^2 + 3}{2}, \qquad \text{and} \qquad f^{-1}(s) = \frac{s^2 + 3}{2}$$

all represent the same function.

Suppose you are given a function that is *not* one-to-one on its entire domain. By restricting the domain to an interval on which the function *is* one-to-one, you can conclude that the new function has an inverse function on the restricted domain.

EXAMPLE 4 Testing Whether a Function Is One-to-One

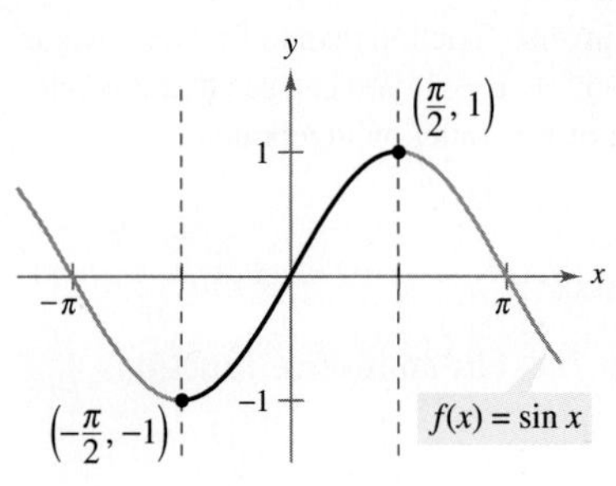

f is one-to-one on the interval $[-\pi/2, \pi/2]$.
Figure 1.41

Show that the sine function

$$f(x) = \sin x$$

is not one-to-one on the entire real line. Then show that f is one-to-one on the closed interval $[-\pi/2, \pi/2]$.

Solution It is clear that f is not one-to-one, because many different x-values yield the same y-value. For instance,

$$\sin(0) = 0 = \sin(\pi).$$

Moreover, from the graph of $f(x) = \sin x$ in Figure 1.41, you can see that when f is restricted to the interval $[-\pi/2, \pi/2]$, then the restricted function *is* one-to-one.

Inverse Trigonometric Functions

From the graphs of the six basic trigonometric functions, you can see that they do not have inverse functions. (Graphs of the six basic trigonometric functions are shown in Appendix D.) The functions that are called "inverse trigonometric functions" are actually inverses of trigonometric functions whose domains have been restricted.

For instance, in Example 4, you saw that the sine function is one-to-one on the interval $[-\pi/2, \pi/2]$ (see Figure 1.42). On this interval, you can define the inverse of the *restricted* sine function to be

$$y = \arcsin x \quad \text{if and only if} \quad \sin y = x$$

where $-1 \le x \le 1$ and $-\pi/2 \le \arcsin x \le \pi/2$. From Figures 1.42 (a) and (b), you can see that you can obtain the graph of $y = \arcsin x$ by reflecting the graph of $y = \sin x$ in the line $y = x$ on the interval $[-\pi/2, \pi/2]$.

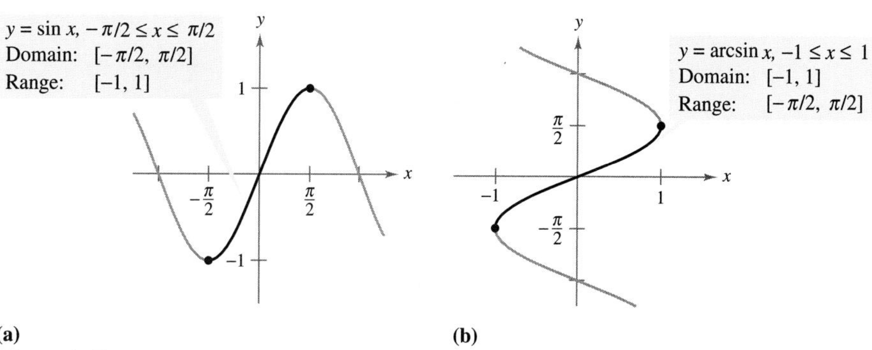

(a) (b)
Figure 1.42

Under suitable restrictions, each of the six trigonometric functions is one-to-one and so has an inverse function, as indicated in the following definition. (The term "iff" is used to represent the phrase "if and only if.")

EXPLORATION

Inverse Secant Function In the definition at the right, the inverse secant function is defined by restricting the domain of the secant function to the intervals

$$\left[0, \frac{\pi}{2}\right) \cup \left(\frac{\pi}{2}, \pi\right].$$

Most other texts and reference books agree with this, but some disagree. What other domains might make sense? Explain your reasoning graphically. Most calculators do not have a key for the inverse secant function. How can you use a calculator to evaluate the inverse secant function?

Definition of Inverse Trigonometric Functions

Function	*Domain*	*Range*
$y = \arcsin x$ iff $\sin y = x$	$-1 \le x \le 1$	$-\frac{\pi}{2} \le y \le \frac{\pi}{2}$
$y = \arccos x$ iff $\cos y = x$	$-1 \le x \le 1$	$0 \le y \le \pi$
$y = \arctan x$ iff $\tan y = x$	$-\infty < x < \infty$	$-\frac{\pi}{2} < y < \frac{\pi}{2}$
$y = \operatorname{arccot} x$ iff $\cot y = x$	$-\infty < x < \infty$	$0 < y < \pi$
$y = \operatorname{arcsec} x$ iff $\sec y = x$	$\lvert x \rvert \ge 1$	$0 \le y \le \pi, \quad y \ne \frac{\pi}{2}$
$y = \operatorname{arccsc} x$ iff $\csc y = x$	$\lvert x \rvert \ge 1$	$-\frac{\pi}{2} \le y \le \frac{\pi}{2}, \quad y \ne 0$

NOTE The term $\arcsin x$ is read as "the arcsine of x" or sometimes "the angle whose sine is x." An alternative notation for the inverse sine function is $\sin^{-1} x$.

The graphs of the six inverse trigonometric functions are shown in Figure 1.43.

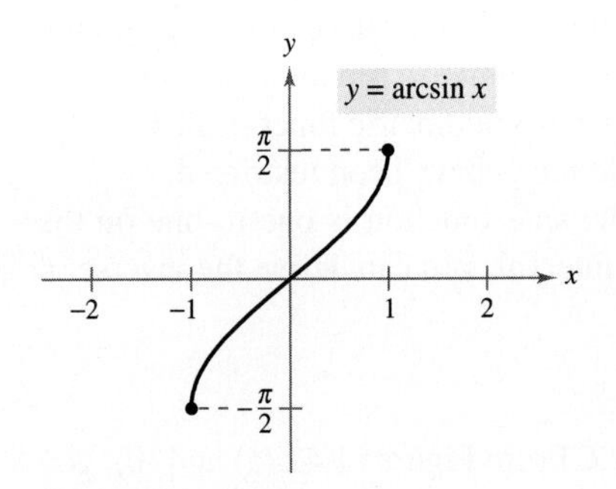

Domain: $[-1, 1]$
Range: $[-\pi/2, \pi/2]$

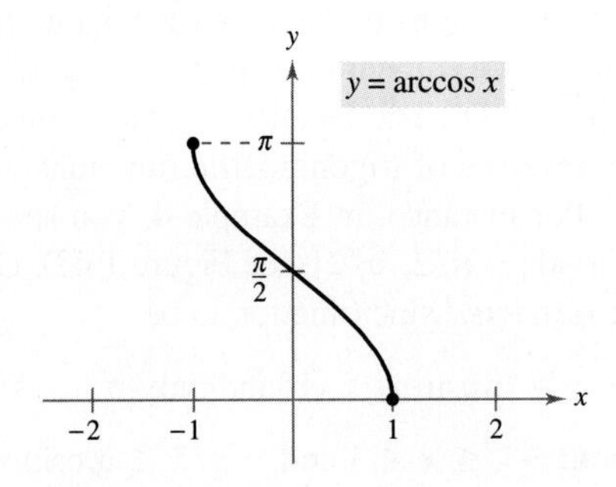

Domain: $[-1, 1]$
Range: $[0, \pi]$

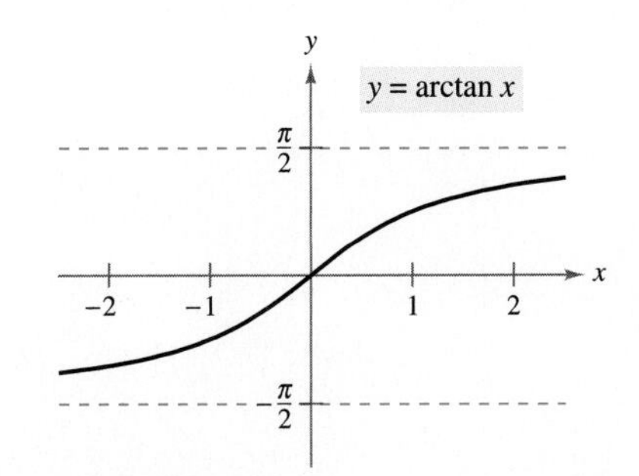

Domain: $(-\infty, \infty)$
Range: $(-\pi/2, \pi/2)$

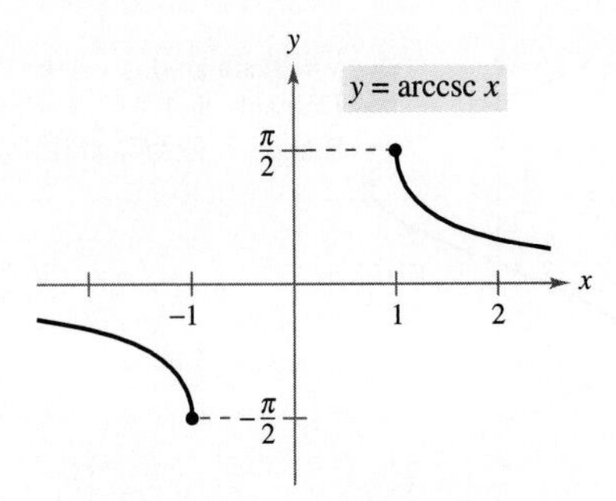

Domain: $(-\infty, -1] \cup [1, \infty)$
Range: $[-\pi/2, 0) \cup (0, \pi/2]$

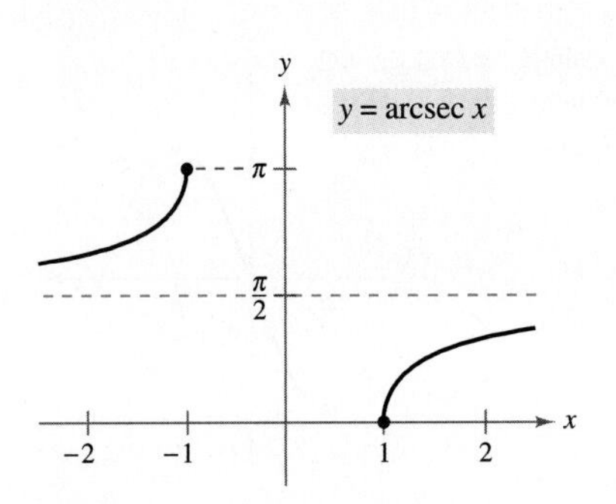

Domain: $(-\infty, -1] \cup [1, \infty)$
Range: $[0, \pi/2) \cup (\pi/2, \pi]$

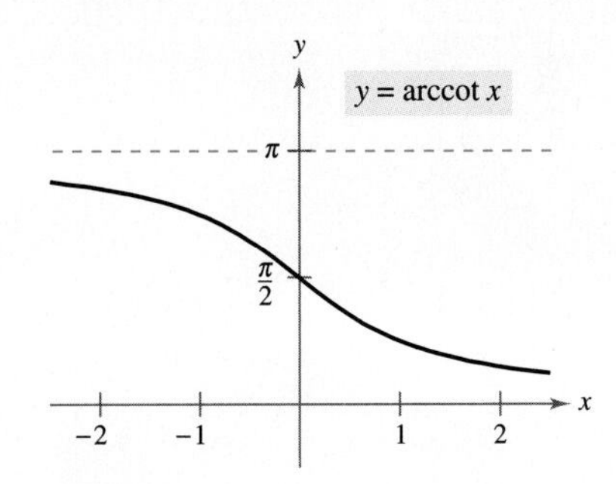

Domain: $(-\infty, \infty)$
Range: $(0, \pi)$

Figure 1.43

EXAMPLE 5 Evaluating Inverse Trigonometric Functions

Evaluate each of the following.

a. $\arcsin\left(-\frac{1}{2}\right)$ **b.** $\arccos 0$ **c.** $\arctan \sqrt{3}$ **d.** $\arcsin(0.3)$

NOTE When evaluating inverse trigonometric functions, remember that they denote *angles in radian measure.*

Solution

a. By definition, $y = \arcsin\left(-\frac{1}{2}\right)$ implies that $\sin y = -\frac{1}{2}$. In the interval $[-\pi/2, \pi/2]$, the correct value of y is $-\pi/6$.

$$\arcsin\left(-\frac{1}{2}\right) = -\frac{\pi}{6}$$

b. By definition, $y = \arccos 0$ implies that $\cos y = 0$. In the interval $[0, \pi]$, you have $y = \pi/2$.

$$\arccos 0 = \frac{\pi}{2}$$

c. By definition, $y = \arctan \sqrt{3}$ implies that $\tan y = \sqrt{3}$. In the interval $(-\pi/2, \pi/2)$, you have $y = \pi/3$.

$$\arctan \sqrt{3} = \frac{\pi}{3}$$

d. Using a calculator set in *radian* mode produces

$$\arcsin(0.3) \approx 0.3047.$$

EXPLORATION

Graph $y = \arccos(\cos x)$ for $-4\pi \le x \le 4\pi$. Why isn't the graph the same as the graph of $y = x$?

Inverse functions have the properties

$$f(f^{-1}(x)) = x \quad \text{and} \quad f^{-1}(f(x)) = x.$$

When applying these properties to inverse trigonometric functions, remember that the trigonometric functions have inverse functions only in restricted domains. For x-values outside these domains, these two properties do not hold. For example, $\arcsin(\sin \pi)$ is equal to 0, not π.

Properties of Inverse Trigonometric Functions

1. If $-1 \le x \le 1$ and $-\pi/2 \le y \le \pi/2$, then

$$\sin(\arcsin x) = x \quad \text{and} \quad \arcsin(\sin y) = y.$$

2. If $-\pi/2 < y < \pi/2$, then

$$\tan(\arctan x) = x \quad \text{and} \quad \arctan(\tan y) = y.$$

3. If $|x| \ge 1$ and $0 \le y < \pi/2$ or $\pi/2 < y \le \pi$, then

$$\sec(\text{arcsec } x) = x \quad \text{and} \quad \text{arcsec}(\sec y) = y.$$

Similar properties hold for the other inverse trigonometric functions.

EXAMPLE 6 Solving an Equation

$\arctan(2x - 3) = \dfrac{\pi}{4}$	Write original equation.
$\tan[\arctan(2x - 3)] = \tan\dfrac{\pi}{4}$	Take tangent of both sides.
$2x - 3 = 1$	$\tan(\arctan x) = x$
$x = 2$	Solve for x.

Some problems in calculus require that you evaluate expressions such as $\cos(\arcsin x)$, as shown in Example 7.

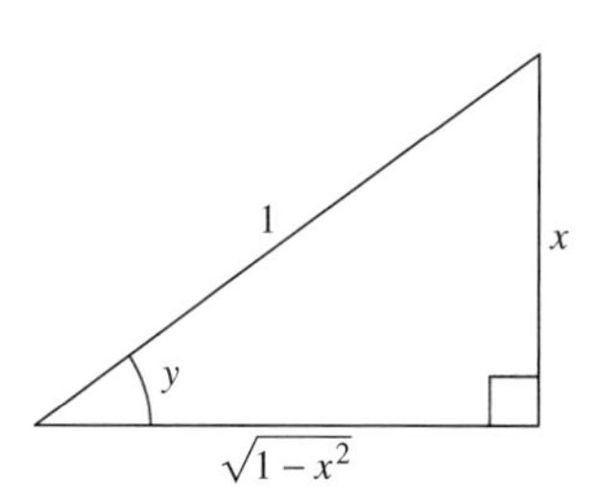

$y = \arcsin x$
Figure 1.44

EXAMPLE 7 Using Right Triangles

a. Given $y = \arcsin x$, where $0 < y < \pi/2$, find $\cos y$.
b. Given $y = \text{arcsec}(\sqrt{5}/2)$, find $\tan y$.

Solution

a. Because $y = \arcsin x$, you know that $\sin y = x$. This relationship between x and y can be represented by a right triangle, as shown in Figure 1.44.

$$\cos y = \cos(\arcsin x) = \frac{\text{adj.}}{\text{hyp.}} = \sqrt{1 - x^2}$$

(This result is also valid for $-\pi/2 < y < 0$.)

b. Use the right triangle shown in Figure 1.45.

$$\tan y = \tan\left[\text{arcsec}\left(\frac{\sqrt{5}}{2}\right)\right] = \frac{\text{opp.}}{\text{adj.}} = \frac{1}{2}$$

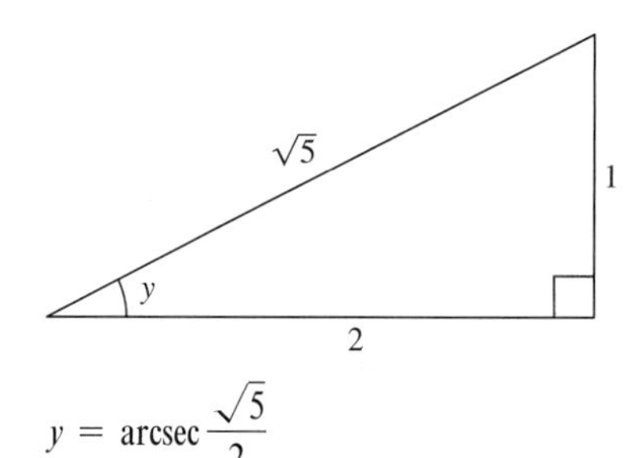

$y = \text{arcsec}\dfrac{\sqrt{5}}{2}$
Figure 1.45

Exercises for Section 1.5

See www.CalcChat.com for worked-out solutions to odd-numbered exercises.

In Exercises 1–8, show that f and g are inverse functions (a) analytically and (b) graphically.

1. $f(x) = 5x + 1,\quad g(x) = (x - 1)/5$
2. $f(x) = 3 - 4x,\quad g(x) = (3 - x)/4$
3. $f(x) = x^3,\quad g(x) = \sqrt[3]{x}$
4. $f(x) = 1 - x^3,\quad g(x) = \sqrt[3]{1 - x}$
5. $f(x) = \sqrt{x - 4},\quad g(x) = x^2 + 4,\ x \geq 0$
6. $f(x) = 16 - x^2,\ x \geq 0;\quad g(x) = \sqrt{16 - x}$
7. $f(x) = 1/x,\quad g(x) = 1/x$
8. $f(x) = \dfrac{1}{1 + x},\ x \geq 0;\quad g(x) = \dfrac{1 - x}{x},\ 0 < x \leq 1$

In Exercises 9–12, match the graph of the function with the graph of its inverse function. [The graphs of the inverse functions are labeled (a), (b), (c), and (d).]

(a)

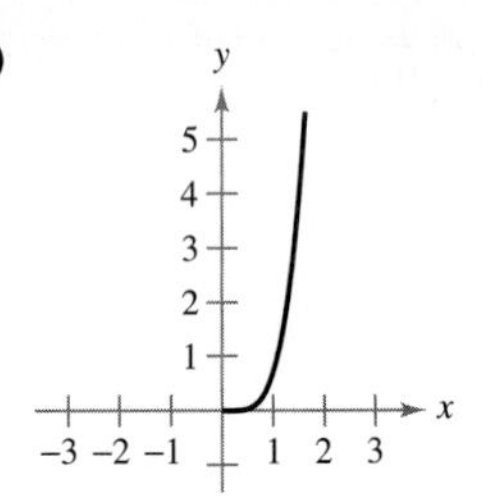

(b)

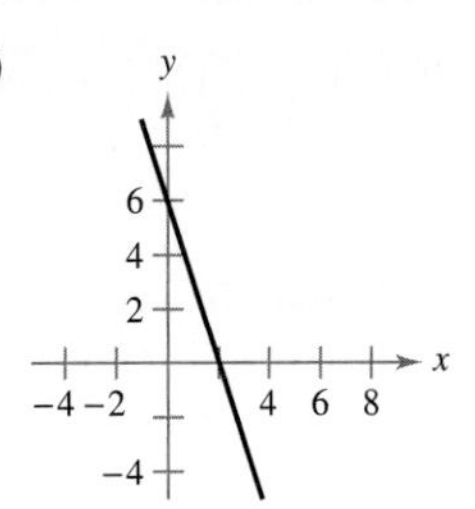

(c)

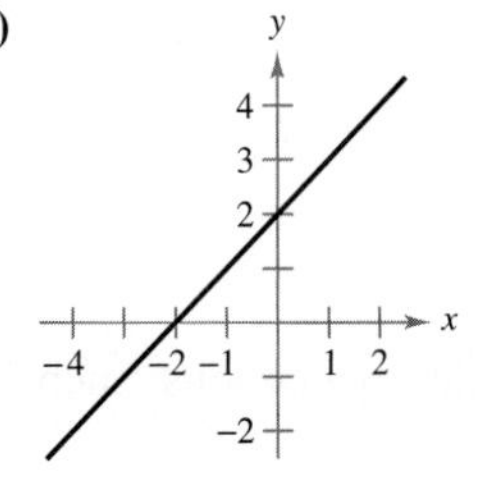

(d)

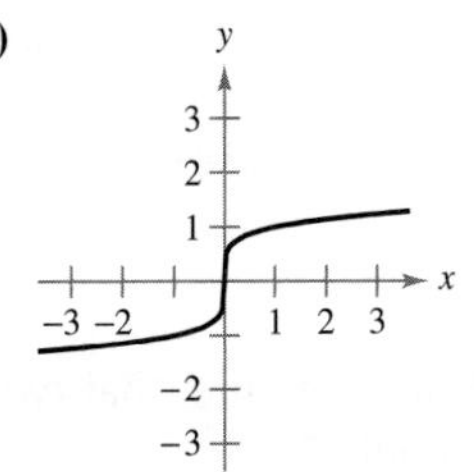

9.

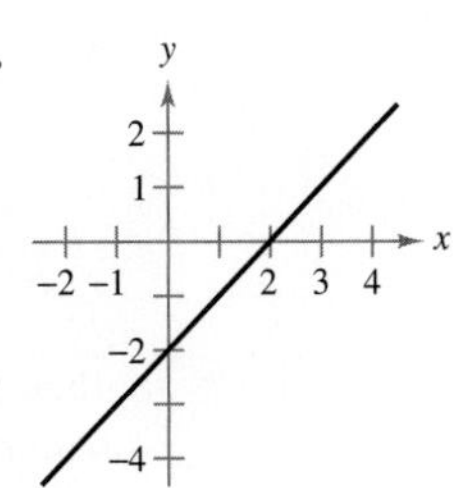

10.

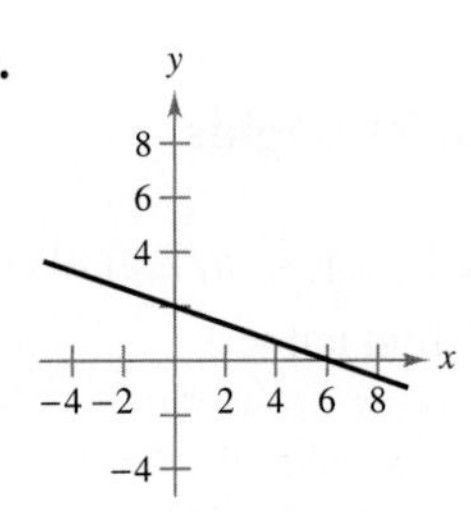

11.

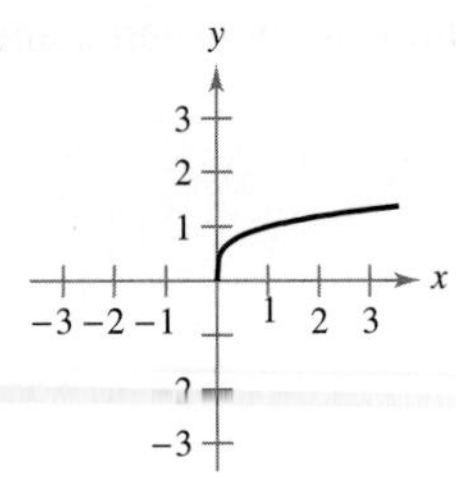

12.

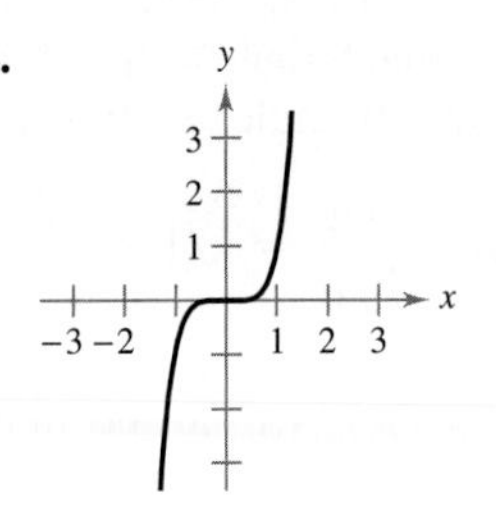

In Exercises 13–16, use the Horizontal Line Test to determine whether the function is one-to-one on its entire domain and therefore has an inverse function. To print an enlarged copy of the graph, go to the website *www.mathgraphs.com*.

13. $f(x) = \frac{3}{4}x + 6$

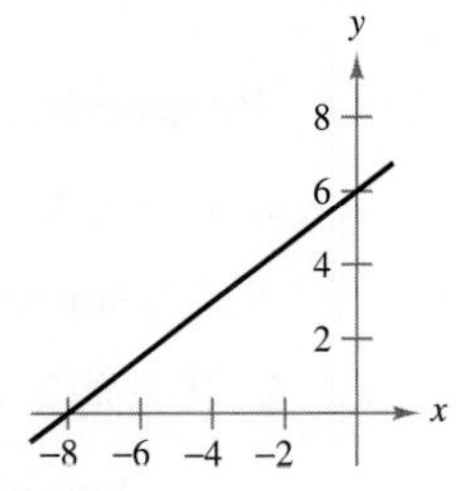

14. $f(x) = 5x - 3$

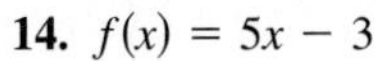

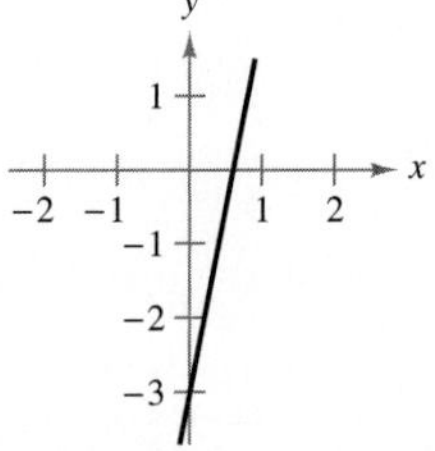

15. $f(\theta) = \sin \theta$

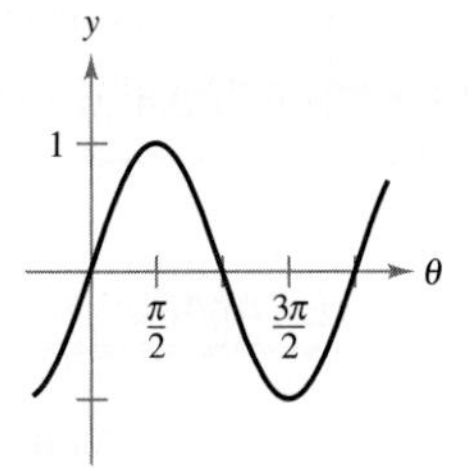

16. $f(x) = \dfrac{x^2}{x^2 + 4}$

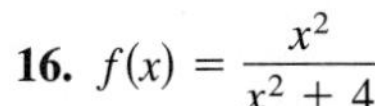

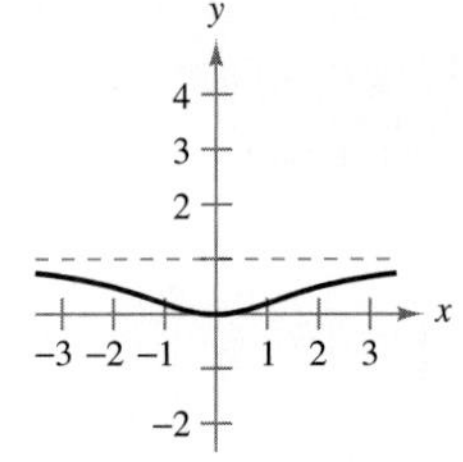

In Exercises 17–22, use a graphing utility to graph the function. Determine whether the function is one-to-one on its entire domain.

17. $h(s) = \dfrac{1}{s - 2} - 3$
18. $f(x) = \dfrac{1}{x^2 + 1}$
19. $g(t) = \dfrac{1}{\sqrt{t^2 + 1}}$
20. $f(x) = 5x\sqrt{x - 1}$
21. $g(x) = (x + 5)^3$
22. $h(x) = |x + 4| - |x - 4|$

In Exercises 23–28, determine whether the function is one-to-one on its entire domain and therefore has an inverse function.

23. $f(x) = (x + a)^3 + b$
24. $f(x) = \sin \dfrac{3x}{2}$
25. $f(x) = \dfrac{x^4}{4} - 2x^2$
26. $f(x) = x^3 - 6x^2 + 12x$
27. $f(x) = 2 - x - x^3$
28. $f(x) = \sqrt[3]{x + 1}$

In Exercises 29–36, find the inverse function of f. Graph (by hand) f and f^{-1}. Describe the relationship between the graphs.

29. $f(x) = 2x - 3$
30. $f(x) = 3x$
31. $f(x) = x^5$
32. $f(x) = x^3 - 1$
33. $f(x) = \sqrt{x}$

34. $f(x) = x^2, \quad x \geq 0$

35. $f(x) = \sqrt{4 - x^2}, \quad x \geq 0$

36. $f(x) = \sqrt{x^2 - 4}, \quad x \geq 2$

In Exercises 37–42, find the inverse function of f. Use a graphing utility to graph f and f^{-1} in the same viewing window. Describe the relationship between the graphs.

37. $f(x) = \sqrt[3]{x - 1}$

38. $f(x) = 3\sqrt[5]{2x - 1}$

39. $f(x) = x^{2/3}, \quad x \geq 0$

40. $f(x) = x^{3/5}$

41. $f(x) = \dfrac{x}{\sqrt{x^2 + 7}}$

42. $f(x) = \dfrac{x + 2}{x}$

In Exercises 43 and 44, use the graph of the function f to complete the table and sketch the graph of f^{-1}. To print an enlarged copy of the graph, go to the website *www.mathgraphs.com*.

43.

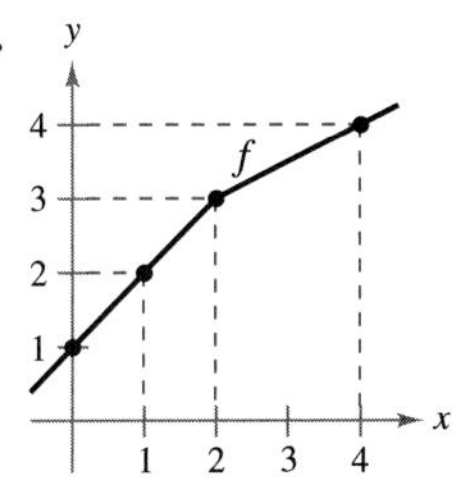

x	1	2	3	4
$f^{-1}(x)$				

44.

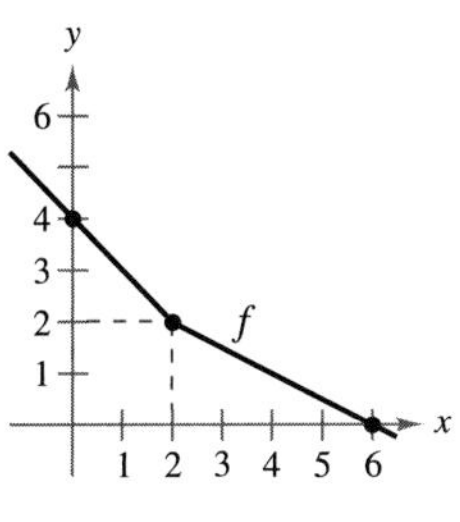

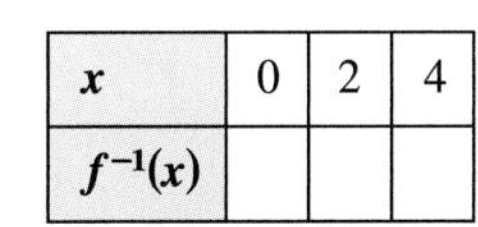

x	0	2	4
$f^{-1}(x)$			

45. ***Cost*** You need 50 pounds of two commodities costing \$1.25 and \$1.60 per pound.

(a) Verify that the total cost is $y = 1.25x + 1.60(50 - x)$, where x is the number of pounds of the less expensive commodity.

(b) Find the inverse function of the cost function. What does each variable represent in the inverse function?

(c) Use the context of the problem to determine the domain of the inverse function.

(d) Determine the number of pounds of the less expensive commodity purchased if the total cost is \$73.

46. ***Temperature*** The formula $C = \frac{5}{9}(F - 32)$, where $F \geq -459.6$, represents the Celsius temperature C as a function of Fahrenheit temperature F.

(a) Find the inverse function of C.

(b) What does the inverse function represent?

(c) Determine the domain of the inverse function.

(d) The temperature is 22°C. What is the corresponding temperature in degrees Fahrenheit?

In Exercises 47 and 48, find f^{-1} over the indicated interval. Use a graphing utility to graph f and f^{-1} in the same viewing window. Describe the relationship between the graphs.

47. $f(x) = \dfrac{x}{x^2 - 4}; \quad (-2, 2)$ **48.** $f(x) = 2 - \dfrac{3}{x^2}; \quad (0, 10)$

Graphical Reasoning **In Exercises 49–52, (a) use a graphing utility to graph the function, (b) use the *drawing* feature of the graphing utility to draw the inverse of the function, and (c) determine whether the graph of the inverse relation is an inverse function. Explain your reasoning.**

49. $f(x) = x^3 + x + 4$

50. $h(x) = x\sqrt{4 - x^2}$

51. $g(x) = \dfrac{3x^2}{x^2 + 1}$

52. $f(x) = \dfrac{4x}{\sqrt{x^2 + 15}}$

In Exercises 53–58, show that f is one-to-one on the indicated interval and therefore has an inverse function on that interval.

	Function	*Interval*
53.	$f(x) = (x - 4)^2$	$[4, \infty)$
54.	$f(x) = \lvert x + 2 \rvert$	$[-2, \infty)$
55.	$f(x) = \dfrac{4}{x^2}$	$(0, \infty)$
56.	$f(x) = \cot x$	$(0, \pi)$
57.	$f(x) = \cos x$	$[0, \pi]$
58.	$f(x) = \sec x$	$\left[0, \dfrac{\pi}{2}\right)$

In Exercises 59–62, determine whether the function is one-to-one. If it is, find its inverse function.

59. $f(x) = \sqrt{x - 2}$

60. $f(x) = -3$

61. $f(x) = \lvert x - 2 \rvert, \quad x \leq 2$

62. $f(x) = ax + b, \quad a \neq 0$

In Exercises 63–66, delete part of the domain so that the function that remains is one-to-one. Find the inverse function of the remaining function and give the domain of the inverse function. (*Note:* There is more than one correct answer.)

63. $f(x) = (x - 3)^2$

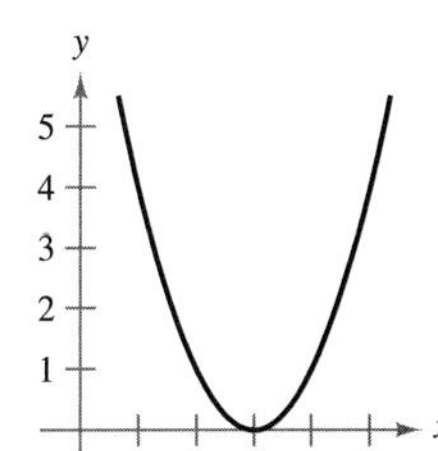

64. $f(x) = 16 - x^4$

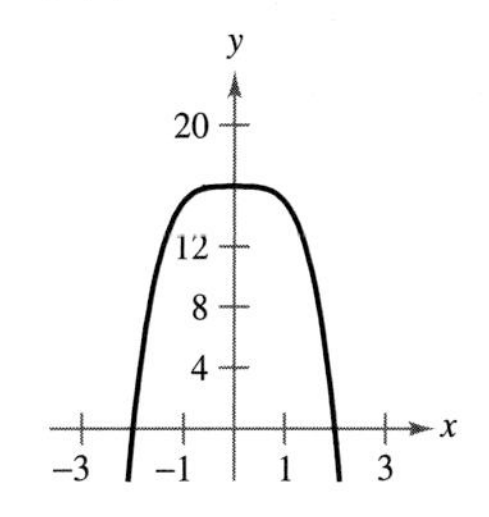

65. $f(x) = |x + 3|$

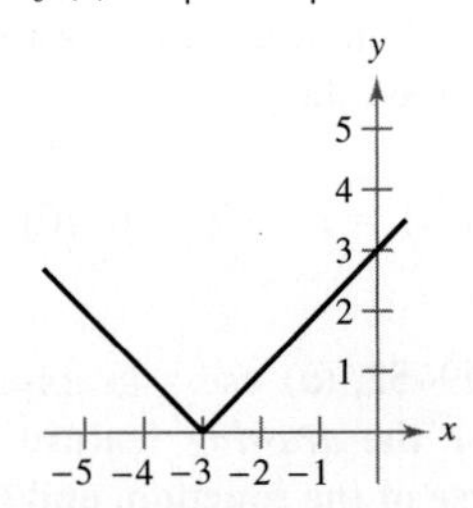

66. $f(x) = |x - 3|$

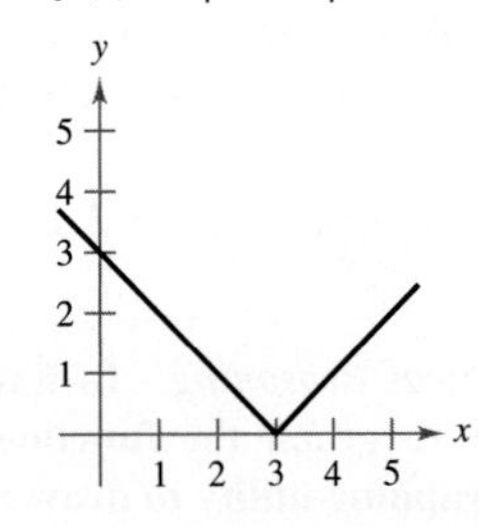

In Exercises 67–72, find $(f^{-1})(a)$ for the function f and real number a.

Function	*Real Number*
67. $f(x) = x^3 + 2x - 1$	$a = 2$
68. $f(x) = 2x^5 + x^3 + 1$	$a = -2$
69. $f(x) = \sin x, \quad -\frac{\pi}{2} \le x \le \frac{\pi}{2}$	$a = \frac{1}{2}$
70. $f(x) = \cos 2x, \quad 0 \le x \le \frac{\pi}{2}$	$a = 1$
71. $f(x) = x^3 - \frac{4}{x}, \quad x > 0$	$a = 6$
72. $f(x) = \sqrt{x - 4}$	$a = 2$

In Exercises 73–76, use the functions $f(x) = \frac{1}{8}x - 3$ and $g(x) = x^3$ to find the indicated value.

73. $(f^{-1} \circ g^{-1})(1)$

74. $(g^{-1} \circ f^{-1})(-3)$

75. $(f^{-1} \circ f^{-1})(6)$

76. $(g^{-1} \circ g^{-1})(-4)$

In Exercises 77–80, use the functions $f(x) = x + 4$ and $g(x) = 2x - 5$ to find the indicated function.

77. $g^{-1} \circ f^{-1}$

78. $f^{-1} \circ g^{-1}$

79. $(f \circ g)^{-1}$

80. $(g \circ f)^{-1}$

In Exercises 81 and 82, (a) use the graph of the function f to determine whether f is one-to-one, (b) state the domain of f^{-1}, and (c) estimate the value of $f^{-1}(2)$.

81.

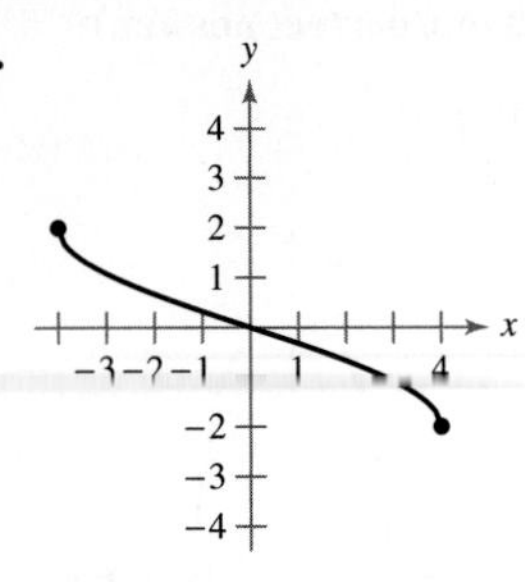

82.

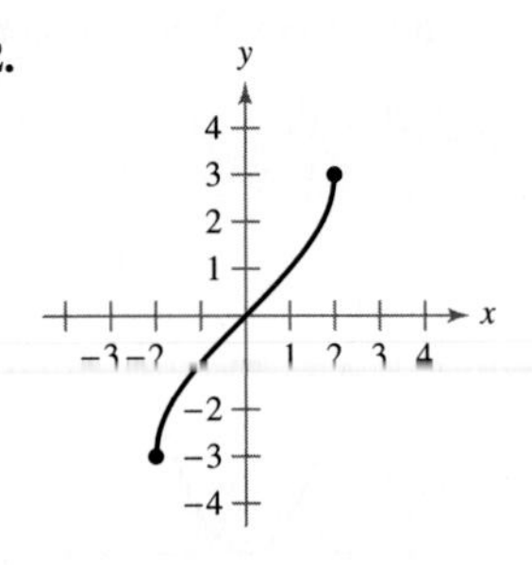

In Exercises 83 and 84, use the graph of the function f to sketch the graph of f^{-1}. To print an enlarged copy of the graph, go to the website *www.mathgraphs.com*.

83.

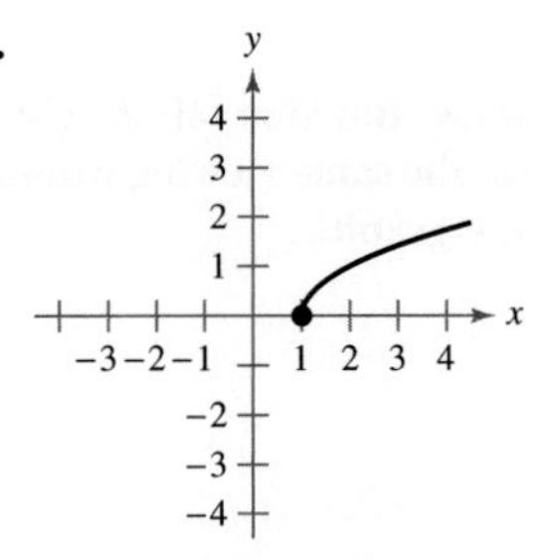

84.

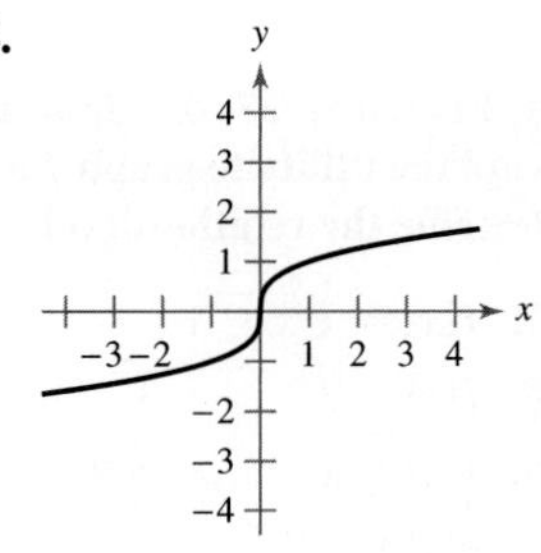

Numerical and Graphical Analysis **In Exercises 85 and 86, (a) use a graphing utility to complete the table, (b) plot the points in the table and graph the function by hand, (c) use a graphing utility to graph the function and compare the result with your hand-drawn graph in part (b), and (d) determine any intercepts and symmetry of the graph.**

x	-1	-0.8	-0.6	-0.4	-0.2	0	0.2	0.4	0.6	0.8	1
y											

85. $y = \arcsin x$

86. $y = \arccos x$

87. Determine the missing coordinates of the points on the graph of the function.

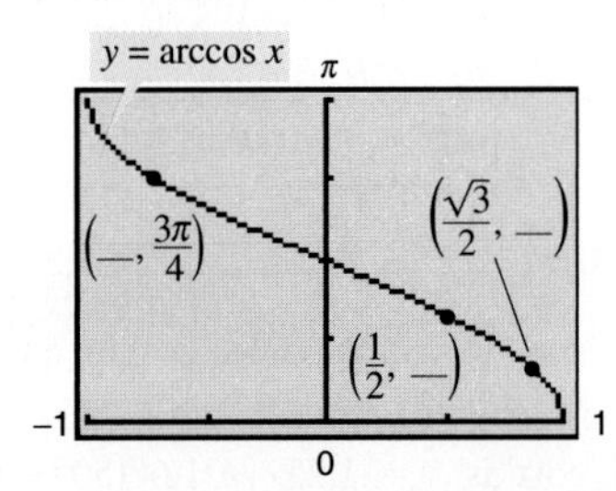

88. Determine the missing coordinates of the points on the graph of the function.

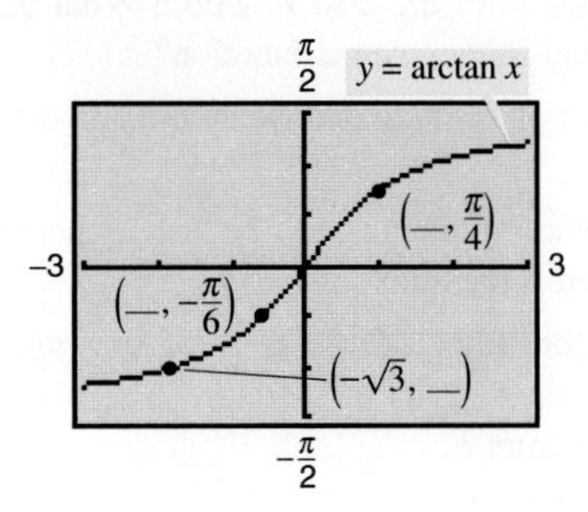

In Exercises 89–96, evaluate the expression without using a calculator.

89. $\arcsin \frac{1}{2}$

90. $\arcsin 0$

91. $\arccos \frac{1}{2}$

92. $\arccos 0$

93. $\arctan \frac{\sqrt{3}}{3}$

94. $\operatorname{arccot}(-\sqrt{3})$

95. $\operatorname{arccsc}(-\sqrt{2})$

96. $\arccos\left(-\frac{\sqrt{3}}{2}\right)$

In Exercises 97–100, use a calculator to approximate the value. Round your answer to two decimal places.

97. $\arccos(-0.8)$

98. $\arcsin(-0.39)$

99. $\operatorname{arcsec} 1.269$

100. $\arctan(-3)$

Writing About Concepts

101. Describe how to find the inverse function of a one-to-one function given by an equation in x and y. Give an example.

102. Describe the relationship between the graph of a function and the graph of its inverse function.

103. Give an example of a function that does *not* have an inverse function.

104. Explain why $\tan \pi = 0$ does not imply that $\arctan 0 = \pi$.

105. Explain why the domains of the trigonometric functions are restricted when finding the inverse trigonometric functions.

106. Explain how to graph $y = \operatorname{arccot} x$ on a graphing utility that does not have the arccotangent function.

In Exercises 107 and 108, use a graphing utility to confirm that f and g are inverse functions. (Remember to restrict the domain of f properly.)

107. $f(x) = \tan x$

$g(x) = \arctan x$

108. $f(x) = \sin x$

$g(x) = \arcsin x$

In Exercises 109 and 110, use the properties of inverse trigonometric functions to evaluate the expression.

109. $\cos[\arccos(-0.1)]$

110. $\arcsin(\sin 3\pi)$

In Exercises 111–116, evaluate the expression without using a calculator. (*Hint:* Make a sketch of a right triangle, as illustrated in Example 7.)

111. (a) $\sin\left(\arcsin \frac{1}{2}\right)$

(b) $\cos\left(\arcsin \frac{1}{2}\right)$

112. (a) $\tan\left(\arccos \frac{\sqrt{2}}{2}\right)$

(b) $\cos\left(\arcsin \frac{5}{13}\right)$

113. (a) $\sin\left(\arctan \frac{3}{4}\right)$

(b) $\sec\left(\arcsin \frac{4}{5}\right)$

114. (a) $\tan(\operatorname{arccot} 2)$

(b) $\cos(\operatorname{arcsec} \sqrt{5})$

115. (a) $\cot\left[\arcsin\left(-\frac{1}{2}\right)\right]$

(b) $\csc\left[\arctan\left(-\frac{5}{12}\right)\right]$

116. (a) $\sec\left[\arctan\left(-\frac{3}{5}\right)\right]$

(b) $\tan\left[\arcsin\left(-\frac{5}{6}\right)\right]$

In Exercises 117–120, solve the equation for x.

117. $\arcsin(3x - \pi) = \frac{1}{2}$

118. $\arctan(2x - 5) = -1$

119. $\arcsin \sqrt{2x} = \arccos \sqrt{x}$

120. $\arccos x = \operatorname{arcsec} x$

In Exercises 121 and 122, find the point of intersection of the graphs of the functions.

121. $y = \arccos x$

$y = \arctan x$

122. $y = \arcsin x$

$y = \arccos x$

In Exercises 123–132, write the expression in algebraic form.

123. $\tan(\arctan x)$

124. $\sin(\arccos x)$

125. $\cos(\arcsin 2x)$

126. $\sec(\arctan 4x)$

127. $\sin(\operatorname{arcsec} x)$

128. $\cos(\operatorname{arccot} x)$

129. $\tan\left(\operatorname{arcsec} \dfrac{x}{3}\right)$

130. $\sec[\arcsin(x - 1)]$

131. $\csc\left(\arctan \dfrac{x}{\sqrt{2}}\right)$

132. $\cos\left(\arcsin \dfrac{x - h}{r}\right)$

In Exercises 133 and 134, fill in the blank.

133. $\arctan \dfrac{9}{x} = \arcsin\left(\square\right), \quad x > 0$

134. $\arcsin \dfrac{\sqrt{36 - x^2}}{6} = \arccos\left(\square\right)$

In Exercises 135 and 136, verify each identity.

135. (a) $\operatorname{arccsc} x = \arcsin \dfrac{1}{x}, \quad |x| \geq 1$

(b) $\arctan x + \arctan \dfrac{1}{x} = \dfrac{\pi}{2}, \quad x > 0$

136. (a) $\arcsin(-x) = -\arcsin x, \quad |x| \leq 1$

(b) $\arccos(-x) = \pi - \arccos x, \quad |x| \leq 1$

In Exercises 137–140, sketch the graph of the function. Use a graphing utility to verify your graph.

137. $f(x) = \arcsin(x - 1)$

138. $f(x) = \arctan x + \dfrac{\pi}{2}$

139. $f(x) = \operatorname{arcsec} 2x$

140. $f(x) = \arccos \dfrac{x}{4}$

141. Prove that if f and g are one-to-one functions, then $(f \circ g)^{-1}(x) = (g^{-1} \circ f^{-1})(x)$.

142. Prove that if f has an inverse function, then $(f^{-1})^{-1} = f$.

143. Prove that if a function has an inverse function, then the inverse function is unique.

144. Prove that a function has an inverse function if and only if it is one-to-one.

True or False? **In Exercises 145–150, determine whether the statement is true or false. If it is false, explain why or give an example that shows it is false.**

145. If f is an even function, then f^{-1} exists.

146. If the inverse function of f exists, then the y-intercept of f is an x-intercept of f^{-1}.

147. $\arcsin^2 x + \arccos^2 x = 1$

148. The range of $y = \arcsin x$ is $[0, \pi]$.

149. If $f(x) = x^n$ where n is odd, then f^{-1} exists.

150. There exists no function f such that $f = f^{-1}$.

151. Prove that

$$\arctan x + \arctan y = \arctan \frac{x + y}{1 - xy}, \quad xy \neq 1.$$

Use this formula to show that

$$\arctan \frac{1}{2} + \arctan \frac{1}{3} = \frac{\pi}{4}.$$

152. ***Think About It*** Use a graphing utility to graph

$$f(x) = \sin x \quad \text{and} \quad g(x) = \arcsin(\sin x).$$

Why isn't the graph of g the line $y = x$?

153. Let $f(x) = a^2 + bx + c$, where $a > 0$ and the domain is all real numbers such that $x \leq -\dfrac{b}{2a}$. Find f^{-1}.

154. Determine conditions on the constants a, b, and c such that the graph of $f(x) = \dfrac{ax + b}{cx - a}$ is symmetric about the line $y = x$.

155. Determine conditions on the constants a, b, c, and d such that $f(x) = \dfrac{ax + b}{cx + d}$ has an inverse function. Then find f^{-1}.

Section 1.6

Exponential and Logarithmic Functions

- **Develop and use properties of exponential functions.**
- **Understand the definition of the number *e*.**
- **Understand the definition of the natural logarithmic function.**
- **Develop and use properties of the natural logarithmic function.**

Exponential Functions

An **exponential function** involves a constant raised to a power, such as $f(x) = 2^x$. You already know how to evaluate 2^x for *rational* values of x. For instance,

$$2^0 = 1, \quad 2^2 = 4, \quad 2^{-1} = \frac{1}{2}, \quad \text{and} \quad 2^{1/2} = \sqrt{2} \approx 1.4142136.$$

For *irrational* values of x, you can define 2^x by considering a sequence of rational numbers that approach x. A full discussion of this process would not be appropriate here, but the general idea is as follows. Suppose you want to define the number $2^{\sqrt{2}}$. Because $\sqrt{2} = 1.414213 \ldots$, you consider the following numbers (which are of the form 2^r, where r is rational).

$$2^1 = 2 < 2^{\sqrt{2}} < 4 = 2^2$$

$$2^{1.4} = 2.639015 \ldots < 2^{\sqrt{2}} < 2.828427 \ldots = 2^{1.5}$$

$$2^{1.41} = 2.657371 \ldots < 2^{\sqrt{2}} < 2.675855 \ldots = 2^{1.42}$$

$$2^{1.414} = 2.664749 \ldots < 2^{\sqrt{2}} < 2.666597 \ldots = 2^{1.415}$$

$$2^{1.4142} = 2.665119 \ldots < 2^{\sqrt{2}} < 2.665303 \ldots = 2^{1.4143}$$

$$2^{1.41421} = 2.665137 \ldots < 2^{\sqrt{2}} < 2.665156 \ldots = 2^{1.41422}$$

$$2^{1.414213} = 2.665143 \ldots < 2^{\sqrt{2}} < 2.665144 \ldots = 2^{1.414214}$$

From these calculations, it seems reasonable to conclude that

$$2^{\sqrt{2}} \approx 2.66514.$$

In practice, you can use a calculator to approximate numbers such as $2^{\sqrt{2}}$.

In general, you can use any positive base a, $a \neq 1$, to define an exponential function. Thus, the exponential function with base a is written as $f(x) = a^x$. Exponential functions, even those with irrational values of x, obey the familiar properties of exponents.

Properties of Exponents

Let a and b be positive real numbers, and let x and y be any real numbers.

1. $a^0 = 1$ **2.** $a^x a^y = a^{x+y}$ **3.** $(a^x)^y = a^{xy}$ **4.** $(ab)^x = a^x b^x$

5. $\dfrac{a^x}{a^y} = a^{x-y}$ **6.** $\left(\dfrac{a}{b}\right)^x = \dfrac{a^x}{b^x}$ **7.** $a^{-x} = \dfrac{1}{a^x}$

EXAMPLE 1 Using Properties of Exponents

a. $(2^2)(2^3) = 2^{2+3} = 2^5$ **b.** $\dfrac{2^2}{2^3} = 2^{2-3} = 2^{-1} = \dfrac{1}{2}$

c. $(3^x)^3 = 3^{3x}$ **d.** $\left(\dfrac{1}{3}\right)^{-x} = (3^{-1})^{-x} = 3^x$

EXAMPLE 2 Sketching Graphs of Exponential Functions

Sketch the graphs of the functions

$$f(x) = 2^x, \quad g(x) = \left(\tfrac{1}{2}\right)^x = 2^{-x}, \quad \text{and} \quad h(x) = 3^x.$$

Solution To sketch the graphs of these functions by hand, you can complete a table of values, plot the corresponding points, and connect the points with smooth curves.

x	-3	-2	-1	0	1	2	3	4
2^x	$\frac{1}{8}$	$\frac{1}{4}$	$\frac{1}{2}$	1	2	4	8	16
2^{-x}	8	4	2	1	$\frac{1}{2}$	$\frac{1}{4}$	$\frac{1}{8}$	$\frac{1}{16}$
3^x	$\frac{1}{27}$	$\frac{1}{9}$	$\frac{1}{3}$	1	3	9	27	81

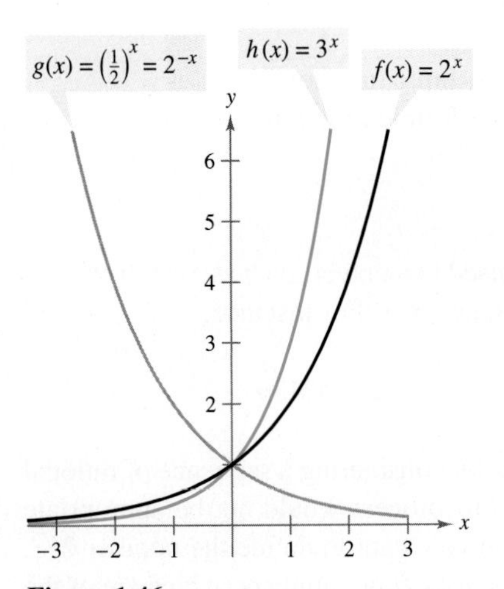

Figure 1.46

Another way to graph these functions is to use a graphing utility. In either case, you should obtain graphs similar to those shown in Figure 1.46.

The shapes of the graphs in Figure 1.46 are typical of the exponential functions a^x and a^{-x} where $a > 1$, as shown in Figure 1.47.

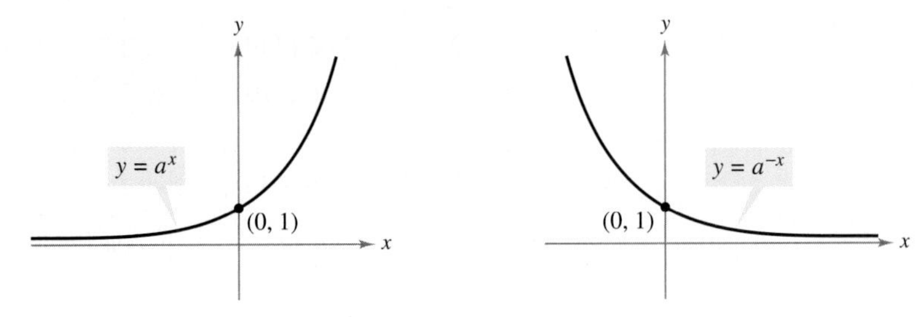

Figure 1.47

> **Properties of Exponential Functions**
>
> Let a be a real number that is greater than 1.
>
> **1.** The domain of $f(x) = a^x$ and $g(x) = a^{-x}$ is $(-\infty, \infty)$.
> **2.** The range of $f(x) = a^x$ and $g(x) = a^{-x}$ is $(0, \infty)$.
> **3.** The y-intercept of $f(x) = a^x$ and $g(x) = a^{-x}$ is $(0, 1)$.
> **4.** The functions $f(x) = a^x$ and $g(x) = a^{-x}$ are one-to-one.

TECHNOLOGY Functions of the form $h(x) = b^{cx}$ have the same types of properties and graphs as functions of the form $f(x) = a^x$ and $g(x) = a^{-x}$. To see why this is true, notice that

$$b^{cx} = (b^c)^x.$$

For instance, $f(x) = 2^{3x}$ can be written as $f(x) = (2^3)^x$ or $f(x) = 8^x$. Try confirming this by graphing $f(x) = 2^{3x}$ and $g(x) = 8^x$ in the same viewing window.

The Number e

In calculus, the natural (or convenient) choice for a base of an exponential number is the irrational number e, whose decimal approximation is

$$e \approx 2.71828182846.$$

This choice may seem anything but natural. However, the convenience of this particular base will become apparent as you continue in this course.

EXAMPLE 3 Investigating the Number e

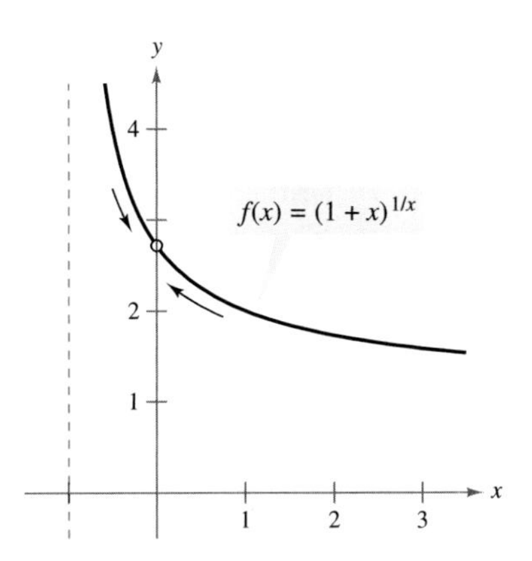

Figure 1.48

Use a graphing utility to graph the function

$$f(x) = (1 + x)^{1/x}.$$

Describe the behavior of the function at values of x that are close to 0.

Solution One way to examine the values of $f(x)$ near 0 is to construct a table.

x	-0.01	-0.001	-0.0001	0.0001	0.001	0.01
$(1 + x)^{1/x}$	2.7320	2.7196	2.7184	2.7181	2.7169	2.7048

From the table, it appears that the closer x gets to 0, the closer $(1 + x)^{1/x}$ gets to e. You can confirm this by graphing the function f, as shown in Figure 1.48. Try using a graphing calculator to obtain this graph. Then zoom in closer and closer to $x = 0$. Although f is not defined when $x = 0$, it is defined for x-values that are arbitrarily close to zero. By zooming in, you can see that the value of $f(x)$ gets closer and closer to $e \approx 2.71828182846$ as x gets closer and closer to 0. Later, when you study limits, you will learn that this result can be written as

$$\lim_{x \to 0} (1 + x)^{1/x} = e$$

which is read as "the limit of $(1 + x)^{1/x}$ as x approaches 0 is e."

EXAMPLE 4 The Graph of the Natural Exponential Function

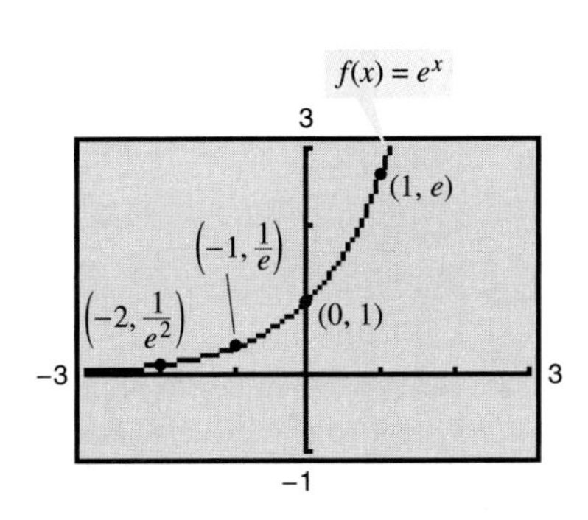

Figure 1.49

Sketch the graph of $f(x) = e^x$.

Solution To sketch the graph by hand, you can complete a table of values.

x	-2	-1	0	1	2
e^x	0.135	0.368	1	2.718	7.389

You can also use a graphing utility to graph the function. From the values in the table, you can see that a good viewing window for the graph is $-3 \le x \le 3$ and $-1 \le y \le 3$, as shown in Figure 1.49.

The Natural Logarithmic Function

Because the natural exponential function $f(x) = e^x$ is one-to-one, it must have an inverse function. Its inverse is called the **natural logarithmic function.** The domain of the natural logarithmic function is the set of positive real numbers.

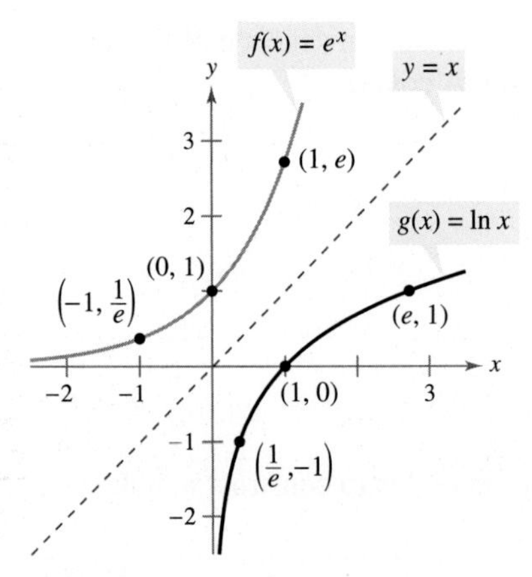

Figure 1.50

> **Definition of the Natural Logarithmic Function**
>
> Let x be a positive real number. The **natural logarithmic function,** denoted by $\ln x$, is defined as follows. ($\ln x$ is read as "el en of x" or "the natural log of x.")
>
> $\ln x = b$ if and only if $e^b = x$.

This definition tells you that a logarithmic equation can be written in an equivalent exponential form, and vice versa. Here are some examples.

Logarithmic Form	*Exponential Form*
$\ln 1 = 0$	$e^0 = 1$
$\ln e = 1$	$e^1 = e$
$\ln e^{-1} = -1$	$e^{-1} = \dfrac{1}{e}$

Because the function $g(x) = \ln x$ is defined to be the inverse of $f(x) = e^x$, it follows that the graph of the natural logarithmic function is a reflection of the graph of the natural exponential function in the line $y = x$, as shown in Figure 1.50. Several other properties of the natural logarithmic function also follow directly from its definition as the inverse of the natural exponential function.

> **Properties of the Natural Logarithmic Function**
>
> **1.** The domain of $g(x) = \ln x$ is $(0, \infty)$.
> **2.** The range of $g(x) = \ln x$ is $(-\infty, \infty)$.
> **3.** The x-intercept of $g(x) = \ln x$ is $(1, 0)$.
> **4.** The function $g(x) = \ln x$ is one-to-one.

Because $f(x) = e^x$ and $g(x) = \ln x$ are inverses of each other, you can conclude that

$$\ln e^x = x \quad \text{and} \quad e^{\ln x} = x.$$

EXPLORATION

The graphing utility screen in Figure 1.51 shows the graph of $y_1 = \ln e^x$ or $y_2 = e^{\ln x}$. Which graph is it? What are the domains of y_1 and y_2? Does $\ln e^x = e^{\ln x}$ for all real values of x? Explain.

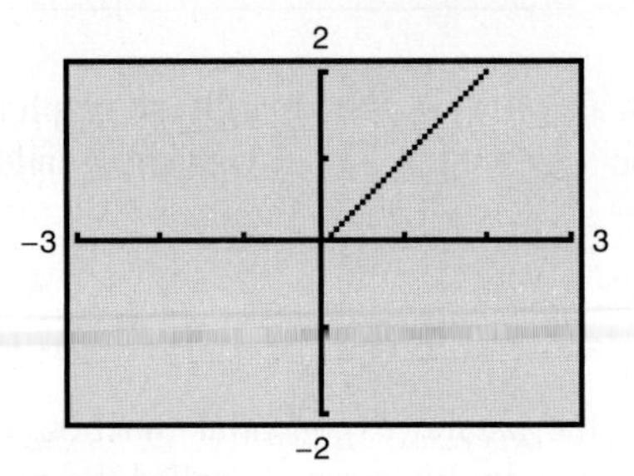

Figure 1.51

Properties of Logarithms

One of the properties of exponents states that when you multiply two exponential functions (having the same base), you add their exponents. For instance,

$$e^x e^y = e^{x+y}.$$

The logarithmic version of this property states that the natural logarithm of the product of two numbers is equal to the sum of the natural logs of the numbers. That is,

$$\ln xy = \ln x + \ln y.$$

This property and the properties dealing with the natural log of a quotient and the natural log of a power are listed here.

Properties of Logarithms

Let x, y, and z be real numbers such that $x > 0$ and $y > 0$.

1. $\ln xy = \ln x + \ln y$

2. $\ln \frac{x}{y} = \ln x - \ln y$

3. $\ln x^z = z \ln x$

EXAMPLE 5 Expanding Logarithmic Expressions

a. $\ln \frac{10}{9} = \ln 10 - \ln 9$ — Property 2

b. $\ln\sqrt{3x+2} = \ln(3x+2)^{1/2}$ — Rewrite with rational exponent.

$= \frac{1}{2}\ln(3x+2)$ — Property 3

c. $\ln \frac{6x}{5} = \ln(6x) - \ln 5$ — Property 2

$= \ln 6 + \ln x - \ln 5$ — Property 1

d.
$$\begin{aligned}\ln \frac{(x^2+3)^2}{x\sqrt[3]{x^2+1}} &= \ln(x^2+3)^2 - \ln\left(x\sqrt[3]{x^2+1}\right)\\ &= 2\ln(x^2+3) - [\ln x + \ln(x^2+1)^{1/3}]\\ &= 2\ln(x^2+3) - \ln x - \ln(x^2+1)^{1/3}\\ &= 2\ln(x^2+3) - \ln x - \frac{1}{3}\ln(x^2+1)\end{aligned}$$

When using the properties of logarithms to rewrite logarithmic functions, you must check to see whether the domain of the rewritten function is the same as the domain of the original function. For instance, the domain of $f(x) = \ln x^2$ is all real numbers except $x = 0$, and the domain of $g(x) = 2\ln x$ is all positive real numbers.

TECHNOLOGY Try using a graphing utility to compare the graphs of

$$f(x) = \ln x^2 \quad \text{and} \quad g(x) = 2\ln x.$$

Which of the graphs in Figure 1.52 is the graph of f? Which is the graph of g?

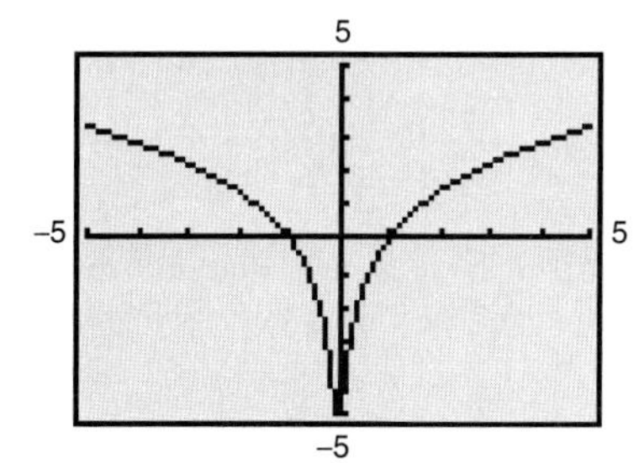

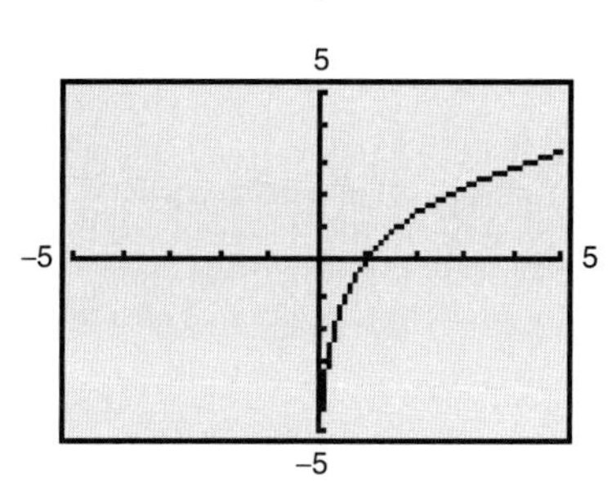

Figure 1.52

EXAMPLE 6 Solving Exponential and Logarithmic Equations

Solve (a) $7 = e^{x+1}$ and (b) $\ln(2x - 3) = 5$.

Solution

a.

$7 = e^{x+1}$	Write original equation.
$\ln 7 = \ln(e^{x+1})$	Take natural log of each side.
$\ln 7 = x + 1$	Apply inverse property.
$-1 + \ln 7 = x$	Solve for x.
$0.946 \approx x$	Use a calculator.

b.

$\ln(2x - 3) = 5$	Write original equation.
$e^{\ln(2x-3)} = e^5$	Exponentiate each side.
$2x - 3 = e^5$	Apply inverse property.
$x = \frac{1}{2}(e^5 + 3)$	Solve for x.
$x \approx 75.707$	Use a calculator.

Exercises for Section 1.6

See www.CalcChat.com for worked-out solutions to odd-numbered exercises.

In Exercises 1 and 2, evaluate the expressions.

1. (a) $25^{3/2}$ (b) $81^{1/2}$ (c) 3^{-2} (d) $27^{-1/3}$
2. (a) $64^{1/3}$ (b) 5^{-4} (c) $\left(\frac{1}{8}\right)^{1/3}$ (d) $\left(\frac{1}{4}\right)^3$

In Exercises 3–6, use the properties of exponents to simplify the expressions.

3. (a) $(5^2)(5^3)$ (b) $(5^2)(5^{-3})$
   (c) $\dfrac{5^3}{25^2}$ (d) $\left(\dfrac{1}{4}\right)^2 2^6$
4. (a) $(2^2)^3$ (b) $(5^4)^{1/2}$
   (c) $[(27^{-1})(27^{2/3})]^3$ (d) $(25^{3/2})(3^2)$
5. (a) $e^2(e^4)$ (b) $(e^3)^4$
   (c) $(e^3)^{-2}$ (d) $\dfrac{e^5}{e^3}$
6. (a) $\left(\dfrac{1}{e}\right)^{-2}$ (b) $\left(\dfrac{e^5}{e^2}\right)^{-1}$
   (c) e^0 (d) $\dfrac{1}{e^{-3}}$

In Exercises 7–16, solve for x.

7. $3^x = 81$
8. $5^{x+1} = 125$
9. $\left(\frac{1}{3}\right)^{x-1} = 27$
10. $\left(\frac{1}{5}\right)^{2x} = 625$
11. $4^3 = (x + 2)^3$
12. $18^2 = (5x - 7)^2$
13. $x^{3/4} = 8$
14. $(x + 3)^{4/3} = 16$
15. $e^{-2x} = e^5$
16. $e^x = 1$

In Exercises 17 and 18, compare the given number with the number e. Is the number less than or greater than e?

17. $\left(1 + \dfrac{1}{1{,}000{,}000}\right)^{1{,}000{,}000}$
18. $1 + 1 + \frac{1}{2} + \frac{1}{6} + \frac{1}{24} + \frac{1}{120} + \frac{1}{720} + \frac{1}{5040}$

In Exercises 19–28, sketch the graph of the function.

19. $y = 3^x$
20. $y = 3^{x-1}$
21. $y = \left(\frac{1}{3}\right)^x$
22. $y = 2^{-x^2}$
23. $f(x) = 3^{-x^2}$
24. $f(x) = 3^{|x|}$
25. $h(x) = e^{x-2}$
26. $g(x) = -e^{x/2}$
27. $y = e^{-x^2}$
28. $y = e^{-x/4}$

29. Use a graphing utility to graph $f(x) = e^x$ and the given function in the same viewing window. How are the two graphs related?
    (a) $g(x) = e^{x-2}$
    (b) $h(x) = -\frac{1}{2}e^x$
    (c) $q(x) = e^{-x} + 3$

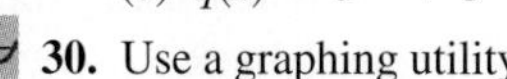

30. Use a graphing utility to graph the function. Describe the shape of the graph for very large and very small values of x.
    (a) $f(x) = \dfrac{8}{1 + e^{-0.5x}}$
    (b) $g(x) = \dfrac{8}{1 + e^{-0.5/x}}$

In Exercises 31–34, match the equation with the correct graph. Assume that a and C are positive real numbers. [The graphs are labeled (a), (b), (c), and (d).]

(a)

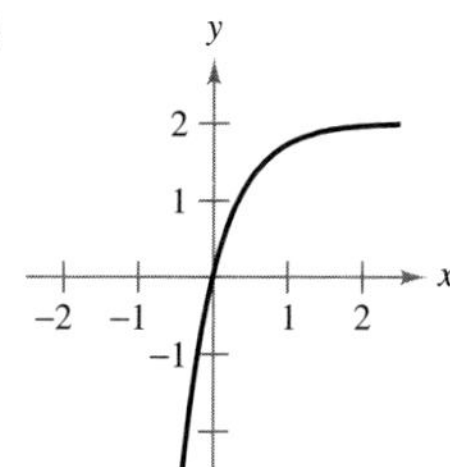

(b)

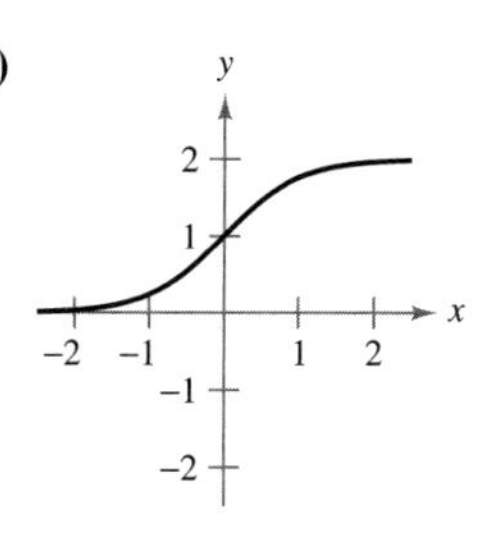

(c)

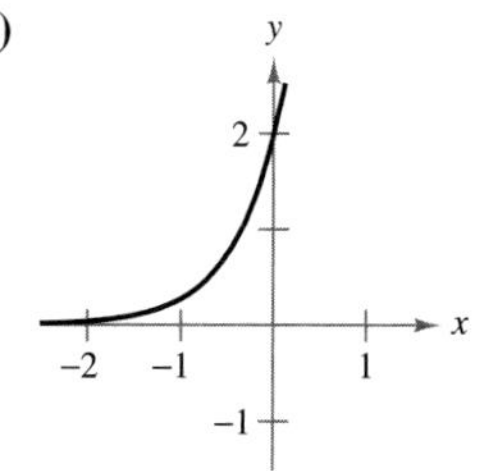

(d)

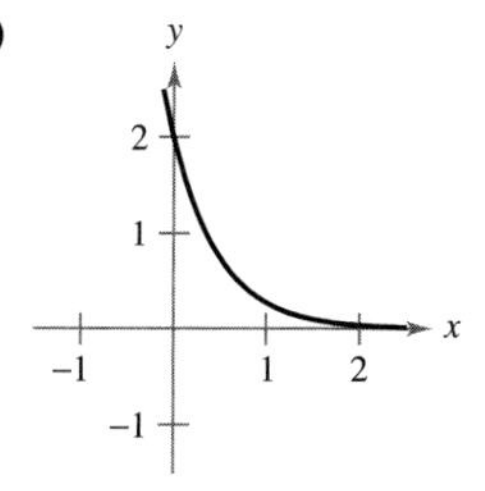

31. $y = Ce^{ax}$

32. $y = Ce^{-ax}$

33. $y = C(1 - e^{-ax})$

34. $y = C/(1 + e^{-ax})$

In Exercises 35–38, match the function with its graph. [The graphs are labeled (a), (b), (c), and (d).]

(a)

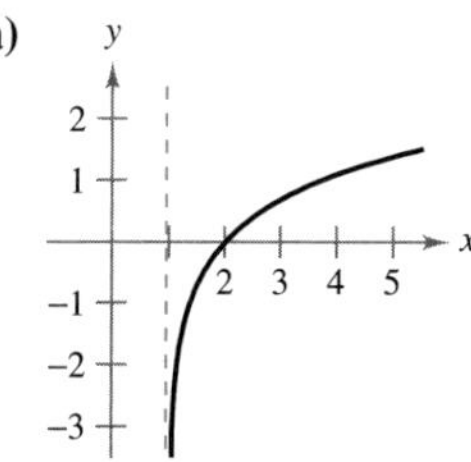

(b)

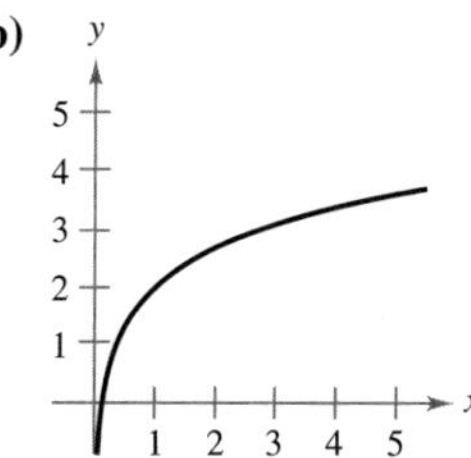

(c)

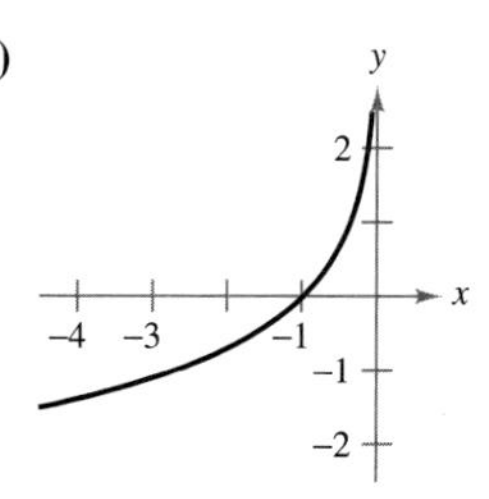

(d)

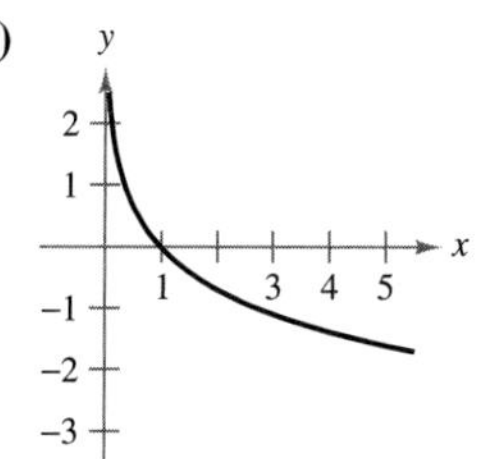

35. $f(x) = \ln x + 2$

36. $f(x) = -\ln x$

37. $f(x) = \ln(x - 1)$

38. $f(x) = -\ln(-x)$

In Exercises 39 and 40, find the exponential function $y = Ca^x$ that fits the graph.

39.

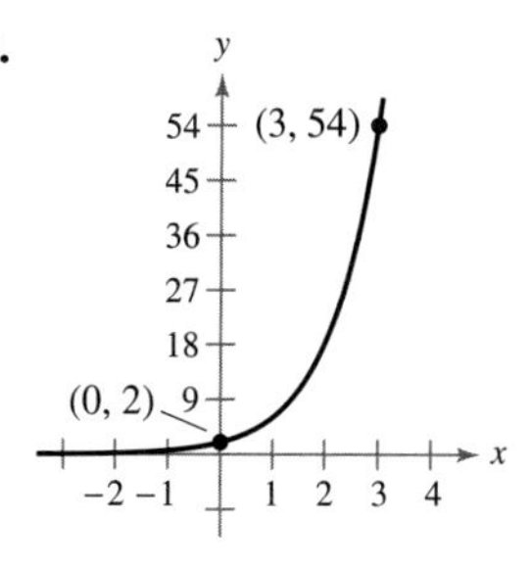

40.

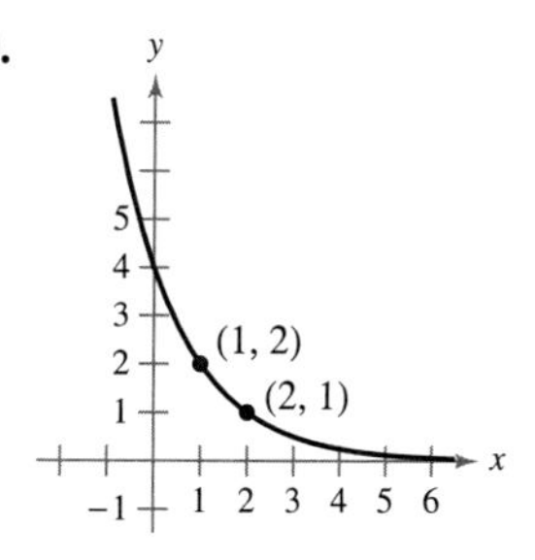

In Exercises 41–44, write the exponential equation as a logarithmic equation, or vice versa.

41. $e^0 = 1$

42. $e^{-2} = 0.1353\ldots$

43. $\ln 2 = 0.6931\ldots$

44. $\ln 0.5 = -0.6931\ldots$

In Exercises 45–50, sketch the graph of the function and state its domain.

45. $f(x) = 3 \ln x$

46. $f(x) = -2 \ln x$

47. $f(x) = \ln 2x$

48. $f(x) = \ln|x|$

49. $f(x) = \ln(x - 1)$

50. $f(x) = 2 + \ln x$

In Exercises 51–54, show that the functions f and g are inverses of each other by graphing them in the same viewing window.

51. $f(x) = e^{2x}$, $g(x) = \ln\sqrt{x}$

52. $f(x) = e^{x/3}$, $g(x) = \ln x^3$

53. $f(x) = e^x - 1$, $g(x) = \ln(x + 1)$

54. $f(x) = e^{x-1}$, $g(x) = 1 + \ln x$

In Exercises 55–58, (a) find the inverse of the function, (b) use a graphing utility to graph f and f^{-1} in the same viewing window, and (c) verify that $f^{-1}(f(x)) = x$ and $f(f^{-1}(x)) = x$.

55. $f(x) = e^{4x-1}$

56. $f(x) = 3e^{-x}$

57. $f(x) = 2\ln(x - 1)$

58. $f(x) = 3 + \ln(2x)$

In Exercises 59–64, apply the inverse properties of $\ln x$ and e^x to simplify the given expression.

59. $\ln e^{x^2}$

60. $\ln e^{2x-1}$

61. $e^{\ln(5x+2)}$

62. $-1 + \ln e^{2x}$

63. $e^{\ln\sqrt{x}}$

64. $-8 + e^{\ln x^3}$

In Exercises 65 and 66, use the properties of logarithms to approximate the indicated logarithms, given that $\ln 2 \approx 0.6931$ and $\ln 3 \approx 1.0986$.

65. (a) $\ln 6$ (b) $\ln \frac{2}{3}$ (c) $\ln 81$ (d) $\ln \sqrt{3}$

66. (a) $\ln 0.25$ (b) $\ln 24$ (c) $\ln \sqrt[3]{12}$ (d) $\ln \frac{1}{72}$

Writing About Concepts

67. In your own words, state the properties of the natural logarithmic function.

68. Explain why $\ln e^x = x$.

69. In your own words, state the properties of the natural exponential function.

70. The table of values below was obtained by evaluating a function. Determine which of the statements may be true and which must be false, and explain why.

(a) y is an exponential function of x.

(b) y is a logarithmic function of x.

(c) x is an exponential function of y.

(d) y is a linear function of x.

x	1	2	8
y	0	1	3

In Exercises 71–80, use the properties of logarithms to expand the logarithmic expression.

71. $\ln \frac{2}{3}$

72. $\ln \sqrt{2^3}$

73. $\ln \dfrac{xy}{z}$

74. $\ln(xyz)$

75. $\ln \frac{1}{5}$

76. $\ln \sqrt[3]{z+1}$

77. $\ln\left(\dfrac{x^2-1}{x^3}\right)^3$

78. $\ln z(z-1)^2$

79. $\ln(3e^2)$

80. $\ln \dfrac{1}{e}$

In Exercises 81–86, write the expression as the logarithm of a single quantity.

81. $\ln(x-2) - \ln(x+2)$

82. $3 \ln x + 2 \ln y - 4 \ln z$

83. $\frac{1}{3}[2 \ln(x+3) + \ln x - \ln(x^2-1)]$

84. $2[\ln x - \ln(x+1) - \ln(x-1)]$

85. $2 \ln 3 - \frac{1}{2} \ln(x^2+1)$

86. $\frac{3}{2}[\ln(x^2+1) - \ln(x+1) - \ln(x-1)]$

In Exercises 87–90, solve for x accurate to three decimal places.

87. (a) $e^{\ln x} = 4$

(b) $\ln e^{2x} = 3$

88. (a) $e^{\ln 2x} = 12$

(b) $\ln e^{-x} = 0$

89. (a) $\ln x = 2$

(b) $e^x = 4$

90. (a) $\ln x^2 = 8$

(b) $e^{-2x} = 5$

In Exercises 91–94, solve the inequality for x.

91. $e^x > 5$

92. $e^{1-x} < 6$

93. $-2 < \ln x < 0$

94. $1 < \ln x < 100$

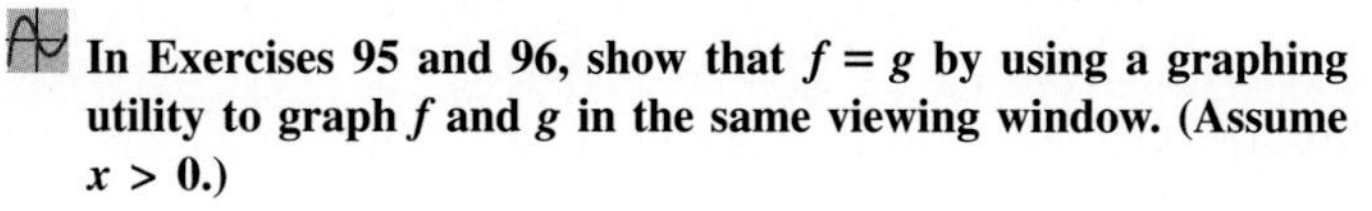

In Exercises 95 and 96, show that $f = g$ by using a graphing utility to graph f and g in the same viewing window. (Assume $x > 0$.)

95. $f(x) = \ln(x^2/4)$

$g(x) = 2 \ln x - \ln 4$

96. $f(x) = \ln \sqrt{x(x^2+1)}$

$g(x) = \frac{1}{2}[\ln x + \ln(x^2+1)]$

97. Prove that $\ln(x/y) = \ln x - \ln y$, $x > 0, y > 0$.

98. Prove that $\ln x^y = y \ln x$.

99. Graph the functions

$f(x) = 6^x$ and $g(x) = x^6$

in the same viewing window. Where do these graphs intersect? As x increases, which function grows more rapidly?

100. Graph the functions

$f(x) = \ln x$ and $g(x) = x^{1/4}$

in the same viewing window. Where do these graphs intersect? As x increases, which function grows more rapidly?

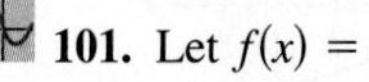

101. Let $f(x) = \ln\left(x + \sqrt{x^2+1}\right)$.

(a) Use a graphing utility to graph f and determine its domain.

(b) Show that f is an odd function.

(c) Find the inverse function of f.

Review Exercises for Chapter 1

See www.CalcChat.com for worked-out solutions to odd-numbered exercises.

In Exercises 1–4, find the intercepts (if any).

1. $y = 2x - 3$

2. $y = (x - 1)(x - 3)$

3. $y = \dfrac{x - 1}{x - 2}$

4. $xy = 4$

In Exercises 5 and 6, check for symmetry with respect to both axes and to the origin.

5. $x^2y - x^2 + 4y = 0$

6. $y = x^4 - x^2 + 3$

In Exercises 7–14, sketch the graph of the equation.

7. $y = \frac{1}{2}(-x + 3)$

8. $4x - 2y = 6$

9. $-\frac{1}{3}x + \frac{5}{6}y = 1$

10. $0.02x + 0.15y = 0.25$

11. $y = 7 - 6x - x^2$

12. $y = 6x - x^2$

13. $y = \sqrt{5 - x}$

14. $y = |x - 4| - 4$

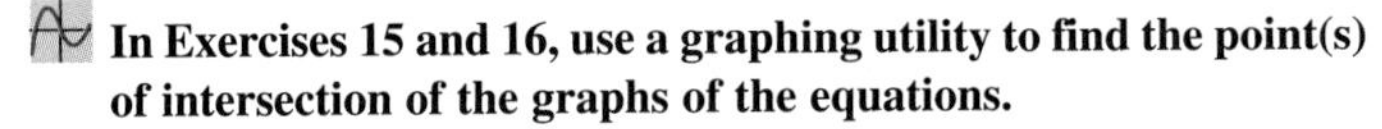

In Exercises 15 and 16, use a graphing utility to find the point(s) of intersection of the graphs of the equations.

15. $3x - 4y = 8$
$x + y = 5$

16. $x - y + 1 = 0$
$y - x^2 = 7$

In Exercises 17 and 18, plot the points and find the slope of the line passing through the points.

17. $\left(\frac{3}{2}, 1\right), \left(5, \frac{5}{2}\right)$

18. $(7, -1), (7, 12)$

In Exercises 19 and 20, use the concept of slope to find t such that the three points are collinear.

19. $(-2, 5), (0, t), (1, 1)$

20. $(-3, 3), (t, -1), (8, 6)$

In Exercises 21–24, find an equation of the line that passes through the point with the indicated slope. Sketch the line.

21. $(0, -5), \quad m = \frac{3}{2}$

22. $(-2, 6), \quad m = 0$

23. $(-3, 0), \quad m = -\frac{2}{3}$

24. $(5, 4), \quad m$ is undefined.

25. Find the equations of the lines passing through $(-2, 4)$ and having the following characteristics.

(a) Slope of $\frac{7}{16}$

(b) Parallel to the line $5x - 3y = 3$

(c) Passing through the origin

(d) Parallel to the y-axis

26. Find the equations of the lines passing through $(1, 3)$ and having the following characteristics.

(a) Slope of $-\frac{2}{3}$

(b) Perpendicular to the line $x + y = 0$

(c) Passing through the point $(2, 4)$

(d) Parallel to the x-axis

27. ***Rate of Change*** The purchase price of a new machine is \$12,500, and its value will decrease by \$850 per year. Use this information to write a linear equation that gives the value V of the machine t years after it is purchased. Find its value at the end of 3 years.

28. ***Break-Even Analysis*** A contractor purchases a piece of equipment for \$36,500 that costs an average of \$9.25 per hour for fuel and maintenance. The equipment operator is paid \$13.50 per hour, and customers are charged \$30 per hour.

(a) Write an equation for the cost C of operating this equipment for t hours.

(b) Write an equation for the revenue R derived from t hours of use.

(c) Find the break-even point for this equipment by finding the time at which $R = C$.

In Exercises 29–32, sketch the graph of the equation and use the Vertical Line Test to determine whether the equation expresses y as a function of x.

29. $x - y^2 = 0$

30. $x^2 - y = 0$

31. $y = x^2 - 2x$

32. $x = 9 - y^2$

33. Evaluate (if possible) the function $f(x) = 1/x$ at the specified values of the independent variable, and simplify the results.

(a) $f(0)$ (b) $\dfrac{f(1 + \Delta x) - f(1)}{\Delta x}$

34. Evaluate (if possible) the function at each value of the independent variable.

$$f(x) = \begin{cases} x^2 + 2, & x < 0 \\ |x - 2|, & x \ge 0 \end{cases}$$

(a) $f(-4)$ (b) $f(0)$ (c) $f(1)$

35. Find the domain and range of each function.

(a) $y = \sqrt{36 - x^2}$ (b) $y = \dfrac{7}{2x - 10}$ (c) $y = \begin{cases} x^2, & x < 0 \\ 2 - x, & x \ge 0 \end{cases}$

36. Given $f(x) = 1 - x^2$ and $g(x) = 2x + 1$, find the following.

(a) $f(x) - g(x)$ (b) $f(x)g(x)$ (c) $g(f(x))$

37. Sketch (on the same set of coordinate axes) a graph of f for $c = -2, 0,$ and 2.

(a) $f(x) = x^3 + c$

(b) $f(x) = (x - c)^3$

(c) $f(x) = (x - 2)^3 + c$

(d) $f(x) = cx^3$

38. Use a graphing utility to graph $f(x) = x^3 - 3x^2$. Use the graph to write a formula for the function g shown in the figure. To print an enlarged copy of the graph, go to the website *www.mathgraphs.com.*

(a)

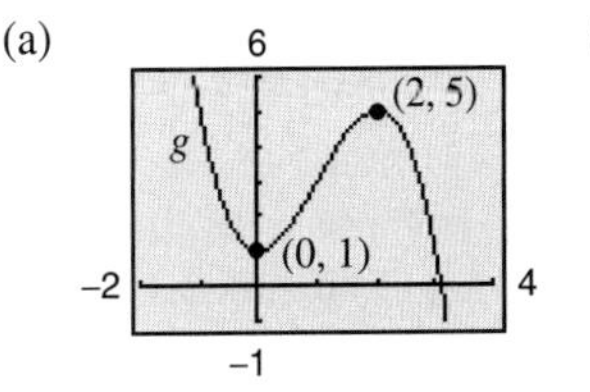

(b)

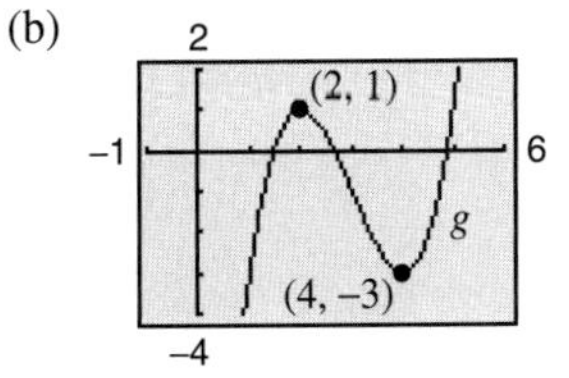

39. ***Think About It*** What is the minimum degree of the polynomial function whose graph approximates the given graph? What sign must the leading coefficient have?

(a)

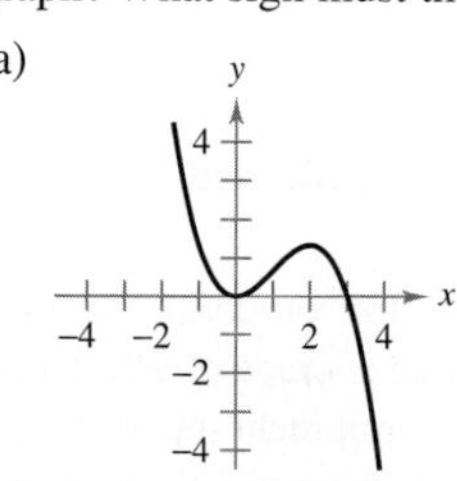

(b)

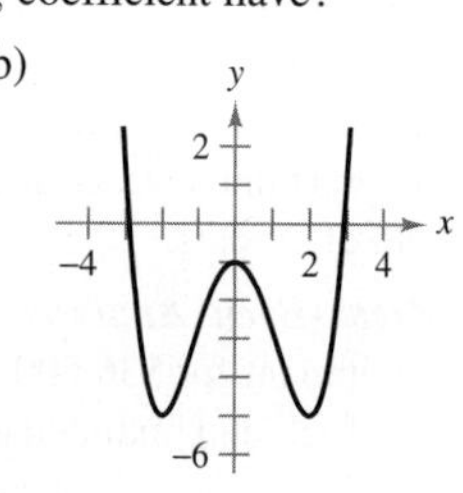

(c)

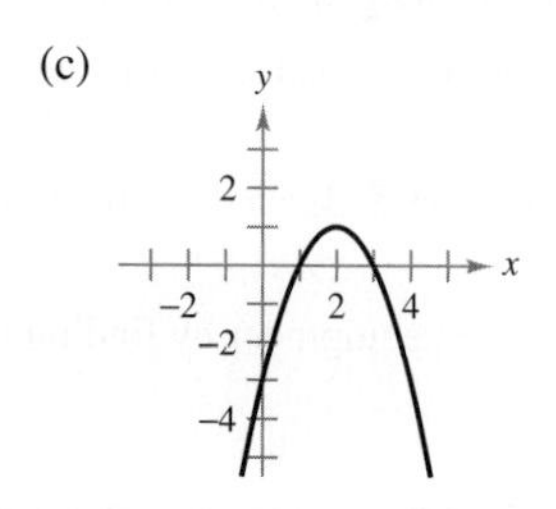

(d)

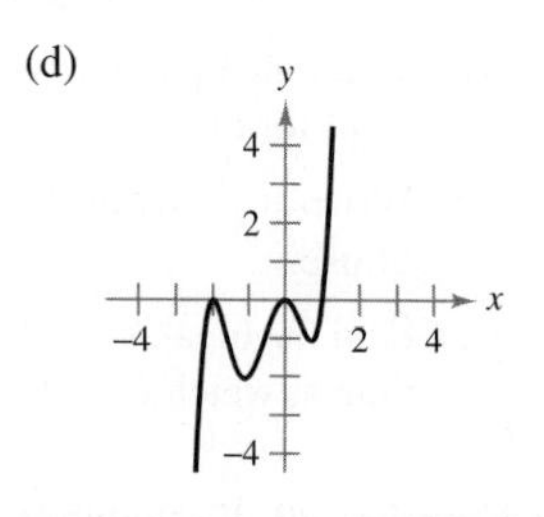

40. ***Writing*** The following graphs give the profits P for two small companies over a period p of 2 years. Create a story to describe the behavior of each profit function for some hypothetical product the company produces.

(a)

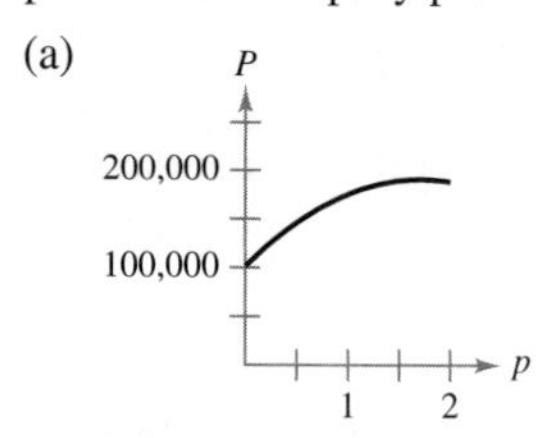

(b)

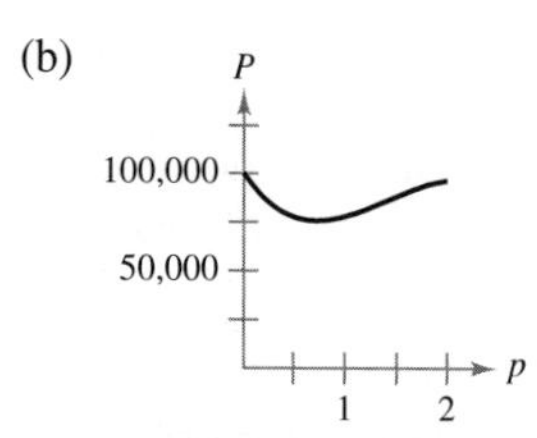

41. ***Harmonic Motion*** The motion of an oscillating weight suspended by a spring was measured by a motion detector. The data collected and the approximate maximum (positive and negative) displacements from equilibrium are shown in the figure. The displacement y is measured in feet and the time t is measured in seconds.

(a) Is y a function of t? Explain.

(b) Approximate the amplitude and period of the oscillations.

(c) Find a model for the data.

(d) Use a graphing utility to graph the model in part (c). Compare the result with the data in the figure.

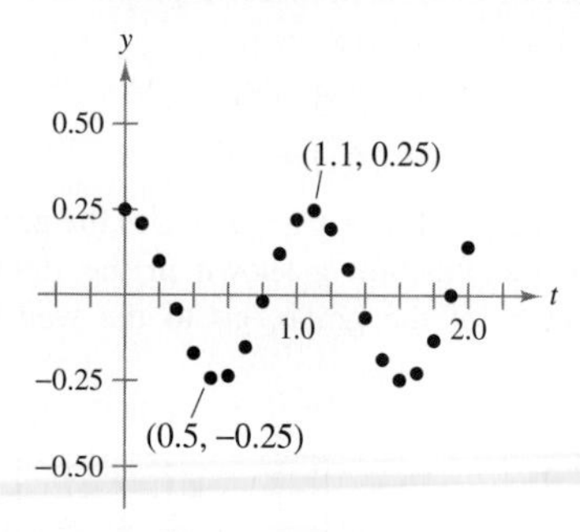

42. ***Stress Test*** A machine part was tested by bending it x centimeters 10 times per minute until the time y (in hours) of failure. The results are recorded in the table.

x	3	6	9	12	15	18	21	24	27	30
y	61	56	53	55	48	35	36	33	44	23

Table for 42

(a) Use the regression capabilities of a graphing utility to find a linear model for the data.

(b) Use a graphing utility to plot the data and graph the model.

(c) Use the graph to determine whether there may have been an error made in conducting one of the tests or in recording the results. If so, eliminate the erroneous point and find a model for the revised data.

In Exercises 43–48, (a) find the inverse of the function, (b) use a graphing utility to graph f and f^{-1} in the same viewing window, and (c) verify that $f^{-1}(f(x)) = x$ and $f(f^{-1}(x)) = x$.

43. $f(x) = \frac{1}{2}x - 3$

44. $f(x) = 5x - 7$

45. $f(x) = \sqrt{x + 1}$

46. $f(x) = x^3 + 2$

47. $f(x) = \sqrt[3]{x + 1}$

48. $f(x) = x^2 - 5, \quad x \geq 0$

In Exercises 49 and 50, sketch the graph of the function by hand.

49. $f(x) = 2 \arctan(x + 3)$

50. $h(x) = -3 \arcsin 2x$

In Exercises 51 and 52, evaluate the expression without using a calculator. (*Hint:* Make a sketch of a right triangle.)

51. $\sin\left(\arcsin \frac{1}{2}\right)$

52. $\tan(\operatorname{arccot} 2)$

In Exercises 53 and 54, sketch the graph of the function by hand.

53. $f(x) = \ln x + 3$

54. $f(x) = \ln(x - 3)$

In Exercises 55 and 56, use the properties of logarithms to expand the logarithmic function.

55. $\ln \sqrt[5]{\dfrac{4x^2 - 1}{4x^2 + 1}}$

56. $\ln[(x^2 + 1)(x - 1)]$

In Exercises 57 and 58, write the expression as the logarithm of a single quantity.

57. $\ln 3 + \frac{1}{3} \ln(4 - x^2) - \ln x$

58. $3[\ln x - 2 \ln(x^2 + 1)] + 2 \ln 5$

In Exercises 59 and 60, solve the equation for x.

59. $\ln \sqrt{x + 1} = 2$

60. $\ln x + \ln(x - 3) = 0$

In Exercises 61 and 62, (a) find the inverse function of f, (b) use a graphing utility to graph f and f^{-1} in the same viewing window, and (c) verify that $f^{-1}(f(x)) = x$ and $f(f^{-1}(x)) = x$.

61. $f(x) = \ln \sqrt{x}$

62. $f(x) = e^{1-x}$

In Exercises 63 and 64, sketch the graph of the function by hand.

63. $y = e^{-x/2}$

64. $y = 4e^{-x^2}$

P.S. Problem Solving

See www.CalcChat.com for worked-out solutions to odd-numbered exercises.

1. Consider the circle $x^2 + y^2 - 6x - 8y = 0$, as shown in the figure.

(a) Find the center and radius of the circle.

(b) Find an equation of the tangent line to the circle at the point $(0, 0)$.

(c) Find an equation of the tangent line to the circle at the point $(6, 0)$.

(d) Where do the two tangent lines intersect?

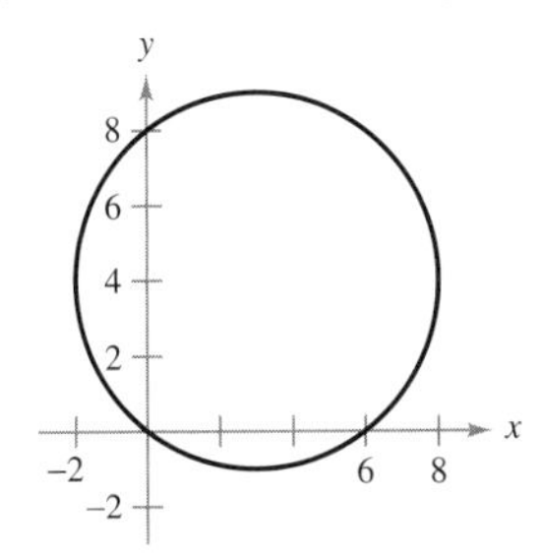

Figure for 1

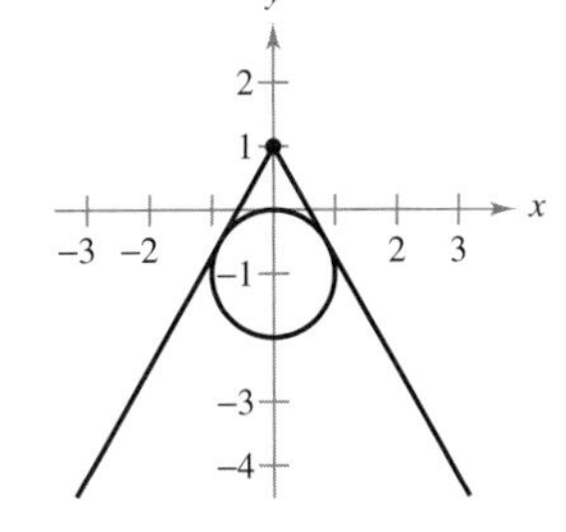

Figure for 2

2. There are two tangent lines from the point $(0, 1)$ to the circle $x^2 + (y + 1)^2 = 1$ (see figure). Find equations of these two lines by using the fact that each tangent line intersects the circle in *exactly* one point.

3. The Heaviside function $H(x)$ is widely used in engineering applications.

$$H(x) = \begin{cases} 1, & x \geq 0 \\ 0, & x < 0 \end{cases}$$

Sketch the graph of the Heaviside function and the graphs of the following functions by hand.

(a) $H(x) - 2$ (b) $H(x - 2)$ (c) $-H(x)$

(d) $H(-x)$ (e) $\frac{1}{2}H(x)$ (f) $-H(x - 2) + 2$

Institute of Electrical Engineers, London

OLIVER HEAVISIDE (1850–1925)

Heaviside was a British mathematician and physicist who contributed to the field of applied mathematics, especially applications of mathematics to electrical engineering. The *Heaviside function* is a classic type of "on-off" function that has applications to electricity and computer science.

4. Consider the graph of the function f shown below. Use this graph to sketch the graphs of the following functions. To print an enlarged copy of the graph, go to the website *www.mathgraphs.com.*

(a) $f(x + 1)$ (b) $f(x) + 1$ (c) $2f(x)$ (d) $f(-x)$

(e) $-f(x)$ (f) $|f(x)|$ (g) $f(|x|)$

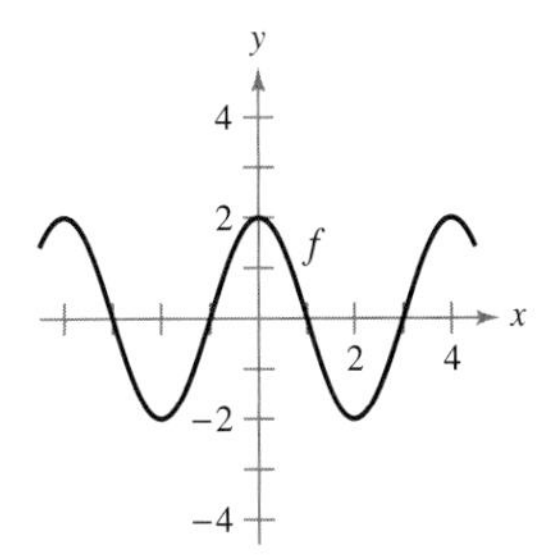

5. A rancher plans to fence a rectangular pasture adjacent to a river. The rancher has 100 meters of fence, and no fencing is needed along the river (see figure).

(a) Write the area A of the pasture as a function of x, the length of the side parallel to the river. What is the domain of A?

(b) Graph the area function $A(x)$ and estimate the dimensions that yield the maximum amount of area for the pasture.

(c) Find the dimensions that yield the maximum amount of area for the pasture by completing the square.

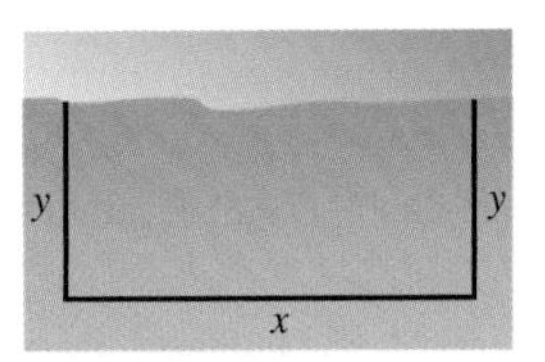

Figure for 5

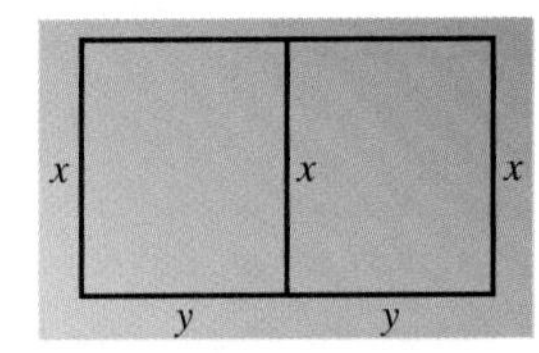

Figure for 6

6. A rancher has 300 feet of fence to enclose two adjacent pastures.

(a) Write the total area A of the two pastures as a function of x (see figure). What is the domain of A?

(b) Graph the area function and estimate the dimensions that yield the maximum amount of area for the pastures.

(c) Find the dimensions that yield the maximum amount of area for the pastures by completing the square.

7. You are in a boat 2 miles from the nearest point on the coast. You are to go to a point Q located 3 miles down the coast and 1 mile inland (see figure). You can row at 2 miles per hour and walk at 4 miles per hour. Write the total time T of the trip as a function of x.

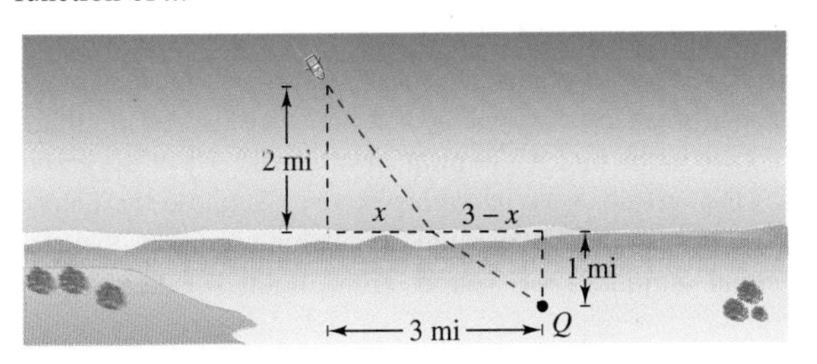

8. Graph the function $f(x) = e^x - e^{-x}$. From the graph the function appears to be one-to-one. Assuming that the function has an inverse, find $f^{-1}(x)$.

9. One of the fundamental themes of calculus is to find the slope of the tangent line to a curve at a point. To see how this can be done, consider the point $(2, 4)$ on the graph of $f(x) = x^2$.

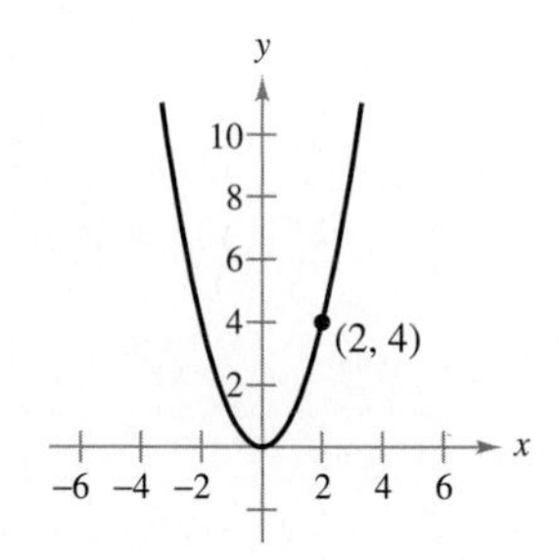

(a) Find the slope of the line joining $(2, 4)$ and $(3, 9)$. Is the slope of the tangent line at $(2, 4)$ greater than or less than this number?

(b) Find the slope of the line joining $(2, 4)$ and $(1, 1)$. Is the slope of the tangent line at $(2, 4)$ greater than or less than this number?

(c) Find the slope of the line joining $(2, 4)$ and $(2.1, 4.41)$. Is the slope of the tangent line at $(2, 4)$ greater than or less than this number?

(d) Find the slope of the line joining $(2, 4)$ and $(2 + h, f(2 + h))$ in terms of the nonzero number h. Verify that $h = 1, -1$, and 0.1 yield the solutions to parts (a)–(c) above.

(e) What is the slope of the tangent line at $(2, 4)$? Explain how you arrived at your answer.

10. Sketch the graph of the function $f(x) = \sqrt{x}$ and label the point $(4, 2)$ on the graph.

(a) Find the slope of the line joining $(4, 2)$ and $(9, 3)$. Is the slope of the tangent line at $(4, 2)$ greater than or less than this number?

(b) Find the slope of the line joining $(4, 2)$ and $(1, 1)$. Is the slope of the tangent line at $(4, 2)$ greater than or less than this number?

(c) Find the slope of the line joining $(4, 2)$ and $(4.41, 2.1)$. Is the slope of the tangent line at $(4, 2)$ greater than or less than this number?

(d) Find the slope of the line joining $(4, 2)$ and $(4 + h, f(4 + h))$ in terms of the nonzero number h.

(e) What is the slope of the tangent line at the point $(4, 2)$? Explain how you arrived at your answer.

11. A large room contains two speakers that are 3 meters apart. The sound intensity I of one speaker is twice that of the other, as shown in the figure. (To print an enlarged copy of the graph, go to the website *www.mathgraphs.com.*) Suppose the listener is free to move about the room to find those positions that receive equal amounts of sound from both speakers. Such a location satisfies two conditions: (1) the sound intensity at the listener's position is directly proportional to the sound level of a source, and (2) the sound intensity is inversely proportional to the square of the distance from the source.

(a) Find the points on the x-axis that receive equal amounts of sound from both speakers.

(b) Find and graph the equation of all locations (x, y) where one could stand and receive equal amounts of sound from both speakers.

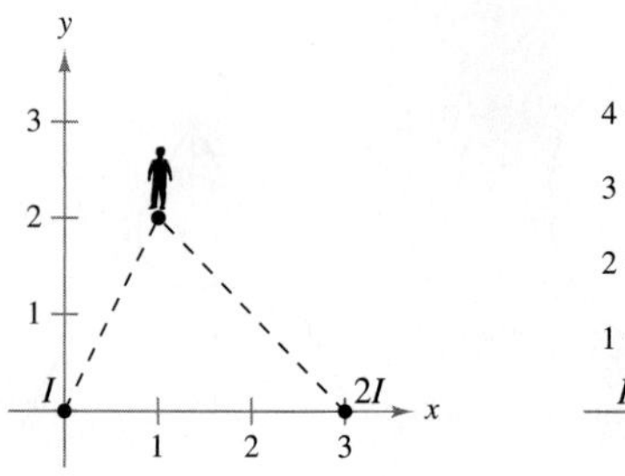

Figure for 11

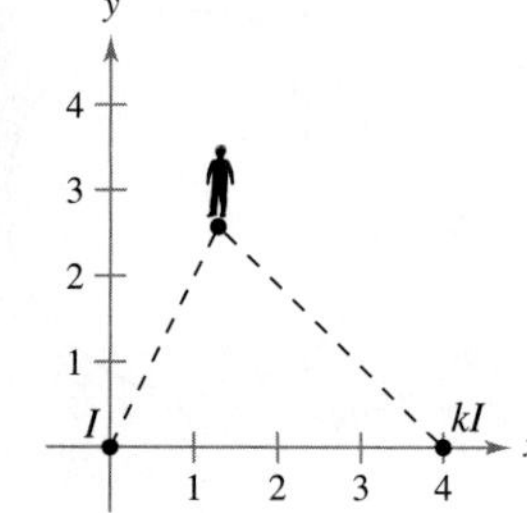

Figure for 12

12. Suppose the speakers in Exercise 11 are 4 meters apart and the sound intensity of one speaker is k times that of the other, as shown in the figure. To print an enlarged copy of the graph, go to the website *www.mathgraphs.com.*

(a) Find the equation of all locations (x, y) where one could stand and receive equal amounts of sound from both speakers.

(b) Graph the equation for the case $k = 3$.

(c) Describe the set of locations of equal sound as k becomes very large.

13. Let d_1 and d_2 be the distances from the point (x, y) to the points $(-1, 0)$ and $(1, 0)$, respectively, as shown in the figure. Show that the equation of the graph of all points (x, y) satisfying $d_1 d_2 = 1$ is $(x^2 + y^2)^2 = 2(x^2 - y^2)$. This curve is called a **lemniscate.** Graph the lemniscate and identify three points on the graph.

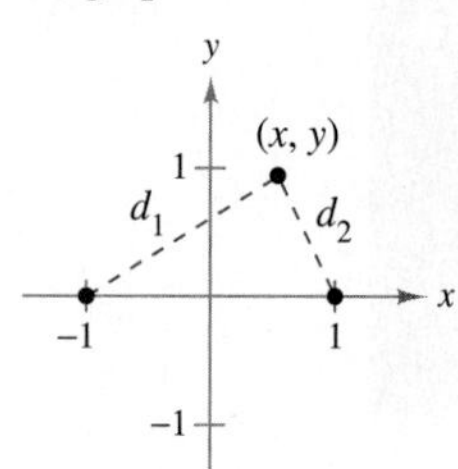

14. Let $f(x) = \dfrac{1}{1 - x}$.

(a) What are the domain and range of f?

(b) Find the composition $f(f(x))$. What is the domain of this function?

(c) Find $f(f(f(x)))$. What is the domain of this function?

(d) Graph $f(f(f(x)))$. Is the graph a line? Why or why not?

2 Limits and Their Properties

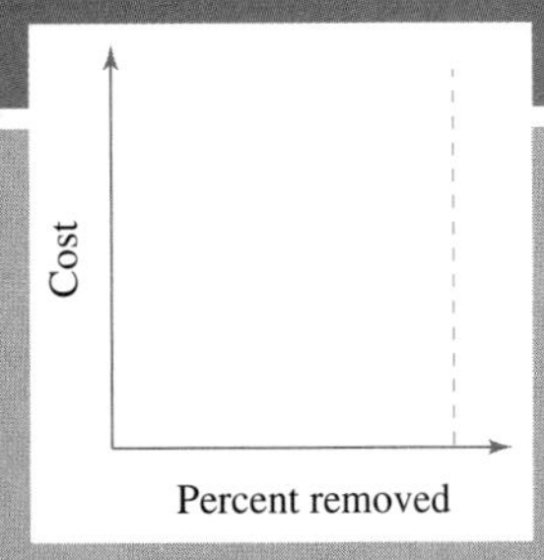

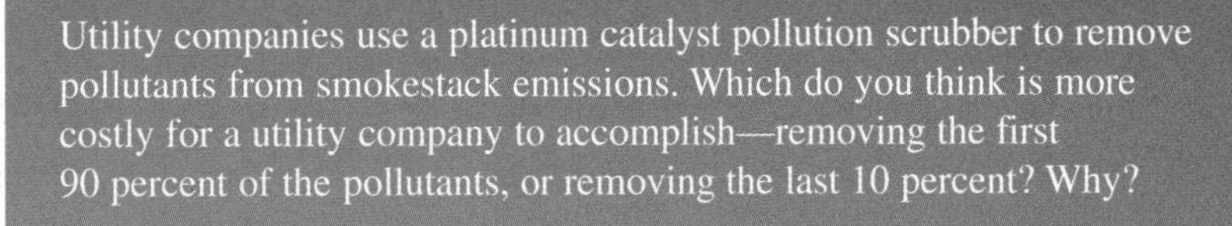

Utility companies use a platinum catalyst pollution scrubber to remove pollutants from smokestack emissions. Which do you think is more costly for a utility company to accomplish—removing the first 90 percent of the pollutants, or removing the last 10 percent? Why?

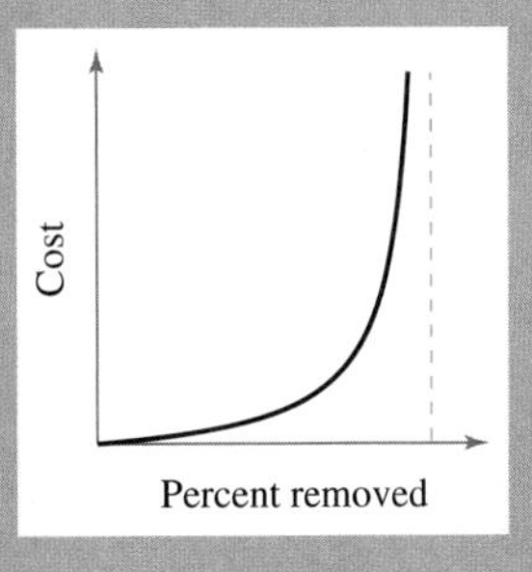

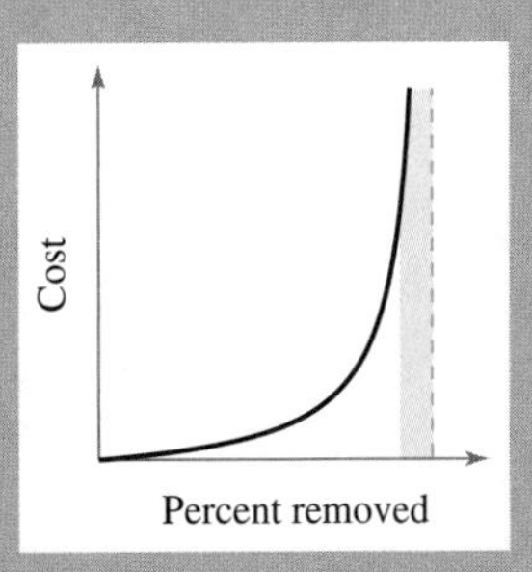

The limit process is a fundamental concept of calculus. One technique you can use to estimate a limit is to graph the function and then determine the behavior of the graph as the independent variable approaches a specific value. In Chapter 2, you will learn how to find limits of functions analytically, graphically, and numerically.

Jeremy Walker/Getty Images

Section 2.1 A Preview of Calculus

- **Understand what calculus is and how it compares with precalculus.**
- **Understand that the tangent line problem is basic to calculus.**
- **Understand that the area problem is also basic to calculus.**

STUDY TIP As you progress through this course, remember that learning calculus is just one of your goals. Your most important goal is to learn how to use calculus to model and solve real-life problems. Here are a few problem-solving strategies that may help you.

- Be sure you understand the question. What is given? What are you asked to find?
- Outline a plan. There are many approaches you could use: look for a pattern, solve a simpler problem, work backwards, draw a diagram, use technology, or any of many other approaches.
- Complete your plan. Be sure to answer the question. Verbalize your answer. For example, rather than writing the answer as $x = 4.6$, it would be better to write the answer as "The area of the region is 4.6 square meters."
- Look back at your work. Does your answer make sense? Is there a way you can check the reasonableness of your answer?

What Is Calculus?

Calculus is the mathematics of change—velocities and accelerations. Calculus is also the mathematics of tangent lines, slopes, areas, volumes, arc lengths, centroids, curvatures, and a variety of other concepts that have enabled scientists, engineers, and economists to model real-life situations.

Although precalculus mathematics deals with velocities, accelerations, tangent lines, slopes, and so on, there is a fundamental difference between precalculus mathematics and calculus. Precalculus mathematics is more static, whereas calculus is more dynamic. Here are some examples.

- An object traveling at a constant velocity can be analyzed with precalculus mathematics. To analyze the velocity of an accelerating object, you need calculus.
- The slope of a line can be analyzed with precalculus mathematics. To analyze the slope of a curve, you need calculus.
- A tangent line to a circle can be analyzed with precalculus mathematics. To analyze a tangent line to a general graph, you need calculus.
- The area of a rectangle can be analyzed with precalculus mathematics. To analyze the area under a general curve, you need calculus.

Each of these situations involves the same general strategy—the reformulation of precalculus mathematics through the use of a limit process. So, one way to answer the question "What is calculus?" is to say that calculus is a "limit machine" that involves three stages. The first stage is precalculus mathematics, such as finding the slope of a line or the area of a rectangle. The second stage is the limit process, and the third stage is a new calculus formulation, such as a derivative or an integral.

Precalculus mathematics ⇨ Limit process ⇨ Calculus

GRACE CHISHOLM YOUNG (1868–1944)

Grace Chisholm Young received her degree in mathematics from Girton College in Cambridge, England. Her early work was published under the name of William Young, her husband. Between 1914 and 1916, Grace Young published work on the foundations of calculus that won her the Gamble Prize from Girton College.

Some students try to learn calculus as if it were simply a collection of new formulas. This is unfortunate. If you reduce calculus to the memorization of differentiation and integration formulas, you will miss a great deal of understanding, self-confidence, and satisfaction.

On the following two pages, some familiar precalculus concepts coupled with their calculus counterparts are listed. Throughout the text, your goal should be to learn how precalculus formulas and techniques are used as building blocks to produce the more general calculus formulas and techniques. Don't worry if you are unfamiliar with some of the "old formulas" listed on the following two pages—you will be reviewing all of them.

As you proceed through this text, we suggest that you come back to this discussion repeatedly. Try to keep track of where you are relative to the three stages involved in the study of calculus. For example, the first three chapters break down as shown.

Chapter 1: Preparation for Calculus	Precalculus
Chapter 2: Limits and Their Properties	Limit process
Chapter 3: Differentiation	Calculus

Without Calculus	With Differential Calculus
Value of $f(x)$ when $x = c$	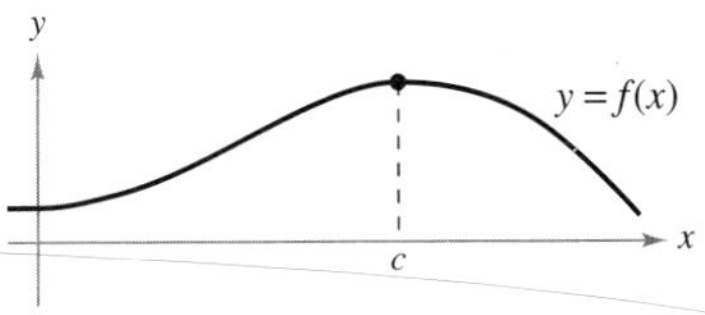 Limit of $f(x)$ as x approaches c 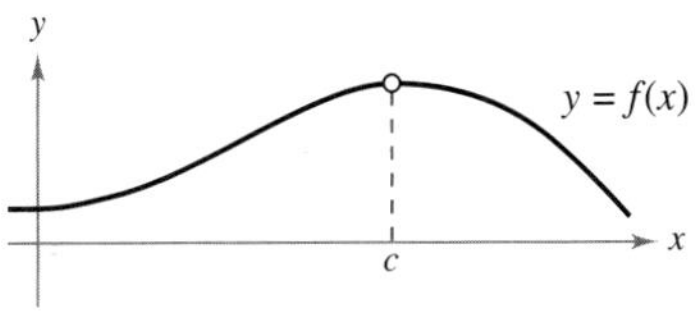
Slope of a line	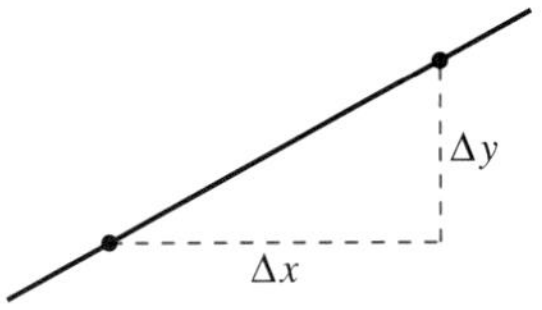 Slope of a curve 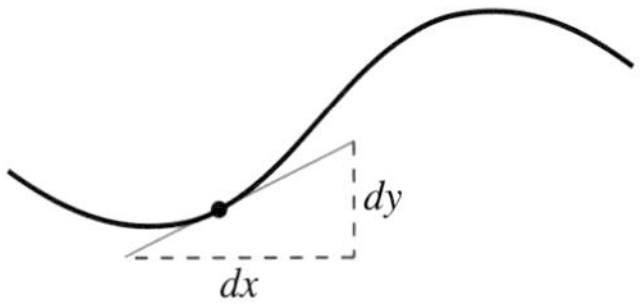
Secant line to a curve	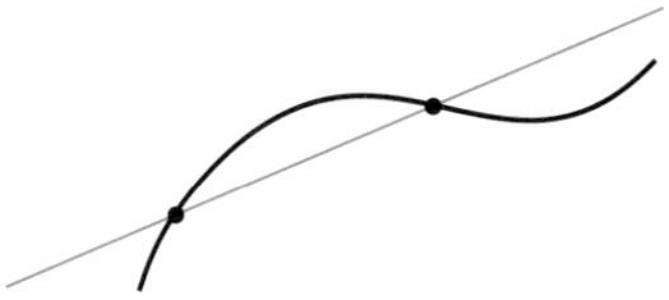 Tangent line to a curve 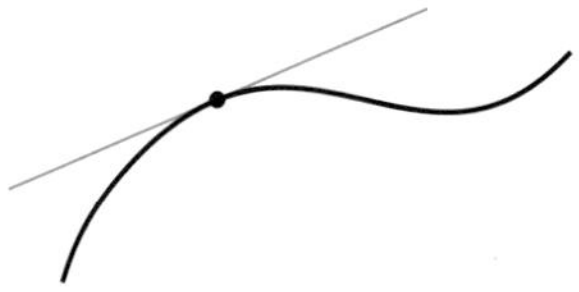
Average rate of change between $t = a$ and $t = b$	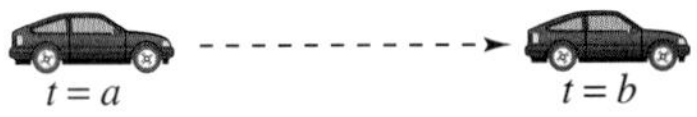 Instantaneous rate of change at $t = c$ 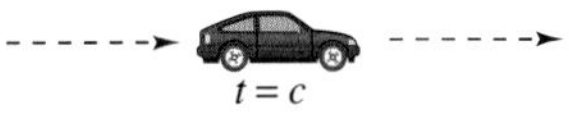
Curvature of a circle	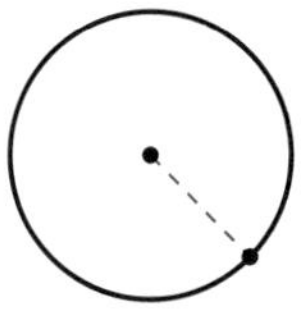 Curvature of a curve 
Height of a curve when $x = c$	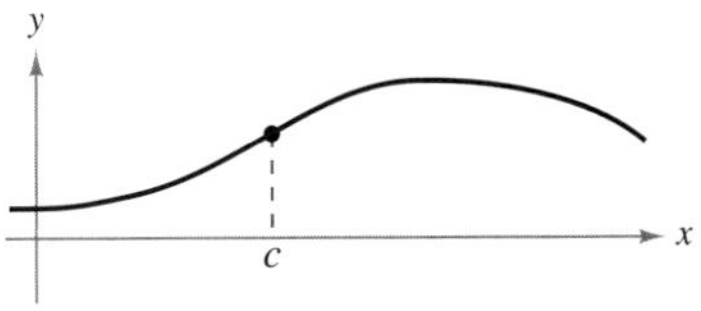 Maximum height of a curve on an interval 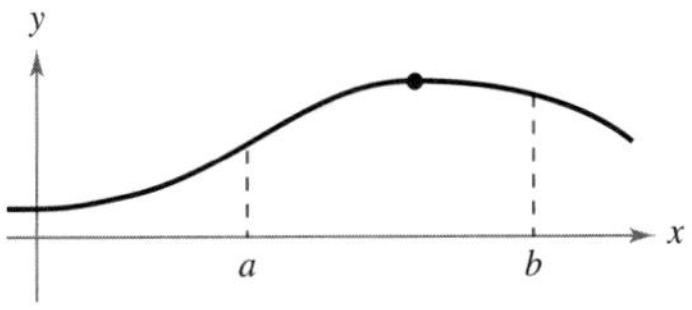
Tangent plane to a sphere	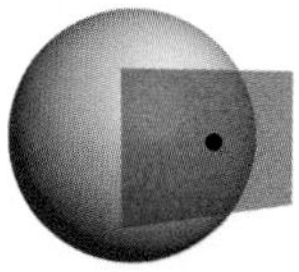 Tangent plane to a surface 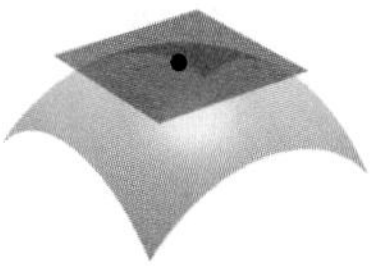
Direction of motion along a straight line	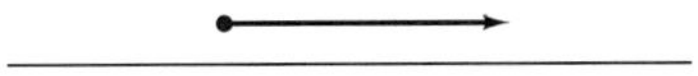 Direction of motion along a curved line 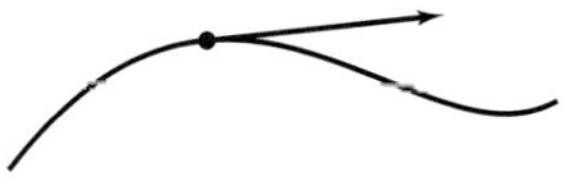

Without Calculus		With Integral Calculus	
Area of a rectangle		Area under a curve	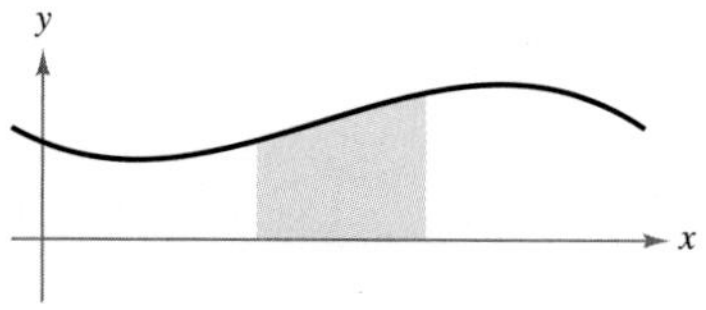 y, x
Work done by a constant force	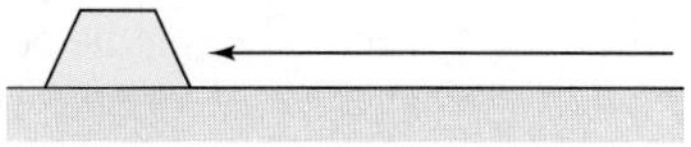	Work done by a variable force	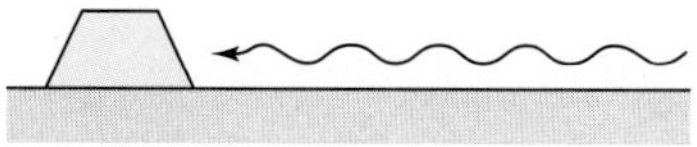
Center of a rectangle	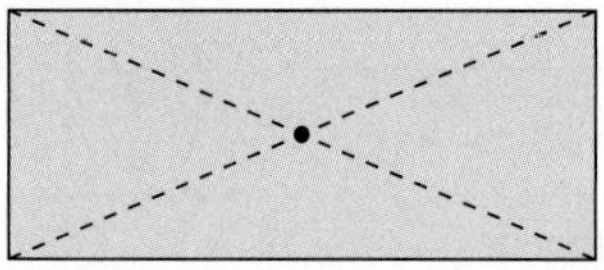	Centroid of a region	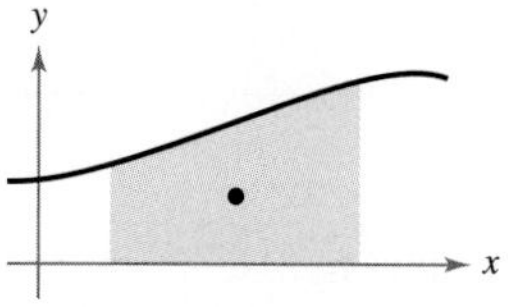 y, x
Length of a line segment	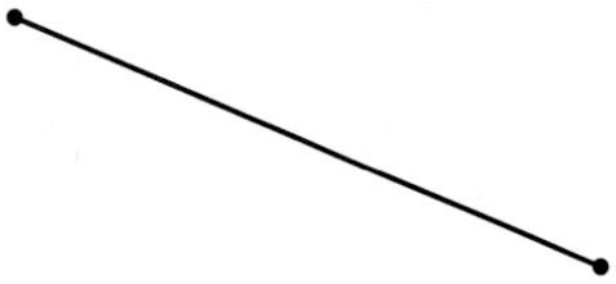	Length of an arc	
Surface area of a cylinder		Surface area of a solid of revolution	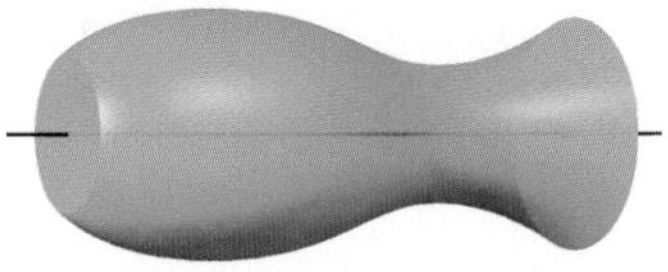
Mass of a solid of constant density		Mass of a solid of variable density	
Volume of a rectangular solid		Volume of a region under a surface	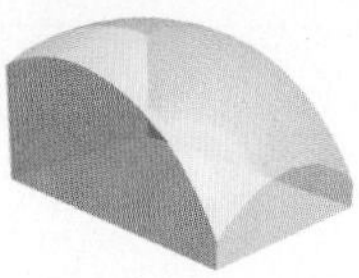
Sum of a finite number of terms	$a_1 + a_2 + \cdots + a_n = S$	Sum of an infinite number of terms	$a_1 + a_2 + a_3 + \cdots = S$

The Tangent Line Problem

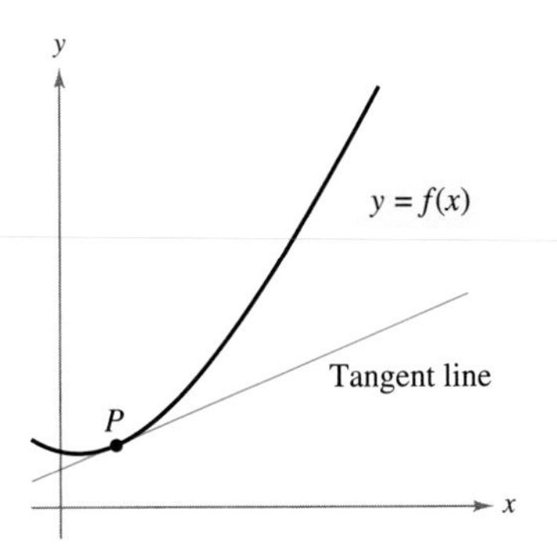

The tangent line to the graph of f at P
Figure 2.1

The notion of a limit is fundamental to the study of calculus. The following brief descriptions of two classic problems in calculus—*the tangent line problem* and *the area problem*—should give you some idea of the way limits are used in calculus.

In the tangent line problem, you are given a function f and a point P on its graph and are asked to find an equation of the tangent line to the graph at point P, as shown in Figure 2.1.

Except for cases involving a vertical tangent line, the problem of finding the **tangent line** at a point P is equivalent to finding the *slope* of the tangent line at P. You can approximate this slope by using a line through the point of tangency and a second point on the curve, as shown in Figure 2.2(a). Such a line is called a **secant line.** If $P(c, f(c))$ is the point of tangency and

$$Q(c + \Delta x, f(c + \Delta x))$$

is a second point on the graph of f, then the slope of the secant line through these two points is given by

$$m_{\text{sec}} = \frac{f(c + \Delta x) - f(c)}{c + \Delta x - c} = \frac{f(c + \Delta x) - f(c)}{\Delta x}.$$

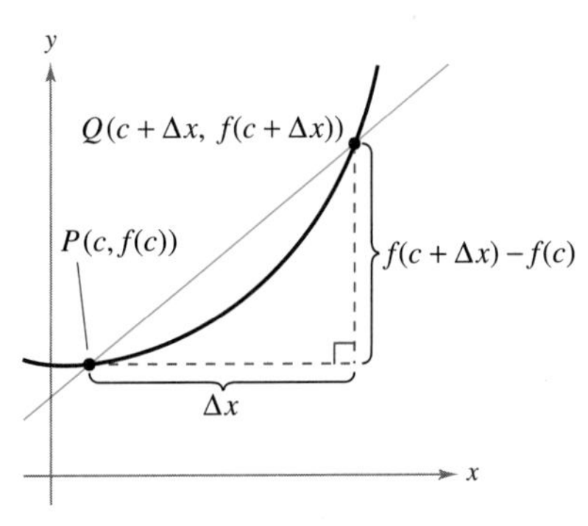

(a) The secant line through $(c, f(c))$ and $(c + \Delta x, f(c + \Delta x))$

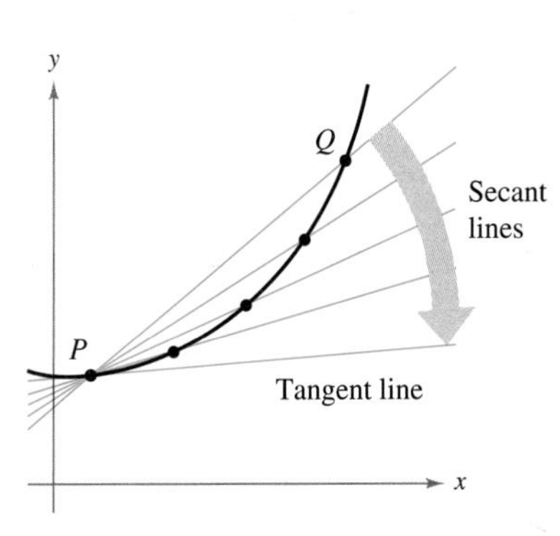

(b) As Q approaches P, the secant lines approach the tangent line.

Figure 2.2

As point Q approaches point P, the slope of the secant line approaches the slope of the tangent line, as shown in Figure 2.2(b). When such a "limiting position" exists, the slope of the tangent line is said to be the **limit** of the slope of the secant line. (Much more will be said about this important problem in Chapter 3.)

EXPLORATION

The following points lie on the graph of $f(x) = x^2$.

$Q_1(1.5, f(1.5))$, $Q_2(1.1, f(1.1))$, $Q_3(1.01, f(1.01))$,
$Q_4(1.001, f(1.001))$, $Q_5(1.0001, f(1.0001))$

Each successive point gets closer to the point $P(1, 1)$. Find the slope of the secant line through Q_1 and P, Q_2 and P, and so on. Graph these secant lines on a graphing utility. Then use your results to estimate the slope of the tangent line to the graph of f at the point P.

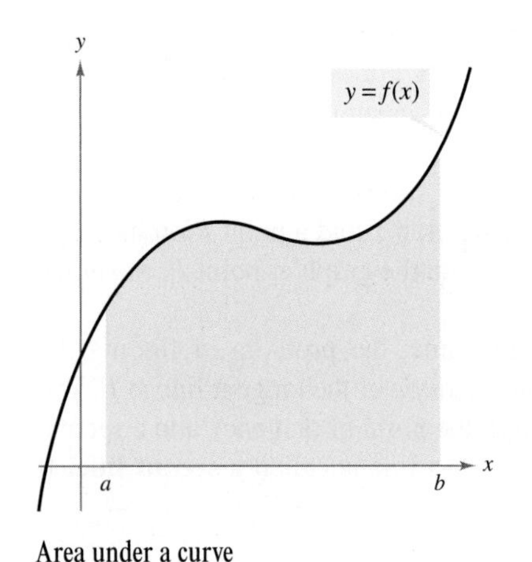

Area under a curve
Figure 2.3

The Area Problem

In the tangent line problem, you saw how the limit process can be applied to the slope of a line to find the slope of a general curve. A second classic problem in calculus is finding the area of a plane region that is bounded by the graphs of functions. This problem can also be solved with a limit process. In this case, the limit process is applied to the area of a rectangle to find the area of a general region.

As a simple example, consider the region bounded by the graph of the function $y = f(x)$, the x-axis, and the vertical lines $x = a$ and $x = b$, as shown in Figure 2.3. You can approximate the area of the region with several rectangular regions, as shown in Figure 2.4. As you increase the number of rectangles, the approximation tends to become better and better because the amount of area missed by the rectangles decreases. Your goal is to determine the limit of the sum of the areas of the rectangles as the number of rectangles increases without bound.

Historical Note

In one of the most astounding events ever to occur in mathematics, it was discovered that the tangent line problem and the area problem are closely related. This discovery led to the birth of calculus. You will learn about the relationship between these two problems when you study the Fundamental Theorem of Calculus in Chapter 5.

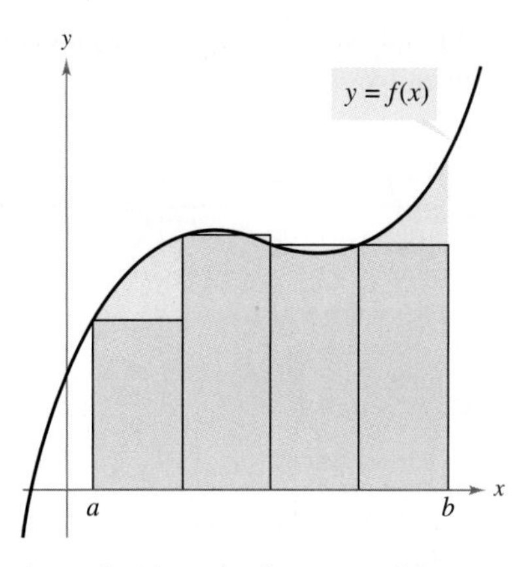

Approximation using four rectangles
Figure 2.4

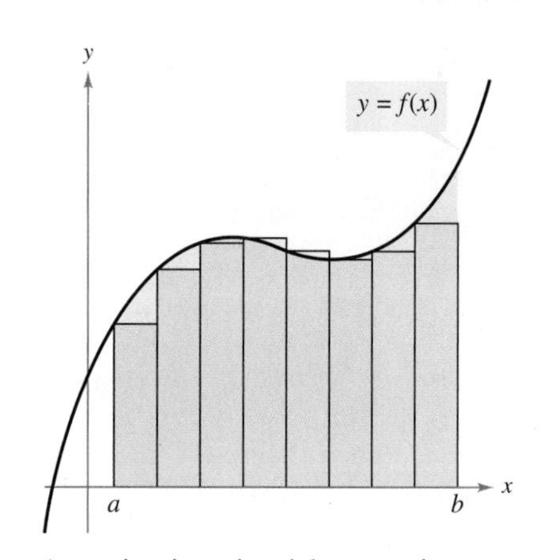

Approximation using eight rectangles

EXPLORATION

Consider the region bounded by the graphs of $f(x) = x^2$, $y = 0$, and $x = 1$, as shown in part (a) of the figure. The area of the region can be approximated by two sets of rectangles—one set inscribed within the region and the other set circumscribed over the region, as shown in parts (b) and (c). Find the sum of the areas of each set of rectangles. Then use your results to approximate the area of the region.

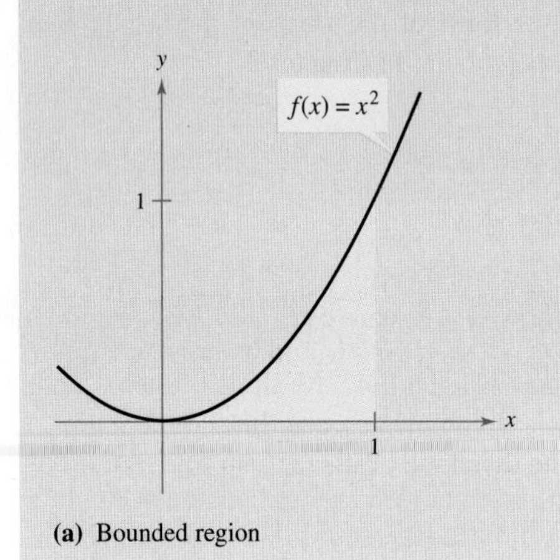

(a) Bounded region

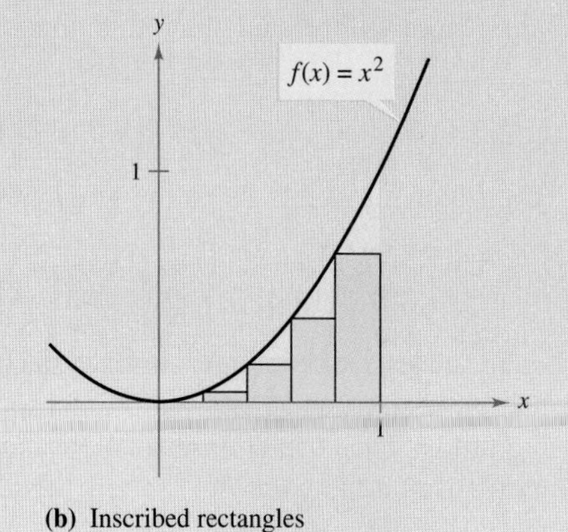

(b) Inscribed rectangles

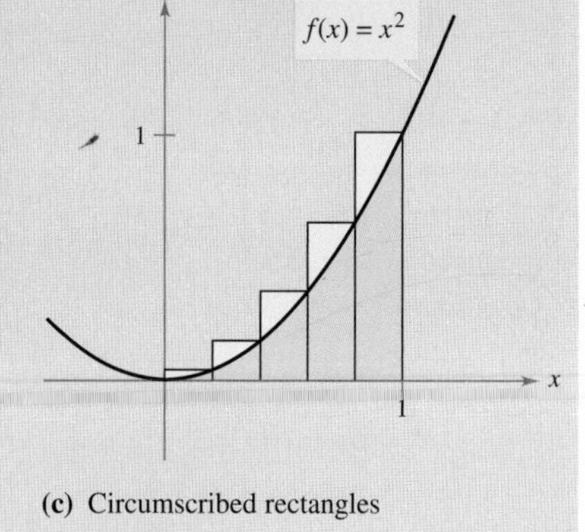

(c) Circumscribed rectangles

Exercises for Section 2.1

See www.CalcChat.com for worked-out solutions to odd-numbered exercises.

In Exercises 1–6, decide whether the problem can be solved using precalculus, or whether calculus is required. If the problem can be solved using precalculus, solve it. If the problem seems to require calculus, explain your reasoning and use a graphical or numerical approach to estimate the solution.

1. Find the distance traveled in 15 seconds by an object traveling at a constant velocity of 20 feet per second.

2. Find the distance traveled in 15 seconds by an object moving with a velocity of $v(t) = 20 + 7\cos t$ feet per second.

3. A bicyclist is riding on a path modeled by the function $f(x) = 0.04(8x - x^2)$, where x and $f(x)$ are measured in miles. Find the rate of change of elevation when $x = 2$.

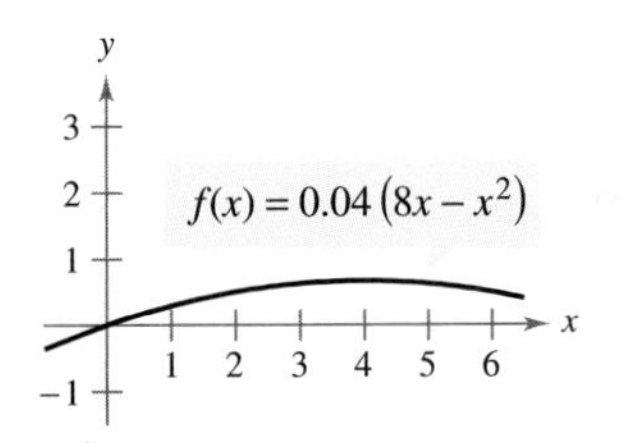

Figure for 3

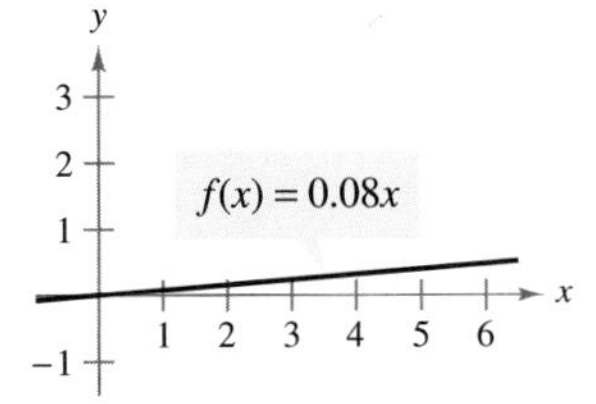

Figure for 4

4. A bicyclist is riding on a path modeled by the function $f(x) = 0.08x$, where x and $f(x)$ are measured in miles. Find the rate of change of elevation when $x = 2$.

5. Find the area of the shaded region.

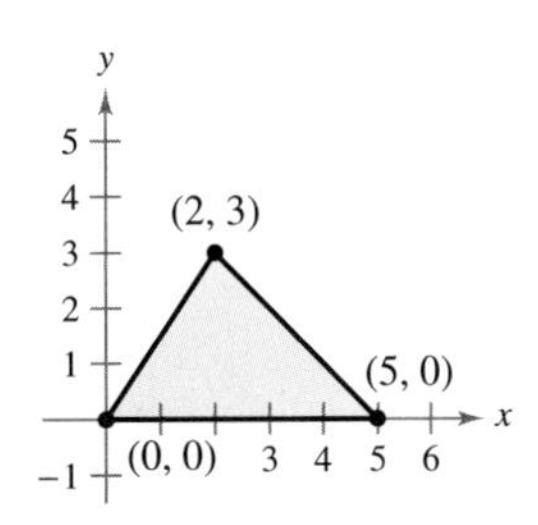

Figure for 5

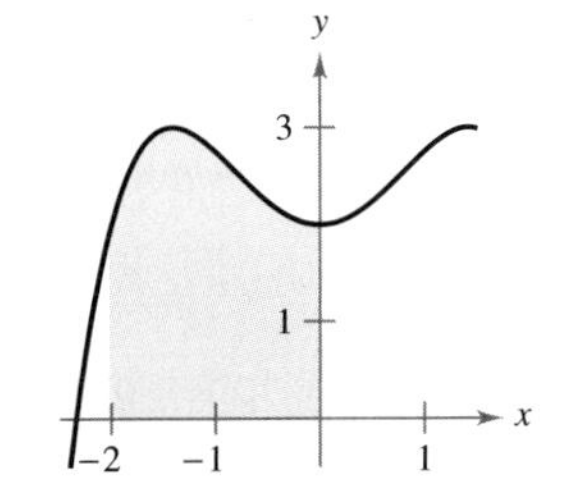

Figure for 6

6. Find the area of the shaded region.

7. ***Secant Lines*** Consider the function $f(x) = 4x - x^2$ and the point $P(1, 3)$ on the graph of f.

(a) Graph f and the secant lines passing through $P(1, 3)$ and $Q(x, f(x))$ for x-values of 2, 1.5, and 0.5.

(b) Find the slope of each secant line.

(c) Use the results of part (b) to estimate the slope of the tangent line of f at $P(1, 3)$. Describe how to improve your approximation of the slope.

8. ***Secant Lines*** Consider the function $f(x) = \sqrt{x}$ and the point $P(4, 2)$ on the graph of f.

(a) Graph f and the secant lines passing through $P(4, 2)$ and $Q(x, f(x))$ for x-values of 1, 3, and 5.

(b) Find the slope of each secant line.

(c) Use the results of part (b) to estimate the slope of the tangent line of f at $P(4, 2)$. Describe how to improve your approximation of the slope.

9. (a) Use the rectangles in each graph to approximate the area of the region bounded by $y = 5/x$, $y = 0$, $x = 1$, and $x = 5$.

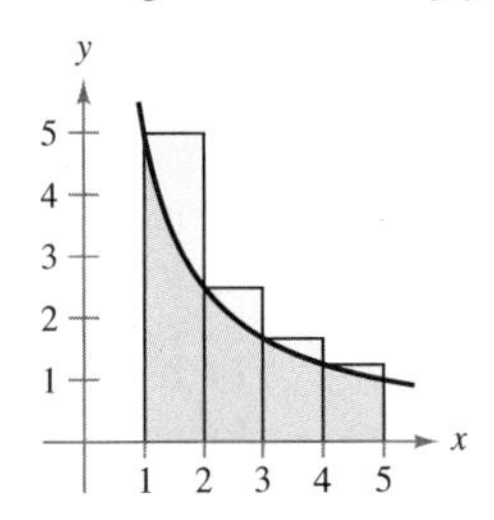

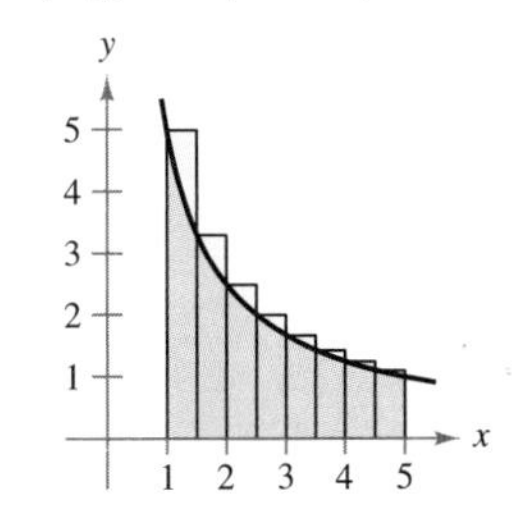

(b) Describe how you could continue this process to obtain a more accurate approximation of the area.

10. (a) Use the rectangles in each graph to approximate the area of the region bounded by $y = \sin x$, $y = 0$, $x = 0$, and $x = \pi$.

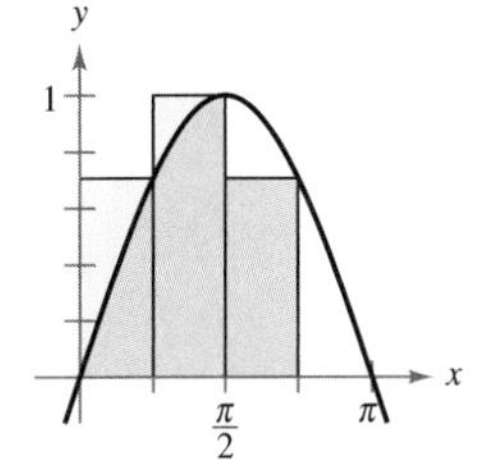

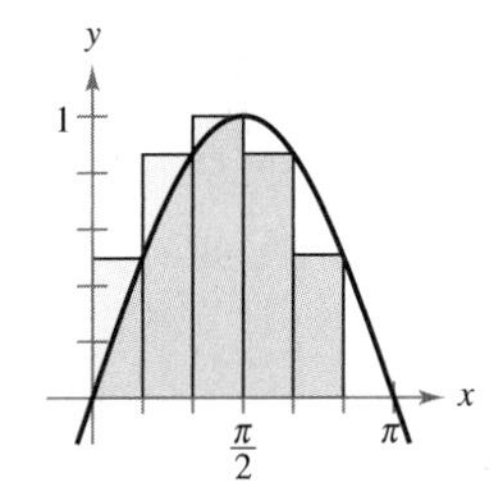

(b) Describe how you could continue this process to obtain a more accurate approximation of the area.

Writing About Concepts

11. Consider the length of the graph of $f(x) = 5/x$ from $(1, 5)$ to $(5, 1)$.

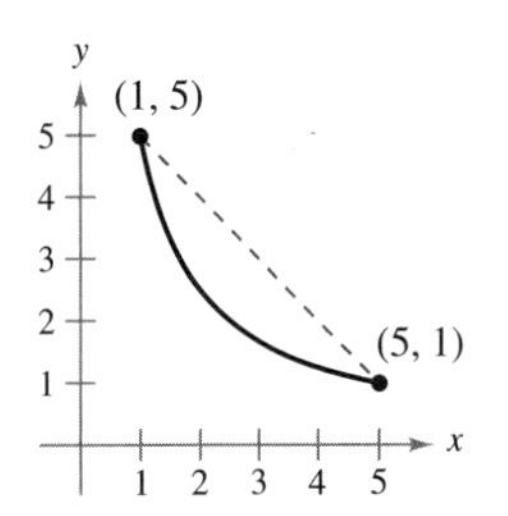

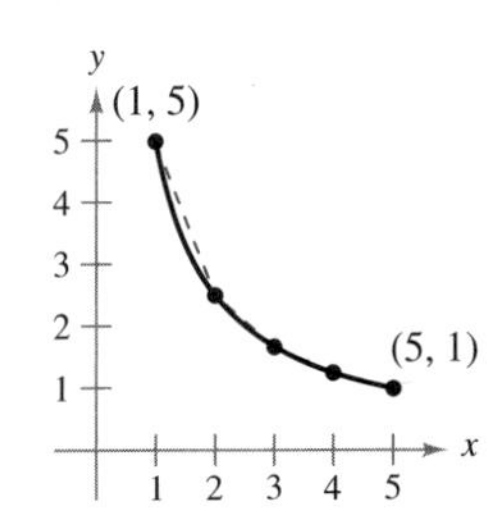

(a) Approximate the length of the curve by finding the distance between its two endpoints, as shown in the first figure.

(b) Approximate the length of the curve by finding the sum of the lengths of four line segments, as shown in the second figure.

(c) Describe how you could continue this process to obtain a more accurate approximation of the length of the curve.

Section 2.2 Finding Limits Graphically and Numerically

- **Estimate a limit using a numerical or graphical approach.**
- **Learn different ways that a limit can fail to exist.**
- **Study and use a formal definition of a limit.**

An Introduction to Limits

Suppose you are asked to sketch the graph of the function f given by

$$f(x) = \frac{x^3 - 1}{x - 1}, \qquad x \neq 1.$$

For all values other than $x = 1$, you can use standard curve-sketching techniques. However, at $x = 1$, it is not clear what to expect. To get an idea of the behavior of the graph of f near $x = 1$, you can use two sets of x-values—one set that approaches 1 from the left and one that approaches 1 from the right, as shown in the table.

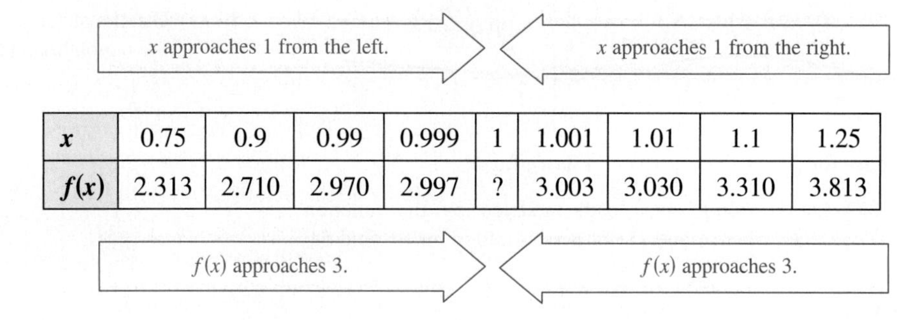

x approaches 1 from the left.
x approaches 1 from the right.
$f(x)$ approaches 3.
$f(x)$ approaches 3.

x	0.75	0.9	0.99	0.999	1	1.001	1.01	1.1	1.25
$f(x)$	2.313	2.710	2.970	2.997	?	3.003	3.030	3.310	3.813

The limit of $f(x)$ as x approaches 1 is 3.
Figure 2.5

The graph of f is a parabola that has a gap at the point $(1, 3)$, as shown in Figure 2.5. Although x cannot equal 1, you can move arbitrarily close to 1, and as a result $f(x)$ moves arbitrarily close to 3. Using limit notation, you can write

$$\lim_{x \to 1} f(x) = 3. \qquad \text{This is read as "the limit of } f(x) \text{ as } x \text{ approaches 1 is 3."}$$

This discussion leads to an informal description of a limit. If $f(x)$ becomes arbitrarily close to a single number L as x approaches c from either side, the **limit** of $f(x)$, as x approaches c, is L. This limit is written as

$$\lim_{x \to c} f(x) = L.$$

EXPLORATION

The discussion above gives an example of how you can estimate a limit *numerically* by constructing a table and *graphically* by drawing a graph. Estimate the following limit numerically by completing the table.

$$\lim_{x \to 2} \frac{x^2 - 3x + 2}{x - 2}$$

x	1.75	1.9	1.99	1.999	2	2.001	2.01	2.1	2.25
$f(x)$	?	?	?	?	?	?	?	?	?

Then use a graphing utility to estimate the limit graphically.

EXAMPLE 1 Estimating a Limit Numerically

Evaluate the function $f(x) = x/(\sqrt{x+1} - 1)$ at several points near $x = 0$ and use the results to estimate the limit

$$\lim_{x \to 0} \frac{x}{\sqrt{x+1} - 1}.$$

Solution The table lists the values of $f(x)$ for several x-values near 0.

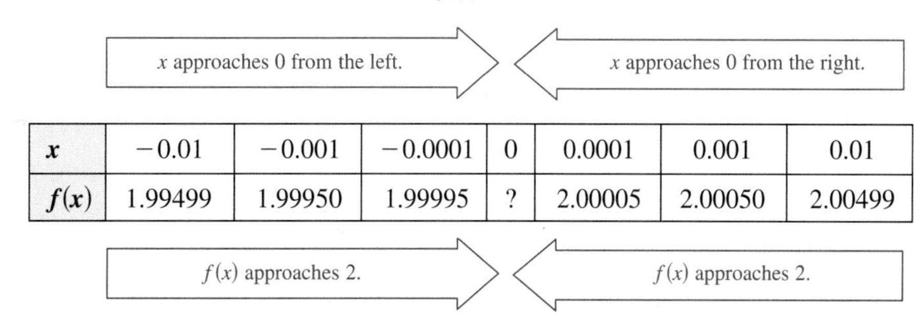

x	-0.01	-0.001	-0.0001	0	0.0001	0.001	0.01
$f(x)$	1.99499	1.99950	1.99995	?	2.00005	2.00050	2.00499

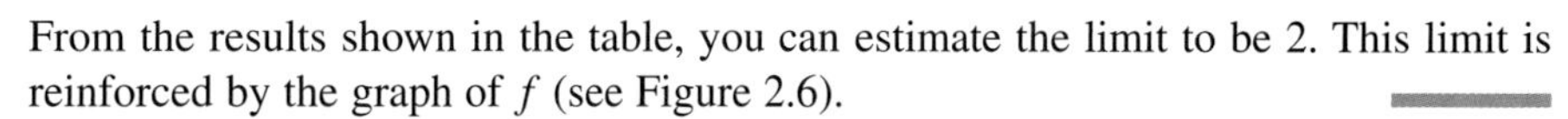

From the results shown in the table, you can estimate the limit to be 2. This limit is reinforced by the graph of f (see Figure 2.6).

The limit of $f(x)$ as x approaches 0 is 2.
Figure 2.6

In Example 1, note that the function is undefined at $x = 0$, and yet $f(x)$ appears to be approaching a limit as x approaches 0. This often happens, and it is important to realize that *the existence or nonexistence of $f(x)$ at $x = c$ has no bearing on the existence of the limit of $f(x)$ as x approaches c.*

EXAMPLE 2 Finding a Limit

Find the limit of $f(x)$ as x approaches 2, where f is defined as

$$f(x) = \begin{cases} 1, & x \neq 2 \\ 0, & x = 2. \end{cases}$$

Solution Because $f(x) = 1$ for all x other than $x = 2$, you can conclude that the limit is 1, as shown in Figure 2.7. So, you can write

$$\lim_{x \to 2} f(x) = 1.$$

The fact that $f(2) = 0$ has no bearing on the existence or value of the limit as x approaches 2. For instance, if the function were defined as

$$f(x) = \begin{cases} 1, & x \neq 2 \\ 2, & x = 2 \end{cases}$$

the limit would be the same.

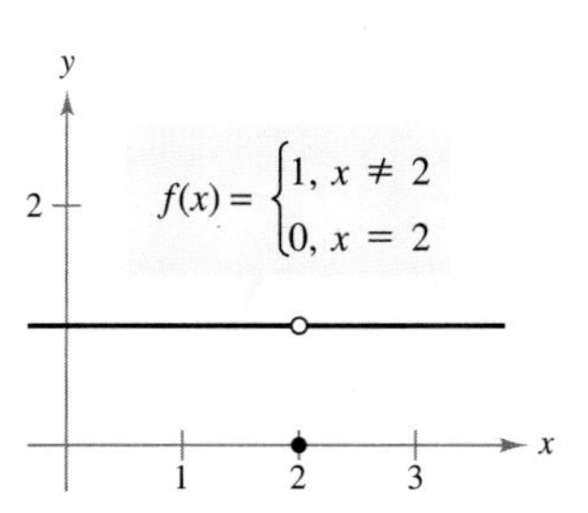

The limit of $f(x)$ as x approaches 2 is 1.
Figure 2.7

So far in this section, you have been estimating limits numerically and graphically. Each of these approaches produces an estimate of the limit. In Section 2.3, you will study analytic techniques for evaluating limits. Throughout the course, try to develop a habit of using this three-pronged approach to problem solving.

1. Numerical approach — Construct a table of values.
2. Graphical approach — Draw a graph by hand or using technology.
3. Analytic approach — Use algebra or calculus.

Limits That Fail to Exist

In the next three examples you will examine some limits that fail to exist.

EXAMPLE 3 Behavior That Differs from the Right and Left

Show that the limit does not exist.

$$\lim_{x \to 0} \frac{|x|}{x}$$

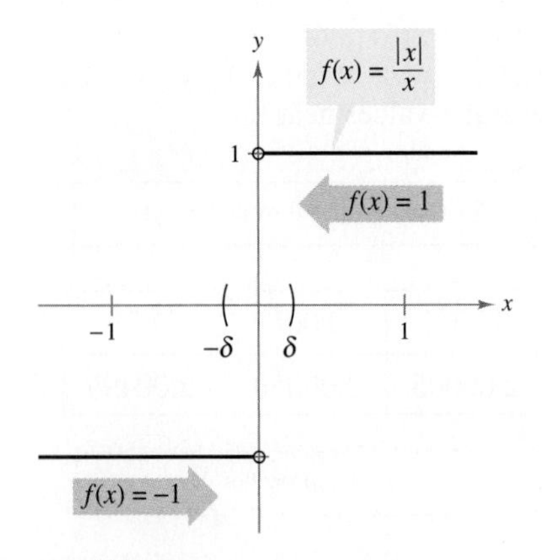

$\lim_{x \to 0} f(x)$ does not exist.
Figure 2.8

Solution Consider the graph of the function $f(x) = |x|/x$. From Figure 2.8, you can see that for positive x-values

$$\frac{|x|}{x} = 1, \qquad x > 0$$

and for negative x-values

$$\frac{|x|}{x} = -1, \qquad x < 0.$$

This means that no matter how close x gets to 0, there will be both positive and negative x-values that yield $f(x) = 1$ and $f(x) = -1$. Specifically, if δ (the lowercase Greek letter *delta*) is a positive number, then for x-values satisfying the inequality $0 < |x| < \delta$, you can classify the values of $|x|/x$ as shown.

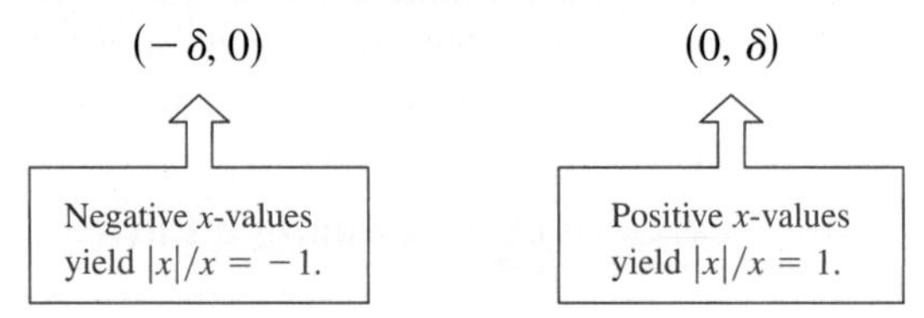

This implies that the limit does not exist.

EXAMPLE 4 Unbounded Behavior

Discuss the existence of the limit

$$\lim_{x \to 0} \frac{1}{x^2}.$$

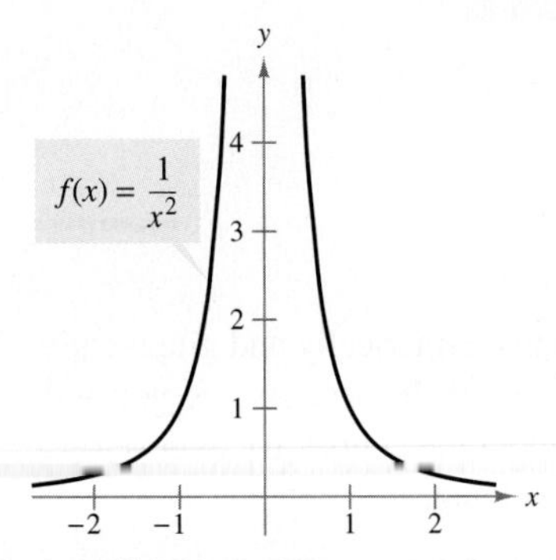

$\lim_{x \to 0} f(x)$ does not exist.
Figure 2.9

Solution Let $f(x) = 1/x^2$. In Figure 2.9, you can see that as x approaches 0 from either the right or the left, $f(x)$ increases without bound. This means that by choosing x close enough to 0, you can force $f(x)$ to be as large as you want. For instance, $f(x)$ will be larger than 100 if you choose x that is within $\frac{1}{10}$ of 0. That is,

$$0 < |x| < \frac{1}{10} \implies f(x) = \frac{1}{x^2} > 100.$$

Similarly, you can force $f(x)$ to be larger than 1,000,000, as follows.

$$0 < |x| < \frac{1}{1000} \implies f(x) = \frac{1}{x^2} > 1{,}000{,}000$$

Because $f(x)$ is not approaching a real number L as x approaches 0, you can conclude that the limit does not exist.

EXAMPLE 5 Oscillating Behavior

Discuss the existence of the limit $\lim_{x \to 0} \sin \frac{1}{x}$.

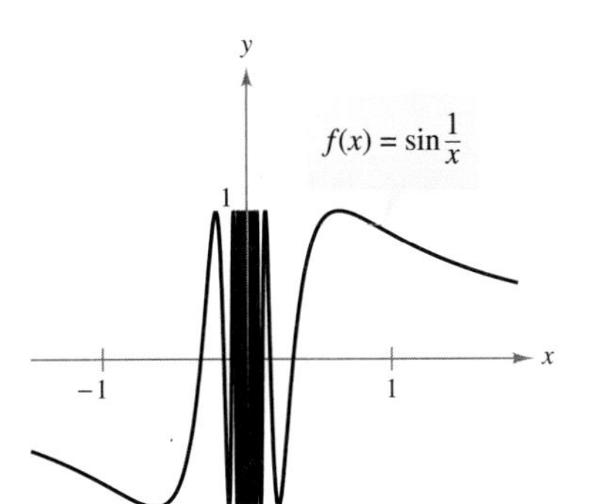

$\lim_{x \to 0} f(x)$ does not exist.
Figure 2.10

Solution Let $f(x) = \sin(1/x)$. In Figure 2.10, you can see that as x approaches 0, $f(x)$ oscillates between -1 and 1. So, the limit does not exist because no matter how small you choose δ, it is possible to choose x_1 and x_2 within δ units of 0 such that $\sin(1/x_1) = 1$ and $\sin(1/x_2) = -1$, as shown in the table.

x	$2/\pi$	$2/3\pi$	$2/5\pi$	$2/7\pi$	$2/9\pi$	$2/11\pi$	$x \to 0$
$\sin(1/x)$	1	-1	1	-1	1	-1	Limit does not exist.

Common Types of Behavior Associated with Nonexistence of a Limit

1. $f(x)$ approaches a different number from the right side of c than it approaches from the left side.
2. $f(x)$ increases or decreases without bound as x approaches c.
3. $f(x)$ oscillates between two fixed values as x approaches c.

There are many other interesting functions that have unusual limit behavior. An often cited one is the *Dirichlet function*

$$f(x) = \begin{cases} 0, & \text{if } x \text{ is rational.} \\ 1, & \text{if } x \text{ is irrational.} \end{cases}$$

Because this function has *no limit* at any real number c, it is *not continuous* at any real number c. You will study continuity more closely in Section 2.4.

The Granger Collection

PETER GUSTAV DIRICHLET (1805–1859)

In the early development of calculus, the definition of a function was much more restricted than it is today, and "functions" such as the Dirichlet function would not have been considered. The modern definition of function was given by the German mathematician Peter Gustav Dirichlet.

TECHNOLOGY PITFALL When you use a graphing utility to investigate the behavior of a function near the x-value at which you are trying to evaluate a limit, remember that you can't always trust the pictures that graphing utilities draw. For instance, if you use a graphing utility to graph the function in Example 5 over an interval containing 0, you will most likely obtain an incorrect graph such as that shown in Figure 2.11. The reason that a graphing utility can't show the correct graph is that the graph has infinitely many oscillations over any interval that contains 0.

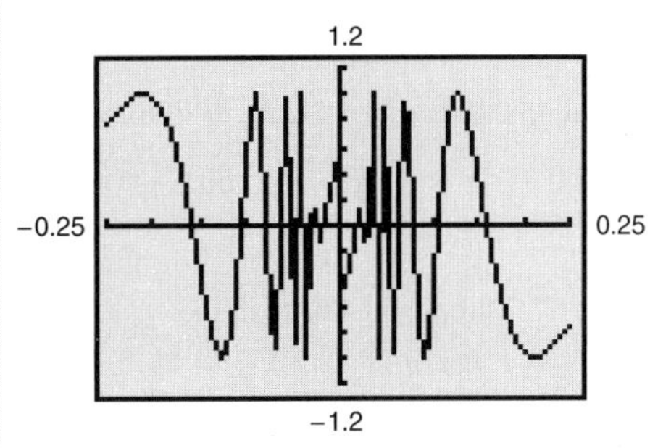

Incorrect graph of $f(x) = \sin(1/x)$
Figure 2.11

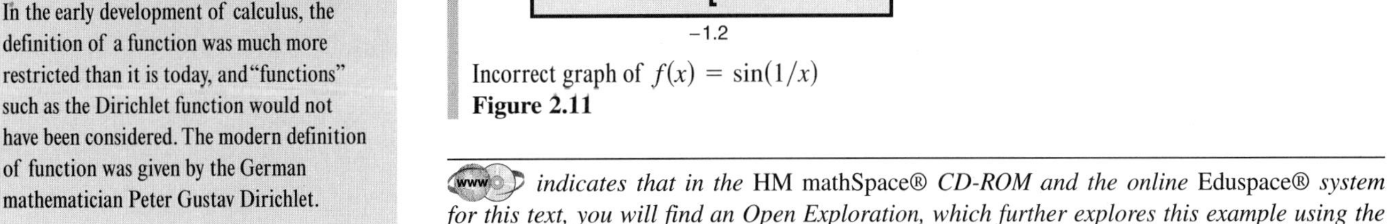

indicates that in the HM mathSpace® *CD-ROM and the online* Eduspace® *system for this text, you will find an Open Exploration, which further explores this example using the computer algebra systems* Maple, Mathcad, Mathematica, *and* Derive.

A Formal Definition of Limit

Let's take another look at the informal description of a limit. If $f(x)$ becomes arbitrarily close to a single number L as x approaches c from either side, then the limit of $f(x)$ as x approaches c is L, written as

$$\lim_{x \to c} f(x) = L.$$

At first glance, this description looks fairly technical. Even so, it is informal because exact meanings have not yet been given to the two phrases

"$f(x)$ becomes arbitrarily close to L"

and

"x approaches c."

The first person to assign mathematically rigorous meanings to these two phrases was Augustin-Louis Cauchy. His **ε-δ definition of limit** is the standard used today.

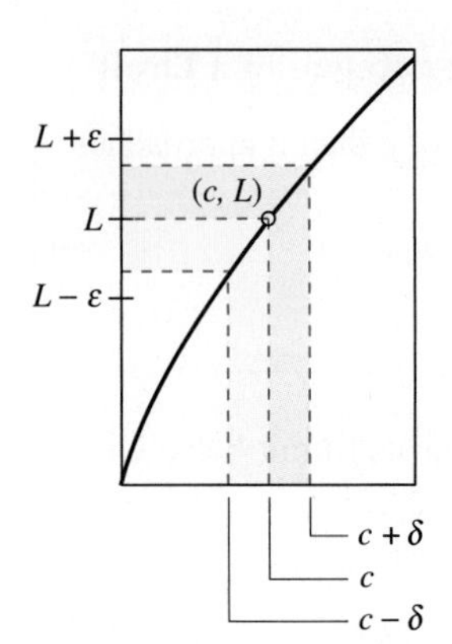

The ε-δ definition of the limit of $f(x)$ as x approaches c
Figure 2.12

In Figure 2.12, let ε (the lowercase Greek letter *epsilon*) represent a (small) positive number. Then the phrase "$f(x)$ becomes arbitrarily close to L" means that $f(x)$ lies in the interval $(L - \varepsilon, L + \varepsilon)$. Using absolute value, you can write this as

$$|f(x) - L| < \varepsilon.$$

Similarly, the phrase "x approaches c" means that there exists a positive number δ such that x lies in either the interval $(c - \delta, c)$ or the interval $(c, c + \delta)$. This fact can be concisely expressed by the double inequality

$$0 < |x - c| < \delta.$$

The first inequality

$0 < |x - c|$ The distance between x and c is more than 0.

expresses the fact that $x \neq c$. The second inequality

$|x - c| < \delta$ x is within δ units of c.

states that x is within a distance δ of c.

Definition of Limit

Let f be a function defined on an open interval containing c (except possibly at c) and let L be a real number. The statement

$$\lim_{x \to c} f(x) = L$$

means that for each $\varepsilon > 0$ there exists a $\delta > 0$ such that if

$$0 < |x - c| < \delta, \quad \text{then} \quad |f(x) - L| < \varepsilon.$$

NOTE Throughout this text, the expression

$$\lim_{x \to c} f(x) = L$$

implies two statements—the limit exists *and* the limit is L.

FOR FURTHER INFORMATION For more on the introduction of rigor to calculus, see "Who Gave You the Epsilon? Cauchy and the Origins of Rigorous Calculus" by Judith V. Grabiner in *The American Mathematical Monthly*. To view this article, go to the website *www.matharticles.com*.

Some functions do not have limits as $x \to c$, but those that do cannot have two different limits as $x \to c$. That is, *if the limit of a function exists, it is unique* (see Exercise 71).

The next three examples should help you develop a better understanding of the ε-δ definition of a limit.

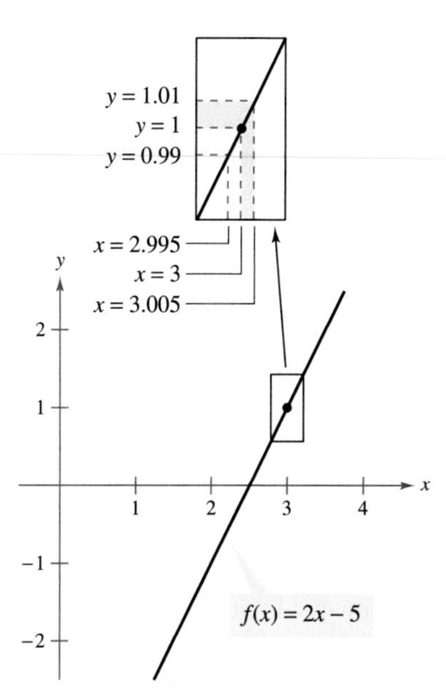

The limit of $f(x)$ as x approaches 3 is 1.
Figure 2.13

EXAMPLE 6 Finding a δ for a Given ε

Given the limit

$$\lim_{x\to 3} (2x - 5) = 1$$

find δ such that $|(2x - 5) - 1| < 0.01$ whenever $0 < |x - 3| < \delta$.

Solution In this problem, you are working with a given value of ε—namely, $\varepsilon = 0.01$. To find an appropriate δ, notice that

$$|(2x - 5) - 1| = |2x - 6| = 2|x - 3|.$$

Because the inequality $|(2x - 5) - 1| < 0.01$ is equivalent to $2|x - 3| < 0.01$, you can choose $\delta = \frac{1}{2}(0.01) = 0.005$. This choice works because

$$0 < |x - 3| < 0.005$$

implies that

$$|(2x - 5) - 1| = 2|x - 3| < 2(0.005) = 0.01$$

as shown in Figure 2.13.

NOTE In Example 6, note that 0.005 is the *largest* value of δ that will guarantee $|(2x - 5) - 1| < 0.01$ whenever $0 < |x - 3| < \delta$. Any *smaller* positive value of δ would, of course, also work.

In Example 6, you found a δ-value for a *given* ε. This does not prove the existence of the limit. To do that, you must prove that you can find a δ for any ε, as shown in the next example.

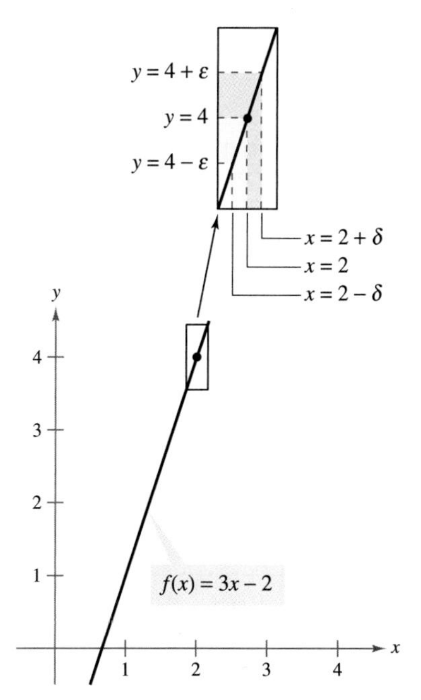

The limit of $f(x)$ as x approaches 2 is 4.
Figure 2.14

EXAMPLE 7 Using the ε-δ Definition of a Limit

Use the ε-δ definition of a limit to prove that

$$\lim_{x\to 2} (3x - 2) = 4.$$

Solution You must show that for each $\varepsilon > 0$, there exists a $\delta > 0$ such that $|(3x - 2) - 4| < \varepsilon$ whenever $0 < |x - 2| < \delta$. Because your choice of δ depends on ε, you need to establish a connection between the absolute values $|(3x - 2) - 4|$ and $|x - 2|$.

$$|(3x - 2) - 4| = |3x - 6| = 3|x - 2|$$

So, for a given $\varepsilon > 0$, you can choose $\delta = \varepsilon/3$. This choice works because

$$0 < |x - 2| < \delta = \frac{\varepsilon}{3}$$

implies that

$$|(3x - 2) - 4| = 3|x - 2| < 3\left(\frac{\varepsilon}{3}\right) = \varepsilon$$

as shown in Figure 2.14.

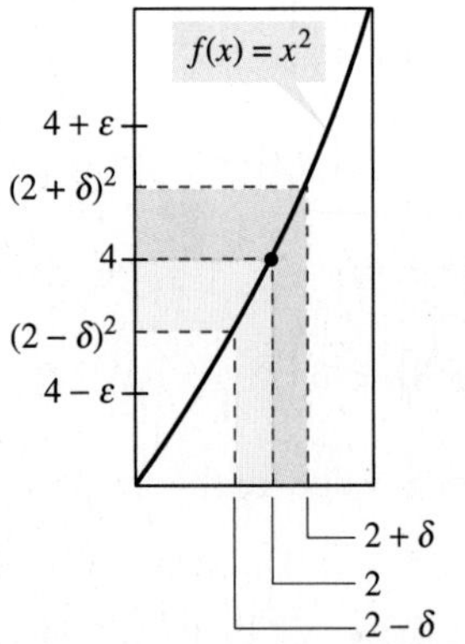

EXAMPLE 8 Using the ε-δ Definition of a Limit

Use the ε-δ definition of a limit to prove that

$$\lim_{x \to 2} x^2 = 4.$$

Solution You must show that for each $\varepsilon > 0$, there exists a $\delta > 0$ such that

$$|x^2 - 4| < \varepsilon \text{ whenever } 0 < |x - 2| < \delta.$$

To find an appropriate δ, begin by writing $|x^2 - 4| = |x - 2||x + 2|$. For all x in the interval $(1, 3)$, you know that $|x + 2| < 5$. So, letting δ be the minimum of $\varepsilon/5$ and 1, it follows that, whenever $0 < |x - 2| < \delta$, you have

$$|x^2 - 4| = |x - 2||x + 2| < \left(\frac{\varepsilon}{5}\right)(5) = \varepsilon$$

as shown in Figure 2.15.

The limit of $f(x)$ as x approaches 2 is 4.
Figure 2.15

Throughout this chapter you will use the ε-δ definition of a limit primarily to prove theorems about limits and to establish the existence or nonexistence of particular types of limits. For *finding* limits, you will learn techniques that are easier to use than the ε-δ definition of a limit.

Exercises for Section 2.2

See www.CalcChat.com for worked-out solutions to odd-numbered exercises.

In Exercises 1–10, complete the table and use the result to estimate the limit. Use a graphing utility to graph the function to confirm your result.

1. $\displaystyle\lim_{x \to 2} \frac{x - 2}{x^2 - x - 2}$

x	1.9	1.99	1.999	2.001	2.01	2.1
$f(x)$						

2. $\displaystyle\lim_{x \to 2} \frac{x - 2}{x^2 - 4}$

x	1.9	1.99	1.999	2.001	2.01	2.1
$f(x)$						

3. $\displaystyle\lim_{x \to 3} \frac{[1/(x + 1)] - (1/4)}{x - 3}$

x	2.9	2.99	2.999	3.001	3.01	3.1
$f(x)$						

4. $\displaystyle\lim_{x \to -3} \frac{\sqrt{1 - x} - 2}{x + 3}$

x	−3.1	−3.01	−3.001	−2.999	−2.99	−2.9
$f(x)$						

5. $\displaystyle\lim_{x \to 0} \frac{\sin x}{x}$

x	−0.1	−0.01	−0.001	0.001	0.01	0.1
$f(x)$						

6. $\displaystyle\lim_{x \to 0} \frac{\cos x - 1}{x}$

x	−0.1	−0.01	−0.001	0.001	0.01	0.1
$f(x)$						

7. $\displaystyle\lim_{x \to 0} \frac{e^x - 1}{x}$

x	−0.1	−0.01	−0.001	0.001	0.01	0.1
$f(x)$						

8. $\displaystyle\lim_{x \to 0} \frac{4}{1 + e^{1/x}}$

x	−0.1	−0.01	−0.001	0.001	0.01	0.1
$f(x)$						

9. $\lim_{x \to 0} \frac{\ln(x + 1)}{x}$

x	-0.1	-0.01	-0.001	0.001	0.01	0.1
$f(x)$						

10. $\lim_{x \to 2} \frac{\ln x - \ln 2}{x - 2}$

x	1.9	1.99	1.999	2.001	2.01	2.1
$f(x)$						

In Exercises 11–20, use the graph to find the limit (if it exists). If the limit does not exist, explain why.

11. $\lim_{x \to 3} (4 - x)$

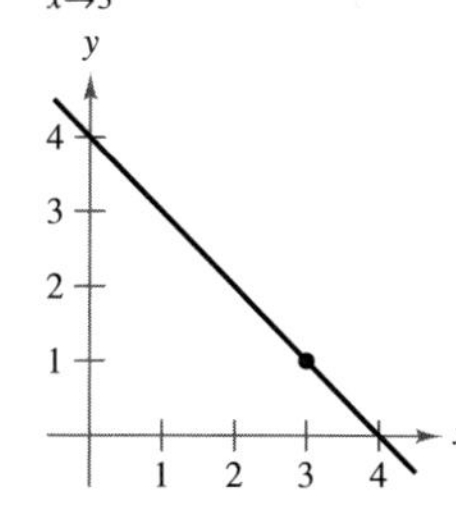

12. $\lim_{x \to 1} (x^2 + 2)$

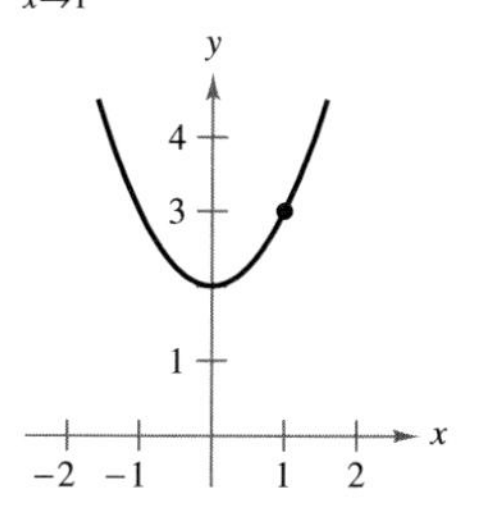

13. $\lim_{x \to 3} \frac{|x - 3|}{x - 3}$

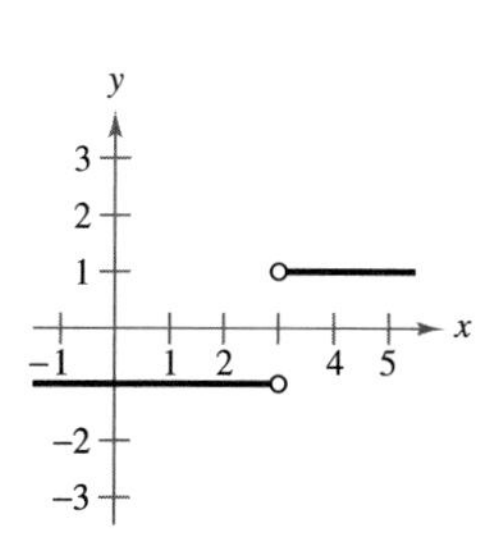

14. $\lim_{x \to 1} f(x)$

$$f(x) = \begin{cases} x^2 + 3, & x \neq 1 \\ 1, & x = 1 \end{cases}$$

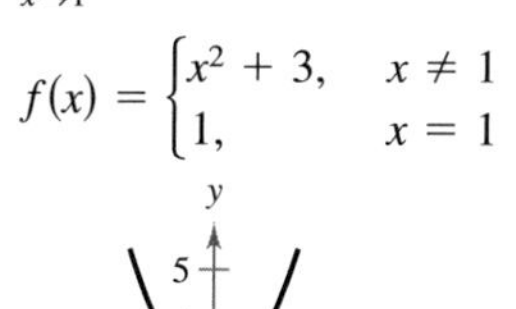

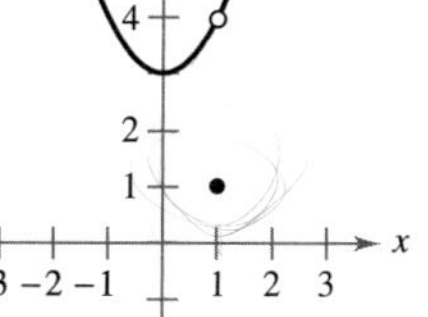

15. $\lim_{x \to 1} \sqrt[3]{x} \ln|x - 2|$

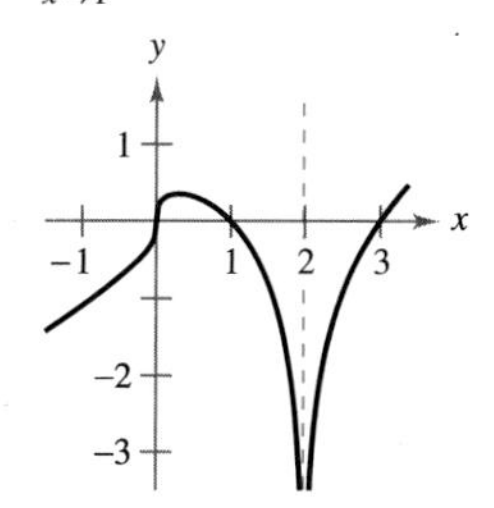

16. $\lim_{x \to 0} \frac{4}{2 + e^{1/x}}$

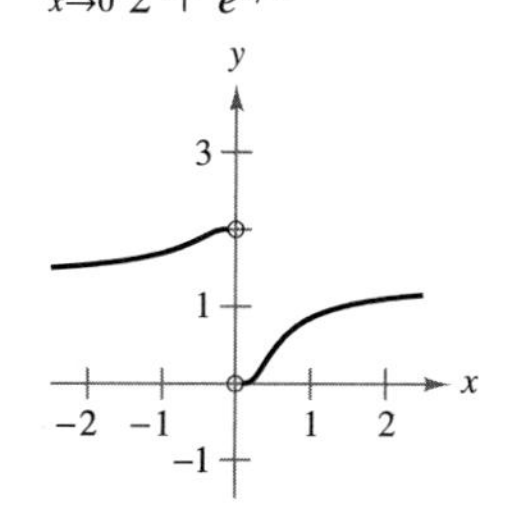

17. $\lim_{x \to \pi/2} \tan x$

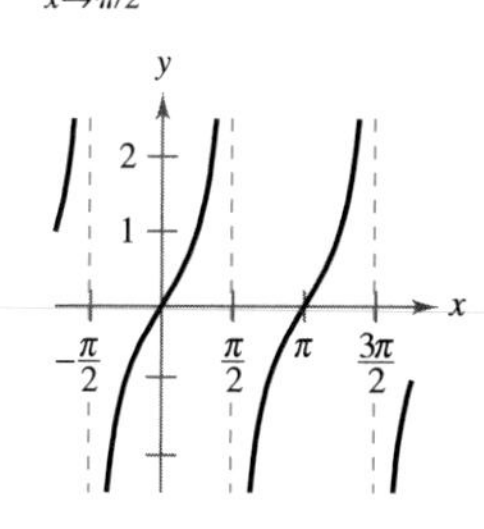

18. $\lim_{x \to 0} 2 \cos \frac{1}{x}$

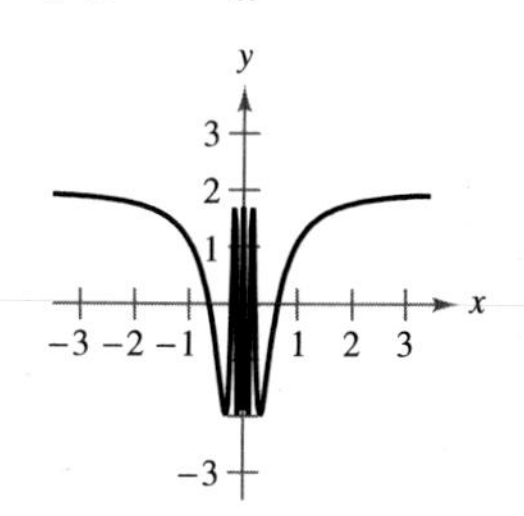

19. $\lim_{x \to 0} \sec x$

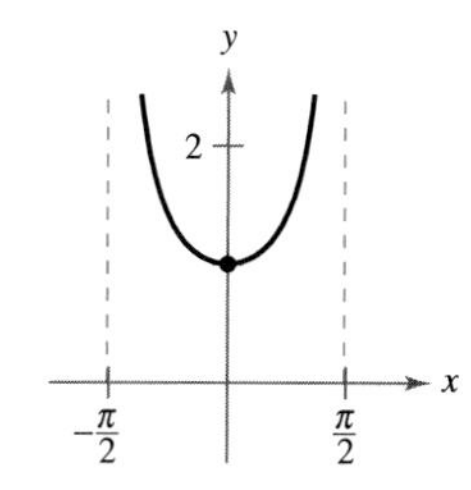

20. $\lim_{x \to 2} \frac{1}{x - 2}$

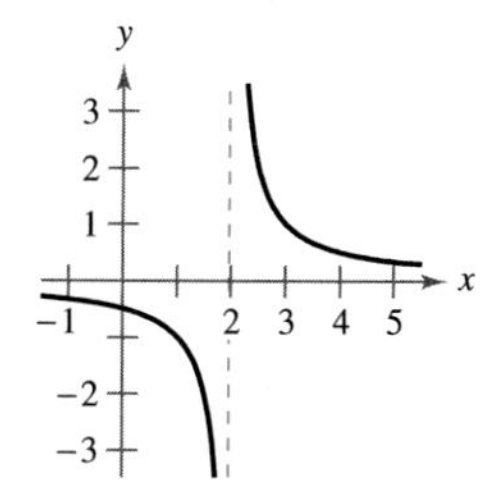

In Exercises 21 and 22, use the graph of the function f to decide whether the value of the given quantity exists. If it does, find it. If not, explain why.

21. (a) $f(1)$

(b) $\lim_{x \to 1} f(x)$

(c) $f(4)$

(d) $\lim_{x \to 4} f(x)$

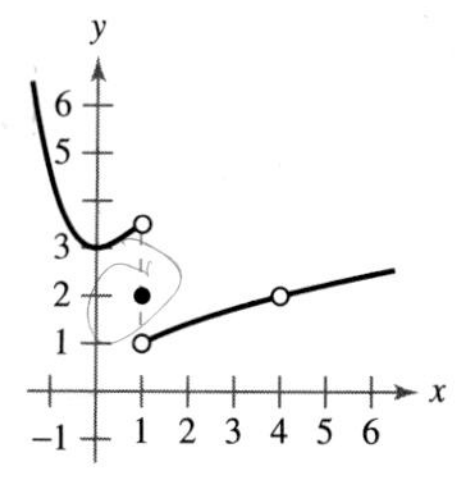

22. (a) $f(-2)$

(b) $\lim_{x \to -2} f(x)$

(c) $f(0)$

(d) $\lim_{x \to 0} f(x)$

(e) $f(2)$

(f) $\lim_{x \to 2} f(x)$

(g) $f(4)$

(h) $\lim_{x \to 4} f(x)$

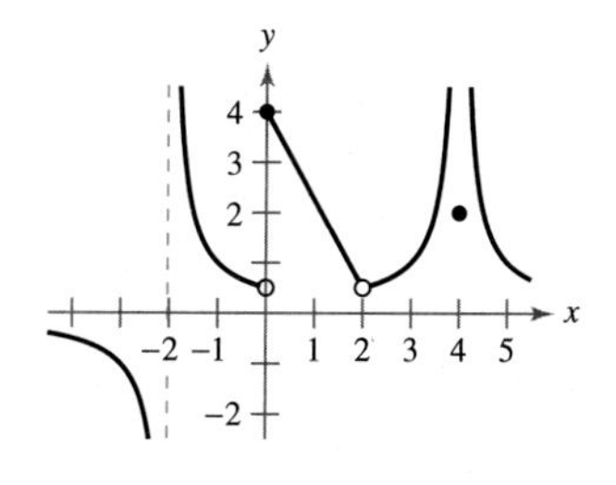

In Exercises 23 and 24, use the graph of f to identify the values of c for which $\lim_{x \to c} f(x)$ exists.

23.

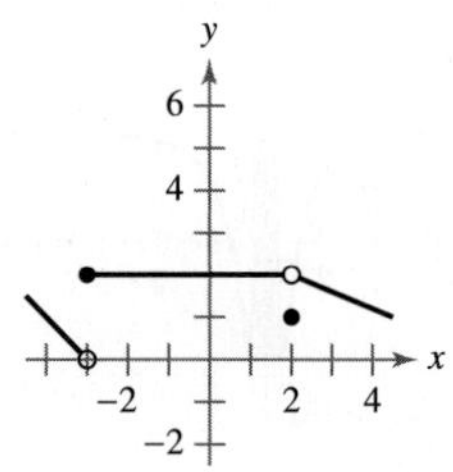

24.

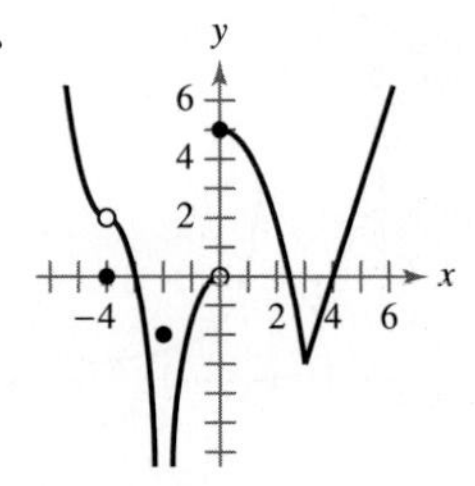

In Exercises 25 and 26, sketch the graph of f. Then identify the values of c for which $\lim_{x \to c} f(x)$ exists.

25. $f(x) = \begin{cases} x^2, & x \le 2 \\ 8 - 2x, & 2 < x < 4 \\ 4, & x \ge 4 \end{cases}$

26. $f(x) = \begin{cases} \sin x, & x < 0 \\ 1 - \cos x, & 0 \le x \le \pi \\ \cos x, & x > \pi \end{cases}$

In Exercises 27 and 28, sketch a graph of a function f that satisfies the given values. (There are many correct answers.)

27. $f(0)$ is undefined.

$\lim_{x \to 0} f(x) = 4$

$f(2) = 6$

$\lim_{x \to 2} f(x) = 3$

28. $f(-2) = 0$

$f(2) = 0$

$\lim_{x \to -2} f(x) = 0$

$\lim_{x \to 2} f(x)$ does not exist.

29. ***Modeling Data*** The cost of a telephone call between two cities is \$0.75 for the first minute and \$0.50 for each additional minute or fraction thereof. A formula for the cost is given by

$$C(t) = 0.75 - 0.50[\![-(t - 1)]\!]$$

where t is the time in minutes.

(*Note:* $[\![x]\!]$ = greatest integer n such that $n \le x$. For example, $[\![3.2]\!] = 3$ and $[\![-1.6]\!] = -2$.)

(a) Use a graphing utility to graph the cost function for $0 < t \le 5$.

(b) Use the graph to complete the table and observe the behavior of the function as t approaches 3.5. Use the graph and the table to find

$$\lim_{t \to 3.5} C(t).$$

t	3	3.3	3.4	3.5	3.6	3.7	4
C				?			

(c) Use the graph to complete the table and observe the behavior of the function as t approaches 3.

t	2	2.5	2.9	3	3.1	3.5	4
C				?			

Does the limit of $C(t)$ as t approaches 3 exist? Explain.

30. Repeat Exercise 29 for

$$C(t) = 0.35 - 0.12[\![-(t - 1)]\!].$$

31. The graph of $f(x) = x + 1$ is shown in the figure. Find δ such that if $0 < |x - 2| < \delta$, then $|f(x) - 3| < 0.4$.

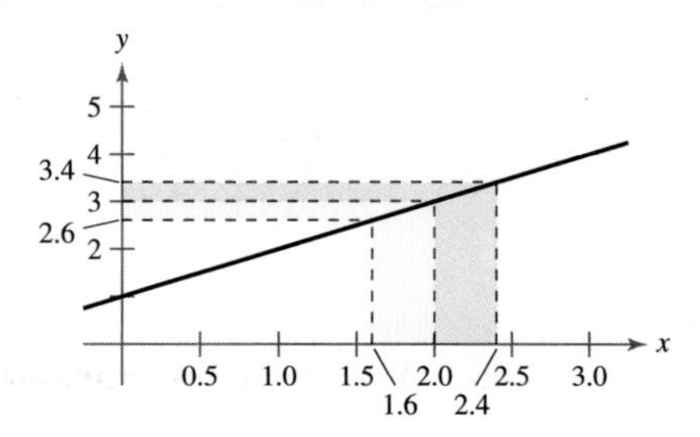

32. The graph of

$$f(x) = \frac{1}{x - 1}$$

is shown in the figure. Find δ such that if $0 < |x - 2| < \delta$, then $|f(x) - 1| < 0.01$.

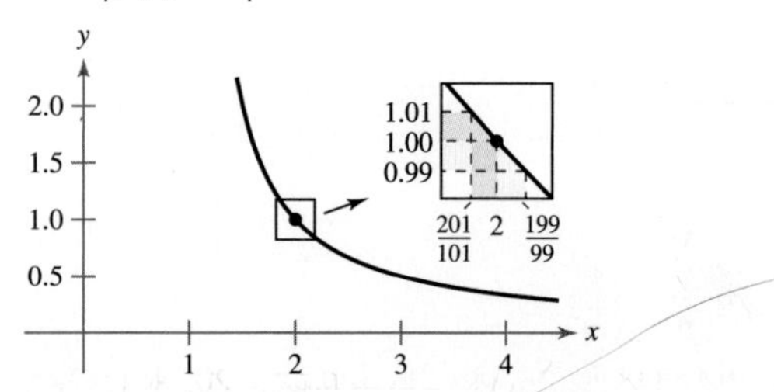

33. The graph of

$$f(x) = 2 - \frac{1}{x}$$

is shown in the figure. Find δ such that if $0 < |x - 1| < \delta$, then $|f(x) - 1| < 0.1$.

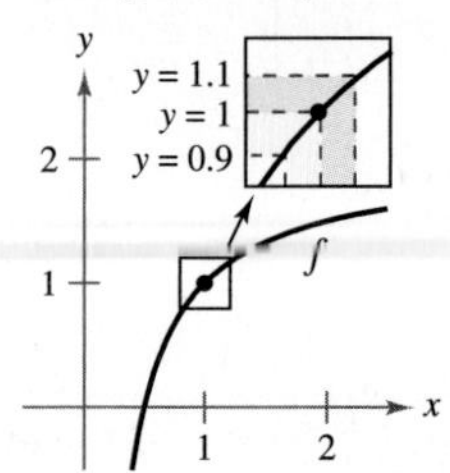

34. The graph of

$f(x) = x^2 - 1$

is shown in the figure. Find δ such that if $0 < |x - 2| < \delta$, then $|f(x) - 3| < 0.2$.

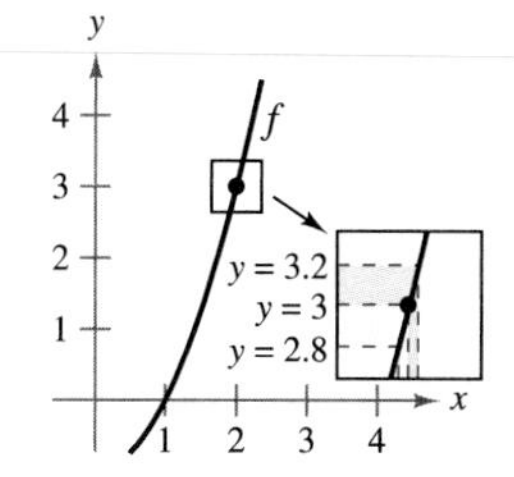

In Exercises 35–38, find the limit L. Then find $\delta > 0$ such that $|f(x) - L| < 0.01$ whenever $0 < |x - c| < \delta$.

35. $\lim_{x \to 2} (3x + 2)$

36. $\lim_{x \to 4} \left(4 - \frac{x}{2}\right)$

37. $\lim_{x \to 2} (x^2 - 3)$

38. $\lim_{x \to 5} (x^2 + 4)$

In Exercises 39–50, find the limit L. Then use the ε-δ definition to prove that the limit is L.

39. $\lim_{x \to 2} (x + 3)$

40. $\lim_{x \to -3} (2x + 5)$

41. $\lim_{x \to -4} \left(\frac{1}{2}x - 1\right)$

42. $\lim_{x \to 1} \left(\frac{2}{3}x + 9\right)$

43. $\lim_{x \to 6} 3$

44. $\lim_{x \to 2} (-1)$

45. $\lim_{x \to 0} \sqrt[3]{x}$

46. $\lim_{x \to 4} \sqrt{x}$

47. $\lim_{x \to -2} |x - 2|$

48. $\lim_{x \to 3} |x - 3|$

49. $\lim_{x \to 1} (x^2 + 1)$

50. $\lim_{x \to -3} (x^2 + 3x)$

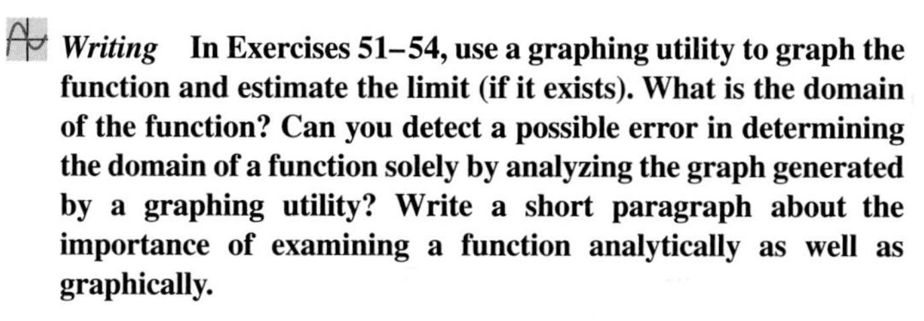

Writing **In Exercises 51–54, use a graphing utility to graph the function and estimate the limit (if it exists). What is the domain of the function? Can you detect a possible error in determining the domain of a function solely by analyzing the graph generated by a graphing utility? Write a short paragraph about the importance of examining a function analytically as well as graphically.**

51. $f(x) = \dfrac{\sqrt{x + 5} - 3}{x - 4}$

$\lim_{x \to 4} f(x)$

52. $f(x) = \dfrac{x - 3}{x^2 - 4x + 3}$

$\lim_{x \to 3} f(x)$

53. $f(x) = \dfrac{x - 9}{\sqrt{x} - 3}$

$\lim_{x \to 9} f(x)$

54. $f(x) = \dfrac{e^{x/2} - 1}{x}$

$\lim_{x \to 0} f(x)$

Writing About Concepts

55. Write a brief description of the meaning of the notation $\lim_{x \to 8} f(x) = 25$.

56. If $f(2) = 4$, can you conclude anything about the limit of $f(x)$ as x approaches 2? Explain your reasoning.

57. If the limit of $f(x)$ as x approaches 2 is 4, can you conclude anything about $f(2)$? Explain your reasoning.

58. Identify three types of behavior associated with the nonexistence of a limit. Illustrate each type with a graph of a function.

59. ***Jewelry*** A jeweler resizes a ring so that its inner circumference is 6 centimeters.

(a) What is the radius of the ring?

(b) If the ring's inner circumference can vary between 5.5 centimeters and 6.5 centimeters, how can the radius vary?

(c) Use the ε-δ definition of a limit to describe this situation. Identify ε and δ.

60. ***Sports*** A sporting goods manufacturer designs a golf ball with a volume of 2.48 cubic inches.

(a) What is the radius of the golf ball?

(b) If the ball's volume can vary between 2.45 cubic inches and 2.51 cubic inches, how can the radius vary?

(c) Use the ε-δ definition of a limit to describe this situation. Identify ε and δ.

61. Consider the function $f(x) = (1 + x)^{1/x}$. Estimate the limit

$$\lim_{x \to 0} (1 + x)^{1/x}$$

by evaluating f at x-values near 0. Sketch the graph of f.

62. Consider the function

$$f(x) = \frac{|x + 1| - |x - 1|}{x}.$$

Estimate

$$\lim_{x \to 0} \frac{|x + 1| - |x - 1|}{x}$$

by evaluating f at x-values near 0. Sketch the graph of f.

The symbol *indicates an exercise in which you are instructed to use graphing technology or a symbolic computer algebra system. The solutions of other exercises may also be facilitated by use of appropriate technology.*

63. ***Graphical Analysis*** The statement

$$\lim_{x\to 2} \frac{x^2 - 4}{x - 2} = 4$$

means that for each $\varepsilon > 0$ there corresponds a $\delta > 0$ such that if $0 < |x - 2| < \delta$, then

$$\left|\frac{x^2 - 4}{x - 2} - 4\right| < \varepsilon.$$

If $\varepsilon = 0.001$, then

$$\left|\frac{x^2 - 4}{x - 2} - 4\right| < 0.001.$$

Use a graphing utility to graph each side of this inequality. Use the *zoom* feature to find an interval $(2 - \delta, 2 + \delta)$ such that the graph of the left side is below the graph of the right side of the inequality.

64. ***Graphical Analysis*** The statement

$$\lim_{x\to 3} \frac{x^2 - 3x}{x - 3}$$

means that for each $\varepsilon > 0$ there corresponds a $\delta > 0$ such that if $0 < |x - 3| < \delta$, then

$$\left|\frac{x^2 - 3x}{x - 3} - 3\right| < \varepsilon.$$

If $\varepsilon = 0.001$, then

$$\left|\frac{x^2 - 3x}{x - 3} - 3\right| < 0.001.$$

Use a graphing utility to graph each side of this inequality. Use the *zoom* feature to find an interval $(3 - \delta, 3 + \delta)$ such that the graph of the left side is below the graph of the right side of the inequality.

True or False? **In Exercises 65–68, determine whether the statement is true or false. If it is false, explain why or give an example that shows it is false.**

65. If f is undefined at $x = c$, then the limit of $f(x)$ as x approaches c does not exist.

66. If the limit of $f(x)$ as x approaches c is 0, then there must exist a number k such that $f(k) < 0.001$.

67. If $f(c) = L$, then $\lim_{x\to c} f(x) = L$.

68. If $\lim_{x\to c} f(x) = L$, then $f(c) = L$.

69. Consider the function $f(x) = \sqrt{x}$.

(a) Is $\lim_{x\to 0.25} \sqrt{x} = 0.5$ a true statement? Explain.

(b) Is $\lim_{x\to 0} \sqrt{x} = 0$ a true statement? Explain.

70. ***Writing*** The definition of limit on page 72 requires that f is a function defined on an open interval containing c, except possibly at c. Why is this requirement necessary?

71. Prove that if the limit of $f(x)$ as $x \to c$ exists, then the limit must be unique. [*Hint:* Let

$$\lim_{x\to c} f(x) = L_1 \quad \text{and} \quad \lim_{x\to c} f(x) = L_2$$

and prove that $L_1 = L_2$.]

72. Consider the line $f(x) = mx + b$, where $m \neq 0$. Use the ε-δ definition of a limit to prove that $\lim_{x\to c} f(x) = mc + b$.

73. Prove that $\lim_{x\to c} f(x) = L$ is equivalent to $\lim_{x\to c} [f(x) - L] = 0$.

74. (a) Given that

$$\lim_{x\to 0} (3x + 1)(3x - 1)x^2 + 0.01 = 0.01$$

prove that there exists an open interval (a, b) containing 0 such that $(3x + 1)(3x - 1)x^2 + 0.01 > 0$ for all $x \neq 0$ in (a, b).

(b) Given that $\lim_{x\to c} g(x) = L$, where $L > 0$, prove that there exists an open interval (a, b) containing c such that $g(x) > 0$ for all $x \neq c$ in (a, b).

75. ***Programming*** Use the programming capabilities of a graphing utility to write a program for approximating $\lim_{x\to c} f(x)$.

Assume the program will be applied only to functions whose limits exist as x approaches c. Let $y_1 = f(x)$ and generate two lists whose entries form the ordered pairs

$$(c \pm [0.1]^n, f(c \pm [0.1]^n))$$

for $n = 0, 1, 2, 3$, and 4.

76. ***Programming*** Use the program you created in Exercise 75 to approximate the limit

$$\lim_{x\to 4} \frac{x^2 - x - 12}{x - 4}.$$

Putnam Exam Challenge

77. Inscribe a rectangle of base b and height h and an isosceles triangle of base b in a circle of radius one as shown. For what value of h do the rectangle and triangle have the same area?

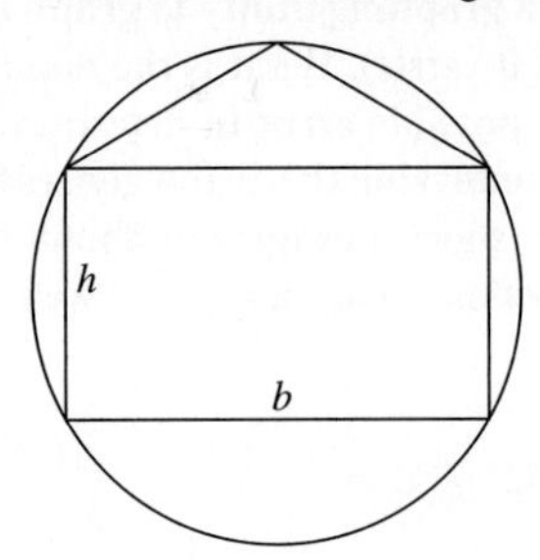

78. A right circular cone has base of radius 1 and height 3. A cube is inscribed in the cone so that one face of the cube is contained in the base of the cone. What is the side-length of the cube?

These problems were composed by the Committee on the Putnam Prize Competition.

Section 2.3 Evaluating Limits Analytically

- **Evaluate a limit using properties of limits.**
- **Develop and use a strategy for finding limits.**
- **Evaluate a limit using dividing out and rationalizing techniques.**
- **Evaluate a limit using the Squeeze Theorem.**

Properties of Limits

In Section 2.2, you learned that the limit of $f(x)$ as x approaches c does not depend on the value of f at $x = c$. It may happen, however, that the limit is precisely $f(c)$. In such cases, the limit can be evaluated by **direct substitution.** That is,

$$\lim_{x \to c} f(x) = f(c). \quad \text{Substitute } c \text{ for } x.$$

Such *well-behaved* functions are **continuous at c.** You will examine this concept more closely in Section 2.4.

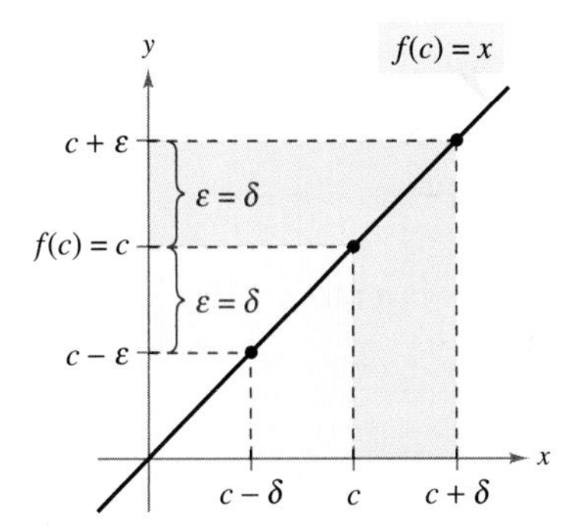

Figure 2.16

THEOREM 2.1 Some Basic Limits

Let b and c be real numbers and let n be a positive integer.

1. $\lim_{x \to c} b = b$ **2.** $\lim_{x \to c} x = c$ **3.** $\lim_{x \to c} x^n = c^n$

Proof To prove Property 2 of Theorem 2.1, you need to show that for each $\varepsilon > 0$ there exists a $\delta > 0$ such that $|x - c| < \varepsilon$ whenever $0 < |x - c| < \delta$. Because the second inequality is a stricter version of the first, you can simply choose $\delta = \varepsilon$, as shown in Figure 2.16. This completes the proof. (Proofs of the other properties of limits in this section are listed in Appendix A or are discussed in the exercises.)

NOTE When you encounter new notations or symbols in mathematics, be sure you know how the notations are read. For instance, the limit in Example 1(c) is read as "the limit of x^2 as x approaches 2 is 4."

EXAMPLE 1 **Evaluating Basic Limits**

a. $\lim_{x \to 2} 3 = 3$ **b.** $\lim_{x \to -4} x = -4$ **c.** $\lim_{x \to 2} x^2 = 2^2 = 4$

THEOREM 2.2 Properties of Limits

Let b and c be real numbers, let n be a positive integer, and let f and g be functions with the following limits.

$$\lim_{x \to c} f(x) = L \quad \text{and} \quad \lim_{x \to c} g(x) = K$$

1. Scalar multiple: $\lim_{x \to c} [b f(x)] = bL$

2. Sum or difference: $\lim_{x \to c} [f(x) \pm g(x)] = L \pm K$

3. Product: $\lim_{x \to c} [f(x)g(x)] = LK$

4. Quotient: $\lim_{x \to c} \dfrac{f(x)}{g(x)} = \dfrac{L}{K}$, provided $K \neq 0$

5. Power: $\lim_{x \to c} [f(x)]^n = L^n$

EXAMPLE 2 The Limit of a Polynomial

$$
\begin{aligned}
\lim_{x\to 2}(4x^2+3) &= \lim_{x\to 2} 4x^2 + \lim_{x\to 2} 3 && \text{Property 2} \\
&= 4\left(\lim_{x\to 2} x^2\right) + \lim_{x\to 2} 3 && \text{Property 1} \\
&= 4(2^2) + 3 && \text{Example 1} \\
&= 19 && \text{Simplify.}
\end{aligned}
$$

In Example 2, note that the limit (as $x \to 2$) of the *polynomial function* $p(x) = 4x^2 + 3$ is simply the value of p at $x = 2$.

$$\lim_{x\to 2} p(x) = p(2) = 4(2^2) + 3 = 19$$

This *direct substitution* property is valid for all polynomial and rational functions with nonzero denominators.

THEOREM 2.3 Limits of Polynomial and Rational Functions

If p is a polynomial function and c is a real number, then

$$\lim_{x\to c} p(x) = p(c).$$

If r is a rational function given by $r(x) = p(x)/q(x)$ and c is a real number such that $q(c) \neq 0$, then

$$\lim_{x\to c} r(x) = r(c) = \frac{p(c)}{q(c)}.$$

EXAMPLE 3 The Limit of a Rational Function

Find the limit: $\lim_{x\to 1} \dfrac{x^2 + x + 2}{x + 1}$.

Solution Because the denominator is not 0 when $x = 1$, you can apply Theorem 2.3 to obtain

$$\lim_{x\to 1} \frac{x^2 + x + 2}{x + 1} = \frac{1^2 + 1 + 2}{1 + 1} = \frac{4}{2} = 2.$$

Polynomial functions and rational functions are two of the three basic types of algebraic functions. The following theorem deals with the limit of the third type of algebraic function—one that involves a radical. See Appendix A for a proof of this theorem.

THEOREM 2.4 The Limit of a Function Involving a Radical

Let n be a positive integer. The following limit is valid for all c if n is odd, and is valid for $c > 0$ if n is even.

$$\lim_{x\to c} \sqrt[n]{x} = \sqrt[n]{c}$$

THE SQUARE ROOT SYMBOL

The first use of a symbol to denote the square root can be traced to the sixteenth century. Mathematicians first used the symbol $\surd$, which had only two strokes. This symbol was chosen because it resembled a lowercase *r*, to stand for the Latin word *radix*, meaning root.

NOTE Your goal in this section is to become familiar with limits that can be evaluated by direct substitution. In the following library of elementary functions, what are the values of c for which

$$\lim_{x\to c} f(x) = f(c)?$$

Polynomial function:

$$f(x) = a_n x^n + \cdots + a_1 x + a_0$$

Rational function: (p and q are polynomials):

$$f(x) = \frac{p(x)}{q(x)}$$

Trigonometric functions:

$$f(x) = \sin x, \quad f(x) = \cos x$$

$$f(x) = \tan x, \quad f(x) = \cot x$$

$$f(x) = \sec x, \quad f(x) = \csc x$$

Exponential functions:

$$f(x) = a^x, \quad f(x) = e^x$$

Natural logarithmic function:

$$f(x) = \ln x$$

The following theorem greatly expands your ability to evaluate limits because it shows how to analyze the limit of a composite function. See Appendix A for a proof of this theorem.

THEOREM 2.5 The Limit of a Composite Function

If f and g are functions such that $\lim_{x \to c} g(x) = L$ and $\lim_{x \to L} f(x) = f(L)$, then

$$\lim_{x \to c} f(g(x)) = f\left(\lim_{x \to c} g(x)\right) = f(L).$$

EXAMPLE 4 The Limit of a Composite Function

Because

$$\lim_{x \to 0} (x^2 + 4) = 0^2 + 4 = 4 \qquad \text{and} \qquad \lim_{x \to 4} \sqrt{x} = 2$$

it follows that

$$\lim_{x \to 0} \sqrt{x^2 + 4} = \sqrt{4} = 2.$$

You have seen that the limits of many algebraic functions can be evaluated by direct substitution. The basic transcendental functions (trigonometric, exponential, and logarithmic) also possess this desirable quality, as shown in the next theorem (presented without proof).

THEOREM 2.6 Limits of Transcendental Functions

Let c be a real number in the domain of the given trigonometric function.

1. $\lim_{x \to c} \sin x = \sin c$
2. $\lim_{x \to c} \cos x = \cos c$
3. $\lim_{x \to c} \tan x = \tan c$
4. $\lim_{x \to c} \cot x = \cot c$
5. $\lim_{x \to c} \sec x = \sec c$
6. $\lim_{x \to c} \csc x = \csc c$
7. $\lim_{x \to c} a^x = a^c, \ (a > 0)$
8. $\lim_{x \to c} \ln x = \ln c$

EXAMPLE 5 Limits of Transcendental Functions

a. $\lim_{x \to 0} \sin x = \sin(0) = 0$

b. $\lim_{x \to 2} (2 + \ln x) = 2 + \ln 2$

c. $\lim_{x \to \pi} (x \cos x) = \left(\lim_{x \to \pi} x\right)\left(\lim_{x \to \pi} \cos x\right) = \pi \cos(\pi) = -\pi$

d. $\lim_{x \to 0} \dfrac{\tan x}{x^2 + 1} = \dfrac{\lim_{x \to 0} \tan x}{\lim_{x \to 0} x^2 + 1} = \dfrac{0}{0^2 + 1} = 0$

e. $\lim_{x \to -1} xe^x = \left(\lim_{x \to -1} x\right)\left(\lim_{x \to -1} e^x\right) = (-1)(e^{-1}) = -e^{-1}$

f. $\lim_{x \to e} \ln x^3 = \lim_{x \to e} 3 \ln x = 3(1) = 3$

A Strategy for Finding Limits

On the previous three pages, you studied several types of functions whose limits can be evaluated by direct substitution. This knowledge, together with the following theorem, can be used to develop a strategy for finding limits. A proof of this theorem is given in Appendix A.

> **THEOREM 2.7 Functions That Agree at All But One Point**
>
> Let c be a real number and let $f(x) = g(x)$ for all $x \neq c$ in an open interval containing c. If the limit of $g(x)$ as x approaches c exists, then the limit of $f(x)$ also exists and
>
> $$\lim_{x\to c} f(x) = \lim_{x\to c} g(x).$$

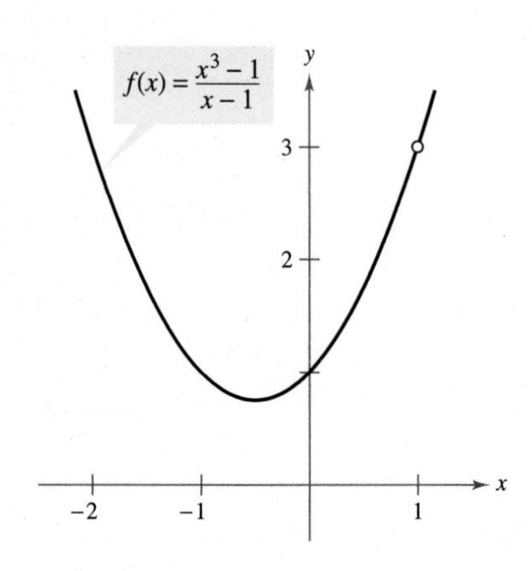

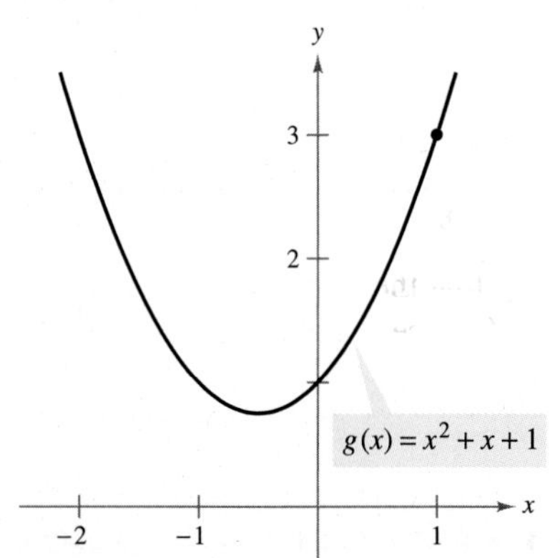

f and *g* agree at all but one point.
Figure 2.17

EXAMPLE 6 Finding the Limit of a Function

Find the limit: $\displaystyle\lim_{x\to 1} \frac{x^3-1}{x-1}$.

Solution Let $f(x) = (x^3 - 1)/(x - 1)$. By factoring and dividing out like factors, you can rewrite f as

$$f(x) = \frac{\cancel{(x-1)}(x^2+x+1)}{\cancel{(x-1)}} = x^2 + x + 1 = g(x), \qquad x \neq 1.$$

So, for all x-values other than $x = 1$, the functions f and g agree, as shown in Figure 2.17. Because $\lim_{x\to 1} g(x)$ exists, you can apply Theorem 2.7 to conclude that f and g have the same limit at $x = 1$.

$$\begin{aligned}
\lim_{x\to 1}\frac{x^3-1}{x-1} &= \lim_{x\to 1}\frac{(x-1)(x^2+x+1)}{x-1} && \text{Factor.}\\
&= \lim_{x\to 1}\frac{\cancel{(x-1)}(x^2+x+1)}{\cancel{x-1}} && \text{Divide out like factors.}\\
&= \lim_{x\to 1}(x^2+x+1) && \text{Apply Theorem 2.7.}\\
&= 1^2+1+1 && \text{Use direct substitution.}\\
&= 3 && \text{Simplify.}
\end{aligned}$$

STUDY TIP When applying this strategy for finding a limit, remember that some functions do not have a limit (as x approaches c). For instance, the following limit does not exist.

$$\lim_{x\to 1}\frac{x^3+1}{x-1}$$

> **A Strategy for Finding Limits**
>
> 1. Learn to recognize which limits can be evaluated by direct substitution. (These limits are listed in Theorems 2.1 through 2.6.)
> 2. If the limit of $f(x)$ as x approaches c *cannot* be evaluated by direct substitution, try to find a function g that agrees with f for all x other than $x = c$. [Choose g such that the limit of $g(x)$ *can* be evaluated by direct substitution.]
> 3. Apply Theorem 2.7 to conclude *analytically* that
>
> $$\lim_{x\to c} f(x) = \lim_{x\to c} g(x) = g(c).$$
> 4. Use a *graph* or *table* to reinforce your conclusion.

Dividing Out and Rationalizing Techniques

Two techniques for finding limits analytically are shown in Examples 7 and 8. The first technique involves dividing out common factors, and the second technique involves rationalizing the numerator of a fractional expression.

EXAMPLE 7 Dividing Out Technique

Find the limit: $\displaystyle\lim_{x\to-3}\frac{x^2+x-6}{x+3}$.

Solution Although you are taking the limit of a rational function, you *cannot* apply Theorem 2.3 because the limit of the denominator is 0.

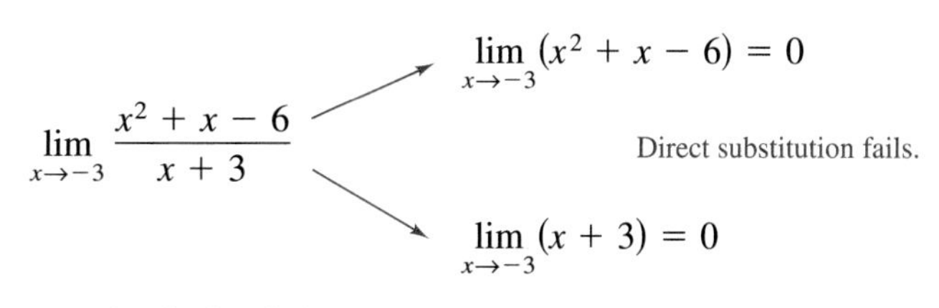

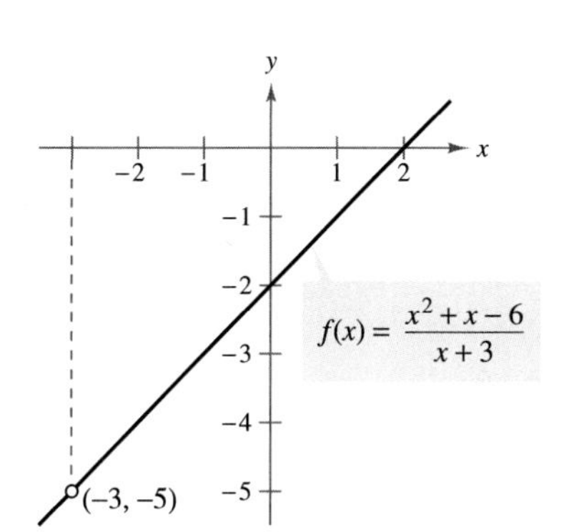

f is undefined when $x=-3$.
Figure 2.18

Because the limit of the numerator is also 0, the numerator and denominator have a *common factor* of $(x+3)$. So, for all $x\neq-3$, you can divide out this factor to obtain

$$f(x)=\frac{x^2+x-6}{x+3}=\frac{\cancel{(x+3)}(x-2)}{\cancel{x+3}}=x-2=g(x), \qquad x\neq-3.$$

Using Theorem 2.7, it follows that

$$\lim_{x\to-3}\frac{x^2+x-6}{x+3}=\lim_{x\to-3}(x-2) \qquad \text{Apply Theorem 2.7.}$$
$$=-5. \qquad \text{Use direct substitution.}$$

This result is shown graphically in Figure 2.18. Note that the graph of the function f coincides with the graph of the function $g(x)=x-2$, except that the graph of f has a gap at the point $(-3,-5)$.

NOTE In the solution of Example 7, be sure you see the usefulness of the Factor Theorem of Algebra. This theorem states that if c is a zero of a polynomial function, $(x-c)$ is a factor of the polynomial. So, if you apply direct substitution to a rational function and obtain

$$r(c)=\frac{p(c)}{q(c)}=\frac{0}{0}$$

you can conclude that $(x-c)$ must be a common factor to both $p(x)$ and $q(x)$.

In Example 7, direct substitution produced the meaningless fractional form $0/0$. An expression such as $0/0$ is called an **indeterminate form** because you cannot (from the form alone) determine the limit. When you try to evaluate a limit and encounter this form, remember that you must rewrite the fraction so that the new denominator does not have 0 as its limit. One way to do this is to *divide out common factors*, as shown in Example 7. A second way is to *rationalize the numerator*, as shown in Example 8.

−3 − δ −5 + ε −3 + δ Glitch near (−3, −5) −5 − ε

Incorrect graph of f
Figure 2.19

TECHNOLOGY PITFALL Because the graphs of

$$f(x)=\frac{x^2+x-6}{x+3} \qquad \text{and} \qquad g(x)=x-2$$

differ only at the point $(-3,-5)$, a standard graphing utility setting may not distinguish clearly between these graphs. However, because of the pixel configuration and rounding error of a graphing utility, it may be possible to find screen settings that distinguish between the graphs. Specifically, by repeatedly zooming in near the point $(-3,-5)$ on the graph of f, your graphing utility may show glitches or irregularities that do not exist on the actual graph. (See Figure 2.19.) By changing the screen settings on your graphing utility, you may obtain the correct graph of f.

EXAMPLE 8 Rationalizing Technique

Find the limit: $\lim\limits_{x \to 0} \dfrac{\sqrt{x+1}-1}{x}$.

Solution By direct substitution, you obtain the indeterminate form 0/0.

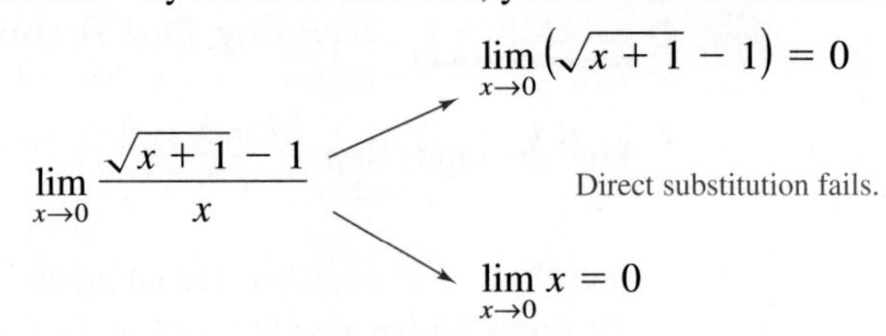

In this case, you can rewrite the fraction by rationalizing the numerator.

$$
\begin{aligned}
\frac{\sqrt{x+1}-1}{x} &= \left(\frac{\sqrt{x+1}-1}{x}\right)\left(\frac{\sqrt{x+1}+1}{\sqrt{x+1}+1}\right) \\
&= \frac{(x+1)-1}{x(\sqrt{x+1}+1)} \\
&= \frac{\cancel{x}}{\cancel{x}(\sqrt{x+1}+1)} \\
&= \frac{1}{\sqrt{x+1}+1}, \quad x \neq 0
\end{aligned}
$$

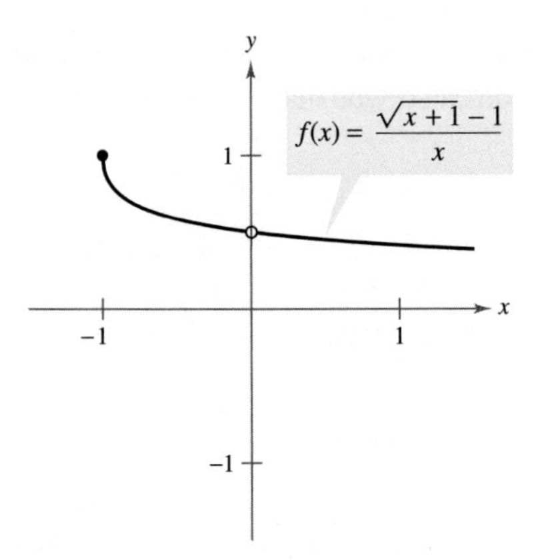

The limit of $f(x)$ as x approaches 0 is $\frac{1}{2}$.
Figure 2.20

Now, using Theorem 2.7, you can evaluate the limit as shown.

$$
\begin{aligned}
\lim_{x \to 0} \frac{\sqrt{x+1}-1}{x} &= \lim_{x \to 0} \frac{1}{\sqrt{x+1}+1} \\
&= \frac{1}{1+1} \\
&= \frac{1}{2}
\end{aligned}
$$

A table or a graph can reinforce your conclusion that the limit is $\frac{1}{2}$. (See Figure 2.20.)

x	−0.25	−0.1	−0.01	−0.001	0	0.001	0.01	0.1	0.25
$f(x)$	0.5359	0.5132	0.5013	0.5001	?	0.4999	0.4988	0.4881	0.4721

$f(x)$ approaches 0.5. $f(x)$ approaches 0.5.

NOTE The rationalizing technique for evaluating limits is based on multiplication by a convenient form of 1. In Example 8, the convenient form is

$$1 = \frac{\sqrt{x+1}+1}{\sqrt{x+1}+1}.$$

The Squeeze Theorem

The next theorem concerns the limit of a function that is squeezed between two other functions, each of which has the same limit at a given x-value, as shown in Figure 2.21. (The proof of this theorem is given in Appendix A.)

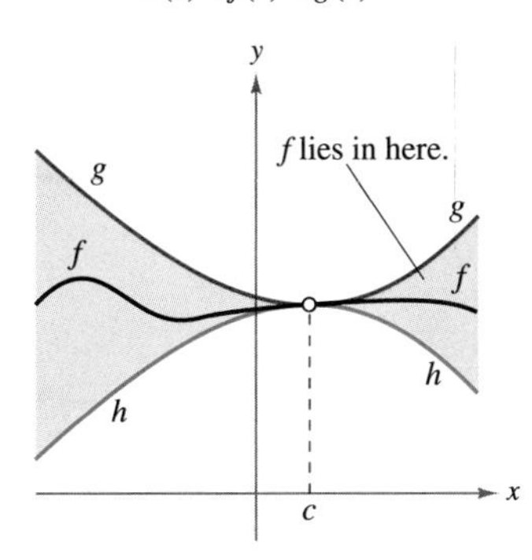

The Squeeze Theorem
Figure 2.21

THEOREM 2.8 The Squeeze Theorem

If $h(x) \le f(x) \le g(x)$ for all x in an open interval containing c, except possibly at c itself, and if

$$\lim_{x\to c} h(x) = L = \lim_{x\to c} g(x)$$

then $\lim_{x\to c} f(x)$ exists and is equal to L.

FOR FURTHER INFORMATION For more information on the function $f(x) = (\sin x)/x$, see the article "The Function $(\sin x)/x$" by William B. Gearhart and Harris S. Shultz in *The College Mathematics Journal.* To view this article, go to the website *www.matharticles.com.*

You can see the usefulness of the Squeeze Theorem in the proof of Theorem 2.9.

THEOREM 2.9 Three Special Limits

1. $\lim_{x\to 0} \dfrac{\sin x}{x} = 1$ **2.** $\lim_{x\to 0} \dfrac{1-\cos x}{x} = 0$ **3.** $\lim_{x\to 0} (1+x)^{1/x} = e$

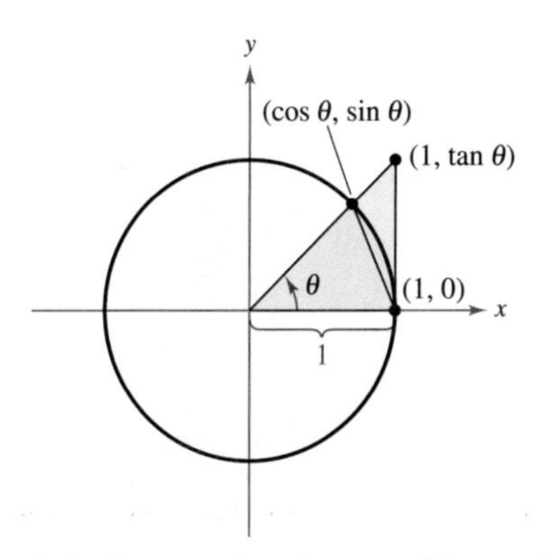

A circular sector is used to prove Theorem 2.9.
Figure 2.22

Proof To avoid the confusion of two different uses of x, the proof of the first limit is presented using the variable θ, where θ is an acute positive angle *measured in radians*. Figure 2.22 shows a circular sector that is squeezed between two triangles.

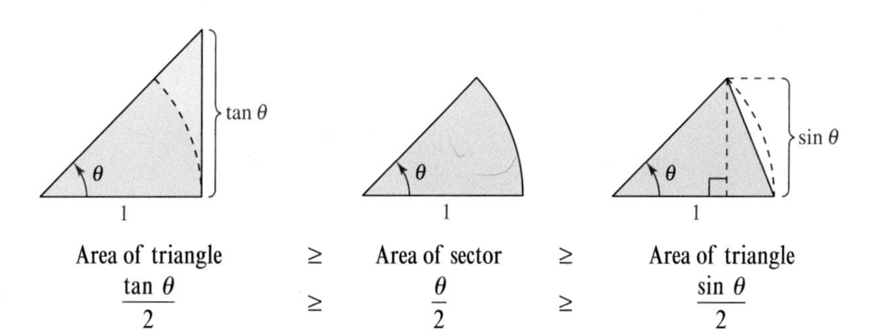

Area of triangle	$\ge$	Area of sector	$\ge$	Area of triangle
$\dfrac{\tan\theta}{2}$	$\ge$	$\dfrac{\theta}{2}$	$\ge$	$\dfrac{\sin\theta}{2}$

Multiplying each expression by $2/\sin\theta$ produces

$$\frac{1}{\cos\theta} \ge \frac{\theta}{\sin\theta} \ge 1$$

and taking reciprocals and reversing the inequalities yields

$$\cos\theta \le \frac{\sin\theta}{\theta} \le 1.$$

Because $\cos\theta = \cos(-\theta)$ and $(\sin\theta)/\theta = [\sin(-\theta)]/(-\theta)$, you can conclude that this inequality is valid for *all* nonzero θ in the open interval $(-\pi/2, \pi/2)$. Finally, because $\lim_{\theta\to 0} \cos\theta = 1$ and $\lim_{\theta\to 0} 1 = 1$, you can apply the Squeeze Theorem to conclude that $\lim_{\theta\to 0} (\sin\theta)/\theta = 1$. The proof of the second limit is left as an exercise (see Exercise 126). Recall from Section 1.6 that the third limit is actually the definition of the number e.

NOTE The third limit of Theorem 2.9 will be used in Chapter 3 in the development of the formula for the derivative of the exponential function $f(x) = e^x$.

EXAMPLE 9 A Limit Involving a Trigonometric Function

Find the limit: $\lim_{x \to 0} \dfrac{\tan x}{x}$.

Solution Direct substitution yields the indeterminate form $0/0$. To solve this problem, you can write $\tan x$ as $(\sin x)/(\cos x)$ and obtain

$$\lim_{x \to 0} \frac{\tan x}{x} = \lim_{x \to 0} \left(\frac{\sin x}{x}\right)\left(\frac{1}{\cos x}\right).$$

Now, because

$$\lim_{x \to 0} \frac{\sin x}{x} = 1 \quad \text{and} \quad \lim_{x \to 0} \frac{1}{\cos x} = 1$$

you can obtain

$$\begin{aligned} \lim_{x \to 0} \frac{\tan x}{x} &= \left(\lim_{x \to 0} \frac{\sin x}{x}\right)\left(\lim_{x \to 0} \frac{1}{\cos x}\right) \\ &= (1)(1) \\ &= 1. \end{aligned}$$

(See Figure 2.23.)

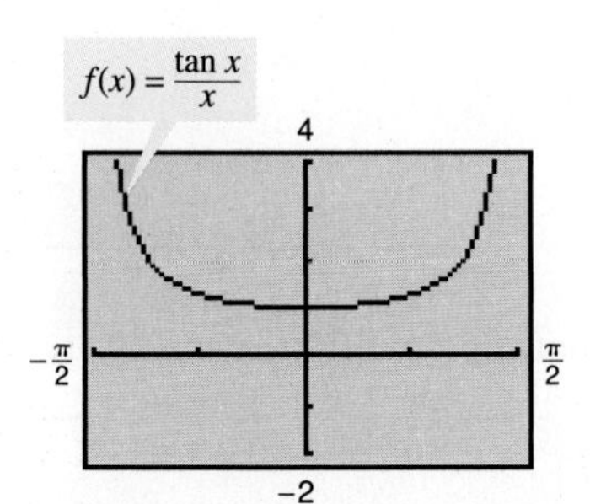

The limit of $f(x)$ as x approaches 0 is 1.
Figure 2.23

EXAMPLE 10 A Limit Involving a Trigonometric Function

Find the limit: $\lim_{x \to 0} \dfrac{\sin 4x}{x}$.

Solution Direct substitution yields the indeterminate form $0/0$. To solve this problem, you can rewrite the limit as

$$\lim_{x \to 0} \frac{\sin 4x}{x} = 4\left(\lim_{x \to 0} \frac{\sin 4x}{4x}\right).$$

Now, by letting $y = 4x$ and observing that $x \to 0$ if and only if $y \to 0$, you can write

$$\begin{aligned} \lim_{x \to 0} \frac{\sin 4x}{x} &= 4\left(\lim_{x \to 0} \frac{\sin 4x}{4x}\right) \\ &= 4\left(\lim_{y \to 0} \frac{\sin y}{y}\right) \\ &= 4(1) \\ &= 4. \end{aligned}$$

(See Figure 2.24.)

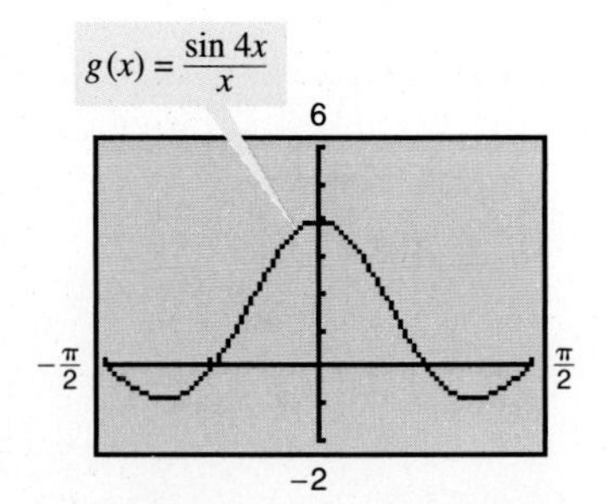

The limit of $g(x)$ as x approaches 0 is 4.
Figure 2.24

TECHNOLOGY Try using a graphing utility to confirm the limits in the examples and exercise set. For instance, Figures 2.23 and 2.24 show the graphs of

$$f(x) = \frac{\tan x}{x} \quad \text{and} \quad g(x) = \frac{\sin 4x}{x}.$$

Note that the first graph appears to contain the point $(0, 1)$ and the second graph appears to contain the point $(0, 4)$, which lends support to the conclusions obtained in Examples 9 and 10.

Exercises for Section 2.3

See www.CalcChat.com for worked-out solutions to odd-numbered exercises.

In Exercises 1–4, use a graphing utility to graph the function and visually estimate the limits.

1. $h(x) = x^2 - 5x$

(a) $\lim_{x\to 5} h(x)$

(b) $\lim_{x\to -1} h(x)$

2. $g(x) = \dfrac{12(\sqrt{x} - 3)}{x - 9}$

(a) $\lim_{x\to 4} g(x)$

(b) $\lim_{x\to 0} g(x)$

3. $f(x) = x \cos x$

(a) $\lim_{x\to 0} f(x)$

(b) $\lim_{x\to \pi/3} f(x)$

4. $f(t) = t|t - 4|$

(a) $\lim_{t\to 4} f(t)$

(b) $\lim_{t\to -1} f(t)$

In Exercises 5–32, find the limit.

5. $\lim_{x\to 2} x^4$

6. $\lim_{x\to -2} x^5$

7. $\lim_{x\to 0} (2x - 1)$

8. $\lim_{x\to -3} (3x + 2)$

9. $\lim_{x\to -3} (2x^2 + 4x + 1)$

10. $\lim_{x\to 1} (3x^3 - 4x^2 + 3)$

11. $\lim_{x\to 2} \dfrac{1}{x}$

12. $\lim_{x\to -3} \dfrac{2}{x + 2}$

13. $\lim_{x\to 1} \dfrac{x - 3}{x^2 + 4}$

14. $\lim_{x\to 3} \dfrac{2x - 5}{x + 3}$

15. $\lim_{x\to 7} \dfrac{5x}{\sqrt{x + 2}}$

16. $\lim_{x\to 3} \dfrac{\sqrt{x + 1}}{x - 4}$

17. $\lim_{x\to 3} \sqrt{x + 1}$

18. $\lim_{x\to 4} \sqrt[3]{x + 23}$

19. $\lim_{x\to \pi/2} \sin x$

20. $\lim_{x\to \pi} \tan x$

21. $\lim_{x\to 2} \cos\left(\dfrac{\pi x}{3}\right)$

22. $\lim_{x\to 1} \sin\left(\dfrac{\pi x}{2}\right)$

23. $\lim_{x\to 0} \sec 2x$

24. $\lim_{x\to \pi} \cos 5x$

25. $\lim_{x\to 5\pi/6} \sin x$

26. $\lim_{x\to 5\pi/3} \cos x$

27. $\lim_{x\to 3} \tan\left(\dfrac{\pi x}{4}\right)$

28. $\lim_{x\to 7} \sec\left(\dfrac{\pi x}{6}\right)$

29. $\lim_{x\to 0} e^x \cos 2x$

30. $\lim_{x\to 0} e^{-x} \sin \pi x$

31. $\lim_{x\to 1} (\ln 3x + e^x)$

32. $\lim_{x\to 1} \ln\left(\dfrac{x}{e^x}\right)$

In Exercises 33–36, find the limits.

33. $f(x) = 5 - x,\ g(x) = x^3$

(a) $\lim_{x\to 1} f(x)$ (b) $\lim_{x\to 4} g(x)$ (c) $\lim_{x\to 1} g(f(x))$

34. $f(x) = x + 7,\ g(x) = x^2$

(a) $\lim_{x\to -3} f(x)$ (b) $\lim_{x\to 4} g(x)$ (c) $\lim_{x\to -3} g(f(x))$

35. $f(x) = 4 - x^2,\ g(x) = \sqrt{x + 1}$

(a) $\lim_{x\to 1} f(x)$ (b) $\lim_{x\to 3} g(x)$ (c) $\lim_{x\to 1} g(f(x))$

36. $f(x) = 2x^2 - 3x + 1,\ g(x) = \sqrt[3]{x + 6}$

(a) $\lim_{x\to 4} f(x)$ (b) $\lim_{x\to 21} g(x)$ (c) $\lim_{x\to 4} g(f(x))$

In Exercises 37–40, use the information to evaluate the limits.

37. $\lim_{x\to c} f(x) = 2$

$\lim_{x\to c} g(x) = 3$

(a) $\lim_{x\to c} [5g(x)]$

(b) $\lim_{x\to c} [f(x) + g(x)]$

(c) $\lim_{x\to c} [f(x)g(x)]$

(d) $\lim_{x\to c} \dfrac{f(x)}{g(x)}$

38. $\lim_{x\to c} f(x) = \frac{3}{2}$

$\lim_{x\to c} g(x) = \frac{1}{2}$

(a) $\lim_{x\to c} [4f(x)]$

(b) $\lim_{x\to c} [f(x) + g(x)]$

(c) $\lim_{x\to c} [f(x)g(x)]$

(d) $\lim_{x\to c} \dfrac{f(x)}{g(x)}$

39. $\lim_{x\to c} f(x) = 4$

(a) $\lim_{x\to c} [f(x)]^3$

(b) $\lim_{x\to c} \sqrt{f(x)}$

(c) $\lim_{x\to c} [3f(x)]$

(d) $\lim_{x\to c} [f(x)]^{3/2}$

40. $\lim_{x\to c} f(x) = 27$

(a) $\lim_{x\to c} \sqrt[3]{f(x)}$

(b) $\lim_{x\to c} \dfrac{f(x)}{18}$

(c) $\lim_{x\to c} [f(x)]^2$

(d) $\lim_{x\to c} [f(x)]^{2/3}$

In Exercises 41–44, use the graph to determine the limit visually (if it exists). Write a simpler function that agrees with the given function at all but one point.

41. $g(x) = \dfrac{-2x^2 + x}{x}$

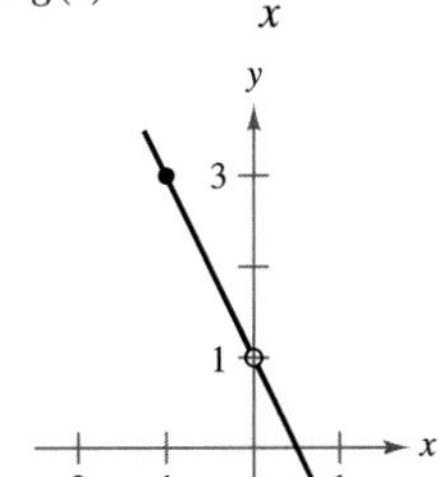

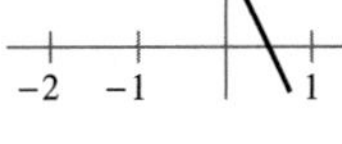

(a) $\lim_{x\to 0} g(x)$

(b) $\lim_{x\to -1} g(x)$

42. $h(x) = \dfrac{x^2 - 3x}{x}$

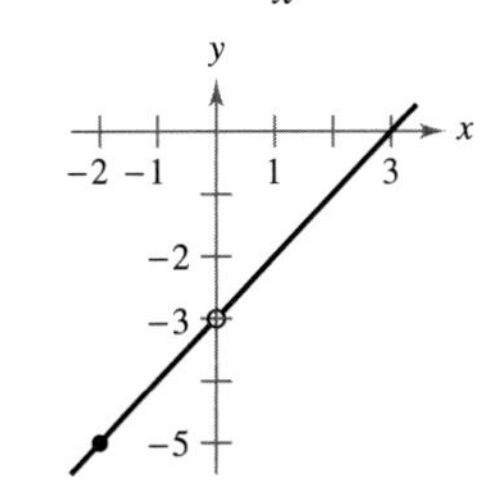

(a) $\lim_{x\to -2} h(x)$

(b) $\lim_{x\to 0} h(x)$

43. $g(x) = \dfrac{x^3 - x}{x - 1}$

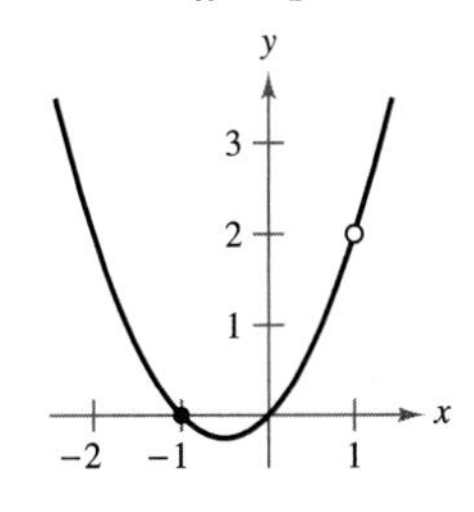

(a) $\lim_{x\to 1} g(x)$

(b) $\lim_{x\to -1} g(x)$

44. $f(x) = \dfrac{x}{x^2 - x}$

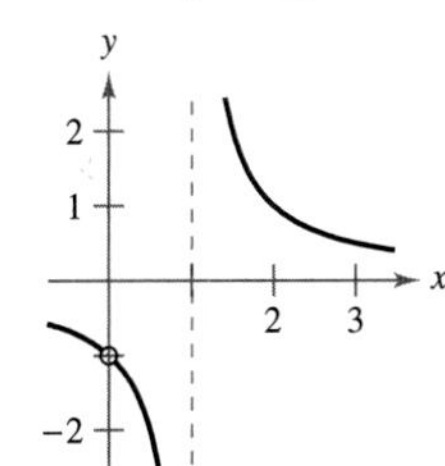

(a) $\lim_{x\to 1} f(x)$

(b) $\lim_{x\to 0} f(x)$

In Exercises 45–50, find the limit of the function (if it exists). Write a simpler function that agrees with the given function at all but one point. Use a graphing utility to confirm your result.

45. $\lim\limits_{x\to-1} \dfrac{x^2-1}{x+1}$

46. $\lim\limits_{x\to-1} \dfrac{2x^2-x-3}{x+1}$

47. $\lim\limits_{x\to2} \dfrac{x^3-8}{x-2}$

48. $\lim\limits_{x\to-1} \dfrac{x^3+1}{x+1}$

49. $\lim\limits_{x\to-4} \dfrac{(x+4)\ln(x+6)}{x^2-16}$

50. $\lim\limits_{x\to0} \dfrac{e^{2x}-1}{e^x-1}$

In Exercises 51–64, find the limit (if it exists).

51. $\lim\limits_{x\to5} \dfrac{x-5}{x^2-25}$

52. $\lim\limits_{x\to3} \dfrac{3-x}{x^2-9}$

53. $\lim\limits_{x\to-3} \dfrac{x^2+x-6}{x^2-9}$

54. $\lim\limits_{x\to3} \dfrac{x^2-x-6}{x^2-5x+6}$

55. $\lim\limits_{x\to0} \dfrac{\sqrt{x+5}-\sqrt{5}}{x}$

56. $\lim\limits_{x\to0} \dfrac{\sqrt{3+x}-\sqrt{3}}{x}$

57. $\lim\limits_{x\to4} \dfrac{\sqrt{x+5}-3}{x-4}$

58. $\lim\limits_{x\to3} \dfrac{\sqrt{x+1}-2}{x-3}$

59. $\lim\limits_{x\to0} \dfrac{[1/(3+x)]-(1/3)}{x}$

60. $\lim\limits_{x\to0} \dfrac{[1/(x+4)]-(1/4)}{x}$

61. $\lim\limits_{\Delta x\to0} \dfrac{2(x+\Delta x)-2x}{\Delta x}$

62. $\lim\limits_{\Delta x\to0} \dfrac{(x+\Delta x)^2-x^2}{\Delta x}$

63. $\lim\limits_{\Delta x\to0} \dfrac{(x+\Delta x)^2-2(x+\Delta x)+1-(x^2-2x+1)}{\Delta x}$

64. $\lim\limits_{\Delta x\to0} \dfrac{(x+\Delta x)^3-x^3}{\Delta x}$

Graphical, Numerical, and Analytic Analysis **In Exercises 65–68, use a graphing utility to graph the function and estimate the limit. Use a table to reinforce your conclusion. Then find the limit by analytic methods.**

65. $\lim\limits_{x\to0} \dfrac{\sqrt{x+2}-\sqrt{2}}{x}$

66. $\lim\limits_{x\to16} \dfrac{4-\sqrt{x}}{x-16}$

67. $\lim\limits_{x\to0} \dfrac{[1/(2+x)]-(1/2)}{x}$

68. $\lim\limits_{x\to2} \dfrac{x^5-32}{x-2}$

In Exercises 69–82, determine the limit of the transcendental function (if it exists).

69. $\lim\limits_{x\to0} \dfrac{\sin x}{5x}$

70. $\lim\limits_{x\to0} \dfrac{5(1-\cos x)}{x}$

71. $\lim\limits_{x\to0} \dfrac{\sin x(1-\cos x)}{2x^2}$

72. $\lim\limits_{\theta\to0} \dfrac{\cos\theta\tan\theta}{\theta}$

73. $\lim\limits_{x\to0} \dfrac{\sin^2 x}{x}$

74. $\lim\limits_{x\to0} \dfrac{2\tan^2 x}{x}$

75. $\lim\limits_{h\to0} \dfrac{(1-\cos h)^2}{h}$

76. $\lim\limits_{\phi\to\pi} \phi\sec\phi$

77. $\lim\limits_{x\to\pi/2} \dfrac{\cos x}{\cot x}$

78. $\lim\limits_{x\to\pi/4} \dfrac{1-\tan x}{\sin x-\cos x}$

79. $\lim\limits_{x\to0} \dfrac{1-e^{-x}}{e^x-1}$

80. $\lim\limits_{x\to0} \dfrac{4(e^{2x}-1)}{e^x-1}$

81. $\lim\limits_{t\to0} \dfrac{\sin 3t}{2t}$

82. $\lim\limits_{x\to0} \dfrac{\sin 2x}{\sin 3x}$ $\left[\textit{Hint: } \text{Find } \lim\limits_{x\to0}\left(\dfrac{2\sin 2x}{2x}\right)\left(\dfrac{3x}{3\sin 3x}\right).\right]$

Graphical, Numerical, and Analytic Analysis **In Exercises 83–88, use a graphing utility to graph the function and estimate the limit. Use a table to reinforce your conclusion. Then find the limit by analytic methods.**

83. $\lim\limits_{t\to0} \dfrac{\sin 3t}{t}$

84. $\lim\limits_{x\to0} \dfrac{\cos x-1}{2x^2}$

85. $\lim\limits_{x\to0} \dfrac{\sin x^2}{x}$

86. $\lim\limits_{x\to0} \dfrac{\sin x}{\sqrt[3]{x}}$

87. $\lim\limits_{x\to1} \dfrac{\ln x}{x-1}$

88. $\lim\limits_{x\to\ln 2} \dfrac{e^{3x}-8}{e^{2x}-4}$

In Exercises 89–92, find $\lim\limits_{\Delta x\to0} \dfrac{f(x+\Delta x)-f(x)}{\Delta x}$.

89. $f(x)=2x+3$

90. $f(x)=\sqrt{x}$

91. $f(x)=\dfrac{4}{x}$

92. $f(x)=x^2-4x$

In Exercises 93 and 94, use the Squeeze Theorem to find $\lim\limits_{x\to c} f(x)$.

93. $c=0$

$4-x^2\le f(x)\le 4+x^2$

94. $c=a$

$b-|x-a|\le f(x)\le b+|x-a|$

In Exercises 95–100, use a graphing utility to graph the given function and the equations $y=|x|$ and $y=-|x|$ in the same viewing window. Using the graphs to visually observe the Squeeze Theorem, find $\lim\limits_{x\to0} f(x)$.

95. $f(x)=x\cos x$

96. $f(x)=|x\sin x|$

97. $f(x)=|x|\sin x$

98. $f(x)=|x|\cos x$

99. $f(x)=x\sin\dfrac{1}{x}$

100. $h(x)=x\cos\dfrac{1}{x}$

Writing About Concepts

101. In the context of finding limits, discuss what is meant by two functions that agree at all but one point.

102. Give an example of two functions that agree at all but one point.

103. What is meant by an indeterminate form?

104. In your own words, explain the Squeeze Theorem.

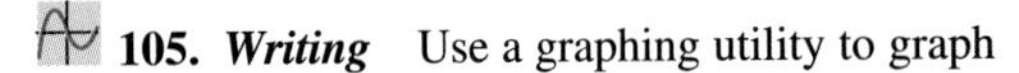

105. ***Writing*** Use a graphing utility to graph

$$f(x) = x, \quad g(x) = \sin x, \quad \text{and} \quad h(x) = \frac{\sin x}{x}$$

in the same viewing window. Compare the magnitudes of $f(x)$ and $g(x)$ when x is "close to" 0. Use the comparison to write a short paragraph explaining why $\lim_{x \to 0} h(x) = 1$.

106. ***Writing*** Use a graphing utility to graph

$$f(x) = x, \; g(x) = \sin^2 x, \; \text{and } h(x) = \frac{\sin^2 x}{x}$$

in the same viewing window. Compare the magnitudes of $f(x)$ and $g(x)$ when x is "close to" 0. Use the comparison to write a short paragraph explaining why $\lim_{x \to 0} h(x) = 0$.

Free-Falling Object **In Exercises 107 and 108, use the position function $s(t) = -16t^2 + 1000$, which gives the height (in feet) of an object that has fallen for t seconds from a height of 1000 feet. The velocity at time $t = a$ seconds is given by**

$$\lim_{t \to a} \frac{s(a) - s(t)}{a - t}.$$

107. If a construction worker drops a wrench from a height of 1000 feet, how fast will the wrench be falling after 5 seconds?

108. If a construction worker drops a wrench from a height of 1000 feet, when will the wrench hit the ground? At what velocity will the wrench impact the ground?

Free-Falling Object **In Exercises 109 and 110, use the position function $s(t) = -4.9t^2 + 150$, which gives the height (in meters) of an object that has fallen from a height of 150 meters. The velocity at time $t = a$ seconds is given by**

$$\lim_{t \to a} \frac{s(a) - s(t)}{a - t}.$$

109. Find the velocity of the object when $t = 3$.

110. At what velocity will the object impact the ground?

111. Find two functions f and g such that $\lim_{x \to 0} f(x)$ and $\lim_{x \to 0} g(x)$ do not exist, but $\lim_{x \to 0} [f(x) + g(x)]$ does exist.

112. Prove that if $\lim_{x \to c} f(x)$ exists and $\lim_{x \to c} [f(x) + g(x)]$ does not exist, then $\lim_{x \to c} g(x)$ does not exist.

113. Prove Property 1 of Theorem 2.1.

114. Prove Property 3 of Theorem 2.1. (You may use Property 3 of Theorem 2.2.)

115. Prove Property 1 of Theorem 2.2.

116. Prove that if $\lim_{x \to c} f(x) = 0$, then $\lim_{x \to c} |f(x)| = 0$.

117. Prove that if $\lim_{x \to c} f(x) = 0$ and $|g(x)| \le M$ for a fixed number M and all $x \ne c$, then $\lim_{x \to c} f(x)g(x) = 0$.

118. (a) Prove that if $\lim_{x \to c} |f(x)| = 0$, then $\lim_{x \to c} f(x) = 0$.

(*Note:* This is the converse of Exercise 116.)

(b) Prove that if $\lim_{x \to c} f(x) = L$, then $\lim_{x \to c} |f(x)| = |L|$.

[*Hint:* Use the inequality $\| f(x)| - |L\| \le |f(x) - L|$.]

True or False? **In Exercises 119–124, determine whether the statement is true or false. If it is false, explain why or give an example that shows it is false.**

119. $\lim_{x \to 0} \frac{|x|}{x} = 1$

120. $\lim_{x \to \pi} \frac{\sin x}{x} = 1$

121. If $f(x) = g(x)$ for all real numbers other than $x = 0$, and $\lim_{x \to 0} f(x) = L$, then $\lim_{x \to 0} g(x) = L$.

122. If $\lim_{x \to c} f(x) = L$, then $f(c) = L$.

123. $\lim_{x \to 2} f(x) = 3$, where $f(x) = \begin{cases} 3, & x \le 2 \\ 0, & x > 2 \end{cases}$

124. If $f(x) < g(x)$ for all $x \ne a$, then $\lim_{x \to a} f(x) < \lim_{x \to a} g(x)$.

125. ***Think About It*** Find a function f to show that the converse of Exercise 118(b) is not true. [*Hint:* Find a function f such that $\lim_{x \to c} |f(x)| = |L|$ but $\lim_{x \to c} f(x)$ does not exist.]

126. Prove the second part of Theorem 2.9 by proving that

$$\lim_{x \to 0} \frac{1 - \cos x}{x} = 0.$$

127. Let $f(x) = \begin{cases} 0, & \text{if } x \text{ is rational} \\ 1, & \text{if } x \text{ is irrational} \end{cases}$

and

$$g(x) = \begin{cases} 0, & \text{if } x \text{ is rational} \\ x, & \text{if } x \text{ is irrational.} \end{cases}$$

Find (if possible) $\lim_{x \to 0} f(x)$ and $\lim_{x \to 0} g(x)$.

128. ***Graphical Reasoning*** Consider $f(x) = \dfrac{\sec x - 1}{x^2}$.

(a) Find the domain of f.

(b) Use a graphing utility to graph f. Is the domain of f obvious from the graph? If not, explain.

(c) Use the graph of f to approximate $\lim_{x \to 0} f(x)$.

(d) Confirm the answer in part (c) analytically.

129. ***Approximation***

(a) Find $\lim_{x \to 0} \dfrac{1 - \cos x}{x^2}$.

(b) Use the result in part (a) to derive the approximation $\cos x \approx 1 - \frac{1}{2}x^2$ for x near 0.

(c) Use the result in part (b) to approximate $\cos(0.1)$.

(d) Use a calculator to approximate $\cos(0.1)$ to four decimal places. Compare the result with part (c).

130. ***Think About It*** When using a graphing utility to generate a table to approximate $\lim_{x \to 0} [(\sin x)/x]$, a student concluded that the limit was 0.01745 rather than 1. Determine the probable cause of the error.

Section 2.4 Continuity and One-Sided Limits

- **Determine continuity at a point and continuity on an open interval.**
- **Determine one-sided limits and continuity on a closed interval.**
- **Use properties of continuity.**
- **Understand and use the Intermediate Value Theorem.**

Continuity at a Point and on an Open Interval

In mathematics, the term *continuous* has much the same meaning as it has in everyday usage. To say that a function f is continuous at $x = c$ means that there is no interruption in the graph of f at c. That is, its graph is unbroken at c and there are no holes, jumps, or gaps. Figure 2.25 identifies three values of x at which the graph of f is *not* continuous. At all other points in the interval (a, b), the graph of f is uninterrupted and **continuous.**

> **EXPLORATION**
>
> Informally, you might say that a function is *continuous* on an open interval if its graph can be drawn with a pencil without lifting the pencil from the paper. Use a graphing utility to graph each function on the given interval. From the graphs, which functions would you say are continuous on the interval? Do you think you can trust the results you obtained graphically? Explain your reasoning.
>
> | *Function* | *Interval* |
> |---|---|
> | **a.** $y = x^2 + 1$ | $(-3, 3)$ |
> | **b.** $y = \dfrac{1}{x - 2}$ | $(-3, 3)$ |
> | **c.** $y = \dfrac{\sin x}{x}$ | $(-\pi, \pi)$ |
> | **d.** $y = \dfrac{x^2 - 4}{x + 2}$ | $(-3, 3)$ |
> | **e.** $y = \begin{cases} 2x - 4, & x \le 0 \\ x + 1, & x > 0 \end{cases}$ | $(-3, 3)$ |

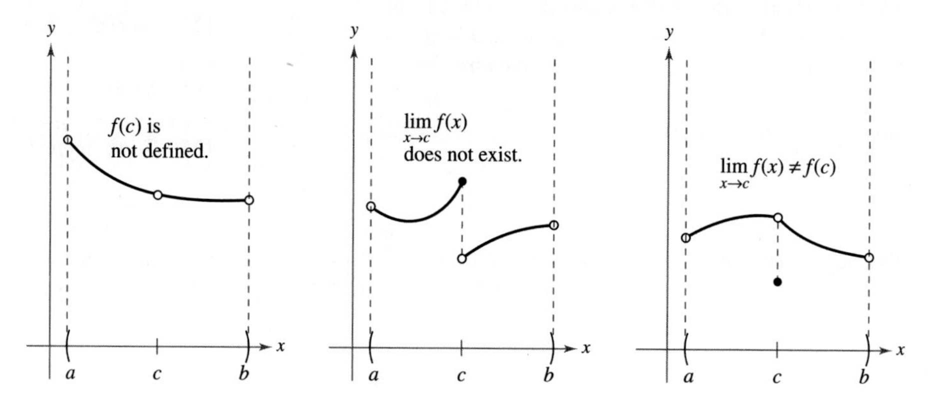

Three conditions exist for which the graph of f is not continuous at $x = c$.
Figure 2.25

In Figure 2.25, it appears that continuity at $x = c$ can be destroyed by any one of the following conditions.

1. The function is not defined at $x = c$.
2. The limit of $f(x)$ does not exist at $x = c$.
3. The limit of $f(x)$ exists at $x = c$, but it is not equal to $f(c)$.

If *none* of the above three conditions is true, the function f is called **continuous at c,** as indicated in the following important definition.

Definition of Continuity

Continuity at a Point: A function f is **continuous at c** if the following three conditions are met.

1. $f(c)$ is defined.
2. $\lim_{x \to c} f(x)$ exists.
3. $\lim_{x \to c} f(x) = f(c)$.

Continuity on an Open Interval: A function is **continuous on an open interval (a, b)** if it is continuous at each point in the interval. A function that is continuous on the entire real line $(-\infty, \infty)$ is **everywhere continuous.**

FOR FURTHER INFORMATION For more information on the concept of continuity, see the article "Leibniz and the Spell of the Continuous" by Hardy Grant in *The College Mathematics Journal*. To view this article, go to the website *www.matharticles.com*.

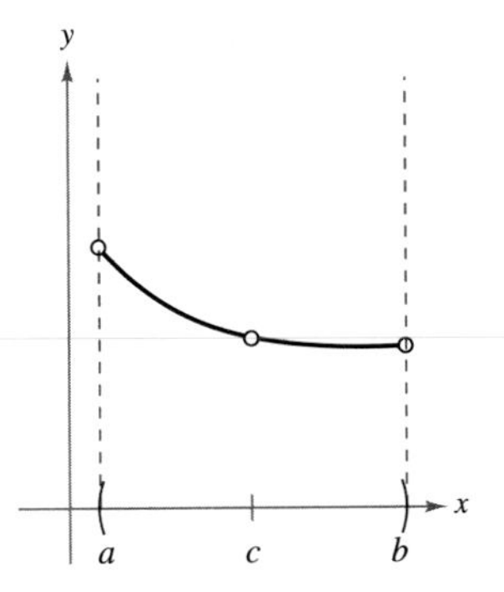

(a) Removable discontinuity

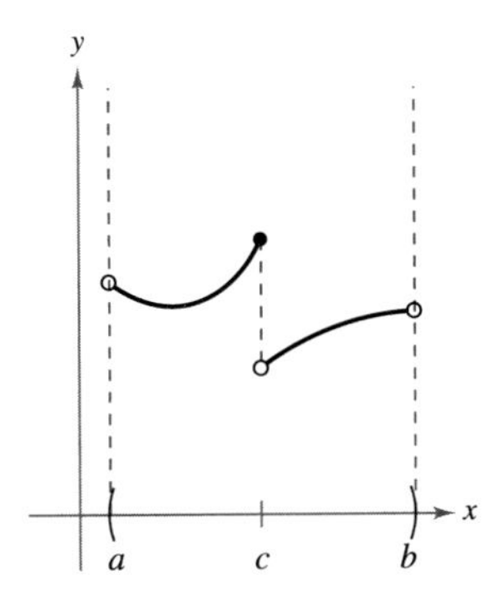

(b) Nonremovable discontinuity

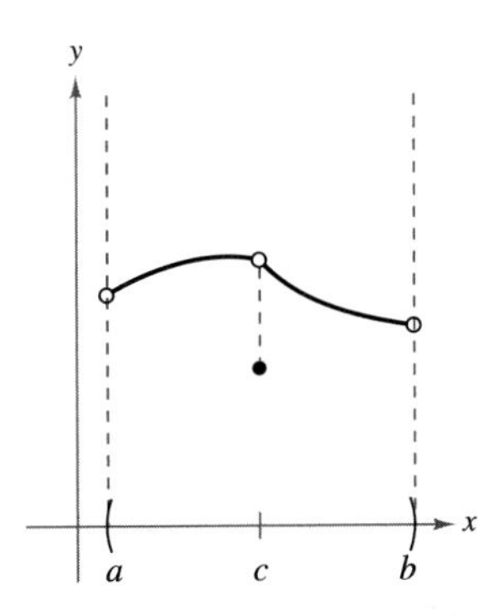

(c) Removable discontinuity
Figure 2.26

Consider an open interval I that contains a real number c. If a function f is defined on I (except possibly at c), and f is not continuous at c, then f is said to have a **discontinuity** at c. Discontinuities fall into two categories: **removable** and **nonremovable.** A discontinuity at c is called removable if f can be made continuous by appropriately defining (or redefining) $f(c)$. For instance, the functions shown in Figure 2.26(a) and (c) have removable discontinuities at c, and the function shown in Figure 2.26(b) has a nonremovable discontinuity at c.

EXAMPLE 1 Continuity of a Function

Discuss the continuity of each function.

a. $f(x) = \dfrac{1}{x}$ **b.** $g(x) = \dfrac{x^2 - 1}{x - 1}$ **c.** $h(x) = \begin{cases} x + 1, & x \le 0 \\ e^x, & x > 0 \end{cases}$ **d.** $y = \sin x$

Solution

a. The domain of f is all nonzero real numbers. From Theorem 2.3, you can conclude that f is continuous at every x-value in its domain. At $x = 0$, f has a nonremovable discontinuity, as shown in Figure 2.27(a). In other words, there is no way to define $f(0)$ so as to make the function continuous at $x = 0$.

b. The domain of g is all real numbers except $x = 1$. From Theorem 2.3, you can conclude that g is continuous at every x-value in its domain. At $x = 1$, the function has a removable discontinuity, as shown in Figure 2.27(b). If $g(1)$ is defined as 2, the "newly defined" function is continuous for all real numbers.

c. The domain of h is all real numbers. The function h is continuous on $(-\infty, 0)$ and $(0, \infty)$, and, because $\lim\limits_{x \to 0} h(x) = 1$, h is continuous on the entire real number line, as shown in Figure 2.27(c).

d. The domain of y is all real numbers. From Theorem 2.6, you can conclude that the function is continuous on its entire domain, $(-\infty, \infty)$, as shown in Figure 2.27(d).

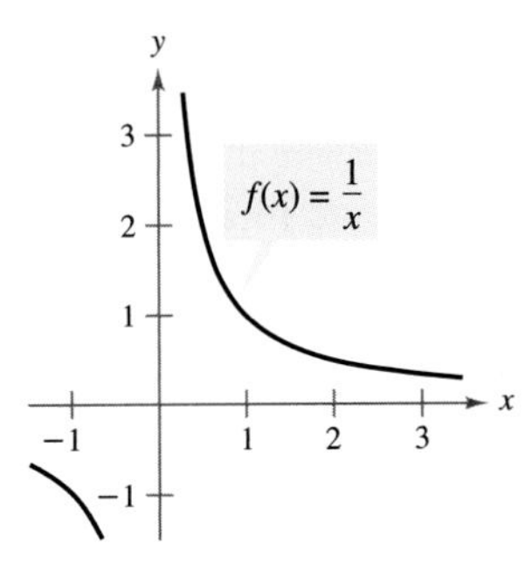

(a) Nonremovable discontinuity at $x = 0$

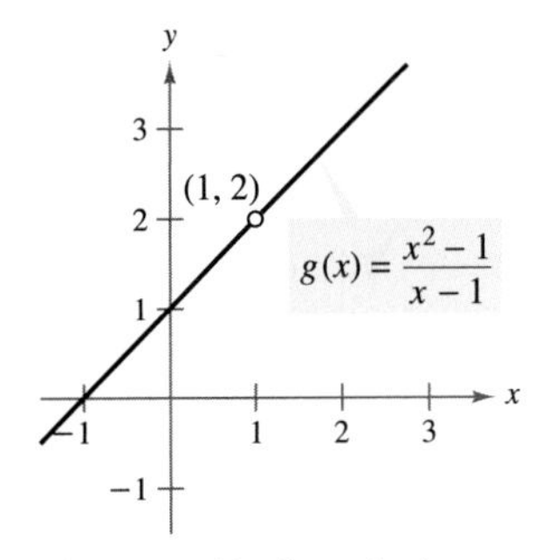

(b) Removable discontinuity at $x = 1$

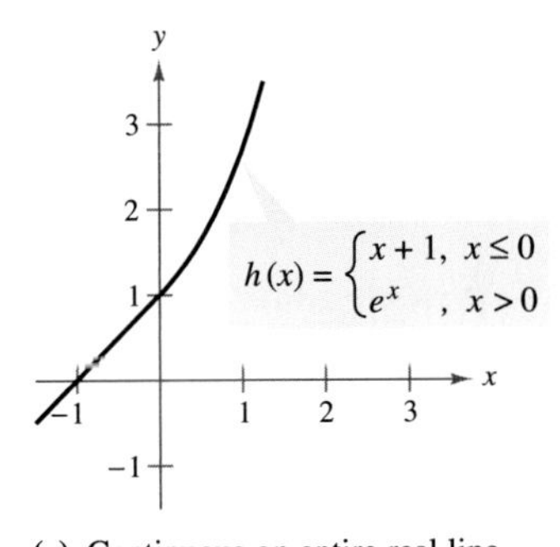

(c) Continuous on entire real line
Figure 2.27

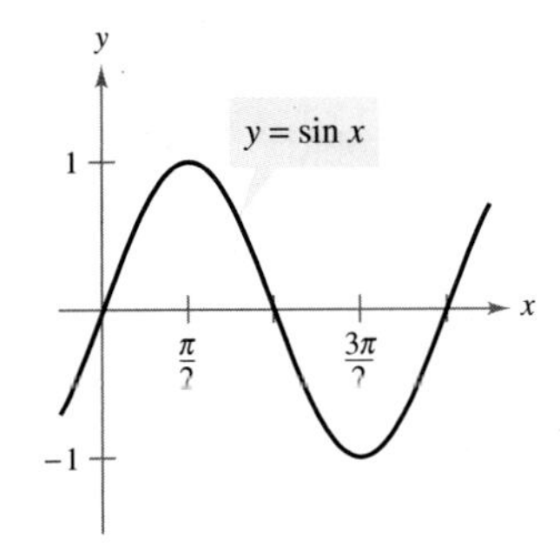

(d) Continuous on entire real line

STUDY TIP Some people may refer to the function in Example 1(a) as "discontinuous." We have found that this terminology can be confusing. Rather than saying the function is discontinuous, we prefer to say that it has a discontinuity at $x = 0$.

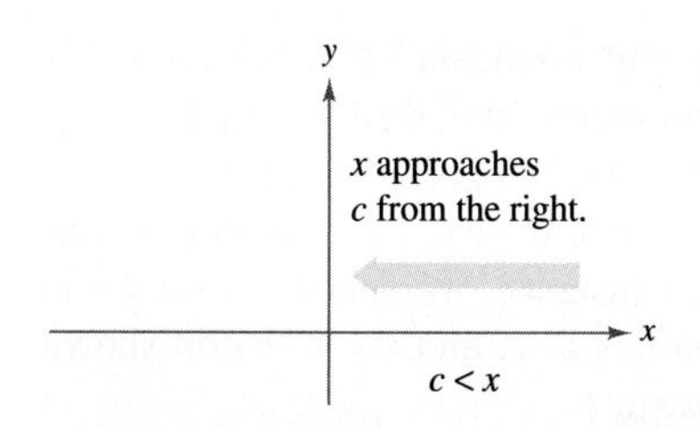

(a) Limit from right

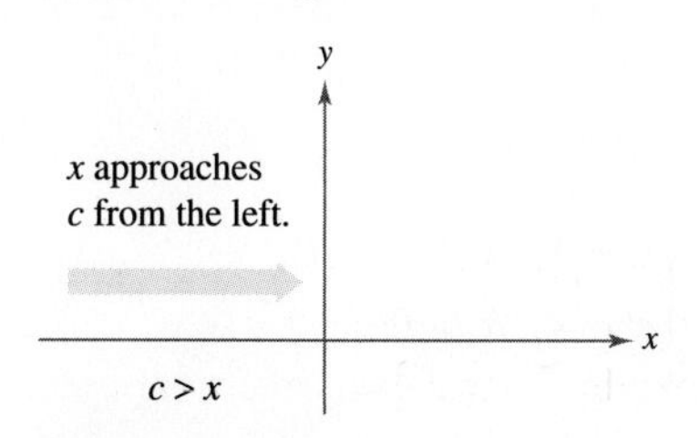

(b) Limit from left
Figure 2.28

One-Sided Limits and Continuity on a Closed Interval

To understand continuity on a closed interval, you first need to look at a different type of limit called a **one-sided limit.** For example, the **limit from the right** means that x approaches c from values greater than c [see Figure 2.28(a)]. This limit is denoted as

$$\lim_{x \to c^+} f(x) = L. \qquad \text{Limit from the right}$$

Similarly, the **limit from the left** means that x approaches c from values less than c [see Figure 2.28(b)]. This limit is denoted as

$$\lim_{x \to c^-} f(x) = L. \qquad \text{Limit from the left}$$

One-sided limits are useful in taking limits of functions involving radicals. For instance, if n is an even integer,

$$\lim_{x \to 0^+} \sqrt[n]{x} = 0.$$

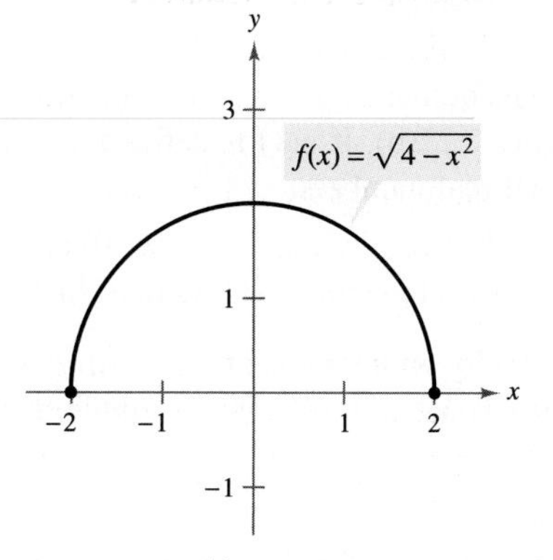

The limit of $f(x)$ as x approaches -2 from the right is 0.
Figure 2.29

EXAMPLE 2 A One-Sided Limit

Find the limit of $f(x) = \sqrt{4 - x^2}$ as x approaches -2 from the right.

Solution As shown in Figure 2.29, the limit as x approaches -2 from the right is

$$\lim_{x \to -2^+} \sqrt{4 - x^2} = 0.$$

One-sided limits can be used to investigate the behavior of **step functions.** One common type of step function is the **greatest integer function** $[\![x]\!]$, defined by

$$[\![x]\!] = \text{greatest integer } n \text{ such that } n \le x. \qquad \text{Greatest integer function}$$

For instance, $[\![2.5]\!] = 2$ and $[\![-2.5]\!] = -3$.

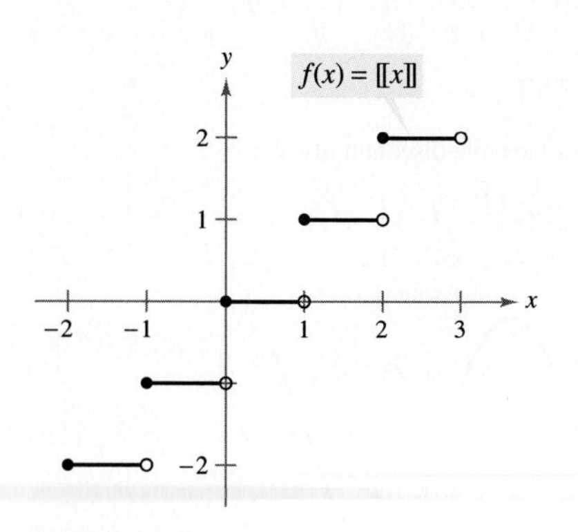

Greatest integer function
Figure 2.30

EXAMPLE 3 The Greatest Integer Function

Find the limit of the greatest integer function $f(x) = [\![x]\!]$ as x approaches 0 from the left and from the right.

Solution As shown in Figure 2.30, the limit as x approaches 0 *from the left* is given by

$$\lim_{x \to 0^-} [\![x]\!] = -1$$

and the limit as x approaches 0 *from the right* is given by

$$\lim_{x \to 0^+} [\![x]\!] = 0.$$

The greatest integer function has a discontinuity at zero because the left and right limits at zero are different. By similar reasoning, you can see that the greatest integer function has a discontinuity at any integer n.

When the limit from the left is not equal to the limit from the right, the (two-sided) limit *does not exist*. The next theorem makes this more explicit. The proof of this theorem follows directly from the definition of a one-sided limit.

THEOREM 2.10 The Existence of a Limit

Let f be a function and let c and L be real numbers. The limit of $f(x)$ as x approaches c is L if and only if

$$\lim_{x \to c^-} f(x) = L \quad \text{and} \quad \lim_{x \to c^+} f(x) = L.$$

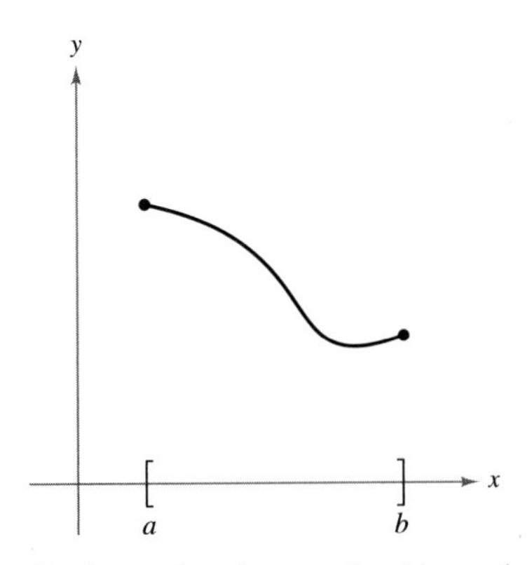

Continuous function on a closed interval
Figure 2.31

The concept of a one-sided limit allows you to extend the definition of continuity to closed intervals. Basically, a function is continuous on a closed interval if it is continuous in the interior of the interval and exhibits one-sided continuity at the endpoints. This is stated formally as follows.

Definition of Continuity on a Closed Interval

A function f is **continuous on the closed interval $[a, b]$** if it is continuous on the open interval (a, b) and

$$\lim_{x \to a^+} f(x) = f(a) \quad \text{and} \quad \lim_{x \to b^-} f(x) = f(b).$$

The function f is **continuous from the right** at a and **continuous from the left** at b (see Figure 2.31).

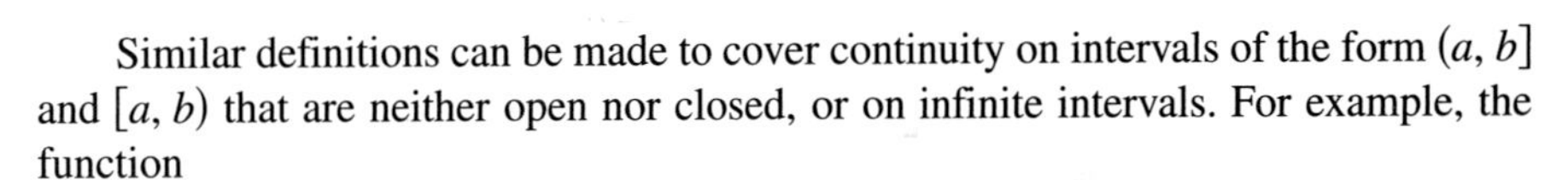

Similar definitions can be made to cover continuity on intervals of the form $(a, b]$ and $[a, b)$ that are neither open nor closed, or on infinite intervals. For example, the function

$$f(x) = \sqrt{x}$$

is continuous on the infinite interval $[0, \infty)$, and the function

$$g(x) = \sqrt{2 - x}$$

is continuous on the infinite interval $(-\infty, 2]$.

EXAMPLE 4 Continuity on a Closed Interval

Discuss the continuity of $f(x) = \sqrt{1 - x^2}$.

Solution The domain of f is the closed interval $[-1, 1]$. At all points in the open interval $(-1, 1)$, the continuity of f follows from Theorems 2.4 and 2.5. Moreover, because

$$\lim_{x \to -1^+} \sqrt{1 - x^2} = 0 = f(-1) \qquad \text{Continuous from the right}$$

and

$$\lim_{x \to 1^-} \sqrt{1 - x^2} = 0 = f(1) \qquad \text{Continuous from the left}$$

you can conclude that f is continuous on the closed interval $[-1, 1]$, as shown in Figure 2.32.

f is continuous on $[-1, 1]$.
Figure 2.32

The next example shows how a one-sided limit can be used to determine the value of absolute zero on the Kelvin scale.

EXAMPLE 5 Charles's Law and Absolute Zero

On the Kelvin scale, *absolute zero* is the temperature 0 K. Although temperatures of approximately 0.0001 K have been produced in laboratories, absolute zero has never been attained. In fact, evidence suggests that absolute zero *cannot* be attained. How did scientists determine that 0 K is the "lower limit" of the temperature of matter? What is absolute zero on the Celsius scale?

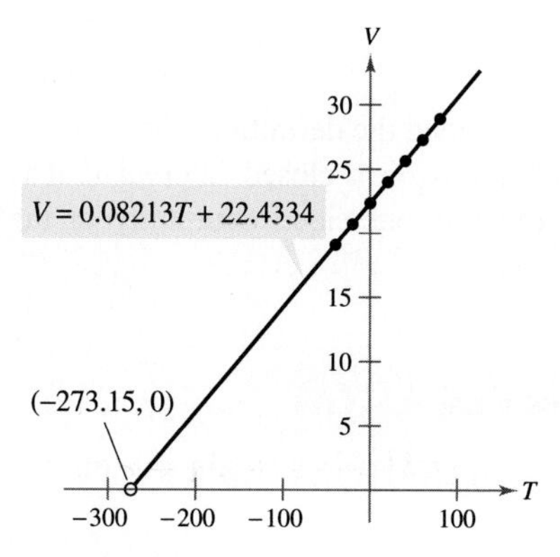

The volume of hydrogen gas depends on its temperature.
Figure 2.33

Solution The determination of absolute zero stems from the work of the French physicist Jacques Charles (1746–1823). Charles discovered that the volume of gas at a constant pressure increases linearly with the temperature of the gas. The table illustrates this relationship between volume and temperature. In the table, one mole of hydrogen is held at a constant pressure of one atmosphere. The volume V is measured in liters and the temperature T is measured in degrees Celsius.

T	−40	−20	0	20	40	60	80
V	19.1482	20.7908	22.4334	24.0760	25.7186	27.3612	29.0038

The points represented by the table are shown in Figure 2.33. Moreover, by using the points in the table, you can determine that T and V are related by the linear equation

$$V = 0.08213T + 22.4334 \quad \text{or} \quad T = \frac{V - 22.4334}{0.08213}.$$

By reasoning that the volume of the gas can approach 0 (but never equal or go below 0), you can determine that the "least possible temperature" is given by

$$\lim_{V \to 0^+} T = \lim_{V \to 0^+} \frac{V - 22.4334}{0.08213}$$

$$= \frac{0 - 22.4334}{0.08213} \qquad \text{Use direct substitution.}$$

$$\approx -273.15.$$

So, absolute zero on the Kelvin scale (0 K) is approximately $-273.15°$ on the Celsius scale.

University of Colorado at Boulder, Office of News Services

In 1995, physicists Carl Wieman and Eric Cornell of the University of Colorado at Boulder used lasers and evaporation to produce a supercold gas in which atoms overlap. This gas is called a Bose-Einstein condensate. "We get to within a billionth of a degree of absolute zero," reported Wieman. (*Source:* Time *magazine, April 10, 2000*)

The following table shows the temperatures in Example 5, converted to the Fahrenheit scale. Try repeating the solution shown in Example 5 using these temperatures and volumes. Use the result to find the value of absolute zero on the Fahrenheit scale.

T	−40	−4	32	68	104	140	176
V	19.1482	20.7908	22.4334	24.0760	25.7186	27.3612	29.0038

NOTE Charles's Law for gases (assuming constant pressure) can be stated as

$$V = RT \qquad \text{Charles's Law}$$

where V is volume, R is constant, and T is temperature. In the statement of this law, what property must the temperature scale have?

AUGUSTIN-LOUIS CAUCHY (1789–1857)

The concept of a continuous function was first introduced by Augustin-Louis Cauchy in 1821. The definition given in his text *Cours d'Analyse* stated that indefinite small changes in y were the result of indefinite small changes in x. "... $f(x)$ will be called a *continuous* function if ... the numerical values of the difference $f(x + \alpha) - f(x)$ decrease indefinitely with those of α ...."

Properties of Continuity

In Section 2.3, you studied several properties of limits. Each of those properties yields a corresponding property pertaining to the continuity of a function. For instance, Theorem 2.11 follows directly from Theorem 2.2.

THEOREM 2.11 Properties of Continuity

If b is a real number and f and g are continuous at $x = c$, then the following functions are also continuous at c.

1. Scalar multiple: bf
2. Sum and difference: $f \pm g$
3. Product: fg
4. Quotient: $\dfrac{f}{g}$, if $g(c) \neq 0$

The following types of functions are continuous at every point in their domains.

1. Polynomial: $p(x) = a_n x^n + a_{n-1}x^{n-1} + \cdots + a_1 x + a_0$
2. Rational: $r(x) = \dfrac{p(x)}{q(x)}, \quad q(x) \neq 0$
3. Radical: $f(x) = \sqrt[n]{x}$
4. Trigonometric: $\sin x, \cos x, \tan x, \cot x, \sec x, \csc x$
5. Exponential and logarithmic: $f(x) = a^x, f(x) = e^x, f(x) = \ln x$

By combining Theorem 2.11 with this summary, you can conclude that a wide variety of elementary functions are continuous at every point in their domains.

EXAMPLE 6 Applying Properties of Continuity

By Theorem 2.11, it follows that each of the following functions is continuous at every point in its domain.

$$f(x) = x + e^x, \qquad f(x) = 3\tan x, \qquad f(x) = \frac{x^2 + 1}{\cos x}$$

For instance, the first function is continuous at every real number because the functions $y = x$ and $y = e^x$ are continuous at every real number and the sum of continuous functions is continuous.

The next theorem, which is a consequence of Theorem 2.5, allows you to determine the continuity of *composite* functions such as

$$f(x) = \sin 3x, \qquad f(x) = \sqrt{x^2 + 1}, \qquad f(x) = \tan\frac{1}{x}.$$

NOTE One consequence of Theorem 2.12 is that if f and g satisfy the given conditions, you can determine the limit of $f(g(x))$ as x approaches c to be

$$\lim_{x \to c} f(g(x)) = f(g(c)).$$

THEOREM 2.12 Continuity of a Composite Function

If g is continuous at c and f is continuous at $g(c)$, then the composite function given by $(f \circ g)(x) = f(g(x))$ is continuous at c.

EXAMPLE 7 Testing for Continuity

Describe the interval(s) on which each function is continuous.

a. $f(x) = \tan x$ **b.** $g(x) = \begin{cases} \sin \dfrac{1}{x}, & x \neq 0 \\ 0, & x = 0 \end{cases}$ **c.** $h(x) = \begin{cases} x \sin \dfrac{1}{x}, & x \neq 0 \\ 0, & x = 0 \end{cases}$

Solution

a. The tangent function $f(x) = \tan x$ is undefined at

$$x = \frac{\pi}{2} + n\pi, \qquad n \text{ is an integer.}$$

At all other points it is continuous. So, $f(x) = \tan x$ is continuous on the open intervals

$$\ldots, \left(-\frac{3\pi}{2}, -\frac{\pi}{2}\right), \left(-\frac{\pi}{2}, \frac{\pi}{2}\right), \left(\frac{\pi}{2}, \frac{3\pi}{2}\right), \ldots$$

as shown in Figure 2.34(a).

b. Because $y = 1/x$ is continuous except at $x = 0$ and the sine function is continuous for all real values of x, it follows that $y = \sin(1/x)$ is continuous at all real values except $x = 0$. At $x = 0$, the limit of $g(x)$ does not exist (see Example 5, Section 2.2). So, g is continuous on the intervals $(-\infty, 0)$ and $(0, \infty)$, as shown in Figure 2.34(b).

c. This function is similar to that in part (b) except that the oscillations are damped by the factor x. Using the Squeeze Theorem, you obtain

$$-|x| \leq x \sin \frac{1}{x} \leq |x|, \qquad x \neq 0$$

and you can conclude that

$$\lim_{x \to 0} h(x) = 0.$$

So, h is continuous on the entire real number line, as shown in Figure 2.34(c).

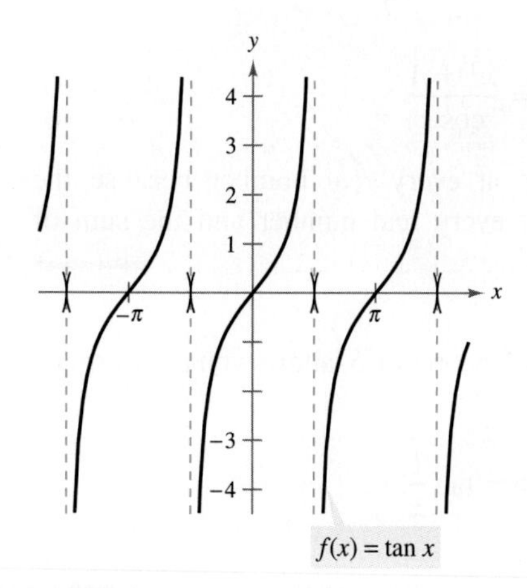

(a) f is continuous on each open interval in its domain.

(b) g is continuous on $(-\infty, 0)$ and $(0, \infty)$.

(c) h is continuous on the entire real number line.

Figure 2.34

The Intermediate Value Theorem

Theorem 2.13 is an important theorem concerning the behavior of functions that are continuous on a closed interval.

THEOREM 2.13 Intermediate Value Theorem

If f is continuous on the closed interval $[a, b]$ and k is any number between $f(a)$ and $f(b)$, then there is at least one number c in $[a, b]$ such that

$$f(c) = k.$$

NOTE The Intermediate Value Theorem tells you that at least one c exists, but it does not give a method for finding c. Such theorems are called **existence theorems.** By referring to a text on advanced calculus, you will find that a proof of this theorem is based on a property of real numbers called *completeness*. The Intermediate Value Theorem states that for a continuous function f, if x takes on all values between a and b, $f(x)$ must take on all values between $f(a)$ and $f(b)$.

As a simple example of this theorem, consider a person's height. Suppose that a girl is 5 feet tall on her thirteenth birthday and 5 feet 7 inches tall on her fourteenth birthday. Then, for any height h between 5 feet and 5 feet 7 inches, there must have been a time t when her height was exactly h. This seems reasonable because human growth is continuous and a person's height does not abruptly change from one value to another.

The Intermediate Value Theorem guarantees the existence of *at least one* number c in the closed interval $[a, b]$. There may, of course, be more than one number c such that $f(c) = k$, as shown in Figure 2.35. A function that is not continuous does not necessarily possess the intermediate value property. For example, the graph of the function shown in Figure 2.36 jumps over the horizontal line given by $y = k$, and for this function there is no value of c in $[a, b]$ such that $f(c) = k$.

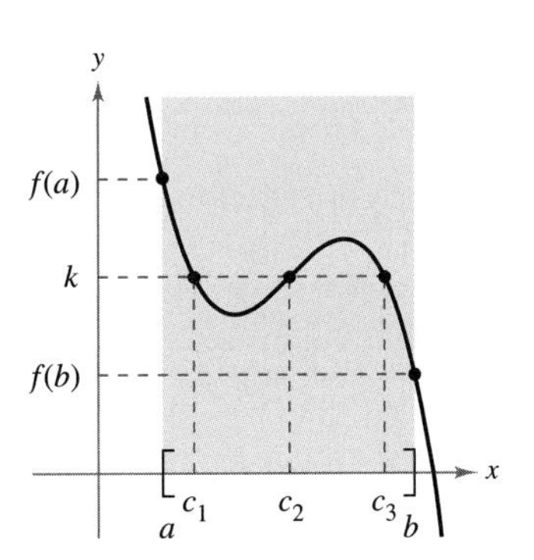

f is continuous on $[a, b]$.
[There exist three c's such that $f(c) = k$.]
Figure 2.35

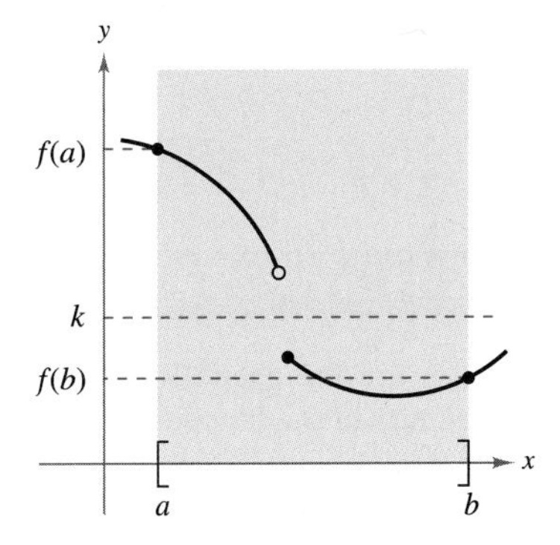

f is not continuous on $[a, b]$.
[There are no c's such that $f(c) = k$.]
Figure 2.36

The Intermediate Value Theorem often can be used to locate the zeros of a function that is continuous on a closed interval. Specifically, if f is continuous on $[a, b]$ and $f(a)$ and $f(b)$ differ in sign, the Intermediate Value Theorem guarantees the existence of at least one zero of f in the closed interval $[a, b]$.

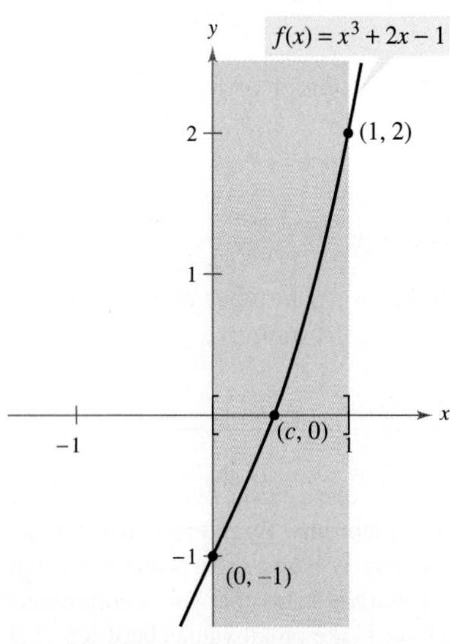

f is continuous on $[0, 1]$ with $f(0) < 0$ and $f(1) > 0$.
Figure 2.37

EXAMPLE 8 An Application of the Intermediate Value Theorem

Use the Intermediate Value Theorem to show that the polynomial function $f(x) = x^3 + 2x - 1$ has a zero in the interval $[0, 1]$.

Solution Note that f is continuous on the closed interval $[0, 1]$. Because

$$f(0) = 0^3 + 2(0) - 1 = -1 \quad \text{and} \quad f(1) = 1^3 + 2(1) - 1 = 2$$

it follows that $f(0) < 0$ and $f(1) > 0$. You can therefore apply the Intermediate Value Theorem to conclude that there must be some c in $[0, 1]$ such that

$$f(c) = 0 \qquad f \text{ has a zero in the closed interval } [0, 1].$$

as shown in Figure 2.37.

The **bisection method** for approximating the real zeros of a continuous function is similar to the method used in Example 8. If you know that a zero exists in the closed interval $[a, b]$, the zero must lie in the interval $[a, (a + b)/2]$ or $[(a + b)/2, b]$. From the sign of $f([a + b]/2)$, you can determine which interval contains the zero. By repeatedly bisecting the interval, you can "close in" on the zero of the function.

TECHNOLOGY You can also use the *zoom* feature of a graphing utility to approximate the real zeros of a continuous function. By repeatedly zooming in on the point where the graph crosses the x-axis, and adjusting the x-axis scale, you can approximate the zero of the function to any desired accuracy. The zero of $x^3 + 2x - 1$ is approximately 0.453, as shown in Figure 2.38.

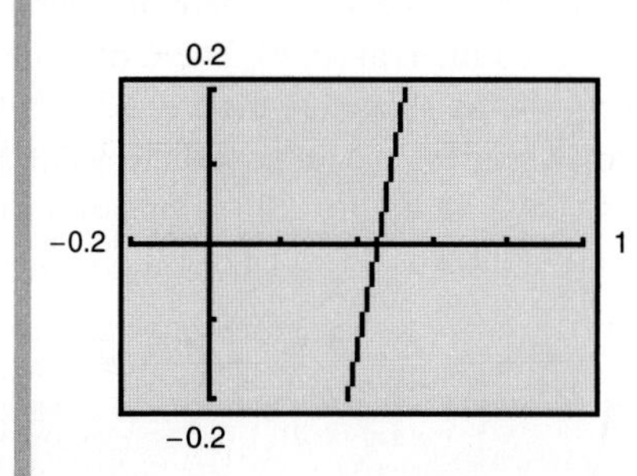

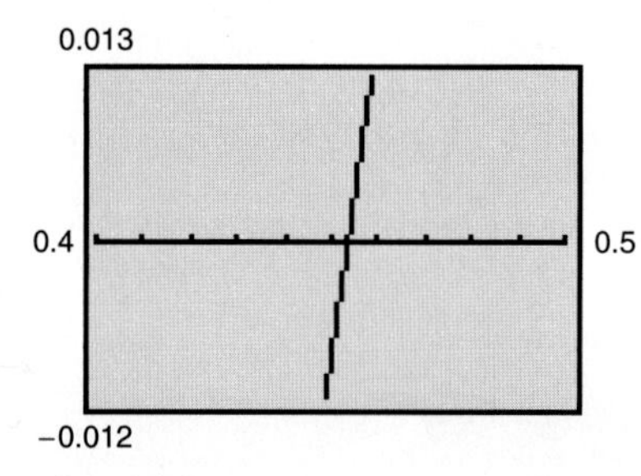

Figure 2.38 Zooming in on the zero of $f(x) = x^3 + 2x - 1$

Exercises for Section 2.4

See www.CalcChat.com for worked-out solutions to odd-numbered exercises.

In Exercises 1–6, use the graph to determine the limit, and discuss the continuity of the function.

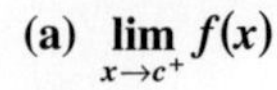

(a) $\lim\limits_{x \to c^+} f(x)$ (b) $\lim\limits_{x \to c^-} f(x)$ (c) $\lim\limits_{x \to c} f(x)$

1.

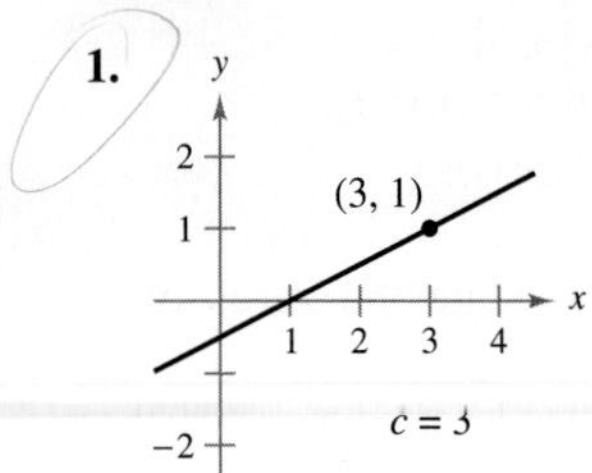

2.

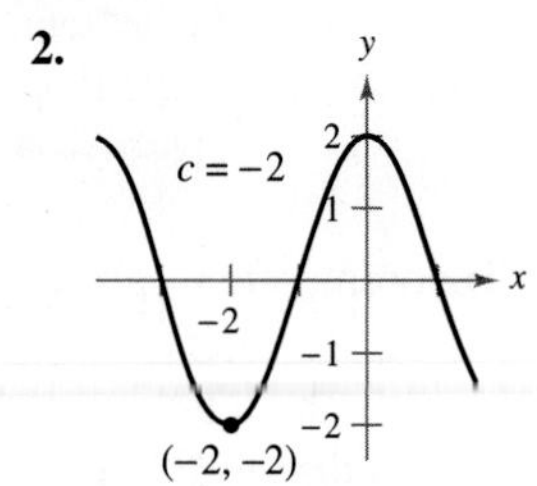

3.

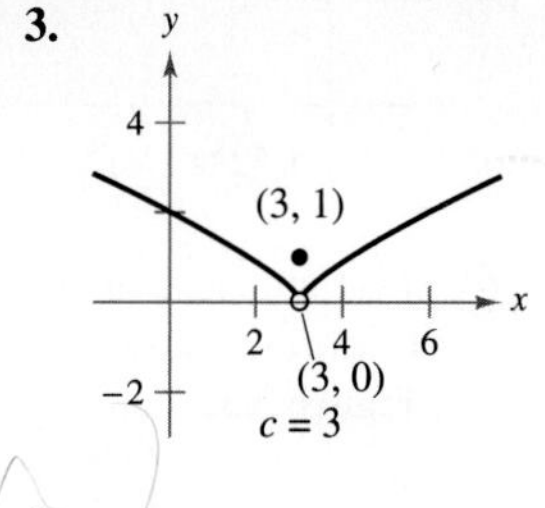

4.

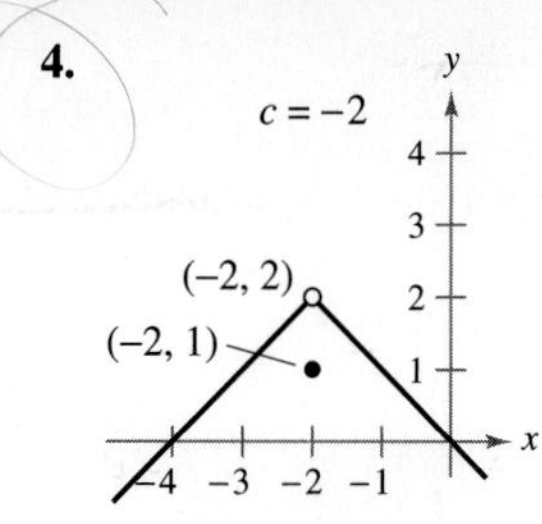

5.

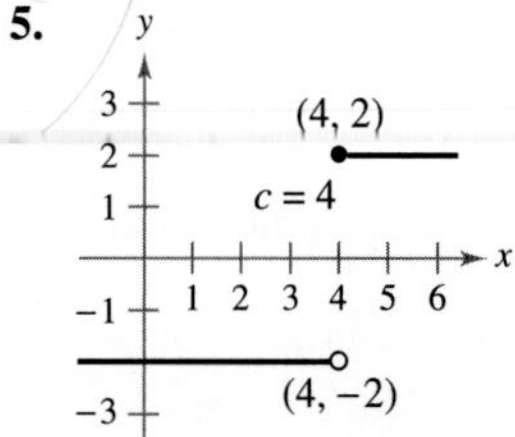

6.

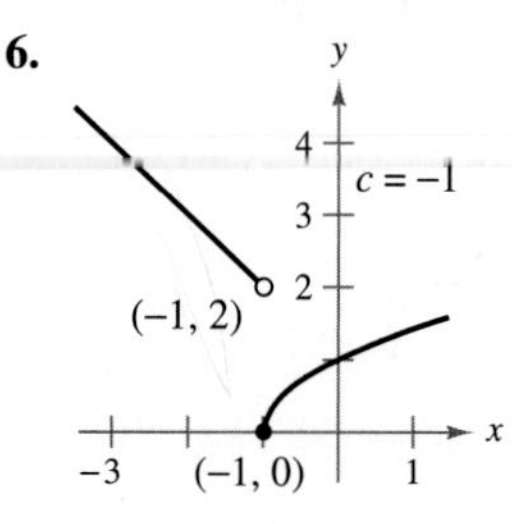

In Exercises 7–28, find the limit (if it exists). If it does not exist, explain why.

7. $\lim_{x\to 5^+} \frac{x-5}{x^2-25}$

8. $\lim_{x\to 2^+} \frac{2-x}{x^2-4}$

9. $\lim_{x\to -3^-} \frac{x}{\sqrt{x^2-9}}$

10. $\lim_{x\to 4^-} \frac{\sqrt{x}-2}{x-4}$

11. $\lim_{x\to 0^-} \frac{|x|}{x}$

12. $\lim_{x\to 3^+} \frac{|x-3|}{x-3}$

13. $\lim_{\Delta x\to 0^-} \frac{\frac{1}{x+\Delta x}-\frac{1}{x}}{\Delta x}$

14. $\lim_{\Delta x\to 0^+} \frac{(x+\Delta x)^2 + x + \Delta x - (x^2+x)}{\Delta x}$

15. $\lim_{x\to 3^-} f(x)$, where $f(x) = \begin{cases} \frac{x+2}{2}, & x \le 3 \\ \frac{12-2x}{3}, & x > 3 \end{cases}$

16. $\lim_{x\to 2} f(x)$, where $f(x) = \begin{cases} x^2-4x+6, & x < 2 \\ -x^2+4x-2, & x \ge 2 \end{cases}$

17. $\lim_{x\to 1} f(x)$, where $f(x) = \begin{cases} x^3+1, & x < 1 \\ x+1, & x \ge 1 \end{cases}$

18. $\lim_{x\to 1^+} f(x)$, where $f(x) = \begin{cases} x, & x \le 1 \\ 1-x, & x > 1 \end{cases}$

19. $\lim_{x\to\pi} \cot x$

20. $\lim_{x\to\pi/2} \sec x$

21. $\lim_{x\to 4^-} (3[\![x]\!] - 5)$

22. $\lim_{x\to 3^+} (3x - [\![x]\!])$

23. $\lim_{x\to 3} (2 - [\![-x]\!])$

24. $\lim_{x\to 1} \left(1 - \left[\!\!\left[-\frac{x}{2}\right]\!\!\right]\right)$

25. $\lim_{x\to 3^+} \ln(x-3)$

26. $\lim_{x\to 6^-} \ln(6-x)$

27. $\lim_{x\to 2^-} \ln[x^2(3-x)]$

28. $\lim_{x\to 5^+} \ln \frac{x}{\sqrt{x-4}}$

In Exercises 29–32, discuss the continuity of each function.

29. $f(x) = \frac{1}{x^2-4}$

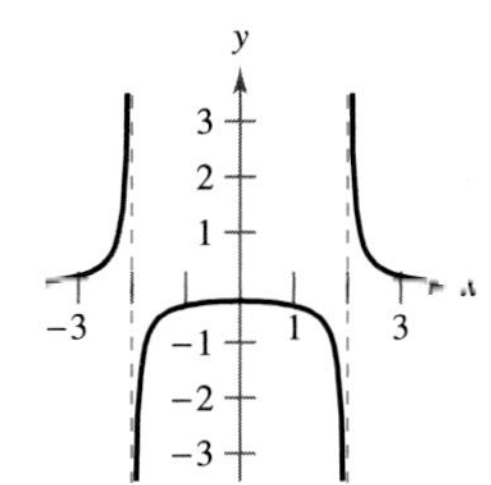

30. $f(x) = \frac{x^2-1}{x+1}$

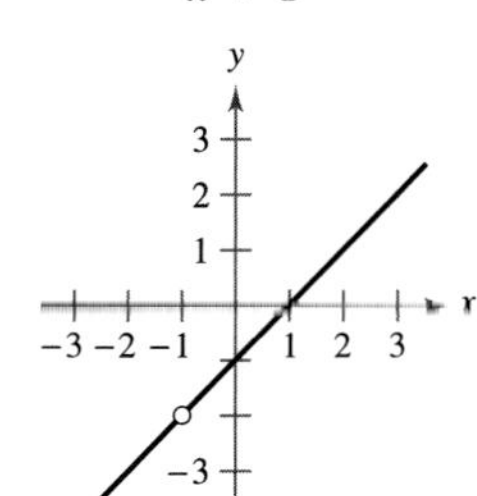

31. $f(x) = \frac{1}{2}[\![x]\!] + x$

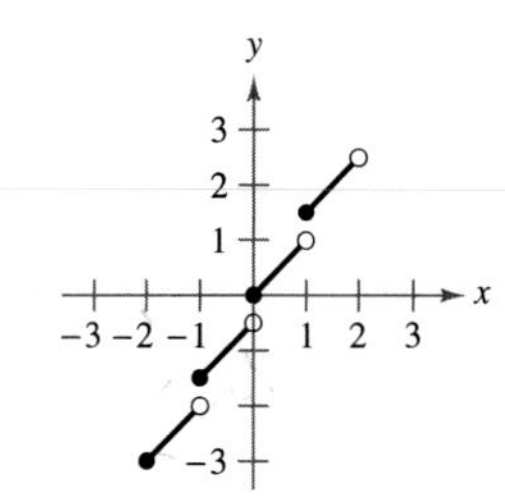

32. $f(x) = \begin{cases} x, & x < 1 \\ 2, & x = 1 \\ 2x-1, & x > 1 \end{cases}$

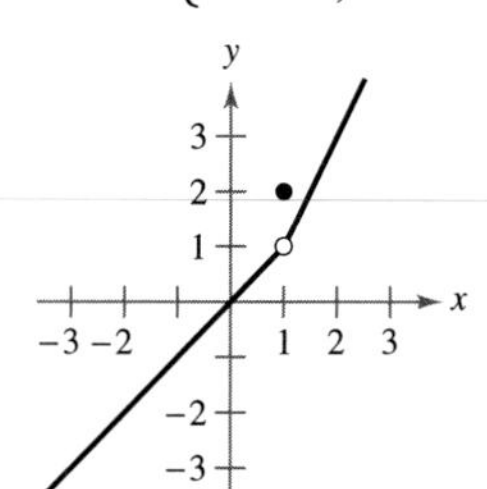

In Exercises 33–36, discuss the continuity of the function on the closed interval.

Function	*Interval*
33. $g(x) = \sqrt{25-x^2}$	$[-5, 5]$
34. $f(t) = 2 - \sqrt{9-t^2}$	$[-2, 2]$
35. $f(x) = \begin{cases} 3-x, & x \le 0 \\ 3+\frac{1}{2}x, & x > 0 \end{cases}$	$[-1, 4]$
36. $g(x) = \frac{1}{x^2-4}$	$[-1, 2]$

In Exercises 37–60, find the x-values (if any) at which f is not continuous. Which of the discontinuities are removable?

37. $f(x) = x^2 - 2x + 1$

38. $f(x) = \frac{1}{x^2+1}$

39. $f(x) = 3x - \cos x$

40. $f(x) = \cos \frac{\pi x}{4}$

41. $f(x) = \frac{x}{x^2-x}$

42. $f(x) = \frac{x}{x^2-1}$

43. $f(x) = \frac{x}{x^2+1}$

44. $f(x) = \frac{x-3}{x^2-9}$

45. $f(x) = \frac{x+2}{x^2-3x-10}$

46. $f(x) = \frac{x-1}{x^2+x-2}$

47. $f(x) = \frac{|x+2|}{x+2}$

48. $f(x) = \frac{|x-3|}{x-3}$

49. $f(x) = \begin{cases} x, & x \le 1 \\ x^2, & x > 1 \end{cases}$

50. $f(x) = \begin{cases} -2x+3, & x < 1 \\ x^2, & x \ge 1 \end{cases}$

51. $f(x) = \begin{cases} \frac{1}{2}x+1, & x \le 2 \\ 3-x, & x > 2 \end{cases}$

52. $f(x) = \begin{cases} -2x, & x \le 2 \\ x^2-4x+1, & x > 2 \end{cases}$

53. $f(x) = \begin{cases} \tan \frac{\pi x}{4}, & |x| < 1 \\ x, & |x| \ge 1 \end{cases}$

54. $f(x) = \begin{cases} \csc \frac{\pi x}{6}, & |x-3| \le 2 \\ 2, & |x-3| > 2 \end{cases}$

55. $f(x) = \begin{cases} \ln(x+1), & x \ge 0 \\ 1-x^2, & x < 0 \end{cases}$

56. $f(x) = \begin{cases} 10-3e^{5-x}, & x > 5 \\ 10-\frac{3}{5}x, & x \le 5 \end{cases}$

57. $f(x) = \csc 2x$

58. $f(x) = \tan \dfrac{\pi x}{4}$

59. $f(x) = [\![x - 1]\!]$

60. $f(x) = 3 - [\![x]\!]$

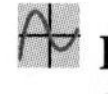

In Exercises 61 and 62, use a graphing utility to graph the function. From the graph, estimate

$$\lim_{x \to 0^+} f(x) \quad \text{and} \quad \lim_{x \to 0^-} f(x).$$

Is the function continuous on the entire real number line? Explain.

61. $f(x) = \dfrac{|x^2 - 4|x}{x + 2}$

62. $f(x) = \dfrac{|x^2 + 4x|(x + 2)}{x + 4}$

In Exercises 63–66, find the constants *a* and *b* such that the function is continuous on the entire real number line.

63. $f(x) = \begin{cases} x^3, & x \le 2 \\ ax^2, & x > 2 \end{cases}$

64. $g(x) = \begin{cases} \dfrac{4 \sin x}{x}, & x < 0 \\ a - 2x, & x \ge 0 \end{cases}$

65. $f(x) = \begin{cases} 2, & x \le -1 \\ ax + b, & -1 < x < 3 \\ -2, & x \ge 3 \end{cases}$

66. $g(x) = \begin{cases} \dfrac{x^2 - a^2}{x - a}, & x \ne a \\ 8, & x = a \end{cases}$

In Exercises 67–70, discuss the continuity of the composite function $h(x) = f(g(x))$.

67. $f(x) = x^2$
$g(x) = x - 1$

68. $f(x) = \dfrac{1}{\sqrt{x}}$
$g(x) = x - 1$

69. $f(x) = \dfrac{1}{x - 6}$
$g(x) = x^2 + 5$

70. $f(x) = \sin x$
$g(x) = x^2$

In Exercises 71–74, use a graphing utility to graph the function. Use the graph to determine any *x*-values at which the function is not continuous.

71. $f(x) = [\![x]\!] - x$

72. $h(x) = \dfrac{1}{x^2 - x - 2}$

73. $g(x) = \begin{cases} 2x - 4, & x \le 3 \\ x^2 - 2x, & x > 3 \end{cases}$

74. $f(x) = \begin{cases} \dfrac{\cos x - 1}{x}, & x < 0 \\ 5x, & x \ge 0 \end{cases}$

In Exercises 75–78, describe the interval(s) on which the function is continuous.

75. $f(x) = \dfrac{x}{x^2 + 1}$

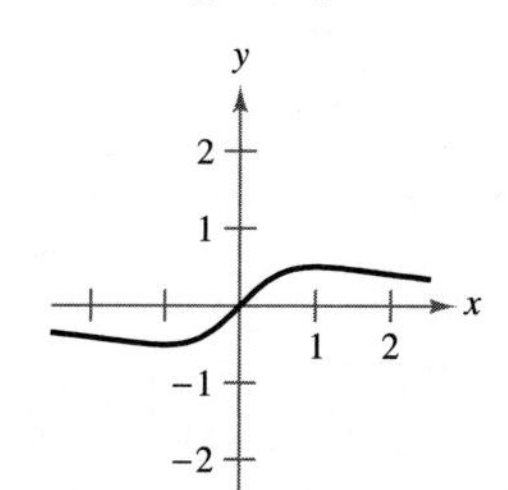

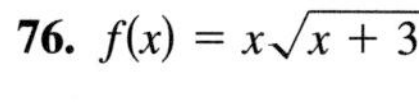

76. $f(x) = x\sqrt{x + 3}$

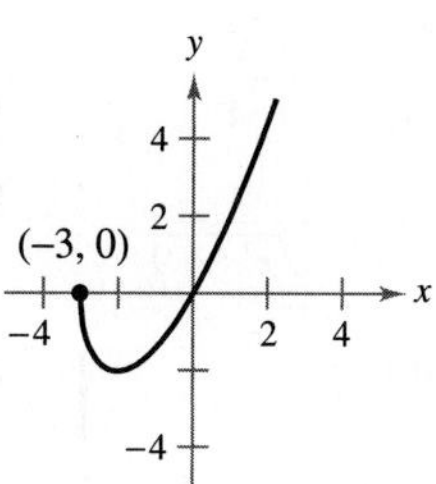

77. $f(x) = \sec \dfrac{\pi x}{4}$

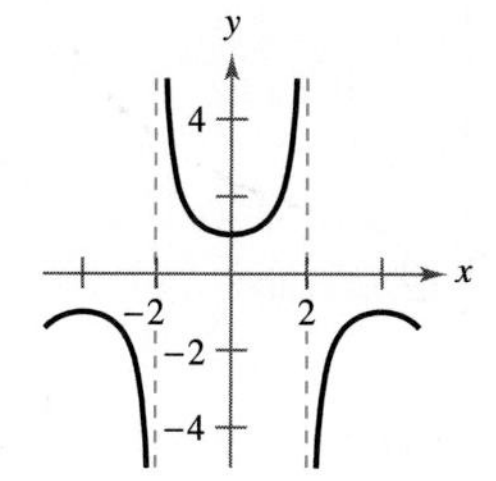

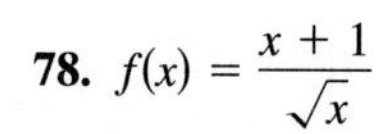

78. $f(x) = \dfrac{x + 1}{\sqrt{x}}$

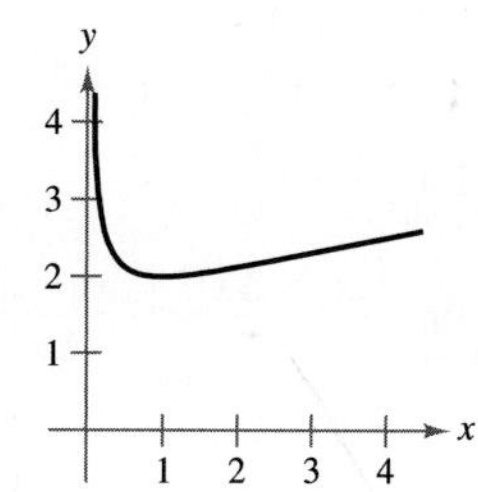

Writing **In Exercises 79–82, use a graphing utility to graph the function on the interval $[-4, 4]$. Does the graph of the function appear continuous on this interval? Is the function continuous on $[-4, 4]$? Write a short paragraph about the importance of examining a function analytically as well as graphically.**

79. $f(x) = \dfrac{\sin x}{x}$

80. $f(x) = \dfrac{x^3 - 8}{x - 2}$

81. $f(x) = \dfrac{\ln(x^2 + 1)}{x}$

82. $f(x) = \dfrac{e^{-x} + 1}{e^x - 1}$

Writing **In Exercises 83–86, explain why the function has a zero in the given interval.**

Function	*Interval*
83. $f(x) = x^2 - 4x + 3$	$[2, 4]$
84. $f(x) = x^3 + 3x - 2$	$[0, 1]$
85. $h(x) = -2e^{-x/2} \cos 2x$	$\left[0, \dfrac{\pi}{2}\right]$
86. $g(t) = (t^3 + 2t - 2) \ln(t^2 + 4)$	$[0, 1]$

In Exercises 87–90, use the Intermediate Value Theorem and a graphing utility to approximate the zero of the function in the interval [0, 1]. Repeatedly "zoom in" on the graph of the function to approximate the zero accurate to two decimal places. Use the *zero* or *root* feature of the graphing utility to approximate the zero accurate to four decimal places.

87. $f(x) = x^3 + x - 1$

88. $f(x) = x^3 + 3x - 3$

89. $g(t) = 2 \cos t - 3t$

90. $h(\theta) = 1 + \theta - 3 \tan \theta$

In Exercises 91–94, verify that the Intermediate Value Theorem applies to the indicated interval and find the value of c guaranteed by the theorem.

91. $f(x) = x^2 + x - 1, \quad [0, 5], \quad f(c) = 11$

92. $f(x) = x^2 - 6x + 8, \quad [0, 3], \quad f(c) = 0$

93. $f(x) = x^3 - x^2 + x - 2, \quad [0, 3], \quad f(c) = 4$

94. $f(x) = \dfrac{x^2 + x}{x - 1}, \quad \left[\dfrac{5}{2}, 4\right], \quad f(c) = 6$

Writing About Concepts

95. State how continuity is destroyed at $x = c$ for each of the following.

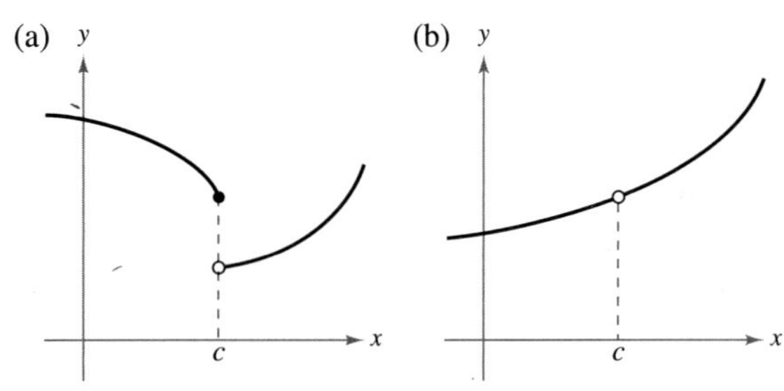

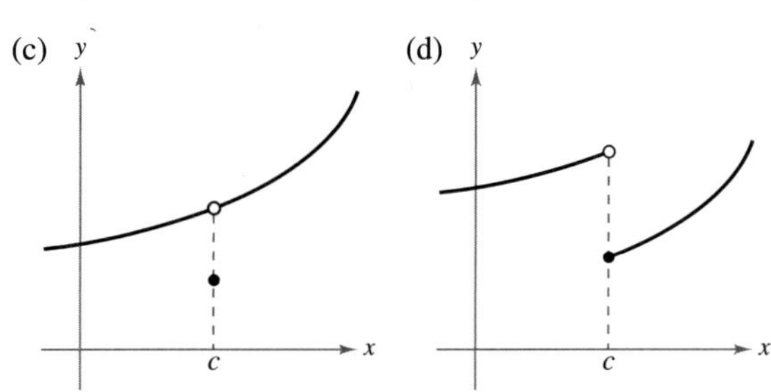

96. Describe the difference between a discontinuity that is removable and one that is nonremovable. In your explanation, give examples of the following.

(a) A function with a nonremovable discontinuity at $x = 2$

(b) A function with a removable discontinuity at $x = -2$

(c) A function that has both of the characteristics described in parts (a) and (b)

97. Sketch the graph of any function f such that

$$\lim_{x \to 3^+} f(x) = 1 \quad \text{and} \quad \lim_{x \to 3^-} f(x) = 0.$$

Is the function continuous at $x = 3$? Explain.

98. If the functions f and g are continuous for all real x, is $f + g$ always continuous for all real x? Is f/g always continuous for all real x? If either is not continuous, give an example to verify your conclusion.

True or False? **In Exercises 99–102, determine whether the statement is true or false. If it is false, explain why or give an example that shows it is false.**

99. If $\lim_{x \to c} f(x) = L$ and $f(c) = L$, then f is continuous at c.

100. If $f(x) = g(x)$ for $x \neq c$ and $f(c) \neq g(c)$, then either f or g is not continuous at c.

101. A rational function can have infinitely many x-values at which it is not continuous.

102. The function $f(x) = |x - 1|/(x - 1)$ is continuous on $(-\infty, \infty)$.

103. ***Swimming Pool*** Every day you dissolve 28 ounces of chlorine in a swimming pool. The graph shows the amount of chlorine $f(t)$ in the pool after t days.

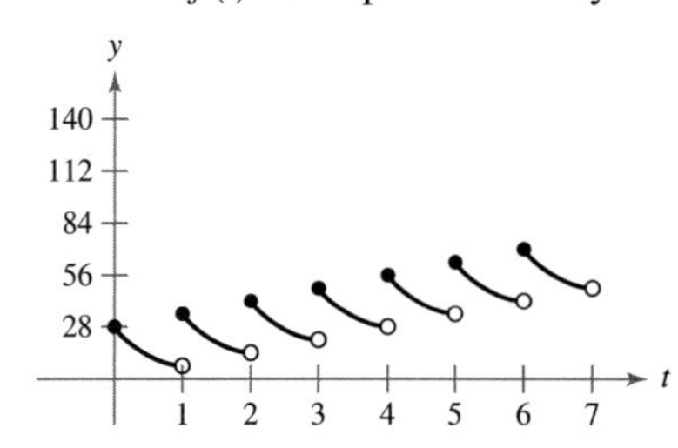

Estimate and interpret $\lim_{t \to 4^-} f(t)$ and $\lim_{t \to 4^+} f(t)$.

104. ***Think About It*** Describe how the functions

$f(x) = 3 + [\![x]\!]$

and

$g(x) = 3 - [\![-x]\!]$

differ.

105. ***Telephone Charges*** A dial-direct long distance call between two cities costs \$1.04 for the first 2 minutes and \$0.36 for each additional minute or fraction thereof. Use the greatest integer function to write the cost C of a call in terms of time t (in minutes). Sketch the graph of this function and discuss its continuity.

106. ***Inventory Management*** The number of units in inventory in a small company is given by

$$N(t) = 25\left(2\left[\!\!\left[\frac{t + 2}{2}\right]\!\!\right] - t\right)$$

where t is the time in months. Sketch the graph of this function and discuss its continuity. How often must this company replenish its inventory?

107. ***Déjà Vu*** At 8:00 A.M. on Saturday a man begins running up the side of a mountain to his weekend campsite (see figure on next page). On Sunday morning at 8:00 A.M. he runs back down the mountain. It takes him 20 minutes to run up, but only 10 minutes to run down. At some point on the way down, he realizes that he passed the same place at exactly the same time on Saturday. Prove that he is correct. [*Hint:* Let $s(t)$ and $r(t)$ be the position functions for the runs up and down, and apply the Intermediate Value Theorem to the function $f(t) = s(t) - r(t)$.]

Figure for 107

108. ***Volume*** Use the Intermediate Value Theorem to show that for all spheres with radii in the interval $[1, 5]$, there is one with a volume of 275 cubic centimeters.

109. Prove that if f is continuous and has no zeros on $[a, b]$, then either

$$f(x) > 0 \text{ for all } x \text{ in } [a, b] \text{ or } f(x) < 0 \text{ for all } x \text{ in } [a, b].$$

110. Show that the Dirichlet function

$$f(x) = \begin{cases} 0, & \text{if } x \text{ is rational} \\ 1, & \text{if } x \text{ is irrational} \end{cases}$$

is not continuous at any real number.

111. Show that the function

$$f(x) = \begin{cases} 0, & \text{if } x \text{ is rational} \\ kx, & \text{if } x \text{ is irrational} \end{cases}$$

is continuous only at $x = 0$. (Assume that k is any nonzero real number.)

112. The **signum function** is defined by

$$\operatorname{sgn}(x) = \begin{cases} -1, & x < 0 \\ 0, & x = 0 \\ 1, & x > 0. \end{cases}$$

Sketch a graph of $\operatorname{sgn}(x)$ and find the following (if possible).

(a) $\lim_{x \to 0^-} \operatorname{sgn}(x)$ (b) $\lim_{x \to 0^+} \operatorname{sgn}(x)$ (c) $\lim_{x \to 0} \operatorname{sgn}(x)$

113. ***Modeling Data*** After an object falls for t seconds, the speed S (in feet per second) of the object is recorded in the table.

t	0	5	10	15	20	25	30
S	0	48.2	53.5	55.2	55.9	56.2	56.3

(a) Create a line graph of the data.

(b) Does there appear to be a limiting speed of the object? If there is a limiting speed, identify a possible cause.

114. ***Creating Models*** A swimmer crosses a pool of width b by swimming in a straight line from $(0, 0)$ to $(2b, b)$. (See figure.)

(a) Let f be a function defined as the y-coordinate of the point on the long side of the pool that is nearest the swimmer at any given time during the swimmer's path across the pool. Determine the function f and sketch its graph. Is it continuous? Explain.

(b) Let g be the minimum distance between the swimmer and the long sides of the pool. Determine the function g and sketch its graph. Is it continuous? Explain.

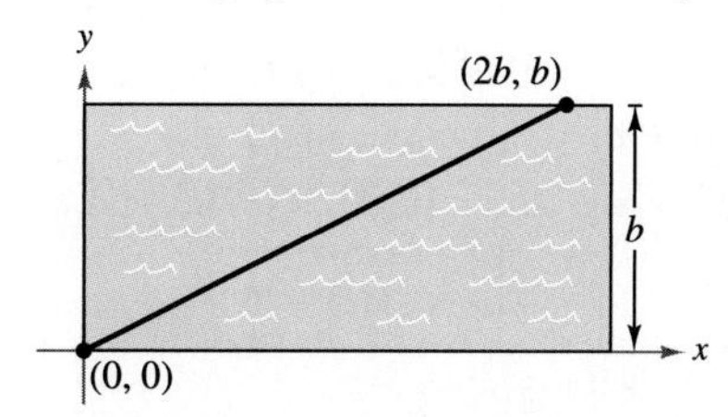

115. Find all values of c such that f is continuous on $(-\infty, \infty)$.

$$f(x) = \begin{cases} 1 - x^2, & x \le c \\ x, & x > c \end{cases}$$

116. Prove that for any real number y there exists x in $(-\pi/2, \pi/2)$ such that $\tan x = y$.

117. Let $f(x) = \left(\sqrt{x + c^2} - c\right)/x$, $c > 0$. What is the domain of f? How can you define f at $x = 0$ in order for f to be continuous there?

118. Prove that if $\lim_{\Delta x \to 0} f(c + \Delta x) = f(c)$, then f is continuous at c.

119. Discuss the continuity of the function $h(x) = x[\![x]\!]$.

120. (a) Let $f_1(x)$ and $f_2(x)$ be continuous on the closed interval $[a, b]$. If $f_1(a) < f_2(a)$ and $f_1(b) > f_2(b)$, prove that there exists c between a and b such that $f_1(c) = f_2(c)$.

(b) Show that there exists c in $\left[0, \frac{\pi}{2}\right]$ such that $\cos x = x$. Use a graphing utility to approximate c to three decimal places.

121. ***Think About It*** Consider the function

$$f(x) = \frac{4}{1 + 2^{4/x}}.$$

(a) What is the domain of the function?

(b) Use a graphing utility to graph the function.

(c) Determine $\lim_{x \to 0-} f(x)$ and $\lim_{x \to 0+} f(x)$.

(d) Use your knowledge of the exponential function to explain the behavior of f near $x = 0$.

Putnam Exam Challenge

122. Prove or disprove: if x and y are real numbers with $y \ge 0$ and $y(y + 1) \le (x + 1)^2$, then $y(y - 1) \le x^2$.

123. Determine all polynomials $P(x)$ such that $P(x^2 + 1) = (P(x))^2 + 1$ and $P(0) = 0$.

These problems were composed by the Committee on the Putnam Prize Competition.

Section 2.5 Infinite Limits

- **Determine infinite limits from the left and from the right.**
- **Find and sketch the vertical asymptotes of the graph of a function.**

Infinite Limits

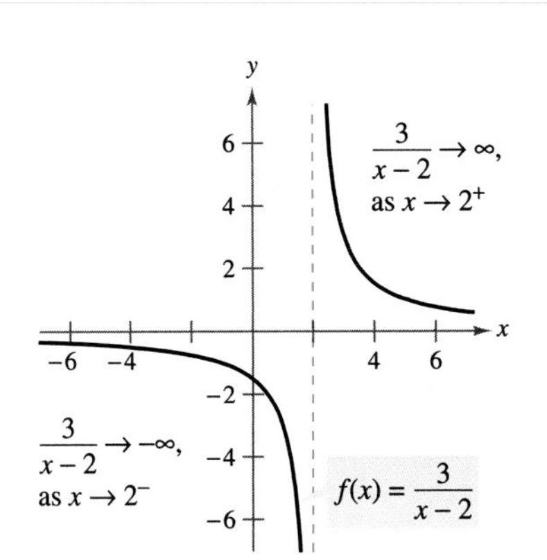

$f(x)$ increases and decreases without bound as x approaches 2.
Figure 2.39

Let f be the function given by

$$f(x) = \frac{3}{x-2}.$$

From Figure 2.39 and the table, you can see that $f(x)$ *decreases without bound* as x approaches 2 from the left, and $f(x)$ *increases without bound* as x approaches 2 from the right. This behavior is denoted as

$$\lim_{x \to 2^-} \frac{3}{x-2} = -\infty \qquad f(x) \text{ decreases without bound as } x \text{ approaches 2 from the left.}$$

and

$$\lim_{x \to 2^+} \frac{3}{x-2} = \infty. \qquad f(x) \text{ increases without bound as } x \text{ approaches 2 from the right.}$$

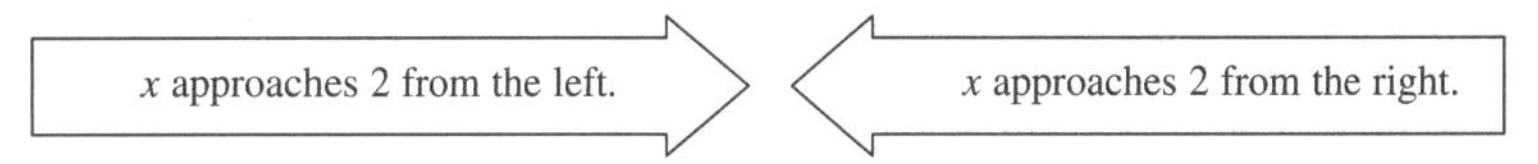

x	1.5	1.9	1.99	1.999	2	2.001	2.01	2.1	2.5
$f(x)$	−6	−30	−300	−3000	?	3000	300	30	6

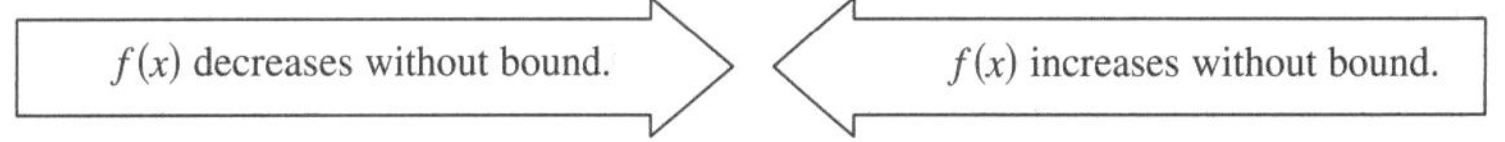

A limit in which $f(x)$ increases or decreases without bound as x approaches c is called an **infinite limit.**

Definition of Infinite Limits

Let f be a function that is defined at every real number in some open interval containing c (except possibly at c itself). The statement

$$\lim_{x \to c} f(x) = \infty$$

means that for each $M > 0$ there exists a $\delta > 0$ such that $f(x) > M$ whenever $0 < |x - c| < \delta$ (see Figure 2.40). Similarly, the statement

$$\lim_{x \to c} f(x) = -\infty$$

means that for each $N < 0$ there exists a $\delta > 0$ such that $f(x) < N$ whenever $0 < |x - c| < \delta$.

To define the **infinite limit from the left,** replace $0 < |x - c| < \delta$ by $c - \delta < x < c$. To define the **infinite limit from the right,** replace $0 < |x - c| < \delta$ by $c < x < c + \delta$.

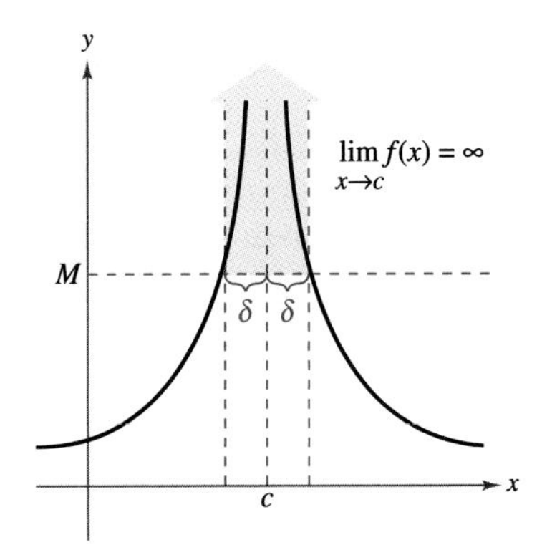

Infinite limits
Figure 2.40

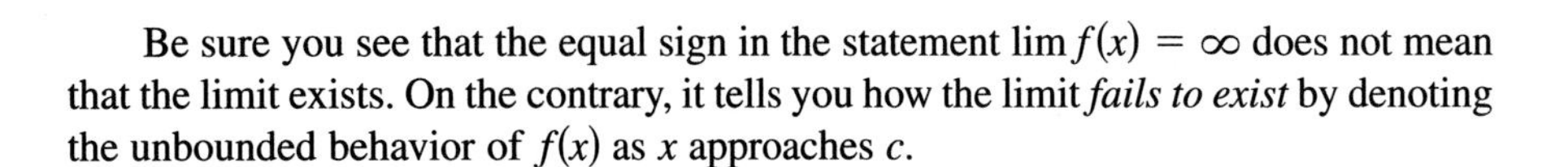

Be sure you see that the equal sign in the statement $\lim f(x) = \infty$ does not mean that the limit exists. On the contrary, it tells you how the limit *fails to exist* by denoting the unbounded behavior of $f(x)$ as x approaches c.

EXPLORATION

Use a graphing utility to graph each function. For each function, analytically find the single real number c that is not in the domain. Then graphically find the limit of $f(x)$ as x approaches c from the left and from the right.

a. $f(x) = \dfrac{3}{x - 4}$ **b.** $f(x) = \dfrac{1}{2 - x}$

c. $f(x) = \dfrac{2}{(x - 3)^2}$ **d.** $f(x) = \dfrac{-3}{(x + 2)^2}$

EXAMPLE 1 Determining Infinite Limits from a Graph

Use Figure 2.41 to determine the limit of each function as x approaches 1 from the left and from the right.

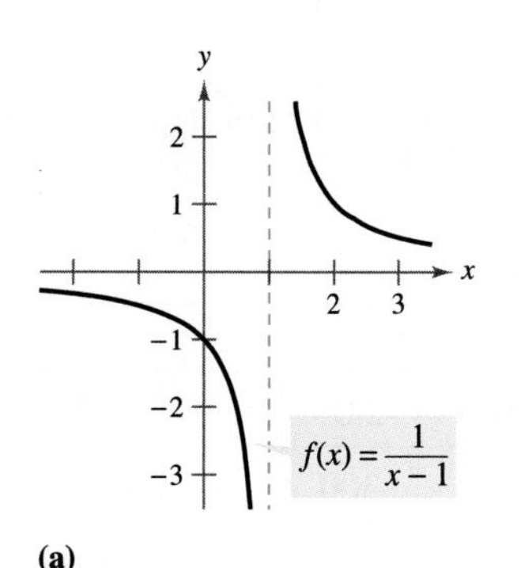

(a)

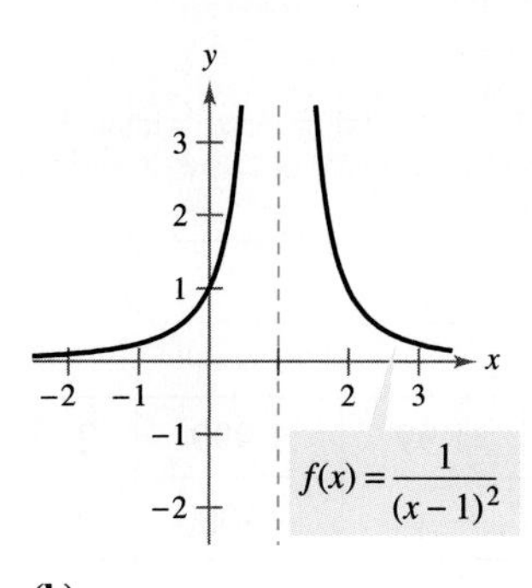

(b)

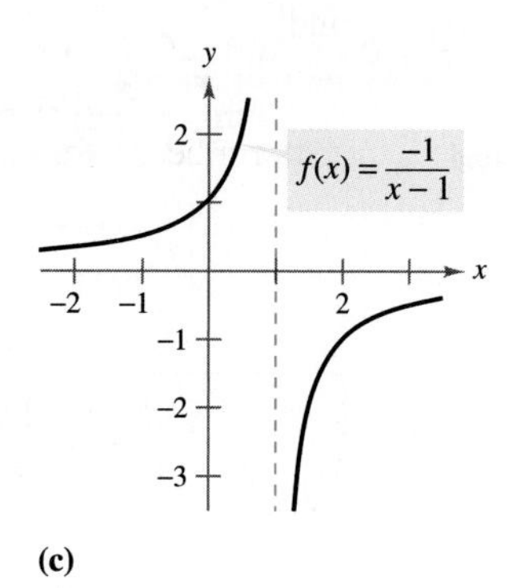

(c)

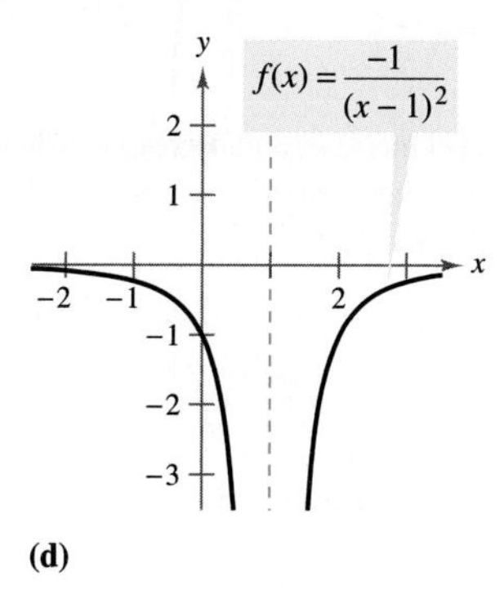

(d)

Figure 2.41 Each graph has an asymptote at $x = 1$.

Solution

a. $\displaystyle\lim_{x\to 1^-} \frac{1}{x - 1} = -\infty$ and $\displaystyle\lim_{x\to 1^+} \frac{1}{x - 1} = \infty$

b. $\displaystyle\lim_{x\to 1} \frac{1}{(x - 1)^2} = \infty$ Limit from each side is ∞.

c. $\displaystyle\lim_{x\to 1^-} \frac{-1}{x - 1} = \infty$ and $\displaystyle\lim_{x\to 1^+} \frac{-1}{x - 1} = -\infty$

d. $\displaystyle\lim_{x\to 1} \frac{-1}{(x - 1)^2} = -\infty$ Limit from each side is $-\infty$.

Vertical Asymptotes

If it were possible to extend the graphs in Figure 2.41 toward positive and negative infinity, you would see that each graph becomes arbitrarily close to the vertical line $x = 1$. This line is a **vertical asymptote** of the graph of f. (You will study other types of asymptotes in Sections 4.5 and 4.6.)

NOTE If a function f has a vertical asymptote at $x = c$, then f is *not continuous* at c.

Definition of a Vertical Asymptote

If $f(x)$ approaches infinity (or negative infinity) as x approaches c from the right or the left, then the line $x = c$ is a **vertical asymptote** of the graph of f.

In Example 1, note that each of the functions is a *quotient* and that the vertical asymptote occurs at a number where the denominator is 0 (and the numerator is not 0). The next theorem generalizes this observation. (A proof of this theorem is given in Appendix A.)

THEOREM 2.14 Vertical Asymptotes

Let f and g be continuous on an open interval containing c. If $f(c) \neq 0$, $g(c) = 0$, and there exists an open interval containing c such that $g(x) \neq 0$ for all $x \neq c$ in the interval, then the graph of the function given by

$$h(x) = \frac{f(x)}{g(x)}$$

has a vertical asymptote at $x = c$.

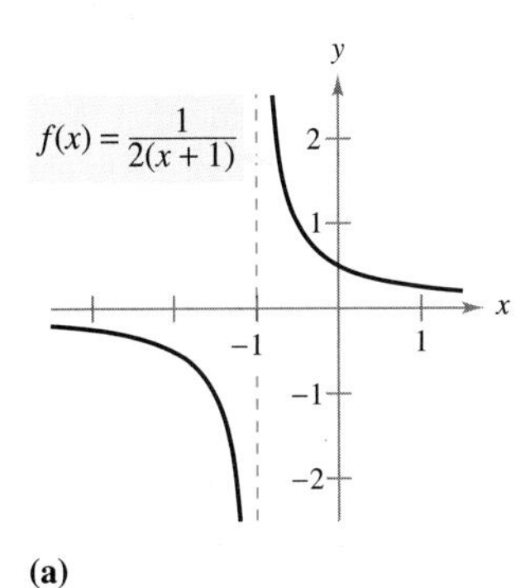

(a)

$f(x) = \dfrac{x^2+1}{x^2-1}$

(b)

EXAMPLE 2 Finding Vertical Asymptotes

Determine all vertical asymptotes of the graph of each function.

a. $f(x) = \dfrac{1}{2(x+1)}$ **b.** $f(x) = \dfrac{x^2+1}{x^2-1}$ **c.** $f(x) = \cot x$

Solution

a. When $x = -1$, the denominator of

$$f(x) = \frac{1}{2(x+1)}$$

is 0 and the numerator is not 0. So, by Theorem 2.14, you can conclude that $x = -1$ is a vertical asymptote, as shown in Figure 2.42(a).

b. By factoring the denominator as

$$f(x) = \frac{x^2+1}{x^2-1} = \frac{x^2+1}{(x-1)(x+1)}$$

you can see that the denominator is 0 at $x = -1$ and $x = 1$. Moreover, because the numerator is not 0 at these two points, you can apply Theorem 2.14 to conclude that the graph of f has two vertical asymptotes, as shown in Figure 2.42(b).

c. By writing the cotangent function in the form

$$f(x) = \cot x = \frac{\cos x}{\sin x}$$

you can apply Theorem 2.14 to conclude that vertical asymptotes occur at all values of x such that $\sin x = 0$ and $\cos x \neq 0$, as shown in Figure 2.42(c). So, the graph of this function has infinitely many vertical asymptotes. These asymptotes occur when $x = n\pi$, where n is an integer.

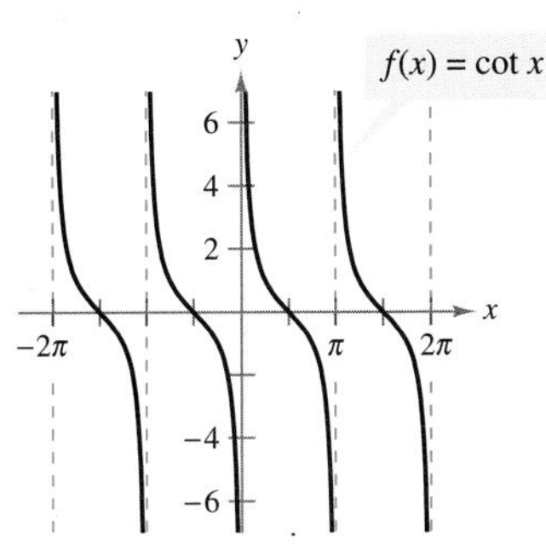

(c)

Functions with vertical asymptotes
Figure 2.42

Theorem 2.14 requires that the value of the numerator at $x = c$ be nonzero. If both the numerator and the denominator are 0 at $x = c$, you obtain the *indeterminate form* $0/0$, and you cannot determine the limit behavior at $x = c$ without further investigation, as illustrated in Example 3.

EXAMPLE 3 A Rational Function with Common Factors

Determine all vertical asymptotes of the graph of

$$f(x) = \frac{x^2 + 2x - 8}{x^2 - 4}.$$

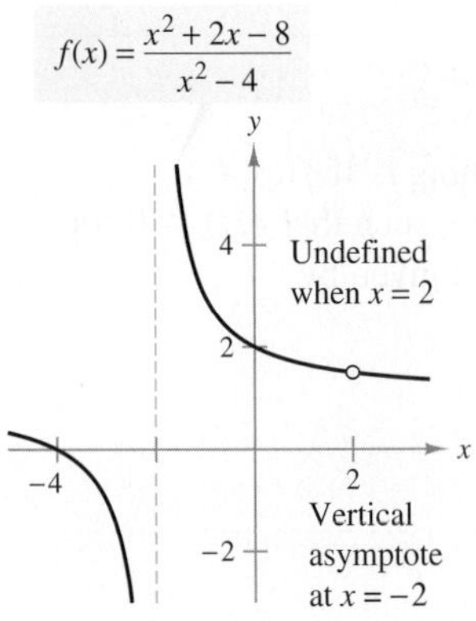

Solution Begin by simplifying the expression, as shown.

$$\begin{aligned} f(x) &= \frac{x^2 + 2x - 8}{x^2 - 4} \\ &= \frac{(x + 4)(x - 2)}{(x + 2)(x - 2)} \\ &= \frac{x + 4}{x + 2}, \quad x \neq 2 \end{aligned}$$

$f(x) = \dfrac{x^2 + 2x - 8}{x^2 - 4}$

Undefined when $x = 2$

Vertical asymptote at $x = -2$

$f(x)$ increases and decreases without bound as x approaches -2.
Figure 2.43

At all x-values other than $x = 2$, the graph of f coincides with the graph of $g(x) = (x + 4)/(x + 2)$. So, you can apply Theorem 2.14 to g to conclude that there is a vertical asymptote at $x = -2$, as shown in Figure 2.43. From the graph, you can see that

$$\lim_{x \to -2^-} \frac{x^2 + 2x - 8}{x^2 - 4} = -\infty \quad \text{and} \quad \lim_{x \to -2^+} \frac{x^2 + 2x - 8}{x^2 - 4} = \infty.$$

Note that $x = 2$ is *not* a vertical asymptote. Rather, $x = 2$ is a removable discontinuity.

EXAMPLE 4 Determining Infinite Limits

Find each limit.

$$\lim_{x \to 1^-} \frac{x^2 - 3x}{x - 1} \quad \text{and} \quad \lim_{x \to 1^+} \frac{x^2 - 3x}{x - 1}$$

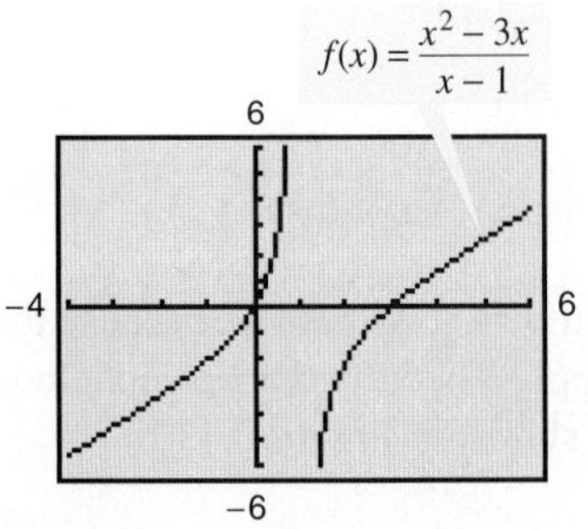

Solution Because the denominator is 0 when $x = 1$ (and the numerator is not zero), you know that the graph of

$$f(x) = \frac{x^2 - 3x}{x - 1}$$

$f(x) = \dfrac{x^2 - 3x}{x - 1}$

f has a vertical asymptote at $x = 1$.
Figure 2.44

has a vertical asymptote at $x = 1$. This means that each of the given limits is either ∞ or $-\infty$. A graphing utility can help determine the result. From the graph of f shown in Figure 2.44, you can see that the graph approaches ∞ from the left of $x = 1$ and approaches $-\infty$ from the right of $x = 1$. So, you can conclude that

$$\lim_{x \to 1^-} \frac{x^2 - 3x}{x - 1} = \infty \qquad \text{The limit from the left is infinity.}$$

and

$$\lim_{x \to 1^+} \frac{x^2 - 3x}{x - 1} = -\infty. \qquad \text{The limit from the right is negative infinity.}$$

TECHNOLOGY PITFALL When using a graphing calculator or graphing software, be careful to interpret correctly the graph of a function with a vertical asymptote—graphing utilities often have difficulty drawing this type of graph correctly.

THEOREM 2.15 Properties of Infinite Limits

Let c and L be real numbers and let f and g be functions such that

$$\lim_{x\to c} f(x) = \infty \quad \text{and} \quad \lim_{x\to c} g(x) = L.$$

1. Sum or difference: $\lim_{x\to c} [f(x) \pm g(x)] = \infty$
2. Product: $\lim_{x\to c} [f(x)g(x)] = \infty, \quad L > 0$
   $\lim_{x\to c} [f(x)g(x)] = -\infty, \quad L < 0$
3. Quotient: $\lim_{x\to c} \dfrac{g(x)}{f(x)} = 0$

Similar properties hold for one-sided limits and for functions for which the limit of $f(x)$ as x approaches c is $-\infty$.

NOTE With a graphing utility, you can confirm that the natural logarithmic function has a vertical asymptote at $x = 0$. (See Figure 2.45.) This implies that

$$\lim_{x\to 0+} \ln x = -\infty.$$

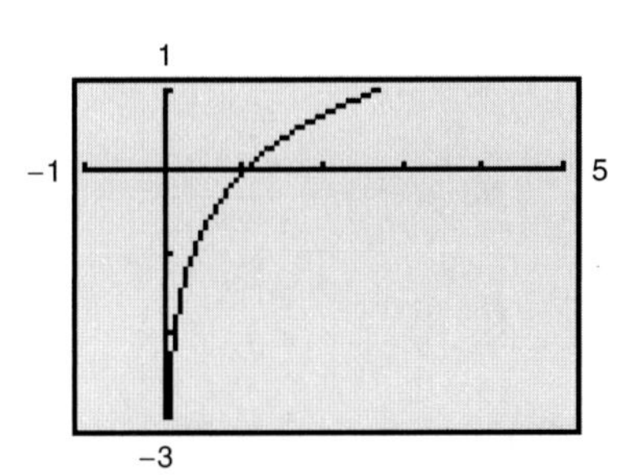

Figure 2.45

Proof To show that the limit of $f(x) + g(x)$ is infinite, choose $M > 0$. You then need to find $\delta > 0$ such that

$$[f(x) + g(x)] > M$$

whenever $0 < |x - c| < \delta$. For simplicity's sake, you can assume L is positive and let $M_1 = M + 1$. Because the limit of $f(x)$ is infinite, there exists δ_1 such that $f(x) > M_1$ whenever $0 < |x - c| < \delta_1$. Also, because the limit of $g(x)$ is L, there exists δ_2 such that $|g(x) - L| < 1$ whenever $0 < |x - c| < \delta_2$. By letting δ be the smaller of δ_1 and δ_2, you can conclude that $0 < |x - c| < \delta$ implies $f(x) > M + 1$ and $|g(x) - L| < 1$. The second of these two inequalities implies that $g(x) > L - 1$, and, adding this to the first inequality, you can write

$$f(x) + g(x) > (M + 1) + (L - 1) = M + L > M.$$

So, you can conclude that

$$\lim_{x\to c} [f(x) + g(x)] = \infty.$$

The proofs of the remaining properties are left as exercises (see Exercise 78).

EXAMPLE 5 Determining Limits

a. Because $\lim_{x\to 0} 1 = 1$ and $\lim_{x\to 0} \dfrac{1}{x^2} = \infty$, you can write

$$\lim_{x\to 0} \left(1 + \frac{1}{x^2}\right) = \infty. \qquad \text{Property 1, Theorem 2.15}$$

b. Because $\lim_{x\to 1^-} (x^2 + 1) = 2$ and $\lim_{x\to 1^-} (\cot \pi x) = -\infty$, you can write

$$\lim_{x\to 1^-} \frac{x^2 + 1}{\cot \pi x} = 0. \qquad \text{Property 3, Theorem 2.15}$$

c. Because $\lim_{x\to 0^+} 3 = 3$ and $\lim_{x\to 0^+} \ln x = -\infty$, you can write

$$\lim_{x\to 0^+} 3 \ln x = -\infty. \qquad \text{Property 2, Theorem 2.15}$$

Exercises for Section 2.5

See www.CalcChat.com for worked-out solutions to odd-numbered exercises.

In Exercises 1–4, determine whether $f(x)$ approaches ∞ or $-\infty$ as x approaches -2 from the left and from the right.

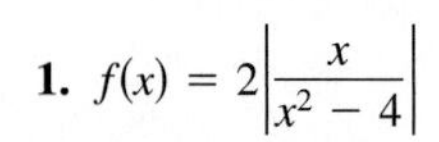

1. $f(x) = 2\left|\dfrac{x}{x^2 - 4}\right|$

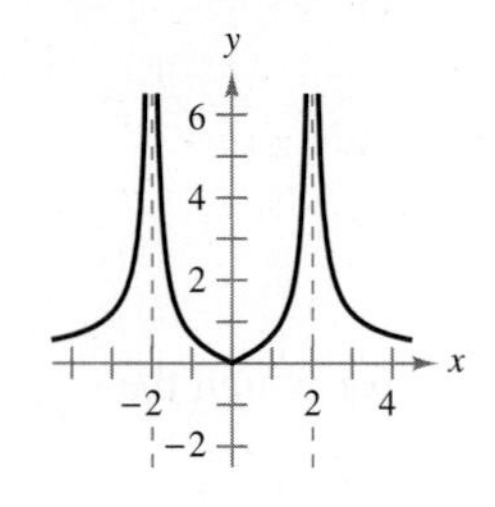

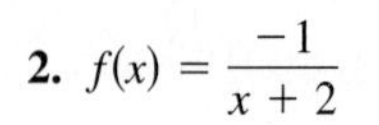

2. $f(x) = \dfrac{-1}{x + 2}$

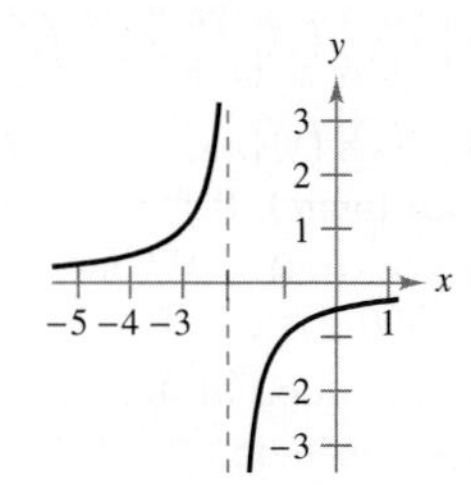

3. $f(x) = \tan \dfrac{\pi x}{4}$

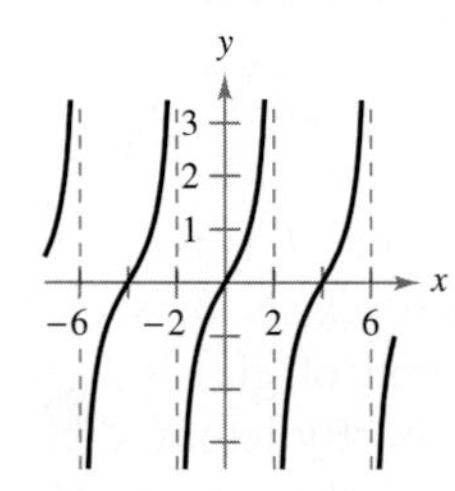

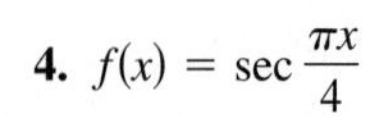

4. $f(x) = \sec \dfrac{\pi x}{4}$

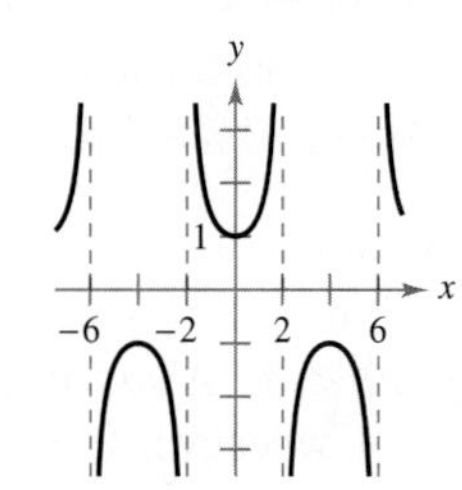

Numerical and Graphical Analysis **In Exercises 5–8, determine whether $f(x)$ approaches ∞ or $-\infty$ as x approaches -3 from the left and from the right by completing the table. Use a graphing utility to graph the function to confirm your answer.**

x	-3.5	-3.1	-3.01	-3.001
$f(x)$				

x	-2.999	-2.99	-2.9	-2.5
$f(x)$				

5. $f(x) = \dfrac{1}{x^2 - 9}$

6. $f(x) = \dfrac{x}{x^2 - 9}$

7. $f(x) = \dfrac{x^2}{x^2 - 9}$

8. $f(x) = \sec \dfrac{\pi x}{6}$

In Exercises 9–32, find the vertical asymptotes (if any) of the function.

9. $f(x) = \dfrac{1}{x^2}$

10. $f(x) = \dfrac{4}{(x - 2)^3}$

11. $h(x) = \dfrac{x^2 - 2}{x^2 - x - 2}$

12. $g(x) = \dfrac{2 + x}{x^2(1 - x)}$

13. $f(x) = \dfrac{x^2}{x^2 - 4}$

14. $f(x) = \dfrac{-4x}{x^2 + 4}$

15. $g(t) = \dfrac{t - 1}{t^2 + 1}$

16. $h(s) = \dfrac{2s - 3}{s^2 - 25}$

17. $f(x) = \tan 2x$

18. $f(x) = \sec \pi x$

19. $T(t) = 1 - \dfrac{4}{t^2}$

20. $g(x) = \dfrac{\frac{1}{2}x^3 - x^2 - 4x}{3x^2 - 6x - 24}$

21. $f(x) = \dfrac{x}{x^2 + x - 2}$

22. $f(x) = \dfrac{4x^2 + 4x - 24}{x^4 - 2x^3 - 9x^2 + 18x}$

23. $g(x) = \dfrac{x^3 + 1}{x + 1}$

24. $h(x) = \dfrac{x^2 - 4}{x^3 - 2x^2 + x - 2}$

25. $f(x) = \dfrac{e^{-2x}}{x - 1}$

26. $g(x) = xe^{-2x}$

27. $h(t) = \dfrac{\ln(t^2 + 1)}{t + 2}$

28. $f(z) = \ln(z^2 - 4)$

29. $f(x) = \dfrac{1}{e^x - 1}$

30. $f(x) = \ln(x + 3)$

31. $s(t) = \dfrac{t}{\sin t}$

32. $g(\theta) = \dfrac{\tan \theta}{\theta}$

In Exercises 33–38, determine whether the function has a vertical asymptote or a removable discontinuity at $x = -1$. Graph the function using a graphing utility to confirm your answer.

33. $f(x) = \dfrac{x^2 - 1}{x + 1}$

34. $f(x) = \dfrac{x^2 - 6x - 7}{x + 1}$

35. $f(x) = \dfrac{x^2 + 1}{x + 1}$

36. $f(x) = \dfrac{\sin(x + 1)}{x + 1}$

37. $f(x) = \dfrac{e^{2(x+1)} - 1}{e^{x+1} - 1}$

38. $f(x) = \dfrac{\ln(x^2 + 1)}{x + 1}$

In Exercises 39–54, find the limit.

39. $\displaystyle\lim_{x \to 2^+} \frac{x - 3}{x - 2}$

40. $\displaystyle\lim_{x \to 1^+} \frac{x(2 + x)}{1 - x}$

41. $\displaystyle\lim_{x \to 3^+} \frac{x^2}{x^2 - 9}$

42. $\displaystyle\lim_{x \to 4^-} \frac{x^2}{x^2 + 16}$

43. $\displaystyle\lim_{x \to -3^-} \frac{x^2 + 2x - 3}{x^2 + x - 6}$

44. $\displaystyle\lim_{x \to (-1/2)^+} \frac{6x^2 + x - 1}{4x^2 - 4x - 3}$

45. $\displaystyle\lim_{x \to 1} \frac{x^2 - x}{(x^2 + 1)(x - 1)}$

46. $\displaystyle\lim_{x \to 3} \frac{x - 2}{x^2}$

47. $\displaystyle\lim_{x \to 0^-} \left(1 + \frac{1}{x}\right)$

48. $\displaystyle\lim_{x \to 0^-} \left(x^2 - \frac{2}{x}\right)$

49. $\displaystyle\lim_{x \to 0^+} \frac{2}{\sin x}$

50. $\displaystyle\lim_{x \to (\pi/2)^+} \frac{-2}{\cos x}$

51. $\displaystyle\lim_{x \to (\pi/2)^-} \ln|\cos x|$

52. $\displaystyle\lim_{x \to 0+} e^{-0.5x} \sin x$

53. $\displaystyle\lim_{x \to 1/2} x \sec \pi x$

54. $\displaystyle\lim_{x \to 1/2} x^2 \tan \pi x$

In Exercises 55–58, use a graphing utility to graph the function and determine the one-sided limit.

55. $f(x) = \dfrac{x^2 + x + 1}{x^3 - 1}$

$\displaystyle\lim_{x \to 1^+} f(x)$

56. $f(x) = \dfrac{x^3 - 1}{x^2 + x + 1}$

$\displaystyle\lim_{x \to 1^-} f(x)$

57. $f(x) = \dfrac{1}{x^2 - 25}$

$\lim_{x \to 5^-} f(x)$

58. $f(x) = \sec \dfrac{\pi x}{6}$

$\lim_{x \to 3^+} f(x)$

Writing About Concepts

59. In your own words, describe the meaning of an infinite limit. Is ∞ a real number?

60. In your own words, describe what is meant by an asymptote of a graph.

61. Write a rational function with vertical asymptotes at $x = 6$ and $x = -2$, and with a zero at $x = 3$.

62. Does every rational function have a vertical asymptote? Explain.

63. Use the graph of the function f (see figure) to sketch the graph of $g(x) = 1/f(x)$ on the interval $[-2, 3]$. To print an enlarged copy of the graph, go to the website *www.mathgraphs.com*.

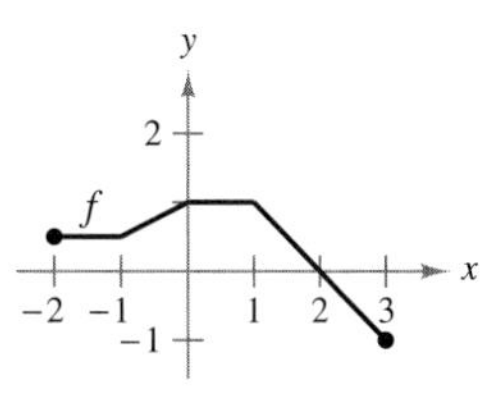

64. ***Boyle's Law*** For a quantity of gas at a constant temperature, the pressure P is inversely proportional to the volume V. Find the limit of P as $V \to 0^+$.

65. ***Rate of Change*** A patrol car is parked 50 feet from a long warehouse (see figure). The revolving light on top of the car turns at a rate of $\frac{1}{2}$ revolution per second. The rate r at which the light beam moves along the wall is

$r = 50\pi \sec^2 \theta$ ft/sec.

(a) Find r when θ is $\pi/6$.

(b) Find r when θ is $\pi/3$.

(c) Find the limit of r as $\theta \to (\pi/2)^-$.

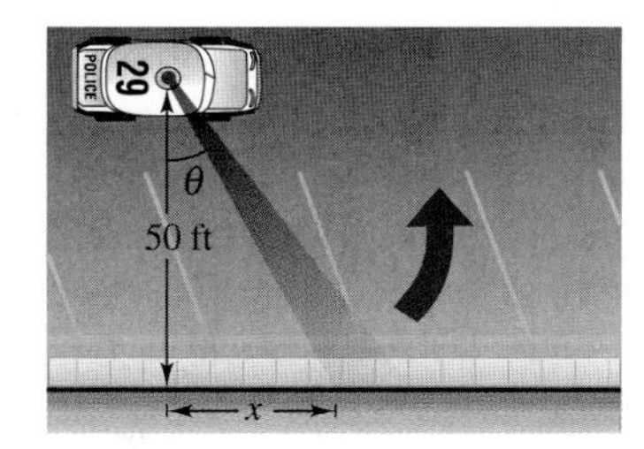

66. ***Illegal Drugs*** The cost in millions of dollars for a governmental agency to seize $x\%$ of an illegal drug is

$$C = \frac{528x}{100 - x}, \quad 0 \le x < 100.$$

(a) Find the cost of seizing 25% of the drug.

(b) Find the cost of seizing 50% of the drug.

(c) Find the cost of seizing 75% of the drug.

(d) Find the limit of C as $x \to 100^-$ and interpret its meaning.

67. ***Relativity*** According to the theory of relativity, the mass m of a particle depends on its velocity v. That is,

$$m = \frac{m_0}{\sqrt{1 - (v^2/c^2)}}$$

where m_0 is the mass when the particle is at rest and c is the speed of light. Find the limit of the mass as v approaches c^-.

68. ***Rate of Change*** A 25-foot ladder is leaning against a house (see figure). If the base of the ladder is pulled away from the house at a rate of 2 feet per second, the top will move down the wall at a rate r of

$$r = \frac{2x}{\sqrt{625 - x^2}} \text{ ft/sec}$$

where x is the distance between the ladder base and the house.

(a) Find r when x is 7 feet.

(b) Find r when x is 15 feet.

(c) Find the limit of r as $x \to 25^-$.

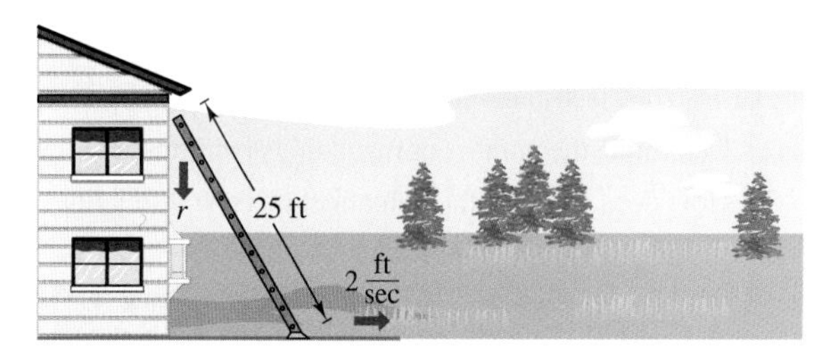

69. ***Average Speed*** On a trip of d miles to another city, a truck driver's average speed was x miles per hour. On the return trip, the average speed was y miles per hour. The average speed for the round trip was 50 miles per hour.

(a) Verify that $y = \dfrac{25x}{x - 25}$. What is the domain?

(b) Complete the table.

x	30	40	50	60
y				

Are the values of y different than you expected? Explain.

(c) Find the limit of y as $x \to 25^+$ and interpret its meaning.

70. ***Numerical and Graphical Analysis*** Use a graphing utility to complete the table for each function and graph each function to estimate the limit. What is the value of the limit when the power on x in the denominator is greater than 3?

x	1	0.5	0.2	0.1	0.01	0.001	0.0001
$f(x)$							

(a) $\lim_{x \to 0^+} \dfrac{x - \sin x}{x}$

(b) $\lim_{x \to 0^+} \dfrac{x - \sin x}{x^2}$

(c) $\lim_{x \to 0^+} \dfrac{x - \sin x}{x^3}$

(d) $\lim_{x \to 0^+} \dfrac{x - \sin x}{x^4}$

71. ***Numerical and Graphical Analysis*** Consider the shaded region outside the sector of a circle of radius 10 meters and inside a right triangle (see figure).

(a) Write the area $A = f(\theta)$ of the region as a function of θ. Determine the domain of the function.

(b) Use a graphing utility to complete the table and graph the function over the appropriate domain.

θ	0.3	0.6	0.9	1.2	1.5
$f(\theta)$					

(c) Find the limit of A as $\theta \to (\pi/2)^-$.

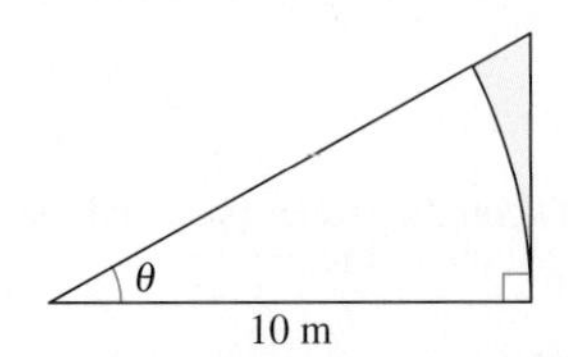

72. ***Numerical and Graphical Reasoning*** A crossed belt connects a 20-centimeter pulley (10-cm radius) on an electric motor with a 40-centimeter pulley (20-cm radius) on a saw arbor (see figure). The electric motor runs at 1700 revolutions per minute.

(a) Determine the number of revolutions per minute of the saw.

(b) How does crossing the belt affect the saw in relation to the motor?

(c) Let L be the total length of the belt. Write L as a function of ϕ, where ϕ is measured in radians. What is the domain of the function? (*Hint:* Add the lengths of the straight sections of the belt and the length of the belt around each pulley.)

(d) Use a graphing utility to complete the table.

ϕ	0.3	0.6	0.9	1.2	1.5
L					

(e) Use a graphing utility to graph the function over the appropriate domain.

(f) Find $\lim_{\phi \to (\pi/2)^-} L$. Use a geometric argument as the basis of a second method of finding this limit.

(g) Find $\lim_{\phi \to 0^+} L$.

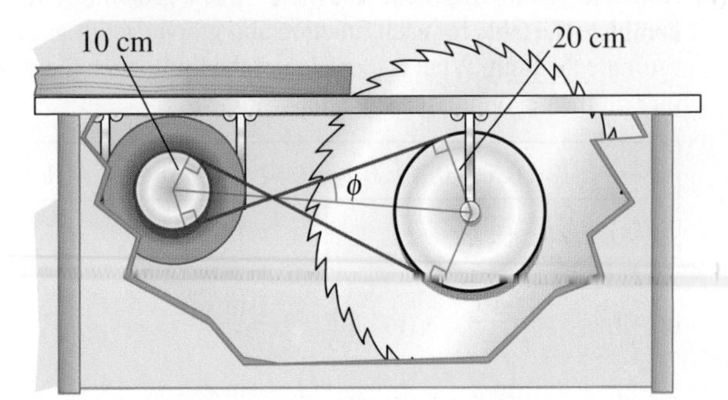

True or False? **In Exercises 73–76, determine whether the statement is true or false. If it is false, explain why or give an example that shows it is false.**

73. If $p(x)$ is a polynomial, then the graph of the function given by $f(x) = \dfrac{p(x)}{x-1}$ has a vertical asymptote at $x = 1$.

74. The graph of a rational function has at least one vertical asymptote.

75. The graphs of polynomial functions have no vertical asymptotes.

76. If f has a vertical asymptote at $x = 0$, then f is undefined at $x = 0$.

77. Find functions f and g such that $\lim_{x \to c} f(x) = \infty$ and $\lim_{x \to c} g(x) = \infty$, but $\lim_{x \to c} [f(x) - g(x)] \neq 0$.

78. Prove the remaining properties of Theorem 2.15.

79. Prove that if $\lim_{x \to c} f(x) = \infty$, then $\lim_{x \to c} \dfrac{1}{f(x)} = 0$.

80. Prove that if $\lim_{x \to c} \dfrac{1}{f(x)} = 0$, then $\lim_{x \to c} f(x)$ does not exist.

Infinite Limits **In Exercises 81 and 82, use the ε-δ definition of infinite limits to prove the statement.**

81. $\lim_{x \to 3^+} \dfrac{1}{x-3} = \infty$

82. $\lim_{x \to 4^-} \dfrac{1}{x-4} = -\infty$

Section Project: Graphs and Limits of Trigonometric Functions

Recall from Theorem 2.9 that the limit of $f(x) = (\sin x)/x$ as x approaches 0 is 1.

(a) Use a graphing utility to graph the function f on the interval $-\pi \leq x \leq \pi$. Explain how this graph helps confirm that $\lim_{x \to 0} \dfrac{\sin x}{x} = 1$.

(b) Explain how you could use a table of values to confirm the value of this limit numerically.

(c) Graph $g(x) = \sin x$ by hand. Sketch a tangent line at the point $(0, 0)$ and visually estimate the slope of this tangent line.

(d) Let $(x, \sin x)$ be a point on the graph of g near $(0, 0)$, and write a formula for the slope of the secant line joining $(x, \sin x)$ and $(0, 0)$. Evaluate this formula for $x = 0.1$ and $x = 0.01$. Then find the exact slope of the tangent line to g at the point $(0, 0)$.

(e) Sketch the graph of the cosine function $h(x) = \cos x$. What is the slope of the tangent line at the point $(0, 1)$? Use limits to find this slope analytically.

(f) Find the slope of the tangent line to $k(x) = \tan x$ at $(0, 0)$.

Review Exercises for Chapter 2

See www.CalcChat.com for worked-out solutions to odd-numbered exercises.

In Exercises 1 and 2, determine whether the problem can be solved using precalculus, or if calculus is required. If the problem can be solved using precalculus, solve it. If the problem seems to require calculus, explain your reasoning. Use a graphical or numerical approach to estimate the solution.

1. Find the distance between the points $(1, 1)$ and $(3, 9)$ along the curve $y = x^2$.

2. Find the distance between the points $(1, 1)$ and $(3, 9)$ along the line $y = 4x - 3$.

In Exercises 3–6, complete the table and use the result to estimate the limit. Use a graphing utility to graph the function to confirm your result.

x	-0.1	-0.01	-0.001	0.001	0.01	0.1
$f(x)$						

3. $\displaystyle\lim_{x\to 0} \frac{[4/(x+2)] - 2}{x}$

4. $\displaystyle\lim_{x\to 0} \frac{4\left(\sqrt{x+2} - \sqrt{2}\right)}{x}$

5. $\displaystyle\lim_{x\to 0} \frac{20(e^{x/2} - 1)}{x - 1}$

6. $\displaystyle\lim_{x\to 0} \frac{\ln(x+5) - \ln 5}{x}$

In Exercises 7–10, use the graph to determine each limit.

7. $h(x) = \dfrac{x^2 - 2x}{x}$

(a) $\displaystyle\lim_{x\to 0} h(x)$ (b) $\displaystyle\lim_{x\to -1} h(x)$

8. $g(x) = \dfrac{3x}{x - 2}$

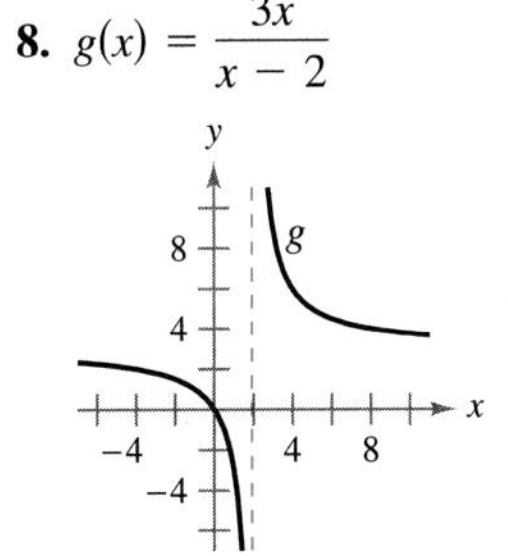

(a) $\displaystyle\lim_{x\to 2} g(x)$ (b) $\displaystyle\lim_{x\to 0} g(x)$

9. $f(t) = \dfrac{\ln(t + 2)}{t}$

(a) $\displaystyle\lim_{t\to 0} f(t)$ (b) $\displaystyle\lim_{t\to -1} f(t)$

10. $g(x) = e^{-x/2} \sin \pi x$

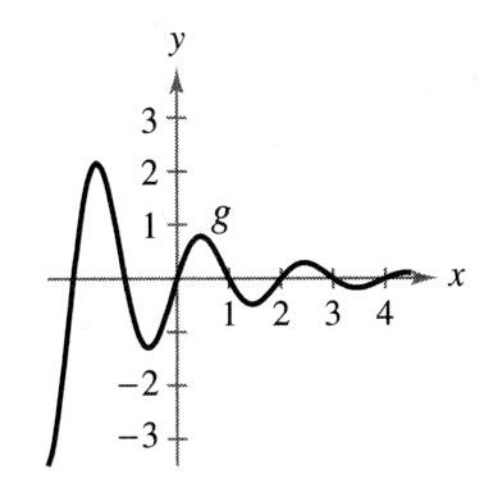

(a) $\displaystyle\lim_{x\to 0} g(x)$ (b) $\displaystyle\lim_{x\to 2} g(x)$

In Exercises 11–14, find the limit L. Then use the ε-δ definition to prove that the limit is L.

11. $\displaystyle\lim_{x\to 1} (3 - x)$

12. $\displaystyle\lim_{x\to 9} \sqrt{x}$

13. $\displaystyle\lim_{x\to 2} (x^2 - 3)$

14. $\displaystyle\lim_{x\to 5} 9$

In Exercises 15–30, find the limit (if it exists).

15. $\displaystyle\lim_{t\to 4} \sqrt{t + 2}$

16. $\displaystyle\lim_{y\to 4} 3|y - 1|$

17. $\displaystyle\lim_{t\to -2} \frac{t + 2}{t^2 - 4}$

18. $\displaystyle\lim_{t\to 3} \frac{t^2 - 9}{t - 3}$

19. $\displaystyle\lim_{x\to 4} \frac{\sqrt{x} - 2}{x - 4}$

20. $\displaystyle\lim_{x\to 0} \frac{\sqrt{4 + x} - 2}{x}$

21. $\displaystyle\lim_{x\to 0} \frac{[1/(x + 1)] - 1}{x}$

22. $\displaystyle\lim_{s\to 0} \frac{\left(1/\sqrt{1 + s}\right) - 1}{s}$

23. $\displaystyle\lim_{x\to -5} \frac{x^3 + 125}{x + 5}$

24. $\displaystyle\lim_{x\to -2} \frac{x^2 - 4}{x^3 + 8}$

25. $\displaystyle\lim_{x\to 0} \frac{1 - \cos x}{\sin x}$

26. $\displaystyle\lim_{x\to \pi/4} \frac{4x}{\tan x}$

27. $\displaystyle\lim_{\Delta x\to 0} \frac{\sin[(\pi/6) + \Delta x] - (1/2)}{\Delta x}$

[*Hint:* $\sin(\theta + \phi) = \sin\theta\cos\phi + \cos\theta\sin\phi$]

28. $\displaystyle\lim_{\Delta x\to 0} \frac{\cos(\pi + \Delta x) + 1}{\Delta x}$

[*Hint:* $\cos(\theta + \phi) = \cos\theta\cos\phi - \sin\theta\sin\phi$]

29. $\displaystyle\lim_{x\to 1} e^{x-1} \sin \frac{\pi x}{2}$

30. $\displaystyle\lim_{x\to 2} \frac{\ln(x - 1)^2}{\ln(x - 1)}$

In Exercises 31 and 32, evaluate the limit given that $\lim_{x\to c} f(x) = -\frac{3}{4}$ and $\lim_{x\to c} g(x) = \frac{2}{3}$.

31. $\displaystyle\lim_{x\to c} [f(x)g(x)]$

32. $\displaystyle\lim_{x\to c} [f(x) + 2g(x)]$

Numerical, Graphical, and Analytic Analysis **In Exercises 33 and 34, consider**

$$\lim_{x\to 1^+} f(x).$$

(a) Complete the table to estimate the limit.

(b) Use a graphing utility to graph the function and use the graph to estimate the limit.

(c) Rationalize the numerator to find the exact value of the limit analytically.

x	1.1	1.01	1.001	1.0001
$f(x)$				

33. $f(x) = \dfrac{\sqrt{2x + 1} - \sqrt{3}}{x - 1}$

34. $f(x) = \dfrac{1 - \sqrt[3]{x}}{x - 1}$

[*Hint:* $a^3 - b^3 = (a - b)(a^2 + ab + b^2)$]

Free-Falling Object **In Exercises 35 and 36, use the position function**

$$s(t) = -4.9t^2 + 200$$

which gives the height (in meters) of an object that has fallen from a height of 200 meters. The velocity at time $t = a$ seconds is given by

$$\lim_{t \to a} \frac{s(a) - s(t)}{a - t}.$$

35. Find the velocity of the object when $t = 4$.

36. At what velocity will the object impact the ground?

In Exercises 37–42, find the limit (if it exists). If the limit does not exist, explain why.

37. $\lim_{x \to 3^-} \frac{|x - 3|}{x - 3}$

38. $\lim_{x \to 4} [\![x - 1]\!]$

39. $\lim_{x \to 2} f(x)$, where $f(x) = \begin{cases} (x - 2)^2, & x \le 2 \\ 2 - x, & x > 2 \end{cases}$

40. $\lim_{x \to 1^+} g(x)$, where $g(x) = \begin{cases} \sqrt{1 - x}, & x \le 1 \\ x + 1, & x > 1 \end{cases}$

41. $\lim_{t \to 1} h(t)$, where $h(t) = \begin{cases} t^3 + 1, & t < 1 \\ \frac{1}{2}(t + 1), & t \ge 1 \end{cases}$

42. $\lim_{s \to -2} f(s)$, where $f(s) = \begin{cases} -s^2 - 4s - 2, & s \le -2 \\ s^2 + 4s + 6, & s > -2 \end{cases}$

In Exercises 43–54, determine the intervals on which the function is continuous.

43. $f(x) = [\![x + 3]\!]$

44. $f(x) = \frac{3x^2 - x - 2}{x - 1}$

45. $f(x) = \begin{cases} \frac{3x^2 - x - 2}{x - 1}, & x \ne 1 \\ 0, & x = 1 \end{cases}$

46. $f(x) = \begin{cases} 5 - x, & x \le 2 \\ 2x - 3, & x > 2 \end{cases}$

47. $f(x) = \frac{1}{(x - 2)^2}$

48. $f(x) = \sqrt{\frac{x + 1}{x}}$

49. $f(x) = \frac{3}{x + 1}$

50. $f(x) = \frac{x + 1}{2x + 2}$

51. $f(x) = \csc \frac{\pi x}{2}$

52. $f(x) = \tan 2x$

53. $g(x) = 2e^{[\![x]\!]/4}$

54. $h(x) = 5 \ln|x - 3|$

55. Determine the value of c such that the function is continuous on the entire real number line.

$$f(x) = \begin{cases} x + 3, & x \le 2 \\ cx + 6, & x > 2 \end{cases}$$

56. Determine the values of b and c such that the function is continuous on the entire real number line.

$$f(x) = \begin{cases} x + 1, & 1 < x < 3 \\ x^2 + bx + c, & |x - 2| \ge 1 \end{cases}$$

57. Use the Intermediate Value Theorem to show that

$$f(x) = 2x^3 - 3$$

has a zero in the interval $[1, 2]$.

58. ***Delivery Charges*** The cost of sending an overnight package from New York to Atlanta is \$9.80 for the first pound and \$2.50 for each additional pound or fraction thereof. Use the greatest integer function to create a model for the cost C of overnight delivery of a package weighing x pounds. Use a graphing utility to graph the function and discuss its continuity.

59. ***Compound Interest*** A sum of \$5000 is deposited in a savings plan that pays 12% interest compounded semiannually. The account balance after t years is given by $A = 5000(1.06)^{[\![2t]\!]}$. Use a graphing utility to graph the function and discuss its continuity.

60. Let $f(x) = \sqrt{x(x - 1)}$.

(a) Find the domain of f.

(b) Find $\lim_{x \to 0^-} f(x)$.

(c) Find $\lim_{x \to 1^+} f(x)$.

In Exercises 61–66, find the vertical asymptotes (if any) of the function.

61. $g(x) = 1 + \frac{2}{x}$

62. $h(x) = \frac{4x}{4 - x^2}$

63. $f(x) = \frac{8}{(x - 10)^2}$

64. $f(x) = \csc \pi x$

65. $g(x) = \ln(9 - x^2)$

66. $f(x) = 10e^{-2/x}$

In Exercises 67–78, find the one-sided limit.

67. $\lim_{x \to -2^-} \frac{2x^2 + x + 1}{x + 2}$

68. $\lim_{x \to (1/2)^+} \frac{x}{2x - 1}$

69. $\lim_{x \to -1^+} \frac{x + 1}{x^3 + 1}$

70. $\lim_{x \to -1^-} \frac{x + 1}{x^4 - 1}$

71. $\lim_{x \to 1^-} \frac{x^2 + 2x + 1}{x - 1}$

72. $\lim_{x \to -1^+} \frac{x^2 - 2x + 1}{x + 1}$

73. $\lim_{x \to 0^+} \frac{\sin 4x}{5x}$

74. $\lim_{x \to 0^+} \frac{\sec x}{x}$

75. $\lim_{x \to 0^+} \frac{\csc 2x}{x}$

76. $\lim_{x \to 0^-} \frac{\cos^2 x}{x}$

77. $\lim_{x \to 0^+} \ln(\sin x)$

78. $\lim_{x \to 0^-} 12e^{-2/x}$

79. The function f is defined as follows.

$$f(x) = \frac{\tan 2x}{x}, \quad x \ne 0$$

(a) Find $\lim_{x \to 0} \frac{\tan 2x}{x}$ (if it exists).

(b) Can the function f be defined at $x = 0$ such that it is continuous at $x = 0$?

P.S. Problem Solving

See www.CalcChat.com for worked-out solutions to odd-numbered exercises.

1. Let $P(x, y)$ be a point on the parabola $y = x^2$ in the first quadrant. Consider the triangle $\triangle PAO$ formed by P, $A(0, 1)$, and the origin $O(0, 0)$, and the triangle $\triangle PBO$ formed by P, $B(1, 0)$, and the origin.

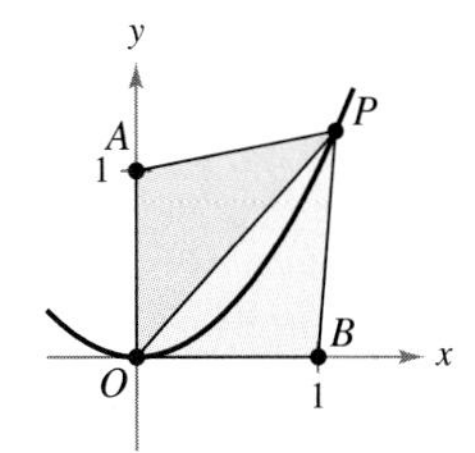

(a) Write the perimeter of each triangle in terms of x.

(b) Let $r(x)$ be the ratio of the perimeters of the two triangles,

$$r(x) = \frac{\text{Perimeter } \triangle PAO}{\text{Perimeter } \triangle PBO}.$$

Complete the table.

x	4	2	1	0.1	0.01
Perimeter $\triangle PAO$					
Perimeter $\triangle PBO$					
$r(x)$					

(c) Calculate $\lim_{x \to 0^+} r(x)$.

2. Let $P(x, y)$ be a point on the parabola $y = x^2$ in the first quadrant. Consider the triangle $\triangle PAO$ formed by P, $A(0, 1)$, and the origin $O(0, 0)$, and the triangle $\triangle PBO$ formed by P, $B(1, 0)$, and the origin.

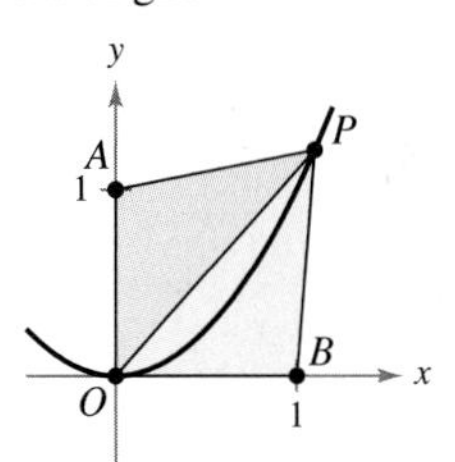

(a) Write the area of each triangle in terms of x.

(b) Let $a(x)$ be the ratio of the areas of the two triangles,

$$a(x) = \frac{\text{Area } \triangle PBO}{\text{Area } \triangle PAO}.$$

Complete the table.

x	4	2	1	0.1	0.01
Area $\triangle PAO$					
Area $\triangle PBO$					
$a(x)$					

(c) Calculate $\lim_{x \to 0^+} a(x)$.

3. (a) Find the area of a regular hexagon inscribed in a circle of radius 1. How close is this area to that of the circle?

(b) Find the area A_n of an n-sided regular polygon inscribed in a circle of radius 1. Write your answer as a function of n.

(c) Complete the table.

n	6	12	24	48	96
A_n					

(d) What number does A_n approach as n gets larger and larger?

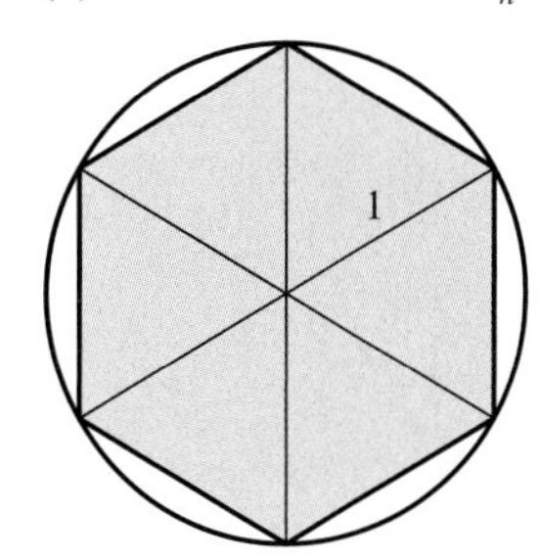

Figure for 3

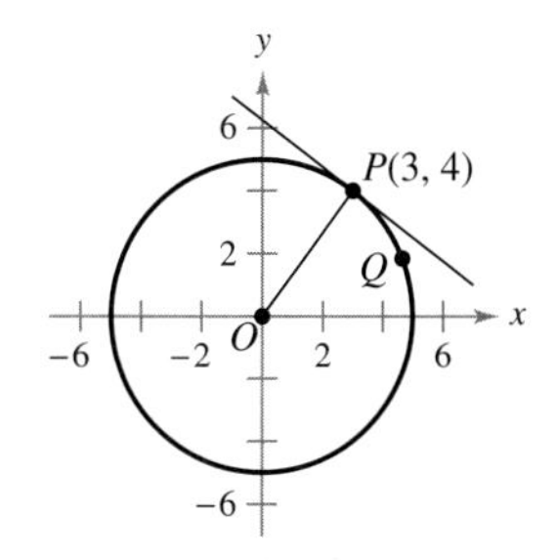

Figure for 4

4. Let $P(3, 4)$ be a point on the circle $x^2 + y^2 = 25$.

(a) What is the slope of the line joining P and $O(0, 0)$?

(b) Find an equation of the tangent line to the circle at P.

(c) Let $Q(x, y)$ be another point on the circle in the first quadrant. Find the slope m_x of the line joining P and Q in terms of x.

(d) Calculate $\lim_{x \to 3} m_x$. How does this number relate to your answer in part (b)?

5. Let $P(5, -12)$ be a point on the circle $x^2 + y^2 = 169$.

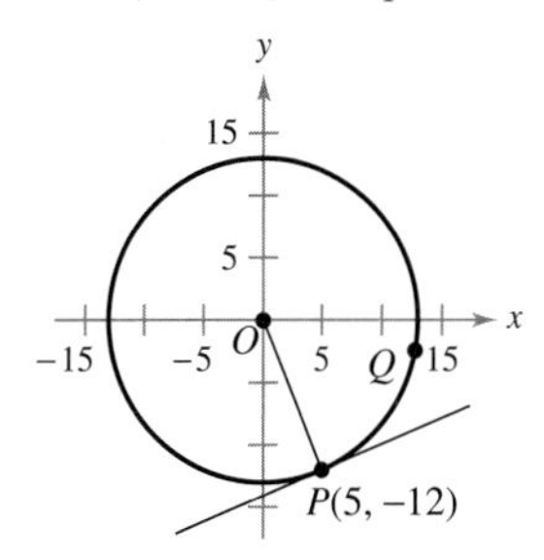

(a) What is the slope of the line joining P and $O(0, 0)$?

(b) Find an equation of the tangent line to the circle at P.

(c) Let $Q(x, y)$ be another point on the circle in the fourth quadrant. Find the slope m_x of the line joining P and Q in terms of x.

(d) Calculate $\lim_{x \to 5} m_x$. How does this number relate to your answer in part (b)?

6. Find the values of the constants a and b such that

$$\lim_{x \to 0} \frac{\sqrt{a + bx} - \sqrt{3}}{x} = \sqrt{3}.$$

7. Consider the function $f(x) = \dfrac{\sqrt{3 + x^{1/3}} - 2}{x - 1}$.

(a) Find the domain of f.

(b) Use a graphing utility to graph the function.

(c) Calculate $\lim_{x\to -27^+} f(x)$.

(d) Calculate $\lim_{x\to 1} f(x)$.

8. Determine all values of the constant a such that the following function is continuous for all real numbers.

$$f(x) = \begin{cases} \dfrac{ax}{\tan x}, & x \geq 0 \\ a^2 - 2, & x < 0 \end{cases}$$

9. Consider the graphs of the four functions g_1, g_2, g_3, and g_4.

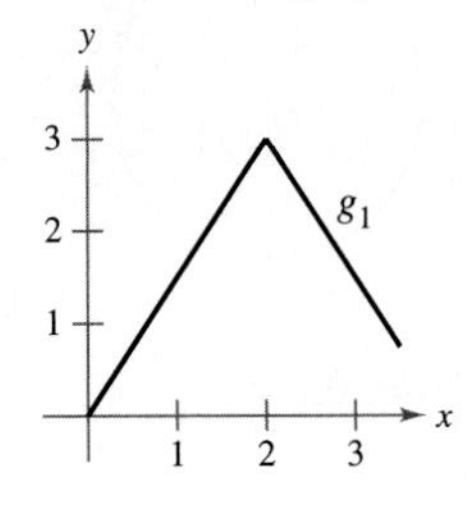

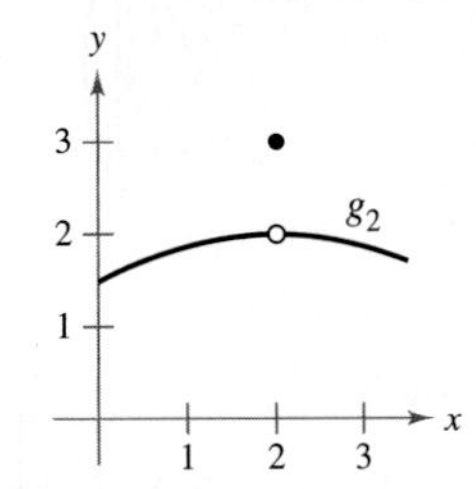

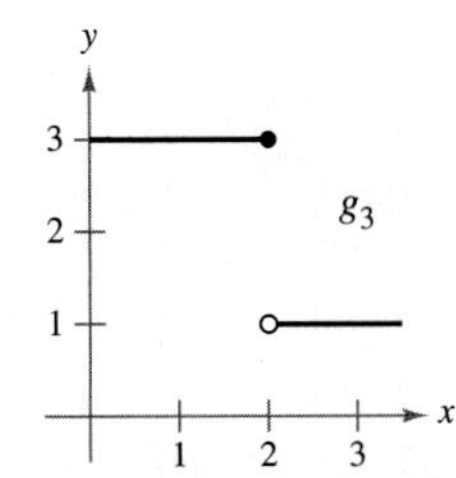

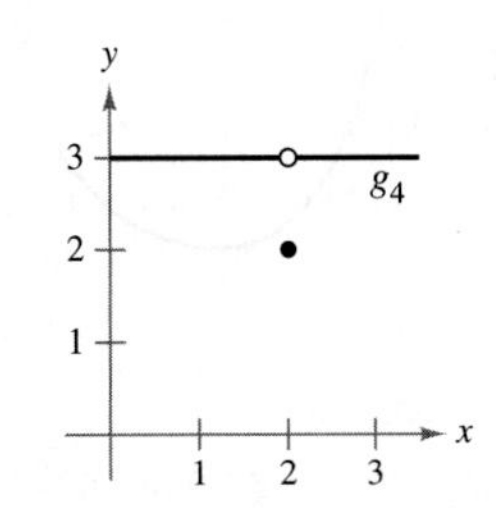

For each given condition of the function f, which of the graphs could be the graph of f?

(a) $\lim_{x\to 2} f(x) = 3$

(b) f is continuous at 2.

(c) $\lim_{x\to 2^-} f(x) = 3$

10. Sketch the graph of the function $f(x) = \left[\!\left[\dfrac{1}{x} \right]\!\right]$.

(a) Evaluate $f\left(\frac{1}{4}\right)$, $f(3)$, and $f(1)$.

(b) Evaluate the limits $\lim_{x\to 1^-} f(x)$, $\lim_{x\to 1^+} f(x)$, $\lim_{x\to 0^-} f(x)$, and $\lim_{x\to 0^+} f(x)$.

(c) Discuss the continuity of the function.

11. Sketch the graph of the function $f(x) = [\![x]\!] + [\![-x]\!]$.

(a) Evaluate $f(1)$, $f(0)$, $f\left(\frac{1}{2}\right)$, and $f(-2.7)$.

(b) Evaluate the limits $\lim_{x\to 1^-} f(x)$, $\lim_{x\to 1^+} f(x)$, and $\lim_{x\to \frac{1}{2}} f(x)$.

(c) Discuss the continuity of the function.

12. To escape Earth's gravitational field, a rocket must be launched with an initial velocity called the **escape velocity.** A rocket launched from the surface of Earth has velocity v (in miles per second) given by

$$v = \sqrt{\frac{2GM}{r} + v_0^2 - \frac{2GM}{R}} \approx \sqrt{\frac{192{,}000}{r} + v_0^2 - 48}$$

where v_0 is the initial velocity, r is the distance from the rocket to the center of Earth, G is the gravitational constant, M is the mass of Earth, and R is the radius of Earth (approximately 4000 miles).

(a) Find the value of v_0 for which you obtain an infinite limit for r as v tends to zero. This value of v_0 is the escape velocity for Earth.

(b) A rocket launched from the surface of the moon has velocity v (in miles per second) given by

$$v = \sqrt{\frac{1920}{r} + v_0^2 - 2.17}.$$

Find the escape velocity for the moon.

(c) A rocket launched from the surface of a planet has velocity v (in miles per second) given by

$$v = \sqrt{\frac{10{,}600}{r} + v_0^2 - 6.99}.$$

Find the escape velocity for this planet. Is the mass of this planet larger or smaller than that of Earth? (Assume that the mean density of this planet is the same as that of Earth.)

13. For positive numbers $a < b$, the **pulse function** is defined as

$$P_{a,b}(x) = H(x - a) - H(x - b) = \begin{cases} 0, & x < a \\ 1, & a \leq x < b \\ 0, & x \geq b \end{cases}$$

where $H(x) = \begin{cases} 1, & x \geq 0 \\ 0, & x < 0 \end{cases}$ is the Heaviside function.

(a) Sketch the graph of the pulse function.

(b) Find the following limits:

(i) $\lim_{x\to a^+} P_{a,b}(x)$ (ii) $\lim_{x\to a^-} P_{a,b}(x)$

(iii) $\lim_{x\to b^+} P_{a,b}(x)$ (iv) $\lim_{x\to b^-} P_{a,b}(x)$

(c) Discuss the continuity of the pulse function.

(d) Why is

$$U(x) = \frac{1}{b - a} P_{a,b}(x)$$

called the **unit** pulse function?

14. Let a be a nonzero constant. Prove that if $\lim_{x\to 0} f(x) = L$, then $\lim_{x\to 0} f(ax) = L$. Show by means of an example that a must be nonzero.

3 Differentiation

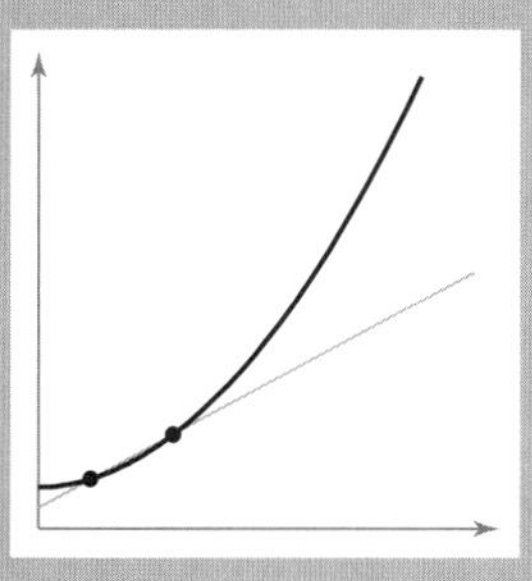

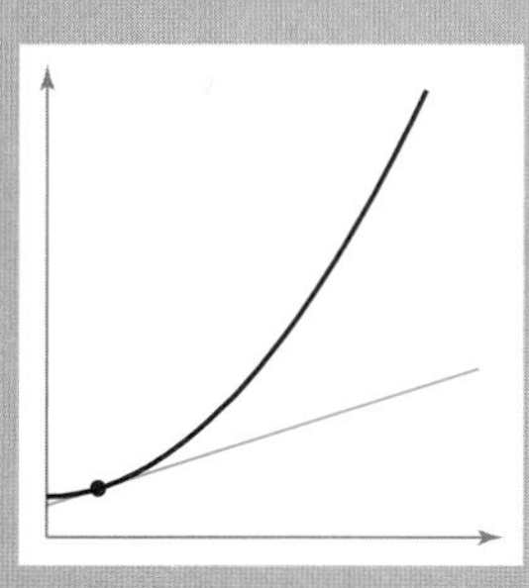

To approximate the slope of a tangent line to a graph at a given point, find the slope of the secant line through the given point and a second point on the graph. As the second point approaches the given point, the approximation tends to become more accurate. In Section 3.1, you will use limits to find slopes of tangent lines to graphs. This process is called *differentiation*.

You pump air at a steady rate into a deflated balloon until the balloon bursts. Does the diameter of the balloon change faster when you first start pumping the air, or just before the balloon bursts? Why?

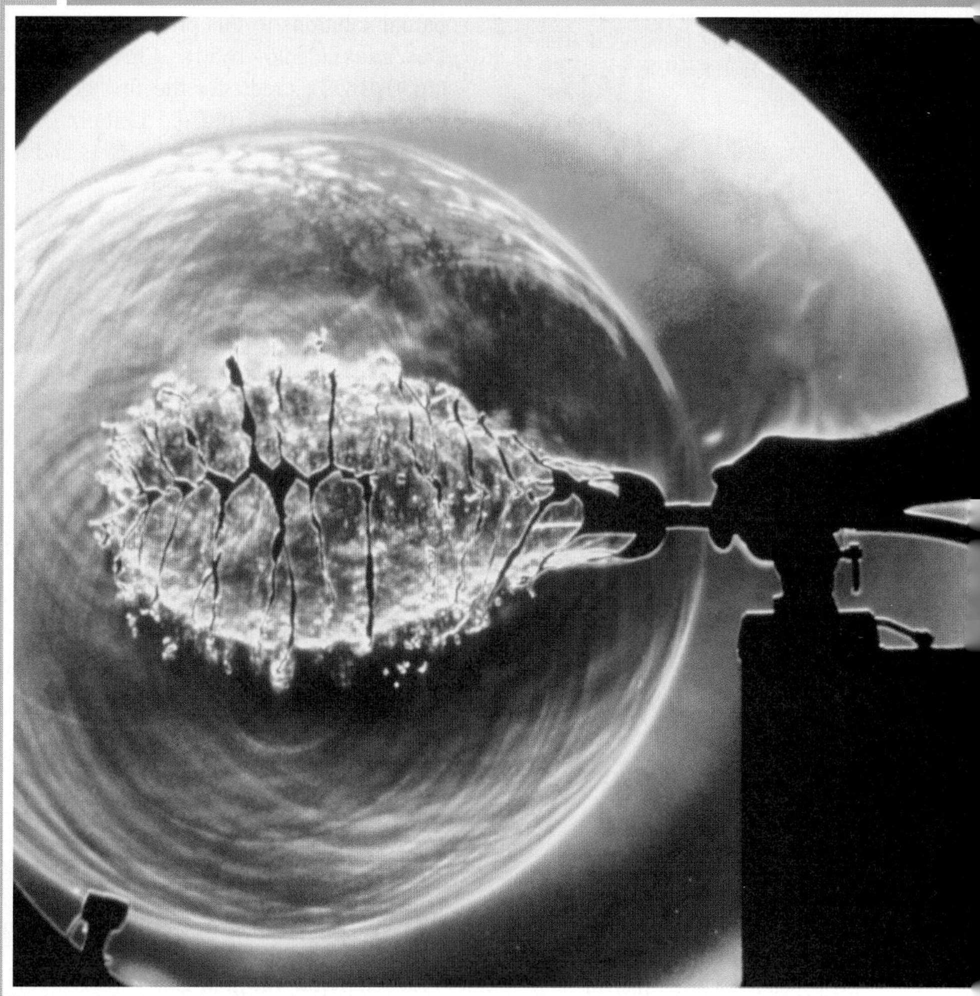

Dr. Gary Settles/SPL/Photo Researchers

Section 3.1 The Derivative and the Tangent Line Problem

- **Find the slope of the tangent line to a curve at a point.**
- **Use the limit definition to find the derivative of a function.**
- **Understand the relationship between differentiability and continuity.**

ISAAC NEWTON (1642–1727)
In addition to his work in calculus, Newton made revolutionary contributions to physics, including the Law of Universal Gravitation and his three laws of motion.

The Tangent Line Problem

Calculus grew out of four major problems that European mathematicians were working on during the seventeenth century.

1. The tangent line problem (Section 2.1 and this section)
2. The velocity and acceleration problem (Sections 3.2 and 3.3)
3. The minimum and maximum problem (Section 4.1)
4. The area problem (Sections 2.1 and 5.2)

Each problem involves the notion of a limit, and calculus can be introduced with any of the four problems.

A brief introduction to the tangent line problem is given in Section 2.1. Although partial solutions to this problem were given by Pierre de Fermat (1601–1665), René Descartes (1596–1650), Christian Huygens (1629–1695), and Isaac Barrow (1630–1677), credit for the first general solution is usually given to Isaac Newton (1642–1727) and Gottfried Leibniz (1646–1716). Newton's work on this problem stemmed from his interest in optics and light refraction.

What does it mean to say that a line is tangent to a curve at a point? For a circle, the tangent line at a point P is the line that is perpendicular to the radial line at point P, as shown in Figure 3.1.

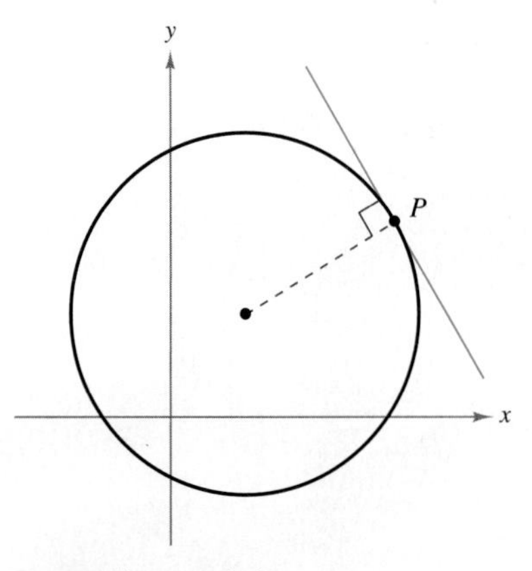

Tangent line to a circle
Figure 3.1

For a general curve, however, the problem is more difficult. For example, how would you define the tangent lines shown in Figure 3.2? You might say that a line is tangent to a curve at a point P if it touches, but does not cross, the curve at point P. This definition would work for the first curve shown in Figure 3.2, but not for the second. *Or* you might say that a line is tangent to a curve if the line touches or intersects the curve at exactly one point. This definition would work for a circle but not for more general curves, as the third curve in Figure 3.2 shows.

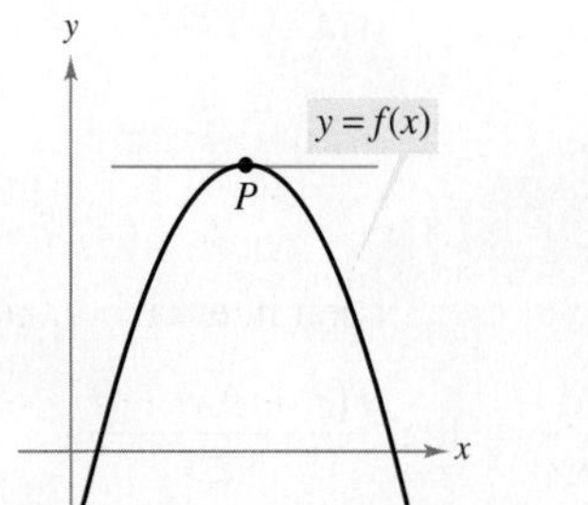

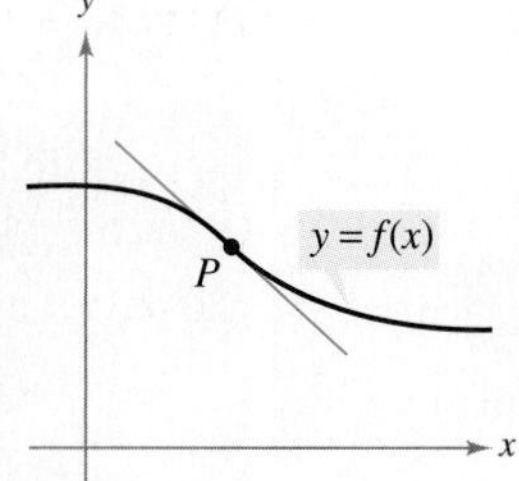

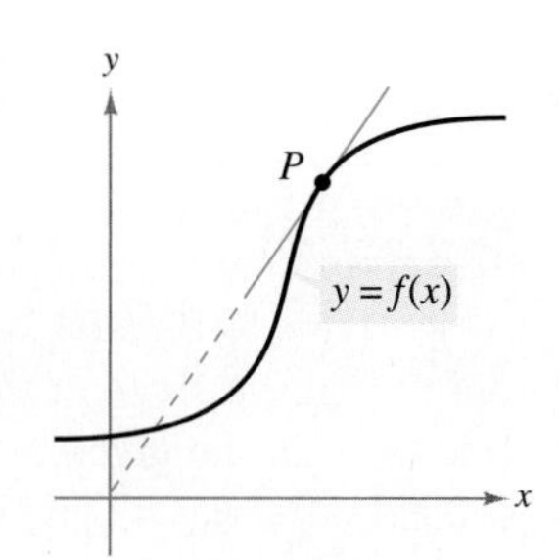

Tangent line to a curve at a point
Figure 3.2

For Further Information For more information on the crediting of mathematical discoveries to the first "discoverer," see the article "Mathematical Firsts—Who Done It?" by Richard H. Williams and Roy D. Mazzagatti in *Mathematics Teacher*. To view this article, go to the website *www.matharticles.com*.

EXPLORATION

Identifying a Tangent Line Use a graphing utility to graph the function $f(x) = 2x^3 - 4x^2 + 3x - 5$. On the same screen, graph $y = x - 5$, $y = 2x - 5$, and $y = 3x - 5$. Which of these lines, if any, appears to be tangent to the graph of f at the point $(0, -5)$? Explain your reasoning.

Essentially, the problem of finding the tangent line at a point P boils down to the problem of finding the *slope* of the tangent line at point P. You can approximate this slope using a **secant line*** through the point of tangency and a second point on the curve, as shown in Figure 3.3. If $(c, f(c))$ is the point of tangency and $(c + \Delta x, f(c + \Delta x))$ is a second point on the graph of f, the slope of the secant line through the two points is given by substitution into the slope formula

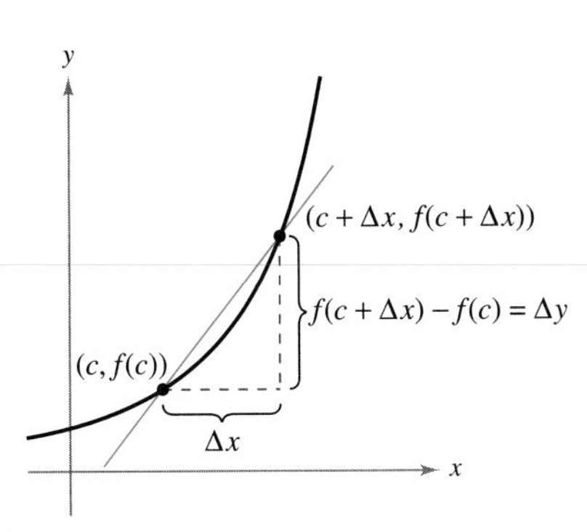

The secant line through $(c, f(c))$ and $(c + \Delta x, f(c + \Delta x))$
Figure 3.3

$$m = \frac{y_2 - y_1}{x_2 - x_1}$$

$$m_{\text{sec}} = \frac{f(c + \Delta x) - f(c)}{(c + \Delta x) - c} \qquad \frac{\text{Change in } y}{\text{Change in } x}$$

$$m_{\text{sec}} = \frac{f(c + \Delta x) - f(c)}{\Delta x}. \qquad \text{Slope of secant line}$$

The right-hand side of this equation is a **difference quotient.** The denominator Δx is the **change in x,** and the numerator $\Delta y = f(c + \Delta x) - f(c)$ is the **change in y.**

The beauty of this procedure is that you can obtain more and more accurate approximations of the slope of the tangent line by choosing points closer and closer to the point of tangency, as shown in Figure 3.4.

The Tangent Line Problem

In 1637, mathematician René Descartes stated this about the tangent line problem:

"And I dare say that this is not only the most useful and general problem in geometry that I know, but even that I ever desire to know."

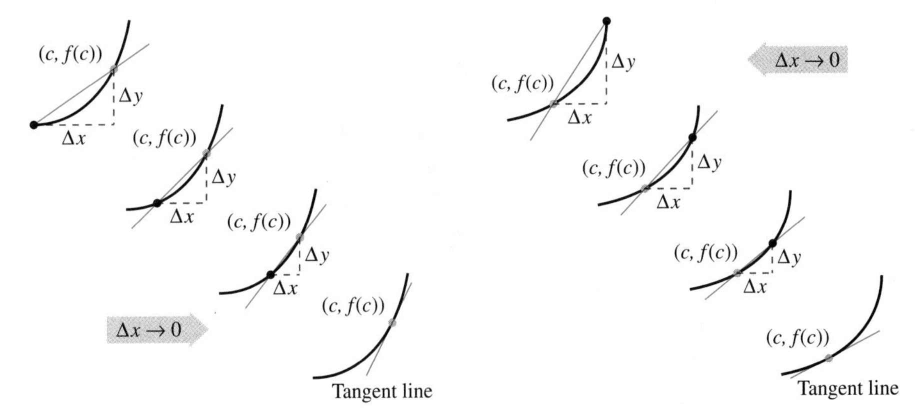

Tangent line approximations
Figure 3.4

Definition of Tangent Line with Slope m

If f is defined on an open interval containing c, and if the limit

$$\lim_{\Delta x \to 0} \frac{\Delta y}{\Delta x} = \lim_{\Delta x \to 0} \frac{f(c + \Delta x) - f(c)}{\Delta x} = m$$

exists, then the line passing through $(c, f(c))$ with slope m is the **tangent line** to the graph of f at the point $(c, f(c))$.

The slope of the tangent line to the graph of f at the point $(c, f(c))$ is also called the **slope of the graph of f at $x = c$.**

---

* *This use of the word* secant *comes from the Latin* secare, *meaning to cut, and is not a reference to the trigonometric function of the same name.*

EXAMPLE 1 The Slope of the Graph of a Linear Function

Find the slope of the graph of

$$f(x) = 2x - 3$$

at the point $(2, 1)$.

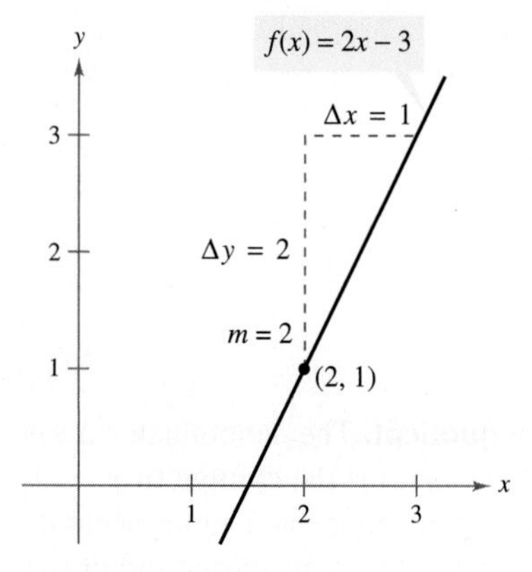

The slope of f at $(2, 1)$ is $m = 2$.
Figure 3.5

Solution To find the slope of the graph of f when $c = 2$, you can apply the definition of the slope of a tangent line, as shown.

$$\begin{aligned}
\lim_{\Delta x \to 0} \frac{f(2 + \Delta x) - f(2)}{\Delta x} &= \lim_{\Delta x \to 0} \frac{[2(2 + \Delta x) - 3] - [2(2) - 3]}{\Delta x} \\
&= \lim_{\Delta x \to 0} \frac{4 + 2\Delta x - 3 - 4 + 3}{\Delta x} \\
&= \lim_{\Delta x \to 0} \frac{2\cancel{\Delta x}}{\cancel{\Delta x}} \\
&= \lim_{\Delta x \to 0} 2 \\
&= 2
\end{aligned}$$

The slope of f at $(c, f(c)) = (2, 1)$ is $m = 2$, as shown in Figure 3.5.

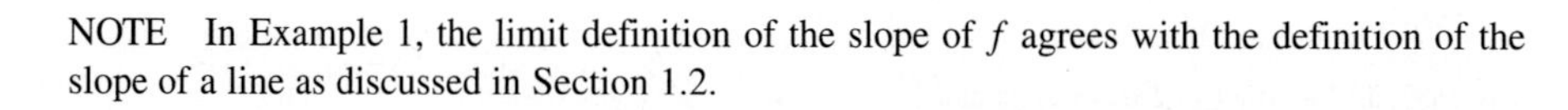

NOTE In Example 1, the limit definition of the slope of f agrees with the definition of the slope of a line as discussed in Section 1.2.

The graph of a linear function has the same slope at any point. This is not true of nonlinear functions, as shown in the following example.

EXAMPLE 2 Tangent Lines to the Graph of a Nonlinear Function

Find the slopes of the tangent lines to the graph of

$$f(x) = x^2 + 1$$

at the points $(0, 1)$ and $(-1, 2)$, as shown in Figure 3.6.

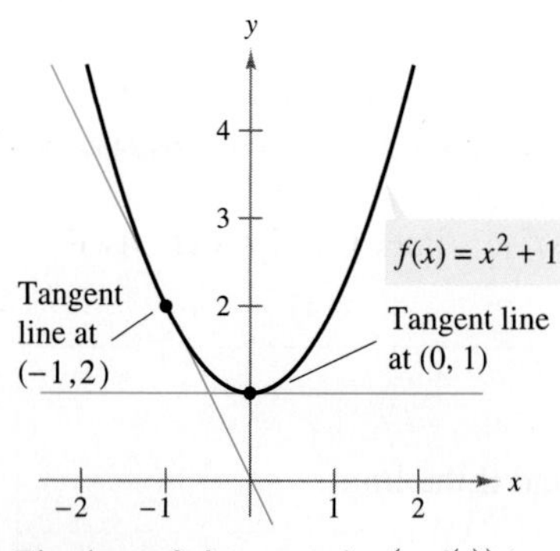

The slope of f at any point $(c, f(c))$ is $m = 2c$.
Figure 3.6

Solution Let $(c, f(c))$ represent an arbitrary point on the graph of f. Then the slope of the tangent line at $(c, f(c))$ is given by

$$\begin{aligned}
\lim_{\Delta x \to 0} \frac{f(c + \Delta x) - f(c)}{\Delta x} &= \lim_{\Delta x \to 0} \frac{[(c + \Delta x)^2 + 1] - (c^2 + 1)}{\Delta x} \\
&= \lim_{\Delta x \to 0} \frac{c^2 + 2c(\Delta x) + (\Delta x)^2 + 1 - c^2 - 1}{\Delta x} \\
&= \lim_{\Delta x \to 0} \frac{2c(\Delta x) + (\Delta x)^2}{\Delta x} \\
&= \lim_{\Delta x \to 0} (2c + \Delta x) \\
&= 2c.
\end{aligned}$$

So, the slope at *any* point $(c, f(c))$ on the graph of f is $m = 2c$. At the point $(0, 1)$, the slope is $m = 2(0) = 0$, and at $(-1, 2)$, the slope is $m = 2(-1) = -2$.

NOTE In Example 2, note that c is held constant in the limit process (as $\Delta x \to 0$).

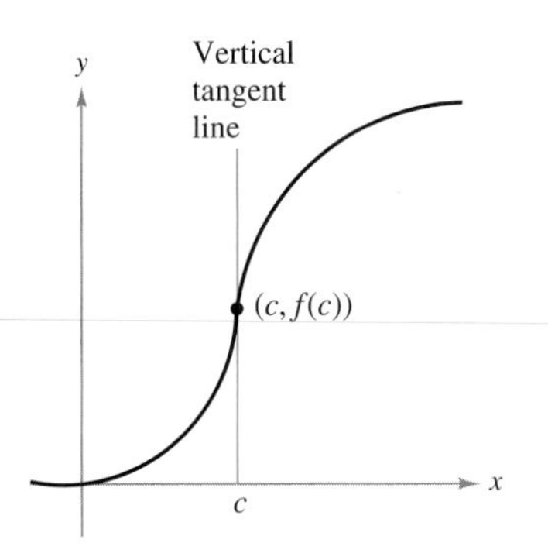

The graph of f has a vertical tangent line at $(c, f(c))$.
Figure 3.7

The definition of a tangent line to a curve does not cover the possibility of a vertical tangent line. For vertical tangent lines, you can use the following definition. If f is continuous at c and

$$\lim_{\Delta x \to 0} \frac{f(c + \Delta x) - f(c)}{\Delta x} = \infty \quad \text{or} \quad \lim_{\Delta x \to 0} \frac{f(c + \Delta x) - f(c)}{\Delta x} = -\infty$$

the vertical line $x = c$ passing through $(c, f(c))$ is a **vertical tangent line** to the graph of f. For example, the function shown in Figure 3.7 has a vertical tangent line at $(c, f(c))$. If the domain of f is the closed interval $[a, b]$, you can extend the definition of a vertical tangent line to include the endpoints by considering continuity and limits from the right (for $x = a$) and from the left (for $x = b$).

The Derivative of a Function

You have now arrived at a crucial point in the study of calculus. The limit used to define the slope of a tangent line is also used to define one of the two fundamental operations of calculus—**differentiation.**

Definition of the Derivative of a Function

The **derivative** of f at x is given by

$$f'(x) = \lim_{\Delta x \to 0} \frac{f(x + \Delta x) - f(x)}{\Delta x}$$

provided the limit exists. For all x for which this limit exists, f' is a function of x.

Be sure you see that the derivative of a function of x is also a function of x. This "new" function gives the slope of the tangent line to the graph of f at the point $(x, f(x))$, provided that the graph has a tangent line at this point.

The process of finding the derivative of a function is called **differentiation.** A function is **differentiable** at x if its derivative exists at x and is **differentiable on an open interval (a, b)** if it is differentiable at every point in the interval.

In addition to $f'(x)$, which is read as "f prime of x," other notations are used to denote the derivative of $y = f(x)$. The most common are

$$f'(x), \quad \frac{dy}{dx}, \quad y', \quad \frac{d}{dx}[f(x)], \quad D_x[y]. \qquad \text{Notation for derivatives}$$

The notation dy/dx is read as "the derivative of y *with respect to* x." Using limit notation, you can write

$$\begin{aligned} \frac{dy}{dx} &= \lim_{\Delta x \to 0} \frac{\Delta y}{\Delta x} \\ &= \lim_{\Delta x \to 0} \frac{f(x + \Delta x) - f(x)}{\Delta x} \\ &= f'(x). \end{aligned}$$

EXAMPLE 3 Finding the Derivative by the Limit Process

Find the derivative of $f(x) = x^3 + 2x$.

Solution

$$
\begin{aligned}
f'(x) &= \lim_{\Delta x \to 0} \frac{f(x + \Delta x) - f(x)}{\Delta x} && \text{Definition of derivative} \\
&= \lim_{\Delta x \to 0} \frac{(x + \Delta x)^3 + 2(x + \Delta x) - (x^3 + 2x)}{\Delta x} \\
&= \lim_{\Delta x \to 0} \frac{x^3 + 3x^2\Delta x + 3x(\Delta x)^2 + (\Delta x)^3 + 2x + 2\Delta x - x^3 - 2x}{\Delta x} \\
&= \lim_{\Delta x \to 0} \frac{3x^2\Delta x + 3x(\Delta x)^2 + (\Delta x)^3 + 2\Delta x}{\Delta x} \\
&= \lim_{\Delta x \to 0} \frac{\cancel{\Delta x}[3x^2 + 3x\Delta x + (\Delta x)^2 + 2]}{\cancel{\Delta x}} \\
&= \lim_{\Delta x \to 0} [3x^2 + 3x\Delta x + (\Delta x)^2 + 2] \\
&= 3x^2 + 2
\end{aligned}
$$

STUDY TIP When using the definition to find a derivative of a function, the key is to rewrite the difference quotient so that Δx does not occur as a factor of the denominator.

Remember that the derivative of a function f is itself a function, which can be used to find the slope of the tangent line at the point $(x, f(x))$ on the graph of f.

EXAMPLE 4 Using the Derivative to Find the Slope at a Point

Find $f'(x)$ for $f(x) = \sqrt{x}$. Then find the slope of the graph of f at the points $(1, 1)$ and $(4, 2)$. Discuss the behavior of f at $(0, 0)$.

Solution Use the procedure for rationalizing numerators, as discussed in Section 2.3.

$$
\begin{aligned}
f'(x) &= \lim_{\Delta x \to 0} \frac{f(x + \Delta x) - f(x)}{\Delta x} && \text{Definition of derivative} \\
&= \lim_{\Delta x \to 0} \frac{\sqrt{x + \Delta x} - \sqrt{x}}{\Delta x} \\
&= \lim_{\Delta x \to 0} \left(\frac{\sqrt{x + \Delta x} - \sqrt{x}}{\Delta x}\right)\left(\frac{\sqrt{x + \Delta x} + \sqrt{x}}{\sqrt{x + \Delta x} + \sqrt{x}}\right) \\
&= \lim_{\Delta x \to 0} \frac{(x + \Delta x) - x}{\Delta x\left(\sqrt{x + \Delta x} + \sqrt{x}\right)} \\
&= \lim_{\Delta x \to 0} \frac{\cancel{\Delta x}}{\cancel{\Delta x}\left(\sqrt{x + \Delta x} + \sqrt{x}\right)} \\
&= \lim_{\Delta x \to 0} \frac{1}{\sqrt{x + \Delta x} + \sqrt{x}} \\
&= \frac{1}{2\sqrt{x}}, \quad x > 0
\end{aligned}
$$

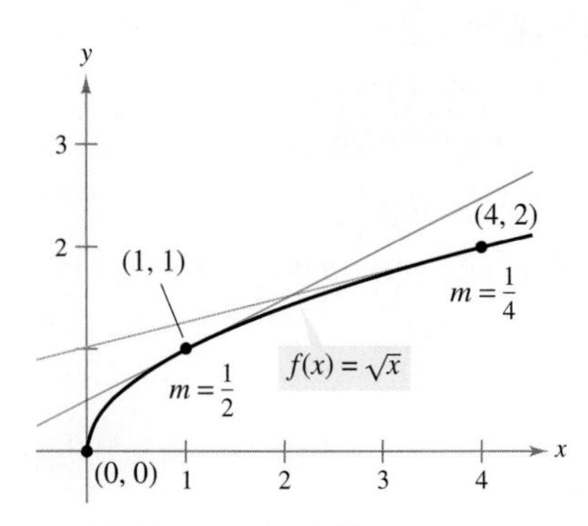

The slope of f at $(x, f(x))$, $x > 0$, is $m = 1/\left(2\sqrt{x}\right)$.
Figure 3.8

At the point $(1, 1)$, the slope is $f'(1) = \frac{1}{2}$. At the point $(4, 2)$, the slope is $f'(4) = \frac{1}{4}$. See Figure 3.8. At the point $(0, 0)$, the slope is undefined. Moreover, the graph of f has a vertical tangent line at $(0, 0)$.

---

indicates that in the HM mathSpace® *CD-ROM and the online* Eduspace® *system for this text, you will find an Open Exploration, which further explores this example using the computer algebra systems* Maple, Mathcad, Mathematica, *and* Derive.

In many applications, it is convenient to use a variable other than x as the independent variable, as shown in Example 5.

EXAMPLE 5 Finding the Derivative of a Function

Find the derivative with respect to t for the function $y = 2/t$.

Solution Considering $y = f(t)$, you obtain

$$\frac{dy}{dt} = \lim_{\Delta t \to 0} \frac{f(t + \Delta t) - f(t)}{\Delta t} \qquad \text{Definition of derivative}$$

$$= \lim_{\Delta t \to 0} \frac{\dfrac{2}{t + \Delta t} - \dfrac{2}{t}}{\Delta t} \qquad f(t + \Delta t) = 2/(t + \Delta t) \text{ and } f(t) = 2/t$$

$$= \lim_{\Delta t \to 0} \frac{\dfrac{2t - 2(t + \Delta t)}{t(t + \Delta t)}}{\Delta t} \qquad \text{Combine fractions in numerator.}$$

$$= \lim_{\Delta t \to 0} \frac{-2\Delta t}{\Delta t(t)(t + \Delta t)} \qquad \text{Divide out common factor of } \Delta t.$$

$$= \lim_{\Delta t \to 0} \frac{-2}{t(t + \Delta t)} \qquad \text{Simplify.}$$

$$= -\frac{2}{t^2}. \qquad \text{Evaluate limit as } \Delta t \to 0.$$

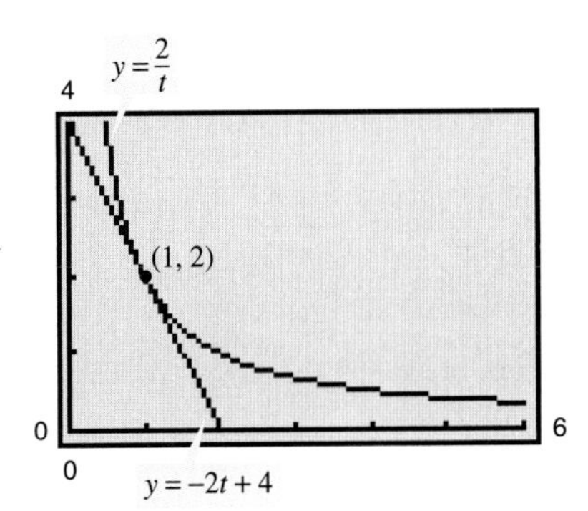

At the point $(1, 2)$, the line $y = -2t + 4$ is tangent to the graph of $y = 2/t$.
Figure 3.9

TECHNOLOGY A graphing utility can be used to reinforce the result given in Example 5. For instance, using the formula $dy/dt = -2/t^2$, you know that the slope of the graph of $y = 2/t$ at the point $(1, 2)$ is $m = -2$. This implies that an equation of the tangent line to the graph at $(1, 2)$ is

$$y - 2 = -2(t - 1) \quad \text{or} \quad y = -2t + 4$$

as shown in Figure 3.9.

Differentiability and Continuity

The following alternative limit form of the derivative is useful in investigating the relationship between differentiability and continuity. The derivative of f at c is

$$f'(c) = \lim_{x \to c} \frac{f(x) - f(c)}{x - c} \qquad \text{Alternative form of derivative}$$

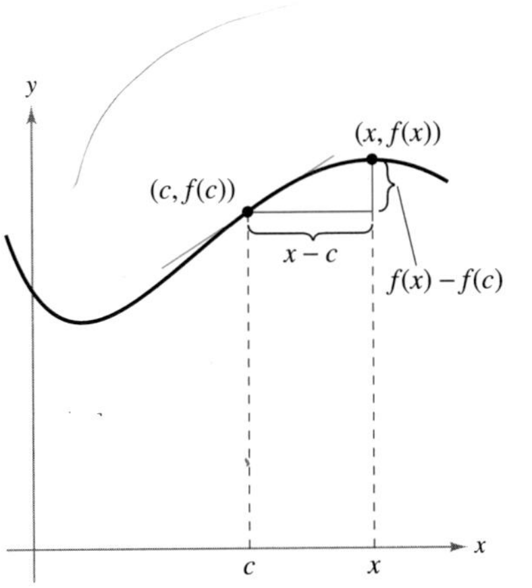

provided this limit exists (see Figure 3.10). (A proof of the equivalence of this form is given in Appendix A.) Note that the existence of the limit in this alternative form requires that the one-sided limits

$$\lim_{x \to c^-} \frac{f(x) - f(c)}{x - c} \quad \text{and} \quad \lim_{x \to c^+} \frac{f(x) - f(c)}{x - c}$$

exist and are equal. These one-sided limits are called the **derivatives from the left and from the right,** respectively. It follows that f is **differentiable on the closed interval $[a, b]$** if it is differentiable on (a, b) and if the derivative from the right at a and the derivative from the left at b both exist.

y
(x, f(x))
(c, f(c))
x − c
f(x) − f(c)
c
x
x

As x approaches c, the secant line approaches the tangent line.
Figure 3.10

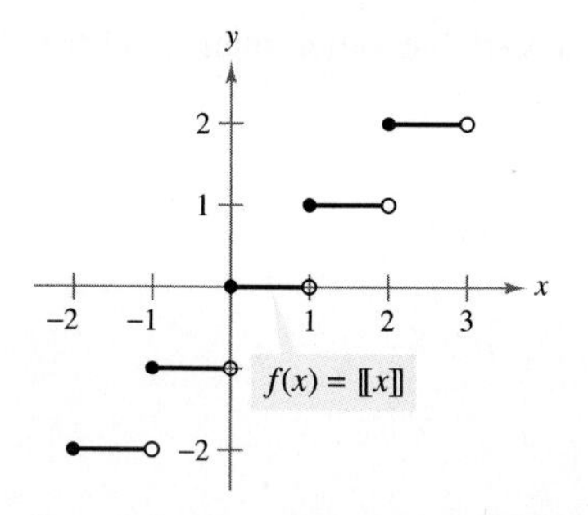

The greatest integer function is not differentiable at $x = 0$, because it is not continuous at $x = 0$.
Figure 3.11

If a function is not continuous at $x = c$, it is also not differentiable at $x = c$. For instance, the greatest integer function

$$f(x) = [\![x]\!]$$

is not continuous at $x = 0$, and so it is not differentiable at $x = 0$ (see Figure 3.11). You can verify this by observing that

$$\lim_{x\to 0^-} \frac{f(x) - f(0)}{x - 0} = \lim_{x\to 0^-} \frac{[\![x]\!] - 0}{x} = \infty \qquad \text{Derivative from the left}$$

and

$$\lim_{x\to 0^+} \frac{f(x) - f(0)}{x - 0} = \lim_{x\to 0^+} \frac{[\![x]\!] - 0}{x} = 0. \qquad \text{Derivative from the right}$$

Although it is true that differentiability implies continuity (as we will show in Theorem 3.1), the converse is not true. That is, it is possible for a function to be continuous at $x = c$ and *not* differentiable at $x = c$. Examples 6 and 7 illustrate this possibility.

EXAMPLE 6 A Graph with a Sharp Turn

The function

$$f(x) = |x - 2|$$

shown in Figure 3.12 is continuous at $x = 2$. But, the one-sided limits

$$\lim_{x\to 2^-} \frac{f(x) - f(2)}{x - 2} = \lim_{x\to 2^-} \frac{|x - 2| - 0}{x - 2} = -1 \qquad \text{Derivative from the left}$$

and

$$\lim_{x\to 2^+} \frac{f(x) - f(2)}{x - 2} = \lim_{x\to 2^+} \frac{|x - 2| - 0}{x - 2} = 1 \qquad \text{Derivative from the right}$$

are not equal. So, f is not differentiable at $x = 2$ and the graph of f does not have a tangent line at the point $(2, 0)$.

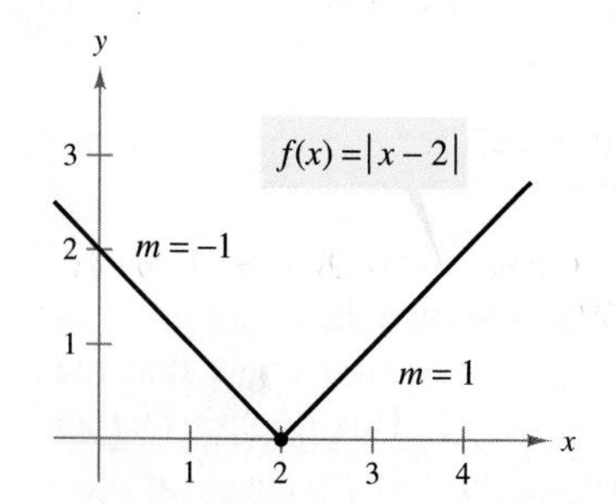

f is not differentiable at $x = 2$, because the derivatives from the left and from the right are not equal.
Figure 3.12

EXAMPLE 7 A Graph with a Vertical Tangent Line

The function

$$f(x) = x^{1/3}$$

is continuous at $x = 0$, as shown in Figure 3.13. But, because the limit

$$\begin{aligned}\lim_{x\to 0} \frac{f(x) - f(0)}{x - 0} &= \lim_{x\to 0} \frac{x^{1/3} - 0}{x} \\ &= \lim_{x\to 0} \frac{1}{x^{2/3}} \\ &= \infty\end{aligned}$$

is infinite, you can conclude that the tangent line is vertical at $x = 0$. So, f is not differentiable at $x = 0$.

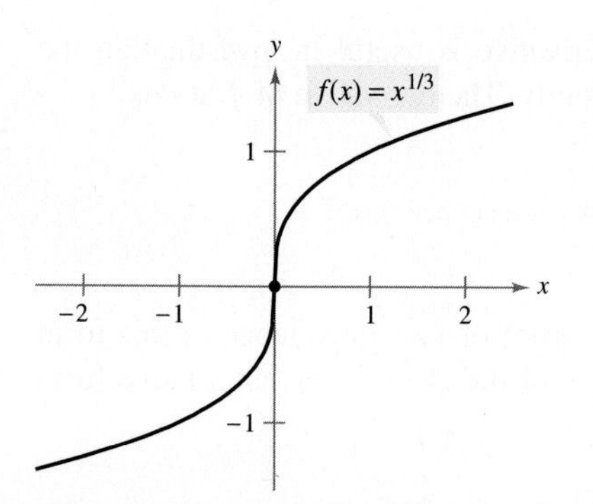

f is not differentiable at $x = 0$, because f has a vertical tangent line at $x = 0$.
Figure 3.13

From Examples 6 and 7, you can see that a function is not differentiable at a point at which its graph has a sharp turn *or* a vertical tangent line.

TECHNOLOGY Some graphing utilities, such as *Derive*, *Maple*, *Mathcad*, *Mathematica*, and the *TI-89*, perform symbolic differentiation. Others perform *numerical differentiation* by finding values of derivatives using the formula

$$f'(x) \approx \frac{f(x + \Delta x) - f(x - \Delta x)}{2\Delta x}$$

where Δx is a small number such as 0.001. Can you see any problems with this definition? For instance, using this definition, what is the value of the derivative of $f(x) = |x|$ when $x = 0$?

THEOREM 3.1 Differentiability Implies Continuity

If f is differentiable at $x = c$, then f is continuous at $x = c$.

Proof You can prove that f is continuous at $x = c$ by showing that $f(x)$ approaches $f(c)$ as $x \to c$. To do this, use the differentiability of f at $x = c$ and consider the following limit.

$$\begin{aligned}
\lim_{x \to c} [f(x) - f(c)] &= \lim_{x \to c} \left[(x - c)\left(\frac{f(x) - f(c)}{x - c}\right)\right] \\
&= \left[\lim_{x \to c} (x - c)\right]\left[\lim_{x \to c} \frac{f(x) - f(c)}{x - c}\right] \\
&= (0)[f'(c)] \\
&= 0
\end{aligned}$$

Because the difference $f(x) - f(c)$ approaches zero as $x \to c$, you can conclude that $\lim_{x \to c} f(x) = f(c)$. So, f is continuous at $x = c$.

You can summarize the relationship between continuity and differentiability as follows

1. If a function is differentiable at $x = c$, then it is continuous at $x = c$. So, differentiability implies continuity.
2. It is possible for a function to be continuous at $x = c$ and not be differentiable at $x = c$. So, continuity does not imply differentiability.

Exercises for Section 3.1

See www.CalcChat.com for worked-out solutions to odd-numbered exercises.

In Exercises 1 and 2, estimate the slope of the graph at the points (x_1, y_1) and (x_2, y_2).

1. (a)

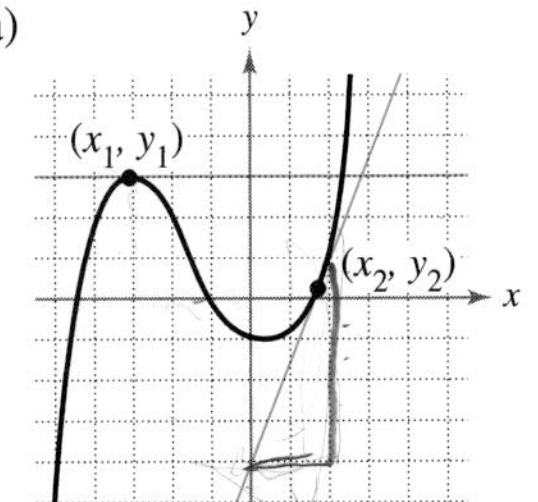

(b)

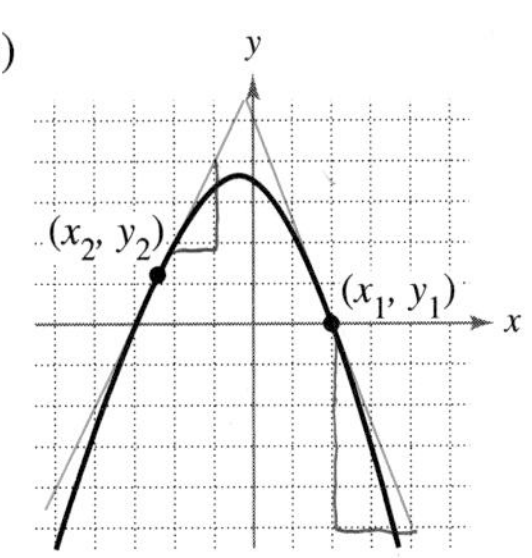

2. (a)

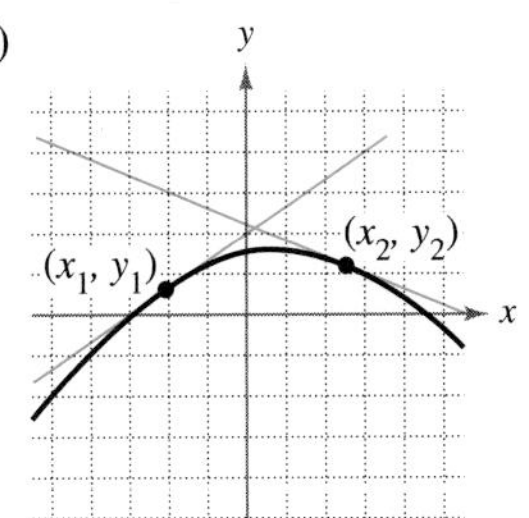

(b)

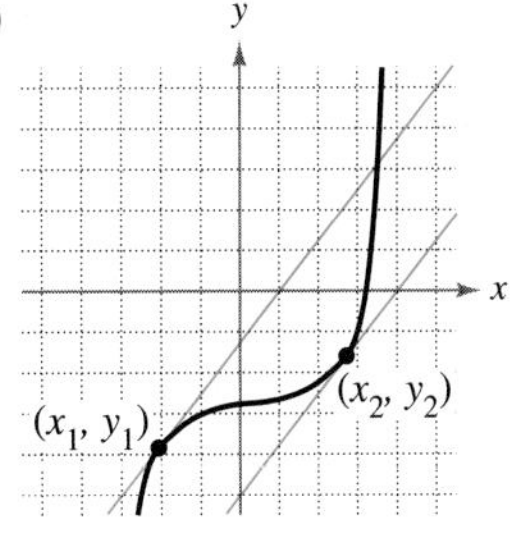

In Exercises 3 and 4, use the graph shown in the figure. To print an enlarged copy of the graph, go to the website *www.mathgraphs.com*.

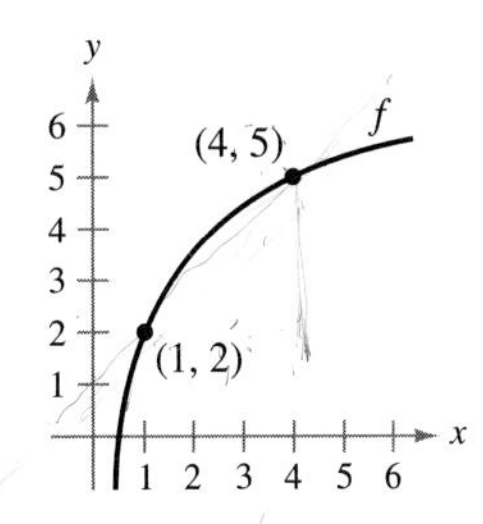

3. Identify or sketch each of the quantities on the figure.

(a) $f(1)$ and $f(4)$ (b) $f(4) - f(1)$

(c) $y = \dfrac{f(4) - f(1)}{4 - 1}(x - 1) + f(1)$

4. Insert the proper inequality symbol (< or >) between the given quantities.

(a) $\dfrac{f(4) - f(1)}{4 - 1} \quad \blacksquare \quad \dfrac{f(4) - f(3)}{4 - 3}$

(b) $\dfrac{f(4) - f(1)}{4 - 1} \quad \blacksquare \quad f'(1)$

In Exercises 5–10, find the slope of the tangent line to the graph of the function at the given point.

5. $f(x) = 3 - 2x, \quad (-1, 5)$
6. $g(x) = \frac{3}{2}x + 1, \quad (-2, -2)$
7. $g(x) = x^2 - 4, \quad (1, -3)$
8. $g(x) = 5 - x^2, \quad (2, 1)$
9. $f(t) = 3t - t^2, \quad (0, 0)$
10. $h(t) = t^2 + 3, \quad (-2, 7)$

In Exercises 11–24, find the derivative by the limit process.

11. $f(x) = 3$
12. $g(x) = -5$
13. $f(x) = -5x$
14. $f(x) = 3x + 2$
15. $h(s) = 3 + \frac{2}{3}s$
16. $f(x) = 9 - \frac{1}{2}x$
17. $f(x) = 2x^2 + x - 1$
18. $f(x) = 1 - x^2$
19. $f(x) = x^3 - 12x$
20. $f(x) = x^3 + x^2$
21. $f(x) = \dfrac{1}{x - 1}$
22. $f(x) = \dfrac{1}{x^2}$
23. $f(x) = \sqrt{x + 1}$
24. $f(x) = \dfrac{4}{\sqrt{x}}$

In Exercises 25–32, (a) find an equation of the tangent line to the graph of f at the given point, (b) use a graphing utility to graph the function and its tangent line at the point, and (c) use the *derivative* feature of a graphing utility to confirm your results.

25. $f(x) = x^2 + 1, \quad (2, 5)$
26. $f(x) = x^2 + 2x + 1, \quad (-3, 4)$
27. $f(x) = x^3, \quad (2, 8)$
28. $f(x) = x^3 + 1, \quad (1, 2)$
29. $f(x) = \sqrt{x}, \quad (1, 1)$
30. $f(x) = \sqrt{x - 1}, \quad (5, 2)$
31. $f(x) = x + \dfrac{4}{x}, \quad (4, 5)$
32. $f(x) = \dfrac{1}{x + 1}, \quad (0, 1)$

In Exercises 33–36, find an equation of the line that is tangent to the graph of f *and* parallel to the given line.

Function	*Line*
33. $f(x) = x^3$	$3x - y + 1 = 0$
34. $f(x) = x^3 + 2$	$3x - y - 4 = 0$
35. $f(x) = \dfrac{1}{\sqrt{x}}$	$x + 2y - 6 = 0$
36. $f(x) = \dfrac{1}{\sqrt{x - 1}}$	$x + 2y + 7 = 0$

In Exercises 37–40, the graph of f is given. Select the graph of f'.

37.

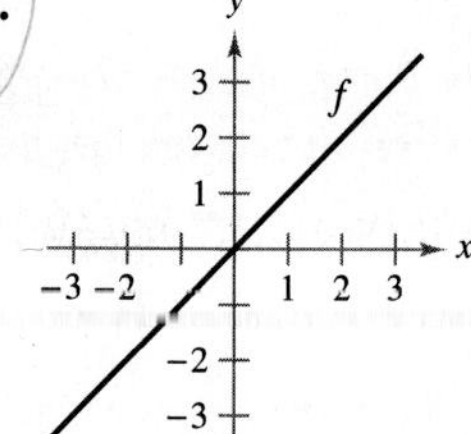

38.

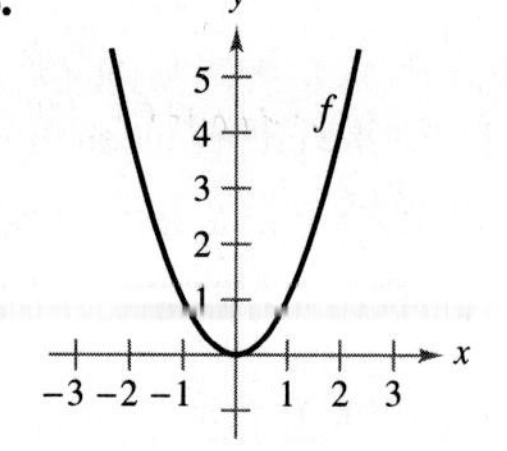

39.

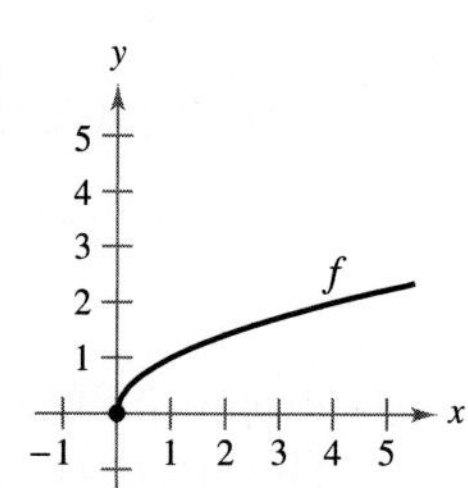

40.

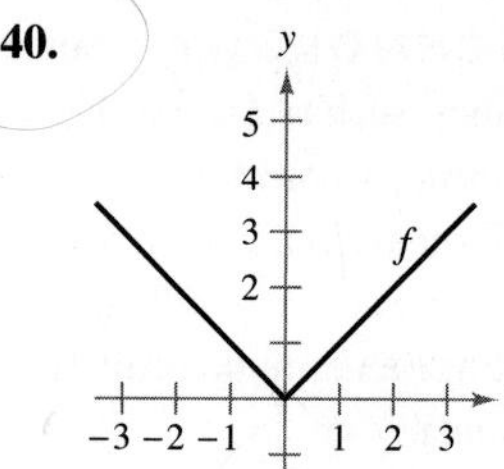

(a)

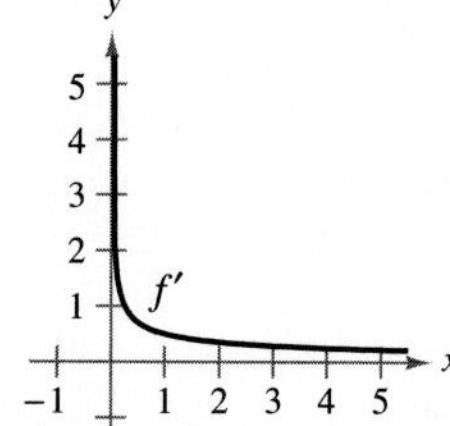

(b)

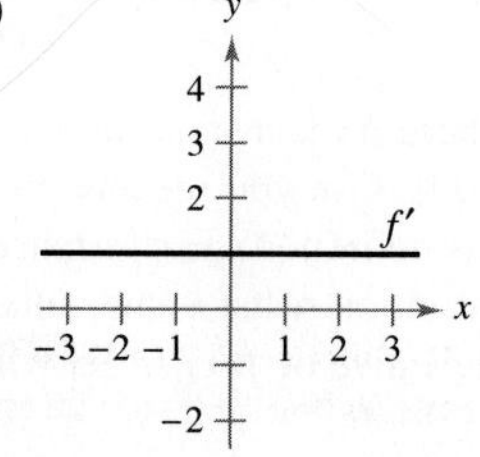

(c)

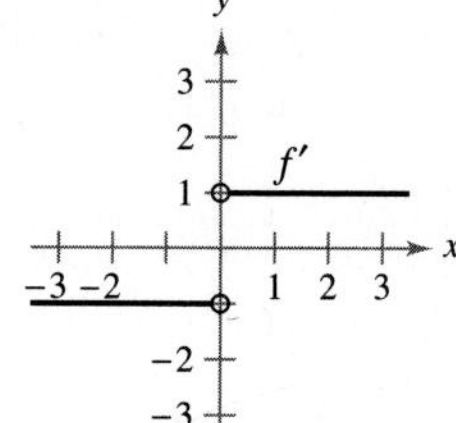

(d)

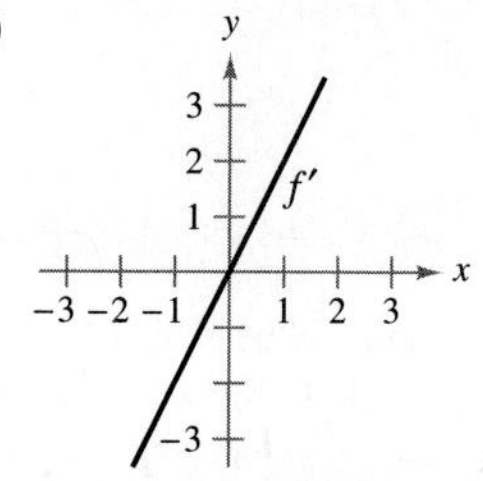

41. The tangent line to the graph of $y = g(x)$ at the point $(5, 2)$ passes through the point $(9, 0)$. Find $g(5)$ and $g'(5)$.

42. The tangent line to the graph of $y = h(x)$ at the point $(-1, 4)$ passes through the point $(3, 6)$. Find $h(-1)$ and $h'(-1)$.

Writing About Concepts

In Exercises 43–46, sketch the graph of f'. Explain how you found your answer.

43.

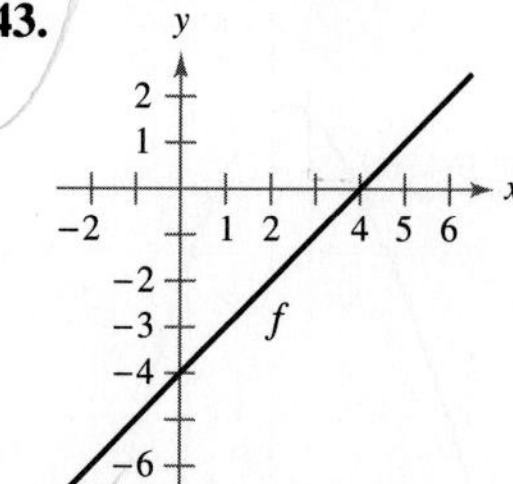

44.

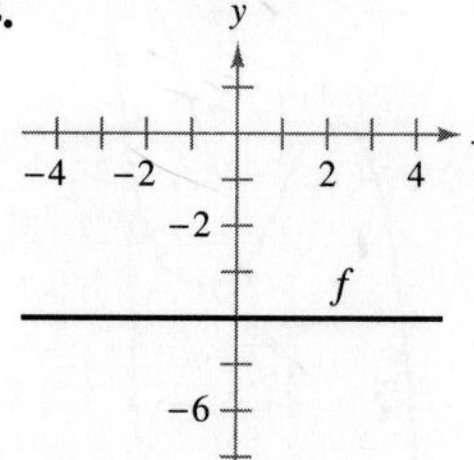

45.

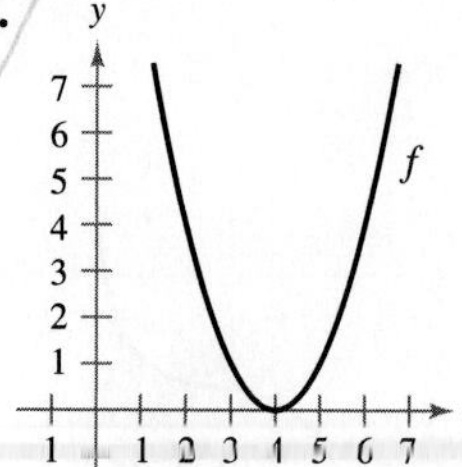

46.

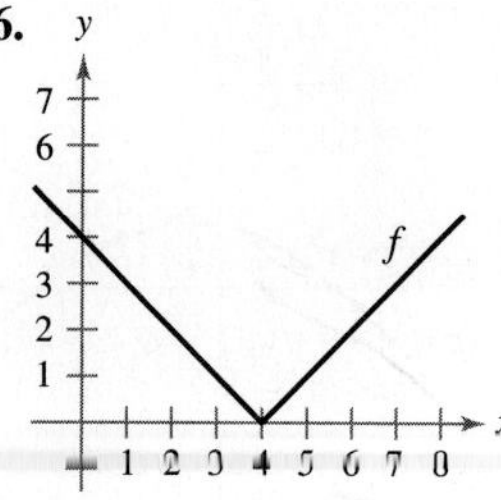

47. Sketch a graph of a function whose derivative is always negative.

Writing About Concepts *(continued)*

48. Sketch a graph of a function whose derivative is always positive.

In Exercises 49–52, the limit represents $f'(c)$ for a function f and a number c. Find f and c.

49. $\lim_{\Delta x \to 0} \dfrac{[5 - 3(1 + \Delta x)] - 2}{\Delta x}$

50. $\lim_{\Delta x \to 0} \dfrac{(-2 + \Delta x)^3 + 8}{\Delta x}$

51. $\lim_{x \to 6} \dfrac{-x^2 + 36}{x - 6}$

52. $\lim_{x \to 9} \dfrac{2\sqrt{x} - 6}{x - 9}$

In Exercises 53–55, identify a function f that has the following characteristics. Then sketch the function.

53. $f(0) = 2$; $f'(x) = -3, -\infty < x < \infty$

54. $f(0) = 4; f'(0) = 0$; $f'(x) < 0$ for $x < 0$; $f'(x) > 0$ for $x > 0$

55. $f(0) = 0; f'(0) = 0; f'(x) > 0$ if $x \neq 0$

56. Assume that $f'(c) = 3$. Find $f'(-c)$ if (a) f is an odd function and (b) f is an even function.

In Exercises 57 and 58, find equations of the two tangent lines to the graph of f that pass through the indicated point.

57. $f(x) = 4x - x^2$

58. $f(x) = x^2$

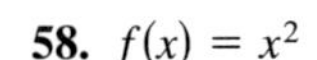

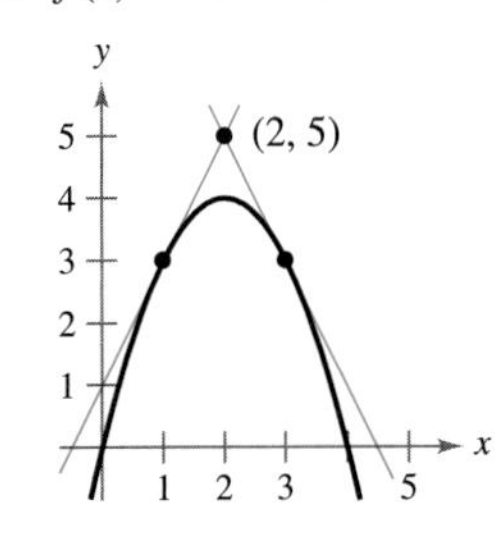

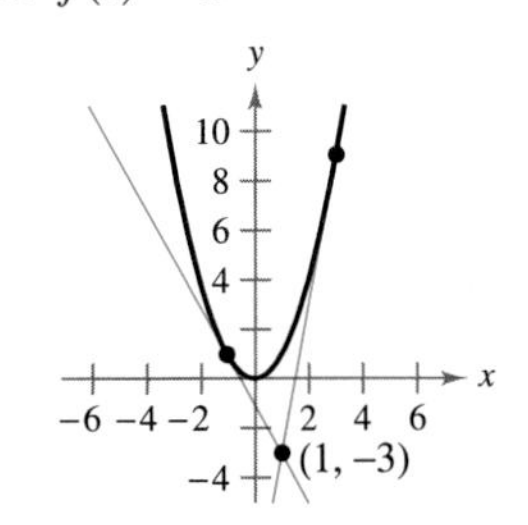

59. ***Graphical Reasoning*** The figure shows the graph of g'.

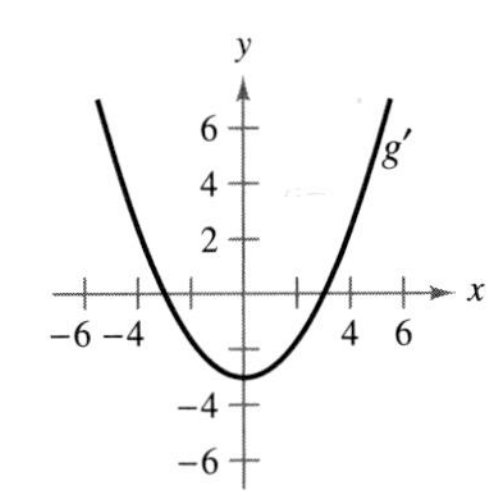

(a) $g'(0) =$ ▭ (b) $g'(3) =$ ▭

(c) What can you conclude about the graph of g knowing that $g'(1) = -\frac{8}{3}$?

(d) What can you conclude about the graph of g knowing that $g'(-4) = \frac{7}{3}$?

(e) Is $g(6) - g(4)$ positive or negative? Explain.

(f) Is it possible to find $g(2)$ from the graph? Explain.

60. ***Graphical Reasoning*** Use a graphing utility to graph each function and its tangent lines at $x = -1, x = 0$, and $x = 1$. Based on the results, determine whether the slopes of tangent lines to the graph of a function at different values of x are always distinct.

(a) $f(x) = x^2$ (b) $g(x) = x^3$

Graphical, Numerical, and Analytic Analysis **In Exercises 61 and 62, use a graphing utility to graph f on the interval $[-2, 2]$. Complete the table by graphically estimating the slope of the graph at the indicated points. Then evaluate the slopes analytically and compare your results with those obtained graphically.**

x	-2	-1.5	-1	-0.5	0	0.5	1	1.5	2
$f(x)$									
$f'(x)$									

61. $f(x) = \frac{1}{4}x^3$

62. $f(x) = \frac{1}{2}x^2$

Graphical Reasoning **In Exercises 63 and 64, use a graphing utility to graph the functions f and g in the same viewing window where**

$$g(x) = \frac{f(x + 0.01) - f(x)}{0.01}.$$

Label the graphs and describe the relationship between them.

63. $f(x) = 2x - x^2$

64. $f(x) = 3\sqrt{x}$

In Exercises 65 and 66, evaluate $f(2)$ and $f(2.1)$ and use the results to approximate $f'(2)$.

65. $f(x) = x(4 - x)$

66. $f(x) = \frac{1}{4}x^3$

Graphical Reasoning **In Exercises 67 and 68, use a graphing utility to graph the function and its derivative in the same viewing window. Label the graphs and describe the relationship between them.**

67. $f(x) = \dfrac{1}{\sqrt{x}}$

68. $f(x) = \dfrac{x^3}{4} - 3x$

Writing **In Exercises 69 and 70, consider the functions f and $S_{\Delta x}$ where**

$$S_{\Delta x}(x) = \frac{f(2 + \Delta x) - f(2)}{\Delta x}(x - 2) + f(2).$$

(a) Use a graphing utility to graph f and $S_{\Delta x}$ in the same viewing window for $\Delta x = 1, 0.5$, and 0.1.

(b) Give a written description of the graphs of S for the different values of Δx in part (a).

69. $f(x) = 4 - (x - 3)^2$

70. $f(x) = x + \dfrac{1}{x}$

In Exercises 71–80, use the alternative form of the derivative to find the derivative at $x = c$ (if it exists).

71. $f(x) = x^2 - 1,\ c = 2$

72. $g(x) = x(x - 1),\ c = 1$

73. $f(x) = x^3 + 2x^2 + 1,\ c = -2$

74. $f(x) = x^3 + 2x,\ c = 1$

75. $g(x) = \sqrt{|x|},\ c = 0$

76. $f(x) = 1/x,\ c = 3$

77. $f(x) = (x - 6)^{2/3},\ c = 6$

78. $g(x) = (x + 3)^{1/3},\ c = -3$

79. $h(x) = |x + 5|,\ c = -5$

80. $f(x) = |x - 4|,\ c = 4$

In Exercises 81–86, describe the x-values at which f is differentiable.

81. $f(x) = \dfrac{1}{x + 1}$

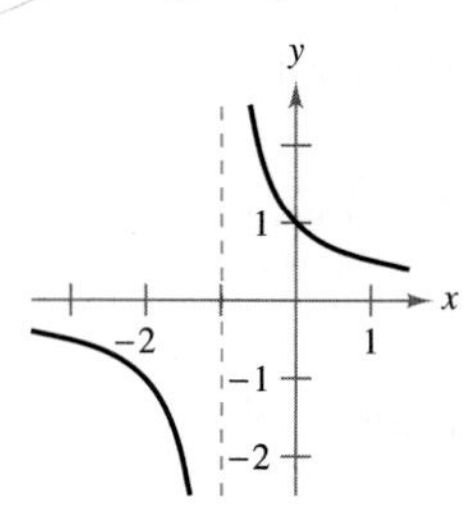

82. $f(x) = |x^2 - 9|$

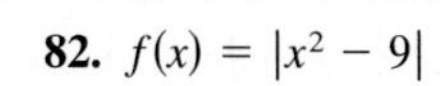

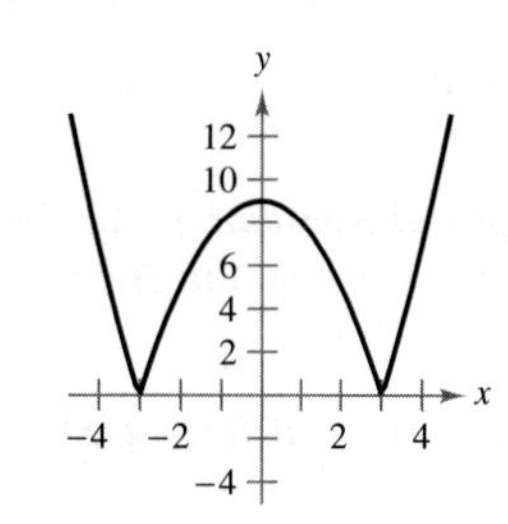

83. $f(x) = (x - 3)^{2/3}$

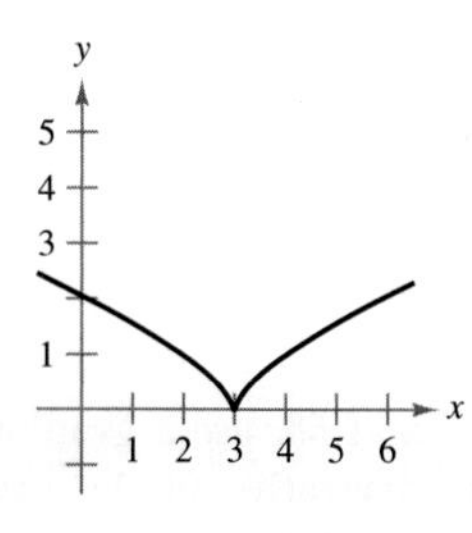

84. $f(x) = \dfrac{x^2}{x^2 - 4}$

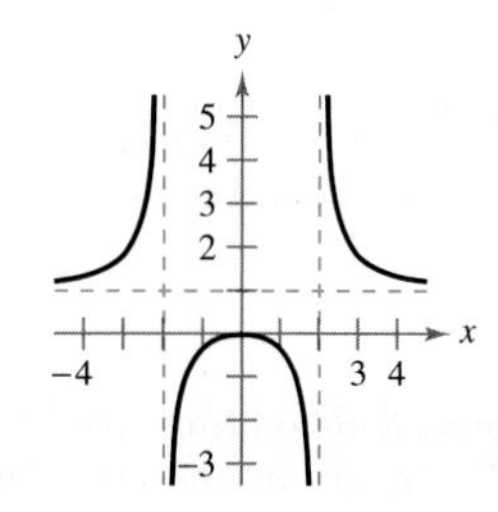

85. $f(x) = \sqrt{x - 1}$

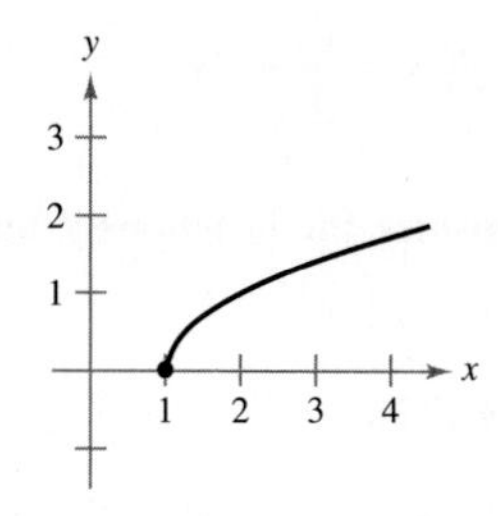

86. $f(x) = \begin{cases} x^2 - 4, & x \le 0 \\ 4 - x^2, & x > 0 \end{cases}$

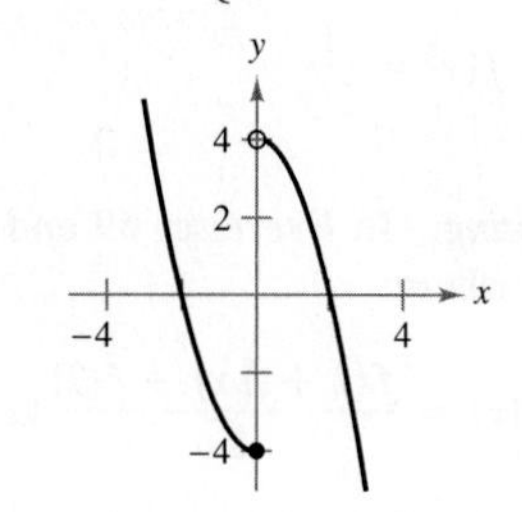

Graphical Analysis **In Exercises 87–90, use a graphing utility to find the x-values at which f is differentiable.**

87. $f(x) = |x + 3|$

88. $f(x) = \dfrac{2x}{x - 1}$

89. $f(x) = x^{2/5}$

90. $f(x) = \begin{cases} x^3 - 3x^2 + 3x, & x \le 1 \\ x^2 - 2x, & x > 1 \end{cases}$

In Exercises 91–94, find the derivatives from the left and from the right at $x = 1$ (if they exist). Is the function differentiable at $x = 1$?

91. $f(x) = |x - 1|$

92. $f(x) = \sqrt{1 - x^2}$

93. $f(x) = \begin{cases} (x - 1)^3, & x \le 1 \\ (x - 1)^2, & x > 1 \end{cases}$

94. $f(x) = \begin{cases} x, & x \le 1 \\ x^2, & x > 1 \end{cases}$

In Exercises 95 and 96, determine whether the function is differentiable at $x = 2$.

95. $f(x) = \begin{cases} x^2 + 1, & x \le 2 \\ 4x - 3, & x > 2 \end{cases}$

96. $f(x) = \begin{cases} \frac{1}{2}x + 1, & x < 2 \\ \sqrt{2x}, & x \ge 2 \end{cases}$

97. *Graphical Reasoning* A line with slope m passes through the point $(0, 4)$ and has the equation $y = mx + 4$.

(a) Write the distance d between the line and the point $(3, 1)$ as a function of m.

(b) Use a graphing utility to graph the function d in part (a). Based on the graph, is the function differentiable at every value of m? If not, where is it not differentiable?

98. *Conjecture* Consider the functions $f(x) = x^2$ and $g(x) = x^3$.

(a) Graph f and f' on the same set of axes.

(b) Graph g and g' on the same set of axes.

(c) Identify a pattern between f and g and their respective derivatives. Use the pattern to make a conjecture about $h'(x)$ if $h(x) = x^n$, where n is an integer and $n \ge 2$.

(d) Find $f'(x)$ if $f(x) = x^4$. Compare the result with the conjecture in part (c). Is this a proof of your conjecture? Explain.

True or False? **In Exercises 99–102, determine whether the statement is true or false. If it is false, explain why or give an example that shows it is false.**

99. The slope of the tangent line to the differentiable function f at the point $(2, f(2))$ is $\dfrac{f(2 + \Delta x) - f(2)}{\Delta x}$.

100. If a function is continuous at a point, then it is differentiable at that point.

101. If a function has derivatives from both the right and the left at a point, then it is differentiable at that point.

102. If a function is differentiable at a point, then it is continuous at that point.

103. Let $f(x) = \begin{cases} x \sin \frac{1}{x}, & x \ne 0 \\ 0, & x = 0 \end{cases}$ and $g(x) = \begin{cases} x^2 \sin \frac{1}{x}, & x \ne 0 \\ 0, & x = 0 \end{cases}$.

Show that f is continuous, but not differentiable, at $x = 0$. Show that g is differentiable at 0, and find $g'(0)$.

104. *Writing* Use a graphing utility to graph the two functions $f(x) = x^2 + 1$ and $g(x) = |x| + 1$ in the same viewing window. Use the *zoom* and *trace* features to analyze the graphs near the point $(0, 1)$. What do you observe? Which function is differentiable at this point? Write a short paragraph describing the geometric significance of differentiability at a point.

Section 3.2 Basic Differentiation Rules and Rates of Change

- **Find the derivative of a function using the Constant Rule.**
- **Find the derivative of a function using the Power Rule.**
- **Find the derivative of a function using the Constant Multiple Rule.**
- **Find the derivative of a function using the Sum and Difference Rules.**
- **Find the derivative of the sine, cosine, and exponential functions.**
- **Use derivatives to find rates of change.**

The Constant Rule

In Section 3.1 you used the limit definition to find derivatives. In this and the next two sections, you will be introduced to several "differentiation rules" that allow you to find derivatives without the *direct* use of the limit definition.

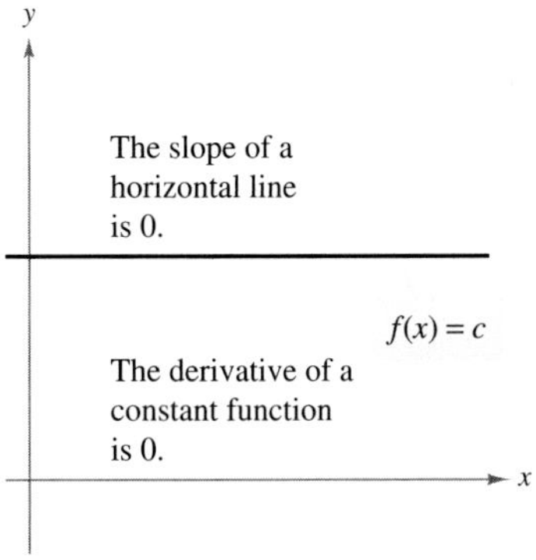

y
The slope of a horizontal line is 0.
$f(x) = c$
The derivative of a constant function is 0.
x

The Constant Rule
Figure 3.14

NOTE In Figure 3.14, note that the Constant Rule is equivalent to saying that the slope of a horizontal line is 0. This demonstrates the relationship between slope and derivative.

THEOREM 3.2 The Constant Rule

The derivative of a constant function is 0. That is, if c is a real number, then

$$\frac{d}{dx}[c] = 0.$$

Proof Let $f(x) = c$. Then, by the limit definition of the derivative,

$$\begin{aligned}\frac{d}{dx}[c] &= f'(x) \\ &= \lim_{\Delta x \to 0} \frac{f(x + \Delta x) - f(x)}{\Delta x} \\ &= \lim_{\Delta x \to 0} \frac{c - c}{\Delta x} \\ &= \lim_{\Delta x \to 0} 0 \\ &= 0.\end{aligned}$$

EXAMPLE 1 Using the Constant Rule

	Function	*Derivative*
a.	$y = 7$	$\dfrac{dy}{dx} = 0$
b.	$f(x) = 0$	$f'(x) = 0$
c.	$s(t) = -3$	$s'(t) = 0$
d.	$y = k\pi^2$, k is constant	$y' = 0$

EXPLORATION

Writing a Conjecture Use the definition of the derivative given in Section 3.1 to find the derivative of each of the following. What patterns do you see? Use your results to write a conjecture about the derivative of $f(x) = x^n$.

a. $f(x) = x^1$ **b.** $f(x) = x^2$ **c.** $f(x) = x^3$
d. $f(x) = x^4$ **e.** $f(x) = x^{1/2}$ **f.** $f(x) = x^{-1}$

The Power Rule

Before proving the next rule, review the procedure for expanding a binomial.

$$(x + \Delta x)^2 = x^2 + 2x\Delta x + (\Delta x)^2$$
$$(x + \Delta x)^3 = x^3 + 3x^2\Delta x + 3x(\Delta x)^2 + (\Delta x)^3$$
$$(x + \Delta x)^4 = x^4 + 4x^3\Delta x + 6x^2(\Delta x)^2 + 4x(\Delta x)^3 + (\Delta x)^4$$

The general binomial expansion for a positive integer n is

$$(x + \Delta x)^n = x^n + nx^{n-1}(\Delta x) + \underbrace{\frac{n(n-1)x^{n-2}}{2}(\Delta x)^2 + \cdots + (\Delta x)^n}_{(\Delta x)^2 \text{ is a factor of these terms.}}.$$

This binomial expansion is used in proving a special case of the Power Rule.

THEOREM 3.3 The Power Rule

If n is a rational number, then the function $f(x) = x^n$ is differentiable and

$$\frac{d}{dx}[x^n] = nx^{n-1}.$$

For f to be differentiable at $x = 0$, n must be a number such that x^{n-1} is defined on an interval containing 0.

Proof If n is a positive integer greater than 1, then the binomial expansion produces the following.

$$\frac{d}{dx}[x^n] = \lim_{\Delta x \to 0} \frac{(x + \Delta x)^n - x^n}{\Delta x}$$

$$= \lim_{\Delta x \to 0} \frac{x^n + nx^{n-1}(\Delta x) + \frac{n(n-1)x^{n-2}}{2}(\Delta x)^2 + \cdots + (\Delta x)^n - x^n}{\Delta x}$$

$$= \lim_{\Delta x \to 0} \left[nx^{n-1} + \frac{n(n-1)x^{n-2}}{2}(\Delta x) + \cdots + (\Delta x)^{n-1} \right]$$

$$= nx^{n-1} + 0 + \cdots + 0$$

$$= nx^{n-1}$$

This proves the case for which n is a positive integer greater than 1. We leave it to you to prove the case for $n = 1$. Example 7 in Section 3.3 proves the case for which n is a negative integer. In Exercise 91 in Section 3.5, you are asked to prove the case for which n is rational.

When using the Power Rule, the case for which $n = 1$ is best thought of as a separate differentiation rule. That is,

$$\frac{d}{dx}[x] = 1. \qquad \text{Power Rule when } n = 1$$

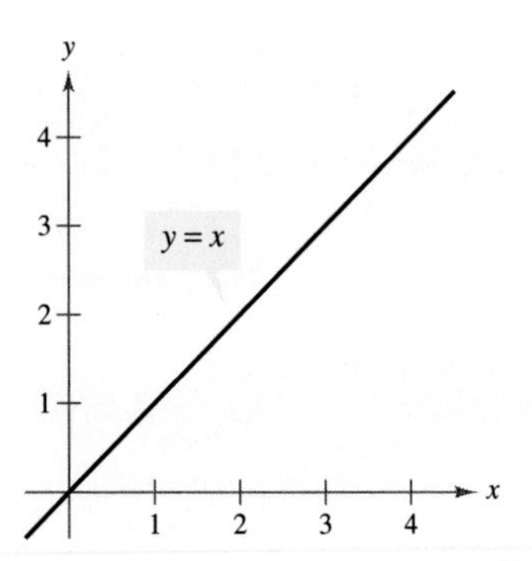

The slope of the line $y = x$ is 1.
Figure 3.15

This rule is consistent with the fact that the slope of the line $y = x$ is 1, as shown in Figure 3.15.

EXAMPLE 2 Using the Power Rule

Function	*Derivative*
a. $f(x) = x^3$	$f'(x) = 3x^2$
b. $g(x) = \sqrt[3]{x}$	$g'(x) = \dfrac{d}{dx}[x^{1/3}] = \dfrac{1}{3}x^{-2/3} = \dfrac{1}{3x^{2/3}}$
c. $y = \dfrac{1}{x^2}$	$\dfrac{dy}{dx} = \dfrac{d}{dx}[x^{-2}] = (-2)x^{-3} = -\dfrac{2}{x^3}$

In Example 2(c), note that *before* differentiating, $1/x^2$ was rewritten as x^{-2}. Rewriting is the first step in *many* differentiation problems.

Given:		Rewrite:		Differentiate:		Simplify:
$y = \dfrac{1}{x^2}$	⇨	$y = x^{-2}$	⇨	$\dfrac{dy}{dx} = (-2)x^{-3}$	⇨	$\dfrac{dy}{dx} = -\dfrac{2}{x^3}$

EXAMPLE 3 Finding the Slope of a Graph

Find the slope of the graph of $f(x) = x^4$ when

a. $x = -1$ **b.** $x = 0$ **c.** $x = 1$.

Solution The derivative of f is $f'(x) = 4x^3$.

a. When $x = -1$, the slope is $f'(-1) = 4(-1)^3 = -4$. Slope is negative.

b. When $x = 0$, the slope is $f'(0) = 4(0)^3 = 0$. Slope is zero.

c. When $x = 1$, the slope is $f'(1) = 4(1)^3 = 4$. Slope is positive.

In Figure 3.16, note that the slope of the graph is negative at the point $(-1, 1)$, the slope is zero at the point $(0, 0)$, and the slope is positive at the point $(1, 1)$.

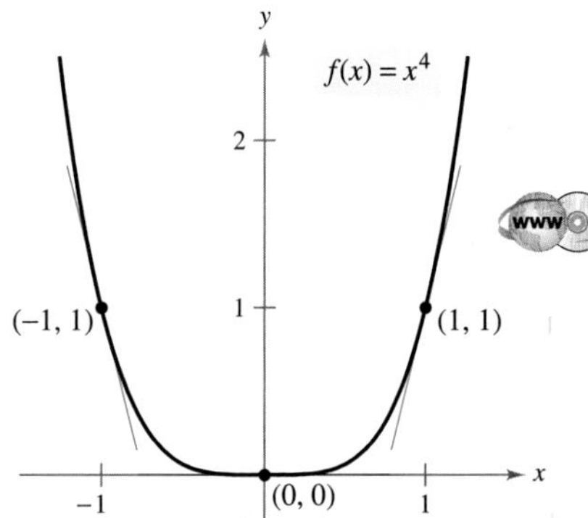

The slope of a graph at a point is the value of the derivative at that point.
Figure 3.16

EXAMPLE 4 Finding an Equation of a Tangent Line

Find an equation of the tangent line to the graph of $f(x) = x^2$ when $x = -2$.

Solution To find the *point* on the graph of f, evaluate the original function at $x = -2$.

$$(-2, f(-2)) = (-2, 4) \qquad \text{Point on graph}$$

To find the *slope* of the graph when $x = -2$, evaluate the derivative, $f'(x) = 2x$, at $x = -2$.

$$m = f'(-2) = -4 \qquad \text{Slope of graph at } (-2, 4)$$

Now, using the point-slope form of the equation of a line, you can write

$$y - y_1 = m(x - x_1) \qquad \text{Point-slope form}$$

$$y - 4 = -4[x - (-2)] \qquad \text{Substitute for } y_1, m, \text{ and } x_1.$$

$$y = -4x - 4. \qquad \text{Simplify.}$$

(See Figure 3.17.)

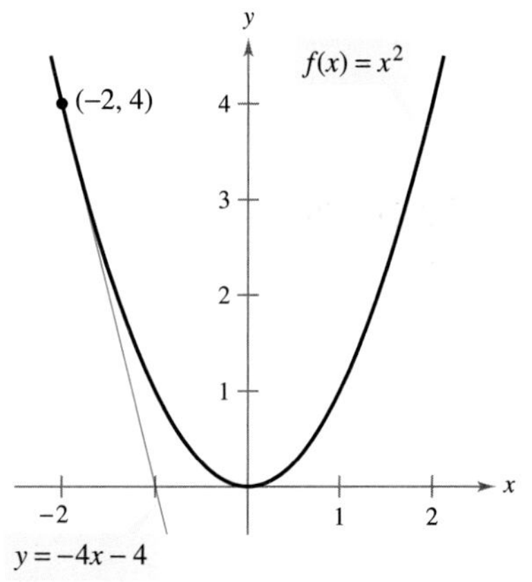

The line $y = -4x - 4$ is tangent to the graph of $f(x) = x^2$ at the point $(-2, 4)$.
Figure 3.17

The Constant Multiple Rule

THEOREM 3.4 The Constant Multiple Rule

If f is a differentiable function and c is a real number, then cf is also differentiable and $\frac{d}{dx}[cf(x)] = cf'(x)$.

Proof

$$\begin{aligned}\frac{d}{dx}[cf(x)] &= \lim_{\Delta x \to 0} \frac{cf(x + \Delta x) - cf(x)}{\Delta x} && \text{Definition of derivative}\\ &= \lim_{\Delta x \to 0} c\left[\frac{f(x + \Delta x) - f(x)}{\Delta x}\right]\\ &= c\left[\lim_{\Delta x \to 0} \frac{f(x + \Delta x) - f(x)}{\Delta x}\right]\\ &= cf'(x)\end{aligned}$$

Informally, the Constant Multiple Rule states that constants can be factored out of the differentiation process, even if the constants appear in the denominator.

$$\frac{d}{dx}[cf(x)] = c\frac{d}{dx}[f(x)] = cf'(x)$$

$$\frac{d}{dx}\left[\frac{f(x)}{c}\right] = \frac{d}{dx}\left[\left(\frac{1}{c}\right)f(x)\right] = \left(\frac{1}{c}\right)\frac{d}{dx}[f(x)] = \left(\frac{1}{c}\right)f'(x)$$

EXAMPLE 5 Using the Constant Multiple Rule

Function	*Derivative*
a. $y = \frac{2}{x}$	$\frac{dy}{dx} = \frac{d}{dx}[2x^{-1}] = 2\frac{d}{dx}[x^{-1}] = 2(-1)x^{-2} = -\frac{2}{x^2}$
b. $f(t) = \frac{4t^2}{5}$	$f'(t) = \frac{d}{dt}\left[\frac{4}{5}t^2\right] = \frac{4}{5}\frac{d}{dt}[t^2] = \frac{4}{5}(2t) = \frac{8}{5}t$
c. $y = 2\sqrt{x}$	$\frac{dy}{dx} = \frac{d}{dx}[2x^{1/2}] = 2\left(\frac{1}{2}x^{-1/2}\right) = x^{-1/2} = \frac{1}{\sqrt{x}}$
d. $y = \frac{1}{2\sqrt[3]{x^2}}$	$\frac{dy}{dx} = \frac{d}{dx}\left[\frac{1}{2}x^{-2/3}\right] = \frac{1}{2}\left(-\frac{2}{3}\right)x^{-5/3} = -\frac{1}{3x^{5/3}}$
e. $y = -\frac{3x}{2}$	$y' = \frac{d}{dx}\left[-\frac{3}{2}x\right] = -\frac{3}{2}(1) = -\frac{3}{2}$

The Constant Multiple Rule and the Power Rule can be combined into one rule. The combination rule is

$$D_x[cx^n] = cnx^{n-1}.$$

EXAMPLE 6 Using Parentheses When Differentiating

Original Function	*Rewrite*	*Differentiate*	*Simplify*
a. $y = \dfrac{5}{2x^3}$	$y = \dfrac{5}{2}(x^{-3})$	$y' = \dfrac{5}{2}(-3x^{-4})$	$y' = -\dfrac{15}{2x^4}$
b. $y = \dfrac{5}{(2x)^3}$	$y = \dfrac{5}{8}(x^{-3})$	$y' = \dfrac{5}{8}(-3x^{-4})$	$y' = -\dfrac{15}{8x^4}$
c. $y = \dfrac{7}{3x^{-2}}$	$y = \dfrac{7}{3}(x^2)$	$y' = \dfrac{7}{3}(2x)$	$y' = \dfrac{14x}{3}$
d. $y = \dfrac{7}{(3x)^{-2}}$	$y = 63(x^2)$	$y' = 63(2x)$	$y' = 126x$

The Sum and Difference Rules

THEOREM 3.5 The Sum and Difference Rules

The sum (or difference) of two differentiable functions f and g is itself differentiable. Moreover, the derivative of $f + g$ (or $f - g$) is the sum (or difference) of the derivatives of f and g.

$$\frac{d}{dx}[f(x) + g(x)] = f'(x) + g'(x) \qquad \text{Sum Rule}$$

$$\frac{d}{dx}[f(x) - g(x)] = f'(x) - g'(x) \qquad \text{Difference Rule}$$

Proof A proof of the Sum Rule follows from Theorem 2.2. (The Difference Rule can be proved in a similar way.)

$$\begin{aligned}
\frac{d}{dx}[f(x) + g(x)] &= \lim_{\Delta x \to 0} \frac{[f(x + \Delta x) + g(x + \Delta x)] - [f(x) + g(x)]}{\Delta x} \\
&= \lim_{\Delta x \to 0} \frac{f(x + \Delta x) + g(x + \Delta x) - f(x) - g(x)}{\Delta x} \\
&= \lim_{\Delta x \to 0} \left[\frac{f(x + \Delta x) - f(x)}{\Delta x} + \frac{g(x + \Delta x) - g(x)}{\Delta x}\right] \\
&= \lim_{\Delta x \to 0} \frac{f(x + \Delta x) - f(x)}{\Delta x} + \lim_{\Delta x \to 0} \frac{g(x + \Delta x) - g(x)}{\Delta x} \\
&= f'(x) + g'(x)
\end{aligned}$$

EXPLORATION

Use a graphing utility to graph the function

$$f(x) = \frac{\sin(x + \Delta x) - \sin x}{\Delta x}$$

for $\Delta x = 0.01$. What does this function represent? Compare this graph with that of the cosine function. What do you think the derivative of the sine function equals?

The Sum and Difference Rules can be extended to any finite number of functions. For instance, if $F(x) = f(x) + g(x) - h(x)$, then $F'(x) = f'(x) + g'(x) - h'(x)$.

EXAMPLE 7 Using the Sum and Difference Rules

Function	*Derivative*
a. $f(x) = x^3 - 4x + 5$	$f'(x) = 3x^2 - 4$
b. $g(x) = -\dfrac{x^4}{2} + 3x^3 - 2x$	$g'(x) = -2x^3 + 9x^2 - 2$

FOR FURTHER INFORMATION For the outline of a geometric proof of the derivatives of the sine and cosine functions, see the article "The Spider's Spacewalk Derivation of sin′ and cos′" by Tim Hesterberg in *The College Mathematics Journal*. To view this article, go to the website *www.matharticles.com*.

Derivatives of Sine and Cosine Functions

In Section 2.3, you studied the following limits.

$$\lim_{\Delta x \to 0} \frac{\sin \Delta x}{\Delta x} = 1 \quad \text{and} \quad \lim_{\Delta x \to 0} \frac{1 - \cos \Delta x}{\Delta x} = 0$$

These two limits can be used to prove differentiation rules for the sine and cosine functions. (The derivatives of the other four trigonometric functions are discussed in Section 3.3.)

THEOREM 3.6 Derivatives of Sine and Cosine Functions

$$\frac{d}{dx}[\sin x] = \cos x \qquad \frac{d}{dx}[\cos x] = -\sin x$$

Proof

$$\frac{d}{dx}[\sin x] = \lim_{\Delta x \to 0} \frac{\sin(x + \Delta x) - \sin x}{\Delta x} \qquad \text{Definition of derivative}$$

$$= \lim_{\Delta x \to 0} \frac{\sin x \cos \Delta x + \cos x \sin \Delta x - \sin x}{\Delta x}$$

$$= \lim_{\Delta x \to 0} \frac{\cos x \sin \Delta x - (\sin x)(1 - \cos \Delta x)}{\Delta x}$$

$$= \lim_{\Delta x \to 0} \left[(\cos x)\left(\frac{\sin \Delta x}{\Delta x}\right) - (\sin x)\left(\frac{1 - \cos \Delta x}{\Delta x}\right)\right]$$

$$= \cos x \left(\lim_{\Delta x \to 0} \frac{\sin \Delta x}{\Delta x}\right) - \sin x \left(\lim_{\Delta x \to 0} \frac{1 - \cos \Delta x}{\Delta x}\right)$$

$$= (\cos x)(1) - (\sin x)(0)$$

$$= \cos x$$

This differentiation rule is shown graphically in Figure 3.18. Note that for each x, the *slope* of the sine curve is equal to the value of the cosine. The proof of the second rule is left as an exercise (see Exercise 116).

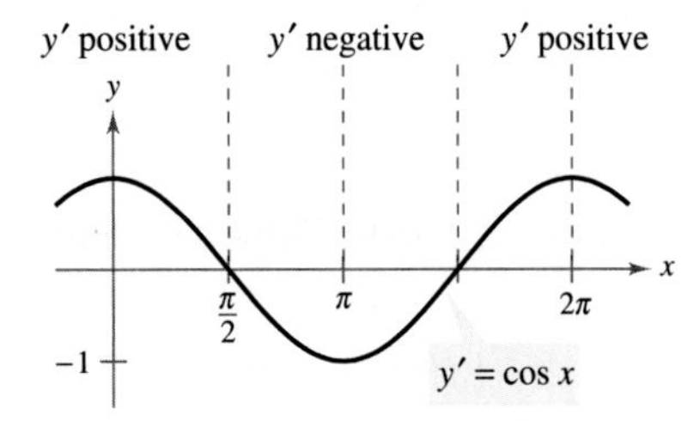

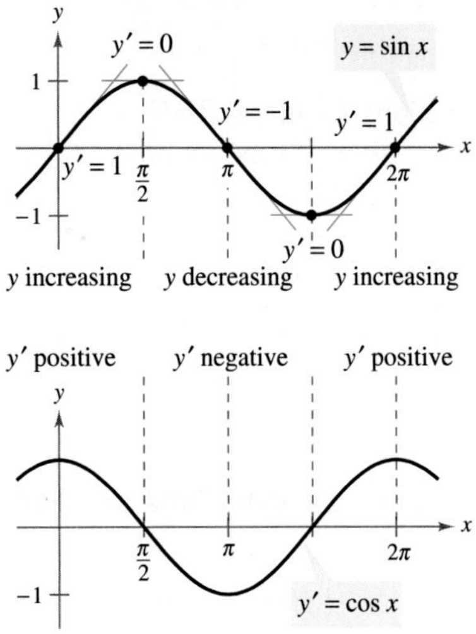

The derivative of the sine function is the cosine function.
Figure 3.18

EXAMPLE 8 Derivatives Involving Sines and Cosines

Function	*Derivative*
a. $y = 2 \sin x$	$y' = 2 \cos x$
b. $y = \dfrac{\sin x}{2} = \dfrac{1}{2} \sin x$	$y' = \dfrac{1}{2} \cos x = \dfrac{\cos x}{2}$
c. $y = x + \cos x$	$y' = 1 - \sin x$

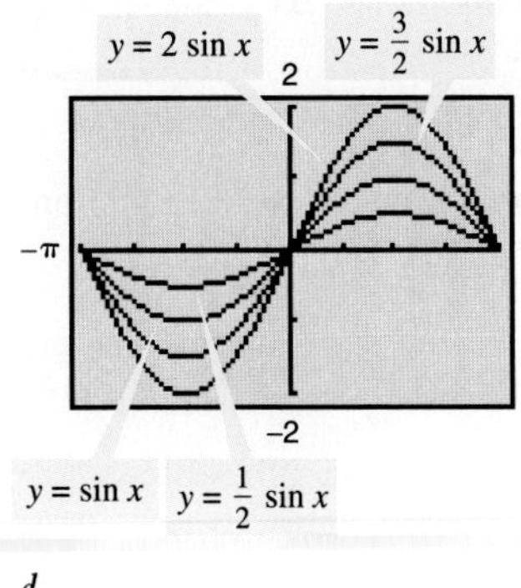

TECHNOLOGY A graphing utility can provide insight into the interpretation of a derivative. For instance, Figure 3.19 shows the graphs of

$$y = a \sin x$$

for $a = \frac{1}{2}, 1, \frac{3}{2}$, and 2. Estimate the slope of each graph at the point $(0, 0)$. Then verify your estimates analytically by evaluating the derivative of each function when $x = 0$.

$\frac{d}{dx}[a \sin x] = a \cos x$
Figure 3.19

EXPLORATION

Use a graphing utility to graph the function

$$f(x) = \frac{e^{x+\Delta x} - e^x}{\Delta x}$$

for $\Delta x = 0.01$. What does this function represent? Compare this graph with that of the exponential function. What do you think the derivative of the exponential function equals?

Derivatives of Exponential Functions

One of the most intriguing (and useful) characteristics of the natural exponential function is that *it is its own derivative*. Consider the following.

Let $f(x) = e^x$.

$$\begin{aligned} f'(x) &= \lim_{\Delta x \to 0} \frac{f(x + \Delta x) - f(x)}{\Delta x} \\ &= \lim_{\Delta x \to 0} \frac{e^{x+\Delta x} - e^x}{\Delta x} \\ &= \lim_{\Delta x \to 0} \frac{e^x(e^{\Delta x} - 1)}{\Delta x} \end{aligned}$$

The definition of e

$$\lim_{\Delta x \to 0} (1 + \Delta x)^{1/\Delta x} = e$$

tells you that for small values of Δx, you have $e \approx (1 + \Delta x)^{1/\Delta x}$, which implies that $e^{\Delta x} \approx 1 + \Delta x$. Replacing $e^{\Delta x}$ by this approximation produces the following.

$$\begin{aligned} f'(x) &= \lim_{\Delta x \to 0} \frac{e^x[e^{\Delta x} - 1]}{\Delta x} \\ &= \lim_{\Delta x \to 0} \frac{e^x[(1 + \Delta x) - 1]}{\Delta x} \\ &= \lim_{\Delta x \to 0} \frac{e^x \Delta x}{\Delta x} \\ &= e^x \end{aligned}$$

This result is stated in the next theorem.

STUDY TIP The key to the formula for the derivative of $f(x) = e^x$ is the limit

$$\lim_{x \to 0} (1 + x)^{1/x} = e.$$

This important limit was introduced on page 51 and formalized later on page 85. It is used to conclude that for $\Delta x \approx 0$,

$$(1 + \Delta x)^{1/\Delta x} \approx e.$$

THEOREM 3.7 Derivative of the Natural Exponential Function

$$\frac{d}{dx}[e^x] = e^x$$

You can interpret Theorem 3.7 graphically by saying that the slope of the graph of $f(x) = e^x$ at any point (x, e^x) is equal to the y-coordinate of the point, as shown in Figure 3.20.

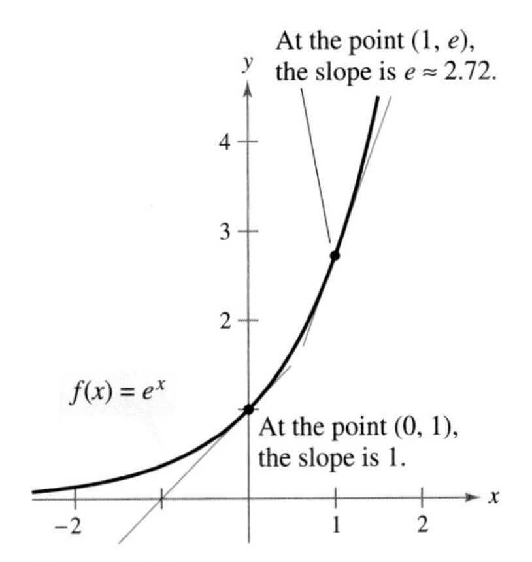

Figure 3.20

EXAMPLE 9 Derivatives of Exponential Functions

Find the derivative of each function.

a. $f(x) = 3e^x$ **b.** $f(x) = x^2 + e^x$ **c.** $f(x) = \sin x - e^x$

Solution

a. $f'(x) = 3\frac{d}{dx}[e^x] = 3e^x$

b. $f'(x) = \frac{d}{dx}[x^2] + \frac{d}{dx}[e^x] = 2x + e^x$

c. $f'(x) = \frac{d}{dx}[\sin x] - \frac{d}{dx}[e^x] = \cos x - e^x$

Rates of Change

You have seen how the derivative is used to determine slope. The derivative can also be used to determine the rate of change of one variable with respect to another. Applications involving rates of change occur in a wide variety of fields. A few examples are population growth rates, production rates, water flow rates, velocity, and acceleration.

A common use for rate of change is to describe the motion of an object moving in a straight line. In such problems, it is customary to use either a horizontal or a vertical line with a designated origin to represent the line of motion. On such lines, movement to the right (or upward) is considered to be in the positive direction, and movement to the left (or downward) is considered to be in the negative direction.

The function s that gives the position (relative to the origin) of an object as a function of time t is called a **position function.** If, over a period of time Δt, the object changes its position by the amount $\Delta s = s(t + \Delta t) - s(t)$, then, by the familiar formula

$$\text{Rate} = \frac{\text{distance}}{\text{time}}$$

the **average velocity** is

$$\frac{\text{Change in distance}}{\text{Change in time}} = \frac{\Delta s}{\Delta t}. \qquad \text{Average velocity}$$

EXAMPLE 10 Finding Average Velocity of a Falling Object

If a billiard ball is dropped from a height of 100 feet, its height s at time t is given by the position function

$$s = -16t^2 + 100 \qquad \text{Position function}$$

where s is measured in feet and t is measured in seconds. Find the average velocity over each of the following time intervals.

a. $[1, 2]$ **b.** $[1, 1.5]$ **c.** $[1, 1.1]$

Solution

a. For the interval $[1, 2]$, the object falls from a height of $s(1) = -16(1)^2 + 100 = 84$ feet to a height of $s(2) = -16(2)^2 + 100 = 36$ feet. The average velocity is

$$\frac{\Delta s}{\Delta t} = \frac{36 - 84}{2 - 1} = \frac{-48}{1} = -48 \text{ feet per second.}$$

b. For the interval $[1, 1.5]$, the object falls from a height of 84 feet to a height of 64 feet. The average velocity is

$$\frac{\Delta s}{\Delta t} = \frac{64 - 84}{1.5 - 1} = \frac{-20}{0.5} = -40 \text{ feet per second.}$$

c. For the interval $[1, 1.1]$, the object falls from a height of 84 feet to a height of 80.64 feet. The average velocity is

$$\frac{\Delta s}{\Delta t} = \frac{80.64 - 84}{1.1 - 1} = \frac{-3.36}{0.1} = -33.6 \text{ feet per second.}$$

Note that the average velocities are *negative*, indicating that the object is moving downward.

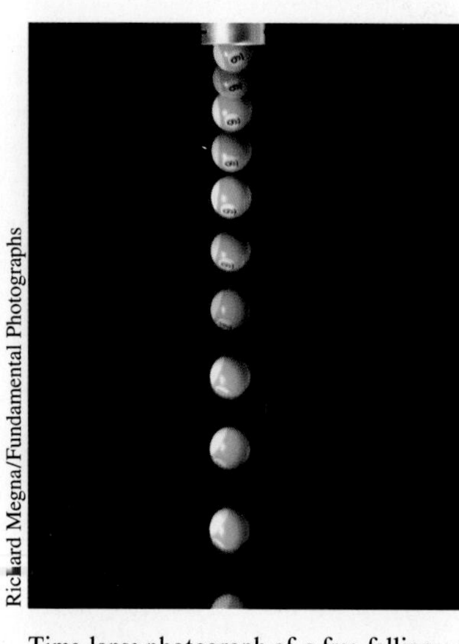

Time-lapse photograph of a free-falling billiard ball

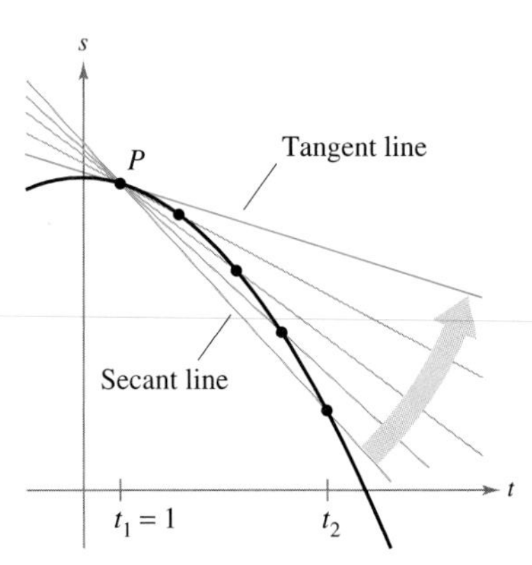

The average velocity between t_1 and t_2 is the slope of the secant line, and the instantaneous velocity at t_1 is the slope of the tangent line.
Figure 3.21

Suppose that in Example 10 you wanted to find the *instantaneous* velocity (or simply the velocity) of the object when $t = 1$. Just as you can approximate the slope of the tangent line by calculating the slope of the secant line, you can approximate the velocity at $t = 1$ by calculating the average velocity over a small interval $[1, 1 + \Delta t]$ (see Figure 3.21). By taking the limit as Δt approaches zero, you obtain the velocity when $t = 1$. Try doing this—you will find that the velocity when $t = 1$ is -32 feet per second.

In general, if $s = s(t)$ is the position function for an object moving along a straight line, the **velocity** of the object at time t is

$$v(t) = \lim_{\Delta t \to 0} \frac{s(t + \Delta t) - s(t)}{\Delta t} = s'(t).$$ Velocity function

In other words, the velocity function is the derivative of the position function. Velocity can be negative, zero, or positive. The **speed** of an object is the absolute value of its velocity. Speed cannot be negative.

The position of a free-falling object (neglecting air resistance) under the influence of gravity can be represented by the equation

$$s(t) = \frac{1}{2}gt^2 + v_0 t + s_0$$ Position function

where s_0 is the initial height of the object, v_0 is the initial velocity of the object, and g is the acceleration due to gravity. On Earth, the value of g is approximately -32 feet per second per second or -9.8 meters per second per second.

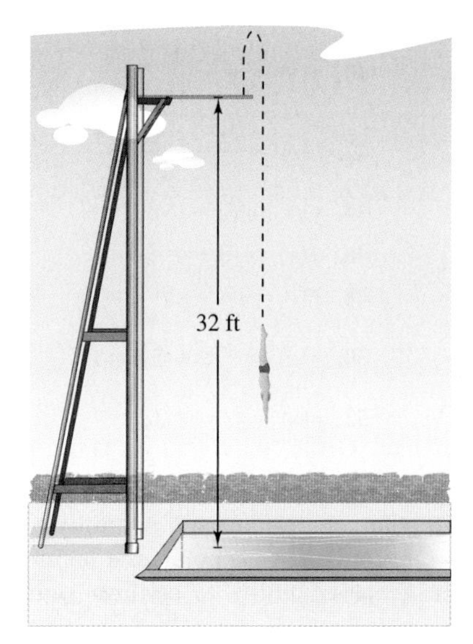

Velocity is positive when an object is rising and negative when an object is falling.
Figure 3.22

NOTE In Figure 3.22, note that the diver moves upward for the first half-second because the velocity is positive for $0 < t < \frac{1}{2}$. When the velocity is 0, the diver has reached the maximum height of the dive.

EXAMPLE 11 Using the Derivative to Find Velocity

At time $t = 0$, a diver jumps from a platform diving board that is 32 feet above the water (see Figure 3.22). The position of the diver is given by

$$s(t) = -16t^2 + 16t + 32$$ Position function

where s is measured in feet and t is measured in seconds.

a. When does the diver hit the water?

b. What is the diver's velocity at impact?

Solution

a. To find the time t when the diver hits the water, let $s = 0$ and solve for t.

$$-16t^2 + 16t + 32 = 0$$ Set position function equal to 0.

$$-16(t + 1)(t - 2) = 0$$ Factor.

$$t = -1 \text{ or } 2$$ Solve for t.

Because $t \geq 0$, choose the positive value to conclude that the diver hits the water at $t = 2$ seconds.

b. The velocity at time t is given by the derivative $s'(t) = -32t + 16$. So, the velocity at time $t = 2$ is

$$s'(2) = -32(2) + 16 = -48 \text{ feet per second.}$$

Exercises for Section 3.2

See www.CalcChat.com for worked-out solutions to odd-numbered exercises.

In Exercises 1 and 2, use the graph to estimate the slope of the tangent line to $y = x^n$ at the point $(1, 1)$. Verify your answer analytically. To print an enlarged copy of the graph, go to the website *www.mathgraphs.com*.

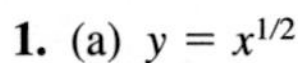

1. (a) $y = x^{1/2}$ (b) $y = x^3$

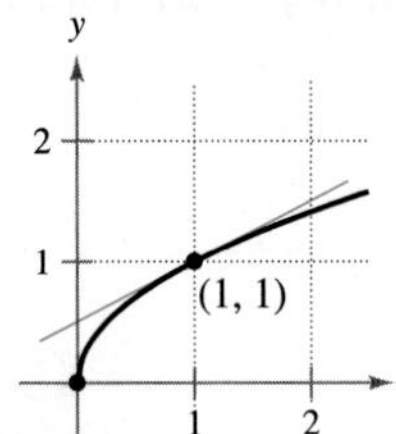

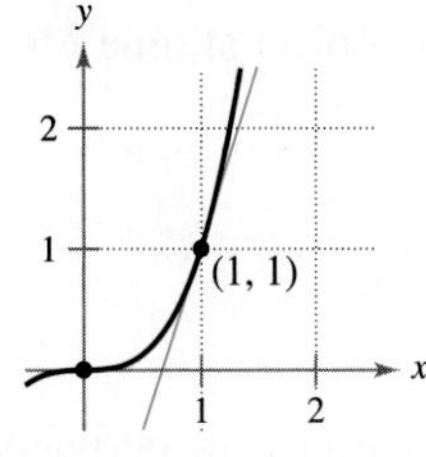

2. (a) $y = x^{-1/2}$ (b) $y = x^{-1}$

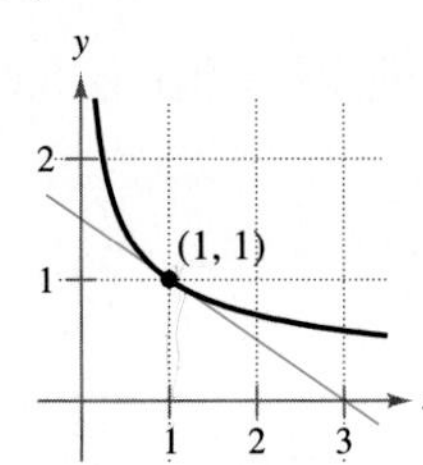

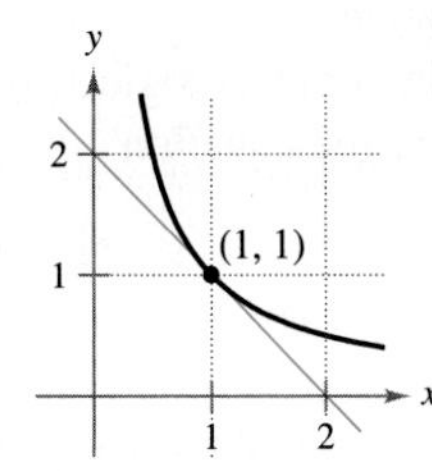

In Exercises 3–24, find the derivative of the function.

3. $y = 8$

4. $f(x) = -6$

5. $y = x^6$

6. $y = \dfrac{1}{x^8}$

7. $f(x) = \sqrt[5]{x}$

8. $g(x) = \sqrt[6]{x}$

9. $f(x) = x + 1$

10. $g(x) = 3x - 1$

11. $f(t) = -2t^2 + 3t - 6$

12. $y = t^2 + 2t - 3$

13. $g(x) = x^2 + 4x^3$

14. $y = 8 - x^3$

15. $s(t) = t^3 - 2t + 4$

16. $f(x) = 2x^3 - 4x^2 + 3x$

17. $f(x) = 6x - 5e^x$

18. $h(t) = t^3 + 2e^t$

19. $y = \dfrac{\pi}{2}\sin\theta - \cos\theta$

20. $g(t) = \pi\cos t$

21. $y = x^2 - \frac{1}{2}\cos x$

22. $y = 5 + \sin x$

23. $y = \frac{1}{2}e^x - 3\sin x$

24. $y = \frac{3}{4}e^x + 2\cos x$

In Exercises 25–30, complete the table using Example 6 as a model.

Original Function	Rewrite	Differentiate	Simplify
25. $y = \dfrac{5}{2x^2}$			
26. $y = \dfrac{4}{3x^2}$			
27. $y = \dfrac{3}{(2x)^3}$			
28. $y = \dfrac{\pi}{(5x)^2}$			
29. $y = \dfrac{\sqrt{x}}{x}$			
30. $y = \dfrac{4}{x^{-3}}$			

In Exercises 31–38, find the slope of the graph of the function at the indicated point. Use the *derivative* feature of a graphing utility to confirm your results.

Function	Point
31. $f(x) = \dfrac{3}{x^2}$	$(1, 3)$
32. $f(t) = 3 - \dfrac{3}{5t}$	$\left(\frac{3}{5}, 3\right)$
33. $f(x) = -\frac{1}{2} + \frac{7}{5}x^3$	$\left(0, -\frac{1}{2}\right)$
34. $f(x) = 3(5 - x)^2$	$(5, 0)$
35. $f(\theta) = 4\sin\theta - \theta$	$(0, 0)$
36. $g(t) = 2 + 3\cos t$	$(\pi, -1)$
37. $f(t) = \frac{3}{4}e^t$	$\left(0, \frac{3}{4}\right)$
38. $g(x) = -4e^x$	$(1, -4e)$

In Exercises 39–52, find the derivative of the function.

39. $g(t) = t^2 - \dfrac{4}{t^3}$

40. $f(x) = x + \dfrac{1}{x^2}$

41. $f(x) = \dfrac{x^3 - 3x^2 + 4}{x^2}$

42. $h(x) = \dfrac{2x^2 - 3x + 1}{x}$

43. $y = x(x^2 + 1)$

44. $y = 3x(6x - 5x^2)$

45. $f(x) = \sqrt{x} - 6\sqrt[3]{x}$

46. $f(x) = \sqrt[3]{x} + \sqrt[5]{x}$

47. $h(s) = s^{4/5} - s^{2/3}$

48. $f(t) = t^{2/3} - t^{1/3} + 4$

49. $f(x) = 6\sqrt{x} + 5\cos x$

50. $f(x) = \dfrac{2}{\sqrt[3]{x}} + 5\cos x$

51. $f(x) = x^{-2} - 2e^x$

52. $g(x) = \sqrt{x} - 3e^x$

In Exercises 53–56, (a) find an equation of the tangent line to the graph of f at the given point, (b) use a graphing utility to graph the function and its tangent line at the point, and (c) use the *derivative* feature of a graphing utility to confirm your results.

Function	Point
53. $y = x^4 - x$	$(-1, 2)$
54. $f(x) = \dfrac{2}{\sqrt[4]{x^3}}$	$(1, 2)$
55. $g(x) = x + e^x$	$(0, 1)$
56. $h(t) = \sin t + \frac{1}{2}e^t$	$\left(\pi, \frac{1}{2}e^\pi\right)$

In Exercises 57–62, determine the point(s) (if any) at which the graph of the function has a horizontal tangent line.

57. $y = x^4 - 8x^2 + 2$

58. $y = \dfrac{1}{x^2}$

59. $y = x + \sin x, \quad 0 \le x < 2\pi$

60. $y = \sqrt{3}x + 2\cos x, \quad 0 \le x < 2\pi$

61. $y = -4x + e^x$

62. $y = x + 4e^x$

In Exercises 63–66, find k such that the line is tangent to the graph of the function.

	Function	*Line*
63.	$f(x) = x^2 - kx$	$y = 4x - 9$
64.	$f(x) = k - x^2$	$y = -4x + 7$
65.	$f(x) = \dfrac{k}{x}$	$y = -\dfrac{3}{4}x + 3$
66.	$f(x) = k\sqrt{x}$	$y = x + 4$

Writing About Concepts

67. Use the graph of f to answer each question. To print an enlarged copy of the graph, go to the website *www.mathgraphs.com.*

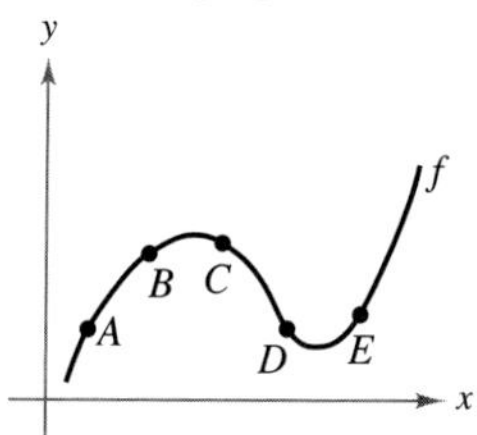

(a) Between which two consecutive points is the average rate of change of the function greatest?

(b) Is the average rate of change of the function between A and B greater than or less than the instantaneous rate of change at B?

(c) Sketch a tangent line to the graph between C and D such that the slope of the tangent line is the same as the average rate of change of the function between C and D.

68. Sketch the graph of a function f such that $f' > 0$ for all x and the rate of change of the function is decreasing.

In Exercises 69 and 70, the relationship between f and g is given. Explain the relationship between f' and g'.

69. $g(x) = f(x) + 6$

70. $g(x) = -5f(x)$

Writing About Concepts *(continued)*

In Exercises 71 and 72, the graphs of a function f and its derivative f' are shown on the same set of coordinate axes. Label the graphs as f or f' and write a short paragraph stating the criteria used in making the selection. To print an enlarged copy of the graph, go to the website *www.mathgraphs.com.*

71.

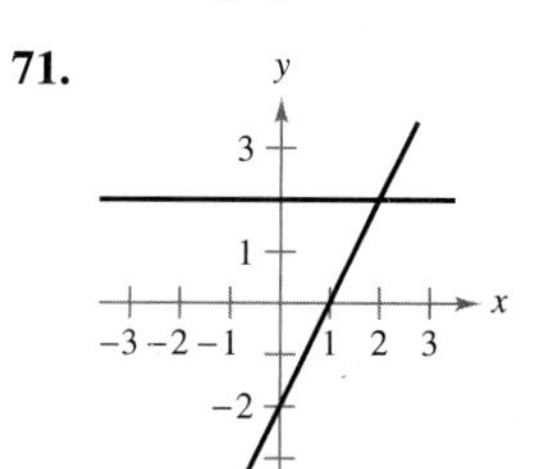

72.

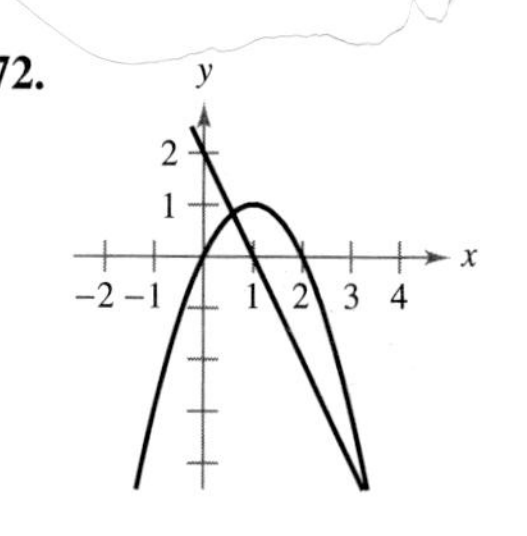

73. Sketch the graphs of $y = x^2$ and $y = -x^2 + 6x - 5$, and sketch the two lines that are tangent to both graphs. Find equations of these lines.

74. Show that the graphs of the two equations $y = x$ and $y = 1/x$ have tangent lines that are perpendicular to each other at their point of intersection.

75. Show that the graph of the function

$f(x) = 3x + \sin x + 2$

does not have a horizontal tangent line.

76. Show that the graph of the function

$f(x) = x^5 + 3x^3 + 5x$

does not have a tangent line with a slope of 3.

In Exercises 77 and 78, find an equation of the tangent line to the graph of the function f through the point (x_0, y_0) not on the graph. To find the point of tangency (x, y) on the graph of f, solve the equation

$$f'(x) = \frac{y_0 - y}{x_0 - x}.$$

77. $f(x) = \sqrt{x}$

$(x_0, y_0) = (-4, 0)$

78. $f(x) = \dfrac{2}{x}$

$(x_0, y_0) = (5, 0)$

79. *Linear Approximation* Use a graphing utility (in square mode) to zoom in on the graph of $f(x) = 4 - \frac{1}{2}x^2$ to approximate $f'(1)$. Use the derivative to find $f'(1)$.

80. *Linear Approximation* Use a graphing utility (in square mode) to zoom in on the graph of $f(x) = 4\sqrt{x} + 1$ to approximate $f'(4)$. Use the derivative to find $f'(4)$.

81. ***Linear Approximation*** Consider the function $f(x) = x^{3/2}$ with the solution point $(4, 8)$.

(a) Use a graphing utility to obtain the graph of f. Use the *zoom* feature to obtain successive magnifications of the graph in the neighborhood of the point $(4, 8)$. After zooming in a few times, the graph should appear nearly linear. Use the *trace* feature to determine the coordinates of a point near $(4, 8)$. Find an equation of the secant line $S(x)$ through the two points.

(b) Find the equation of the line

$$T(x) = f'(4)(x - 4) + f(4)$$

tangent to the graph of f passing through the given point. Why are the linear functions S and T nearly the same?

(c) Use a graphing utility to graph f and T on the same set of coordinate axes. Note that T is a good approximation of f when x is close to 4. What happens to the accuracy of the approximation as you move farther away from the point of tangency?

(d) Demonstrate the conclusion in part (c) by completing the table.

Δx	-3	-2	-1	-0.5	-0.1	0
$f(4 + \Delta x)$						
$T(4 + \Delta x)$						

Δx	0.1	0.5	1	2	3
$f(4 + \Delta x)$					
$T(4 + \Delta x)$					

82. ***Linear Approximation*** Repeat Exercise 81 for the function $f(x) = x^3$, where $T(x)$ is the line tangent to the graph at the point $(1, 1)$. Explain why the accuracy of the linear approximation decreases more rapidly than in Exercise 81.

True or False? **In Exercises 83–88, determine whether the statement is true or false. If it is false, explain why or give an example that shows it is false.**

83. If $f'(x) = g'(x)$, then $f(x) = g(x)$.

84. If $f(x) = g(x) + c$, then $f'(x) = g'(x)$.

85. If $y = \pi^2$, then $dy/dx = 2\pi$.

86. If $y = x/\pi$, then $dy/dx = 1/\pi$.

87. If $g(x) = 3f(x)$, then $g'(x) = 3f'(x)$.

88. If $f(x) = 1/x^n$, then $f'(x) = 1/(nx^{n-1})$.

In Exercises 89–92, find the average rate of change of the function over the given interval. Compare this average rate of change with the instantaneous rates of change at the endpoints of the interval.

89. $f(x) = \dfrac{-1}{x}$, $[1, 2]$

90. $f(x) = \cos x$, $\left[0, \dfrac{\pi}{3}\right]$

91. $g(x) = x^2 + e^x$, $[0, 1]$

92. $h(x) = x^3 - \frac{1}{2}e^x$, $[0, 2]$

Vertical Motion **In Exercises 93 and 94, use the position function $s(t) = -16t^2 + v_0 t + s_0$ for free-falling objects.**

93. A silver dollar is dropped from the top of a building that is 1362 feet tall.

(a) Determine the position and velocity functions for the coin.

(b) Determine the average velocity on the interval $[1, 2]$.

(c) Find the instantaneous velocities when $t = 1$ and $t = 2$.

(d) Find the time required for the coin to reach ground level.

(e) Find the velocity of the coin at impact.

94. A ball is thrown straight down from the top of a 220-foot building with an initial velocity of -22 feet per second. What is its velocity after 3 seconds? What is its velocity after falling 108 feet?

Vertical Motion **In Exercises 95 and 96, use the position function $s(t) = -4.9t^2 + v_0 t + s_0$ for free-falling objects.**

95. A projectile is shot upward from the surface of Earth with an initial velocity of 120 meters per second. What is its velocity after 5 seconds? After 10 seconds?

96. To estimate the height of a building, a stone is dropped from the top of the building into a pool of water at ground level. How high is the building if the splash is seen 6.8 seconds after the stone is dropped?

Think About It **In Exercises 97 and 98, the graph of a position function is shown. It represents the distance in miles that a person drives during a 10-minute trip to work. Make a sketch of the corresponding velocity function.**

97.

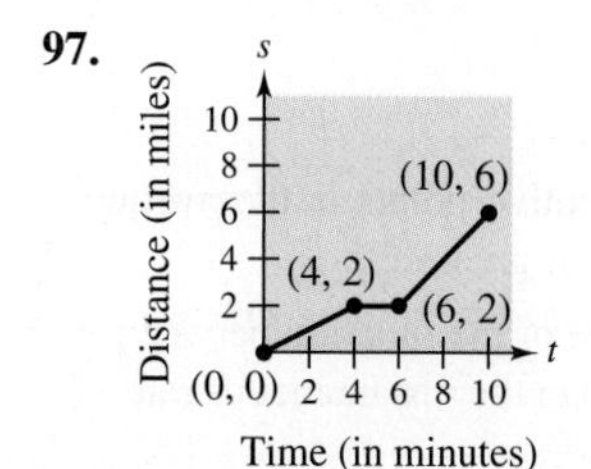

98.

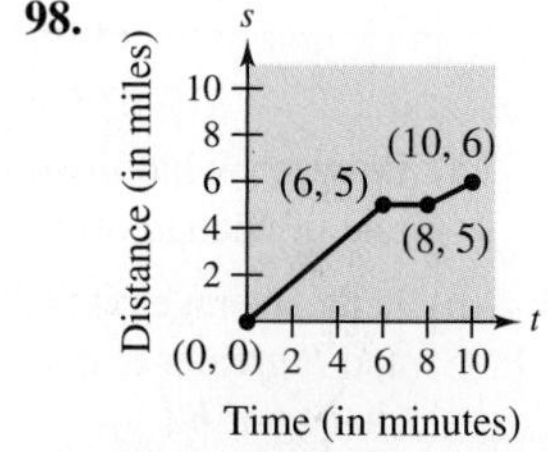

Think About It **In Exercises 99 and 100, the graph of a velocity function is shown. It represents the velocity in miles per hour during a 10-minute drive to work. Make a sketch of the corresponding position function.**

99.

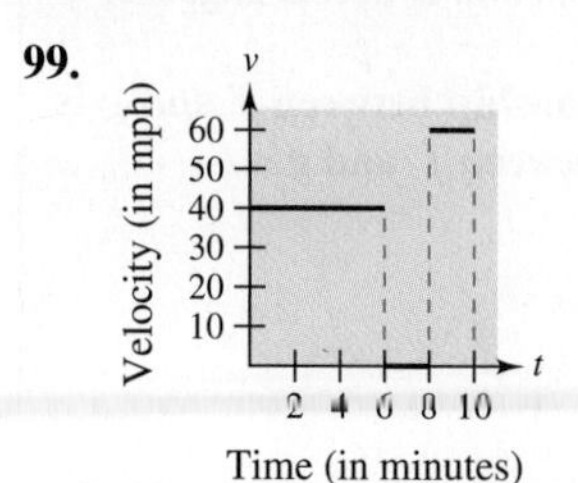

100.

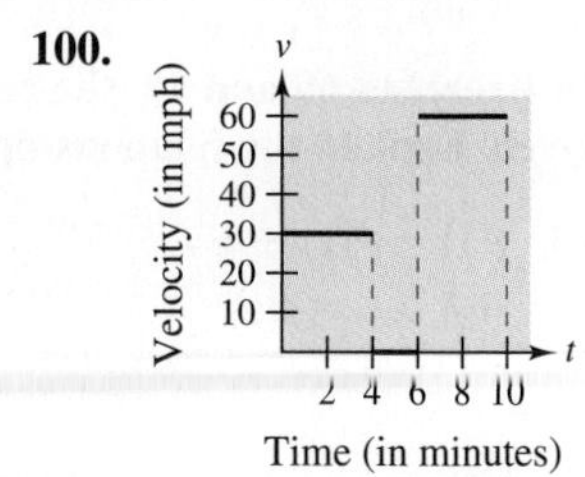

101. *Modeling Data* The stopping distance of an automobile, on dry, level pavement, traveling at a speed v (kilometers per hour) is the distance R (meters) the car travels during the reaction time of the driver plus the distance B (meters) the car travels after the brakes are applied (see figure). The table shows the results of an experiment.

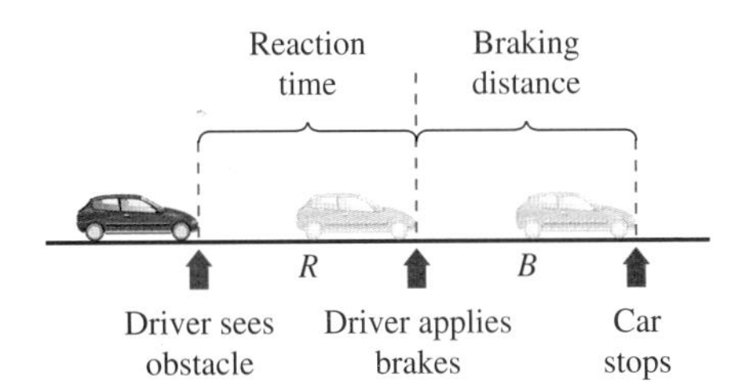

Speed, v	20	40	60	80	100
Reaction Time Distance, R	8.3	16.7	25.0	33.3	41.7
Braking Time Distance, B	2.3	9.0	20.2	35.8	55.9

(a) Use the regression capabilities of a graphing utility to find a linear model for reaction time distance.

(b) Use the regression capabilities of a graphing utility to find a quadratic model for braking distance.

(c) Determine the polynomial giving the total stopping distance T.

(d) Use a graphing utility to graph the functions R, B, and T in the same viewing window.

(e) Find the derivative of T and the rates of change of the total stopping distance for $v = 40$, $v = 80$, and $v = 100$.

(f) Use the results of this exercise to draw conclusions about the total stopping distance as speed increases.

102. *Fuel Cost* A car is driven 15,000 miles a year and gets x miles per gallon. Assume that the average fuel cost is \$1.55 per gallon. Find the annual cost of fuel C as a function of x and use this function to complete the table.

x	10	15	20	25	30	35	40
C							
dC/dx							

Who would benefit more from a one-mile-per-gallon increase in fuel efficiency—the driver of a car that gets 15 miles per gallon or the driver of a car that gets 35 miles per gallon? Explain.

103. *Volume* The volume of a cube with sides of length s is given by $V = s^3$. Find the rate of change of the volume with respect to s when $s = 4$ centimeters.

104. *Area* The area of a square with sides of length s is given by $A = s^2$. Find the rate of change of the area with respect to s when $s = 4$ meters.

105. *Velocity* Verify that the average velocity over the time interval $[t_0 - \Delta t, t_0 + \Delta t]$ is the same as the instantaneous velocity at $t = t_0$ for the position function

$$s(t) = -\tfrac{1}{2}at^2 + c.$$

106. *Inventory Management* The annual inventory cost C for a manufacturer is

$$C = \frac{1{,}008{,}000}{Q} + 6.3Q$$

where Q is the order size when the inventory is replenished. Find the change in annual cost when Q is increased from 350 to 351, and compare this with the instantaneous rate of change when $Q = 350$.

107. *Writing* The number of gallons N of regular unleaded gasoline sold by a gasoline station at a price of p dollars per gallon is given by $N = f(p)$.

(a) Describe the meaning of $f'(1.479)$.

(b) Is $f'(1.479)$ usually positive or negative? Explain.

108. *Newton's Law of Cooling* This law states that the rate of change of the temperature of an object is proportional to the difference between the object's temperature T and the temperature T_a of the surrounding medium. Write an equation for this law.

109. Find an equation of the parabola $y = ax^2 + bx + c$ that passes through $(0, 1)$ and is tangent to the line $y = x - 1$ at $(1, 0)$.

110. Let (a, b) be an arbitrary point on the graph of $y = 1/x$, $x > 0$. Prove that the area of the triangle formed by the tangent line through (a, b) and the coordinate axes is 2.

111. Find the tangent line(s) to the curve $y = x^3 - 9x$ through the point $(1, -9)$.

112. Find the equation(s) of the tangent line(s) to the parabola $y = x^2$ through the given point.

(a) $(0, a)$ (b) $(a, 0)$

Are there any restrictions on the constant a?

In Exercises 113 and 114, find *a* and *b* such that *f* is differentiable everywhere.

113. $f(x) = \begin{cases} ax^3, & x \le 2 \\ x^2 + b, & x > 2 \end{cases}$

114. $f(x) = \begin{cases} \cos x, & x < 0 \\ ax + b, & x \ge 0 \end{cases}$

115. Where are the functions $f_1(x) = |\sin x|$ and $f_2(x) = \sin|x|$ differentiable?

116. Prove that $\dfrac{d}{dx}[\cos x] = -\sin x$.

FOR FURTHER INFORMATION For a geometric interpretation of the derivatives of trigonometric functions, see the article "Sines and Cosines of the Times" by Victor J. Katz in *Math Horizons*. To view this article, go to the website *www.matharticles.com*.

Section 3.3 Product and Quotient Rules and Higher-Order Derivatives

- **Find the derivative of a function using the Product Rule.**
- **Find the derivative of a function using the Quotient Rule.**
- **Find the derivative of a trigonometric function.**
- **Find a higher-order derivative of a function.**

The Product Rule

In Section 3.2 you learned that the derivative of the sum of two functions is simply the sum of their derivatives. The rules for the derivatives of the product and quotient of two functions are not as simple.

NOTE A version of the Product Rule that some people prefer is

$$\frac{d}{dx}[f(x)g(x)] = f'(x)g(x) + f(x)g'(x).$$

The advantage of this form is that it generalizes easily to products involving three or more factors.

THEOREM 3.8 The Product Rule

The product of two differentiable functions f and g is itself differentiable. Moreover, the derivative of fg is the first function times the derivative of the second, plus the second function times the derivative of the first.

$$\frac{d}{dx}[f(x)g(x)] = f(x)g'(x) + g(x)f'(x)$$

Proof Some mathematical proofs, such as the proof of the Sum Rule, are straightforward. Others involve clever steps that may appear unmotivated to a reader. This proof involves such a step—subtracting and adding the same quantity—which is shown in color.

$$\begin{aligned}
\frac{d}{dx}[f(x)g(x)] &= \lim_{\Delta x \to 0} \frac{f(x + \Delta x)g(x + \Delta x) - f(x)g(x)}{\Delta x} \\
&= \lim_{\Delta x \to 0} \frac{f(x + \Delta x)g(x + \Delta x) - f(x + \Delta x)g(x) + f(x + \Delta x)g(x) - f(x)g(x)}{\Delta x} \\
&= \lim_{\Delta x \to 0} \left[f(x + \Delta x)\frac{g(x + \Delta x) - g(x)}{\Delta x} + g(x)\frac{f(x + \Delta x) - f(x)}{\Delta x} \right] \\
&= \lim_{\Delta x \to 0} \left[f(x + \Delta x)\frac{g(x + \Delta x) - g(x)}{\Delta x} \right] + \lim_{\Delta x \to 0} \left[g(x)\frac{f(x + \Delta x) - f(x)}{\Delta x} \right] \\
&= \lim_{\Delta x \to 0} f(x + \Delta x) \cdot \lim_{\Delta x \to 0} \frac{g(x + \Delta x) - g(x)}{\Delta x} + \lim_{\Delta x \to 0} g(x) \cdot \lim_{\Delta x \to 0} \frac{f(x + \Delta x) - f(x)}{\Delta x} \\
&= f(x)g'(x) + g(x)f'(x)
\end{aligned}$$

Note that $\lim_{\Delta x \to 0} f(x + \Delta x) = f(x)$ because f is given to be differentiable and therefore is continuous.

The Product Rule can be extended to cover products involving more than two factors. For example, if f, g, and h are differentiable functions of x, then

$$\frac{d}{dx}[f(x)g(x)h(x)] = f'(x)g(x)h(x) + f(x)g'(x)h(x) + f(x)g(x)h'(x).$$

For instance, the derivative of $y = x^2 \sin x \cos x$ is

$$\begin{aligned}
\frac{dy}{dx} &= 2x \sin x \cos x + x^2 \cos x \cos x + x^2 \sin x(-\sin x) \\
&= 2x \sin x \cos x + x^2(\cos^2 x - \sin^2 x).
\end{aligned}$$

THE PRODUCT RULE

When Leibniz originally wrote a formula for the Product Rule, he was motivated by the expression

$$(x + dx)(y + dy) - xy$$

from which he subtracted $dx\,dy$ (as being negligible) and obtained the differential form $x\,dy + y\,dx$. This derivation resulted in the traditional form of the Product Rule. (*Source:* The History of Mathematics *by David M. Burton*)

The derivative of a product of two functions is not (in general) given by the product of the derivatives of the two functions. To see this, try comparing the product of the derivatives of $f(x) = 3x - 2x^2$ and $g(x) = 5 + 4x$ with the derivative in Example 1.

EXAMPLE 1 Using the Product Rule

Find the derivative of $h(x) = (3x - 2x^2)(5 + 4x)$.

Solution

$$\begin{aligned}
h'(x) &= \overbrace{(3x - 2x^2)}^{\text{First}} \overbrace{\frac{d}{dx}[5 + 4x]}^{\text{Derivative of second}} + \overbrace{(5 + 4x)}^{\text{Second}} \overbrace{\frac{d}{dx}[3x - 2x^2]}^{\text{Derivative of first}} && \text{Apply Product Rule.} \\
&= (3x - 2x^2)(4) + (5 + 4x)(3 - 4x) \\
&= (12x - 8x^2) + (15 - 8x - 16x^2) \\
&= -24x^2 + 4x + 15
\end{aligned}$$

In Example 1, you have the option of finding the derivative with or without the Product Rule. To find the derivative without the Product Rule, you can write

$$\begin{aligned}
D_x[(3x - 2x^2)(5 + 4x)] &= D_x[-8x^3 + 2x^2 + 15x] \\
&= -24x^2 + 4x + 15.
\end{aligned}$$

In the next example, you must use the Product Rule.

EXAMPLE 2 Using the Product Rule

Find the derivative of $y = xe^x$.

Solution

$$\begin{aligned}
\frac{d}{dx}[xe^x] &= x\frac{d}{dx}[e^x] + e^x\frac{d}{dx}[x] && \text{Apply Product Rule.} \\
&= xe^x + e^x(1) \\
&= e^x(x + 1)
\end{aligned}$$

EXAMPLE 3 Using the Product Rule

Find the derivative of $y = 2x \cos x - 2 \sin x$.

Solution

$$\begin{aligned}
\frac{dy}{dx} &= \overbrace{(2x)\left(\frac{d}{dx}[\cos x]\right) + (\cos x)\left(\frac{d}{dx}[2x]\right)}^{\text{Product Rule}} - \overbrace{2\frac{d}{dx}[\sin x]}^{\text{Constant Multiple Rule}} \\
&= (2x)(-\sin x) + (\cos x)(2) - 2(\cos x) \\
&= -2x \sin x
\end{aligned}$$

NOTE In Example 3, notice that you use the Product Rule when both factors of the product are variable, and you use the Constant Multiple Rule when one of the factors is a constant.

The Quotient Rule

THEOREM 3.9 The Quotient Rule

The quotient f/g of two differentiable functions f and g is itself differentiable at all values of x for which $g(x) \neq 0$. Moreover, the derivative of f/g is given by the denominator times the derivative of the numerator minus the numerator times the derivative of the denominator, all divided by the square of the denominator.

$$\frac{d}{dx}\left[\frac{f(x)}{g(x)}\right] = \frac{g(x)f'(x) - f(x)g'(x)}{[g(x)]^2}, \qquad g(x) \neq 0$$

Proof As with the proof of Theorem 3.8, the key to this proof is subtracting and adding the same quantity.

$$\frac{d}{dx}\left[\frac{f(x)}{g(x)}\right] = \lim_{\Delta x \to 0} \frac{\dfrac{f(x+\Delta x)}{g(x+\Delta x)} - \dfrac{f(x)}{g(x)}}{\Delta x} \qquad \text{Definition of derivative}$$

$$= \lim_{\Delta x \to 0} \frac{g(x)f(x+\Delta x) - f(x)g(x+\Delta x)}{\Delta x g(x)g(x+\Delta x)}$$

$$= \lim_{\Delta x \to 0} \frac{g(x)f(x+\Delta x) - f(x)g(x) + f(x)g(x) - f(x)g(x+\Delta x)}{\Delta x g(x)g(x+\Delta x)}$$

$$= \frac{\displaystyle\lim_{\Delta x \to 0} \frac{g(x)[f(x+\Delta x) - f(x)]}{\Delta x} - \lim_{\Delta x \to 0} \frac{f(x)[g(x+\Delta x) - g(x)]}{\Delta x}}{\displaystyle\lim_{\Delta x \to 0} [g(x)g(x+\Delta x)]}$$

$$= \frac{g(x)\left[\displaystyle\lim_{\Delta x \to 0} \frac{f(x+\Delta x) - f(x)}{\Delta x}\right] - f(x)\left[\displaystyle\lim_{\Delta x \to 0} \frac{g(x+\Delta x) - g(x)}{\Delta x}\right]}{\displaystyle\lim_{\Delta x \to 0} [g(x)g(x+\Delta x)]}$$

$$= \frac{g(x)f'(x) - f(x)g'(x)}{[g(x)]^2}$$

Note that $\lim_{\Delta x \to 0} g(x + \Delta x) = g(x)$ because g is given to be differentiable and therefore is continuous.

TECHNOLOGY Graphing utilities can be used to compare the graph of a function with the graph of its derivative. For instance, in Figure 3.23, the graph of the function in Example 4 appears to have two points that have horizontal tangent lines. What are the values of y' at these two points?

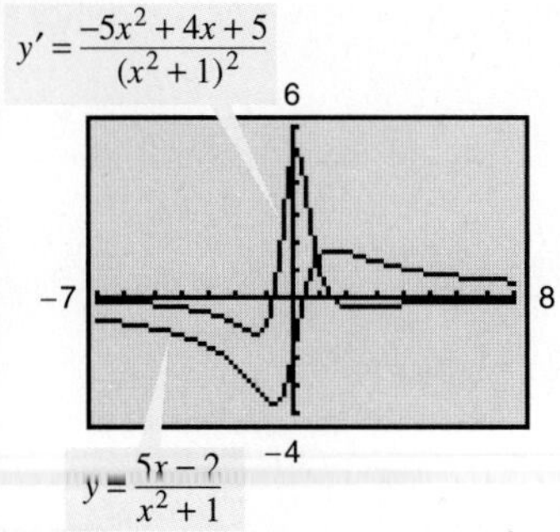

$y' = \dfrac{-5x^2 + 4x + 5}{(x^2 + 1)^2}$

$y = \dfrac{5x - 2}{x^2 + 1}$

Graphical comparison of a function and its derivative
Figure 3.23

EXAMPLE 4 Using the Quotient Rule

Find the derivative of $y = \dfrac{5x - 2}{x^2 + 1}$.

Solution

$$\frac{d}{dx}\left[\frac{5x - 2}{x^2 + 1}\right] = \frac{(x^2 + 1)\dfrac{d}{dx}[5x - 2] - (5x - 2)\dfrac{d}{dx}[x^2 + 1]}{(x^2 + 1)^2} \qquad \text{Apply Quotient Rule.}$$

$$= \frac{(x^2 + 1)(5) - (5x - 2)(2x)}{(x^2 + 1)^2}$$

$$= \frac{(5x^2 + 5) - (10x^2 - 4x)}{(x^2 + 1)^2}$$

$$= \frac{-5x^2 + 4x + 5}{(x^2 + 1)^2}$$

Note the use of parentheses in Example 4. A liberal use of parentheses is recommended for *all* types of differentiation problems. For instance, with the Quotient Rule, it is a good idea to enclose all factors and derivatives in parentheses, and to pay special attention to the subtraction required in the numerator.

When differentiation rules were introduced in the preceding section, the need for rewriting *before* differentiating was emphasized. The next example illustrates this point with the Quotient Rule.

EXAMPLE 5 Rewriting Before Differentiating

Find an equation of the tangent line to the graph of $f(x) = \dfrac{3 - (1/x)}{x + 5}$ at $(-1, 1)$.

Solution Begin by rewriting the function.

$$f(x) = \frac{3 - (1/x)}{x + 5} \qquad \text{Write original function.}$$

$$= \frac{x\left(3 - \frac{1}{x}\right)}{x(x + 5)} \qquad \text{Multiply numerator and denominator by } x.$$

$$= \frac{3x - 1}{x^2 + 5x} \qquad \text{Rewrite.}$$

$$f'(x) = \frac{(x^2 + 5x)(3) - (3x - 1)(2x + 5)}{(x^2 + 5x)^2} \qquad \text{Apply Quotient Rule.}$$

$$= \frac{(3x^2 + 15x) - (6x^2 + 13x - 5)}{(x^2 + 5x)^2}$$

$$= \frac{-3x^2 + 2x + 5}{(x^2 + 5x)^2} \qquad \text{Simplify.}$$

To find the slope at $(-1, 1)$, evaluate $f'(-1)$.

$$f'(-1) = 0 \qquad \text{Slope of graph at } (-1, 1)$$

Then, using the point-slope form of the equation of a line, you can determine that the equation of the tangent line at $(-1, 1)$ is $y = 1$. See Figure 3.24.

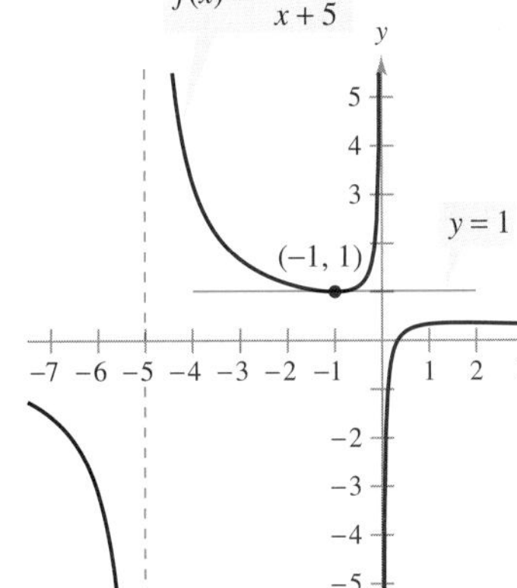

The line $y = 1$ is tangent to the graph of $f(x)$ at the point $(-1, 1)$.
Figure 3.24

Not every quotient needs to be differentiated by the Quotient Rule. For example, each quotient in the next example can be considered as the product of a constant times a function of x. In such cases it is more convenient to use the Constant Multiple Rule.

EXAMPLE 6 Using the Constant Multiple Rule

	Original Function	*Rewrite*	*Differentiate*	*Simplify*
a.	$y = \dfrac{x^2 + 3x}{6}$	$y = \dfrac{1}{6}(x^2 + 3x)$	$y' = \dfrac{1}{6}(2x + 3)$	$y' = \dfrac{2x + 3}{6}$
b.	$y = \dfrac{5x^4}{8}$	$y = \dfrac{5}{8}x^4$	$y' = \dfrac{5}{8}(4x^3)$	$y' = \dfrac{5}{2}x^3$
c.	$y = \dfrac{-3(3x - 2x^2)}{7x}$	$y = -\dfrac{3}{7}(3 - 2x)$	$y' = -\dfrac{3}{7}(-2)$	$y' = \dfrac{6}{7}$
d.	$y = \dfrac{9}{5x^2}$	$y = \dfrac{9}{5}(x^{-2})$	$y' = \dfrac{9}{5}(-2x^{-3})$	$y' = -\dfrac{18}{5x^3}$

NOTE To see the benefit of using the Constant Multiple Rule for some quotients, try using the Quotient Rule to differentiate the functions in Example 6—you should obtain the same results but with more work.

Lt Col Brown is the Best !!

In Section 3.2, the Power Rule was proved only for the case where the exponent n is a positive integer greater than 1. The next example extends the proof to include negative integer exponents.

EXAMPLE 7 Proof of the Power Rule (Negative Integer Exponents)

If n is a negative integer, there exists a positive integer k such that $n = -k$. So, by the Quotient Rule, you can write

$$\frac{d}{dx}[x^n] = \frac{d}{dx}\left[\frac{1}{x^k}\right]$$

$$= \frac{x^k(0) - (1)(kx^{k-1})}{(x^k)^2} \qquad \text{Quotient Rule and Power Rule}$$

$$= \frac{0 - kx^{k-1}}{x^{2k}}$$

$$= -kx^{-k-1}$$

$$= nx^{n-1}. \qquad n = -k$$

So, the Power Rule

$$D_x[x^n] = nx^{n-1} \qquad \text{Power Rule}$$

is valid for any integer. In Exercise 91 in Section 3.5, you are asked to prove the case for which n is any rational number.

Derivatives of Trigonometric Functions

Knowing the derivatives of the sine and cosine functions, you can use the Quotient Rule to find the derivatives of the four remaining trigonometric functions.

THEOREM 3.10 Derivatives of Trigonometric Functions

$$\frac{d}{dx}[\tan x] = \sec^2 x \qquad \frac{d}{dx}[\cot x] = -\csc^2 x$$

$$\frac{d}{dx}[\sec x] = \sec x \tan x \qquad \frac{d}{dx}[\csc x] = -\csc x \cot x$$

Proof Considering $\tan x = (\sin x)/(\cos x)$ and applying the Quotient Rule, you obtain

$$\frac{d}{dx}[\tan x] = \frac{(\cos x)(\cos x) - (\sin x)(-\sin x)}{\cos^2 x} \qquad \text{Apply Quotient Rule.}$$

$$= \frac{\cos^2 x + \sin^2 x}{\cos^2 x}$$

$$= \frac{1}{\cos^2 x}$$

$$= \sec^2 x.$$

The proofs of the other three parts of the theorem are left as an exercise (see Exercise 93).

EXAMPLE 8 Differentiating Trigonometric Functions

NOTE Because of trigonometric identities, the derivative of a trigonometric function can take many forms. This presents a challenge when you are trying to match your answers to those given in the back of the text.

	Function	*Derivative*
a.	$y = x - \tan x$	$\dfrac{dy}{dx} = 1 - \sec^2 x$
b.	$y = x \sec x$	$y' = x(\sec x \tan x) + (\sec x)(1)$ $= (\sec x)(1 + x \tan x)$

EXAMPLE 9 Different Forms of a Derivative

Differentiate both forms of $y = \dfrac{1 - \cos x}{\sin x} = \csc x - \cot x.$

Solution

First form:

$$y = \frac{1 - \cos x}{\sin x}$$

$$y' = \frac{(\sin x)(\sin x) - (1 - \cos x)(\cos x)}{\sin^2 x}$$

$$= \frac{\sin^2 x + \cos^2 x - \cos x}{\sin^2 x}$$

$$= \frac{1 - \cos x}{\sin^2 x}$$

Second form:

$$y = \csc x - \cot x$$

$$y' = -\csc x \cot x + \csc^2 x$$

To verify that the two derivatives are equal, you can write

$$\frac{1 - \cos x}{\sin^2 x} = \frac{1}{\sin^2 x} - \left(\frac{1}{\sin x}\right)\left(\frac{\cos x}{\sin x}\right)$$

$$= \csc^2 x - \csc x \cot x.$$

The summary below shows that much of the work in obtaining a simplified form of a derivative occurs *after* differentiating. Note that two characteristics of a simplified form are the absence of negative exponents and the combining of like terms.

	f′(x) After Differentiating	***f′(x) After Simplifying***
Example 1	$(3x - 2x^2)(4) + (5 + 4x)(3 - 4x)$	$-24x^2 + 4x + 15$
Example 3	$(2x)(-\sin x) + (\cos x)(2) - 2(\cos x)$	$-2x \sin x$
Example 4	$\dfrac{(x^2 + 1)(5) - (5x - 2)(2x)}{(x^2 + 1)^2}$	$\dfrac{-5x^2 + 4x + 5}{(x^2 + 1)^2}$
Example 5	$\dfrac{(x^2 + 5x)(3) - (3x - 1)(2x + 5)}{(x^2 + 5x)^2}$	$\dfrac{-3x^2 + 2x + 5}{(x^2 + 5x)^2}$
Example 9	$\dfrac{(\sin x)(\sin x) - (1 - \cos x)(\cos x)}{\sin^2 x}$	$\dfrac{1 - \cos x}{\sin^2 x}$

EXPLORATION

For which of the functions

$y = e^x, \quad y = \dfrac{1}{e^x}$

$y = \sin x, \quad y = \cos x$

are the following equations true?

a. $y = y'$ b. $y = y''$

c. $y = y'''$ d. $y = y^{(4)}$

Without determining the actual derivative, is $y = y^{(8)}$ for $y = \sin x$ true? What conclusion can you draw from this?

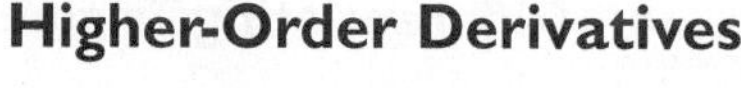

Higher-Order Derivatives

Just as you can obtain a velocity function by differentiating a position function, you can obtain an **acceleration** function by differentiating a velocity function. Another way of looking at this is that you can obtain an acceleration function by differentiating a position function *twice*.

$$s(t) \qquad \text{Position function}$$
$$v(t) = s'(t) \qquad \text{Velocity function}$$
$$a(t) = v'(t) = s''(t) \qquad \text{Acceleration function}$$

The function given by $a(t)$ is the **second derivative** of $s(t)$ and is denoted by $s''(t)$.

The second derivative is an example of a **higher-order derivative.** You can define derivatives of any positive integer order. For instance, the **third derivative** is the derivative of the second derivative. Higher-order derivatives are denoted as follows.

First derivative:	y',	$f'(x)$,	$\dfrac{dy}{dx}$,	$\dfrac{d}{dx}[f(x)]$,	$D_x[y]$
Second derivative:	y'',	$f''(x)$,	$\dfrac{d^2y}{dx^2}$,	$\dfrac{d^2}{dx^2}[f(x)]$,	$D_x^2[y]$
Third derivative:	y''',	$f'''(x)$,	$\dfrac{d^3y}{dx^3}$,	$\dfrac{d^3}{dx^3}[f(x)]$,	$D_x^3[y]$
Fourth derivative:	$y^{(4)}$,	$f^{(4)}(x)$,	$\dfrac{d^4y}{dx^4}$,	$\dfrac{d^4}{dx^4}[f(x)]$,	$D_x^4[y]$
⋮					
nth derivative:	$y^{(n)}$,	$f^{(n)}(x)$,	$\dfrac{d^ny}{dx^n}$,	$\dfrac{d^n}{dx^n}[f(x)]$,	$D_x^n[y]$

THE MOON

The moon's mass is 7.349×10^{22} kilograms, and Earth's mass is 5.976×10^{24} kilograms. The moon's radius is 1737 kilometers, and Earth's radius is 6378 kilometers. Because the gravitational force on the surface of a planet is directly proportional to its mass and inversely proportional to the square of its radius, the ratio of the gravitational force on Earth to the gravitational force on the moon is

$$\frac{(5.976 \times 10^{24})/6378^2}{(7.349 \times 10^{22})/1737^2} \approx 6.03.$$

EXAMPLE 10 Finding the Acceleration Due to Gravity

Because the moon has no atmosphere, a falling object on the moon encounters no air resistance. In 1971, astronaut David Scott demonstrated that a feather and a hammer fall at the same rate on the moon. The position function for each of these falling objects is given by

$$s(t) = -0.81t^2 + 2$$

where $s(t)$ is the height in meters and t is the time in seconds. What is the ratio of Earth's gravitational force to the moon's?

Solution To find the acceleration, differentiate the position function twice.

$$s(t) = -0.81t^2 + 2 \qquad \text{Position function}$$
$$s'(t) = -1.62t \qquad \text{Velocity function}$$
$$s''(t) = -1.62 \qquad \text{Acceleration function}$$

So, the acceleration due to gravity on the moon is -1.62 meters per second per second. Because the acceleration due to gravity on Earth is -9.8 meters per second per second, the ratio of Earth's gravitational force to the moon's is

$$\frac{\text{Earth's gravitational force}}{\text{Moon's gravitational force}} = \frac{-9.8}{-1.62} \approx 6.05.$$

Exercises for Section 3.3

See www.CalcChat.com for worked-out solutions to odd-numbered exercises.

In Exercises 1–6, use the Product Rule to differentiate the function.

1. $g(x) = (x^2 + 1)(x^2 - 2x)$
2. $f(x) = (6x + 5)(x^3 - 2)$
3. $h(t) = \sqrt[3]{t}(t^2 + 4)$
4. $g(s) = \sqrt{s}(4 - s^2)$
5. $f(x) = x^3 \cos x$
6. $g(x) = \sqrt{x} \sin x$

In Exercises 7–12, use the Quotient Rule to differentiate the function.

7. $f(x) = \dfrac{x}{x^2 + 1}$
8. $g(t) = \dfrac{t^2 + 2}{2t - 7}$
9. $h(x) = \dfrac{\sqrt[3]{x}}{x^3 + 1}$
10. $h(s) = \dfrac{s}{\sqrt{s} - 1}$
11. $g(x) = \dfrac{\sin x}{x^2}$
12. $f(t) = \dfrac{\cos t}{t^3}$

In Exercises 13–20, find $f'(x)$ and $f'(c)$.

	Function	*Value of c*
13.	$f(x) = (x^3 - 3x)(2x^2 + 3x + 5)$	$c = 0$
14.	$f(x) = (x^2 - 2x + 1)(x^3 - 1)$	$c = 1$
15.	$f(x) = \dfrac{x^2 - 4}{x - 3}$	$c = 1$
16.	$f(x) = \dfrac{x + 1}{x - 1}$	$c = 2$
17.	$f(x) = x \cos x$	$c = \dfrac{\pi}{4}$
18.	$f(x) = \dfrac{\sin x}{x}$	$c = \dfrac{\pi}{6}$
19.	$f(x) = e^x \sin x$	$c = 0$
20.	$f(x) = \dfrac{\cos x}{e^x}$	$c = 0$

In Exercises 21–26, complete the table without using the Quotient Rule (see Example 6).

	Function	*Rewrite*	*Differentiate*	*Simplify*
21.	$y = \dfrac{x^2 + 2x}{3}$			
22.	$y = \dfrac{5x^2 - 3}{4}$			
23.	$y = \dfrac{7}{3x^3}$			
24.	$y = \dfrac{5}{4x^2}$			
25.	$y = \dfrac{4x^{3/2}}{x}$			
26.	$y = \dfrac{3x^2 - 5}{7}$			

In Exercises 27–40, find the derivative of the algebraic function.

27. $f(x) = \dfrac{3 - 2x - x^2}{x^2 - 1}$
28. $f(x) = \dfrac{x^3 + 3x + 2}{x^2 + 1}$
29. $f(x) = x\left(1 - \dfrac{4}{x + 3}\right)$
30. $f(x) = x^4\left(1 - \dfrac{2}{x + 1}\right)$
31. $f(x) = \dfrac{2x + 5}{\sqrt{x}}$
32. $f(x) = \sqrt[3]{x}\left(\sqrt{x} + 3\right)$
33. $h(s) = (s^3 - 2)^2$
34. $h(x) = (x^2 + 1)^2$
35. $f(x) = \dfrac{2 - \frac{1}{x}}{x - 3}$
36. $g(x) = x^2\left(\dfrac{2}{x} - \dfrac{1}{x + 1}\right)$
37. $f(x) = (3x^3 + 4x)(x - 5)(x + 1)$
38. $f(x) = (x^2 - x)(x^2 + 1)(x^2 + x + 1)$
39. $f(x) = \dfrac{x^2 + c^2}{x^2 - c^2}$, c is a constant
40. $f(x) = \dfrac{c^2 - x^2}{c^2 + x^2}$, c is a constant

In Exercises 41–58, find the derivative of the transcendental function.

41. $f(t) = t^2 \sin t$
42. $f(\theta) = (\theta + 1) \cos \theta$
43. $f(t) = \dfrac{\cos t}{t}$
44. $f(x) = \dfrac{\sin x}{x}$
45. $f(x) = -e^x + \tan x$
46. $y = e^x - \cot x$
47. $g(t) = \sqrt[4]{t} + 8 \sec t$
48. $h(s) = \dfrac{1}{s} - 10 \csc s$
49. $y = \dfrac{3(1 - \sin x)}{2 \cos x}$
50. $y = \dfrac{\sec x}{x}$
51. $y = -\csc x - \sin x$
52. $y = x \cos x + \sin x$
53. $f(x) = x^2 \tan x$
54. $f(x) = 2 \sin x \cos x$
55. $y = 2x \sin x + x^2 e^x$
56. $h(x) = 2e^x \cos x$
57. $y = \dfrac{e^x}{4\sqrt{x}}$
58. $y = \dfrac{2e^x}{x^2 + 1}$

In Exercises 59–62, use a computer algebra system to differentiate the function.

59. $g(x) = \left(\dfrac{x + 1}{x + 2}\right)(2x - 5)$
60. $f(x) = \left(\dfrac{x^2 - x - 3}{x^2 + 1}\right)(x^2 + x + 1)$
61. $g(\theta) = \dfrac{\theta}{1 - \sin \theta}$
62. $f(\theta) = \dfrac{\sin \theta}{1 - \cos \theta}$

In Exercises 63–66, evaluate the derivative of the function at the indicated point. Use a graphing utility to verify your result.

	Function	*Point*
63.	$y = \dfrac{1 + \csc x}{1 - \csc x}$	$\left(\dfrac{\pi}{6}, -3\right)$

Function	*Point*
64. $f(x) = \tan x \cot x$	$(1, 1)$
65. $h(t) = \dfrac{\sec t}{t}$	$\left(\pi, -\dfrac{1}{\pi}\right)$
66. $f(x) = \sin x(\sin x + \cos x)$	$\left(\dfrac{\pi}{4}, 1\right)$

In Exercises 67–72, (a) find an equation of the tangent line to the graph of f at the given point, (b) use a graphing utility to graph the function and its tangent line at the point, and (c) use the *derivative* feature of a graphing utility to confirm your results.

Function	*Point*
67. $f(x) = (x^3 - 3x + 1)(x + 2)$	$(1, -3)$
68. $f(x) = \dfrac{(x - 1)}{(x + 1)}$	$\left(2, \dfrac{1}{3}\right)$
69. $f(x) = \tan x$	$\left(\dfrac{\pi}{4}, 1\right)$
70. $f(x) = \sec x$	$\left(\dfrac{\pi}{3}, 2\right)$
71. $f(x) = (x - 1)e^x$	$(1, 0)$
72. $f(x) = \dfrac{e^x}{x + 4}$	$\left(0, \dfrac{1}{4}\right)$

Famous Curves **In Exercises 73–76, find an equation of the tangent line to the graph at the given point. (The graphs in Exercises 73 and 74 are called *witches of Agnesi*. The graphs in Exercises 75 and 76 are called *serpentines*.)**

73.

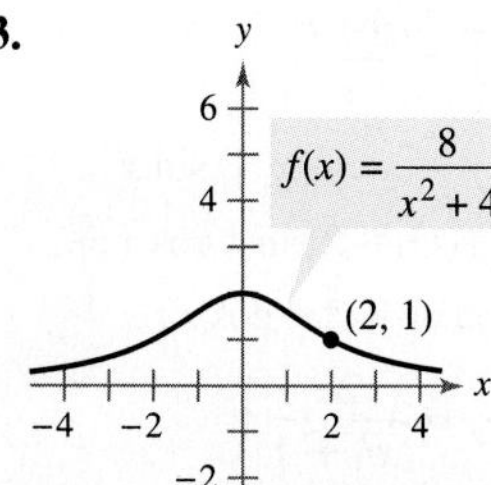

74.

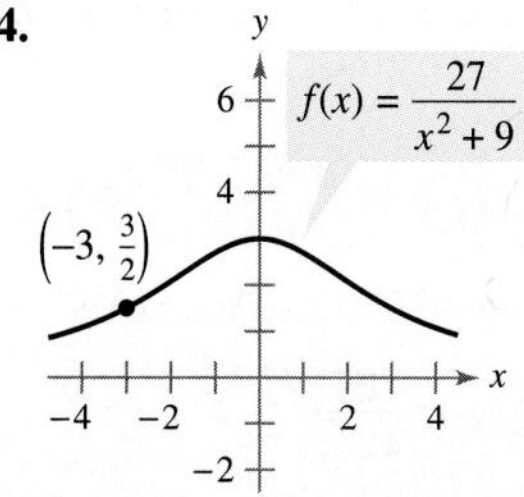

75.

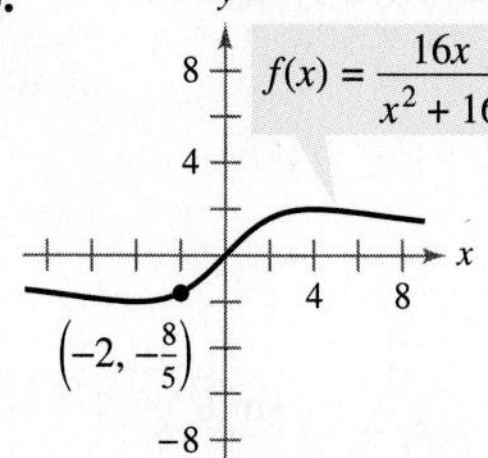

76.

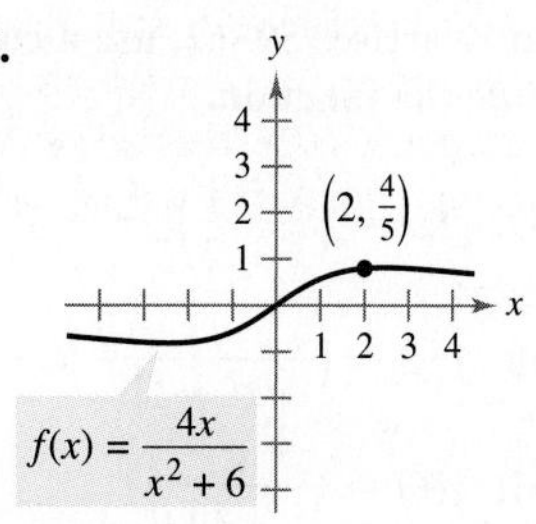

In Exercises 77–80, determine the point(s) at which the graph of the function has a horizontal tangent.

77. $f(x) = \dfrac{x^2}{x - 1}$ **78.** $f(x) = \dfrac{x^2}{x^2 + 1}$

79. $g(x) = \dfrac{8(x - 2)}{e^x}$ **80.** $f(x) = e^x \sin x, \quad [0, \pi]$

81. ***Tangent Lines*** Find equations of the tangent lines to the graph of $f(x) = \dfrac{x + 1}{x - 1}$ that are parallel to the line $2y + x = 6$. Then graph the function and the tangent lines.

82. ***Tangent Lines*** Find equations of the tangent lines to the graph of $f(x) = \dfrac{x}{x - 1}$ that pass through the point $(-1, 5)$. Then graph the function and the tangent lines.

In Exercises 83 and 84, verify that $f'(x) = g'(x)$, and explain the relationship between f and g.

83. $f(x) = \dfrac{3x}{x + 2}, \quad g(x) = \dfrac{5x + 4}{x + 2}$

84. $f(x) = \dfrac{\sin x - 3x}{x}, \quad g(x) = \dfrac{\sin x + 2x}{x}$

In Exercises 85 and 86, use the graphs of f and g. Let $p(x) = f(x)g(x)$ and $q(x) = \dfrac{f(x)}{g(x)}$.

85. 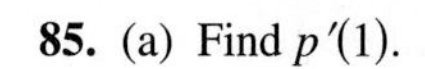
(a) Find $p'(1)$.
(b) Find $q'(4)$.

86. (a) Find $p'(4)$.
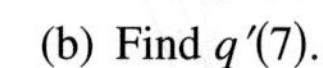
(b) Find $q'(7)$.

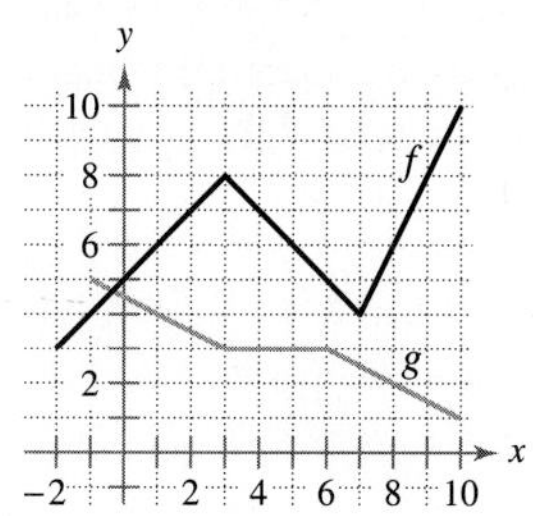

87. ***Area*** The length of a rectangle is given by $2t + 1$ and its height is $\sqrt{t}$, where t is time in seconds and the dimensions are in centimeters. Find the rate of change of the area with respect to time.

88. ***Volume*** The radius of a right circular cylinder is given by $\sqrt{t + 2}$ and its height is $\frac{1}{2}\sqrt{t}$, where t is time in seconds and the dimensions are in inches. Find the rate of change of the volume with respect to time.

89. ***Inventory Replenishment*** The ordering and transportation cost C for the components used in manufacturing a product is

$$C = \frac{375{,}000 + 6x^2}{x}, \quad x \geq 1$$

where C is measured in dollars and x is the order size. Find the rate of change of C with respect to x when (a) $x = 200$, (b) $x = 250$, and (c) $x = 300$. Interpret the meaning of these values.

90. ***Boyle's Law*** This law states that if the temperature of a gas remains constant, its pressure is inversely proportional to its volume. Use the derivative to show that the rate of change of the pressure is inversely proportional to the square of the volume.

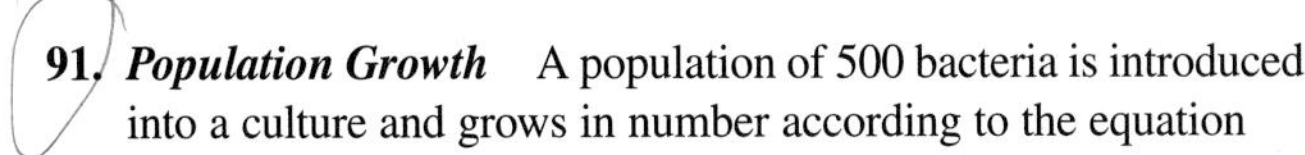

91. ***Population Growth*** A population of 500 bacteria is introduced into a culture and grows in number according to the equation

$$P(t) = 500\left(1 + \frac{4t}{50 + t^2}\right)$$

where t is measured in hours. Find the rate at which the population is growing when $t = 2$.

92. ***Gravitational Force*** Newton's Law of Universal Gravitation states that the force F between two masses, m_1 and m_2, is

$$F = \frac{Gm_1m_2}{d^2}$$

where G is a constant and d is the distance between the masses. Find an equation that gives the instantaneous rate of change of F with respect to d. (Assume m_1 and m_2 represent moving points.)

93. Prove the following differentiation rules.

(a) $\frac{d}{dx}[\sec x] = \sec x \tan x$ (b) $\frac{d}{dx}[\csc x] = -\csc x \cot x$

(c) $\frac{d}{dx}[\cot x] = -\csc^2 x$

94. ***Rate of Change*** Determine whether there exist any values of x in the interval $[0, 2\pi)$ such that the rate of change of $f(x) = \sec x$ and the rate of change of $g(x) = \csc x$ are equal.

95. ***Modeling Data*** The table shows the numbers n (in thousands) of motor homes sold in the United States and the retail values v (in billions of dollars) of these motor homes for the years 1996 through 2001. The year is represented by t, with $t = 6$ corresponding to 1996. *(Source: Recreation Vehicle Industry Association)*

Year, *t*	6	7	8	9	10	11
n	247.5	254.5	292.7	321.2	300.1	256.8
v	6.3	6.9	8.4	10.4	9.5	8.6

(a) Use a graphing utility to find cubic models for the number of motor homes sold $n(t)$ and the total retail value $v(t)$ of the motor homes.

(b) Graph each model found in part (a).

(c) Find $A = v(t)/n(t)$, then graph A. What does this function represent?

(d) Interpret $A'(t)$ in the context of these data.

96. ***Satellites*** When satellites observe Earth, they can scan only part of Earth's surface. Some satellites have sensors that can measure the angle θ shown in the figure. Let h represent the satellite's distance from Earth's surface and let r represent Earth's radius.

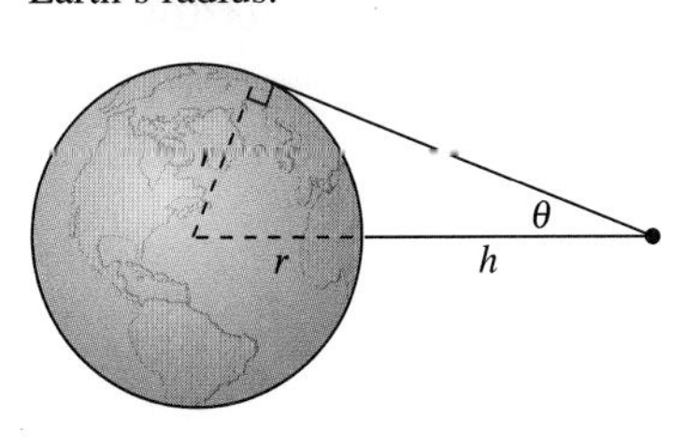

(a) Show that $h = r(\csc \theta - 1)$.

(b) Find the rate at which h is changing with respect to θ when $\theta = 30°$. (Assume $r = 3960$ miles.)

In Exercises 97–104, find the second derivative of the function.

97. $f(x) = 4x^{3/2}$

98. $f(x) = x + 32x^{-2}$

99. $f(x) = \frac{x}{x - 1}$

100. $f(x) = \frac{x^2 + 2x - 1}{x}$

101. $f(x) = 3 \sin x$

102. $f(x) = \sec x$

103. $g(x) = \frac{e^x}{x}$

104. $h(t) = e^t \sin t$

In Exercises 105–108, find the given higher-order derivative.

	Given	*Find*
105.	$f'(x) = x^2$	$f''(x)$
106.	$f''(x) = 2 - \frac{2}{x}$	$f'''(x)$
107.	$f'''(x) = 2\sqrt{x}$	$f^{(4)}(x)$
108.	$f^{(4)}(x) = 2x + 1$	$f^{(6)}(x)$

Writing About Concepts

109. Sketch the graph of a differentiable function f such that $f(2) = 0$, $f' < 0$ for $-\infty < x < 2$, and $f' > 0$ for $2 < x < \infty$.

110. Sketch the graph of a differentiable function f such that $f > 0$ and $f' < 0$ for all real numbers x.

In Exercises 111–114, use the given information to find $f'(2)$.

$g(2) = 3$ and $g'(2) = -2$

$h(2) = -1$ and $h'(2) = 4$

111. $f(x) = 2g(x) + h(x)$

112. $f(x) = 4 - h(x)$

113. $f(x) = \frac{g(x)}{h(x)}$

114. $f(x) = g(x)h(x)$

In Exercises 115 and 116, the graphs of f, f', and f'' are shown on the same set of coordinate axes. Which is which? Explain your reasoning. To print an enlarged copy of the graph, go to the website *www.mathgraphs.com*.

115.

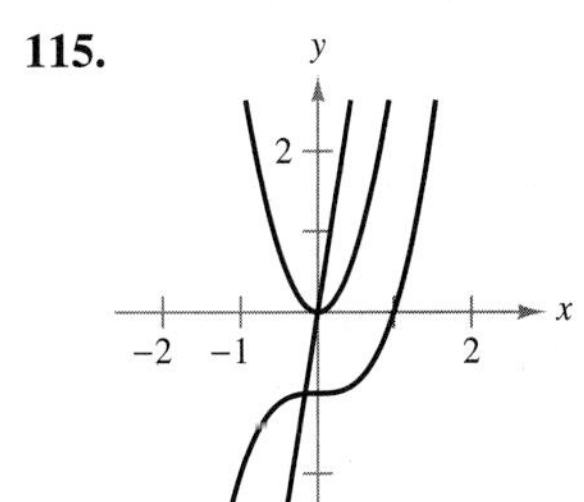

116.

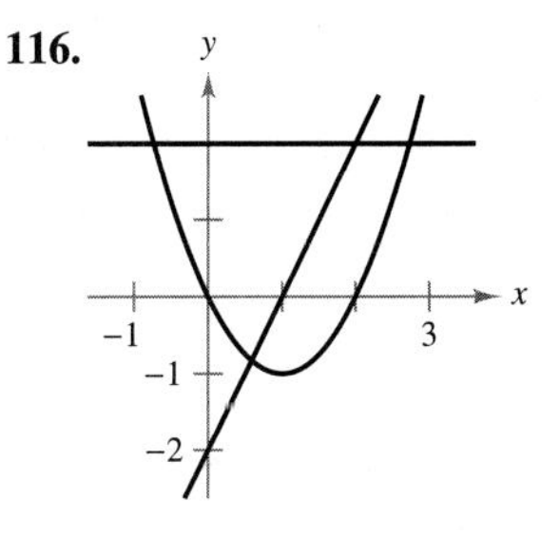

In Exercises 117–120, the graph of f is shown. Sketch the graphs of f' and f''. To print an enlarged copy of the graph, go to the website *www.mathgraphs.com*.

117. **118.**

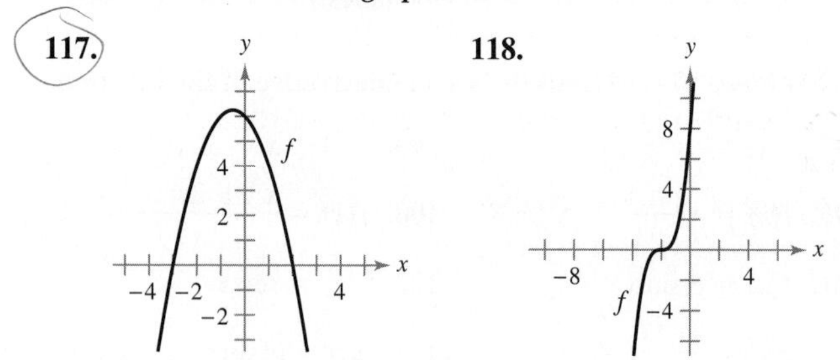

119. **120.**

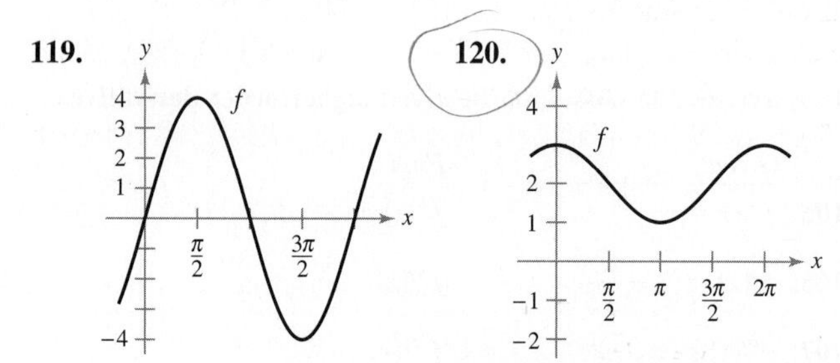

121. ***Acceleration*** The velocity of an object in meters per second is $v(t) = 36 - t^2$, $0 \le t \le 6$. Find the velocity and acceleration of the object when $t = 3$. What can be said about the speed of the object when the velocity and acceleration have opposite signs?

122. ***Particle Motion*** The figure shows the graphs of the position, velocity, and acceleration functions of a particle.

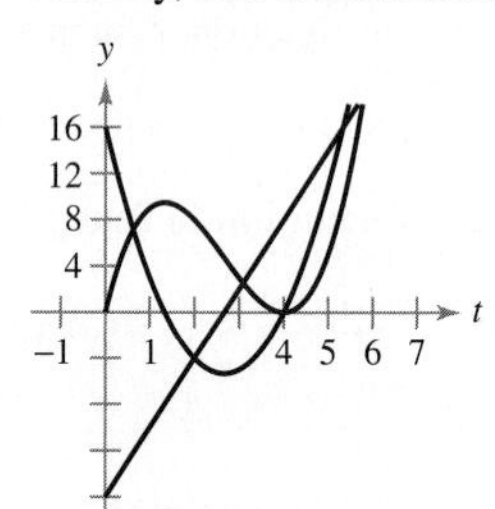

(a) Copy the graphs of the functions shown. Identify each graph. Explain your reasoning. To print an enlarged copy of the graph, go to the website *www.mathgraphs.com*.

(b) On your sketch, identify when the particle speeds up and when it slows down. Explain your reasoning.

Finding a Pattern **In Exercises 123 and 124, develop a general rule for $f^{(n)}(x)$ given $f(x)$.**

123. $f(x) = x^n$ **124.** $f(x) = \dfrac{1}{x}$

125. ***Finding a Pattern*** Consider the function $f(x) = g(x)h(x)$.

(a) Use the Product Rule to generate rules for finding $f''(x)$, $f'''(x)$, and $f^{(4)}(x)$.

(b) Use the results in part (a) to write a general rule for $f^{(n)}(x)$.

126. ***Finding a Pattern*** Develop a general rule for $[xf(x)]^{(n)}$, where f is a differentiable function of x.

Differential Equations **In Exercises 127–130, verify that the function satisfies the differential equation.**

	Function	*Differential Equation*
127.	$y = \dfrac{1}{x}, x > 0$	$x^3 y'' + 2x^2 y' = 0$
128.	$y = 2x^3 - 6x + 10$	$-y''' - xy'' - 2y' = -24x^2$
129.	$y = 2 \sin x + 3$	$y'' + y = 3$
130.	$y = 3 \cos x + \sin x$	$y'' + y = 0$

Linear and Quadratic Approximations **The linear and quadratic approximations of a function f at $x = a$ are**

$$P_1(x) = f'(a)(x - a) + f(a) \text{ and}$$

$$P_2(x) = \tfrac{1}{2}f''(a)(x - a)^2 + f'(a)(x - a) + f(a).$$

In Exercises 131 and 132, (a) find the specified linear and quadratic approximations of f, (b) use a graphing utility to graph f and the approximations, (c) determine whether P_1 or P_2 is the better approximation, and (d) state how the accuracy changes as you move farther from $x = a$.

131. $f(x) = \ln x$, $a = 1$ **132.** $f(x) = e^x$, $a = 0$

True or False? **In Exercises 133–138, determine whether the statement is true or false. If it is false, explain why or give an example that shows it is false.**

133. If $y = f(x)g(x)$, then $dy/dx = f'(x)g'(x)$.

134. If $y = (x + 1)(x + 2)(x + 3)(x + 4)$, then $d^5y/dx^5 = 0$.

135. If $f'(c)$ and $g'(c)$ are zero and $h(x) = f(x)g(x)$, then $h'(c) = 0$.

136. If $f(x)$ is an nth-degree polynomial, then $f^{(n+1)}(x) = 0$.

137. The second derivative represents the rate of change of the first derivative.

138. If the velocity of an object is constant, then its acceleration is zero.

139. Find a second-degree polynomial $f(x) = ax^2 + bx + c$ such that its graph has a tangent line with slope 10 at the point $(2, 7)$ and an x-intercept at $(1, 0)$.

140. Consider the third-degree polynomial

$$f(x) = ax^3 + bx^2 + cx + d, \quad a \neq 0.$$

Determine conditions for a, b, c, and d if the graph of f has (a) no horizontal tangents, (b) exactly one horizontal tangent, and (c) exactly two horizontal tangents. Give an example for each case.

141. Find the derivative of $f(x) = x|x|$. Does $f''(0)$ exist?

142. ***Think About It*** Let f and g be functions whose first and second derivatives exist on an interval I. Which of the following formulas is (are) true?

(a) $fg'' - f''g = (fg' - f'g)'$

(b) $fg'' + f''g = (fg)''$

Section 3.4 The Chain Rule

- **Find the derivative of a composite function using the Chain Rule.**
- **Find the derivative of a function using the General Power Rule.**
- **Simplify the derivative of a function using algebra.**
- **Find the derivative of a transcendental function using the Chain Rule.**
- **Find the derivative of a function involving the natural logarithmic function.**
- **Define and differentiate exponential functions that have bases other than e.**

The Chain Rule

This text has yet to discuss one of the most powerful differentiation rules—the **Chain Rule.** This rule deals with composite functions and adds a surprising versatility to the rules discussed in the two previous sections. For example, compare the following functions. Those on the left can be differentiated without the Chain Rule, and those on the right are best done with the Chain Rule.

Without the Chain Rule	*With the Chain Rule*
$y = x^2 + 1$	$y = \sqrt{x^2 + 1}$
$y = \sin x$	$y = \sin 6x$
$y = 3x + 2$	$y = (3x + 2)^5$
$y = e^x + \tan x$	$y = e^{5x} + \tan x^2$

Basically, the Chain Rule states that if y changes dy/du times as fast as u, and u changes du/dx times as fast as x, then y changes $(dy/du)(du/dx)$ times as fast as x.

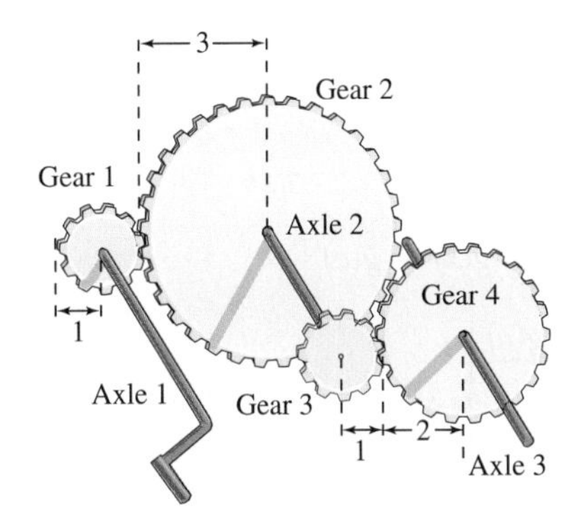

Axle 1: y revolutions per minute
Axle 2: u revolutions per minute
Axle 3: x revolutions per minute
Figure 3.25

EXAMPLE 1 The Derivative of a Composite Function

A set of gears is constructed, as shown in Figure 3.25, such that the second and third gears are on the same axle. As the first axle revolves, it drives the second axle, which in turn drives the third axle. Let y, u, and x represent the numbers of revolutions per minute of the first, second, and third axles, respectively. Find dy/du, du/dx, and dy/dx, and show that

$$\frac{dy}{dx} = \frac{dy}{du} \cdot \frac{du}{dx}.$$

Solution Because the circumference of the second gear is three times that of the first, the first axle must make three revolutions to turn the second axle once. Similarly, the second axle must make two revolutions to turn the third axle once, and you can write

$$\frac{dy}{du} = 3 \quad \text{and} \quad \frac{du}{dx} = 2.$$

Combining these two results, you know that the first axle must make six revolutions to turn the third axle once. So, you can write

$$\frac{dy}{dx} = \boxed{\text{Rate of change of first axle with respect to second axle}} \cdot \boxed{\text{Rate of change of second axle with respect to third axle}}$$

$$= \frac{dy}{du} \cdot \frac{du}{dx} = 3 \cdot 2 = 6 = \boxed{\text{Rate of change of first axle with respect to third axle}}$$

In other words, the rate of change of y with respect to x is the product of the rate of change of y with respect to u and the rate of change of u with respect to x.

EXPLORATION

Using the Chain Rule Each of the following functions can be differentiated using rules that you studied in Sections 3.2 and 3.3. For each function, find the derivative using those rules. Then find the derivative using the Chain Rule. Compare your results. Which method is simpler?

a. $\dfrac{2}{3x + 1}$

b. $(x + 2)^3$

c. $\sin 2x$

Example 1 illustrates a simple case of the Chain Rule. The general rule is stated below.

THEOREM 3.11 The Chain Rule

If $y = f(u)$ is a differentiable function of u and $u = g(x)$ is a differentiable function of x, then $y = f(g(x))$ is a differentiable function of x and

$$\frac{dy}{dx} = \frac{dy}{du} \cdot \frac{du}{dx}$$

or, equivalently,

$$\frac{d}{dx}[f(g(x))] = f'(g(x))g'(x).$$

Proof Let $h(x) = f(g(x))$. Then, using the alternative form of the derivative, you need to show that, for $x = c$,

$$h'(c) = f'(g(c))g'(c).$$

An important consideration in this proof is the behavior of g as x approaches c. A problem occurs if there are values of x, other than c, such that $g(x) = g(c)$. Appendix A shows how to use the differentiability of f and g to overcome this problem. For now, assume that $g(x) \neq g(c)$ for values of x other than c. In the proofs of the Product Rule and the Quotient Rule, the same quantity was added and subtracted to obtain the desired form. This proof uses a similar technique—multiplying and dividing by the same (nonzero) quantity. Note that because g is differentiable, it is also continuous, and it follows that $g(x) \rightarrow g(c)$ as $x \rightarrow c$.

$$\begin{aligned}
h'(c) &= \lim_{x \to c} \frac{f(g(x)) - f(g(c))}{x - c} \\
&= \lim_{x \to c} \left[\frac{f(g(x)) - f(g(c))}{g(x) - g(c)} \cdot \frac{g(x) - g(c)}{x - c}\right], \quad g(x) \neq g(c) \\
&= \left[\lim_{x \to c} \frac{f(g(x)) - f(g(c))}{g(x) - g(c)}\right]\left[\lim_{x \to c} \frac{g(x) - g(c)}{x - c}\right] \\
&= f'(g(c))g'(c)
\end{aligned}$$

When applying the Chain Rule, it is helpful to think of the composite function $f \circ g$ as having two parts—an inner part and an outer part.

Outer function

$$y = f(g(x)) = f(u)$$

Inner function

The derivative of $y = f(u)$ is the derivative of the outer function (at the inner function u) *times* the derivative of the inner function.

$$y' = f'(u) \cdot u'$$

Derivative of outer function

Derivative of inner function

EXAMPLE 2 Decomposition of a Composite Function

$y = f(g(x))$	$u = g(x)$	$y = f(u)$
a. $y = \dfrac{1}{x+1}$	$u = x + 1$	$y = \dfrac{1}{u}$
b. $y = \sin 2x$	$u = 2x$	$y = \sin u$
c. $y = \sqrt{3x^2 - x + 1}$	$u = 3x^2 - x + 1$	$y = \sqrt{u}$
d. $y = \tan^2 x$	$u = \tan x$	$y = u^2$

EXAMPLE 3 Using the Chain Rule

Find dy/dx for $y = (x^2 + 1)^3$.

Solution For this function, you can consider the inside function to be $u = x^2 + 1$. By the Chain Rule, you obtain

$$\frac{dy}{dx} = \underbrace{3(x^2+1)^2}_{\frac{dy}{du}}\underbrace{(2x)}_{\frac{du}{dx}} = 6x(x^2+1)^2.$$

STUDY TIP You could also solve the problem in Example 3 without using the Chain Rule by observing that

$$y = x^6 + 3x^4 + 3x^2 + 1$$

and

$$y' = 6x^5 + 12x^3 + 6x.$$

Verify that this is the same result as the derivative in Example 3. Which method would you use to find

$$\frac{d}{dx}(x^2 + 1)^{50}?$$

The General Power Rule

The function in Example 3 is an example of one of the most common types of composite functions, $y = [u(x)]^n$. The rule for differentiating such functions is called the **General Power Rule,** and it is a special case of the Chain Rule.

THEOREM 3.12 The General Power Rule

If $y = [u(x)]^n$, where u is a differentiable function of x and n is a rational number, then

$$\frac{dy}{dx} = n[u(x)]^{n-1}\frac{du}{dx}$$

or, equivalently,

$$\frac{d}{dx}[u^n] = nu^{n-1}u'.$$

Proof Because $y = u^n$, you apply the Chain Rule to obtain

$$\frac{dy}{dx} = \left(\frac{dy}{du}\right)\left(\frac{du}{dx}\right) = \frac{d}{du}[u^n]\frac{du}{dx}.$$

By the (Simple) Power Rule in Section 3.2, you have $D_u[u^n] = nu^{n-1}$, and it follows that

$$\frac{dy}{dx} = n[u(x)]^{n-1}\frac{du}{dx}.$$

EXAMPLE 4 Applying the General Power Rule

Find the derivative of $f(x) = (3x - 2x^2)^3$.

Solution Let $u = 3x - 2x^2$. Then

$$f(x) = (3x - 2x^2)^3 = u^3$$

and, by the General Power Rule, the derivative is

$$f'(x) = \underbrace{3}_{n}\underbrace{(3x - 2x^2)^2}_{u^{n-1}}\underbrace{\frac{d}{dx}[3x - 2x^2]}_{u'} \quad \text{Apply General Power Rule.}$$

$$= 3(3x - 2x^2)^2(3 - 4x). \quad \text{Differentiate } 3x - 2x^2.$$

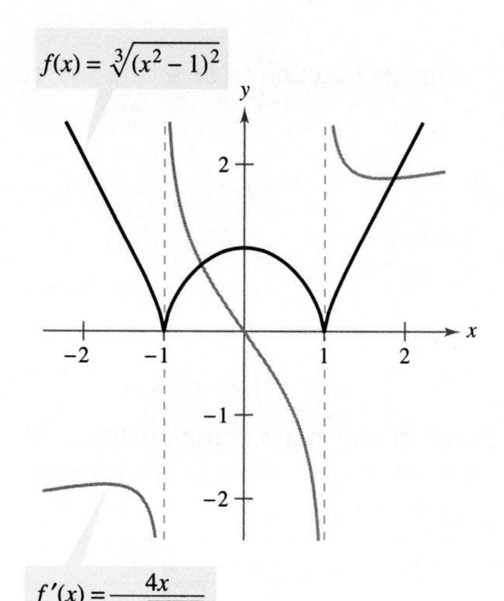

The derivative of f is 0 at $x = 0$ and is undefined at $x = \pm 1$.
Figure 3.26

EXAMPLE 5 Differentiating Functions Involving Radicals

Find all points on the graph of $f(x) = \sqrt[3]{(x^2 - 1)^2}$ for which $f'(x) = 0$ and those for which $f'(x)$ does not exist.

Solution Begin by rewriting the function as

$$f(x) = (x^2 - 1)^{2/3}.$$

Then, applying the General Power Rule (with $u = x^2 - 1$) produces

$$f'(x) = \underbrace{\frac{2}{3}}_{n}\underbrace{(x^2 - 1)^{-1/3}}_{u^{n-1}}\underbrace{(2x)}_{u'} \quad \text{Apply General Power Rule.}$$

$$= \frac{4x}{3\sqrt[3]{x^2 - 1}}. \quad \text{Write in radical form.}$$

So, $f'(x) = 0$ when $x = 0$ and $f'(x)$ does not exist when $x = \pm 1$, as shown in Figure 3.26.

EXAMPLE 6 Differentiating Quotients with Constant Numerators

Differentiate $g(t) = \dfrac{-7}{(2t - 3)^2}$.

Solution Begin by rewriting the function as

$$g(t) = -7(2t - 3)^{-2}.$$

Then, applying the General Power Rule produces

$$g'(t) = \underbrace{(-7)}_{\text{Constant Multiple Rule}}\underbrace{(-2)}_{n}\underbrace{(2t - 3)^{-3}}_{u^{n-1}}\underbrace{(2)}_{u'} \quad \text{Apply General Power Rule.}$$

$$= 28(2t - 3)^{-3} \quad \text{Simplify.}$$

$$= \frac{28}{(2t - 3)^3}. \quad \text{Write with positive exponent.}$$

NOTE Try differentiating the function in Example 6 using the Quotient Rule. You should obtain the same result, but using the Quotient Rule is less efficient than using the General Power Rule.

Simplifying Derivatives

The next three examples illustrate some techniques for simplifying the "raw derivatives" of functions involving products, quotients, and composites.

EXAMPLE 7 Simplifying by Factoring Out the Least Powers

$$
\begin{aligned}
f(x) &= x^2\sqrt{1-x^2} && \text{Original function}\\
&= x^2(1-x^2)^{1/2} && \text{Rewrite.}\\
f'(x) &= x^2\frac{d}{dx}\left[(1-x^2)^{1/2}\right] + (1-x^2)^{1/2}\frac{d}{dx}[x^2] && \text{Product Rule}\\
&= x^2\left[\frac{1}{2}(1-x^2)^{-1/2}(-2x)\right] + (1-x^2)^{1/2}(2x) && \text{General Power Rule}\\
&= -x^3(1-x^2)^{-1/2} + 2x(1-x^2)^{1/2} && \text{Simplify.}\\
&= x(1-x^2)^{-1/2}[-x^2(1) + 2(1-x^2)] && \text{Factor.}\\
&= \frac{x(2-3x^2)}{\sqrt{1-x^2}} && \text{Simplify.}
\end{aligned}
$$

EXAMPLE 8 Simplifying the Derivative of a Quotient

TECHNOLOGY Symbolic differentiation utilities are capable of differentiating very complicated functions. Often, however, the result is given in unsimplified form. If you have access to such a utility, use it to find the derivatives of the functions given in Examples 7, 8, and 9. Then compare the results with those given on this page.

$$
\begin{aligned}
f(x) &= \frac{x}{\sqrt[3]{x^2+4}} && \text{Original function}\\
&= \frac{x}{(x^2+4)^{1/3}} && \text{Rewrite.}\\
f'(x) &= \frac{(x^2+4)^{1/3}(1) - x(1/3)(x^2+4)^{-2/3}(2x)}{(x^2+4)^{2/3}} && \text{Quotient Rule}\\
&= \frac{1}{3}(x^2+4)^{-2/3}\left[\frac{3(x^2+4) - (2x^2)(1)}{(x^2+4)^{2/3}}\right] && \text{Factor.}\\
&= \frac{x^2+12}{3(x^2+4)^{4/3}} && \text{Simplify.}
\end{aligned}
$$

EXAMPLE 9 Simplifying the Derivative of a Power

$$
\begin{aligned}
y &= \left(\frac{3x-1}{x^2+3}\right)^2 && \text{Original function}\\
y' &= \underbrace{2}_{n}\underbrace{\left(\frac{3x-1}{x^2+3}\right)}_{u^{n-1}}\underbrace{\frac{d}{dx}\left[\frac{3x-1}{x^2+3}\right]}_{u'} && \text{General Power Rule}\\
&= \left[\frac{2(3x-1)}{x^2+3}\right]\left[\frac{(x^2+3)(3) - (3x-1)(2x)}{(x^2+3)^2}\right] && \text{Quotient Rule}\\
&= \frac{2(3x-1)(3x^2+9-6x^2+2x)}{(x^2+3)^3} && \text{Multiply.}\\
&= \frac{2(3x-1)(-3x^2+2x+9)}{(x^2+3)^3} && \text{Simplify.}
\end{aligned}
$$

Trigonometric Functions and the Chain Rule

The "Chain Rule versions" of the derivatives of the six trigonometric functions and the natural exponential function are as follows.

$$\frac{d}{dx}[\sin u] = (\cos u)\,u' \qquad \frac{d}{dx}[\cos u] = -(\sin u)\,u'$$

$$\frac{d}{dx}[\tan u] = (\sec^2 u)\,u' \qquad \frac{d}{dx}[\cot u] = -(\csc^2 u)\,u'$$

$$\frac{d}{dx}[\sec u] = (\sec u \tan u)\,u' \qquad \frac{d}{dx}[\csc u] = -(\csc u \cot u)\,u'$$

$$\frac{d}{dx}[e^u] = e^u u'$$

EXAMPLE 10 Applying the Chain Rule to Transcendental Functions

NOTE Be sure that you understand the mathematical conventions regarding parentheses and trigonometric functions. For instance, in Example 10(a), $\sin 2x$ is written to mean $\sin(2x)$.

a. $y = \sin \overbrace{2x}^{u}$ $\qquad y' = \overbrace{\cos 2x}^{\cos u}\,\overbrace{\frac{d}{dx}[2x]}^{u'} = (\cos 2x)(2) = 2\cos 2x$

b. $y = \cos\overbrace{(x-1)}^{u}$ $\qquad y' = \overbrace{-\sin(x-1)}^{-\sin u}\,\overbrace{\frac{d}{dx}[x-1]}^{u'} = -\sin(x-1)$

c. $y = e^{\overbrace{3x}^{u}}$ $\qquad y' = \overbrace{e^{3x}}^{e^u}\,\overbrace{\frac{d}{dx}[3x]}^{u'} = 3e^{3x}$

EXAMPLE 11 Parentheses and Trigonometric Functions

a. $y = \cos 3x^2 = \cos(3x^2)$ $\qquad y' = (-\sin 3x^2)(6x) = -6x \sin 3x^2$

b. $y = (\cos 3)x^2$ $\qquad y' = (\cos 3)(2x) = 2x \cos 3$

c. $y = \cos(3x)^2 = \cos(9x^2)$ $\qquad y' = (-\sin 9x^2)(18x) = -18x \sin 9x^2$

d. $y = \cos^2 x = (\cos x)^2$ $\qquad y' = 2(\cos x)(-\sin x) = -2\cos x \sin x$

To find the derivative of a function of the form $k(x) = f(g(h(x)))$, you need to apply the Chain Rule twice, as shown in Example 12.

EXAMPLE 12 Repeated Application of the Chain Rule

$f(t) = \sin^3 4t$ — Original function

$= (\sin 4t)^3$ — Rewrite.

$f'(t) = 3(\sin 4t)^2 \dfrac{d}{dt}[\sin 4t]$ — Apply Chain Rule once.

$= 3(\sin 4t)^2(\cos 4t)\dfrac{d}{dt}[4t]$ — Apply Chain Rule a second time.

$= 3(\sin 4t)^2(\cos 4t)(4)$

$= 12\sin^2 4t \cos 4t$ — Simplify.

The Derivative of the Natural Logarithmic Function

Up to this point in the text, derivatives of algebraic functions have been algebraic and derivatives of transcendental functions have been transcendental. The next theorem looks at an unusual situation in which the derivative of a transcendental function is algebraic. Specifically, the derivative of the natural logarithmic function is the algebraic function $1/x$.

EXPLORATION

Use the *table* feature of a graphing utility to display the values of $f(x) = \ln x$ and its derivative for $x = 0, 1, 2, 3, \ldots$. What do these values tell you about the derivative of the natural logarithmic function?

THEOREM 3.13 Derivative of the Natural Logarithmic Function

Let u be a differentiable function of x.

1. $\dfrac{d}{dx}[\ln x] = \dfrac{1}{x}, \quad x > 0$

2. $\dfrac{d}{dx}[\ln u] = \dfrac{1}{u}\dfrac{du}{dx} = \dfrac{u'}{u}, \quad u > 0$

Proof To prove the first part, let $y = \ln x$, which implies that $e^y = x$. Differentiating both sides of this equation produces the following.

$$y = \ln x$$
$$e^y = x$$
$$\frac{d}{dx}[e^y] = \frac{d}{dx}[x]$$
$$e^y \frac{dy}{dx} = 1 \qquad \text{Chain Rule}$$
$$\frac{dy}{dx} = \frac{1}{e^y}$$
$$\frac{dy}{dx} = \frac{1}{x}$$

The second part of the theorem can be obtained by applying the Chain Rule to the first part.

EXAMPLE 13 Differentiation of Logarithmic Functions

a. $\dfrac{d}{dx}[\ln(2x)] = \dfrac{u'}{u} = \dfrac{2}{2x} = \dfrac{1}{x} \qquad u = 2x$

b. $\dfrac{d}{dx}[\ln(x^2 + 1)] = \dfrac{u'}{u} = \dfrac{2x}{x^2 + 1} \qquad u = x^2 + 1$

c. $\dfrac{d}{dx}[x \ln x] = x\left(\dfrac{d}{dx}[\ln x]\right) + (\ln x)\left(\dfrac{d}{dx}[x]\right) \qquad$ Product Rule

$= x\left(\dfrac{1}{x}\right) + (\ln x)(1)$

$= 1 + \ln x$

d. $\dfrac{d}{dx}[(\ln x)^3] = 3(\ln x)^2 \dfrac{d}{dx}[\ln x] \qquad$ Chain Rule

$= 3(\ln x)^2 \dfrac{1}{x}$

The Granger Collection

JOHN NAPIER (1550–1617)

Logarithms were invented by the Scottish mathematician John Napier. Although he did not introduce the *natural* logarithmic function, it is sometimes called the *Napterian* logarithm.

John Napier used logarithmic properties to simplify *calculations* involving products, quotients, and powers. Of course, given the availability of calculators, there is now little need for this particular application of logarithms. However, there is great value in using logarithmic properties to simplify *differentiation* involving products, quotients, and powers.

EXAMPLE 14 Logarithmic Properties as Aids to Differentiation

Differentiate $f(x) = \ln\sqrt{x+1}$.

Solution Because

$$f(x) = \ln\sqrt{x+1} = \ln(x+1)^{1/2} = \frac{1}{2}\ln(x+1) \qquad \text{Rewrite before differentiating.}$$

you can write

$$f'(x) = \frac{1}{2}\left(\frac{1}{x+1}\right) = \frac{1}{2(x+1)}. \qquad \text{Differentiate.}$$

EXAMPLE 15 Logarithmic Properties as Aids to Differentiation

Differentiate $f(x) = \ln\dfrac{x(x^2+1)^2}{\sqrt{2x^3-1}}$.

NOTE In Examples 14 and 15, be sure that you see the benefit of applying logarithmic properties *before* differentiation. Consider, for instance, the difficulty of direct differentiation of the function given in Example 15.

Solution

$$f(x) = \ln\frac{x(x^2+1)^2}{\sqrt{2x^3-1}} \qquad \text{Write original function.}$$

$$= \ln x + 2\ln(x^2+1) - \frac{1}{2}\ln(2x^3-1) \qquad \text{Rewrite before differentiating.}$$

$$f'(x) = \frac{1}{x} + 2\left(\frac{2x}{x^2+1}\right) - \frac{1}{2}\left(\frac{6x^2}{2x^3-1}\right) \qquad \text{Differentiate.}$$

$$= \frac{1}{x} + \frac{4x}{x^2+1} - \frac{3x^2}{2x^3-1} \qquad \text{Simplify.}$$

Because the natural logarithm is undefined for negative numbers, you will often encounter expressions of the form $\ln|u|$. Theorem 3.14 states that you can differentiate functions of the form $y = \ln|u|$ as if the absolute value sign were not present.

THEOREM 3.14 Derivative Involving Absolute Value

If u is a differentiable function of x such that $u \neq 0$, then

$$\frac{d}{dx}[\ln|u|] = \frac{u'}{u}.$$

Proof If $u > 0$, then $|u| = u$, and the result follows from Theorem 3.13. If $u < 0$, then $|u| = -u$, and you have

$$\frac{d}{dx}[\ln|u|] = \frac{d}{dx}[\ln(-u)] = \frac{-u'}{-u} = \frac{u'}{u}.$$

Bases Other than e

The **base** of the natural exponential function is e. This "natural" base can be used to assign a meaning to a general base a.

Definition of Exponential Function to Base a

If a is a positive real number $(a \neq 1)$ and x is any real number, then the **exponential function to the base a** is denoted by a^x and is defined by

$$a^x = e^{(\ln a)x}.$$

If $a = 1$, then $y = 1^x = 1$ is a constant function.

Logarithmic functions to bases other than e can be defined in much the same way as exponential functions to other bases are defined.

Definition of Logarithmic Function to Base a

If a is a positive real number $(a \neq 1)$ and x is any positive real number, then the **logarithmic function to the base a** is denoted by $\log_a x$ and is defined as

$$\log_a x = \frac{1}{\ln a} \ln x.$$

To differentiate exponential and logarithmic functions to other bases, you have two options: (1) use the definitions of a^x and $\log_a x$ and differentiate using the rules for the natural exponential and logarithmic functions, or (2) use the following differentiation rules for bases other than e.

NOTE These differentiation rules are similar to those for the natural exponential function and the natural logarithmic function. In fact, they differ only by the constant factors $\ln a$ and $1/\ln a$. This points out one reason why, for calculus, e is the most convenient base.

THEOREM 3.15 Derivatives for Bases Other than e

Let a be a positive real number $(a \neq 1)$ and let u be a differentiable function of x.

1. $\dfrac{d}{dx}[a^x] = (\ln a)a^x$
2. $\dfrac{d}{dx}[a^u] = (\ln a)a^u \dfrac{du}{dx}$
3. $\dfrac{d}{dx}[\log_a x] = \dfrac{1}{(\ln a)x}$
4. $\dfrac{d}{dx}[\log_a u] = \dfrac{1}{(\ln a)u}\dfrac{du}{dx}$

Proof By definition, $a^x = e^{(\ln a)x}$. Therefore, you can prove the first rule by letting $u = (\ln a)x$ and differentiating with base e to obtain

$$\frac{d}{dx}[a^x] = \frac{d}{dx}[e^{(\ln a)x}] = e^u \frac{du}{dx} = e^{(\ln a)x}(\ln a) = (\ln a)a^x.$$

To prove the third rule, you can write

$$\frac{d}{dx}[\log_a x] = \frac{d}{dx}\left[\frac{1}{\ln a} \ln x\right] = \frac{1}{\ln a}\left(\frac{1}{x}\right) = \frac{1}{(\ln a)x}.$$

The second and fourth rules are simply the Chain Rule versions of the first and third rules.

EXAMPLE 16 Differentiating Functions to Other Bases

Find the derivative of each of the following.

a. $y = 2^x$ **b.** $y = 2^{3x}$ **c.** $y = \log_{10} \cos x$

Solution

a. $y' = \frac{d}{dx}[2^x] = (\ln 2)2^x$

b. $y' = \frac{d}{dx}[2^{3x}] = (\ln 2)2^{3x}(3) = (3 \ln 2)2^{3x}$

Try writing 2^{3x} as 8^x and differentiating to see that you obtain the same result.

c. $y' = \frac{d}{dx}[\log_{10} \cos x] = \frac{-\sin x}{(\ln 10) \cos x} = -\frac{1}{\ln 10} \tan x$

STUDY TIP To become skilled at differentiation, you should memorize each rule. As an aid to memorization, note that the cofunctions (cosine, cotangent, and cosecant) require a negative sign as part of their derivatives.

This section conludes with a summary of the differentiation rules studied so far.

Summary of Differentiation Rules

General Differentiation Rules

Let u and v be differentiable functions of x.

Constant Rule:

$\frac{d}{dx}[c] = 0$

(Simple) Power Rule:

$\frac{d}{dx}[x^n] = nx^{n-1}$ $\frac{d}{dx}[x] = 1$

Constant Multiple Rule:

$\frac{d}{dx}[cu] = cu'$

Sum or Difference Rule:

$\frac{d}{dx}[u \pm v] = u' \pm v'$

Product Rule:

$\frac{d}{dx}[uv] = uv' + vu'$

Quotient Rule:

$\frac{d}{dx}\left[\frac{u}{v}\right] = \frac{vu' - uv'}{v^2}$

Chain Rule:

$\frac{d}{dx}[f(u)] = f'(u)\, u'$

General Power Rule:

$\frac{d}{dx}[u^n] = nu^{n-1}u'$

Derivatives of Trigonometric Functions

$\frac{d}{dx}[\sin x] = \cos x$ $\frac{d}{dx}[\tan x] = \sec^2 x$ $\frac{d}{dx}[\sec x] = \sec x \tan x$

$\frac{d}{dx}[\cos x] = -\sin x$ $\frac{d}{dx}[\cot x] = -\csc^2 x$ $\frac{d}{dx}[\csc x] = -\csc x \cot x$

Derivatives of Exponential and Logarithmic Functions

$\frac{d}{dx}[e^x] = e^x$ $\frac{d}{dx}[\ln x] = \frac{1}{x}$

$\frac{d}{dx}[a^x] = (\ln a)a^x$ $\frac{d}{dx}[\log_a x] = \frac{1}{(\ln a)x}$

Exercises for Section 3.4

See www.CalcChat.com for worked-out solutions to odd-numbered exercises.

In Exercises 1–8, complete the table using Example 2 as a model.

$y = f(g(x))$	$u = g(x)$	$y = f(u)$
1. $y = (6x - 5)^4$		
2. $y = \dfrac{1}{\sqrt{x + 2}}$		
3. $y = \sqrt{x^2 - 1}$		
4. $y = 3\tan(\pi x^2)$		
5. $y = \csc^3 x$		
6. $y = \cos\dfrac{3x}{2}$		
7. $y = e^{-2x}$		
8. $y = (\ln x)^3$		

In Exercises 9–36, find the derivative of the function.

9. $y = (2x - 7)^3$
10. $y = (2x^3 + 1)^2$
11. $g(x) = 3(4 - 9x)^4$
12. $y = 3(5 - x^2)^5$
13. $f(x) = (9 - x^2)^{2/3}$
14. $f(t) = (9t + 7)^{2/3}$
15. $f(t) = \sqrt{1 - t}$
16. $g(x) = \sqrt{5 - 3x}$
17. $y = \sqrt[3]{9x^2 + 4}$
18. $g(x) = \sqrt{x^2 - 2x + 1}$
19. $y = 2\sqrt[4]{4 - x^2}$
20. $f(x) = -3\sqrt[4]{2 - 9x}$
21. $y = \dfrac{1}{x - 2}$
22. $s(t) = \dfrac{1}{t^2 + 3t - 1}$
23. $f(t) = \left(\dfrac{1}{t - 3}\right)^2$
24. $y = -\dfrac{8}{(t + 3)^3}$
25. $y = \dfrac{1}{\sqrt{x + 2}}$
26. $g(t) = \sqrt{\dfrac{1}{t^2 - 2}}$
27. $f(x) = x^2(x - 2)^4$
28. $f(x) = x(3x - 7)^3$
29. $y = x\sqrt{1 - x^2}$
30. $y = \frac{1}{2}x^2\sqrt{16 - x^2}$
31. $y = \dfrac{x}{\sqrt{x^2 + 1}}$
32. $y = \dfrac{x}{\sqrt{x^4 + 2}}$
33. $g(x) = \left(\dfrac{x + 5}{x^2 + 2}\right)^2$
34. $h(t) = \left(\dfrac{t^2}{t^3 + 2}\right)^2$
35. $f(v) = \left(\dfrac{1 - 2v}{1 + v}\right)^3$
36. $g(x) = \left(\dfrac{3x^2 - 1}{2x + 5}\right)^3$

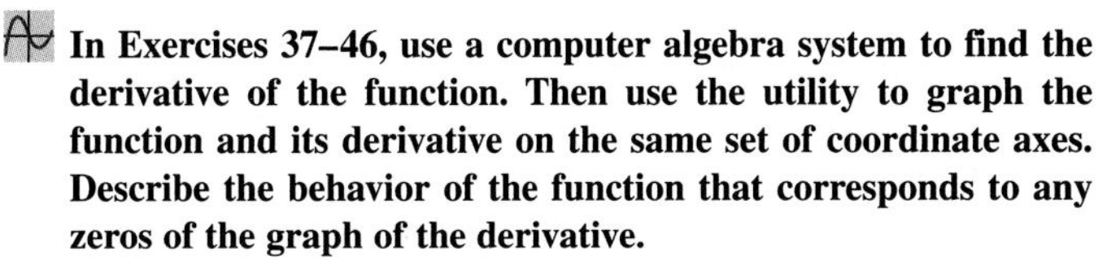

In Exercises 37–46, use a computer algebra system to find the derivative of the function. Then use the utility to graph the function and its derivative on the same set of coordinate axes. Describe the behavior of the function that corresponds to any zeros of the graph of the derivative.

37. $y = \dfrac{\sqrt{x} + 1}{x^2 + 1}$
38. $y = \sqrt{\dfrac{2x}{x + 1}}$
39. $g(t) = \dfrac{3t^2}{\sqrt{t^2 + 2t - 1}}$
40. $f(x) = \sqrt{x}(2 - x)^2$
41. $y = \sqrt{\dfrac{x + 1}{x}}$
42. $y = (t^2 - 9)\sqrt{t + 2}$
43. $s(t) = \dfrac{-2(2 - t)\sqrt{1 + t}}{3}$
44. $g(x) = \sqrt{x - 1} + \sqrt{x + 1}$
45. $y = \dfrac{\cos \pi x + 1}{x}$
46. $y = x^2 \tan\dfrac{1}{x}$

In Exercises 47 and 48, find the slope of the tangent line to the sine function at the origin. Compare this value with the number of complete cycles in the interval $[0, 2\pi]$. What can you conclude about the slope of the sine function $\sin ax$ at the origin?

47. (a)

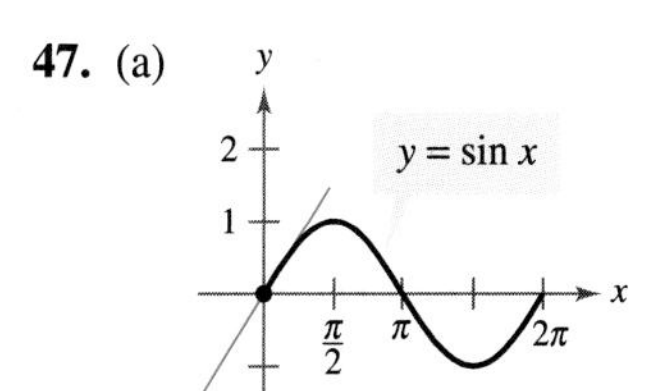

(b)

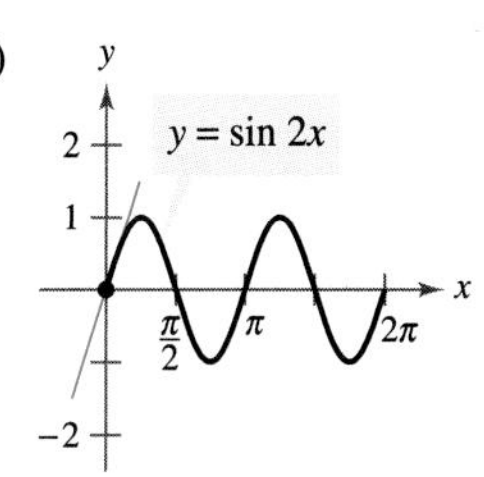

48. (a)

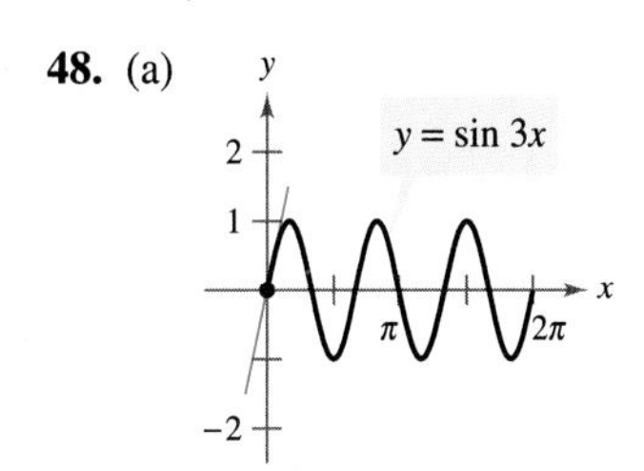

(b)

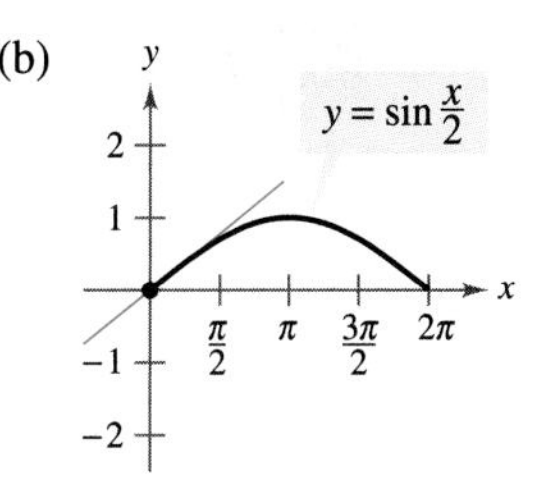

In Exercises 49 and 50, find the slope of the tangent line to the graph of the function at the point $(0, 1)$.

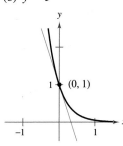

49. (a) $y = e^{3x}$

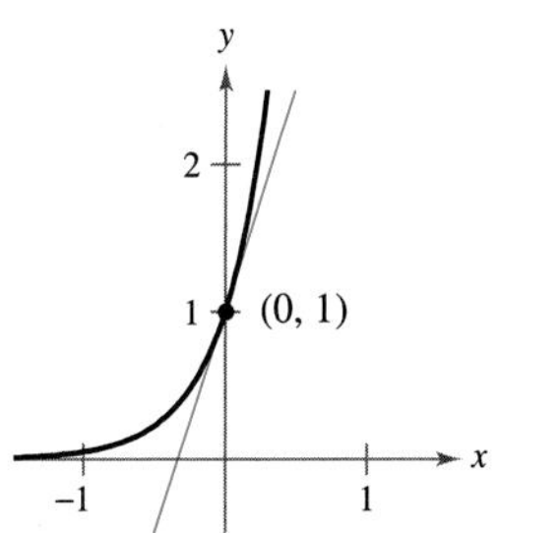

(b) $y = e^{-3x}$

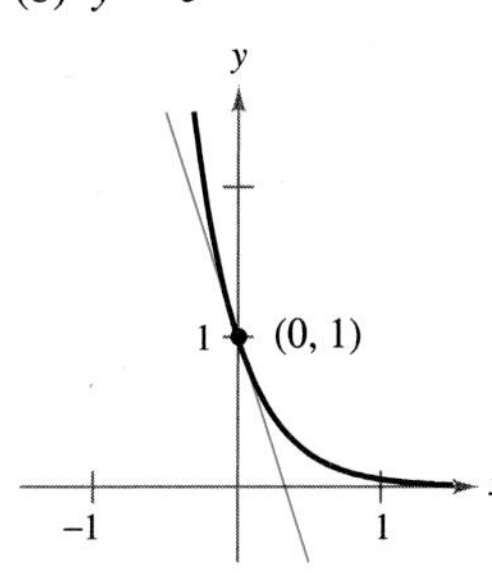

50. (a) $y = e^{2x}$

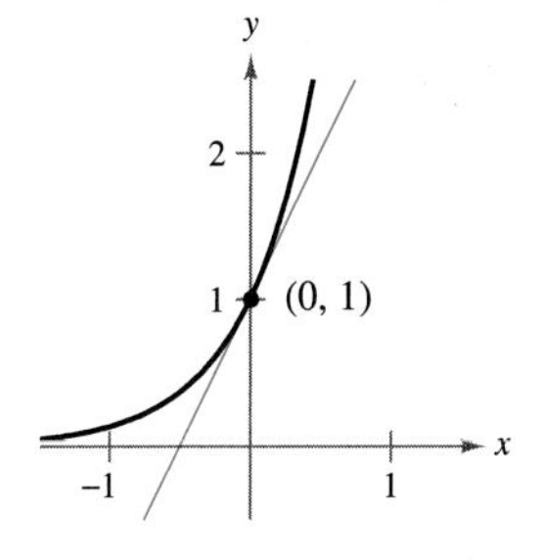

(b) $y = e^{-2x}$

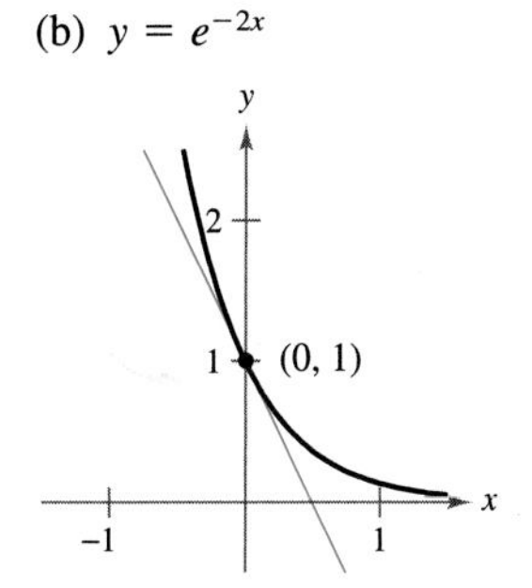

In Exercises 51–54, find the slope of the tangent line to the graph of the logarithmic function at the point (1, 0).

51. $y = \ln x^3$

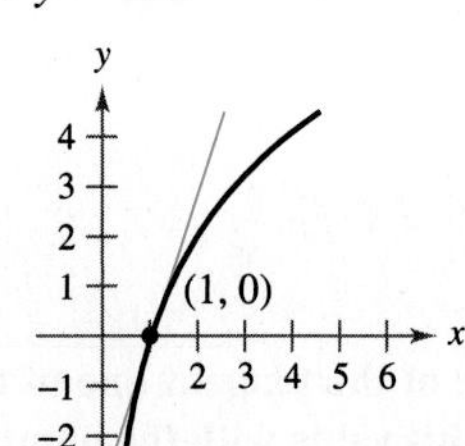

52. $y = \ln x^{3/2}$

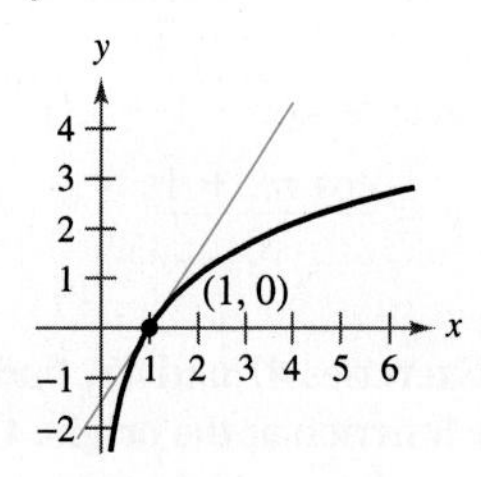

53. $y = \ln x^2$

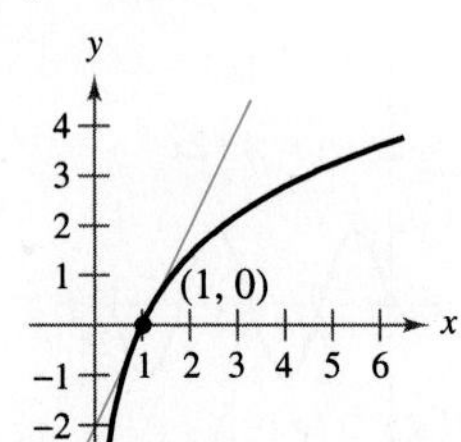

54. $y = \ln x^{1/2}$

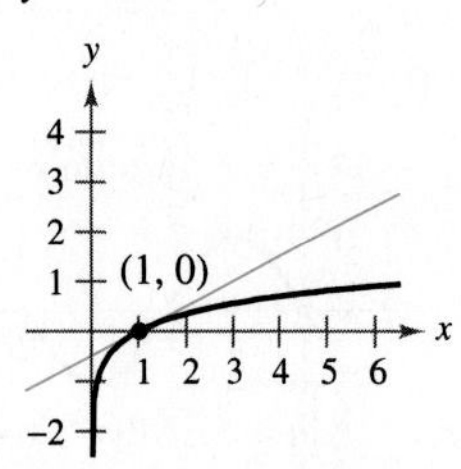

In Exercises 55–100, find the derivative of the function.

55. $y = \cos 3x$
56. $y = \sin \pi x$
57. $g(x) = 3 \tan 4x$
58. $h(x) = \sec x^3$
59. $f(\theta) = \frac{1}{4} \sin^2 2\theta$
60. $g(t) = 5 \cos^3 \pi t$
61. $y = \sqrt{x} + \frac{1}{4} \sin(2x)^2$
62. $y = 3x - 5 \cos(2x)^2$
63. $y = \sin(\cos x)$
64. $y = \sin \sqrt[3]{x} + \sqrt[3]{\sin x}$
65. $f(x) = e^{2x}$
66. $y = e^{-x^2}$
67. $y = e^{\sqrt{x}}$
68. $y = x^2 e^{-x}$
69. $g(t) = (e^{-t} + e^t)^3$
70. $g(t) = e^{-3/t^2}$
71. $y = \ln(e^{x^2})$
72. $y = \ln\left(\frac{1 + e^x}{1 - e^x}\right)$
73. $y = \frac{2}{e^x + e^{-x}}$
74. $y = \frac{e^x - e^{-x}}{2}$
75. $y = x^2 e^x - 2xe^x + 2e^x$
76. $y = xe^x - e^x$
77. $f(x) = e^{-x} \ln x$
78. $f(x) = e^3 \ln x$
79. $y = e^x(\sin x + \cos x)$
80. $y = \ln e^x$
81. $g(x) = \ln x^2$
82. $h(x) = \ln(2x^2 + 3)$
83. $y = (\ln x)^4$
84. $y = x \ln x$
85. $y = \ln\left(x\sqrt{x^2 - 1}\right)$
86. $y = \ln \sqrt{x^2 - 9}$
87. $f(x) = \ln\left(\frac{x}{x^2 + 1}\right)$
88. $f(x) = \ln\left(\frac{2x}{x + 3}\right)$
89. $g(t) = \frac{\ln t}{t^2}$
90. $h(t) = \frac{\ln t}{t}$
91. $y = \ln \sqrt{\frac{x + 1}{x - 1}}$
92. $y = \ln \sqrt[3]{\frac{x - 2}{x + 2}}$
93. $y = \frac{-\sqrt{x^2 + 1}}{x} + \ln\left(x + \sqrt{x^2 + 1}\right)$
94. $y = \frac{-\sqrt{x^2 + 4}}{2x^2} - \frac{1}{4} \ln\left(\frac{2 + \sqrt{x^2 + 4}}{x}\right)$
95. $y = \ln|\sin x|$
96. $y = \ln|\csc x|$
97. $y = \ln\left|\frac{\cos x}{\cos x - 1}\right|$
98. $y = \ln|\sec x + \tan x|$
99. $y = \ln\left|\frac{-1 + \sin x}{2 + \sin x}\right|$
100. $y = \ln\sqrt{1 + \sin^2 x}$

In Exercises 101–106, find the second derivative of the function.

101. $f(x) = 2(x^2 - 1)^3$
102. $f(x) = \frac{1}{x - 2}$
103. $f(x) = \sin x^2$
104. $f(x) = \sec^2 \pi x$
105. $f(x) = (3 + 2x)e^{-3x}$
106. $g(x) = \sqrt{x} + e^x \ln x$

In Exercises 107–114, evaluate the derivative of the function at the indicated point. Use a graphing utility to verify your result.

	Function	*Point*
107.	$s(t) = \sqrt{t^2 + 2t + 8}$	$(2, 4)$
108.	$y = \sqrt[5]{3x^3 + 4x}$	$(2, 2)$
109.	$f(x) = \frac{3}{x^3 - 4}$	$\left(-1, -\frac{3}{5}\right)$
110.	$f(x) = \frac{1}{(x^2 - 3x)^2}$	$\left(4, \frac{1}{16}\right)$
111.	$f(t) = \frac{3t + 2}{t - 1}$	$(0, -2)$
112.	$f(x) = \frac{x + 1}{2x - 3}$	$(2, 3)$
113.	$y = 37 - \sec^3(2x)$	$(0, 36)$
114.	$y = \frac{1}{x} + \sqrt{\cos x}$	$\left(\frac{\pi}{2}, \frac{2}{\pi}\right)$

In Exercises 115–122, (a) find an equation of the tangent line to the graph of f at the indicated point, (b) use a graphing utility to graph the function and its tangent line at the point, and (c) use the derivative feature of a graphing utility to confirm your results.

	Function	*Point*
115.	$f(x) = \sqrt{3x^2 - 2}$	$(3, 5)$
116.	$f(x) = \frac{1}{3}x\sqrt{x^2 + 5}$	$(2, 2)$
117.	$f(x) = \sin 2x$	$(\pi, 0)$
118.	$y = \cos 3x$	$\left(\frac{\pi}{4}, -\frac{\sqrt{2}}{2}\right)$
119.	$y = 2 \tan^3 x$	$\left(\frac{\pi}{4}, 2\right)$
120.	$f(x) = \tan^2 x$	$\left(\frac{\pi}{4}, 1\right)$
121.	$y = 4 - x^2 - \ln\left(\frac{1}{2}x + 1\right)$	$(0, 4)$
122.	$y = 2e^{1 - x^2}$	$(1, 2)$

In Exercises 123–138, find the derivative of the function.

123. $f(x) = 4^x$

124. $g(x) = 5^{-x}$

125. $y = 5^{x-2}$

126. $y = x(6^{-2x})$

127. $g(t) = t^2 2^t$

128. $f(t) = \dfrac{3^{2t}}{t}$

129. $h(\theta) = 2^{-\theta} \cos \pi\theta$

130. $g(\alpha) = 5^{-\alpha/2} \sin 2\alpha$

131. $y = \log_3 x$

132. $y = \log_{10} 2x$

133. $f(x) = \log_2 \dfrac{x^2}{x-1}$

134. $h(x) = \log_3 \dfrac{x\sqrt{x-1}}{2}$

135. $y = \log_5 \sqrt{x^2 - 1}$

136. $y = \log_{10} \dfrac{x^2 - 1}{x}$

137. $g(t) = \dfrac{10 \log_4 t}{t}$

138. $f(t) = t^{3/2} \log_2 \sqrt{t+1}$

Writing About Concepts

In Exercises 139–142, the graphs of a function f and its derivative f' are shown. Label the graphs as f or f' and write a short paragraph stating the criteria used in making the selection. To print an enlarged copy of the graph, go to the website *www.mathgraphs.com*.

139.

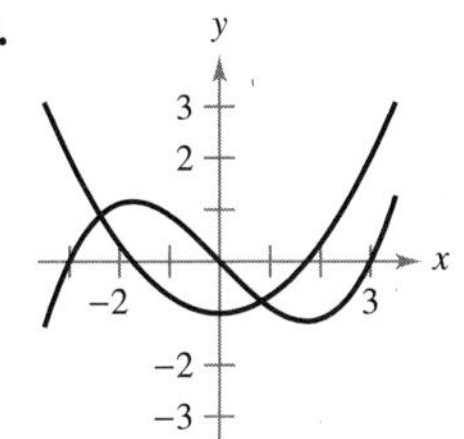

140.

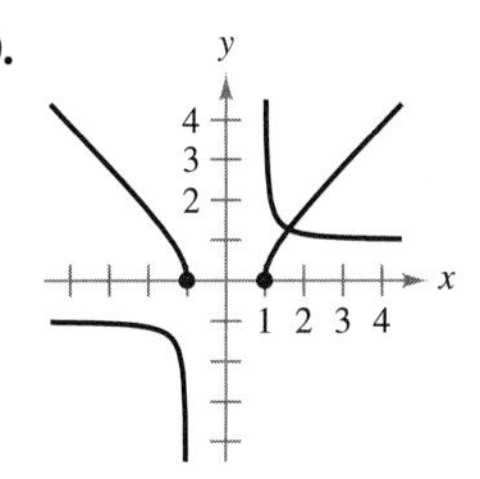

141.

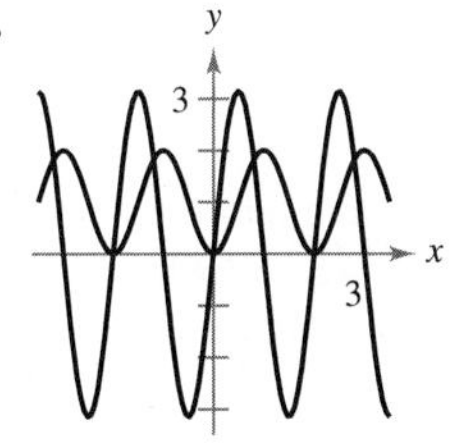

142.

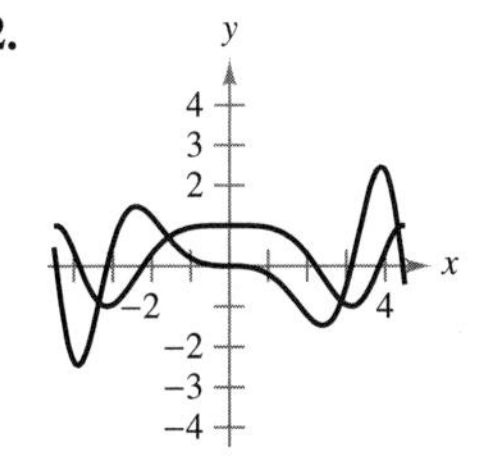

In Exercises 143 and 144, the relationship between f and g is given. Explain the relationship between f' and g'.

143. $g(x) = f(3x)$

144. $g(x) = f(x^2)$

145. Given that $g(5) = -3$, $g'(5) = 6$, $h(5) = 3$, and $h'(5) = -2$, find $f'(5)$ (if possible) for each of the following. If it is not possible, state what additional information is required.

(a) $f(x) = g(x)h(x)$

(b) $f(x) = g(h(x))$

(c) $f(x) = \dfrac{g(x)}{h(x)}$

(d) $f(x) = [g(x)]^3$

146. (a) Find the derivative of the function $g(x) = \sin^2 x + \cos^2 x$ in two ways.

(b) For $f(x) = \sec^2 x$ and $g(x) = \tan^2 x$, show that

$f'(x) = g'(x).$

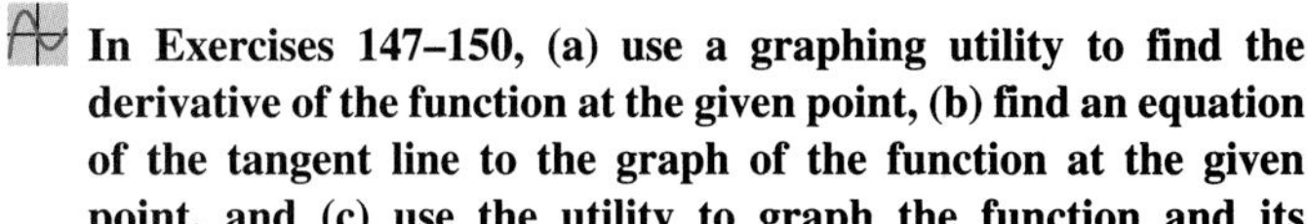

In Exercises 147–150, (a) use a graphing utility to find the derivative of the function at the given point, (b) find an equation of the tangent line to the graph of the function at the given point, and (c) use the utility to graph the function and its tangent line in the same viewing window.

147. $g(t) = \dfrac{3t^2}{\sqrt{t^2 + 2t - 1}}, \quad \left(\dfrac{1}{2}, \dfrac{3}{2}\right)$

148. $f(x) = \sqrt{x}(2 - x)^2, \quad (4, 8)$

149. $s(t) = \dfrac{(4 - 2t)\sqrt{1+t}}{3}, \quad \left(0, \dfrac{4}{3}\right)$

150. $y = (t^2 - 9)\sqrt{t+2}, \quad (2, -10)$

Famous Curves **In Exercises 151 and 152, find an equation of the tangent line to the graph at the given point. Then use a graphing utility to graph the function and its tangent line in the same viewing window.**

151. Top half of circle

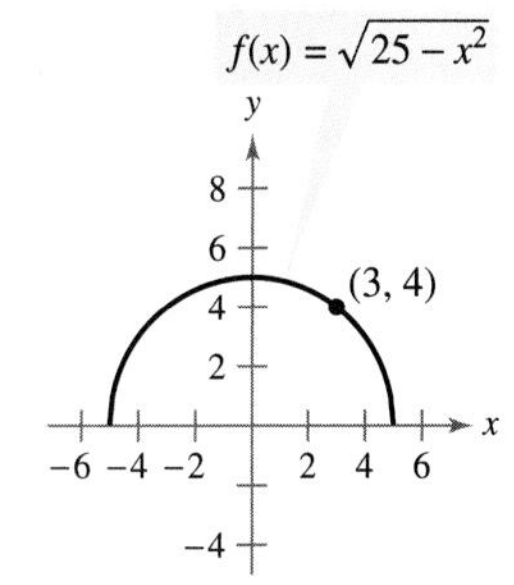

152. Bullet-nose curve

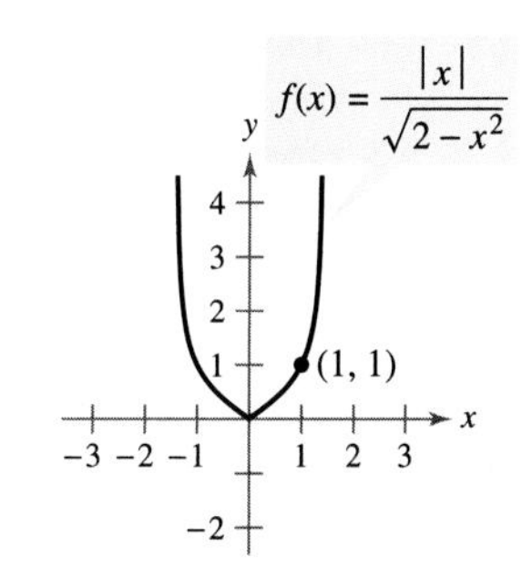

153. ***Horizontal Tangent Line*** Determine the point(s) in the interval $(0, 2\pi)$ at which the graph of $f(x) = 2\cos x + \sin 2x$ has a horizontal tangent line.

154. ***Horizontal Tangent Line*** Determine the point(s) at which the graph of $f(x) = \dfrac{x}{\sqrt{2x-1}}$ has a horizontal tangent line.

In Exercises 155–158, evaluate the second derivative of the function at the given point. Use a computer algebra system to verify your result.

155. $h(x) = \frac{1}{9}(3x + 1)^3, \quad \left(1, \frac{64}{9}\right)$

156. $f(x) = \dfrac{1}{\sqrt{x+4}}, \quad \left(0, \dfrac{1}{2}\right)$

157. $f(x) = \cos(x^2), \quad (0, 1)$

158. $g(t) = \tan 2t, \quad \left(\dfrac{\pi}{6}, \sqrt{3}\right)$

159. ***Doppler Effect*** The frequency F of a fire truck siren heard by a stationary observer is

$$F = \frac{132{,}400}{331 \pm v}$$

where $\pm v$ represents the velocity of the accelerating fire truck in meters per second. Find the rate of change of F with respect to v when

(a) the fire truck is approaching at a velocity of 30 meters per second (use $-v$).

(b) the fire truck is moving away at a velocity of 30 meters per second (use $+v$).

160. ***Harmonic Motion*** The displacement from equilibrium of an object in harmonic motion on the end of a spring is

$$y = \tfrac{1}{3}\cos 12t - \tfrac{1}{4}\sin 12t$$

where y is measured in feet and t is the time in seconds. Determine the position and velocity of the object when $t = \pi/8$.

161. ***Pendulum*** A 15-centimeter pendulum moves according to the equation $\theta = 0.2\cos 8t$, where θ is the angular displacement from the vertical in radians and t is the time in seconds. Determine the maximum angular displacement and the rate of change of θ when $t = 3$ seconds.

162. ***Wave Motion*** A buoy oscillates in simple harmonic motion $y = A\cos\omega t$ as waves move past it. The buoy moves a total of 3.5 feet (vertically) from its low point to its high point. It returns to its high point every 10 seconds.

(a) Write an equation describing the motion of the buoy if it is at its high point at $t = 0$.

(b) Determine the velocity of the buoy as a function of t.

163. ***Circulatory System*** The speed S of blood that is r centimeters from the center of an artery is

$$S = C(R^2 - r^2)$$

where C is a constant, R is the radius of the artery, and S is measured in centimeters per second. Suppose a drug is administered and the artery begins to dilate at a rate of dR/dt. At a constant distance r, find the rate at which S changes with respect to t for $C = 1.76 \times 10^5$, $R = 1.2 \times 10^{-2}$, and $dR/dt = 10^{-5}$.

164. ***Modeling Data*** The normal daily maximum temperatures T (in degrees Fahrenheit) for Denver, Colorado, are shown in the table. *(Source: National Oceanic and Atmospheric Administration)*

Month	Jan	Feb	Mar	Apr	May	Jun
Temperature	43.2	47.2	53.7	60.9	70.5	82.1

Month	Jul	Aug	Sep	Oct	Nov	Dec
Temperature	88.0	86.0	77.4	66.0	51.5	44.1

(a) Use a graphing utility to plot the data and find a model for the data of the form

$$T(t) = a + b\sin(\pi t/6 - c)$$

where T is the temperature and t is the time in months, with $t = 1$ corresponding to January.

(b) Use a graphing utility to graph the model. How well does the model fit the data?

(c) Find T' and use a graphing utility to graph the derivative.

(d) Based on the graph of the derivative, during what times does the temperature change most rapidly? Most slowly? Do your answers agree with your observations of the temperature changes? Explain.

165. ***Volume*** Air is being pumped into a spherical balloon so that the radius is increasing at the rate of $dr/dt = 3$ inches per second. What is the rate of change of the volume of the balloon, in cubic inches per second, when $r = 8$ inches? [*Hint:* $V = \frac{4}{3}\pi r^3$]

166. ***Think About It*** The table shows some values of the derivative of an unknown function f. Complete the table by finding (if possible) the derivative of each transformation of f.

(a) $g(x) = f(x) - 2$ (b) $h(x) = 2f(x)$

(c) $r(x) = f(-3x)$ (d) $s(x) = f(x + 2)$

x	-2	-1	0	1	2	3
$f'(x)$	4	$\frac{2}{3}$	$-\frac{1}{3}$	-1	-2	-4
$g'(x)$						
$h'(x)$						
$r'(x)$						
$s'(x)$						

167. ***Modeling Data*** The table shows the temperature T (°F) at which water boils at selected pressures p (pounds per square inch). *(Source: Standard Handbook of Mechanical Engineers)*

p	5	10	14.696 (1 atm)	20
T	162.24°	193.21°	212.00°	227.96°

p	30	40	60	80	100
T	250.33°	267.25°	292.71°	312.03°	327.81°

A model that approximates the data is

$$T = 87.97 + 34.96\ln p + 7.91\sqrt{p}.$$

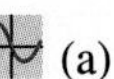
(a) Use a graphing utility to plot the data and graph the model.

(b) Find the rate of change of T with respect to p when $p = 10$ and $p = 70$.

168. ***Depreciation*** After t years, the value of a car purchased for $20,000 is

$V(t) = 20{,}000\left(\frac{3}{4}\right)^t.$

(a) Use a graphing utility to graph the function and determine the value of the car 2 years after it was purchased.

(b) Find the rate of change of V with respect to t when $t = 1$ and $t = 4$.

169. ***Inflation*** If the annual rate of inflation averages 5% over the next 10 years, the approximate cost C of goods or services during any year in that decade is $C(t) = P(1.05)^t$, where t is the time in years and P is the present cost.

(a) If the price of an oil change for your car is presently $24.95, estimate the price 10 years from now.

(b) Find the rate of change of C with respect to t when $t = 1$ and $t = 8$.

(c) Verify that the rate of change of C is proportional to C. What is the constant of proportionality?

170. ***Finding a Pattern*** Consider the function $f(x) = \sin \beta x$, where β is a constant.

(a) Find the first-, second-, third-, and fourth-order derivatives of the function.

(b) Verify that the function and its second derivative satisfy the equation $f''(x) + \beta^2 f(x) = 0$.

(c) Use the results in part (a) to write general rules for the even- and odd-order derivatives

$f^{(2k)}(x)$ and $f^{(2k-1)}(x)$.

[*Hint:* $(-1)^k$ is positive if k is even and negative if k is odd.]

171. ***Conjecture*** Let f be a differentiable function of period p.

(a) Is the function f' periodic? Verify your answer.

(b) Consider the function $g(x) = f(2x)$. Is the function $g'(x)$ periodic? Verify your answer.

172. ***Think About It*** Let $r(x) = f(g(x))$ and $s(x) = g(f(x))$, where f and g are shown in the figure. Find (a) $r'(1)$ and (b) $s'(4)$.

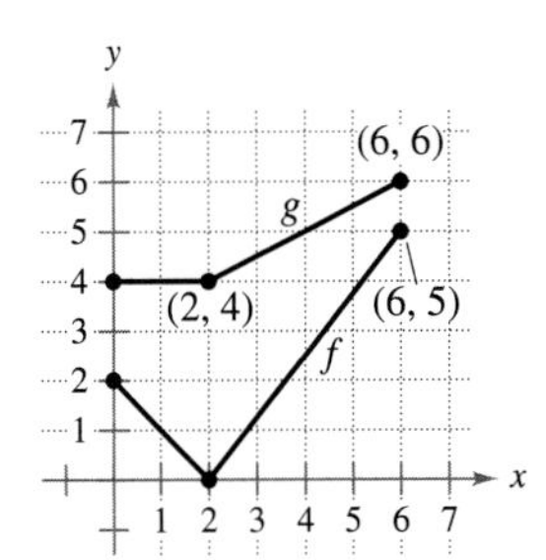

173. (a) Show that the derivative of an odd function is even. That is, if $f(-x) = -f(x)$, then $f'(-x) = f'(x)$.

(b) Show that the derivative of an even function is odd. That is, if $f(-x) = f(x)$, then $f'(-x) = -f'(x)$.

174. Let u be a differentiable function of x. Use the fact that $|u| = \sqrt{u^2}$ to prove that

$$\frac{d}{dx}[|u|] = u'\frac{u}{|u|}, \quad u \neq 0.$$

In Exercises 175–178, use the result of Exercise 174 to find the derivative of the function.

175. $g(x) = |2x - 3|$

176. $f(x) = |x^2 - 4|$

177. $h(x) = |x| \cos x$

178. $f(x) = |\sin x|$

Linear and Quadratic Approximations **The linear and quadratic approximations of a function f at $x = a$ are**

$$P_1(x) = f'(a)(x - a) + f(a) \text{ and}$$

$$P_2(x) = \tfrac{1}{2}f''(a)(x - a)^2 + f'(a)(x - a) + f(a).$$

In Exercises 179–182, (a) find the specified linear and quadratic approximations of f, (b) use a graphing utility to graph f and the approximations, (c) determine whether P_1 or P_2 is the better approximation, and (d) state how the accuracy changes as you move farther from $x = a$.

179. $f(x) = \tan \dfrac{\pi x}{4}$, $a = 1$

180. $f(x) = \sec 2x$, $a = \dfrac{\pi}{6}$

181. $f(x) = e^{-x^2/2}$, $a = 0$

182. $f(x) = x \ln x$, $a = 1$

True or False? **In Exercises 183–185, determine whether the statement is true or false. If it is false, explain why or give an example that shows it is false.**

183. If $y = (1 - x)^{1/2}$, then $y' = \frac{1}{2}(1 - x)^{-1/2}$.

184. If $f(x) = \sin^2(2x)$, then $f'(x) = 2(\sin 2x)(\cos 2x)$.

185. If y is a differentiable function of u, u is a differentiable function of v, and v is a differentiable function of x, then

$$\frac{dy}{dx} = \frac{dy}{du}\frac{du}{dv}\frac{dv}{dx}.$$

Putnam Exam Challenge

186. Let $f(x) = a_1 \sin x + a_2 \sin 2x + \cdot \cdot \cdot + a_n \sin nx$, where $a_1, a_2, \ldots, a_n$ are real numbers and where n is a positive integer. Given that $|f(x)| \leq |\sin x|$ for all real x, prove that $|a_1 + 2a_2 + \cdot \cdot \cdot + na_n| \leq 1$.

187. Let k be a fixed positive integer. The nth derivative of $\dfrac{1}{x^k - 1}$ has the form

$$\frac{P_n(x)}{(x^k - 1)^{n+1}}$$

where $P_n(x)$ is a polynomial. Find $P_n(1)$.

These problems were composed by the Committee on the Putnam Prize Competition.

Section 3.5 Implicit Differentiation

- **Distinguish between functions written in implicit form and explicit form.**
- **Use implicit differentiation to find the derivative of a function.**
- **Find derivatives of functions using logarithmic differentiation.**

Implicit and Explicit Functions

Up to this point in the text, most functions have been expressed in **explicit form.** For example, in the equation

$$y = 3x^2 - 5 \qquad \text{Explicit form}$$

the variable y is explicitly written as a function of x. Some functions, however, are only *implied* by an equation. For instance, the function $y = 1/x$ is defined **implicitly** by the equation $xy = 1$. Suppose you were asked to find dy/dx for this equation. You could begin by writing y explicitly as a function of x and then differentiating.

Implicit Form	*Explicit Form*	*Derivative*
$xy = 1$	$y = \frac{1}{x} = x^{-1}$	$\frac{dy}{dx} = -x^{-2} = -\frac{1}{x^2}$

This strategy works whenever you can solve for the function explicitly. You cannot, however, use this procedure when you are unable to solve for y as a function of x. For instance, how would you find dy/dx for the equation $x^2 - 2y^3 + 4y = 2$, where it is very difficult to express y as a function of x explicitly? To do this, you can use **implicit differentiation.**

To understand how to find dy/dx implicitly, you must realize that the differentiation is taking place *with respect to x*. This means that when you differentiate terms involving x alone, you can differentiate as usual. However, when you differentiate terms involving y, you must apply the Chain Rule, because you are assuming that y is defined implicitly as a differentiable function of x.

EXPLORATION

Graphing an Implicit Equation How could you use a graphing utility to sketch the graph of the equation

$$x^2 - 2y^3 + 4y = 2?$$

Here are two possible approaches.

a. Solve the equation for x. Switch the roles of x and y and graph the two resulting equations. The combined graphs will show a 90° rotation of the graph of the original equation.

b. Set the graphing utility to *parametric* mode and graph the equations

$$x = -\sqrt{2t^3 - 4t + 2}$$
$$y = t$$

and

$$x = \sqrt{2t^3 - 4t + 2}$$
$$y = t.$$

From either of these two approaches, can you decide whether the graph has a tangent line at the point $(0, 1)$? Explain your reasoning.

EXAMPLE 1 Differentiating with Respect to x

a. $\frac{d}{dx}[x^3] = 3x^2$ — Variables agree: Use Simple Power Rule.

(Variables agree)

b. $\frac{d}{dx}[\underbrace{y^3}_{u^n}] = \underbrace{3y^2}_{nu^{n-1}}\underbrace{\frac{dy}{dx}}_{u'}$ — Variables disagree: Use Chain Rule.

(Variables disagree)

c. $\frac{d}{dx}[x + 3y] = 1 + 3\frac{dy}{dx}$ — Chain Rule: $\frac{d}{dx}[3y] = 3y'$

d. $\frac{d}{dx}[xy^2] = x\frac{d}{dx}[y^2] + y^2\frac{d}{dx}[x]$ — Product Rule

$= x\left(2y\frac{dy}{dx}\right) + y^2(1)$ — Chain Rule

$= 2xy\frac{dy}{dx} + y^2$ — Simplify.

Implicit Differentiation

Guidelines for Implicit Differentiation

1. Differentiate both sides of the equation *with respect to x.*
2. Collect all terms involving dy/dx on the left side of the equation and move all other terms to the right side of the equation.
3. Factor dy/dx out of the left side of the equation.
4. Solve for dy/dx by dividing both sides of the equation by the left-hand factor that does not contain dy/dx.

EXAMPLE 2 Implicit Differentiation

Find dy/dx given that $y^3 + y^2 - 5y - x^2 = -4$.

Solution

NOTE In Example 2, note that implicit differentiation can produce an expression for dy/dx that contains both x and y.

1. Differentiate both sides of the equation with respect to x.

$$\frac{d}{dx}[y^3 + y^2 - 5y - x^2] = \frac{d}{dx}[-4]$$

$$\frac{d}{dx}[y^3] + \frac{d}{dx}[y^2] - \frac{d}{dx}[5y] - \frac{d}{dx}[x^2] = \frac{d}{dx}[-4]$$

$$3y^2\frac{dy}{dx} + 2y\frac{dy}{dx} - 5\frac{dy}{dx} - 2x = 0$$

2. Collect the dy/dx terms on the left side of the equation.

$$3y^2\frac{dy}{dx} + 2y\frac{dy}{dx} - 5\frac{dy}{dx} = 2x$$

3. Factor dy/dx out of the left side of the equation.

$$\frac{dy}{dx}(3y^2 + 2y - 5) = 2x$$

4. Solve for dy/dx by dividing by $(3y^2 + 2y - 5)$.

$$\frac{dy}{dx} = \frac{2x}{3y^2 + 2y - 5}$$

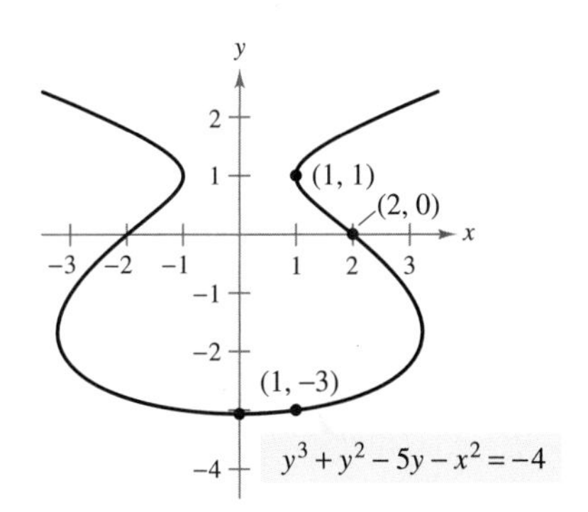

Point on Graph	*Slope of Graph*
$(2, 0)$	$-\frac{4}{5}$
$(1, -3)$	$\frac{1}{8}$
$x = 0$	0
$(1, 1)$	Undefined

The implicit equation

$$y^3 + y^2 - 5y - x^2 = -4$$

has the derivative

$$\frac{dy}{dx} = \frac{2x}{3y^2 + 2y - 5}.$$

Figure 3.27

To see how you can use an *implicit derivative*, consider the graph shown in Figure 3.27. From the graph, you can see that y is not a function of x. Even so, the derivative found in Example 2 gives a formula for the slope of the tangent line at a point on this graph. The slopes at several points on the graph are shown below the graph.

TECHNOLOGY With most graphing utilities, it is easy to graph an equation that explicitly represents y as a function of x. Graphing other equations, however, can require some ingenuity. For instance, to graph the equation given in Example 2, use a graphing utility, set in *parametric* mode, to graph the parametric representations $x = \sqrt{t^3 + t^2 - 5t + 4}$, $y = t$, and $x = -\sqrt{t^3 + t^2 - 5t + 4}$, $y = t$, for $-5 \le t \le 5$. How does the result compare with the graph shown in Figure 3.27?

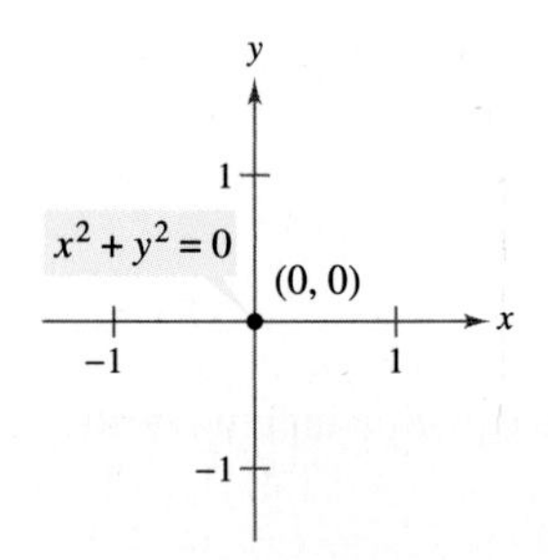

(a)

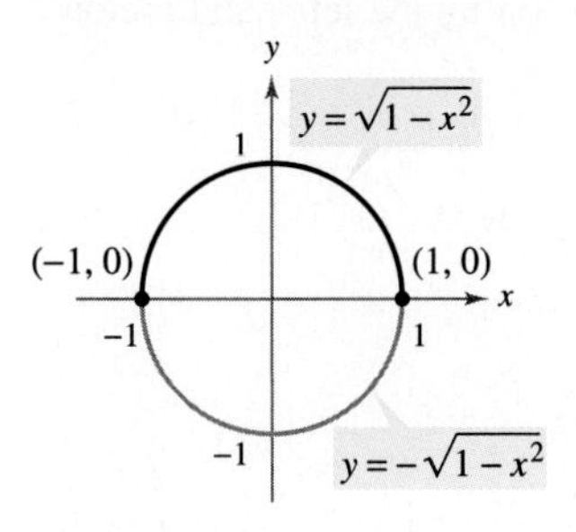

(b)

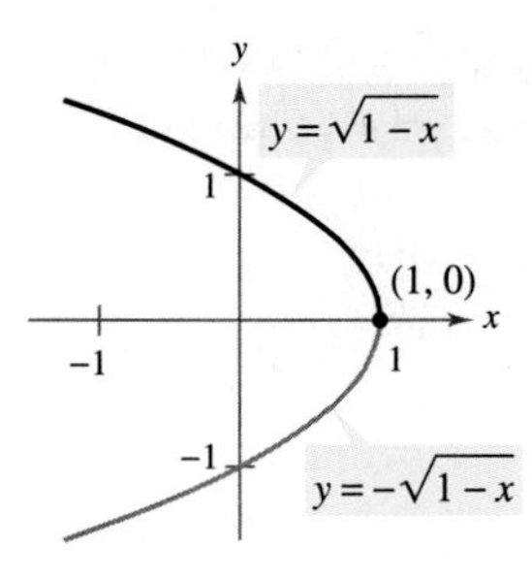

(c)

Some graph segments can be represented by differentiable functions.

Figure 3.28

It is meaningless to solve for dy/dx in an equation that has no solution points. (For example, $x^2 + y^2 = -4$ has no solution points.) If, however, a segment of a graph can be represented by a differentiable function, dy/dx will have meaning as the slope at each point on the segment. Recall that a function is not differentiable at (1) points with vertical tangents and (2) points at which the function is not continuous.

EXAMPLE 3 Representing a Graph by Differentiable Functions

If possible, represent y as a differentiable function of x (see Figure 3.28).

a. $x^2 + y^2 = 0$ **b.** $x^2 + y^2 = 1$ **c.** $x + y^2 = 1$

Solution

a. The graph of this equation is a single point. So, the equation does not define y as a differentiable function of x.

b. The graph of this equation is the unit circle, centered at $(0, 0)$. The upper semicircle is given by the differentiable function

$$y = \sqrt{1 - x^2}, \quad -1 < x < 1$$

and the lower semicircle is given by the differentiable function

$$y = -\sqrt{1 - x^2}, \quad -1 < x < 1.$$

At the points $(-1, 0)$ and $(1, 0)$, the slope of the graph is undefined.

c. The upper half of this parabola is given by the differentiable function

$$y = \sqrt{1 - x}, \quad x < 1$$

and the lower half of this parabola is given by the differentiable function

$$y = -\sqrt{1 - x}, \quad x < 1.$$

At the point $(1, 0)$, the slope of the graph is undefined.

EXAMPLE 4 Finding the Slope of a Graph Implicitly

Determine the slope of the tangent line to the graph of

$$x^2 + 4y^2 = 4$$

at the point $(\sqrt{2}, -1/\sqrt{2})$. See Figure 3.29.

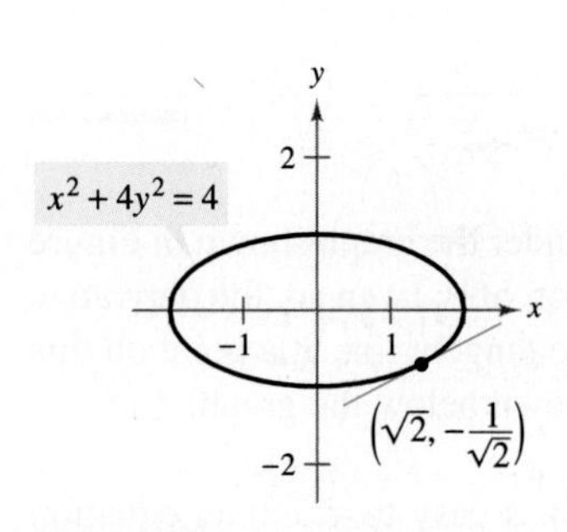

Figure 3.29

Solution

$$x^2 + 4y^2 = 4 \qquad \text{Write original equation.}$$

$$2x + 8y\frac{dy}{dx} = 0 \qquad \text{Differentiate with respect to } x.$$

$$\frac{dy}{dx} = \frac{-2x}{8y} = \frac{-x}{4y} \qquad \text{Solve for } \frac{dy}{dx}.$$

So, at $(\sqrt{2}, -1/\sqrt{2})$, the slope is

$$\frac{dy}{dx} = \frac{-\sqrt{2}}{-4/\sqrt{2}} = \frac{1}{2}. \qquad \text{Evaluate } \frac{dy}{dx} \text{ when } x = \sqrt{2} \text{ and } y = -\frac{1}{\sqrt{2}}.$$

NOTE To see the benefit of implicit differentiation, try doing Example 4 using the explicit function $y = -\frac{1}{2}\sqrt{4 - x^2}$.

EXAMPLE 5 Finding the Slope of a Graph Implicitly

Determine the slope of the graph of $3(x^2 + y^2)^2 = 100xy$ at the point $(3, 1)$.

Solution

$$\frac{d}{dx}[3(x^2 + y^2)^2] = \frac{d}{dx}[100xy]$$

$$3(2)(x^2 + y^2)\left(2x + 2y\frac{dy}{dx}\right) = 100\left[x\frac{dy}{dx} + y(1)\right]$$

$$12y(x^2 + y^2)\frac{dy}{dx} - 100x\frac{dy}{dx} = 100y - 12x(x^2 + y^2)$$

$$[12y(x^2 + y^2) - 100x]\frac{dy}{dx} = 100y - 12x(x^2 + y^2)$$

$$\frac{dy}{dx} = \frac{100y - 12x(x^2 + y^2)}{-100x + 12y(x^2 + y^2)}$$

$$= \frac{25y - 3x(x^2 + y^2)}{-25x + 3y(x^2 + y^2)}$$

At the point $(3, 1)$, the slope of the graph is

$$\frac{dy}{dx} = \frac{25(1) - 3(3)(3^2 + 1^2)}{-25(3) + 3(1)(3^2 + 1^2)} = \frac{25 - 90}{-75 + 30} = \frac{-65}{-45} = \frac{13}{9}$$

as shown in Figure 3.30. This graph is called a **lemniscate.**

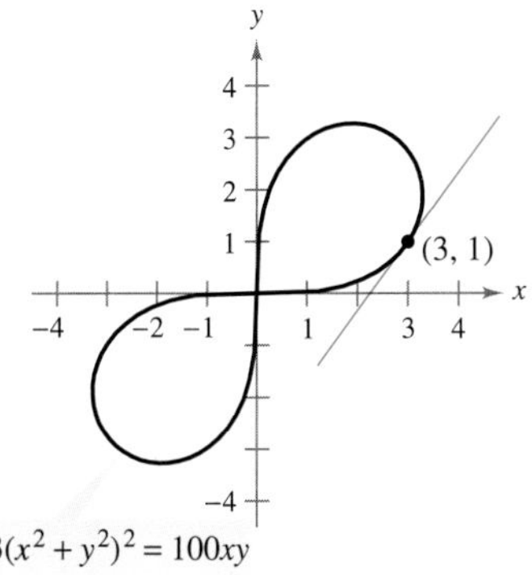

$3(x^2 + y^2)^2 = 100xy$

Lemniscate

Figure 3.30

EXAMPLE 6 Determining a Differentiable Function

Find dy/dx implicitly for the equation $\sin y = x$. Then find the largest interval of the form $-a < y < a$ on which y is a differentiable function of x (see Figure 3.31).

Solution

$$\frac{d}{dx}[\sin y] = \frac{d}{dx}[x]$$

$$\cos y\frac{dy}{dx} = 1$$

$$\frac{dy}{dx} = \frac{1}{\cos y}$$

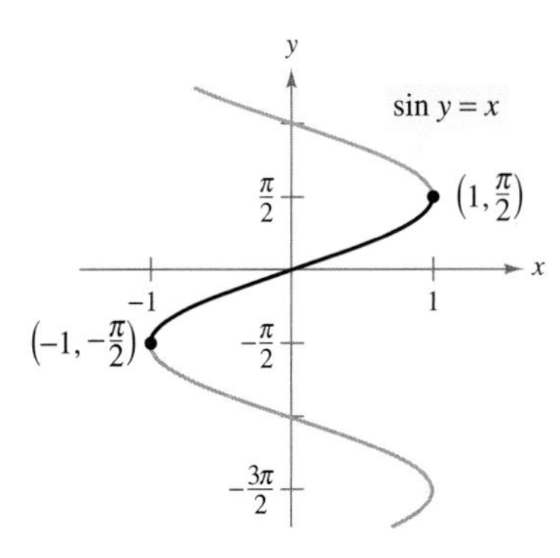

The derivative is $\dfrac{dy}{dx} = \dfrac{1}{\sqrt{1 - x^2}}$.

Figure 3.31

The largest interval about the origin for which y is a differentiable function of x is $-\pi/2 < y < \pi/2$. To see this, note that $\cos y$ is positive for all y in this interval and is 0 at the endpoints. If you restrict y to the interval $-\pi/2 < y < \pi/2$, you should be able to write dy/dx explicitly as a function of x. To do this, you can use

$$\cos y = \sqrt{1 - \sin^2 y}$$

$$= \sqrt{1 - x^2}, \quad -\frac{\pi}{2} < y < \frac{\pi}{2}$$

and conclude that

$$\frac{dy}{dx} = \frac{1}{\sqrt{1 - x^2}}.$$

ISAAC BARROW (1630–1677)

The graph in Example 8 is called the **kappa curve** because it resembles the Greek letter kappa, κ. The general solution for the tangent line to this curve was discovered by the English mathematician Isaac Barrow. Newton was Barrow's student, and they corresponded frequently regarding their work in the early development of calculus.

With implicit differentiation, the form of the derivative often can be simplified (as in Example 6) by an appropriate use of the *original* equation. A similar technique can be used to find and simplify higher-order derivatives obtained implicitly.

EXAMPLE 7 Finding the Second Derivative Implicitly

Given $x^2 + y^2 = 25$, find $\dfrac{d^2y}{dx^2}$.

Solution Differentiating each term with respect to x produces

$$2x + 2y\frac{dy}{dx} = 0$$

$$2y\frac{dy}{dx} = -2x$$

$$\frac{dy}{dx} = \frac{-2x}{2y} = -\frac{x}{y}.$$

Differentiating a second time with respect to x yields

$$\frac{d^2y}{dx^2} = -\frac{(y)(1) - (x)(dy/dx)}{y^2} \qquad \text{Quotient Rule}$$

$$= -\frac{y - (x)(-x/y)}{y^2} \qquad \text{Substitute } -x/y \text{ for } \frac{dy}{dx}.$$

$$= -\frac{y^2 + x^2}{y^3} \qquad \text{Simplify.}$$

$$= -\frac{25}{y^3}. \qquad \text{Substitute 25 for } x^2 + y^2.$$

EXAMPLE 8 Finding a Tangent Line to a Graph

Find the tangent line to the graph given by $x^2(x^2 + y^2) = y^2$ at the point $\left(\sqrt{2}/2, \sqrt{2}/2\right)$, as shown in Figure 3.32.

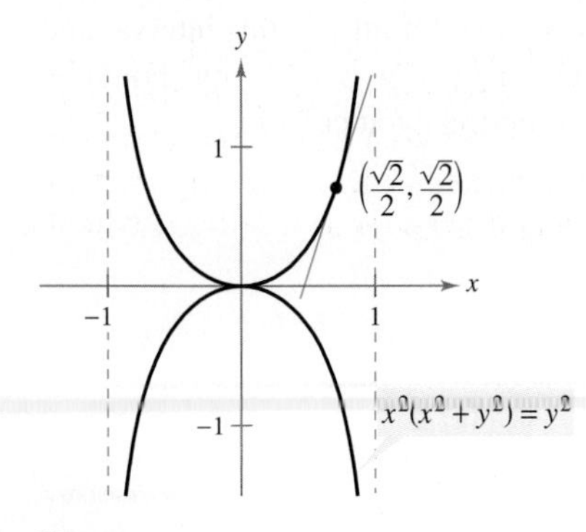

Kappa curve
Figure 3.32

Solution By rewriting and differentiating implicitly, you obtain

$$x^4 + x^2y^2 - y^2 = 0$$

$$4x^3 + x^2\left(2y\frac{dy}{dx}\right) + 2xy^2 - 2y\frac{dy}{dx} = 0$$

$$2y(x^2 - 1)\frac{dy}{dx} = -2x(2x^2 + y^2)$$

$$\frac{dy}{dx} = \frac{x(2x^2 + y^2)}{y(1 - x^2)}.$$

At the point $\left(\sqrt{2}/2, \sqrt{2}/2\right)$, the slope is

$$\frac{dy}{dx} = \frac{\left(\sqrt{2}/2\right)[2(1/2) + (1/2)]}{\left(\sqrt{2}/2\right)[1 - (1/2)]} = \frac{3/2}{1/2} = 3$$

and the equation of the tangent line at this point is

$$y - \frac{\sqrt{2}}{2} = 3\left(x - \frac{\sqrt{2}}{2}\right)$$

$$y = 3x - \sqrt{2}.$$

Logarithmic Differentiation

On occasion, it is convenient to use logarithms as aids in differentiating nonlogarithmic functions. This procedure is called **logarithmic differentiation.**

EXAMPLE 9 Logarithmic Differentiation

Find the derivative of $y = \frac{(x-2)^2}{\sqrt{x^2+1}}, x \neq 2$.

Solution Note that $y > 0$ and so $\ln y$ is defined. Begin by taking the natural logarithms of both sides of the equation. Then apply logarithmic properties and differentiate implicitly. Finally, solve for y'.

$$\ln y = \ln \frac{(x-2)^2}{\sqrt{x^2+1}} \qquad \text{Take ln of both sides.}$$

$$\ln y = 2\ln(x-2) - \frac{1}{2}\ln(x^2+1) \qquad \text{Logarithmic properties}$$

$$\frac{y'}{y} = 2\left(\frac{1}{x-2}\right) - \frac{1}{2}\left(\frac{2x}{x^2+1}\right) \qquad \text{Differentiate.}$$

$$= \frac{2}{x-2} - \frac{x}{x^2+1} \qquad \text{Simplify.}$$

$$y' = y\left(\frac{2}{x-2} - \frac{x}{x^2+1}\right) \qquad \text{Solve for } y'.$$

$$= \frac{(x-2)^2}{\sqrt{x^2+1}}\left[\frac{x^2+2x+2}{(x-2)(x^2+1)}\right] \qquad \text{Substitute for } y.$$

$$= \frac{(x-2)(x^2+2x+2)}{(x^2+1)^{3/2}} \qquad \text{Simplify.}$$

Exercises for Section 3.5

See www.CalcChat.com for worked-out solutions to odd-numbered exercises.

In Exercises 1–20, find dy/dx by implicit differentiation.

1. $x^2 + y^2 = 36$
2. $x^2 - y^2 = 81$
3. $x^{1/2} + y^{1/2} = 9$
4. $x^3 + y^3 = 8$
5. $x^3 - xy + y^2 = 4$
6. $x^2y + y^2x = -3$
7. $xe^y - 10x + 3y = 0$
8. $e^{xy} + x^2 - y^2 = 10$
9. $x^3y^3 - y = x$
10. $\sqrt{xy} = x - 2y$
11. $x^3 - 2x^2y + 3xy^2 = 38$
12. $2 \sin x \cos y = 1$
13. $\sin x + 2 \cos 2y = 1$
14. $(\sin \pi x + \cos \pi y)^2 = 2$
15. $\sin x = x(1 + \tan y)$
16. $\cot y = x - y$
17. $y = \sin(xy)$
18. $x = \sec \frac{1}{y}$
19. $x^2 - 3 \ln y + y^2 = 10$
20. $\ln xy + 5x = 30$

In Exercises 21–24, (a) find two explicit functions by solving the equation for y in terms of x, (b) sketch the graph of the equation and label the parts given by the corresponding explicit functions, (c) differentiate the explicit functions, and (d) find dy/dx implicitly and show that the result is equivalent to that of part (c).

21. $x^2 + y^2 = 16$
22. $x^2 + y^2 - 4x + 6y + 9 = 0$
23. $9x^2 + 16y^2 = 144$
24. $4y^2 - x^2 = 4$

In Exercises 25–34, find dy/dx by implicit differentiation and evaluate the derivative at the indicated point.

Equation	*Point*
25. $xy = 4$	$(-4, -1)$
26. $x^3 - y^2 = 0$	$(1, 1)$
27. $y^2 = \dfrac{x^2 - 9}{x^2 + 9}$	$(3, 0)$
28. $(x + y)^3 = x^3 + y^3$	$(-1, 1)$
29. $x^{2/3} + y^{2/3} = 5$	$(8, 1)$
30. $x^3 + y^3 = 2xy$	$(1, 1)$
31. $\tan(x + y) = x$	$(0, 0)$
32. $x \cos y = 1$	$\left(2, \dfrac{\pi}{3}\right)$
33. $3e^{xy} - x = 0$	$(3, 0)$
34. $y^2 = \ln x$	$(e, 1)$

Famous Curves **In Exercises 35–38, find the slope of the tangent line to the graph at the indicated point.**

35. Witch of Agnesi:

$(x^2 + 4)y = 8$

Point: $(2, 1)$

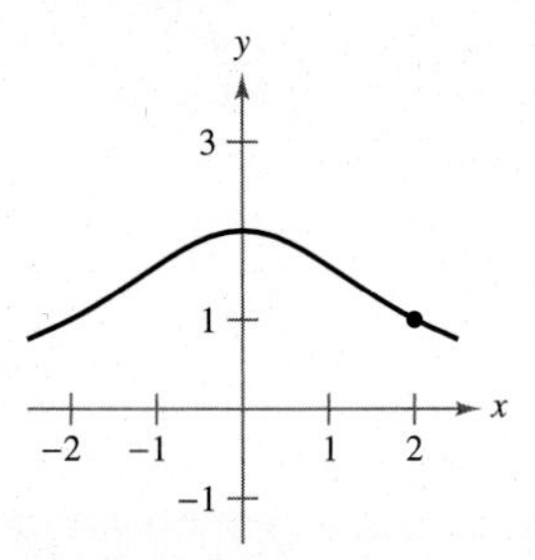

36. Cissoid:

$(4 - x)y^2 = x^3$

Point: $(2, 2)$

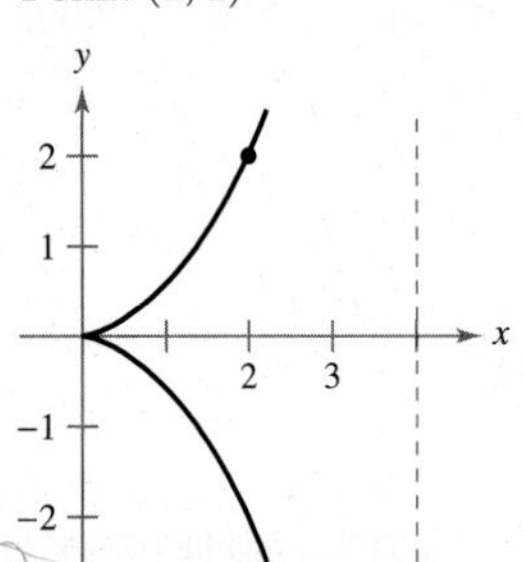

37. Bifolium:

$(x^2 + y^2)^2 = 4x^2y$

Point: $(1, 1)$

38. Folium of Descartes:

$x^3 + y^3 - 6xy = 0$

Point: $\left(\frac{4}{3}, \frac{8}{3}\right)$

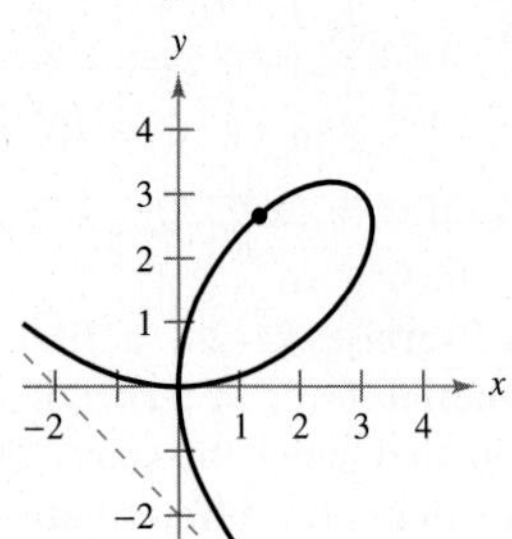

Famous Curves **In Exercises 39–46, find an equation of the tangent line to the graph at the given point. To print an enlarged copy of the graph, go to the website *www.mathgraphs.com*.**

39. Parabola

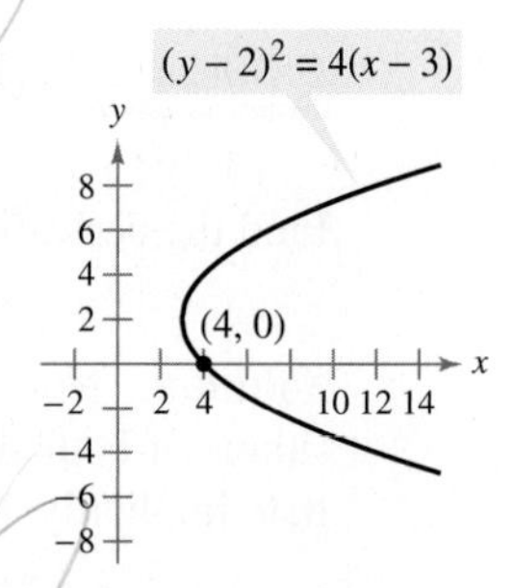

40. Circle

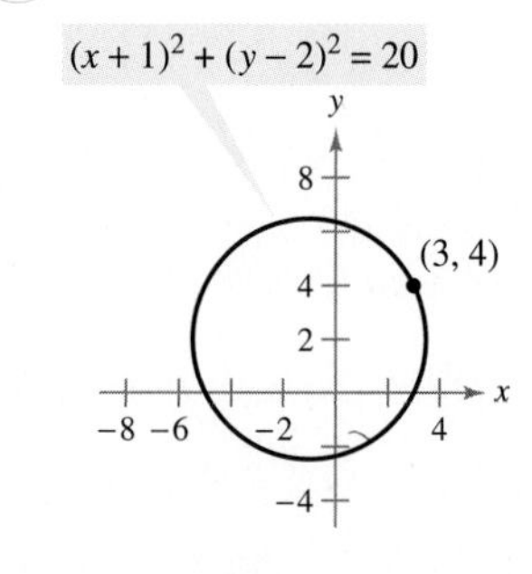

41. Rotated hyperbola

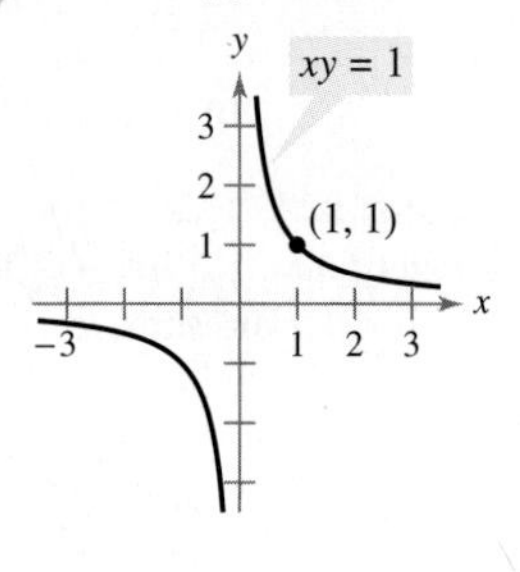

42. Rotated ellipse

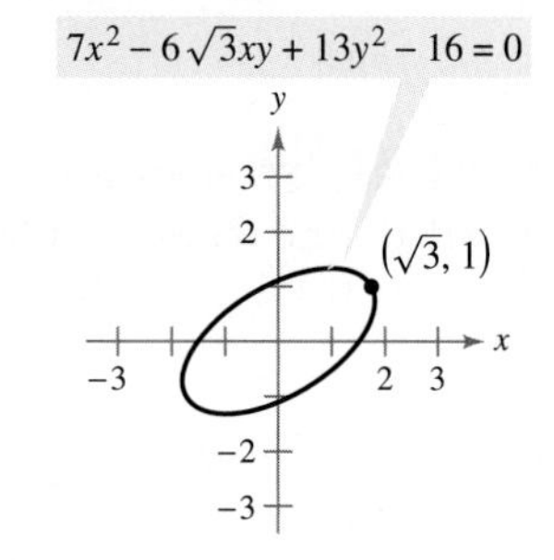

43. Cruciform

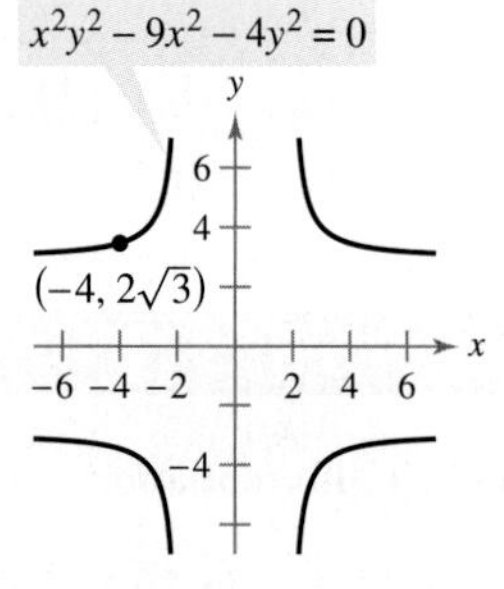

44. Astroid

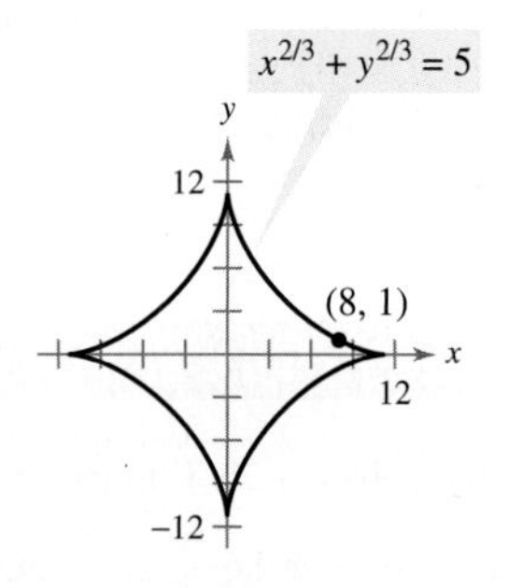

45. Lemniscate

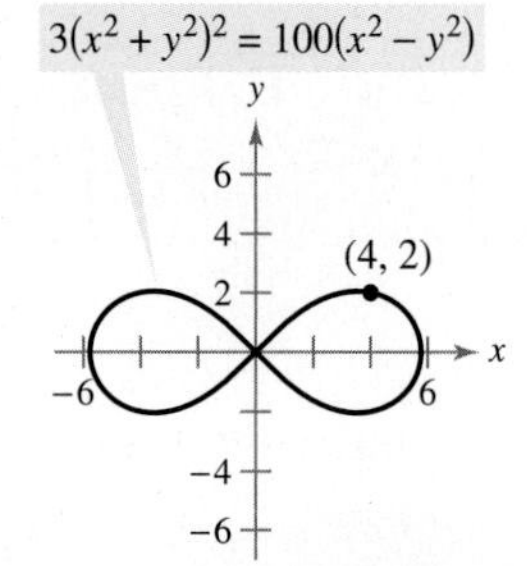

46. Kappa curve

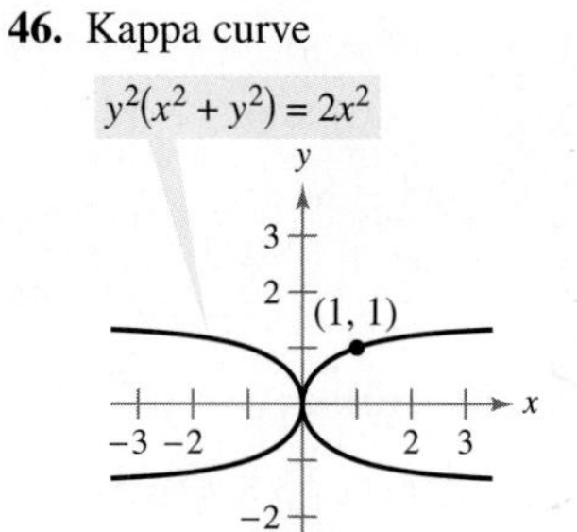

47. (a) Use implicit differentiation to find an equation of the tangent line to the ellipse $\dfrac{x^2}{2} + \dfrac{y^2}{8} = 1$ at $(1, 2)$.

(b) Show that the equation of the tangent line to the ellipse $\dfrac{x^2}{a^2} + \dfrac{y^2}{b^2} = 1$ at (x_0, y_0) is $\dfrac{x_0x}{a^2} + \dfrac{y_0y}{b^2} = 1$.

48. (a) Use implicit differentiation to find an equation of the tangent line to the hyperbola $\frac{x^2}{6} - \frac{y^2}{8} = 1$ at $(3, -2)$.

(b) Show that the equation of the tangent line to the hyperbola $\frac{x^2}{a^2} - \frac{y^2}{b^2} = 1$ at (x_0, y_0) is $\frac{x_0 x}{a^2} - \frac{y_0 y}{b^2} = 1$.

In Exercises 49 and 50, find dy/dx implicitly and find the largest interval of the form $-a < y < a$ or $0 < y < a$ such that y is a differentiable function of x. Write dy/dx as a function of x.

49. $\tan y = x$ **50.** $\cos y = x$

In Exercises 51–56, find d^2y/dx^2 in terms of x and y.

51. $x^2 + y^2 = 36$ **52.** $x^2y^2 - 2x = 3$

53. $x^2 - y^2 = 16$ **54.** $1 - xy = x - y$

55. $y^2 = x^3$ **56.** $y^2 = 4x$

In Exercises 57 and 58, use a graphing utility to graph the equation. Find an equation of the tangent line to the graph at the given point and graph the tangent line in the same viewing window.

57. $\sqrt{x} + \sqrt{y} = 4,\ (9, 1)$ **58.** $y^2 = \frac{x - 1}{x^2 + 1},\ \left(2, \frac{\sqrt{5}}{5}\right)$

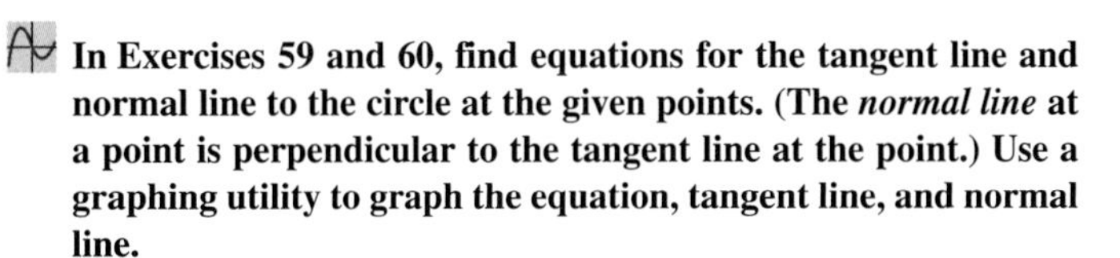

In Exercises 59 and 60, find equations for the tangent line and normal line to the circle at the given points. (The *normal line* at a point is perpendicular to the tangent line at the point.) Use a graphing utility to graph the equation, tangent line, and normal line.

59. $x^2 + y^2 = 25$
$(4, 3), (-3, 4)$

60. $x^2 + y^2 = 9$
$(0, 3), (2, \sqrt{5})$

61. Show that the normal line at any point on the circle $x^2 + y^2 = r^2$ passes through the origin.

62. Two circles of radius 4 are tangent to the graph of $y^2 = 4x$ at the point $(1, 2)$. Find equations of these two circles.

In Exercises 63 and 64, find the points at which the graph of the equation has a vertical or horizontal tangent line.

63. $25x^2 + 16y^2 + 200x - 160y + 400 = 0$

64. $4x^2 + y^2 - 8x + 4y + 4 = 0$

In Exercises 65–74, find dy/dx using logarithmic differentiation.

65. $y = x\sqrt{x^2 - 1}$

66. $y = \sqrt{(x - 1)(x - 2)(x - 3)}$

67. $y = \frac{x^2\sqrt{3x - 2}}{(x - 1)^2}$ **68.** $y = \sqrt{\frac{x^2 - 2}{x^2 + 2}}$

69. $y = \frac{x(x - 1)^{3/2}}{\sqrt{x + 1}}$ **70.** $y = \frac{(x + 1)(x + 2)}{(x - 1)(x - 2)}$

71. $y = x^{2/x}$ **72.** $y = x^{x-1}$

73. $y = (x - 2)^{x+1}$ **74.** $y = (1 + x)^{1/x}$

Orthogonal Trajectories **In Exercises 75–78, use a graphing utility to sketch the intersecting graphs of the equations and show that they are orthogonal. [Two graphs are *orthogonal* if at their point(s) of intersection their tangent lines are perpendicular to each other.]**

75. $2x^2 + y^2 = 6$
$y^2 = 4x$

76. $y^2 = x^3$
$2x^2 + 3y^2 = 5$

77. $x + y = 0$
$x = \sin y$

78. $x^3 = 3(y - 1)$
$x(3y - 29) = 3$

Orthogonal Trajectories **In Exercises 79 and 80, verify that the two families of curves are orthogonal, where C and K are real numbers. Use a graphing utility to graph the two families for two values of C and two values of K.**

79. $xy = C,\quad x^2 - y^2 = K$ **80.** $x^2 + y^2 = C^2,\quad y = Kx$

In Exercises 81–84, differentiate (a) with respect to x (y is a function of x) and (b) with respect to t (x and y are functions of t).

81. $2y^2 - 3x^4 = 0$ **82.** $x^2 - 3xy^2 + y^3 = 10$

83. $\cos \pi y - 3 \sin \pi x = 1$ **84.** $4 \sin x \cos y = 1$

Writing About Concepts

85. Describe the difference between the explicit form of a function and an implicit equation. Give an example of each.

86. In your own words, state the guidelines for implicit differentiation.

87. ***Orthogonal Trajectories*** The figure below shows the topographic map carried by a group of hikers. The hikers are in a wooded area on top of the hill shown on the map and they decide to follow a path of steepest descent (orthogonal trajectories to the contours on the map). Draw their routes if they start from point A and if they start from point B. If their goal is to reach the road along the top of the map, which starting point should they use? To print an enlarged copy of the map, go to the website *www.mathgraphs.com*.

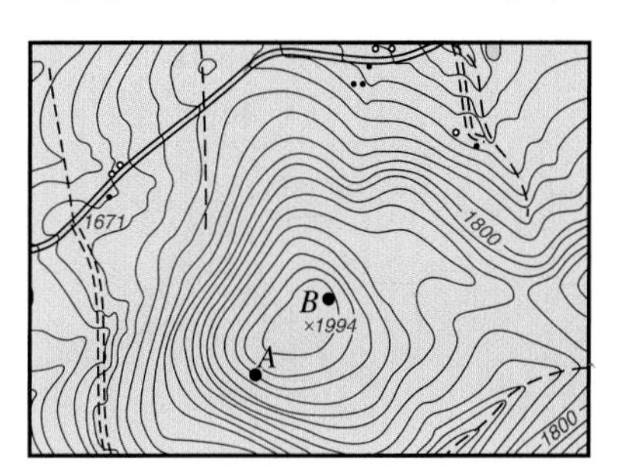

88. ***Weather Map*** The weather map shows several *isobars*—curves that represent areas of constant air pressure. Three high pressures H and one low pressure L are shown on the map. Given that wind speed is greatest along the orthogonal trajectories of the isobars, use the map to determine the areas having high wind speed.

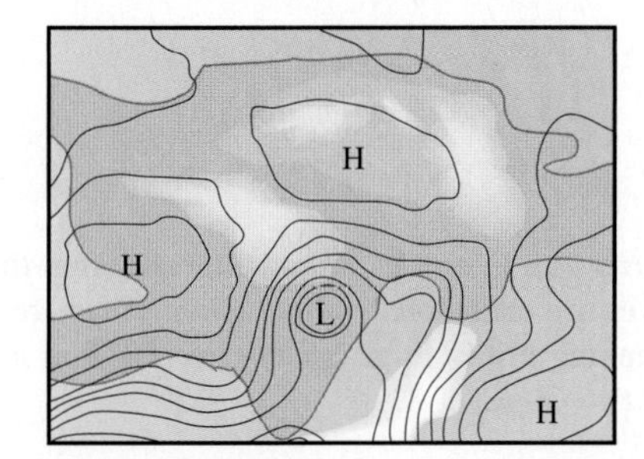

89. Consider the equation $x^4 = 4(4x^2 - y^2)$.

(a) Use a graphing utility to graph the equation.

(b) Find and graph the four tangent lines to the curve for $y = 3$.

(c) Find the exact coordinates of the point of intersection of the two tangent lines in the first quadrant.

90. Let L be any tangent line to the curve $\sqrt{x} + \sqrt{y} = \sqrt{c}$. Show that the sum of the x- and y-intercepts of L is c.

91. Prove (Theorem 3.3) that

$$\frac{d}{dx}[x^n] = nx^{n-1}$$

for the case in which n is a rational number. (*Hint:* Write $y = x^{p/q}$ in the form $y^q = x^p$ and differentiate implicitly. Assume that p and q are integers, where $q > 0$.)

92. ***Slope*** Find all points on the circle $x^2 + y^2 = 25$ where the slope is $\frac{3}{4}$.

93. ***Horizontal Tangent*** Determine the point(s) at which the graph of $y^4 = y^2 - x^2$ has a horizontal tangent.

94. ***Tangent Lines*** Find equations of both tangent lines to the ellipse $\frac{x^2}{4} + \frac{y^2}{9} = 1$ that passes through the point $(4, 0)$.

95. ***Normals to a Parabola*** The graph shows the normal lines from the point $(2, 0)$ to the graph of the parabola $x = y^2$. How many normal lines are there from the point $(x_0, 0)$ to the graph of the parabola if (a) $x_0 = \frac{1}{4}$, (b) $x_0 = \frac{1}{2}$, and (c) $x_0 = 1$? For what value of x_0 are two of the normal lines perpendicular to each other?

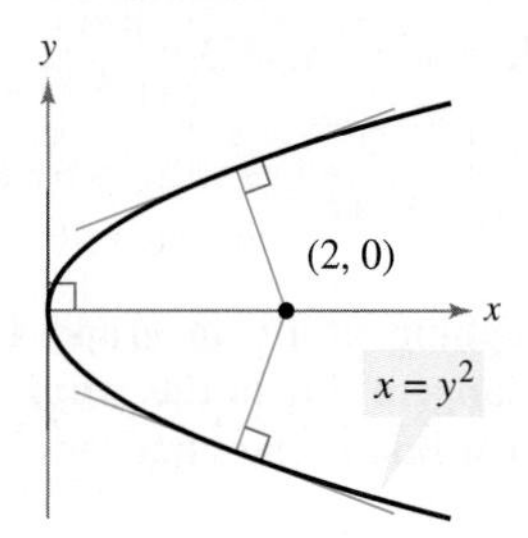

96. ***Normal Lines*** (a) Find an equation of the normal line to the ellipse

$$\frac{x^2}{32} + \frac{y^2}{8} = 1$$

at the point $(4, 2)$. (b) Use a graphing utility to graph the ellipse and the normal line. (c) At what other point does the normal line intersect the ellipse?

Section Project: Optical Illusions

In each graph below, an optical illusion is created by having lines intersect a family of curves. In each case, the lines appear to be curved. Find the value of dy/dx for the given values of x and y.

(a) Circles: $x^2 + y^2 = C^2$

$x = 3, y = 4, C = 5$

(b) Hyperbolas: $xy = C$

$x = 1, y = 4, C = 4$

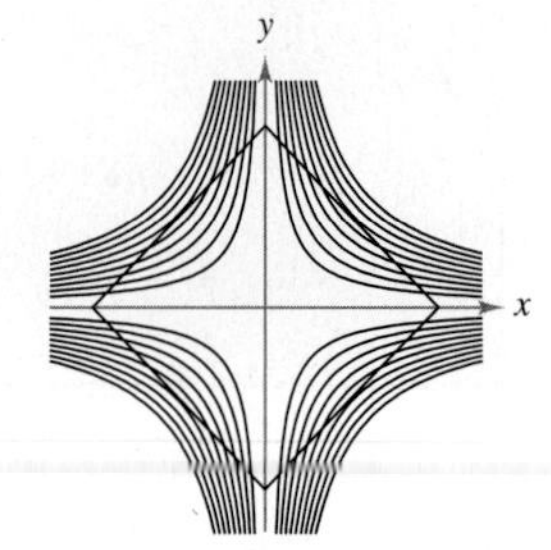

(c) Lines: $ax = by$

$x = \sqrt{3}, y = 3,$

$a = \sqrt{3}, b = 1$

(d) Cosine curves: $y = C \cos x$

$x = \frac{\pi}{3}, y = \frac{1}{3}, C = \frac{2}{3}$

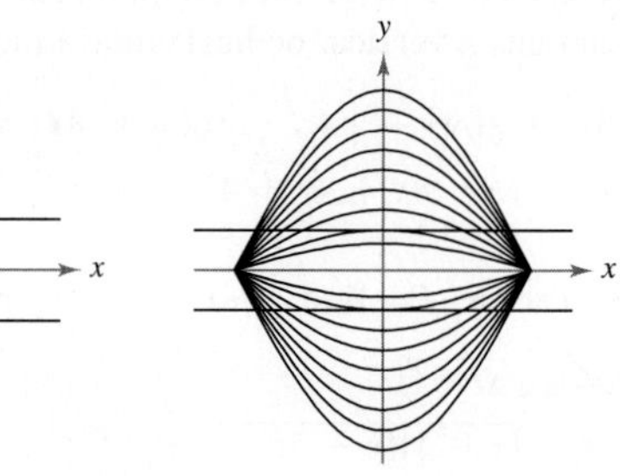

FOR FURTHER INFORMATION For more information on the mathematics of optical illusions, see the article "Descriptive Models for Perception of Optical Illusions" by David A. Smith in *The UMAP Journal*.

Section 3.6 Derivatives of Inverse Functions

- **Find the derivative of an inverse function.**
- **Differentiate an inverse trigonometric function.**
- **Review the basic differentiation formulas for elementary functions.**

Derivative of an Inverse Function

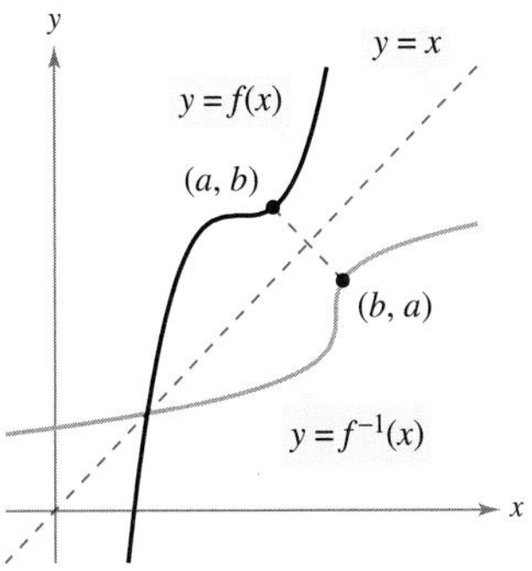

The graph of f^{-1} is a reflection of the graph of f in the line $y = x$.
Figure 3.33

The next two theorems discuss the derivative of an inverse function. The reasonableness of Theorem 3.16 follows from the reflective property of inverse functions, as shown in Figure 3.33. Proofs of the two theorems are given in Appendix A.

THEOREM 3.16 Continuity and Differentiability of Inverse Functions

Let f be a function whose domain is an interval I. If f has an inverse function, then the following statements are true.

1. If f is continuous on its domain, then f^{-1} is continuous on its domain.
2. If f is differentiable on an interval containing c and $f'(c) \neq 0$, then f^{-1} is differentiable at $f(c)$.

THEOREM 3.17 The Derivative of an Inverse Function

Let f be a function that is differentiable on an interval I. If f has an inverse function g, then g is differentiable at any x for which $f'(g(x)) \neq 0$. Moreover,

$$g'(x) = \frac{1}{f'(g(x))}, \quad f'(g(x)) \neq 0.$$

EXAMPLE 1 Evaluating the Derivative of an Inverse Function

Let $f(x) = \frac{1}{4}x^3 + x - 1$.

a. What is the value of $f^{-1}(x)$ when $x = 3$?

b. What is the value of $(f^{-1})'(x)$ when $x = 3$?

Solution Notice that f is one-to-one and therefore has an inverse function.

a. Because $f(2) = 3$, you know that $f^{-1}(3) = 2$.

b. Because the function f is differentiable and has an inverse function, you can apply Theorem 3.17 (with $g = f^{-1}$) to write

$$(f^{-1})'(3) = \frac{1}{f'(f^{-1}(3))} = \frac{1}{f'(2)}.$$

Moreover, using $f'(x) = \frac{3}{4}x^2 + 1$, you can conclude that

$$(f^{-1})'(3) = \frac{1}{f'(2)} = \frac{1}{\frac{3}{4}(2^2) + 1} = \frac{1}{4}.$$

(See Figure 3.34.)

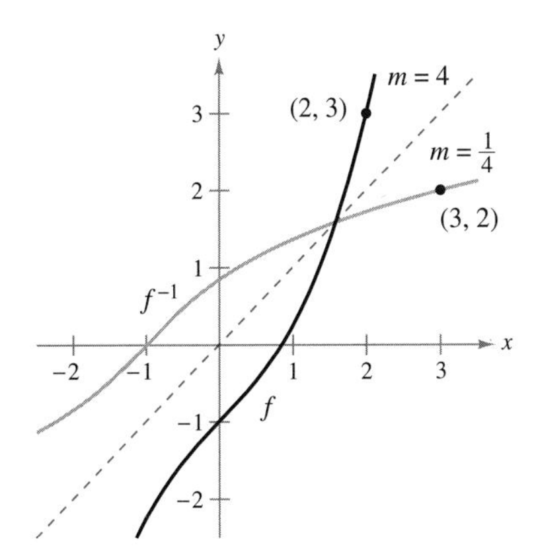

The graphs of the inverse functions f and f^{-1} have reciprocal slopes at points (a, b) and (b, a).
Figure 3.34

In Example 1, note that at the point $(2, 3)$ the slope of the graph of f is 4 and at the point $(3, 2)$ the slope of the graph of f^{-1} is $\frac{1}{4}$ (see Figure 3.34). This reciprocal relationship (which follows from Theorem 3.17) is sometimes written as

$$\frac{dy}{dx} = \frac{1}{dx/dy}.$$

EXAMPLE 2 Graphs of Inverse Functions Have Reciprocal Slopes

Let $f(x) = x^2$ (for $x \geq 0$) and let $f^{-1}(x) = \sqrt{x}$. Show that the slopes of the graphs of f and f^{-1} are reciprocals at each of the following points.

a. $(2, 4)$ and $(4, 2)$ **b.** $(3, 9)$ and $(9, 3)$

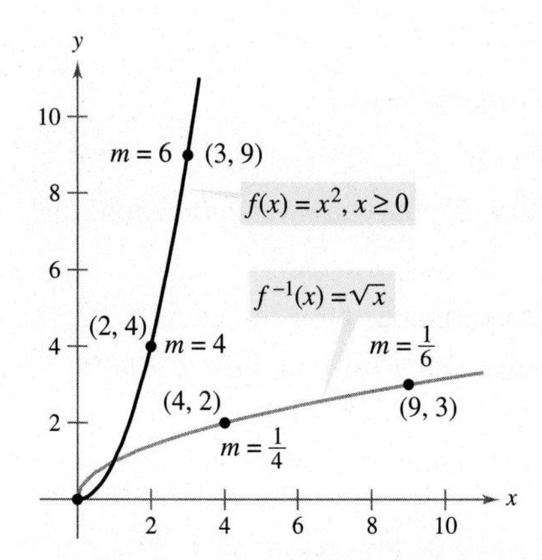

At $(0, 0)$, the derivative of f is 0 and the derivative of f^{-1} does not exist.
Figure 3.35

Solution The derivatives of f and f^{-1} are $f'(x) = 2x$ and $(f^{-1})'(x) = \dfrac{1}{2\sqrt{x}}$.

a. At $(2, 4)$, the slope of the graph of f is $f'(2) = 2(2) = 4$. At $(4, 2)$, the slope of the graph of f^{-1} is

$$(f^{-1})'(4) = \frac{1}{2\sqrt{4}} = \frac{1}{2(2)} = \frac{1}{4}.$$

b. At $(3, 9)$, the slope of the graph of f is $f'(3) = 2(3) = 6$. At $(9, 3)$, the slope of the graph of f^{-1} is

$$(f^{-1})'(9) = \frac{1}{2\sqrt{9}} = \frac{1}{2(3)} = \frac{1}{6}.$$

So, in both cases, the slopes are reciprocals, as shown in Figure 3.35.

When determining the derivative of an inverse function, you have two options: (1) you can apply Theorem 3.17, or (2) you can use implicit differentiation. The first approach is illustrated in Example 3, and the second in the proof of Theorem 3.18.

EXAMPLE 3 Finding the Derivative of an Inverse Function

Find the derivative of the inverse tangent function.

Solution Let $f(x) = \tan x$, $-\pi/2 < x < \pi/2$. Then let $g(x) = \arctan x$ be the inverse tangent function. To find the derivative of $g(x)$, use the fact that $f'(x) = \sec^2 x = \tan^2 x + 1$, and apply Theorem 3.17 as follows.

$$g'(x) = \frac{1}{f'(g(x))} = \frac{1}{f'(\arctan x)} = \frac{1}{[\tan(\arctan x)]^2 + 1} = \frac{1}{x^2 + 1}$$

Derivatives of Inverse Trigonometric Functions

In Section 3.4, you saw that the derivative of the *transcendental* function $f(x) = \ln x$ is the *algebraic* function $f'(x) = 1/x$. You will now see that the derivatives of the inverse trigonometric functions also are algebraic (even though the inverse trigonometric functions are themselves transcendental).

The following theorem lists the derivatives of the six inverse trigonometric functions. Note that the derivatives of arccos u, arccot u, and arccsc u are the *negatives* of the derivatives of arcsin u, arctan u, and arcsec u, respectively.

THEOREM 3.18 Derivatives of Inverse Trigonometric Functions

Let u be a differentiable function of x.

$$\frac{d}{dx}[\arcsin u] = \frac{u'}{\sqrt{1-u^2}} \qquad \frac{d}{dx}[\arccos u] = \frac{-u'}{\sqrt{1-u^2}}$$

$$\frac{d}{dx}[\arctan u] = \frac{u'}{1+u^2} \qquad \frac{d}{dx}[\text{arccot } u] = \frac{-u'}{1+u^2}$$

$$\frac{d}{dx}[\text{arcsec } u] = \frac{u'}{|u|\sqrt{u^2-1}} \qquad \frac{d}{dx}[\text{arccsc } u] = \frac{-u'}{|u|\sqrt{u^2-1}}$$

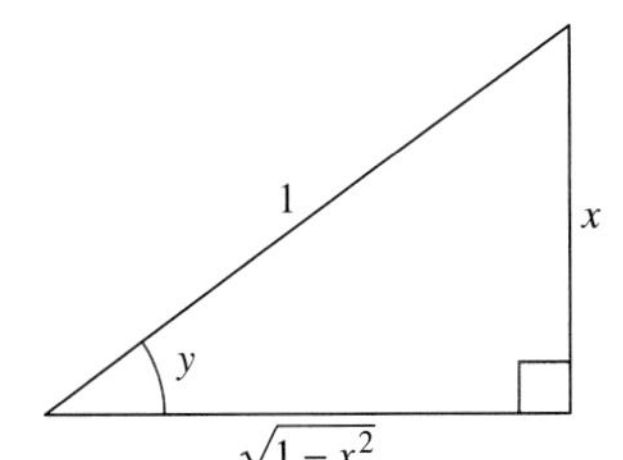

$y = \arcsin x$
Figure 3.36

Proof Let $y = \arcsin x$, $-\pi/2 \le y \le \pi/2$ (see Figure 3.36). So, $\sin y = x$, and you can use implicit differentiation as follows.

$$\sin y = x$$

$$(\cos y)\left(\frac{dy}{dx}\right) = 1$$

$$\frac{dy}{dx} = \frac{1}{\cos y} = \frac{1}{\sqrt{1-\sin^2 y}} = \frac{1}{\sqrt{1-x^2}}$$

Because u is a differentiable function of x, you can use the Chain Rule to write

$$\frac{d}{dx}[\arcsin u] = \frac{u'}{\sqrt{1-u^2}}, \quad \text{where } u' = \frac{du}{dx}.$$

Proofs of the other differentiation rules are left as exercises (see Exercise 71).

TECHNOLOGY If your graphing utility does not have the arcsecant function, you can obtain its graph using

$$f(x) = \text{arcsec } x = \arccos \frac{1}{x}.$$

There is no common agreement on the definition of $\text{arcsec } x$ (or $\text{arccsc } x$) for negative values of x. When we defined the range of the arcsecant, we chose to preserve the reciprocal identity $\text{arcsec } x = \arccos(1/x)$. For example, to evaluate $\text{arcsec}(-2)$, you can write

$$\text{arcsec}(-2) = \arccos(-0.5) \approx 2.09.$$

One of the consequences of the definition of the inverse secant function given in this text is that its graph has a positive slope at every x-value in its domain. This accounts for the absolute value sign in the formula for the derivative of $\text{arcsec } x$.

EXPLORATION

Suppose that you want to find a linear approximation to the graph of the function in Example 4. You decide to use the tangent line at the origin, as shown below. Use a graphing utility to describe an interval about the origin where the tangent line is within 0.01 unit of the graph of the function. What might a person mean by saying that the original function is "locally linear"?

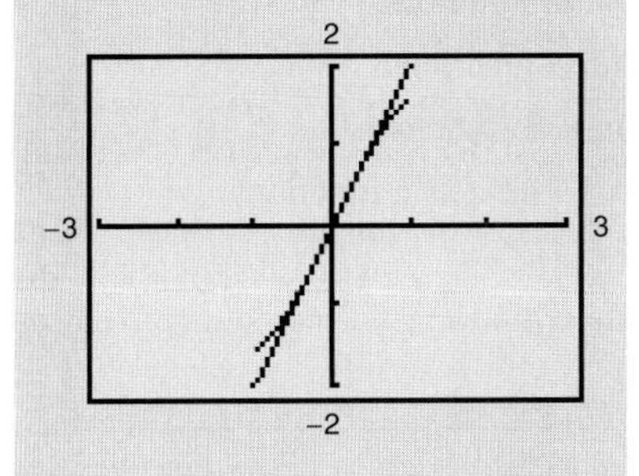

EXAMPLE 4 A Derivative That Can Be Simplified

Differentiate $y = \arcsin x + x\sqrt{1-x^2}$.

Solution

$$\begin{aligned} y' &= \frac{1}{\sqrt{1-x^2}} + x\left(\frac{1}{2}\right)(-2x)(1-x^2)^{-1/2} + \sqrt{1-x^2} \\ &= \frac{1}{\sqrt{1-x^2}} - \frac{x^2}{\sqrt{1-x^2}} + \sqrt{1-x^2} \\ &= \sqrt{1-x^2} + \sqrt{1-x^2} \\ &= 2\sqrt{1-x^2} \end{aligned}$$

EXAMPLE 5 Differentiating Inverse Trigonometric Functions

a. $\dfrac{d}{dx}[\arcsin(2x)] = \dfrac{2}{\sqrt{1-(2x)^2}}$ $\quad u = 2x$

$= \dfrac{2}{\sqrt{1-4x^2}}$

b. $\dfrac{d}{dx}[\arctan(3x)] = \dfrac{3}{1+(3x)^2}$ $\quad u = 3x$

$= \dfrac{3}{1+9x^2}$

c. $\dfrac{d}{dx}[\arcsin\sqrt{x}] = \dfrac{(1/2)x^{-1/2}}{\sqrt{1-x}}$ $\quad u = \sqrt{x}$

$= \dfrac{1}{2\sqrt{x}\sqrt{1-x}}$

$= \dfrac{1}{2\sqrt{x-x^2}}$

d. $\dfrac{d}{dx}[\operatorname{arcsec} e^{2x}] = \dfrac{2e^{2x}}{e^{2x}\sqrt{(e^{2x})^2-1}}$ $\quad u = e^{2x}$

$= \dfrac{2e^{2x}}{e^{2x}\sqrt{e^{4x}-1}}$

$= \dfrac{2}{\sqrt{e^{4x}-1}}$

In part (d), the absolute value sign is not necessary because $e^{2x} > 0$.

Review of Basic Differentiation Rules

In the 1600s, Europe was ushered into the scientific age by such great thinkers as Descartes, Galileo, Huygens, Newton, and Kepler. These men believed that nature is governed by basic laws—laws that can, for the most part, be written in terms of mathematical equations. One of the most influential publications of this period—*Dialogue on the Great World Systems*, by Galileo Galilei—has become a classic description of modern scientific thought.

As mathematics has developed during the past few hundred years, a small number of elementary functions has proven sufficient for modeling most* phenomena in physics, chemistry, biology, engineering, economics, and a variety of other fields. An **elementary function** is a function from the following list or one that can be formed as the sum, product, quotient, or composition of functions in the list.

Algebraic Functions	*Transcendental Functions*
Polynomial functions	Logarithmic functions
Rational functions	Exponential functions
Functions involving radicals	Trigonometric functions
	Inverse trigonometric functions

With the differentiation rules introduced so far in the text, you can differentiate any elementary function. For convenience, these differentiation rules are summarized on the next page.

GALILEO GALILEI (1564–1642)

Galileo's approach to science departed from the accepted Aristotelian view that nature had describable *qualities*, such as "fluidity" and "potentiality." He chose to describe the physical world in terms of measurable *quantities*, such as time, distance, force, and mass.

* *Some important functions used in engineering and science (such as Bessel functions and gamma functions) are not elementary functions.*

Basic Differentiation Rules for Elementary Functions

1. $\frac{d}{dx}[cu] = cu'$	**2.** $\frac{d}{dx}[u \pm v] = u' \pm v'$	**3.** $\frac{d}{dx}[uv] = uv' + vu'$
4. $\frac{d}{dx}\left[\frac{u}{v}\right] = \frac{vu' - uv'}{v^2}$	**5.** $\frac{d}{dx}[c] = 0$	**6.** $\frac{d}{dx}[u^n] = nu^{n-1}u'$
7. $\frac{d}{dx}[x] = 1$	**8.** $\frac{d}{dx}[\lvert u \rvert] = \frac{u}{\lvert u \rvert}(u'), \quad u \neq 0$	**9.** $\frac{d}{dx}[\ln u] = \frac{u'}{u}$
10. $\frac{d}{dx}[e^u] = e^u u'$	**11.** $\frac{d}{dx}[\log_a u] = \frac{u'}{(\ln a)u}$	**12.** $\frac{d}{dx}[a^u] = (\ln a)a^u u'$
13. $\frac{d}{dx}[\sin u] = (\cos u)u'$	**14.** $\frac{d}{dx}[\cos u] = -(\sin u)u'$	**15.** $\frac{d}{dx}[\tan u] = (\sec^2 u)u'$
16. $\frac{d}{dx}[\cot u] = -(\csc^2 u)u'$	**17.** $\frac{d}{dx}[\sec u] = (\sec u \tan u)u'$	**18.** $\frac{d}{dx}[\csc u] = -(\csc u \cot u)u'$
19. $\frac{d}{dx}[\arcsin u] = \frac{u'}{\sqrt{1 - u^2}}$	**20.** $\frac{d}{dx}[\arccos u] = \frac{-u'}{\sqrt{1 - u^2}}$	**21.** $\frac{d}{dx}[\arctan u] = \frac{u'}{1 + u^2}$
22. $\frac{d}{dx}[\text{arccot } u] = \frac{-u'}{1 + u^2}$	**23.** $\frac{d}{dx}[\text{arcsec } u] = \frac{u'}{\lvert u \rvert \sqrt{u^2 - 1}}$	**24.** $\frac{d}{dx}[\text{arccsc } u] = \frac{-u'}{\lvert u \rvert \sqrt{u^2 - 1}}$

Exercises for Section 3.6

See www.CalcChat.com for worked-out solutions to odd-numbered exercises.

In Exercises 1–6, find $(f^{-1})'(a)$ for the function f and real number a.

	Function	Real Number
1.	$f(x) = x^3 + 2x - 1$	$a = 2$
2.	$f(x) = \frac{1}{27}(x^5 + 2x^3)$	$a = -11$
3.	$f(x) = \sin x, \quad -\frac{\pi}{2} \leq x \leq \frac{\pi}{2}$	$a = \frac{1}{2}$
4.	$f(x) = \cos 2x, \quad 0 \leq x \leq \frac{\pi}{2}$	$a = 1$
5.	$f(x) = x^3 - \frac{4}{x}$	$a = 6$
6.	$f(x) = \sqrt{x - 4}$	$a = 2$

In Exercises 7–10, show that the slopes of the graphs of f and f^{-1} are reciprocals at the indicated points.

	Function	Point
7.	$f(x) = x^3$	$\left(\frac{1}{2}, \frac{1}{8}\right)$
	$f^{-1}(x) = \sqrt[3]{x}$	$\left(\frac{1}{8}, \frac{1}{2}\right)$
8.	$f(x) = 3 - 4x$	$(1, -1)$
	$f^{-1}(x) = \frac{3 - x}{4}$	$(-1, 1)$
9.	$f(x) = \sqrt{x - 4}$	$(5, 1)$
	$f^{-1}(x) = x^2 + 4, \quad x \geq 0$	$(1, 5)$
10.	$f(x) = \frac{4}{1 + x^2}, \quad x \geq 0$	$(1, 2)$
	$f^{-1}(x) = \sqrt{\frac{4 - x}{x}}$	$(2, 1)$

In Exercises 11–14, (a) find an equation of the tangent line to the graph of f at the indicated point and (b) use a graphing utility to graph the function and its tangent line at the point.

	Function	Point
11.	$f(x) = \arcsin 2x$	$\left(\frac{\sqrt{2}}{4}, \frac{\pi}{4}\right)$
12.	$f(x) = \arctan x$	$\left(-1, -\frac{\pi}{4}\right)$
13.	$f(x) = \arccos x^2$	$\left(0, \frac{\pi}{2}\right)$
14.	$f(x) = \text{arcsec } x$	$\left(\sqrt{2}, \frac{\pi}{4}\right)$

In Exercises 15–18, find dy/dx at the indicated point for the equation.

15. $x = y^3 - 7y^2 + 2, \ (-4, 1)$

16. $x = 2\ln(y^2 - 3), \ (0, 4)$

17. $x \arctan x = e^y, \ \left(1, \ln\frac{\pi}{4}\right)$

18. $\arcsin xy = \frac{2}{3} \arctan 2x, \ \left(\frac{1}{2}, 1\right)$

In Exercises 19–44, find the derivative of the function.

19. $f(x) = 2\arcsin(x - 1)$ **20.** $f(t) = \arcsin t^2$

21. $g(x) = 3\arccos(x/2)$ **22.** $f(x) = \operatorname{arcsec} 3x$

23. $f(x) = \arctan(x/a)$ **24.** $f(x) = \arctan\sqrt{x}$

25. $g(x) = \dfrac{\arcsin 3x}{x}$ **26.** $h(x) = x^2 \arctan x$

27. $g(x) = \dfrac{\arccos x}{x + 1}$ **28.** $g(x) = e^x \arcsin x$

29. $h(x) = \operatorname{arccot} 6x$ **30.** $f(x) = \operatorname{arcsec} 2x$

31. $h(t) = \sin(\arccos t)$ **32.** $f(x) = \arcsin x + \arccos x$

33. $y = x\arccos x - \sqrt{1 - x^2}$

34. $y = 2\ln(t^2 + 4) - \arctan\dfrac{t}{2}$

35. $y = \dfrac{1}{2}\left(\dfrac{1}{2}\ln\dfrac{x + 1}{x - 1} + \arctan x\right)$

36. $y = \dfrac{1}{2}\left[x\sqrt{4 - x^2} + 4\arcsin\left(\dfrac{x}{2}\right)\right]$

37. $g(t) = \tan(\arcsin t)$ **38.** $f(x) = \operatorname{arcsec} x + \operatorname{arccsc} x$

39. $y = x\arcsin x + \sqrt{1 - x^2}$

40. $y = x\arctan 2x - \frac{1}{4}\ln(1 + 4x^2)$

41. $y = 8\arcsin\dfrac{x}{4} - \dfrac{x\sqrt{16 - x^2}}{2}$

42. $y = 25\arcsin\dfrac{x}{5} - x\sqrt{25 - x^2}$

43. $y = \arctan x + \dfrac{x}{1 + x^2}$ **44.** $y = \arctan\dfrac{x}{2} - \dfrac{1}{2(x^2 + 4)}$

In Exercises 45–50, find an equation of the tangent line to the graph of the function at the given point.

45. $y = 2\arcsin x$

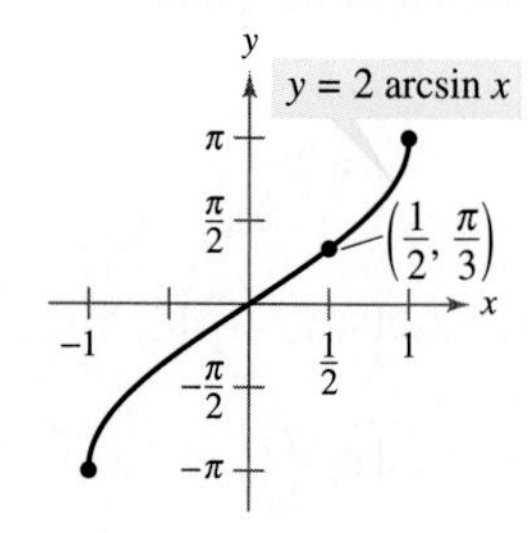

46. $y = \dfrac{1}{2}\arccos x$

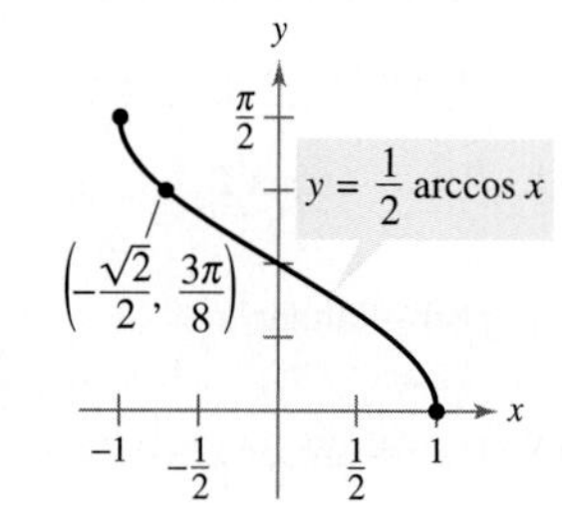

47. $y = \arctan\dfrac{x}{2}$

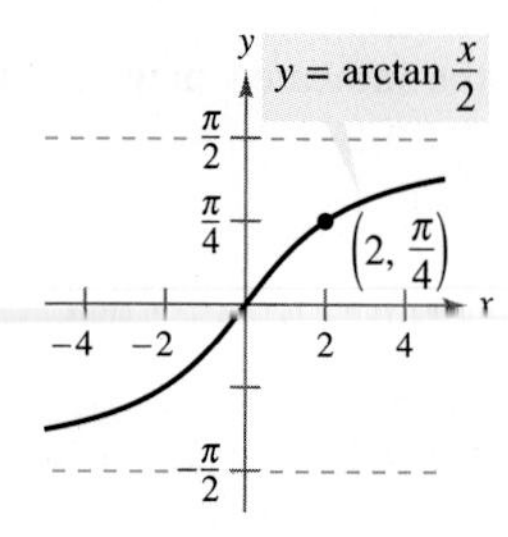

48. $y = \operatorname{arcsec} 4x$

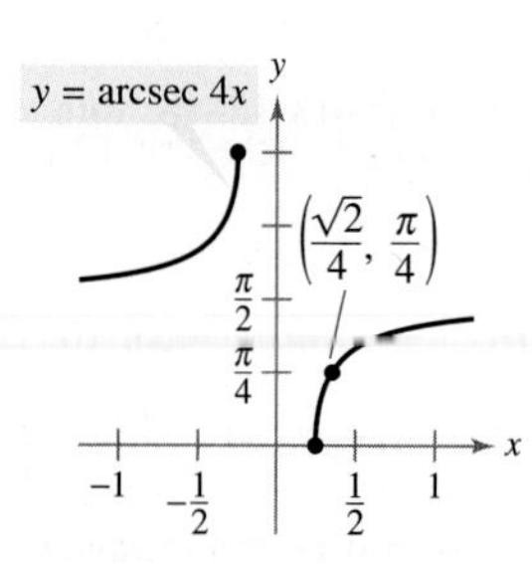

49. $y = 4x\arccos(x - 1)$

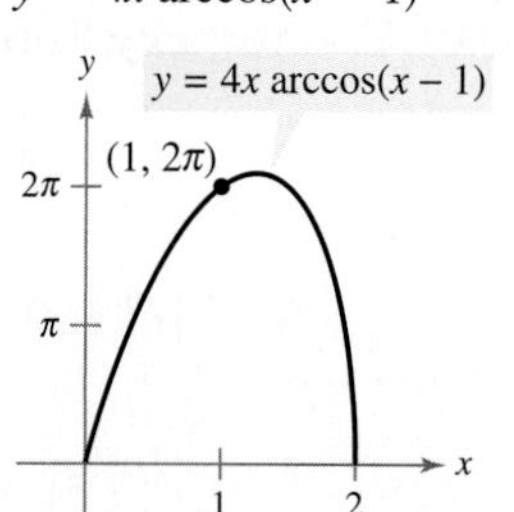

50. $y = 3x\arcsin x$

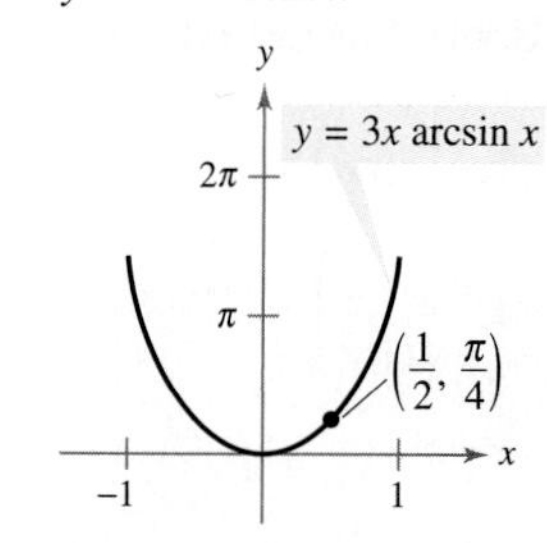

51. Find equations of all tangent lines to the graph of $f(x) = \arccos x$ that have slope -2.

52. Find an equation of the tangent line to the graph of $g(x) = \arctan x$ when $x = 1$.

Linear and Quadratic Approximations **In Exercises 53–56, use a computer algebra system to find the linear approximation**

$$P_1(x) = f(a) + f'(a)(x - a)$$

and the quadratic approximation

$$P_2(x) = f(a) + f'(a)(x - a) + \frac{1}{2}f''(a)(x - a)^2$$

to the function f at $x = a$. Sketch the graph of the function and its linear and quadratic approximations.

53. $f(x) = \arcsin x,\quad a = \frac{1}{2}$ **54.** $f(x) = \arctan x,\quad a = 1$

55. $f(x) = \arctan x,\quad a = 0$ **56.** $f(x) = \arccos x,\quad a = 0$

Implicit Differentiation **In Exercises 57–60, find an equation of the tangent line to the graph of the equation at the given point.**

57. $x^2 + x\arctan y = y - 1,\quad \left(-\dfrac{\pi}{4}, 1\right)$

58. $\arctan(xy) = \arcsin(x + y),\quad (0, 0)$

59. $\arcsin x + \arcsin y = \dfrac{\pi}{2},\quad \left(\dfrac{\sqrt{2}}{2}, \dfrac{\sqrt{2}}{2}\right)$

60. $\arctan(x + y) = y^2 + \dfrac{\pi}{4},\quad (1, 0)$

Writing About Concepts

In Exercises 61 and 62, the derivative of the function has the same sign for all x in its domain, but the function is not one-to-one. Explain.

61. $f(x) = \tan x$ **62.** $f(x) = \dfrac{x}{x^2 - 4}$

63. State the theorem that gives the method for finding the derivative of an inverse function.

64. Are the derivatives of the inverse trigonometric functions algebraic or transcendental functions? List the derivatives of the inverse trigonometric functions.

65. ***Angular Rate of Change*** An airplane flies at an altitude of 5 miles toward a point directly over an observer. Consider θ and x as shown in the figure.

(a) Write θ as a function of x.

(b) The speed of the plane is 400 miles per hour. Find $d\theta/dt$ when $x = 10$ miles and $x = 3$ miles.

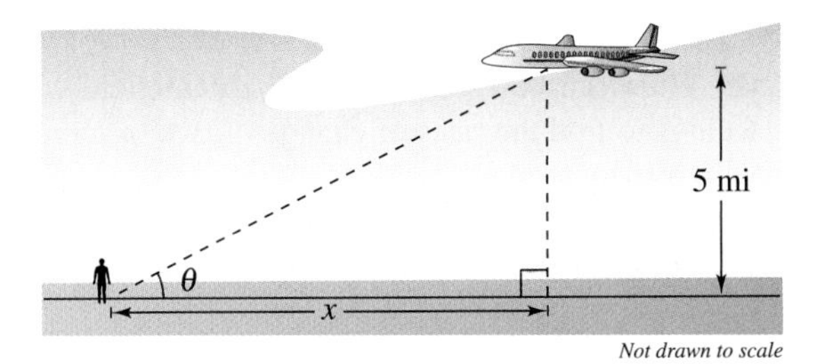

66. ***Writing*** Repeat Exercise 65 if the altitude of the plane is 3 miles and describe how the altitude affects the rate of change of θ.

67. ***Angular Rate of Change*** In a free-fall experiment, an object is dropped from a height of 256 feet. A camera on the ground 500 feet from the point of impact records the fall of the object (see figure).

(a) Find the position function giving the height of the object at time t, assuming the object is released at time $t = 0$. At what time will the object reach ground level?

(b) Find the rates of change of the angle of elevation of the camera when $t = 1$ and $t = 2$.

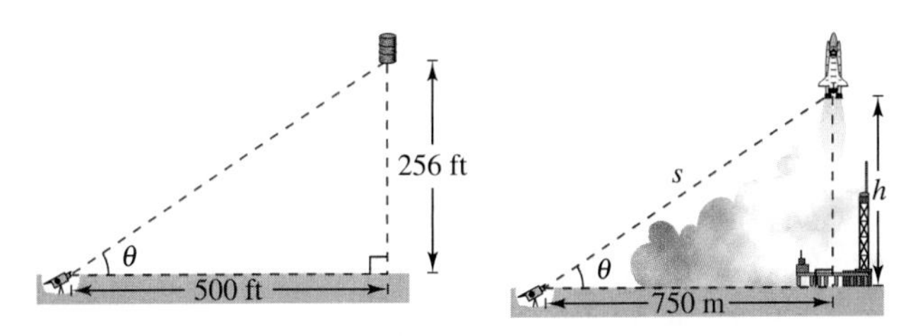

Figure for 67 **Figure for 68**

68. ***Angular Rate of Change*** A television camera at ground level is filming the lift-off of a space shuttle at a point 750 meters from the launch pad. Let θ be the angle of elevation of the shuttle and let s be the distance between the camera and the shuttle (see figure). Write θ as a function of s for the period of time when the shuttle is moving vertically. Differentiate the result to find $d\theta/dt$ in terms of s and ds/dt.

69. ***Angular Rate of Change*** An observer is standing 300 feet from the point at which a balloon is released. The balloon rises at a rate of 5 feet per second. How fast is the angle of elevation of the observer's line of sight increasing when the balloon is 100 feet high?

70. ***Farm Workers*** The table gives the number of farm workers (in millions) in the United States for selected years from 1910 to 2000. *(Source: U.S. Department of Agriculture)*

Year	1910	1920	1930	1940	1950
Workers	13.56	13.43	12.50	10.98	9.93

Year	1960	1970	1980	1990	2000
Workers	7.06	4.52	3.73	2.91	2.95

A model for the data is

$$y = 24.760 - 0.361t + 0.001t^2 - 79.564 \text{ arccot } t$$

where t is time in years, with $t = 0$ corresponding to the year 1900, and y is the number of workers in millions.

(a) Use a graphing utility to plot the data and graph the model.

(b) Find the rate of change of the number of workers when $t = 20$ and $t = 60$.

71. Verify each differentiation formula.

(a) $\dfrac{d}{dx}[\arctan u] = \dfrac{u'}{1 + u^2}$

(b) $\dfrac{d}{dx}[\text{arcsec } u] = \dfrac{u'}{|u|\sqrt{u^2 - 1}}$

(c) $\dfrac{d}{dx}[\arccos u] = \dfrac{-u'}{\sqrt{1 - u^2}}$

(d) $\dfrac{d}{dx}[\text{arccot } u] = \dfrac{-u'}{1 + u^2}$

(e) $\dfrac{d}{dx}[\text{arccsc } u] = \dfrac{-u'}{|u|\sqrt{u^2 - 1}}$

72. ***Existence of an Inverse*** Determine the values of k such that the function $f(x) = kx + \sin x$ has an inverse function.

True or False? **In Exercises 73 and 74, determine whether the statement is true or false. If it is false, explain why or give an example that shows it is false.**

73. The slope of the graph of the inverse tangent function is positive for all x.

74. $\dfrac{d}{dx}[\arctan(\tan x)] = 1$ for all x in the domain.

75. Prove that $\arcsin x = \arctan\left(\dfrac{x}{\sqrt{1 - x^2}}\right)$, $|x| < 1$.

76. Prove that $\arccos x = \dfrac{\pi}{2} - \arctan\left(\dfrac{x}{\sqrt{1 - x^2}}\right)$, $|x| < 1$.

77. Some calculus textbooks define the inverse secant function using the range $[0, \pi/2) \cup [\pi, 3\pi/2)$.

(a) Sketch the graph of $y = \text{arcsec } x$ using this range.

(b) Show that $y' = \dfrac{1}{x\sqrt{x^2 - 1}}$.

78. Compare the graphs of $y_1 = \sin(\arcsin x)$ and $y_2 = \arcsin(\sin x)$. What are the domains and ranges of y_1 and y_2?

79. Show that the function

$$f(x) = \arcsin \frac{x - 2}{2} - 2 \arcsin\left(\frac{\sqrt{x}}{2}\right)$$

is constant for $0 \le x \le 4$.

80. Use a graphing utility to graph the functions $y_1 = \arctan x$ and $y_2 = \arctan e^x$. Compare the functions and explain why the ranges are not the same.

Section 3.7 Related Rates

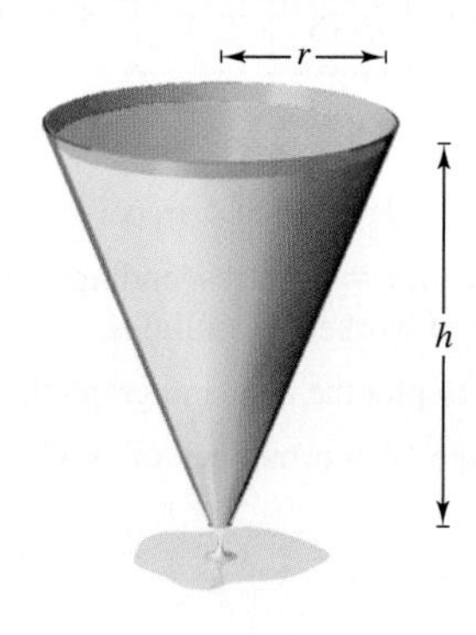

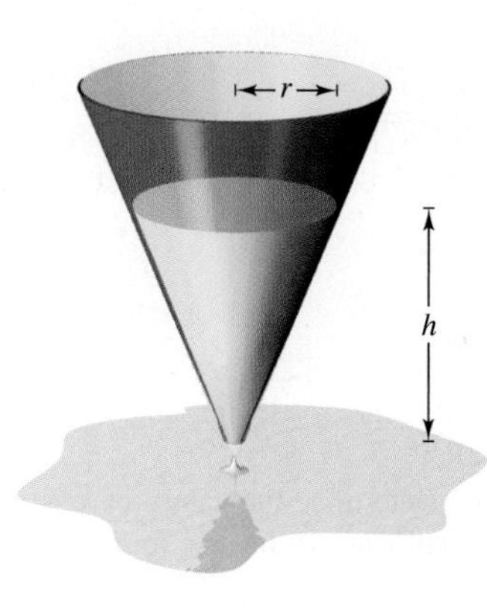

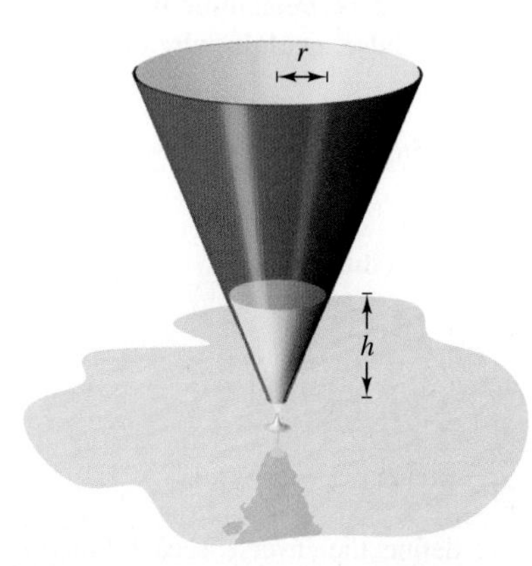

Volume is related to radius and height.
Figure 3.37

- **Find a related rate.**
- **Use related rates to solve real-life problems.**

Finding Related Rates

You have seen how the Chain Rule can be used to find dy/dx implicitly. Another important use of the Chain Rule is to find the rates of change of two or more related variables that are changing with respect to *time*.

For example, when water is drained out of a conical tank (see Figure 3.37), the volume V, the radius r, and the height h of the water level are all functions of time t. Knowing that these variables are related by the equation

$$V = \frac{\pi}{3} r^2 h \qquad \text{Original equation}$$

you can differentiate implicitly with respect to t to obtain the **related-rate** equation

$$\frac{d}{dt}(V) = \frac{d}{dt}\left(\frac{\pi}{3} r^2 h\right)$$

$$\frac{dV}{dt} = \frac{\pi}{3}\left[r^2 \frac{dh}{dt} + h\left(2r\frac{dr}{dt}\right)\right] \qquad \text{Differentiate with respect to } t.$$

$$= \frac{\pi}{3}\left(r^2 \frac{dh}{dt} + 2rh\frac{dr}{dt}\right).$$

From this equation, you can see that the rate of change of V is related to the rates of change of both h and r.

EXPLORATION

Finding a Related Rate In the conical tank shown in Figure 3.37, suppose that the height is changing at a rate of -0.2 foot per minute and the radius is changing at a rate of -0.1 foot per minute. What is the rate of change of the volume when the radius is $r = 1$ foot and the height is $h = 2$ feet? Does the rate of change of the volume depend on the values of r and h? Explain.

EXAMPLE 1 Two Rates That Are Related

Suppose x and y are both differentiable functions of t and are related by the equation $y = x^2 + 3$. Find dy/dt when $x = 1$, given that $dx/dt = 2$ when $x = 1$.

Solution Using the Chain Rule, you can differentiate both sides of the equation *with respect to t*.

$$y = x^2 + 3 \qquad \text{Write original equation.}$$

$$\frac{d}{dt}[y] = \frac{d}{dt}[x^2 + 3] \qquad \text{Differentiate with respect to } t.$$

$$\frac{dy}{dt} = 2x\frac{dx}{dt} \qquad \text{Chain Rule}$$

When $x = 1$ and $dx/dt = 2$, you have

$$\frac{dy}{dt} = 2(1)(2) = 4.$$

FOR FURTHER INFORMATION To learn more about the history of related-rate problems, see the article "The Lengthening Shadow: The Story of Related Rates" by Bill Austin, Don Barry, and David Berman in *Mathematics Magazine*. To view this article, go to the website *www.matharticles.com*.

Problem Solving with Related Rates

In Example 1, you were *given* an equation that related the variables x and y and were asked to find the rate of change of y when $x = 1$.

Equation: $y = x^2 + 3$

Given rate: $\dfrac{dx}{dt} = 2$ when $x = 1$

Find: $\dfrac{dy}{dt}$ when $x = 1$

In each of the remaining examples in this section, you must *create* a mathematical model from a verbal description.

EXAMPLE 2 Ripples in a Pond

W. Cody/Corbis

Total area increases as the outer radius increases.
Figure 3.38

A pebble is dropped into a calm pond, causing ripples in the form of concentric circles, as shown in Figure 3.38. The radius r of the outer ripple is increasing at a constant rate of 1 foot per second. When the radius is 4 feet, at what rate is the total area A of the disturbed water changing?

Solution The variables r and A are related by $A = \pi r^2$. The rate of change of the radius r is $dr/dt = 1$.

Equation: $A = \pi r^2$

Given rate: $\dfrac{dr}{dt} = 1$

Find: $\dfrac{dA}{dt}$ when $r = 4$

With this information, you can proceed as in Example 1.

$$\frac{d}{dt}[A] = \frac{d}{dt}[\pi r^2] \qquad \text{Differentiate with respect to } t.$$

$$\frac{dA}{dt} = 2\pi r \frac{dr}{dt} \qquad \text{Chain Rule}$$

$$\frac{dA}{dt} = 2\pi(4)(1) = 8\pi \qquad \text{Substitute 4 for } r \text{ and 1 for } dr/dt.$$

When the radius is 4 feet, the area is changing at a rate of 8π square feet per second.

NOTE When using these guidelines, be sure you perform Step 3 before Step 4. Substituting the known values of the variables before differentiating will produce an inappropriate derivative.

Guidelines For Solving Related-Rate Problems

1. Identify all *given* quantities and quantities *to be determined*. Make a sketch and label the quantities.
2. Write an equation involving the variables whose rates of change either are given or are to be determined.
3. Using the Chain Rule, implicitly differentiate both sides of the equation *with respect to time t*.
4. *After* completing Step 3, substitute into the resulting equation all known values for the variables and their rates of change. Then solve for the required rate of change.

The table below lists examples of mathematical models involving rates of change. For instance, the rate of change in the first example is the velocity of a car.

Verbal Statement	Mathematical Model
The velocity of a car after traveling for 1 hour is 50 miles per hour.	$x =$ distance traveled $\frac{dx}{dt} = 50$ when $t = 1$
Water is being pumped into a swimming pool at a rate of 10 cubic meters per hour.	$V =$ volume of water in pool $\frac{dV}{dt} = 10\ \text{m}^3/\text{hr}$
A gear is revolving at a rate of 25 revolutions per minute (1 revolution $= 2\pi$ radians).	$\theta =$ angle of revolution $\frac{d\theta}{dt} = 25(2\pi)$ rad/min

EXAMPLE 3 An Inflating Balloon

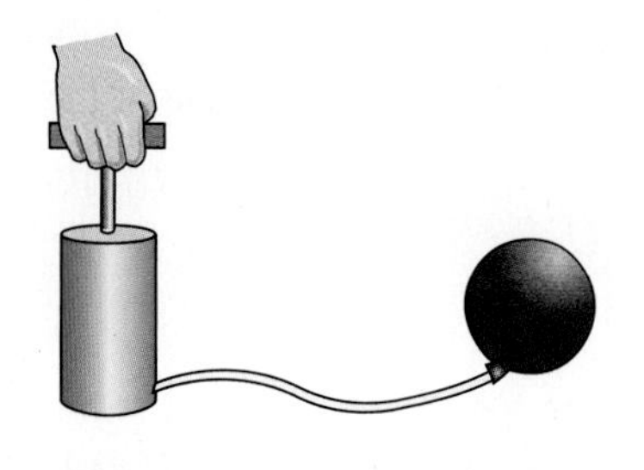

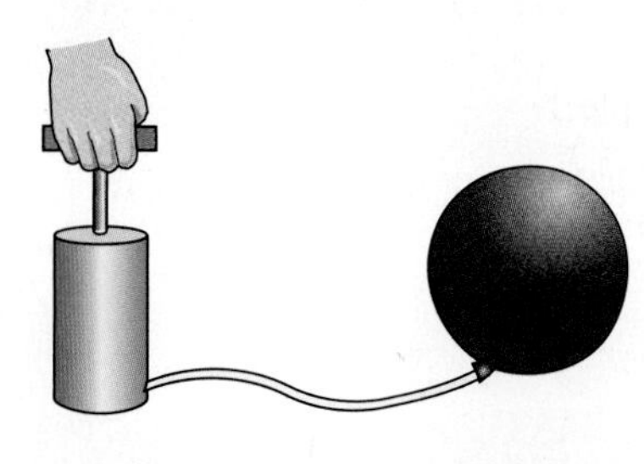

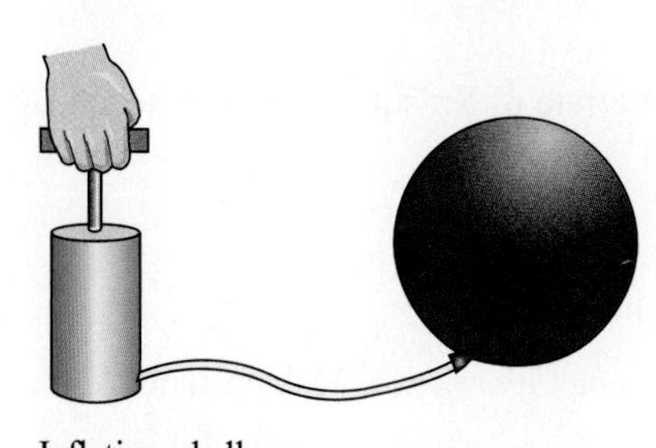

Inflating a balloon
Figure 3.39

Air is being pumped into a spherical balloon (see Figure 3.39) at a rate of 4.5 cubic feet per minute. Find the rate of change of the radius when the radius is 2 feet.

Solution Let V be the volume of the balloon and let r be its radius. Because the volume is increasing at a rate of 4.5 cubic feet per minute, you know that at time t the rate of change of the volume is $dV/dt = \frac{9}{2}$. So, the problem can be stated as shown.

Given rate: $\frac{dV}{dt} = \frac{9}{2}$ (constant rate)

Find: $\frac{dr}{dt}$ when $r = 2$

To find the rate of change of the radius, you must find an equation that relates the radius r to the volume V.

Equation: $V = \frac{4}{3}\pi r^3$ Volume of a sphere

Differentiating both sides of the equation with respect to t produces

$$\frac{dV}{dt} = 4\pi r^2 \frac{dr}{dt}$$ Differentiate with respect to t.

$$\frac{dr}{dt} = \frac{1}{4\pi r^2}\left(\frac{dV}{dt}\right).$$ Solve for dr/dt.

Finally, when $r = 2$, the rate of change of the radius is

$$\frac{dr}{dt} = \frac{1}{16\pi}\left(\frac{9}{2}\right) \approx 0.09 \text{ foot per minute.}$$

In Example 3, note that the volume is increasing at a *constant* rate but the radius is increasing at a *variable* rate. Just because two rates are related does not mean that they are proportional. In this particular case, the radius is growing more and more slowly as t increases. Do you see why?

EXAMPLE 4 The Speed of an Airplane Tracked by Radar

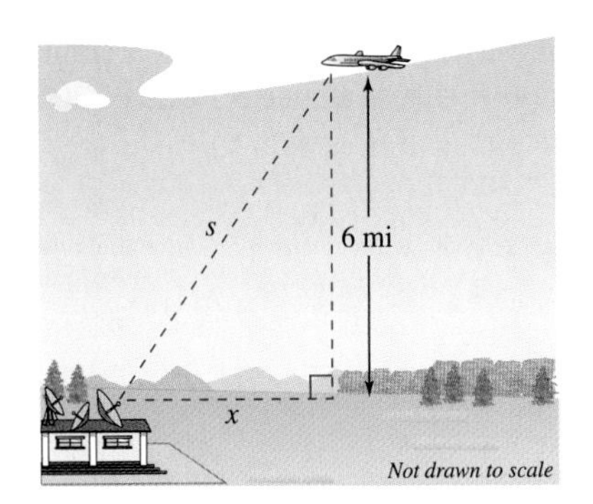

An airplane is flying at an altitude of 6 miles, s miles from the station.
Figure 3.40

An airplane is flying on a flight path that will take it directly over a radar tracking station, as shown in Figure 3.40. If s is decreasing at a rate of 400 miles per hour when $s = 10$ miles, what is the speed of the plane?

Solution Let x be the horizontal distance from the station, as shown in Figure 3.40. Notice that when $s = 10$, $x = \sqrt{10^2 - 36} = 8$.

Given rate: $ds/dt = -400$ when $s = 10$

Find: dx/dt when $s = 10$ and $x = 8$

You can find the velocity of the plane as shown.

Equation: $x^2 + 6^2 = s^2$ — Pythagorean Theorem

$$2x\frac{dx}{dt} = 2s\frac{ds}{dt}$$ Differentiate with respect to t.

$$\frac{dx}{dt} = \frac{s}{x}\left(\frac{ds}{dt}\right)$$ Solve for dx/dt.

$$\frac{dx}{dt} = \frac{10}{8}(-400)$$ Substitute for s, x, and ds/dt.

$$= -500 \text{ miles per hour}$$ Simplify.

Because the velocity is -500 miles per hour, the *speed* is 500 miles per hour.

EXAMPLE 5 A Changing Angle of Elevation

Find the rate of change of the angle of elevation of the camera shown in Figure 3.41 at 10 seconds after lift-off.

Solution Let θ be the angle of elevation, as shown in Figure 3.41. When $t = 10$, the height s of the rocket is $s = 50t^2 = 50(10)^2 = 5000$ feet.

Given rate: $ds/dt = 100t =$ velocity of rocket

Find: $d\theta/dt$ when $t = 10$ and $s = 5000$

Using Figure 3.41, you can relate s and θ by the equation $\tan\theta = s/2000$.

Equation: $$\tan\theta = \frac{s}{2000}$$ See Figure 3.41.

$$(\sec^2\theta)\frac{d\theta}{dt} = \frac{1}{2000}\left(\frac{ds}{dt}\right)$$ Differentiate with respect to t.

$$\frac{d\theta}{dt} = \cos^2\theta\,\frac{100t}{2000}$$ Substitute $100t$ for ds/dt.

$$= \left(\frac{2000}{\sqrt{s^2 + 2000^2}}\right)^2\frac{100t}{2000}$$ $\cos\theta = 2000/\sqrt{s^2 + 2000^2}$

When $t = 10$ and $s = 5000$, you have

$$\frac{d\theta}{dt} = \frac{2000(100)(10)}{5000^2 + 2000^2} = \frac{2}{29} \text{ radian per second.}$$

So, when $t = 10$, θ is changing at a rate of $\frac{2}{29}$ radian per second.

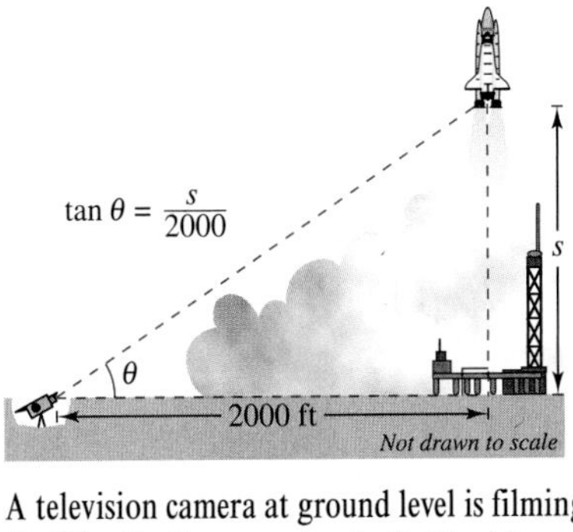

A television camera at ground level is filming the lift-off of a space shuttle that is rising vertically according to the position equation $s = 50t^2$, where s is measured in feet and t is measured in seconds. The camera is 2000 feet from the launch pad.
Figure 3.41

EXAMPLE 6 The Velocity of a Piston

In the engine shown in Figure 3.42, a 7-inch connecting rod is fastened to a crank of radius 3 inches. The crankshaft rotates counterclockwise at a constant rate of 200 revolutions per minute. Find the velocity of the piston when $\theta = \pi/3$.

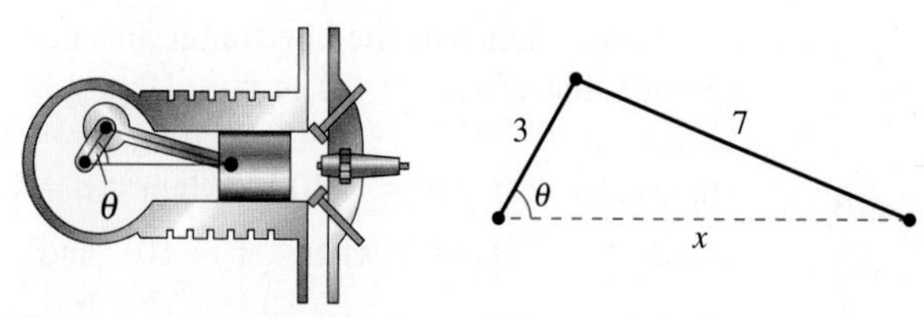

The velocity of a piston is related to the angle of the crankshaft.
Figure 3.42

Solution Label the distances as shown in Figure 3.42. Because a complete revolution corresponds to 2π radians, it follows that $d\theta/dt = 200(2\pi) = 400\pi$ radians per minute.

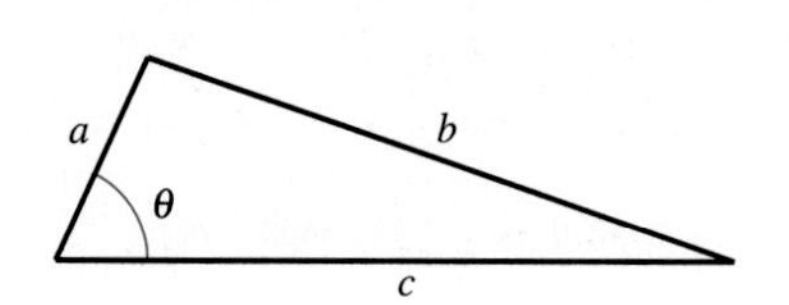

Law of Cosines:
$b^2 = a^2 + c^2 - 2ac\cos\theta$
Figure 3.43

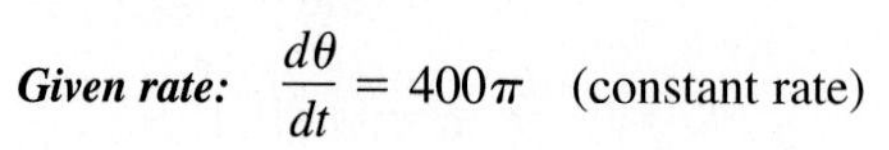

Given rate: $\dfrac{d\theta}{dt} = 400\pi$ (constant rate)

Find: $\dfrac{dx}{dt}$ when $\theta = \dfrac{\pi}{3}$

You can use the Law of Cosines (Figure 3.43) to find an equation that relates x and θ.

Equation:

$$7^2 = 3^2 + x^2 - 2(3)(x)\cos\theta$$

$$0 = 2x\frac{dx}{dt} - 6\left(-x\sin\theta\frac{d\theta}{dt} + \cos\theta\frac{dx}{dt}\right)$$

$$(6\cos\theta - 2x)\frac{dx}{dt} = 6x\sin\theta\frac{d\theta}{dt}$$

$$\frac{dx}{dt} = \frac{6x\sin\theta}{6\cos\theta - 2x}\left(\frac{d\theta}{dt}\right)$$

When $\theta = \pi/3$, you can solve for x as shown.

$$7^2 = 3^2 + x^2 - 2(3)(x)\cos\frac{\pi}{3}$$

$$49 = 9 + x^2 - 6x\left(\frac{1}{2}\right)$$

$$0 = x^2 - 3x - 40$$

$$0 = (x - 8)(x + 5)$$

$$x = 8 \qquad \text{Choose positive solution.}$$

So, when $x = 8$ and $\theta = \pi/3$, the velocity of the piston is

$$\frac{dx}{dt} = \frac{6(8)(\sqrt{3}/2)}{6(1/2) - 16}(400\pi)$$

$$= \frac{9600\pi\sqrt{3}}{-13}$$

$$\approx -4018 \text{ inches per minute.}$$

NOTE The velocity in Example 6 is negative because x represents a distance that is decreasing.

Exercises for Section 3.7

See www.CalcChat.com for worked-out solutions to odd-numbered exercises.

In Exercises 1–4, assume that x and y are both differentiable functions of t and find the required values of dy/dt and dx/dt.

	Equation	*Find*	*Given*
1.	$y = \sqrt{x}$	(a) $\dfrac{dy}{dt}$ when $x = 4$	$\dfrac{dx}{dt} = 3$
		(b) $\dfrac{dx}{dt}$ when $x = 25$	$\dfrac{dy}{dt} = 2$
2.	$y = 2(x^2 - 3x)$	(a) $\dfrac{dy}{dt}$ when $x = 3$	$\dfrac{dx}{dt} = 2$
		(b) $\dfrac{dx}{dt}$ when $x = 1$	$\dfrac{dy}{dt} = 5$
3.	$xy = 4$	(a) $\dfrac{dy}{dt}$ when $x = 8$	$\dfrac{dx}{dt} = 10$
		(b) $\dfrac{dx}{dt}$ when $x = 1$	$\dfrac{dy}{dt} = -6$
4.	$x^2 + y^2 = 25$	(a) $\dfrac{dy}{dt}$ when $x = 3, y = 4$	$\dfrac{dx}{dt} = 8$
		(b) $\dfrac{dx}{dt}$ when $x = 4, y = 3$	$\dfrac{dy}{dt} = -2$

In Exercises 5–8, a point is moving along the graph of the given function such that dx/dt is 2 centimeters per second. Find dy/dt for the given values of x.

	Function	*Values of x*		
5.	$y = x^2 + 1$	(a) $x = -1$	(b) $x = 0$	(c) $x = 1$
6.	$y = \dfrac{1}{1 + x^2}$	(a) $x = -2$	(b) $x = 0$	(c) $x = 2$
7.	$y = \tan x$	(a) $x = -\dfrac{\pi}{3}$	(b) $x = -\dfrac{\pi}{4}$	(c) $x = 0$
8.	$y = \sin x$	(a) $x = \dfrac{\pi}{6}$	(b) $x = \dfrac{\pi}{4}$	(c) $x = \dfrac{\pi}{3}$

Writing About Concepts

In Exercises 9 and 10, use the graph of f to (a) determine whether dy/dt is positive or negative given that dx/dt is negative, and (b) determine whether dx/dt is positive or negative given that dy/dt is positive.

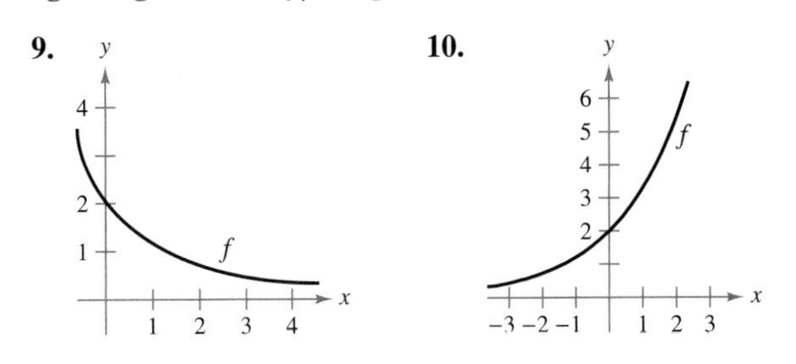

11. Consider the linear function $y = ax + b$. If x changes at a constant rate, does y change at a constant rate? If so, does it change at the same rate as x? Explain.

Writing About Concepts *(continued)*

12. In your own words, state the guidelines for solving related-rate problems.

13. Find the rate of change of the distance between the origin and a moving point on the graph of $y = x^2 + 1$ if $dx/dt = 2$ centimeters per second.

14. Find the rate of change of the distance between the origin and a moving point on the graph of $y = \sin x$ if $dx/dt = 2$ centimeters per second.

15. ***Area*** The radius r of a circle is increasing at a rate of 3 centimeters per minute. Find the rate of change of the area when (a) $r = 6$ centimeters and (b) $r = 24$ centimeters.

16. ***Area*** Let A be the area of a circle of radius r that is changing with respect to time. If dr/dt is constant, is dA/dt constant? Explain.

17. ***Area*** The included angle of the two sides of constant equal length s of an isosceles triangle is θ.

(a) Show that the area of the triangle is given by $A = \frac{1}{2}s^2 \sin\theta$.

(b) If θ is increasing at the rate of $\frac{1}{2}$ radian per minute, find the rates of change of the area when $\theta = \pi/6$ and $\theta = \pi/3$.

(c) Explain why the rate of change of the area of the triangle is not constant even though $d\theta/dt$ is constant.

18. ***Volume*** The radius r of a sphere is increasing at a rate of 2 inches per minute.

(a) Find the rate of change of the volume when $r = 6$ inches and $r = 24$ inches.

(b) Explain why the rate of change of the volume of the sphere is not constant even though dr/dt is constant.

19. ***Volume*** A hemispherical water tank with radius 6 meters is filled to a depth of h meters. The volume of water in the tank is given by $V = \frac{1}{3}\pi h(108 - h^2)$, $0 < h < 6$. If water is being pumped into the tank at the rate of 3 cubic meters per minute, find the rate of change of the depth of the water when $h = 2$ meters.

20. ***Volume*** All edges of a cube are expanding at a rate of 3 centimeters per second. How fast is the volume changing when each edge is (a) 1 centimeter and (b) 10 centimeters?

21. ***Surface Area*** The conditions are the same as in Exercise 20. Determine how fast the *surface area* is changing when each edge is (a) 1 centimeter and (b) 10 centimeters.

22. ***Volume*** The formula for the volume of a cone is $V = \frac{1}{3}\pi r^2 h$. Find the rate of change of the volume if dr/dt is 2 inches per minute and $h = 3r$ when (a) $r = 6$ inches and (b) $r = 24$ inches.

23. ***Volume*** At a sand and gravel plant, sand is falling off a conveyor and onto a conical pile at a rate of 10 cubic feet per minute. The diameter of the base of the cone is approximately three times the altitude. At what rate is the height of the pile changing when the pile is 15 feet high?

24. *Depth* A conical tank (with vertex down) is 10 feet across the top and 12 feet deep. If water is flowing into the tank at a rate of 10 cubic feet per minute, find the rate of change of the depth of the water when the water is 8 feet deep.

25. *Depth* A swimming pool is 12 meters long, 6 meters wide, 1 meter deep at the shallow end, and 3 meters deep at the deep end (see figure). Water is being pumped into the pool at $\frac{1}{4}$ cubic meter per minute, and there is 1 meter of water at the deep end.

(a) What percent of the pool is filled?

(b) At what rate is the water level rising?

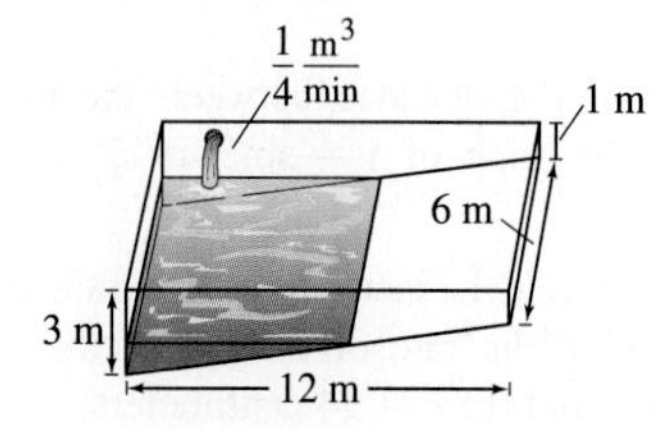

Figure for 25

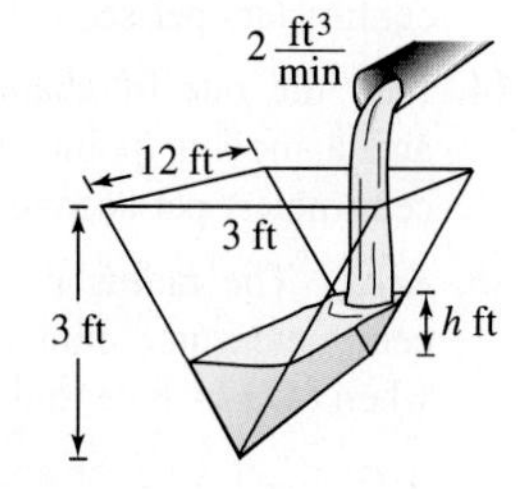

Figure for 26

26. *Depth* A trough is 12 feet long and 3 feet across the top (see figure). Its ends are isosceles triangles with altitudes of 3 feet.

(a) If water is being pumped into the trough at 2 cubic feet per minute, how fast is the water level rising when h is 1 foot deep?

(b) If the water is rising at a rate of $\frac{3}{8}$ inch per minute when $h = 2$, determine the rate at which water is being pumped into the trough.

27. *Moving Ladder* A ladder 25 feet long is leaning against the wall of a house (see figure). The base of the ladder is pulled away from the wall at a rate of 2 feet per second.

(a) How fast is the top of the ladder moving down the wall when its base is 7 feet, 15 feet, and 24 feet from the wall?

(b) Consider the triangle formed by the side of the house, the ladder, and the ground. Find the rate at which the area of the triangle is changing when the base of the ladder is 7 feet from the wall.

(c) Find the rate at which the angle between the ladder and the wall of the house is changing when the base of the ladder is 7 feet from the wall.

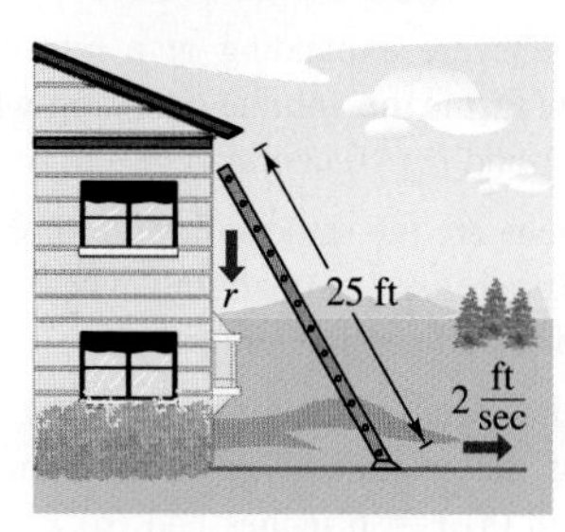

Figure for 27

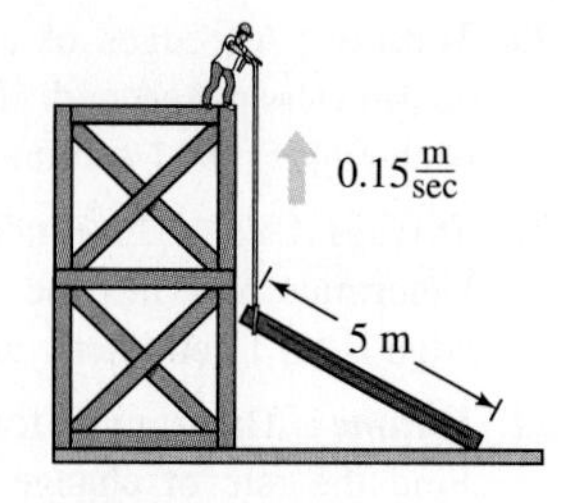

Figure for 28

FOR FURTHER INFORMATION For more information on the mathematics of moving ladders, see the article "The Falling Ladder Paradox" by Paul Scholten and Andrew Simoson in *The College Mathematics Journal*. To view this article, go to the website *www.matharticles.com*.

28. *Construction* A construction worker pulls a five-meter plank up the side of a building under construction by means of a rope tied to one end of the plank (see figure). Assume the opposite end of the plank follows a path perpendicular to the wall of the building and the worker pulls the rope at a rate of 0.15 meter per second. How fast is the end of the plank sliding along the ground when it is 2.5 meters from the wall of the building?

29. *Construction* A winch at the top of a 12-meter building pulls a pipe of the same length to a vertical position, as shown in the figure. The winch pulls in rope at a rate of -0.2 meter per second. Find the rate of vertical change and the rate of horizontal change at the end of the pipe when $y = 6$.

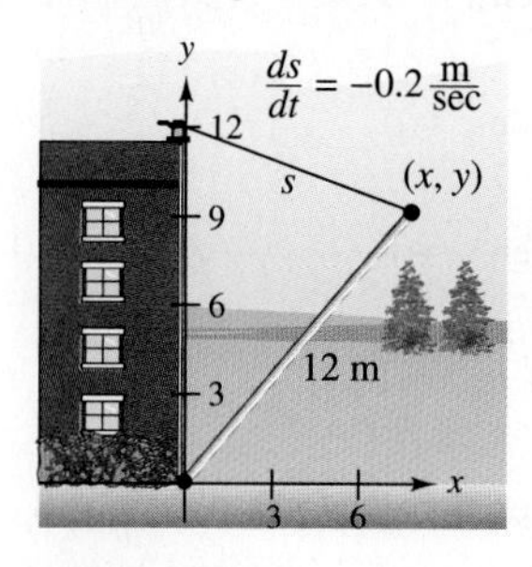

Figure for 29

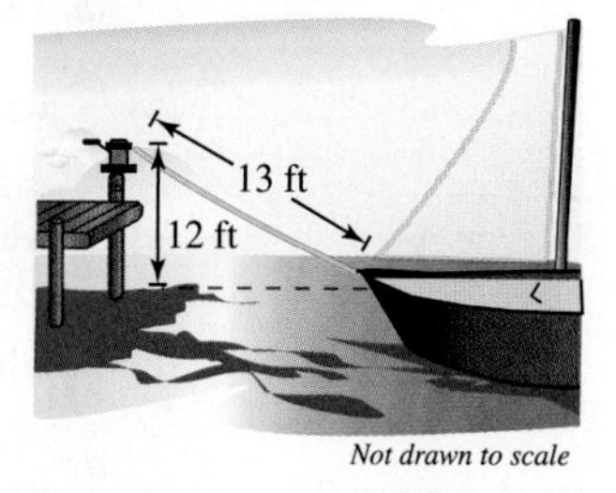

Figure for 30

30. *Boating* A boat is pulled into a dock by means of a winch 12 feet above the deck of the boat (see figure).

(a) The winch pulls in rope at a rate of 4 feet per second. Determine the speed of the boat when there is 13 feet of rope out. What happens to the speed of the boat as it gets closer to the dock?

(b) Suppose the boat is moving at a constant rate of 4 feet per second. Determine the speed at which the winch pulls in rope when there is a total of 13 feet of rope out. What happens to the speed at which the winch pulls in rope as the boat gets closer to the dock?

31. *Air Traffic Control* An air traffic controller spots two planes at the same altitude converging on a point as they fly at right angles to each other (see figure). One plane is 150 miles from the point moving at 450 miles per hour. The other plane is 200 miles from the point moving at 600 miles per hour.

(a) At what rate is the distance between the planes decreasing?

(b) How much time does the air traffic controller have to get one of the planes on a different flight path?

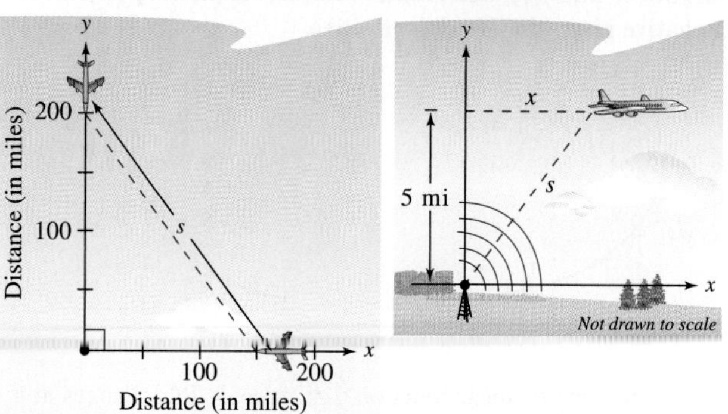

Figure for 31

Figure for 32

32. ***Air Traffic Control*** An airplane is flying at an altitude of 5 miles and passes directly over a radar antenna (see figure on previous page). When the plane is 10 miles away ($s = 10$), the radar detects that the distance s is changing at a rate of 240 miles per hour. What is the speed of the plane?

33. ***Baseball*** A baseball diamond has the shape of a square with sides 90 feet long (see figure). A player running from second base to third base at a speed of 28 feet per second is 30 feet from third base. At what rate is the player's distance s from home plate changing?

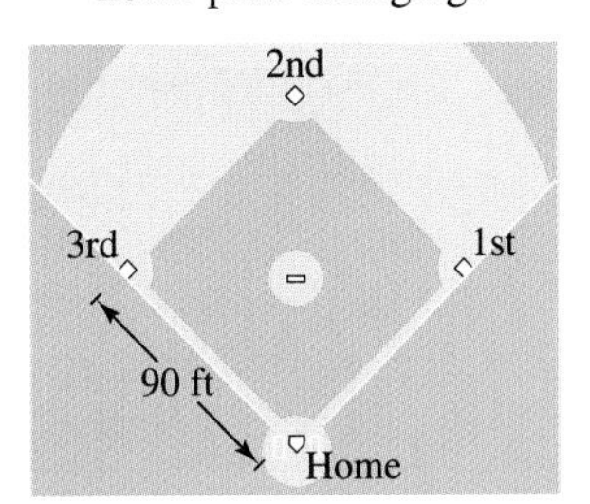

Figure for 33 and 34

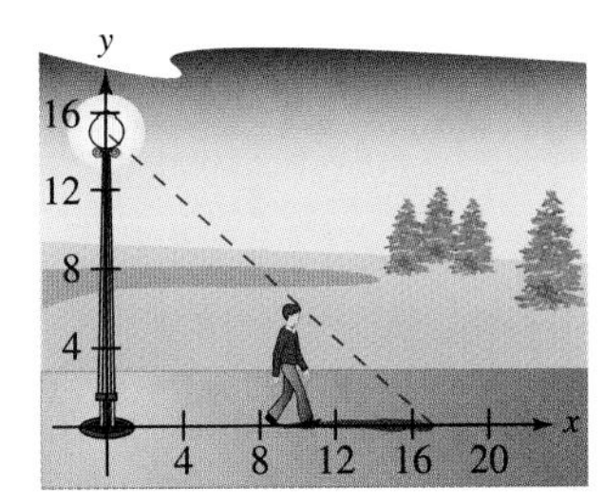

Figure for 35

34. ***Baseball*** For the baseball diamond in Exercise 33, suppose the player is running from first to second at a speed of 28 feet per second. Find the rate at which the distance from home plate is changing when the player is 30 feet from second base.

35. ***Shadow Length*** A man 6 feet tall walks at a rate of 5 feet per second away from a light that is 15 feet above the ground (see figure). When he is 10 feet from the base of the light,

(a) at what rate is the tip of his shadow moving?

(b) at what rate is the length of his shadow changing?

36. ***Shadow Length*** Repeat Exercise 35 for a man 6 feet tall walking at a rate of 5 feet per second *toward* a light that is 20 feet above the ground (see figure).

Figure for 36

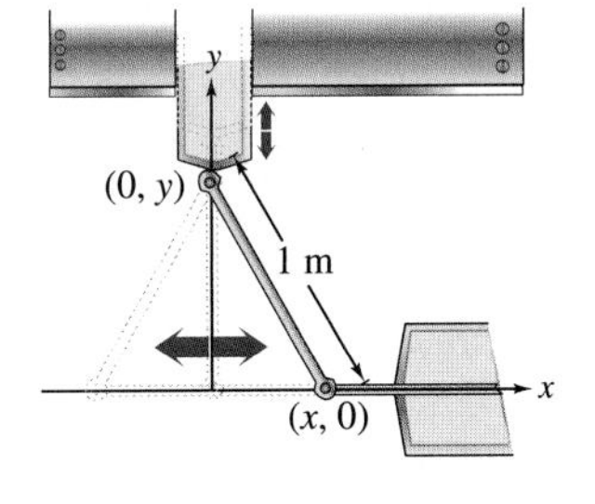

Figure for 37

37. ***Machine Design*** The endpoints of a movable rod of length 1 meter have coordinates $(x, 0)$ and $(0, y)$ (see figure). The position of the end of the rod on the x-axis is

$$x(t) = \frac{1}{2} \sin \frac{\pi t}{6}$$

where t is the time in seconds.

(a) Find the time of one complete cycle of the rod.

(b) What is the lowest point reached by the end of the rod on the y-axis?

(c) Find the speed of the y-axis endpoint when the x-axis endpoint is $\left(\frac{1}{4}, 0\right)$.

38. ***Machine Design*** Repeat Exercise 37 for a position function of $x(t) = \frac{3}{5} \sin \pi t$. Use the point $\left(\frac{3}{10}, 0\right)$ for part (c).

39. ***Evaporation*** As a spherical raindrop falls, it reaches a layer of dry air and begins to evaporate at a rate that is proportional to its surface area ($S = 4\pi r^2$). Show that the radius of the raindrop decreases at a constant rate.

40. ***Electricity*** The combined electrical resistance R of R_1 and R_2, connected in parallel, is given by

$$\frac{1}{R} = \frac{1}{R_1} + \frac{1}{R_2}$$

where R, R_1, and R_2 are measured in ohms. R_1 and R_2 are increasing at rates of 1 and 1.5 ohms per second, respectively. At what rate is R changing when $R_1 = 50$ ohms and $R_2 = 75$ ohms?

41. ***Adiabatic Expansion*** When a certain polyatomic gas undergoes adiabatic expansion, its pressure p and volume V satisfy the equation $pV^{1.3} = k$, where k is a constant. Find the relationship between the related rates dp/dt and dV/dt.

42. ***Roadway Design*** Cars on a certain roadway travel on a circular arc of radius r. In order not to rely on friction alone to overcome the centrifugal force, the road is banked at an angle of magnitude θ from the horizontal (see figure). The banking angle must satisfy the equation $rg \tan \theta = v^2$, where v is the velocity of the cars and $g = 32$ feet per second per second is the acceleration due to gravity. Find the relationship between the related rates dv/dt and $d\theta/dt$.

43. ***Angle of Elevation*** A balloon rises at a rate of 3 meters per second from a point on the ground 30 meters from an observer. Find the rate of change of the angle of elevation of the balloon from the observer when the balloon is 30 meters above the ground.

44. ***Angle of Elevation*** A fish is reeled in at a rate of 1 foot per second from a point 10 feet above the water (see figure). At what rate is the angle between the line and the water changing when there is a total of 25 feet of line out?

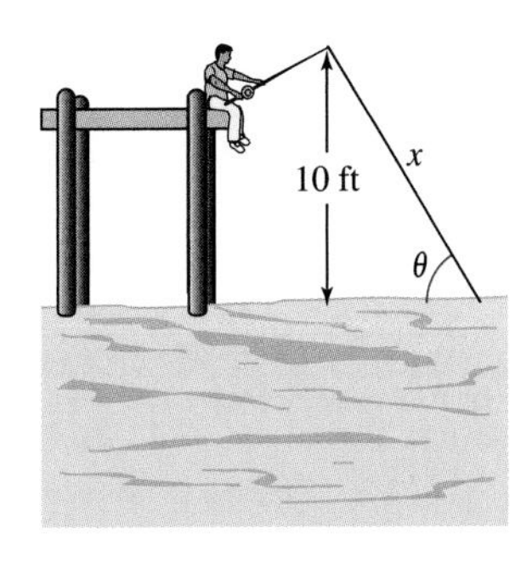

45. ***Relative Humidity*** When the dewpoint is 65° Fahrenheit, the relative humidity H is

$$H = \frac{4347}{400{,}000{,}000} e^{369{,}444/(50t + 19{,}793)}$$

where t is the temperature in degrees Fahrenheit.

(a) Determine the relative humidity when $t = 65°$ and $t = 80°$.

(b) At 10 A.M., the temperature is 75° and increasing at the rate of 2° per hour. Find the rate at which the relative humidity is changing.

46. ***Linear vs. Angular Speed*** A patrol car is parked 50 feet from a long warehouse (see figure). The revolving light on top of the car turns at a rate of 30 revolutions per minute. How fast is the light beam moving along the wall when the beam makes angles of (a) $\theta = 30°$, (b) $\theta = 60°$, and (c) $\theta = 70°$ with the line perpendicular from the light to the wall?

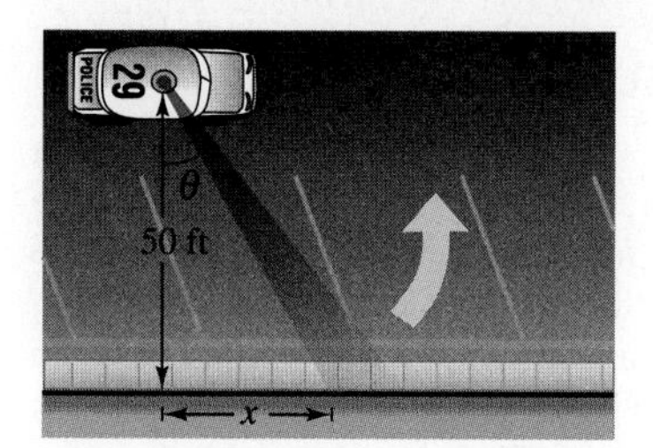

Figure for 46

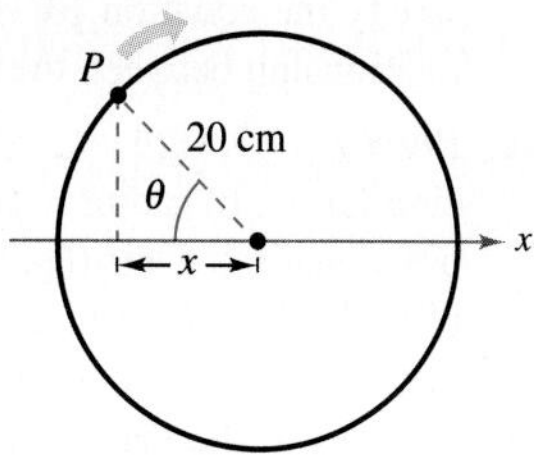

Figure for 47

47. ***Linear vs. Angular Speed*** A wheel of radius 20 centimeters revolves at a rate of 10 revolutions per second. A dot is painted at a point P on the rim of the wheel (see figure).

(a) Find dx/dt as a function of θ.

(b) Use a graphing utility to graph the function in part (a).

(c) When is the absolute value of the rate of change of x greatest? When is it least?

(d) Find dx/dt when $\theta = 30°$ and $\theta = 60°$.

48. ***Flight Control*** An airplane is flying in still air with an airspeed of 240 miles per hour. If it is climbing at an angle of 22°, find the rate at which it is gaining altitude.

49. ***Security Camera*** A security camera is centered 50 feet above a 100-foot hallway (see figure). It is easiest to design the camera with a constant angular rate of rotation, but this results in a variable rate at which the images of the surveillance area are recorded. So, it is desirable to design a system with a variable rate of rotation and a constant rate of movement of the scanning beam along the hallway. Find a model for the variable rate of rotation if $|dx/dt| = 2$ feet per second.

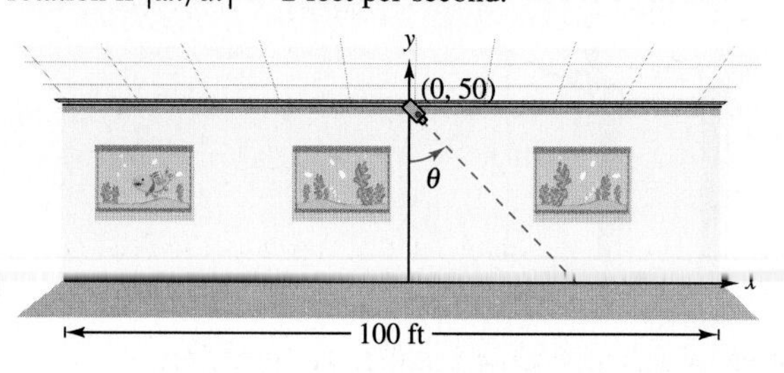

50. ***Think About It*** Describe the relationship between the rate of change of y and the rate of change of x in each expression. Assume all variables and derivatives are positive.

(a) $\dfrac{dy}{dt} = 3\dfrac{dx}{dt}$ (b) $\dfrac{dy}{dt} = x(L - x)\dfrac{dx}{dt}, \quad 0 \le x \le L$

51. ***Angle of Elevation*** An airplane flies at an altitude of 5 miles toward a point directly over an observer (see figure). The speed of the plane is 600 miles per hour. Find the rates at which the angle of elevation θ is changing when the angle is (a) $\theta = 30°$, (b) $\theta = 60°$, and (c) $\theta = 75°$.

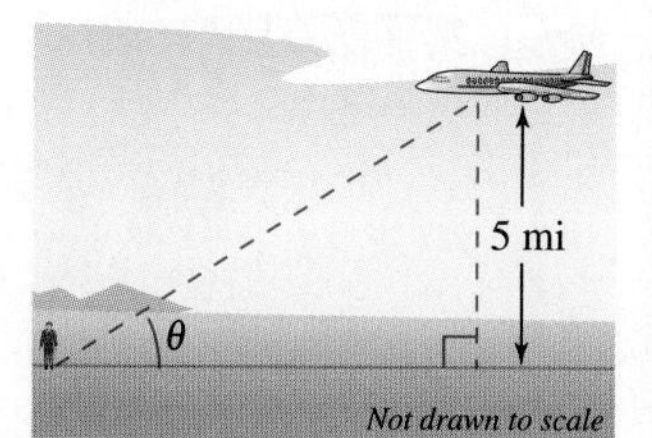

Figure for 51

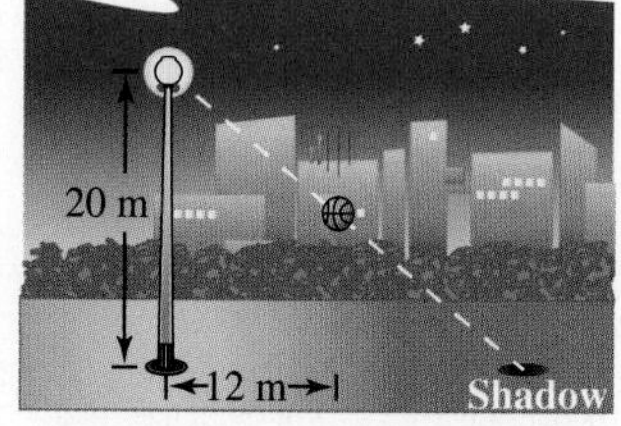

Figure for 52

52. ***Moving Shadow*** A ball is dropped from a height of 20 meters, 12 meters away from the top of a 20-meter lamppost (see figure). The ball's shadow, caused by the light at the top of the lamppost, is moving along the level ground. How fast is the shadow moving 1 second after the ball is released? *(Submitted by Dennis Gittinger, St. Philips College, San Antonio, TX)*

Acceleration **In Exercises 53 and 54, find the acceleration of the specified object. (*Hint:* Recall that if a variable is changing at a constant rate, its acceleration is zero.)**

53. Find the acceleration of the top of the ladder described in Exercise 27 when the base of the ladder is 7 feet from the wall.

54. Find the acceleration of the boat in Exercise 30(a) when there is a total of 13 feet of rope out.

55. ***Modeling Data*** The table shows the numbers (in millions) of single women (never married) s and married women m in the civilian work force in the United States for the years 1993 through 2001. *(Source: U.S. Bureau of Labor Statistics)*

Year	1993	1994	1995	1996	1997	1998	1999	2000	2001
s	15.0	15.3	15.5	15.8	16.5	17.1	17.6	17.8	18.0
m	32.0	32.9	33.4	33.6	33.8	33.9	34.4	34.6	34.7

(a) Use the regression capabilities of a graphing utility to find a model of the form $m(s) = as^3 + bs^2 + cs + d$ for the data, where t is the time in years, with $t = 3$ corresponding to 1993.

(b) Find dm/dt. Then use the model to estimate dm/dt for $t = 10$ if it is predicted that the number of single women in the work force will increase at the rate of 0.75 million per year.

Section 3.8 Newton's Method

- **Approximate a zero of a function using Newton's Method.**

Newton's Method

In this section you will study a technique for approximating the real zeros of a function. The technique is called **Newton's Method,** and it uses tangent lines to approximate the graph of the function near its x-intercepts.

To see how Newton's Method works, consider a function f that is continuous on the interval $[a, b]$ and differentiable on the interval (a, b). If $f(a)$ and $f(b)$ differ in sign, then, by the Intermediate Value Theorem, f must have at least one zero in the interval (a, b). Suppose you estimate this zero to occur at

$$x = x_1 \qquad \text{First estimate}$$

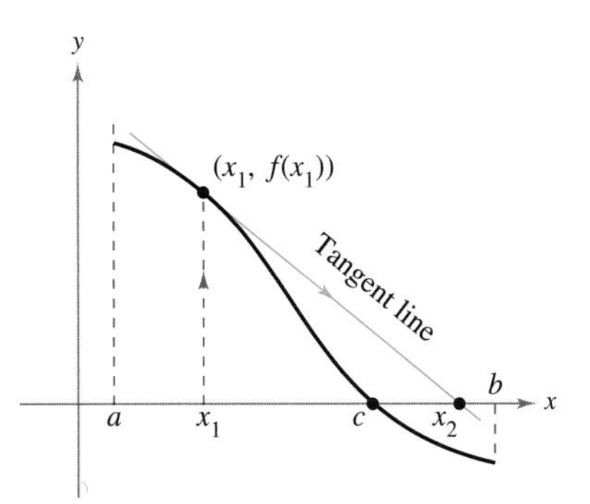

(a)

as shown in Figure 3.44(a). Newton's Method is based on the assumption that the graph of f and the tangent line at $(x_1, f(x_1))$ both cross the x-axis at *about* the same point. Because you can easily calculate the x-intercept for this tangent line, you can use it as a second (and, usually, better) estimate for the zero of f. The tangent line passes through the point $(x_1, f(x_1))$ with a slope of $f'(x_1)$. In point-slope form, the equation of the tangent line is therefore

$$y - f(x_1) = f'(x_1)(x - x_1)$$
$$y = f'(x_1)(x - x_1) + f(x_1).$$

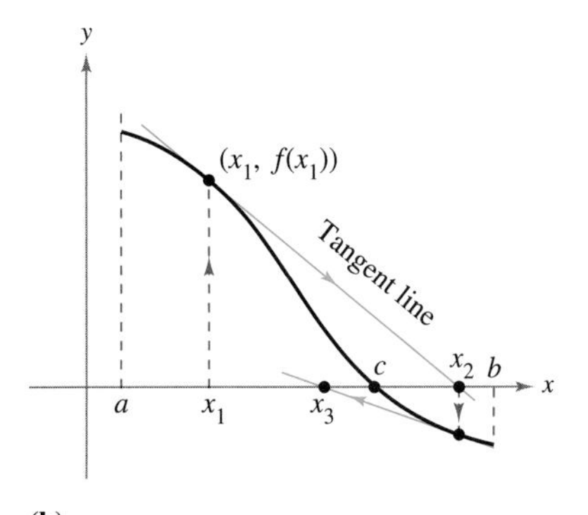

(b)
The x-intercept of the tangent line approximates the zero of f.
Figure 3.44

Letting $y = 0$ and solving for x produces

$$x = x_1 - \frac{f(x_1)}{f'(x_1)}.$$

So, from the initial estimate x_1 you obtain a new estimate

$$x_2 = x_1 - \frac{f(x_1)}{f'(x_1)}. \qquad \text{Second estimate [see Figure 3.44(b)]}$$

You can improve on x_2 and calculate yet a third estimate

$$x_3 = x_2 - \frac{f(x_2)}{f'(x_2)}. \qquad \text{Third estimate}$$

Repeated application of this process is called Newton's Method.

> **Newton's Method**
>
> Isaac Newton first described the method for approximating the real zeros of a function in his text *Method of Fluxions*. Although the book was written in 1671, it was not published until 1736. Meanwhile, in 1690, Joseph Raphson (1648–1715) published a paper describing a method for approximating the real zeros of a function that was very similar to Newton's. For this reason, the method is often referred to as the Newton-Raphson method.

Newton's Method for Approximating the Zeros of a Function

Let $f(c) = 0$, where f is differentiable on an open interval containing c. Then, to approximate c, use the following steps.

1. Make an initial estimate x_1 that is close to c. (A graph is helpful.)
2. Determine a new approximation
$$x_{n+1} = x_n - \frac{f(x_n)}{f'(x_n)}.$$
3. If $|x_n - x_{n+1}|$ is within the desired accuracy, let x_{n+1} serve as the final approximation. Otherwise, return to Step 2 and calculate a new approximation.

Each successive application of this procedure is called an **iteration.**

NOTE For many functions, just a few iterations of Newton's Method will produce approximations having very small errors, as shown in Example 1.

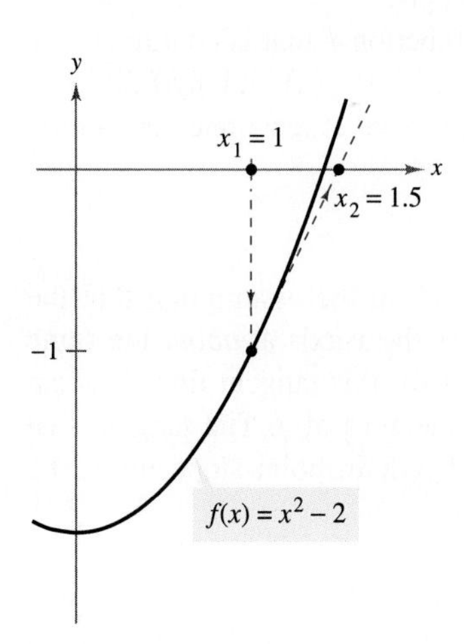

The first iteration of Newton's Method
Figure 3.45

EXAMPLE 1 Using Newton's Method

Calculate three iterations of Newton's Method to approximate a zero of $f(x) = x^2 - 2$. Use $x_1 = 1$ as the initial guess.

Solution Because $f(x) = x^2 - 2$, you have $f'(x) = 2x$, and the iterative process is given by the formula

$$x_{n+1} = x_n - \frac{f(x_n)}{f'(x_n)} = x_n - \frac{x_n^2 - 2}{2x_n}.$$

The calculations for three iterations are shown in the table.

n	x_n	$f(x_n)$	$f'(x_n)$	$\frac{f(x_n)}{f'(x_n)}$	$x_n - \frac{f(x_n)}{f'(x_n)}$
1	1.000000	−1.000000	2.000000	−0.500000	1.500000
2	1.500000	0.250000	3.000000	0.083333	1.416667
3	1.416667	0.006945	2.833334	0.002451	1.414216
4	1.414216				

Of course, in this case you know that the two zeros of the function are $\pm\sqrt{2}$. To six decimal places, $\sqrt{2} = 1.414214$. So, after only three iterations of Newton's Method, you have obtained an approximation that is within 0.000002 of an actual root. The first iteration of this process is shown in Figure 3.45.

EXAMPLE 2 Using Newton's Method

Use Newton's Method to approximate the zeros of

$$f(x) = e^x + x.$$

Continue the iterations until two successive approximations differ by less than 0.0001.

Solution Begin by sketching a graph of f, as shown in Figure 3.46. From the graph, you can observe that the function has only one zero, which occurs near $x = -0.6$. Next, differentiate f and form the iterative formula

$$x_{n+1} = x_n - \frac{f(x_n)}{f'(x_n)} = x_n - \frac{e^{x_n} + x_n}{e^{x_n} + 1}.$$

The calculations are shown in the table.

n	x_n	$f(x_n)$	$f'(x_n)$	$\frac{f(x_n)}{f'(x_n)}$	$x_n - \frac{f(x_n)}{f'(x_n)}$
1	−0.60000	−0.05119	1.54881	−0.03305	−0.56695
2	−0.56695	0.00030	1.56725	0.00019	−0.56714
3	−0.56714	0.00000	1.56714	0.00000	−0.56714
4	0.56714				

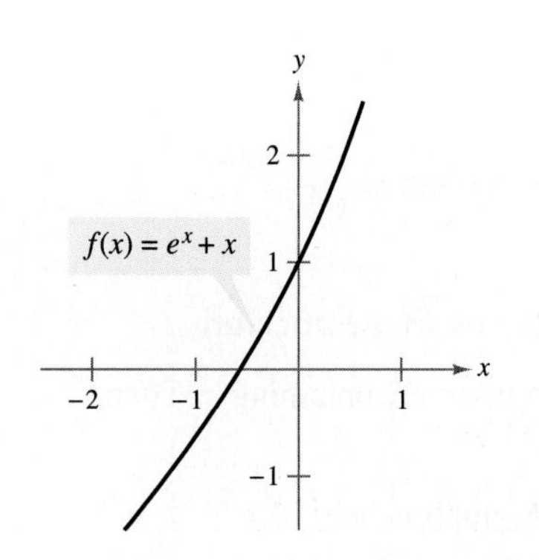

After three iterations of Newton's Method, the zero of f is approximated to the desired accuracy.
Figure 3.46

Because two successive approximations differ by less than the required 0.0001, you can estimate the zero of f to be -0.56714.

When, as in Examples 1 and 2, the approximations approach a limit, the sequence $x_1, x_2, x_3, \ldots, x_n, \ldots$ is said to **converge.** Moreover, if the limit is c, it can be shown that c must be a zero of f.

Newton's Method does not always yield a convergent sequence. One way it can fail to do so is shown in Figure 3.47. Because Newton's Method involves division by $f'(x_n)$, it is clear that the method will fail if the derivative is zero for any x_n in the sequence. When you encounter this problem, you can usually overcome it by choosing a different value for x_1. Another way Newton's Method can fail is shown in the next example.

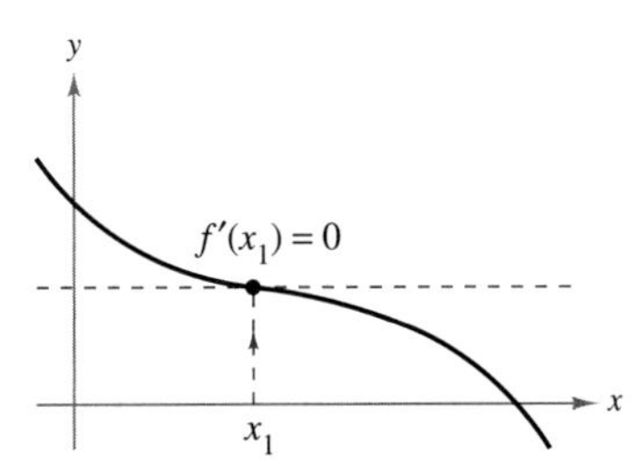

Newton's Method fails to converge if $f'(x_n) = 0$.
Figure 3.47

EXAMPLE 3 An Example in Which Newton's Method Fails

The function $f(x) = x^{1/3}$ is not differentiable at $x = 0$. Show that Newton's Method fails to converge using $x_1 = 0.1$.

Solution Because $f'(x) = \frac{1}{3}x^{-2/3}$, the iterative formula is

$$\begin{aligned} x_{n+1} &= x_n - \frac{f(x_n)}{f'(x_n)} \\ &= x_n - \frac{x_n^{1/3}}{\frac{1}{3}x_n^{-2/3}} \\ &= x_n - 3x_n \\ &= -2x_n. \end{aligned}$$

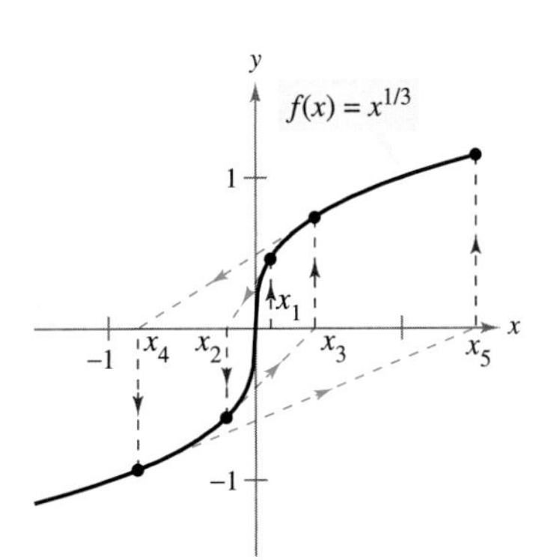

Newton's Method fails to converge for every x-value other than the actual zero of f.
Figure 3.48

The calculations are shown in the table. This table and Figure 3.48 indicate that x_n continues to increase in magnitude as $n \to \infty$, and so the limit of the sequence does not exist.

n	x_n	$f(x_n)$	$f'(x_n)$	$\frac{f(x_n)}{f'(x_n)}$	$x_n - \frac{f(x_n)}{f'(x_n)}$
1	0.10000	0.46416	1.54720	0.30000	−0.20000
2	−0.20000	−0.58480	0.97467	−0.60000	0.40000
3	0.40000	0.73681	0.61401	1.20000	−0.80000
4	−0.80000	−0.92832	0.38680	−2.40000	1.60000

NOTE In Example 3, the initial estimate $x_1 = 0.1$ fails to produce a convergent sequence. Try showing that Newton's Method also fails for every other choice of x_1 (other than the actual zero).

It can be shown that a condition sufficient to produce convergence of Newton's Method to a zero of f is that

$$\left|\frac{f(x)f''(x)}{[f'(x)]^2}\right| < 1$$

Condition for convergence

on an open interval containing the zero. For instance, in Example 1 this test would yield $f(x) = x^2 - 2, f'(x) = 2x, f''(x) = 2$, and

$$\left|\frac{f(x)f''(x)}{[f'(x)]^2}\right| = \left|\frac{(x^2-2)(2)}{4x^2}\right| = \left|\frac{1}{2} - \frac{1}{x^2}\right|.$$

Example 1

On the interval $(1, 3)$, this quantity is less than 1 and therefore the convergence of Newton's Method is guaranteed. On the other hand, in Example 3, you have $f(x) = x^{1/3}, f'(x) = \frac{1}{3}x^{-2/3}, f''(x) = -\frac{2}{9}x^{-5/3}$, and

$$\left|\frac{f(x)f''(x)}{[f'(x)]^2}\right| = \left|\frac{x^{1/3}(-2/9)(x^{-5/3})}{(1/9)(x^{-4/3})}\right| = 2$$

Example 3

which is not less than 1 for any value of x, so you cannot conclude that Newton's Method will converge.

Niels Henrik Abel (1802–1829)

Evariste Galois (1811–1832)

Although the lives of both Abel and Galois were brief, their work in the fields of analysis and abstract algebra was far-reaching.

Algebraic Solutions of Polynomial Equations

The zeros of some functions, such as

$$f(x) = x^3 - 2x^2 - x + 2$$

can be found by simple algebraic techniques, such as factoring. The zeros of other functions, such as

$$f(x) = x^3 - x + 1$$

cannot be found by *elementary* algebraic methods. This particular function has only one real zero, and by using more advanced algebraic techniques you can determine the zero to be

$$x = -\sqrt[3]{\frac{3-\sqrt{23/3}}{6}} - \sqrt[3]{\frac{3+\sqrt{23/3}}{6}}.$$

Because the *exact* solution is written in terms of square roots and cube roots, it is called a **solution by radicals.**

NOTE Try approximating the real zero of $f(x) = x^3 - x + 1$ and compare your result with the exact solution shown above.

The determination of radical solutions of a polynomial equation is one of the fundamental problems of algebra. The earliest such result is the Quadratic Formula, which dates back at least to Babylonian times. The general formula for the zeros of a cubic function was developed much later. In the sixteenth century an Italian mathematician, Jerome Cardan, published a method for finding radical solutions to cubic and quartic equations. Then, for 300 years, the problem of finding a general quintic formula remained open. Finally, in the nineteenth century, the problem was answered independently by two young mathematicians. Niels Henrik Abel, a Norwegian mathematician, and Evariste Galois, a French mathematician, proved that it is not possible to solve a *general* fifth- (or higher-) degree polynomial equation by radicals. Of course, you can solve particular fifth-degree equations such as $x^5 - 1 = 0$, but Abel and Galois were able to show that no general *radical* solution exists.

Exercises for Section 3.8

See www.CalcChat.com for worked-out solutions to odd-numbered exercises.

In Exercises 1–4, complete two iterations of Newton's Method for the function using the given initial guess.

1. $f(x) = x^2 - 3, \quad x_1 = 1.7$
2. $f(x) = 2x^2 - 3, \quad x_1 = 1$
3. $f(x) = \sin x, \quad x_1 = 3$
4. $f(x) = \tan x, \quad x_1 = 0.1$

In Exercises 5–16, approximate the zero(s) of the function. Use Newton's Method and continue the process until two successive approximations differ by less than 0.001. Then find the zero(s) using a graphing utility and compare the results.

5. $f(x) = x^3 + x - 1$
6. $f(x) = x^5 + x - 1$
7. $f(x) = 3\sqrt{x - 1} - x$
8. $f(x) = x - 2\sqrt{x + 1}$
9. $f(x) = x - e^{-x}$
10. $f(x) = x - 3 + \ln x$
11. $f(x) = x^3 + 3$
12. $f(x) = 3 - 2x^3$
13. $f(x) = x^3 - 3.9x^2 + 4.79x - 1.881$
14. $f(x) = \frac{1}{2}x^4 - 3x - 3$
15. $f(x) = x + \sin(x + 1)$
16. $f(x) = x^3 - \cos x$

In Exercises 17–24, apply Newton's Method to approximate the x-value(s) of the indicated point(s) of intersection of the two graphs. Continue the process until two successive approximations differ by less than 0.001. [*Hint:* Let $h(x) = f(x) - g(x)$.]

17. $f(x) = 2x + 1$
$g(x) = \sqrt{x + 4}$

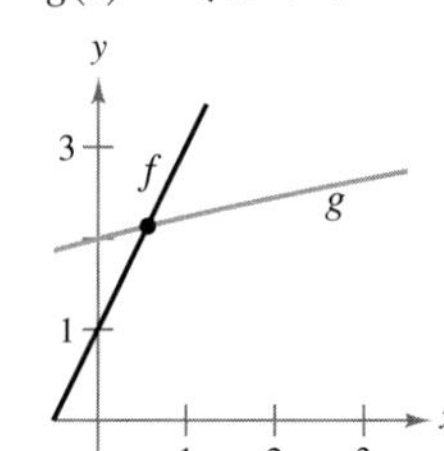

18. $f(x) = 3 - x$
$g(x) = 1/(x^2 + 1)$

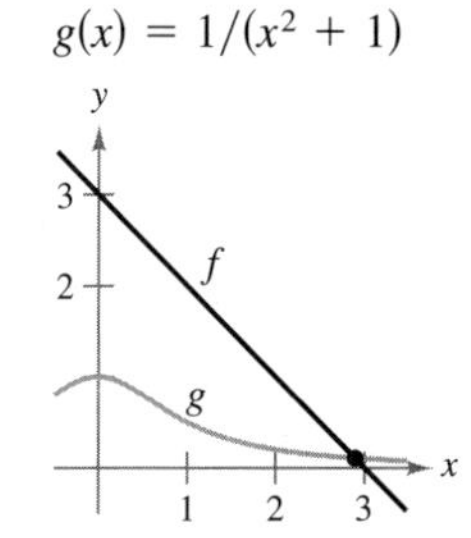

19. $f(x) = x$
$g(x) = \tan x$

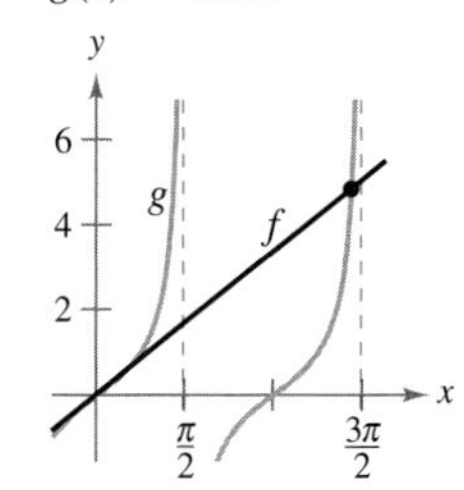

20. $f(x) = x^2$
$g(x) = \cos x$

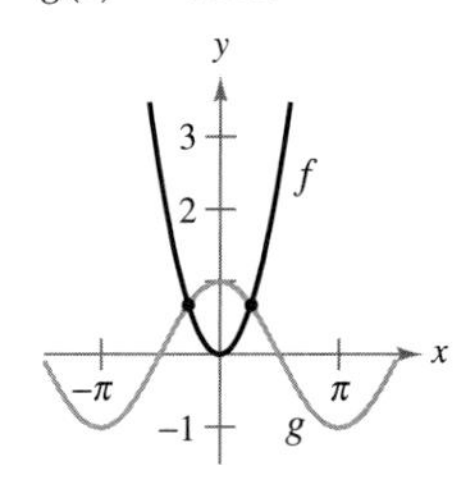

21. $f(x) = -x$
$g(x) = \ln x$

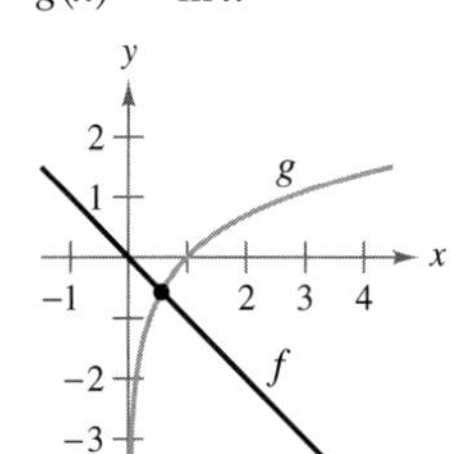

22. $f(x) = 2 - x^2$
$g(x) = e^{x/2}$

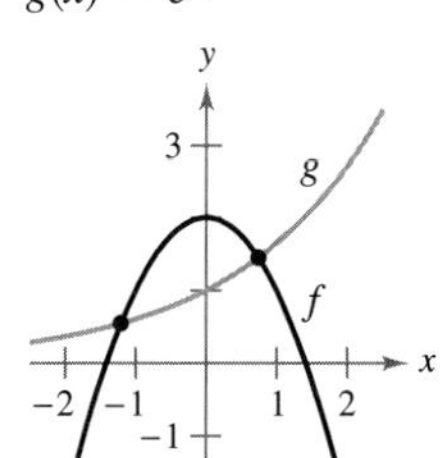

23. $f(x) = \arccos x$
$g(x) = \arctan x$

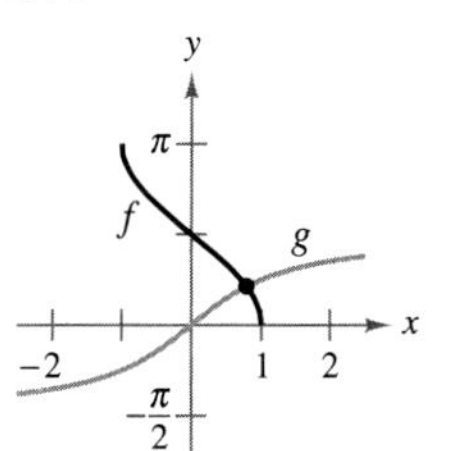

24. $f(x) = 1 - x$
$g(x) = \arcsin x$

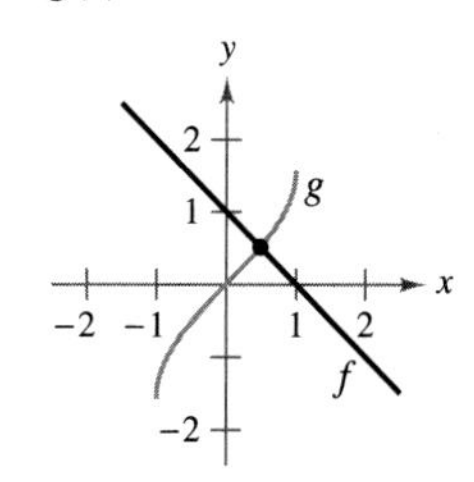

25. ***Mechanic's Rule*** The Mechanic's Rule for approximating $\sqrt{a}$, $a > 0$, is

$$x_{n+1} = \frac{1}{2}\left(x_n + \frac{a}{x_n}\right), \quad n = 1, 2, 3 \ldots$$

where x_1 is an approximation of $\sqrt{a}$.

(a) Use Newton's Method and the function $f(x) = x^2 - a$ to derive the Mechanic's Rule.

(b) Use the Mechanic's Rule to approximate $\sqrt{5}$ and $\sqrt{7}$ to three decimal places.

26. (a) Use Newton's Method and the function $f(x) = x^n - a$ to obtain a general rule for approximating $x = \sqrt[n]{a}$.

(b) Use the general rule found in part (a) to approximate $\sqrt[4]{6}$ and $\sqrt[3]{15}$ to three decimal places.

In Exercises 27–30, apply Newton's Method using the given initial guess, and explain why the method fails.

27. $y = 2x^3 - 6x^2 + 6x - 1, \quad x_1 = 1$
28. $y = 4x^3 - 12x^2 + 12x - 3, \quad x_1 = \frac{3}{2}$

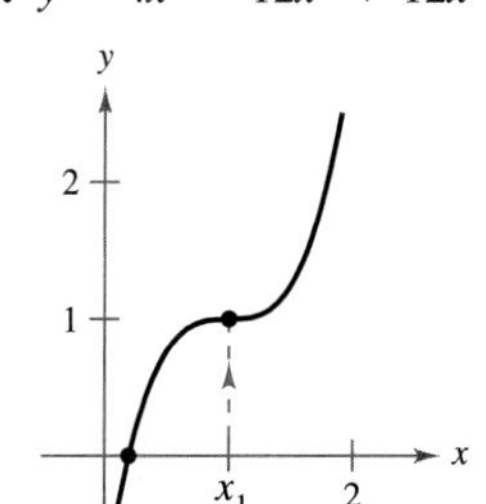

Figure for 27

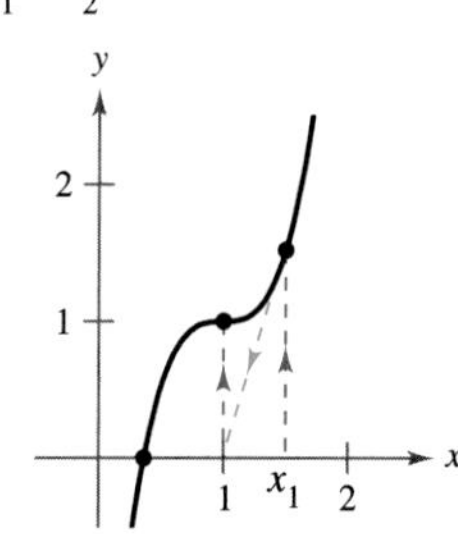

Figure for 28

29. $f(x) = -x^3 + 6x^2 - 10x + 6, \quad x_1 = 2$
30. $f(x) = 2\sin x + \cos 2x, \quad x_1 = \dfrac{3\pi}{2}$

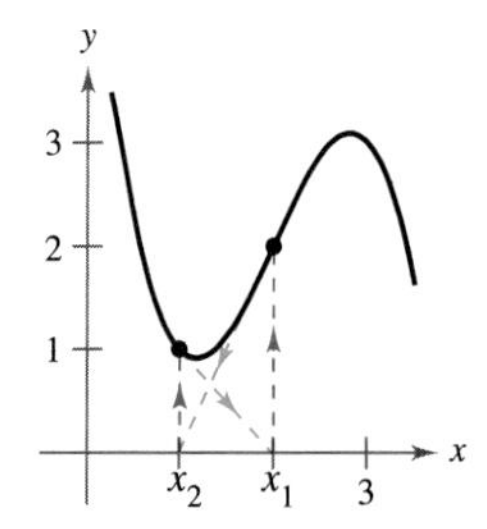

Figure for 29

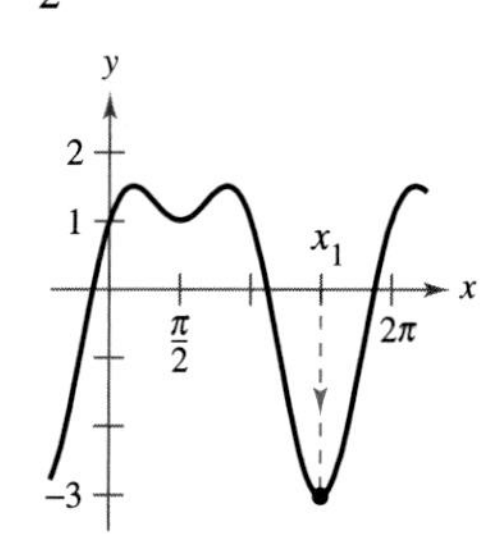

Figure for 30

Writing About Concepts

31. In your own words and using a sketch, describe Newton's Method for approximating the zeros of a function.

32. Under what conditions will Newton's Method fail?

Fixed Point **In Exercises 33–36, approximate the fixed point of the function to two decimal places. [A *fixed point* x_0 of a function f is a value of x such that $f(x_0) = x_0$.]**

33. $f(x) = \cos x$

34. $f(x) = \cot x, \quad 0 < x < \pi$

35. $f(x) = e^{x/10}$

36. $f(x) = -\ln x$

37. ***Writing*** Consider the function $f(x) = x^3 - 3x^2 + 3$.

(a) Use a graphing utility to graph f.

(b) Use Newton's Method with $x_1 = 1$ as an initial guess.

(c) Repeat part (b) using $x_1 = \frac{1}{4}$ as an initial guess and observe that the result is different.

(d) To understand why the results in parts (b) and (c) are different, sketch the tangent lines to the graph of f at the points $(1, f(1))$ and $\left(\frac{1}{4}, f\left(\frac{1}{4}\right)\right)$. Find the x-intercept of each tangent line and compare the intercepts with the first iteration of Newton's Method using the respective initial guesses.

(e) Write a short paragraph summarizing how Newton's Method works. Use the results of this exercise to describe why it is important to select the initial guess carefully.

38. ***Writing*** Repeat the steps in Exercise 37 for the function $f(x) = \sin x$ with initial guesses of $x_1 = 1.8$ and $x_1 = 3$.

39. Use Newton's Method to show that the equation $x_{n+1} = x_n(2 - ax_n)$ can be used to approximate $1/a$ if x_1 is an initial guess for the reciprocal of a. Note that this method of approximating reciprocals uses only the operations of multiplication and subtraction. [*Hint:* Consider $f(x) = (1/x) - a$.]

40. Use the result of Exercise 39 to approximate the indicated reciprocal to three decimal places.

(a) $\frac{1}{3}$ (b) $\frac{1}{11}$

41. ***Advertising Costs*** A company that produces portable CD players estimates that the profit for selling a particular model is

$$P = -76x^3 + 4830x^2 - 320{,}000, \quad 0 \le x \le 60$$

where P is the profit in dollars and x is the advertising expense in 10,000s of dollars (see figure). According to this model, find the smaller of two advertising amounts that yield a profit P of \$2,500,000.

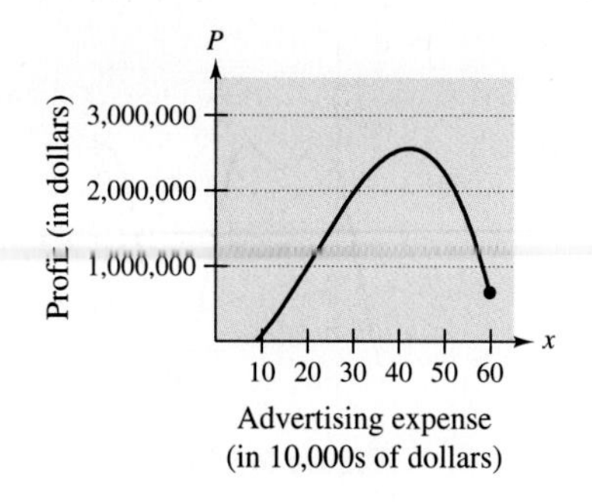

42. ***Engine Power*** The torque produced by a compact automobile engine is approximated by the model

$$T = 0.808x^3 - 17.974x^2 + 71.248x + 110.843, \quad 1 \le x \le 5$$

where T is the torque in foot-pounds and x is the engine speed in thousands of revolutions per minute (see figure). Approximate the two engine speeds that yield a torque T of 170 foot-pounds.

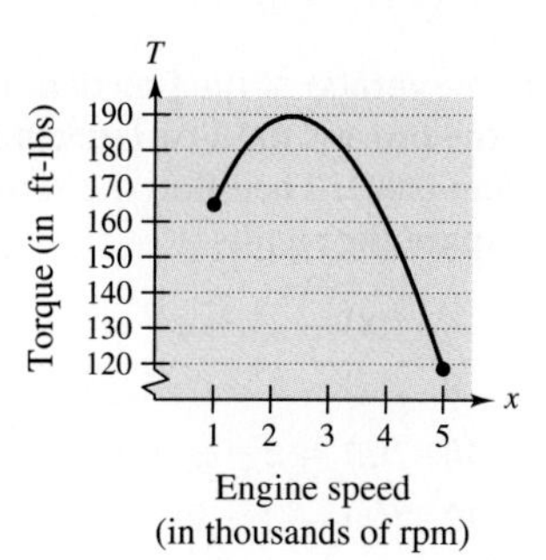

True or False? **In Exercises 43–46, determine whether the statement is true or false. If it is false, explain why or give an example that shows it is false.**

43. The zeros of $f(x) = p(x)/q(x)$ coincide with the zeros of $p(x)$.

44. If the coefficients of a polynomial function are all positive, then the polynomial has no positive zeros.

45. If $f(x)$ is a cubic polynomial such that $f'(x)$ is never zero, then any initial guess will force Newton's Method to converge to the zero of f.

46. The roots of $\sqrt{f(x)} = 0$ coincide with the roots of $f(x) = 0$.

In Exercises 47 and 48, *write a computer program or use a spreadsheet* to find the zeros of a function using Newton's Method. Approximate the zeros of the function accurate to three decimal places. The output should be a table with the following headings.

$$n, \quad x_n, \quad f(x_n), \quad f'(x_n), \quad \frac{f(x_n)}{f'(x_n)}, \quad x_n - \frac{f(x_n)}{f'(x_n)}$$

47. $f(x) = \frac{1}{4}x^3 - 3x^2 + \frac{3}{4}x - 2$

48. $f(x) = \sqrt{4 - x^2}\sin(x - 2)$

49. ***Tangent Lines*** The graph of $f(x) = -\sin x$ has infinitely many tangent lines that pass through the origin. Use Newton's Method to approximate the slope of the tangent line having the greatest slope to three decimal places.

50. Consider the function $f(x) = 2x^3 - 20x^2 - 12x - 24$.

(a) Use a graphing utility to determine the number of zeros of f.

(b) Use Newton's Method with an initial estimate of $x_1 = 2$ to approximate the zero of f to four decimal places.

(c) Repeat part (b) using initial estimates of $x_1 = 10$ and $x_1 = 100$.

(d) Discuss the results of parts (b) and (c). What can you conclude?

Review Exercises for Chapter 3

See www.CalcChat.com for worked-out solutions to odd-numbered exercises.

In Exercises 1–4, find the derivative of the function by using the definition of the derivative.

1. $f(x) = x^2 - 2x + 3$
2. $f(x) = \dfrac{x+1}{x-1}$
3. $f(x) = \sqrt{x} + 1$
4. $f(x) = 2/x$

In Exercises 5 and 6, describe the x-values at which f is differentiable.

5. $f(x) = (x+1)^{2/3}$

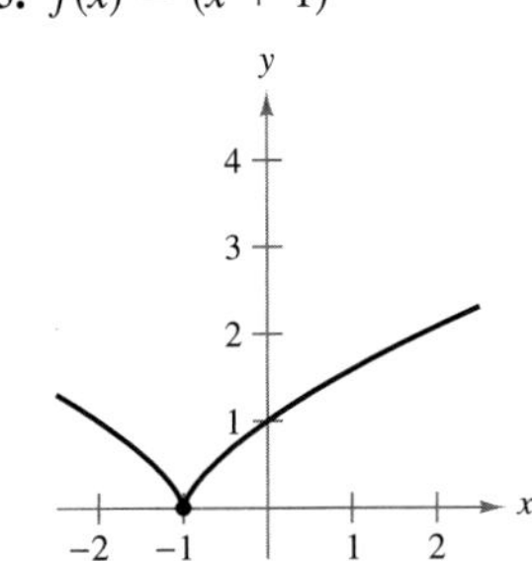

6. $f(x) = 4x/(x+3)$

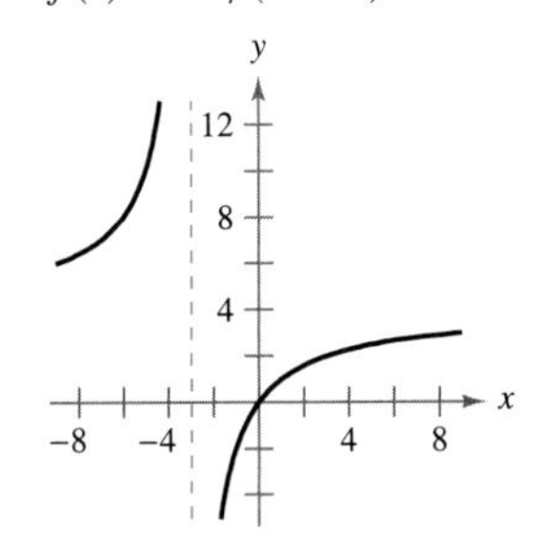

7. Sketch the graph of $f(x) = 4 - |x - 2|$.
   (a) Is f continuous at $x = 2$?
   (b) Is f differentiable at $x = 2$? Explain.

8. Sketch the graph of $f(x) = \begin{cases} x^2 + 4x + 2, & x < -2 \\ 1 - 4x - x^2, & x \geq -2. \end{cases}$
   (a) Is f continuous at $x = -2$?
   (b) Is f differentiable at $x = -2$? Explain.

In Exercises 9 and 10, find the slope of the tangent line to the graph of the function at the specified point.

9. $g(x) = \dfrac{2}{3}x^2 - \dfrac{x}{6}, \quad \left(-1, \dfrac{5}{6}\right)$
10. $h(x) = \dfrac{3x}{8} - 2x^2, \quad \left(-2, -\dfrac{35}{4}\right)$

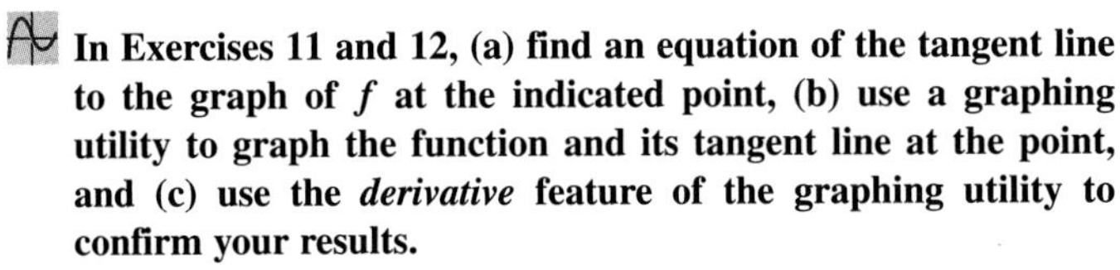

In Exercises 11 and 12, (a) find an equation of the tangent line to the graph of f at the indicated point, (b) use a graphing utility to graph the function and its tangent line at the point, and (c) use the *derivative* feature of the graphing utility to confirm your results.

11. $f(x) = x^3 - 1, \quad (-1, -2)$
12. $f(x) = \dfrac{2}{x+1}, \quad (0, 2)$

In Exercises 13 and 14, use the alternative form of the derivative to find the derivative at $x = c$ (if it exists).

13. $g(x) = x^2(x - 1), \quad c = 2$
14. $f(x) = \dfrac{1}{x+1}, \quad c = 2$

Writing **In Exercises 15 and 16, the figure shows the graphs of a function and its derivative. Label the graphs as f or f' and write a short paragraph stating the criteria used in making the selection. To print an enlarged copy of the graph, go to the website *www.mathgraphs.com*.**

15.

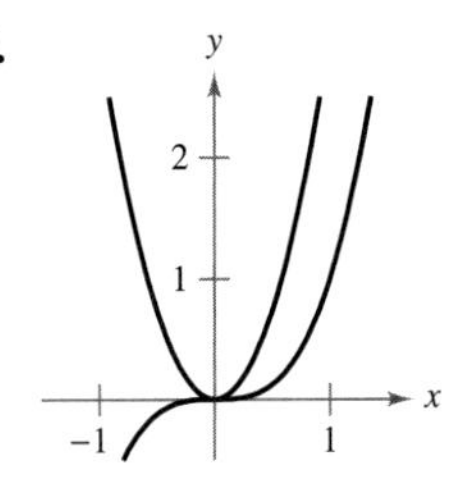

16.

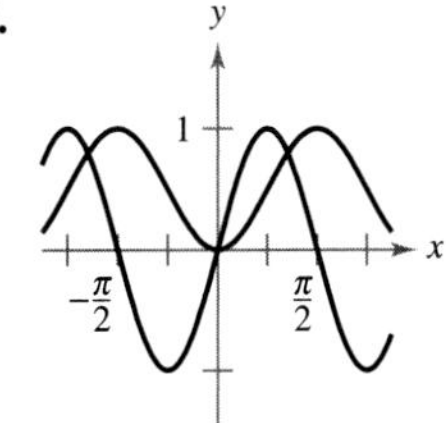

In Exercises 17–32, find the derivative of the function.

17. $y = 25$
18. $y = -12$
19. $f(x) = x^8$
20. $g(x) = x^{12}$
21. $h(t) = 3t^4$
22. $f(t) = -8t^5$
23. $f(x) = x^3 - 3x^2$
24. $g(s) = 4s^4 - 5s^2$
25. $h(x) = 6\sqrt{x} + 3\sqrt[3]{x}$
26. $f(x) = x^{1/2} - x^{-1/2}$
27. $g(t) = \dfrac{2}{3t^2}$
28. $h(x) = \dfrac{2}{(3x)^2}$
29. $f(\theta) = 2\theta - 3\sin\theta$
30. $g(\alpha) = 4\cos\alpha + 6$
31. $f(t) = 3\cos t - 4e^t$
32. $g(s) = \frac{5}{3}\sin s - 2e^s$

33. ***Vibrating String*** When a guitar string is plucked, it vibrates with a frequency of $F = 200\sqrt{T}$, where F is measured in vibrations per second and the tension T is measured in pounds. Find the rate of change of F when (a) $T = 4$ and (b) $T = 9$.

34. ***Vertical Motion*** A ball is dropped from a height of 100 feet. One second later, another ball is dropped from a height of 75 feet. Which ball hits the ground first?

35. ***Vertical Motion*** To estimate the height of a building, a weight is dropped from the top of the building into a pool at ground level. How high is the building if the splash is seen 9.2 seconds after the weight is dropped?

36. ***Vertical Motion*** A bomb is dropped from an airplane at an altitude of 14,400 feet. How long will it take for the bomb to reach the ground? (Because of the motion of the plane, the fall will not be vertical, but the time will be the same as that for a vertical fall.) The plane is moving at 600 miles per hour. How far will the bomb move horizontally after it is released from the plane?

37. ***Projectile Motion*** A thrown ball follows a path described by $y = x - 0.02x^2$.
    (a) Sketch a graph of the path.
    (b) Find the total horizontal distance the ball was thrown.
    (c) At what x-value does the ball reach its maximum height? (Use the symmetry of the path.)
    (d) Find an equation that gives the instantaneous rate of change of the height of the ball with respect to the horizontal change. Evaluate the equation at $x = 0, 10, 25, 30$, and 50.
    (e) What is the instantaneous rate of change of the height when the ball reaches its maximum height?

38. ***Projectile Motion*** The path of a projectile thrown at an angle of 45° with level ground is

$$y = x - \frac{32}{v_0^2}(x^2)$$

where the initial velocity is v_0 feet per second.

(a) Find the x-coordinate of the point where the projectile strikes the ground. Use the symmetry of the path of the projectile to locate the x-coordinate of the point where the projectile reaches its maximum height.

(b) What is the instantaneous rate of change of the height when the projectile is at its maximum height?

(c) Show that doubling the initial velocity of the projectile multiplies both the maximum height and the range by a factor of 4.

(d) Find the maximum height and range of a projectile thrown with an initial velocity of 70 feet per second. Use a graphing utility to sketch the path of the projectile.

39. ***Horizontal Motion*** The position function of a particle moving along the x-axis is

$$x(t) = t^2 - 3t + 2$$

for $-\infty < t < \infty$.

(a) Find the velocity of the particle.

(b) Find the open t-interval(s) in which the particle is moving to the left.

(c) Find the position of the particle when the velocity is 0.

(d) Find the speed of the particle when the position is 0.

40. ***Modeling Data*** The speed of a car in miles per hour and its stopping distance in feet are recorded in the table.

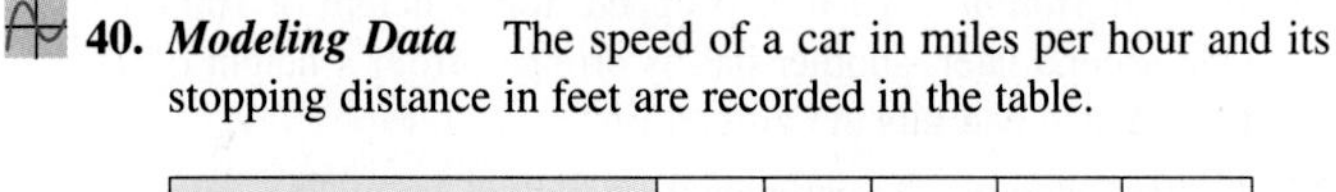

Speed (x)	20	30	40	50	60
Stopping Distance (y)	25	55	105	188	300

(a) Use the regression capabilities of a graphing utility to find a quadratic model for the data.

(b) Use a graphing utility to plot the data and graph the model.

(c) Use a graphing utility to graph dy/dx.

(d) Use the model to approximate the stopping distance at a speed of 65 miles per hour.

(e) Use the graphs in parts (b) and (c) to explain the change in stopping distance as the speed increases.

In Exercises 41–57, find the derivative of the function.

41. $f(x) = (3x^2 + 7)(x^2 - 2x + 3)$

42. $g(x) = (x^3 - 3x)(x + 2)$

43. $h(x) = \sqrt{x}\sin x$

44. $f(t) = t^3 \cos t$

45. $f(x) = \dfrac{2x^3 - 1}{x^2}$

46. $f(x) = \dfrac{x + 1}{x - 1}$

47. $f(x) = \dfrac{x^2 + x - 1}{x^2 - 1}$

48. $f(x) = \dfrac{6x - 5}{x^2 + 1}$

49. $f(x) = \dfrac{1}{4 - 3x^2}$

50. $f(x) = \dfrac{9}{3x^2 - 2x}$

51. $y = \dfrac{x^2}{\cos x}$

52. $y = \dfrac{\sin x}{x^2}$

53. $y = 3x^2 \sec x$

54. $y = 2x - x^2 \tan x$

55. $y = -x \tan x$

56. $y = \dfrac{1 + \sin x}{1 - \sin x}$

57. $y = 4xe^x$

58. ***Acceleration*** The velocity of an object in meters per second is $v(t) = 36 - t^2$, $0 \le t \le 6$. Find the velocity and acceleration of the object when $t = 4$.

In Exercises 59–62, find the second derivative of the function.

59. $g(t) = t^3 - 3t + 2$

60. $f(x) = 12\sqrt[4]{x}$

61. $f(\theta) = 3 \tan \theta$

62. $h(t) = 4 \sin t - 5 \cos t$

In Exercises 63 and 64, show that the function satisfies the equation.

	Function	*Equation*
63.	$y = 2 \sin x + 3 \cos x$	$y'' + y = 0$
64.	$y = \dfrac{10 - \cos x}{x}$	$xy' + y = \sin x$

65. ***Rate of Change*** Determine whether there exist any values of x in the interval $[0, 2\pi)$ such that the rate of change of $f(x) = \sec x$ and the rate of change of $g(x) = \csc x$ are equal.

66. ***Volume*** The radius of a right circular cylinder is given by $\sqrt{t + 2}$ and its height is $\frac{1}{2}\sqrt{t}$, where t is time in seconds and the dimensions are in inches. Find the rate of change of the volume with respect to time.

In Exercises 67–96, find the derivative of the function.

67. $f(x) = \sqrt{1 - x^3}$

68. $f(x) = \sqrt[3]{x^2 - 1}$

69. $h(x) = \left(\dfrac{x - 3}{x^2 + 1}\right)^2$

70. $f(x) = \left(x^2 + \dfrac{1}{x}\right)^5$

71. $f(s) = (s^2 - 1)^{5/2}(s^3 + 5)$

72. $h(\theta) = \dfrac{\theta}{(1 - \theta)^3}$

73. $y = 3\cos(3x + 1)$

74. $y = 1 - \cos 2x + 2\cos^2 x$

75. $y = \frac{1}{2}\csc 2x$

76. $y = \csc 3x + \cot 3x$

77. $y = \dfrac{x}{2} - \dfrac{\sin 2x}{4}$

78. $y = \dfrac{\sec^7 x}{7} - \dfrac{\sec^5 x}{5}$

79. $y = \dfrac{2}{3}\sin^{3/2} x - \dfrac{2}{7}\sin^{7/2} x$

80. $f(x) = \dfrac{3x}{\sqrt{x^2 + 1}}$

81. $y = \dfrac{\sin \pi x}{x + 2}$

82. $y = \dfrac{\cos(x - 1)}{x - 1}$

83. $g(t) = t^2 e^{t/4}$

84. $h(z) = e^{-z^2/2}$

85. $y = \sqrt{e^{2x} + e^{-2x}}$

86. $y = 3e^{-3/t}$

87. $g(x) = \dfrac{x^2}{e^x}$

88. $f(\theta) = \dfrac{1}{2}e^{\sin 2\theta}$

89. $g(x) = \ln \sqrt{x}$

90. $h(x) = \ln \dfrac{x(x - 1)}{x - 2}$

91. $f(x) = x\sqrt{\ln x}$

92. $f(x) = \ln[x(x^2 - 2)^{2/3}]$

93. $y = \dfrac{1}{b^2}\left[\ln(a + bx) + \dfrac{a}{a + bx}\right]$

94. $y = \dfrac{1}{b^2}[a + bx - a\ln(a + bx)]$

95. $y = -\dfrac{1}{a}\ln\dfrac{a + bx}{x}$

96. $y = -\dfrac{1}{ax} + \dfrac{b}{a^2}\ln\dfrac{a + bx}{x}$

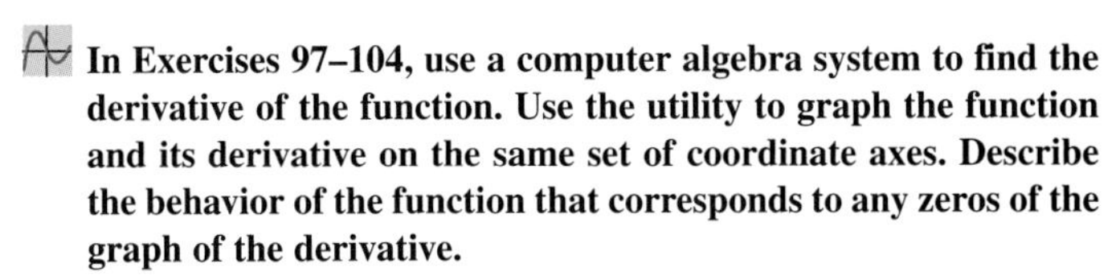

In Exercises 97–104, use a computer algebra system to find the derivative of the function. Use the utility to graph the function and its derivative on the same set of coordinate axes. Describe the behavior of the function that corresponds to any zeros of the graph of the derivative.

97. $f(t) = t^2(t - 1)^5$

98. $f(x) = [(x - 2)(x + 4)]^2$

99. $g(x) = \dfrac{2x}{\sqrt{x + 1}}$

100. $g(x) = x\sqrt{x^2 + 1}$

101. $f(t) = \sqrt{t + 1}\,\sqrt[3]{t + 1}$

102. $y = \sqrt{3x}(x + 2)^3$

103. $y = \tan\sqrt{1 - x}$

104. $y = 2\csc^3(\sqrt{x})$

In Exercises 105–108, find the second derivative of the function.

105. $y = 2x^2 + \sin 2x$

106. $y = \dfrac{1}{x} + \tan x$

107. $f(x) = \cot x$

108. $y = \sin^2 x$

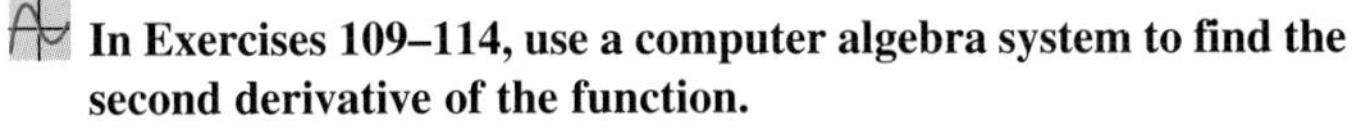

In Exercises 109–114, use a computer algebra system to find the second derivative of the function.

109. $f(t) = \dfrac{t}{(1 - t)^2}$

110. $g(x) = \dfrac{6x - 5}{x^2 + 1}$

111. $g(\theta) = \tan 3\theta - \sin(\theta - 1)$

112. $h(x) = x\sqrt{x^2 - 1}$

113. $g(x) = x^3 \ln x$

114. $f(x) = 6x^2 e^{-x/3}$

115. ***Refrigeration*** The temperature T of food put in a freezer is

$$T = \frac{700}{t^2 + 4t + 10}$$

where t is the time in hours. Find the rate of change of T with respect to t at each of the following times.

(a) $t = 1$ (b) $t = 3$ (c) $t = 5$ (d) $t = 10$

116. ***Fluid Flow*** The emergent velocity v of a liquid flowing from a hole in the bottom of a tank is given by $v = \sqrt{2gh}$, where g is the acceleration due to gravity (32 feet per second per second) and h is the depth of the liquid in the tank. Find the rate of change of v with respect to h when (a) $h = 9$ and (b) $h = 4$. (Note that $g = +32$ feet per second per second. The sign of g depends on how a problem is modeled. In this case, letting g be negative would produce an imaginary value for v.)

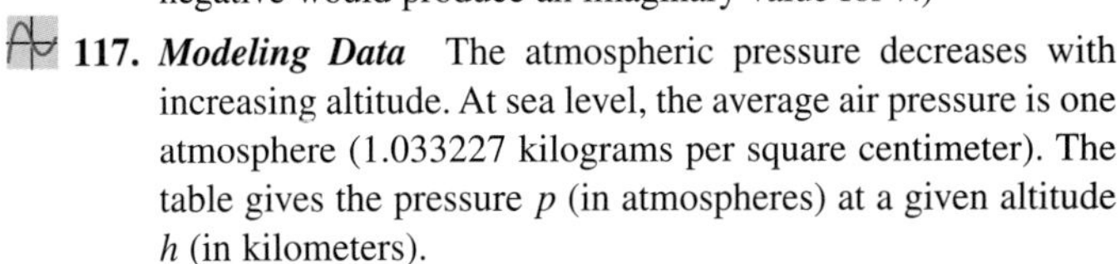

117. ***Modeling Data*** The atmospheric pressure decreases with increasing altitude. At sea level, the average air pressure is one atmosphere (1.033227 kilograms per square centimeter). The table gives the pressure p (in atmospheres) at a given altitude h (in kilometers).

h	0	5	10	15	20	25
p	1	0.55	0.25	0.12	0.06	0.02

(a) Use a graphing utility to find a model of the form $p = a + b\ln h$ for the data. Explain why the result is an error message.

(b) Use a graphing utility to find the logarithmic model $h = a + b\ln p$ for the data.

(c) Use a graphing utility to plot the data and graph the logarithmic model.

(d) Use the model to estimate the altitude at which the pressure is 0.75 atmosphere.

(e) Use the model to estimate the pressure at an altitude of 13 kilometers.

(f) Find the rate of change of pressure when $h = 5$ and $h = 20$. Interpret the results in the context of the problem.

118. ***Tractrix*** A person walking along a dock drags a boat by a 10-meter rope. The boat travels along a path known as a *tractrix* (see figure). The equation of this path is

$$y = 10\ln\left(\frac{10 + \sqrt{100 - x^2}}{x}\right) - \sqrt{100 - x^2}.$$

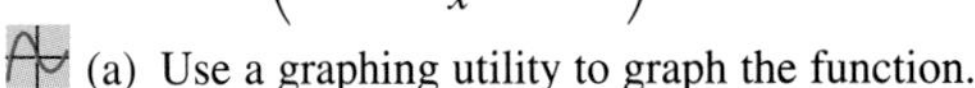

(a) Use a graphing utility to graph the function.

(b) What is the slope of the path when $x = 5$ and $x = 9$?

(c) What does the slope of the path approach as $x \to 10$?

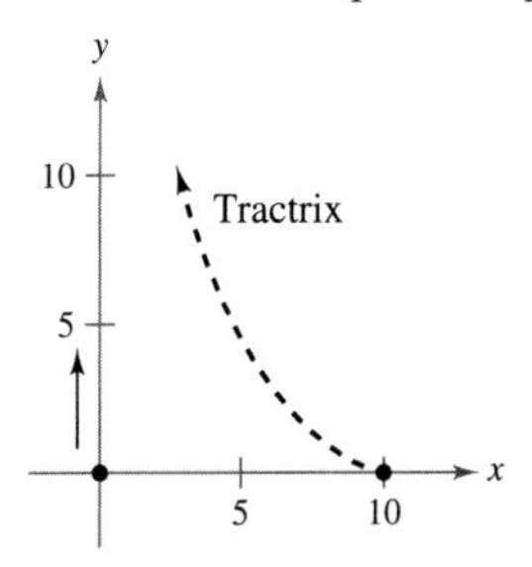

In Exercises 119–126, use implicit differentiation to find dy/dx.

119. $x^2 + 3xy + y^3 = 10$

120. $x^2 + 9y^2 - 4x + 3y = 0$

121. $\cos x^2 = xe^y$

122. $ye^x + xe^y = xy$

123. $y\sqrt{x} - x\sqrt{y} = 16$

124. $y^2 = (x - y)(x^2 + y)$

125. $x\sin y = y\cos x$

126. $\cos(x + y) = x$

In Exercises 127–130, find the equations of the tangent line and the normal line to the graph of the equation at the indicated point. Use a graphing utility to graph the equation, the tangent line, and the normal line.

127. $x^2 + y^2 = 20,\quad (2, 4)$

128. $x^2 - y^2 = 16,\quad (5, 3)$

129. $y\ln x + y^2 = 0,\quad (e, -1)$

130. $\ln(x + y) = x,\quad (0, 1)$

In Exercises 131 and 132, use logarithmic differentiation to find dy/dx.

131. $y = \dfrac{x\sqrt{x^2 + 1}}{x + 4}$

132. $y = \dfrac{(2x + 1)^3(x^2 - 1)^2}{x + 3}$

In Exercise 133–136, find $(f^{-1})'(a)$ for the function f and real number a.

	Function	*Real number*
133.	$f(x) = x^3 + 2$	$a = -1$
134.	$f(x) = x\sqrt{x - 3}$	$a = 4$
135.	$f(x) = \tan x, \quad -\dfrac{\pi}{4} \le x \le \dfrac{\pi}{4}$	$a = \dfrac{\sqrt{3}}{3}$
136.	$f(x) = \ln x$	$a = 0$

In Exercises 137–142, find the derivative of the function.

137. $y = \tan(\arcsin x)$

138. $y = \arctan(x^2 - 1)$

139. $y = x \operatorname{arcsec} x$

140. $y = \frac{1}{2} \arctan e^{2x}$

141. $y = x(\arcsin x)^2 - 2x + 2\sqrt{1 - x^2} \arcsin x$

142. $y = \sqrt{x^2 - 4} - 2 \operatorname{arcsec} \dfrac{x}{2}, \quad 2 < x < 4$

143. A point moves along the curve $y = \sqrt{x}$ in such a way that the y-value is increasing at a rate of 2 units per second. At what rate is x changing for each of the following values?

(a) $x = \frac{1}{2}$ (b) $x = 1$ (c) $x = 4$

144. ***Surface Area*** The edges of a cube are expanding at a rate of 5 centimeters per second. How fast is the surface area changing when each edge is 4.5 centimeters?

145. ***Changing Depth*** The cross section of a 5-meter trough is an isosceles trapezoid with a 2-meter lower base, a 3-meter upper base, and an altitude of 2 meters. Water is running into the trough at a rate of 1 cubic meter per minute. How fast is the water level rising when the water is 1 meter deep?

146. ***Linear and Angular Velocity*** A rotating beacon is located 1 kilometer off a straight shoreline (see figure). If the beacon rotates at a rate of 3 revolutions per minute, how fast (in kilometers per hour) does the beam of light appear to be moving to a viewer who is $\frac{1}{2}$ kilometer down the shoreline?

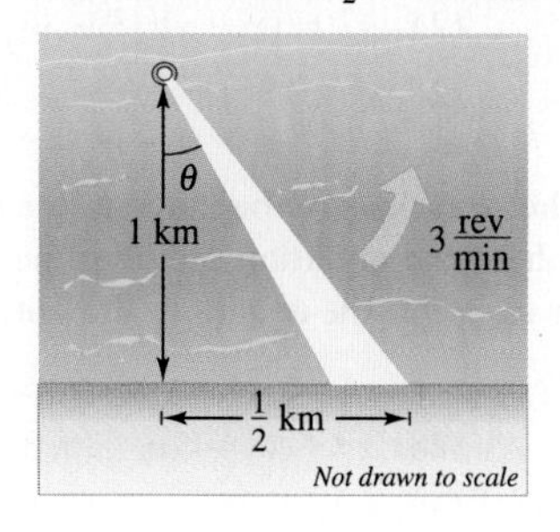

147. ***Moving Shadow*** A sandbag is dropped from a balloon at a height of 60 meters when the angle of elevation to the sun is 30° (see figure). Find the rate at which the shadow of the sandbag is traveling along the ground when the sandbag is at a height of 35 meters. [*Hint:* The position of the sandbag is given by $s(t) = 60 - 4.9t^2$.]

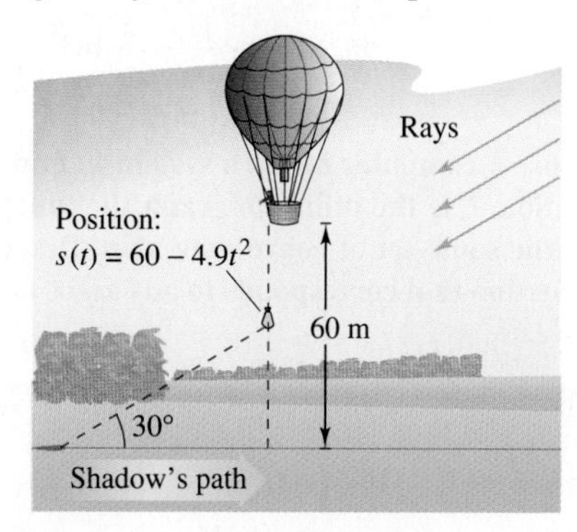

148. ***Geometry*** Consider the rectangle shown in the figure.

(a) Find the area of the rectangle as a function of x.

(b) Find the rate of change of the area when $x = 4$ centimeters if $dx/dt = 4$ centimeters per minute.

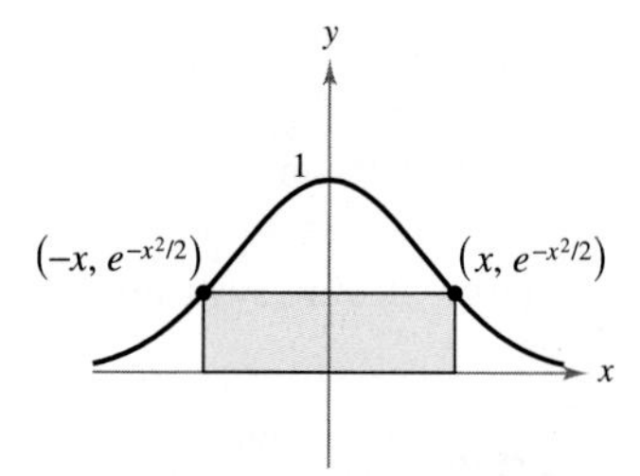

In Exercises 149–152, use Newton's Method to approximate any real zeros of the function accurate to three decimal places. Use the root-finding capabilities of a graphing utility to verify your results.

149. $f(x) = x^3 - 3x - 1$

150. $f(x) = x^3 + 2x + 1$

151. $g(x) = xe^x - 4$

152. $f(x) = 3 - x \ln x$

In Exercises 153 and 154, use Newton's Method to approximate, to three decimal places, the x-values of any points of intersection of the graphs of the equations. Use a graphing utility to verify your results.

153. $y = x^4$
$y = x + 3$

154. $y = \sin \pi x$
$y = 1 - x$

P.S. Problem Solving

See www.CalcChat.com for worked-out solutions to odd-numbered exercises.

1. Consider the graph of the parabola $y = x^2$.

(a) Find the radius r of the largest possible circle centered on the y-axis that is tangent to the parabola at the origin, as indicated in the figure. This circle is called the **circle of curvature** (see Section 12.5). Use a graphing utility to graph the circle and parabola in the same viewing window.

(b) Find the center $(0, b)$ of the circle of radius 1 centered on the y-axis that is tangent to the parabola at two points, as indicated in the figure. Use a graphing utility to graph the circle and parabola in the same viewing window.

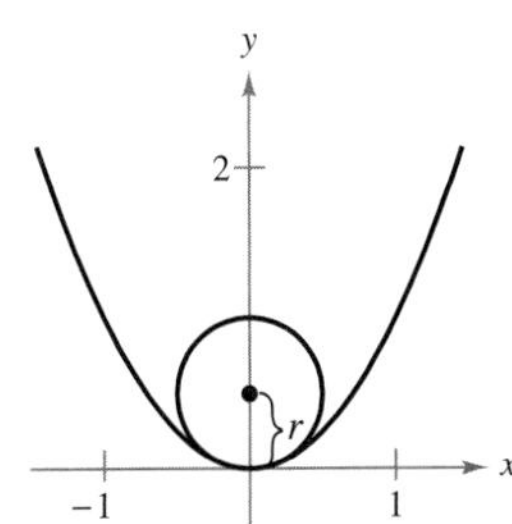

Figure for 1(a)

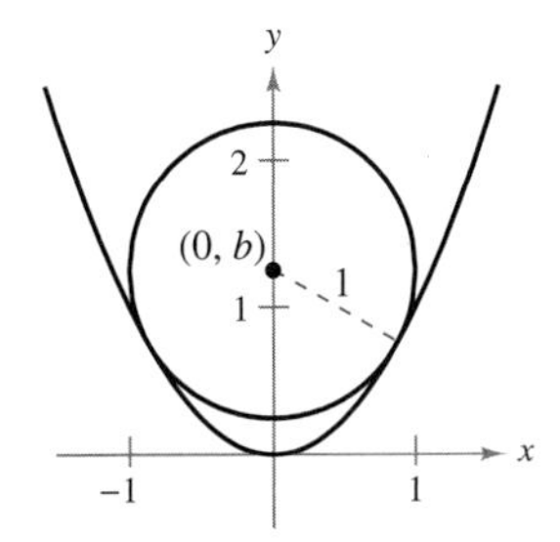

Figure for 1(b)

2. Graph the two parabolas $y = x^2$ and $y = -x^2 + 2x - 5$ in the same coordinate plane. Find equations of the two lines simultaneously tangent to both parabolas.

3. (a) Find the polynomial $P_1(x) = a_0 + a_1x$ whose value and slope agree with the value and slope of $f(x) = \cos x$ at the point $x = 0$.

(b) Find the polynomial $P_2(x) = a_0 + a_1x + a_2x^2$ whose value and first two derivatives agree with the value and first two derivatives of $f(x) = \cos x$ at the point $x = 0$. This polynomial is called the second-degree **Taylor polynomial** of $f(x) = \cos x$ at $x = 0$.

(c) Complete the table comparing the values of f and P_2. What do you observe?

x	-1.0	-0.1	-0.001	0	0.001	0.1	1.0
$\cos x$							
$P_2(x)$							

(d) Find the third-degree Taylor polynomial of $f(x) = \sin x$ at $x = 0$.

4. (a) Find an equation of the tangent line to the parabola $y = x^2$ at the point $(2, 4)$.

(b) Find an equation of the normal line to $y = x^2$ at the point $(2, 4)$. (The normal line is perpendicular to the tangent line.) Where does this line intersect the parabola a second time?

(c) Find equations of the tangent line and normal line to $y = x^2$ at the point $(0, 0)$.

(d) Prove that for any point $(a, b) \neq (0, 0)$ on the parabola $y = x^2$, the normal line intersects the graph a second time.

5. Find a third-degree polynomial $p(x)$ that is tangent to the line $y = 14x - 13$ at the point $(1, 1)$, and tangent to the line $y = -2x - 5$ at the point $(-1, -3)$.

6. Find a function of the form $f(x) = a + b \cos cx$ that is tangent to the line $y = 1$ at the point $(0, 1)$, and tangent to the line

$$y = x + \frac{3}{2} - \frac{\pi}{4}$$

at the point $\left(\frac{\pi}{4}, \frac{3}{2}\right)$.

7. The graph of the **eight curve,**

$$x^4 = a^2(x^2 - y^2),\ a \neq 0,$$

is shown below.

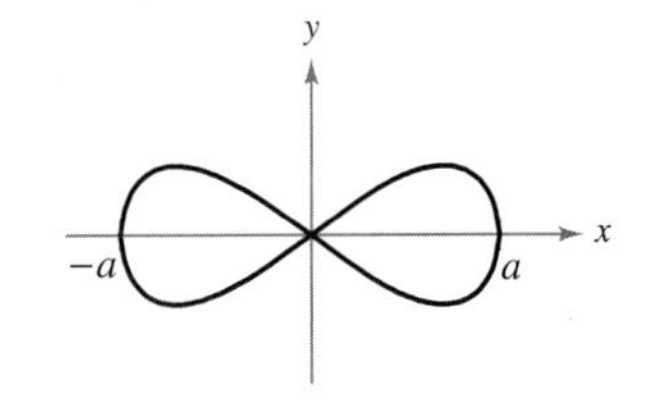

(a) Explain how you could use a graphing utility to obtain the graph of this curve.

(b) Use a graphing utility to graph the curve for various values of the constant a. Describe how a affects the shape of the curve.

(c) Determine the points on the curve where the tangent line is horizontal.

8. The graph of the **pear-shaped quartic,**

$$b^2y^2 = x^3(a - x),\ a, b > 0,$$

is shown below.

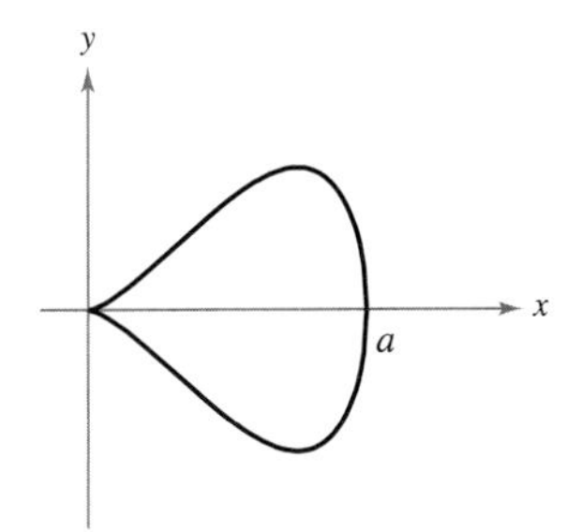

(a) Explain how you could use a graphing utility to obtain the graph of this curve.

(b) Use a graphing utility to graph the curve for various values of the constants a and b. Describe how a and b affect the shape of the curve.

(c) Determine the points on the curve where the tangent line is horizontal.

9. A man 6 feet tall walks at a rate of 5 feet per second toward a streetlight that is 30 feet high (see figure). The man's 3-foot-tall child follows at the same speed, but 10 feet behind the man. At times, the shadow behind the child is caused by the man, and at other times, by the child.

(a) Suppose the man is 90 feet from the streetlight. Show that the man's shadow extends beyond the child's shadow.

(b) Suppose the man is 60 feet from the streetlight. Show that the child's shadow extends beyond the man's shadow.

(c) Determine the distance d from the man to the streetlight at which the tips of the two shadows are exactly the same distance from the streetlight.

(d) Determine how fast the tip of the shadow is moving as a function of x, the distance between the man and the streetlight. Discuss the continuity of this shadow speed function.

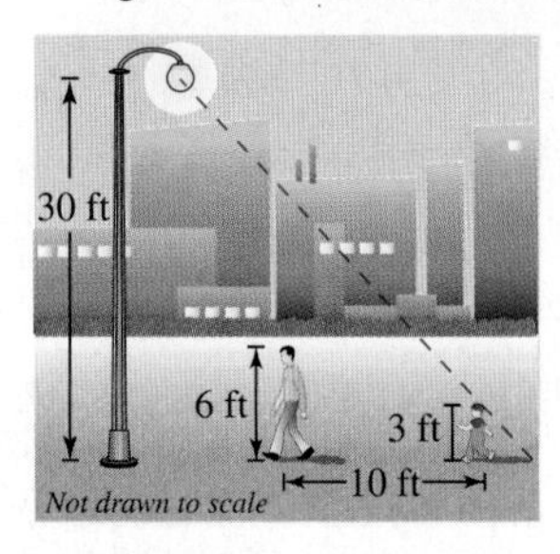

Figure for 9

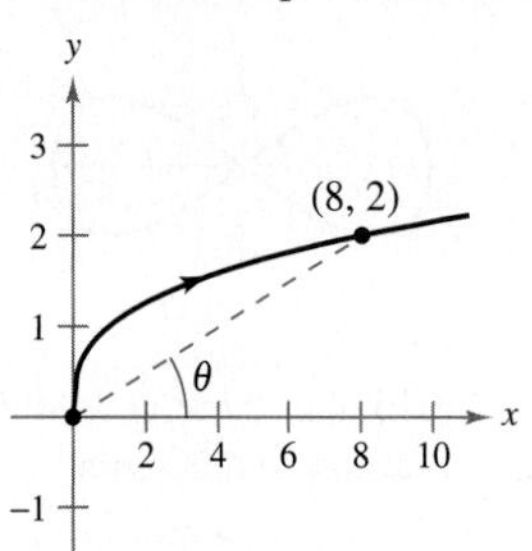

Figure for 10

10. A particle is moving along the graph of $y = \sqrt[3]{x}$ (see figure). When $x = 8$, the y-component of its position is increasing at the rate of 1 centimeter per second.

(a) How fast is the x-component changing at this moment?

(b) How fast is the distance from the origin changing at this moment?

(c) How fast is the angle of inclination θ changing at this moment?

11. Let L be the tangent line to the graph of the function $y = \ln x$ at the point (a, b). Show that the distance between b and c is always equal to 1.

12. Let L be the tangent line to the graph of the function $y = e^x$ at the point (a, b). Show that the distance between a and c is always equal to 1.

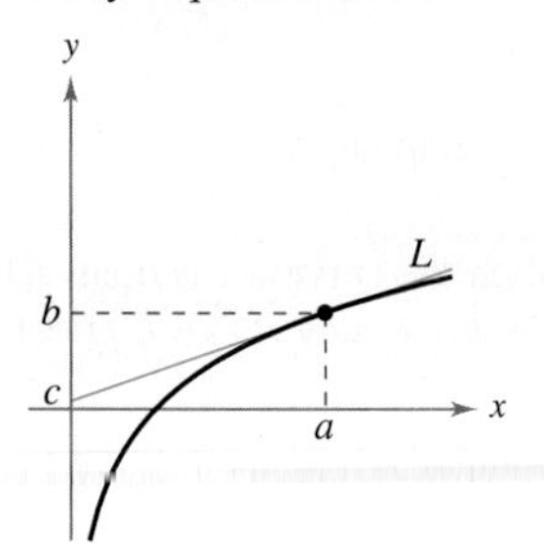

Figure for 11

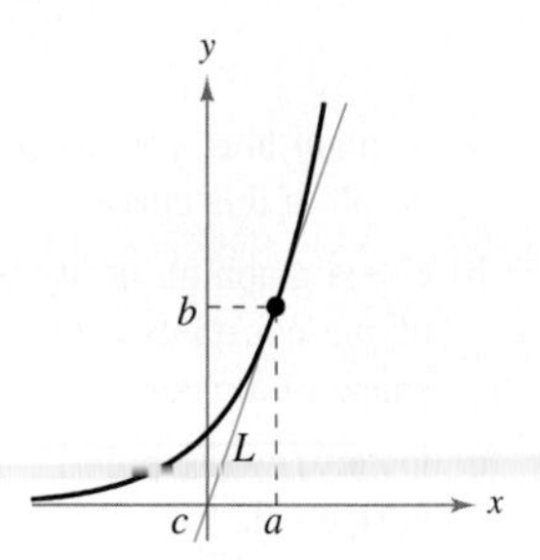

Figure for 12

13. The fundamental limit

$$\lim_{x\to 0} \frac{\sin x}{x} = 1$$

assumes that x is measured in radians. What happens if we assume that x is measured in degrees instead of radians?

(a) Set your calculator to degree mode and complete the table.

z **(in degrees)**	0.1	0.01	0.0001
$\dfrac{\sin z}{z}$			

(b) Use the table to estimate $\lim_{z\to 0} \dfrac{\sin z}{z}$ for z in degrees. What is the exact value of this limit? (*Hint:* $180° = \pi$ radians)

(c) Use the limit definition of the derivative to find $\dfrac{d}{dz} \sin z$ for z in degrees.

(d) Define the new functions $S(z) = \sin(cz)$ and $C(z) = \cos(cz)$, where $c = \pi/180$. Find $S(90)$ and $C(180)$. Use the Chain Rule to calculate $\dfrac{d}{dz}S(z)$.

(e) Explain why calculus is made easier by using radians instead of degrees.

14. An astronaut standing on the moon throws a rock into the air. The height of the rock is $s = -\frac{27}{10}t^2 + 27t + 6$, where s is measured in feet and t is measured in seconds.

(a) Find expressions for the velocity and acceleration of the rock.

(b) Find the time when the rock is at its highest point by finding the time when the velocity is zero. What is the rock's height at this time?

(c) How does the acceleration of the rock compare with the acceleration due to gravity on Earth?

15. If a is the acceleration of an object, the *jerk* j is defined by $j = a'(t)$.

(a) Use this definition to give a physical interpretation of j.

(b) The figure shows the graphs of the position, velocity, acceleration, and jerk functions of a vehicle. Identify each graph and explain your reasoning.

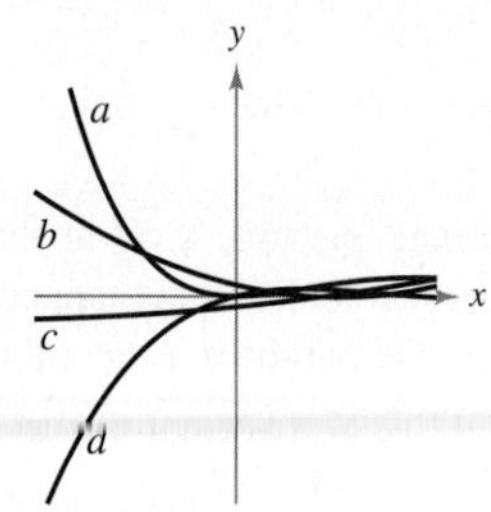

4 Applications of Differentiation

When a glassblower removes a glowing "blob" of molten glass from a kiln, its temperature is about 1700°F. At first, the molten glass cools rapidly. Then, as the temperature of the glass approaches room temperature, it cools more and more slowly. Will the temperature of the glass ever actually reach room temperature? Why?

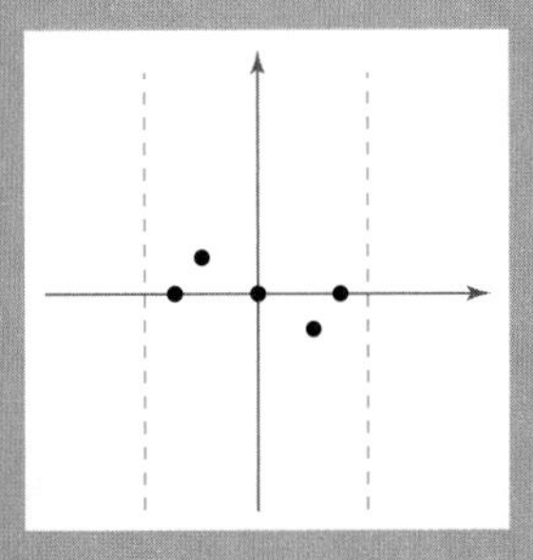

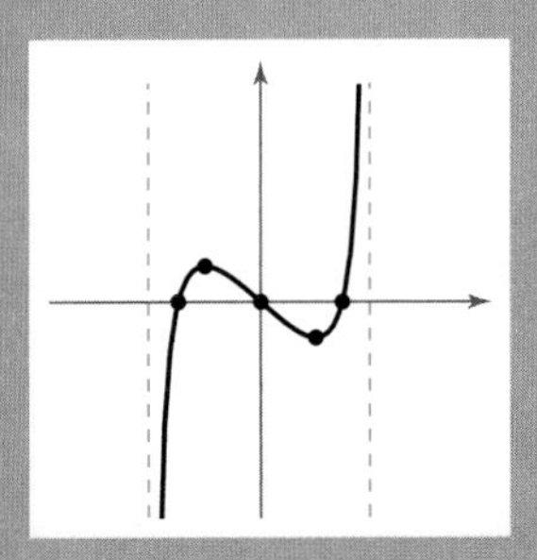

In Chapter 4, you will use calculus to analyze graphs of functions. For example, you can use the derivative of a function to determine the function's maximum and minimum values. You can use limits to identify any asymptotes of the function's graph. In Section 4.6, you will combine these techniques to sketch the graph of a function.

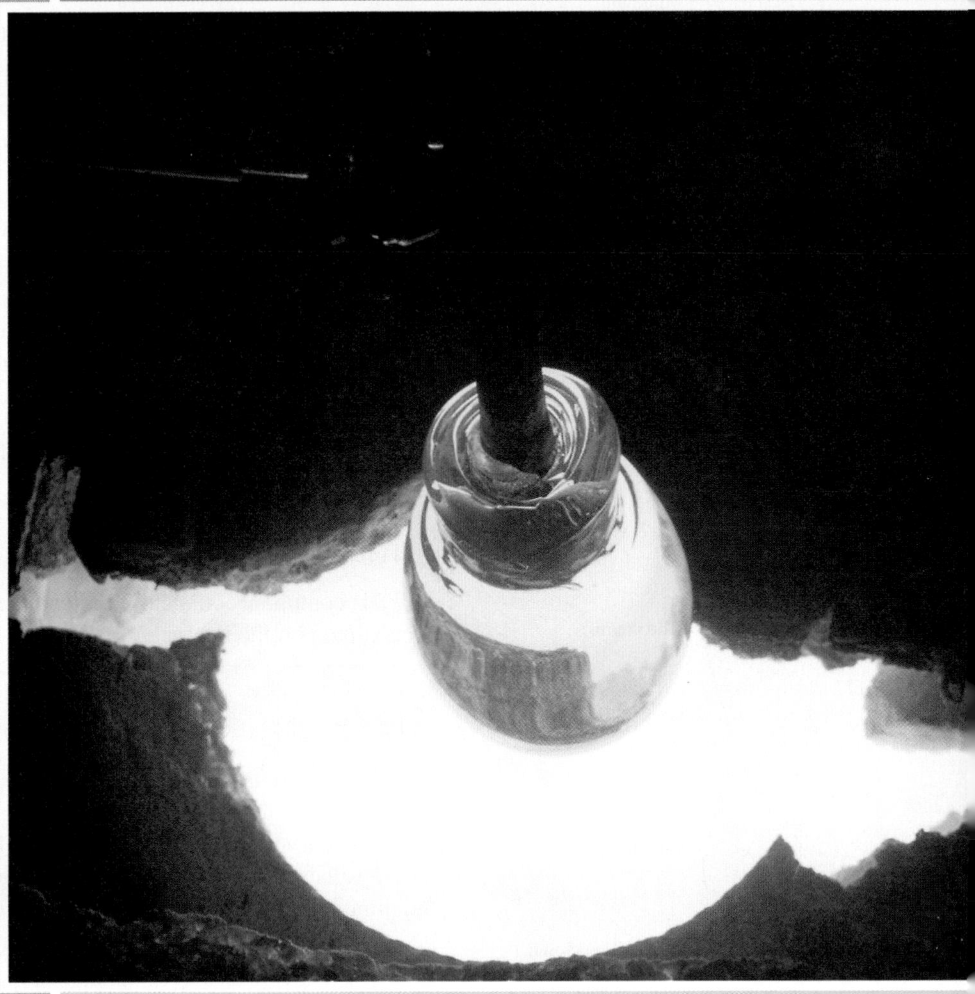

www.shawnolson.net/a/507

Section 4.1 Extrema on an Interval

- **Understand the definition of extrema of a function on an interval.**
- **Understand the definition of relative extrema of a function on an open interval.**
- **Find extrema on a closed interval.**

Extrema of a Function

In calculus, much effort is devoted to determining the behavior of a function f on an interval I. Does f have a maximum value on I? Does it have a minimum value? Where is the function increasing? Where is it decreasing? In this chapter you will learn how derivatives can be used to answer these questions. You will also see why these questions are important in real-life applications.

(a) f is continuous, $[-1, 2]$ is closed.

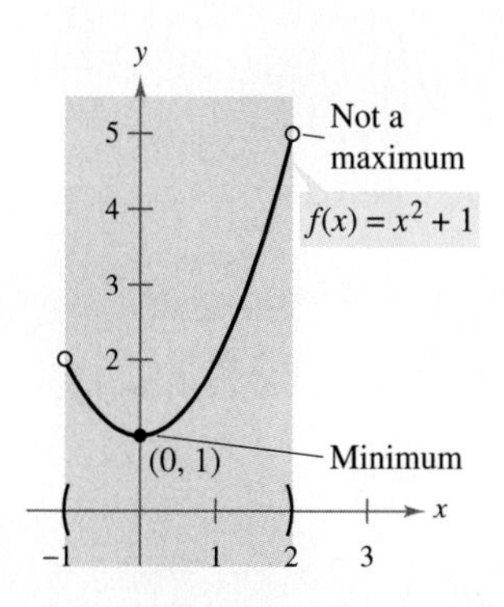

(b) f is continuous, $(-1, 2)$ is open.

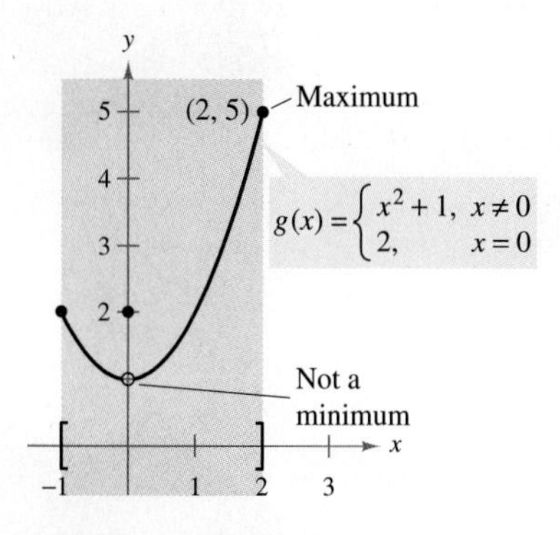

(c) g is not continuous, $[-1, 2]$ is closed.
Extrema can occur at interior points or endpoints of an interval. Extrema that occur at the endpoints are called **endpoint extrema.**
Figure 4.1

Definition of Extrema

Let f be defined on an interval I containing c.

1. $f(c)$ is the **minimum of f on I** if $f(c) \leq f(x)$ for all x in I.
2. $f(c)$ is the **maximum of f on I** if $f(c) \geq f(x)$ for all x in I.

The minimum and maximum of a function on an interval are the **extreme values,** or **extrema** (the singular form of extrema is extremum), of the function on the interval. The minimum and maximum of a function on an interval are also called the **absolute minimum** and **absolute maximum** on the interval.

A function need not have a minimum or a maximum on an interval. For instance, in Figure 4.1(a) and (b), you can see that the function $f(x) = x^2 + 1$ has both a minimum and a maximum on the closed interval $[-1, 2]$, but does not have a maximum on the open interval $(-1, 2)$. Moreover, in Figure 4.1(c), you can see that continuity (or the lack of it) can affect the existence of an extremum on the interval. This suggests the following theorem. (Although the Extreme Value Theorem is intuitively plausible, a proof of this theorem is not within the scope of this text.)

THEOREM 4.1 The Extreme Value Theorem

If f is continuous on a closed interval $[a, b]$, then f has both a minimum and a maximum on the interval.

EXPLORATION

Finding Minimum and Maximum Values The Extreme Value Theorem (like the Intermediate Value Theorem) is an *existence theorem* because it tells of the existence of minimum and maximum values but does not show how to find these values. Use the extreme-value capability of a graphing utility to find the minimum and maximum values of each of the following functions. In each case, do you think the x-values are exact or approximate? Explain your reasoning.

a. $f(x) = x^2 - 4x + 5$ on the closed interval $[-1, 3]$
b. $f(x) = x^3 - 2x^2 - 3x - 2$ on the closed interval $[-1, 3]$

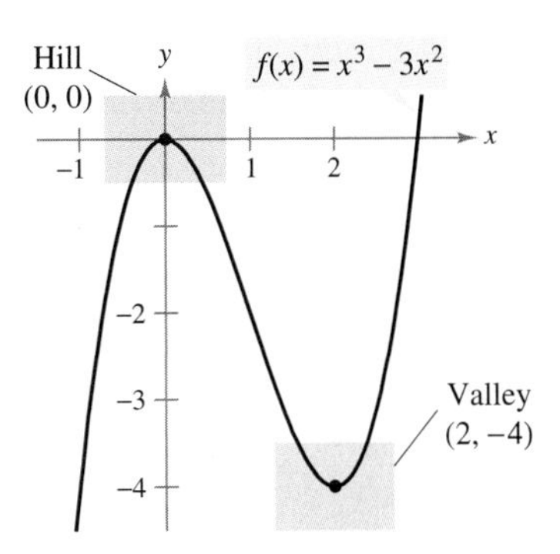

f has a relative maximum at $(0, 0)$ and a relative minimum at $(2, -4)$.
Figure 4.2

Relative Extrema and Critical Numbers

In Figure 4.2, the graph of $f(x) = x^3 - 3x^2$ has a **relative maximum** at the point $(0, 0)$ and a **relative minimum** at the point $(2, -4)$. Informally, you can think of a relative maximum as occurring on a "hill" on the graph, and a relative minimum as occurring in a "valley" on the graph. Such a hill and valley can occur in two ways. If the hill (or valley) is smooth and rounded, the graph has a horizontal tangent line at the high point (or low point). If the hill (or valley) is sharp and peaked, the graph represents a function that is not differentiable at the high point (or low point).

Definition of Relative Extrema

1. If there is an open interval containing c on which $f(c)$ is a maximum, then $f(c)$ is called a **relative maximum** of f, or you can say that f has a **relative maximum at $(c, f(c))$.**
2. If there is an open interval containing c on which $f(c)$ is a minimum, then $f(c)$ is called a **relative minimum** of f, or you can say that f has a **relative minimum at $(c, f(c))$.**

The plural of relative maximum is relative maxima, and the plural of relative minimum is relative minima.

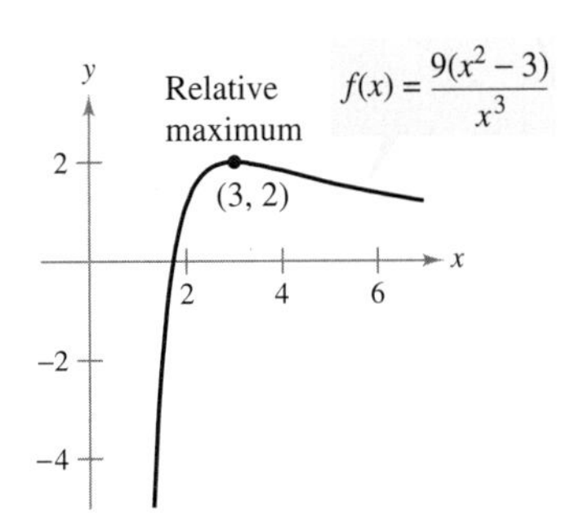

(a) $f'(3) = 0$

Example 1 examines the derivatives of functions at *given* relative extrema. (Much more is said about *finding* the relative extrema of a function in Section 4.3.)

EXAMPLE 1 The Value of the Derivative at Relative Extrema

Find the value of the derivative at each of the relative extrema shown in Figure 4.3.

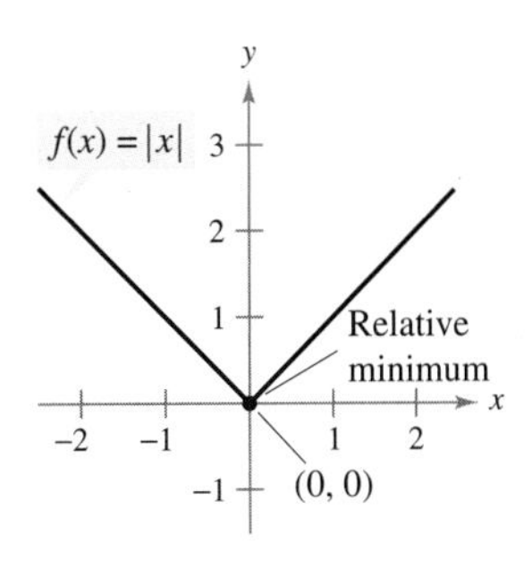

(b) $f'(0)$ does not exist.

Solution

a. The derivative of $f(x) = \dfrac{9(x^2 - 3)}{x^3}$ is

$$f'(x) = \frac{x^3(18x) - (9)(x^2 - 3)(3x^2)}{(x^3)^2} \qquad \text{Differentiate using Quotient Rule.}$$

$$= \frac{9(9 - x^2)}{x^4}. \qquad \text{Simplify.}$$

At the point $(3, 2)$, the value of the derivative is $f'(3) = 0$ [see Figure 4.3(a)].

b. At $x = 0$, the derivative of $f(x) = |x|$ *does not exist* because the following one-sided limits differ [see Figure 4.3(b)].

$$\lim_{x \to 0^-} \frac{f(x) - f(0)}{x - 0} = \lim_{x \to 0^-} \frac{|x|}{x} = -1 \qquad \text{Limit from the left}$$

$$\lim_{x \to 0^+} \frac{f(x) - f(0)}{x - 0} = \lim_{x \to 0^+} \frac{|x|}{x} = 1 \qquad \text{Limit from the right}$$

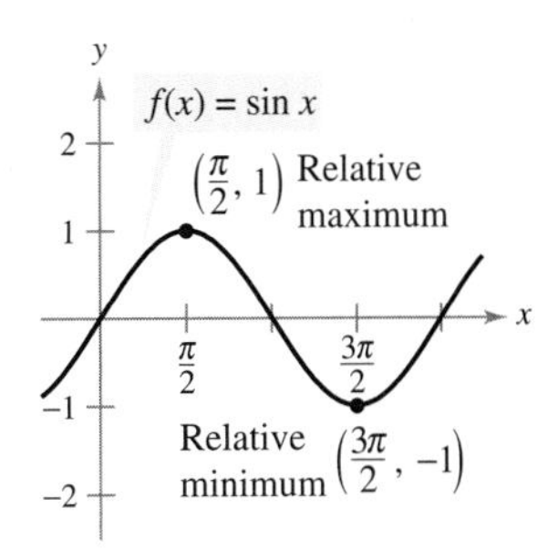

(c) $f'\left(\frac{\pi}{2}\right) = 0;\ f'\left(\frac{3\pi}{2}\right) = 0$
Figure 4.3

c. The derivative of $f(x) = \sin x$ is

$$f'(x) = \cos x.$$

At the point $(\pi/2, 1)$, the value of the derivative is $f'(\pi/2) = \cos(\pi/2) = 0$. At the point $(3\pi/2, -1)$, the value of the derivative is $f'(3\pi/2) = \cos(3\pi/2) = 0$ [see Figure 4.3(c)].

Note in Example 1 that at each relative extremum, the derivative is either zero or does not exist. The x-values at these special points are called **critical numbers.** Figure 4.4 illustrates the two types of critical numbers.

TECHNOLOGY Use a graphing utility to examine the graphs of the following four functions. Only one of the functions has critical numbers. Which is it?

$f(x) = e^x$
$f(x) = \ln x$
$f(x) = \sin x$
$f(x) = \tan x$

Definition of a Critical Number

Let f be defined at c. If $f'(c) = 0$ or if f is not differentiable at c, then c is a **critical number** of f.

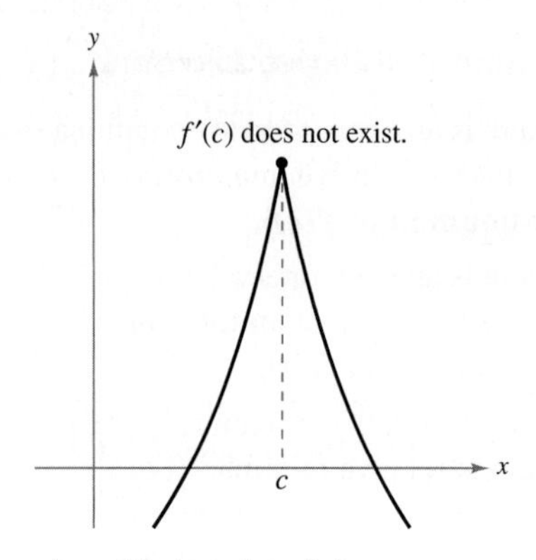

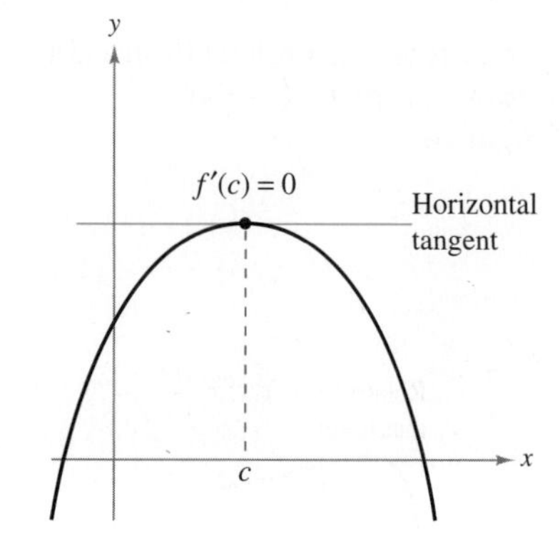

c is a critical number of f.
Figure 4.4

THEOREM 4.2 Relative Extrema Occur Only at Critical Numbers

If f has a relative minimum or relative maximum at $x = c$, then c is a critical number of f.

PIERRE DE FERMAT (1601–1665)

For Fermat, who was trained as a lawyer, mathematics was more of a hobby than a profession. Nevertheless, Fermat made many contributions to analytic geometry, number theory, calculus, and probability. In letters to friends, he wrote of many of the fundamental ideas of calculus, long before Newton or Leibniz. For instance, the theorem at the right is sometimes attributed to Fermat.

Proof

Case 1: If f is *not* differentiable at $x = c$, then, by definition, c is a critical number of f and the theorem is valid.

Case 2: If f is differentiable at $x = c$, then $f'(c)$ must be positive, negative, or 0. Suppose $f'(c)$ is positive. Then

$$f'(c) = \lim_{x \to c} \frac{f(x) - f(c)}{x - c} > 0$$

which implies that there exists an interval (a, b) containing c such that

$$\frac{f(x) - f(c)}{x - c} > 0, \text{ for all } x \neq c \text{ in } (a, b).$$ [See Exercise 74(b), Section 2.2.]

Because this quotient is positive, the signs of the denominator and numerator must agree. This produces the following inequalities for x-values in the interval (a, b).

Left of c: $x < c$ and $f(x) < f(c)$ ⇨ $f(c)$ is not a relative minimum

Right of c: $x > c$ and $f(x) > f(c)$ ⇨ $f(c)$ is not a relative maximum

So, the assumption that $f'(c) > 0$ contradicts the hypothesis that $f(c)$ is a relative extremum. Assuming that $f'(c) < 0$ produces a similar contradiction, you are left with only one possibility—namely, $f'(c) = 0$. So, by definition, c is a critical number of f and the theorem is valid.

Finding Extrema on a Closed Interval

Theorem 4.2 states that the relative extrema of a function can occur *only* at the critical numbers of the function. Knowing this, you can use the following guidelines to find extrema on a closed interval.

Guidelines for Finding Extrema on a Closed Interval

To find the extrema of a continuous function f on a closed interval $[a, b]$, use the following steps.

1. Find the critical numbers of f in (a, b).
2. Evaluate f at each critical number in (a, b).
3. Evaluate f at each endpoint of $[a, b]$.
4. The least of these values is the minimum. The greatest is the maximum.

The next three examples show how to apply these guidelines. Be sure you see that finding the critical numbers of the function is only part of the procedure. Evaluating the function at the critical numbers *and* the endpoints is the other part.

EXAMPLE 2 Finding Extrema on a Closed Interval

Find the extrema of $f(x) = 3x^4 - 4x^3$ on the interval $[-1, 2]$.

Solution Begin by differentiating the function.

$$f(x) = 3x^4 - 4x^3 \quad \text{Write original function.}$$
$$f'(x) = 12x^3 - 12x^2 \quad \text{Differentiate.}$$

To find the critical numbers of f, you must find all x-values for which $f'(x) = 0$ and all x-values for which $f'(x)$ does not exist.

$$f'(x) = 12x^3 - 12x^2 = 0 \quad \text{Set } f'(x) \text{ equal to 0.}$$
$$12x^2(x - 1) = 0 \quad \text{Factor.}$$
$$x = 0, 1 \quad \text{Critical numbers}$$

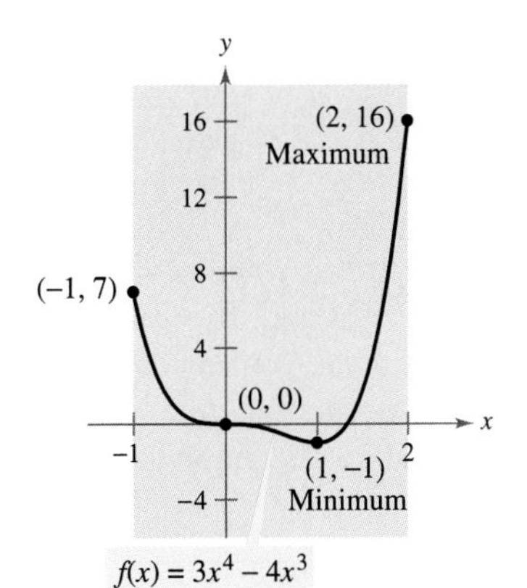

On the closed interval $[-1, 2]$, f has a minimum at $(1, -1)$ and a maximum at $(2, 16)$.
Figure 4.5

Because f' is defined for all x, you can conclude that these are the only critical numbers of f. By evaluating f at these two critical numbers and at the endpoints of $[-1, 2]$, you can determine that the maximum is $f(2) = 16$ and the minimum is $f(1) = -1$, as shown in the table. The graph of f is shown in Figure 4.5.

Left Endpoint	Critical Number	Critical Number	Right Endpoint
$f(-1) = 7$	$f(0) = 0$	$f(1) = -1$ Minimum	$f(2) = 16$ Maximum

In Figure 4.5, note that the critical number $x = 0$ does not yield a relative minimum or a relative maximum. This tells you that the converse of Theorem 4.2 is not true. In other words, *the critical numbers of a function need not produce relative extrema.*

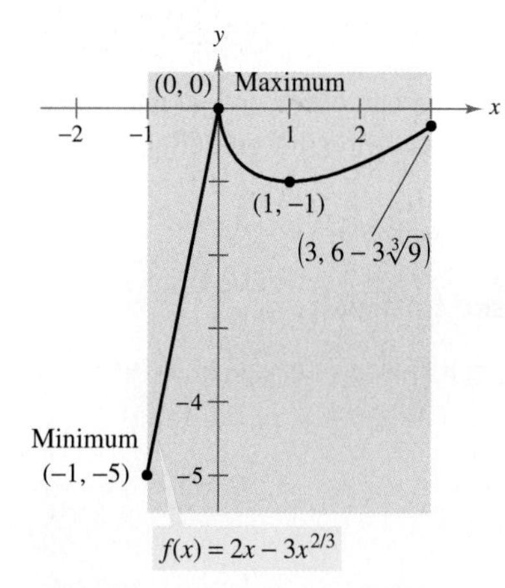

On the closed interval $[-1, 3]$, f has a minimum at $(-1, -5)$ and a maximum at $(0, 0)$.
Figure 4.6

EXAMPLE 3 Finding Extrema on a Closed Interval

Find the extrema of $f(x) = 2x - 3x^{2/3}$ on the interval $[-1, 3]$.

Solution Begin by differentiating the function.

$$f(x) = 2x - 3x^{2/3} \qquad \text{Write original function.}$$

$$f'(x) = 2 - \frac{2}{x^{1/3}} = 2\left(\frac{x^{1/3} - 1}{x^{1/3}}\right) \qquad \text{Differentiate.}$$

From this derivative, you can see that the function has two critical numbers in the interval $[-1, 3]$. The number 1 is a critical number because $f'(1) = 0$, and the number 0 is a critical number because $f'(0)$ does not exist. By evaluating f at these two numbers and at the endpoints of the interval, you can conclude that the minimum is $f(-1) = -5$ and the maximum is $f(0) = 0$, as shown in the table. The graph of f is shown in Figure 4.6.

Left Endpoint	Critical Number	Critical Number	Right Endpoint
$f(-1) = -5$ Minimum	$f(0) = 0$ Maximum	$f(1) = -1$	$f(3) = 6 - 3\sqrt[3]{9} \approx -0.24$

EXAMPLE 4 Finding Extrema on a Closed Interval

Find the extrema of $f(x) = 2 \sin x - \cos 2x$ on the interval $[0, 2\pi]$.

Solution This function is differentiable for all real x, so you can find all critical numbers by differentiating the function and setting $f'(x)$ equal to zero, as shown.

$$f(x) = 2 \sin x - \cos 2x \qquad \text{Write original function.}$$

$$f'(x) = 2 \cos x + 2 \sin 2x = 0 \qquad \text{Set } f'(x) \text{ equal to 0.}$$

$$2 \cos x + 4 \cos x \sin x = 0 \qquad \sin 2x = 2 \cos x \sin x$$

$$2(\cos x)(1 + 2 \sin x) = 0 \qquad \text{Factor.}$$

In the interval $[0, 2\pi]$, the factor $\cos x$ is zero when $x = \pi/2$ and when $x = 3\pi/2$. The factor $(1 + 2 \sin x)$ is zero when $x = 7\pi/6$ and when $x = 11\pi/6$. By evaluating f at these four critical numbers and at the endpoints of the interval, you can conclude that the maximum is $f(\pi/2) = 3$ and the minimum occurs at *two* points, $f(7\pi/6) = -3/2$ and $f(11\pi/6) = -3/2$, as shown in the table. The graph is shown in Figure 4.7.

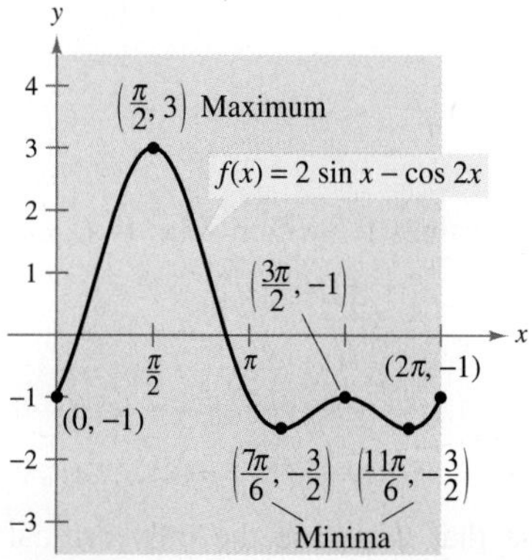

On the closed interval $[0, 2\pi]$, f has minima at $(7\pi/6, -3/2)$ and $(11\pi/6, -3/2)$ and a maximum at $(\pi/2, 3)$.
Figure 4.7

Left Endpoint	Critical Number	Critical Number	Critical Number	Critical Number	Right Endpoint
$f(0) = -1$	$f\left(\frac{\pi}{2}\right) = 3$ Maximum	$f\left(\frac{7\pi}{6}\right) = -\frac{3}{2}$ Minimum	$f\left(\frac{3\pi}{2}\right) = -1$	$f\left(\frac{11\pi}{6}\right) = -\frac{3}{2}$ Minimum	$f(2\pi) = -1$

indicates that in the HM mathSpace® *CD-ROM and the online* Eduspace® *system for this text, you will find an Open Exploration, which further explores this example using the computer algebra systems* Maple, Mathcad, Mathematica, *and* Derive.

Exercises for Section 4.1

See www.CalcChat.com for worked-out solutions to odd-numbered exercises.

In Exercises 1 and 2, decide whether each labeled point is an absolute maximum, an absolute minimum, a relative maximum, a relative minimum, or none of these.

1.

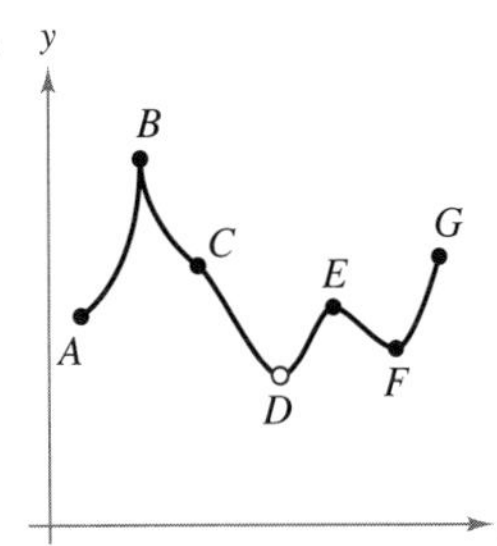

2.

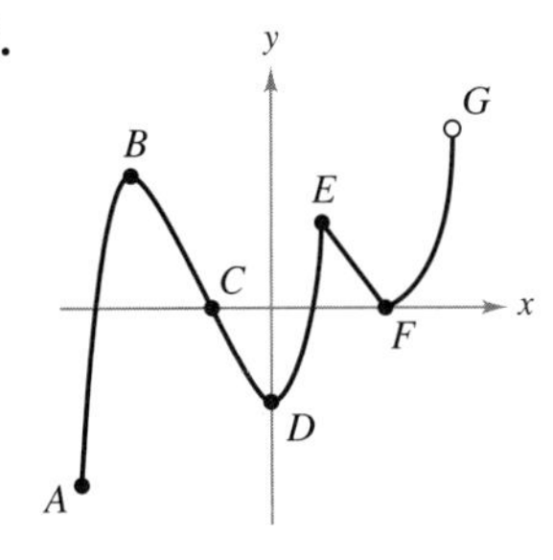

In Exercises 3–8, find the value of the derivative (if it exists) at each indicated extremum.

3. $f(x) = \dfrac{x^2}{x^2 + 4}$

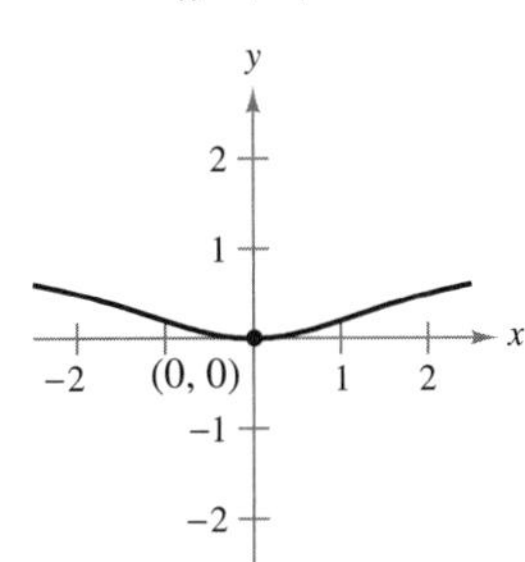

4. $f(x) = \cos \dfrac{\pi x}{2}$

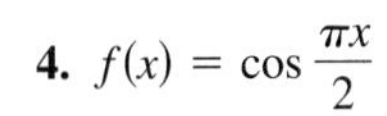

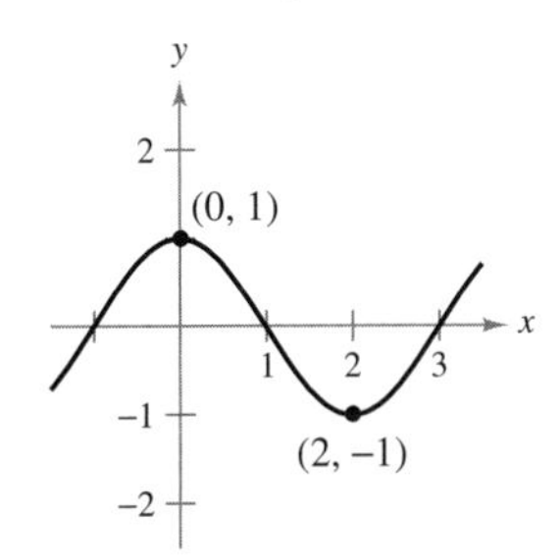

5. $f(x) = x + \dfrac{27}{2x^2}$

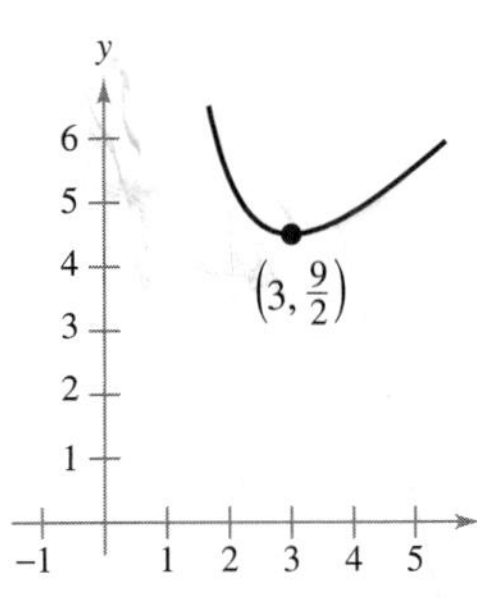

6. $f(x) = -3x\sqrt{x + 1}$

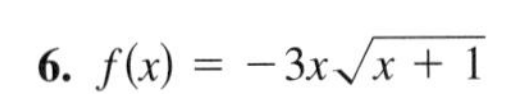

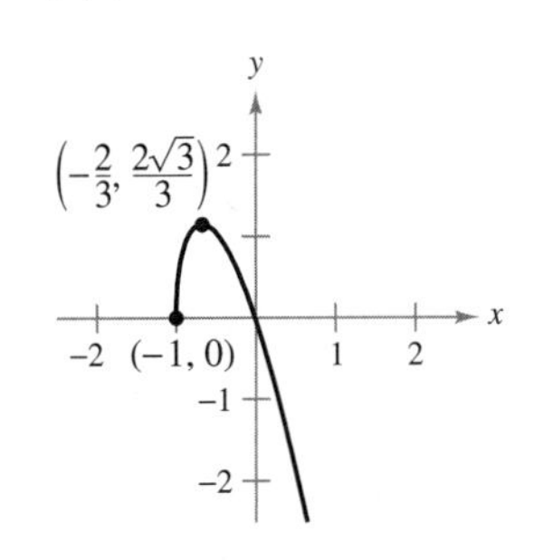

7. $f(x) = (x + 2)^{2/3}$

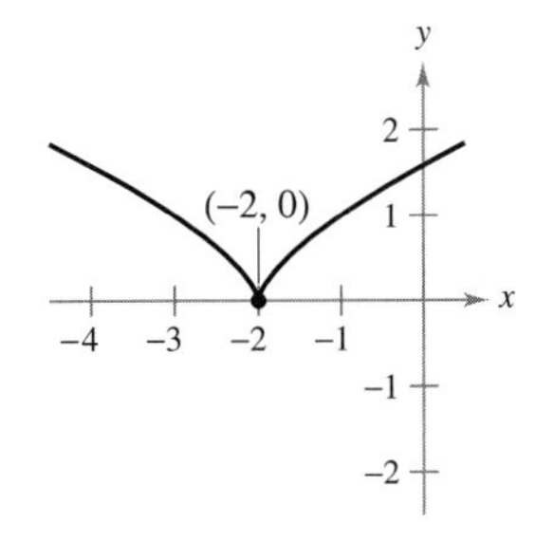

8. $f(x) = 4 - |x|$

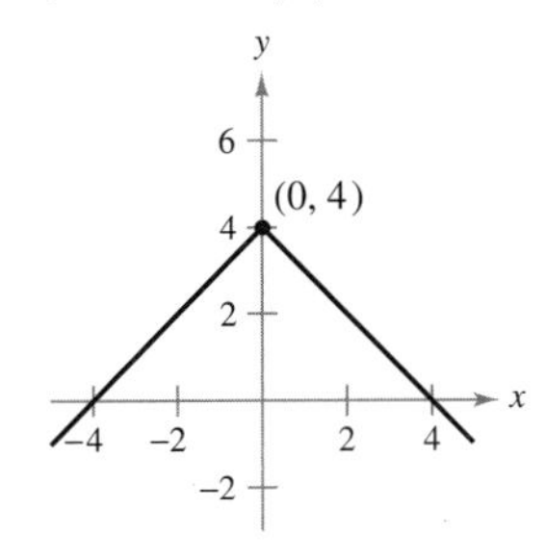

In Exercises 9–12, approximate the critical numbers of the function shown in the graph. Determine whether the function has a relative maximum, a relative minimum, an absolute maximum, an absolute minimum, or none of these at each critical number on the interval shown.

9.

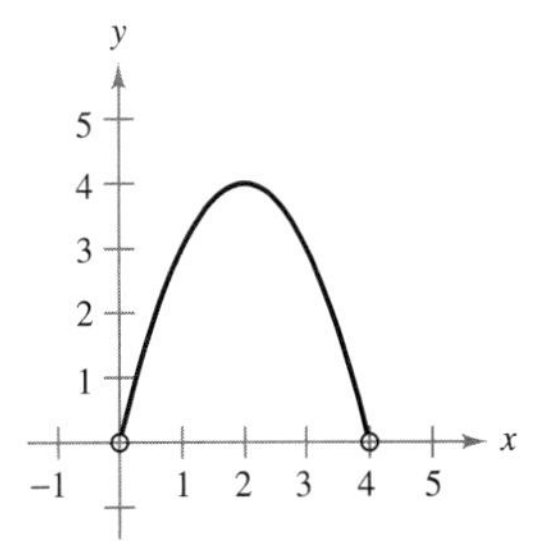

10.

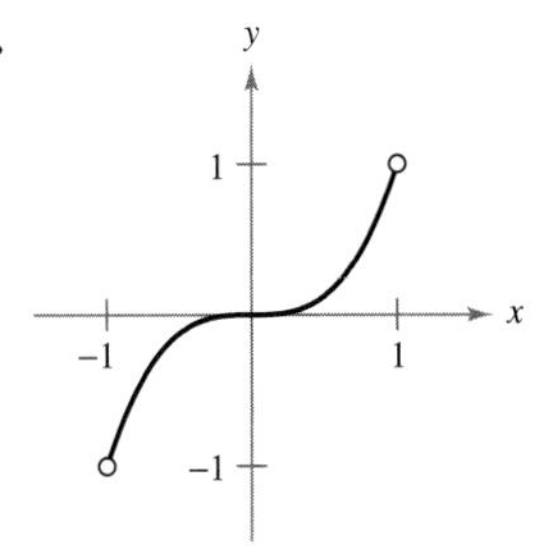

11.

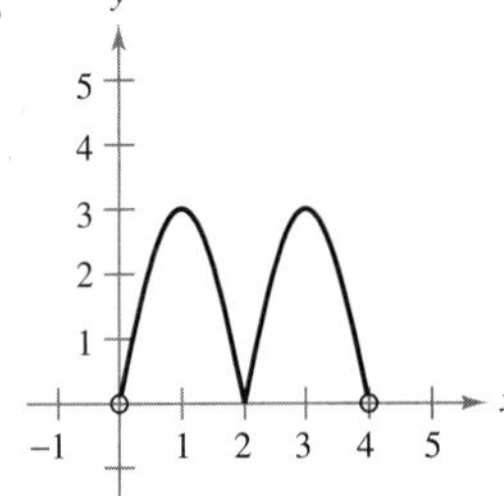

12.

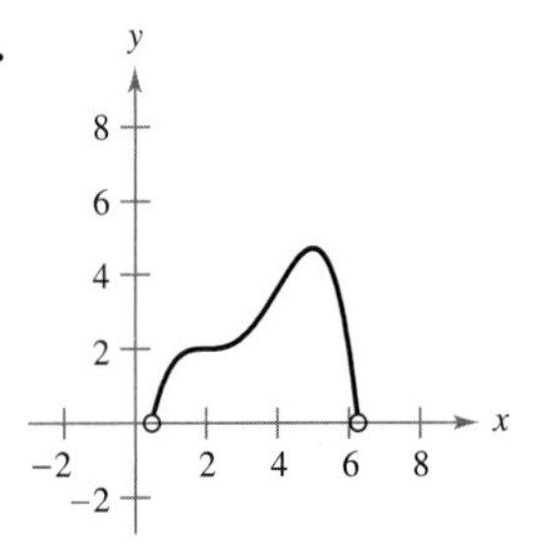

In Exercises 13–20, find any critical numbers of the function.

13. $f(x) = x^2(x - 3)$

14. $g(x) = x^2(x^2 - 4)$

15. $g(t) = t\sqrt{4 - t},\ t < 3$

16. $f(x) = \dfrac{4x}{x^2 + 1}$

17. $h(x) = \sin^2 x + \cos x$
$0 < x < 2\pi$

18. $f(\theta) = 2 \sec \theta + \tan \theta$
$0 < \theta < 2\pi$

19. $f(x) = x^2 \log_2(x^2 + 1)$

20. $g(x) = 4x^2(3^x)$

In Exercises 21–38, locate the absolute extrema of the function on the closed interval.

21. $f(x) = 2(3 - x),\ [-1, 2]$

22. $f(x) = \dfrac{2x + 5}{3},\ [0, 5]$

23. $f(x) = -x^2 + 3x,\ [0, 3]$

24. $f(x) = x^2 + 2x - 4,\ [-1, 1]$

25. $f(x) = x^3 - \dfrac{3}{2}x^2,\ [-1, 2]$

26. $f(x) = x^3 - 12x,\ [0, 4]$

27. $y = 3x^{2/3} - 2x,\ [-1, 1]$

28. $g(x) = \sqrt[3]{x},\ [-1, 1]$

29. $g(t) = \dfrac{t^2}{t^2 + 3},\ [-1, 1]$

30. $y = 3 - |t - 3|,\ [-1, 5]$

31. $h(s) = \dfrac{1}{s - 2},\ [0, 1]$

32. $h(t) = \dfrac{t}{t - 2},\ [3, 5]$

33. $y = e^x \sin x,\ [0, \pi]$

34. $y = x \ln(x + 3),\ [0, 3]$

35. $f(x) = \cos \pi x,\ \left[0, \dfrac{1}{6}\right]$

36. $g(x) = \sec x,\ \left[-\dfrac{\pi}{6}, \dfrac{\pi}{3}\right]$

37. $y = \dfrac{4}{x} + \tan\left(\dfrac{\pi x}{8}\right),\ [1, 2]$

38. $y = x^2 - 2 - \cos x,\ [-1, 3]$

In Exercises 39 and 40, locate the absolute extrema of the function (if any exist) over each interval.

39. $f(x) = 2x - 3$

(a) $[0, 2]$ (b) $[0, 2)$

(c) $(0, 2]$ (d) $(0, 2)$

40. $f(x) = \sqrt{4 - x^2}$

(a) $[-2, 2]$ (b) $[-2, 0)$

(c) $(-2, 2)$ (d) $[1, 2)$

In Exercises 41–46, use a graphing utility to graph the function. Locate the absolute extrema of the function on the given interval.

Function	*Interval*
41. $f(x) = \begin{cases} 2x + 2, & 0 \le x \le 1 \\ 4x^2, & 1 < x \le 3 \end{cases}$	$[0, 3]$
42. $f(x) = \begin{cases} 2 - x^2, & 1 \le x < 3 \\ 2 - 3x, & 3 \le x \le 5 \end{cases}$	$[1, 5]$
43. $f(x) = \dfrac{3}{x - 1}$	$(1, 4]$
44. $f(x) = \dfrac{2}{2 - x}$	$[0, 2)$
45. $f(x) = x^4 - 2x^3 + x + 1$	$[-1, 3]$
46. $f(x) = \sqrt{x} + \cos \dfrac{x}{2}$	$[0, 2\pi]$

In Exercises 47–52, (a) use a computer algebra system to graph the function and approximate any absolute extrema on the indicated interval. (b) Use the utility to find any critical numbers, and use them to find any absolute extrema not located at the endpoints. Compare the results with those in part (a).

Function	*Interval*
47. $f(x) = 3.2x^5 + 5x^3 - 3.5x$	$[0, 1]$
48. $f(x) = \dfrac{4}{3}x\sqrt{3 - x}$	$[0, 3]$
49. $f(x) = (x^2 - 2x)\ln(x + 3)$	$[0, 3]$
50. $f(x) = \sqrt{x + 4}\, e^{x^2/10}$	$[-2, 2]$
51. $f(x) = 2x \arctan(x - 1)$	$[0, 2]$
52. $f(x) = (x - 4) \arcsin \dfrac{x}{4}$	$[-2, 4]$

In Exercises 53–56, use a computer algebra system to find the maximum value of $|f''(x)|$ on the closed interval. (This value is used in the error estimate for the Trapezoidal Rule, as discussed in Section 5.6.)

Function	*Interval*	*Function*	*Interval*
53. $f(x) = \sqrt{1 + x^3}$	$[0, 2]$	54. $f(x) = \dfrac{1}{x^2 + 1}$	$\left[\dfrac{1}{2}, 3\right]$
55. $f(x) = e^{-x^2/2}$	$[0, 1]$	56. $f(x) = x\ln(x + 1)$	$[0, 2]$

In Exercises 57 and 58, use a computer algebra system to find the maximum value of $|f^{4}(x)|$ on the closed interval. (This value is used in the error estimate for Simpson's Rule, as discussed in Section 5.6.)

Function	*Interval*	*Function*	*Interval*
57. $f(x) = (x + 1)^{2/3}$	$[0, 2]$	58. $f(x) = \dfrac{1}{x^2 + 1}$	$[-1, 1]$

59. Explain why the function $f(x) = \tan x$ has a maximum on $[0, \pi/4]$ but not on $[0, \pi]$.

60. ***Writing*** Write a short paragraph explaining why a continuous function on an open interval may not have a maximum or minimum. Illustrate your explanation with a sketch of the graph of such a function.

Writing About Concepts

In Exercises 61 and 62, graph a function on the interval $[-2, 5]$ having the given characteristics.

61. Absolute maximum at $x = -2$
Absolute minimum at $x = 1$
Relative maximum at $x = 3$

62. Relative minimum at $x = -1$
Critical number at $x = 0$, but no extrema
Absolute maximum at $x = 2$
Absolute minimum at $x = 5$

In Exercises 63–66, determine from the graph whether f has a minimum in the open interval (a, b).

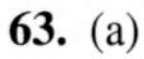

63. (a)

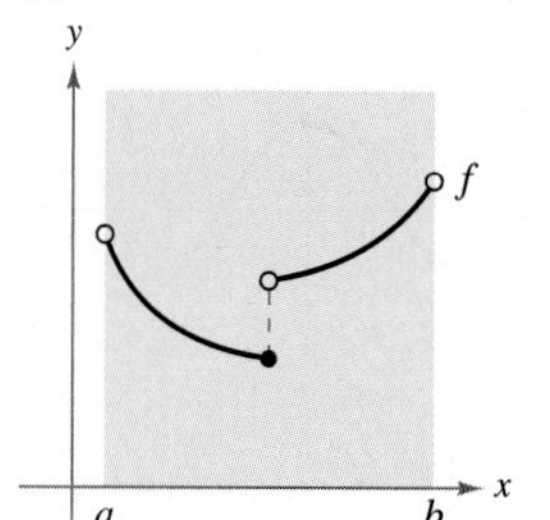

(b)

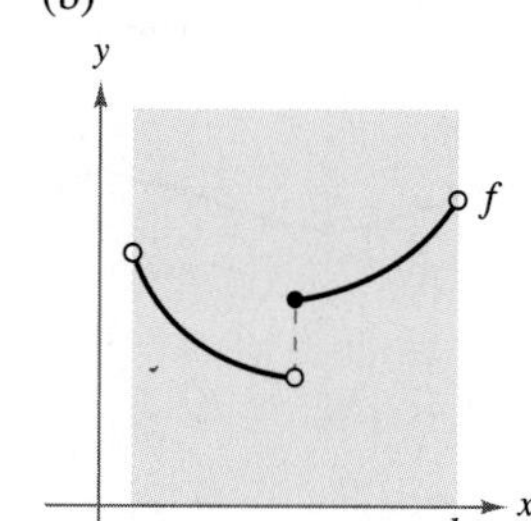

64. (a)

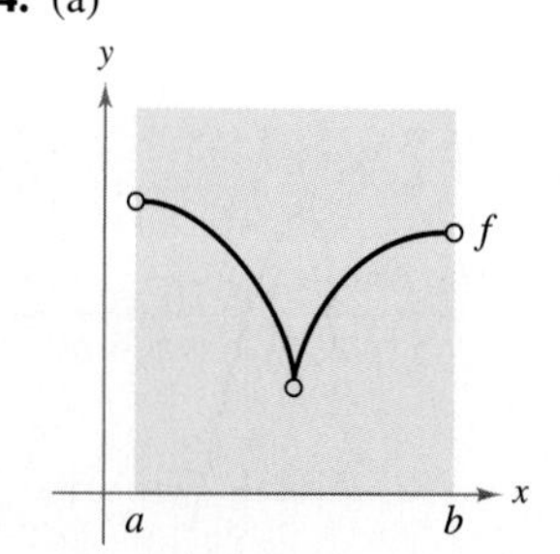

(b)

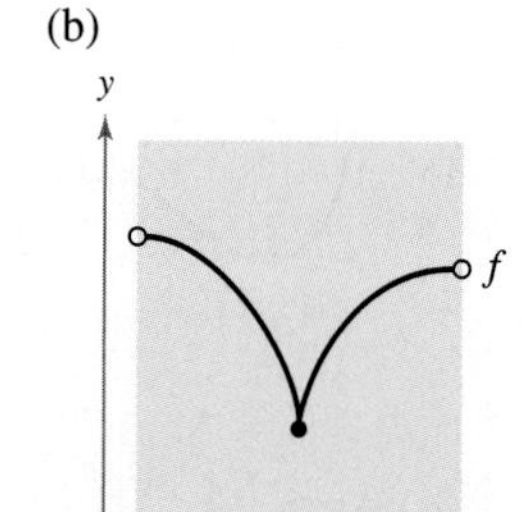

65. (a)

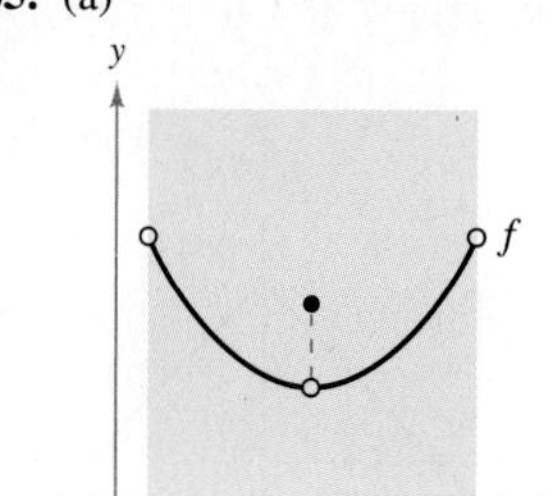

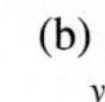

(b)

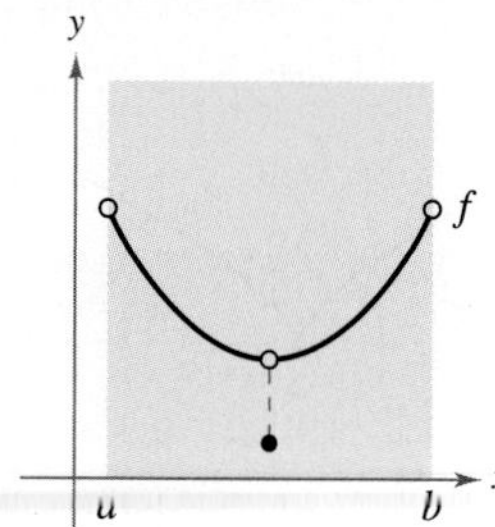

Writing About Concepts (continued)

66. (a) (b)

y y

f f

a b x a b x

67. ***Lawn Sprinkler*** A lawn sprinkler is constructed in such a way that $d\theta/dt$ is constant, where θ ranges between 45° and 135° (see figure). The distance the water travels horizontally is

$$x = \frac{v^2 \sin 2\theta}{32}, \quad 45° \le \theta \le 135°$$

where v is the speed of the water. Find dx/dt and explain why this lawn sprinkler does not water evenly. What part of the lawn receives the most water?

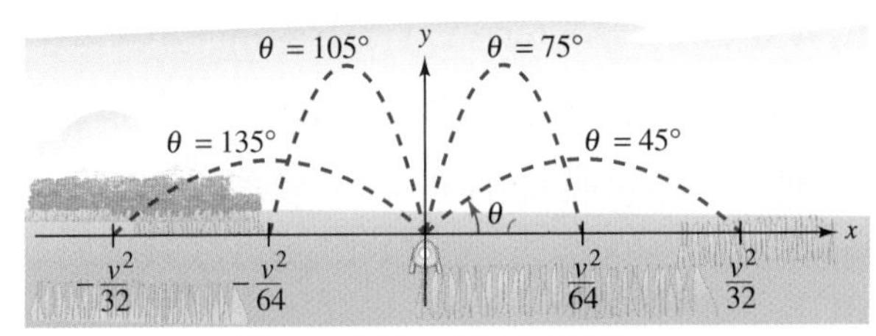

Water sprinkler: $45° \le \theta \le 135°$

FOR FURTHER INFORMATION For more information on the "calculus of lawn sprinklers," see the article "Design of an Oscillating Sprinkler" by Bart Braden in *Mathematics Magazine*. To view this article, go to the website *www.matharticles.com.*

68. ***Honeycomb*** The surface area of a cell in a honeycomb is

$$S = 6hs + \frac{3s^2}{2}\left(\frac{\sqrt{3} - \cos\theta}{\sin\theta}\right)$$

where h and s are positive constants and θ is the angle at which the upper faces meet the altitude of the cell (see figure). Find the angle θ $(\pi/6 \le \theta \le \pi/2)$ that minimizes the surface area S.

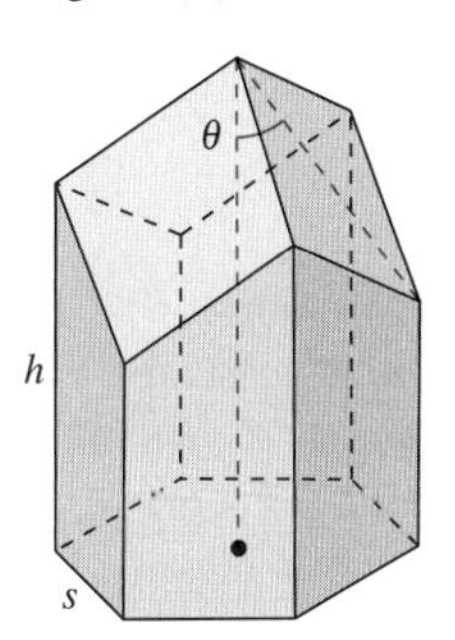

FOR FURTHER INFORMATION For more information on the geometric structure of a honeycomb cell, see the article "The Design of Honeycombs" by Anthony L. Peressini in UMAP Module 502, published by COMAP, Inc., Suite 210, 57 Bedford Street, Lexington, MA.

True or False? **In Exercises 69–72, determine whether the statement is true or false. If it is false, explain why or give an example that shows it is false.**

69. The maximum of a function that is continuous on a closed interval can occur at two different values in the interval.

70. If a function is continuous on a closed interval, then it must have a minimum on the interval.

71. If $x = c$ is a critical number of the function f, then it is also a critical number of the function $g(x) = f(x) + k$, where k is a constant.

72. If $x = c$ is a critical number of the function f, then it is also a critical number of the function $g(x) = f(x - k)$, where k is a constant.

73. Let the function f be differentiable on an interval I containing c. If f has a maximum value at $x = c$, show that $-f$ has a minimum value at $x = c$.

74. Consider the cubic function $f(x) = ax^3 + bx^2 + cx + d$ where $a \ne 0$. Show that f can have zero, one, or two critical numbers and give an example of each case.

75. ***Highway Design*** In order to build a highway, it is necessary to fill a section of a valley where the grades (slopes) of the sides are 9% and 6% (see figure). The top of the filled region will have the shape of a parabolic arc that is tangent to the two slopes at the points A and B. The horizontal distance between the points A and B is 1000 feet.

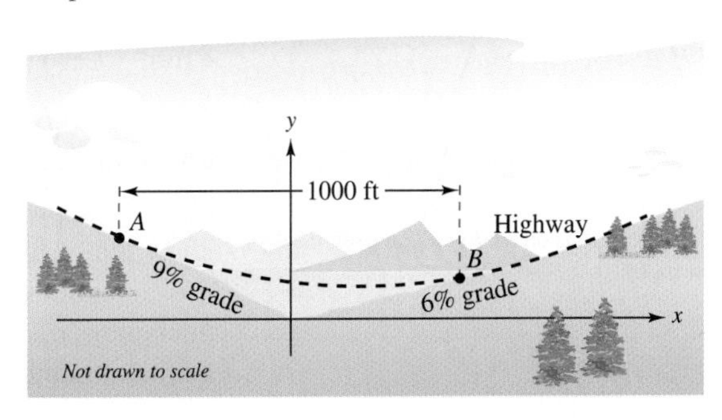

(a) Find a quadratic function $y = ax^2 + bx + c$, $-500 \le x \le 500$, that describes the top of the filled region.

(b) Construct a table giving the depths d of the fill for $x = -500, -400, -300, -200, -100, 0, 100, 200, 300, 400$, and 500.

(c) What will be the lowest point on the completed highway? Will it be directly over the point where the two hillsides come together?

Section 4.2 Rolle's Theorem and the Mean Value Theorem

- **Understand and use Rolle's Theorem.**
- **Understand and use the Mean Value Theorem.**

Rolle's Theorem

> **ROLLE'S THEOREM**
>
> French mathematician Michel Rolle first published the theorem that bears his name in 1691. Before this time, however, Rolle was one of the most vocal critics of calculus, stating that it gave erroneous results and was based on unsound reasoning. Later in life, Rolle came to see the usefulness of calculus.

The Extreme Value Theorem (Section 4.1) states that a continuous function on a closed interval $[a, b]$ must have both a minimum and a maximum on the interval. Both of these values, however, can occur at the endpoints. **Rolle's Theorem,** named after the French mathematician Michel Rolle (1652–1719), gives conditions that guarantee the existence of an extreme value in the *interior* of a closed interval.

EXPLORATION

Extreme Values in a Closed Interval Sketch a rectangular coordinate plane on a piece of paper. Label the points $(1, 3)$ and $(5, 3)$. Using a pencil or pen, draw the graph of a differentiable function f that starts at $(1, 3)$ and ends at $(5, 3)$. Is there at least one point on the graph for which the derivative is zero? Would it be possible to draw the graph so that there *isn't* a point for which the derivative is zero? Explain your reasoning.

THEOREM 4.3 Rolle's Theorem

Let f be continuous on the closed interval $[a, b]$ and differentiable on the open interval (a, b). If

$$f(a) = f(b)$$

then there is at least one number c in (a, b) such that $f'(c) = 0$.

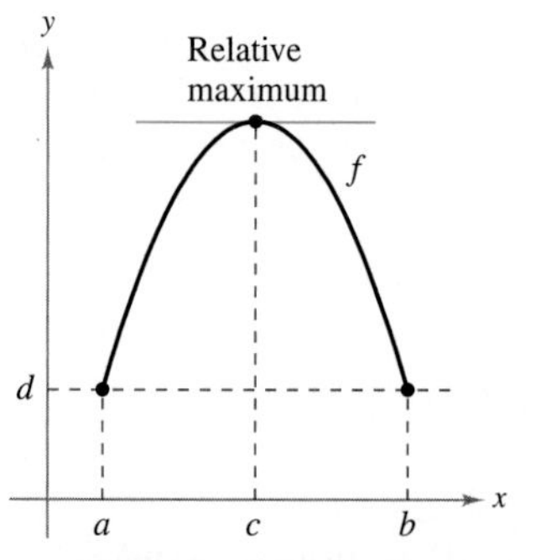

(a) f is continuous on $[a, b]$ and differentiable on (a, b).

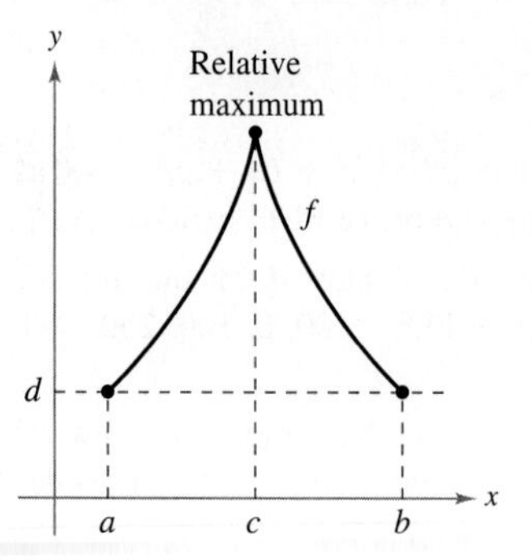

(b) f is continuous on $[a, b]$.
Figure 4.8

Proof Let $f(a) = d = f(b)$.

Case 1: If $f(x) = d$ for all x in $[a, b]$, then f is constant on the interval and, by Theorem 3.2, $f'(x) = 0$ for all x in (a, b).

Case 2: Suppose $f(x) > d$ for some x in (a, b). By the Extreme Value Theorem, you know that f has a maximum at some c in the interval. Moreover, because $f(c) > d$, this maximum does not occur at either endpoint. So, f has a maximum in the *open* interval (a, b). This implies that $f(c)$ is a *relative* maximum and, by Theorem 4.2, c is a critical number of f. Finally, because f is differentiable at c, you can conclude that $f'(c) = 0$.

Case 3: If $f(x) < d$ for some x in (a, b), you can use an argument similar to that in Case 2, but involving the minimum instead of the maximum.

From Rolle's Theorem, you can see that if a function f is continuous on $[a, b]$ and differentiable on (a, b), and if $f(a) = f(b)$, then there must be at least one x-value between a and b at which the graph of f has a horizontal tangent, as shown in Figure 4.8(a). If the differentiability requirement is dropped from Rolle's Theorem, f will still have a critical number in (a, b), but it may not yield a horizontal tangent. Such a case is shown in Figure 4.8(b).

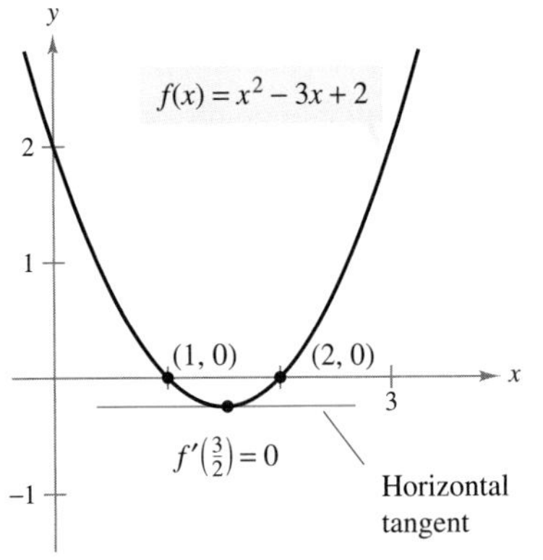

The x-value for which $f'(x) = 0$ is between the two x-intercepts.
Figure 4.9

EXAMPLE 1 Illustrating Rolle's Theorem

Find the two x-intercepts of

$$f(x) = x^2 - 3x + 2$$

and show that $f'(x) = 0$ at some point between the two x-intercepts.

Solution Note that f is differentiable on the entire real number line. Setting $f(x)$ equal to 0 produces

$$x^2 - 3x + 2 = 0 \qquad \text{Set } f(x) \text{ equal to 0.}$$
$$(x - 1)(x - 2) = 0. \qquad \text{Factor.}$$

So, $f(1) = f(2) = 0$, and from Rolle's Theorem you know that there *exists* at least one c in the interval $(1, 2)$ such that $f'(c) = 0$. To *find* such a c, you can solve the equation

$$f'(x) = 2x - 3 = 0 \qquad \text{Set } f'(x) \text{ equal to 0.}$$

and determine that $f'(x) = 0$ when $x = \frac{3}{2}$. Note that the x-value lies in the open interval $(1, 2)$, as shown in Figure 4.9.

Rolle's Theorem states that if f satisfies the conditions of the theorem, there must be *at least* one point between a and b at which the derivative is 0. There may of course be more than one such point, as shown in the next example.

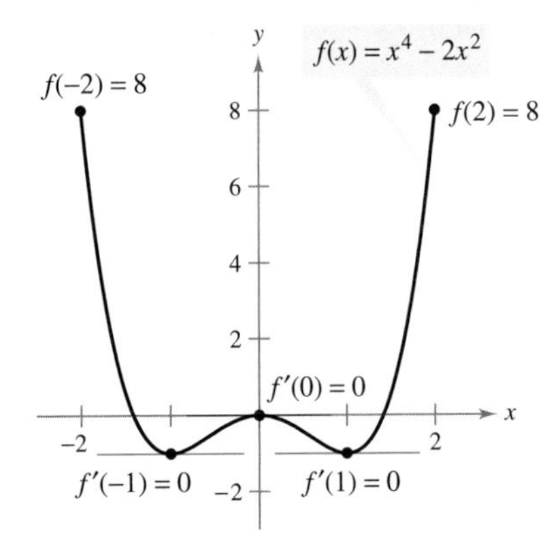

$f'(x) = 0$ for more than one x-value in the interval $(-2, 2)$.
Figure 4.10

EXAMPLE 2 Illustrating Rolle's Theorem

Let $f(x) = x^4 - 2x^2$. Find all values of c in the interval $(-2, 2)$ such that $f'(c) = 0$.

Solution To begin, note that the function satisfies the conditions of Rolle's Theorem. That is, f is continuous on the interval $[-2, 2]$ and differentiable on the interval $(-2, 2)$. Moreover, because $f(-2) = f(2) = 8$, you can conclude that there exists at least one c in $(-2, 2)$ such that $f'(c) = 0$. Setting the derivative equal to 0 produces

$$f'(x) = 4x^3 - 4x = 0 \qquad \text{Set } f'(x) \text{ equal to 0.}$$
$$4x(x - 1)(x + 1) = 0 \qquad \text{Factor.}$$
$$x = 0, 1, -1. \qquad x\text{-values for which } f'(x) = 0$$

So, in the interval $(-2, 2)$, the derivative is zero at three different values of x, as shown in Figure 4.10.

TECHNOLOGY PITFALL A graphing utility can be used to indicate whether the points on the graphs in Examples 1 and 2 are relative minima or relative maxima of the functions. When using a graphing utility, however, you should keep in mind that it can give misleading pictures of graphs. For example, use a graphing utility to graph

$$f(x) = 1 - (x - 1)^2 - \frac{1}{1000(x - 1)^{1/7} + 1}.$$

With most viewing windows, it appears that the function has a maximum of 1 when $x = 1$ (see Figure 4.11). By evaluating the function at $x = 1$, however, you can see that $f(1) = 0$. To determine the behavior of this function near $x = 1$, you need to examine the graph analytically to get the complete picture.

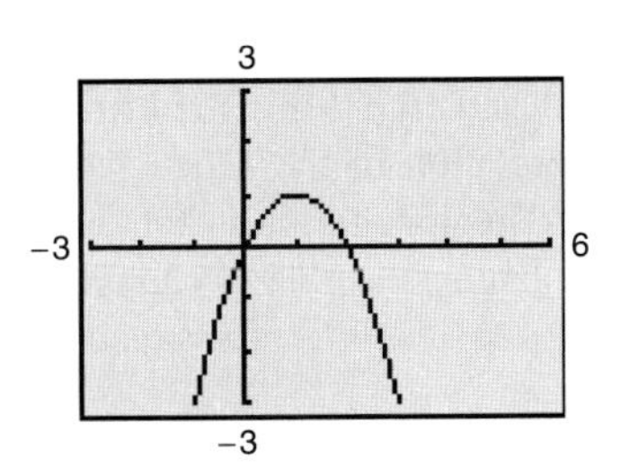

Figure 4.11

The Mean Value Theorem

Rolle's Theorem can be used to prove another theorem—the **Mean Value Theorem.**

> **THEOREM 4.4 The Mean Value Theorem**
>
> If f is continuous on the closed interval $[a, b]$ and differentiable on the open interval (a, b), then there exists a number c in (a, b) such that
>
> $$f'(c) = \frac{f(b) - f(a)}{b - a}.$$

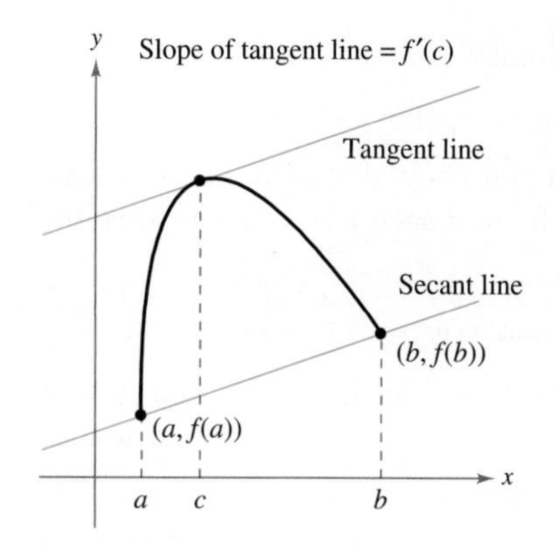

Figure 4.12

Proof Refer to Figure 4.12. The equation of the secant line containing the points $(a, f(a))$ and $(b, f(b))$ is

$$y = \left[\frac{f(b) - f(a)}{b - a}\right](x - a) + f(a).$$

Let $g(x)$ be the difference between $f(x)$ and y. Then

$$\begin{aligned} g(x) &= f(x) - y \\ &= f(x) - \left[\frac{f(b) - f(a)}{b - a}\right](x - a) - f(a). \end{aligned}$$

By evaluating g at a and b, you can see that $g(a) = 0 = g(b)$. Because f is continuous on $[a, b]$, it follows that g is also continuous on $[a, b]$. Furthermore, because f is differentiable, g is also differentiable, and you can apply Rolle's Theorem to the function g. So, there exists a number c in (a, b) such that $g'(c) = 0$, which implies that

$$\begin{aligned} 0 &= g'(c) \\ &= f'(c) - \frac{f(b) - f(a)}{b - a}. \end{aligned}$$

So, there exists a number c in (a, b) such that

$$f'(c) = \frac{f(b) - f(a)}{b - a}.$$

JOSEPH-LOUIS LAGRANGE (1736–1813)

The Mean Value Theorem was first proved by the famous mathematician Joseph-Louis Lagrange. Born in Italy, Lagrange held a position in the court of Frederick the Great in Berlin for 20 years. Afterward, he moved to France, where he met emperor Napoleon Bonaparte, who is quoted as saying, "Lagrange is the lofty pyramid of the mathematical sciences."

NOTE The "mean" in the Mean Value Theorem refers to the mean (or average) rate of change of f in the interval $[a, b]$.

Although the Mean Value Theorem can be used directly in problem solving, it is used more often to prove other theorems. In fact, some people consider this to be the most important theorem in calculus—it is closely related to the Fundamental Theorem of Calculus discussed in Chapter 5. For now, you can get an idea of the versatility of this theorem by looking at the results stated in Exercises 83–91 in this section.

The Mean Value Theorem has implications for both basic interpretations of the derivative. Geometrically, the theorem guarantees the existence of a tangent line that is parallel to the secant line through the points $(a, f(a))$ and $(b, f(b))$, as shown in Figure 4.12. Example 3 illustrates this geometric interpretation of the Mean Value Theorem. In terms of rates of change, the Mean Value Theorem implies that there must be a point in the open interval (a, b) at which the instantaneous rate of change is equal to the average rate of change over the interval $[a, b]$. This is illustrated in Example 4.

EXAMPLE 3 Finding a Tangent Line

Given $f(x) = 5 - (4/x)$, find all values of c in the open interval $(1, 4)$ such that

$$f'(c) = \frac{f(4) - f(1)}{4 - 1}.$$

Solution The slope of the secant line through $(1, f(1))$ and $(4, f(4))$ is

$$\frac{f(4) - f(1)}{4 - 1} = \frac{4 - 1}{4 - 1} = 1.$$

Because f satisfies the conditions of the Mean Value Theorem, there exists at least one number c in $(1, 4)$ such that $f'(c) = 1$. Solving the equation $f'(x) = 1$ yields

$$f'(x) = \frac{4}{x^2} = 1$$

which implies that $x = \pm 2$. So, in the interval $(1, 4)$, you can conclude that $c = 2$, as shown in Figure 4.13.

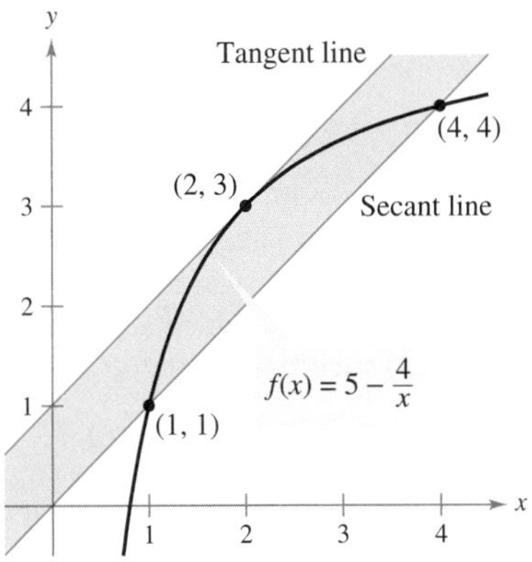

The tangent line at $(2, 3)$ is parallel to the secant line through $(1, 1)$ and $(4, 4)$.
Figure 4.13

EXAMPLE 4 Finding an Instantaneous Rate of Change

Two stationary patrol cars equipped with radar are 5 miles apart on a highway, as shown in Figure 4.14. As a truck passes the first patrol car, its speed is clocked at 55 miles per hour. Four minutes later, when the truck passes the second patrol car, its speed is clocked at 50 miles per hour. Prove that the truck must have exceeded the speed limit (of 55 miles per hour) at some time during the 4 minutes.

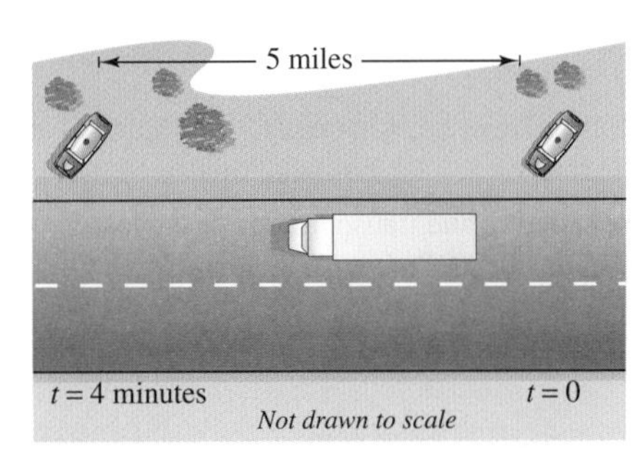

At some time t, the instantaneous velocity is equal to the average velocity over 4 minutes.
Figure 4.14

Solution Let $t = 0$ be the time (in hours) when the truck passes the first patrol car. The time when the truck passes the second patrol car is

$$t = \frac{4}{60} = \frac{1}{15} \text{ hour.}$$

By letting $s(t)$ represent the distance (in miles) traveled by the truck, you have $s(0) = 0$ and $s(\frac{1}{15}) = 5$. So, the average velocity of the truck over the five-mile stretch of highway is

$$\begin{aligned}\text{Average velocity} &= \frac{s(1/15) - s(0)}{(1/15) - 0} \\ &= \frac{5}{1/15} = 75 \text{ miles per hour.}\end{aligned}$$

Assuming that the position function is differentiable, you can apply the Mean Value Theorem to conclude that the truck must have been traveling at a rate of 75 miles per hour sometime during the 4 minutes.

A useful alternative form of the Mean Value Theorem is as follows: If f is continuous on $[a, b]$ and differentiable on (a, b), then there exists a number c in (a, b) such that

$$f(b) = f(a) + (b - a)f'(c). \qquad \text{Alternative form of Mean Value Theorem}$$

NOTE When doing the exercises for this section, keep in mind that polynomial functions, rational functions, and transcendental functions are differentiable at all points in their domains.

Exercises for Section 4.2

See www.CalcChat.com for worked-out solutions to odd-numbered exercises.

In Exercises 1–4, explain why Rolle's Theorem does not apply to the function even though there exist a and b such that $f(a) = f(b)$.

1. $f(x) = 1 - |x - 1|$

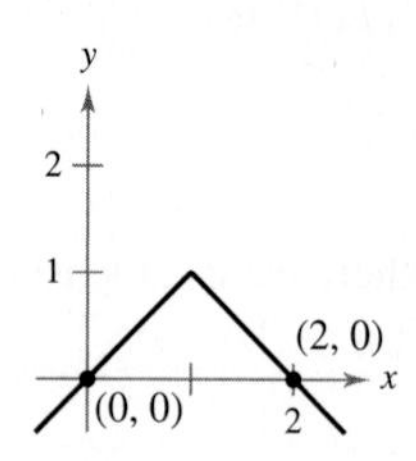

2. $f(x) = \cot \frac{x}{2}$

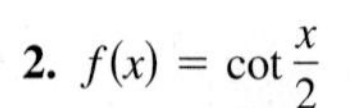

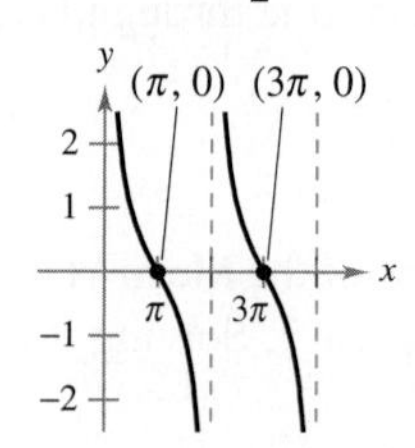

3. $f(x) = \left|\frac{1}{x}\right|$, $[-1, 1]$

4. $f(x) = \sqrt{(2 - x^{2/3})^3}$, $[-1, 1]$

In Exercises 5–8, find the two x-intercepts of the function f and show that $f'(x) = 0$ at some point between the two x-intercepts.

5. $f(x) = x^2 - x - 2$
6. $f(x) = x(x - 3)$
7. $f(x) = x\sqrt{x + 4}$
8. $f(x) = -3x\sqrt{x + 1}$

Rolle's Theorem **In Exercises 9 and 10, the graph of f is shown. Apply Rolle's Theorem and find all values of c such that $f'(c) = 0$ at some point between the labeled intercepts.**

9. $f(x) = x^2 + 3x - 4$

10.

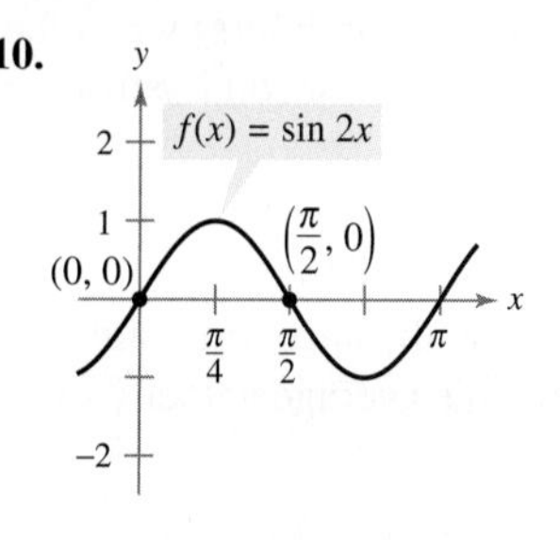

In Exercises 11–26, determine whether Rolle's Theorem can be applied to f on the closed interval $[a, b]$. If Rolle's Theorem can be applied, find all values of c in the open interval (a, b) such that $f'(c) = 0$.

11. $f(x) = x^2 - 2x, [0, 2]$
12. $f(x) = x^2 - 5x + 4, [1, 4]$
13. $f(x) = (x - 1)(x - 2)(x - 3), [1, 3]$
14. $f(x) = (x - 3)(x + 1)^2, [-1, 3]$
15. $f(x) = x^{2/3} - 1, [-8, 8]$
16. $f(x) = 3 - |x - 3|, [0, 6]$
17. $f(x) = \dfrac{x^2 - 2x - 3}{x + 2}, [-1, 3]$
18. $f(x) = \dfrac{x^2 - 1}{x}, [-1, 1]$
19. $f(x) = (x^2 - 2x)e^x, [0, 2]$
20. $f(x) = x - 2\ln x, [1, 3]$
21. $f(x) = \sin x, [0, 2\pi]$
22. $f(x) = \cos x, [0, 2\pi]$
23. $f(x) = \dfrac{6x}{\pi} - 4\sin^2 x, \left[0, \dfrac{\pi}{6}\right]$
24. $f(x) = \cos 2x, \left[-\dfrac{\pi}{12}, \dfrac{\pi}{6}\right]$
25. $f(x) = \tan x, [0, \pi]$
26. $f(x) = \sec x, \left[-\dfrac{\pi}{4}, \dfrac{\pi}{4}\right]$

In Exercises 27–32, use a graphing utility to graph the function on the closed interval $[a, b]$. Determine whether Rolle's Theorem can be applied to f on the interval and, if so, find all values of c in the open interval (a, b) such that $f'(c) = 0$.

27. $f(x) = |x| - 1, \quad [-1, 1]$
28. $f(x) = x - x^{1/3}, \quad [0, 1]$
29. $f(x) = 4x - \tan \pi x, \quad \left[-\frac{1}{4}, \frac{1}{4}\right]$
30. $f(x) = \dfrac{x}{2} - \sin \dfrac{\pi x}{6}, \quad [-1, 0]$
31. $f(x) = 2 + \arcsin(x^2 - 1), \quad [-1, 1]$
32. $f(x) = 2 + (x^2 - 4x)(2^{-x/4}), \quad [0, 4]$

33. ***Vertical Motion*** The height of a ball t seconds after it is thrown upward from a height of 32 feet and with an initial velocity of 48 feet per second is $f(t) = -16t^2 + 48t + 32$.
    (a) Verify that $f(1) = f(2)$.
    (b) According to Rolle's Theorem, what must be the velocity at some time in the interval $(1, 2)$? Find that time.

34. ***Reorder Costs*** The ordering and transportation cost C of components used in a manufacturing process is approximated by

$$C(x) = 10\left(\frac{1}{x} + \frac{x}{x + 3}\right)$$

where C is measured in thousands of dollars and x is the order size in hundreds.
    (a) Verify that $C(3) = C(6)$.
    (b) According to Rolle's Theorem, the rate of change of cost must be 0 for some order size in the interval $(3, 6)$. Find that order size.

In Exercises 35 and 36, copy the graph and sketch the secant line to the graph through the points $(a, f(a))$ and $(b, f(b))$. Then sketch any tangent lines to the graph for each value of c guaranteed by the Mean Value Theorem. To print an enlarged copy of the graph, go to the website *www.mathgraphs.com*.

35.

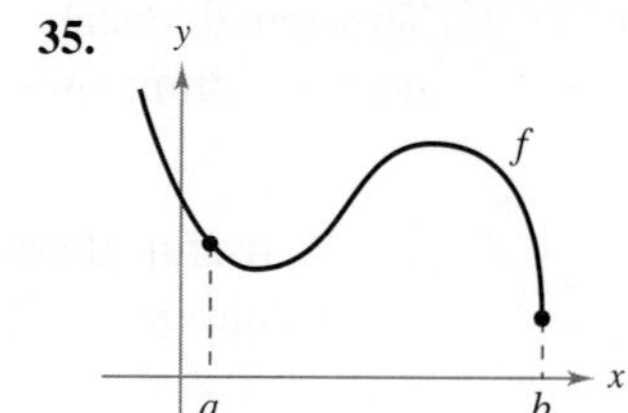

36.

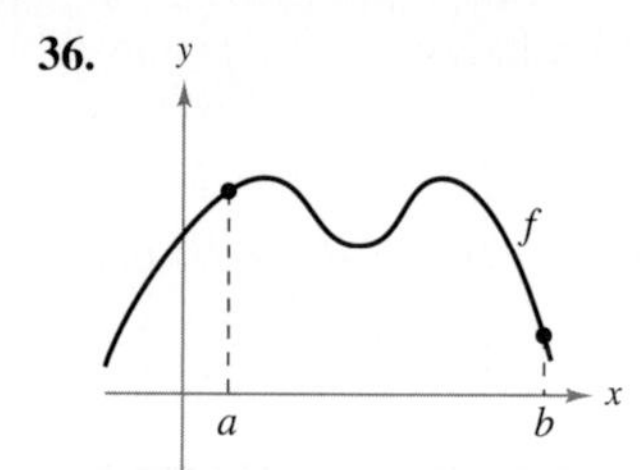

Writing **In Exercises 37–40, explain why the Mean Value Theorem does not apply to the function f on the interval [0, 6].**

37.

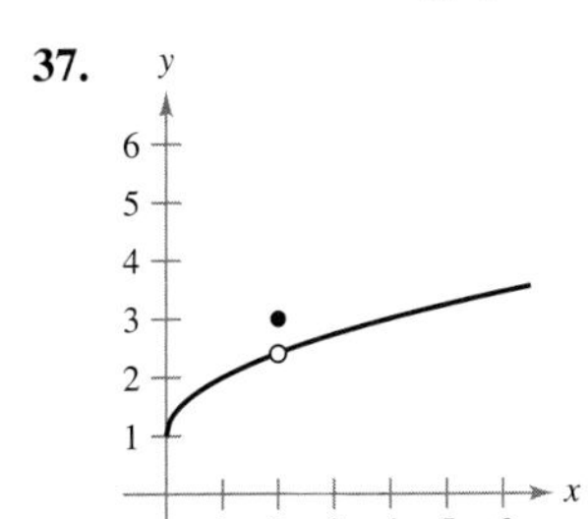

38.

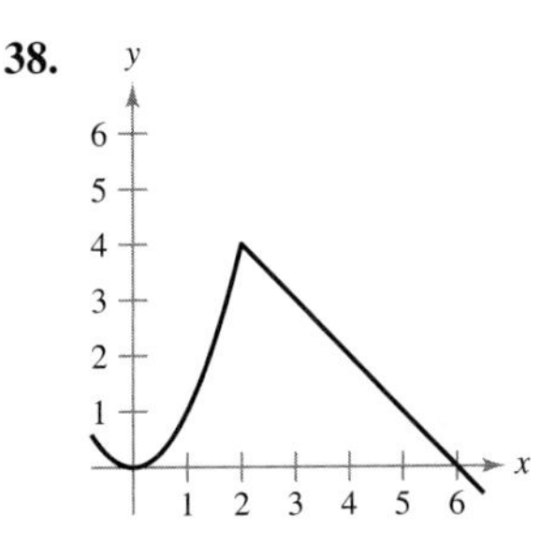

39. $f(x) = \dfrac{1}{x-3}$

40. $f(x) = |x - 3|$

41. ***Mean Value Theorem*** Consider the graph of the function $f(x) = x^2 + 1$. (a) Find the equation of the secant line joining the points $(-1, 2)$ and $(2, 5)$. (b) Use the Mean Value Theorem to determine a point c in the interval $(-1, 2)$ such that the tangent line at c is parallel to the secant line. (c) Find the equation of the tangent line through c. (d) Use a graphing utility to graph f, the secant line, and the tangent line.

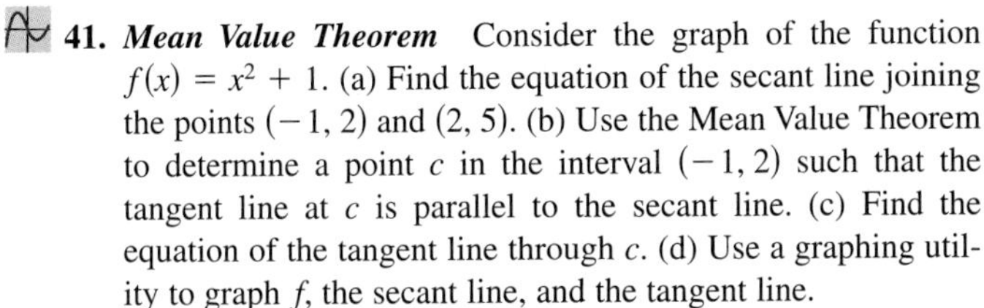

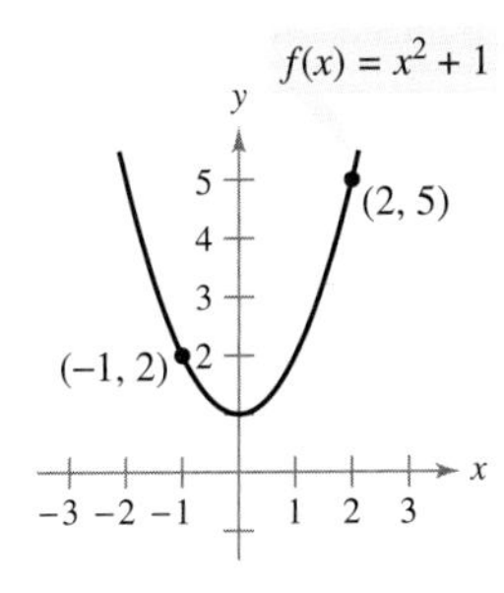

Figure for 41

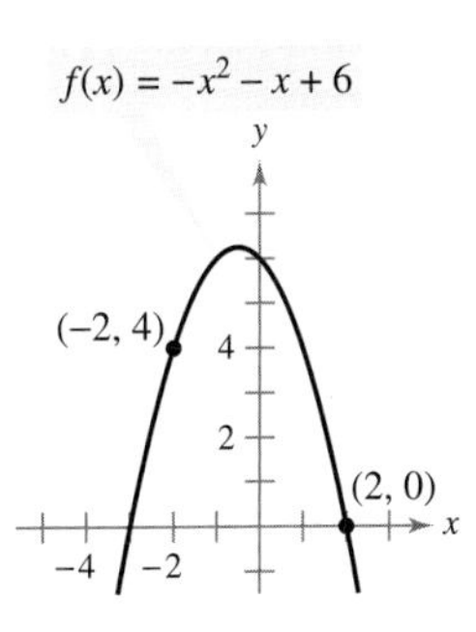

Figure for 42

42. ***Mean Value Theorem*** Consider the graph of the function $f(x) = -x^2 - x + 6$. (a) Find the equation of the secant line joining the points $(-2, 4)$ and $(2, 0)$. (b) Use the Mean Value Theorem to determine a point c in the interval $(-2, 2)$ such that the tangent line at c is parallel to the secant line. (c) Find the equation of the tangent line through c. (d) Use a graphing utility to graph f, the secant line, and the tangent line.

In Exercises 43–52, determine whether the Mean Value Theorem can be applied to f on the closed interval $[a, b]$. If the Mean Value Theorem can be applied, find all values of c in the open interval (a, b) such that $f'(c) = \dfrac{f(b) - f(a)}{b - a}$.

43. $f(x) = x^2,\ [-2, 1]$

44. $f(x) = x(x^2 - x - 2),\ [-1, 1]$

45. $f(x) = x^{2/3},\ [0, 1]$

46. $f(x) = \dfrac{x+1}{x},\ \left[\frac{1}{2}, 2\right]$

47. $f(x) = \sqrt{2 - x},\ [-7, 2]$

48. $f(x) = x^3,\ [0, 1]$

49. $f(x) = \sin x,\ [0, \pi]$

50. $f(x) = 2\sin x + \sin 2x,\ [0, \pi]$

51. $f(x) = x \log_2 x,\ [1, 2]$

52. $f(x) = \arctan(1 - x),\ [0, 1]$

In Exercises 53–58, use a graphing utility to (a) graph the function f on the given interval, (b) find and graph the secant line through points on the graph of f at the endpoints of the given interval, and (c) find and graph any tangent lines to the graph of f that are parallel to the secant line.

53. $f(x) = \dfrac{x}{x+1},\ \left[-\frac{1}{2}, 2\right]$

54. $f(x) = x - 2\sin x,\ [-\pi, \pi]$

55. $f(x) = \sqrt{x},\ [1, 9]$

56. $f(x) = -x^4 + 4x^3 + 8x^2 + 5,\ [0, 5]$

57. $f(x) = 2e^{x/4}\cos\dfrac{\pi x}{4},\ [0, 2]$

58. $f(x) = \ln|\sec \pi x|,\ \left[0, \frac{1}{4}\right]$

Writing About Concepts

59. Let f be continuous on $[a, b]$ and differentiable on (a, b). If there exists c in (a, b) such that $f'(c) = 0$, does it follow that $f(a) = f(b)$? Explain.

60. Let f be continuous on the closed interval $[a, b]$ and differentiable on the open interval (a, b). Also, suppose that $f(a) = f(b)$ and that c is a real number in the interval such that $f'(c) = 0$. Find an interval for the function g over which Rolle's Theorem can be applied, and find the corresponding critical number of g (k is a constant).

(a) $g(x) = f(x) + k$ (b) $g(x) = f(x - k)$

(c) $g(x) = f(kx)$

61. The function

$$f(x) = \begin{cases} 0, & x = 0 \\ 1 - x, & 0 < x \le 1 \end{cases}$$

is differentiable on $(0, 1)$ and satisfies $f(0) = f(1)$. However, its derivative is never zero on $(0, 1)$. Does this contradict Rolle's Theorem? Explain.

62. Can you find a function f such that $f(-2) = -2$, $f(2) = 6$, and $f'(x) < 1$ for all x? Why or why not?

63. ***Speed*** A plane begins its takeoff at 2:00 P.M. on a 2500-mile flight. The plane arrives at its destination at 7:30 P.M. Explain why there are at least two times during the flight when the speed of the plane is 400 miles per hour.

64. ***Temperature*** When an object is removed from a furnace and placed in an environment with a constant temperature of 90°F, its core temperature is 1500°F. Five hours later the core temperature is 390°F. Explain why there must exist a time in the interval when the temperature is decreasing at a rate of 222°F per hour.

65. ***Velocity*** Two bicyclists begin a race at 8:00 A.M. They both finish the race 2 hours and 15 minutes later. Prove that at some time during the race, the bicyclists are traveling at the same velocity.

66. ***Acceleration*** At 9:13 A.M., a sports car is traveling 35 miles per hour. Two minutes later, the car is traveling 85 miles per hour. Prove that at some time during this two-minute interval, the car's acceleration is exactly 1500 miles per hour squared.

67. ***Graphical Reasoning*** The figure shows two parts of the graph of a continuous differentiable function f on $[-10, 4]$. The derivative f' is also continuous. To print an enlarged copy of the graph, go to the website *www.mathgraphs.com.*

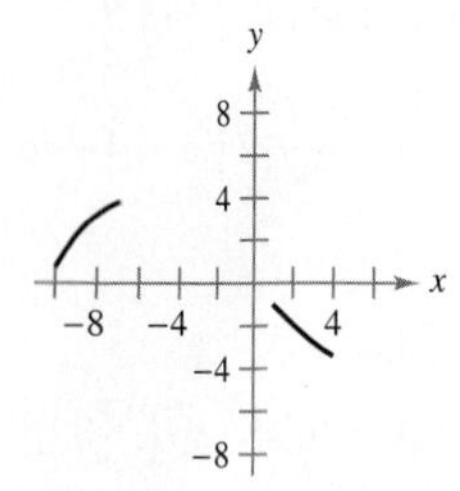

(a) Explain why f must have at least one zero in $[-10, 4]$.

(b) Explain why f' must also have at least one zero in the interval $[-10, 4]$. What are these zeros called?

(c) Make a possible sketch of the function with one zero of f' on the interval $[-10, 4]$.

(d) Make a possible sketch of the function with two zeros of f' on the interval $[-10, 4]$.

(e) Were the conditions of continuity of f and f' necessary to do parts (a) through (d)? Explain.

68. Consider the function $f(x) = 3\cos^2\left(\frac{\pi x}{2}\right)$.

(a) Use a graphing utility to graph f and f'.

(b) Is f a continuous function? Is f' a continuous function?

(c) Does Rolle's Theorem apply on the interval $[-1, 1]$? Does it apply on the interval $[1, 2]$? Explain.

(d) Evaluate, if possible, $\lim_{x\to 3^-} f'(x)$ and $\lim_{x\to 3^+} f'(x)$.

Think About It **In Exercises 69 and 70, sketch the graph of an arbitrary function f that satisfies the given condition but does not satisfy the conditions of the Mean Value Theorem on the interval $[-5, 5]$.**

69. f is continuous on $[-5, 5]$.

70. f is not continuous on $[-5, 5]$.

In Exercises 71 and 72, use the Intermediate Value Theorem and Rolle's Theorem to prove that the equation has exactly one real solution.

71. $x^5 + x^3 + x + 1 = 0$ **72.** $2x - 2 - \cos x = 0$

73. Determine the values of a, b, and c such that the function f satisfies the hypotheses of the Mean Value Theorem on the interval $[0, 3]$.

$$f(x) = \begin{cases} 1, & x = 0 \\ ax + b, & 0 < x \le 1 \\ x^2 + 4x + c, & 1 < x \le 3 \end{cases}$$

74. Determine the values of a, b, c, and d such that the function f satisfies the hypotheses of the Mean Value Theorem on the interval $[-1, 2]$.

$$f(x) = \begin{cases} a, & x = -1 \\ 2, & -1 < x \le 0 \\ bx^2 + c, & 0 < x \le 1 \\ dx + 4, & 1 < x \le 2 \end{cases}$$

Differential Equations **In Exercises 75–78, find a function f that has the derivative $f'(x)$ and whose graph passes through the given point. Explain your reasoning.**

75. $f'(x) = 0$, $(2, 5)$ **76.** $f'(x) = 4$, $(0, 1)$

77. $f'(x) = 2x$, $(1, 0)$ **78.** $f'(x) = 2x + 3$, $(1, 0)$

True or False? **In Exercises 79–82, determine whether the statement is true or false. If it is false, explain why or give an example that shows it is false.**

79. The Mean Value Theorem can be applied to $f(x) = 1/x$ on the interval $[-1, 1]$.

80. If the graph of a function has three x-intercepts, then it must have at least two points at which its tangent line is horizontal.

81. If the graph of a polynomial function has three x-intercepts, then it must have at least two points at which its tangent line is horizontal.

82. If $f'(x) = 0$ for all x in the domain of f, then f is a constant function.

83. Prove that if $a > 0$ and n is any positive integer, then the polynomial function $p(x) = x^{2n+1} + ax + b$ cannot have two real roots.

84. Prove that if $f'(x) = 0$ for all x in an interval (a, b), then f is constant on (a, b).

85. Let $p(x) = Ax^2 + Bx + C$. Prove that for any interval $[a, b]$, the value c guaranteed by the Mean Value Theorem is the midpoint of the interval.

86. (a) Let $f(x) = x^2$ and $g(x) = -x^3 + x^2 + 3x + 2$. Then $f(-1) = g(-1)$ and $f(2) = g(2)$. Show that there is at least one value c in the interval $(-1, 2)$ where the tangent line to f at $(c, f(c))$ is parallel to the tangent line to g at $(c, g(c))$. Identify c.

(b) Let f and g be differentiable functions on $[a, b]$ where $f(a) = g(a)$ and $f(b) = g(b)$. Show that there is at least one value c in the interval (a, b) where the tangent line to f at $(c, f(c))$ is parallel to the tangent line to g at $(c, g(c))$.

87. Prove that if f is differentiable on $(-\infty, \infty)$ and $f'(x) < 1$ for all real numbers, then f has at most one fixed point. A fixed point of a function f is a real number c such that $f(c) = c$.

88. Use the result of Exercise 87 to show that $f(x) = \frac{1}{2}\cos x$ has at most one fixed point.

89. Prove that $|\cos a - \cos b| \le |a - b|$ for all a and b.

90. Prove that $|\sin a - \sin b| \le |a - b|$ for all a and b.

91. Let $0 < a < b$. Use the Mean Value Theorem to show that

$$\sqrt{b} - \sqrt{a} < \frac{b - a}{2\sqrt{a}}.$$

Section 4.3 Increasing and Decreasing Functions and the First Derivative Test

- **Determine intervals on which a function is increasing or decreasing.**
- **Apply the First Derivative Test to find relative extrema of a function.**

Increasing and Decreasing Functions

In this section you will learn how derivatives can be used to *classify* relative extrema as either relative minima or relative maxima. First, it is important to define increasing and decreasing functions.

Definitions of Increasing and Decreasing Functions

A function f is **increasing** on an interval if for any two numbers x_1 and x_2 in the interval, $x_1 < x_2$ implies $f(x_1) < f(x_2)$.

A function f is **decreasing** on an interval if for any two numbers x_1 and x_2 in the interval, $x_1 < x_2$ implies $f(x_1) > f(x_2)$.

y
$x = a$
$x = b$
f
Decreasing
Increasing
Constant
x
$f'(x) < 0$
$f'(x) = 0$
$f'(x) > 0$

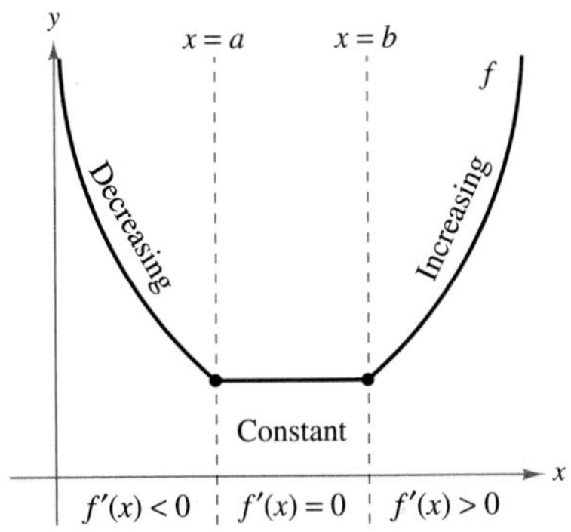

The derivative is related to the slope of a function.
Figure 4.15

A function is increasing if, *as x moves to the right*, its graph moves up, and is decreasing if its graph moves down. For example, the function in Figure 4.15 is decreasing on the interval $(-\infty, a)$, is constant on the interval (a, b), and is increasing on the interval (b, ∞). As shown in Theorem 4.5 below, a positive derivative implies that the function is increasing; a negative derivative implies that the function is decreasing; and a zero derivative on an entire interval implies that the function is constant on that interval.

THEOREM 4.5 Test for Increasing and Decreasing Functions

Let f be a function that is continuous on the closed interval $[a, b]$ and differentiable on the open interval (a, b).

1. If $f'(x) > 0$ for all x in (a, b), then f is increasing on $[a, b]$.
2. If $f'(x) < 0$ for all x in (a, b), then f is decreasing on $[a, b]$.
3. If $f'(x) = 0$ for all x in (a, b), then f is constant on $[a, b]$.

Proof To prove the first case, assume that $f'(x) > 0$ for all x in the interval (a, b) and let $x_1 < x_2$ be any two points in the interval. By the Mean Value Theorem, you know that there exists a number c such that $x_1 < c < x_2$, and

$$f'(c) = \frac{f(x_2) - f(x_1)}{x_2 - x_1}.$$

Because $f'(c) > 0$ and $x_2 - x_1 > 0$, you know that

$$f(x_2) - f(x_1) > 0$$

which implies that $f(x_1) < f(x_2)$. So, f is increasing on the interval. The second case has a similar proof (see Exercise 107), and the third case was given as Exercise 84 in Section 4.2.

NOTE The conclusions in the first two cases of Theorem 4.5 are valid even if $f'(x) = 0$ at a finite number of x-values in (a, b).

EXAMPLE 1 Intervals on Which f Is Increasing or Decreasing

Find the open intervals on which $f(x) = x^3 - \frac{3}{2}x^2$ is increasing or decreasing.

Solution Note that f is differentiable on the entire real number line. To determine the critical numbers of f, set $f'(x)$ equal to zero.

$$f(x) = x^3 - \frac{3}{2}x^2 \qquad \text{Write original function.}$$

$$f'(x) = 3x^2 - 3x = 0 \qquad \text{Differentiate and set } f'(x) \text{ equal to 0.}$$

$$3(x)(x-1) = 0 \qquad \text{Factor.}$$

$$x = 0, 1 \qquad \text{Critical numbers}$$

Figure 4.16

Because there are no points for which f' does not exist, you can conclude that $x = 0$ and $x = 1$ are the only critical numbers. The table summarizes the testing of the three intervals determined by these two critical numbers.

Interval	$-\infty < x < 0$	$0 < x < 1$	$1 < x < \infty$
Test Value	$x = -1$	$x = \frac{1}{2}$	$x = 2$
Sign of $f'(x)$	$f'(-1) = 6 > 0$	$f'(\frac{1}{2}) = -\frac{3}{4} < 0$	$f'(2) = 6 > 0$
Conclusion	Increasing	Decreasing	Increasing

So, f is increasing on the intervals $(-\infty, 0)$ and $(1, \infty)$ and decreasing on the interval $(0, 1)$, as shown in Figure 4.16.

Example 1 gives you one example of how to find intervals on which a function is increasing or decreasing. The guidelines below summarize the steps followed in the example.

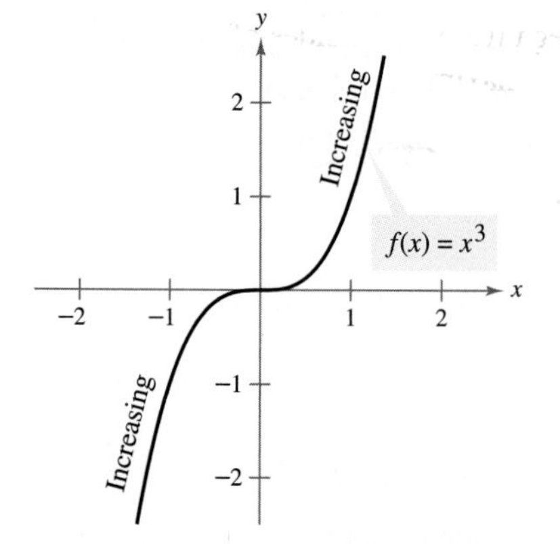

(a) Strictly monotonic function

Guidelines for Finding Intervals on Which a Function Is Increasing or Decreasing

Let f be continuous on the interval (a, b). To find the open intervals on which f is increasing or decreasing, use the following steps.

1. Locate the critical numbers of f in (a, b), and use these numbers to determine test intervals.
2. Determine the sign of $f'(x)$ at one test value in each of the intervals.
3. Use Theorem 4.5 to determine whether f is increasing or decreasing on each interval.

These guidelines are also valid if the interval (a, b) is replaced by an interval of the form $(-\infty, b)$, (a, ∞), or $(-\infty, \infty)$.

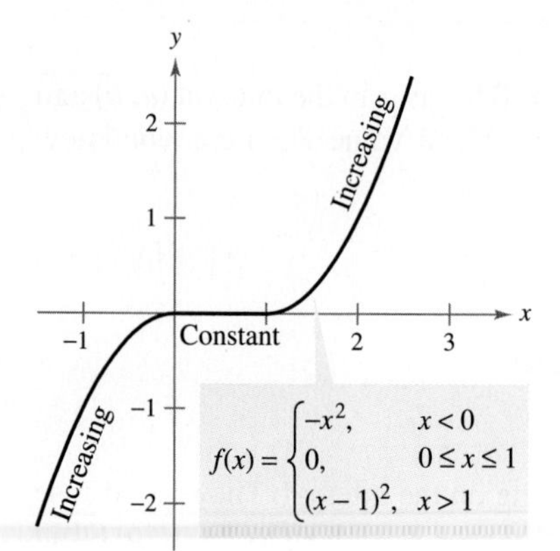

(b) Not strictly monotonic

Figure 4.17

A function is **strictly monotonic** on an interval if it is either increasing on the entire interval or decreasing on the entire interval. For instance, the function $f(x) = x^3$ is strictly monotonic on the entire real number line because it is increasing on the entire real number line, as shown in Figure 4.17(a). The function shown in Figure 4.17(b) is not strictly monotonic on the entire real number line because it is constant on the interval $[0, 1]$.

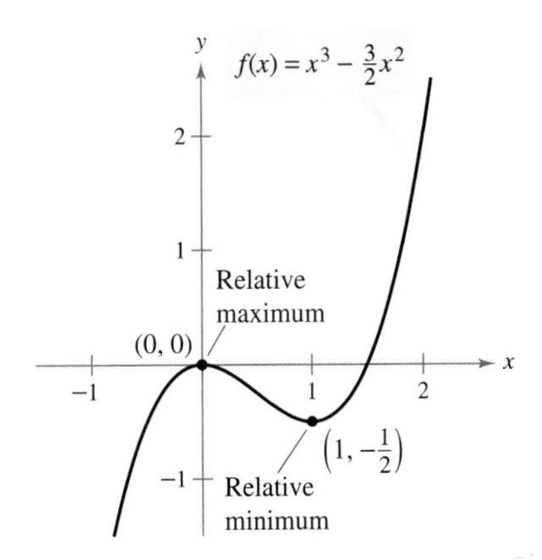

Relative extrema of f
Figure 4.18

The First Derivative Test

After you have determined the intervals on which a function is increasing or decreasing, it is not difficult to locate the relative extrema of the function. For instance, in Figure 4.18 (from Example 1), the function

$$f(x) = x^3 - \frac{3}{2}x^2$$

has a relative maximum at the point $(0, 0)$ because f is increasing immediately to the left of $x = 0$ and decreasing immediately to the right of $x = 0$. Similarly, f has a relative minimum at the point $\left(1, -\frac{1}{2}\right)$ because f is decreasing immediately to the left of $x = 1$ and increasing immediately to the right of $x = 1$. The following theorem, called the First Derivative Test, makes this more explicit.

THEOREM 4.6 The First Derivative Test

Let c be a critical number of a function f that is continuous on an open interval I containing c. If f is differentiable on the interval, except possibly at c, then $f(c)$ can be classified as follows.

1. If $f'(x)$ changes from negative to positive at c, then f has a *relative minimum* at $(c, f(c))$.
2. If $f'(x)$ changes from positive to negative at c, then f has a *relative maximum* at $(c, f(c))$.
3. If $f'(x)$ is positive on both sides of c or negative on both sides of c, then $f(c)$ is neither a relative minimum nor a relative maximum.

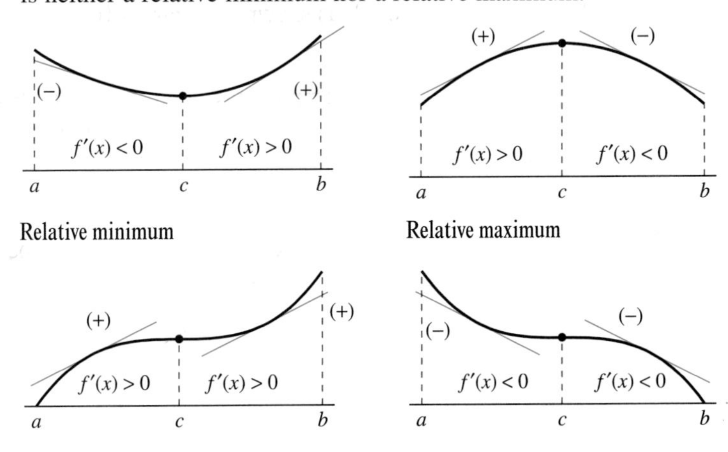

Proof Assume that $f'(x)$ changes from negative to positive at c. Then there exist a and b in I such that

$$f'(x) < 0 \text{ for all } x \text{ in } (a, c)$$

and

$$f'(x) > 0 \text{ for all } x \text{ in } (c, b).$$

By Theorem 4.5, f is decreasing on $[a, c]$ and increasing on $[c, b]$. So, $f(c)$ is a minimum of f on the open interval (a, b) and, consequently, a relative minimum of f. This proves the first case of the theorem. The second case can be proved in a similar way (see Exercise 108).

EXAMPLE 2 Applying the First Derivative Test

Find the relative extrema of the function $f(x) = \frac{1}{2}x - \sin x$ in the interval $(0, 2\pi)$.

Solution Note that f is continuous on the interval $(0, 2\pi)$. To determine the critical numbers of f in this interval, set $f'(x)$ equal to 0.

$$f'(x) = \frac{1}{2} - \cos x = 0 \qquad \text{Set } f'(x) \text{ equal to 0.}$$

$$\cos x = \frac{1}{2}$$

$$x = \frac{\pi}{3}, \frac{5\pi}{3} \qquad \text{Critical numbers}$$

Because there are no points for which f' does not exist, you can conclude that $x = \pi/3$ and $x = 5\pi/3$ are the only critical numbers. The table summarizes the testing of the three intervals determined by these two critical numbers.

Interval	$0 < x < \frac{\pi}{3}$	$\frac{\pi}{3} < x < \frac{5\pi}{3}$	$\frac{5\pi}{3} < x < 2\pi$
Test Value	$x = \frac{\pi}{4}$	$x = \pi$	$x = \frac{7\pi}{4}$
Sign of $f'(x)$	$f'\left(\frac{\pi}{4}\right) < 0$	$f'(\pi) > 0$	$f'\left(\frac{7\pi}{4}\right) < 0$
Conclusion	Decreasing	Increasing	Decreasing

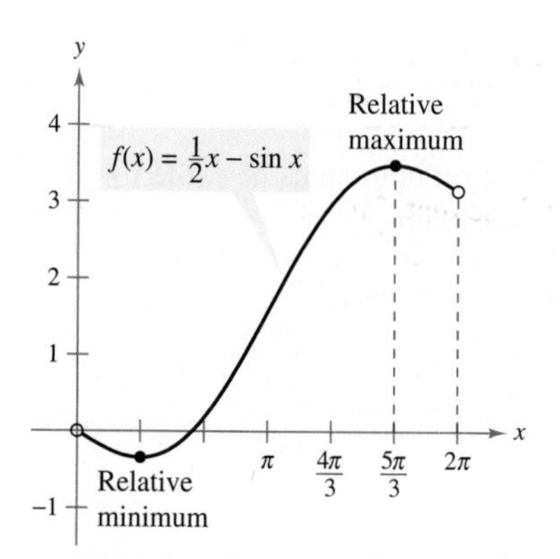

A relative minimum occurs where f changes from decreasing to increasing, and a relative maximum occurs where f changes from increasing to decreasing.
Figure 4.19

By applying the First Derivative Test, you can conclude that f has a relative minimum at the point where

$$x = \frac{\pi}{3} \qquad x\text{-value where relative minimum occurs}$$

and a relative maximum at the point where

$$x = \frac{5\pi}{3} \qquad x\text{-value where relative maximum occurs}$$

as shown in Figure 4.19.

EXPLORATION

Comparing Graphical and Analytic Approaches From Section 4.2, you know that, *by itself*, a graphing utility can give misleading information about the relative extrema of a graph. *Used in conjunction with an analytic approach*, however, a graphing utility can provide a good way to reinforce your conclusions. Try using a graphing utility to graph the function in Example 2. Then use the *zoom* and *trace* features to estimate the relative extrema. How close are your graphical approximations?

Note that in Examples 1 and 2 the given functions are differentiable on the entire real number line. For such functions, the only critical numbers are those for which $f'(x) = 0$. Example 3 concerns a function that has two types of critical numbers—those for which $f'(x) = 0$ and those for which f is not differentiable.

EXAMPLE 3 Applying the First Derivative Test

Find the relative extrema of

$$f(x) = (x^2 - 4)^{2/3}.$$

Solution Begin by noting that f is continuous on the entire real number line. The derivative of f

$$f'(x) = \frac{2}{3}(x^2 - 4)^{-1/3}(2x) \qquad \text{General Power Rule}$$

$$= \frac{4x}{3(x^2 - 4)^{1/3}} \qquad \text{Simplify.}$$

is 0 when $x = 0$ and does not exist when $x = \pm 2$. So, the critical numbers are $x = -2$, $x = 0$, and $x = 2$. The table summarizes the testing of the four intervals determined by these three critical numbers.

Interval	$-\infty < x < -2$	$-2 < x < 0$	$0 < x < 2$	$2 < x < \infty$
Test Value	$x = -3$	$x = -1$	$x = 1$	$x = 3$
Sign of $f'(x)$	$f'(-3) < 0$	$f'(-1) > 0$	$f'(1) < 0$	$f'(3) > 0$
Conclusion	Decreasing	Increasing	Decreasing	Increasing

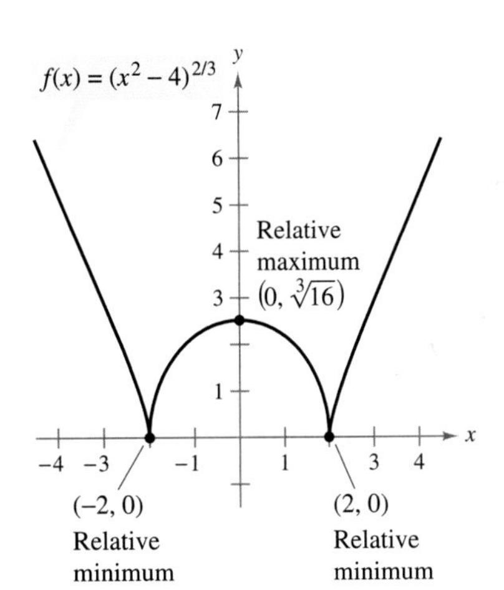

You can apply the First Derivative Test to find relative extrema.
Figure 4.20

By applying the First Derivative Test, you can conclude that f has a relative minimum at the point $(-2, 0)$, a relative maximum at the point $(0, \sqrt[3]{16})$, and another relative minimum at the point $(2, 0)$, as shown in Figure 4.20.

TECHNOLOGY PITFALL When using a graphing utility to graph a function involving radicals or rational exponents, be sure you understand the way the utility evaluates radical expressions. For instance, even though

$$f(x) = (x^2 - 4)^{2/3}$$

and

$$g(x) = [(x^2 - 4)^2]^{1/3}$$

are the same algebraically, some graphing utilities distinguish between these two functions. Which of the graphs shown in Figure 4.21 is incorrect? Why did the graphing utility produce an incorrect graph?

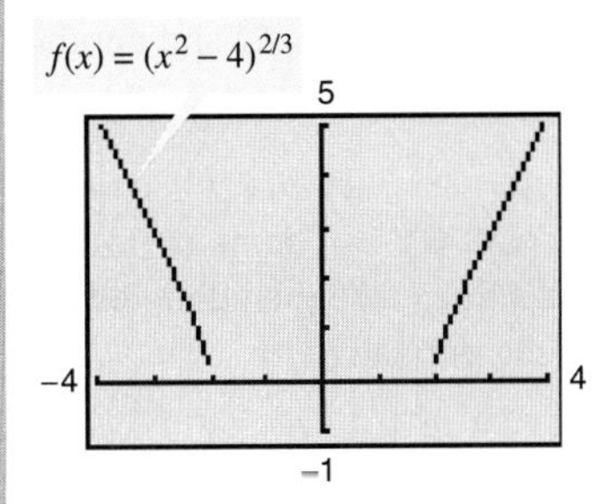

$g(x) = [(x^2 - 4)^2]^{1/3}$

Which graph is incorrect?
Figure 4.21

When using the First Derivative Test, be sure to consider the domain of the function. For instance, in the next example, the function

$$f(x) = \frac{x^4 + 1}{x^2}$$

is not defined when $x = 0$. This x-value must be used with the critical numbers to determine the test intervals.

EXAMPLE 4 Applying the First Derivative Test

Find the relative extrema of $f(x) = \dfrac{x^4 + 1}{x^2}$.

Solution

$f(x) = x^2 + x^{-2}$ — Rewrite original function.

$f'(x) = 2x - 2x^{-3}$ — Differentiate.

$= 2x - \dfrac{2}{x^3}$ — Rewrite with positive exponent.

$= \dfrac{2(x^4 - 1)}{x^3}$ — Simplify.

$= \dfrac{2(x^2 + 1)(x - 1)(x + 1)}{x^3}$ — Factor.

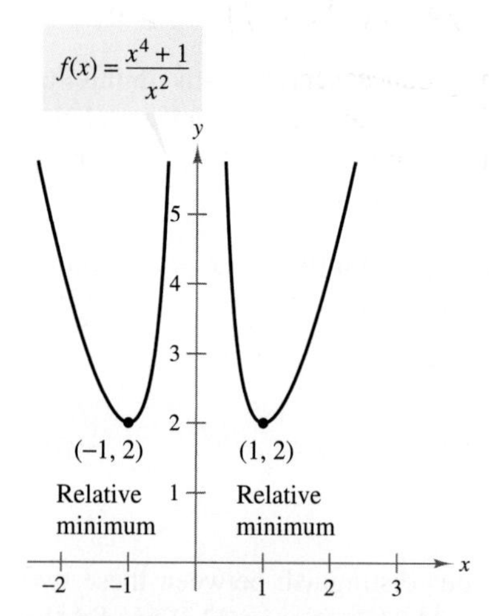

x-values that are not in the domain of f, as well as critical numbers, determine test intervals for f'.
Figure 4.22

So, $f'(x)$ is zero at $x = \pm 1$. Moreover, because $x = 0$ is not in the domain of f, you should use this x-value along with the critical numbers to determine the test intervals.

$x = \pm 1$ — Critical numbers, $f'(\pm 1) = 0$

$x = 0$ — 0 is not in the domain of f.

The table summarizes the testing of the four intervals determined by these three x-values.

Interval	$-\infty < x < -1$	$-1 < x < 0$	$0 < x < 1$	$1 < x < \infty$
Test Value	$x = -2$	$x = -\frac{1}{2}$	$x = \frac{1}{2}$	$x = 2$
Sign of $f'(x)$	$f'(-2) < 0$	$f'(-\frac{1}{2}) > 0$	$f'(\frac{1}{2}) < 0$	$f'(2) > 0$
Conclusion	Decreasing	Increasing	Decreasing	Increasing

By applying the First Derivative Test, you can conclude that f has one relative minimum at the point $(-1, 2)$ and another at the point $(1, 2)$, as shown in Figure 4.22.

TECHNOLOGY The most difficult step in applying the First Derivative Test is finding the values for which the derivative is equal to 0. For instance, the values of x for which the derivative of

$$f(x) = \frac{x^4 + 1}{x^2 + 1}$$

is equal to zero are $x = 0$ and $x = \pm\sqrt{\sqrt{2} - 1}$. If you have access to technology that can perform symbolic differentiation and solve equations, use it to apply the First Derivative Test to this function.

EXAMPLE 5 The Path of a Projectile

Michael Kevin Daly/Corbis

If a projectile is propelled from ground level and air resistance is neglected, the object will travel farthest with an initial angle of 45°. If, however, the projectile is propelled from a point above ground level, the angle that yields a maximum horizontal distance is not 45° (see Example 5).

Neglecting air resistance, the path of a projectile that is propelled at an angle θ is

$$y = \frac{g \sec^2 \theta}{2v_0^2}x^2 + (\tan \theta)x + h, \quad 0 \le \theta \le \frac{\pi}{2}$$

where y is the height, x is the horizontal distance, g is the acceleration due to gravity, v_0 is the initial velocity, and h is the initial height. (This equation is derived in Section 12.3.) Let $g = -32$ feet per second per second, $v_0 = 24$ feet per second, and $h = 9$ feet. What value of θ will produce a maximum horizontal distance?

Solution To find the distance the projectile travels, let $y = 0$ and use the Quadratic Formula to solve for x.

$$\frac{g \sec^2 \theta}{2v_0^2}x^2 + (\tan \theta)x + h = 0$$

$$\frac{-32 \sec^2 \theta}{2(24^2)}x^2 + (\tan \theta)x + 9 = 0$$

$$-\frac{\sec^2 \theta}{36}x^2 + (\tan \theta)x + 9 = 0$$

$$x = \frac{-\tan \theta \pm \sqrt{\tan^2 \theta + \sec^2 \theta}}{-\sec^2 \theta/18}$$

$$x = 18 \cos \theta\left(\sin \theta + \sqrt{\sin^2 \theta + 1}\right), \quad x \ge 0$$

At this point, you need to find the value of θ that produces a maximum value of x. Applying the First Derivative Test by hand would be very tedious. Using technology to solve the equation $dx/d\theta = 0$, however, eliminates most of the messy computations. The result is that the maximum value of x occurs when

$$\theta \approx 0.61548 \text{ radian, or } 35.3°.$$

This conclusion is reinforced by sketching the path of the projectile for different values of θ, as shown in Figure 4.23. Of the three paths shown, note that the distance traveled is greatest for $\theta = 35°$.

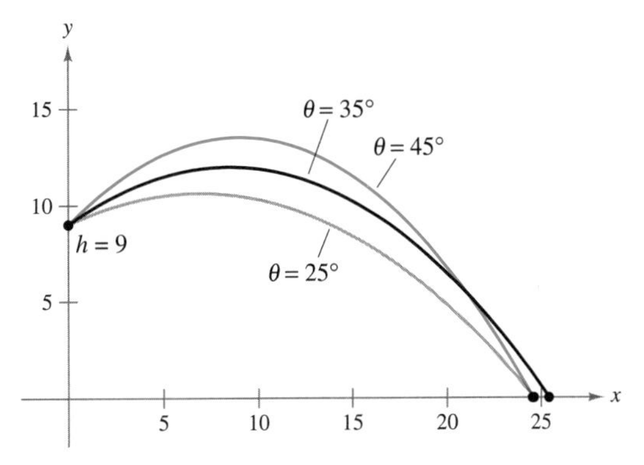

The path of a projectile with initial angle θ
Figure 4.23

NOTE A computer simulation of this example is given in the *HM mathSpace®* CD-ROM and the online *Eduspace®* system for this text. Using that simulation, you can experimentally discover that the maximum value of x occurs when $\theta \approx 35.3°$.

Exercises for Section 4.3

See www.CalcChat.com for worked-out solutions to odd-numbered exercises.

In Exercises 1 and 2, use the graph of f to find (a) the largest open interval on which f is increasing, and (b) the largest open interval on which f is decreasing.

1. 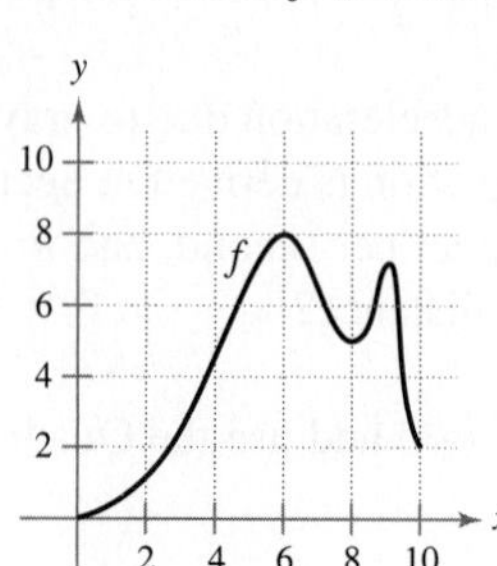

2. 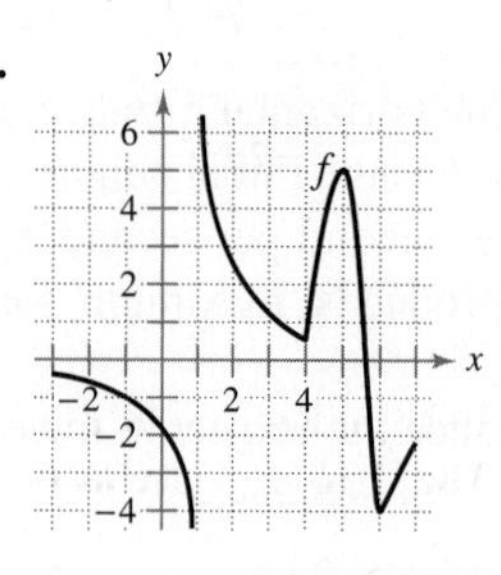

In Exercises 3–16, identify the open intervals on which the function is increasing or decreasing.

3. $f(x) = x^2 - 6x + 8$

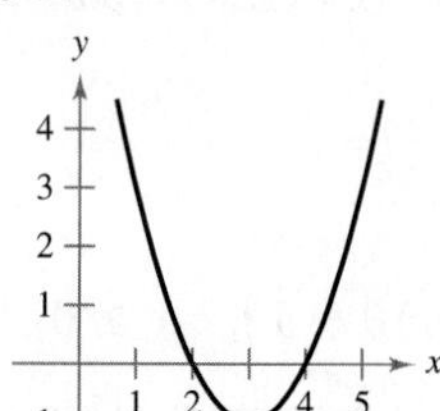

4. $y = -(x + 1)^2$

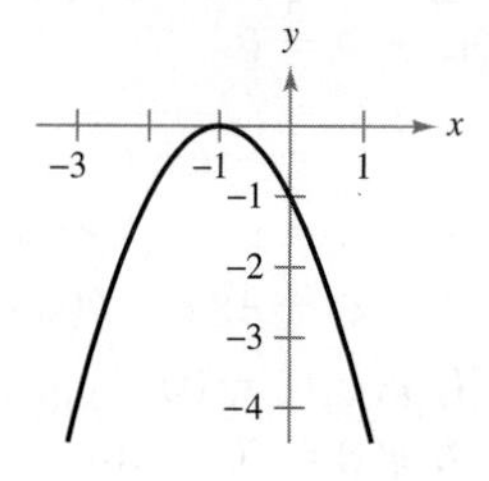

5. $y = \dfrac{x^3}{4} - 3x$

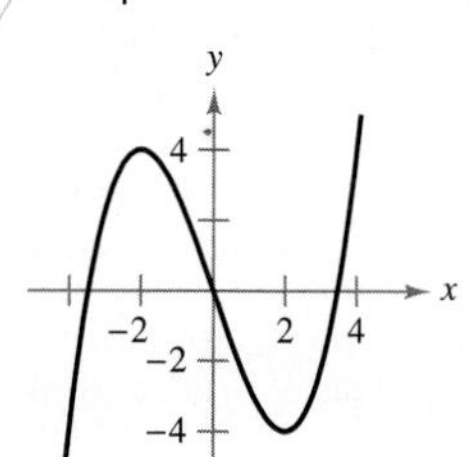

6. $f(x) = x^4 - 2x^2$

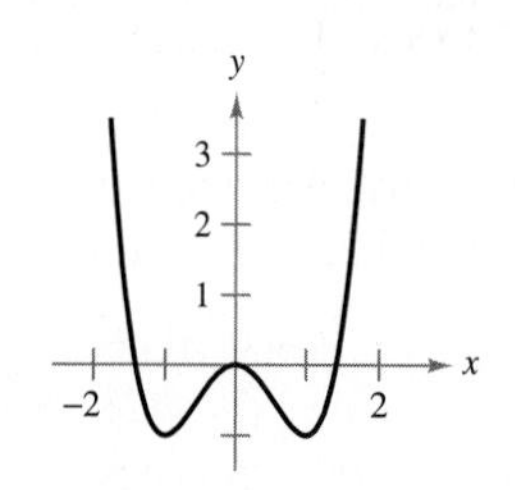

7. $f(x) = \sin x + 2, 0 < x < 2\pi$

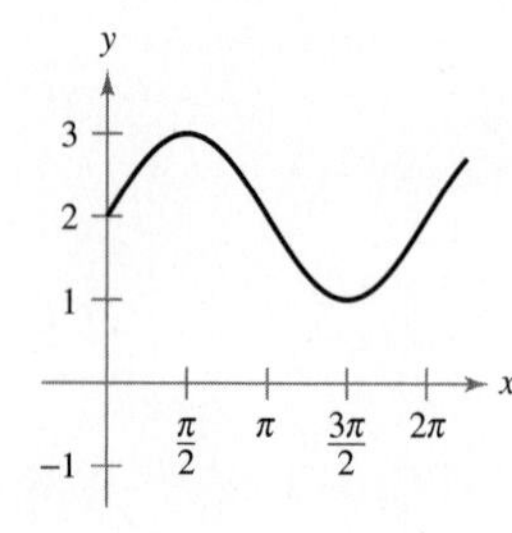

8. $h(x) = \cos \dfrac{x}{2}, 0 < x < 2\pi$

9. $f(x) = \dfrac{1}{x^2}$

10. $y = \dfrac{x^2}{x + 1}$

11. $g(x) = x^2 - 2x - 8$

12. $h(x) = 27x - x^3$

13. $y = x\sqrt{16 - x^2}$

14. $y = x + \dfrac{4}{x}$

15. $y = x - 2 \cos x, \quad 0 < x < 2\pi$

16. $f(x) = \cos^2 x - \cos x, \quad 0 < x < 2\pi$

In Exercises 17–46, find the critical numbers of f (if any). Find the open intervals on which the function is increasing or decreasing and locate all relative extrema. Use a graphing utility to confirm your results.

17. $f(x) = x^2 - 6x$
18. $f(x) = x^2 + 8x + 10$
19. $f(x) = -2x^2 + 4x + 3$
20. $f(x) = -(x^2 + 8x + 12)$
21. $f(x) = 2x^3 + 3x^2 - 12x$
22. $f(x) = x^3 - 6x^2 + 15$
23. $f(x) = x^2(3 - x)$
24. $f(x) = (x + 2)^2(x - 1)$
25. $f(x) = \dfrac{x^5 - 5x}{5}$
26. $f(x) = x^4 - 32x + 4$
27. $f(x) = x^{1/3} + 1$
28. $f(x) = x^{2/3} - 4$
29. $f(x) = (x - 1)^{2/3}$
30. $f(x) = (x - 1)^{1/3}$
31. $f(x) = 5 - |x - 5|$
32. $f(x) = |x + 3| - 1$
33. $f(x) = x + \dfrac{1}{x}$
34. $f(x) = \dfrac{x}{x + 1}$
35. $f(x) = \dfrac{x^2}{x^2 - 9}$
36. $f(x) = \dfrac{x + 3}{x^2}$
37. $f(x) = \dfrac{x^2 - 2x + 1}{x + 1}$
38. $f(x) = \dfrac{x^2 - 3x - 4}{x - 2}$
39. $f(x) = (3 - x)e^{x-3}$
40. $f(x) = (x - 1)e^x$
41. $f(x) = 4(x - \arcsin x)$
42. $f(x) = x \arctan x$
43. $g(x) = (x)3^{-x}$
44. $f(x) = 2^{x^2-3}$
45. $f(x) = x - \log_4 x$
46. $f(x) = \dfrac{x^3}{3} - \ln x$

In Exercises 47–54, consider the function on the interval $(0, 2\pi)$. For each function, (a) find the open interval(s) on which the function is increasing or decreasing, (b) apply the First Derivative Test to identify all relative extrema, and (c) use a graphing utility to confirm your results.

47. $f(x) = \dfrac{x}{2} + \cos x$
48. $f(x) = \sin x \cos x$
49. $f(x) = \sin x + \cos x$
50. $f(x) = x + 2 \sin x$
51. $f(x) = \cos^2(2x)$
52. $f(x) = \sqrt{3} \sin x + \cos x$
53. $f(x) = \sin^2 x + \sin x$
54. $f(x) = \dfrac{\sin x}{1 + \cos^2 x}$

In Exercises 55–60, (a) use a computer algebra system to differentiate the function, (b) sketch the graphs of f and f' on the same set of coordinate axes over the indicated interval, (c) find the critical numbers of f in the open interval, and (d) find the interval(s) on which f' is positive and the interval(s) on which it is negative. Compare the behavior of f and the sign of f'.

55. $f(x) = 2x\sqrt{9 - x^2}, [-3, 3]$
56. $f(x) = 10\left(5 - \sqrt{x^2 - 3x + 16}\right), [0, 5]$

57. $f(t) = t^2 \sin t, [0, 2\pi]$

58. $f(x) = \dfrac{x}{2} + \cos \dfrac{x}{2}, [0, 4\pi]$

59. $f(x) = \frac{1}{2}(x^2 - \ln x), (0, 3]$

60. $f(x) = (4 - x^2)e^x, [0, 2]$

In Exercises 61 and 62, use symmetry, extrema, and zeros to sketch the graph of f. How do the functions f and g differ? Explain.

61. $f(x) = \dfrac{x^5 - 4x^3 + 3x}{x^2 - 1}, \quad g(x) = x(x^2 - 3)$

62. $f(t) = \cos^2 t - \sin^2 t, \quad g(t) = 1 - 2\sin^2 t, \quad (-2, 2)$

Think About It **In Exercises 63–68, the graph of f is shown in the figure. Sketch a graph of the derivative of f. To print an enlarged copy of the graph, go to the website *www.mathgraphs.com*.**

63.

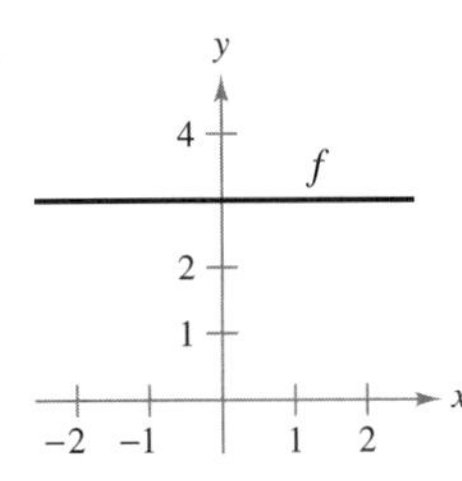

64.

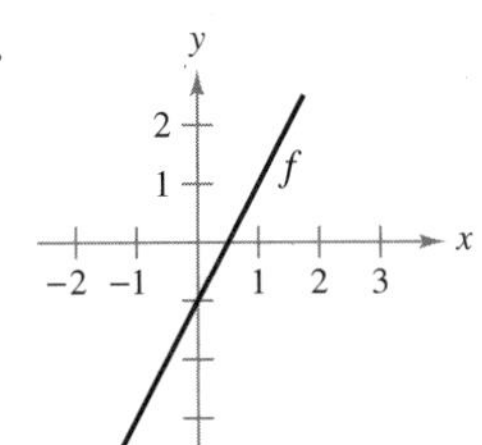

65.

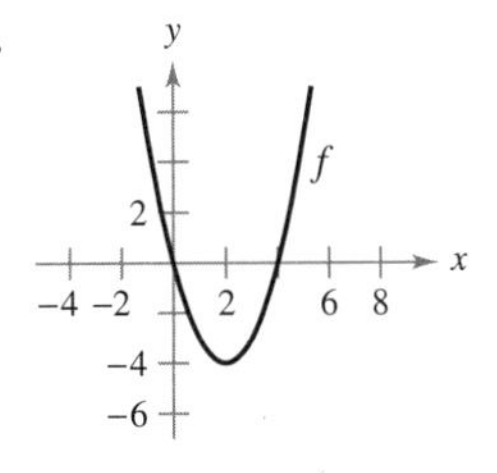

66.

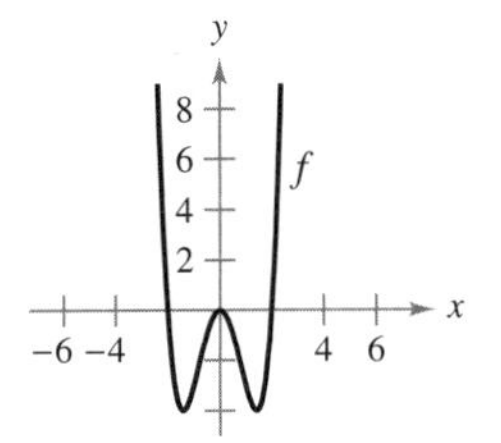

67.

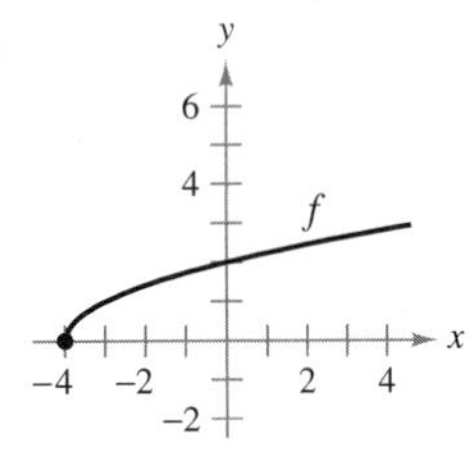

68.

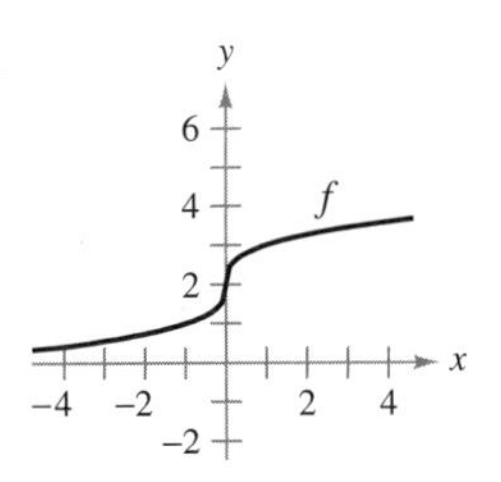

In Exercises 69–72, use the graph of f' to (a) identify the interval(s) on which f is increasing or decreasing, and (b) estimate the values of x at which f has a relative maximum or minimum.

69.

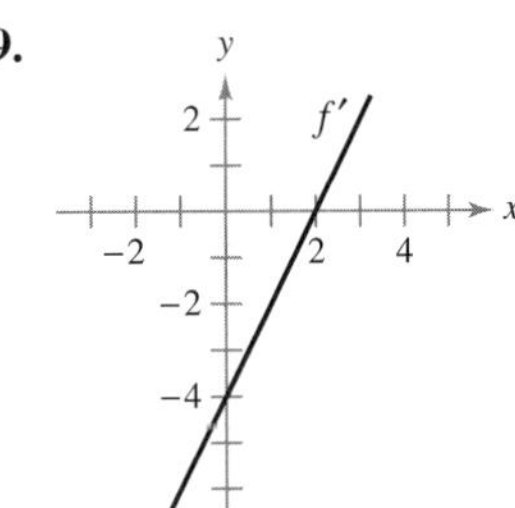

70.

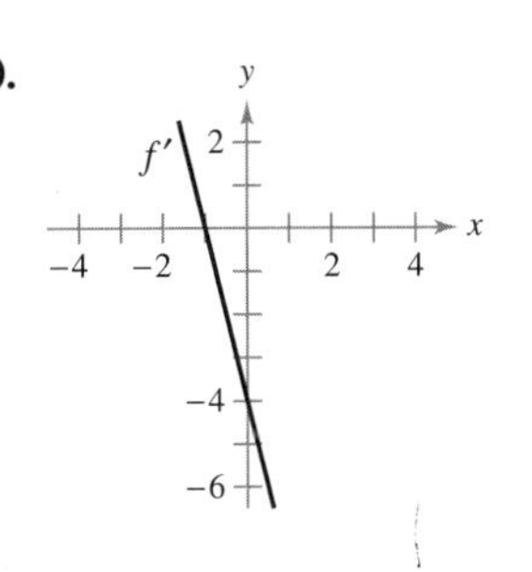

71. 72.

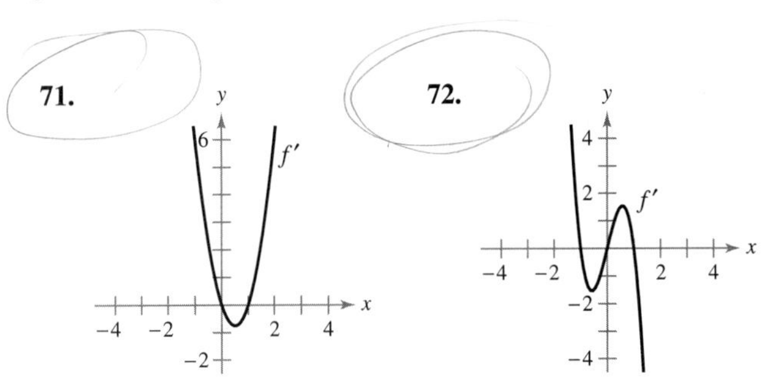

Writing About Concepts

In Exercises 73–78, assume that f is differentiable for all x. The signs of f' are as follows.

$f'(x) > 0$ on $(-\infty, -4)$

$f'(x) < 0$ on $(-4, 6)$

$f'(x) > 0$ on $(6, \infty)$

Supply the appropriate inequality for the indicated value of c.

	Function	*Sign of $g'(c)$*
73.	$g(x) = f(x) + 5$	$g'(0)$ ▢ 0
74.	$g(x) = 3f(x) - 3$	$g'(-5)$ ▢ 0
75.	$g(x) = -f(x)$	$g'(-6)$ ▢ 0
76.	$g(x) = -f(x)$	$g'(0)$ ▢ 0
77.	$g(x) = f(x - 10)$	$g'(0)$ ▢ 0
78.	$g(x) = f(x - 10)$	$g'(8)$ ▢ 0

79. Sketch the graph of the arbitrary function f such that

$$f'(x)\begin{cases} > 0, & x < 4 \\ \text{undefined}, & x = 4. \\ < 0, & x > 4 \end{cases}$$

80. A differentiable function f has one critical number at $x = 5$. Identify the relative extrema of f at the critical number if $f'(4) = -2.5$ and $f'(6) = 3$.

81. ***Think About It*** The function f is differentiable on the interval $[-1, 1]$. The table shows the values of f' for selected values of x. Sketch the graph of f, approximate the critical numbers, and identify the relative extrema.

x	-1	-0.75	-0.50	-0.25
$f'(x)$	-10	-3.2	-0.5	0.8

x	0	0.25	0.50	0.75	1
$f'(x)$	5.6	3.6	-0.2	-6.7	-20.1

82. ***Think About It*** The function f is differentiable on the interval $[0, \pi]$. The table shows the values of f' for selected values of x. Sketch the graph of f, approximate the critical numbers, and identify the relative extrema.

x	0	$\pi/6$	$\pi/4$	$\pi/3$	$\pi/2$
$f'(x)$	3.14	−0.23	−2.45	−3.11	0.69

x	$2\pi/3$	$3\pi/4$	$5\pi/6$	π
$f'(x)$	3.00	1.37	−1.14	−2.84

83. ***Rolling a Ball Bearing*** A ball bearing is placed on an inclined plane and begins to roll. The angle of elevation of the plane is θ. The distance (in meters) the ball bearing rolls in t seconds is $s(t) = 4.9(\sin\theta)t^2$.

(a) Determine the speed of the ball bearing after t seconds.

(b) Complete the table and use it to determine the value of θ that produces the maximum speed at a particular time.

θ	0	$\pi/4$	$\pi/3$	$\pi/2$	$2\pi/3$	$3\pi/4$	π
$s'(t)$							

84. ***Numerical, Graphical, and Analytic Analysis*** The concentration C of a chemical in the bloodstream t hours after injection into muscle tissue is

$$C(t) = \frac{3t}{27 + t^3}, \quad t \geq 0.$$

(a) Complete the table and use it to approximate the time when the concentration is greatest.

t	0	0.5	1	1.5	2	2.5	3
$C(t)$							

(b) Use a graphing utility to graph the concentration function and use the graph to approximate the time when the concentration is greatest.

(c) Use calculus to determine analytically the time when the concentration is greatest.

85. ***Numerical, Graphical, and Analytic Analysis*** Consider the functions $f(x) = x$ and $g(x) = \sin x$ on the interval $(0, \pi)$.

(a) Complete the table and make a conjecture about which is the greater function on the interval $(0, \pi)$.

x	0.5	1	1.5	2	2.5	3
$f(x)$						
$g(x)$						

(b) Use a graphing utility to graph the functions and use the graphs to make a conjecture about which is the greater function on the interval $(0, \pi)$.

(c) Prove that $f(x) > g(x)$ on the interval $(0, \pi)$. [*Hint:* Show that $h'(x) > 0$ where $h = f - g$.]

86. ***Numerical, Graphical, and Analytic Analysis*** Consider the functions $f(x) = x$ and $g(x) = \tan x$ on the interval $(0, \pi/2)$.

(a) Complete the table and make a conjecture about which is the greater function on the interval $(0, \pi/2)$.

x	0.25	0.5	0.75	1	1.25	1.5
$f(x)$						
$g(x)$						

(b) Use a graphing utility to graph the functions and use the graphs to make a conjecture about which is the greater function on the interval $(0, \pi/2)$.

(c) Prove that $f(x) < g(x)$ on the interval $(0, \pi/2)$. [*Hint:* Show that $h'(x) > 0$, where $h = g - f$.]

87. ***Trachea Contraction*** Coughing forces the trachea (windpipe) to contract, which affects the velocity v of the air passing through the trachea. The velocity of the air during coughing is

$$v = k(R - r)r^2, \quad 0 \leq r < R$$

where k is constant, R is the normal radius of the trachea, and r is the radius during coughing. What radius will produce the maximum air velocity?

88. ***Profit*** The profit P (in dollars) made by a fast-food restaurant selling x hamburgers is

$$P = 40{,}000(e^{-x} - 1) - 3x + 850\sqrt{x}, \quad 0 \leq x \leq 35{,}000.$$

Find the open intervals on which P is increasing or decreasing.

89. ***Modeling Data*** The end-of-year assets for the Medicare Hospital Insurance Trust Fund (in billions of dollars) for the years 1995 through 2001 are shown.

1995: 130.3; 1996: 124.9; 1997: 115.6; 1998: 120.4;

1999: 141.4; 2000: 177.5; 2001: 208.7

(Source: U.S. Centers for Medicare and Medicaid Services)

(a) Use the regression capabilities of a graphing utility to find a model of the form $M = at^2 + bt + c$ for the data. (Let $t = 5$ represent 1995.)

(b) Use a graphing utility to plot the data and graph the model.

(c) Analytically find the minimum of the model and compare the result with the actual data.

90. ***Modeling Data*** The number of bankruptcies (in thousands) for the years 1988 through 2001 are shown.

1988: 594.6; 1989: 643.0; 1990: 725.5; 1991: 880.4;

1992: 972.5; 1993: 918.7; 1994: 845.3; 1995: 858.1;

1996: 1042.1; 1997: 1317.0; 1998: 1429.5;

1999: 1392.0; 2000: 1277.0; 2001: 1386.6

(Source: Administrative Office of the U.S. Courts)

(a) Use the regression capabilities of a graphing utility to find a model of the form $B = at^4 + bt^3 + ct^2 + dt + e$ for the data. (Let $t = 8$ represent 1988.)

(b) Use a graphing utility to plot the data and graph the model.

(c) Find the maximum of the model and compare the result with the actual data.

Motion Along a Line **In Exercises 91–94, the function $s(t)$ describes the motion of a particle moving along a line. For each function, (a) find the velocity function of the particle at any time $t \geq 0$, (b) identify the time interval(s) when the particle is moving in a positive direction, (c) identify the time interval(s) when the particle is moving in a negative direction, and (d) identify the time(s) when the particle changes its direction.**

91. $s(t) = 6t - t^2$

92. $s(t) = t^2 - 7t + 10$

93. $s(t) = t^3 - 5t^2 + 4t$

94. $s(t) = t^3 - 20t^2 + 128t - 280$

Motion Along a Line **In Exercises 95 and 96, the graph shows the position of a particle moving along a line. Describe how the particle's position changes with respect to time.**

95.

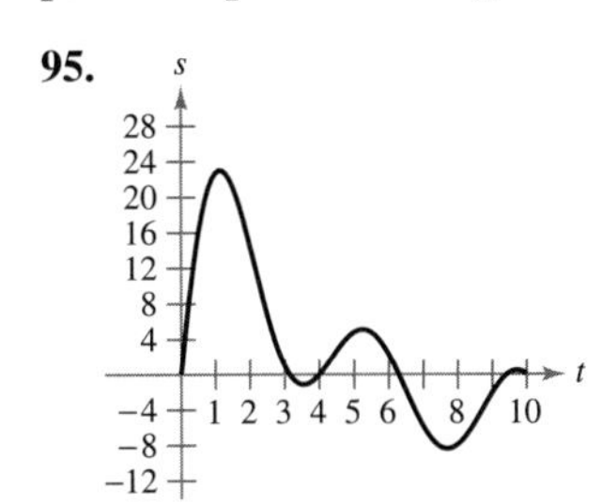

96.

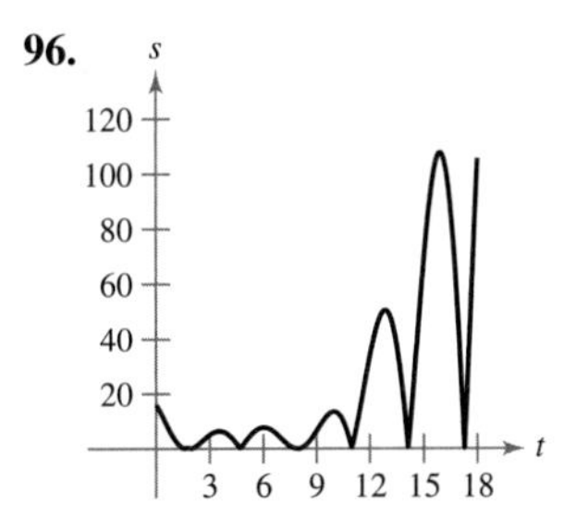

Creating Polynomial Functions **In Exercises 97–100, find a polynomial function**

$$f(x) = a_n x^n + a_{n-1} x^{n-1} + \cdot \cdot \cdot + a_2 x^2 + a_1 x + a_0$$

that has only the specified extrema. (a) Determine the minimum degree of the function and give the criteria you used in determining the degree. (b) Using the fact that the coordinates of the extrema are solution points of the function, and that the x-coordinates are critical numbers, determine a system of linear equations whose solution yields the coefficients of the required function. (c) Use a graphing utility to solve the system of equations and determine the function. (d) Use a graphing utility to confirm your result graphically.

97. Relative minimum: $(0, 0)$; Relative maximum: $(2, 2)$

98. Relative minimum: $(0, 0)$; Relative maximum: $(4, 1000)$

99. Relative minima: $(0, 0)$, $(4, 0)$
Relative maximum: $(2, 4)$

100. Relative minimum: $(1, 2)$
Relative maxima: $(-1, 4)$, $(3, 4)$

True or False? **In Exercises 101–106, determine whether the statement is true or false. If it is false, explain why or give an example that shows it is false.**

101. The sum of two increasing functions is increasing.

102. The product of two increasing functions is increasing.

103. Every nth-degree polynomial has $(n - 1)$ critical numbers.

104. An nth-degree polynomial has at most $(n - 1)$ critical numbers.

105. There is a relative maximum or minimum at each critical number.

106. The relative maxima of the function f are $f(1) = 4$ and $f(3) = 10$. So, f has at least one minimum for some x in the interval $(1, 3)$.

107. Prove the second case of Theorem 4.5.

108. Prove the second case of Theorem 4.6.

109. Let $x > 0$ and $n > 1$ be real numbers. Prove that $(1 + x)^n > 1 + nx$.

110. Use the definitions of increasing and decreasing functions to prove that $f(x) = x^3$ is increasing on $(-\infty, \infty)$.

111. Use the definitions of increasing and decreasing functions to prove that $f(x) = 1/x$ is decreasing on $(0, \infty)$.

Section Project: Rainbows

Rainbows are formed when light strikes raindrops and is reflected and refracted, as shown in the figure. (This figure shows a cross section of a spherical raindrop.) The Law of Refraction states that $(\sin \alpha)/(\sin \beta) = k$, where $k \approx 1.33$ (for water). The angle of deflection is given by $D = \pi + 2\alpha - 4\beta$.

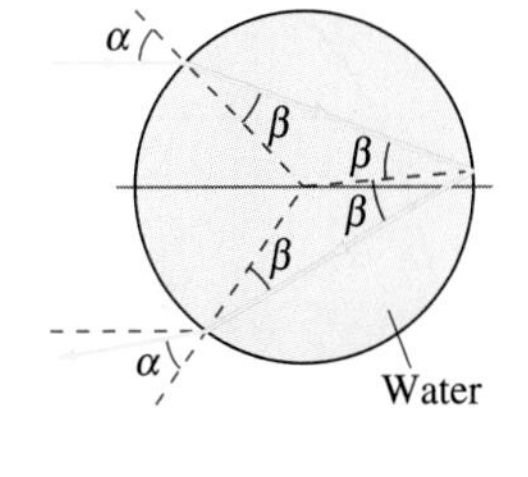

(a) Use a graphing utility to graph

$$D = \pi + 2\alpha - 4 \sin^{-1}(1/k \sin \alpha),$$

$$0 \leq \alpha \leq \pi/2.$$

(b) Prove that the minimum angle of deflection occurs when

$$\cos \alpha = \sqrt{\frac{k^2 - 1}{3}}.$$

For water, what is the minimum angle of deflection, D_{min}? (The angle $\pi - D_{\text{min}}$ is called the *rainbow angle*.) What value of α produces this minimum angle? (A ray of sunlight that strikes a raindrop at this angle, α, is called a *rainbow ray*.)

FOR FURTHER INFORMATION For more information about the mathematics of rainbows, see the article "Somewhere Within the Rainbow" by Steven Janke in *The UMAP Journal.*

Section 4.4 Concavity and the Second Derivative Test

- **Determine intervals on which a function is concave upward or concave downward.**
- **Find any points of inflection of the graph of a function.**
- **Apply the Second Derivative Test to find relative extrema of a function.**

Concavity

You have already seen that locating the intervals on which a function f increases or decreases helps to describe its graph. In this section, you will see how locating the intervals on which f' increases or decreases can be used to determine where the graph of f is *curving upward* or *curving downward.*

Definition of Concavity

Let f be differentiable on an open interval I. The graph of f is **concave upward** on I if f' is increasing on the interval and **concave downward** on I if f' is decreasing on the interval.

The following graphical interpretation of concavity is useful. (See Appendix A for a proof of these results.)

1. Let f be differentiable on an open interval I. If the graph of f is concave *upward* on I, then the graph of f lies *above* all of its tangent lines on I. [See Figure 4.24(a).]
2. Let f be differentiable on an open interval I. If the graph of f is concave *downward* on I, then the graph of f lies *below* all of its tangent lines on I. [See Figure 4.24(b).]

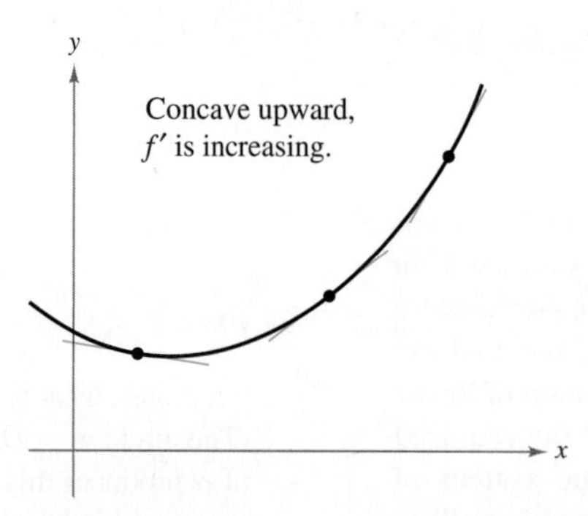

(a) The graph of f lies above its tangent lines.

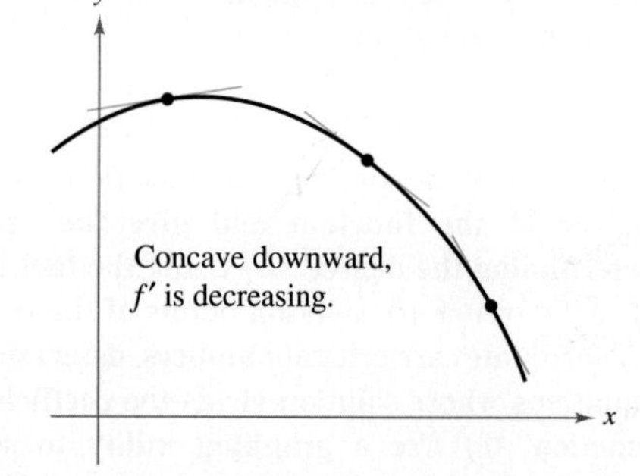

(b) The graph of f lies below its tangent lines.

Figure 4.24

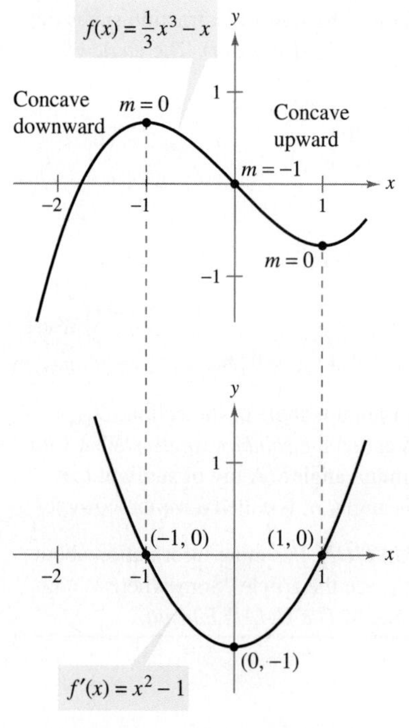

The concavity of f is related to the slope of its derivative.
Figure 4.25

To find the open intervals on which the graph of a function f is concave upward or downward, you need to find the intervals on which f' is increasing or decreasing. For instance, the graph of

$$f(x) = \tfrac{1}{3}x^3 - x$$

is concave downward on the open interval $(-\infty, 0)$ because $f'(x) = x^2 - 1$ is decreasing there. (See Figure 4.25.) Similarly, the graph of f is concave upward on the interval $(0, \infty)$ because f' is increasing on $(0, \infty)$.

The following theorem shows how to use the *second* derivative of a function f to determine intervals on which the graph of f is concave upward or downward. A proof of this theorem follows directly from Theorem 4.5 and the definition of concavity.

THEOREM 4.7 Test for Concavity

Let f be a function whose second derivative exists on an open interval I.

1. If $f''(x) > 0$ for all x in I, then the graph of f is concave upward in I.
2. If $f''(x) < 0$ for all x in I, then the graph of f is concave downward in I.

Note that a third case of Theorem 4.7 could be that if $f''(x) = 0$ for all x in I, then f is linear. Note, however, that concavity is not defined for a line. In other words, a straight line is neither concave upward nor concave downward.

To apply Theorem 4.7, first locate the x-values at which $f''(x) = 0$ or f'' does not exist. Second, use these x-values to determine test intervals. Finally, test the sign of $f''(x)$ in each of the test intervals.

EXAMPLE 1 Determining Concavity

Determine the open intervals on which the graph of

$$f(x) = e^{-x^2/2}$$

is concave upward or downward.

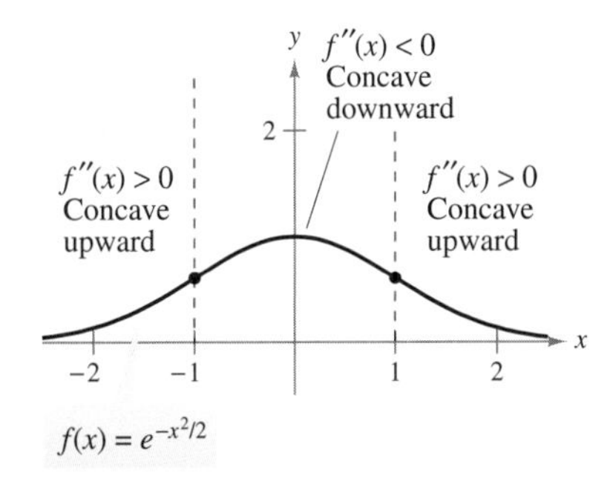

From the sign of f'' you can determine the concavity of the graph of f.
Figure 4.26

Solution Begin by observing that f is continuous on the entire real number line. Next, find the second derivative of f.

$$f'(x) = -xe^{-x^2/2} \qquad \text{First derivative}$$

$$f''(x) = (-x)(-x)e^{-x^2/2} + e^{-x^2/2}(-1) \qquad \text{Differentiate.}$$

$$= e^{-x^2/2}(x^2 - 1) \qquad \text{Second derivative}$$

Because $f''(x) = 0$ when $x = \pm 1$ and f'' is defined on the entire real number line, you should test f'' in the intervals $(-\infty, -1)$, $(-1, 1)$, and $(1, \infty)$. The results are shown in the table and in Figure 4.26.

Interval	$-\infty < x < -1$	$-1 < x < 1$	$1 < x < \infty$
Test Value	$x = -2$	$x = 0$	$x = 2$
Sign of $f''(x)$	$f''(-2) > 0$	$f''(0) < 0$	$f''(2) > 0$
Conclusion	Concave upward	Concave downward	Concave upward

NOTE The function in Example 1 is similar to the normal probability density function, whose general form is

$$f(x) = \frac{1}{\sigma\sqrt{2\pi}} e^{-x^2/2\sigma^2}$$

where σ is the standard deviation (σ is the lowercase Greek letter sigma). This "bell-shaped" curve is concave downward on the interval $(-\sigma, \sigma)$.

The function given in Example 1 is continuous on the entire real number line. If there are x-values at which the function is not continuous, these values should be used along with the points at which $f''(x) = 0$ or $f''(x)$ does not exist to form the test intervals.

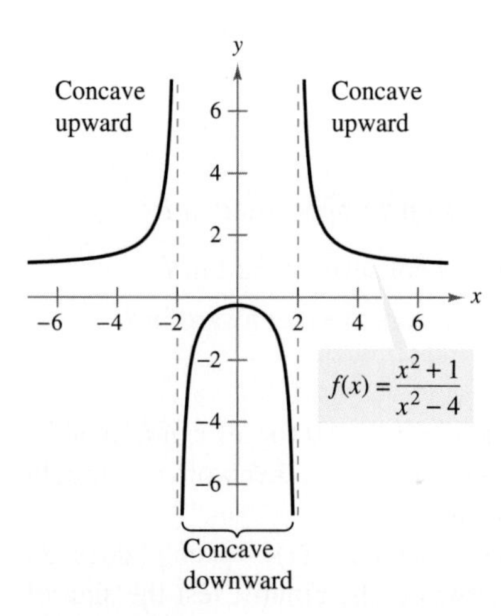

Figure 4.27

EXAMPLE 2 Determining Concavity

Determine the open intervals on which the graph of $f(x) = \dfrac{x^2 + 1}{x^2 - 4}$ is concave upward or downward.

Solution Differentiating twice produces the following.

$$f(x) = \frac{x^2 + 1}{x^2 - 4} \qquad \text{Write original function.}$$

$$f'(x) = \frac{(x^2 - 4)(2x) - (x^2 + 1)(2x)}{(x^2 - 4)^2} \qquad \text{Differentiate.}$$

$$= \frac{-10x}{(x^2 - 4)^2} \qquad \text{First derivative}$$

$$f''(x) = \frac{(x^2 - 4)^2(-10) - (-10x)(2)(x^2 - 4)(2x)}{(x^2 - 4)^4} \qquad \text{Differentiate.}$$

$$= \frac{10(3x^2 + 4)}{(x^2 - 4)^3} \qquad \text{Second derivative}$$

There are no points at which $f''(x) = 0$, but at $x = \pm 2$ the function f is not continuous, so test for concavity in the intervals $(-\infty, -2)$, $(-2, 2)$, and $(2, \infty)$, as shown in the table. The graph of f is shown in Figure 4.27.

Interval	$-\infty < x < -2$	$-2 < x < 2$	$2 < x < \infty$
Test Value	$x = -3$	$x = 0$	$x = 3$
Sign of $f''(x)$	$f''(-3) > 0$	$f''(0) < 0$	$f''(3) > 0$
Conclusion	Concave upward	Concave downward	Concave upward

Points of Inflection

The graph in Figure 4.26 has two points at which the concavity changes. If the tangent line to the graph exists at such a point, that point is a **point of inflection.** Three types of points of inflection are shown in Figure 4.28.

Definition of Point of Inflection

Let f be a function that is continuous on an open interval and let c be a point in the interval. If the graph of f has a tangent line at this point $(c, f(c))$, then this point is a **point of inflection** of the graph of f if the concavity of f changes from upward to downward (or downward to upward) at the point.

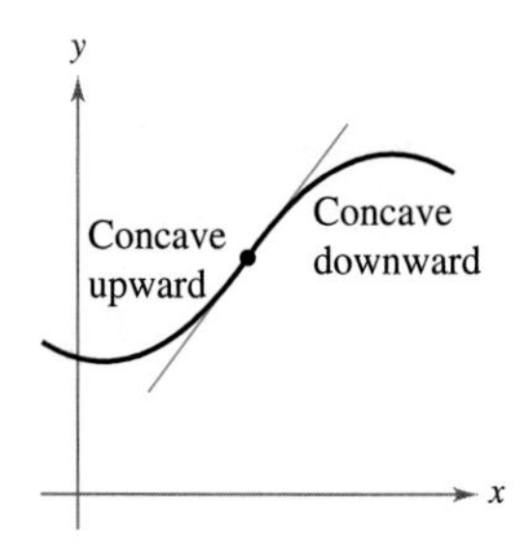

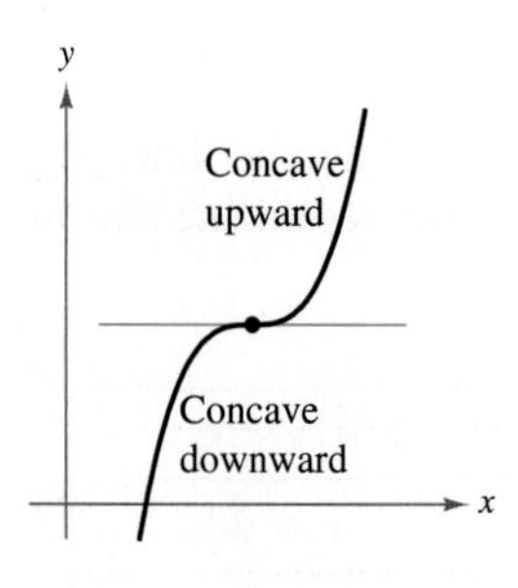

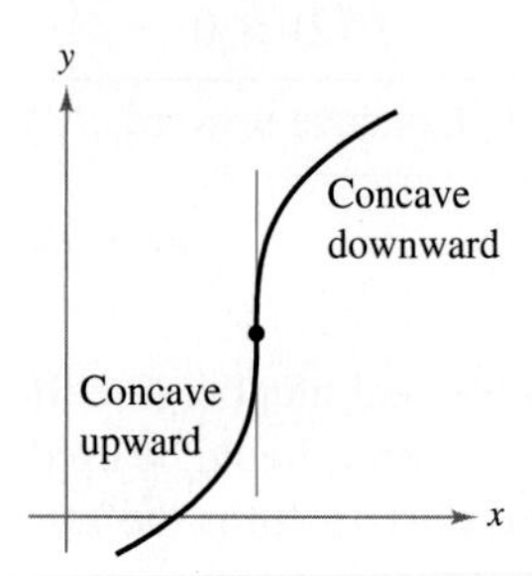

The concavity of f changes at a point of inflection. Note that a graph crosses its tangent line at a point of inflection.
Figure 4.28

NOTE The definition of *point of inflection* given in this book requires that the tangent line exists at the point of inflection. Some books do not require this. For instance, we do not consider the function

$$f(x) = \begin{cases} x^3, & x < 0 \\ x^2 + 2x, & x \geq 0 \end{cases}$$

to have a point of inflection at the origin, even though the concavity of the graph changes from concave downward to concave upward.

To locate *possible* points of inflection, you can determine the values of x for which $f''(x) = 0$ or $f''(x)$ does not exist. This is similar to the procedure for locating relative extrema of f.

THEOREM 4.8 Points of Inflection

If $(c, f(c))$ is a point of inflection of the graph of f, then either $f''(c) = 0$ or f'' does not exist at $x = c$.

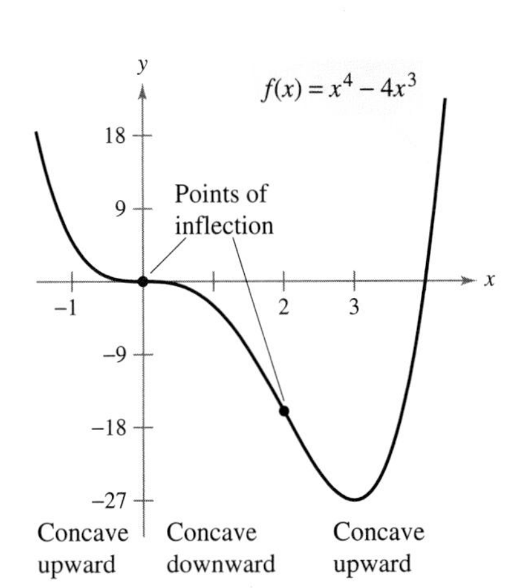

Points of inflection can occur where $f''(x) = 0$ or f'' does not exist.
Figure 4.29

EXAMPLE 3 Finding Points of Inflection

Determine the points of inflection and discuss the concavity of the graph of $f(x) = x^4 - 4x^3$.

Solution Differentiating twice produces the following.

$$f(x) = x^4 - 4x^3 \qquad \text{Write original function.}$$

$$f'(x) = 4x^3 - 12x^2 \qquad \text{Find first derivative.}$$

$$f''(x) = 12x^2 - 24x = 12x(x - 2) \qquad \text{Find second derivative.}$$

Setting $f''(x) = 0$, you can determine that the possible points of inflection occur at $x = 0$ and $x = 2$. By testing the intervals determined by these x-values, you can conclude that they both yield points of inflection. A summary of this testing is shown in the table, and the graph of f is shown in Figure 4.29.

Interval	$-\infty < x < 0$	$0 < x < 2$	$2 < x < \infty$
Test Value	$x = -1$	$x = 1$	$x = 3$
Sign of $f''(x)$	$f''(-1) > 0$	$f''(1) < 0$	$f''(3) > 0$
Conclusion	Concave upward	Concave downward	Concave upward

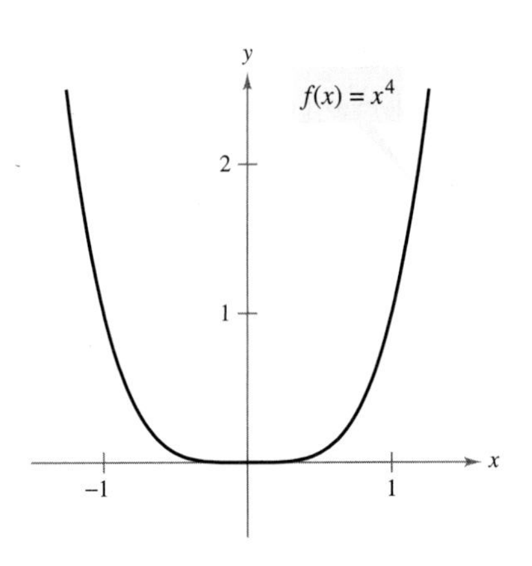

$f''(0) = 0$, but $(0, 0)$ is not a point of inflection.
Figure 4.30

The converse of Theorem 4.8 is not generally true. That is, it is possible for the second derivative to be 0 at a point that is *not* a point of inflection. For instance, the graph of $f(x) = x^4$ is shown in Figure 4.30. The second derivative is 0 when $x = 0$, but the point $(0, 0)$ is not a point of inflection because the graph of f is concave upward on both intervals $-\infty < x < 0$ and $0 < x < \infty$.

EXPLORATION

Consider a general cubic function of the form

$$f(x) = ax^3 + bx^2 + cx + d.$$

You know that the value of d has a bearing on the location of the graph but has no bearing on the value of the first derivative at given values of x. Graphically, this is true because changes in the value of d shift the graph up or down but do not change its basic shape. Use a graphing utility to graph several cubics with different values of c. Then give a graphical explanation of why changes in c do not affect the values of the second derivative.

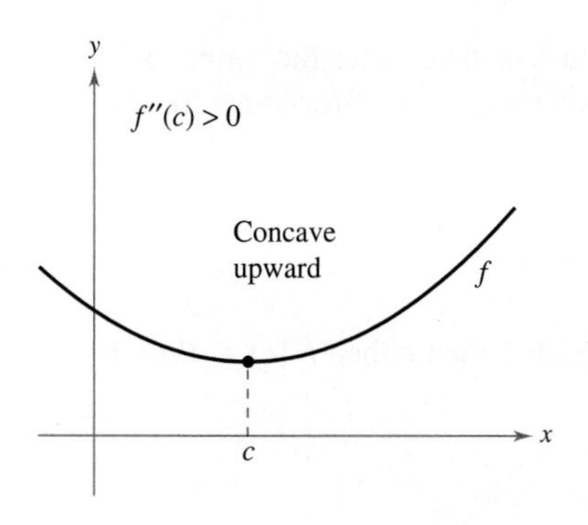

If $f'(c) = 0$ and $f''(c) > 0$, $f(c)$ is a relative minimum.

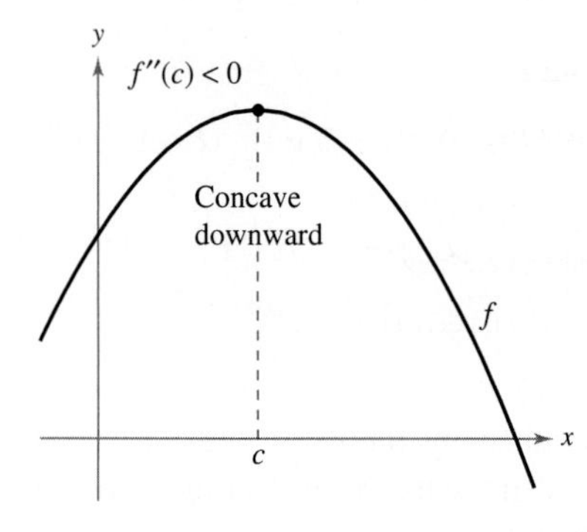

If $f'(c) = 0$ and $f''(c) < 0$, $f(c)$ is a relative maximum.
Figure 4.31

The Second Derivative Test

In addition to testing for concavity, the second derivative can be used to perform a simple test for relative maxima and minima. The test is based on the fact that if the graph of a function f is concave upward on an open interval containing c, and $f'(c) = 0$, $f(c)$ must be a relative minimum of f. Similarly, if the graph of a function f is concave downward on an open interval containing c, and $f'(c) = 0$, $f(c)$ must be a relative maximum of f (see Figure 4.31).

THEOREM 4.9 Second Derivative Test

Let f be a function such that $f'(c) = 0$ and the second derivative of f exists on an open interval containing c.

1. If $f''(c) > 0$, then $f(c)$ is a relative minimum.

2. If $f''(c) < 0$, then $f(c)$ is a relative maximum.

If $f''(c) = 0$, the test fails. That is, f may have a relative maximum, a relative minimum, or neither. In such cases, you can use the First Derivative Test.

Proof If $f'(c) = 0$ and $f''(c) > 0$, there exists an open interval I containing c for which

$$\frac{f'(x) - f'(c)}{x - c} = \frac{f'(x)}{x - c} > 0$$

for all $x \neq c$ in I. If $x < c$, then $x - c < 0$ and $f'(x) < 0$. Also, if $x > c$, then $x - c > 0$ and $f'(x) > 0$. So, $f'(x)$ changes from negative to positive at c, and the First Derivative Test implies that $f(c)$ is a relative minimum. A proof of the second case is left to you.

EXAMPLE 4 Using the Second Derivative Test

Find the relative extrema for $f(x) = -3x^5 + 5x^3$.

Solution Begin by finding the critical numbers of f.

$$f'(x) = -15x^4 + 15x^2 = 15x^2(1 - x^2) = 0 \quad \text{Set } f'(x) \text{ equal to 0.}$$

$$x = -1, 0, 1 \quad \text{Critical numbers}$$

Using

$$f''(x) = -60x^3 + 30x = 30(-2x^3 + x)$$

you can apply the Second Derivative Test as shown below.

Point	$(-1, -2)$	$(1, 2)$	$(0, 0)$
Sign of $f''(x)$	$f''(-1) > 0$	$f''(1) < 0$	$f''(0) = 0$
Conclusion	Relative minimum	Relative maximum	Test fails

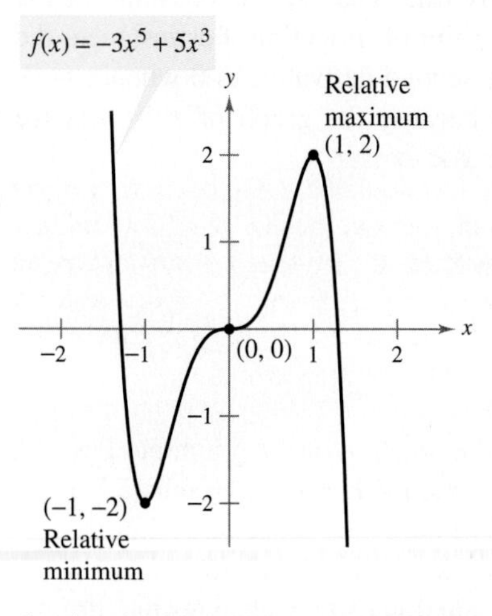

$(0, 0)$ is neither a relative minimum nor a relative maximum.
Figure 4.32

Because the Second Derivative Test fails at $(0, 0)$, you can use the First Derivative Test and observe that f increases to the left and right of $x = 0$. So, $(0, 0)$ is neither a relative minimum nor a relative maximum (even though the graph has a horizontal tangent line at this point). The graph of f is shown in Figure 4.32.

Exercises for Section 4.4

See www.CalcChat.com for worked-out solutions to odd-numbered exercises.

In Exercises 1–10, determine the open intervals on which the graph is concave upward or concave downward.

1. $y = x^2 - x - 2$

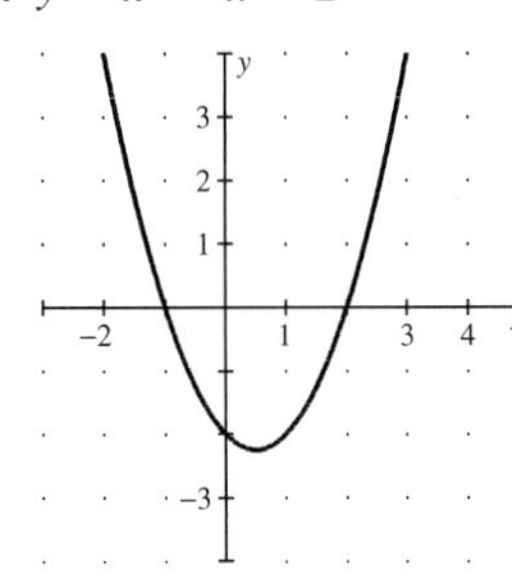

Generated by Derive

2. $y = -x^3 + 3x^2 - 2$

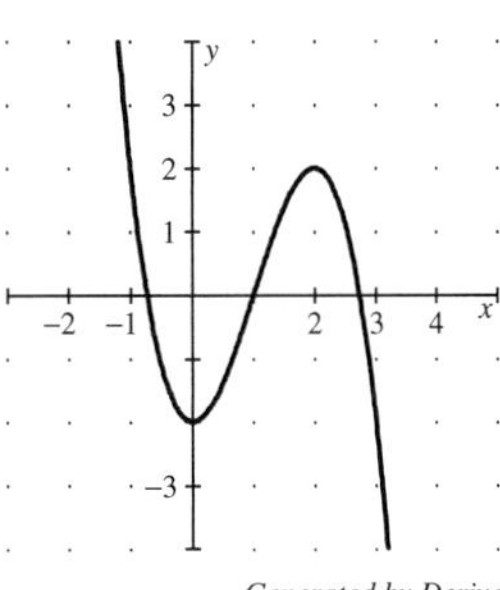

Generated by Derive

3. $f(x) = \dfrac{24}{x^2 + 12}$

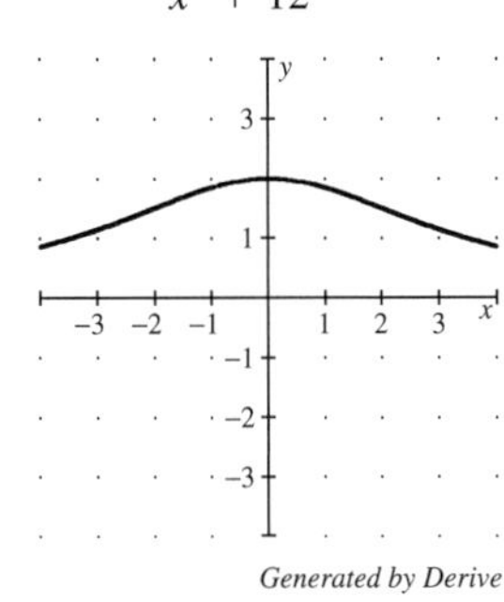

Generated by Derive

4. $f(x) = \dfrac{x^2 - 1}{2x + 1}$

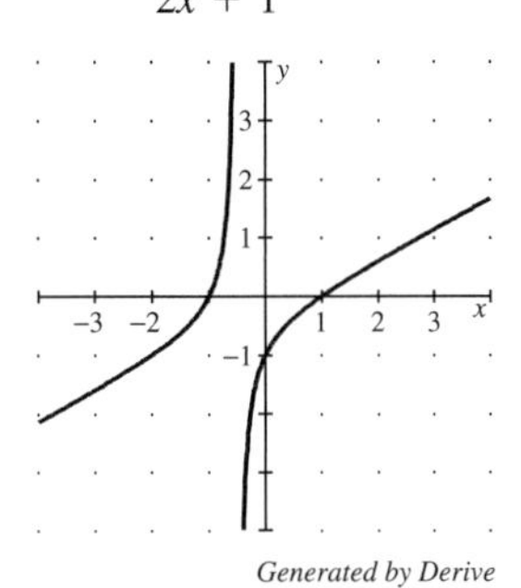

Generated by Derive

5. $f(x) = \dfrac{x^2 + 1}{x^2 - 1}$

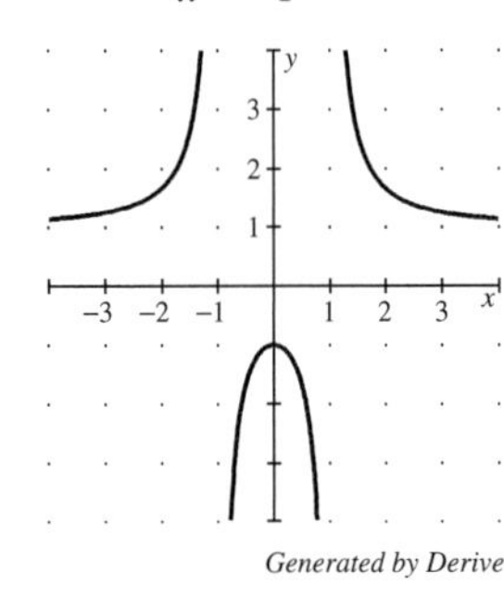

Generated by Derive

6. $y = \dfrac{-3x^5 + 40x^3 + 135x}{270}$

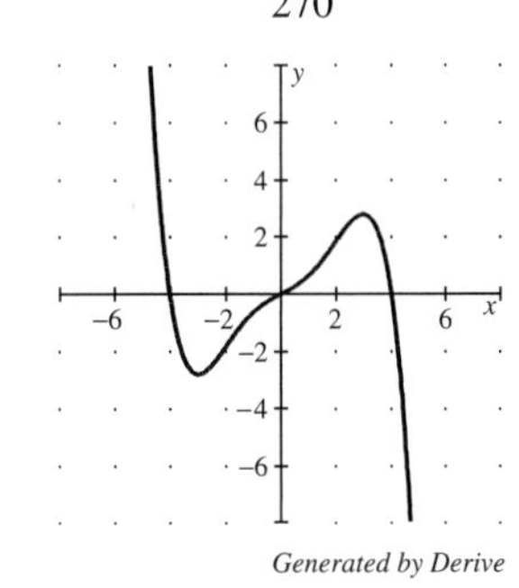

Generated by Derive

7. $g(x) = 3x^2 - x^3$
8. $h(x) = x^5 - 5x + 2$
9. $y = 2x - \tan x, \quad \left(-\dfrac{\pi}{2}, \dfrac{\pi}{2}\right)$
10. $y = x + \dfrac{2}{\sin x}, \quad (-\pi, \pi)$

In Exercises 11–28, find the points of inflection and discuss the concavity of the graph of the function.

11. $f(x) = x^3 - 6x^2 + 12x$
12. $f(x) = 2x^3 - 3x^2 - 12x + 5$
13. $f(x) = \frac{1}{4}x^4 - 2x^2$
14. $f(x) = 2x^4 - 8x + 3$
15. $f(x) = x(x - 4)^3$
16. $f(x) = x^3(x - 2)$
17. $f(x) = x\sqrt{x + 3}$
18. $f(x) = x\sqrt{x + 1}$
19. $f(x) = \dfrac{x}{x^2 + 1}$
20. $f(x) = \dfrac{x + 1}{\sqrt{x}}$
21. $f(x) = \sin \dfrac{x}{2}, \quad [0, 4\pi]$
22. $f(x) = 2 \csc \dfrac{3x}{2}, \quad (0, 2\pi)$
23. $f(x) = \sec\left(x - \dfrac{\pi}{2}\right), \quad (0, 4\pi)$
24. $f(x) = \sin x + \cos x, \quad [0, 2\pi]$
25. $f(x) = 2 \sin x + \sin 2x, \quad [0, 2\pi]$
26. $f(x) = x + 2 \cos x, \quad [0, 2\pi]$
27. $y = x - \ln x$
28. $y = \frac{1}{2}(e^x - e^{-x})$

In Exercises 29–54, find all relative extrema. Use the Second Derivative Test where applicable.

29. $f(x) = x^4 - 4x^3 + 2$
30. $f(x) = x^2 + 3x - 8$
31. $f(x) = (x - 5)^2$
32. $f(x) = -(x - 5)^2$
33. $f(x) = x^3 - 3x^2 + 3$
34. $f(x) = x^3 - 9x^2 + 27x$
35. $g(x) = x^2(6 - x)^3$
36. $g(x) = -\frac{1}{8}(x + 2)^2(x - 4)^2$
37. $f(x) = x^{2/3} - 3$
38. $f(x) = \sqrt{x^2 + 1}$
39. $f(x) = x + \dfrac{4}{x}$
40. $f(x) = \dfrac{x}{x - 1}$
41. $f(x) = \cos x - x, \quad [0, 4\pi]$
42. $f(x) = 2 \sin x + \cos 2x, \quad [0, 2\pi]$
43. $y = \dfrac{1}{2}x^2 - \ln x$
44. $y = x \ln x$
45. $y = \dfrac{x}{\ln x}$
46. $y = x^2 \ln \dfrac{x}{4}$
47. $f(x) = \dfrac{e^x + e^{-x}}{2}$
48. $g(x) = \dfrac{1}{\sqrt{2\pi}} e^{-(x-3)^2/2}$
49. $f(x) = x^2 e^{-x}$
50. $f(x) = xe^{-x}$
51. $f(x) = 8x(4^{-x})$
52. $y = x^2 \log_3 x$
53. $f(x) = \operatorname{arcsec} x - x$
54. $f(x) = \arcsin x - 2x$

In Exercises 55–58, use a computer algebra system to analyze the function over the given interval. (a) Find the first and second derivatives of the function. (b) Find any relative extrema and points of inflection. (c) Graph f, f', and f'' on the same set of coordinate axes and state the relationship between the behavior of f and the signs of f' and f''.

55. $f(x) = 0.2x^2(x - 3)^3, \quad [-1, 4]$
56. $f(x) = x^2\sqrt{6 - x^2}, \quad \left[-\sqrt{6}, \sqrt{6}\right]$
57. $f(x) = \sin x - \frac{1}{3} \sin 3x + \frac{1}{5} \sin 5x, \quad [0, \pi]$
58. $f(x) = \sqrt{2x} \sin x, \quad [0, 2\pi]$

Writing About Concepts

59. Consider a function f such that f' is increasing. Sketch graphs of f for (a) $f' < 0$ and (b) $f' > 0$.

60. Consider a function f such that f' is decreasing. Sketch graphs of f for (a) $f' < 0$ and (b) $f' > 0$.

61. Sketch the graph of a function f that does *not* have a point of inflection at $(c, f(c))$ even though $f''(c) = 0$.

62. S represents weekly sales of a product. What can be said of S' and S'' for each of the following?

(a) The rate of change of sales is increasing.

(b) Sales are increasing at a slower rate.

(c) The rate of change of sales is constant.

(d) Sales are steady.

(e) Sales are declining, but at a slower rate.

(f) Sales have bottomed out and have started to rise.

In Exercises 63–66, the graph of f is shown. Graph f, f', and f'' on the same set of coordinate axes. To print an enlarged copy of the graph, go to the website *www.mathgraphs.com*.

63.

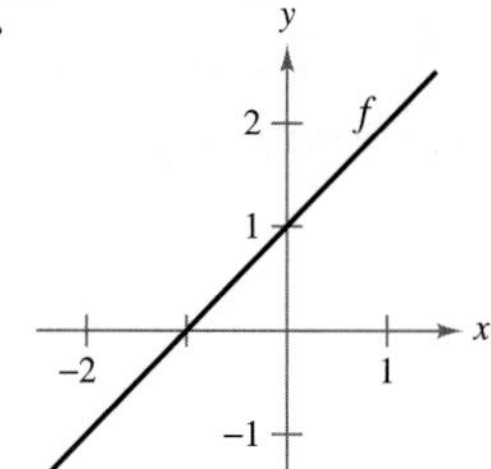

64.

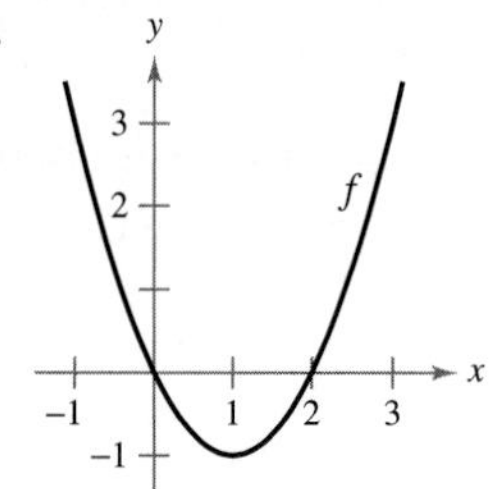

65.

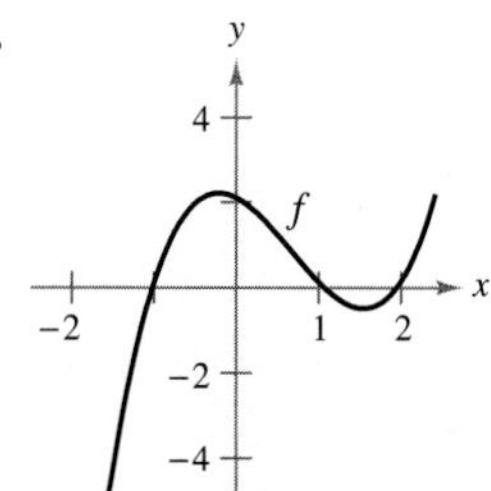

66.

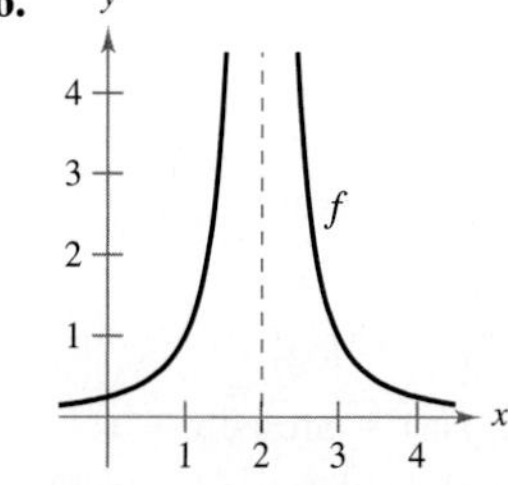

Think About It **In Exercises 67–70, sketch the graph of a function f having the given characteristics.**

67. $f(2) = f(4) = 0$
$f(3)$ is defined.
$f'(x) < 0$ if $x < 3$
$f'(3)$ does not exist.
$f'(x) > 0$ if $x > 3$
$f''(x) < 0, x \neq 3$

68. $f(0) = f(2) = 0$
$f'(x) > 0$ if $x < 1$
$f'(1) = 0$
$f'(x) < 0$ if $x > 1$
$f''(x) < 0$

69. $f(2) = f(4) = 0$
$f'(x) > 0$ if $x < 3$
$f'(3)$ does not exist.
$f'(x) < 0$ if $x > 3$
$f''(x) > 0, x \neq 3$

70. $f(0) = f(2) = 0$
$f'(x) < 0$ if $x < 1$
$f'(1) = 0$
$f'(x) > 0$ if $x > 1$
$f''(x) > 0$

71. ***Conjecture*** Consider the function $f(x) = (x - 2)^n$.

(a) Use a graphing utility to graph f for $n = 1, 2, 3$, and 4. Use the graphs to make a conjecture about the relationship between n and any inflection points of the graph of f.

(b) Verify your conjecture in part (a).

72. (a) Graph $f(x) = \sqrt[3]{x}$ and identify the inflection point.

(b) Does $f''(x)$ exist at the inflection point? Explain.

In Exercises 73 and 74, find a, b, c, and d such that the cubic $f(x) = ax^3 + bx^2 + cx + d$ satisfies the given conditions.

73. Relative maximum: $(3, 3)$
Relative minimum: $(5, 1)$
Inflection point: $(4, 2)$

74. Relative maximum: $(2, 4)$
Relative minimum: $(4, 2)$
Inflection point: $(3, 3)$

75. ***Aircraft Glide Path*** A small aircraft starts its descent from an altitude of 1 mile, 4 miles west of the runway (see figure).

(a) Find the cubic $f(x) = ax^3 + bx^2 + cx + d$ on the interval $[-4, 0]$ that describes a smooth glide path for the landing.

(b) The function in part (a) models the glide path of the plane. When would the plane be descending at the most rapid rate?

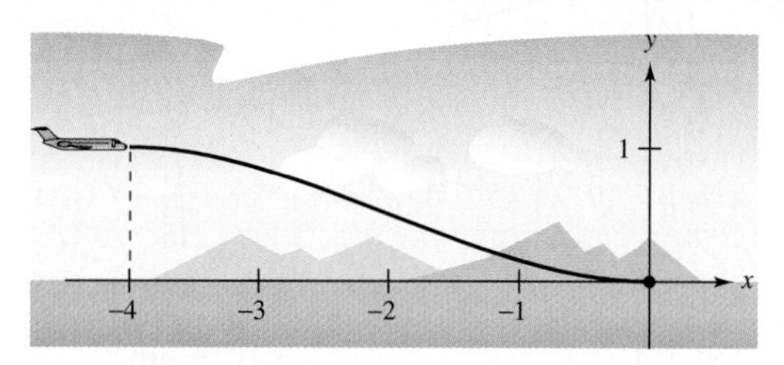

FOR FURTHER INFORMATION For more information on this type of modeling, see the article "How Not to Land at Lake Tahoe!" by Richard Barshinger in *The American Mathematical Monthly*. To view this article, go to the website *www.matharticles.com*.

76. ***Highway Design*** A section of highway connecting two hillsides with grades of 6% and 4% is to be built between two points that are separated by a horizontal distance of 2000 feet (see figure). At the point where the two hillsides come together, there is a 50-foot difference in elevation.

(a) Design a section of highway connecting the hillsides modeled by the function $f(x) = ax^3 + bx^2 + cx + d$ $(-1000 \le x \le 1000)$. At the points A and B, the slope of the model must match the grade of the hillside.

(b) Use a graphing utility to graph the model.

(c) Use a graphing utility to graph the derivative of the model.

(d) Determine the grade at the steepest part of the transitional section of the highway.

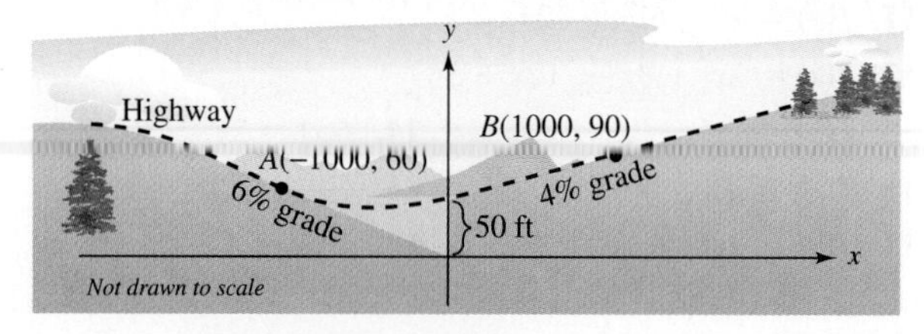

77. *Beam Deflection* The deflection D of a beam of length L is $D = 2x^4 - 5Lx^3 + 3L^2x^2$, where x is the distance from one end of the beam. Find the value of x that yields the maximum deflection.

78. *Specific Gravity* A model for the specific gravity of water S is

$$S = \frac{5.755}{10^8}T^3 - \frac{8.521}{10^6}T^2 + \frac{6.540}{10^5}T + 0.99987, \; 0 < T < 25$$

where T is the water temperature in degrees Celsius.

(a) Use a computer algebra system to find the coordinates of the maximum value of the function.

(b) Sketch a graph of the function over the specified domain. (Use a setting in which $0.996 \le S \le 1.001$.)

(c) Estimate the specific gravity of water when $T = 20°$.

79. *Average Cost* A manufacturer has determined that the total cost C of operating a factory is $C = 0.5x^2 + 15x + 5000$, where x is the number of units produced. At what level of production will the average cost per unit be minimized? (The average cost per unit is C/x.)

80. *Modeling Data* The average typing speed S (words per minute) of a typing student after t weeks of lessons is shown in the table.

t	5	10	15	20	25	30
S	38	56	79	90	93	94

A model for the data is $S = \dfrac{100t^2}{65 + t^2}$, $t > 0$.

(a) Use a graphing utility to plot the data and graph the model.

(b) Use the second derivative to determine the concavity of S. Compare the result with the graph in part (a).

(c) What is the sign of the first derivative for $t > 0$? By combining this information with the concavity of the model, what inferences can be made about the typing speed as t increases?

Linear and Quadratic Approximations **In Exercises 81–84, use a graphing utility to graph the function. Then graph the linear and quadratic approximations**

$$P_1(x) = f(a) + f'(a)(x - a)$$

and

$$P_2(x) = f(a) + f'(a)(x - a) + \tfrac{1}{2}f''(a)(x - a)^2$$

in the same viewing window. Compare the values of f, P_1, and P_2 and their first derivatives at $x = a$. How do the approximations change as you move farther away from $x = a$?

Function	*Value of a*
81. $f(x) = 2(\sin x + \cos x)$	$a = \dfrac{\pi}{4}$
82. $f(x) = 2(\sin x + \cos x)$	$a = 0$
83. $f(x) = \arctan x$	$a = -1$
84. $f(x) = \dfrac{\sqrt{x}}{x - 1}$	$a = 2$

85. Use a graphing utility to graph $y = x \sin(1/x)$. Show that the graph is concave downward to the right of $x = 1/\pi$.

86. Show that the point of inflection of $f(x) = x(x - 6)^2$ lies midway between the relative extrema of f.

87. Prove that every cubic function with three distinct real zeros has a point of inflection whose x-coordinate is the average of the three zeros.

88. Show that the cubic polynomial $p(x) = ax^3 + bx^2 + cx + d$ has exactly one point of inflection (x_0, y_0), where

$$x_0 = \frac{-b}{3a} \quad \text{and} \quad y_0 = \frac{2b^3}{27a^2} - \frac{bc}{3a} + d.$$

Use this formula to find the point of inflection of

$p(x) = x^3 - 3x^2 + 2$.

True or False? **In Exercises 89–94, determine whether the statement is true or false. If it is false, explain why or give an example that shows it is false.**

89. The graph of every cubic polynomial has precisely one point of inflection.

90. The graph of $f(x) = 1/x$ is concave downward for $x < 0$ and concave upward for $x > 0$, and thus it has a point of inflection at $x = 0$.

91. The maximum value of $y = 3\sin x + 2\cos x$ is 5.

92. The maximum slope of the graph of $y = \sin(bx)$ is b.

93. If $f'(c) > 0$, then f is concave upward at $x = c$.

94. If $f''(2) = 0$, then the graph of f must have a point of inflection at $x = 2$.

In Exercises 95 and 96, let f and g represent differentiable functions such that $f'' \neq 0$ and $g'' \neq 0$.

95. Show that if f and g are concave upward on the interval (a, b), then $f + g$ is also concave upward on (a, b).

96. Prove that if f and g are positive, increasing, and concave upward on the interval (a, b), then fg is also concave upward on (a, b).

Section 4.5 Limits at Infinity

- **Determine (finite) limits at infinity.**
- **Determine the horizontal asymptotes, if any, of the graph of a function.**
- **Determine infinite limits at infinity.**

Limits at Infinity

This section discusses the "end behavior" of a function on an *infinite* interval. Consider the graph of

$$f(x) = \frac{3x^2}{x^2 + 1}$$

as shown in Figure 4.33. Graphically, you can see that the values of $f(x)$ appear to approach 3 as x increases without bound or decreases without bound. You can come to the same conclusions numerically, as shown in the table.

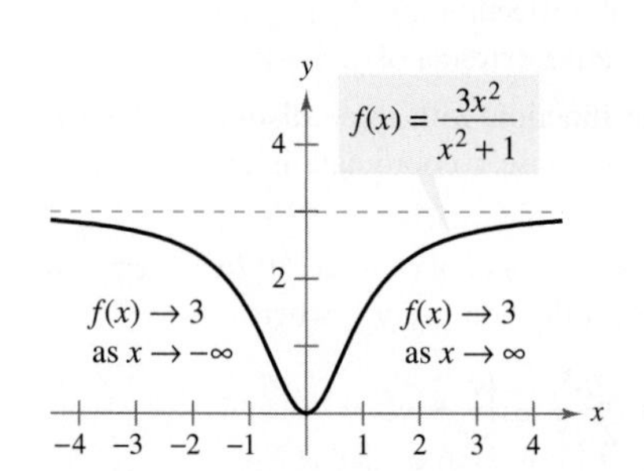

The limit of $f(x)$ as x approaches $-\infty$ or ∞ is 3.
Figure 4.33

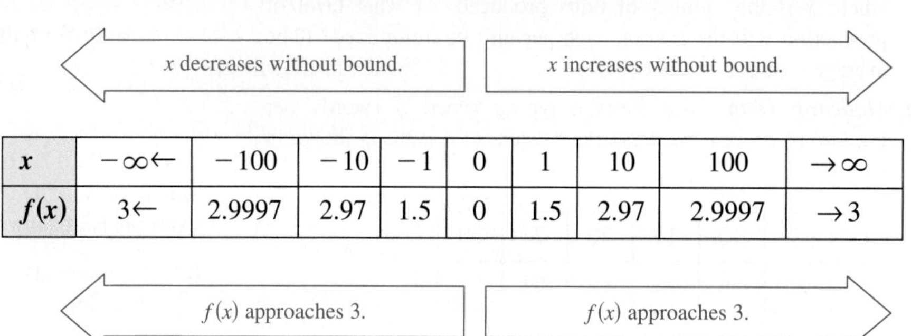

x	$-\infty \leftarrow$	-100	-10	-1	0	1	10	100	$\to \infty$
$f(x)$	$3 \leftarrow$	2.9997	2.97	1.5	0	1.5	2.97	2.9997	$\to 3$

The table suggests that the value of $f(x)$ approaches 3 as x increases without bound $(x \to \infty)$. Similarly, $f(x)$ approaches 3 as x decreases without bound $(x \to -\infty)$. These **limits at infinity** are denoted by

$$\lim_{x \to -\infty} f(x) = 3 \quad \text{Limit at negative infinity}$$

and

$$\lim_{x \to \infty} f(x) = 3. \quad \text{Limit at positive infinity}$$

NOTE The statement $\lim_{x \to -\infty} f(x) = L$ or $\lim_{x \to \infty} f(x) = L$ means that the limit exists *and* the limit is equal to L.

To say that a statement is true as x increases *without bound* means that for some (large) real number M, the statement is true for *all* x in the interval $\{x: x > M\}$. The following definition uses this concept.

Definition of Limits at Infinity

Let L be a real number.

1. The statement $\lim_{x \to \infty} f(x) = L$ means that for each $\varepsilon > 0$ there exists an $M > 0$ such that $|f(x) - L| < \varepsilon$ whenever $x > M$.
2. The statement $\lim_{x \to -\infty} f(x) = L$ means that for each $\varepsilon > 0$ there exists an $N < 0$ such that $|f(x) - L| < \varepsilon$ whenever $x < N$.

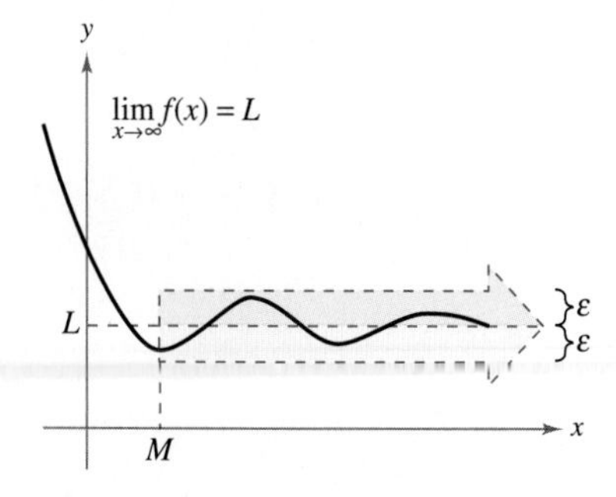

$f(x)$ is within ε units of L as $x \to \infty$.
Figure 4.34

The definition of a limit at infinity is shown in Figure 4.34. In this figure, note that for a given positive number ε there exists a positive number M such that, for $x > M$, the graph of f will lie between the horizontal lines given by $y = L + \varepsilon$ and $y = L - \varepsilon$.

EXPLORATION

Use a graphing utility to graph

$$f(x) = \frac{2x^2 + 4x - 6}{3x^2 + 2x - 16}.$$

Describe all the important features of the graph. Can you find a single viewing window that shows all of these features clearly? Explain your reasoning.

What are the horizontal asymptotes of the graph? How far to the right do you have to move on the graph so that the graph is within 0.001 unit of its horizontal asymptote? Explain your reasoning.

Horizontal Asymptotes

In Figure 4.34, the graph of f approaches the line $y = L$ as x increases without bound. The line $y = L$ is called a **horizontal asymptote** of the graph of f.

Definition of a Horizontal Asymptote

The line $y = L$ is a **horizontal asymptote** of the graph of f if

$$\lim_{x \to -\infty} f(x) = L$$

or

$$\lim_{x \to \infty} f(x) = L.$$

Note that from this definition, it follows that the graph of a *function* of x can have at most two horizontal asymptotes—one to the right and one to the left.

Limits at infinity have many of the same properties of limits discussed in Section 2.3. For example, if $\lim_{x \to \infty} f(x)$ and $\lim_{x \to \infty} g(x)$ both exist, then

$$\lim_{x \to \infty} [f(x) + g(x)] = \lim_{x \to \infty} f(x) + \lim_{x \to \infty} g(x)$$

and

$$\lim_{x \to \infty} [f(x)g(x)] = \left[\lim_{x \to \infty} f(x)\right]\left[\lim_{x \to \infty} g(x)\right].$$

Similar properties hold for limits at $-\infty$.

When evaluating limits at infinity, the following theorem is helpful. (A proof of part 1 of this theorem is given in Appendix A.)

THEOREM 4.10 Limits at Infinity

1. If r is a positive rational number and c is any real number, then

$$\lim_{x \to \infty} \frac{c}{x^r} = 0 \quad \text{and} \quad \lim_{x \to -\infty} \frac{c}{x^r} = 0.$$

The second limit is valid only if x^r is defined when $x < 0$.

2. $\lim_{x \to -\infty} e^x = 0 \quad$ and $\quad \lim_{x \to \infty} e^{-x} = 0$

EXAMPLE 1 Evaluating a Limit at Infinity

a. $\displaystyle \lim_{x \to \infty} \left(5 - \frac{2}{x^2}\right) = \lim_{x \to \infty} 5 - \lim_{x \to \infty} \frac{2}{x^2}$ Property of limits

$= 5 - 0$

$= 5$

b. $\displaystyle \lim_{x \to \infty} \frac{3}{e^x} = \lim_{x \to \infty} 3e^{-x}$

$\displaystyle = 3 \lim_{x \to \infty} e^{-x}$ Property of limits

$= 3(0)$

$= 0$

EXAMPLE 2 Evaluating a Limit at Infinity

Find the limit: $\lim_{x\to\infty} \dfrac{2x-1}{x+1}$.

Solution Note that both the numerator and the denominator approach infinity as x approaches infinity.

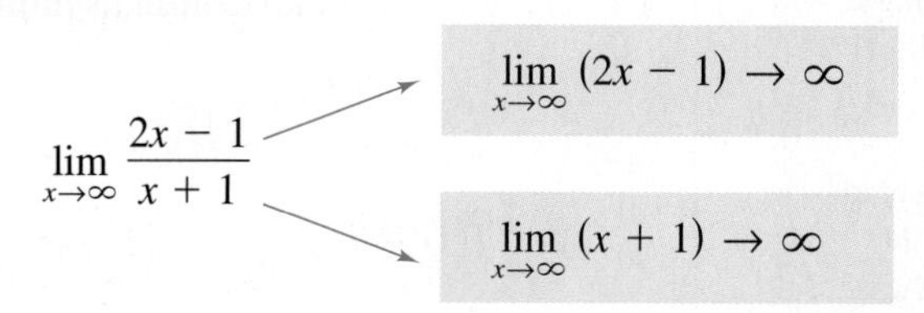

This results in $\frac{\infty}{\infty}$, an **indeterminate form.** To resolve this problem, you can divide both the numerator and the denominator by x. After dividing, the limit may be evaluated as follows.

NOTE When you encounter an indeterminate form such as the one in Example 2, you should divide the numerator and denominator by the highest power of x in the *denominator*.

$$\lim_{x\to\infty} \frac{2x-1}{x+1} = \lim_{x\to\infty} \frac{\frac{2x-1}{x}}{\frac{x+1}{x}} \qquad \text{Divide numerator and denominator by } x.$$

$$= \lim_{x\to\infty} \frac{2-\frac{1}{x}}{1+\frac{1}{x}} \qquad \text{Simplify.}$$

$$= \frac{\lim\limits_{x\to\infty} 2 - \lim\limits_{x\to\infty} \frac{1}{x}}{\lim\limits_{x\to\infty} 1 + \lim\limits_{x\to\infty} \frac{1}{x}} \qquad \text{Take limits of numerator and denominator.}$$

$$= \frac{2-0}{1+0} \qquad \text{Apply Theorem 4.10.}$$

$$= 2$$

So, the line $y = 2$ is a horizontal asymptote to the right. By taking the limit as $x \to -\infty$, you can see that $y = 2$ is also a horizontal asymptote to the left. The graph of the function is shown in Figure 4.35.

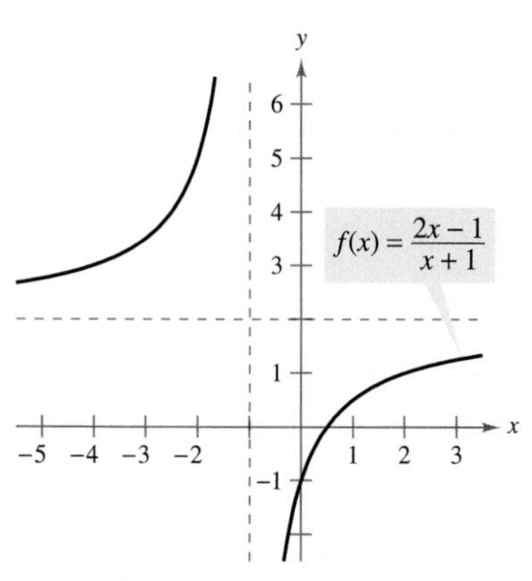

$y = 2$ is a horizontal asymptote.
Figure 4.35

TECHNOLOGY You can test the reasonableness of the limit found in Example 2 by evaluating $f(x)$ for a few large positive values of x. For instance,

$$f(100) \approx 1.9703, \quad f(1000) \approx 1.9970, \quad \text{and} \quad f(10{,}000) \approx 1.9997.$$

Another way to test the reasonableness of the limit is to use a graphing utility. For instance, in Figure 4.36, the graph of

$$f(x) = \frac{2x-1}{x+1}$$

is shown with the horizontal line $y = 2$. Note that as x increases, the graph of f moves closer and closer to its horizontal asymptote.

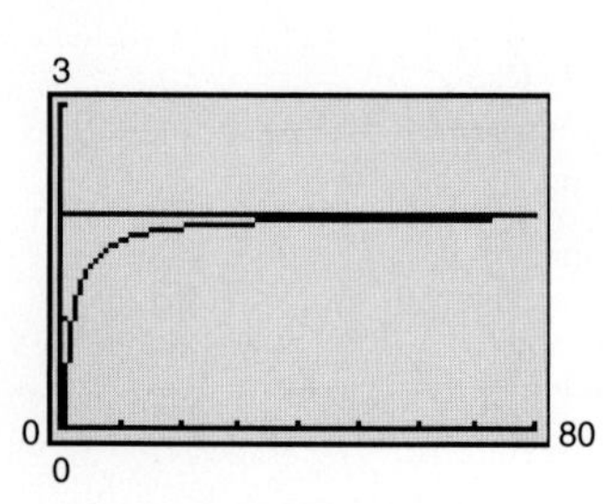

As x increases, the graph of f moves closer and closer to the line $y = 2$.
Figure 4.36

EXAMPLE 3 A Comparison of Three Rational Functions

Find each limit.

a. $\displaystyle\lim_{x\to\infty} \frac{2x + 5}{3x^2 + 1}$ **b.** $\displaystyle\lim_{x\to\infty} \frac{2x^2 + 5}{3x^2 + 1}$ **c.** $\displaystyle\lim_{x\to\infty} \frac{2x^3 + 5}{3x^2 + 1}$

Solution In each case, attempting to evaluate the limit produces the indeterminate form ∞/∞.

a. Divide both the numerator and the denominator by x^2.

$$\lim_{x\to\infty} \frac{2x + 5}{3x^2 + 1} = \lim_{x\to\infty} \frac{(2/x) + (5/x^2)}{3 + (1/x^2)} = \frac{0 + 0}{3 + 0} = \frac{0}{3} = 0$$

b. Divide both the numerator and the denominator by x^2.

$$\lim_{x\to\infty} \frac{2x^2 + 5}{3x^2 + 1} = \lim_{x\to\infty} \frac{2 + (5/x^2)}{3 + (1/x^2)} = \frac{2 + 0}{3 + 0} = \frac{2}{3}$$

c. Divide both the numerator and the denominator by x^2.

$$\lim_{x\to\infty} \frac{2x^3 + 5}{3x^2 + 1} = \lim_{x\to\infty} \frac{2x + (5/x^2)}{3 + (1/x^2)} = \frac{\infty}{3}$$

You can conclude that the limit *does not exist* because the numerator increases without bound while the denominator approaches 3.

The Granger Collection

MARIA GAETANA AGNESI (1718–1799)

Agnesi was one of a handful of women to receive credit for significant contributions to mathematics before the twentieth century. In her early twenties, she wrote the first text that included both differential and integral calculus. By age 30, she was an honorary member of the faculty at the University of Bologna.

Guidelines for Finding Limits at $\pm\infty$ of Rational Functions

1. If the degree of the numerator is *less than* the degree of the denominator, then the limit of the rational function is 0.
2. If the degree of the numerator is *equal to* the degree of the denominator, then the limit of the rational function is the ratio of the leading coefficients.
3. If the degree of the numerator is *greater than* the degree of the denominator, then the limit of the rational function does not exist.

Use these guidelines to check the results in Example 3. These limits seem reasonable when you consider that for large values of x, the highest-power term of the rational function is the most "influential" in determining the limit. For instance, the limit as x approaches infinity of the function

$$f(x) = \frac{1}{x^2 + 1}$$

is 0 because the denominator overpowers the numerator as x increases or decreases without bound, as shown in Figure 4.37.

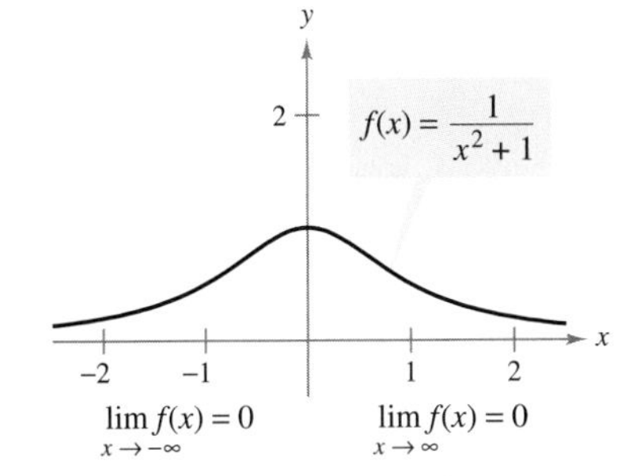

f has a horizontal asymptote at $y = 0$.
Figure 4.37

The function shown in Figure 4.37 is a special case of a type of curve studied by the Italian mathematician Maria Gaetana Agnesi. The general form of this function is

$$f(x) = \frac{8a^3}{x^2 + 4a^2} \qquad \text{Witch of Agnesi}$$

and, through a mistranslation of the Italian word *vertéré*, the curve has come to be known as the Witch of Agnesi. Agnesi's work with this curve first appeared in a comprehensive text on calculus that was published in 1748.

FOR FURTHER INFORMATION For more information on the contributions of women to mathematics, see the article "Why Women Succeed in Mathematics" by Mona Fabricant, Sylvia Svitak, and Patricia Clark Kenschaft in *Mathematics Teacher*. To view this article, go to the website *www.matharticles.com*.

In Figure 4.37, you can see that the function

$$f(x) = \frac{1}{x^2 + 1}$$

approaches the same horizontal asymptote to the right and to the left. This is always true of rational functions. Functions that are not rational, however, may approach different horizontal asymptotes to the right and to the left. A common example of such a function is the **logistic function** shown in the next example.

EXAMPLE 4 A Function with Two Horizontal Asymptotes

Show that the *logistic function*

$$f(x) = \frac{1}{1 + e^{-x}}$$

has different horizontal asymptotes to the left and to the right.

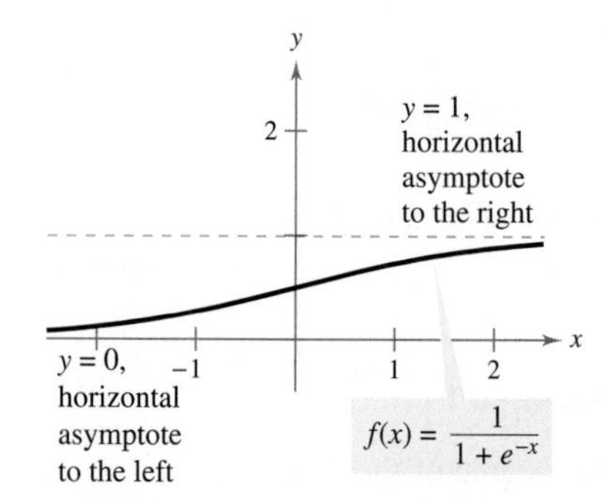

Functions that are not rational may have different right and left horizontal asymptotes.
Figure 4.38

Solution To begin, try using a graphing utility to graph the function. From Figure 4.38 it appears that

$$y = 0 \quad \text{and} \quad y = 1$$

are horizontal asymptotes to the left and to the right, respectively. The following table shows the same results numerically.

x	-10	-5	-2	-1	1	2	5	10
$f(x)$	0.000	0.007	0.119	0.269	0.731	0.881	0.9933	1.0000

Finally, you can obtain the same results analytically, as follows.

$$\lim_{x\to\infty} \frac{1}{1 + e^{-x}} = \frac{\lim_{x\to\infty} 1}{\lim_{x\to\infty} (1 + e^{-x})} = \frac{1}{1 + 0} = 1 \qquad y = 1 \text{ is a horizontal asymptote to the right.}$$

The denominator approaches infinity as x approaches negative infinity. So, the quotient approaches 0 and thus the limit is 0.

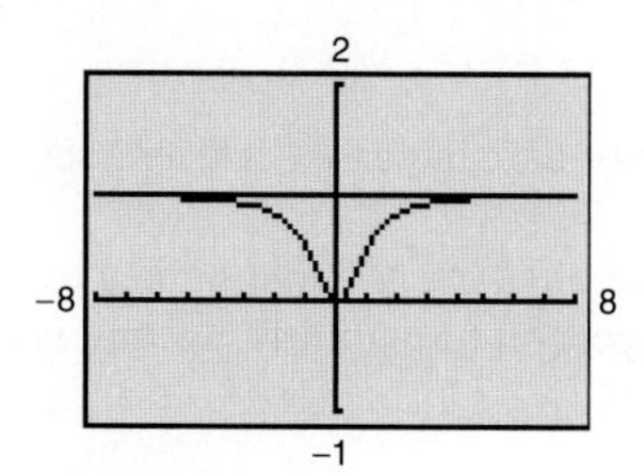

The horizontal asymptote appears to be the line $y = 1$ but it is actually the line $y = 2$.
Figure 4.39

TECHNOLOGY PITFALL If you use a graphing utility to help estimate a limit, be sure that you also confirm the estimate analytically—the pictures shown by a graphing utility can be misleading. For instance, Figure 4.39 shows one view of the graph of

$$y = \frac{2x^3 + 1000x^2 + x}{x^3 + 1000x^2 + x + 1000}.$$

From this view, one could be convinced that the graph has $y = 1$ as a horizontal asymptote. An analytical approach shows that the horizontal asymptote is actually $y = 2$. Confirm this by enlarging the viewing window on the graphing utility.

In Section 2.3 (Example 9), you saw how the Squeeze Theorem can be used to evaluate limits involving trigonometric functions. This theorem is also valid for limits at infinity.

EXAMPLE 5 Limits Involving Trigonometric Functions

Find each limit.

a. $\lim\limits_{x\to\infty} \sin x$ **b.** $\lim\limits_{x\to\infty} \dfrac{\sin x}{x}$

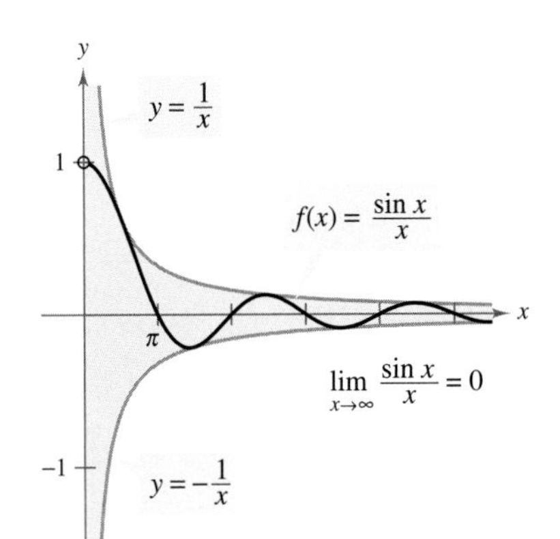

As x increases without bound, $f(x)$ approaches 0.
Figure 4.40

Solution

a. As x approaches infinity, the sine function oscillates between 1 and -1. So, this limit does not exist.

b. Because $-1 \le \sin x \le 1$, it follows that for $x > 0$,

$$-\frac{1}{x} \le \frac{\sin x}{x} \le \frac{1}{x}$$

where $\lim\limits_{x\to\infty} (-1/x) = 0$ and $\lim\limits_{x\to\infty} (1/x) = 0$. So, by the Squeeze Theorem, you can obtain

$$\lim_{x\to\infty} \frac{\sin x}{x} = 0$$

as shown in Figure 4.40.

EXAMPLE 6 Oxygen Level in a Pond

Suppose that $f(t)$ measures the level of oxygen in a pond, where $f(t) = 1$ is the normal (unpolluted) level and the time t is measured in weeks. When $t = 0$, organic waste is dumped into the pond, and as the waste material oxidizes, the level of oxygen in the pond is

$$f(t) = \frac{t^2 - t + 1}{t^2 + 1}.$$

What percent of the normal level of oxygen exists in the pond after 1 week? After 2 weeks? After 10 weeks? What is the limit as t approaches infinity?

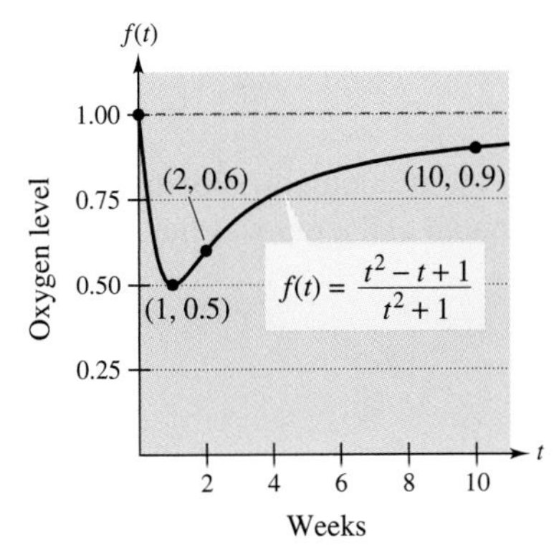

The level of oxygen in a pond approaches the normal level of 1 as t approaches ∞.
Figure 4.41

Solution When $t = 1$, 2, and 10, the levels of oxygen are as shown.

$$f(1) = \frac{1^2 - 1 + 1}{1^2 + 1} = \frac{1}{2} = 50\% \qquad \text{1 week}$$

$$f(2) = \frac{2^2 - 2 + 1}{2^2 + 1} = \frac{3}{5} = 60\% \qquad \text{2 weeks}$$

$$f(10) = \frac{10^2 - 10 + 1}{10^2 + 1} = \frac{91}{101} \approx 90.1\% \qquad \text{10 weeks}$$

To find the limit as t approaches infinity, divide the numerator and the denominator by t^2 to obtain

$$\lim_{t\to\infty} \frac{t^2 - t + 1}{t^2 + 1} = \lim_{t\to\infty} \frac{1 - (1/t) + (1/t^2)}{1 + (1/t^2)} = \frac{1 - 0 + 0}{1 + 0} = 1 = 100\%.$$

See Figure 4.41.

Infinite Limits at Infinity

Many functions do not approach a finite limit as x increases (or decreases) without bound. For instance, no polynomial function has a finite limit at infinity. The following definition is used to describe the behavior of polynomial and other functions at infinity.

NOTE Determining whether a function has an infinite limit at infinity is useful in analyzing the "end behavior" of its graph. You will see examples of this in Section 4.6 on curve sketching.

Definition of Infinite Limits at Infinity

Let f be a function defined on the interval (a, ∞).

1. The statement $\lim_{x\to\infty} f(x) = \infty$ means that for each positive number M, there is a corresponding number $N > 0$ such that $f(x) > M$ whenever $x > N$.
2. The statement $\lim_{x\to\infty} f(x) = -\infty$ means that for each negative number M, there is a corresponding number $N > 0$ such that $f(x) < M$ whenever $x > N$.

Similar definitions can be given for the statements $\lim_{x\to-\infty} f(x) = \infty$ and $\lim_{x\to-\infty} f(x) = -\infty$.

EXAMPLE 7 Finding Infinite Limits at Infinity

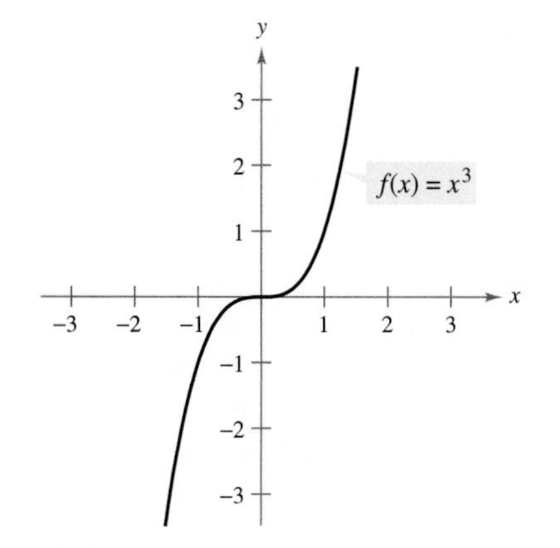

Figure 4.42

Find each limit.

a. $\lim_{x\to\infty} x^3$ **b.** $\lim_{x\to-\infty} x^3$

Solution

a. As x increases without bound, x^3 also increases without bound. So, you can write $\lim_{x\to\infty} x^3 = \infty$.

b. As x decreases without bound, x^3 also decreases without bound. So, you can write $\lim_{x\to-\infty} x^3 = -\infty$.

The graph of $f(x) = x^3$ in Figure 4.42 illustrates these two results. These results agree with the Leading Coefficient Test for polynomial functions as described in Section 1.3.

EXAMPLE 8 Finding Infinite Limits at Infinity

Find each limit.

a. $\lim_{x\to\infty} \dfrac{2x^2 - 4x}{x + 1}$ **b.** $\lim_{x\to-\infty} \dfrac{2x^2 - 4x}{x + 1}$

Solution One way to evaluate each of these limits is to use long division to rewrite the improper rational function as the sum of a polynomial and a rational function.

$$\text{a. } \lim_{x\to\infty} \frac{2x^2 - 4x}{x + 1} = \lim_{x\to\infty}\left(2x - 6 + \frac{6}{x + 1}\right) = \infty$$

$$\text{b. } \lim_{x\to-\infty} \frac{2x^2 - 4x}{x + 1} = \lim_{x\to-\infty}\left(2x - 6 + \frac{6}{x + 1}\right) = -\infty$$

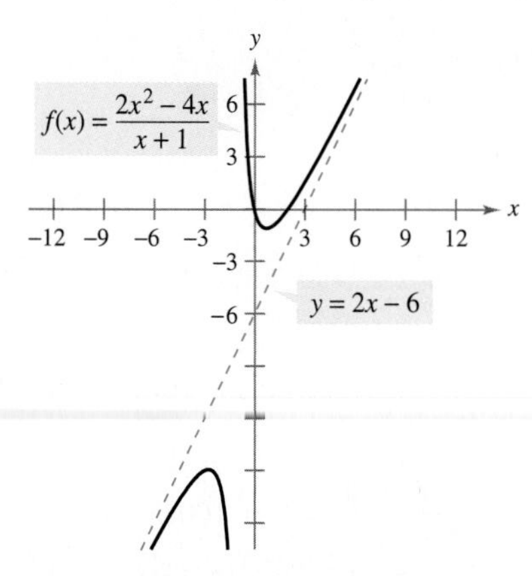

Figure 4.43

The statements above can be interpreted as saying that as x approaches $\pm\infty$, the function $f(x) = (2x^2 - 4x)/(x + 1)$ behaves like the function $g(x) = 2x - 6$. In Section 4.6, you will see that this is graphically described by saying that the line $y = 2x - 6$ is a slant asymptote of the graph of f, as shown in Figure 4.43.

Exercises for Section 4.5

See www.CalcChat.com for worked-out solutions to odd-numbered exercises.

In Exercises 1 and 2, describe in your own words what the statement means.

1. $\lim_{x\to\infty} f(x) = 4$

2. $\lim_{x\to-\infty} f(x) = 2$

In Exercises 3–8, match the function with one of the graphs [(a), (b), (c), (d), (e), or (f)] using horizontal asymptotes as an aid.

(a)

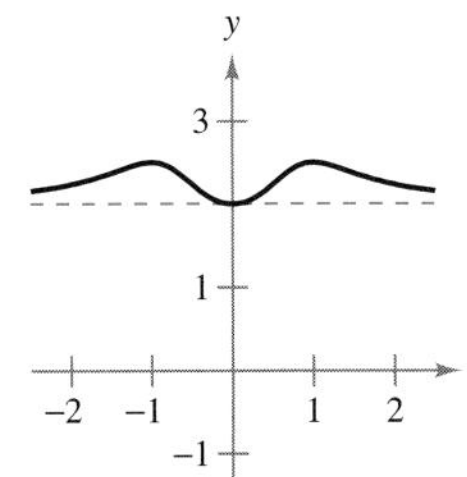

(b)

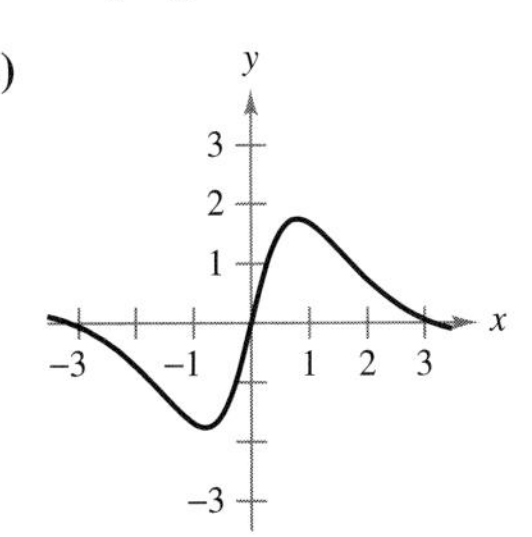

(c)

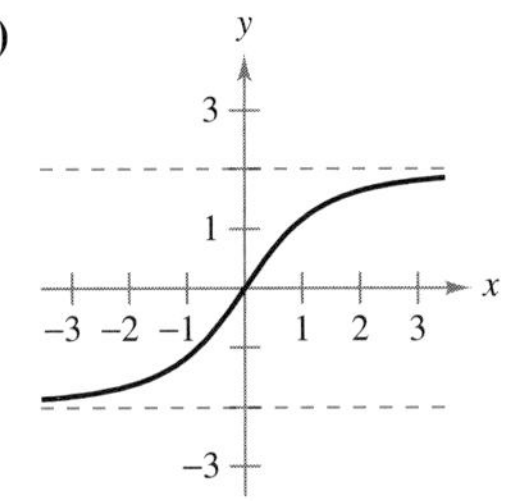

(d)

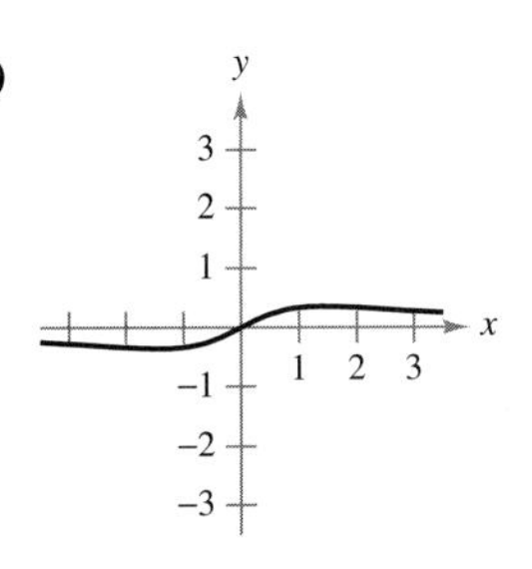

(e)

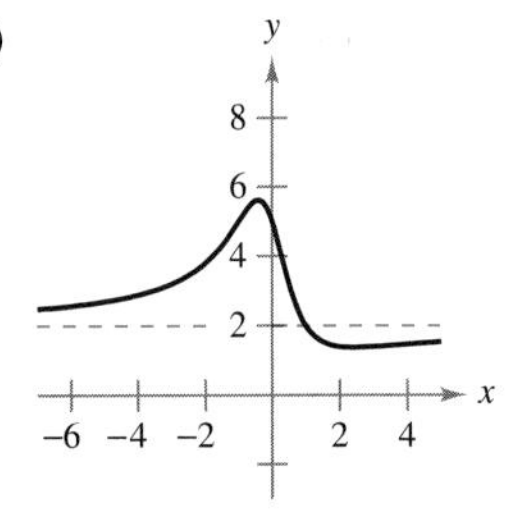

(f)

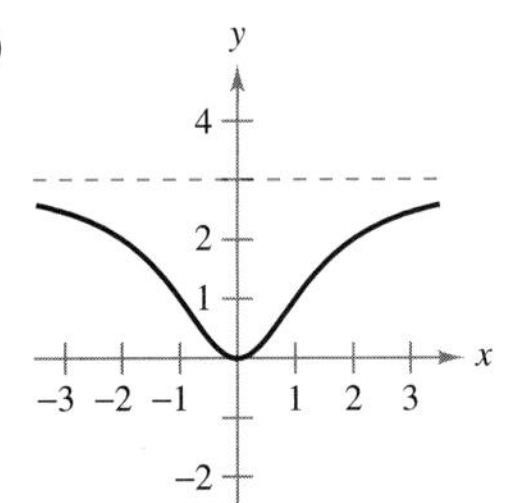

3. $f(x) = \dfrac{3x^2}{x^2 + 2}$

4. $f(x) = \dfrac{2x}{\sqrt{x^2 + 2}}$

5. $f(x) = \dfrac{x}{x^2 + 2}$

6. $f(x) = 2 + \dfrac{x^2}{x^4 + 1}$

7. $f(x) = \dfrac{4 \sin x}{x^2 + 1}$

8. $f(x) = \dfrac{2x^2 - 3x + 5}{x^2 + 1}$

Numerical and Graphical Analysis **In Exercises 9–14, use a graphing utility to complete the table and estimate the limit as x approaches infinity. Then use a graphing utility to graph the function and estimate the limit graphically.**

x	10^0	10^1	10^2	10^3	10^4	10^5	10^6
$f(x)$							

9. $f(x) = \dfrac{4x + 3}{2x - 1}$

10. $f(x) = \dfrac{2x^2}{x + 1}$

11. $f(x) = \dfrac{-6x}{\sqrt{4x^2 + 5}}$

12. $f(x) = \dfrac{8x}{\sqrt{x^2 - 3}}$

13. $f(x) = 5 - \dfrac{1}{x^2 + 1}$

14. $f(x) = 4 + \dfrac{3}{x^2 + 2}$

In Exercises 15 and 16, find $\lim_{x\to\infty} h(x)$, if possible.

15. $f(x) = 5x^3 - 3x^2 + 10$

   (a) $h(x) = \dfrac{f(x)}{x^2}$ (b) $h(x) = \dfrac{f(x)}{x^3}$ (c) $h(x) = \dfrac{f(x)}{x^4}$

16. $f(x) = 5x^2 - 3x + 7$

   (a) $h(x) = \dfrac{f(x)}{x}$ (b) $h(x) = \dfrac{f(x)}{x^2}$ (c) $h(x) = \dfrac{f(x)}{x^3}$

In Exercises 17–20, find each limit, if possible.

17. (a) $\lim_{x\to\infty} \dfrac{x^2 + 2}{x^3 - 1}$

   (b) $\lim_{x\to\infty} \dfrac{x^2 + 2}{x^2 - 1}$

   (c) $\lim_{x\to\infty} \dfrac{x^2 + 2}{x - 1}$

18. (a) $\lim_{x\to\infty} \dfrac{3 - 2x}{3x^3 - 1}$

   (b) $\lim_{x\to\infty} \dfrac{3 - 2x}{3x - 1}$

   (c) $\lim_{x\to\infty} \dfrac{3 - 2x^2}{3x - 1}$

19. (a) $\lim_{x\to\infty} \dfrac{5 - 2x^{3/2}}{3x^2 - 4}$

   (b) $\lim_{x\to\infty} \dfrac{5 - 2x^{3/2}}{3x^{3/2} - 4}$

   (c) $\lim_{x\to\infty} \dfrac{5 - 2x^{3/2}}{3x - 4}$

20. (a) $\lim_{x\to\infty} \dfrac{5x^{3/2}}{4x^2 + 1}$

   (b) $\lim_{x\to\infty} \dfrac{5x^{3/2}}{4x^{3/2} + 1}$

   (c) $\lim_{x\to\infty} \dfrac{5x^{3/2}}{4\sqrt{x} + 1}$

In Exercises 21–34, find the limit.

21. $\lim_{x\to\infty} \dfrac{2x - 1}{3x + 2}$

22. $\lim_{x\to\infty} \dfrac{3x^3 + 2}{9x^3 - 2x^2 + 7}$

23. $\lim_{x\to\infty} \dfrac{x}{x^2 - 1}$

24. $\lim_{x\to\infty} \left(4 + \dfrac{3}{x}\right)$

25. $\lim_{x\to-\infty} \dfrac{6x^2}{x + 3}$

26. $\lim_{x\to-\infty} \left(\dfrac{1}{2}x - \dfrac{4}{x^2}\right)$

27. $\lim_{x\to-\infty} \dfrac{x}{\sqrt{x^2 - x}}$

28. $\lim_{x\to-\infty} \dfrac{x}{\sqrt{x^2 + 1}}$

29. $\lim_{x\to-\infty} \dfrac{2x + 1}{\sqrt{x^2 - x}}$

30. $\lim_{x\to-\infty} \dfrac{-3x + 1}{\sqrt{x^2 + x}}$

31. $\lim_{x\to\infty} \dfrac{\sin 2x}{x}$

32. $\lim_{x\to\infty} \dfrac{3(x - \cos x)}{x}$

33. $\lim_{x\to\infty} \dfrac{1}{2x + \sin x}$

34. $\lim_{x\to\infty} \cos \dfrac{1}{x}$

35. $\lim_{x\to\infty} (2 - 5e^{-x})$

36. $\lim_{x\to-\infty} (2 + 5e^{x})$

37. $\lim_{x\to-\infty} \dfrac{3}{1 + 2e^x}$

38. $\lim_{x\to\infty} \dfrac{8}{4 - 10^{-x/2}}$

39. $\lim_{x\to\infty} \log_{10}(1 + 10^{-x})$

40. $\lim_{x\to\infty} \left[\dfrac{5}{2} + \ln\left(\dfrac{x^2 + 1}{x^2}\right)\right]$

41. $\lim_{t\to\infty} \left(\dfrac{5}{t} - \arctan t\right)$

42. $\lim_{u\to\infty} \operatorname{arcsec}(u + 1)$

In Exercises 43–46, use a graphing utility to graph the function and identify any horizontal asymptotes.

43. $f(x) = \dfrac{|x|}{x + 1}$

44. $f(x) = \dfrac{|3x + 2|}{x - 2}$

45. $f(x) = \dfrac{3x}{\sqrt{x^2 + 2}}$

46. $f(x) = \dfrac{\sqrt{9x^2 - 2}}{2x + 1}$

In Exercises 47 and 48, find the limit. (*Hint:* Let $x = 1/t$ and find the limit as $t \to 0^+$.)

47. $\lim\limits_{x \to \infty} x \sin \dfrac{1}{x}$

48. $\lim\limits_{x \to \infty} x \tan \dfrac{1}{x}$

In Exercises 49–54, find the limit. (*Hint:* Treat the expression as a fraction whose denominator is 1, and rationalize the numerator.) Use a graphing utility to verify your result.

49. $\lim\limits_{x \to -\infty} \left(x + \sqrt{x^2 + 3}\right)$

50. $\lim\limits_{x \to \infty} \left(2x - \sqrt{4x^2 + 1}\right)$

51. $\lim\limits_{x \to \infty} \left(x - \sqrt{x^2 + x}\right)$

52. $\lim\limits_{x \to -\infty} \left(3x + \sqrt{9x^2 - x}\right)$

53. $\lim\limits_{x \to \infty} \left(4x - \sqrt{16x^2 - x}\right)$

54. $\lim\limits_{x \to -\infty} \left(\dfrac{x}{2} + \sqrt{\dfrac{1}{4}x^2 + x}\right)$

Numerical, Graphical, and Analytic Analysis **In Exercises 55–58, use a graphing utility to complete the table and estimate the limit as x approaches infinity. Then use a graphing utility to graph the function and estimate the limit. Finally, find the limit analytically and compare your results with the estimates.**

x	10^0	10^1	10^2	10^3	10^4	10^5	10^6
$f(x)$							

55. $f(x) = x - \sqrt{x(x - 1)}$

56. $f(x) = x^2 - x\sqrt{x(x - 1)}$

57. $f(x) = x \sin \dfrac{1}{2x}$

58. $f(x) = \dfrac{x + 1}{x\sqrt{x}}$

Writing About Concepts

59. The graph of a function f is shown below. To print an enlarged copy of the graph, go to the website *www.mathgraphs.com*.

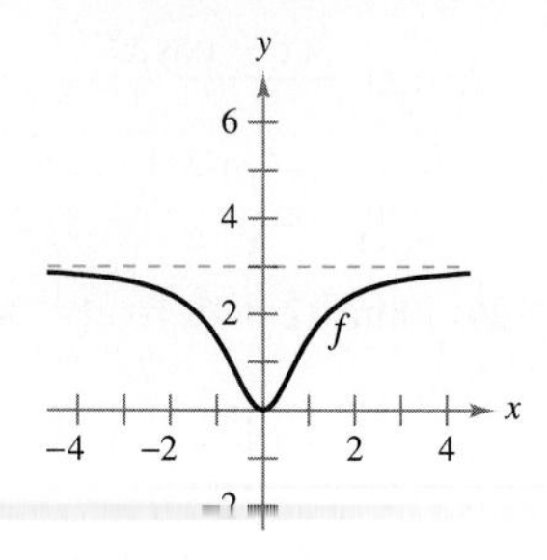

(a) Sketch f'.

(b) Use the graphs to estimate $\lim\limits_{x \to \infty} f(x)$ and $\lim\limits_{x \to \infty} f'(x)$.

(c) Explain the answers you gave in part (b).

Writing About Concepts *(continued)*

60. Sketch a graph of a differentiable function f that satisfies the following conditions and has $x = 2$ as its only critical number.

$f'(x) < 0$ for $x < 2$ $\qquad$ $f'(x) > 0$ for $x > 2$

$\lim\limits_{x \to -\infty} f(x) = \lim\limits_{x \to \infty} f(x) = 6$

61. Is it possible to sketch a graph of a function that satisfies the conditions of Exercise 60 and has *no* points of inflection? Explain.

62. If f is a continuous function such that $\lim\limits_{x \to \infty} f(x) = 5$, find, if possible, $\lim\limits_{x \to -\infty} f(x)$ for each specified condition.

(a) The graph of f is symmetric to the y-axis.

(b) The graph of f is symmetric to the origin.

In Exercises 63–80, sketch the graph of the equation. Look for extrema, intercepts, symmetry, and asymptotes as necessary. Use a graphing utility to verify your result.

63. $y = \dfrac{2 + x}{1 - x}$

64. $y = \dfrac{x - 3}{x - 2}$

65. $y = \dfrac{x}{x^2 - 4}$

66. $y = \dfrac{2x}{9 - x^2}$

67. $y = \dfrac{x^2}{x^2 + 9}$

68. $y = \dfrac{x^2}{x^2 - 9}$

69. $y = \dfrac{2x^2}{x^2 - 4}$

70. $y = \dfrac{2x^2}{x^2 + 4}$

71. $xy^2 = 4$

72. $x^2y = 4$

73. $y = \dfrac{2x}{1 - x}$

74. $y = \dfrac{2x}{1 - x^2}$

75. $y = 2 - \dfrac{3}{x^2}$

76. $y = 1 + \dfrac{1}{x}$

77. $y = 3 + \dfrac{2}{x}$

78. $y = 4\left(1 - \dfrac{1}{x^2}\right)$

79. $y = \dfrac{x^3}{\sqrt{x^2 - 4}}$

80. $y = \dfrac{x}{\sqrt{x^2 - 4}}$

In Exercises 81–92, use a computer algebra system to analyze the graph of the function. Label any extrema and/or asymptotes that exist.

81. $f(x) = 5 - \dfrac{1}{x^2}$

82. $f(x) = \dfrac{x^2}{x^2 - 1}$

83. $f(x) = \dfrac{x}{x^2 - 4}$

84. $f(x) = \dfrac{1}{x^2 - x - 2}$

85. $f(x) = \dfrac{x - 2}{x^2 - 4x + 3}$

86. $f(x) = \dfrac{x + 1}{x^2 + x + 1}$

87. $f(x) = \dfrac{3x}{\sqrt{4x^2 + 1}}$

88. $g(x) = \dfrac{2x}{\sqrt{3x^2 + 1}}$

89. $g(x) = \sin\left(\dfrac{x}{x-2}\right), \quad x > 3$ **90.** $f(x) = \dfrac{2 \sin 2x}{x}$

91. $f(x) = 2 + (x^2 - 3)e^{-x}$ **92.** $f(x) = \dfrac{10 \ln x}{x^2\sqrt{x}}$

In Exercises 93 and 94, (a) use a graphing utility to graph f and g in the same viewing window, (b) verify algebraically that f and g represent the same function, and (c) zoom out sufficiently far so that the graph appears as a line. What equation does this line appear to have? (Note that the points at which the function is not continuous are not readily seen when you zoom out.)

93. $f(x) = \dfrac{x^3 - 3x^2 + 2}{x(x-3)}$

$g(x) = x + \dfrac{2}{x(x-3)}$

94. $f(x) = -\dfrac{x^3 - 2x^2 + 2}{2x^2}$

$g(x) = -\dfrac{1}{2}x + 1 - \dfrac{1}{x^2}$

95. ***Average Cost*** A business has a cost of $C = 0.5x + 500$ for producing x units. The average cost per unit is $\overline{C} = \dfrac{C}{x}$. Find the limit of $\overline{C}$ as x approaches infinity.

96. ***Engine Efficiency*** The efficiency of an internal combustion engine is

$$\text{Efficiency (\%)} = 100\left[1 - \frac{1}{(v_1/v_2)^c}\right]$$

where v_1/v_2 is the ratio of the uncompressed gas to the compressed gas and c is a positive constant dependent on the engine design. Find the limit of the efficiency as the compression ratio approaches infinity.

97. ***Physics*** Newton's First Law of Motion and Einstein's Special Theory of Relativity differ concerning a particle's behavior as its velocity approaches the speed of light c. Functions N and E represent the predicted velocity v with respect to time t for a particle accelerated by a constant force. Write a limit statement that describes each theory.

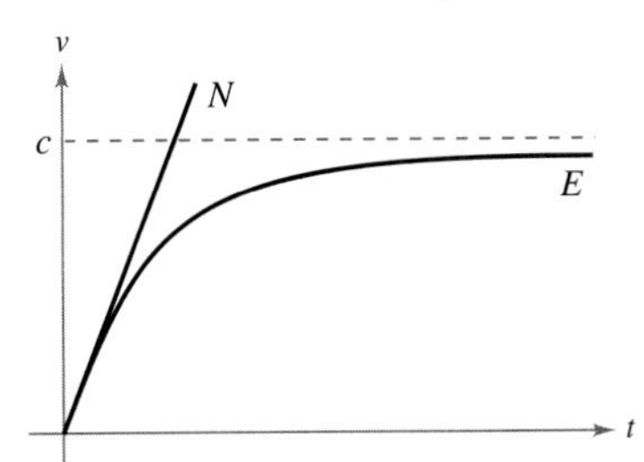

98. ***Temperature*** The graph shows the temperature T (in degrees Fahrenheit) of an apple pie t seconds after it is removed from an oven and placed on a cooling rack.

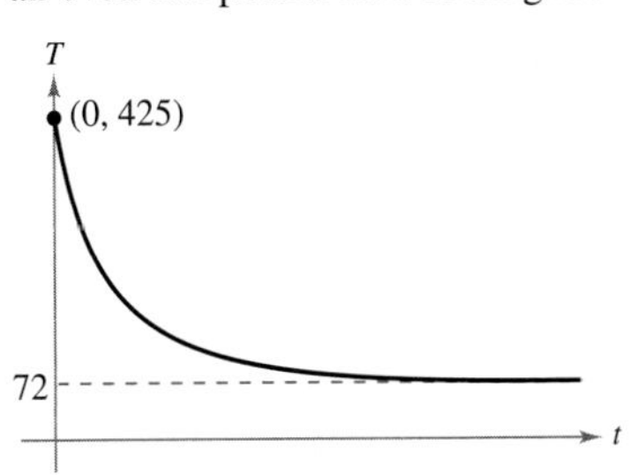

(a) Find $\lim_{t \to 0^+} T$. What does this limit represent?

(b) Find $\lim_{t \to \infty} T$. What does this limit represent?

99. ***Modeling Data*** A heat probe is attached to the heat exchanger of a heating system. The temperature T (in degrees Celsius) is recorded t seconds after the furnace is started. The results for the first 2 minutes are recorded in the table.

t	0	15	30	45	60	75	90	105	120
T	25.2°	36.9°	45.5°	51.4°	56.0°	59.6°	62.0°	64.0°	65.2°

(a) Use the regression capabilities of a graphing utility to find a model of the form $T_1 = at^2 + bt + c$ for the data.

(b) Use a graphing utility to graph T_1.

(c) A rational model for the data is $T_2 = \dfrac{1451 + 86t}{58 + t}$. Use a graphing utility to graph the model.

(d) Find $T_1(0)$ and $T_2(0)$.

(e) Find $\lim_{t \to \infty} T_2$.

(f) Interpret the result in part (e) in the context of the problem. Is it possible to do this type of analysis using T_1? Explain.

100. ***Modeling Data*** A container contains 5 liters of a 25% brine solution. The table shows the concentrations C of the mixture after adding x liters of a 75% brine solution to the container.

x	0	0.5	1	1.5	2	2.5	3	3.5	4
C	0.25	0.295	0.333	0.365	0.393	0.417	0.438	0.456	0.472

(a) Use the regression features of a graphing utility to find a model of the form $C_1 = ax^2 + bx + c$ for the data.

(b) Use a graphing utility to graph C_1.

(c) A rational model for these data is $C_2 = \dfrac{5 + 3x}{20 + 4x}$. Use a graphing utility to graph C_2.

(d) Find $\lim_{x \to \infty} C_1$ and $\lim_{x \to \infty} C_2$. Which model do you think best represents the concentration of the mixture? Explain.

(e) What is the limiting concentration?

101. ***Timber Yield*** The yield V (in millions of cubic feet per acre) for a stand of timber at age t (in years) is $V = 7.1e^{(-48.1)/t}$.

(a) Find the limiting volume of wood per acre as t approaches infinity.

(b) Find the rates at which the yield is changing when $t = 20$ years and $t = 60$ years.

102. *Learning Theory* In a group project in learning theory, a mathematical model for the proportion P of correct responses after n trials was found to be

$$P = \frac{0.83}{1 + e^{-0.2n}}.$$

(a) Find the limiting proportion of correct responses as n approaches infinity.

(b) Find the rates at which P is changing after $n = 3$ trials and $n = 10$ trials.

103. *Writing* Consider the function $f(x) = \dfrac{2}{1 + e^{1/x}}$.

(a) Use a graphing utility to graph f.

(b) Write a short paragraph explaining why the graph has a horizontal asymptote at $y = 1$ and why the function has a nonremovable discontinuity at $x = 0$.

104. *Writing* In your own words, state the guidelines for finding the limit of a rational function. Give examples.

105. A line with slope m passes through the point $(0, 4)$.

(a) Write the distance d between the line and the point $(3, 1)$ as a function of m.

(b) Use a graphing utility to graph the equation in part (a).

(c) Find $\lim_{m\to\infty} d(m)$ and $\lim_{m\to-\infty} d(m)$. Interpret the results geometrically.

106. A line with slope m passes through the point $(0, -2)$.

(a) Write the distance d between the line and the point $(4, 2)$ as a function of m.

(b) Use a graphing utility to graph the equation in part (a).

(c) Find $\lim_{m\to\infty} d(m)$ and $\lim_{m\to-\infty} d(m)$. Interpret the results geometrically.

107. The graph of $f(x) = \dfrac{2x^2}{x^2 + 2}$ is shown.

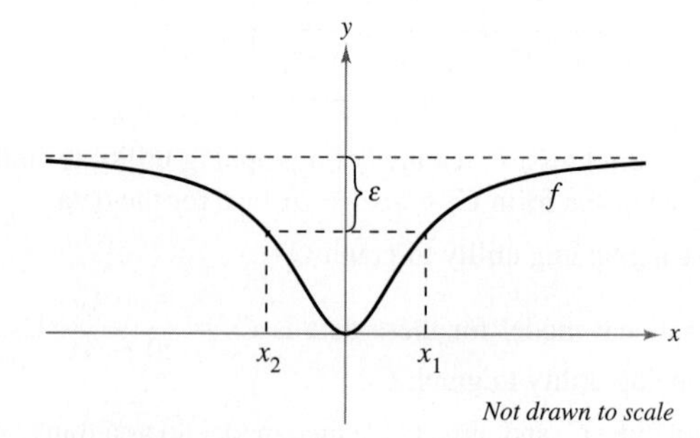

Not drawn to scale

(a) Find $L = \lim_{x\to\infty} f(x)$.

(b) Determine x_1 and x_2 in terms of ε.

(c) Determine M, where $M > 0$, such that $|f(x) - L| < \varepsilon$ for $x > M$.

(d) Determine N, where $N < 0$, such that $|f(x) - L| < \varepsilon$ for $x < N$.

108. The graph of $f(x) = \dfrac{6x}{\sqrt{x^2 + 2}}$ is shown.

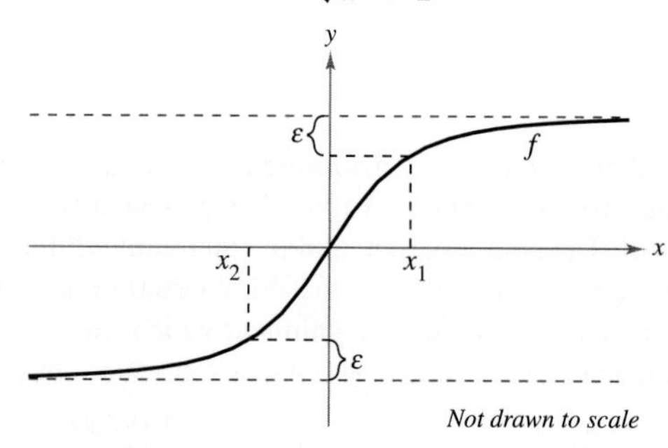

Not drawn to scale

(a) Find $L = \lim_{x\to\infty} f(x)$ and $K = \lim_{x\to-\infty} f(x)$.

(b) Determine x_1 and x_2 in terms of ε.

(c) Determine M, where $M > 0$, such that $|f(x) - L| < \varepsilon$ for $x > M$.

(d) Determine N, where $N < 0$, such that $|f(x) - K| < \varepsilon$ for $x < N$.

109. Consider $\lim_{x\to\infty} \dfrac{3x}{\sqrt{x^2 + 3}}$. Use the definition of limits at infinity to find values of M that correspond to (a) $\varepsilon = 0.5$ and (b) $\varepsilon = 0.1$.

110. Consider $\lim_{x\to-\infty} \dfrac{3x}{\sqrt{x^2 + 3}}$. Use the definition of limits at infinity to find values of N that correspond to (a) $\varepsilon = 0.5$ and (b) $\varepsilon = 0.1$.

In Exercises 111–114, use the definition of limits at infinity to prove the limit.

111. $\lim_{x\to\infty} \dfrac{1}{x^2} = 0$

112. $\lim_{x\to\infty} \dfrac{2}{\sqrt{x}} = 0$

113. $\lim_{x\to-\infty} \dfrac{1}{x^3} = 0$

114. $\lim_{x\to-\infty} \dfrac{1}{x - 2} = 0$

115. Prove that if $p(x) = a_n x^n + \cdots + a_1 x + a_0$ and $q(x) = b_m x^m + \cdots + b_1 x + b_0$ $(a_n \neq 0, b_m \neq 0)$, then

$$\lim_{x\to\infty} \frac{p(x)}{q(x)} = \begin{cases} 0, & n < m \\ \dfrac{a_n}{b_m}, & n = m. \\ \pm\infty, & n > m \end{cases}$$

116. Use the definition of infinite limits at infinity to prove that $\lim_{x\to\infty} x^3 = \infty$.

True or False? **In Exercises 117 and 118, determine whether the statement is true or false. If it is false, explain why or give an example that shows it is false.**

117. If $f'(x) > 0$ for all real numbers x, then f increases without bound.

118. If $f''(x) < 0$ for all real numbers x, then f decreases without bound.

Section 4.6 A Summary of Curve Sketching

- **Analyze and sketch the graph of a function.**

Analyzing the Graph of a Function

It would be difficult to overstate the importance of using graphs in mathematics. Descartes's introduction of analytic geometry contributed significantly to the rapid advances in calculus that began during the mid-seventeenth century. In the words of Lagrange, "As long as algebra and geometry traveled separate paths, their advance was slow and their applications limited. But when these two sciences joined company, they drew from each other fresh vitality and thenceforth marched on at a rapid pace toward perfection."

So far, you have studied several concepts that are useful in analyzing the graph of a function.

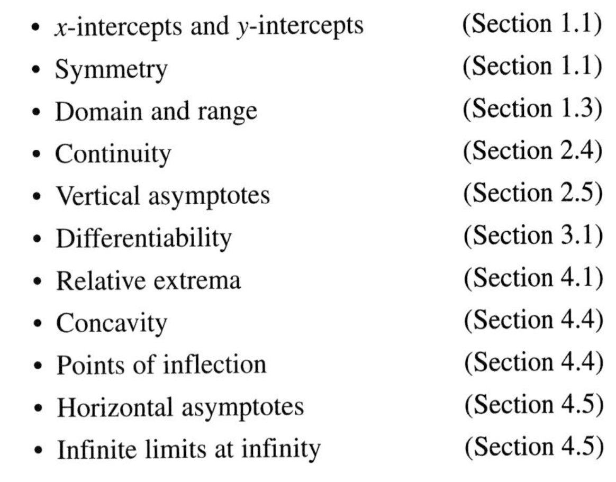

- x-intercepts and y-intercepts (Section 1.1)
- Symmetry (Section 1.1)
- Domain and range (Section 1.3)
- Continuity (Section 2.4)
- Vertical asymptotes (Section 2.5)
- Differentiability (Section 3.1)
- Relative extrema (Section 4.1)
- Concavity (Section 4.4)
- Points of inflection (Section 4.4)
- Horizontal asymptotes (Section 4.5)
- Infinite limits at infinity (Section 4.5)

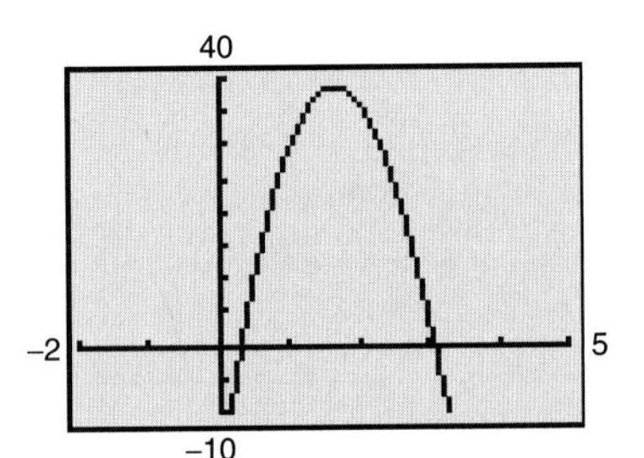

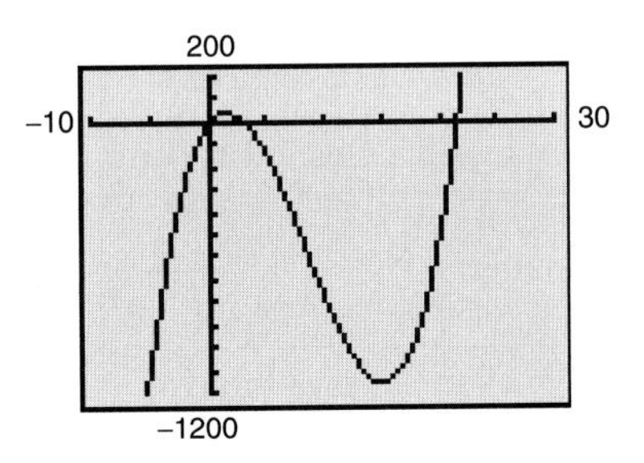

Different viewing windows for the graph of $f(x) = x^3 - 25x^2 + 74x - 20$
Figure 4.44

When you are sketching the graph of a function, either by hand or with a graphing utility, remember that normally you cannot show the *entire* graph. The decision as to which part of the graph you choose to show is often crucial. For instance, which of the viewing windows in Figure 4.44 better represents the graph of

$$f(x) = x^3 - 25x^2 + 74x - 20?$$

By seeing both views, it is clear that the second viewing window gives a more complete representation of the graph. But would a third viewing window reveal other interesting portions of the graph? To answer this, you need to use calculus to interpret the first and second derivatives. Here are some guidelines for determining a good viewing window for the graph of a function.

Guidelines for Analyzing the Graph of a Function

1. Determine the domain and range of the function.
2. Determine the intercepts, asymptotes, and symmetry of the graph.
3. Locate the x-values for which $f'(x)$ and $f''(x)$ either are zero or do not exist. Use the results to determine relative extrema and points of inflection.

NOTE In these guidelines, note the importance of *algebra* (as well as calculus) for solving the equations $f(x) = 0$, $f'(x) = 0$, and $f''(x) = 0$.

EXAMPLE 1 Sketching the Graph of a Rational Function

Analyze and sketch the graph of $f(x) = \dfrac{2(x^2 - 9)}{x^2 - 4}$.

Solution

First derivative:	$f'(x) = \dfrac{20x}{(x^2 - 4)^2}$
Second derivative:	$f''(x) = \dfrac{-20(3x^2 + 4)}{(x^2 - 4)^3}$
x-intercepts:	$(-3, 0), (3, 0)$
y-intercept:	$\left(0, \frac{9}{2}\right)$
Vertical asymptotes:	$x = -2, x = 2$
Horizontal asymptote:	$y = 2$
Critical number:	$x = 0$
Possible points of inflection:	None
Domain:	All real numbers except $x = \pm 2$
Symmetry:	With respect to y-axis
Test intervals:	$(-\infty, -2), (-2, 0), (0, 2), (2, \infty)$

The table shows how the test intervals are used to determine several characteristics of the graph. The graph of f is shown in Figure 4.45.

	$f(x)$	$f'(x)$	$f''(x)$	Characteristic of Graph
$-\infty < x < -2$		$-$	$-$	Decreasing, concave downward
$x = -2$	Undef.	Undef.	Undef.	Vertical asymptote
$-2 < x < 0$		$-$	$+$	Decreasing, concave upward
$x = 0$	$\frac{9}{2}$	0	$+$	Relative minimum
$0 < x < 2$		$+$	$+$	Increasing, concave upward
$x = 2$	Undef.	Undef.	Undef.	Vertical asymptote
$2 < x < \infty$		$+$	$-$	Increasing, concave downward

Using calculus, you can be certain that you have determined all characteristics of the graph of f.
Figure 4.45

FOR FURTHER INFORMATION For more information on the use of technology to graph rational functions, see the article "Graphs of Rational Functions for Computer Assisted Calculus" by Stan Byrd and Terry Walters in *The College Mathematics Journal*. To view this article, go to the website *www.matharticles.com*.

Be sure you understand all of the implications of creating a table such as that shown in Example 1. Because of the use of calculus, you can *be sure* that the graph has no relative extrema or points of inflection other than those shown in Figure 4.45.

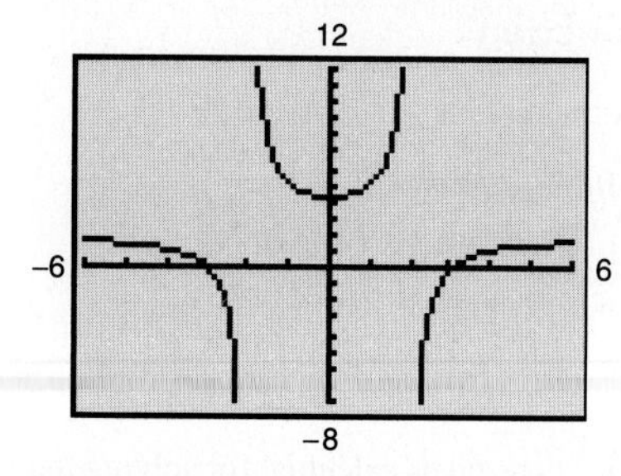

By not using calculus you may overlook important characteristics of the graph of g.
Figure 4.46

TECHNOLOGY PITFALL Without using the type of analysis outlined in Example 1, it is easy to obtain an incomplete view of a graph's basic characteristics. For instance, Figure 4.46 shows a view of the graph of

$$g(x) = \frac{2(x^2 - 9)(x - 20)}{(x^2 - 4)(x - 21)}.$$

From this view, it appears that the graph of g is about the same as the graph of f shown in Figure 4.45. The graphs of these two functions, however, differ significantly. Try enlarging the viewing window to see the differences.

EXAMPLE 2 Sketching the Graph of a Rational Function

Analyze and sketch the graph of $f(x) = \dfrac{x^2 - 2x + 4}{x - 2}$.

Solution

First derivative: $f'(x) = \dfrac{x(x-4)}{(x-2)^2}$

Second derivative: $f''(x) = \dfrac{8}{(x-2)^3}$

x-intercepts: None

y-intercept: $(0, -2)$

Vertical asymptote: $x = 2$

Horizontal asymptotes: None

End behavior: $\lim_{x\to-\infty} f(x) = -\infty, \ \lim_{x\to\infty} f(x) = \infty$

Critical numbers: $x = 0, x = 4$

Possible points of inflection: None

Domain: All real numbers except $x = 2$

Test intervals: $(-\infty, 0), (0, 2), (2, 4), (4, \infty)$

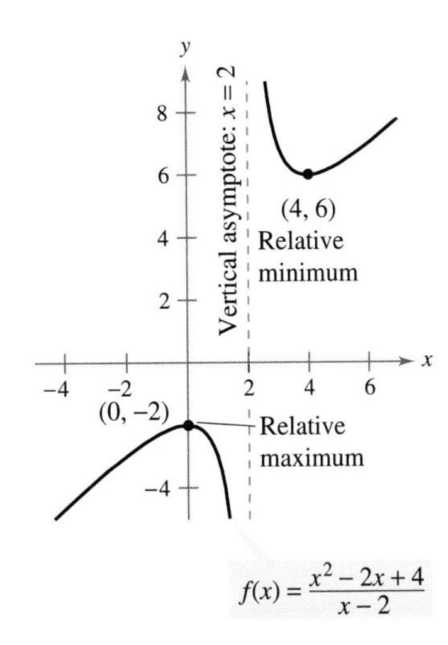

Figure 4.47

The analysis of the graph of f is shown in the table, and the graph is shown in Figure 4.47.

	$f(x)$	$f'(x)$	$f''(x)$	Characteristic of Graph
$-\infty < x < 0$		$+$	$-$	Increasing, concave downward
$x = 0$	-2	0	$-$	Relative maximum
$0 < x < 2$		$-$	$-$	Decreasing, concave downward
$x = 2$	Undef.	Undef.	Undef.	Vertical asymptote
$2 < x < 4$		$-$	$+$	Decreasing, concave upward
$x = 4$	6	0	$+$	Relative minimum
$4 < x < \infty$		$+$	$+$	Increasing, concave upward

Although the graph of the function in Example 2 has no horizontal asymptote, it does have a slant asymptote. The graph of a rational function (having no common factors and whose denominator is of degree 1 or greater) has a **slant asymptote** if the degree of the numerator exceeds the degree of the denominator by exactly 1. To find the slant asymptote, use long division to rewrite the rational function as the sum of a first-degree polynomial and another rational function.

$$f(x) = \frac{x^2 - 2x + 4}{x - 2} \qquad \text{Write original equation.}$$

$$= x + \frac{4}{x - 2} \qquad \text{Rewrite using long division.}$$

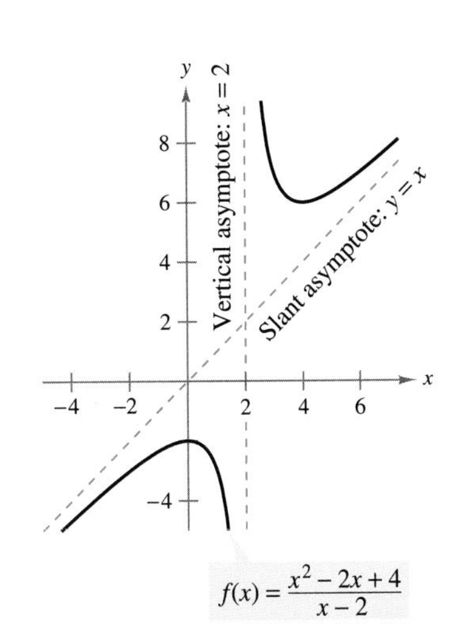

A slant asymptote
Figure 4.48

In Figure 4.48, note that the graph of f approaches the slant asymptote $y = x$ as x approaches $-\infty$ or ∞.

EXAMPLE 3 Sketching the Graph of a Logistic Function

Analyze and sketch the graph of the *logistic function* $f(x) = \dfrac{1}{1 + e^{-x}}$.

Solution

$$f'(x) = \frac{e^{-x}}{(1 + e^{-x})^2} \qquad \text{and} \qquad f''(x) = -\frac{e^{-x}(e^{-x} - 1)}{(1 + e^{-x})^3}$$

The graph has only one intercept, $\left(0, \frac{1}{2}\right)$. It has no vertical asymptotes, but it has two horizontal asymptotes: $y = 1$ (to the right) and $y = 0$ (to the left). The function has no critical numbers and one possible point of inflection (at $x = 0$). The domain of the function is all real numbers. The analysis of the graph of f is shown in the table, and the graph is shown in Figure 4.49.

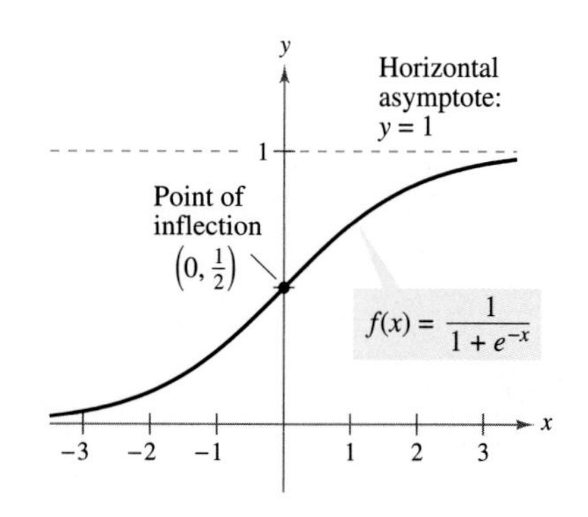

Figure 4.49

	$f(x)$	$f'(x)$	$f''(x)$	Characteristic of Graph
$-\infty < x < 0$		+	+	Increasing, concave upward
$x = 0$	$\frac{1}{2}$	$\frac{1}{4}$	0	Point of inflection
$0 < x < \infty$		+	−	Increasing, concave downward

EXAMPLE 4 Sketching the Graph of a Radical Function

Analyze and sketch the graph of $f(x) = 2x^{5/3} - 5x^{4/3}$.

Solution

$$f'(x) = \frac{10}{3}x^{1/3}(x^{1/3} - 2) \qquad f''(x) = \frac{20(x^{1/3} - 1)}{9x^{2/3}}$$

The function has two intercepts: $(0, 0)$ and $\left(\frac{125}{8}, 0\right)$. There are no horizontal or vertical asymptotes. The function has two critical numbers ($x = 0$ and $x = 8$) and two possible points of inflection ($x = 0$ and $x = 1$). The domain is all real numbers. The analysis of the graph of f is shown in the table, and the graph is shown in Figure 4.50.

Figure 4.50

	$f(x)$	$f'(x)$	$f''(x)$	Characteristic of Graph
$-\infty < x < 0$		+	−	Increasing, concave downward
$x = 0$	0	0	Undef.	Relative maximum
$0 < x < 1$		−	−	Decreasing, concave downward
$x = 1$	-3	−	0	Point of inflection
$1 < x < 8$		−	+	Decreasing, concave upward
$x = 8$	-16	0	+	Relative minimum
$8 < x < \infty$		+	+	Increasing, concave upward

EXAMPLE 5 Sketching the Graph of a Polynomial Function

Analyze and sketch the graph of $f(x) = x^4 - 12x^3 + 48x^2 - 64x$.

Solution Begin by factoring to obtain

$$\begin{aligned} f(x) &= x^4 - 12x^3 + 48x^2 - 64x \\ &= x(x - 4)^3. \end{aligned}$$

Then, using the factored form of $f(x)$, you can perform the following analysis.

First derivative:	$f'(x) = 4(x - 1)(x - 4)^2$
Second derivative:	$f''(x) = 12(x - 4)(x - 2)$
x-intercepts:	$(0, 0), (4, 0)$
y-intercept:	$(0, 0)$
Vertical asymptotes:	None
Horizontal asymptotes:	None
End behavior:	$\lim_{x \to -\infty} f(x) = \infty, \lim_{x \to \infty} f(x) = \infty$
Critical numbers:	$x = 1, x = 4$
Possible points of inflection:	$x = 2, x = 4$
Domain:	All real numbers
Test intervals:	$(-\infty, 1), (1, 2), (2, 4), (4, \infty)$

The analysis of the graph of f is shown in the table, and the graph is shown in Figure 4.51(a). Using a computer algebra system such as *Derive* [see Figure 3.51(b)] can help you verify your analysis.

	$f(x)$	$f'(x)$	$f''(x)$	Characteristic of Graph
$-\infty < x < 1$		$-$	$+$	Decreasing, concave upward
$x = 1$	-27	0	$+$	Relative minimum
$1 < x < 2$		$+$	$+$	Increasing, concave upward
$x = 2$	-16	$+$	0	Point of inflection
$2 < x < 4$		$+$	$-$	Increasing, concave downward
$x = 4$	0	0	0	Point of inflection
$4 < x < \infty$		$+$	$+$	Increasing, concave upward

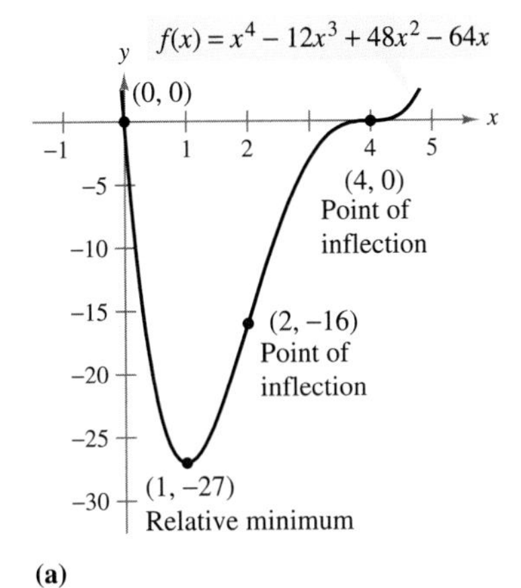

(a)

Generated by Derive

(b)

A polynomial function of even degree must have at least one relative extremum.

Figure 4.51

The fourth-degree polynomial function in Example 5 has one relative minimum and no relative maxima. In general, a polynomial function of degree n can have *at most* $n - 1$ relative extrema, and *at most* $n - 2$ points of inflection. Moreover, polynomial functions of even degree must have *at least* one relative extremum.

Remember from the Leading Coefficient Test described in Section 1.3 that the "end behavior" of the graph of a polynomial function is determined by its leading coefficient and its degree. For instance, because the polynomial in Example 5 has a positive leading coefficient, the graph rises to the right. Moreover, because the degree is even, the graph also rises to the left.

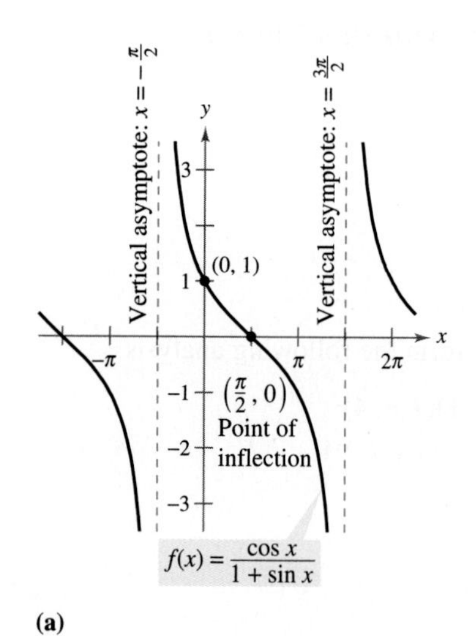

(a)

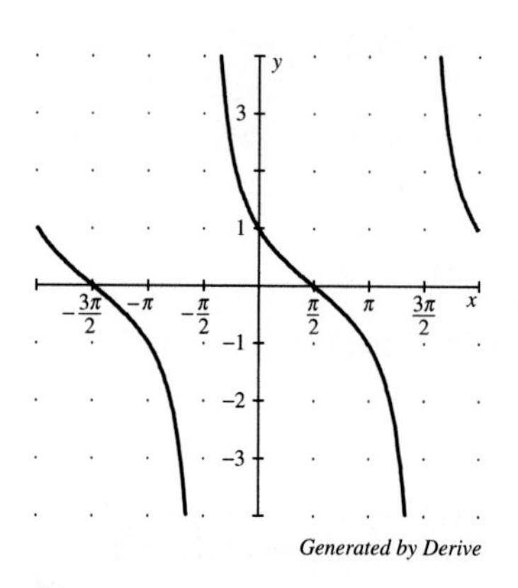

(b)

Figure 4.52

EXAMPLE 6 Sketching the Graph of a Trigonometric Function

Analyze and sketch the graph of $f(x) = \dfrac{\cos x}{1 + \sin x}$.

Solution Because the function has a period of 2π, you can restrict the analysis of the graph to any interval of length 2π. For convenience, choose $(-\pi/2, 3\pi/2)$.

First derivative: $f'(x) = -\dfrac{1}{1 + \sin x}$

Second derivative: $f''(x) = \dfrac{\cos x}{(1 + \sin x)^2}$

Period: 2π

x-intercept: $\left(\dfrac{\pi}{2}, 0\right)$

y-intercept: $(0, 1)$

Vertical asymptotes: $x = -\dfrac{\pi}{2}, x = \dfrac{3\pi}{2}$ See Note below.

Horizontal asymptotes: None

Critical numbers: None

Possible points of inflection: $x = \dfrac{\pi}{2}$

Domain: All real numbers except $x = \dfrac{3 + 4n}{2}\pi$

Test intervals: $\left(-\dfrac{\pi}{2}, \dfrac{\pi}{2}\right), \left(\dfrac{\pi}{2}, \dfrac{3\pi}{2}\right)$

The analysis of the graph of f on the interval $(-\pi/2, 3\pi/2)$ is shown in the table, and the graph is shown in Figure 4.52(a). Compare this with the graph generated by the computer algebra system *Derive* in Figure 4.52(b).

	$f(x)$	$f'(x)$	$f''(x)$	Characteristic of Graph
$x = -\dfrac{\pi}{2}$	Undef.	Undef.	Undef.	Vertical asymptote
$-\dfrac{\pi}{2} < x < \dfrac{\pi}{2}$		$-$	$+$	Decreasing, concave upward
$x = \dfrac{\pi}{2}$	0	$-\dfrac{1}{2}$	0	Point of inflection
$\dfrac{\pi}{2} < x < \dfrac{3\pi}{2}$		$-$	$-$	Decreasing, concave downward
$x = \dfrac{3\pi}{2}$	Undef.	Undef.	Undef.	Vertical asymptote

NOTE By substituting $-\pi/2$ or $3\pi/2$ into the function, you obtain the form $0/0$. This is called an indeterminate form and you will study this in Section 8.7. To determine that the function has vertical asymptotes at these two values, you can rewrite the function as follows.

$$f(x) = \frac{\cos x}{1 + \sin x} = \frac{(\cos x)(1 - \sin x)}{(1 + \sin x)(1 - \sin x)} = \frac{(\cos x)(1 - \sin x)}{\cos^2 x} = \frac{1 - \sin x}{\cos x}$$

In this form, it is clear that the graph of f has vertical asymptotes when $x = -\pi/2$ and $3\pi/2$.

EXAMPLE 7 Analyzing an Inverse Trigonometric Graph

Analyze the graph of $y = (\arctan x)^2$.

Solution From the derivative

$$y' = 2(\arctan x)\left(\frac{1}{1+x^2}\right) = \frac{2 \arctan x}{1+x^2}$$

you can see that the only critical number is $x = 0$. By the First Derivative Test, this value corresponds to a relative minimum. From the second derivative

$$y'' = \frac{(1+x^2)\left(\frac{2}{1+x^2}\right) - (2 \arctan x)(2x)}{(1+x^2)^2} = \frac{2(1 - 2x \arctan x)}{(1+x^2)^2}$$

it follows that points of inflection occur when $2x \arctan x = 1$. Using Newton's Method, these points occur when $x \approx \pm 0.765$. Finally, because

$$\lim_{x \to \pm\infty} (\arctan x)^2 = \frac{\pi^2}{4}$$

it follows that the graph has a horizontal asymptote at $y = \pi^2/4$. The graph is shown in Figure 4.53.

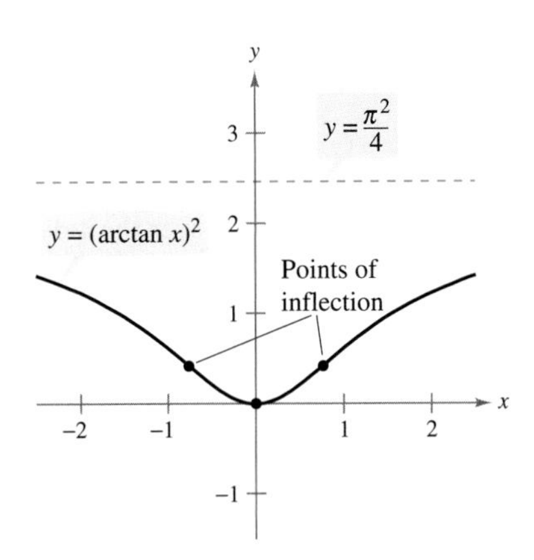

The graph of $y = (\arctan x)^2$ has a horizontal asymptote at $y = \pi^2/4$.
Figure 4.53

Exercises for Section 4.6

See www.CalcChat.com for worked-out solutions to odd-numbered exercises.

In Exercises 1–4, match the graph of f in the left column with that of its derivative in the right column.

Graph of f

1.

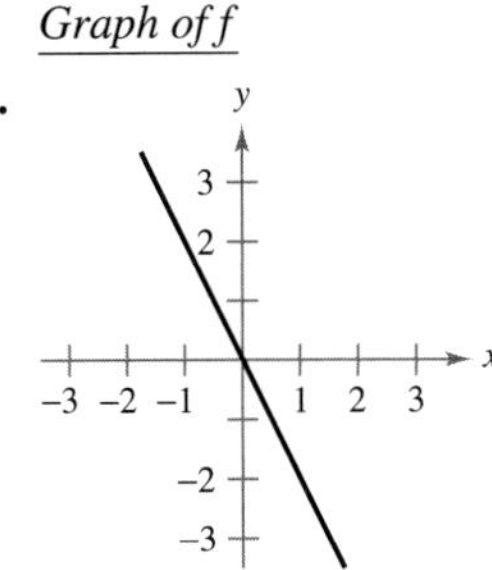

2.

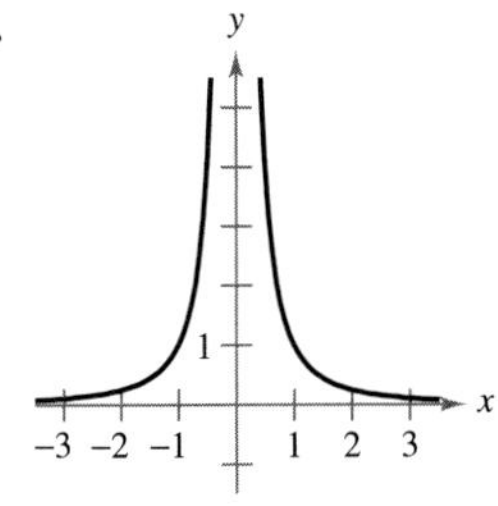

3.

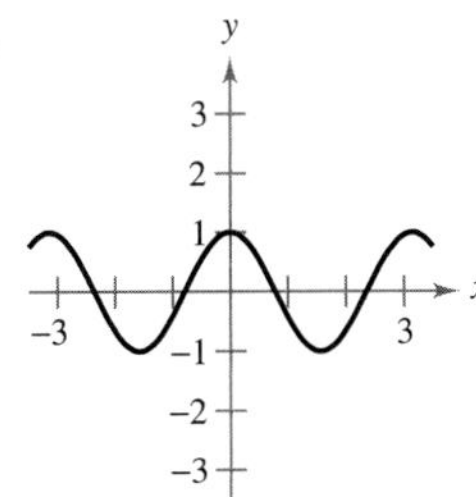

4.

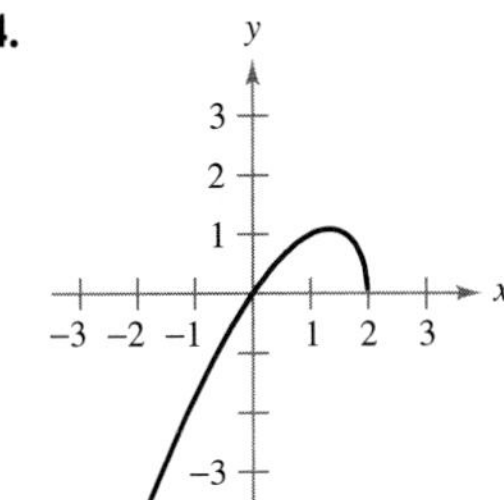

Graph of f'

(a)

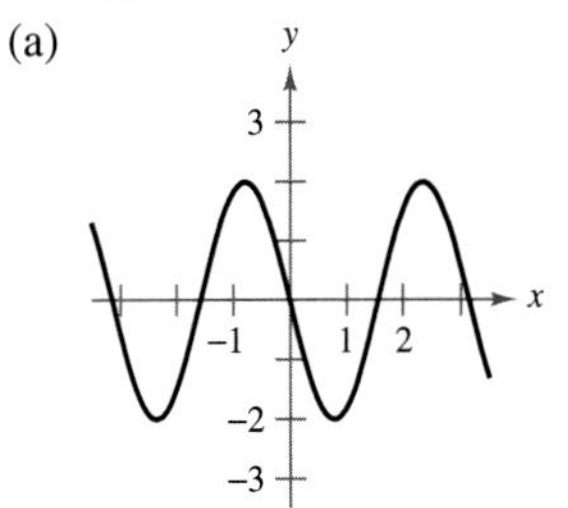

(b)

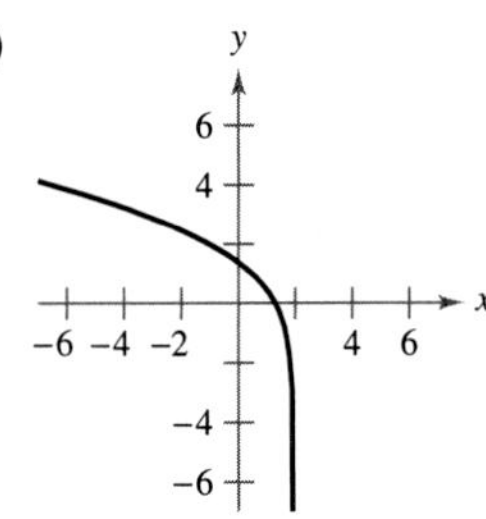

(c)

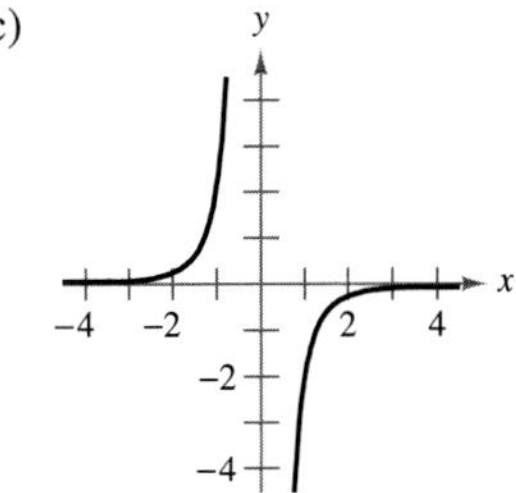

(d)

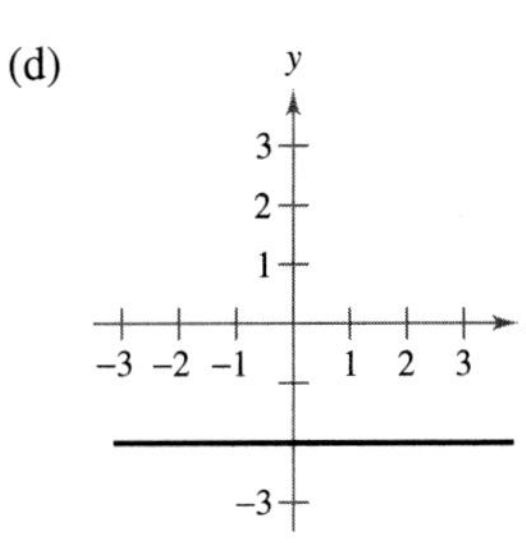

5. ***Graphical Reasoning*** The graph of f is shown in the figure.
   (a) For which values of x is $f'(x)$ zero? Positive? Negative?
   (b) For which values of x is $f''(x)$ zero? Positive? Negative?
   (c) On what interval is f' an increasing function?
   (d) For which value of x is $f'(x)$ minimum? For this value of x, how does the rate of change of f compare with the rate of change of f for other values of x? Explain.

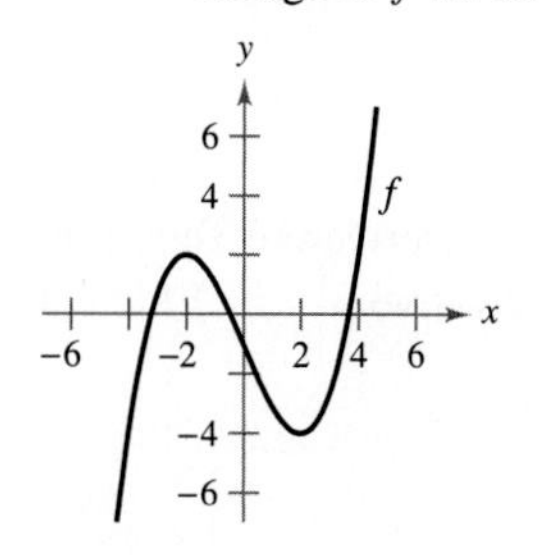

Figure for 5

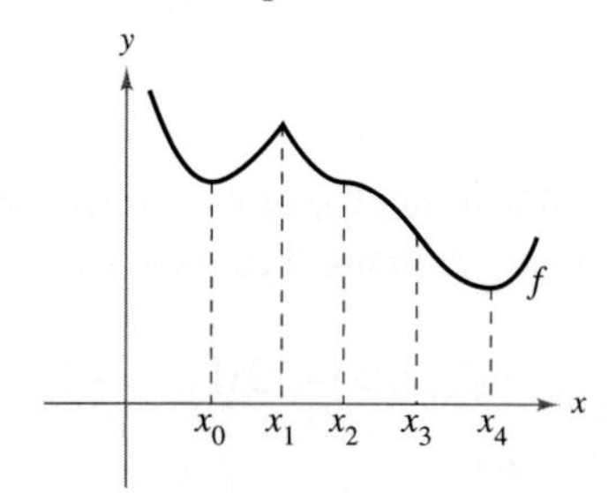

Figure for 6

6. ***Graphical Reasoning*** Identify the real numbers x_0, x_1, x_2, x_3, and x_4 in the figure such that each of the following is true.
   (a) $f'(x) = 0$
   (b) $f''(x) = 0$
   (c) $f'(x)$ does not exist.
   (d) f has a relative maximum.
   (e) f has a point of inflection.

In Exercises 7–50, analyze and sketch a graph of the function. Label any intercepts, relative extrema, points of inflection, and asymptotes. Use a graphing utility to verify your results.

7. $y = \dfrac{x^2}{x^2 + 3}$
8. $y = \dfrac{x}{x^2 + 1}$
9. $y = \dfrac{1}{x - 2} - 3$
10. $y = \dfrac{x^2 + 1}{x^2 - 9}$
11. $y = \dfrac{2x}{x^2 - 1}$
12. $f(x) = \dfrac{x + 2}{x}$

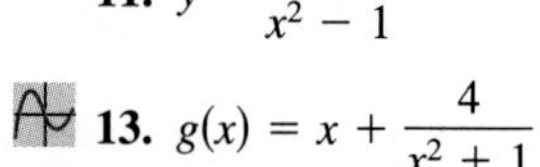

13. $g(x) = x + \dfrac{4}{x^2 + 1}$
14. $f(x) = x + \dfrac{32}{x^2}$
15. $f(x) = \dfrac{x^2 + 1}{x}$
16. $f(x) = \dfrac{x^3}{x^2 - 4}$
17. $y = \dfrac{x^2 - 6x + 12}{x - 4}$
18. $y = \dfrac{2x^2 - 5x + 5}{x - 2}$
19. $y = x\sqrt{4 - x}$
20. $g(x) = x\sqrt{9 - x}$
21. $h(x) = x\sqrt{9 - x^2}$
22. $y = x\sqrt{16 - x^2}$
23. $y = 3x^{2/3} - 2x$
24. $y = 3(x - 1)^{2/3} - (x - 1)^2$
25. $y = x^3 - 3x^2 + 3$
26. $y = -\frac{1}{3}(x^3 - 3x + 2)$
27. $y = 2 - x - x^3$
28. $f(x) = \frac{1}{3}(x - 1)^3 + 2$
29. $f(x) = 3x^3 - 9x + 1$
30. $f(x) = (x + 1)(x - 2)(x - 5)$
31. $y = 3x^4 + 4x^3$
32. $y = 3x^4 - 6x^2 + \frac{5}{3}$
33. $f(x) = x^4 - 4x^3 + 16x$
34. $f(x) = x^4 - 8x^3 + 18x^2 - 16x + 5$
35. $y = x^5 - 5x$
36. $y = (x - 1)^5$
37. $y = |2x - 3|$
38. $y = |x^2 - 6x + 5|$
39. $f(x) = e^{3x}(2 - x)$
40. $f(x) = -2 + e^{3x}(4 - 2x)$
41. $g(t) = \dfrac{10}{1 + 4e^{-t}}$
42. $h(x) = \dfrac{10}{2 + 3e^{-x/2}}$
43. $y = (x - 1)\ln(x - 1)$
44. $y = \frac{1}{24}x^3 - \ln x$
45. $g(x) = 6\arcsin\left(\dfrac{x - 2}{2}\right)^2$
46. $h(x) = 7\arctan(x + 1) - \ln(x^2 + 2x + 2)$
47. $f(x) = \dfrac{x}{2^{x-2}}$
48. $h(t) = (3 - t)3^t$
49. $g(x) = \log_4(x - x^2)$
50. $f(x) = \log_2|x^2 - 4x|$

In Exercises 51–58, sketch a graph of the function over the indicated interval. Use a graphing utility to verify your graph.

	Function	*Interval*
51.	$y = \sin x - \frac{1}{18}\sin 3x$	$0 \le x \le 2\pi$
52.	$y = \cos x - \frac{1}{2}\cos 2x$	$0 \le x \le 2\pi$
53.	$y = 2x - \tan x$	$-\frac{\pi}{2} < x < \frac{\pi}{2}$
54.	$y = 2(x - 2) + \cot x$	$0 < x < \pi$
55.	$y = 2(\csc x + \sec x)$	$0 < x < \frac{\pi}{2}$
56.	$y = \sec^2\left(\frac{\pi x}{8}\right) - 2\tan\left(\frac{\pi x}{8}\right) - 1$	$-3 < x < 3$
57.	$g(x) = x\tan x$	$-\frac{3\pi}{2} < x < \frac{3\pi}{2}$
58.	$g(x) = x\cot x$	$-2\pi < x < 2\pi$

In Exercises 59–64, use a computer algebra system to analyze and graph the function. Identify any relative extrema, points of inflection, and asymptotes.

59. $f(x) = \dfrac{20x}{x^2 + 1} - \dfrac{1}{x}$
60. $f(x) = 5\left(\dfrac{1}{x - 4} - \dfrac{1}{x + 2}\right)$
61. $f(x) = \dfrac{x}{\sqrt{x^2 + 7}}$
62. $f(x) = \dfrac{4x}{\sqrt{x^2 + 15}}$
63. $y = \dfrac{x}{2} + \ln\left(\dfrac{x}{x + 3}\right)$
64. $y = \dfrac{3x}{2}(1 + 4e^{-x/3})$

Think About It **In Exercises 65–68, create a function whose graph has the indicated characteristics. (There is more than one correct answer.)**

65. Vertical asymptote: $x = 5$
    Horizontal asymptote: $y = 0$
66. Vertical asymptote: $x = -3$
    Horizontal asymptote: None
67. Vertical asymptote: $x = 5$
    Slant asymptote: $y = 3x + 2$
68. Vertical asymptote: $x = 0$
    Slant asymptote: $y = -x$

Writing About Concepts

In Exercises 69 and 70, the graphs of f, f', and f'' are shown on the same set of coordinate axes. Which is which? Explain your reasoning. To print an enlarged copy of the graph, go to the website *www.mathgraphs.com*.

69.

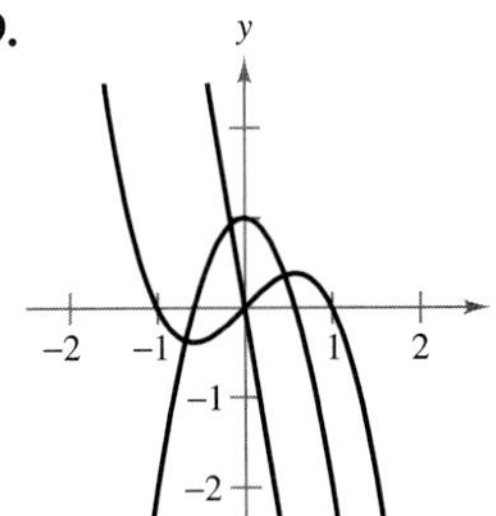

70.

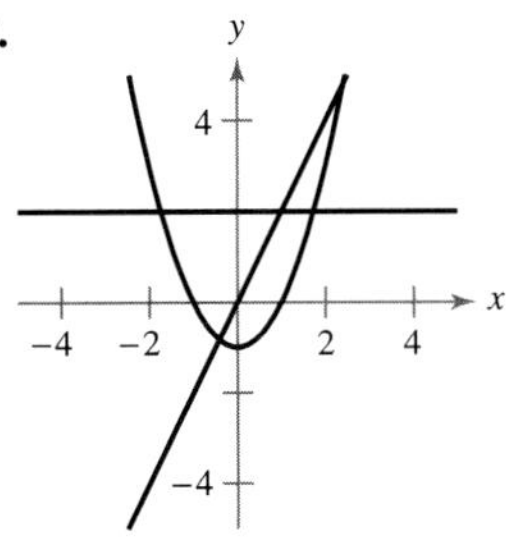

In Exercises 71–74, use the graph of f' to sketch a graph of f and the graph of f''. To print an enlarged copy of the graph, go to the website *www.mathgraphs.com*.

71.

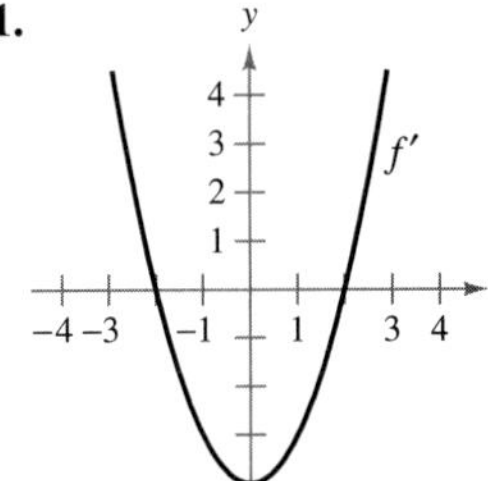

72.

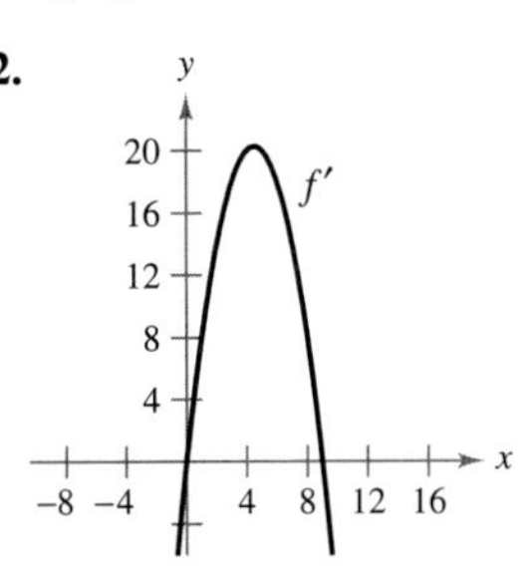

73.

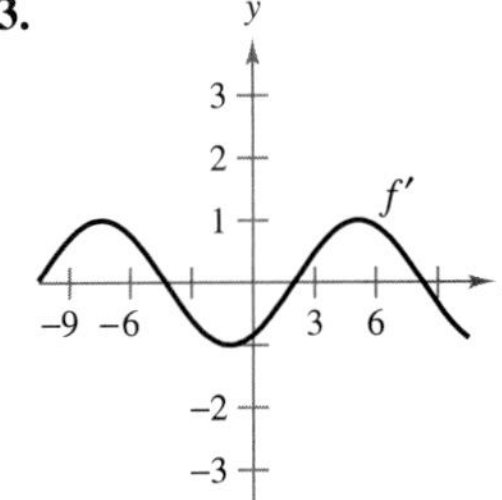

74.

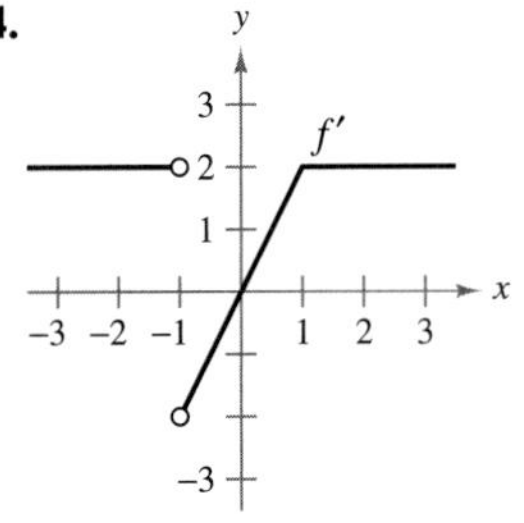

(Submitted by Bill Fox, Moberly Area Community College, Moberly, MO)

75. Suppose $f'(t) < 0$ for all t in the interval $(2, 8)$. Explain why $f(3) > f(5)$.

76. Suppose $f(0) = 3$ and $2 \le f'(x) \le 4$ for all x in the interval $[-5, 5]$. Determine the greatest and least possible values of $f(2)$.

In Exercises 77–80, use a graphing utility to graph the function. Use the graph to determine whether it is possible for the graph of a function to cross its horizontal asymptote. Do you think it is possible for the graph of a function to cross its vertical asymptote? Why or why not?

77. $f(x) = \dfrac{4(x-1)^2}{x^2 - 4x + 5}$

78. $g(x) = \dfrac{3x^4 - 5x + 3}{x^4 + 1}$

79. $h(x) = \dfrac{\sin 2x}{x}$

80. $f(x) = \dfrac{\cos 3x}{4x}$

Writing **In Exercises 81 and 82, use a graphing utility to graph the function. Explain why there is no vertical asymptote when a superficial examination of the function may indicate that there should be one.**

81. $h(x) = \dfrac{6 - 2x}{3 - x}$

82. $g(x) = \dfrac{x^2 + x - 2}{x - 1}$

Writing **In Exercises 83–86, use a graphing utility to graph the function and determine the slant asymptote of the graph. Zoom out repeatedly and describe how the graph on the display appears to change. Why does this occur?**

83. $f(x) = -\dfrac{x^2 - 3x - 1}{x - 2}$

84. $g(x) = \dfrac{2x^2 - 8x - 15}{x - 5}$

85. $f(x) = \dfrac{x^3}{x^2 + 1}$

86. $h(x) = \dfrac{-x^3 + x^2 + 4}{x^2}$

87. ***Graphical Reasoning*** Consider the function

$$f(x) = \frac{\cos^2 \pi x}{\sqrt{x^2 + 1}}, \quad 0 < x < 4.$$

(a) Use a computer algebra system to graph the function and use the graph to approximate the critical numbers visually.

(b) Use a computer algebra system to find f' and approximate the critical numbers. Are the results the same as the visual approximation in part (a)? Explain.

88. ***Graphical Reasoning*** Consider the function

$$f(x) = \tan(\sin \pi x).$$

(a) Use a graphing utility to graph the function.

(b) Identify any symmetry of the graph.

(c) Is the function periodic? If so, what is the period?

(d) Identify any extrema on $(-1, 1)$.

(e) Use a graphing utility to determine the concavity of the graph on $(0, 1)$.

89. ***Graphical Reasoning*** Consider the function

$$f(x) = \frac{ax}{(x - b)^2}.$$

(a) Determine the effect on the graph of f if $b \ne 0$ and a is varied. Consider cases where a is positive and a is negative.

(b) Determine the effect on the graph of f if $a \ne 0$ and b is varied.

90. Consider the function $f(x) = \frac{1}{2}(ax)^2 - (ax), \quad a \ne 0$.

(a) Determine the changes (if any) in the intercepts, extrema, and concavity of the graph of f when a is varied.

(b) In the same viewing window, use a graphing utility to graph the function for four different values of a.

91. *Investigation* Consider the function $f(x) = \dfrac{3x^n}{x^4 + 1}$ for nonnegative integer values of n.

(a) Discuss the relationship between the value of n and the symmetry of the graph.

(b) For which values of n will the x-axis be the horizontal asymptote?

(c) For which value of n will $y = 3$ be the horizontal asymptote?

(d) What is the asymptote of the graph when $n = 5$?

(e) Use a graphing utility to graph f for the indicated values of n in the table. Use the graph to determine the number of extrema M and the number of inflection points N of the graph.

n	0	1	2	3	4	5
M						
N						

92. *Investigation* Let $P(x_0, y_0)$ be an arbitrary point on the graph of f such that $f'(x_0) \neq 0$, as shown in the figure. Verify each statement.

(a) The x-intercept of the tangent line is $\left(x_0 - \dfrac{f(x_0)}{f'(x_0)}, 0\right)$.

(b) The y-intercept of the tangent line is $(0, f(x_0) - x_0 f'(x_0))$.

(c) The x-intercept of the normal line is $(x_0 + f(x_0)f'(x_0), 0)$.

(d) The y-intercept of the normal line is $\left(0, y_0 + \dfrac{x_0}{f'(x_0)}\right)$.

(e) $|BC| = \left|\dfrac{f(x_0)}{f'(x_0)}\right|$ (f) $|PC| = \left|\dfrac{f(x_0)\sqrt{1 + [f'(x_0)]^2}}{f'(x_0)}\right|$

(g) $|AB| = |f(x_0)f'(x_0)|$

(h) $|AP| = |f(x_0)|\sqrt{1 + [f'(x_0)]^2}$

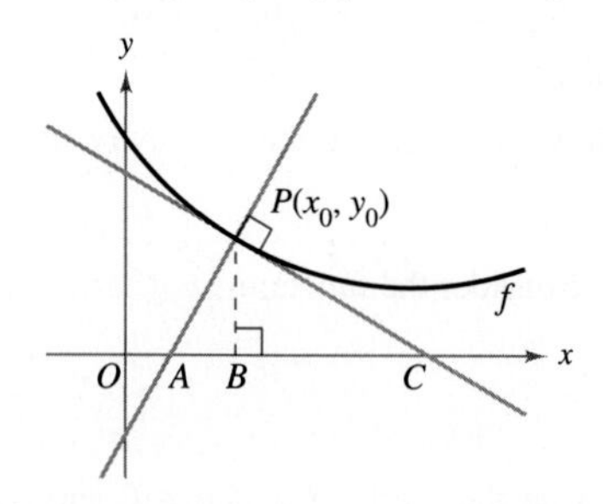

93. *Modeling Data* The data in the table show the number N of bacteria in a culture at time t, where t is measured in days.

t	1	2	3	4	5	6	7	8
N	25	200	804	1756	2296	2434	2467	2473

A model for these data is given by

$$N = \frac{24{,}670 - 35{,}153t + 13{,}250t^2}{100 - 39t + 7t^2}, \quad 1 \leq t \leq 8.$$

(a) Use a graphing utility to plot the data and graph the model.

(b) Use the model to estimate the number of bacteria when $t = 10$.

(c) Approximate the day when the number of bacteria is greatest.

(d) Use a computer algebra system to determine the time when the rate of increase in the number of bacteria is greatest.

(e) Find $\lim_{t \to \infty} N(t)$.

94. *Modeling Data* A meteorologist measures the atmospheric pressure P (in kilograms per square meter) at altitude h (in kilometers). The data are shown below.

h	0	5	10	15	20
P	10,332	5583	2376	1240	517

(a) Use a graphing utility to plot the points $(h, \ln P)$. Use the regression capabilities of the graphing utility to find a linear model for the revised data points.

(b) The line in part (a) has the form $\ln P = ah + b$. Write the equation in exponential form.

(c) Use a graphing utility to plot the original data and graph the exponential model in part (b).

(d) Find the rate of change of the pressure when $h = 5$ and $h = 18$.

95. Let f be a function that is positive and differentiable on the entire real number line. Let $g(x) = \ln f(x)$.

(a) If g is increasing, must f be increasing? Explain.

(b) If the graph of f is concave upward, must the graph of g be concave upward? Explain.

96. *Conjecture* Use a graphing utility to graph f and g in the same viewing window and determine which is increasing at the faster rate for "large" values of x. What can you conclude about the rate of growth of the natural logarithmic function?

(a) $f(x) = \ln x,\ g(x) = \sqrt{x}$ (b) $f(x) = \ln x,\ g(x) = \sqrt[4]{x}$

Slant Asymptotes **In Exercises 97 and 98, the graph of the function has two slant asymptotes. Identify each slant asymptote. Then graph the function and its asymptotes.**

97. $y = \sqrt{4 + 16x^2}$ **98.** $y = \sqrt{x^2 + 6x}$

Putnam Exam Challenge

99. Let $f(x)$ be defined for $a \leq x \leq b$. Assuming appropriate properties of continuity and derivability, prove for $a < x < b$ that

$$\frac{\dfrac{f(x) - f(a)}{x - a} - \dfrac{f(b) - f(a)}{b - a}}{x - b} = \frac{1}{2}f''(\beta)$$

where β is some number between a and b.

This problem was composed by the Committee on the Putnam Prize Competition.

Section 4.7 Optimization Problems

- **Solve applied minimum and maximum problems.**

Applied Minimum and Maximum Problems

One of the most common applications of calculus involves the determination of minimum and maximum values. Consider how frequently you hear or read terms such as greatest profit, least cost, least time, greatest voltage, optimum size, least size, greatest strength, and greatest distance. Before outlining a general problem-solving strategy for such problems, let's look at an example.

EXAMPLE 1 Finding Maximum Volume

A manufacturer wants to design an open box having a square base and a surface area of 108 square inches, as shown in Figure 4.54. What dimensions will produce a box with maximum volume?

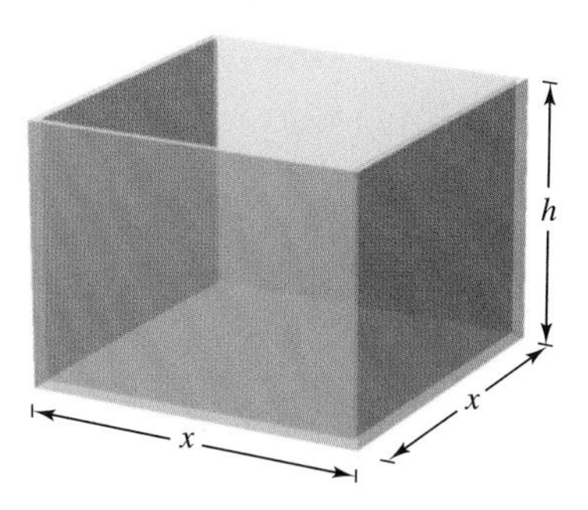

Open box with square base:
$S = x^2 + 4xh = 108$
Figure 4.54

Solution Because the box has a square base, its volume is

$$V = x^2h. \qquad \text{Primary equation}$$

This equation is called the **primary equation** because it gives a formula for the quantity to be optimized. The surface area of the box is

$$S = (\text{area of base}) + (\text{area of four sides})$$
$$S = x^2 + 4xh = 108. \qquad \text{Secondary equation}$$

Because V is to be maximized, you want to write V as a function of just one variable. To do this, you can solve the equation $x^2 + 4xh = 108$ for h in terms of x to obtain $h = (108 - x^2)/(4x)$. Substituting into the primary equation produces

$$V = x^2h \qquad \text{Function of two variables}$$
$$= x^2\left(\frac{108 - x^2}{4x}\right) \qquad \text{Substitute for } h.$$
$$= 27x - \frac{x^3}{4}. \qquad \text{Function of one variable}$$

Before finding which x-value will yield a maximum value of V, you should determine the *feasible domain.* That is, what values of x make sense in this problem? You know that $V \geq 0$. You also know that x must be nonnegative and that the area of the base $(A = x^2)$ is at most 108. So, the feasible domain is

$$0 \leq x \leq \sqrt{108}. \qquad \text{Feasible domain}$$

To maximize V, find the critical numbers of the volume function on the interval $\left[0, \sqrt{108}\right]$.

$$\frac{dV}{dx} = 27 - \frac{3x^2}{4} = 0 \qquad \text{Set derivative equal to 0.}$$
$$3x^2 = 108 \qquad \text{Simplify.}$$
$$x = \pm 6 \qquad \text{Critical numbers}$$

So, the critical numbers are $x = \pm 6$. You do not need to consider $x = -6$ because it is outside the domain. Evaluating V at the critical number $x = 6$ and at the endpoints of the domain produces $V(0) = 0$, $V(6) = 108$, and $V\left(\sqrt{108}\right) = 0$. So, V is maximum when $x = 6$ and the dimensions of the box are $6 \times 6 \times 3$ inches.

TECHNOLOGY You can verify your answer by using a graphing utility to graph the volume function

$$V = 27x - \frac{x^3}{4}.$$

Use a viewing window in which $0 \leq x \leq \sqrt{108} \approx 10.4$ and $0 \leq y \leq 120$ and the *trace* feature to determine the maximum value of V.

In Example 1, you should realize that there are infinitely many open boxes having 108 square inches of surface area. To begin solving the problem, you might ask yourself which basic shape would seem to yield a maximum volume. Should the box be tall, squat, or nearly cubical?

You might even try calculating a few volumes, as shown in Figure 4.55, to see if you can get a better feeling for what the optimum dimensions should be. Remember that you are not ready to begin solving a problem until you have clearly identified what the problem is.

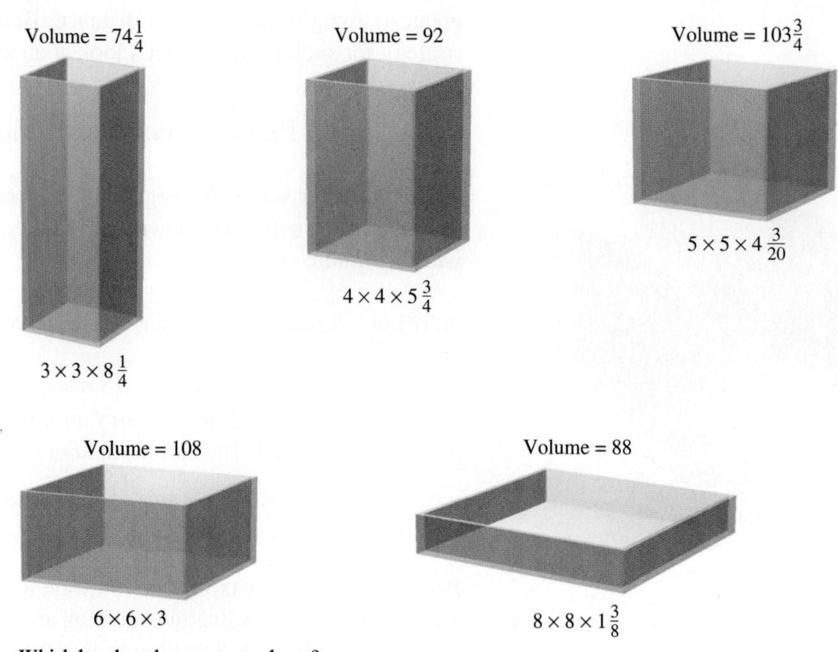

Which box has the greatest volume?
Figure 4.55

Example 1 illustrates the following guidelines for solving applied minimum and maximum problems.

Guidelines for Solving Applied Minimum and Maximum Problems

1. Identify all *given* quantities and quantities *to be determined*. If possible, make a sketch.
2. Write a **primary equation** for the quantity that is to be maximized or minimized. (A review of several useful formulas from geometry is presented inside the back cover.)
3. Reduce the primary equation to one having a *single independent variable*. This may involve the use of **secondary equations** relating the independent variables of the primary equation.
4. Determine the feasible domain of the primary equation. That is, determine the values for which the stated problem makes sense.
5. Determine the desired maximum or minimum value by the calculus techniques discussed in Sections 4.1 through 4.4.

NOTE When performing Step 5, recall that to determine the maximum or minimum value of a continuous function f on a closed interval, you should compare the values of f at its critical numbers with the values of f at the endpoints of the interval.

EXAMPLE 2 Finding Minimum Distance

Which points on the graph of $y = 4 - x^2$ are closest to the point $(0, 2)$?

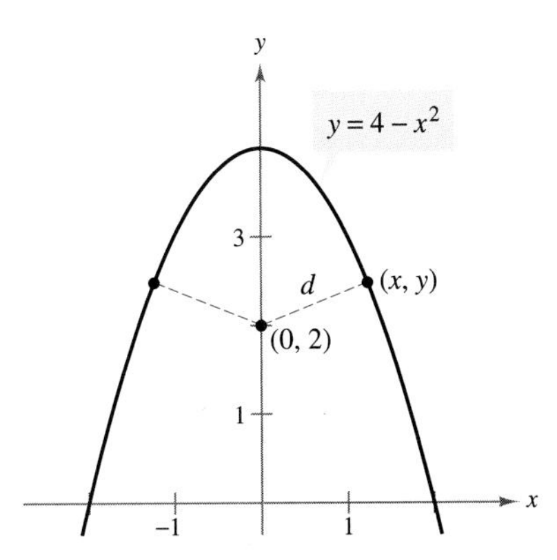

The quantity to be minimized is distance: $d = \sqrt{(x - 0)^2 + (y - 2)^2}$.
Figure 4.56

Solution Figure 4.56 shows that there are two points at a minimum distance from the point $(0, 2)$. The distance between the point $(0, 2)$ and a point (x, y) on the graph of $y = 4 - x^2$ is given by

$$d = \sqrt{(x - 0)^2 + (y - 2)^2}. \qquad \text{Primary equation}$$

Using the secondary equation $y = 4 - x^2$, you can rewrite the primary equation as

$$d = \sqrt{x^2 + (4 - x^2 - 2)^2} = \sqrt{x^4 - 3x^2 + 4}.$$

Because d is smallest when the expression inside the radical is smallest, you need only find the critical numbers of $f(x) = x^4 - 3x^2 + 4$. Note that the domain of f is the entire real number line. So, there are no endpoints of the domain to consider. Moreover, setting $f'(x)$ equal to 0 yields

$$f'(x) = 4x^3 - 6x = 2x(2x^2 - 3) = 0$$

$$x = 0, \sqrt{\frac{3}{2}}, -\sqrt{\frac{3}{2}}.$$

The First Derivative Test verifies that $x = 0$ yields a relative maximum, whereas both $x = \sqrt{3/2}$ and $x = -\sqrt{3/2}$ yield a minimum distance. So, the closest points are $\left(\sqrt{3/2}, 5/2\right)$ and $\left(-\sqrt{3/2}, 5/2\right)$.

EXAMPLE 3 Finding Minimum Area

A rectangular page is to contain 24 square inches of print. The margins at the top and bottom of the page are to be $1\frac{1}{2}$ inches, and the margins on the left and right are to be 1 inch (see Figure 4.57). What should the dimensions of the page be so that the least amount of paper is used?

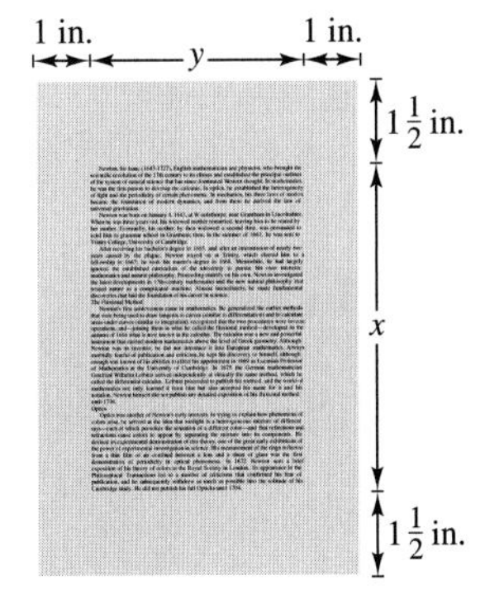

The quantity to be minimized is area: $A = (x + 3)(y + 2)$.
Figure 4.57

Solution Let A be the area to be minimized.

$$A = (x + 3)(y + 2) \qquad \text{Primary equation}$$

The printed area inside the margins is given by

$$24 = xy. \qquad \text{Secondary equation}$$

Solving this equation for y produces $y = 24/x$. Substitution into the primary equation produces

$$A = (x + 3)\left(\frac{24}{x} + 2\right) = 30 + 2x + \frac{72}{x}. \qquad \text{Function of one variable}$$

Because x must be positive, you are interested only in values of A for $x > 0$. To find the critical numbers, differentiate with respect to x.

$$\frac{dA}{dx} = 2 - \frac{72}{x^2} = 0 \quad \Longrightarrow \quad x^2 = 36$$

So, the critical numbers are $x = \pm 6$. You do not have to consider $x = -6$ because it is outside the domain. The First Derivative Test confirms that A is a minimum when $x = 6$. So, $y = \frac{24}{6} = 4$ and the dimensions of the page should be $x + 3 = 9$ inches by $y + 2 = 6$ inches.

EXAMPLE 4 Finding Minimum Length

Two posts, one 12 feet high and the other 28 feet high, stand 30 feet apart. They are to be stayed by two wires, attached to a single stake, running from ground level to the top of each post. Where should the stake be placed to use the least amount of wire?

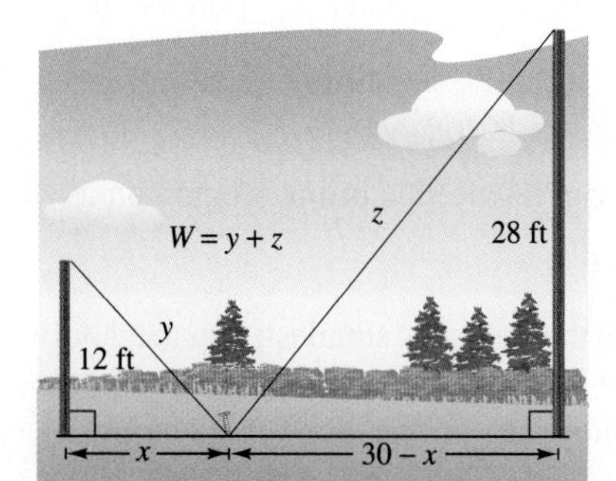

The quantity to be minimized is length. From the diagram, you can see that x varies between 0 and 30.
Figure 4.58

Solution Let W be the wire length to be minimized. Using Figure 4.58, you can write

$$W = y + z. \qquad \text{Primary equation}$$

In this problem, rather than solving for y in terms of z (or vice versa), you can solve for both y and z in terms of a third variable x, as shown in Figure 4.58. From the Pythagorean Theorem, you obtain

$$x^2 + 12^2 = y^2$$
$$(30 - x)^2 + 28^2 = z^2$$

which implies that

$$y = \sqrt{x^2 + 144}$$
$$z = \sqrt{x^2 - 60x + 1684}.$$

So, W is given by

$$\begin{aligned} W &= y + z \\ &= \sqrt{x^2 + 144} + \sqrt{x^2 - 60x + 1684}, \qquad 0 \le x \le 30. \end{aligned}$$

Differentiating W with respect to x yields

$$\frac{dW}{dx} = \frac{x}{\sqrt{x^2 + 144}} + \frac{x - 30}{\sqrt{x^2 - 60x + 1684}}.$$

By letting $dW/dx = 0$, you obtain

$$\begin{aligned} \frac{x}{\sqrt{x^2 + 144}} + \frac{x - 30}{\sqrt{x^2 - 60x + 1684}} &= 0 \\ x\sqrt{x^2 - 60x + 1684} &= (30 - x)\sqrt{x^2 + 144} \\ x^2(x^2 - 60x + 1684) &= (30 - x)^2(x^2 + 144) \\ x^4 - 60x^3 + 1684x^2 &= x^4 - 60x^3 + 1044x^2 - 8640x + 129{,}600 \\ 640x^2 + 8640x - 129{,}600 &= 0 \\ 320(x - 9)(2x + 45) &= 0 \\ x &= 9, -22.5. \end{aligned}$$

Because $x = -22.5$ is not in the domain and

$$W(0) \approx 53.04, \qquad W(9) = 50, \qquad \text{and} \qquad W(30) \approx 60.31$$

you can conclude that the wire should be staked at 9 feet from the 12-foot pole.

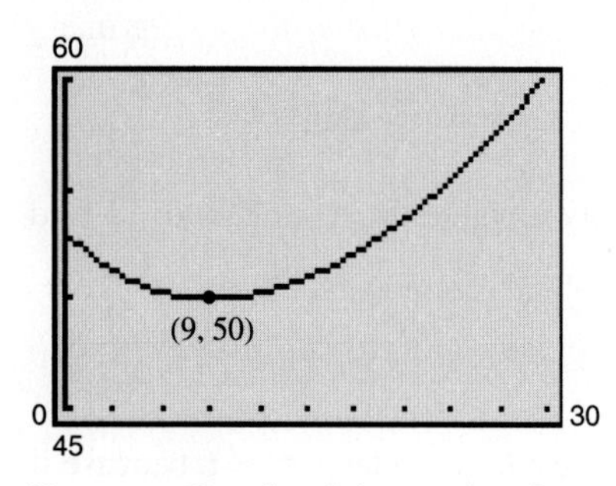

You can confirm the minimum value of W with a graphing utility.
Figure 4.59

TECHNOLOGY From Example 4, you can see that applied optimization problems can involve a lot of algebra. If you have access to a graphing utility, you can confirm that $x = 9$ yields a minimum value of W by graphing

$$W = \sqrt{x^2 + 144} + \sqrt{x^2 - 60x + 1684}$$

as shown in Figure 4.59.

In each of the first four examples, the extreme value occurred at a critical number. Although this happens often, remember that an extreme value can also occur at an endpoint of an interval, as shown in Example 5.

EXAMPLE 5 An Endpoint Maximum

Four feet of wire is to be used to form a square and a circle. How much of the wire should be used for the square and how much should be used for the circle to enclose the maximum total area?

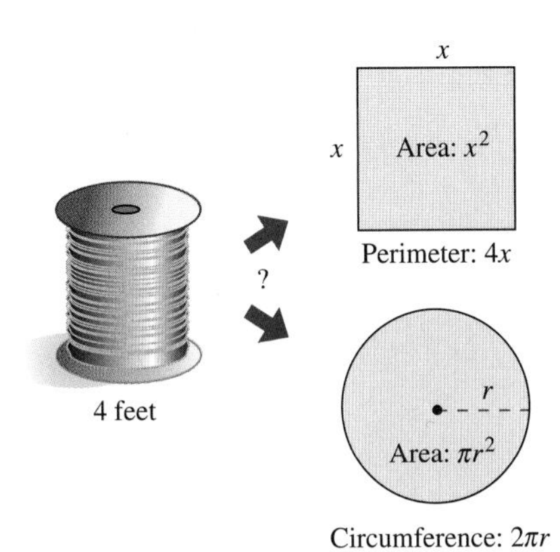

The quantity to be maximized is area: $A = x^2 + \pi r^2$.
Figure 4.60

Solution The total area (see Figure 4.60) is given by

$$A = (\text{area of square}) + (\text{area of circle})$$
$$A = x^2 + \pi r^2. \qquad \text{Primary equation}$$

Because the total length of wire is 4 feet, you obtain

$$4 = (\text{perimeter of square}) + (\text{circumference of circle})$$
$$4 = 4x + 2\pi r.$$

So, $r = 2(1 - x)/\pi$, and by substituting into the primary equation you have

$$\begin{aligned} A &= x^2 + \pi\left[\frac{2(1-x)}{\pi}\right]^2 \\ &= x^2 + \frac{4(1-x)^2}{\pi} \\ &= \frac{1}{\pi}[(\pi + 4)x^2 - 8x + 4]. \end{aligned}$$

The feasible domain is $0 \le x \le 1$ restricted by the square's perimeter. Because

$$\frac{dA}{dx} = \frac{2(\pi + 4)x - 8}{\pi}$$

the only critical number in $(0, 1)$ is $x = 4/(\pi + 4) \approx 0.56$. So, using

$$A(0) \approx 1.273, \quad A(0.56) \approx 0.56, \quad \text{and} \quad A(1) = 1$$

you can conclude that the maximum area occurs when $x = 0$. That is, *all* the wire is used for the circle.

EXPLORATION

What would the answer be if Example 5 asked for the dimensions needed to enclose the *minimum* total area?

Let's review the primary equations developed in the first five examples. As applications go, these five examples are fairly simple, and yet the resulting primary equations are quite complicated.

$$V = 27x - \frac{x^3}{4} \qquad W = \sqrt{x^2 + 144} + \sqrt{x^2 - 60x + 1684}$$

$$d = \sqrt{x^4 - 3x^2 + 4} \qquad A = \frac{1}{\pi}[(\pi + 4)x^2 - 8x + 4]$$

$$A = 30 + 2x + \frac{72}{x}$$

You must expect that real-life applications often involve equations that are *at least as complicated* as these five. Remember that one of the main goals of this course is to learn to use calculus to analyze equations that initially seem formidable.

The camera should be 2.236 feet from the painting to maximize the angle β.
Figure 4.61

EXAMPLE 6 Maximizing an Angle

A photographer is taking a picture of a 4-foot painting hung in an art gallery. The camera lens is 1 foot below the lower edge of the painting, as shown in Figure 4.61. How far should the camera be from the painting to maximize the angle subtended by the camera lens?

Solution In Figure 4.61, let β be the angle to be maximized.

$$\beta = \theta - \alpha \qquad \text{Primary equation}$$

From Figure 4.61, you can see that $\cot\theta = \dfrac{x}{5}$ and $\cot\alpha = \dfrac{x}{1}$. Therefore, $\theta = \operatorname{arccot}\dfrac{x}{5}$ and $\alpha = \operatorname{arccot} x$. So,

$$\beta = \operatorname{arccot}\frac{x}{5} - \operatorname{arccot} x.$$

Differentiating β with respect to x produces

$$\begin{aligned}\frac{d\beta}{dx} &= \frac{-1/5}{1 + (x^2/25)} - \frac{-1}{1 + x^2}\\ &= \frac{-5}{25 + x^2} + \frac{1}{1 + x^2}\\ &= \frac{4(5 - x^2)}{(25 + x^2)(1 + x^2)}.\end{aligned}$$

Because $d\beta/dx = 0$ when $x = \sqrt{5}$, you can conclude from the First Derivative Test that this distance yields a maximum value of β. So, the distance is $x \approx 2.236$ feet and the angle is $\beta \approx 0.7297$ radian $\approx 41.81°$.

EXAMPLE 7 Finding a Maximum Revenue

The demand function for a product is modeled by

$$p = 56e^{-0.000012x} \qquad \text{Demand function}$$

where p is the price per unit (in dollars) and x is the number of units. What price will yield a maximum revenue?

Solution Substituting for p (from the demand function) produces the revenue function

$$R = xp = 56xe^{-0.000012x}. \qquad \text{Primary equation}$$

The rate of change of revenue R with respect to the number of units sold x is called the *marginal revenue*. Setting the marginal revenue equal to zero,

$$\frac{dR}{dx} = 56x(e^{-0.000012x})(-0.000012) + e^{-0.000012x}(56) = 0$$

yields $x \approx 83{,}333$ units. From this, you can conclude that the maximum revenue occurs when the price is

$$p = 56e^{-0.000012(83{,}333)} \approx \$20.60.$$

So, a price of about \$20.60 will yield a maximum revenue (see Figure 4.62).

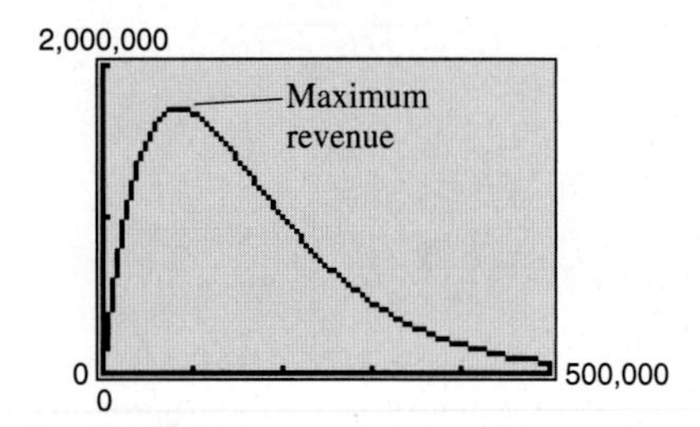

Figure 4.62

Exercises for Section 4.7

See www.CalcChat.com for worked-out solutions to odd-numbered exercises.

1. ***Numerical, Graphical, and Analytic Analysis*** Find two positive numbers whose sum is 110 and whose product is a maximum.

   (a) Analytically complete six rows of a table such as the one below. (The first two rows are shown.)

First Number x	Second Number	Product P
10	$110 - 10$	$10(110 - 10) = 1000$
20	$110 - 20$	$20(110 - 20) = 1800$

   (b) Use a graphing utility to generate additional rows of the table. Use the table to estimate the solution. (*Hint:* Use the *table* feature of the graphing utility.)

   (c) Write the product P as a function of x.

   (d) Use a graphing utility to graph the function in part (c) and estimate the solution from the graph.

   (e) Use calculus to find the critical number of the function in part (c). Then find the two numbers.

2. ***Numerical, Graphical, and Analytic Analysis*** An open box of maximum volume is to be made from a square piece of material, 24 inches on a side, by cutting equal squares from the corners and turning up the sides (see figure).

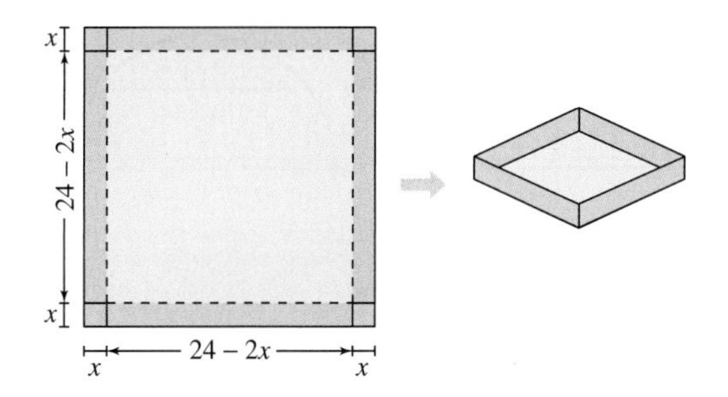

   (a) Analytically complete six rows of a table such as the one below. (The first two rows are shown.) Use the table to guess the maximum volume.

Height	Length and Width	Volume
1	$24 - 2(1)$	$1[24 - 2(1)]^2 = 484$
2	$24 - 2(2)$	$2[24 - 2(2)]^2 = 800$

   (b) Write the volume V as a function of x.

   (c) Use calculus to find the critical number of the function in part (b) and find the maximum value.

   (d) Use a graphing utility to graph the function in part (b) and verify the maximum volume from the graph.

In Exercises 3–8, find two positive numbers that satisfy the given requirements.

3. The sum is S and the product is a maximum.
4. The product is 192 and the sum is a minimum.
5. The product is 192 and the sum of the first plus three times the second is a minimum.
6. The second number is the reciprocal of the first and the sum is a minimum.
7. The sum of the first and twice the second is 100 and the product is a maximum.
8. The sum of the first number squared and the second is 27 and the product is a maximum.

In Exercises 9 and 10, find the length and width of a rectangle that has the given perimeter and a maximum area.

9. Perimeter: 100 meters
10. Perimeter: P units

In Exercises 11 and 12, find the length and width of a rectangle that has the given area and a minimum perimeter.

11. Area: 64 square feet
12. Area: A square centimeters

In Exercises 13–16, find the point on the graph of the function that is closest to the given point.

	Function	*Point*		*Function*	*Point*
13.	$f(x) = \sqrt{x}$	$(4, 0)$	14.	$f(x) = \sqrt{x - 8}$	$(2, 0)$
15.	$f(x) = x^2$	$\left(2, \frac{1}{2}\right)$	16.	$f(x) = (x + 1)^2$	$(5, 3)$

17. ***Chemical Reaction*** In an autocatalytic chemical reaction, the product formed is a catalyst for the reaction. If Q_0 is the amount of the original substance and x is the amount of catalyst formed, the rate of chemical reaction is

$$\frac{dQ}{dx} = kx(Q_0 - x).$$

   For what value of x will the rate of chemical reaction be greatest?

18. ***Traffic Control*** On a given day, the flow rate F (in cars per hour) on a congested roadway is

$$F = \frac{v}{22 + 0.02v^2}$$

   where v is the speed of the traffic in miles per hour. What speed will maximize the flow rate on the road?

19. ***Area*** A farmer plans to fence a rectangular pasture adjacent to a river. The pasture must contain 180,000 square meters in order to provide enough grass for the herd. What dimensions would require the least amount of fencing if no fencing is needed along the river?

20. *Maximum Area* A rancher has 200 feet of fencing with which to enclose two adjacent rectangular corrals (see figure). What dimensions should be used so that the enclosed area will be a maximum?

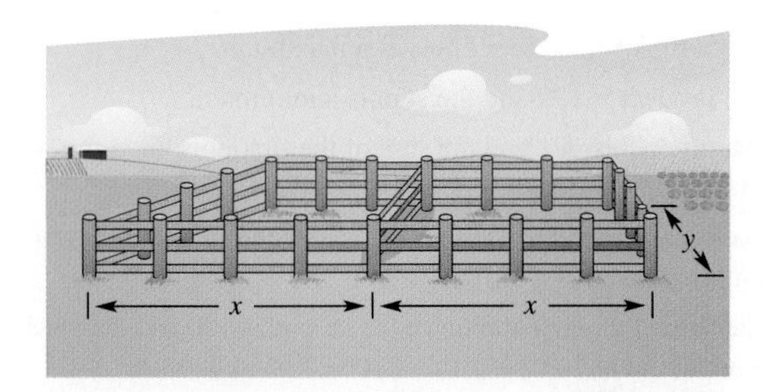

21. *Maximum Volume*

(a) Verify that each of the rectangular solids shown in the figure has a surface area of 150 square inches.

(b) Find the volume of each solid.

(c) Determine the dimensions of a rectangular solid (with a square base) of maximum volume if its surface area is 150 square inches.

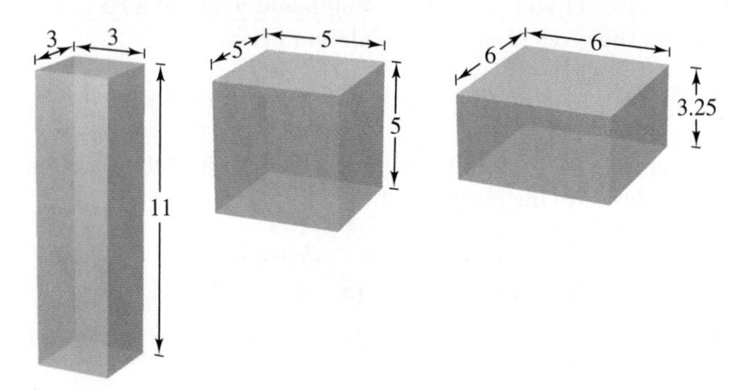

22. *Maximum Volume* Determine the dimensions of a rectangular solid (with a square base) with maximum volume if its surface area is 337.5 square centimeters.

23. *Maximum Area* A Norman window is constructed by adjoining a semicircle to the top of an ordinary rectangular window (see figure). Find the dimensions of a Norman window of maximum area if the total perimeter is 16 feet.

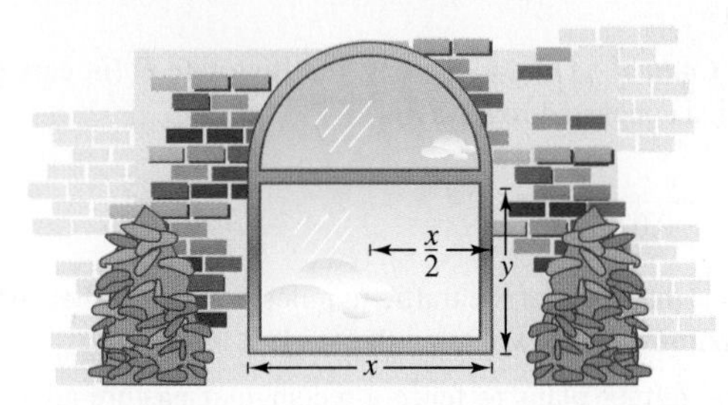

24. *Maximum Area* A rectangle is bounded by the x- and y-axes and the graph of $y = (6 - x)/2$ (see figure). What length and width should the rectangle have so that its area is a maximum?

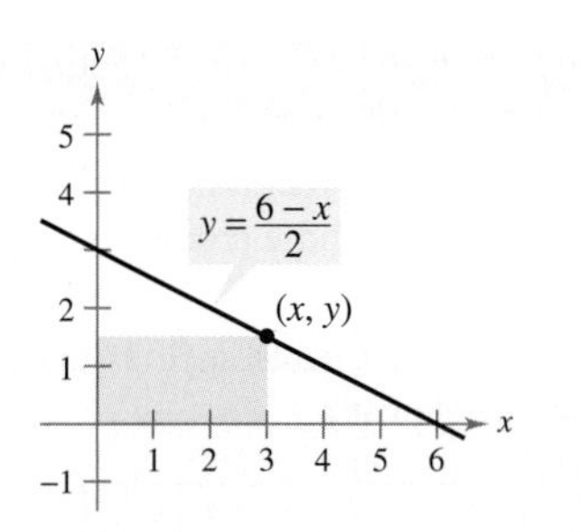

Figure for 24

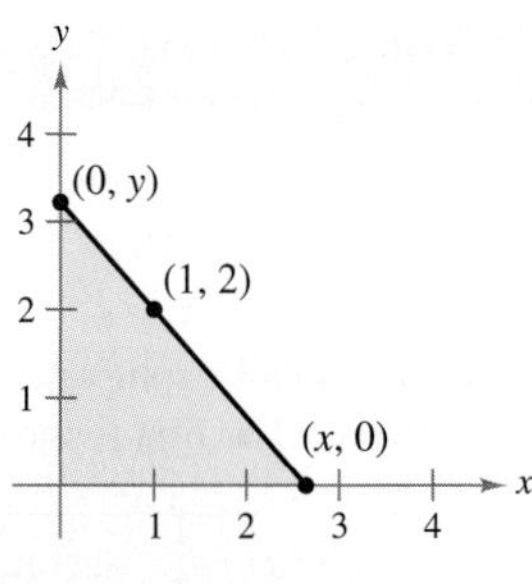

Figure for 25

25. *Minimum Length* A right triangle is formed in the first quadrant by the x- and y-axes and a line through the point $(1, 2)$ (see figure).

(a) Write the length L of the hypotenuse as a function of x.

(b) Use a graphing utility to approximate x graphically such that the length of the hypotenuse is a minimum.

(c) Find the vertices of the triangle such that its area is a minimum.

26. *Maximum Area* Find the area of the largest isosceles triangle that can be inscribed in a circle of radius 4 (see figure).

(a) Solve by writing the area as a function of h.

(b) Solve by writing the area as a function of α.

(c) Identify the type of triangle of maximum area.

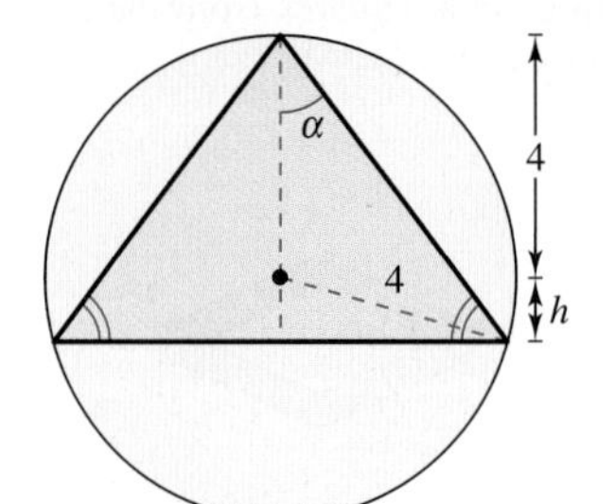

Figure for 26

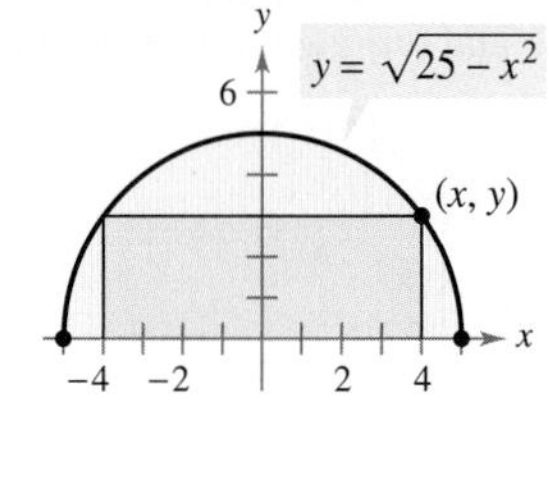

Figure for 27

27. *Maximum Area* A rectangle is bounded by the x-axis and the semicircle $y = \sqrt{25 - x^2}$ (see figure). What length and width should the rectangle have so that its area is a maximum?

28. *Area* Find the dimensions of the largest rectangle that can be inscribed in a semicircle of radius r (see Exercise 27).

29. *Area* A rectangular page is to contain 30 square inches of print. The margins on each side are 1 inch. Find the dimensions of the page such that the least amount of paper is used.

30. *Area* A rectangular page is to contain 36 square inches of print. The margins on each side are to be $1\frac{1}{2}$ inches. Find the dimensions of the page such that the least amount of paper is used.

31. ***Numerical, Graphical, and Analytic Analysis*** An exercise room consists of a rectangle with a semicircle on each end. A 200-meter running track runs around the outside of the room.

(a) Draw a figure to represent the problem. Let x and y represent the length and width of the rectangle.

(b) Analytically complete six rows of a table such as the one below. (The first two rows are shown.) Use the table to guess the maximum area of the rectangular region.

Length x	Width y	Area
10	$\frac{2}{\pi}(100 - 10)$	$(10)\frac{2}{\pi}(100 - 10) \approx 573$
20	$\frac{2}{\pi}(100 - 20)$	$(20)\frac{2}{\pi}(100 - 20) \approx 1019$

(c) Write the area A as a function of x.

(d) Use calculus to find the critical number of the function in part (c) and find the maximum value.

(e) Use a graphing utility to graph the function in part (c) and verify the maximum area from the graph.

32. ***Numerical, Graphical, and Analytic Analysis*** A right circular cylinder is to be designed to hold 22 cubic inches of a soft drink (approximately 12 fluid ounces).

(a) Analytically complete six rows of a table such as the one below. (The first two rows are shown.)

Radius r	Height	Surface Area
0.2	$\frac{22}{\pi(0.2)^2}$	$2\pi(0.2)\left[0.2 + \frac{22}{\pi(0.2)^2}\right] \approx 220.3$
0.4	$\frac{22}{\pi(0.4)^2}$	$2\pi(0.4)\left[0.4 + \frac{22}{\pi(0.4)^2}\right] \approx 111.0$

(b) Use a graphing utility to generate additional rows of the table. Use the table to estimate the minimum surface area. (*Hint:* Use the *table* feature of the graphing utility.)

(c) Write the surface area S as a function of r.

(d) Use a graphing utility to graph the function in part (c) and estimate the minimum surface area from the graph.

(e) Use calculus to find the critical number of the function in part (c) and find dimensions that will yield the minimum surface area.

33. ***Maximum Volume*** A rectangular package to be sent by a postal service can have a maximum combined length and girth (perimeter of a cross section) of 108 inches (see figure). Find the dimensions of the package of maximum volume that can be sent. (Assume the cross section is square.)

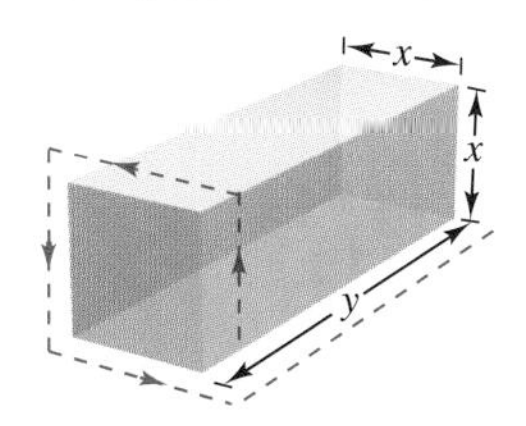

34. ***Maximum Volume*** Rework Exercise 33 for a cylindrical package. (The cross section is circular.)

35. ***Maximum Volume*** Find the volume of the largest right circular cone that can be inscribed in a sphere of radius r.

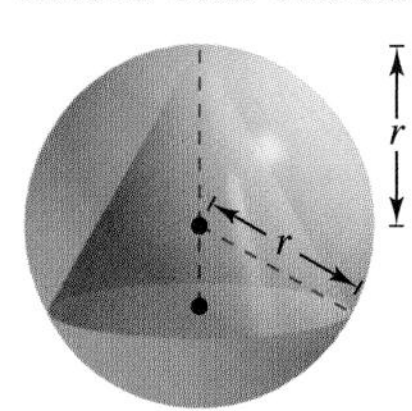

36. ***Maximum Volume*** Find the volume of the largest right circular cylinder that can be inscribed in a sphere of radius r.

Writing About Concepts

37. The perimeter of a rectangle is 20 feet. Of all possible dimensions, the maximum area is 25 square feet when its length and width are both 5 feet. Are there dimensions that yield a minimum area? Explain.

38. A shampoo bottle is a right circular cylinder. Because the surface area of the bottle does not change when it is squeezed, is it true that the volume remains the same? Explain.

39. ***Minimum Surface Area*** A solid is formed by adjoining two hemispheres to the ends of a right circular cylinder. The total volume of the solid is 12 cubic centimeters. Find the radius of the cylinder that produces the minimum surface area.

40. ***Minimum Cost*** An industrial tank of the shape described in Exercise 39 must have a volume of 3000 cubic feet. The hemispherical ends cost twice as much per square foot of surface area as the sides. Find the dimensions that will minimize cost.

41. ***Minimum Area*** The sum of the perimeters of an equilateral triangle and a square is 10. Find the dimensions of the triangle and the square that produce a minimum total area.

42. ***Maximum Area*** Twenty feet of wire is to be used to form two figures. In each of the following cases, how much wire should be used for each figure so that the total enclosed area is a maximum?

(a) Equilateral triangle and square

(b) Square and regular pentagon

(c) Regular pentagon and regular hexagon

(d) Regular hexagon and circle

What can you conclude from this pattern? {*Hint:* The area of a regular polygon with n sides of length x is $A = (n/4)[\cot(\pi/n)]x^2$.}

43. ***Beam Strength*** A wooden beam has a rectangular cross section of height h and width w (see figure on the next page). The strength S of the beam is directly proportional to the width and the square of the height. What are the dimensions of the strongest beam that can be cut from a round log of diameter 24 inches? (*Hint:* $S = kh^2w$, where k is the proportionality constant.)

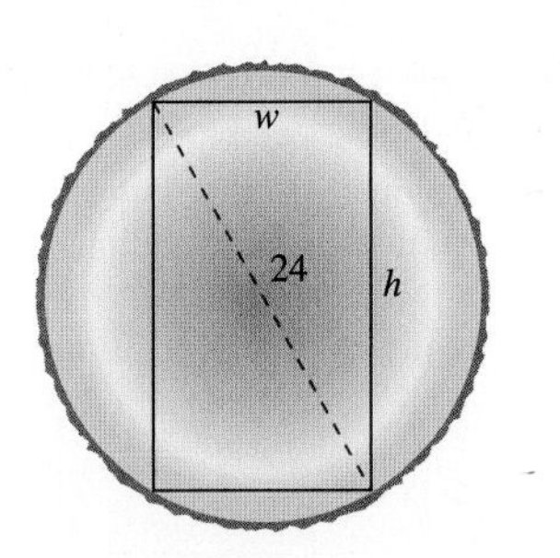

Figure for 43

Figure for 44

44. ***Minimum Length*** Two factories are located at the coordinates $(-x, 0)$ and $(x, 0)$ with their power supply located at $(0, h)$ (see figure). Find y such that the total length of power line from the power supply to the factories is a minimum.

45. ***Projectile Range*** The range R of a projectile fired with an initial velocity v_0 at an angle θ with the horizontal is $R = \dfrac{v_0^2 \sin 2\theta}{g}$, where g is the acceleration due to gravity. Find the angle θ such that the range is a maximum.

46. ***Conjecture*** Consider the functions $f(x) = \frac{1}{2}x^2$ and $g(x) = \frac{1}{16}x^4 - \frac{1}{2}x^2$ on the domain $[0, 4]$.

(a) Use a graphing utility to graph the functions on the specified domain.

(b) Write the vertical distance d between the functions as a function of x and use calculus to find the value of x for which d is maximum.

(c) Find the equations of the tangent lines to the graphs of f and g at the critical number found in part (b). Graph the tangent lines. What is the relationship between the lines?

(d) Make a conjecture about the relationship between tangent lines to the graphs of two functions at the value of x at which the vertical distance between the functions is greatest, and prove your conjecture.

47. ***Illumination*** A light source is located over the center of a circular table of diameter 4 feet (see figure). Find the height h of the light source such that the illumination I at the perimeter of the table is maximum if $I = k(\sin \alpha)/s^2$, where s is the slant height, α is the angle at which the light strikes the table, and k is a constant.

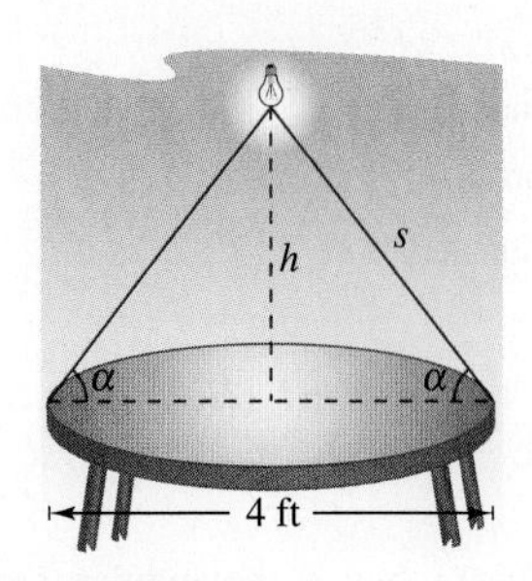

48. ***Illumination*** The illumination from a light source is directly proportional to the strength of the source and inversely proportional to the square of the distance from the source. Two light sources of intensities I_1 and I_2 are d units apart. What point on the line segment joining the two sources has the least illumination?

49. ***Minimum Time*** A man is in a boat 2 miles from the nearest point on the coast. He is to go to a point Q, located 3 miles down the coast and 1 mile inland (see figure). He can row at 2 miles per hour and walk at 4 miles per hour. Toward what point on the coast should he row in order to reach point Q in the least time?

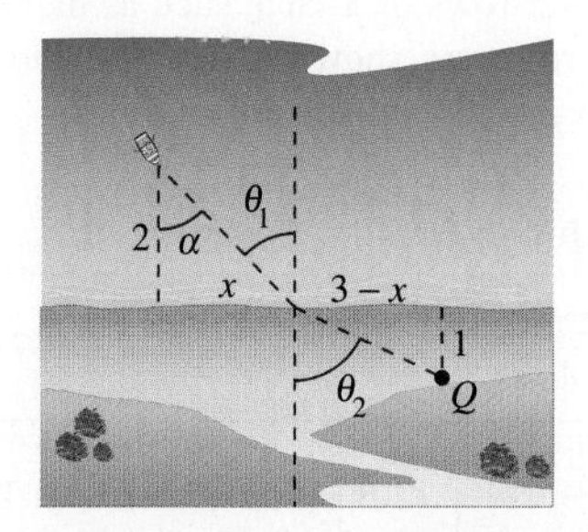

50. ***Minimum Time*** Consider Exercise 49 if the point Q is on the shoreline rather than 1 mile inland.

(a) Write the travel time T as a function of α.

(b) Use the result of part (a) to find the minimum time to reach Q.

(c) The man can row at v_1 miles per hour and walk at v_2 miles per hour. Write the time T as a function of α. Show that the critical number of T depends only on v_1 and v_2 and not on the distances. Explain how this result would be more beneficial to the man than the result of Exercise 49.

(d) Describe how to apply the result of part (c) to minimizing the cost of constructing a power transmission cable that costs c_1 dollars per mile under water and c_2 dollars per mile over land.

51. ***Minimum Time*** The conditions are the same as in Exercise 49 except that the man can row at v_1 miles per hour and walk at v_2 miles per hour. If θ_1 and θ_2 are the magnitudes of the angles, show that the man will reach point Q in the least time when

$$\frac{\sin \theta_1}{v_1} = \frac{\sin \theta_2}{v_2}.$$

52. ***Minimum Time*** When light waves, traveling in a transparent medium, strike the surface of a second transparent medium, they change direction. This change of direction is called *refraction* and is defined by **Snell's Law of Refraction,**

$$\frac{\sin \theta_1}{v_1} = \frac{\sin \theta_2}{v_2}$$

where θ_1 and θ_2 are the magnitudes of the angles shown in the figure and v_1 and v_2 are the velocities of light in the two media. Show that this problem is equivalent to Exercise 51, and that light waves traveling from P to Q follow the path of minimum time.

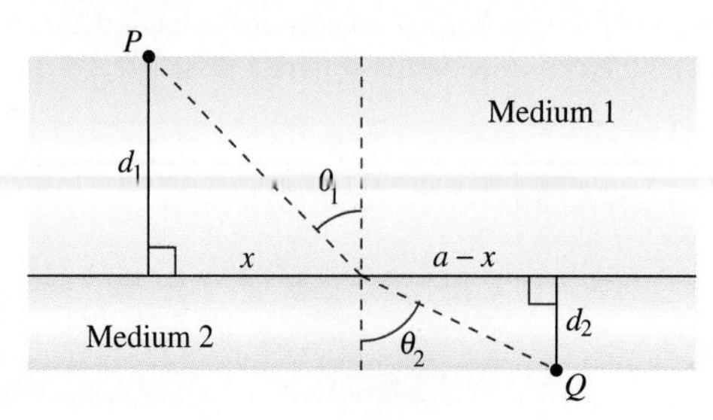

53. Sketch the graph of $f(x) = 2 - 2 \sin x$ on the interval $[0, \pi/2]$.

(a) Find the distance from the origin to the y-intercept and the distance from the origin to the x-intercept.

(b) Write the distance d from the origin to a point on the graph of f as a function of x. Use a graphing utility to graph d and find the minimum distance.

(c) Use calculus and the *zero* or *root* feature of a graphing utility to find the value of x that minimizes the function d on the interval $[0, \pi/2]$. What is the minimum distance?

(Submitted by Tim Chapell, Penn Valley Community College, Kansas City, MO.)

54. *Minimum Cost* An offshore oil well is 2 kilometers off the coast. The refinery is 4 kilometers down the coast. Laying pipe in the ocean is twice as expensive as on land. What path should the pipe follow in order to minimize the cost?

55. *Maximum Volume* A sector with central angle θ is cut from a circle of radius 12 inches (see figure), and the edges of the sector are brought together to form a cone. Find the magnitude of θ such that the volume of the cone is a maximum.

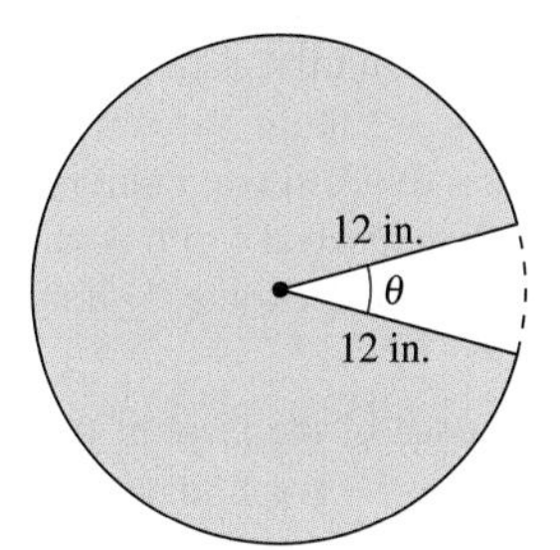

Figure for 55

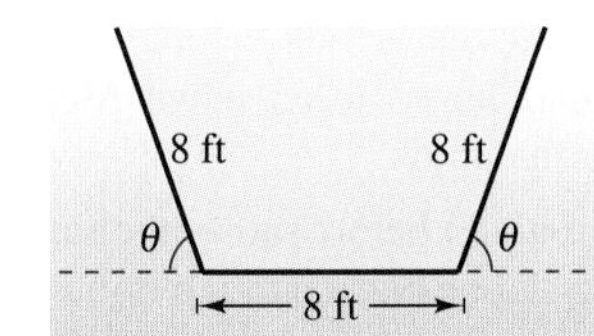

Figure for 56

56. *Numerical, Graphical, and Analytic Analysis* The cross sections of an irrigation canal are isosceles trapezoids of which three sides are 8 feet long (see figure). Determine the angle of elevation θ of the sides such that the area of the cross section is a maximum by completing the following.

(a) Analytically complete six rows of a table such as the one below. (The first two rows are shown.)

Base 1	Base 2	Altitude	Area
8	$8 + 16 \cos 10°$	$8 \sin 10°$	≈ 22.1
8	$8 + 16 \cos 20°$	$8 \sin 20°$	≈ 42.5

(b) Use a graphing utility to generate additional rows of the table and estimate the maximum cross-sectional area. (*Hint:* Use the *table* feature of the graphing utility.)

(c) Write the cross-sectional area A as a function of θ.

(d) Use calculus to find the critical number of the function in part (c) and find the angle that will yield the maximum cross-sectional area.

(e) Use a graphing utility to graph the function in part (c) and verify the maximum cross-sectional area.

57. *Minimum Cost* The ordering and transportation cost C of the components used in manufacturing a product is

$$C = 100\left(\frac{200}{x^2} + \frac{x}{x + 30}\right), \quad x \geq 1$$

where C is measured in thousands of dollars and x is the order size in hundreds. Find the order size that minimizes the cost. (*Hint:* Use the *root* feature of a graphing utility.)

58. *Diminishing Returns* The profit P (in thousands of dollars) for a company spending an amount s (in thousands of dollars) on advertising is

$$P = -\tfrac{1}{10}s^3 + 6s^2 + 400.$$

(a) Find the amount of money the company should spend on advertising in order to obtain a maximum profit.

(b) The *point of diminishing returns* is the point at which the rate of growth of the profit function begins to decline. Find the point of diminishing returns.

59. *Area* Find the area of the largest rectangle that can be inscribed under the curve $y = e^{-x^2}$ in the first and second quadrants.

60. *Population Growth* Fifty elk are introduced into a game preserve. It is estimated that their population will increase according to the model $p(t) = 250/(1 + 4e^{-t/3})$, where t is measured in years. At what rate is the population increasing when $t = 2$? After how many years is the population increasing most rapidly?

61. Verify that the function

$$y = \frac{L}{1 + ae^{-x/b}}, \quad a > 0, b > 0, L > 0$$

increases at the maximum rate when $y = L/2$.

62. *Area* Perform the following steps to find the maximum area of the rectangle shown in the figure.

(a) Solve for c in the equation $f(c) = f(c + x)$.

(b) Use the result in part (a) to write the area A as a function of x. [*Hint*: $A = xf(c)$]

(c) Use a graphing utility to graph the area function. Use the graph to approximate the dimensions of the rectangle of maximum area. Determine the required area.

(d) Use a graphing utility to graph the expression for c found in part (a). Use the graph to approximate

$$\lim_{x\to 0^+} c \quad \text{and} \quad \lim_{x\to\infty} c.$$

Use this result to describe the changes in the dimensions and position of the rectangle for $0 < x < \infty$.

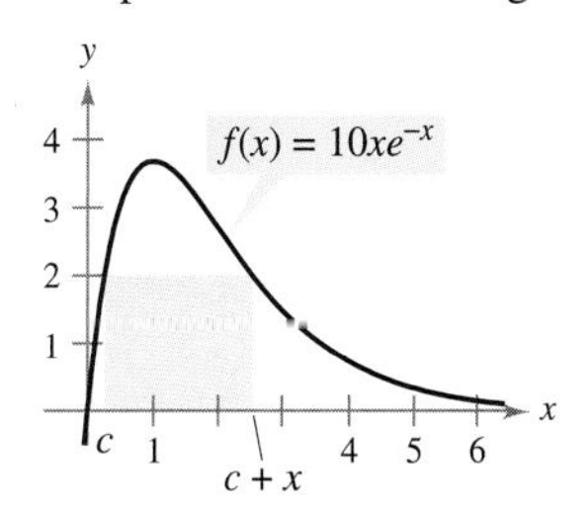

Minimum Distance **In Exercises 63–65, consider a fuel distribution center located at the origin of the rectangular coordinate system (units in miles; see figures). The center supplies three factories with coordinates (4, 1), (5, 6), and (10, 3). A trunk line will run from the distribution center along the line $y = mx$, and feeder lines will run to the three factories. The objective is to find m such that the lengths of the feeder lines are minimized.**

63. Minimize the sum of the squares of the lengths of vertical feeder lines given by

$$S_1 = (4m - 1)^2 + (5m - 6)^2 + (10m - 3)^2.$$

Find the equation for the trunk line by this method and then determine the sum of the lengths of the feeder lines.

64. Minimize the sum of the absolute values of the lengths of vertical feeder lines given by

$$S_2 = |4m - 1| + |5m - 6| + |10m - 3|.$$

Find the equation for the trunk line by this method and then determine the sum of the lengths of the feeder lines. (*Hint:* Use a graphing utility to graph the function S_2 and approximate the required critical number.)

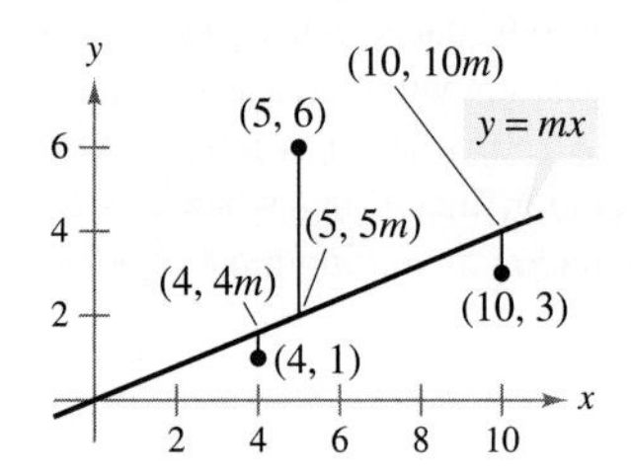

Figure for 63 and 64

Figure for 65

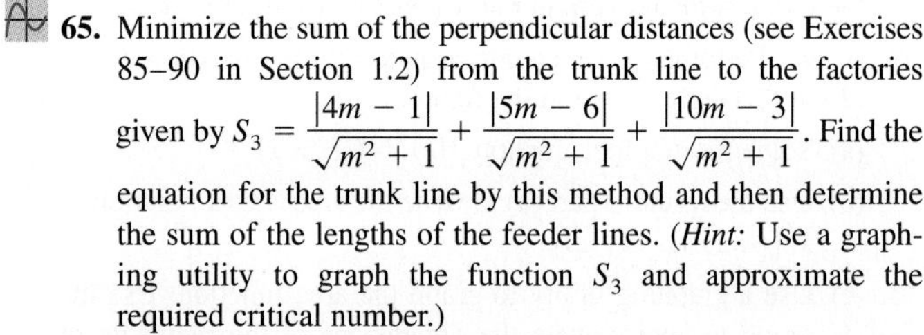

65. Minimize the sum of the perpendicular distances (see Exercises 85–90 in Section 1.2) from the trunk line to the factories given by $S_3 = \dfrac{|4m - 1|}{\sqrt{m^2 + 1}} + \dfrac{|5m - 6|}{\sqrt{m^2 + 1}} + \dfrac{|10m - 3|}{\sqrt{m^2 + 1}}$. Find the equation for the trunk line by this method and then determine the sum of the lengths of the feeder lines. (*Hint:* Use a graphing utility to graph the function S_3 and approximate the required critical number.)

66. *Maximum Area* Consider a symmetric cross inscribed in a circle of radius r (see figure).

(a) Write the area A of the cross as a function of x and find the value of x that maximizes the area.

(b) Write the area A of the cross as a function of θ and find the value of θ that maximizes the area.

(c) Show that the critical numbers of parts (a) and (b) yield the same maximum area. What is that area?

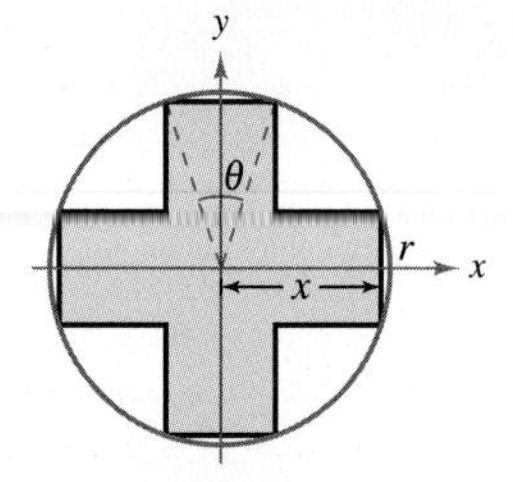

Putnam Exam Challenge

67. Find the maximum value of $f(x) = x^3 - 3x$ on the set of all real numbers x satisfying $x^4 + 36 \le 13x^2$. Explain your reasoning.

68. Find the minimum value of

$$\frac{(x + 1/x)^6 - (x^6 + 1/x^6) - 2}{(x + 1/x)^3 + (x^3 + 1/x^3)}$$

for $x > 0$.

These problems were composed by the Committee on the Putnam Prize Competition.

Section Project: Connecticut River

Whenever the Connecticut River reaches a level of 105 feet above sea level, two Northampton, Massachusetts flood control station operators begin a round-the-clock river watch. Every 2 hours, they check the height of the river, using a scale marked off in tenths of a foot, and record the data in a log book. In the spring of 1996, the flood watch lasted from April 4, when the river reached 105 feet and was rising at 0.2 foot per hour, until April 25, when the level subsided again to 105 feet. Between those dates, their log shows that the river rose and fell several times, at one point coming close to the 115-foot mark. If the river had reached 115 feet, the city would have closed down Mount Tom Road (Route 5, south of Northampton).

The graph below shows the rate of change of the level of the river during one portion of the flood watch. Use the graph to answer each question.

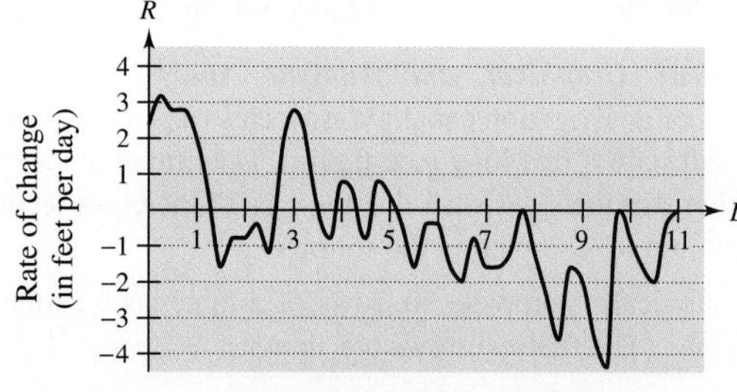

(a) On what date was the river rising most rapidly? How do you know?

(b) On what date was the river falling most rapidly? How do you know?

(c) There were two dates in a row on which the river rose, then fell, then rose again during the course of the day. On which days did this occur, and how do you know?

(d) At 1 minute past midnight, April 14, the river level was 111.0 feet. Estimate the river's height 24 hours later and 48 hours later. Explain how you made your estimates.

(e) The river crested at 114.4 feet. On what date do you think this occurred?

(Submitted by Mary Murphy, Smith College, Northampton, MA)

Section 4.8 Differentials

- **Understand the concept of a tangent line approximation.**
- **Compare the value of the differential, dy, with the actual change in y, Δy.**
- **Estimate a propagated error using a differential.**
- **Find the differential of a function using differentiation formulas.**

> **EXPLORATION**
>
> ***Tangent Line Approximation*** Use a graphing utility to graph
>
> $f(x) = x^2.$
>
> In the same viewing window, graph the tangent line to the graph of f at the point $(1, 1)$. Zoom in twice on the point of tangency. Does your graphing utility distinguish between the two graphs? Use the *trace* feature to compare the two graphs. As the x-values get closer to 1, what can you say about the y-values?

Tangent Line Approximations

Newton's Method (Section 3.8) is an example of the use of a tangent line to a graph to approximate the graph. In this section, you will study other situations in which the graph of a function can be approximated by a straight line.

To begin, consider a function f that is differentiable at c. The equation for the tangent line at the point $(c, f(c))$ is given by

$$y - f(c) = f'(c)(x - c)$$
$$y = f(c) + f'(c)(x - c)$$

and is called the **tangent line approximation** (or **linear approximation**) **of f at c.** Because c is a constant, y is a linear function of x. Moreover, by restricting the values of x to be sufficiently close to c, the values of y can be used as approximations (to any desired accuracy) of the values of the function f. In other words, as $x \to c$, the limit of y is $f(c)$.

EXAMPLE 1 Using a Tangent Line Approximation

Find the tangent line approximation of

$$f(x) = 1 + \sin x$$

at the point $(0, 1)$. Then use a table to compare the y-values of the linear function with those of $f(x)$ on an open interval containing $x = 0$.

Solution The derivative of f is

$$f'(x) = \cos x. \qquad \text{First derivative}$$

So, the equation of the tangent line to the graph of f at the point $(0, 1)$ is

$$y - f(0) = f'(0)(x - 0)$$
$$y - 1 = (1)(x - 0)$$
$$y = 1 + x. \qquad \text{Tangent line approximation}$$

The table compares the values of y given by this linear approximation with the values of $f(x)$ near $x = 0$. Notice that the closer x is to 0, the better the approximation is. This conclusion is reinforced by the graph shown in Figure 4.63.

x	-0.5	-0.1	-0.01	0	0.01	0.1	0.5
$f(x) = 1 + \sin x$	0.521	0.9002	0.9900002	1	1.0099998	1.0998	1.479
$y = 1 + x$	0.5	0.9	0.99	1	1.01	1.1	1.5

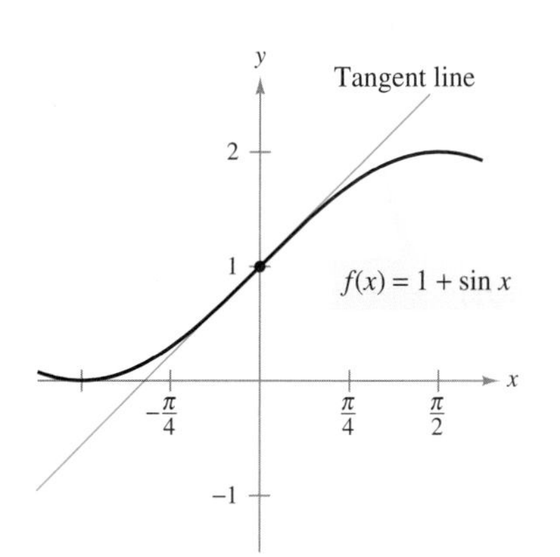

The tangent line approximation of f at the point $(0, 1)$
Figure 4.63

NOTE Be sure you see that this linear approximation of $f(x) = 1 + \sin x$ depends on the point of tangency. At a different point on the graph of f, you would obtain a different tangent line approximation.

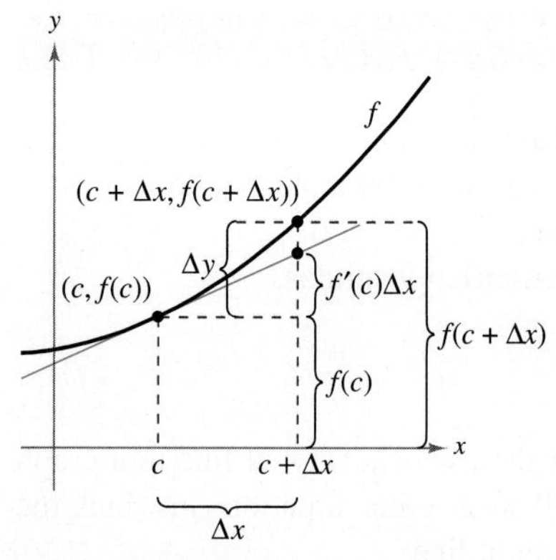

When Δx is small, $\Delta y = f(c + \Delta x) - f(c)$ is approximated by $f'(c)\Delta x$.
Figure 4.64

Differentials

When the tangent line to the graph of f at the point $(c, f(c))$

$$y = f(c) + f'(c)(x - c) \qquad \text{Tangent line at } (c, f(c))$$

is used as an approximation of the graph of f, the quantity $x - c$ is called the change in x, and is denoted by Δx, as shown in Figure 4.64. When Δx is small, the change in y (denoted by Δy) can be approximated as shown.

$$\Delta y = f(c + \Delta x) - f(c) \qquad \text{Actual change in } y$$
$$\approx f'(c)\Delta x \qquad \text{Approximate change in } y$$

For such an approximation, the quantity Δx is traditionally denoted by dx, and is called the **differential of *x*.** The expression $f'(x)\,dx$ is denoted by dy, and is called the **differential of *y*.**

Definition of Differentials

Let $y = f(x)$ represent a function that is differentiable on an open interval containing x. The **differential of *x*** (denoted by dx) is any nonzero real number. The **differential of *y*** (denoted by dy) is

$$dy = f'(x)\,dx.$$

In many types of applications, the differential of y can be used as an approximation of the change in y. That is,

$$\Delta y \approx dy \qquad \text{or} \qquad \Delta y \approx f'(x)dx.$$

EXAMPLE 2 Comparing Δy and dy

Let $y = x^2$. Find dy when $x = 1$ and $dx = 0.01$. Compare this value with Δy for $x = 1$ and $\Delta x = 0.01$.

Solution Because $y = f(x) = x^2$, you have $f'(x) = 2x$, and the differential dy is given by

$$dy = f'(x)\,dx = f'(1)(0.01) = 2(0.01) = 0.02. \qquad \text{Differential of } y$$

Now, using $\Delta x = 0.01$, the change in y is

$$\Delta y = f(x + \Delta x) - f(x) = f(1.01) - f(1) = (1.01)^2 - 1^2 = 0.0201.$$

Figure 4.65 shows the geometric comparison of dy and Δy. Try comparing other values of dy and Δy. You will see that the values become closer to each other as dx (or Δx) approaches 0.

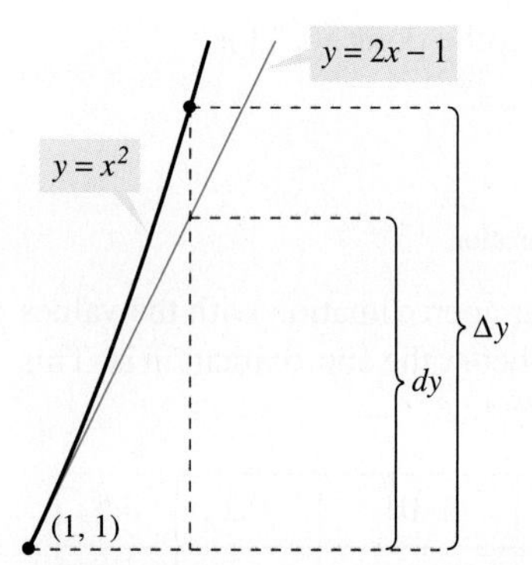

The change in y, Δy, is approximated by the differential of y, dy.
Figure 4.65

In Example 2, the tangent line to the graph of $f(x) = x^2$ at $x = 1$ is

$$y = 2x - 1 \quad \text{or} \quad g(x) = 2x - 1. \qquad \text{Tangent line to the graph of } f \text{ at } x = 1$$

For x-values near 1, this line is close to the graph of f, as shown in Figure 4.65. For instance,

$$f(1.01) = 1.01^2 = 1.0201 \quad \text{and} \quad g(1.01) = 2(1.01) - 1 = 1.02.$$

Error Propagation

Physicists and engineers tend to make liberal use of the approximation of Δy by dy. One way this occurs in practice is in the estimation of errors propagated by physical measuring devices. For example, if you let x represent the measured value of a variable and let $x + \Delta x$ represent the exact value, then Δx is the *error in measurement.* Finally, if the measured value x is used to compute another value $f(x)$, the difference between $f(x + \Delta x)$ and $f(x)$ is the **propagated error.**

$$\underbrace{f(x + \overbrace{\Delta x}^{\text{Measurement error}})}_{\text{Exact value}} - \underbrace{f(x)}_{\text{Measured value}} = \overbrace{\Delta y}^{\text{Propagated error}}$$

EXAMPLE 3 Estimation of Error

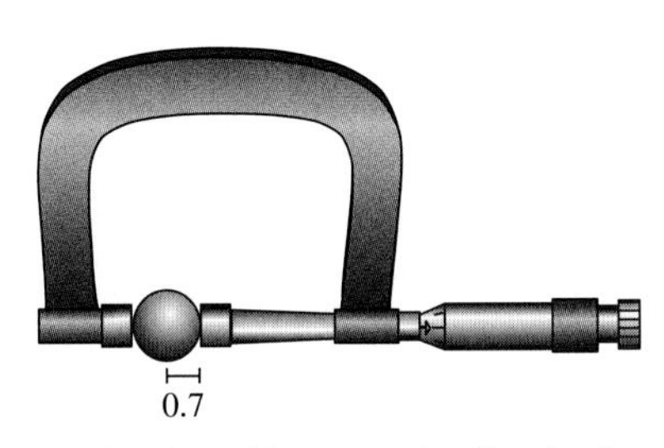

Ball bearing with measured radius that is correct to within 0.01 inch
Figure 4.66

The radius of a ball bearing is measured to be 0.7 inch, as shown in Figure 4.66. If the measurement is correct to within 0.01 inch, estimate the propagated error in the volume V of the ball bearing.

Solution The formula for the volume of a sphere is $V = \frac{4}{3}\pi r^3$, where r is the radius of the sphere. So, you can write

$$r = 0.7 \qquad \text{Measured radius}$$

and

$$-0.01 \le \Delta r \le 0.01. \qquad \text{Possible error}$$

To approximate the propagated error in the volume, differentiate V to obtain $dV/dr = 4\pi r^2$ and write

$$\begin{aligned} \Delta V &\approx dV && \text{Approximate } \Delta V \text{ by } dV. \\ &= 4\pi r^2\, dr \\ &= 4\pi(0.7)^2(\pm 0.01) && \text{Substitute for } r \text{ and } dr. \\ &\approx \pm 0.06158 \text{ in}^3. \end{aligned}$$

So the volume has a propagated error of about 0.06 cubic inch.

Would you say that the propagated error in Example 3 is large or small? The answer is best given in *relative* terms by comparing dV with V. The ratio

$$\begin{aligned} \frac{dV}{V} &= \frac{4\pi r^2\, dr}{\frac{4}{3}\pi r^3} && \text{Ratio of } dV \text{ to } V \\ &= \frac{3\, dr}{r} && \text{Simplify.} \\ &\approx \frac{3}{0.7}(\pm 0.01) && \text{Substitute for } dr \text{ and } r. \\ &\approx \pm 0.0429 \end{aligned}$$

is called the **relative error.** The corresponding **percent error** is approximately 4.29%.

Calculating Differentials

Each of the differentiation rules that you studied in Chapter 3 can be written in **differential form.** For example, suppose u and v are differentiable functions of x. By the definition of differentials, you have

$$du = u'\,dx \quad \text{and} \quad dv = v'\,dx.$$

So, you can write the differential form of the Product Rule as follows.

$$\begin{aligned} d[uv] &= \frac{d}{dx}[uv]\,dx && \text{Differential of } uv \\ &= [uv' + vu']\,dx && \text{Product Rule} \\ &= uv'\,dx + vu'\,dx \\ &= u\,dv + v\,du \end{aligned}$$

Differential Formulas

Let u and v be differentiable functions of x.

Constant multiple: $d[cu] = c\,du$

Sum or difference: $d[u \pm v] = du \pm dv$

Product: $d[uv] = u\,dv + v\,du$

Quotient: $d\left[\frac{u}{v}\right] = \frac{v\,du - u\,dv}{v^2}$

EXAMPLE 4 Finding Differentials

Function	Derivative	Differential
a. $y = x^2$	$\frac{dy}{dx} = 2x$	$dy = 2x\,dx$
b. $y = 2\sin x$	$\frac{dy}{dx} = 2\cos x$	$dy = 2\cos x\,dx$
c. $y = xe^x$	$\frac{dy}{dx} = e^x(x + 1)$	$dy = e^x(x + 1)\,dx$
d. $y = \frac{1}{x}$	$\frac{dy}{dx} = -\frac{1}{x^2}$	$dy = -\frac{dx}{x^2}$

The notation in Example 4 is called the **Leibniz notation** for derivatives and differentials, named after the German mathematician Gottfried Wilhelm Leibniz. The beauty of this notation is that it provides an easy way to remember several important calculus formulas by making it seem as though the formulas were derived from algebraic manipulations of differentials. For instance, in Leibniz notation, the *Chain Rule*

$$\frac{dy}{dx} = \frac{dy}{du}\frac{du}{dx}$$

would appear to be true because the du's divide out. Even though this reasoning is *incorrect*, the notation does help one remember the Chain Rule.

GOTTFRIED WILHELM LEIBNIZ (1646–1716)

Both Leibniz and Newton are credited with creating calculus. It was Leibniz, however, who tried to broaden calculus by developing rules and formal notation. He often spent days choosing an appropriate notation for a new concept.

EXAMPLE 5 Finding the Differential of a Composite Function

$$y = f(x) = \sin 3x \qquad \text{Original function}$$
$$f'(x) = 3 \cos 3x \qquad \text{Apply Chain Rule.}$$
$$dy = f'(x)\,dx = 3 \cos 3x\,dx \qquad \text{Differential form}$$

EXAMPLE 6 Finding the Differential of a Composite Function

$$y = f(x) = (x^2 + 1)^{1/2} \qquad \text{Original function}$$
$$f'(x) = \frac{1}{2}(x^2 + 1)^{-1/2}(2x) = \frac{x}{\sqrt{x^2 + 1}} \qquad \text{Apply Chain Rule.}$$
$$dy = f'(x)\,dx = \frac{x}{\sqrt{x^2 + 1}}\,dx \qquad \text{Differential form}$$

Differentials can be used to approximate function values. To do this for the function given by $y = f(x)$, you use the formula

$$f(x + \Delta x) \approx f(x) + dy = f(x) + f'(x)\,dx$$

which is derived from the approximation $\Delta y = f(x + \Delta x) - f(x) \approx dy$. The key to using this formula is to choose a value for x that makes the calculations easier, as shown in Example 7.

EXAMPLE 7 Approximating Function Values

Use differentials to approximate $\sqrt{16.5}$.

Solution Using $f(x) = \sqrt{x}$, you can write

$$f(x + \Delta x) \approx f(x) + f'(x)\,dx = \sqrt{x} + \frac{1}{2\sqrt{x}}\,dx.$$

Now, choosing $x = 16$ and $dx = 0.5$, you obtain the following approximation.

$$f(x + \Delta x) = \sqrt{16.5} \approx \sqrt{16} + \frac{1}{2\sqrt{16}}(0.5) = 4 + \left(\frac{1}{8}\right)\left(\frac{1}{2}\right) = 4.0625$$

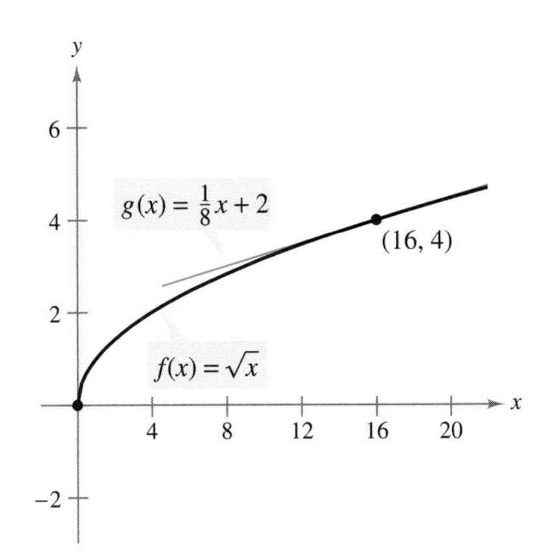

Figure 4.67

The tangent line approximation to $f(x) = \sqrt{x}$ at $x = 16$ is the line $g(x) = \frac{1}{8}x + 2$. For x-values near 16, the graphs of f and g are close together, as shown in Figure 4.67. For instance,

$$f(16.5) = \sqrt{16.5} \approx 4.0620 \quad \text{and} \quad g(16.5) = \frac{1}{8}(16.5) + 2 = 4.0625.$$

In fact, if you use a graphing utility to zoom in near the point of tangency (16, 4), you will see that the two graphs appear to coincide. Notice also that as you move farther away from the point of tangency, the linear approximation is less accurate.

Exercises for Section 4.8

See www.CalcChat.com for worked-out solutions to odd-numbered exercises.

In Exercises 1–6, find the equation of the tangent line T to the graph of f at the indicated point. Use this linear approximation to complete the table.

x	1.9	1.99	2	2.01	2.1
$f(x)$					
$T(x)$					

Function	*Point*
1. $f(x) = x^2$	$(2, 4)$
2. $f(x) = \dfrac{6}{x^2}$	$\left(2, \dfrac{3}{2}\right)$
3. $f(x) = x^5$	$(2, 32)$
4. $f(x) = \sqrt{x}$	$(2, \sqrt{2})$
5. $f(x) = \sin x$	$(2, \sin 2)$
6. $f(x) = \log_2 x$	$(2, 1)$

In Exercises 7–10, use the information to evaluate and compare Δy and dy.

7. $y = \frac{1}{2}x^3$ $\quad x = 2 \quad \Delta x = dx = 0.1$

8. $y = 1 - 2x^2$ $\quad x = 0 \quad \Delta x = dx = -0.1$

9. $y = x^4 + 1$ $\quad x = -1 \quad \Delta x = dx = 0.01$

10. $y = 2x + 1$ $\quad x = 2 \quad \Delta x = dx = 0.01$

In Exercises 11–24, find the differential dy of the given function.

11. $y = 3x^2 - 4$

12. $y = 3x^{2/3}$

13. $y = \dfrac{x + 1}{2x - 1}$

14. $y = \sqrt{9 - x^2}$

15. $y = x\sqrt{1 - x^2}$

16. $y = \sqrt{x} + 1/\sqrt{x}$

17. $y = \ln\sqrt{4 - x^2}$

18. $y = e^{-0.5x}\cos 4x$

19. $y = 2x - \cot^2 x$

20. $y = x\sin x$

21. $y = \dfrac{1}{3}\cos\left(\dfrac{6\pi x - 1}{2}\right)$

22. $y = \dfrac{\sec^2 x}{x^2 + 1}$

23. $y = x \arcsin x$

24. $y = \arctan(x - 2)$

In Exercises 25–28, use differentials and the graph of f to approximate (a) $f(1.9)$ and (b) $f(2.04)$. To print an enlarged copy of the graph, go to the website *www.mathgraphs.com*.

25.

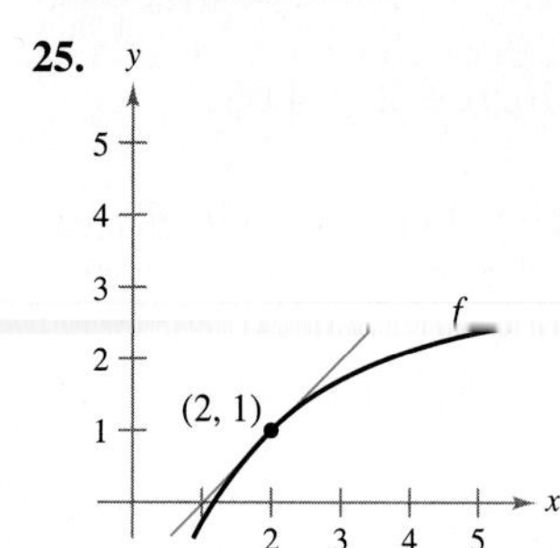

26.

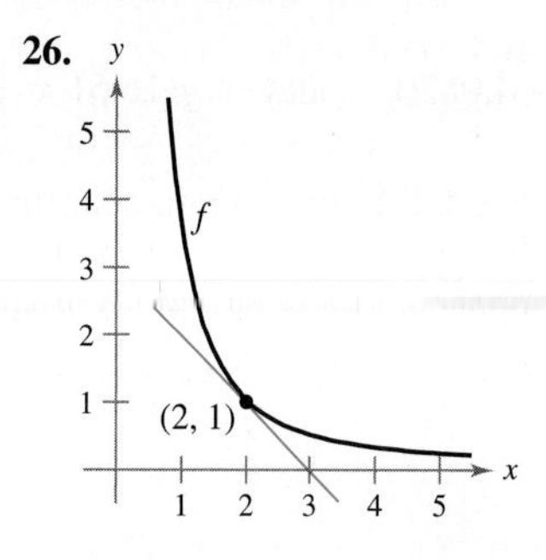

27.

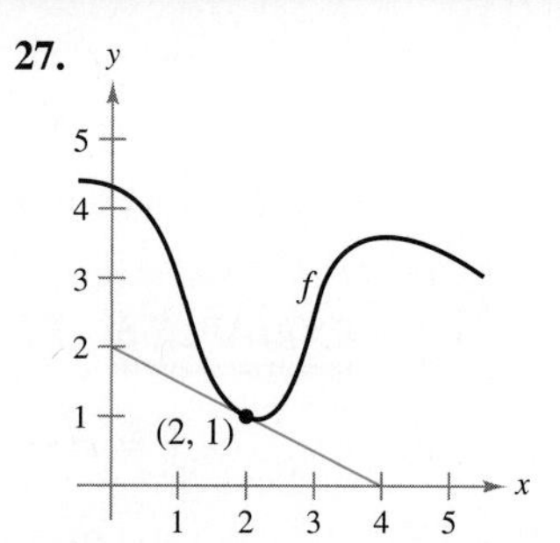

28.

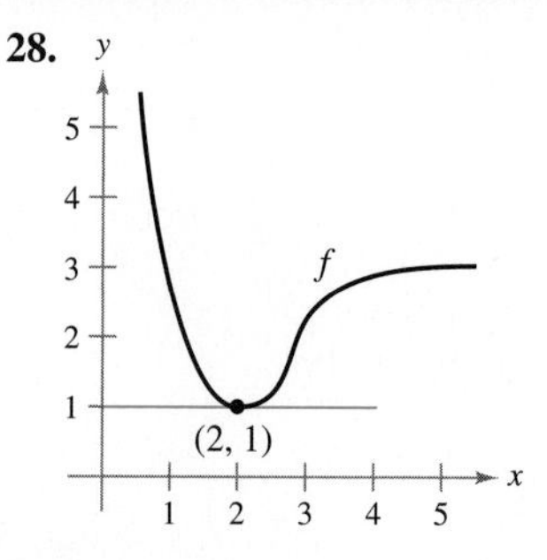

In Exercises 29 and 30, use differentials and the graph of g' to approximate (a) $g(2.93)$ and (b) $g(3.1)$ given that $g(3) = 8$.

29.

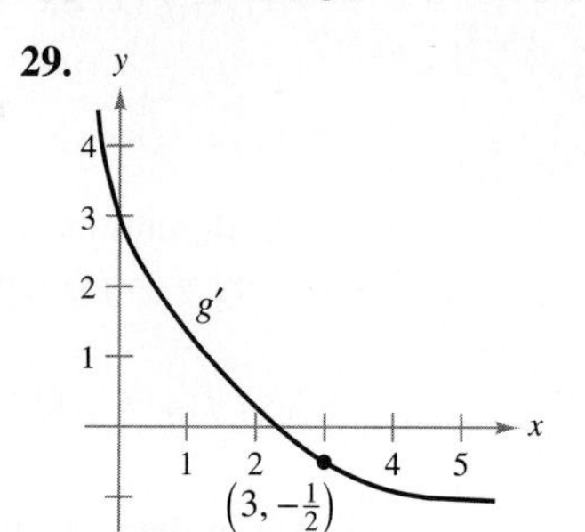

30.

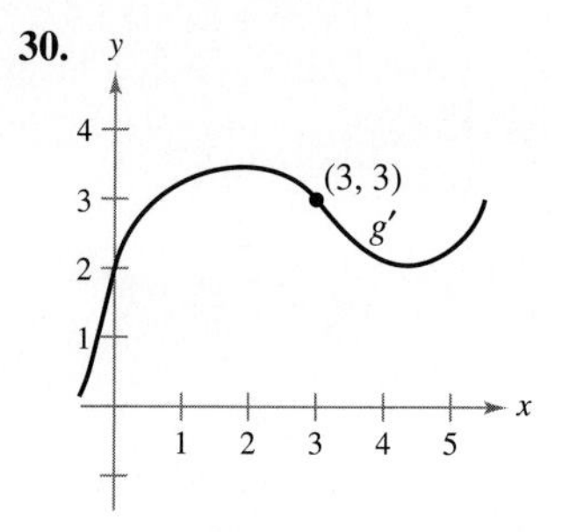

31. ***Area*** The measurement of the side of a square is found to be 12 inches, with a possible error of $\frac{1}{64}$ inch. Use differentials to approximate the possible propagated error in computing the area of the square.

32. ***Area*** The measurements of the base and altitude of a triangle are found to be 36 and 50 centimeters, respectively. The possible error in each measurement is 0.25 centimeter. Use differentials to approximate the possible propagated error in computing the area of the triangle.

33. ***Area*** The measurement of the radius of the end of a log is found to be 14 inches, with a possible error of $\frac{1}{4}$ inch. Use differentials to approximate the possible propagated error in computing the area of the end of the log.

34. ***Volume and Surface Area*** The measurement of the edge of a cube is found to be 12 inches, with a possible error of 0.03 inch. Use differentials to approximate the maximum possible propagated error in computing (a) the volume of the cube and (b) the surface area of the cube.

35. ***Area*** The measurement of a side of a square is found to be 15 centimeters, with a possible error of 0.05 centimeter.

(a) Approximate the percent error in computing the area of the square.

(b) Estimate the maximum allowable percent error in measuring the side if the error in computing the area cannot exceed 2.5%.

36. ***Circumference*** The measurement of the circumference of a circle is found to be 60 centimeters, with a possible error of 1.2 centimeters.

(a) Approximate the percent error in computing the area of the circle.

(b) Estimate the maximum allowable percent error in measuring the circumference if the error in computing the area cannot exceed 3%.

37. ***Volume and Surface Area*** The radius of a sphere is measured to be 6 inches, with a possible error of 0.02 inch. Use differentials to approximate the maximum possible error in calculating (a) the volume of the sphere, (b) the surface area of the sphere, and (c) the relative errors in parts (a) and (b).

38. ***Profit*** The profit P for a company is given by

$$P = (500x - x^2) - \left(\tfrac{1}{2}x^2 - 77x + 3000\right).$$

Approximate the change and percent change in profit as production changes from $x = 115$ to $x = 120$ units.

39. ***Profit*** The profit P for a company is $P = 100xe^{-x/400}$ where x is sales. Approximate the change and percent change in profit as sales change from $x = 115$ to $x = 120$ units.

40. ***Relative Humidity*** When the dewpoint is 65° Fahrenheit, the relative humidity H is modeled by

$$H = \frac{4347}{400{,}000{,}000}e^{369{,}444/(50t + 19{,}793)}$$

where t is the air temperature in degrees Fahrenheit. Use differentials to approximate the change in relative humidity at $t = 72$ for a 1-degree change in the air temperature.

In Exercises 41 and 42, the thickness of the shell is 0.2 centimeter. Use differentials to approximate the volume of the shell.

41. A cylindrical shell with height 40 centimeters and radius 5 centimeters

42. A spherical shell of radius 100 centimeters

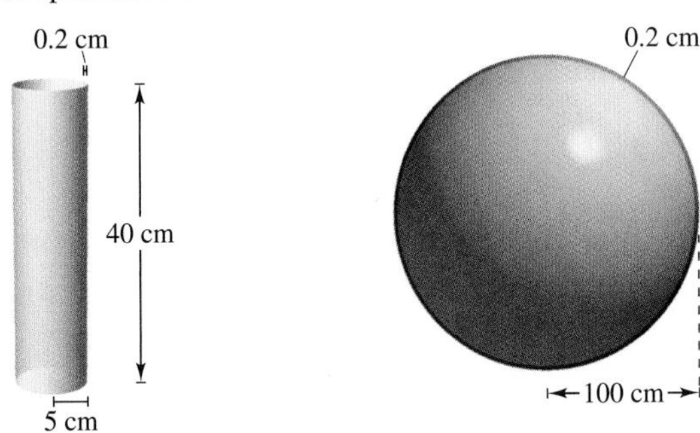

Figure for 41 **Figure for 42**

43. ***Triangle Measurements*** The measurement of one side of a right triangle is found to be 9.5 inches, and the angle opposite that side is 26°45′ with a possible error of 15′.

(a) Approximate the percent error in computing the length of the hypotenuse.

(b) Estimate the maximum allowable percent error in measuring the angle if the error in computing the length of the hypotenuse cannot exceed 2%.

44. ***Ohm's Law*** A current of I amperes passes through a resistor of R ohms. **Ohm's Law** states that the voltage E applied to the resistor is $E = IR$. If the voltage is constant, show that the magnitude of the relative error in R caused by a change in I is equal in magnitude to the relative error in I.

45. ***Projectile Motion*** The range R of a projectile is

$$R = \frac{v_0^2}{32}(\sin 2\theta)$$

where v_0 is the initial velocity in feet per second and θ is the angle of elevation. If $v_0 = 2200$ feet per second and θ is changed from 10° to 11°, use differentials to approximate the change in the range.

46. ***Surveying*** A surveyor standing 50 feet from the base of a large tree measures the angle of elevation to the top of the tree as 71.5°. How accurately must the angle be measured if the percent error in estimating the height of the tree is to be less than 6%?

In Exercises 47–50, use differentials to approximate the value of the expression. Compare your answer with that of a calculator.

47. $\sqrt{99.4}$ **48.** $\sqrt[3]{26}$ **49.** $\sqrt[4]{624}$ **50.** $(2.99)^3$

Writing **In Exercises 51 and 52, give a short explanation of why the approximation is valid.**

51. $\sqrt{4.02} \approx 2 + \frac{1}{4}(0.02)$ **52.** $\tan 0.05 \approx 0 + 1(0.05)$

In Exercises 53–56, verify the tangent line approximation of the function at the given point. Then use a graphing utility to graph the function and its approximation in the same viewing window.

	Function	*Approximation*	*Point*
53.	$f(x) = \sqrt{x + 4}$	$y = 2 + \dfrac{x}{4}$	$(0, 2)$
54.	$f(x) = \sqrt{x}$	$y = \dfrac{1}{2} + \dfrac{x}{2}$	$(1, 1)$
55.	$f(x) = \tan x$	$y = x$	$(0, 0)$
56.	$f(x) = \dfrac{1}{1 - x}$	$y = 1 + x$	$(0, 1)$

Writing About Concepts

57. Describe the change in accuracy of dy as an approximation for Δy when Δx is decreased.

58. When using differentials, what is meant by the terms *propagated error*, *relative error*, and *percent error*?

True or False? **In Exercises 59–62, determine whether the statement is true or false. If it is false, explain why or give an example that shows it is false.**

59. If $y = x + c$, then $dy = dx$.

60. If $y = ax + b$, then $\Delta y/\Delta x = dy/dx$.

61. If y is differentiable, then $\lim_{\Delta x \to 0} (\Delta y - dy) = 0$.

62. If $y = f(x)$, f is increasing and differentiable, and $\Delta x > 0$, then $\Delta y \ge dy$.

Review Exercises for Chapter 4

See www.CalcChat.com for worked-out solutions to odd-numbered exercises.

1. Give the definition of a critical number, and graph a function f showing the different types of critical numbers.

2. Consider the odd function f that is continuous, differentiable, and has the functional values shown in the table.

x	-5	-4	-1	0	2	3	6
$f(x)$	1	3	2	0	-1	-4	0

(a) Determine $f(4)$.

(b) Determine $f(-3)$.

(c) Plot the points and make a possible sketch of the graph of f on the interval $[-6, 6]$. What is the smallest number of critical points in the interval? Explain.

(d) Does there exist at least one real number c in the interval $(-6, 6)$ where $f'(c) = -1$? Explain.

(e) Is it possible that $\lim_{x \to 0} f(x)$ does not exist? Explain.

(f) Is it necessary that $f'(x)$ exists at $x = 2$? Explain.

In Exercises 3 and 4, find the absolute extrema of the function on the closed interval. Use a graphing utility to graph the function over the indicated interval to confirm your results.

3. $g(x) = 2x + 5 \cos x, \quad [0, 2\pi]$

4. $f(x) = \dfrac{x}{\sqrt{x^2 + 1}}, \quad [0, 2]$

In Exercises 5 and 6, determine whether Rolle's Theorem can be applied to f on the closed interval $[a, b]$. If Rolle's Theorem can be applied, find all values of c in the open interval (a, b) such that $f'(c) = 0$.

5. $f(x) = (x - 2)(x + 3)^2, \quad [-3, 2]$

6. $f(x) = |x - 2| - 2, \quad [0, 4]$

7. Consider the function $f(x) = 3 - |x - 4|$.

(a) Graph the function and verify that $f(1) = f(7)$.

(b) Note that $f'(x)$ is not equal to zero for any x in $[1, 7]$. Explain why this does not contradict Rolle's Theorem.

8. Can the Mean Value Theorem be applied to the function $f(x) = 1/x^2$ on the interval $[-2, 1]$? Explain.

In Exercises 9–12, find the point(s) guaranteed by the Mean Value Theorem for the closed interval $[a, b]$.

9. $f(x) = x^{2/3}, \quad [1, 8]$

10. $f(x) = \dfrac{1}{x}, \quad [1, 4]$

11. $f(x) = x - \cos x, \quad \left[-\dfrac{\pi}{2}, \dfrac{\pi}{2}\right]$

12. $f(x) = x \log_2 x, \quad [1, 2]$

13. For the function $f(x) = Ax^2 + Bx + C$, determine the value of c guaranteed by the Mean Value Theorem in the interval $[x_1, x_2]$.

14. Demonstrate the result of Exercise 13 for $f(x) = 2x^2 - 3x + 1$ on the interval $[0, 4]$.

In Exercises 15–20, find the critical numbers (if any) and the open intervals on which the function is increasing or decreasing.

15. $f(x) = (x - 1)^2(x - 3)$

16. $g(x) = (x + 1)^3$

17. $h(x) = \sqrt{x}(x - 3), \quad x > 0$

18. $f(x) = \sin x + \cos x, \quad [0, 2\pi]$

19. $f(t) = (2 - t)2^t$

20. $g(x) = 2x \ln x$

In Exercises 21 and 22, use the First Derivative Test to find any relative extrema of the function. Use a graphing utility to verify your results.

21. $h(t) = \dfrac{1}{4}t^4 - 8t$

22. $g(x) = \dfrac{3}{2} \sin\left(\dfrac{\pi x}{2} - 1\right), \quad [0, 4]$

23. ***Harmonic Motion*** The height of an object attached to a spring is given by the harmonic equation

$$y = \tfrac{1}{3} \cos 12t - \tfrac{1}{4} \sin 12t$$

where y is measured in inches and t is measured in seconds.

(a) Calculate the height and velocity of the object when $t = \pi/8$ second.

(b) Show that the maximum displacement of the object is $\frac{5}{12}$ inch.

(c) Find the period P of y. Also, find the frequency f (number of oscillations per second) if $f = 1/P$.

24. ***Writing*** The general equation giving the height of an oscillating object attached to a spring is

$$y = A \sin \sqrt{\frac{k}{m}}\,t + B \cos \sqrt{\frac{k}{m}}\,t$$

where k is the spring constant and m is the mass of the object.

(a) Show that the maximum displacement of the object is $\sqrt{A^2 + B^2}$.

(b) Show that the object oscillates with a frequency of

$$f = \frac{1}{2\pi}\sqrt{\frac{k}{m}}.$$

In Exercises 25 and 26, determine the points of inflection of the function.

25. $f(x) = x + \cos x, \quad [0, 2\pi]$

26. $f(x) = (x + 2)^2(x - 4)$

In Exercises 27 and 28, use the Second Derivative Test to find all relative extrema.

27. $g(x) = 2x^2(1 - x^2)$

28. $h(t) = t - 4\sqrt{t + 1}$

Think About It **In Exercises 29 and 30, sketch the graph of a function f having the indicated characteristics.**

29. $f(0) = f(6) = 0$
$f'(3) = f'(5) = 0$
$f'(x) > 0$ if $x < 3$
$f'(x) > 0$ if $3 < x < 5$
$f'(x) < 0$ if $x > 5$
$f''(x) < 0$ if $x < 3$ and $x > 4$
$f''(x) > 0,\ 3 < x < 4$

30. $f(0) = 4,\ f(6) = 0$
$f'(x) < 0$ if $x < 2$ and $x > 4$
$f'(2)$ does not exist.
$f'(4) = 0$
$f'(x) > 0$ if $2 < x < 4$
$f''(x) < 0,\ x \neq 2$

31. ***Writing*** A newspaper headline states that "The rate of growth of the national deficit is decreasing." What does this mean? What does it imply about the graph of the deficit as a function of time?

32. ***Inventory Cost*** The cost of inventory depends on the ordering and storage costs according to the inventory model

$$C = \left(\frac{Q}{x}\right)s + \left(\frac{x}{2}\right)r.$$

Determine the order size that will minimize the cost, assuming that sales occur at a constant rate, Q is the number of units sold per year, r is the cost of storing one unit for 1 year, s is the cost of placing an order, and x is the number of units per order.

33. ***Modeling Data*** Outlays for national defense D (in billions of dollars) for selected years from 1970 through 1999 are shown in the table, where t is time in years, with $t = 0$ corresponding to 1970. *(Source: U.S. Office of Management and Budget)*

t	0	5	10	15	20
D	90.4	103.1	155.1	279.0	328.3

t	25	26	27	28	29
D	309.9	302.7	309.8	310.3	320.2

(a) Use the regression capabilities of a graphing utility to fit a model of the form $D = at^4 + bt^3 + ct^2 + dt + e$ to the data.

(b) Use a graphing utility to plot the data and graph the model.

(c) For the years shown in the table, when does the model indicate that the outlay for national defense is at a maximum? When is it at a minimum?

(d) For the years shown in the table, when does the model indicate that the outlay for national defense is increasing at the greatest rate?

34. ***Climb Rate*** The time t (in minutes) for a small plane to climb to an altitude of h feet is

$$t = 50 \log_{10} \frac{18{,}000}{18{,}000 - h}$$

where 18,000 feet is the plane's absolute ceiling.

(a) Determine the domain of the function appropriate for the context of the problem.

(b) Use a graphing utility to graph the time function and identify any asymptotes.

(c) Find the time when the altitude is increasing at the greatest rate.

In Exercises 35–42, find the limit.

35. $\displaystyle\lim_{x\to\infty} \frac{2x^2}{3x^2 + 5}$

36. $\displaystyle\lim_{x\to\infty} \frac{2x}{3x^2 + 5}$

37. $\displaystyle\lim_{x\to-\infty} \frac{3x^2}{x + 5}$

38. $\displaystyle\lim_{x\to-\infty} \frac{\sqrt{x^2 + x}}{-2x}$

39. $\displaystyle\lim_{x\to\infty} \frac{5\cos x}{x}$

40. $\displaystyle\lim_{x\to\infty} \frac{3x}{\sqrt{x^2 + 4}}$

41. $\displaystyle\lim_{x\to-\infty} \frac{6x}{x + \cos x}$

42. $\displaystyle\lim_{x\to-\infty} \frac{x}{2\sin x}$

In Exercises 43–50, find any vertical and horizontal asymptotes of the graph of the function. Use a graphing utility to verify your results.

43. $h(x) = \dfrac{2x + 3}{x - 4}$

44. $g(x) = \dfrac{5x^2}{x^2 + 2}$

45. $f(x) = \dfrac{3}{x} - 2$

46. $f(x) = \dfrac{3x}{\sqrt{x^2 + 2}}$

47. $f(x) = \dfrac{5}{3 + 2e^{-x}}$

48. $g(x) = 30xe^{-2x}$

49. $g(x) = 3\ln(1 + e^{-x/4})$

50. $h(x) = 10\ln\left(\dfrac{x}{x + 1}\right)$

In Exercises 51–54, use a graphing utility to graph the function. Use the graph to approximate any relative extrema or asymptotes.

51. $f(x) = x^3 + \dfrac{243}{x}$

52. $f(x) = |x^3 - 3x^2 + 2x|$

53. $f(x) = \dfrac{x - 1}{1 + 3x^2}$

54. $g(x) = \dfrac{\pi^2}{3} - 4\cos x + \cos 2x$

In Exercises 55–80, analyze and sketch the graph of the function.

55. $f(x) = 4x - x^2$

56. $f(x) = 4x^3 - x^4$

57. $f(x) = x\sqrt{16 - x^2}$

58. $f(x) = (x^2 - 4)^2$

59. $f(x) = (x - 1)^3(x - 3)^2$

60. $f(x) = (x - 3)(x + 2)^3$

61. $f(x) = x^{1/3}(x + 3)^{2/3}$

62. $f(x) = (x - 2)^{1/3}(x + 1)^{2/3}$

63. $f(x) = \dfrac{x+1}{x-1}$

64. $f(x) = \dfrac{2x}{1+x^2}$

65. $f(x) = \dfrac{4}{1+x^2}$

66. $f(x) = \dfrac{x^2}{1+x^4}$

67. $f(x) = x^3 + x + \dfrac{4}{x}$

68. $f(x) = x^2 + \dfrac{1}{x}$

69. $f(x) = |x^2 - 9|$

70. $f(x) = |x - 1| + |x - 3|$

71. $h(x) = (1 - x)e^x$

72. $g(x) = 5xe^{-x^2}$

73. $g(x) = (x + 3)\ln(x + 3)$

74. $h(t) = \dfrac{\ln t}{t^2}$

75. $f(x) = \dfrac{10 \log_4 x}{x}$

76. $g(x) = 100x(3^{-x})$

77. $f(x) = x + \cos x, \quad 0 \le x \le 2\pi$

78. $f(x) = \dfrac{1}{\pi}(2 \sin \pi x - \sin 2\pi x), \quad -1 \le x \le 1$

79. $y = 4x - 6 \arctan x$

80. $y = \dfrac{1}{2}x^2 - \arcsin \dfrac{x}{2}$

81. Find the maximum and minimum points on the graph of

$$x^2 + 4y^2 - 2x - 16y + 13 = 0$$

(a) without using calculus.

(b) using calculus.

82. Consider the function $f(x) = x^n$ for positive integer values of n.

(a) For what values of n does the function have a relative minimum at the origin?

(b) For what values of n does the function have a point of inflection at the origin?

83. ***Minimum Distance*** At noon, ship A is 100 kilometers due east of ship B. Ship A is sailing west at 12 kilometers per hour, and ship B is sailing south at 10 kilometers per hour. At what time will the ships be nearest to each other, and what will this distance be?

84. ***Maximum Area*** Find the dimensions of the rectangle of maximum area, with sides parallel to the coordinate axes, that can be inscribed in the ellipse given by

$$\frac{x^2}{144} + \frac{y^2}{16} = 1.$$

85. ***Minimum Length*** A right triangle in the first quadrant has the coordinate axes as sides, and the hypotenuse passes through the point $(1, 8)$. Find the vertices of the triangle such that the length of the hypotenuse is minimum.

86. ***Minimum Length*** The wall of a building is to be braced by a beam that must pass over a parallel fence 5 feet high and 4 feet from the building. Find the length of the shortest beam that can be used.

87. ***Maximum Area*** Three sides of a trapezoid have the same length s. Of all such possible trapezoids, show that the one of maximum area has a fourth side of length $2s$.

88. ***Maximum Area*** Show that the greatest area of any rectangle inscribed in a triangle is one-half that of the triangle.

89. ***Maximum Length*** Find the length of the longest pipe that can be carried level around a right-angle corner at the intersection of two corridors of widths 4 feet and 6 feet. (Do not use trigonometry.)

90. ***Maximum Length*** Rework Exercise 89, given corridors of widths a meters and b meters.

91. ***Maximum Length*** A hallway of width 6 feet meets a hallway of width 9 feet at right angles. Find the length of the longest pipe that can be carried level around this corner. [*Hint:* If L is the length of the pipe, show that

$$L = 6 \csc \theta + 9 \csc\left(\frac{\pi}{2} - \theta\right)$$

where θ is the angle between the pipe and the wall of the narrower hallway.]

92. ***Maximum Length*** Rework Exercise 91, given that one hallway is of width a meters and the other is of width b meters. Show that the result is the same as in Exercise 90.

Minimum Cost **In Exercises 93 and 94, find the speed v (in miles per hour) that will minimize costs on a 110-mile delivery trip. The cost per hour for fuel is C dollars, and the driver is paid W dollars per hour. (Assume there are no costs other than wages and fuel.)**

93. Fuel cost: $C = \dfrac{v^2}{600}$

Driver: $W = \$5$

94. Fuel cost: $C = \dfrac{v^2}{500}$

Driver: $W = \$7.50$

In Exercises 95 and 96, find the differential dy.

95. $y = x(1 - \cos x)$

96. $y = \sqrt{36 - x^2}$

97. ***Surface Area and Volume*** The diameter of a sphere is measured to be 18 centimeters, with a maximum possible error of 0.05 centimeter. Use differentials to approximate the possible propagated error and percent error in calculating the surface area and the volume of the sphere.

98. ***Demand Function*** A company finds that the demand for its commodity is $p = 75 - \frac{1}{4}x$. If x changes from 7 to 8, find and compare the values of Δp and dp.

P.S. Problem Solving

See www.CalcChat.com for worked-out solutions to odd-numbered exercises.

1. Graph the fourth-degree polynomial $p(x) = x^4 + ax^2 + 1$ for various values of the constant a.

(a) Determine the values of a for which p has exactly one relative minimum.

(b) Determine the values of a for which p has exactly one relative maximum.

(c) Determine the values of a for which p has exactly two relative minima.

(d) Show that the graph of p cannot have exactly two relative extrema.

2. (a) Graph the fourth-degree polynomial $p(x) = ax^4 - 6x^2$ for $a = -3, -2, -1, 0, 1, 2,$ and 3. For what values of the constant a does p have a relative minimum or relative maximum?

(b) Show that p has a relative maximum for all values of the constant a.

(c) Determine analytically the values of a for which p has a relative minimum.

(d) Let $(x, y) = (x, p(x))$ be a relative extremum of p. Show that (x, y) lies on the graph of $y = -3x^2$. Verify this result graphically by graphing $y = -3x^2$ together with the seven curves from part (a).

3. Let $f(x) = \frac{c}{x} + x^2$. Determine all values of the constant c such that f has a relative minimum, but no relative maximum.

4. (a) Let $f(x) = ax^2 + bx + c, a \neq 0$, be a quadratic polynomial. How many points of inflection does the graph of f have?

(b) Let $f(x) = ax^3 + bx^2 + cx + d, a \neq 0$, be a cubic polynomial. How many points of inflection does the graph of f have?

(c) Suppose the function $y = f(x)$ satisfies the equation $\frac{dy}{dx} = ky\left(1 - \frac{y}{L}\right)$ where k and L are positive constants. Show that the graph of f has a point of inflection at the point where $y = \frac{L}{2}$. (This equation is called the **logistic differential equation.**)

5. Prove Darboux's Theorem: Let f be differentiable on the closed interval $[a, b]$ such that $f'(a) = y_1$ and $f'(b) = y_2$. If d lies between y_1 and y_2, then there exists c in (a, b) such that $f'(c) = d$.

6. Let f and g be functions that are continuous on $[a, b]$ and differentiable on (a, b). Prove that if $f(a) = g(a)$ and $g'(x) > f'(x)$ for all x in (a, b), then $g(b) > f(b)$.

7. Prove the following **Extended Mean Value Theorem:** If f and f' are continuous on the closed interval $[a, b]$, and if f'' exists on the open interval (a, b), then there exists a number c in (a, b) such that

$$f(b) = f(a) + f'(a)(b - a) + \frac{1}{2}f''(c)(b - a)^2.$$

8. (a) Let $V = x^3$. Find dV and ΔV. Show that for small values of x, the difference $\Delta V - dV$ is very small in the sense that there exists ε such that $\Delta V - dV = \varepsilon\Delta x$, where $\varepsilon \to 0$ as $\Delta x \to 0$.

(b) Generalize this result by showing that if $y = f(x)$ is a differentiable function, then $\Delta y - dy = \varepsilon\Delta x$, where $\varepsilon \to 0$ as $\Delta x \to 0$.

9. The amount of illumination of a surface is proportional to the intensity of the light source, inversely proportional to the square of the distance from the light source, and proportional to $\sin\theta$, where θ is the angle at which the light strikes the surface. A rectangular room measures 10 feet by 24 feet, with a 10-foot ceiling. Determine the height at which the light should be placed to allow the corners of the floor to receive as much light as possible.

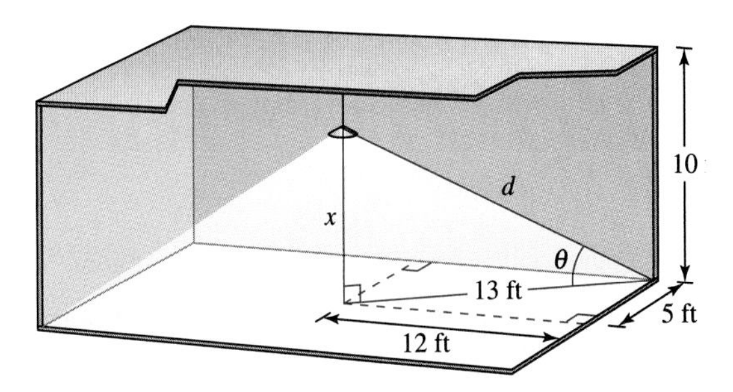

10. Consider a room in the shape of a cube, 4 meters on each side. A bug at point P wants to walk to point Q at the opposite corner, as shown in the figure. Use calculus to determine the shortest path. Can you solve the problem without calculus?

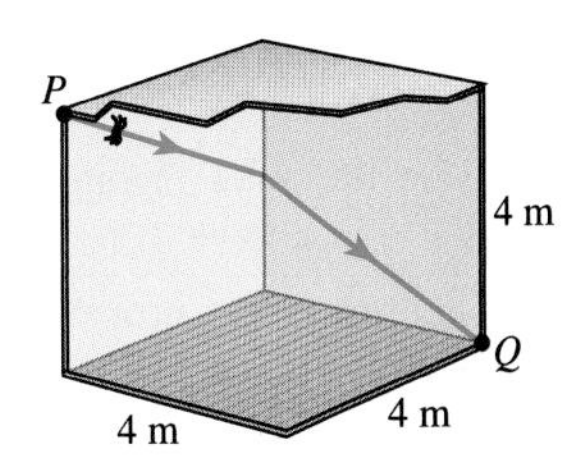

11. The line joining P and Q crosses two parallel lines, as shown in the figure. The point R is d units from P. How far from Q should the point S be so that the sum of the areas of the two shaded triangles is a minimum? So that the sum is a maximum?

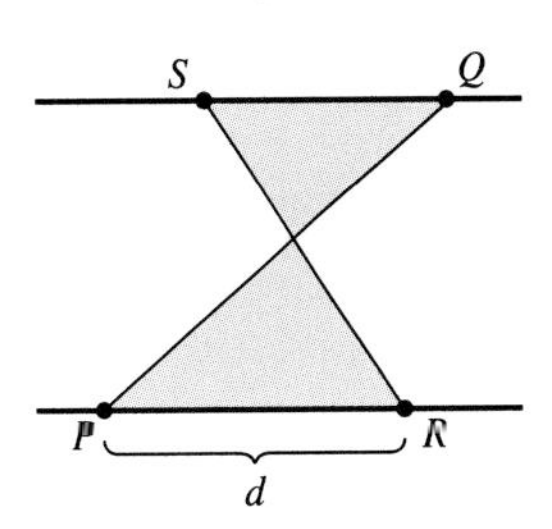

12. The figures show a rectangle, a circle, and a semicircle inscribed in a triangle bounded by the coordinate axes and the first-quadrant portion of the line with intercepts (3, 0) and (0, 4). Find the dimensions of each inscribed figure such that its area is maximum. State whether calculus was helpful in finding the required dimensions. Explain your reasoning.

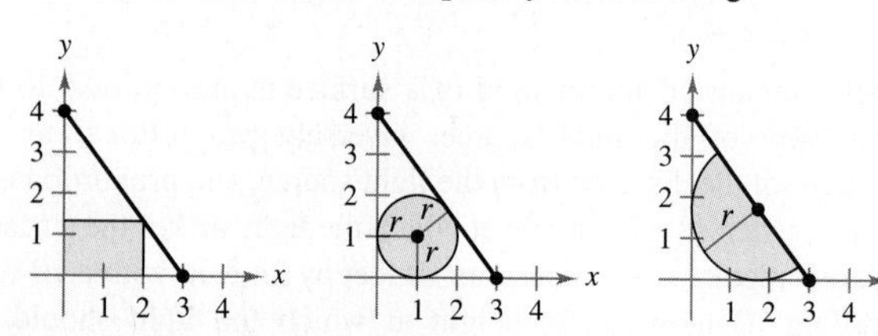

13. (a) Prove that $\lim_{x\to\infty} x^2 = \infty$.

(b) Prove that $\lim_{x\to\infty} \left(\frac{1}{x^2}\right) = 0$.

(c) Let L be a real number. Prove that if $\lim_{x\to\infty} f(x) = L$, then $\lim_{y\to 0^+} f\left(\frac{1}{y}\right) = L$.

14. Find the point on the graph of $y = \dfrac{1}{1+x^2}$ (see figure) where the tangent line has the greatest slope, and the point where the tangent line has the least slope.

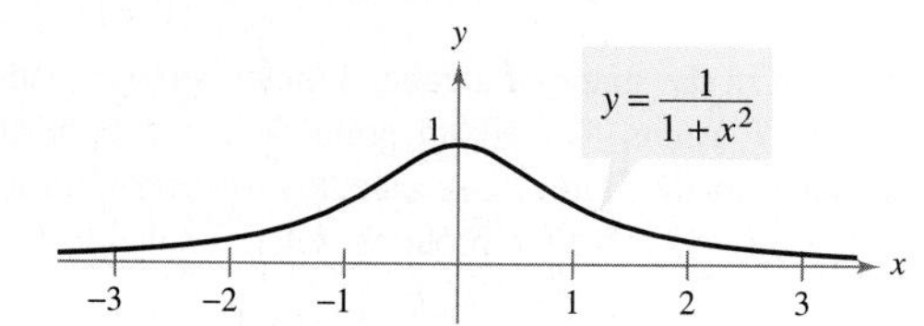

15. (a) Let x be a positive number. Use the *table* feature of a graphing utility to verify that $\sqrt{1+x} < \frac{1}{2}x + 1$.

(b) Use the Mean Value Theorem to prove that $\sqrt{1+x} < \frac{1}{2}x + 1$ for all positive real numbers x.

16. (a) Let x be a positive number. Use the *table* feature of a graphing utility to verify that $\sin x < x$.

(b) Use the Mean Value Theorem to prove that $\sin x < x$ for all positive real numbers x.

17. The police department must determine the speed limit on a bridge such that the flow rate of cars is maximized per unit time. The greater the speed limit, the farther apart the cars must be in order to keep a safe stopping distance. Experimental data on the stopping distance d (in meters) for various speeds v (in kilometers per hour) are shown in the table.

v	20	40	60	80	100
d	5.1	13.7	27.2	44.2	66.4

(a) Convert the speeds v in the table to the speeds s in meters per second. Use the regression capabilities of a graphing utility to find a model of the form $d(s) = as^2 + bs + c$ for the data.

(b) Consider two consecutive vehicles of average length 5.5 meters, traveling at a safe speed on the bridge. Let T be the difference between the times (in seconds) when the front bumpers of the vehicles pass a given point on the bridge. Verify that this difference in times is given by

$$T = \frac{d(s)}{s} + \frac{5.5}{s}.$$

(c) Use a graphing utility to graph the function T and estimate the speed s that minimizes the time between vehicles.

(d) Use calculus to determine the speed that minimizes T. What is the minimum value of T? Convert the required speed to kilometers per hour.

(e) Find the optimal distance between vehicles for the posted speed limit determined in part (d).

18. A legal-sized sheet of paper (8.5 inches by 14 inches) is folded so that corner P touches the opposite 14-inch edge at R. (*Note:* $PQ = \sqrt{C^2 - x^2}$.)

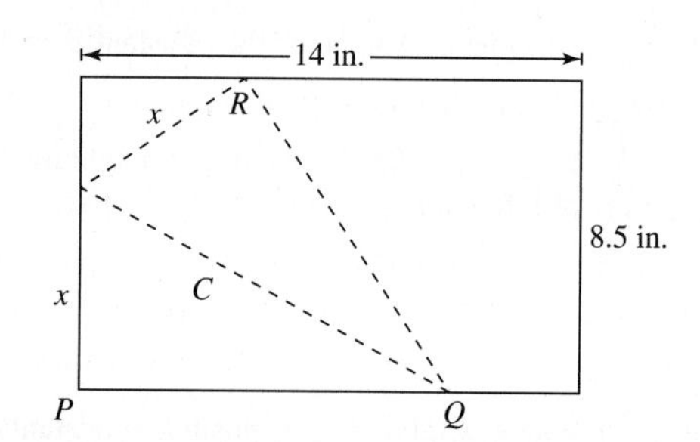

(a) Show that $C^2 = \dfrac{2x^3}{2x - 8.5}$.

(b) What is the domain of C?

(c) Determine the x-value that minimizes C.

(d) Determine the minimum length of C.

19. Let $f(x) = \sin(\ln x)$.

(a) Determine the domain of the function f.

(b) Find two values of x satisfying $f(x) = 1$.

(c) Find two values of x satisfying $f(x) = -1$.

(d) What is the range of the function f?

(e) Calculate $f'(x)$ and use calculus to find the maximum value of f on the interval $[1, 10]$.

(f) Use a graphing utility to graph f in the viewing window $[0, 5] \times [-2, 2]$ and estimate $\lim_{x\to 0^+} f(x)$, if it exists.

(g) Determine $\lim_{x\to 0^+} f(x)$ analytically, if it exists.

5 Integration

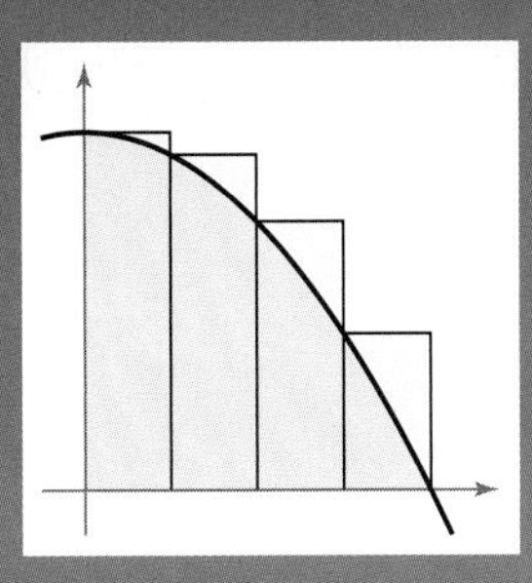

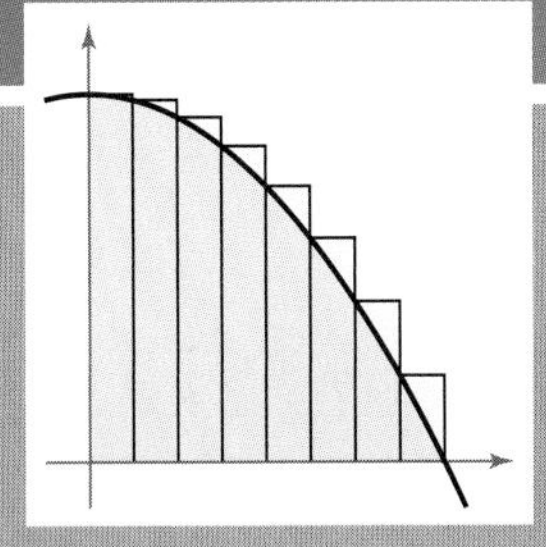

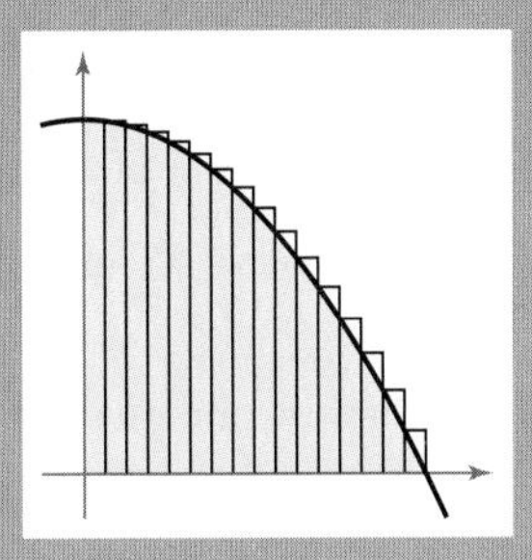

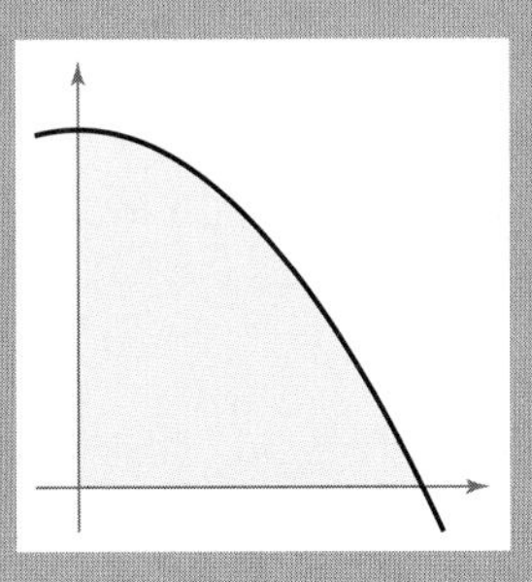

The area of a parabolic region can be approximated by the sum of the areas of rectangles. As you increase the number of rectangles, the approximation tends to become more and more accurate. In Section 5.2, you will learn how the limit process can be used to find the areas of a wide variety of regions. This process is called *integration* and is closely related to differentiation.

This photo of a jet breaking the sound barrier was taken by Ensign John Gay. At the time the photo was taken, was the jet's velocity constant or changing? Why?

Section 5.1 Antiderivatives and Indefinite Integration

- **Write the general solution of a differential equation.**
- **Use indefinite integral notation for antiderivatives.**
- **Use basic integration rules to find antiderivatives.**
- **Find a particular solution of a differential equation.**

EXPLORATION

Finding Antiderivatives For each derivative, describe the original function F.

a. $F'(x) = 2x$

b. $F'(x) = x$

c. $F'(x) = x^2$

d. $F'(x) = \dfrac{1}{x^2}$

e. $F'(x) = \dfrac{1}{x^3}$

f. $F'(x) = \cos x$

What strategy did you use to find F?

Antiderivatives

Suppose you were asked to find a function F whose derivative is $f(x) = 3x^2$. From your knowledge of derivatives, you would probably say that

$$F(x) = x^3 \text{ because } \frac{d}{dx}[x^3] = 3x^2.$$

The function F is an *antiderivative* of f.

Definition of an Antiderivative

A function F is an **antiderivative** of f on an interval I if $F'(x) = f(x)$ for all x in I.

Note that F is called *an* antiderivative of f, rather than *the* antiderivative of f. To see why, observe that

$$F_1(x) = x^3, \quad F_2(x) = x^3 - 5, \quad \text{and} \quad F_3(x) = x^3 + 97$$

are all antiderivatives of $f(x) = 3x^2$. In fact, for any constant C, the function given by $F(x) = x^3 + C$ is an antiderivative of f.

THEOREM 5.1 Representation of Antiderivatives

If F is an antiderivative of f on an interval I, then G is an antiderivative of f on the interval I if and only if G is of the form $G(x) = F(x) + C$, for all x in I, where C is a constant.

Proof The proof of Theorem 5.1 in one direction is straightforward. That is, if $G(x) = F(x) + C$, $F'(x) = f(x)$, and C is a constant, then

$$G'(x) = \frac{d}{dx}[F(x) + C] = F'(x) + 0 = f(x).$$

To prove this theorem in the other direction, assume that G is an antiderivative of f. Define a function H such that

$$H(x) = G(x) - F(x).$$

If H is not constant on the interval I, then there must exist a and b $(a < b)$ in the interval such that $H(a) \neq H(b)$. Moreover, because H is differentiable on (a, b), you can apply the Mean Value Theorem to conclude that there exists some c in (a, b) such that

$$H'(c) = \frac{H(b) - H(a)}{b - a}.$$

Because $H(b) \neq H(a)$, it follows that $H'(c) \neq 0$. However, because $G'(c) = F'(c)$, you know that $H'(c) = G'(c) - F'(c) = 0$, which contradicts the fact that $H'(c) \neq 0$. Consequently, you can conclude that $H(x)$ is a constant, C. So, $G(x) - F(x) = C$ and it follows that $G(x) = F(x) + C$.

Using Theorem 5.1, you can represent the entire family of antiderivatives of a function by adding a constant to a *known* antiderivative. For example, knowing that $D_x[x^2] = 2x$, you can represent the family of *all* antiderivatives of $f(x) = 2x$ by

$$G(x) = x^2 + C \qquad \text{Family of all antiderivatives of } f(x) = 2x$$

where C is a constant. The constant C is called the **constant of integration.** The family of functions represented by G is the **general antiderivative** of f, and $G(x) = x^2 + C$ is the **general solution** of the *differential equation*

$$G'(x) = 2x. \qquad \text{Differential equation}$$

A **differential equation** in x and y is an equation that involves x, y, and derivatives of y. For instance, $y' = 3x$ and $y' = x^2 + 1$ are examples of differential equations.

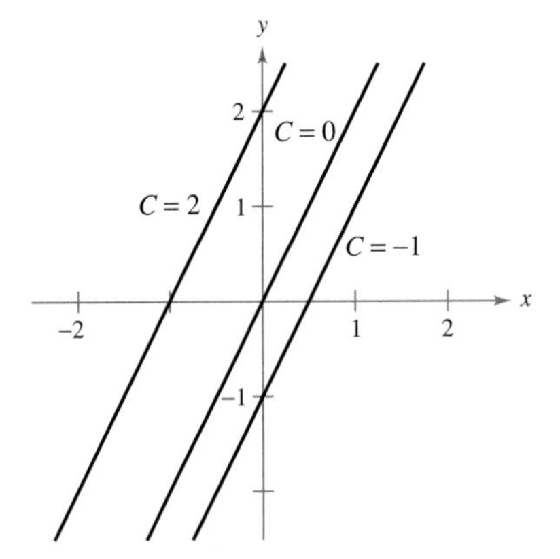

Functions of the form $y = 2x + C$
Figure 5.1

EXAMPLE 1 Solving a Differential Equation

Find the general solution of the differential equation $y' = 2$.

Solution To begin, you need to find a function whose derivative is 2. One such function is

$$y = 2x. \qquad 2x \text{ is } an \text{ antiderivative of } 2.$$

Now, you can use Theorem 5.1 to conclude that the general solution of the differential equation is

$$y = 2x + C. \qquad \text{General solution}$$

The graphs of several functions of the form $y = 2x + C$ are shown in Figure 5.1.

Notation for Antiderivatives

When solving a differential equation of the form

$$\frac{dy}{dx} = f(x)$$

it is convenient to write it in the equivalent differential form

$$dy = f(x)\,dx.$$

The operation of finding all solutions of this equation is called **antidifferentiation** (or **indefinite integration**) and is denoted by an integral sign $\int$. The general solution is denoted by

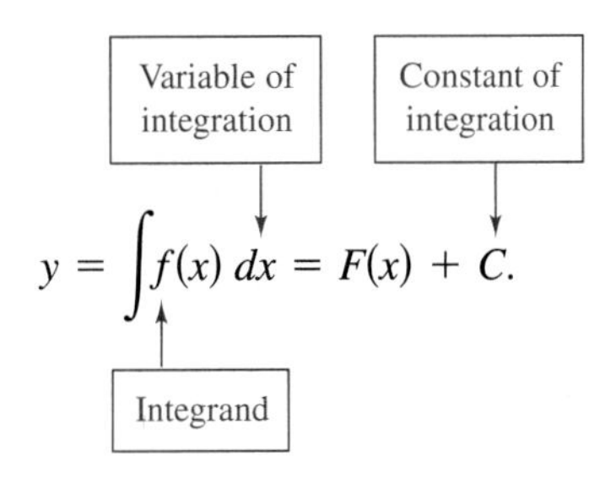

$$y = \int f(x)\,dx = F(x) + C.$$

The expression $\int f(x)dx$ is read as the *antiderivative of f with respect to x.* So, the differential dx serves to identify x as the variable of integration. The term **indefinite integral** is a synonym for antiderivative.

NOTE In this text, the notation $\int f(x)\,dx = F(x) + C$ means that F is an antiderivative of f *on an interval.*

Basic Integration Rules

The inverse nature of integration and differentiation can be verified by substituting $F'(x)$ for $f(x)$ in the indefinite integration definition to obtain

$$\int F'(x)\,dx = F(x) + C.$$

Integration is the "inverse" of differentiation.

Moreover, if $\int f(x)\,dx = F(x) + C$, then

$$\frac{d}{dx}\left[\int f(x)\,dx\right] = f(x).$$

Differentiation is the "inverse" of integration.

NOTE The Power Rule for integration has the restriction that $n \neq -1$. To evaluate $fx^{-1}\,dx$, you must use the natural log rule. (See Exercise 106)

These two equations allow you to obtain integration formulas directly from differentiation formulas, as shown in the following summary.

Basic Integration Rules

Differentiation Formula	*Integration Formula*	
$\frac{d}{dx}[C] = 0$	$\int 0\,dx = C$	
$\frac{d}{dx}[kx] = k$	$\int k\,dx = kx + C$	
$\frac{d}{dx}[kf(x)] = kf'(x)$	$\int kf(x)\,dx = k\int f(x)\,dx$	
$\frac{d}{dx}[f(x) \pm g(x)] = f'(x) \pm g'(x)$	$\int [f(x) \pm g(x)]\,dx = \int f(x)\,dx \pm \int g(x)\,dx$	
$\frac{d}{dx}[x^n] = nx^{n-1}$	$\int x^n\,dx = \frac{x^{n+1}}{n+1} + C, \quad n \neq -1$	Power Rule
$\frac{d}{dx}[\sin x] = \cos x$	$\int \cos x\,dx = \sin x + C$	
$\frac{d}{dx}[\cos x] = -\sin x$	$\int \sin x\,dx = -\cos x + C$	
$\frac{d}{dx}[\tan x] = \sec^2 x$	$\int \sec^2 x\,dx = \tan x + C$	
$\frac{d}{dx}[\sec x] = \sec x \tan x$	$\int \sec x \tan x\,dx = \sec x + C$	
$\frac{d}{dx}[\cot x] = -\csc^2 x$	$\int \csc^2 x\,dx = -\cot x + C$	
$\frac{d}{dx}[\csc x] = -\csc x \cot x$	$\int \csc x \cot x\,dx = -\csc x + C$	
$\frac{d}{dx}[e^x] = e^x$	$\int e^x\,dx = e^x + C$	
$\frac{d}{dx}[a^x] = (\ln a)a^x$	$\int a^x\,dx = \left(\frac{1}{\ln a}\right)a^x + C$	
$\frac{d}{dx}[\ln x] = \frac{1}{x}, \ x > 0$	$\int \frac{1}{x}\,dx = \ln\lvert x\rvert + C$	

EXAMPLE 2 Applying the Basic Integration Rules

Describe the antiderivatives of $3x$.

Solution

$$\int 3x\,dx = 3\int x\,dx \qquad \text{Constant Multiple Rule}$$

$$= 3\int x^1\,dx \qquad \text{Rewrite } x \text{ as } x^1.$$

$$= 3\left(\frac{x^2}{2}\right) + C \qquad \text{Power Rule } (n = 1)$$

$$= \frac{3}{2}x^2 + C \qquad \text{Simplify.}$$

When indefinite integrals are evaluated, a strict application of the basic integration rules tends to produce complicated constants of integration. For instance, in Example 2, you could have written

$$\int 3x\,dx = 3\int x\,dx = 3\left(\frac{x^2}{2} + C\right) = \frac{3}{2}x^2 + 3C.$$

However, because C represents *any* constant, it is both cumbersome and unnecessary to write $3C$ as the constant of integration. So, $\frac{3}{2}x^2 + 3C$ is written in the simpler form $\frac{3}{2}x^2 + C$.

In Example 2, note that the general pattern of integration is similar to that of differentiation.

Original integral ⇨ Rewrite ⇨ Integrate ⇨ Simplify

EXAMPLE 3 Rewriting Before Integrating

TECHNOLOGY Some software programs, such as *Derive*, *Maple*, *Mathcad*, *Mathematica*, and the *TI-89*, are capable of performing integration symbolically. If you have access to such a symbolic integration utility, try using it to evaluate the indefinite integrals in Example 3.

	Original Integral	*Rewrite*	*Integrate*	*Simplify*
a.	$\int \frac{1}{x^3}\,dx$	$\int x^{-3}\,dx$	$\frac{x^{-2}}{-2} + C$	$-\frac{1}{2x^2} + C$
b.	$\int \sqrt{x}\,dx$	$\int x^{1/2}\,dx$	$\frac{x^{3/2}}{3/2} + C$	$\frac{2}{3}x^{3/2} + C$
c.	$\int 2\sin x\,dx$	$2\int \sin x\,dx$	$2(-\cos x) + C$	$-2\cos x + C$
d.	$\int \frac{3}{x}\,dx$	$3\int \frac{1}{x}\,dx$	$3(\ln\lvert x\rvert) + C$	$3\ln\lvert x\rvert + C$

NOTE The properties of logarithms presented on page 53 can be used to rewrite anitderivatives in different forms. For instant, the antiderivative in Example 3(d) can be rewritten as

$$3\ln\lvert x\rvert + C = \ln\lvert x\rvert^3 + C.$$

Remember that you can check your answer to an antidifferentiation problem by differentiating. For instance, in Example 3(b), you can check that $\frac{2}{3}x^{3/2} + C$ is the correct antiderivative by differentiating the answer to obtain

$$D_x\left[\frac{2}{3}x^{3/2} + C\right] = \left(\frac{2}{3}\right)\left(\frac{3}{2}\right)x^{1/2} = \sqrt{x}. \qquad \text{Use differentiation to check antiderivative.}$$

---

indicates that in the HM mathSpace® *CD-ROM and the online* Eduspace® *system for this text, you will find an Open Exploration, which further explores this example using the computer algebra systems* Maple, Mathcad, Mathematica, *and* Derive.

The basic integration rules listed earlier in this section allow you to integrate any polynomial function, as shown in Example 4.

EXAMPLE 4 Integrating Polynomial Functions

a. $\int dx = \int 1\, dx$ — Integrand is understood to be 1.

$= x + C$ — Integrate.

b. $\int (x + 2)\, dx = \int x\, dx + \int 2\, dx$

$= \dfrac{x^2}{2} + C_1 + 2x + C_2$ — Integrate.

$= \dfrac{x^2}{2} + 2x + C$ — $C = C_1 + C_2$

The second line in the solution is usually omitted.

c. $\int (3x^4 - 5x^2 + x)\, dx = 3\left(\dfrac{x^5}{5}\right) - 5\left(\dfrac{x^3}{3}\right) + \dfrac{x^2}{2} + C$ — Integrate.

$= \dfrac{3}{5}x^5 - \dfrac{5}{3}x^3 + \dfrac{1}{2}x^2 + C$ — Simplify.

EXAMPLE 5 Rewriting Before Integrating

$\int \dfrac{x+1}{\sqrt{x}}\, dx = \int \left(\dfrac{x}{\sqrt{x}} + \dfrac{1}{\sqrt{x}}\right) dx$ — Rewrite as two fractions.

$= \int (x^{1/2} + x^{-1/2})\, dx$ — Rewrite with fractional exponents.

$= \dfrac{x^{3/2}}{3/2} + \dfrac{x^{1/2}}{1/2} + C$ — Integrate.

$= \dfrac{2}{3}x^{3/2} + 2x^{1/2} + C$ — Simplify.

$= \dfrac{2}{3}\sqrt{x}(x + 3) + C$ — Factor.

NOTE When integrating quotients, do not integrate the numerator and denominator separately. This is no more valid in integration than it is in differentiation. For instance, in Example 5, be sure you understand that

$$\int \frac{x+1}{\sqrt{x}}\, dx = \frac{2}{3}\sqrt{x}(x+3) + C \text{ is not the same as } \frac{\int (x+1)\, dx}{\int \sqrt{x}\, dx} = \frac{\frac{1}{2}x^2 + x + C_1}{\frac{2}{3}x\sqrt{x} + C_2}.$$

EXAMPLE 6 Rewriting Before Integrating

$\int \dfrac{\sin x}{\cos^2 x}\, dx = \int \left(\dfrac{1}{\cos x}\right)\left(\dfrac{\sin x}{\cos x}\right) dx$ — Rewrite as a product.

$= \int \sec x \tan x\, dx$ — Rewrite using trigonometric identities.

$= \sec x + C$ — Integrate.

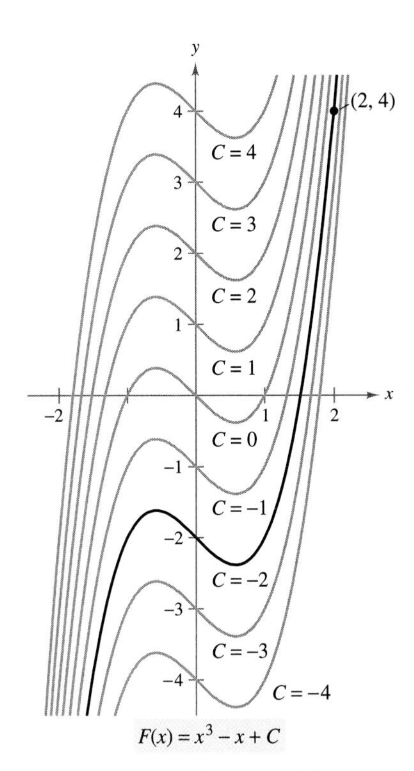

The particular solution that satisfies the initial condition $F(2) = 4$ is $F(x) = x^3 - x - 2$.
Figure 5.2

Initial Conditions and Particular Solutions

You have already seen that the equation $y = \int f(x)\,dx$ has many solutions (each differing from the others by a constant). This means that the graphs of any two antiderivatives of f are vertical translations of each other. For example, Figure 5.2 shows the graphs of several antiderivatives of the form

$$y = \int (3x^2 - 1)\,dx$$

$$= x^3 - x + C \qquad \text{General solution}$$

for various integer values of C. Each of these antiderivatives is a solution of the differential equation

$$\frac{dy}{dx} = 3x^2 - 1.$$

In many applications of integration, you are given enough information to determine a **particular solution.** To do this, you need only know the value of $y = F(x)$ for one value of x. This information is called an **initial condition.** For example, in Figure 5.2, only one curve passes through the point $(2, 4)$. To find this curve, you can use the following information.

$$F(x) = x^3 - x + C \qquad \text{General solution}$$

$$F(2) = 4 \qquad \text{Initial condition}$$

By using the initial condition in the general solution, you can determine that $F(2) = 8 - 2 + C = 4$, which implies that $C = -2$. So, you obtain

$$F(x) = x^3 - x - 2. \qquad \text{Particular solution}$$

EXAMPLE 7 Finding a Particular Solution

Find the general solution of

$$F'(x) = e^x$$

and find the particular solution that satisfies the initial condition $F(0) = 3$.

Solution To find the general solution, integrate to obtain

$$F(x) = \int e^x\,dx$$

$$= e^x + C. \qquad \text{General solution}$$

Using the initial condition $F(0) = 3$, you can solve for C as follows.

$$F(0) = e^0 + C$$

$$3 = 1 + C$$

$$2 = C$$

So, the particular solution, as shown in Figure 5.3, is

$$F(x) = e^x + 2. \qquad \text{Particular solution}$$

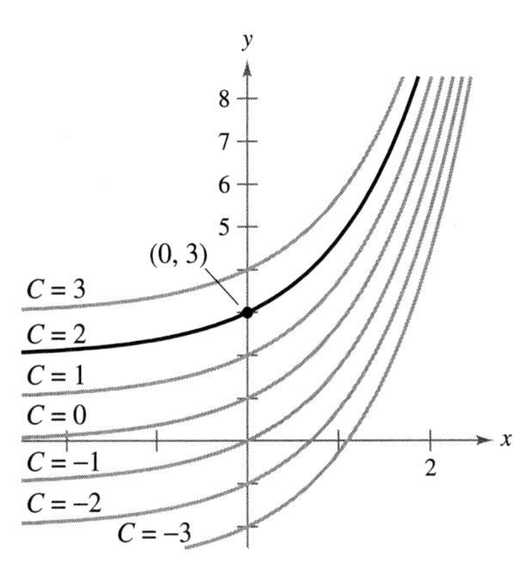

The particular solution that satisfies the initial condition $F(0) = 3$ is $F(x) = e^x + 2$.
Figure 5.3

So far in this section you have been using x as the variable of integration. In applications, it is often convenient to use a different variable. For instance, in the following example involving *time*, the variable of integration is t.

EXAMPLE 8 Solving a Vertical Motion Problem

A ball is thrown upward with an initial velocity of 64 feet per second from an initial height of 80 feet.

a. Find the position function giving the height s as a function of the time t.

b. When does the ball hit the ground?

Solution

a. Let $t = 0$ represent the initial time. The two given initial conditions can be written as follows.

$$s(0) = 80 \qquad \text{Initial height is 80 feet.}$$

$$s'(0) = 64 \qquad \text{Initial velocity is 64 feet per second.}$$

Using -32 feet per second per second as the acceleration due to gravity, you can write

$$s''(t) = -32$$

$$s'(t) = \int s''(t)\, dt$$

$$= \int -32\, dt = -32t + C_1.$$

Using the initial velocity, you obtain $s'(0) = 64 = -32(0) + C_1$, which implies that $C_1 = 64$. Next, by integrating $s'(t)$, you obtain

$$s(t) = \int s'(t)\, dt$$

$$= \int (-32t + 64)\, dt$$

$$= -16t^2 + 64t + C_2.$$

Using the initial height, you obtain

$$s(0) = 80 = -16(0^2) + 64(0) + C_2$$

which implies that $C_2 = 80$. So, the position function is

$$s(t) = -16t^2 + 64t + 80. \qquad \text{See Figure 5.4.}$$

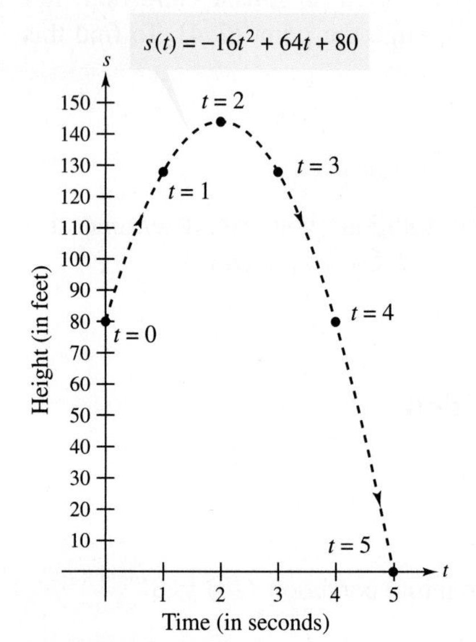

Height of a ball at time t
Figure 5.4

b. Using the position function found in part (a), you can find the time that the ball hits the ground by solving the equation $s(t) = 0$.

$$s(t) = -16t^2 + 64t + 80 = 0$$

$$-16(t + 1)(t - 5) = 0$$

$$t = -1, 5$$

Because t must be positive, you can conclude that the ball hit the ground 5 seconds after it was thrown.

NOTE In Example 8, note that the position function has the form

$$s(t) = \frac{1}{2}gt^2 + v_0t + s_0$$

where $g = -32$, v_0 is the initial velocity, and s_0 is the initial height, as presented in Section 3.2.

Example 8 shows how to use calculus to analyze vertical motion problems in which the acceleration is determined by a gravitational force. You can use a similar strategy to analyze other linear motion problems (vertical or horizontal) in which the acceleration (or deceleration) is the result of some other force, as you will see in Exercises 87–94.

Before you begin the exercise set, be sure you realize that one of the most important steps in integration is *rewriting the integrand* in a form that fits the basic integration rules. To further illustrate this point, here are some additional examples.

Original Integral	*Rewrite*	*Integrate*	*Simplify*
$\int \frac{2}{\sqrt{x}}\, dx$	$2\int x^{-1/2}\, dx$	$2\left(\frac{x^{1/2}}{1/2}\right) + C$	$4x^{1/2} + C$
$\int (t^2 + 1)^2\, dt$	$\int (t^4 + 2t^2 + 1)\, dt$	$\frac{t^5}{5} + 2\left(\frac{t^3}{3}\right) + t + C$	$\frac{1}{5}t^5 + \frac{2}{3}t^3 + t + C$
$\int \frac{x^3 + 3}{x^2}\, dx$	$\int (x + 3x^{-2})\, dx$	$\frac{x^2}{2} + 3\left(\frac{x^{-1}}{-1}\right) + C$	$\frac{1}{2}x^2 - \frac{3}{x} + C$
$\int \sqrt[3]{x}(x - 4)\, dx$	$\int (x^{4/3} - 4x^{1/3})\, dx$	$\frac{x^{7/3}}{7/3} - 4\left(\frac{x^{4/3}}{4/3}\right) + C$	$\frac{3}{7}x^{4/3}(x - 7) + C$

Exercises for Section 5.1

See www.CalcChat.com for worked-out solutions to odd-numbered exercises.

In Exercises 1–4, verify the statement by showing that the derivative of the right side equals the integrand of the left side.

1. $\int \left(-\frac{9}{x^4}\right) dx = \frac{3}{x^3} + C$

2. $\int \left(4x^3 - \frac{1}{x^2}\right) dx = x^4 + \frac{1}{x} + C$

3. $\int (x - 2)(x + 2)\, dx = \frac{1}{3}x^3 - 4x + C$

4. $\int \frac{x^2 - 1}{x^{3/2}}\, dx = \frac{2(x^2 + 3)}{3\sqrt{x}} + C$

In Exercises 5–8, find the general solution of the differential equation and check the result by differentiation.

5. $\frac{dy}{dt} = 3t^2$

6. $\frac{dr}{d\theta} = \pi$

7. $\frac{dy}{dx} = x^{3/2}$

8. $\frac{dy}{dx} = 2x^{-3}$

In Exercises 9–14, complete the table using Example 3 and the examples at the top of this page as a model.

	Original Integral	*Rewrite*	*Integrate*	*Simplify*
9.	$\int \sqrt[3]{x}\, dx$			
10.	$\int \frac{1}{x^2}\, dx$			
11.	$\int \frac{1}{x\sqrt{x}}\, dx$			
12.	$\int x(x^2 + 3)\, dx$			
13.	$\int \frac{1}{2x^3}\, dx$			
14.	$\int \frac{1}{(3x)^2}\, dx$			

In Exercises 15–44, find the indefinite integral and check the result by differentiation.

15. $\int (x + 3)\, dx$

16. $\int (5 - x)\, dx$

17. $\int (x^3 + 5)\, dx$

18. $\int (4x^3 + 6x^2 - 1)\, dx$

19. $\int (x^{3/2} + 2x + 1)\, dx$

20. $\int \left(\sqrt[4]{x^3} + 1\right) dx$

21. $\int \frac{1}{x^3}\, dx$

22. $\int \frac{1}{x^4}\, dx$

23. $\int \frac{x^2 + x + 1}{\sqrt{x}}\, dx$

24. $\int \frac{x^2 + 2x - 3}{x^4}\, dx$

25. $\int (x + 1)(3x - 2)\, dx$

26. $\int (2t^2 - 1)^2\, dt$

27. $\int y^2\sqrt{y}\, dy$

28. $\int (1 + 3t)t^2\, dt$

29. $\int dx$

30. $\int 3\, dt$

31. $\int (2\sin x + 3\cos x)\, dx$

32. $\int (t^2 - \sin t)\, dt$

33. $\int (1 - \csc t \cot t)\, dt$

34. $\int (\theta^2 + \sec^2\theta)\, d\theta$

35. $\int (2\sin x - 5e^x)\, dx$

36. $\int (3x^2 + 2e^x)\, dx$

37. $\int (\sec^2\theta - \sin\theta)\, d\theta$

38. $\int \sec y(\tan y - \sec y)\, dy$

39. $\int (\tan^2 y + 1)\, dy$

40. $\int \frac{\cos x}{1 - \cos^2 x}\, dx$

41. $\int (2x - 4^x)\, dx$

42. $\int (\cos x + 3^x)\, dx$

43. $\int \left(x - \frac{5}{x}\right) dx$

44. $\int \left(\frac{4}{x} + \sec^2 x\right) dx$

In Exercises 45–48, sketch the graphs of the function $g(x) = f(x) + C$ for $C = -2$, $C = 0$, and $C = 3$ on the same set of coordinate axes.

45. $f(x) = \cos x$

46. $f(x) = \sqrt{x}$

47. $f(x) = \ln x$

48. $f(x) = \frac{1}{2}e^x$

In Exercises 49–52, the graph of the derivative of a function is given. Sketch the graphs of *two* functions that have the given derivative. (There is more than one correct answer.) To print an enlarged copy of the graph, go to the website *www.mathgraphs.com*.

49.

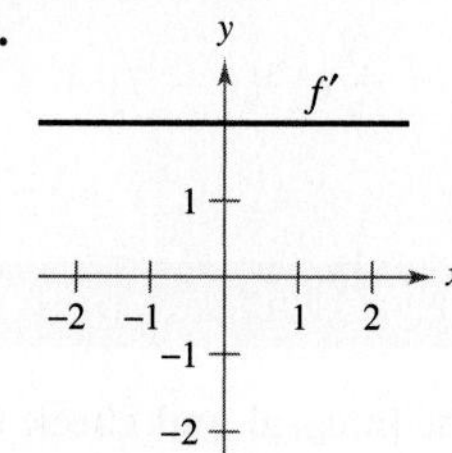

50.

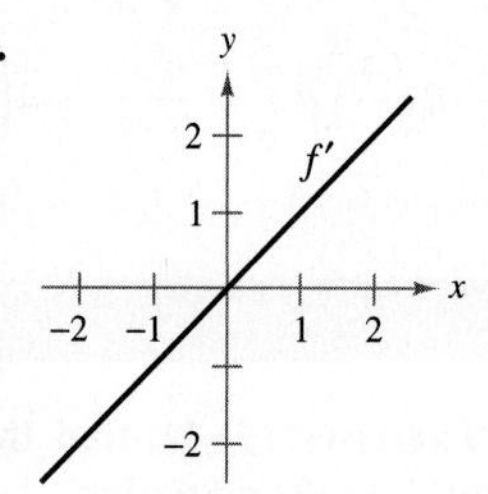

51.

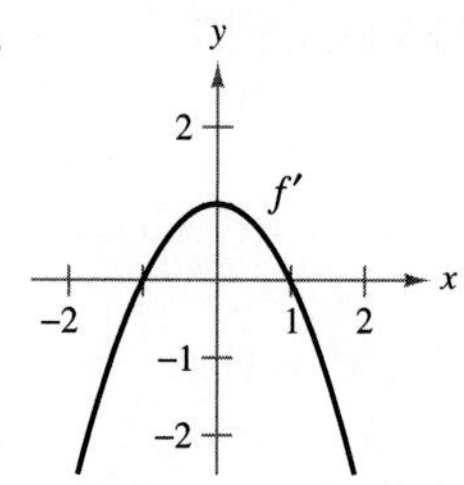

52.

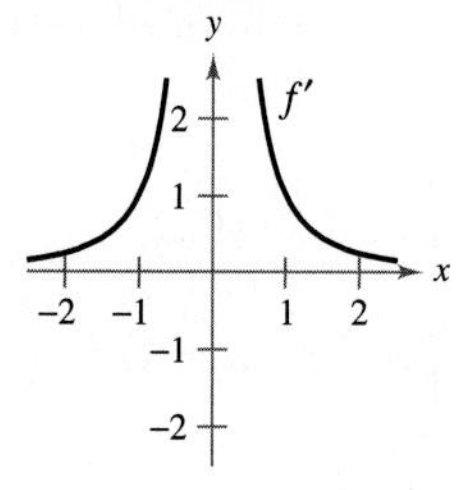

In Exercises 53–56, find the equation for y, given the derivative and the indicated point on the curve.

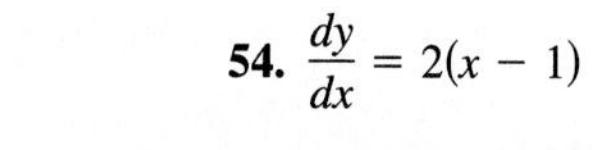

53. $\dfrac{dy}{dx} = 2x - 1$

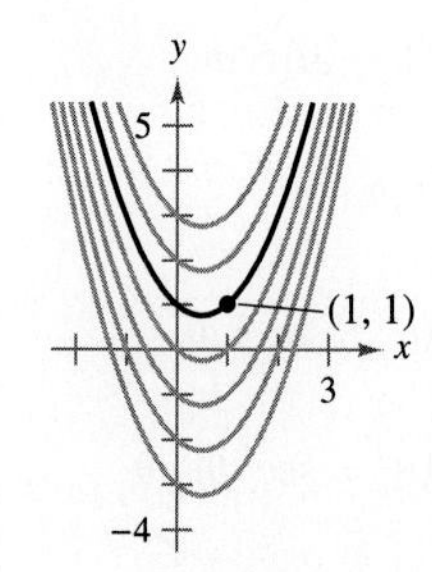

54. $\dfrac{dy}{dx} = 2(x - 1)$

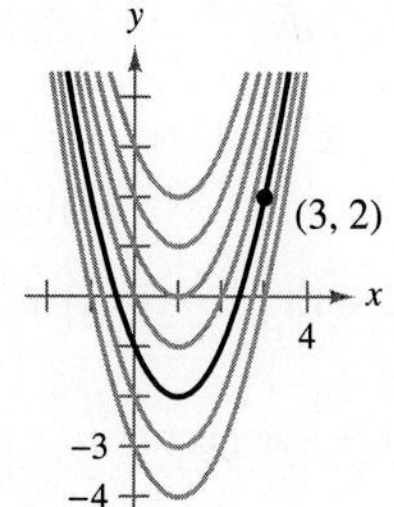

55. $\dfrac{dy}{dx} = \cos x$

(0, 4)

56. $\dfrac{dy}{dx} = \dfrac{3}{x}, \quad x > 0$

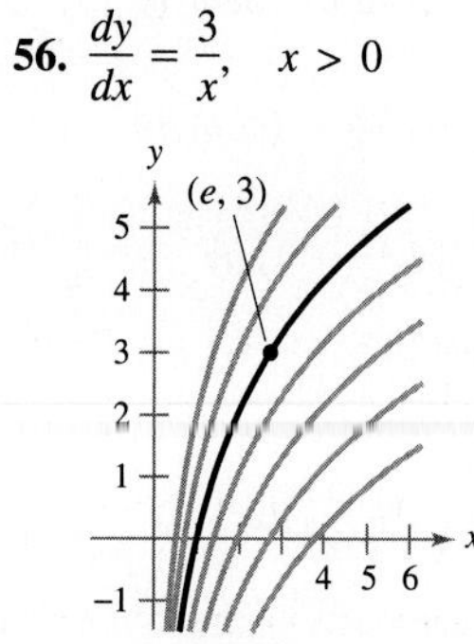

Slope Fields **In Exercises 57–60, a differential equation, a point, and a slope field are given. A *slope field* (or *direction field*) consists of line segments with slopes given by the differential equation. These line segments give a visual perspective of the slopes of the solutions of the differential equation. (a) Sketch two approximate solutions of the differential equation on the slope field, one of which passes through the indicated point. (To print an enlarged copy of the graph, go to the website *www.mathgraphs.com*.) (b) Use integration to find the particular solution of the differential equation and use a graphing utility to graph the solution. Compare the result with the sketches in part (a).**

57. $\dfrac{dy}{dx} = \dfrac{1}{2}x - 1, \quad (4, 2)$

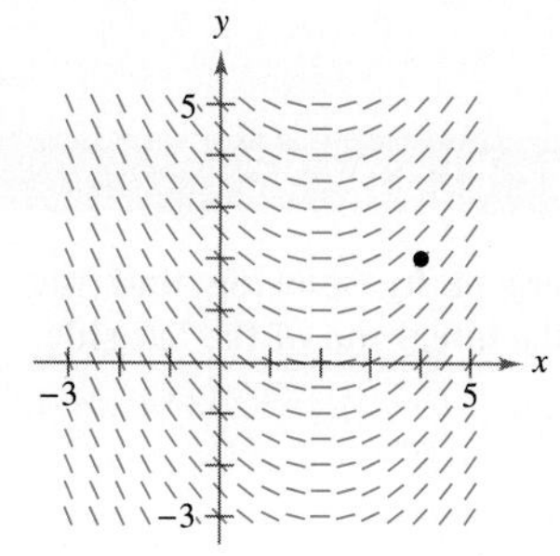

58. $\dfrac{dy}{dx} = x^2 - 1, \quad (-1, 3)$

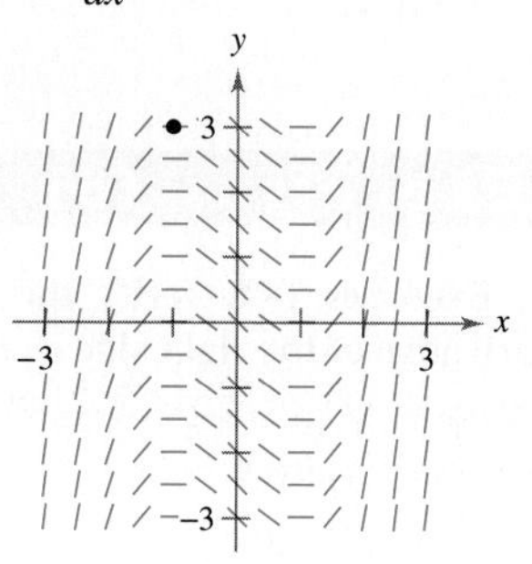

59. $\dfrac{dy}{dx} = \cos x, \ (0, 4)$

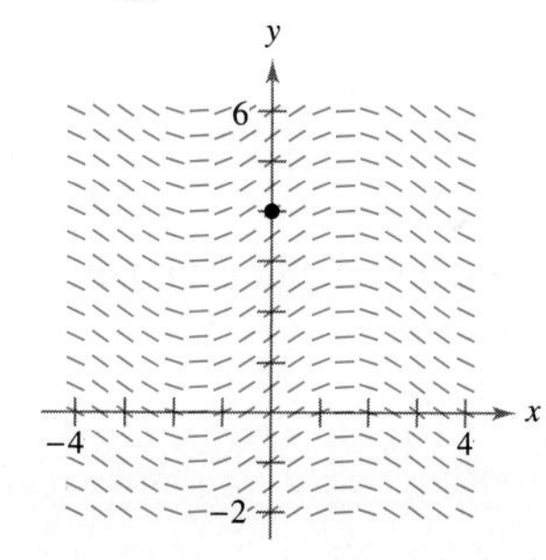

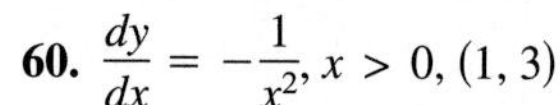

60. $\dfrac{dy}{dx} = -\dfrac{1}{x^2}, \ x > 0, \ (1, 3)$

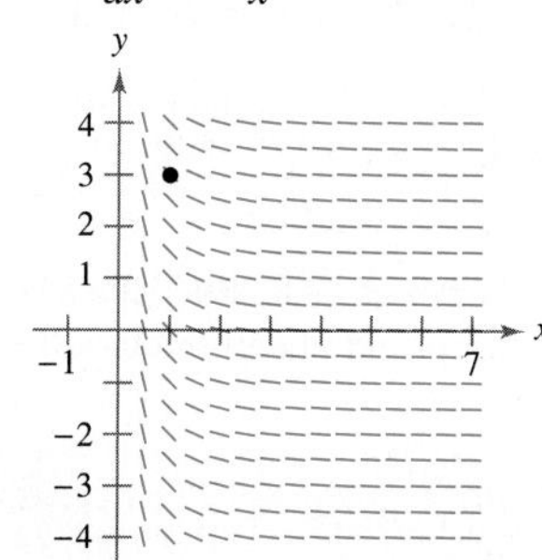

Slope Fields **In Exercises 61 and 62, (a) use a graphing utility to graph a slope field for the differential equation, (b) use integration and the given point to find the particular solution of the differential equation, and (c) graph the solution and the slope field in the same viewing window.**

61. $\dfrac{dy}{dx} = 2x, \ (-2, -2)$

62. $\dfrac{dy}{dx} = 2\sqrt{x}, \ (4, 12)$

In Exercises 63–72, solve the differential equation.

63. $f'(x) = 4x, f(0) = 6$

64. $g'(x) = 6x^2, g(0) = -1$

65. $h'(t) = 8t^3 + 5, h(1) = -4$

66. $f'(s) = 6s - 8s^3, f(2) = 3$

67. $f''(x) = 2, \ f'(2) = 5, \ f(2) = 10$

68. $f''(x) = x^2, \ f'(0) = 6, \ f(0) = 3$

69. $f''(x) = x^{-3/2}, \ f'(4) = 2, \ f(0) = 0$

70. $f''(x) = \sin x, \ f'(0) = 1, \ f(0) = 6$

71. $f''(x) = e^x,\ f'(0) = 2,\ f(0) = 5$

72. $f''(x) = \frac{2}{x^2},\ f'(1) = 4,\ f(1) = 3$

73. ***Tree Growth*** An evergreen nursery usually sells a certain shrub after 6 years of growth and shaping. The growth rate during those 6 years is approximated by $dh/dt = 1.5t + 5$, where t is the time in years and h is the height in centimeters. The seedlings are 12 centimeters tall when planted $(t = 0)$.

(a) Find the height after t years.

(b) How tall are the shrubs when they are sold?

74. ***Population Growth*** The rate of growth dP/dt of a population of bacteria is proportional to the square root of t, where P is the population size and t is the time in days $(0 \le t \le 10)$. That is,

$$\frac{dP}{dt} = k\sqrt{t}.$$

The initial size of the population is 500. After 1 day the population has grown to 600. Estimate the population after 7 days.

Writing About Concepts

75. Use the graph of f' shown in the figure to answer the following, given that $f(0) = -4$.

(a) Approximate the slope of f at $x = 4$. Explain.

(b) Is it possible that $f(2) = -1$? Explain.

(c) Is $f(5) - f(4) > 0$? Explain.

(d) Approximate the value of x where f is maximum. Explain.

(e) Approximate any intervals in which the graph of f is concave upward and any intervals in which it is concave downward. Approximate the x-coordinates of any points of inflection.

(f) Approximate the x-coordinate of the minimum of $f''(x)$.

(g) Sketch an approximate graph of f. To print an enlarged copy of the graph, go to the website *www.mathgraphs.com*.

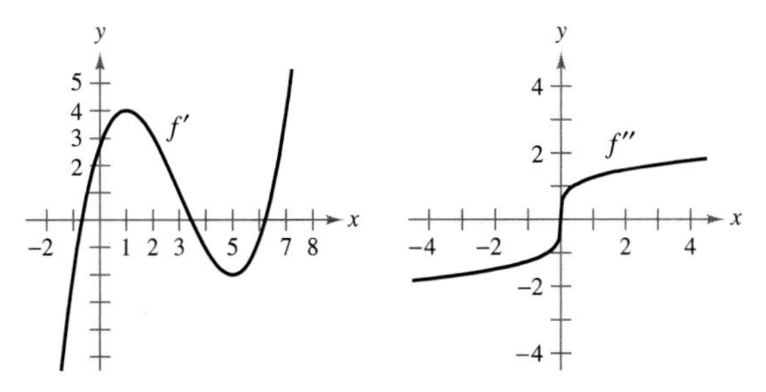

Figure for 75 **Figure for 76**

76. The graphs of f and f' each pass through the origin. Use the graph of f'' shown in the figure to sketch the graphs of f and f'. To print an enlarged copy of the graph, go to the website *www.mathgraphs.com*.

Vertical Motion **In Exercises 77–80, use $a(t) = -32$ feet per second per second as the acceleration due to gravity. (Neglect air resistance.)**

77. A ball is thrown vertically upward from a height of 6 feet with an initial velocity of 60 feet per second. How high will the ball go?

78. Show that the height above the ground of an object thrown upward from a point s_0 feet above the ground with an initial velocity of v_0 feet per second is given by the function

$$f(t) = -16t^2 + v_0 t + s_0.$$

79. With what initial velocity must an object be thrown upward (from ground level) to reach the top of the Washington Monument (approximately 550 feet)?

80. A balloon, rising vertically with a velocity of 8 feet per second, releases a sandbag at the instant it is 64 feet above the ground.

(a) How many seconds after its release will the bag strike the ground?

(b) At what velocity will it hit the ground?

Vertical Motion **In Exercises 81 and 82, use $a(t) = -9.8$ meters per second per second as the acceleration due to gravity. (Neglect air resistance.)**

81. Show that the height above the ground of an object thrown upward from a point s_0 meters above the ground with an initial velocity of v_0 meters per second is given by the function

$$f(t) = -4.9t^2 + v_0 t + s_0.$$

82. The Grand Canyon is 1600 meters deep at its deepest point. A rock is dropped from the rim above this point. Express the height of the rock as a function of the time t in seconds. How long will it take the rock to hit the canyon floor?

83. A baseball is thrown upward from a height of 2 meters with an initial velocity of 10 meters per second. Determine its maximum height.

84. With what initial velocity must an object be thrown upward (from a height of 2 meters) to reach a maximum height of 200 meters?

85. ***Lunar Gravity*** On the moon, the acceleration due to gravity is -1.6 meters per second per second. A stone is dropped from a cliff on the moon and hits the surface of the moon 20 seconds later. How far did it fall? What was its velocity at impact?

86. ***Escape Velocity*** The minimum velocity required for an object to escape Earth's gravitational pull is obtained from the solution of the equation

$$\int v\,dv = -GM\int \frac{1}{y^2}\,dy$$

where v is the velocity of the object projected from Earth, y is the distance from the center of Earth, G is the gravitational constant, and M is the mass of Earth. Show that v and y are related by the equation

$$v^2 = v_0^2 + 2GM\left(\frac{1}{y} - \frac{1}{R}\right)$$

where v_0 is the initial velocity of the object and R is the radius of Earth.

Rectilinear Motion **In Exercises 87–90, consider a particle moving along the x-axis where $x(t)$ is the position of the particle at time t, $x'(t)$ is its velocity, and $x''(t)$ is its acceleration.**

87. $x(t) = t^3 - 6t^2 + 9t - 2, \quad 0 \le t \le 5$

(a) Find the velocity and acceleration of the particle.

(b) Find the open t-intervals on which the particle is moving to the right.

(c) Find the velocity of the particle when the acceleration is 0.

88. Repeat Exercise 87 for the position function

$x(t) = (t - 1)(t - 3)^2, \quad 0 \le t \le 5.$

89. A particle moves along the x-axis at a velocity of $v(t) = 1/\sqrt{t}$, $t > 0$. At time $t = 1$, its position is $x = 4$. Find the acceleration and position functions for the particle.

90. A particle, initially at rest, moves along the x-axis such that its acceleration at time $t > 0$ is given by $a(t) = \cos t$. At the time $t = 0$, its position is $x = 3$.

(a) Find the velocity and position functions for the particle.

(b) Find the values of t for which the particle is at rest.

91. ***Acceleration*** The maker of an automobile advertises that it takes 13 seconds to accelerate from 25 kilometers per hour to 80 kilometers per hour. Assuming constant acceleration, compute the following.

(a) The acceleration in meters per second per second

(b) The distance the car travels during the 13 seconds

92. ***Deceleration*** A car traveling at 45 miles per hour is brought to a stop, at constant deceleration, 132 feet from where the brakes are applied.

(a) How far has the car moved when its speed has been reduced to 30 miles per hour?

(b) How far has the car moved when its speed has been reduced to 15 miles per hour?

(c) Draw the real number line from 0 to 132, and plot the points found in parts (a) and (b). What can you conclude?

93. ***Acceleration*** At the instant the traffic light turns green, a car that has been waiting at an intersection starts with a constant acceleration of 6 feet per second per second. At the same instant, a truck traveling with a constant velocity of 30 feet per second passes the car.

(a) How far beyond its starting point will the car pass the truck?

(b) How fast will the car be traveling when it passes the truck?

94. ***Modeling Data*** The table shows the velocities (in miles per hour) of two cars on an entrance ramp to an interstate highway. The time t is in seconds.

t	0	5	10	15	20	25	30
v_1	0	2.5	7	16	29	45	65
v_2	0	21	38	51	60	64	65

(a) Rewrite the table, converting miles per hour to feet per second.

(b) Use the regression capabilities of a graphing utility to find quadratic models for the data in part (a).

(c) Approximate the distance traveled by each car during the 30 seconds. Explain the difference in the distances.

True or False? **In Exercises 95–100, determine whether the statement is true or false. If it is false, explain why or give an example that shows it is false.**

95. Each antiderivative of an nth-degree polynomial function is an $(n + 1)$th-degree polynomial function.

96. If $p(x)$ is a polynomial function, then p has exactly one antiderivative whose graph contains the origin.

97. If $F(x)$ and $G(x)$ are antiderivatives of $f(x)$, then $F(x) = G(x) + C$.

98. If $f'(x) = g(x)$, then $\int g(x)\, dx = f(x) + C$.

99. $\int f(x)g(x)\, dx = \int f(x)\, dx \int g(x)\, dx$

100. The antiderivative of $f(x)$ is unique.

101. Find a function f such that the graph of f has a horizontal tangent at $(2, 0)$ and $f''(x) = 2x$.

102. The graph of f' is shown. Sketch the graph of f given that f is continuous and $f(0) = 1$.

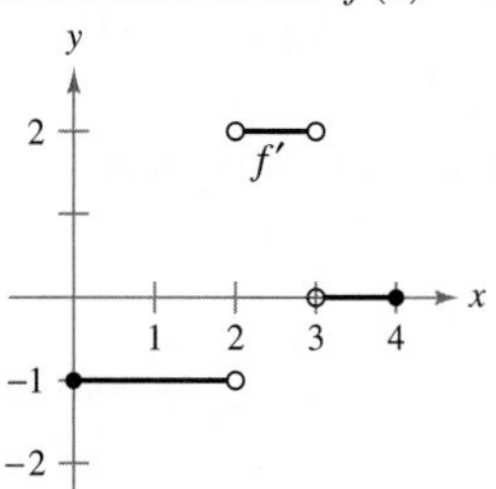

103. If $f'(x) = \begin{cases} 1, & 0 \le x < 2 \\ 3x, & 2 \le x \le 5 \end{cases}$, f is continuous, and $f(1) = 3$, find f. Is f differentiable at $x = 2$?

104. Let $s(x)$ and $c(x)$ be two functions satisfying $s'(x) = c(x)$ and $c'(x) = -s(x)$ for all x. If $s(0) = 0$ and $c(0) = 1$, prove that $[s(x)]^2 + [c(x)]^2 = 1$.

105. ***Verification*** Verify the natural log rule $\displaystyle\int \frac{1}{x}\, dx = \ln|Cx|$, $C \ne 0$, by showing that the derivative of $\ln|Cx|$ is $1/x$.

106. ***Verification*** Verify the natural log rule $\displaystyle\int \frac{1}{x}\, dx = \ln|x| + C$ by showing that the derivative of $\ln|x| + C$ is $1/x$.

Putnam Exam Challenge

107. Suppose f and g are nonconstant, differentiable, real-valued functions on R. Furthermore, suppose that for each pair of real numbers x and y, $f(x + y) = f(x)f(y) - g(x)g(y)$ and $g(x + y) = f(x)g(y) + g(x)f(y)$. If $f'(0) = 0$, prove that $(f(x))^2 + (g(x))^2 = 1$ for all x.

This problem was composed by the Committee on the Putnam Prize Competition.

Section 5.2 Area

- **Use sigma notation to write and evaluate a sum.**
- **Understand the concept of area.**
- **Approximate the area of a plane region.**
- **Find the area of a plane region using limits.**

Sigma Notation

In the preceding section, you studied antidifferentiation. In this section, you will look further into a problem introduced in Section 2.1—that of finding the area of a region in the plane. At first glance, these two ideas may seem unrelated, but you will discover in Section 5.4 that they are closely related by an extremely important theorem called the Fundamental Theorem of Calculus.

This section begins by introducing a concise notation for sums. This notation is called **sigma notation** because it uses the uppercase Greek letter sigma, written as Σ.

Sigma Notation

The sum of n terms $a_1, a_2, a_3, \ldots, a_n$ is written as

$$\sum_{i=1}^{n} a_i = a_1 + a_2 + a_3 + \cdots + a_n$$

where i is the **index of summation,** a_i is the ***i*th term** of the sum, and the **upper and lower bounds of summation** are n and 1.

NOTE The upper and lower bounds must be constant with respect to the index of summation. However, the lower bound doesn't have to be 1. Any integer less than or equal to the upper bound is legitimate.

EXAMPLE 1 Examples of Sigma Notation

a. $\displaystyle\sum_{i=1}^{6} i = 1 + 2 + 3 + 4 + 5 + 6$

b. $\displaystyle\sum_{i=0}^{5} (i + 1) = 1 + 2 + 3 + 4 + 5 + 6$

c. $\displaystyle\sum_{j=3}^{7} j^2 = 3^2 + 4^2 + 5^2 + 6^2 + 7^2$

d. $\displaystyle\sum_{k=1}^{n} \frac{1}{n}(k^2 + 1) = \frac{1}{n}(1^2 + 1) + \frac{1}{n}(2^2 + 1) + \cdots + \frac{1}{n}(n^2 + 1)$

e. $\displaystyle\sum_{i=1}^{n} f(x_i)\,\Delta x = f(x_1)\,\Delta x + f(x_2)\,\Delta x + \cdots + f(x_n)\,\Delta x$

From parts (a) and (b), notice that the same sum can be represented in different ways using sigma notation.

FOR FURTHER INFORMATION For a geometric interpretation of summation formulas, see the article, "Looking at $\sum_{k=1}^{n} k$ and $\sum_{k=1}^{n} k^2$ Geometrically" by Eric Hegblom in *Mathematics Teacher*. To view this article, go to the website *www.matharticles.com.*

Although any variable can be used as the index of summation, i, j, and k are often used. Notice in Example 1 that the index of summation does not appear in the terms of the expanded sum.

THE SUM OF THE FIRST 100 INTEGERS

Carl Friedrich Gauss's (1777–1855) teacher asked him to add all the integers from 1 to 100. When Gauss returned with the correct answer after only a few moments, the teacher could only look at him in astounded silence. This is what Gauss did:

$$\begin{array}{rrrrr} 1\ + & 2\ + & 3\ + & \cdots\ + & 100 \\ 100\ + & 99\ + & 98\ + & \cdots\ + & 1 \\ \hline 101\ + & 101\ + & 101\ + & \cdots\ + & 101 \end{array}$$

$$\frac{100 \times 101}{2} = 5050$$

This is generalized by Theorem 5.2, where

$$\sum_{i=1}^{100} i = \frac{100(101)}{2} = 5050.$$

The following properties of summation can be derived using the associative and commutative properties of addition and the distributive property of multiplication over addition. (In the first property, k is a constant.)

1. $\displaystyle\sum_{i=1}^{n} ka_i = k\sum_{i=1}^{n} a_i$
2. $\displaystyle\sum_{i=1}^{n} (a_i \pm b_i) = \sum_{i=1}^{n} a_i \pm \sum_{i=1}^{n} b_i$

The next theorem lists some useful formulas for sums of powers. A proof of this theorem is given in Appendix A.

THEOREM 5.2 Summation Formulas

1. $\displaystyle\sum_{i=1}^{n} c = cn$
2. $\displaystyle\sum_{i=1}^{n} i = \frac{n(n+1)}{2}$
3. $\displaystyle\sum_{i=1}^{n} i^2 = \frac{n(n+1)(2n+1)}{6}$
4. $\displaystyle\sum_{i=1}^{n} i^3 = \frac{n^2(n+1)^2}{4}$

EXAMPLE 2 Evaluating a Sum

Evaluate $\displaystyle\sum_{i=1}^{n} \frac{i+1}{n^2}$ for $n = 10, 100, 1000$, and $10{,}000$.

Solution Applying Theorem 5.2, you can write

$$\begin{aligned} \sum_{i=1}^{n} \frac{i+1}{n^2} &= \frac{1}{n^2}\sum_{i=1}^{n}(i+1) && \text{Factor constant } 1/n^2 \text{ out of sum.} \\ &= \frac{1}{n^2}\left(\sum_{i=1}^{n} i + \sum_{i=1}^{n} 1\right) && \text{Write as two sums.} \\ &= \frac{1}{n^2}\left[\frac{n(n+1)}{2} + n\right] && \text{Apply Theorem 5.2.} \\ &= \frac{1}{n^2}\left[\frac{n^2+3n}{2}\right] && \text{Simplify.} \\ &= \frac{n+3}{2n}. && \text{Simplify.} \end{aligned}$$

Now you can evaluate the sum by substituting the appropriate values of n, as shown in the table at the left.

n	$\displaystyle\sum_{i=1}^{n} \frac{i+1}{n^2} = \frac{n+3}{2n}$
10	0.65000
100	0.51500
1000	0.50150
10,000	0.50015

In the table, note that the sum appears to approach a limit as n increases. Although the discussion of limits at infinity in Section 4.5 applies to a variable x, where x can be any real number, many of the same results hold true for limits involving the variable n, where n is restricted to positive integer values. So, to find the limit of $(n + 3)/2n$ as n approaches infinity, you can write

$$\lim_{n\to\infty} \frac{n+3}{2n} = \frac{1}{2}.$$

Area

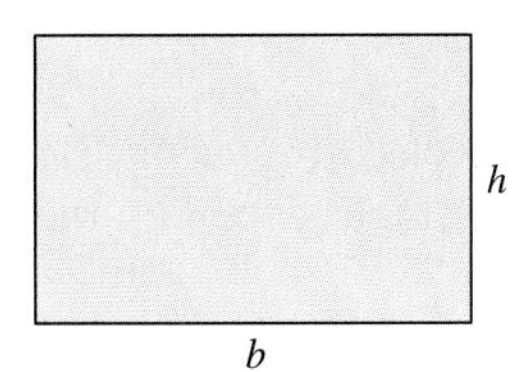

Rectangle: $A = bh$
Figure 5.5

In Euclidean geometry, the simplest type of plane region is a rectangle. Although people often say that the *formula* for the area of a rectangle is $A = bh$, as shown in Figure 5.5, it is actually more proper to say that this is the *definition* of the **area of a rectangle.**

From this definition, you can develop formulas for the areas of many other plane regions. For example, to determine the area of a triangle, you can form a rectangle whose area is twice that of the triangle, as shown in Figure 5.6. Once you know how to find the area of a triangle, you can determine the area of any polygon by subdividing the polygon into triangular regions, as shown in Figure 5.7.

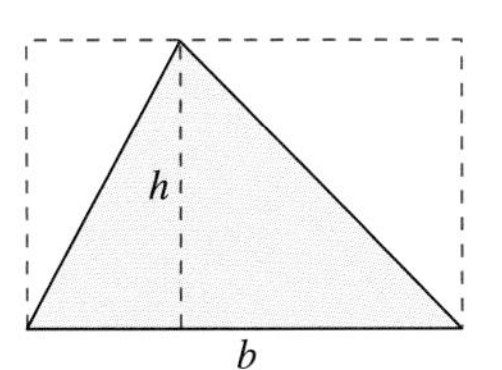

Triangle: $A = \frac{1}{2}bh$
Figure 5.6

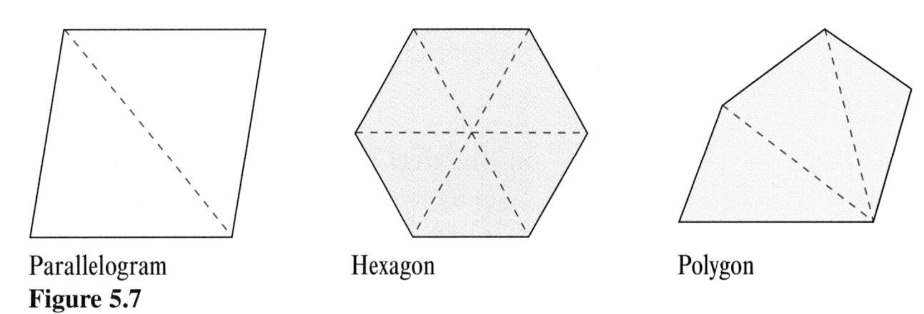

Parallelogram Hexagon Polygon
Figure 5.7

Finding the areas of regions other than polygons is more difficult. The ancient Greeks were able to determine formulas for the areas of some general regions (principally those bounded by conics) by the *exhaustion* method. The clearest description of this method was given by Archimedes. Essentially, the method is a limiting process in which the area is squeezed between two polygons—one inscribed in the region and one circumscribed about the region.

For instance, in Figure 5.8, the area of a circular region is approximated by an n-sided inscribed polygon and an n-sided circumscribed polygon. For each value of n, the area of the inscribed polygon is less than the area of the circle, and the area of the circumscribed polygon is greater than the area of the circle. Moreover, as n increases, the areas of both polygons become better and better approximations of the area of the circle.

ARCHIMEDES (287–212 B.C.)

Archimedes used the method of exhaustion to derive formulas for the areas of ellipses, parabolic segments, and sectors of a spiral. He is considered to have been the greatest applied mathematician of antiquity.

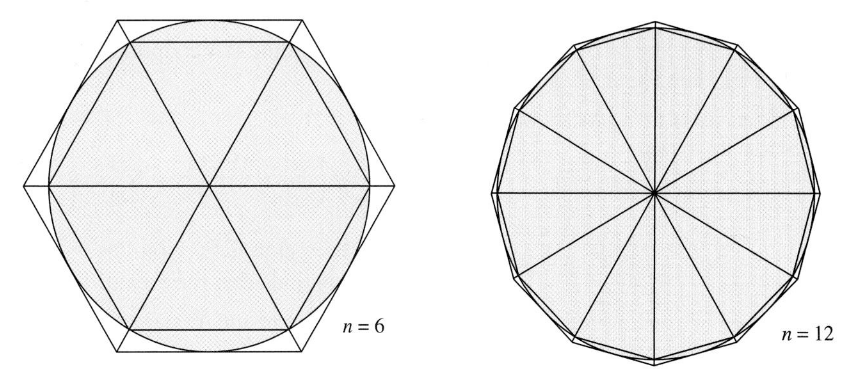

The exhaustion method for finding the area of a circular region
Figure 5.8

FOR FURTHER INFORMATION For an alternative development of the formula for the area of a circle, see the article "Proof Without Words: Area of a Disk is πR^2" by Russell Jay Hendel in *Mathematics Magazine*. To view this article, go to the website *www.matharticles.com*.

A process that is similar to that used by Archimedes to determine the area of a plane region is used in the remaining examples in this section.

The Area of a Plane Region

Recall from Section 2.1 that the origins of calculus are connected to two classic problems: the tangent line problem and the area problem. Example 3 begins the investigation of the area problem.

EXAMPLE 3 Approximating the Area of a Plane Region

Use the five rectangles in Figure 5.9(a) and (b) to find *two* approximations of the area of the region lying between the graph of

$$f(x) = -x^2 + 5$$

and the x-axis between $x = 0$ and $x = 2$.

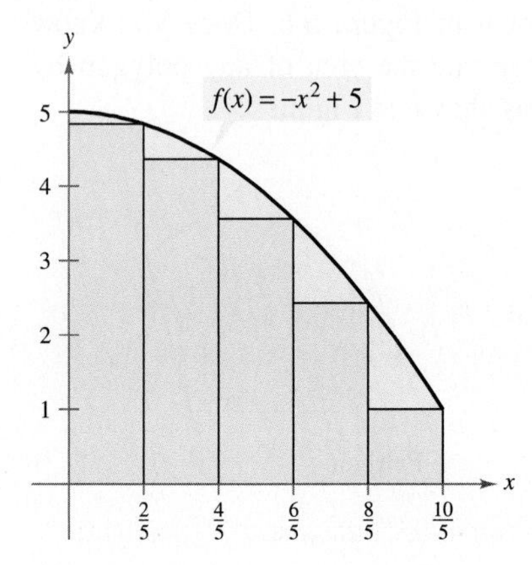

(a) The area of the parabolic region is greater than the area of the rectangles.

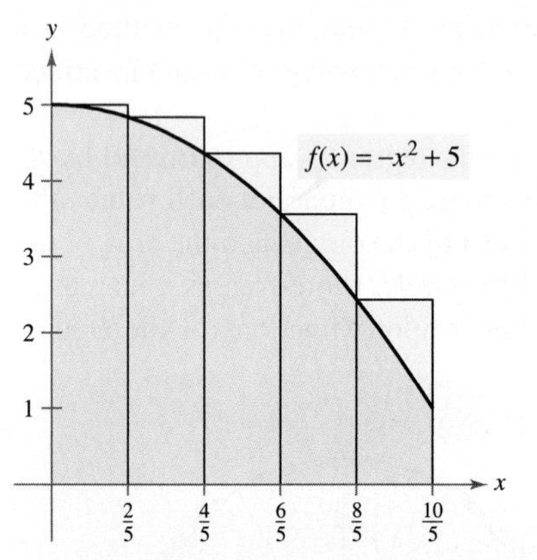

(b) The area of the parabolic region is less than the area of the rectangles.

Figure 5.9

Solution

a. The right endpoints of the five intervals are $\frac{2}{5}i$, where $i = 1, 2, 3, 4, 5$. The width of each rectangle is $\frac{2}{5}$, and the height of each rectangle can be obtained by evaluating f at the right endpoint of each interval.

$$\left[0, \frac{2}{5}\right], \left[\frac{2}{5}, \frac{4}{5}\right], \left[\frac{4}{5}, \frac{6}{5}\right], \left[\frac{6}{5}, \frac{8}{5}\right], \left[\frac{8}{5}, \frac{10}{5}\right]$$

Evaluate f at the right endpoints of these intervals.

The sum of the areas of the five rectangles is

Height Width

$$\sum_{i=1}^{5} f\left(\frac{2i}{5}\right)\left(\frac{2}{5}\right) = \sum_{i=1}^{5}\left[-\left(\frac{2i}{5}\right)^2 + 5\right]\left(\frac{2}{5}\right) = \frac{162}{25} = 6.48.$$

Because each of the five rectangles lies inside the parabolic region, you can conclude that the area of the parabolic region is greater than 6.48.

b. The left endpoints of the five intervals are $\frac{2}{5}(i - 1)$, where $i = 1, 2, 3, 4, 5$. The width of each rectangle is $\frac{2}{5}$, and the height of each rectangle can be obtained by evaluating f at the left endpoint of each interval.

Height Width

$$\sum_{i=1}^{5} f\left(\frac{2i-2}{5}\right)\left(\frac{2}{5}\right) = \sum_{i=1}^{5}\left[-\left(\frac{2i-2}{5}\right)^2 + 5\right]\left(\frac{2}{5}\right) = \frac{202}{25} = 8.08.$$

Because the parabolic region lies within the union of the five rectangular regions, you can conclude that the area of the parabolic region is less than 8.08.

By combining the results in parts (a) and (b), you can conclude that

$$6.48 < (\text{Area of region}) < 8.08.$$

NOTE By increasing the number of rectangles used in Example 3, you can obtain closer and closer approximations of the area of the region. For instance, using 25 rectangles of width $\frac{2}{25}$ each, you can conclude that

$$7.17 < (\text{Area of region}) < 7.49.$$

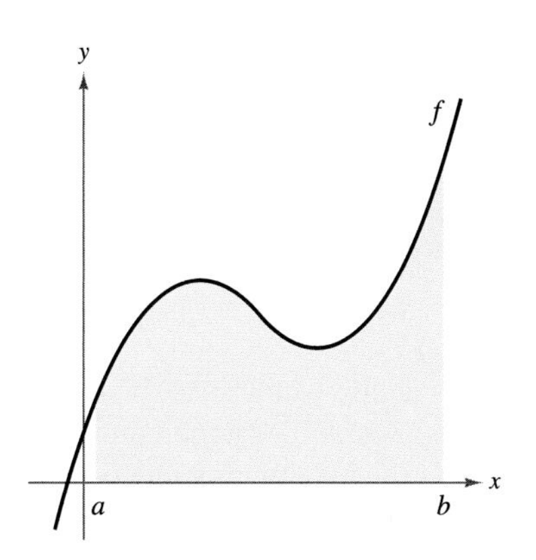

The region under a curve
Figure 5.10

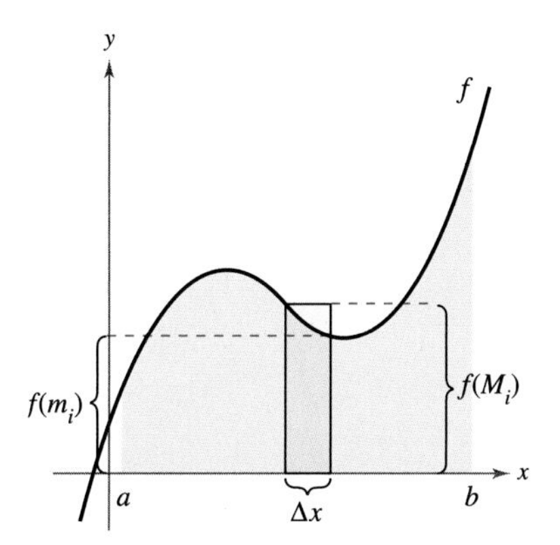

The interval $[a, b]$ is divided into n subintervals of width $\Delta x = \dfrac{b - a}{n}$.
Figure 5.11

Upper and Lower Sums

The procedure used in Example 3 can be generalized as follows. Consider a plane region bounded above by the graph of a nonnegative, continuous function $y = f(x)$, as shown in Figure 5.10. The region is bounded below by the x-axis, and the left and right boundaries of the region are the vertical lines $x = a$ and $x = b$.

To approximate the area of the region, begin by subdividing the interval $[a, b]$ into n subintervals, each of width

$$\Delta x = (b - a)/n$$

as shown in Figure 5.11. The endpoints of the intervals are as follows.

$$\overbrace{a + 0(\Delta x)}^{a = x_0} < \overbrace{a + 1(\Delta x)}^{x_1} < \overbrace{a + 2(\Delta x)}^{x_2} < \cdots < \overbrace{a + n(\Delta x)}^{x_n = b}$$

Because f is continuous, the Extreme Value Theorem guarantees the existence of a minimum and a maximum value of $f(x)$ in *each* subinterval.

$$f(m_i) = \text{Minimum value of } f(x) \text{ in } i\text{th subinterval}$$
$$f(M_i) = \text{Maximum value of } f(x) \text{ in } i\text{th subinterval}$$

Next, define an **inscribed rectangle** lying *inside* the ith subregion and a **circumscribed rectangle** extending *outside* the ith subregion. The height of the ith inscribed rectangle is $f(m_i)$ and the height of the ith circumscribed rectangle is $f(M_i)$. For *each* i, the area of the inscribed rectangle is less than or equal to the area of the circumscribed rectangle.

$$\begin{pmatrix}\text{Area of inscribed}\\ \text{rectangle}\end{pmatrix} = f(m_i)\,\Delta x \le f(M_i)\,\Delta x = \begin{pmatrix}\text{Area of circumscribed}\\ \text{rectangle}\end{pmatrix}$$

The sum of the areas of the inscribed rectangles is called a **lower sum,** and the sum of the areas of the circumscribed rectangles is called an **upper sum.**

$$\text{Lower sum} = s(n) = \sum_{i=1}^{n} f(m_i)\,\Delta x \qquad \text{Area of inscribed rectangles}$$

$$\text{Upper sum} = S(n) = \sum_{i=1}^{n} f(M_i)\,\Delta x \qquad \text{Area of circumscribed rectangles}$$

From Figure 5.12, you can see that the lower sum $s(n)$ is less than or equal to the upper sum $S(n)$. Moreover, the actual area of the region lies between these two sums.

$$s(n) \le (\text{Area of region}) \le S(n)$$

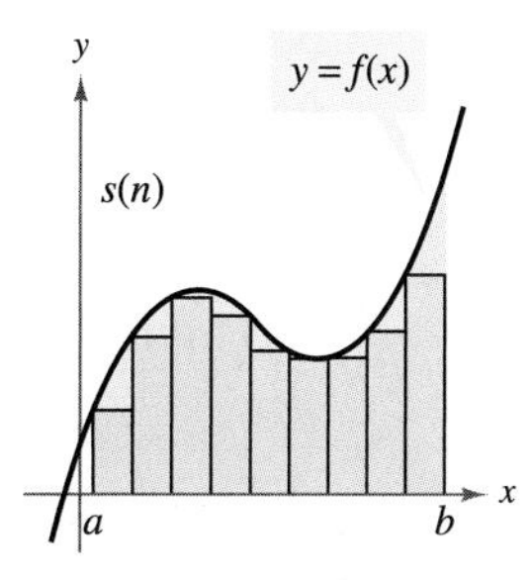

Area of inscribed rectangles is less than area of region.

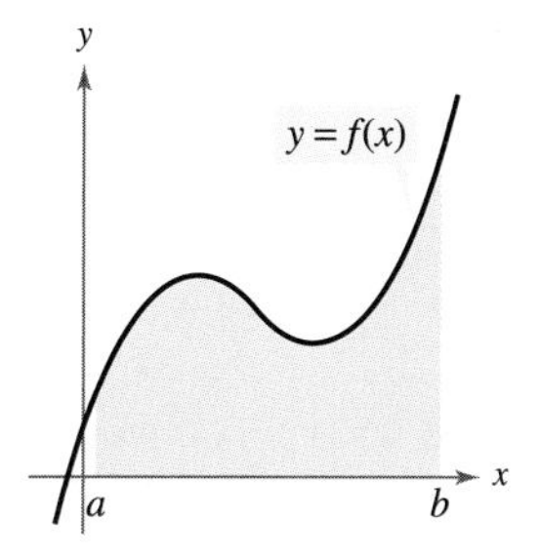

Area of region

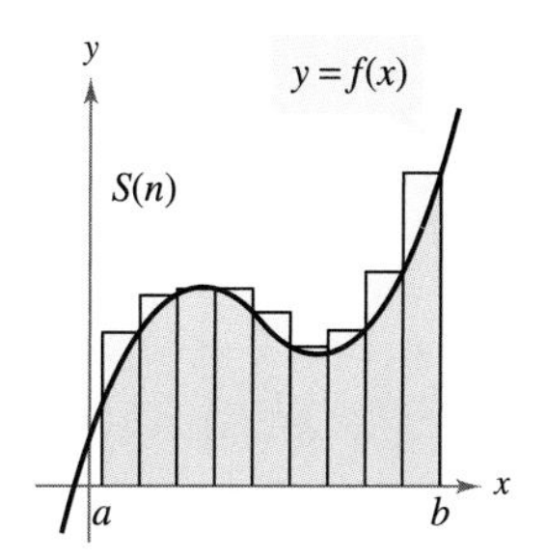

Area of circumscribed rectangles is greater than area of region.

Figure 5.12

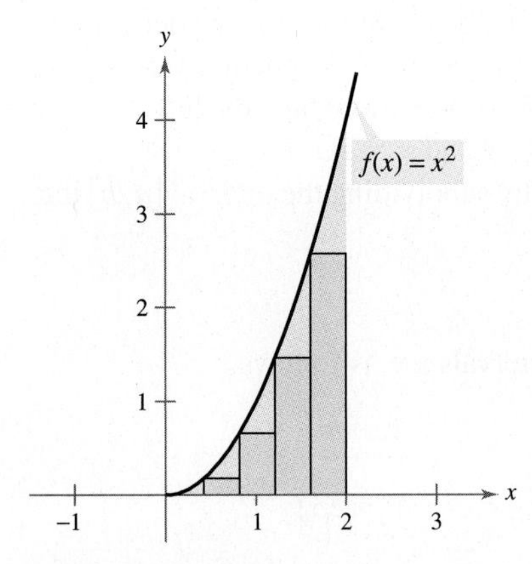

Inscribed rectangles

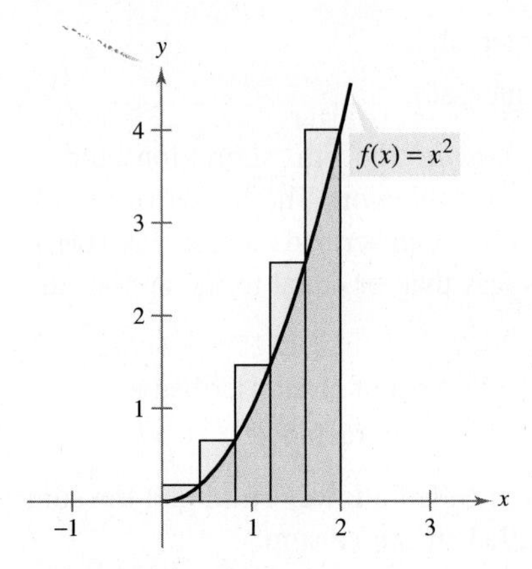

Circumscribed rectangles
Figure 5.13

EXAMPLE 4 Finding Upper and Lower Sums for a Region

Find the upper and lower sums for the region bounded by the graph of $f(x) = x^2$ and the x-axis between $x = 0$ and $x = 2$.

Solution To begin, partition the interval $[0, 2]$ into n subintervals, each of width

$$\Delta x = \frac{b-a}{n} = \frac{2-0}{n} = \frac{2}{n}.$$

Figure 5.13 shows the endpoints of the subintervals and several inscribed and circumscribed rectangles. Because f is increasing on the interval $[0, 2]$, the minimum value on each subinterval occurs at the left endpoint, and the maximum value occurs at the right endpoint.

Left Endpoints	*Right Endpoints*
$m_i = 0 + (i-1)\left(\frac{2}{n}\right) = \frac{2(i-1)}{n}$	$M_i = 0 + i\left(\frac{2}{n}\right) = \frac{2i}{n}$

Using the left endpoints, the lower sum is

$$\begin{aligned}
s(n) = \sum_{i=1}^{n} f(m_i)\,\Delta x &= \sum_{i=1}^{n} f\left[\frac{2(i-1)}{n}\right]\left(\frac{2}{n}\right)\\
&= \sum_{i=1}^{n}\left[\frac{2(i-1)}{n}\right]^2\left(\frac{2}{n}\right)\\
&= \sum_{i=1}^{n}\left(\frac{8}{n^3}\right)(i^2 - 2i + 1)\\
&= \frac{8}{n^3}\left(\sum_{i=1}^{n} i^2 - 2\sum_{i=1}^{n} i + \sum_{i=1}^{n} 1\right)\\
&= \frac{8}{n^3}\left\{\frac{n(n+1)(2n+1)}{6} - 2\left[\frac{n(n+1)}{2}\right] + n\right\}\\
&= \frac{4}{3n^3}(2n^3 - 3n^2 + n)\\
&= \frac{8}{3} - \frac{4}{n} + \frac{4}{3n^2}. \qquad \text{Lower sum}
\end{aligned}$$

Using the right endpoints, the upper sum is

$$\begin{aligned}
S(n) = \sum_{i=1}^{n} f(M_i)\,\Delta x &= \sum_{i=1}^{n} f\left(\frac{2i}{n}\right)\left(\frac{2}{n}\right)\\
&= \sum_{i=1}^{n}\left(\frac{2i}{n}\right)^2\left(\frac{2}{n}\right)\\
&= \sum_{i=1}^{n}\left(\frac{8}{n^3}\right)i^2\\
&= \frac{8}{n^3}\left[\frac{n(n+1)(2n+1)}{6}\right]\\
&= \frac{4}{3n^3}(2n^3 + 3n^2 + n)\\
&= \frac{8}{3} + \frac{4}{n} + \frac{4}{3n^2}. \qquad \text{Upper sum}
\end{aligned}$$

EXPLORATION

For the region given in Example 4, evaluate the lower sum

$$s(n) = \frac{8}{3} - \frac{4}{n} + \frac{4}{3n^2}$$

and the upper sum

$$S(n) = \frac{8}{3} + \frac{4}{n} + \frac{4}{3n^2}$$

for $n = 10$, 100, and 1000. Use your results to determine the area of the region.

Example 4 illustrates some important things about lower and upper sums. First, notice that for any value of n, the lower sum is less than (or equal to) the upper sum.

$$s(n) = \frac{8}{3} - \frac{4}{n} + \frac{4}{3n^2} < \frac{8}{3} + \frac{4}{n} + \frac{4}{3n^2} = S(n)$$

Second, the difference between these two sums lessens as n increases. In fact, if you take the limits as $n \to \infty$, both the upper sum and the lower sum approach $\frac{8}{3}$.

$$\lim_{n\to\infty} s(n) = \lim_{n\to\infty} \left(\frac{8}{3} - \frac{4}{n} + \frac{4}{3n^2}\right) = \frac{8}{3} \qquad \text{Lower sum limit}$$

$$\lim_{n\to\infty} S(n) = \lim_{n\to\infty} \left(\frac{8}{3} + \frac{4}{n} + \frac{4}{3n^2}\right) = \frac{8}{3} \qquad \text{Upper sum limit}$$

The next theorem shows that the equivalence of the limits (as $n \to \infty$) of the upper and lower sums is not mere coincidence. It is true for all functions that are continuous and nonnegative on the closed interval $[a, b]$. The proof of this theorem is best left to a course in advanced calculus.

THEOREM 5.3 Limits of the Lower and Upper Sums

Let f be continuous and nonnegative on the interval $[a, b]$. The limits as $n \to \infty$ of both the lower and upper sums exist and are equal to each other. That is,

$$\begin{aligned}\lim_{n\to\infty} s(n) &= \lim_{n\to\infty} \sum_{i=1}^{n} f(m_i)\,\Delta x \\ &= \lim_{n\to\infty} \sum_{i=1}^{n} f(M_i)\,\Delta x \\ &= \lim_{n\to\infty} S(n)\end{aligned}$$

where $\Delta x = (b - a)/n$ and $f(m_i)$ and $f(M_i)$ are the minimum and maximum values of f on the subinterval.

Because the same limit is attained for both the minimum value $f(m_i)$ and the maximum value $f(M_i)$, it follows from the Squeeze Theorem (Theorem 2.8) that the choice of x in the ith subinterval does not affect the limit. This means that you are free to choose an *arbitrary* x-value in the ith subinterval, as in the following *definition of the area of a region in the plane.*

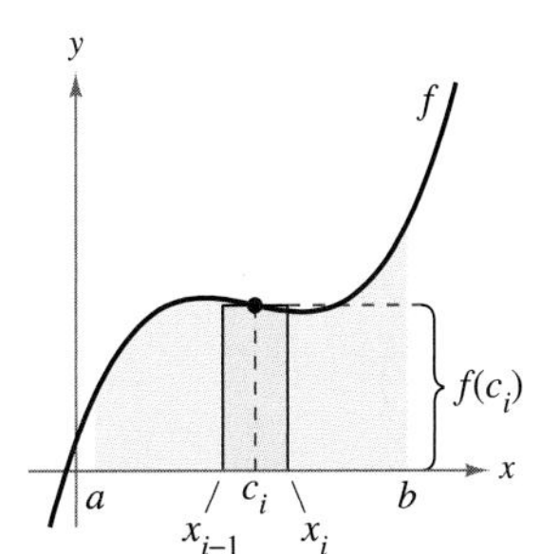

The width of the ith subinterval is $\Delta x = x_i - x_{i-1}$.
Figure 5.14

Definition of the Area of a Region in the Plane

Let f be continuous and nonnegative on the interval $[a, b]$. The area of the region bounded by the graph of f, the x-axis, and the vertical lines $x = a$ and $x = b$ is

$$\text{Area} = \lim_{n\to\infty} \sum_{i=1}^{n} f(c_i)\,\Delta x, \qquad x_{i-1} \le c_i \le x_i$$

where $\Delta x = (b - a)/n$ (see Figure 5.14).

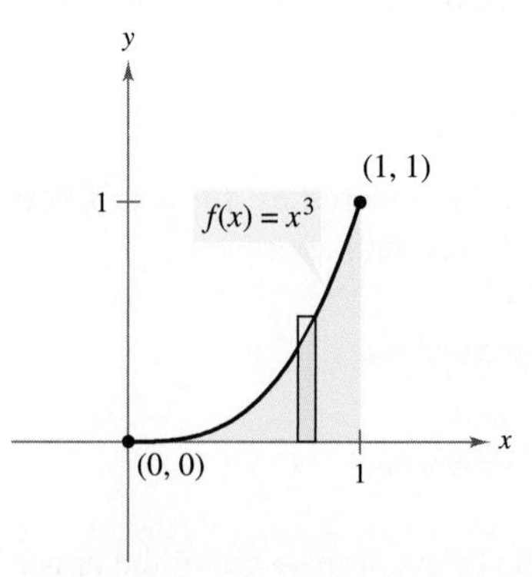

The area of the region bounded by the graph of f, the x-axis, $x = 0$, and $x = 1$ is $\frac{1}{4}$.
Figure 5.15

EXAMPLE 5 Finding Area by the Limit Definition

Find the area of the region bounded by the graph $f(x) = x^3$, the x-axis, and the vertical lines $x = 0$ and $x = 1$, as shown in Figure 5.15.

Solution Begin by noting that f is continuous and nonnegative on the interval $[0, 1]$. Next, partition the interval $[0, 1]$ into n subintervals, each of width $\Delta x = 1/n$. According to the definition of area, you can choose any x-value in the ith subinterval. For this example, the right endpoints $c_i = i/n$ are convenient.

$$\begin{aligned}
\text{Area} = \lim_{n\to\infty} \sum_{i=1}^{n} f(c_i)\,\Delta x &= \lim_{n\to\infty} \sum_{i=1}^{n} \left(\frac{i}{n}\right)^3\left(\frac{1}{n}\right) && \text{Right endpoints: } c_i = \frac{i}{n} \\
&= \lim_{n\to\infty} \frac{1}{n^4}\sum_{i=1}^{n} i^3 \\
&= \lim_{n\to\infty} \frac{1}{n^4}\left[\frac{n^2(n+1)^2}{4}\right] \\
&= \lim_{n\to\infty} \left(\frac{1}{4} + \frac{1}{2n} + \frac{1}{4n^2}\right) \\
&= \frac{1}{4}
\end{aligned}$$

The area of the region is $\frac{1}{4}$.

EXAMPLE 6 Finding Area by the Limit Definition

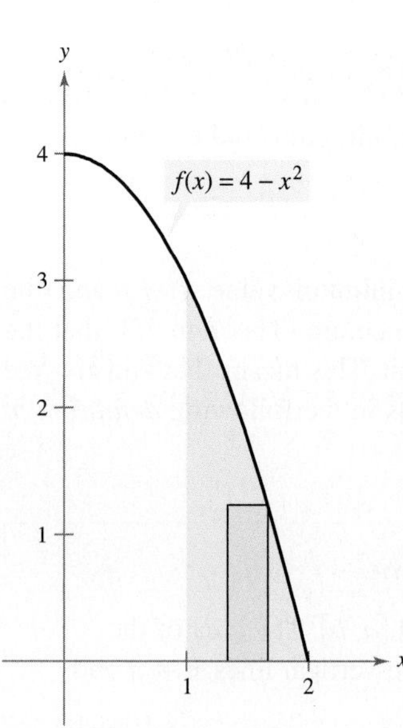

The area of the region bounded by the graph of f, the x-axis, $x = 1$, and $x = 2$ is $\frac{5}{3}$.
Figure 5.16

Find the area of the region bounded by the graph of $f(x) = 4 - x^2$, the x-axis, and the vertical lines $x = 1$ and $x = 2$, as shown in Figure 5.16.

Solution The function f is continuous and nonnegative on the interval $[1, 2]$, so begin by partitioning the interval into n subintervals, each of width $\Delta x = 1/n$. Choosing the right endpoint

$$c_i = a + i\Delta x = 1 + \frac{i}{n} \qquad \text{Right endpoints}$$

of each subinterval, you obtain

$$\begin{aligned}
\text{Area} = \lim_{n\to\infty} \sum_{i=1}^{n} f(c_i)\,\Delta x &= \lim_{n\to\infty} \sum_{i=1}^{n} \left[4 - \left(1 + \frac{i}{n}\right)^2\right]\left(\frac{1}{n}\right) \\
&= \lim_{n\to\infty} \sum_{i=1}^{n} \left(3 - \frac{2i}{n} - \frac{i^2}{n^2}\right)\left(\frac{1}{n}\right) \\
&= \lim_{n\to\infty} \left(\frac{1}{n}\sum_{i=1}^{n} 3 - \frac{2}{n^2}\sum_{i=1}^{n} i - \frac{1}{n^3}\sum_{i=1}^{n} i^2\right) \\
&= \lim_{n\to\infty} \left[3 - \left(1 + \frac{1}{n}\right) - \left(\frac{1}{3} + \frac{1}{2n} + \frac{1}{6n^2}\right)\right] \\
&= 3 - 1 - \frac{1}{3} \\
&= \frac{5}{3}.
\end{aligned}$$

The area of the region is $\frac{5}{3}$.

The last example in this section looks at a region that is bounded by the y-axis (rather than by the x-axis).

EXAMPLE 7 A Region Bounded by the y-axis

Find the area of the region bounded by the graph of $f(y) = y^2$ and the y-axis for $0 \le y \le 1$, as shown in Figure 5.17.

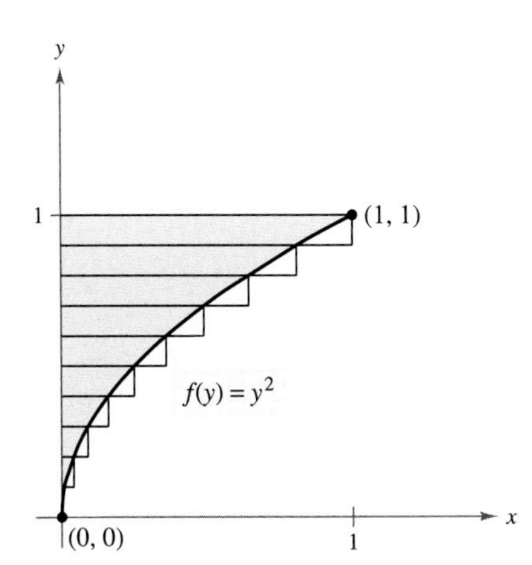

The area of the region bounded by the graph of f and the y-axis for $0 \le y \le 1$ is $\frac{1}{3}$.
Figure 5.17

Solution When f is a continuous, nonnegative function of y, you still can use the same basic procedure shown in Examples 5 and 6. Begin by partitioning the interval $[0, 1]$ into n subintervals, each of width $\Delta y = 1/n$. Then, using the upper endpoints $c_i = i/n$, you obtain

$$\begin{aligned}
\text{Area} &= \lim_{n\to\infty} \sum_{i=1}^{n} f(c_i)\,\Delta y = \lim_{n\to\infty} \sum_{i=1}^{n} \left(\frac{i}{n}\right)^2 \left(\frac{1}{n}\right) \qquad \text{Upper endpoints: } c_i = \frac{i}{n}\\
&= \lim_{n\to\infty} \frac{1}{n^3} \sum_{i=1}^{n} i^2\\
&= \lim_{n\to\infty} \frac{1}{n^3}\left[\frac{n(n+1)(2n+1)}{6}\right]\\
&= \lim_{n\to\infty} \left(\frac{1}{3} + \frac{1}{2n} + \frac{1}{6n^2}\right)\\
&= \frac{1}{3}.
\end{aligned}$$

The area of the region is $\frac{1}{3}$.

Exercises for Section 5.2

See www.CalcChat.com for worked-out solutions to odd-numbered exercises.

In Exercises 1–6, find the sum. Use the summation capabilities of a graphing utility to verify your result.

1. $\displaystyle\sum_{i=1}^{5} (2i + 1)$
2. $\displaystyle\sum_{k=3}^{6} k(k - 2)$
3. $\displaystyle\sum_{k=0}^{4} \frac{1}{k^2 + 1}$
4. $\displaystyle\sum_{j=3}^{5} \frac{1}{j}$
5. $\displaystyle\sum_{k=1}^{4} c$
6. $\displaystyle\sum_{i=1}^{4} [(i - 1)^2 + (i + 1)^3]$

In Exercises 7–14, use sigma notation to write the sum.

7. $\dfrac{1}{3(1)} + \dfrac{1}{3(2)} + \dfrac{1}{3(3)} + \cdots + \dfrac{1}{3(9)}$
8. $\dfrac{5}{1+1} + \dfrac{5}{1+2} + \dfrac{5}{1+3} + \cdots + \dfrac{5}{1+15}$
9. $\left[5\left(\frac{1}{8}\right) + 3\right] + \left[5\left(\frac{2}{8}\right) + 3\right] + \cdots + \left[5\left(\frac{8}{8}\right) + 3\right]$
10. $\left[1 - \left(\frac{1}{4}\right)^2\right] + \left[1 - \left(\frac{2}{4}\right)^2\right] + \cdots + \left[1 - \left(\frac{4}{4}\right)^2\right]$
11. $\left[\left(\frac{2}{n}\right)^3 - \frac{2}{n}\right]\left(\frac{2}{n}\right) + \cdots + \left[\left(\frac{2n}{n}\right)^3 - \frac{2n}{n}\right]\left(\frac{2}{n}\right)$
12. $\left[1 - \left(\frac{2}{n} - 1\right)^2\right]\left(\frac{2}{n}\right) + \cdots + \left[1 - \left(\frac{2n}{n} - 1\right)^2\right]\left(\frac{2}{n}\right)$
13. $\left[2\left(1 + \frac{3}{n}\right)^2\right]\left(\frac{3}{n}\right) + \cdots + \left[2\left(1 + \frac{3n}{n}\right)^2\right]\left(\frac{3}{n}\right)$
14. $\left(\frac{1}{n}\right)\sqrt{1 - \left(\frac{0}{n}\right)^2} + \cdots + \left(\frac{1}{n}\right)\sqrt{1 - \left(\frac{n-1}{n}\right)^2}$

In Exercises 15–20, use the properties of summation and Theorem 5.2 to evaluate the sum. Use the summation capabilities of a graphing utility to verify your result.

15. $\displaystyle\sum_{i=1}^{20} 2i$
16. $\displaystyle\sum_{i=1}^{15} (2i - 3)$
17. $\displaystyle\sum_{i=1}^{20} (i - 1)^2$
18. $\displaystyle\sum_{i=1}^{10} (i^2 - 1)$
19. $\displaystyle\sum_{i=1}^{15} i(i - 1)^2$
20. $\displaystyle\sum_{i=1}^{10} i(i^2 + 1)$

In Exercises 21 and 22, use the summation capabilities of a graphing utility to evaluate the sum. Then use the properties of summation and Theorem 5.2 to verify the sum.

21. $\displaystyle\sum_{i=1}^{20} (i^2 + 3)$
22. $\displaystyle\sum_{i=1}^{15} (i^3 - 2i)$

In Exercises 23–26, bound the area of the shaded region by approximating the upper and lower sums. Use rectangles of width 1.

23.

24.

25.

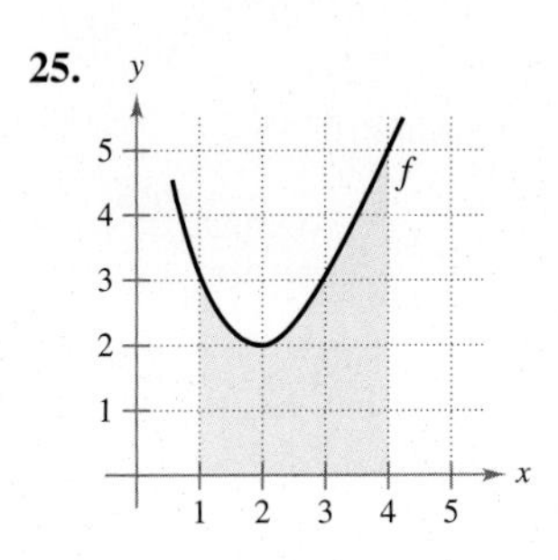

26.

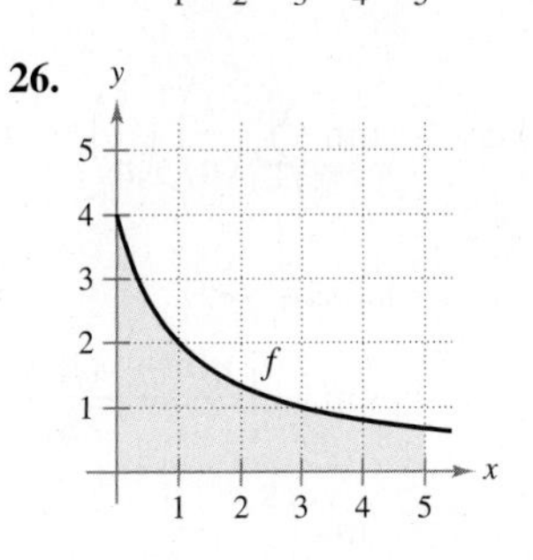

In Exercises 27–30, use upper and lower sums to approximate the area of the region using the given number of subintervals (of equal width).

27. $y = \sqrt{x}$

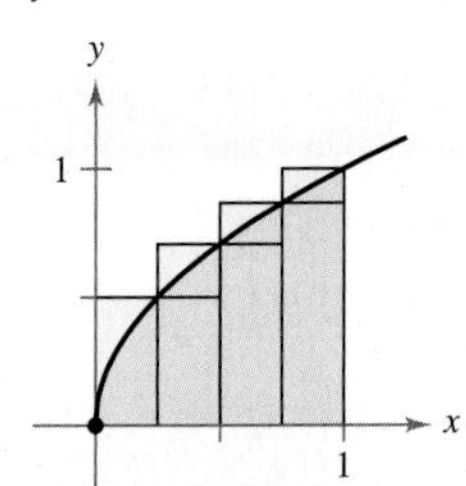

28. $y = 4e^{-x}$

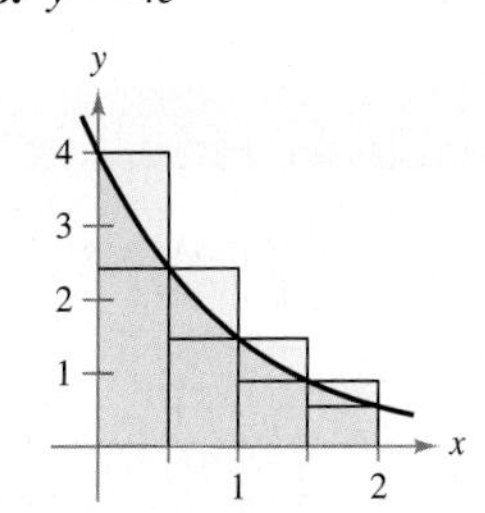

29. $y = \dfrac{1}{x}$

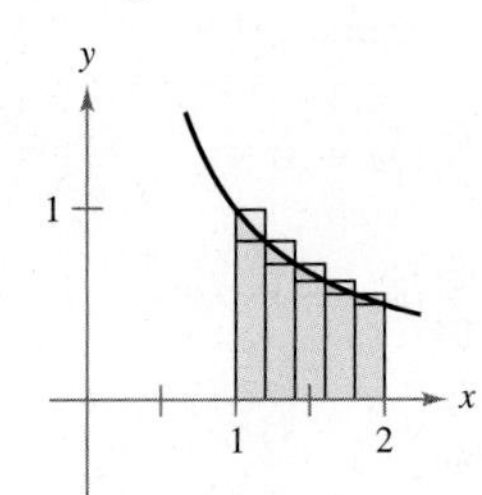

30. $y = \sqrt{1 - x^2}$

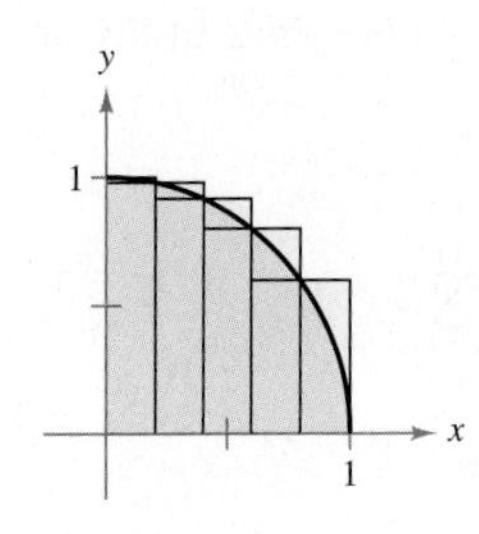

In Exercises 31–34, find the limit of $s(n)$ as $n \to \infty$.

31. $s(n) = \dfrac{81}{n^4}\left[\dfrac{n^2(n+1)^2}{4}\right]$

32. $s(n) = \dfrac{64}{n^3}\left[\dfrac{n(n+1)(2n+1)}{6}\right]$

33. $s(n) = \dfrac{18}{n^2}\left[\dfrac{n(n+1)}{2}\right]$

34. $s(n) = \dfrac{1}{n^2}\left[\dfrac{n(n+1)}{2}\right]$

In Exercises 35–38, use the summation formulas to rewrite the expression without the summation notation. Use the result to find the sum for $n = 10, 100, 1000$, and $10{,}000$.

35. $\displaystyle\sum_{i=1}^{n} \frac{2i+1}{n^2}$

36. $\displaystyle\sum_{j=1}^{n} \frac{4j+1}{n^2}$

37. $\displaystyle\sum_{k=1}^{n} \frac{6k(k-1)}{n^3}$

38. $\displaystyle\sum_{i=1}^{n} \frac{4i^2(i-1)}{n^4}$

In Exercises 39–44, find a formula for the sum of n terms. Use the formula to find the limit as $n \to \infty$.

39. $\displaystyle\lim_{n\to\infty} \sum_{i=1}^{n} \frac{16i}{n^2}$

40. $\displaystyle\lim_{n\to\infty} \sum_{i=1}^{n} \left(\frac{2i}{n}\right)\left(\frac{2}{n}\right)$

41. $\displaystyle\lim_{n\to\infty} \sum_{i=1}^{n} \frac{1}{n^3}(i-1)^2$

42. $\displaystyle\lim_{n\to\infty} \sum_{i=1}^{n} \left(1 + \frac{2i}{n}\right)^2\left(\frac{2}{n}\right)$

43. $\displaystyle\lim_{n\to\infty} \sum_{i=1}^{n} \left(1 + \frac{i}{n}\right)\left(\frac{2}{n}\right)$

44. $\displaystyle\lim_{n\to\infty} \sum_{i=1}^{n} \left(1 + \frac{2i}{n}\right)^3\left(\frac{2}{n}\right)$

45. ***Numerical Reasoning*** Consider a triangle of area 2 bounded by the graphs of $y = x$, $y = 0$, and $x = 2$.

(a) Sketch the region.

(b) Divide the interval $[0, 2]$ into n subintervals of equal width and show that the endpoints are

$$0 < 1\left(\frac{2}{n}\right) < \cdots < (n-1)\left(\frac{2}{n}\right) < n\left(\frac{2}{n}\right).$$

(c) Show that $s(n) = \displaystyle\sum_{i=1}^{n}\left[(i-1)\left(\frac{2}{n}\right)\right]\left(\frac{2}{n}\right)$.

(d) Show that $S(n) = \displaystyle\sum_{i=1}^{n}\left[i\left(\frac{2}{n}\right)\right]\left(\frac{2}{n}\right)$.

(e) Complete the table.

n	5	10	50	100
$s(n)$				
$S(n)$				

(f) Show that $\displaystyle\lim_{n\to\infty} s(n) = \lim_{n\to\infty} S(n) = 2$.

46. ***Numerical Reasoning*** Consider a trapezoid of area 4 bounded by the graphs of $y = x$, $y = 0$, $x = 1$, and $x = 3$.

(a) Sketch the region.

(b) Divide the interval $[1, 3]$ into n subintervals of equal width and show that the endpoints are

$$1 < 1 + 1\left(\frac{2}{n}\right) < \cdots < 1 + (n-1)\left(\frac{2}{n}\right) < 1 + n\left(\frac{2}{n}\right).$$

(c) Show that $s(n) = \displaystyle\sum_{i=1}^{n}\left[1 + (i-1)\left(\frac{2}{n}\right)\right]\left(\frac{2}{n}\right)$.

(d) Show that $S(n) = \displaystyle\sum_{i=1}^{n}\left[1 + i\left(\frac{2}{n}\right)\right]\left(\frac{2}{n}\right)$.

(e) Complete the table.

n	5	10	50	100
$s(n)$				
$S(n)$				

(f) Show that $\displaystyle\lim_{n\to\infty} s(n) = \lim_{n\to\infty} S(n) = 4$.

In Exercises 47–56, use the limit process to find the area of the region between the graph of the function and the x-axis over the given interval. Sketch the region.

47. $y = -2x + 3, \quad [0, 1]$ **48.** $y = 3x - 4, \quad [2, 5]$

49. $y = x^2 + 2, \quad [0, 1]$ **50.** $y = x^2 + 1, \quad [0, 3]$

51. $y = 16 - x^2, \quad [1, 3]$ **52.** $y = 1 - x^2, \quad [-1, 1]$

53. $y = 64 - x^3, \quad [1, 4]$ **54.** $y = 2x - x^3, \quad [0, 1]$

55. $y = x^2 - x^3, \quad [-1, 1]$ **56.** $y = x^2 - x^3, \quad [-1, 0]$

In Exercises 57–62, use the limit process to find the area of the region between the graph of the function and the y-axis over the given y-interval. Sketch the region.

57. $f(y) = 3y, 0 \le y \le 2$ **58.** $g(y) = \frac{1}{2}y, 2 \le y \le 4$

59. $f(y) = y^2, 0 \le y \le 3$ **60.** $f(y) = 4y - y^2, 1 \le y \le 2$

61. $g(y) = 4y^2 - y^3, 1 \le y \le 3$ **62.** $h(y) = y^3 + 1, 1 \le y \le 2$

In Exercises 63–66, use the *Midpoint Rule*

$$\text{Area} \approx \sum_{i=1}^{n} f\left(\frac{x_i + x_{i-1}}{2}\right)\Delta x$$

with $n = 4$ to approximate the area of the region bounded by the graph of the function and the x-axis over the given interval.

63. $f(x) = x^2 + 3, \quad [0, 2]$ **64.** $f(x) = x^2 + 4x, \quad [0, 4]$

65. $f(x) = \tan x, \quad \left[0, \frac{\pi}{4}\right]$ **66.** $f(x) = \sin x, \quad \left[0, \frac{\pi}{2}\right]$

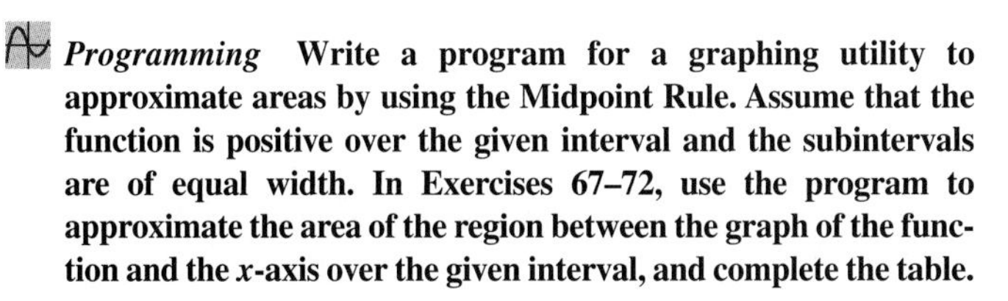

Programming **Write a program for a graphing utility to approximate areas by using the Midpoint Rule. Assume that the function is positive over the given interval and the subintervals are of equal width. In Exercises 67–72, use the program to approximate the area of the region between the graph of the function and the x-axis over the given interval, and complete the table.**

n	4	8	12	16	20
Approximate Area					

67. $f(x) = \sqrt{x}, \quad [0, 4]$ **68.** $f(x) = \dfrac{8}{x^2 + 1}, \quad [2, 6]$

69. $f(x) = \tan\left(\dfrac{\pi x}{8}\right), \quad [1, 3]$ **70.** $f(x) = \cos \sqrt{x}, \quad [0, 2]$

71. $f(x) = \ln x, \quad [1, 5]$ **72.** $f(x) = xe^x, \quad [0, 2]$

Writing About Concepts

Approximation **In Exercises 73 and 74, determine which value best approximates the area of the region between the x-axis and the graph of the function over the given interval. (Make your selection on the basis of a sketch of the region and not by performing calculations.)**

73. $f(x) = 4 - x^2, \quad [0, 2]$

(a) -2 (b) 6 (c) 10 (d) 3 (e) 8

Writing About Concepts *(continued)*

74. $f(x) = \sin \dfrac{\pi x}{4}, \quad [0, 4]$

(a) 3 (b) 1 (c) -2 (d) 8 (e) 6

75. In your own words and using appropriate figures, describe the methods of upper sums and lower sums in approximating the area of a region.

76. Give the definition of the area of a region in the plane.

77. ***Graphical Reasoning*** Consider the region bounded by the graphs of

$$f(x) = \frac{8x}{x + 1},$$

$x = 0$, $x = 4$, and $y = 0$, as shown in the figure. To print an enlarged copy of the graph, go to the website *www.mathgraphs.com*.

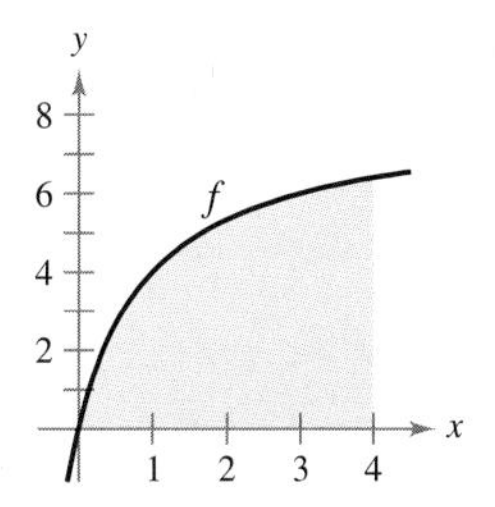

(a) Redraw the figure, and complete and shade the rectangles representing the lower sum when $n = 4$. Find this lower sum.

(b) Redraw the figure, and complete and shade the rectangles representing the upper sum when $n = 4$. Find this upper sum.

(c) Redraw the figure, and complete and shade the rectangles whose heights are determined by the functional values at the midpoint of each subinterval when $n = 4$. Find this sum using the Midpoint Rule.

(d) Verify the following formulas for approximating the area of the region using n subintervals of equal width.

Lower sum: $s(n) = \displaystyle\sum_{i=1}^{n} f\left[(i - 1)\frac{4}{n}\right]\left(\frac{4}{n}\right)$

Upper sum: $S(n) = \displaystyle\sum_{i=1}^{n} f\left[(i)\frac{4}{n}\right]\left(\frac{4}{n}\right)$

Midpoint Rule: $M(n) = \displaystyle\sum_{i=1}^{n} f\left[\left(i - \frac{1}{2}\right)\frac{4}{n}\right]\left(\frac{4}{n}\right)$

(e) Use a graphing utility and the formulas in part (d) to complete the table.

n	4	8	20	100	200
$s(n)$					
$S(n)$					
$M(n)$					

(f) Explain why $s(n)$ increases and $S(n)$ decreases for increasing values of n, as shown in the table in part (e).

78. ***Monte Carlo Method*** The following computer program approximates the area of the region under the graph of a monotonic function and above the x-axis between $x = a$ and $x = b$. Run the program for $a = 0$ and $b = \pi/2$ for several values of N2. Explain why the Monte Carlo Method works. [*Adaptation of Monte Carlo Method program from James M. Sconyers, "Approximation of Area Under a Curve," MATHEMATICS TEACHER 77, no. 2 (February 1984). Copyright © 1984 by the National Council of Teachers of Mathematics. Reprinted with permission.*]

```
10 DEF FNF(X)=SIN(X)
20 A=0
30 B=π/2
40 PRINT "Input Number of Random Points"
50 INPUT N2
60 N1=0
70 IF FNF(A)>FNF(B) THEN YMAX=FNF(A) ELSE
   YMAX=FNF(B)
80 FOR I=1 TO N2
90 X=A+(B-A)*RND(1)
100 Y=YMAX*RND(1)
110 IF Y>=FNF(X) THEN GOTO 130
120 N1=N1+1
130 NEXT I
140 AREA=(N1/N2)*(B-A)*YMAX
150 PRINT "Approximate Area:"; AREA
160 END
```

True or False? **In Exercises 79 and 80, determine whether the statement is true or false. If it is false, explain why or give an example that shows it is false.**

79. The sum of the first n positive integers is $n(n + 1)/2$.

80. If f is continuous and nonnegative on $[a, b]$, then the limits as $n \to \infty$ of its lower sum $s(n)$ and upper sum $S(n)$ both exist and are equal.

81. ***Writing*** Use the figure to write a short paragraph explaining why the formula $1 + 2 + \cdot \cdot \cdot + n = \frac{1}{2}n(n + 1)$ is valid for all positive integers n.

Figure for 81

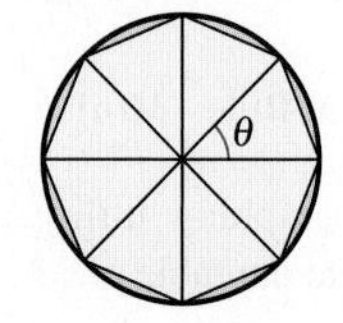

Figure for 82

82. ***Graphical Reasoning*** Consider an n-sided regular polygon inscribed in a circle of radius r. Join the vertices of the polygon to the center of the circle, forming n congruent triangles (see figure).

(a) Determine the central angle θ in terms of n.

(b) Show that the area of each triangle is $\frac{1}{2}r^2 \sin \theta$.

(c) Let A_n be the sum of the areas of the n triangles. Find $\lim_{n \to \infty} A_n$.

83. ***Modeling Data*** The table lists the measurements of a lot bounded by a stream and two straight roads that meet at right angles, where x and y are measured in feet (see figure).

x	0	50	100	150	200	250	300
y	450	362	305	268	245	156	0

(a) Use the regression capabilities of a graphing utility to find a model of the form $y = ax^3 + bx^2 + cx + d$.

(b) Use a graphing utility to plot the data and graph the model.

(c) Use the model in part (a) to estimate the area of the lot.

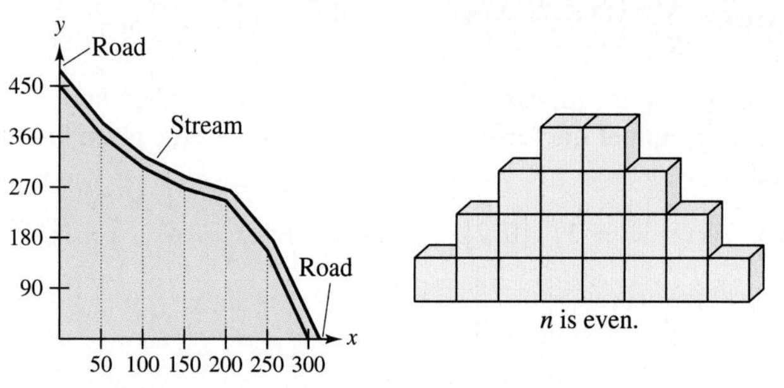

Figure for 83 **Figure for 84**

84. ***Building Blocks*** A child places n cubic building blocks in a row to form the base of a triangular design (see figure). Each successive row contains two fewer blocks than the preceding row. Find a formula for the number of blocks used in the design. (*Hint:* The number of building blocks in the design depends on whether n is odd or even.)

85. Prove each formula by mathematical induction. (You may need to review the method of proof by induction from a precalculus text.)

(a) $\sum_{i=1}^{n} 2i = n(n + 1)$

(b) $\sum_{i=1}^{n} i^3 = \dfrac{n^2(n + 1)^2}{4}$

Putnam Exam Challenge

86. A dart, thrown at random, hits a square target. Assuming that any two parts of the target of equal area are equally likely to be hit, find the probability that the point hit is nearer to the center than to any edge. Write your answer in the form $(a\sqrt{b} + c)/d$, where a, b, c, and d are positive integers.

This problem was composed by the Committee on the Putnam Prize Competition.

Section 5.3 Riemann Sums and Definite Integrals

- **Understand the definition of a Riemann sum.**
- **Evaluate a definite integral using limits.**
- **Evaluate a definite integral using properties of definite integrals.**

Riemann Sums

In the definition of area given in Section 5.2, the partitions have subintervals of *equal width.* This was done only for computational convenience. The following example shows that it is not necessary to have subintervals of equal width.

EXAMPLE 1 A Partition with Subintervals of Unequal Widths

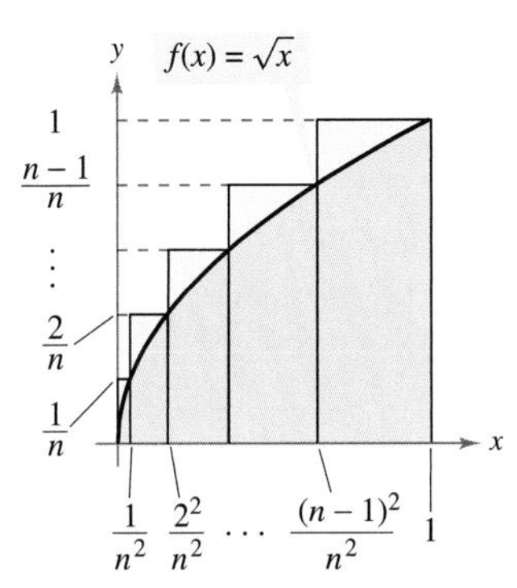

The subintervals do not have equal widths.
Figure 5.18

Consider the region bounded by the graph of $f(x) = \sqrt{x}$ and the x-axis for $0 \le x \le 1$, as shown in Figure 5.18. Evaluate the limit

$$\lim_{n\to\infty} \sum_{i=1}^{n} f(c_i)\,\Delta x_i$$

where c_i is the right endpoint of the partition given by $c_i = i^2/n^2$ and Δx_i is the width of the ith interval.

Solution The width of the ith interval is given by

$$\begin{aligned}\Delta x_i &= \frac{i^2}{n^2} - \frac{(i-1)^2}{n^2}\\ &= \frac{i^2 - i^2 + 2i - 1}{n^2}\\ &= \frac{2i-1}{n^2}.\end{aligned}$$

So, the limit is

$$\begin{aligned}\lim_{n\to\infty} \sum_{i=1}^{n} f(c_i)\,\Delta x_i &= \lim_{n\to\infty} \sum_{i=1}^{n} \sqrt{\frac{i^2}{n^2}}\left(\frac{2i-1}{n^2}\right)\\ &= \lim_{n\to\infty} \frac{1}{n^3} \sum_{i=1}^{n} (2i^2 - i)\\ &= \lim_{n\to\infty} \frac{1}{n^3}\left[2\left(\frac{n(n+1)(2n+1)}{6}\right) - \frac{n(n+1)}{2}\right]\\ &= \lim_{n\to\infty} \frac{4n^3 + 3n^2 - n}{6n^3}\\ &= \frac{2}{3}.\end{aligned}$$

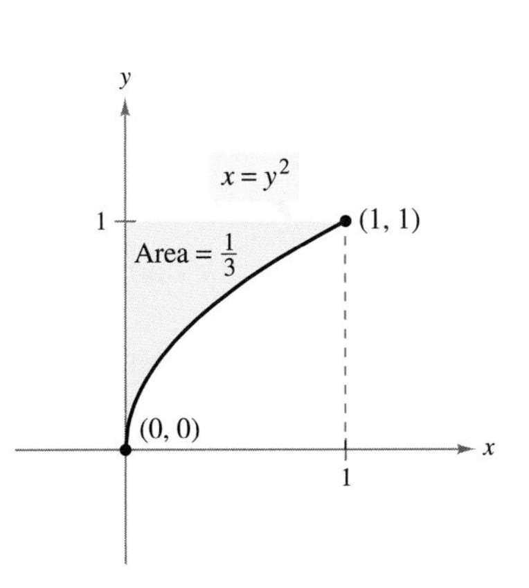

The area of the region bounded by the graph of $x = y^2$ and the y-axis for $0 \le y \le 1$ is $\frac{1}{3}$.
Figure 5.19

From Example 7 in Section 5.2, you know that the region shown in Figure 5.19 has an area of $\frac{1}{3}$. Because the square bounded by $0 \le x \le 1$ and $0 \le y \le 1$ has an area of 1, you can conclude that the area of the region shown in Figure 5.18 has an area of $\frac{2}{3}$. This agrees with the limit found in Example 1, even though that example used a partition having subintervals of unequal widths. The reason this particular partition gave the proper area is that as n increases, the *width of the largest subinterval approaches zero.* This is a key feature of the development of definite integrals.

GEORG FRIEDRICH BERNHARD RIEMANN (1826–1866)

German mathematician Riemann did his most famous work in the areas of non-Euclidean geometry, differential equations, and number theory. It was Riemann's results in physics and mathematics that formed the structure on which Einstein's theory of general relativity is based.

In the preceding section, the limit of a sum was used to define the area of a region in the plane. Finding area by this means is only one of *many* applications involving the limit of a sum. A similar approach can be used to determine quantities as diverse as arc lengths, average values, centroids, volumes, work, and surface areas. The following definition is named after Georg Friedrich Bernhard Riemann. Although the definite integral had been defined and used long before the time of Riemann, he generalized the concept to cover a broader category of functions.

In the following definition of a Riemann sum, note that the function f has no restrictions other than being defined on the interval $[a, b]$. (In the preceding section, the function f was assumed to be continuous and nonnegative because we were dealing with the area under a curve.)

Definition of a Riemann Sum

Let f be defined on the closed interval $[a, b]$, and let Δ be a partition of $[a, b]$ given by

$$a = x_0 < x_1 < x_2 < \cdot \cdot \cdot < x_{n-1} < x_n = b$$

where Δx_i is the width of the ith subinterval $[x_{i-1}, x_i]$. If c_i is *any* point in the ith subinterval, then the sum

$$\sum_{i=1}^{n} f(c_i)\,\Delta x_i, \qquad x_{i-1} \le c_i \le x_i$$

is called a **Riemann sum** of f for the partition Δ.

NOTE The sums in Section 5.2 are examples of Riemann sums, but there are more general Riemann sums than those covered there.

The width of the largest subinterval of a partition Δ is the **norm** of the partition and is denoted by $\|\Delta\|$. If every subinterval is of equal width, the partition is **regular** and the norm is denoted by

$$\|\Delta\| = \Delta x = \frac{b-a}{n}. \qquad \text{Regular partition}$$

For a general partition, the norm is related to the number of subintervals of $[a, b]$ in the following way.

$$\frac{b-a}{\|\Delta\|} \le n \qquad \text{General partition}$$

So, the number of subintervals in a partition approaches infinity as the norm of the partition approaches 0. That is, $\|\Delta\| \to 0$ implies that $n \to \infty$.

The converse of this statement is not true. For example, let Δ_n be the partition of the interval $[0, 1]$ given by

$$0 < \frac{1}{2^n} < \frac{1}{2^{n-1}} < \cdot \cdot \cdot < \frac{1}{8} < \frac{1}{4} < \frac{1}{2} < 1.$$

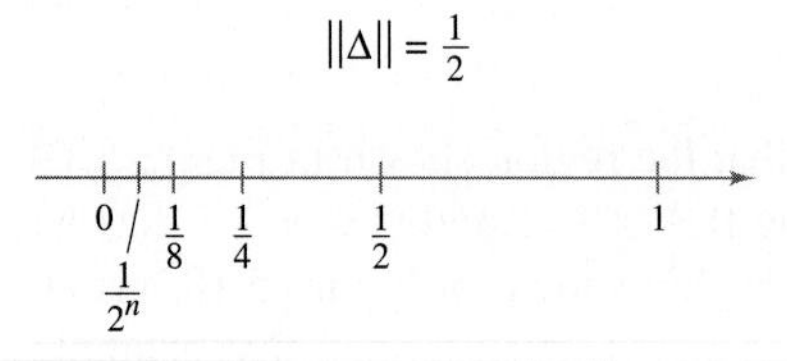

$n \to \infty$ does not imply that $\|\Delta\| \to 0$.
Figure 5.20

As shown in Figure 5.20, for any positive value of n, the norm of the partition Δ_n is $\frac{1}{2}$. So, letting n approach infinity does not force $\|\Delta\|$ to approach 0. In a regular partition, however, the statements $\|\Delta\| \to 0$ and $n \to \infty$ are equivalent.

Definite Integrals

To define the definite integral, consider the following limit.

$$\lim_{\|\Delta\| \to 0} \sum_{i=1}^{n} f(c_i)\,\Delta x_i = L$$

To say that this limit exists means that for $\varepsilon > 0$ there exists a $\delta > 0$ such that for every partition with $\|\Delta\| < \delta$ it follows that

$$\left| L - \sum_{i=1}^{n} f(c_i)\,\Delta x_i \right| < \varepsilon.$$

(This must be true for any choice of c_i in the ith subinterval of Δ.)

FOR FURTHER INFORMATION For insight into the history of the definite integral, see the article "The Evolution of Integration" by A. Shenitzer and J. Steprāns in The *American Mathematical Monthly*. To view this article, go to the website *www.matharticles.com*.

Definition of a Definite Integral

If f is defined on the closed interval $[a, b]$ and the limit

$$\lim_{\|\Delta\| \to 0} \sum_{i=1}^{n} f(c_i)\,\Delta x_i$$

exists (as described above), then f is **integrable** on $[a, b]$ and the limit is denoted by

$$\lim_{\|\Delta\| \to 0} \sum_{i=1}^{n} f(c_i)\,\Delta x_i = \int_a^b f(x)\,dx.$$

The limit is called the **definite integral** of f from a to b. The number a is the **lower limit** of integration, and the number b is the **upper limit** of integration.

It is not a coincidence that the notation used for definite integrals is similar to that used for indefinite integrals. You will see why in the next section when the Fundamental Theorem of Calculus is introduced. For now it is important to see that definite integrals and indefinite integrals are different identities. A definite integral is a *number*, whereas an indefinite integral is a *family of functions*.

A sufficient condition for a function f to be integrable on $[a, b]$ is that it is continuous on $[a, b]$. A proof of this theorem is beyond the scope of this text.

THEOREM 5.4 Continuity Implies Integrability

If a function f is continuous on the closed interval $[a, b]$, then f is integrable on $[a, b]$.

EXPLORATION

The Converse of Theorem 5.4 Is the converse of Theorem 5.4 true? That is, if a function is integrable, does it have to be continuous? Explain your reasoning and give examples.

Describe the relationships among continuity, differentiability, and integrability. Which is the strongest condition? Which is the weakest? Which conditions imply other conditions?

EXAMPLE 2 Evaluating a Definite Integral as a Limit

Evaluate the definite integral $\int_{-2}^{1} 2x\,dx$.

Solution The function $f(x) = 2x$ is integrable on the interval $[-2, 1]$ because it is continuous on $[-2, 1]$. Moreover, the definition of integrability implies that any partition whose norm approaches 0 can be used to determine the limit. For computational convenience, define Δ by subdividing $[-2, 1]$ into n subintervals of equal width

$$\Delta x_i = \Delta x = \frac{b - a}{n} = \frac{3}{n}.$$

Choosing c_i as the right endpoint of each subinterval produces

$$c_i = a + i(\Delta x) = -2 + \frac{3i}{n}.$$

So, the definite integral is given by

$$\begin{aligned}
\int_{-2}^{1} 2x\,dx &= \lim_{\|\Delta\|\to 0} \sum_{i=1}^{n} f(c_i)\,\Delta x_i \\
&= \lim_{n\to\infty} \sum_{i=1}^{n} f(c_i)\,\Delta x \\
&= \lim_{n\to\infty} \sum_{i=1}^{n} 2\left(-2 + \frac{3i}{n}\right)\left(\frac{3}{n}\right) \\
&= \lim_{n\to\infty} \frac{6}{n} \sum_{i=1}^{n} \left(-2 + \frac{3i}{n}\right) \\
&= \lim_{n\to\infty} \frac{6}{n}\left\{-2n + \frac{3}{n}\left[\frac{n(n+1)}{2}\right]\right\} \\
&= \lim_{n\to\infty} \left(-12 + 9 + \frac{9}{n}\right) \\
&= -3.
\end{aligned}$$

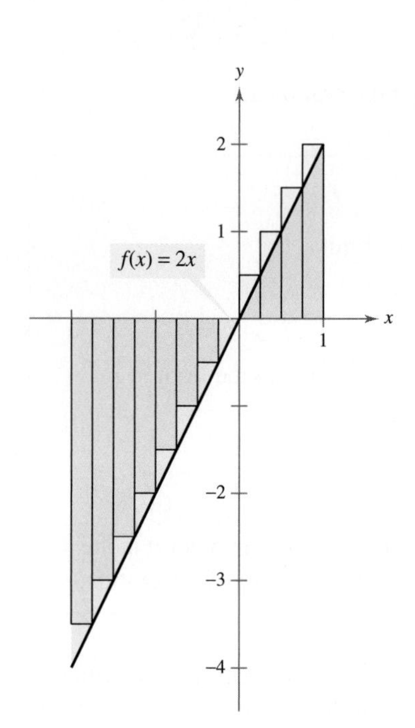

Because the definite integral is negative, it does not represent the area of the region.
Figure 5.21

Because the definite integral in Example 2 is negative, it *cannot* represent the area of the region shown in Figure 5.21. Definite integrals can be positive, negative, or zero. For a definite integral to be interpreted as an area (as defined in Section 5.2), the function f must be continuous and nonnegative on $[a, b]$, as stated in the following theorem. (The proof of this theorem is straightforward—you simply use the definition of area given in Section 5.2.)

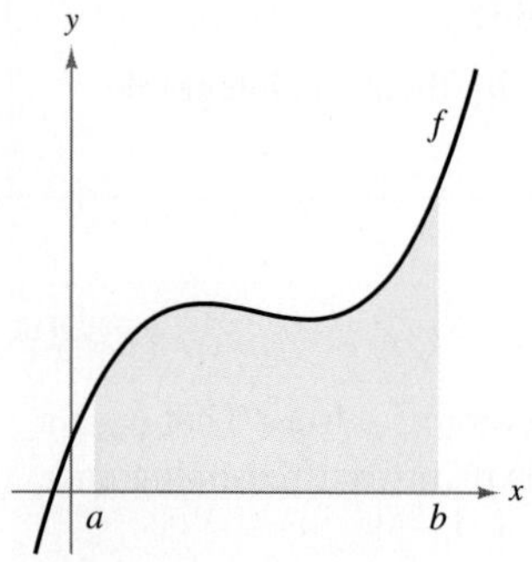

> **THEOREM 5.5 The Definite Integral as the Area of a Region**
>
> If f is continuous and nonnegative on the closed interval $[a, b]$, then the area of the region bounded by the graph of f, the x-axis, and the vertical lines $x = a$ and $x = b$ is given by
>
> $$\text{Area} = \int_a^b f(x)\,dx.$$
>
> (See Figure 5.22.)

You can use a definite integral to find the area of the region bounded by the graph of f, the x-axis, $x = a$, and $x = b$.
Figure 5.22

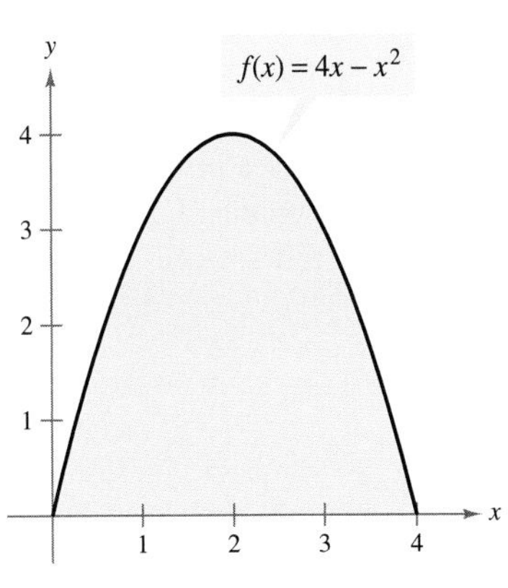

$\text{Area} = \int_0^4 (4x - x^2)\, dx$

Figure 5.23

As an example of Theorem 5.5, consider the region bounded by the graph of

$$f(x) = 4x - x^2$$

and the x-axis, as shown in Figure 5.23. Because f is continuous and nonnegative on the closed interval $[0, 4]$, the area of the region is

$$\text{Area} = \int_0^4 (4x - x^2)\, dx.$$

A straightforward technique for evaluating a definite integral such as this will be discussed in Section 5.4. For now, however, you can evaluate a definite integral in two ways—you can use the limit definition *or* you can check to see whether the definite integral represents the area of a common geometric region such as a rectangle, triangle, or semicircle.

EXAMPLE 3 Areas of Common Geometric Figures

Sketch the region corresponding to each definite integral. Then evaluate each integral using a geometric formula.

a. $\int_1^3 4\, dx$ **b.** $\int_0^3 (x + 2)\, dx$ **c.** $\int_{-2}^2 \sqrt{4 - x^2}\, dx$

Solution A sketch of each region is shown in Figure 5.24.

a. This region is a rectangle of height 4 and width 2.

$$\int_1^3 4\, dx = (\text{Area of rectangle}) = 4(2) = 8$$

b. This region is a trapezoid with an altitude of 3 and parallel bases of lengths 2 and 5. The formula for the area of a trapezoid is $\frac{1}{2}h(b_1 + b_2)$.

$$\int_0^3 (x + 2)\, dx = (\text{Area of trapezoid}) = \frac{1}{2}(3)(2 + 5) = \frac{21}{2}$$

c. This region is a semicircle of radius 2. The formula for the area of a semicircle is $\frac{1}{2}\pi r^2$.

$$\int_{-2}^2 \sqrt{4 - x^2}\, dx = (\text{Area of semicircle}) = \frac{1}{2}\pi(2^2) = 2\pi$$

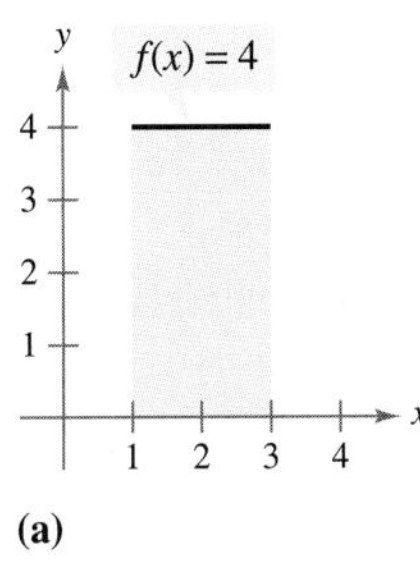

(a)

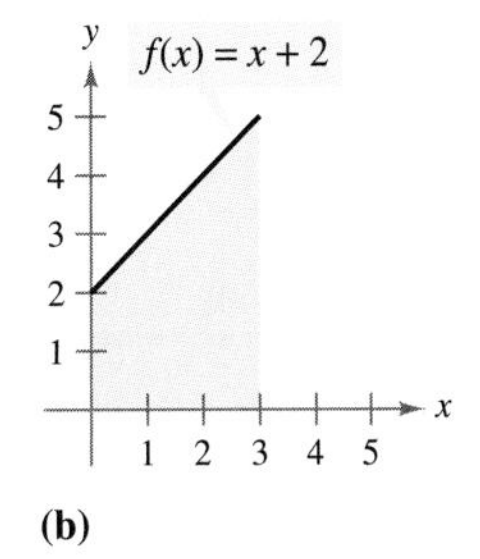

(b)

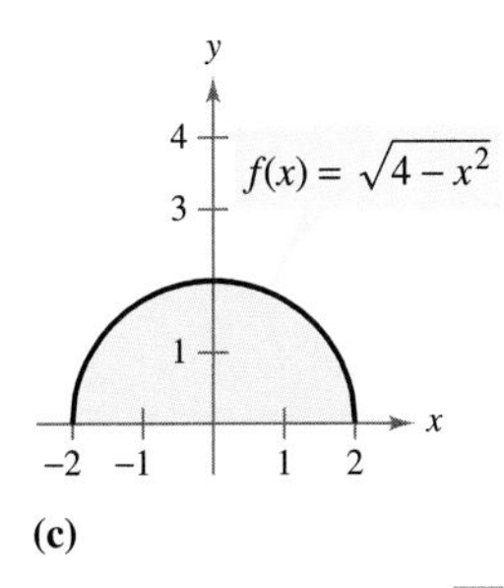

(c)

Figure 5.24

NOTE The variable of integration in a definite integral is sometimes called a *dummy variable* because it can be replaced by any other variable without changing the value of the integral. For instance, the definite integrals

$$\int_0^3 (x + 2)\, dx$$

and

$$\int_0^3 (t + 2)\, dt$$

have the same value.

Properties of Definite Integrals

The definition of the definite integral of f on the interval $[a, b]$ specifies that $a < b$. Now, however, it is convenient to extend the definition to cover cases in which $a = b$ or $a > b$. Geometrically, the following two definitions seem reasonable. For instance, it makes sense to define the area of a region of zero width and finite height to be 0.

Definitions of Two Special Definite Integrals

1. If f is defined at $x = a$, then we define $\int_a^a f(x)\,dx = 0$.

2. If f is integrable on $[a, b]$, then we define $\int_b^a f(x)\,dx = -\int_a^b f(x)\,dx$.

EXAMPLE 4 Evaluating Definite Integrals

a. Because the sine function is defined at $x = \pi$, and the upper and lower limits of integration are equal, you can write

$$\int_\pi^\pi \sin x\,dx = 0.$$

b. The integral $\int_3^0 (x + 2)\,dx$ is the same as that given in Example 3(b) except that the upper and lower limits are interchanged. Because the integral in Example 3(b) has a value of $\frac{21}{2}$, you can write

$$\int_3^0 (x + 2)\,dx = -\int_0^3 (x + 2)\,dx = -\frac{21}{2}.$$

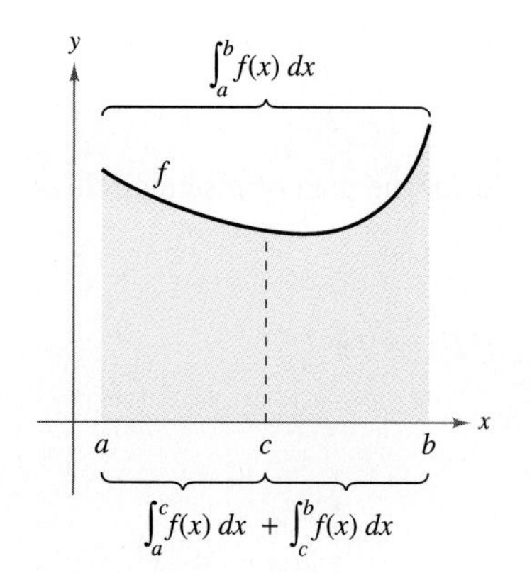

Figure 5.25

In Figure 5.25, the larger region can be divided at $x = c$ into two subregions whose intersection is a line segment. Because the line segment has zero area, it follows that the area of the larger region is equal to the sum of the areas of the two smaller regions.

THEOREM 5.6 Additive Interval Property

If f is integrable on the three closed intervals determined by a, b, and c, then

$$\int_a^b f(x)\,dx = \int_a^c f(x)\,dx + \int_c^b f(x)\,dx.$$

EXAMPLE 5 Using the Additive Interval Property

$$\int_{-1}^1 |x|\,dx = \int_{-1}^0 -x\,dx + \int_0^1 x\,dx \qquad \text{Theorem 5.6}$$

$$= \frac{1}{2} + \frac{1}{2} \qquad \text{Area of a triangle}$$

$$= 1$$

Because the definite integral is defined as the limit of a sum, it inherits the properties of summation given at the top of page 296.

THEOREM 5.7 Properties of Definite Integrals

If f and g are integrable on $[a, b]$ and k is a constant, then the functions of kf and $f \pm g$ are integrable on $[a, b]$, and

1. $\displaystyle\int_a^b kf(x)\,dx = k\int_a^b f(x)\,dx$

2. $\displaystyle\int_a^b [f(x) \pm g(x)]\,dx = \int_a^b f(x)\,dx \pm \int_a^b g(x)\,dx.$

Note that Property 2 of Theorem 5.7 can be extended to cover any finite number of functions. For example,

$$\int_a^b [f(x) + g(x) + h(x)]\,dx = \int_a^b f(x)\,dx + \int_a^b g(x)\,dx + \int_a^b h(x)\,dx.$$

EXAMPLE 6 Evaluation of a Definite Integral

Evaluate $\displaystyle\int_1^3 (-x^2 + 4x - 3)\,dx$ using each of the following values.

$$\int_1^3 x^2\,dx = \frac{26}{3}, \qquad \int_1^3 x\,dx = 4, \qquad \int_1^3 dx = 2$$

Solution

$$\begin{aligned}\int_1^3 (-x^2 + 4x - 3)\,dx &= \int_1^3 (-x^2)\,dx + \int_1^3 4x\,dx + \int_1^3 (-3)\,dx\\ &= -\int_1^3 x^2\,dx + 4\int_1^3 x\,dx - 3\int_1^3 dx\\ &= -\left(\frac{26}{3}\right) + 4(4) - 3(2)\\ &= \frac{4}{3}\end{aligned}$$

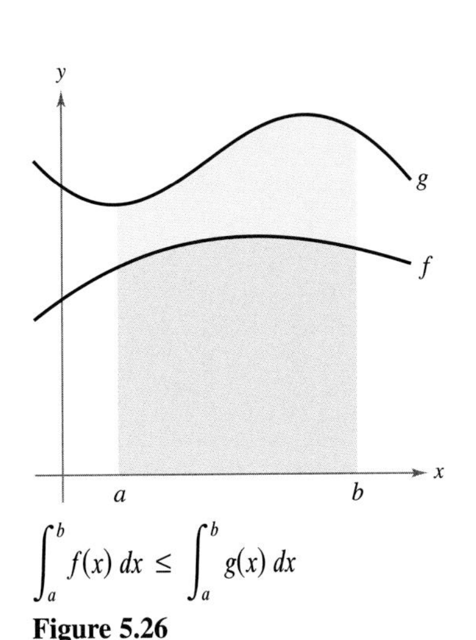

$\displaystyle\int_a^b f(x)\,dx \le \int_a^b g(x)\,dx$

Figure 5.26

If f and g are continuous on the closed interval $[a, b]$ and

$0 \le f(x) \le g(x)$

for $a \le x \le b$, the following properties are true. First, the area of the region bounded by the graph of f and the x-axis (between a and b) must be nonnegative. Second, this area must be less than or equal to the area of the region bounded by the graph of g and the x-axis (between a and b), as shown in Figure 5.26. These two results are generalized in Theorem 5.8. (A proof of this theorem is given in Appendix A.)

THEOREM 5.8 Preservation of Inequality

1. If f is integrable and nonnegative on the closed interval $[a, b]$, then

$$0 \le \int_a^b f(x)\,dx.$$

2. If f and g are integrable on the closed interval $[a, b]$ and $f(x) \le g(x)$ for every x in $[a, b]$, then

$$\int_a^b f(x)\,dx \le \int_a^b g(x)\,dx.$$

Exercises for Section 5.3

See www.CalcChat.com for worked-out solutions to odd-numbered exercises.

In Exercises 1 and 2, use Example 1 as a model to evaluate the limit

$$\lim_{n\to\infty} \sum_{i=1}^{n} f(c_i)\,\Delta x_i$$

over the region bounded by the graphs of the equations.

1. $f(x) = \sqrt{x},\quad y = 0,\quad x = 0,\quad x = 3$
   (*Hint:* Let $c_i = 3i^2/n^2$.)
2. $f(x) = 2\sqrt[3]{x},\quad y = 0,\quad x = 0,\quad x = 1$
   (*Hint:* Let $c_i = i^3/n^3$.)

In Exercises 3–8, evaluate the definite integral by the limit definition.

3. $\displaystyle\int_4^{10} 6\,dx$
4. $\displaystyle\int_{-2}^{3} x\,dx$
5. $\displaystyle\int_{-1}^{1} x^3\,dx$
6. $\displaystyle\int_1^3 3x^2\,dx$
7. $\displaystyle\int_1^2 (x^2 + 1)\,dx$
8. $\displaystyle\int_{-1}^{2} (3x^2 + 2)\,dx$

In Exercises 9–14, write the limit as a definite integral on the interval $[a, b]$, where c_i is any point in the ith subinterval.

	Limit	*Interval*
9.	$\displaystyle\lim_{\|\Delta\|\to 0} \sum_{i=1}^{n} (3c_i + 10)\,\Delta x_i$	$[-1, 5]$
10.	$\displaystyle\lim_{\|\Delta\|\to 0} \sum_{i=1}^{n} 6c_i(4 - c_i)^2\,\Delta x_i$	$[0, 4]$
11.	$\displaystyle\lim_{\|\Delta\|\to 0} \sum_{i=1}^{n} \sqrt{c_i^2 + 4}\,\Delta x_i$	$[0, 3]$
12.	$\displaystyle\lim_{\|\Delta\|\to 0} \sum_{i=1}^{n} \left(\frac{3}{c_i^2}\right)\Delta x_i$	$[1, 3]$
13.	$\displaystyle\lim_{\|\Delta\|\to 0} \sum_{i=1}^{n} \left(1 + \frac{3}{c_i}\right)\Delta x_i$	$[1, 5]$
14.	$\displaystyle\lim_{\|\Delta\|\to 0} \sum_{i=1}^{n} (2^{-c_i}\sin c_i)\,\Delta x_i$	$[0, \pi]$

In Exercises 15–22, set up a definite integral that yields the area of the region. (Do not evaluate the integral.)

15. $f(x) = 3$

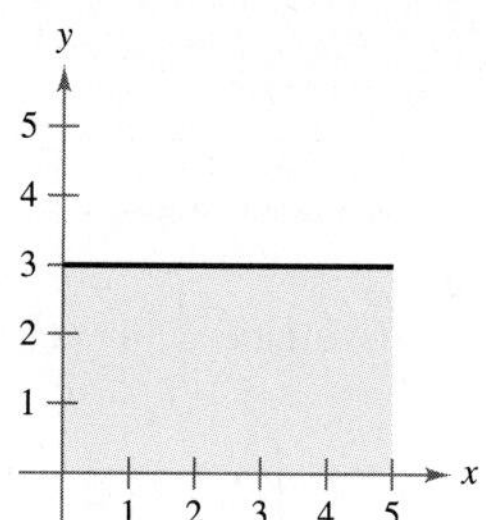

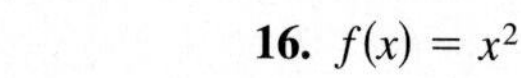

16. $f(x) = x^2$

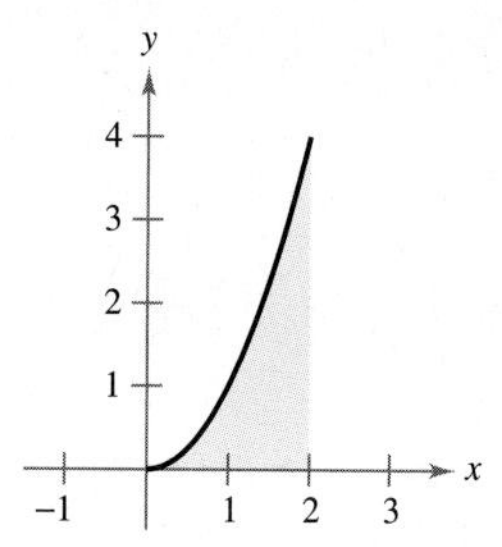

17. $f(x) = \dfrac{2}{x}$

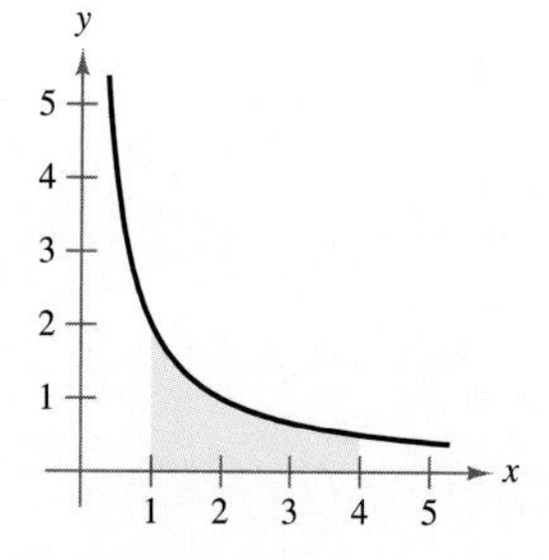

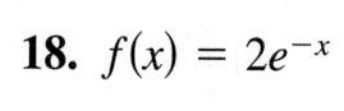

18. $f(x) = 2e^{-x}$

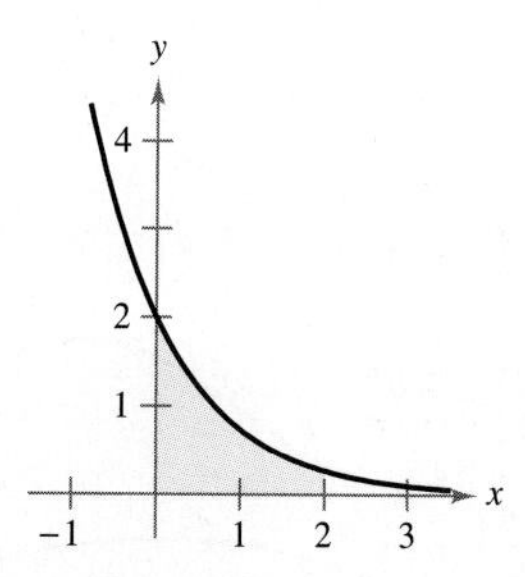

19. $f(x) = \sin x$

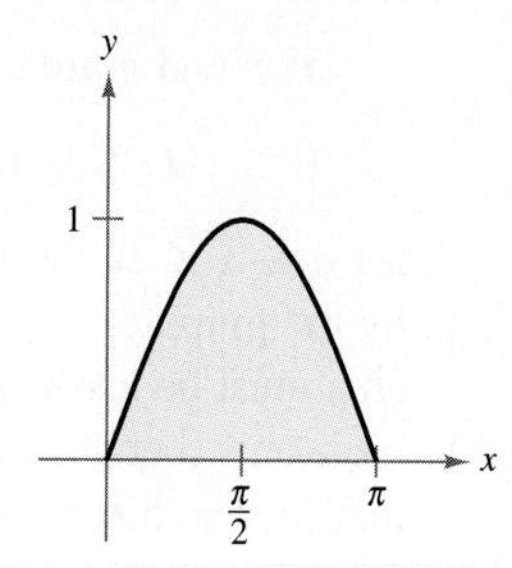

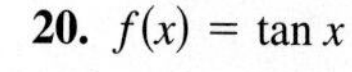

20. $f(x) = \tan x$

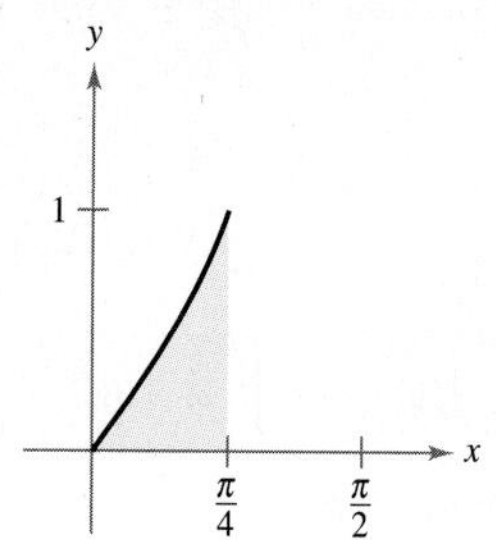

21. $g(y) = y^3$

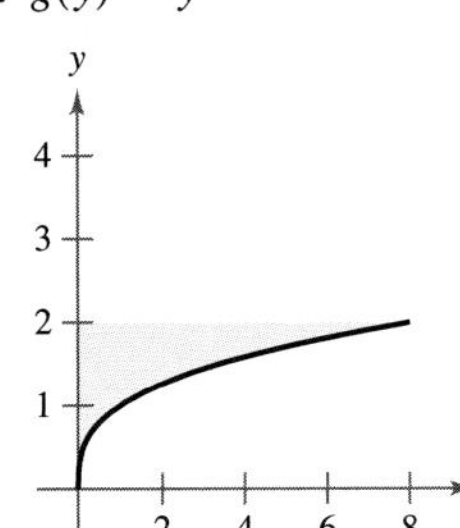

22. $f(y) = (y - 2)^2$

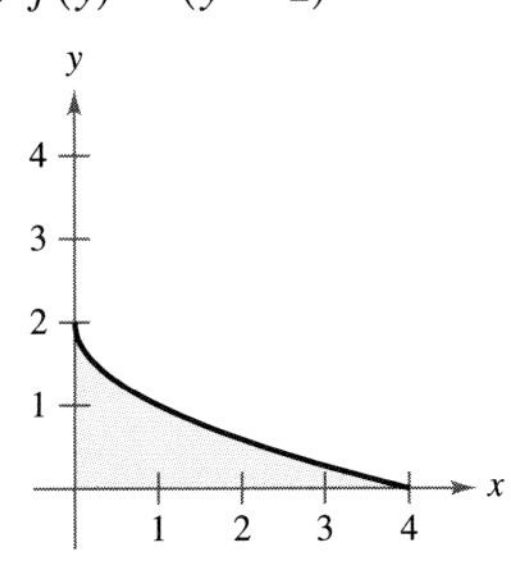

In Exercises 23–32, sketch the region whose area is given by the definite integral. Then use a geometric formula to evaluate the integral ($a > 0, r > 0$).

23. $\int_0^3 4\,dx$ **24.** $\int_{-a}^a 4\,dx$

25. $\int_0^4 x\,dx$ **26.** $\int_0^4 \frac{x}{2}\,dx$

27. $\int_0^2 (2x + 5)\,dx$ **28.** $\int_0^8 (8 - x)\,dx$

29. $\int_{-1}^1 (1 - |x|)\,dx$ **30.** $\int_{-a}^a (a - |x|)\,dx$

31. $\int_{-3}^3 \sqrt{9 - x^2}\,dx$ **32.** $\int_{-r}^r \sqrt{r^2 - x^2}\,dx$

In Exercises 33–40, evaluate the integral using the following values.

$$\int_2^4 x^3\,dx = 60, \quad \int_2^4 x\,dx = 6, \quad \int_2^4 dx = 2$$

33. $\int_4^2 x\,dx$ **34.** $\int_2^2 x^3\,dx$

35. $\int_2^4 4x\,dx$ **36.** $\int_2^4 15\,dx$

37. $\int_2^4 (x - 8)\,dx$ **38.** $\int_2^4 (x^3 + 4)\,dx$

39. $\int_2^4 \left(\frac{1}{2}x^3 - 3x + 2\right)\,dx$ **40.** $\int_2^4 (6 + 2x - x^3)\,dx$

41. Given $\int_0^5 f(x)\,dx = 10$ and $\int_5^7 f(x)\,dx = 3$, evaluate

(a) $\int_0^7 f(x)\,dx.$ (b) $\int_5^0 f(x)\,dx.$

(c) $\int_5^5 f(x)\,dx.$ (d) $\int_0^5 3f(x)\,dx.$

42. Given $\int_0^3 f(x)\,dx = 4$ and $\int_3^6 f(x)\,dx = -1$, evaluate

(a) $\int_0^6 f(x)\,dx.$ (b) $\int_6^3 f(x)\,dx.$

(c) $\int_3^3 f(x)\,dx.$ (d) $\int_3^6 -5f(x)\,dx.$

43. Given $\int_2^6 f(x)\,dx = 10$ and $\int_2^6 g(x)\,dx = -2$, evaluate

(a) $\int_2^6 [f(x) + g(x)]\,dx.$ (b) $\int_2^6 [g(x) - f(x)]\,dx.$

(c) $\int_2^6 2g(x)\,dx.$ (d) $\int_2^6 3f(x)\,dx.$

44. Given $\int_{-1}^1 f(x)\,dx = 0$ and $\int_0^1 f(x)\,dx = 5$, evaluate

(a) $\int_{-1}^0 f(x)\,dx.$ (b) $\int_0^1 f(x)\,dx - \int_{-1}^0 f(x)\,dx.$

(c) $\int_{-1}^1 3f(x)\,dx.$ (d) $\int_0^1 3f(x)\,dx.$

45. Use the table of values to find lower and upper estimates of

$$\int_0^{10} f(x)\,dx.$$

Assume that f is a decreasing function.

x	0	2	4	6	8	10
$f(x)$	32	24	12	-4	-20	-36

46. Use the table of values to estimate

$$\int_0^6 f(x)\,dx.$$

Use three equal subintervals and the (a) left endpoints, (b) right endpoints, and (c) midpoints. If f is an increasing function, how does each estimate compare with the actual value? Explain your reasoning.

x	0	1	2	3	4	5	6
$f(x)$	-6	0	8	18	30	50	80

47. ***Think About It*** The graph of f consists of line segments and a semicircle, as shown in the figure. Evaluate each definite integral by using geometric formulas.

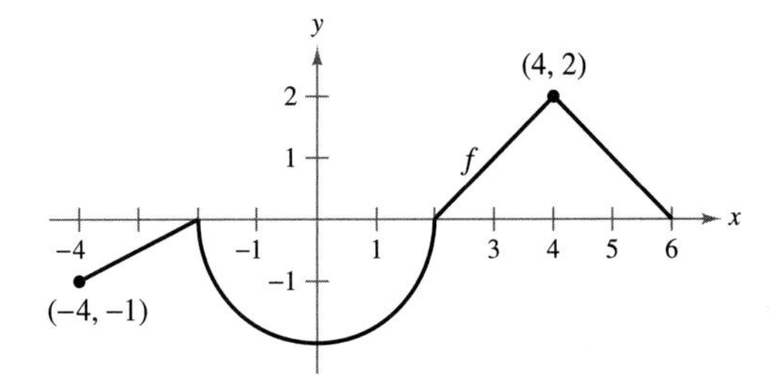

(a) $\int_0^2 f(x)\,dx$ (b) $\int_2^6 f(x)\,dx$

(c) $\int_{-4}^2 f(x)\,dx$ (d) $\int_{-4}^6 f(x)\,dx$

(e) $\int_{-4}^6 |f(x)|\,dx$ (f) $\int_{-4}^6 [f(x) + 2]\,dx$

48. ***Think About It*** The graph of f consists of line segments, as shown in the figure. Evaluate each definite integral by using geometric formulas.

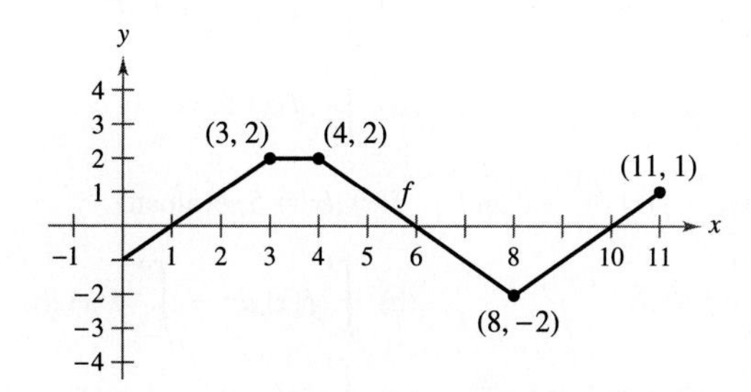

(a) $\int_0^1 -f(x)\,dx$ (b) $\int_3^4 3f(x)\,dx$

(c) $\int_0^7 f(x)\,dx$ (d) $\int_5^{11} f(x)\,dx$

(e) $\int_0^{11} f(x)\,dx$ (f) $\int_4^{10} f(x)\,dx$

49. ***Think About It*** Consider the function f that is continuous on the interval $[-5, 5]$ and for which

$$\int_0^5 f(x)\,dx = 4.$$

Evaluate each integral.

(a) $\int_0^5 [f(x) + 2]\,dx$ (b) $\int_{-2}^3 f(x + 2)\,dx$

(c) $\int_{-5}^5 f(x)\,dx$ (f is even.) (d) $\int_{-5}^5 f(x)\,dx$ (f is odd.)

50. ***Think About It*** A function f is defined below. Use geometric formulas to find $\int_0^8 f(x)\,dx$.

$$f(x) = \begin{cases} 4, & x < 4 \\ x, & x \geq 4 \end{cases}$$

Writing About Concepts

In Exercises 51 and 52, use the figure to fill in the blank with the symbol <, >, or =.

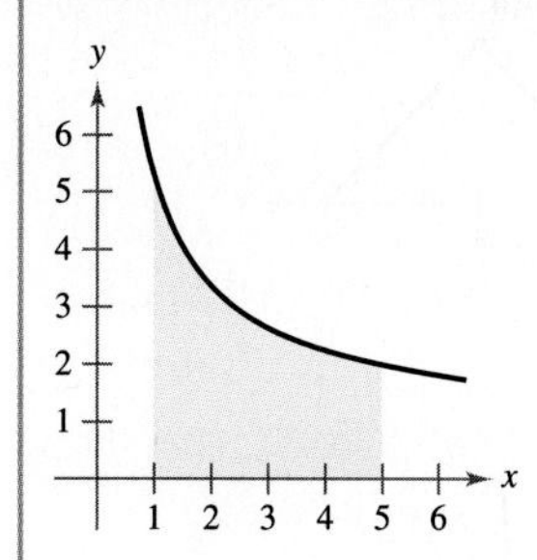

51. The interval $[1, 5]$ is partitioned into n subintervals of equal width Δx, and x_i is the left endpoint of the ith subinterval.

$$\sum_{i=1}^{n} f(x_i)\,\Delta x \quad \square \quad \int_1^5 f(x)\,dx$$

Writing About Concepts *(continued)*

52. The interval $[1, 5]$ is partitioned into n subintervals of equal width Δx, and x_i is the right endpoint of the ith subinterval.

$$\sum_{i=1}^{n} f(x_i)\,\Delta x \quad \square \quad \int_1^5 f(x)\,dx$$

53. Determine whether the function $f(x) = \dfrac{1}{x - 4}$ is integrable on the interval $[3, 5]$. Explain.

54. Give an example of a function that is integrable on the interval $[-1, 1]$, but not continuous on $[-1, 1]$.

In Exercises 55–58, determine which value best approximates the definite integral. Make your selection on the basis of a sketch.

55. $\int_0^4 \sqrt{x}\,dx$

(a) 5 (b) -3 (c) 10 (d) 2 (e) 8

56. $\int_0^{1/2} 4\cos \pi x\,dx$

(a) 4 (b) $\frac{4}{3}$ (c) 16 (d) 2π (e) -6

57. $\int_0^2 2e^{-x^2}\,dx$

(a) $\frac{1}{3}$ (b) 6 (c) 2 (d) 4

58. $\int_1^2 \ln x\,dx$

(a) $\frac{1}{3}$ (b) 1 (c) 4 (d) 3

Programming **Write a program for your graphing utility to approximate a definite integral using the Riemann sum**

$$\sum_{i=1}^{n} f(c_i)\,\Delta x_i$$

where the subintervals are of equal width. The output should give three approximations of the integral where c_i is the left-hand endpoint $L(n)$, midpoint $M(n)$, and right-hand endpoint $R(n)$ of each subinterval. In Exercises 59–64, use the program to approximate the definite integral and complete the table.

n	4	8	12	16	20
$L(n)$					
$M(n)$					
$R(n)$					

59. $\int_0^3 x\sqrt{3 - x}\,dx$

60. $\int_0^3 \dfrac{5}{x^2 + 1}\,dx$

61. $\int_1^3 \dfrac{1}{x}\,dx$

62. $\int_0^4 e^x\,dx$

63. $\int_0^{\pi/2} \sin^2 x \, dx$

64. $\int_0^3 x \sin x \, dx$

True or False? **In Exercises 65–70, determine whether the statement is true or false. If it is false, explain why or give an example that shows it is false.**

65. $\int_a^b [f(x) + g(x)] \, dx = \int_a^b f(x) \, dx + \int_a^b g(x) \, dx$

66. $\int_a^b f(x)g(x) \, dx = \left[\int_a^b f(x) \, dx\right]\left[\int_a^b g(x) \, dx\right]$

67. If the norm of a partition approaches zero, then the number of subintervals approaches infinity.

68. If f is increasing on $[a, b]$, then the minimum value of $f(x)$ on $[a, b]$ is $f(a)$.

69. The value of

$$\int_a^b f(x) \, dx$$

must be positive.

70. The value of

$$\int_2^2 \sin(x^2) \, dx$$

is 0.

71. Find the Riemann sum for $f(x) = x^2 + 3x$ over the interval $[0, 8]$, where $x_0 = 0$, $x_1 = 1$, $x_2 = 3$, $x_3 = 7$, and $x_4 = 8$, and where $c_1 = 1$, $c_2 = 2$, $c_3 = 5$, and $c_4 = 8$.

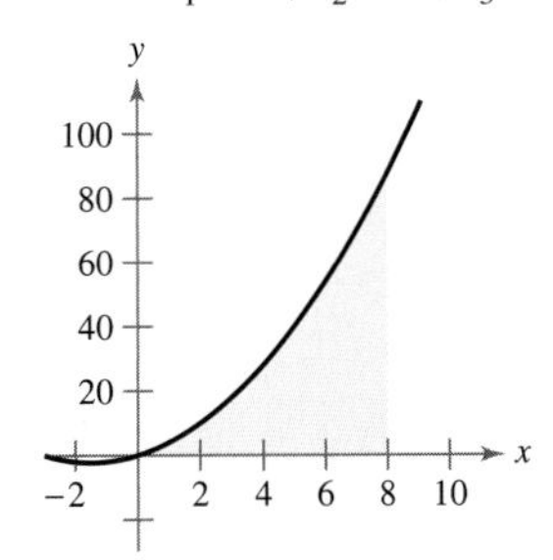

72. Find the Riemann sum for $f(x) = \sin x$ over the interval $[0, 2\pi]$, where $x_0 = 0$, $x_1 = \pi/4$, $x_2 = \pi/3$, $x_3 = \pi$, and $x_4 = 2\pi$, and where $c_1 = \pi/6$, $c_2 = \pi/3$, $c_3 = 2\pi/3$, and $c_4 = 3\pi/2$.

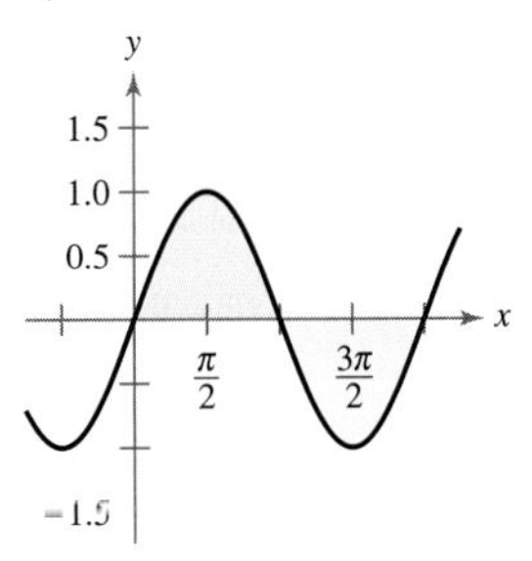

73. Prove that $\int_a^b x \, dx = \frac{b^2 - a^2}{2}$.

74. Prove that $\int_a^b x^2 \, dx = \frac{b^3 - a^3}{3}$.

75. ***Think About It*** Determine whether the Dirichlet function

$$f(x) = \begin{cases} 1, & x \text{ is rational} \\ 0, & x \text{ is irrational} \end{cases}$$

is integrable on the interval $[0, 1]$. Explain.

76. Suppose the function f is defined on $[0, 1]$, as shown in the figure.

$$f(x) = \begin{cases} 0, & x = 0 \\ \frac{1}{x}, & 0 < x \le 1 \end{cases}$$

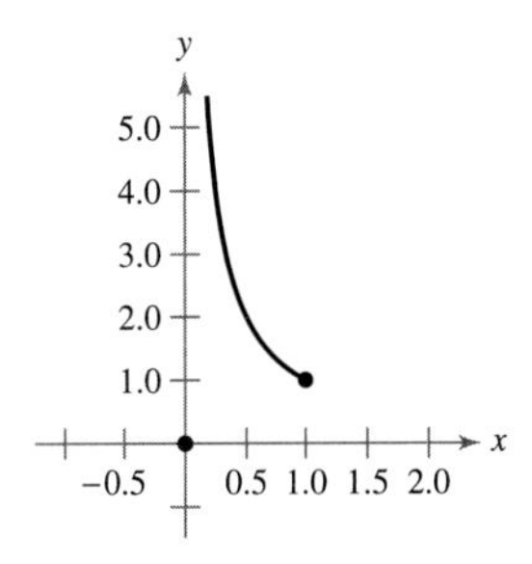

Show that $\int_0^1 f(x) \, dx$ does not exist. Why doesn't this contradict Theorem 5.4?

77. Find the constants a and b that maximize the value of

$$\int_a^b (1 - x^2) \, dx.$$

Explain your reasoning.

78. Evaluate, if possible, the integral $\int_0^2 [\![x]\!] \, dx$.

79. Determine

$$\lim_{n \to \infty} \frac{1}{n^3}[1^2 + 2^2 + 3^2 + \cdots + n^2]$$

by using an appropriate Riemann sum.

Section 5.4 The Fundamental Theorem of Calculus

- **Evaluate a definite integral using the Fundamental Theorem of Calculus.**
- **Understand and use the Mean Value Theorem for Integrals.**
- **Find the average value of a function over a closed interval.**
- **Understand and use the Second Fundamental Theorem of Calculus.**

The Fundamental Theorem of Calculus

You have now been introduced to the two major branches of calculus: differential calculus (introduced with the tangent line problem) and integral calculus (introduced with the area problem). At this point, these two problems might seem unrelated—but there is a very close connection. The connection was discovered independently by Isaac Newton and Gottfried Leibniz and is stated in a theorem that is appropriately called the **Fundamental Theorem of Calculus.**

Informally, the theorem states that differentiation and (definite) integration are inverse operations, in the same sense that division and multiplication are inverse operations. To see how Newton and Leibniz might have anticipated this relationship, consider the approximations shown in Figure 5.27. The slope of the tangent line was defined using the *quotient* $\Delta y/\Delta x$ (the slope of the secant line). Similarly, the area of a region under a curve was defined using the *product* $\Delta y \Delta x$ (the area of a rectangle). So, at least in the primitive approximation stage, the operations of differentiation and definite integration appear to have an inverse relationship in the same sense that division and multiplication are inverse operations. The Fundamental Theorem of Calculus states that the limit processes (used to define the derivative and definite integral) preserve this inverse relationship.

EXPLORATION

Integration and Antidifferentiation Throughout this chapter, you have been using the integral sign to denote an antiderivative (a family of functions) and a definite integral (a number).

Antidifferentiation: $\int f(x)\,dx$

Definite integration: $\int_a^b f(x)\,dx$

The use of this same symbol for both operations makes it appear that they are related. In the early work with calculus, however, it was not known that the two operations were related. Do you think the symbol $\int$ was first applied to antidifferentiation or to definite integration? Explain your reasoning. *(Hint:* The symbol was first used by Leibniz and was derived from the letter *S*.)

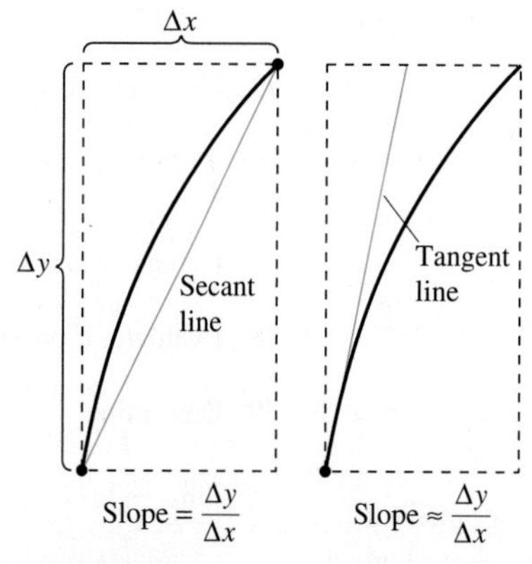

(a) Differentiation

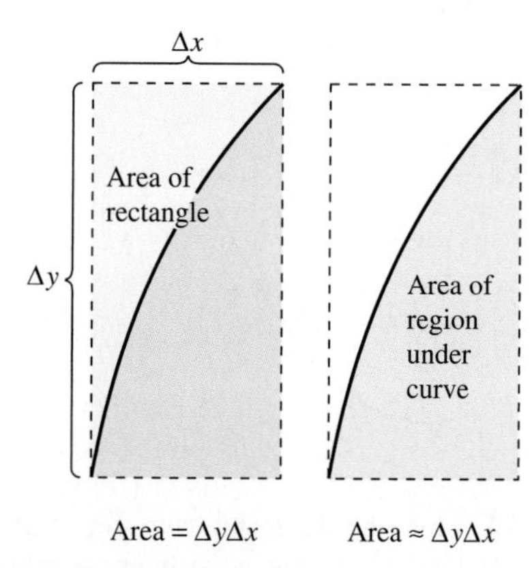

(b) Definite integration

Differentiation and definite integration have an "inverse" relationship.
Figure 5.27

THEOREM 5.9 The Fundamental Theorem of Calculus

If a function f is continuous on the closed interval $[a, b]$ and F is an antiderivative of f on the interval $[a, b]$, then

$$\int_a^b f(x)\,dx = F(b) - F(a).$$

Proof The key to the proof is in writing the difference $F(b) - F(a)$ in a convenient form. Let Δ be the following partition of $[a, b]$.

$$a = x_0 < x_1 < x_2 < \cdot \cdot \cdot < x_{n-1} < x_n = b$$

By pairwise subtraction and addition of like terms, you can write

$$\begin{aligned} F(b) - F(a) &= F(x_n) - F(x_{n-1}) + F(x_{n-1}) - \cdot \cdot \cdot - F(x_1) + F(x_1) - F(x_0) \\ &= \sum_{i=1}^{n} [F(x_i) - F(x_{i-1})]. \end{aligned}$$

By the Mean Value Theorem, you know that there exists a number c_i in the ith subinterval such that

$$F'(c_i) = \frac{F(x_i) - F(x_{i-1})}{x_i - x_{i-1}}.$$

Because $F'(c_i) = f(c_i)$, you can let $\Delta x_i = x_i - x_{i-1}$ and obtain

$$F(b) - F(a) = \sum_{i=1}^{n} f(c_i)\,\Delta x_i.$$

This important equation tells you that by applying the Mean Value Theorem, you can always find a collection of c_i's such that the *constant* $F(b) - F(a)$ is a Riemann sum of f on $[a, b]$. Taking the limit (as $\|\Delta\| \to 0$) produces

$$F(b) - F(a) = \int_a^b f(x)\,dx.$$

The following guidelines can help you understand the use of the Fundamental Theorem of Calculus.

Guidelines for Using the Fundamental Theorem of Calculus

1. *Provided you can find* an antiderivative of f, you now have a way to evaluate a definite integral without having to use the limit of a sum.
2. When applying the Fundamental Theorem of Calculus, the following notation is convenient.

$$\begin{aligned} \int_a^b f(x)\,dx &= F(x)\Big]_a^b \\ &= F(b) - F(a) \end{aligned}$$

For instance, to evaluate $\int_1^3 x^3\,dx$, you can write

$$\int_1^3 x^3\,dx = \frac{x^4}{4}\Big]_1^3 = \frac{3^4}{4} - \frac{1^4}{4} = \frac{81}{4} - \frac{1}{4} = 20.$$

3. It is not necessary to include a constant of integration C in the antiderivative because

$$\begin{aligned} \int_a^b f(x)\,dx &= \Big[F(x) + C\Big]_a^b \\ &= [F(b) + C] - [F(a) + C] \\ &= F(b) - F(a). \end{aligned}$$

EXAMPLE 1 Evaluating a Definite Integral

Evaluate each definite integral.

a. $\displaystyle\int_1^2 (x^2 - 3)\,dx$ **b.** $\displaystyle\int_1^4 3\sqrt{x}\,dx$ **c.** $\displaystyle\int_0^{\pi/4} \sec^2 x\,dx$

Solution

a. $$\int_1^2 (x^2 - 3)\,dx = \left[\frac{x^3}{3} - 3x\right]_1^2 = \left(\frac{8}{3} - 6\right) - \left(\frac{1}{3} - 3\right) = -\frac{2}{3}$$

b. $$\int_1^4 3\sqrt{x}\,dx = 3\int_1^4 x^{1/2}\,dx = 3\left[\frac{x^{3/2}}{3/2}\right]_1^4 = 2(4)^{3/2} - 2(1)^{3/2} = 14$$

c. $$\int_0^{\pi/4} \sec^2 x\,dx = \tan x\Big]_0^{\pi/4} = 1 - 0 = 1$$

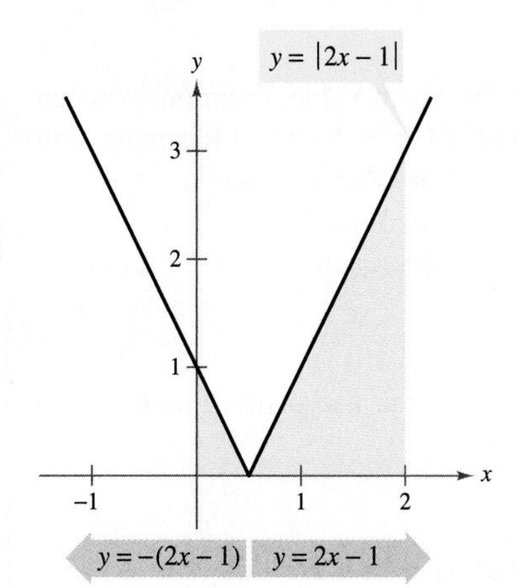

The definite integral of y on $[0, 2]$ is $\frac{5}{2}$.
Figure 5.28

EXAMPLE 2 A Definite Integral Involving Absolute Value

Evaluate $\displaystyle\int_0^2 |2x - 1|\,dx.$

Solution Using Figure 5.28 and the definition of absolute value, you can rewrite the integrand as shown.

$$|2x - 1| = \begin{cases} -(2x - 1), & x < \frac{1}{2} \\ 2x - 1, & x \geq \frac{1}{2} \end{cases}$$

From this, you can rewrite the integral in two parts.

$$\begin{aligned}
\int_0^2 |2x - 1|\,dx &= \int_0^{1/2} -(2x - 1)\,dx + \int_{1/2}^2 (2x - 1)\,dx \\
&= \Big[-x^2 + x\Big]_0^{1/2} + \Big[x^2 - x\Big]_{1/2}^2 \\
&= \left(-\frac{1}{4} + \frac{1}{2}\right) - (0 + 0) + (4 - 2) - \left(\frac{1}{4} - \frac{1}{2}\right) \\
&= \frac{5}{2}
\end{aligned}$$

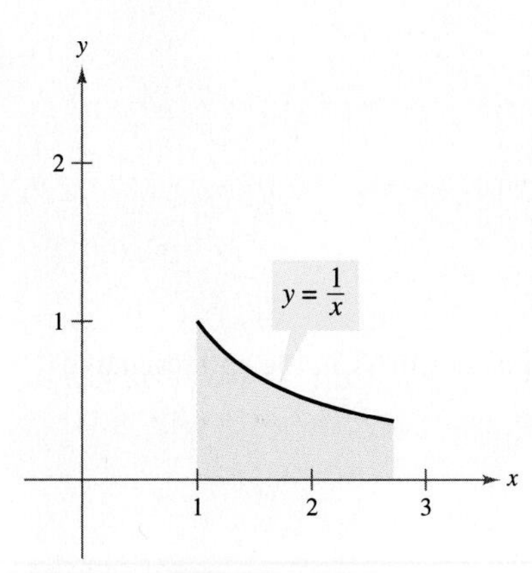

The area of the region bounded by the graph of $y = 1/x$, the x-axis, $x = 1$, and $x = e$ is 1.
Figure 5.29

EXAMPLE 3 Using the Fundamental Theorem to Find Area

Find the area of the region bounded by the graph of $y = 1/x$, the x-axis, and the vertical lines $x = 1$ and $x = e$, as shown in Figure 5.29.

Solution Note that $y > 0$ on the interval $[1, e]$.

$$\begin{aligned}
\text{Area} &= \int_1^e \frac{1}{x}\,dx && \text{Integrate between } x = 1 \text{ and } x = e. \\
&= \Big[\ln x\Big]_1^e && \text{Find antiderivative.} \\
&= (\ln e) - (\ln 1) && \text{Apply Fundamental Theorem of Calculus.} \\
&= 1 && \text{Simplify.}
\end{aligned}$$

The Mean Value Theorem for Integrals

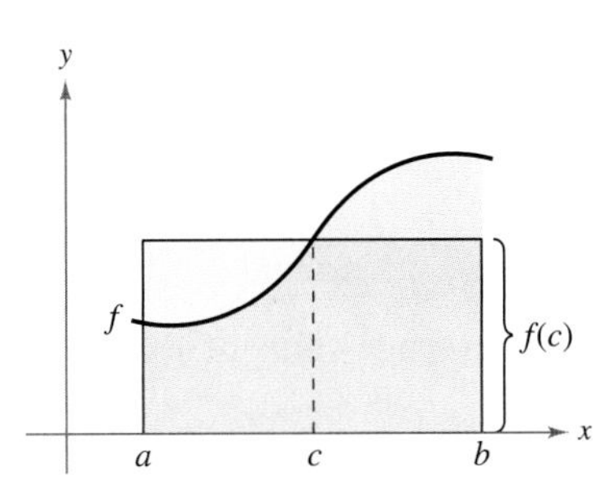

Mean value rectangle:

$$f(c)(b - a) = \int_a^b f(x)\,dx$$

Figure 5.30

In Section 5.2, you saw that the area of a region under a curve is greater than the area of an inscribed rectangle and less than the area of a circumscribed rectangle. The Mean Value Theorem for Integrals states that somewhere "between" the inscribed and circumscribed rectangles there is a rectangle whose area is precisely equal to the area of the region under the curve, as shown in Figure 5.30.

> **THEOREM 5.10 Mean Value Theorem for Integrals**
>
> If f is continuous on the closed interval $[a, b]$, then there exists a number c in the closed interval $[a, b]$ such that
>
> $$\int_a^b f(x)\,dx = f(c)(b - a).$$

Proof

Case 1: If f is constant on the interval $[a, b]$, the theorem is clearly valid because c can be any point in $[a, b]$.

Case 2: If f is not constant on $[a, b]$, then, by the Extreme Value Theorem, you can choose $f(m)$ and $f(M)$ to be the minimum and maximum values of f on $[a, b]$. Because $f(m) \le f(x) \le f(M)$ for all x in $[a, b]$, you can apply Theorem 5.8 to write the following.

$$\int_a^b f(m)\,dx \le \int_a^b f(x)\,dx \le \int_a^b f(M)\,dx \qquad \text{See Figure 5.31.}$$

$$f(m)(b - a) \le \int_a^b f(x)\,dx \le f(M)(b - a)$$

$$f(m) \le \frac{1}{b - a}\int_a^b f(x)\,dx \le f(M)$$

From the third inequality, you can apply the Intermediate Value Theorem to conclude that there exists some c in $[a, b]$ such that

$$f(c) = \frac{1}{b - a}\int_a^b f(x)\,dx \qquad \text{or} \qquad f(c)(b - a) = \int_a^b f(x)\,dx.$$

Inscribed rectangle (less than actual area)

$$\int_a^b f(m)\,dx = f(m)(b - a)$$

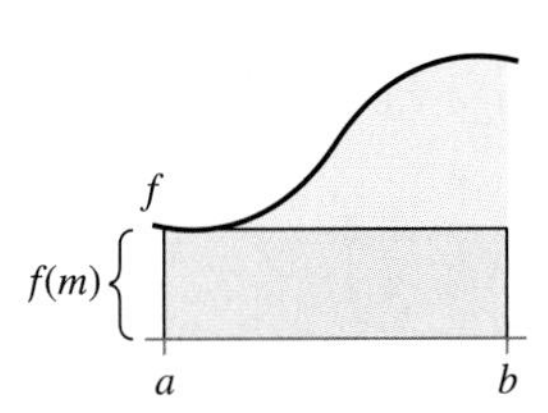

Mean value rectangle (equal to actual area)

$$\int_a^b f(x)\,dx$$

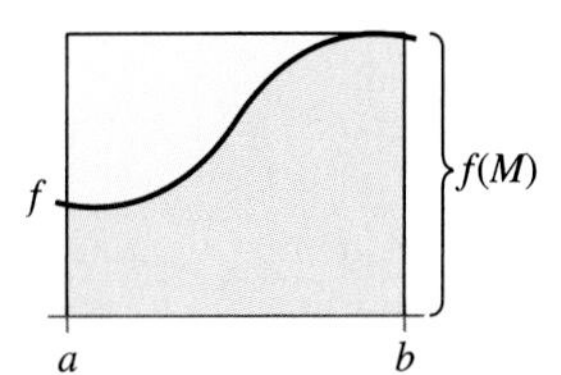

Circumscribed rectangle (greater than actual area)

$$\int_a^b f(M)\,dx = f(M)(b - a)$$

Figure 5.31

NOTE Notice that Theorem 5.10 does not specify how to determine c. It merely guarantees the existence of at least one number c in the interval.

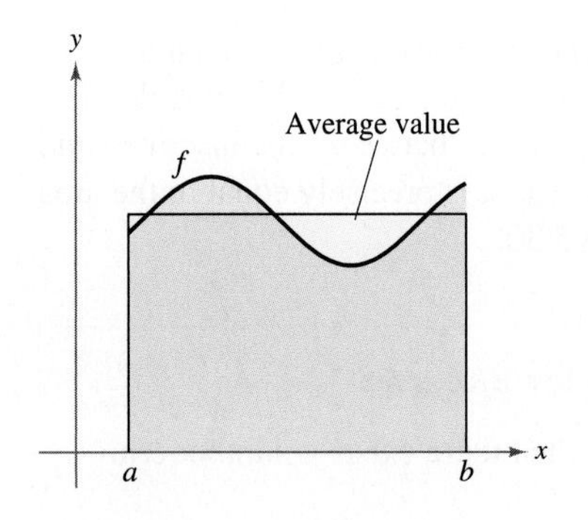

Average value $= \dfrac{1}{b-a}\displaystyle\int_a^b f(x)\,dx$

Figure 5.32

NOTE Notice in Figure 5.32 that the area of the region under the graph of f is equal to the area of the rectangle whose height is the average value.

Average Value of a Function

The value of $f(c)$ given in the Mean Value Theorem for Integrals is called the **average value** of f on the interval $[a, b]$.

> **Definition of the Average Value of a Function on an Interval**
>
> If f is integrable on the closed interval $[a, b]$, then the **average value** of f on the interval is
>
> $$\frac{1}{b-a}\int_a^b f(x)\,dx.$$

To see why the average value of f is defined in this way, suppose that you partition $[a, b]$ into n subintervals of equal width $\Delta x = (b-a)/n$. If c_i is any point in the ith subinterval, the arithmetic average (or mean) of the function values at the c_i's is given by

$$a_n = \frac{1}{n}[f(c_1) + f(c_2) + \cdot\cdot\cdot + f(c_n)].\qquad\text{Average of } f(c_1), \ldots, f(c_n)$$

By multiplying and dividing by $(b-a)$, you can write the average as

$$a_n = \frac{1}{n}\sum_{i=1}^{n} f(c_i)\left(\frac{b-a}{b-a}\right) = \frac{1}{b-a}\sum_{i=1}^{n} f(c_i)\left(\frac{b-a}{n}\right)$$
$$= \frac{1}{b-a}\sum_{i=1}^{n} f(c_i)\,\Delta x.$$

Finally, taking the limit as $n \rightarrow \infty$ produces the average value of f on the interval $[a, b]$, as given in the definition above.

This development of the average value of a function on an interval is only one of many practical uses of definite integrals to represent summation processes. In Chapter 7, you will study other applications, such as volume, arc length, centers of mass, and work.

EXAMPLE 4 Finding the Average Value of a Function

Find the average value of $f(x) = 3x^2 - 2x$ on the interval $[1, 4]$.

Solution The average value is given by

$$\begin{aligned}\frac{1}{b-a}\int_a^b f(x)\,dx &= \frac{1}{3}\int_1^4 (3x^2 - 2x)\,dx\\ &= \frac{1}{3}\Big[x^3 - x^2\Big]_1^4\\ &= \frac{1}{3}[64 - 16 - (1 - 1)]\\ &= \frac{48}{3}\\ &= 16.\end{aligned}$$

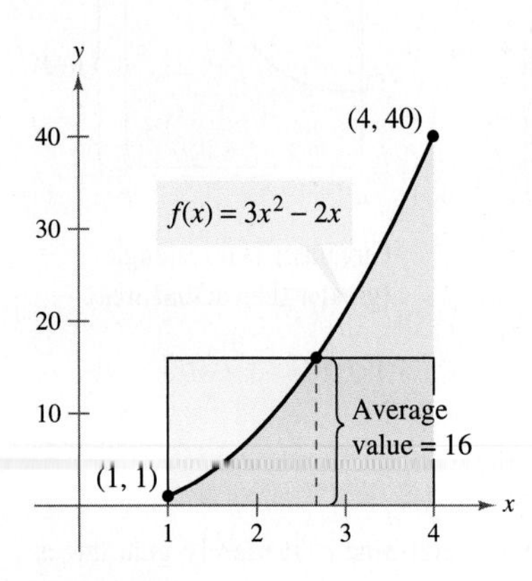

Figure 5.33

(See Figure 5.33.)

The first person to fly at a speed greater than the speed of sound was Charles Yeager. On October 14, 1947, Yeager was clocked at 295.9 meters per second at an altitude of 12.2 kilometers. If Yeager had been flying at an altitude below 11.275 kilometers, this speed would not have "broken the sound barrier." The photo above shows an F-14 *Tomcat*, a supersonic, twin-engine strike fighter. Currently, the *Tomcat* can reach heights of 15.24 kilometers and speeds up to 2 mach (707.78 meters per second).

EXAMPLE 5 The Speed of Sound

At different altitudes in Earth's atmosphere, sound travels at different speeds. The speed of sound $s(x)$ (in meters per second) can be modeled by

$$s(x) = \begin{cases} -4x + 341, & 0 \le x < 11.5 \\ 295, & 11.5 \le x < 22 \\ \frac{3}{4}x + 278.5, & 22 \le x < 32 \\ \frac{3}{2}x + 254.5, & 32 \le x < 50 \\ -\frac{3}{2}x + 404.5, & 50 \le x \le 80 \end{cases}$$

where x is the altitude in kilometers (see Figure 5.34). What is the average speed of sound over the interval $[0, 80]$?

Solution Begin by integrating $s(x)$ over the interval $[0, 80]$. To do this, you can break the integral into five parts.

$$\int_0^{11.5} s(x)\,dx = \int_0^{11.5} (-4x + 341)\,dx = \Big[-2x^2 + 341x\Big]_0^{11.5} = 3657$$

$$\int_{11.5}^{22} s(x)\,dx = \int_{11.5}^{22} (295)\,dx = \Big[295x\Big]_{11.5}^{22} = 3097.5$$

$$\int_{22}^{32} s(x)\,dx = \int_{22}^{32} \left(\tfrac{3}{4}x + 278.5\right)dx = \Big[\tfrac{3}{8}x^2 + 278.5x\Big]_{22}^{32} = 2987.5$$

$$\int_{32}^{50} s(x)\,dx = \int_{32}^{50} \left(\tfrac{3}{2}x + 254.5\right)dx = \Big[\tfrac{3}{4}x^2 + 254.5x\Big]_{32}^{50} = 5688$$

$$\int_{50}^{80} s(x)\,dx = \int_{50}^{80} \left(-\tfrac{3}{2}x + 404.5\right)dx = \Big[-\tfrac{3}{4}x^2 + 404.5x\Big]_{50}^{80} = 9210$$

By adding the values of the five integrals, you have

$$\int_0^{80} s(x)\,dx = 24{,}640.$$

So, the average speed of sound from an altitude of 0 kilometers to an altitude of 80 kilometers is

$$\text{Average speed} = \frac{1}{80}\int_0^{80} s(x)\,dx = \frac{24{,}640}{80} = 308 \text{ meters per second.}$$

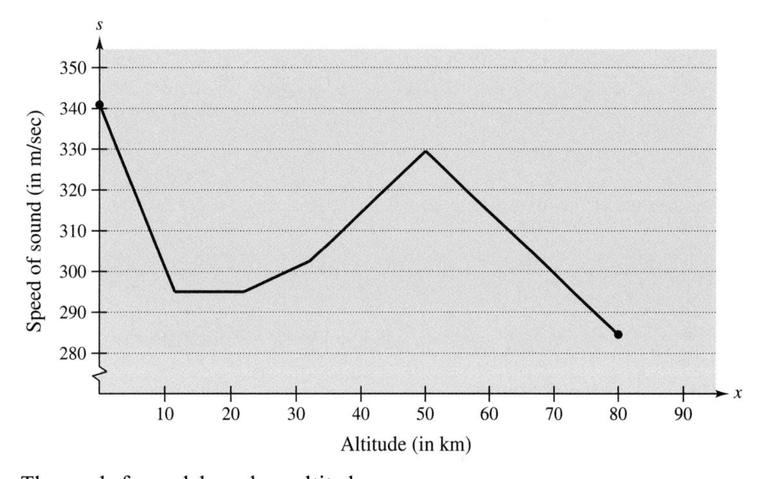

The speed of sound depends on altitude.
Figure 5.34

The Second Fundamental Theorem of Calculus

Earlier you saw that the definite integral of f on the interval $[a, b]$ was defined using the constant b as the upper limit of integration and x as the variable of integration. However, a slightly different situation may arise in which the variable x is used as the upper limit of integration. To avoid the confusion of using x in two different ways, t is temporarily used as the variable of integration. (Remember that the definite integral is *not* a function of its variable of integration.)

The Definite Integral as a Number

Constant

$$\int_a^b f(x)\,dx$$

Constant | f is a function of x.

The Definite Integral as a Function of x

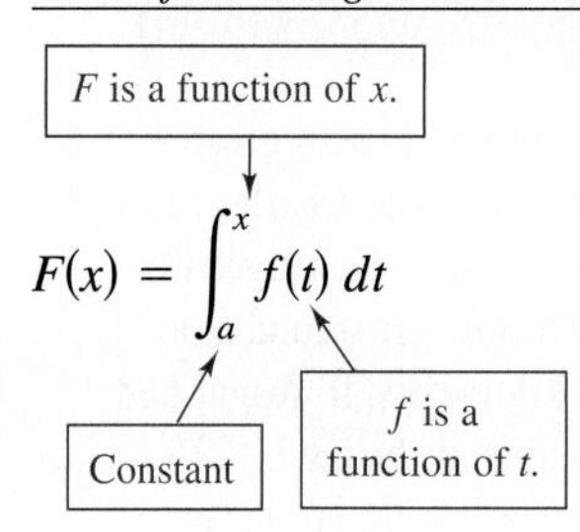

> **EXPLORATION**
>
> Use a graphing utility to graph the function
>
> $$F(x) = \int_0^x \cos t\,dt$$
>
> for $0 \le x \le \pi$. Do you recognize this graph? Explain.

EXAMPLE 6 The Definite Integral as a Function

Evaluate the function

$$F(x) = \int_0^x \cos t\,dt$$

at $x = 0, \pi/6, \pi/4, \pi/3$, and $\pi/2$.

Solution You could evaluate five different definite integrals, one for each of the given upper limits. However, it is much simpler to fix x (as a constant) temporarily and apply the Fundemental Theorem once, to obtain

$$\int_0^x \cos t\,dt = \sin t\Big]_0^x = \sin x - \sin 0 = \sin x.$$

Now, using $F(x) = \sin x$, you can obtain the results shown in Figure 5.35.

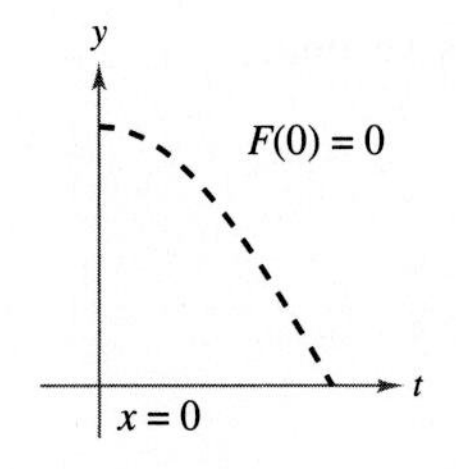

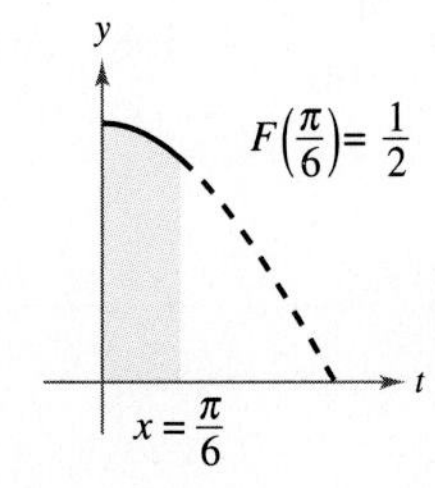

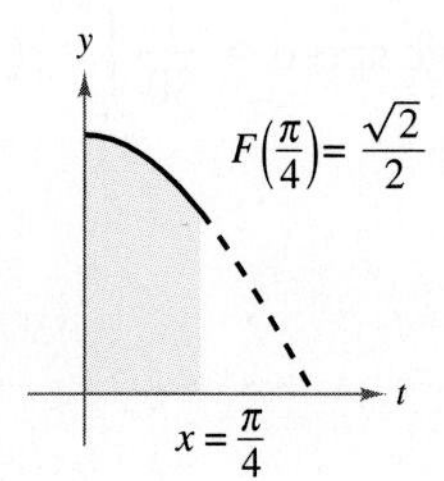

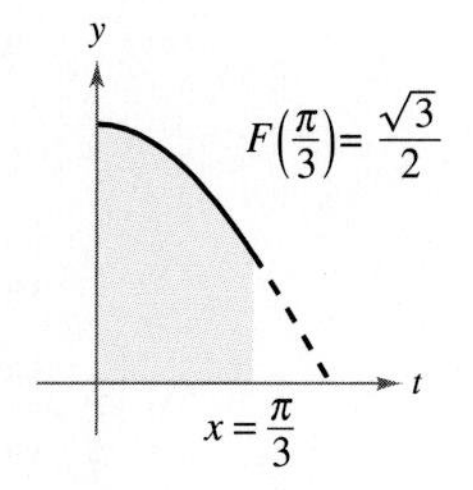

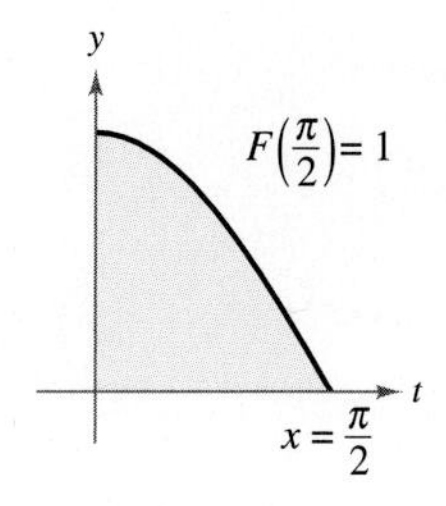

$F(x) = \int_0^x \cos t\,dt$ is the area under the curve $f(t) = \cos t$ from 0 to x.

Figure 5.35

You can think of the function $F(x)$ as *accumulating* the area under the curve $f(t) = \cos t$ from $t = 0$ to $t = x$. For $x = 0$, the area is 0 and $F(0) = 0$. For $x = \pi/2$, $F(\pi/2) = 1$ gives the accumulated area under the cosine curve on the entire interval $[0, \pi/2]$. This interpretation of an integral as an **accumulation function** is used often in applications of integration.

In Example 6, note that the derivative of F is the original integrand (with only the variable changed). That is,

$$\frac{d}{dx}[F(x)] = \frac{d}{dx}[\sin x] = \frac{d}{dx}\left[\int_0^x \cos t\, dt\right] = \cos x.$$

This result is generalized in the following theorem, called the **Second Fundamental Theorem of Calculus.**

THEOREM 5.11 The Second Fundamental Theorem of Calculus

If f is continuous on an open interval I containing a, then, for every x in the interval,

$$\frac{d}{dx}\left[\int_a^x f(t)\, dt\right] = f(x).$$

Proof Begin by defining F as

$$F(x) = \int_a^x f(t)\, dt.$$

Then, by the definition of the derivative, you can write

$$\begin{aligned} F'(x) &= \lim_{\Delta x \to 0} \frac{F(x + \Delta x) - F(x)}{\Delta x} \\ &= \lim_{\Delta x \to 0} \frac{1}{\Delta x}\left[\int_a^{x+\Delta x} f(t)\, dt - \int_a^x f(t)\, dt\right] \\ &= \lim_{\Delta x \to 0} \frac{1}{\Delta x}\left[\int_a^{x+\Delta x} f(t)\, dt + \int_x^a f(t)\, dt\right] \\ &= \lim_{\Delta x \to 0} \frac{1}{\Delta x}\left[\int_x^{x+\Delta x} f(t)\, dt\right]. \end{aligned}$$

From the Mean Value Theorem for Integrals (assuming $\Delta x > 0$), you know there exists a number c in the interval $[x, x + \Delta x]$ such that the integral in the expression above is equal to $f(c)\,\Delta x$. Moreover, because $x \le c \le x + \Delta x$, it follows that $c \to x$ as $\Delta x \to 0$. So, you obtain

$$\begin{aligned} F'(x) &= \lim_{\Delta x \to 0}\left[\frac{1}{\Delta x} f(c)\,\Delta x\right] \\ &= \lim_{\Delta x \to 0} f(c) \\ &= f(x). \end{aligned}$$

A similar argument can be made for $\Delta x < 0$.

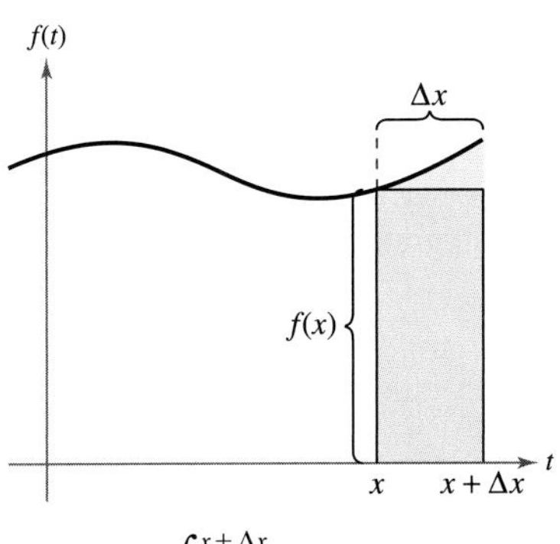

$$f(x)\,\Delta x \approx \int_x^{x+\Delta x} f(t)\, dt$$

Figure 5.36

NOTE Using the area model for definite integrals, you can view the approximation

$$f(x)\,\Delta x \approx \int_x^{x+\Delta x} f(t)\, dt$$

as saying that the area of the rectangle of height $f(x)$ and width Δx is approximately equal to the area of the region lying between the graph of f and the x-axis on the interval $[x, x + \Delta x]$, as shown in Figure 5.36.

Note that the Second Fundamental Theorem of Calculus tells you that if a function is continuous, you can be sure that it has an antiderivative. This antiderivative need not, however, be an elementary function. (Recall the discussion of elementary functions in Section 1.3.)

EXAMPLE 7 Using the Second Fundamental Theorem of Calculus

Evaluate $\dfrac{d}{dx}\left[\displaystyle\int_0^x \sqrt{t^2+1}\,dt\right]$.

Solution Note that $f(t) = \sqrt{t^2+1}$ is continuous on the entire real number line. So, using the Second Fundamental Theorem of Calculus, you can write

$$\frac{d}{dx}\left[\int_0^x \sqrt{t^2+1}\,dt\right] = \sqrt{x^2+1}.$$

The differentiation shown in Example 7 is a straightforward application of the Second Fundamental Theorem of Calculus. The next example shows how this theorem can be combined with the Chain Rule to find the derivative of a function.

EXAMPLE 8 Using the Second Fundamental Theorem of Calculus

Find the derivative of $F(x) = \displaystyle\int_{\pi/2}^{x^3} \cos t\,dt$.

Solution Using $u = x^3$, you can apply the Second Fundamental Theorem of Calculus with the Chain Rule as shown.

$$F'(x) = \frac{dF}{du}\frac{du}{dx} \qquad \text{Chain Rule}$$

$$= \frac{d}{du}[F(x)]\frac{du}{dx} \qquad \text{Definition of } \frac{dF}{du}$$

$$= \frac{d}{du}\left[\int_{\pi/2}^{x^3} \cos t\,dt\right]\frac{du}{dx} \qquad \text{Substitute } \int_{\pi/2}^{x^3} \cos t\,dt \text{ for } F(x).$$

$$= \frac{d}{du}\left[\int_{\pi/2}^{u} \cos t\,dt\right]\frac{du}{dx} \qquad \text{Substitute } u \text{ for } x^3.$$

$$= (\cos u)(3x^2) \qquad \text{Apply Second Fundamental Theorem of Calculus.}$$

$$= (\cos x^3)(3x^2) \qquad \text{Rewrite as function of } x.$$

Because the integrand in Example 8 is easily integrated, you can verify the derivative as follows.

$$F(x) = \int_{\pi/2}^{x^3} \cos t\,dt = \sin t\Big]_{\pi/2}^{x^3} = \sin x^3 - \sin\frac{\pi}{2} = (\sin x^3) - 1$$

In this form, you can apply the Power Rule to verify that the derivative is the same as that obtained in Example 8.

$$F'(x) = (\cos x^3)(3x^2)$$

Exercises for Section 5.4

See www.CalcChat.com for worked-out solutions to odd-numbered exercises.

Graphical Reasoning **In Exercises 1–4, use a graphing utility to graph the integrand. Use the graph to determine whether the definite integral is positive, negative, or zero.**

1. $\int_0^{\pi} \frac{4}{x^2+1}\,dx$
2. $\int_0^{\pi} \cos x\,dx$
3. $\int_{-2}^{2} x\sqrt{x^2+1}\,dx$
4. $\int_{-2}^{2} x\sqrt{2-x}\,dx$

In Exercises 5–26, evaluate the definite integral of the algebraic function. Use a graphing utility to verify your result.

5. $\int_0^1 2x\,dx$
6. $\int_2^7 3\,dv$
7. $\int_{-1}^0 (x-2)\,dx$
8. $\int_2^5 (-3v+4)\,dv$
9. $\int_{-1}^1 (t^2-2)\,dt$
10. $\int_1^3 (3x^2+5x-4)\,dx$
11. $\int_0^1 (2t-1)^2\,dt$
12. $\int_{-1}^1 (t^3-9t)\,dt$
13. $\int_1^2 \left(\frac{3}{x^2}-1\right)dx$
14. $\int_{-2}^{-1} \left(u-\frac{1}{u^2}\right)du$
15. $\int_1^4 \frac{u-2}{\sqrt{u}}\,du$
16. $\int_{-3}^3 v^{1/3}\,dv$
17. $\int_{-1}^1 (\sqrt[3]{t}-2)\,dt$
18. $\int_1^8 \sqrt{\frac{2}{x}}\,dx$
19. $\int_0^1 \frac{x-\sqrt{x}}{3}\,dx$
20. $\int_0^2 (2-t)\sqrt{t}\,dt$
21. $\int_{-1}^0 (t^{1/3}-t^{2/3})\,dt$
22. $\int_{-8}^{-1} \frac{x-x^2}{2\sqrt[3]{x}}\,dx$
23. $\int_0^3 |2x-3|\,dx$
24. $\int_2^5 (3-|x-4|)\,dx$
25. $\int_0^3 |x^2-4|\,dx$
26. $\int_0^4 |x^2-4x+3|\,dx$

In Exercises 27–38, evaluate the definite integral of the transcendental function. Use a graphing utility to verify your result.

27. $\int_0^{\pi} (1+\sin x)\,dx$
28. $\int_0^{\pi/4} \frac{1-\sin^2\theta}{\cos^2\theta}\,d\theta$
29. $\int_{-\pi/6}^{\pi/6} \sec^2 x\,dx$
30. $\int_{\pi/4}^{\pi/2} (2-\csc^2 x)\,dx$
31. $\int_1^e \left(2x-\frac{1}{x}\right)dx$
32. $\int_1^5 \frac{x+1}{x}\,dx$
33. $\int_{-\pi/3}^{\pi/3} 4\sec\theta\tan\theta\,d\theta$
34. $\int_{-\pi/2}^{\pi/2} (2t+\cos t)\,dt$
35. $\int_0^2 (2^x+6)\,dx$
36. $\int_0^3 (t-5^t)\,dt$
37. $\int_{-1}^1 (e^{\theta}+\sin\theta)\,d\theta$
38. $\int_e^{2e} \left(\cos x-\frac{1}{x}\right)dx$

In Exercises 39–44, determine the area of the given region.

39. $y = x - x^2$

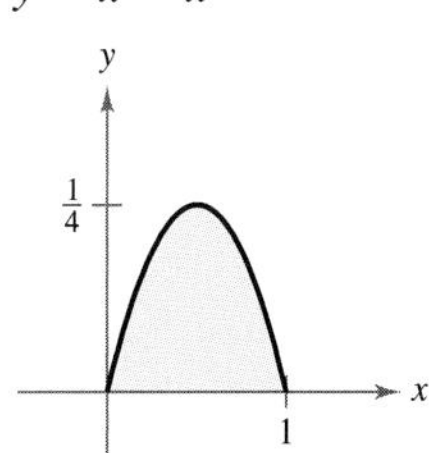

40. $y = 1 - x^4$

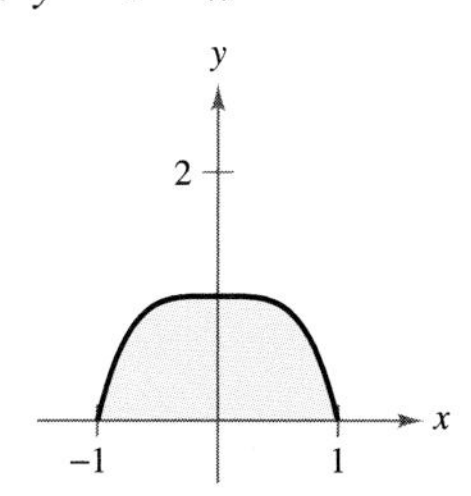

41. $y = (3-x)\sqrt{x}$

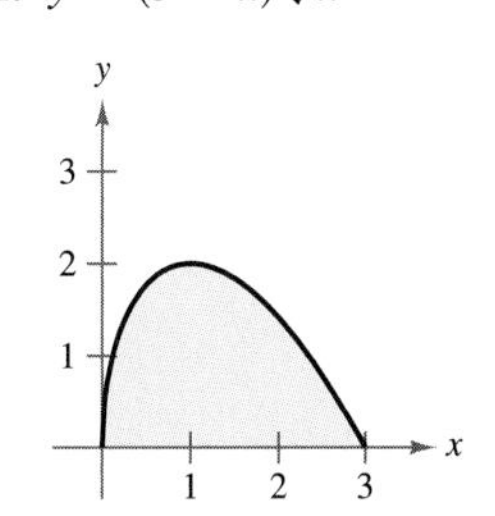

42. $y = \frac{1}{x^2}$

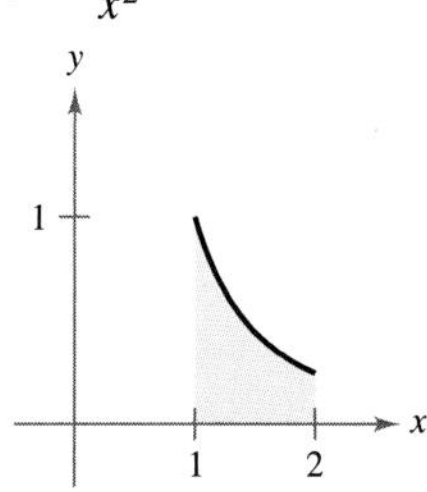

43. $y = \cos x$

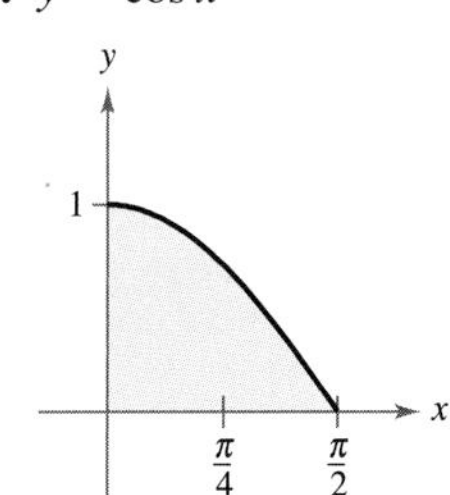

44. $y = x + \sin x$

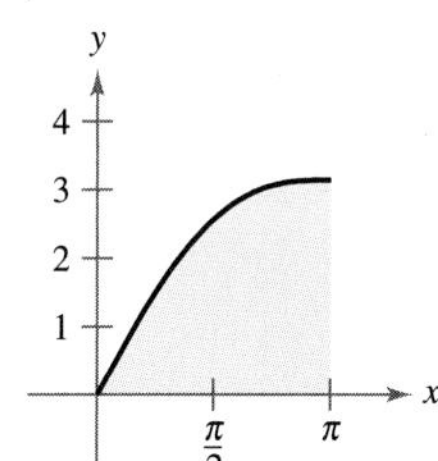

In Exercises 45–50, find the area of the region bounded by the graphs of the equations.

45. $y = 3x^2 + 1,\quad x = 0,\quad x = 2,\quad y = 0$
46. $y = 1 + \sqrt[3]{x},\quad x = 0,\quad x = 8,\quad y = 0$
47. $y = x^3 + x,\quad x = 2,\quad y = 0$
48. $y = -x^2 + 3x,\quad y = 0$
49. $y = \frac{4}{x},\quad x = 1,\quad x = e,\quad y = 0$
50. $y = e^x,\quad x = 0,\quad x = 2,\quad y = 0$

In Exercises 51–56, find the value(s) of *c* guaranteed by the Mean Value Theorem for Integrals for the function over the indicated interval.

	Function	*Interval*
51.	$f(x) = x - 2\sqrt{x}$	$[0, 2]$
52.	$f(x) = 9/x^3$	$[1, 3]$
53.	$f(x) = 2\sec^2 x$	$[-\pi/4, \pi/4]$
54.	$f(x) = \cos x$	$[-\pi/3, \pi/3]$
55.	$f(x) = 5 - \frac{1}{x}$	$[1, 4]$
56.	$f(x) = 10 - 2^x$	$[0, 3]$

In Exercises 57–62, find the average value of the function over the given interval and all values of x in the interval for which the function equals its average value.

Function	*Interval*
57. $f(x) = 4 - x^2$	$[-2, 2]$
58. $f(x) = \dfrac{4(x^2 + 1)}{x^2}$	$[1, 3]$
59. $f(x) = 2e^x$	$[-1, 1]$
60. $f(x) = \dfrac{1}{2x}$	$[1, 4]$
61. $f(x) = \sin x$	$[0, \pi]$
62. $f(x) = \cos x$	$[0, \pi/2]$

63. ***Velocity*** The graph shows the velocity, in feet per second, of a car accelerating from rest. Use the graph to estimate the distance the car travels in 8 seconds.

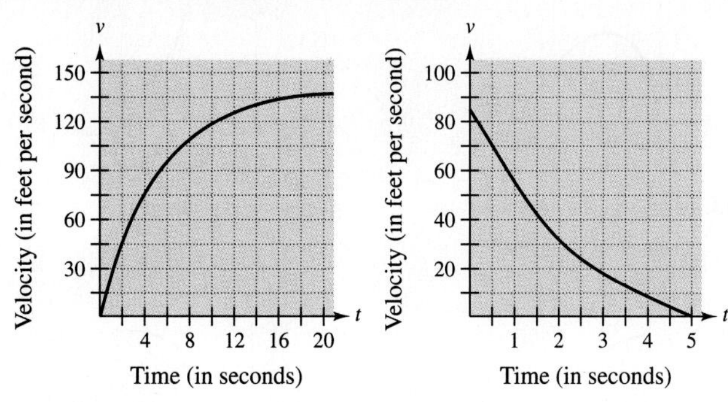

Figure for 63 **Figure for 64**

64. ***Velocity*** The graph shows the velocity of a car as soon as the driver applies the brakes. Use the graph to estimate how far the car travels before it comes to a stop.

Writing About Concepts

65. State the Fundamental Theorem of Calculus.

66. The graph of f is shown in the figure.

(a) Evaluate $\int_1^7 f(x)\,dx$.

(b) Determine the average value of f on the interval $[1, 7]$.

(c) Determine the answers to parts (a) and (b) if the graph is translated two units upward.

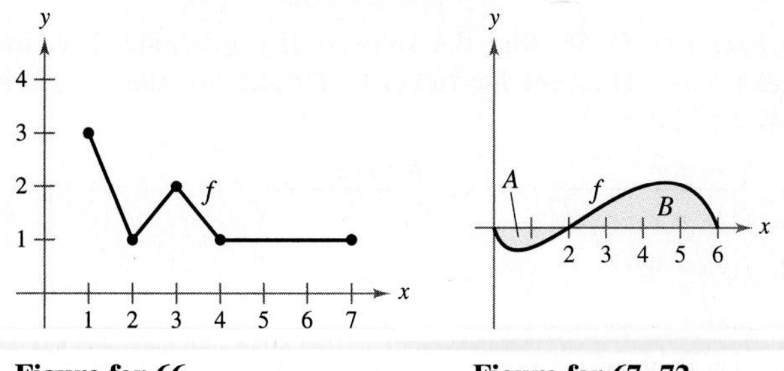

Figure for 66 **Figure for 67–72**

Writing About Concepts *(continued)*

In Exercises 67–72, use the graph of f shown in the figure. The shaded region A has an area of 1.5, and $\int_0^6 f(x)\,dx = 3.5$. Use this information to fill in the blanks.

67. $\displaystyle\int_0^2 f(x)\,dx = \square$

68. $\displaystyle\int_2^6 f(x)\,dx = \square$

69. $\displaystyle\int_0^6 |f(x)|\,dx = \square$

70. $\displaystyle\int_0^2 -2f(x)\,dx = \square$

71. $\displaystyle\int_0^6 [2 + f(x)]\,dx = \square$

72. The average value of f over the interval $[0, 6]$ is $\square$.

73. ***Force*** The force F (in newtons) of a hydraulic cylinder in a press is proportional to the square of $\sec x$, where x is the distance (in meters) that the cylinder is extended in its cycle. The domain of F is $[0, \pi/3]$, and $F(0) = 500$.

(a) Find F as a function of x.

(b) Find the average force exerted by the press over the interval $[0, \pi/3]$.

74. ***Blood Flow*** The velocity v of the flow of blood at a distance r from the central axis of an artery of radius R is

$$v = k(R^2 - r^2)$$

where k is the constant of proportionality. Find the average rate of flow of blood along a radius of the artery. (Use 0 and R as the limits of integration.)

75. ***Respiratory Cycle*** The volume V (in liters) of air in the lungs during a five-second respiratory cycle is approximated by the model

$$V = 0.1729t + 0.1522t^2 - 0.0374t^3$$

where t is the time in seconds. Approximate the average volume of air in the lungs during one cycle.

76. ***Average Sales*** A company fits a model to the monthly sales data of a seasonal product. The model is

$$S(t) = \frac{t}{4} + 1.8 + 0.5\sin\left(\frac{\pi t}{6}\right), \quad 0 \le t \le 24$$

where S is sales (in thousands) and t is time in months.

(a) Use a graphing utility to graph $f(t) = 0.5\sin(\pi t/6)$ for $0 \le t \le 24$. Use the graph to explain why the average value of $f(t)$ is 0 over the interval.

(b) Use a graphing utility to graph $S(t)$ and the line $g(t) = t/4 + 1.8$ in the same viewing window. Use the graph and the result of part (a) to explain why g is called the *trend line*.

77. *Modeling Data* An experimental vehicle is tested on a straight track. It starts from rest, and its velocity v (in meters per second) is recorded in the table every 10 seconds for 1 minute.

t	0	10	20	30	40	50	60
v	0	5	21	40	62	78	83

(a) Use a graphing utility to find a model of the form $v = at^3 + bt^2 + ct + d$ for the data.

(b) Use a graphing utility to plot the data and graph the model.

(c) Use the Fundamental Theorem of Calculus to approximate the distance traveled by the vehicle during the test.

78. *Modeling Data* A department store manager wants to estimate the number of customers that enter the store from noon until closing at 9 P.M. The table shows the number of customers N entering the store during a randomly selected minute each hour from $t - 1$ to t, with $t = 0$ corresponding to noon.

t	1	2	3	4	5	6	7	8	9
N	6	7	9	12	15	14	11	7	2

(a) Draw a histogram of the data.

(b) Estimate the total number of customers entering the store between noon and 9 P.M.

(c) Use the regression capabilities of a graphing utility to find a model of the form $N(t) = at^3 + bt^2 + ct + d$ for the data.

(d) Use a graphing utility to plot the data and graph the model.

(e) Use a graphing utility to evaluate $\int_0^9 N(t)\,dt$, and use the result to estimate the number of customers entering the store between noon and 9 P.M. Compare this with your answer in part (b).

(f) Estimate the average number of customers entering the store per minute between 3 P.M. and 7 P.M.

In Exercises 79–84, find F as a function of x and evaluate it at $x = 2$, $x = 5$, and $x = 8$.

79. $F(x) = \displaystyle\int_0^x (t - 5)\,dt$

80. $F(x) = \displaystyle\int_2^x (t^3 + 2t - 2)\,dt$

81. $F(x) = \displaystyle\int_1^x \frac{10}{v^2}\,dv$

82. $F(x) = \displaystyle\int_2^x -\frac{2}{t^3}\,dt$

83. $F(x) = \displaystyle\int_1^x \cos\theta\,d\theta$

84. $F(x) = \displaystyle\int_0^x \sin\theta\,d\theta$

85. Let $g(x) = \int_0^x f(t)\,dt$, where f is the function whose graph is shown.

(a) Estimate $g(0)$, $g(2)$, $g(4)$, $g(6)$, and $g(8)$.

(b) Find the largest open interval on which g is increasing. Find the largest open interval on which g is decreasing.

(c) Identify any extrema of g.

(d) Sketch a rough graph of g.

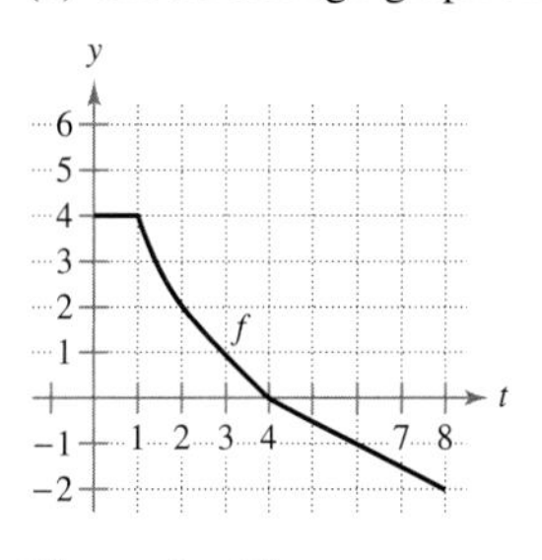

Figure for 85

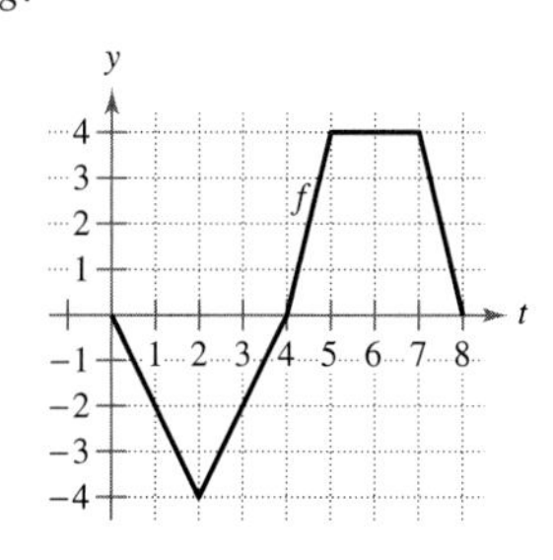

Figure for 86

86. Let $g(x) = \int_0^x f(t)\,dt$, where f is the function whose graph is shown.

(a) Estimate $g(0)$, $g(2)$, $g(4)$, $g(6)$, and $g(8)$.

(b) Find the largest open interval on which g is increasing. Find the largest open interval on which g is decreasing.

(c) Identify any extrema of g.

(d) Sketch a rough graph of g.

In Exercises 87–94, (a) integrate to find F as a function of x and (b) demonstrate the Second Fundamental Theorem of Calculus by differentiating the result in part (a).

87. $F(x) = \displaystyle\int_0^x (t + 2)\,dt$

88. $F(x) = \displaystyle\int_0^x t(t^2 + 1)\,dt$

89. $F(x) = \displaystyle\int_8^x \sqrt[3]{t}\,dt$

90. $F(x) = \displaystyle\int_4^x \sqrt{t}\,dt$

91. $F(x) = \displaystyle\int_{\pi/4}^x \sec^2 t\,dt$

92. $F(x) = \displaystyle\int_{\pi/3}^x \sec t \tan t\,dt$

93. $F(x) = \displaystyle\int_{-1}^x e^t\,dt$

94. $F(x) = \displaystyle\int_1^x \frac{1}{t}\,dt$

In Exercises 95–100, use the Second Fundamental Theorem of Calculus to find $F'(x)$.

95. $F(x) = \displaystyle\int_{-2}^x (t^2 - 2t)\,dt$

96. $F(x) = \displaystyle\int_1^x \frac{t^2}{t^2 + 1}\,dt$

97. $F(x) = \displaystyle\int_{-1}^x \sqrt{t^4 + 1}\,dt$

98. $F(x) = \displaystyle\int_1^x \sqrt[4]{t}\,dt$

99. $F(x) = \displaystyle\int_0^x t\cos t\,dt$

100. $F(x) = \displaystyle\int_0^x \sec^3 t\,dt$

In Exercises 101–106, find $F'(x)$.

101. $F(x) = \displaystyle\int_x^{x+2} (4t + 1)\,dt$

102. $F(x) = \displaystyle\int_{-x}^x t^3\,dt$

103. $F(x) = \displaystyle\int_0^{\sin x} \sqrt{t}\,dt$

104. $F(x) = \displaystyle\int_2^{x^2} \frac{1}{t^3}\,dt$

105. $F(x) = \displaystyle\int_0^{x^3} \sin t^2\,dt$

106. $F(x) = \displaystyle\int_0^{x^2} \sin\theta^2\,d\theta$

107. ***Graphical Analysis*** Approximate the graph of g on the interval $0 \le x \le 4$, where $g(x) = \int_0^x f(t)\,dt$. Identify the x-coordinate of an extremum of g. To print an enlarged copy of the graph, go to the website *www.mathgraphs.com*.

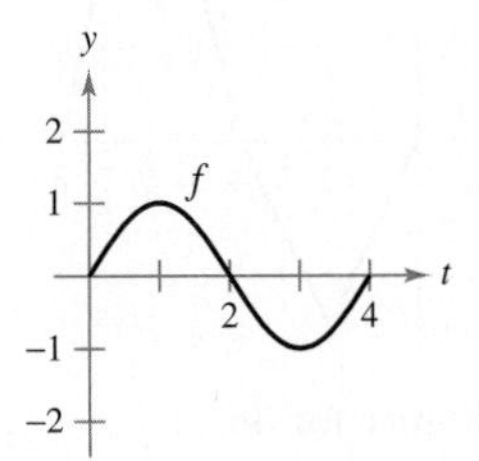

108. ***Area*** The area A between the graph of the function $g(t) = 4 - 4/t^2$ and the t-axis over the interval $[1, x]$ is

$$A(x) = \int_1^x \left(4 - \frac{4}{t^2}\right) dt.$$

(a) Find the horizontal asymptote of the graph of g.

(b) Integrate to find A as a function of x. Does the graph of A have a horizontal asymptote? Explain.

Rectilinear Motion **In Exercises 109–111, consider a particle moving along the x-axis where $x(t)$ is the position of the particle at time t, $x'(t)$ is its velocity, and $\int_a^b |x'(t)|\,dt$ is the distance the particle travels in the interval of time.**

109. The position function is given by $x(t) = t^3 - 6t^2 + 9t - 2$, $0 \le t \le 5$. Find the total distance the particle travels in 5 units of time.

110. Repeat Exercise 109 for the position function given by $x(t) = (t - 1)(t - 3)^2$, $0 \le t \le 5$.

111. A particle moves along the x-axis with velocity $v(t) = 1/\sqrt{t}$, $t > 0$. At time $t = 1$, its position is $x = 4$. Find the total distance traveled by the particle on the interval $1 \le t \le 4$.

112. ***Buffon's Needle Experiment*** A horizontal plane is ruled with parallel lines 2 inches apart. A two-inch needle is tossed randomly onto the plane. The probability that the needle will touch a line is

$$P = \frac{2}{\pi}\int_0^{\pi/2} \sin\theta\,d\theta$$

where θ is the acute angle between the needle and any one of the parallel lines. Find this probability.

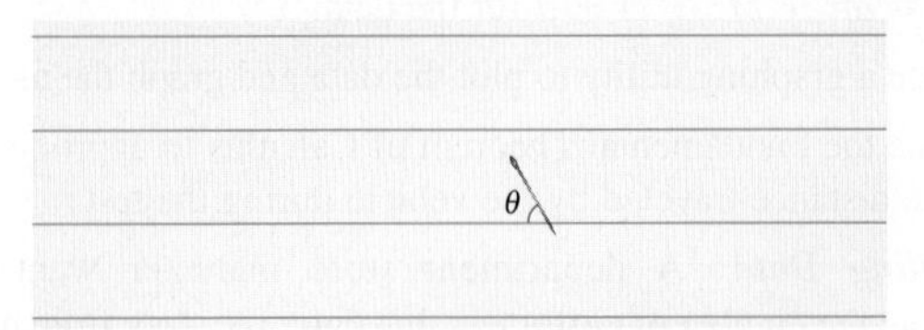

True or False? **In Exercises 113 and 114, determine whether the statement is true or false. If it is false, explain why or give an example that shows it is false.**

113. If $F'(x) = G'(x)$ on the interval $[a, b]$, then $F(b) - F(a) = G(b) - G(a)$.

114. If f is continuous on $[a, b]$, then f is integrable on $[a, b]$.

115. ***Find the Error*** Describe why the statement is incorrect.

$$\int_{-1}^{1} x^{-2}\,dx = \left[-x^{-1}\right]_{-1}^{1} = (-1) - 1 = -2$$

116. Prove that $\dfrac{d}{dx}\left[\int_{u(x)}^{v(x)} f(t)\,dt\right] = f(v(x))v'(x) - f(u(x))u'(x)$.

117. Show that the function

$$f(x) = \int_0^{1/x} \frac{1}{t^2+1}\,dt + \int_0^x \frac{1}{t^2+1}\,dt$$

is constant for $x > 0$.

118. Let $G(x) = \int_0^x \left[s\int_0^s f(t)dt\right] ds$, where f is continuous for all real t. Find (a) $G(0)$, (b) $G'(0)$, (c) $G''(x)$, and (d) $G''(0)$.

Section Project: Demonstrating the Fundamental Theorem

Use a graphing utility to graph the function $y_1 = \sin^2 t$ on the interval $0 \le t \le \pi$. Let $F(x)$ be the following function of x.

$$F(x) = \int_0^x \sin^2 t\,dt$$

(a) Complete the table. Explain why the values of F are increasing.

x	0	$\pi/6$	$\pi/3$	$\pi/2$	$2\pi/3$	$5\pi/6$	π
$F(x)$							

(b) Use the integration capabilities of a graphing utility to graph F.

(c) Use the differentiation capabilities of a graphing utility to graph $F'(x)$. How is this graph related to the graph in part (b)?

(d) Verify that the derivative of $y = (1/2)t - (\sin 2t)/4$ is $\sin^2 t$. Graph y and write a short paragraph about how this graph is related to those in parts (b) and (c).

Section 5.5 Integration by Substitution

- **Use pattern recognition to find an indefinite integral.**
- **Use a change of variables to find an indefinite integral.**
- **Use the General Power Rule for Integration to find an indefinite integral.**
- **Use a change of variables to evaluate a definite integral.**
- **Evaluate a definite integral involving an even or odd function.**

Pattern Recognition

In this section you will study techniques for integrating composite functions. The discussion is split into two parts—*pattern recognition* and *change of variables*. Both techniques involve a ***u*-substitution.** With pattern recognition you perform the substitution mentally, and with change of variables you write the substitution steps.

The role of substitution in integration is comparable to the role of the Chain Rule in differentiation. Recall that for differentiable functions given by $y = F(u)$ and $u = g(x)$, the Chain Rule states that

$$\frac{d}{dx}[F(g(x))] = F'(g(x))g'(x).$$

From the definition of an antiderivative, it follows that

$$\begin{aligned}\int F'(g(x))g'(x)\,dx &= F(g(x)) + C\\ &= F(u) + C.\end{aligned}$$

These results are summarized in the following theorem.

NOTE The statement of Theorem 5.12 doesn't tell how to distinguish between $f(g(x))$ and $g'(x)$ in the integrand. As you become more experienced at integration, your skill in doing this will increase. Of course, part of the key is familiarity with derivatives.

THEOREM 5.12 Antidifferentiation of a Composite Function

Let g be a function whose range is an interval I, and let f be a function that is continuous on I. If g is differentiable on its domain and F is an antiderivative of f on I, then

$$\int f(g(x))g'(x)\,dx = F(g(x)) + C.$$

If $u = g(x)$, then $du = g'(x)\,dx$ and

$$\int f(u)\,du = F(u) + C.$$

STUDY TIP There are several techniques for applying substitution, each differing slightly from the others. However, you should remember that the goal is the same with every technique—*you are trying to find an antiderivative of the integrand.*

EXPLORATION

Recognizing Patterns The integrand in each of the following integrals fits the pattern $f(g(x))g'(x)$. Identify the pattern and use the result to evaluate the integral.

a. $\int 2x(x^2 + 1)^4\,dx$ **b.** $\int 3x^2\sqrt{x^3 + 1}\,dx$ **c.** $\int \sec^2 x(\tan x + 3)\,dx$

The next three integrals are similar to the first three. Show how you can multiply and divide by a constant to evaluate these integrals.

d. $\int x(x^2 + 1)^4\,dx$ **e.** $\int x^2\sqrt{x^3 + 1}\,dx$ **f.** $\int 2\sec^2 x(\tan x + 3)\,dx$

Examples 1 and 2 show how to apply Theorem 5.12 *directly*, by recognizing the presence of $f(g(x))$ and $g'(x)$. Note that the composite function in the integrand has an *outside function* f and an *inside function* g. Moreover, the derivative $g'(x)$ is present as a factor of the integrand.

Outside function

$$\int f(g(x))g'(x)\,dx = F(g(x)) + C$$

Inside function

Derivative of inside function

EXAMPLE 1 Recognizing the $f(g(x))g'(x)$ Pattern

Find $\displaystyle\int (x^2 + 1)^2(2x)\,dx.$

Solution Letting $g(x) = x^2 + 1$, you obtain

$$g'(x) = 2x$$

and

$$f(g(x)) = f(x^2 + 1) = (x^2 + 1)^2.$$

From this, you can recognize that the integrand follows the $f(g(x))g'(x)$ pattern. Using the Power Rule for Integration and Theorem 5.12, you can write

$$\int \overbrace{(x^2 + 1)^2}^{f(g(x))}\overbrace{(2x)}^{g'(x)}\,dx = \frac{1}{3}(x^2 + 1)^3 + C.$$

Try using the Chain Rule to check that the derivative of $\frac{1}{3}(x^2 + 1)^3 + C$ is the integrand of the original integral.

TECHNOLOGY Try using a computer algebra system, such as *Maple, Derive, Mathematica, Mathcad*, or the *TI-89*, to solve the integrals given in Examples 1 and 2. Do you obtain the same antiderivatives that are listed in the examples?

EXAMPLE 2 Recognizing the $f(g(x))g'(x)$ Pattern

Find $\displaystyle\int 5e^{5x}\,dx.$

Solution Letting $g(x) = 5x$, you obtain

$$g'(x) = 5 \quad \text{and} \quad f(g(x)) = f(5x) = e^{5x}.$$

From this, you can recognize that the integrand follows the $f(g(x))g'(x)$ pattern. Using the Exponential Rule for Integration and Theorem 5.12, you can write

$$\int \underbrace{e^{5x}}_{f(g(x))}\underbrace{(5)}_{g'(x)}\,dx = e^{5x} + C.$$

You can check this by differentiating $e^{5x} + C$ to obtain the original integrand.

The integrands in Examples 1 and 2 fit the $f(g(x))g'(x)$ pattern exactly—you only had to recognize the pattern. You can extend this technique considerably with the Constant Multiple Rule

$$\int kf(x)\,dx = k\int f(x)\,dx.$$

Many integrands contain the essential part (the variable part) of $g'(x)$ but are missing a constant multiple. In such cases, you can multiply and divide by the necessary constant multiple, as shown in Example 3.

EXAMPLE 3 Multiplying and Dividing by a Constant

Find $\int x(x^2 + 1)^2\,dx$.

Solution This is similar to the integral given in Example 1, except that the integrand is missing a factor of 2. Recognizing that $2x$ is the derivative of $x^2 + 1$, you can let $g(x) = x^2 + 1$ and supply the $2x$ as follows.

$$\begin{aligned}
\int x(x^2 + 1)^2\,dx &= \int (x^2 + 1)^2\left(\frac{1}{2}\right)(2x)\,dx && \text{Multiply and divide by 2.}\\
&= \frac{1}{2}\int \overbrace{(x^2 + 1)^2}^{f(g(x))}\,\overbrace{(2x)}^{g'(x)}\,dx && \text{Constant Multiple Rule}\\
&= \frac{1}{2}\left[\frac{(x^2 + 1)^3}{3}\right] + C && \text{Integrate.}\\
&= \frac{1}{6}(x^2 + 1)^3 + C && \text{Simplify.}
\end{aligned}$$

In practice, most people would not write as many steps as are shown in Example 3. For instance, you could evaluate the integral by simply writing

$$\begin{aligned}
\int x(x^2 + 1)^2\,dx &= \frac{1}{2}\int (x^2 + 1)^2\,2x\,dx\\
&= \frac{1}{2}\left[\frac{(x^2 + 1)^3}{3}\right] + C\\
&= \frac{1}{6}(x^2 + 1)^3 + C.
\end{aligned}$$

NOTE Be sure you see that the *Constant* Multiple Rule applies only to *constants*. You cannot multiply and divide by a variable and then move the variable outside the integral sign. For instance,

$$\int (x^2 + 1)^2\,dx \neq \frac{1}{2x}\int (x^2 + 1)^2\,(2x)\,dx.$$

After all, if it were legitimate to move variable quantities outside the integral sign, you could move the entire integrand out and simplify the whole process. But the result would be incorrect.

Change of Variables

With a formal **change of variables,** you completely rewrite the integral in terms of u and du (or any other convenient variable). Although this procedure can involve more written steps than the pattern recognition illustrated in Examples 1 to 3, it is useful for complicated integrands. The change of variable technique uses the Leibniz notation for the differential. That is, if $u = g(x)$, then $du = g'(x)\,dx$, and the integral in Theorem 5.12 takes the form

$$\int f(g(x))g'(x)\,dx = \int f(u)\,du = F(u) + C.$$

EXAMPLE 4 Change of Variables

Find $\int \sqrt{2x - 1}\,dx$.

Solution First, let u be the inner function, $u = 2x - 1$. Then calculate the differential du to be $du = 2\,dx$. Now, using $\sqrt{2x - 1} = \sqrt{u}$ and $dx = du/2$, substitute to obtain

$$\int \sqrt{2x-1}\,dx = \int \sqrt{u}\left(\frac{du}{2}\right) \qquad \text{Integral in terms of } u$$

$$= \frac{1}{2}\int u^{1/2}\,du \qquad \text{Constant Multiple Rule}$$

$$= \frac{1}{2}\left(\frac{u^{3/2}}{3/2}\right) + C \qquad \text{Antiderivative in terms of } u$$

$$= \frac{1}{3}u^{3/2} + C \qquad \text{Simplify.}$$

$$= \frac{1}{3}(2x-1)^{3/2} + C. \qquad \text{Antiderivative in terms of } x$$

STUDY TIP Because integration is usually more difficult than differentiation, you should always check your answer to an integration problem by differentiating. For instance, in Example 4 you should differentiate $\frac{1}{3}(2x - 1)^{3/2} + C$ to verify that you obtain the original integrand.

EXAMPLE 5 Change of Variables

Find $\int x\sqrt{2x - 1}\,dx$.

Solution As in the previous example, let $u = 2x - 1$ and obtain $dx = du/2$. Because the integrand contains a factor of x, you must also solve for x in terms of u, as shown.

$$u = 2x - 1 \quad \Rightarrow \quad x = (u + 1)/2 \qquad \text{Solve for } x \text{ in terms of } u.$$

Now, using substitution, you obtain

$$\int x\sqrt{2x-1}\,dx = \int \left(\frac{u+1}{2}\right)u^{1/2}\left(\frac{du}{2}\right)$$

$$= \frac{1}{4}\int (u^{3/2} + u^{1/2})\,du$$

$$= \frac{1}{4}\left(\frac{u^{5/2}}{5/2} + \frac{u^{3/2}}{3/2}\right) + C$$

$$= \frac{1}{10}(2x-1)^{5/2} + \frac{1}{6}(2x-1)^{3/2} + C.$$

To complete the change of variables in Example 5, you solved for x in terms of u. Sometimes this is very difficult. Fortunately it is not always necessary, as shown in the next example.

EXAMPLE 6 Change of Variables

Find $\int \sin^2 3x \cos 3x \, dx$.

Solution Because $\sin^2 3x = (\sin 3x)^2$, you can let $u = \sin 3x$. Then

$$du = (\cos 3x)(3)\, dx.$$

Now, because $\cos 3x \, dx$ is part of the original integral, you can write

$$\frac{du}{3} = \cos 3x \, dx.$$

STUDY TIP When making a change of variables, be sure that your answer is written using the same variables used in the original integrand. For instance, in Example 6, you should not leave your answer as

$$\tfrac{1}{9}u^3 + C$$

but rather, replace u by $\sin 3x$.

Substituting u and $du/3$ in the original integral yields

$$\begin{aligned}\int \sin^2 3x \cos 3x \, dx &= \int u^2 \frac{du}{3} \\ &= \frac{1}{3}\int u^2 \, du \\ &= \frac{1}{3}\left(\frac{u^3}{3}\right) + C \\ &= \frac{1}{9}\sin^3 3x + C.\end{aligned}$$

You can check this by differentiating.

$$\begin{aligned}\frac{d}{dx}\left[\frac{1}{9}\sin^3 3x\right] &= \left(\frac{1}{9}\right)(3)(\sin 3x)^2(\cos 3x)(3) \\ &= \sin^2 3x \cos 3x\end{aligned}$$

Because differentiation produces the original integrand, you know that you have obtained the correct antiderivative.

The steps used for integration by substitution are summarized in the following guidelines.

Guidelines for Making a Change of Variables

1. Choose a substitution $u = g(x)$. Usually, it is best to choose the *inner* part of a composite function, such as a quantity raised to a power.
2. Compute $du = g'(x)\, dx$.
3. Rewrite the integral in terms of the variable u.
4. Find the resulting integral in terms of u.
5. Replace u by $g(x)$ to obtain an antiderivative in terms of x.
6. Check your answer by differentiating.

The General Power Rule for Integration

One of the most common u-substitutions involves quantities in the integrand that are raised to a power. Because of the importance of this type of substitution, it is given a special name—the **General Power Rule for Integration.** A proof of this rule follows directly from the (simple) Power Rule for Integration, together with Theorem 5.12.

THEOREM 5.13 The General Power Rule for Integration

If g is a differentiable function of x, then

$$\int [g(x)]^n g'(x)\, dx = \frac{[g(x)]^{n+1}}{n+1} + C, \qquad n \neq -1.$$

Equivalently, if $u = g(x)$, then

$$\int u^n\, du = \frac{u^{n+1}}{n+1} + C, \qquad n \neq -1.$$

EXAMPLE 7 Substitution and the General Power Rule

a. $$\int 3(3x-1)^4\, dx = \int \overbrace{(3x-1)^4}^{u^4}\overbrace{(3)\, dx}^{du} = \overbrace{\frac{(3x-1)^5}{5}}^{u^5/5} + C$$

b. $$\int (e^x+1)(e^x+x)dx = \int \overbrace{(e^x+x)}^{u^1}\overbrace{(e^x+1)dx}^{du} = \overbrace{\frac{(e^x+x)^2}{2}}^{u^2/2} + C$$

c. $$\int 3x^2\sqrt{x^3-2}\, dx = \int \overbrace{(x^3-2)^{1/2}}^{u^{1/2}}\overbrace{(3x^2)\, dx}^{du} = \overbrace{\frac{(x^3-2)^{3/2}}{3/2}}^{u^{3/2}/(3/2)} + C = \frac{2}{3}(x^3-2)^{3/2} + C$$

d. $$\int \frac{-4x}{(1-2x^2)^2}\, dx = \int \overbrace{(1-2x^2)^{-2}}^{u^{-2}}\overbrace{(-4x)\, dx}^{du} = \overbrace{\frac{(1-2x^2)^{-1}}{-1}}^{u^{-1}/(-1)} + C = -\frac{1}{1-2x^2} + C$$

e. $$\int \cos^2 x \sin x\, dx = -\int \overbrace{(\cos x)^2}^{u^2}\overbrace{(-\sin x)\, dx}^{du} = -\overbrace{\frac{(\cos x)^3}{3}}^{u^3/3} + C$$

EXPLORATION

Suppose you were asked to find one of the following integrals. Which one would you choose? Explain your reasoning.

a. $\int \sqrt{x^3+1}\, dx$ or

$\int x^2\sqrt{x^3+1}\, dx$

b. $\int \tan(3x)\sec^2(3x)\, dx$ or

$\int \tan(3x)\, dx$

Some integrals whose integrands involve quantities raised to powers cannot be found by the General Power Rule. Consider the two integrals

$$\int x(x^2+1)^2\, dx \quad \text{and} \quad \int (x^2+1)^2\, dx.$$

The substitution $u = x^2 + 1$ works in the first integral but not in the second. In the second, the substitution fails because the integrand lacks the factor x needed for du. Fortunately, *for this particular integral*, you can expand the integrand as $(x^2+1)^2 = x^4 + 2x^2 + 1$ and use the (simple) Power Rule to integrate each term.

Change of Variables for Definite Integrals

When using u-substitution with a definite integral, it is often convenient to determine the limits of integration for the variable u rather than to convert the antiderivative back to the variable x and evaluate at the original limits. This change of variables is stated explicitly in the next theorem. The proof follows from Theorem 5.12 combined with the Fundamental Theorem of Calculus.

THEOREM 5.14 Change of Variables for Definite Integrals

If the function $u = g(x)$ has a continuous derivative on the closed interval $[a, b]$ and f is continuous on the range of g, then

$$\int_a^b f(g(x))g'(x)\,dx = \int_{g(a)}^{g(b)} f(u)\,du.$$

EXAMPLE 8 Change of Variables

Evaluate $\displaystyle\int_0^1 x(x^2 + 1)^3\,dx.$

Solution To evaluate this integral, let $u = x^2 + 1$. Then, you obtain

$$u = x^2 + 1 \implies du = 2x\,dx.$$

Before substituting, determine the new upper and lower limits of integration.

Lower Limit	*Upper Limit*
When $x = 0$, $u = 0^2 + 1 = 1$.	When $x = 1$, $u = 1^2 + 1 = 2$.

Now, you can substitute to obtain

$$\int_0^1 x(x^2 + 1)^3\,dx = \frac{1}{2}\int_0^1 (x^2 + 1)^3(2x)\,dx$$

Integration limits for x

$$= \frac{1}{2}\int_1^2 u^3\,du$$

Integration limits for u

$$= \frac{1}{2}\left[\frac{u^4}{4}\right]_1^2$$

$$= \frac{1}{2}\left(4 - \frac{1}{4}\right)$$

$$= \frac{15}{8}.$$

Try rewriting the antiderivative $\frac{1}{2}(u^4/4)$ in terms of the variable x and then evaluate the definite integral at the original limits of integration, as shown.

$$\frac{1}{2}\left[\frac{u^4}{4}\right]_1^2 = \frac{1}{2}\left[\frac{(x^2 + 1)^4}{4}\right]_0^1$$

$$= \frac{1}{2}\left(4 - \frac{1}{4}\right) = \frac{15}{8}$$

Notice that you obtain the same result.

EXAMPLE 9 Change of Variables

Evaluate $A = \int_1^5 \frac{x}{\sqrt{2x-1}}\,dx.$

Solution To evaluate this integral, let $u = \sqrt{2x-1}$. Then, you obtain

$$u^2 = 2x - 1$$
$$u^2 + 1 = 2x$$
$$\frac{u^2+1}{2} = x$$
$$u\,du = dx. \qquad \text{Differentiate each side.}$$

Before substituting, determine the new upper and lower limits of integration.

Lower Limit	*Upper Limit*
When $x = 1$, $u = \sqrt{2-1} = 1$.	When $x = 5$, $u = \sqrt{10-1} = 3$.

Now, substitute to obtain

$$\begin{aligned}\int_1^5 \frac{x}{\sqrt{2x-1}}\,dx &= \int_1^3 \frac{1}{u}\left(\frac{u^2+1}{2}\right)u\,du\\ &= \frac{1}{2}\int_1^3 (u^2+1)\,du\\ &= \frac{1}{2}\left[\frac{u^3}{3}+u\right]_1^3\\ &= \frac{1}{2}\left(9+3-\frac{1}{3}-1\right)\\ &= \frac{16}{3}.\end{aligned}$$

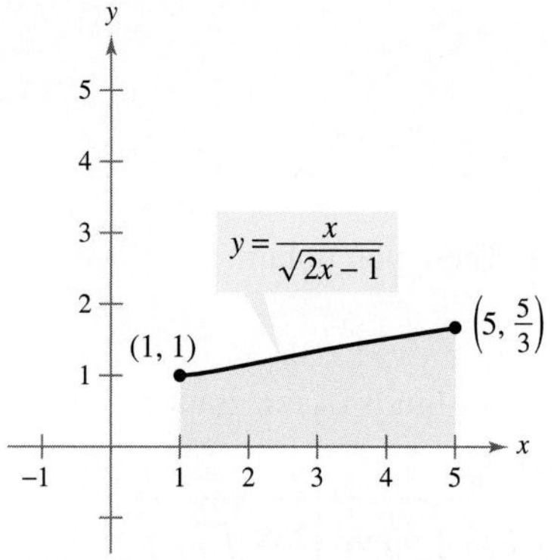

The region before substitution has an area of $\frac{16}{3}$.
Figure 5.37

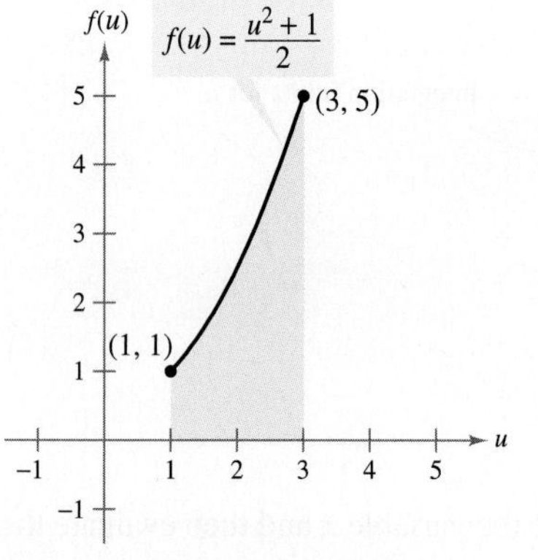

The region after substitution has an area of $\frac{16}{3}$.
Figure 5.38

Geometrically, you can interpret the equation

$$\int_1^5 \frac{x}{\sqrt{2x-1}}\,dx = \int_1^3 \frac{u^2+1}{2}\,du$$

to mean that the two *different* regions shown in Figures 5.37 and 5.38 have the *same* area.

When evaluating definite integrals by substitution, it is possible for the upper limit of integration of the u-variable form to be smaller than the lower limit. If this happens, don't rearrange the limits. Simply evaluate as usual. For example, after substituting $u = \sqrt{1-x}$ in the integral

$$\int_0^1 x^2(1-x)^{1/2}\,dx$$

you obtain $u = \sqrt{1-1} = 0$ when $x = 1$, and $u = \sqrt{1-0} = 1$ when $x = 0$. So, the correct u-variable form of this integral is

$$-2\int_1^0 (1-u^2)^2u^2\,du.$$

Integration of Even and Odd Functions

Even with a change of variables, integration can be difficult. Occasionally, you can simplify the evaluation of a definite integral (over an interval that is symmetric about the y-axis or about the origin) by recognizing the integrand to be an even or odd function (see Figure 5.39).

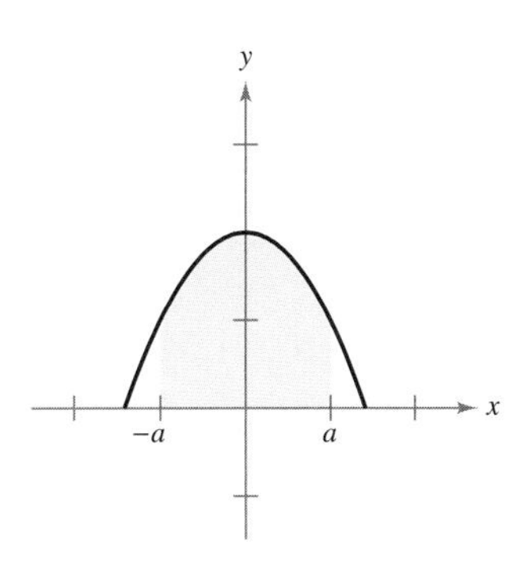

Even function

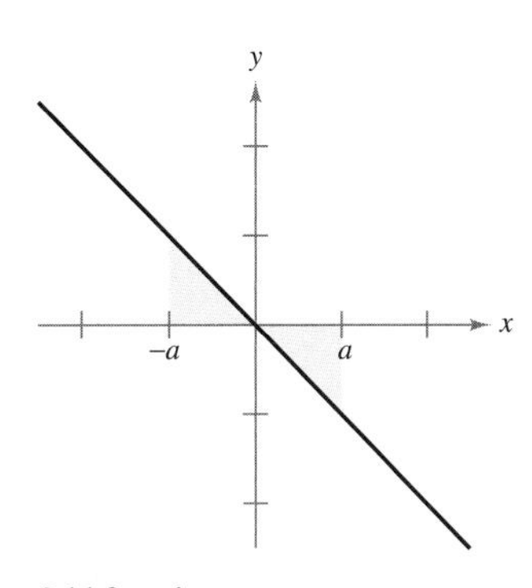

Odd function

Figure 5.39

> **THEOREM 5.15 Integration of Even and Odd Functions**
>
> Let f be integrable on the closed interval $[-a, a]$.
>
> **1.** If f is an *even* function, then $\int_{-a}^{a} f(x)\,dx = 2\int_{0}^{a} f(x)\,dx.$
>
> **2.** If f is an *odd* function, then $\int_{-a}^{a} f(x)\,dx = 0.$

Proof Because f is even, you know that $f(x) = f(-x)$. Using Theorem 5.12 with the substitution $u = -x$ produces

$$\int_{-a}^{0} f(x)\,dx = \int_{a}^{0} f(-u)(-du) = -\int_{a}^{0} f(u)\,du = \int_{0}^{a} f(u)\,du = \int_{0}^{a} f(x)\,dx.$$

Finally, using Theorem 5.6, you obtain

$$\begin{aligned}\int_{-a}^{a} f(x)\,dx &= \int_{-a}^{0} f(x)\,dx + \int_{0}^{a} f(x)\,dx\\ &= \int_{0}^{a} f(x)\,dx + \int_{0}^{a} f(x)\,dx = 2\int_{0}^{a} f(x)\,dx.\end{aligned}$$

This proves the first property. The proof of the second property is left to you (see Exercise 169).

EXAMPLE 10 Integration of an Odd Function

Evaluate $\displaystyle\int_{-\pi/2}^{\pi/2} (\sin^3 x \cos x + \sin x \cos x)\,dx.$

Solution Letting $f(x) = \sin^3 x \cos x + \sin x \cos x$ produces

$$\begin{aligned}f(-x) &= \sin^3(-x)\cos(-x) + \sin(-x)\cos(-x)\\ &= -\sin^3 x \cos x - \sin x \cos x = -f(x).\end{aligned}$$

So, f is an odd function, and because f is symmetric about the origin over $[-\pi/2, \pi/2]$, you can apply Theorem 5.15 to conclude that

$$\int_{-\pi/2}^{\pi/2} (\sin^3 x \cos x + \sin x \cos x)\,dx = 0.$$

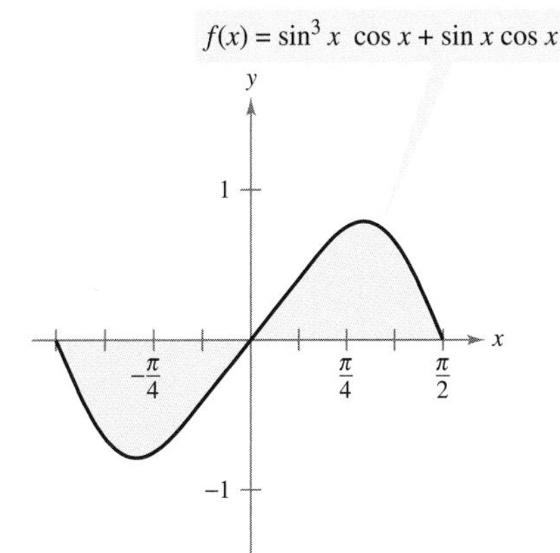

Because f is an odd function,

$$\int_{-\pi/2}^{\pi/2} f(x)\,dx = 0.$$

Figure 5.40

NOTE From Figure 5.40, you can see that the two regions on either side of the y-axis have the same area. However, because one lies below the x-axis and one lies above it, integration produces a cancellation effect. (More will be said about this in Section 7.1.)

Exercises for Section 5.5

See www.CalcChat.com for worked-out solutions to odd-numbered exercises.

In Exercises 1–6, complete the table by identifying u and du for the integral.

$\int f(g(x))g'(x)\,dx$	$u = g(x)$	$du = g'(x)\,dx$
1. $\int (5x^2+1)^2(10x)\,dx$		
2. $\int x^2\sqrt{x^3+1}\,dx$		
3. $\int \dfrac{x}{\sqrt{x^2+1}}\,dx$		
4. $\int \sec 2x \tan 2x\,dx$		
5. $\int \tan^2 x \sec^2 x\,dx$		
6. $\int \dfrac{\cos x}{\sin^3 x}\,dx$		

In Exercises 7–34, find the indefinite integral and check the result by differentiation.

7. $\int (1+2x)^4(2)\,dx$

8. $\int (x^2-9)^3(2x)\,dx$

9. $\int \sqrt{9-x^2}\,(-2x)\,dx$

10. $\int \sqrt[3]{(1-2x^2)}(-4x)\,dx$

11. $\int x^3(x^4+3)^2\,dx$

12. $\int x^2(x^3+5)^4\,dx$

13. $\int x^2(x^3-1)^4\,dx$

14. $\int x(4x^2+3)^3\,dx$

15. $\int t\sqrt{t^2+2}\,dt$

16. $\int t^3\sqrt{t^4+5}\,dt$

17. $\int 5x\sqrt[3]{1-x^2}\,dx$

18. $\int u^2\sqrt{u^3+5}\,du$

19. $\int \dfrac{x}{(1-x^2)^3}\,dx$

20. $\int \dfrac{x^3}{(1+x^4)^2}\,dx$

21. $\int \dfrac{x^2}{(1+x^3)^2}\,dx$

22. $\int \dfrac{x^2}{(9-x^3)^2}\,dx$

23. $\int \dfrac{x}{\sqrt{1-x^2}}\,dx$

24. $\int \dfrac{x^3}{\sqrt{1+x^4}}\,dx$

25. $\int \left(1+\dfrac{1}{t}\right)^3\left(\dfrac{1}{t^2}\right)dt$

26. $\int \left[x^2+\dfrac{1}{(3x)^2}\right]dx$

27. $\int \dfrac{1}{\sqrt{2x}}\,dx$

28. $\int \dfrac{1}{2\sqrt{x}}\,dx$

29. $\int \dfrac{x^2+3x+7}{\sqrt{x}}\,dx$

30. $\int \dfrac{t+2t^2}{\sqrt{t}}\,dt$

31. $\int t^2\left(t-\dfrac{2}{t}\right)dt$

32. $\int \left(\dfrac{t^3}{3}+\dfrac{1}{4t^2}\right)dt$

33. $\int (9-y)\sqrt{y}\,dy$

34. $\int 2\pi y(8-y^{3/2})\,dy$

In Exercises 35–38, solve the differential equation.

35. $\dfrac{dy}{dx} = 4x + \dfrac{4x}{\sqrt{16-x^2}}$

36. $\dfrac{dy}{dx} = \dfrac{10x^2}{\sqrt{1+x^3}}$

37. $\dfrac{dy}{dx} = \dfrac{x+1}{(x^2+2x-3)^2}$

38. $\dfrac{dy}{dx} = \dfrac{x-4}{\sqrt{x^2-8x+1}}$

Slope Fields **In Exercises 39–46, a differential equation, a point, and a slope field are given. A *slope field* consists of line segments with slopes given by the differential equation. These line segments give a visual perspective of the directions of the solutions of the differential equation. (a) Sketch two approximate solutions of the differential equation on the slope field, one of which passes through the given point. (To print an enlarged copy of the graph, go to the website *www.mathgraphs.com*.) (b) Use integration to find the particular solution of the differential equation and use a graphing utility to graph the solution. Compare the result with the sketches in part (a).**

39. $\dfrac{dy}{dx} = x\sqrt{4-x^2}$, $(2, 2)$

40. $\dfrac{dy}{dx} = x^2(x^3-1)^2$, $(1, 0)$

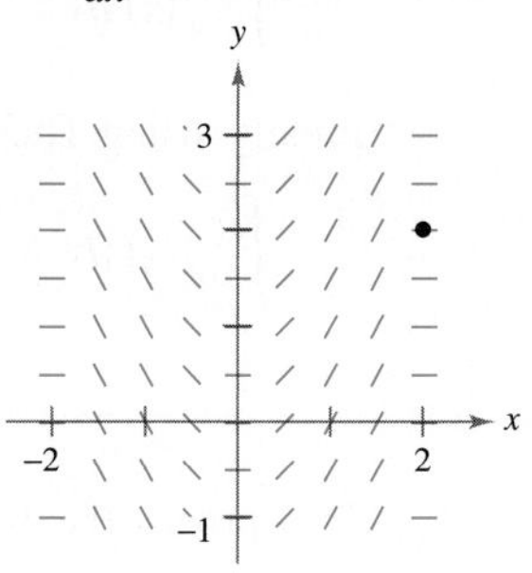

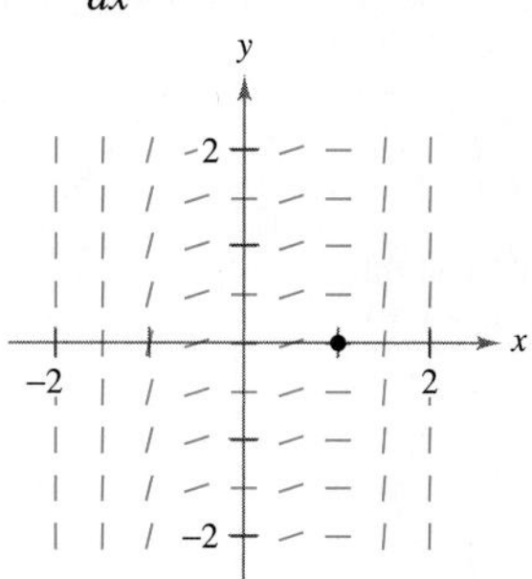

41. $\dfrac{dy}{dx} = x\cos x^2$, $(0, 1)$

42. $\dfrac{dy}{dx} = -2\sec(2x)\tan(2x)$, $(0, -1)$

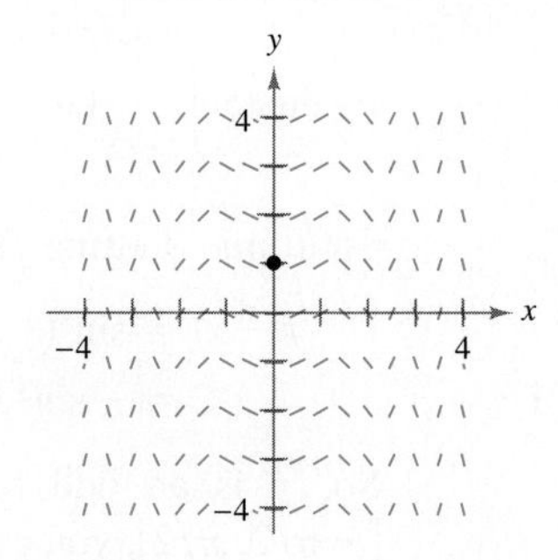

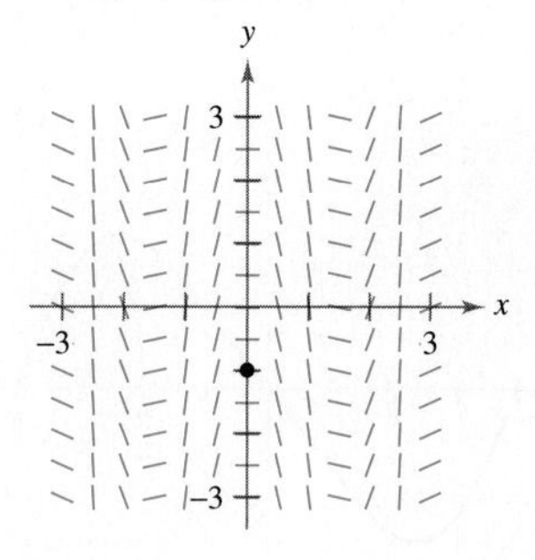

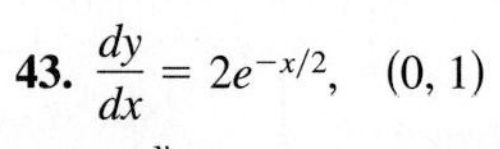

43. $\dfrac{dy}{dx} = 2e^{-x/2}$, $(0, 1)$

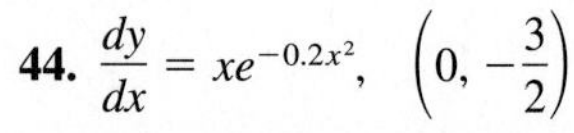

44. $\dfrac{dy}{dx} = xe^{-0.2x^2}$, $\left(0, -\dfrac{3}{2}\right)$

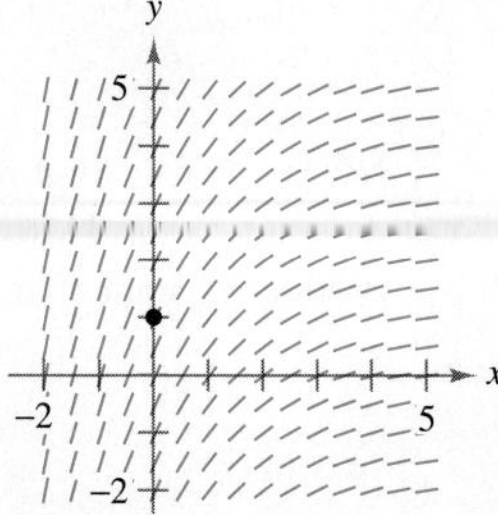

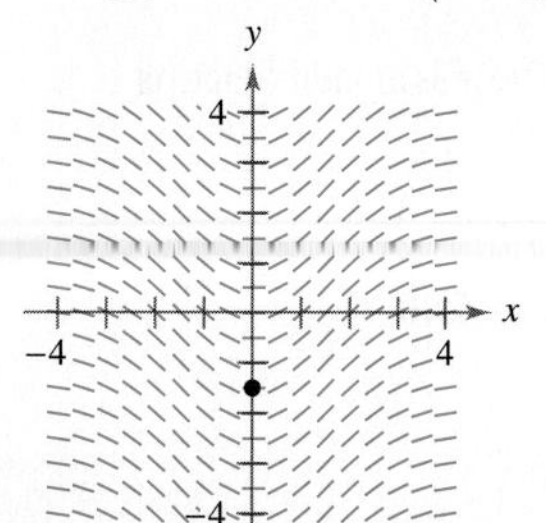

45. $\dfrac{dy}{dx} = e^{x/3}, \quad \left(0, \dfrac{1}{2}\right)$

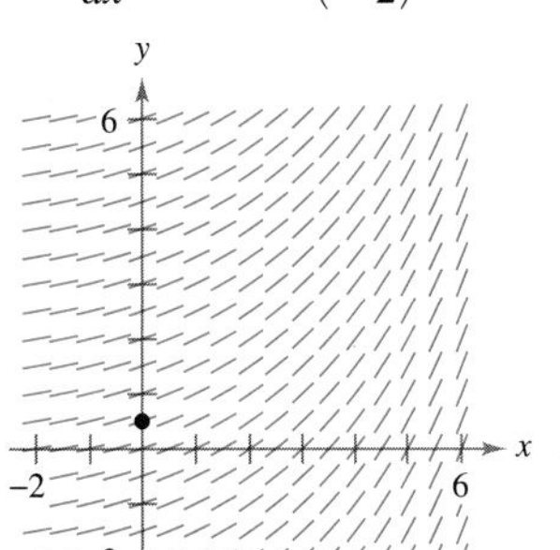

46. $\dfrac{dy}{dx} = e^{\sin x} \cos x, \quad (\pi, 2)$

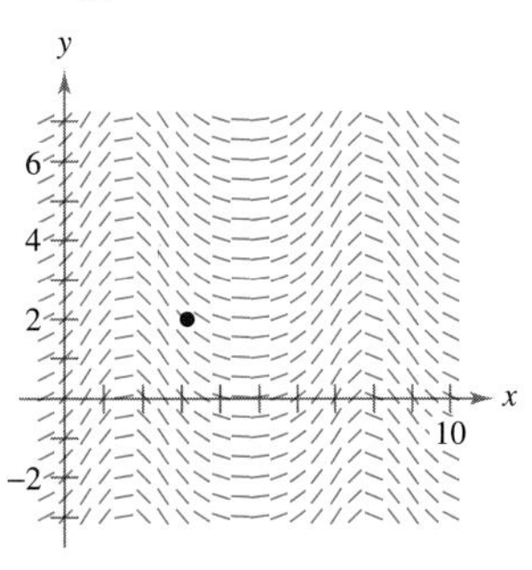

In Exercises 47–78, find the indefinite integral.

47. $\int \pi \sin \pi x \, dx$

48. $\int 4x^3 \sin x^4 \, dx$

49. $\int \sin 2x \, dx$

50. $\int \cos 6x \, dx$

51. $\int \dfrac{1}{\theta^2} \cos \dfrac{1}{\theta} \, d\theta$

52. $\int x \sin x^2 \, dx$

53. $\int e^{5x}(5) \, dx$

54. $\int e^{-x^3}(-3x^2) \, dx$

55. $\int x^2 e^{-x^3} \, dx$

56. $\int (x + 1)e^{x^2+2x} \, dx$

57. $\int \sin 2x \cos 2x \, dx$

58. $\int \sec(2 - x) \tan(2 - x) \, dx$

59. $\int \tan^4 x \sec^2 x \, dx$

60. $\int \sqrt{\tan x} \sec^2 x \, dx$

61. $\int \dfrac{\csc^2 x}{\cot^3 x} \, dx$

62. $\int \dfrac{\sin x}{\cos^3 x} \, dx$

63. $\int \cot^2 x \, dx$

64. $\int \csc^2\left(\dfrac{x}{2}\right) dx$

65. $\int e^x(e^x + 1)^2 \, dx$

66. $\int e^x(1 - 3e^x) \, dx$

67. $\int e^x\sqrt{1 - e^x} \, dx$

68. $\int \dfrac{2e^x - 2e^{-x}}{(e^x + e^{-x})^2} \, dx$

69. $\int \dfrac{5 - e^x}{e^{2x}} \, dx$

70. $\int \dfrac{e^{2x} + 2e^x + 1}{e^x} \, dx$

71. $\int e^{\sin \pi x} \cos \pi x \, dx$

72. $\int e^{\tan 2x} \sec^2 2x \, dx$

73. $\int e^{-x} \sec^2(e^{-x}) \, dx$

74. $\int \ln(e^{2x-1}) \, dx$

75. $\int 3^{x/2} \, dx$

76. $\int 4^{-x} \, dx$

77. $\int x5^{-x^2} \, dx$

78. $\int (3 - x)7^{(3-x)^2} \, dx$

In Exercises 79–84, find an equation for the function f that has the indicated derivative and whose graph passes through the given point.

	Derivative	*Point*
79.	$f'(x) = x\sqrt{4 - x^2}$	$(2, 2)$
80.	$f'(x) = 0.4^{x/3}$	$\left(0, \frac{1}{2}\right)$
81.	$f'(x) = \cos \dfrac{x}{2}$	$(0, 3)$
82.	$f'(x) = \pi \sec \pi x \tan \pi x$	$\left(\frac{1}{3}, 1\right)$
83.	$f'(x) = 2e^{-x/4}$	$(0, 1)$
84.	$f'(x) = x^2e^{-0.2x^3}$	$\left(0, \frac{3}{2}\right)$

In Exercises 85 and 86, find the particular solution of the differential equation that satisfies the initial conditions.

85. $f''(x) = \frac{1}{2}(e^x + e^{-x}), \quad f(0) = 1, f'(0) = 0$

86. $f''(x) = \sin x + e^{2x}, \quad f(0) = \frac{1}{4}, f'(0) = \frac{1}{2}$

In Exercises 87–94, find the indefinite integral by the method shown in Example 5.

87. $\int x\sqrt{x + 2} \, dx$

88. $\int x\sqrt{2x + 3} \, dx$

89. $\int x^2\sqrt{1 - x} \, dx$

90. $\int (x + 1)\sqrt{2 - x} \, dx$

91. $\int \dfrac{x^2 - 1}{\sqrt{2x - 1}} \, dx$

92. $\int \dfrac{2x + 1}{\sqrt{x + 4}} \, dx$

93. $\int \dfrac{-x}{(x + 1) - \sqrt{x + 1}} \, dx$

94. $\int t\sqrt[3]{t - 4} \, dt$

In Exercises 95–112, evaluate the definite integral. Use a graphing utility to verify your result.

95. $\int_{-1}^{1} x(x^2 + 1)^3 \, dx$

96. $\int_{-2}^{4} x^2(x^3 + 8)^2 \, dx$

97. $\int_{1}^{2} 2x^2\sqrt{x^3 + 1} \, dx$

98. $\int_{0}^{2} x\sqrt{4 - x^2} \, dx$

99. $\int_{0}^{4} \dfrac{1}{\sqrt{2x + 1}} \, dx$

100. $\int_{0}^{2} \dfrac{x}{\sqrt{1 + 2x^2}} \, dx$

101. $\int_{0}^{1} e^{-2x} \, dx$

102. $\int_{1}^{2} e^{1-x} \, dx$

103. $\int_{1}^{3} \dfrac{e^{3/x}}{x^2} \, dx$

104. $\int_{0}^{\sqrt{2}} xe^{-(x^2/2)} \, dx$

105. $\int_{1}^{9} \dfrac{1}{\sqrt{x}\left(1 + \sqrt{x}\right)^2} \, dx$

106. $\int_{0}^{2} x\sqrt[3]{4 + x^2} \, dx$

107. $\int_{1}^{2} (x - 1)\sqrt{2 - x} \, dx$

108. $\int_{1}^{5} \dfrac{x}{\sqrt{2x - 1}} \, dx$

109. $\int_{0}^{\pi/2} \cos\left(\dfrac{2x}{3}\right) dx$

110. $\int_{\pi/3}^{\pi/2} (x + \cos x) \, dx$

111. $\int_{-1}^{2} 2^x \, dx$

112. $\int_{-2}^{0} (3^3 - 5^2) \, dx$

Differential Equations **In Exercises 113–116, the graph of a function f is shown. Use the differential equation and the given point to find an equation of the function.**

113. $\dfrac{dy}{dx} = 18x^2(2x^3 + 1)^2$

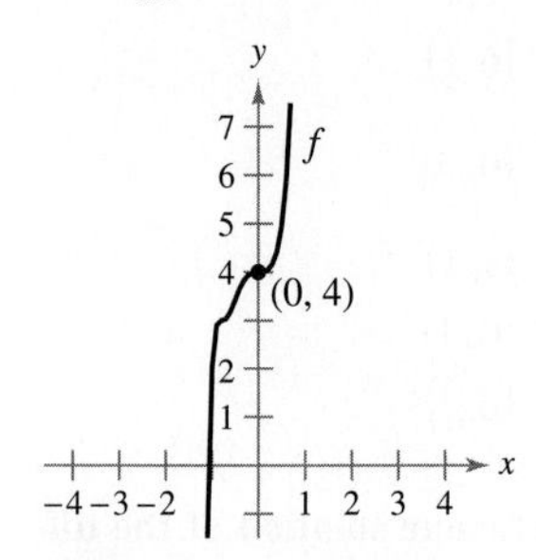

114. $\dfrac{dy}{dx} = \dfrac{-48}{(3x + 5)^3}$

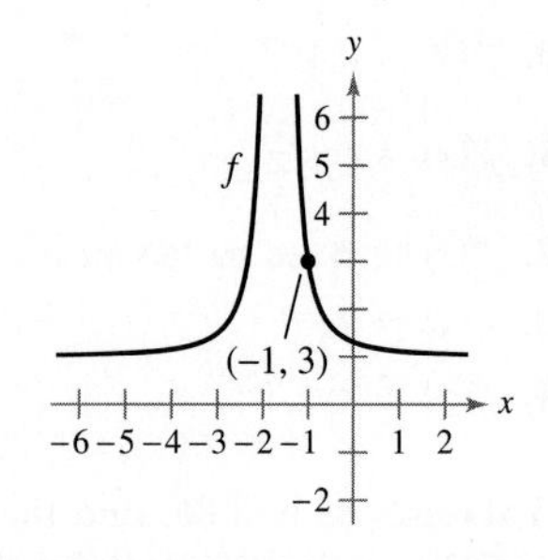

115. $\dfrac{dy}{dx} = \dfrac{2x}{\sqrt{2x^2 - 1}}$

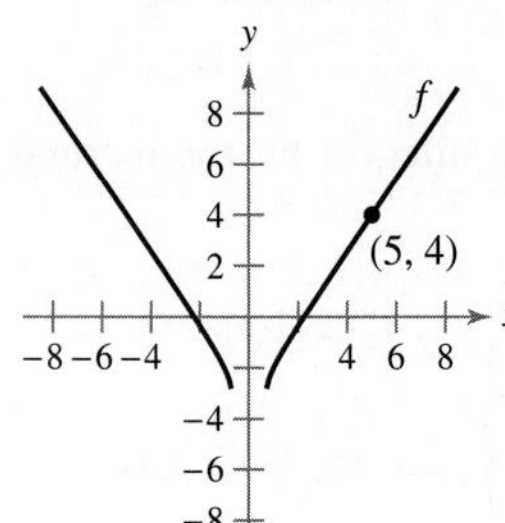

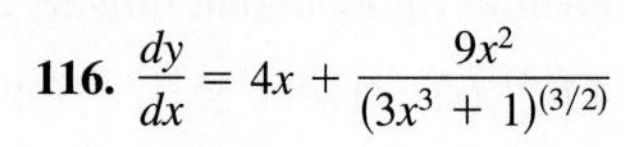

116. $\dfrac{dy}{dx} = 4x + \dfrac{9x^2}{(3x^3 + 1)^{(3/2)}}$

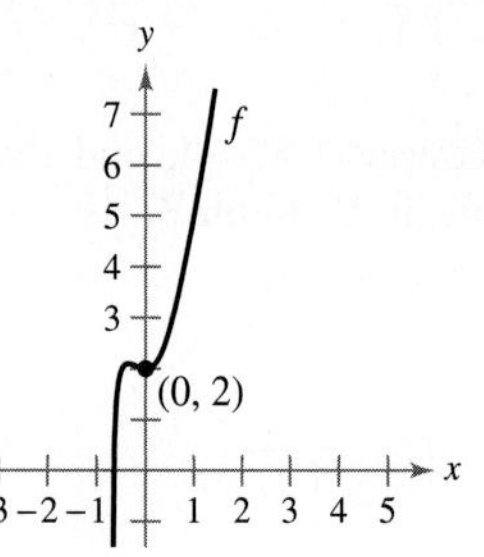

In Exercises 117–122, find the area of the region. Use a graphing utility to verify your result.

117. $\displaystyle\int_0^7 x\sqrt[3]{x + 1}\,dx$

118. $\displaystyle\int_{-2}^6 x^2\sqrt[3]{x + 2}\,dx$

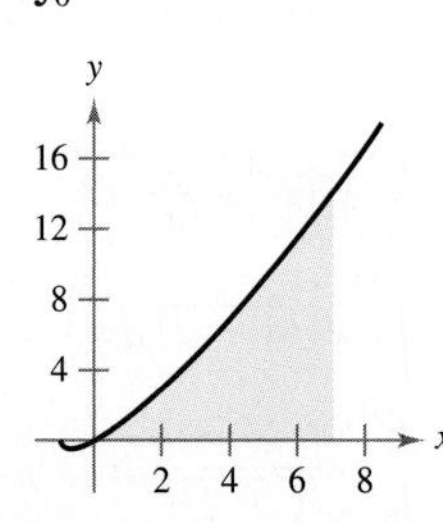

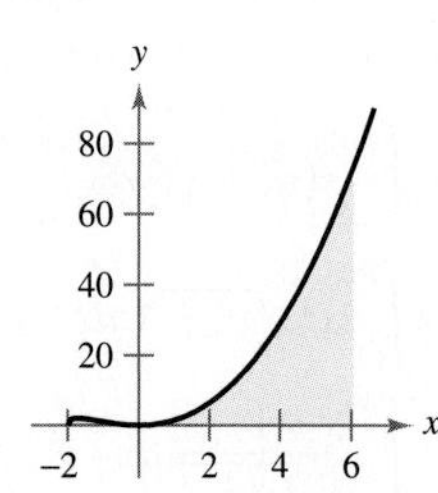

119. $y = 2\sin x + \sin 2x$

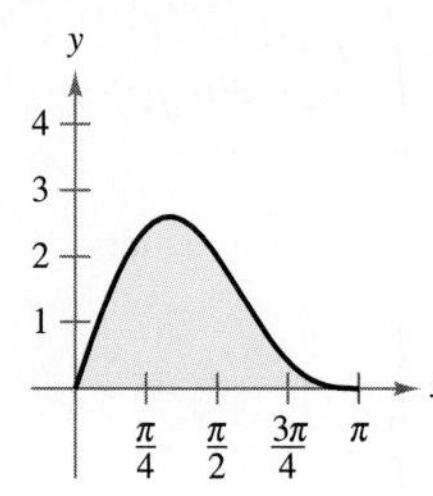

120. $y = \sin x + \cos 2x$

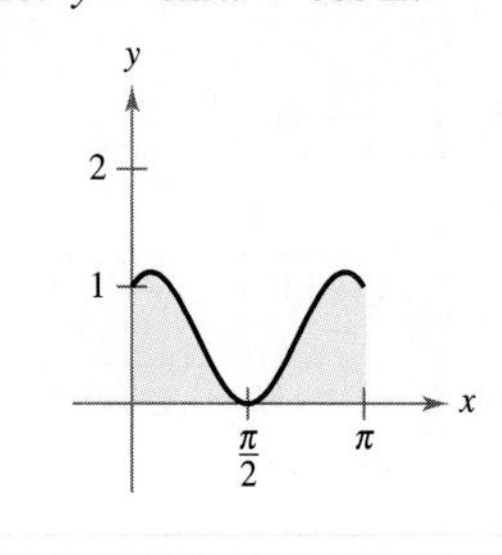

121. $\displaystyle\int_{\pi/2}^{2\pi/3} \sec^2\left(\frac{x}{2}\right) dx$

122. $\displaystyle\int_{\pi/12}^{\pi/4} \csc 2x \cot 2x\,dx$

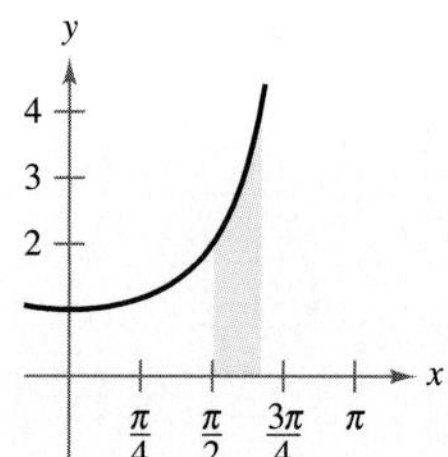

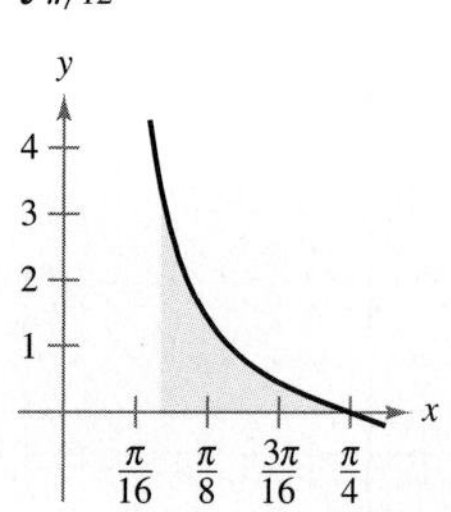

Area **In Exercises 123–126, find the area of the region bounded by the graphs of the equations. Use a graphing utility to graph the region and verify your result.**

123. $y = e^x,\ y = 0,\ x = 0,\ x = 5$

124. $y = e^{-x},\ y = 0,\ x = a,\ x = b$

125. $y = xe^{-x^2/4},\ y = 0,\ x = 0,\ x = \sqrt{6}$

126. $y = e^{-2x} + 2,\ y = 0,\ x = 0,\ x = 2$

In Exercises 127–134, use a graphing utility to evaluate the integral. Graph the region whose area is given by the definite integral.

127. $\displaystyle\int_0^4 \frac{x}{\sqrt{2x + 1}}\,dx$

128. $\displaystyle\int_0^2 x^3\sqrt{x + 2}\,dx$

129. $\displaystyle\int_3^7 x\sqrt{x - 3}\,dx$

130. $\displaystyle\int_1^5 x^2\sqrt{x - 1}\,dx$

131. $\displaystyle\int_0^3 \left(\theta + \cos\frac{\theta}{6}\right) d\theta$

132. $\displaystyle\int_0^{\pi/2} \sin 2x\,dx$

133. $\displaystyle\int_0^{\sqrt{2}} xe^{-x^2/2}\,dx$

134. $\displaystyle\int_0^2 (e^{-2x} + 2)\,dx$

Writing **In Exercises 135 and 136, find the indefinite integral in two ways. Explain any difference in the forms of the answers.**

135. $\displaystyle\int (2x - 1)^2\,dx$

136. $\displaystyle\int \sin x \cos x\,dx$

In Exercises 137–140, evaluate the integral using the properties of even and odd functions as an aid.

137. $\displaystyle\int_{-2}^2 x^2(x^2 + 1)\,dx$

138. $\displaystyle\int_{-\pi/2}^{\pi/2} \sin^2 x \cos x\,dx$

139. $\displaystyle\int_{-2}^2 x(x^2 + 1)^3\,dx$

140. $\displaystyle\int_{-\pi/4}^{\pi/4} \sin x \cos x\,dx$

141. Use $\int_0^2 x^2\,dx = \frac{8}{3}$ to evaluate the definite integrals without using the Fundamental Theorem of Calculus.

(a) $\displaystyle\int_{-2}^0 x^2\,dx$ (b) $\displaystyle\int_{-2}^2 x^2\,dx$

(c) $\displaystyle\int_0^2 -x^2\,dx$ (d) $\displaystyle\int_{-2}^0 3x^2\,dx$

142. Use the symmetry of the graphs of the sine and cosine functions as an aid in evaluating each of the integrals.

(a) $\int_{-\pi/4}^{\pi/4} \sin x\, dx$ (b) $\int_{-\pi/4}^{\pi/4} \cos x\, dx$

(c) $\int_{-\pi/2}^{\pi/2} \cos x\, dx$ (d) $\int_{-\pi/2}^{\pi/2} \sin x \cos x\, dx$

In Exercises 143 and 144, write the integral as the sum of the integral of an odd function and the integral of an even function. Use this simplification to evaluate the integral.

143. $\int_{-4}^{4} (x^3 + 6x^2 - 2x - 3)\, dx$ **144.** $\int_{-\pi}^{\pi} (\sin 3x + \cos 3x)\, dx$

Writing About Concepts

145. Describe why

$$\int x(5 - x^2)^3\, dx \neq \int u^3\, du$$

where $u = 5 - x^2$.

146. Without integrating, explain why

$$\int_{-2}^{2} x(x^2 + 1)^2\, dx = 0.$$

147. ***Cash Flow*** The rate of disbursement dQ/dt of a 2 million dollar federal grant is proportional to the square of $100 - t$. Time t is measured in days $(0 \leq t \leq 100)$, and Q is the amount that remains to be disbursed. Find the amount that remains to be disbursed after 50 days. Assume that all the money will be disbursed in 100 days.

148. ***Depreciation*** The rate of depreciation dV/dt of a machine is inversely proportional to the square of $t + 1$, where V is the value of the machine t years after it was purchased. The initial value of the machine was \$500,000, and its value decreased \$100,000 in the first year. Estimate its value after 4 years.

149. ***Rainfall*** The normal monthly rainfall at the Seattle-Tacoma airport can be approximated by the model

$$R = 3.121 + 2.399 \sin(0.524t + 1.377)$$

where R is measured in inches and t is the time in months, with $t = 1$ corresponding to January. *(Source: U.S. National Oceanic and Atmospheric Administration)*

(a) Determine the extrema of the function over a one-year period.

(b) Use integration to approximate the normal annual rainfall. (*Hint:* Integrate over the interval [0, 12].)

(c) Approximate the average monthly rainfall during the months of October, November, and December.

150. ***Sales*** The sales S (in thousands of units) of a seasonal product are given by the model

$$S = 74.50 + 43.75 \sin \frac{\pi t}{6}$$

where t is the time in months, with $t = 1$ corresponding to January. Find the average sales for each time period.

(a) The first quarter $(0 \leq t \leq 3)$

(b) The second quarter $(3 \leq t \leq 6)$

(c) The entire year $(0 \leq t \leq 12)$

151. ***Water Supply*** A model for the flow rate of water at a pumping station on a given day is

$$R(t) = 53 + 7 \sin\left(\frac{\pi t}{6} + 3.6\right) + 9 \cos\left(\frac{\pi t}{12} + 8.9\right)$$

where $0 \leq t \leq 24$. R is the flow rate in thousands of gallons per hour, and t is the time in hours.

(a) Use a graphing utility to graph the rate function and approximate the maximum flow rate at the pumping station.

(b) Approximate the total volume of water pumped in 1 day.

152. ***Electricity*** The oscillating current in an electrical circuit is

$$I = 2 \sin(60\pi t) + \cos(120\pi t)$$

where I is measured in amperes and t is measured in seconds. Find the average current for each time interval.

(a) $0 \leq t \leq \frac{1}{60}$

(b) $0 \leq t \leq \frac{1}{240}$

(c) $0 \leq t \leq \frac{1}{30}$

Probability **In Exercises 153 and 154, the function**

$$f(x) = kx^n(1 - x)^m, \quad 0 \leq x \leq 1$$

where $n > 0$, $m > 0$, and k is a constant, can be used to represent various probability distributions. If k is chosen such that

$$\int_0^1 f(x)\, dx = 1$$

the probability that x will fall between a and b $(0 \leq a \leq b \leq 1)$ is

$$P_{a,b} = \int_a^b f(x)\, dx.$$

153. The probability that a person will remember between $100a\%$ and $100b\%$ of material learned in an experiment is

$$P_{a,b} = \int_a^b \frac{15}{4} x\sqrt{1 - x}\, dx$$

where x represents the proportion remembered. (See figure on the next page.)

(a) For a randomly chosen individual, what is the probability that he or she will recall between 50% and 75% of the material?

(b) What is the median percent recall? That is, for what value of b is it true that the probability of recalling 0 to $100b\%$ is 0.5?

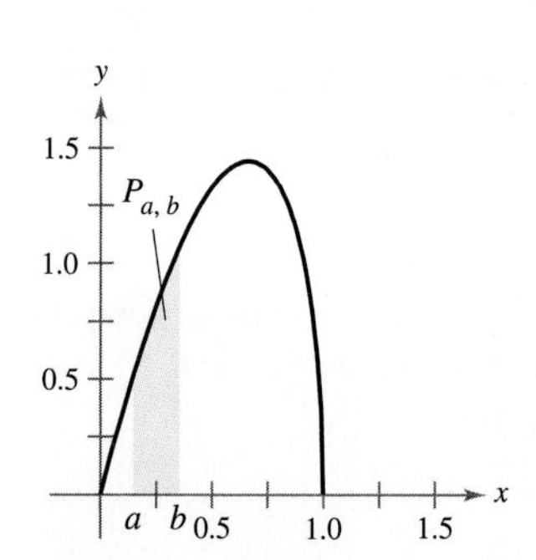

Figure for 153

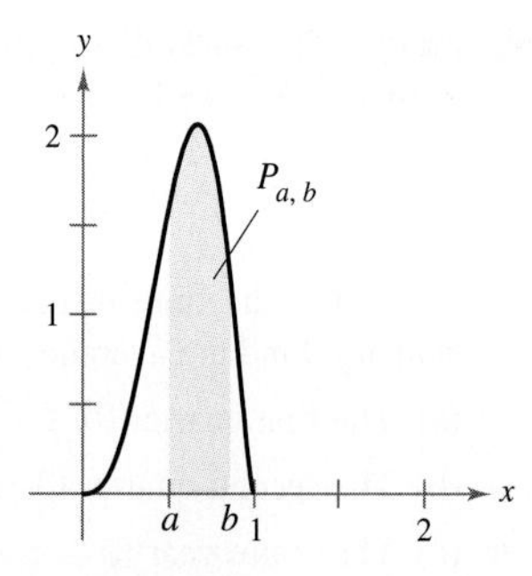

Figure for 154

154. The probability that ore samples taken from a region contain between $100a\%$ and $100b\%$ iron is

$$P_{a,\,b} = \int_a^b \frac{1155}{32} x^3(1-x)^{3/2}\, dx$$

where x represents the proportion of iron. (See figure.) What is the probability that a sample will contain between

(a) 0% and 25% iron?

(b) 50% and 100% iron?

155. ***Probability*** A car battery has an average lifetime of 48 months with a standard deviation of 6 months. The battery lives are normally distributed. The probability that a given battery will last between 48 months and 60 months is

$$0.0665 \int_{48}^{60} e^{-0.0139(t-48)^2}\, dt.$$

Use the integration capabilities of a graphing utility to approximate the integral. Interpret the resulting probability.

156. Given $e^x \ge 1$ for $x \ge 0$, it follows that

$$\int_0^x e^t\, dt \ge \int_0^x 1\, dt.$$

Perform this integration to derive the inequality $e^x \ge 1 + x$ for $x \ge 0$.

157. ***Graphical Analysis*** Consider the functions f and g, where

$$f(x) = 6 \sin x \cos^2 x \quad \text{and} \quad g(t) = \int_0^t f(x)\, dx.$$

(a) Use a graphing utility to graph f and g in the same viewing window.

(b) Explain why g is nonnegative.

(c) Identify the points on the graph of g that correspond to the extrema of f.

(d) Does each of the zeros of f correspond to an extremum of g? Explain.

(e) Consider the function

$$h(t) = \int_{\pi/2}^t f(x)\, dx.$$

Use a graphing utility to graph h. What is the relationship between g and h? Verify your conjecture.

158. Find $\displaystyle\lim_{n\to+\infty} \sum_{i=1}^{n} \frac{\sin(i\pi/n)}{n}$ by evaluating an appropriate definite integral over the interval $[0, 1]$.

159. (a) Show that $\int_0^1 x^2(1-x)^5\, dx = \int_0^1 x^5(1-x)^2\, dx$.

(b) Show that $\int_0^1 x^a(1-x)^b\, dx = \int_0^1 x^b(1-x)^a\, dx$.

160. (a) Show that $\int_0^{\pi/2} \sin^2 x\, dx = \int_0^{\pi/2} \cos^2 x\, dx$.

(b) Show that $\int_0^{\pi/2} \sin^n x\, dx = \int_0^{\pi/2} \cos^n x\, dx$, where n is a positive integer.

True or False? **In Exercises 161–166, determine whether the statement is true or false. If it is false, explain why or give an example that shows it is false.**

161. $\displaystyle\int (2x+1)^2\, dx = \frac{1}{3}(2x+1)^3 + C$

162. $\displaystyle\int x(x^2+1)\, dx = \frac{1}{2}x^2\left(\frac{1}{3}x^3 + x\right) + C$

163. $\displaystyle\int_{-10}^{10} (ax^3 + bx^2 + cx + d)\, dx = 2\int_0^{10} (bx^2 + d)\, dx$

164. $\displaystyle\int_a^b \sin x\, dx = \int_a^{b+2\pi} \sin x\, dx$

165. $\displaystyle 4\int \sin x \cos x\, dx = -\cos 2x + C$

166. $\displaystyle\int \sin^2 2x \cos 2x\, dx = \frac{1}{3}\sin^3 2x + C$

167. Assume that f is continuous everywhere and that c is a constant. Show that

$$\int_{ca}^{cb} f(x)\, dx = c\int_a^b f(cx)\, dx.$$

168. (a) Verify that $\sin u - u\cos u + C = \int u \sin u\, du$.

(b) Use part (a) to show that $\int_0^{\pi^2} \sin\sqrt{x}\, dx = 2\pi$.

169. Complete the proof of Theorem 5.15.

170. Show that if f is continuous on the entire real number line, then

$$\int_a^b f(x+h)\, dx = \int_{a+h}^{b+h} f(x)\, dx.$$

Putnam Exam Challenge

171. If $a_0, a_1, \ldots, a_n$ are real numbers satisfying

$$\frac{a_0}{1} + \frac{a_1}{2} + \cdots + \frac{a_n}{n+1} = 0$$

show that the equation $a_0 + a_1x + a_2x^2 + \cdots + a_nx^n = 0$ has at least one real zero.

172. Find all the continuous positive functions $f(x)$, for $0 \le x \le 1$, such that

$$\int_0^1 f(x)\, dx = 1, \quad \int_0^1 f(x)x\, dx = \alpha, \text{ and } \int_0^1 f(x)x^2\, dx = \alpha^2$$

where α is a real number.

These problems were composed by the Committee on the Putnam Prize Competition.

Section 5.6

Numerical Integration

- **Approximate a definite integral using the Trapezoidal Rule.**
- **Approximate a definite integral using Simpson's Rule.**
- **Analyze the approximate errors in the Trapezoidal Rule and Simpson's Rule.**

The Trapezoidal Rule

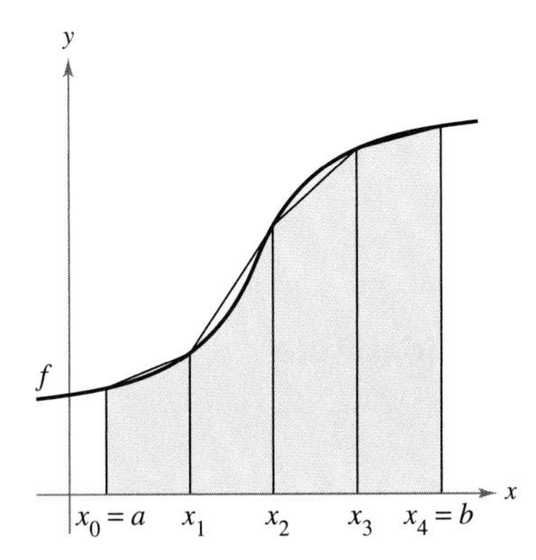

The area of the region can be approximated using four trapezoids.
Figure 5.41

Some elementary functions simply do not have antiderivatives that are elementary functions. For example, there is no elementary function that has any of the following functions as its derivative.

$$\sqrt[3]{x}\sqrt{1-x}, \qquad \sqrt{x}\cos x, \qquad \frac{\cos x}{x}, \qquad \sqrt{1-x^3}, \qquad \sin x^2$$

If you need to evaluate a definite integral involving a function whose antiderivative cannot be found, the Fundamental Theorem of Calculus cannot be applied, and you must resort to an approximation technique. Two such techniques are described in this section.

One way to approximate a definite integral is to use n trapezoids, as shown in Figure 5.41. In the development of this method, assume that f is continuous and positive on the interval $[a, b]$. So, the definite integral

$$\int_a^b f(x)\,dx$$

represents the area of the region bounded by the graph of f and the x-axis, from $x = a$ to $x = b$. First, partition the interval $[a, b]$ into n subintervals, each of width $\Delta x = (b - a)/n$, such that

$$a = x_0 < x_1 < x_2 < \cdot \cdot \cdot < x_n = b.$$

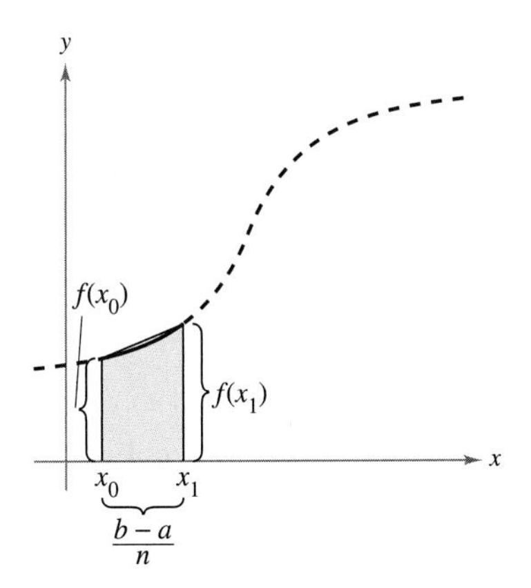

The area of the first trapezoid is $\left[\frac{f(x_0) + f(x_1)}{2}\right]\left(\frac{b-a}{n}\right)$.
Figure 5.42

Then form a trapezoid for each subinterval (see Figure 5.42). The area of the ith trapezoid is

$$\text{Area of } i\text{th trapezoid} = \left[\frac{f(x_{i-1}) + f(x_i)}{2}\right]\left(\frac{b-a}{n}\right).$$

This implies that the sum of the areas of the n trapezoids is

$$\begin{aligned}\text{Area} &= \left(\frac{b-a}{n}\right)\left[\frac{f(x_0) + f(x_1)}{2} + \cdot \cdot \cdot + \frac{f(x_{n-1}) + f(x_n)}{2}\right]\\ &= \left(\frac{b-a}{2n}\right)[f(x_0) + f(x_1) + f(x_1) + f(x_2) + \cdot \cdot \cdot + f(x_{n-1}) + f(x_n)]\\ &= \left(\frac{b-a}{2n}\right)[f(x_0) + 2f(x_1) + 2f(x_2) + \cdot \cdot \cdot + 2f(x_{n-1}) + f(x_n)].\end{aligned}$$

Letting $\Delta x = (b - a)/n$, you can take the limit as $n \to \infty$ to obtain

$$\begin{aligned}&\lim_{n\to\infty}\left(\frac{b-a}{2n}\right)[f(x_0) + 2f(x_1) + \cdot \cdot \cdot + 2f(x_{n-1}) + f(x_n)]\\ &\quad = \lim_{n\to\infty}\left[\frac{[f(a) - f(b)]\,\Delta x}{2} + \sum_{i=1}^{n} f(x_i)\,\Delta x\right]\\ &\quad = \lim_{n\to\infty}\frac{[f(a) - f(b)](b-a)}{2n} + \lim_{n\to\infty}\sum_{i=1}^{n} f(x_i)\,\Delta x\\ &\quad = 0 + \int_a^b f(x)\,dx.\end{aligned}$$

The result is summarized in the following theorem.

THEOREM 5.16 The Trapezoidal Rule

Let f be continuous on $[a, b]$. The Trapezoidal Rule for approximating $\int_a^b f(x)\,dx$ is given by

$$\int_a^b f(x)\,dx \approx \frac{b-a}{2n}[f(x_0) + 2f(x_1) + 2f(x_2) + \cdot \cdot \cdot + 2f(x_{n-1}) + f(x_n)].$$

Moreover, as $n \to \infty$, the right-hand side approaches $\int_a^b f(x)\,dx$.

NOTE Observe that the coefficients in the Trapezoidal Rule have the following pattern.

1 2 2 2 . . . 2 2 1

EXAMPLE 1 Approximation with the Trapezoidal Rule

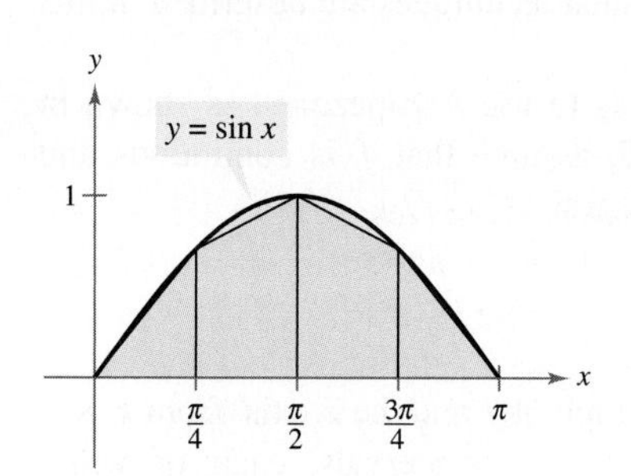

Four subintervals

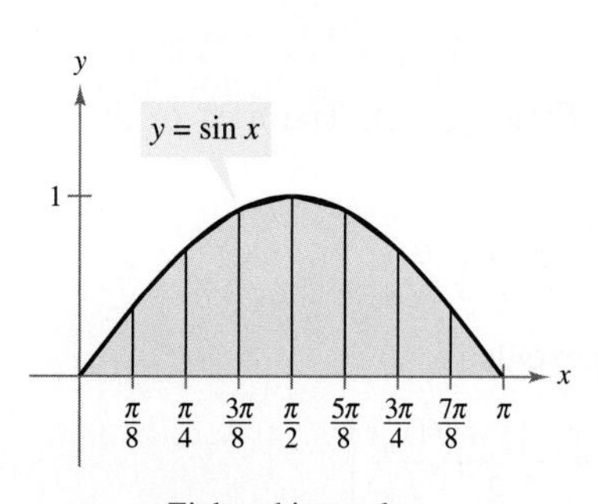

Eight subintervals

Trapezoidal approximations
Figure 5.43

Use the Trapezoidal Rule to approximate

$$\int_0^{\pi} \sin x\,dx.$$

Compare the results for $n = 4$ and $n = 8$, as shown in Figure 5.43.

Solution When $n = 4$, $\Delta x = \pi/4$, and you obtain

$$\int_0^{\pi} \sin x\,dx \approx \frac{\pi}{8}\left(\sin 0 + 2\sin\frac{\pi}{4} + 2\sin\frac{\pi}{2} + 2\sin\frac{3\pi}{4} + \sin \pi\right)$$

$$= \frac{\pi}{8}\left(0 + \sqrt{2} + 2 + \sqrt{2} + 0\right) = \frac{\pi\left(1 + \sqrt{2}\right)}{4} \approx 1.896.$$

When $n = 8$, $\Delta x = \pi/8$, and you obtain

$$\int_0^{\pi} \sin x\,dx \approx \frac{\pi}{16}\left(\sin 0 + 2\sin\frac{\pi}{8} + 2\sin\frac{\pi}{4} + 2\sin\frac{3\pi}{8} + 2\sin\frac{\pi}{2}\right.$$

$$\left. + 2\sin\frac{5\pi}{8} + 2\sin\frac{3\pi}{4} + 2\sin\frac{7\pi}{8} + \sin \pi\right)$$

$$= \frac{\pi}{16}\left(2 + 2\sqrt{2} + 4\sin\frac{\pi}{8} + 4\sin\frac{3\pi}{8}\right) \approx 1.974.$$

For this particular integral, you could have found an antiderivative and determined that the exact area of the region is 2.

TECHNOLOGY Most graphing utilities and computer algebra systems have built-in programs that can be used to approximate the value of a definite integral. Try using such a program to approximate the integral in Example 1. How close is your approximation?

When you use such a program, you need to be aware of its limitations. Often, you are given no indication of the degree of accuracy of the approximation. Other times, you may be given an approximation that is completely wrong. For instance, try using a built-in numerical integration program to evaluate

$$\int_{-1}^{2} \frac{1}{x}\,dx.$$

Your calculator should give an error message. Does it?

It is interesting to compare the Trapezoidal Rule with the Midpoint Rule given in Section 5.2 (Exercises 63–66). For the Trapezoidal Rule, you average the function values at the endpoints of the subintervals, but for the Midpoint Rule you take the function values of the subinterval midpoints.

$$\int_a^b f(x)\,dx \approx \sum_{i=1}^{n} f\left(\frac{x_i + x_{i-1}}{2}\right)\Delta x \qquad \text{Midpoint Rule}$$

$$\int_a^b f(x)\,dx \approx \sum_{i=1}^{n} \left(\frac{f(x_i) + f(x_{i-1})}{2}\right)\Delta x \qquad \text{Trapezoidal Rule}$$

NOTE There are two important points that should be made concerning the Trapezoidal Rule (or the Midpoint Rule). First, the approximation tends to become more accurate as n increases. For instance, in Example 1, if $n = 16$, the Trapezoidal Rule yields an approximation of 1.994. Second, although you could have used the Fundamental Theorem to evaluate the integral in Example 1, this theorem cannot be used to evaluate an integral as simple as $\int_0^\pi \sin x^2\,dx$ because $\sin x^2$ has no elementary antiderivative. Yet, the Trapezoidal Rule can be applied easily to this integral.

Simpson's Rule

One way to view the trapezoidal approximation of a definite integral is to say that on each subinterval, you approximate f by a *first*-degree polynomial. In Simpson's Rule, named after the English mathematician Thomas Simpson (1710–1761), you take this procedure one step further and approximate f by *second*-degree polynomials.

Before presenting Simpson's Rule, a theorem for evaluating integrals of polynomials of degree 2 (or less) is listed.

THEOREM 5.17 Integral of $p(x) = Ax^2 + Bx + C$

If $p(x) = Ax^2 + Bx + C$, then

$$\int_a^b p(x)\,dx = \left(\frac{b-a}{6}\right)\left[p(a) + 4p\left(\frac{a+b}{2}\right) + p(b)\right].$$

Proof

$$\begin{aligned}\int_a^b p(x)\,dx &= \int_a^b (Ax^2 + Bx + C)\,dx\\ &= \left[\frac{Ax^3}{3} + \frac{Bx^2}{2} + Cx\right]_a^b\\ &= \frac{A(b^3 - a^3)}{3} + \frac{B(b^2 - a^2)}{2} + C(b-a)\\ &= \left(\frac{b-a}{6}\right)[2A(a^2 + ab + b^2) + 3B(b+a) + 6C]\end{aligned}$$

By expansion and collection of terms, the expression inside the brackets becomes

$$\underbrace{(Aa^2 + Ba + C)}_{p(a)} + \underbrace{4\left[A\left(\frac{b+a}{2}\right)^2 + B\left(\frac{b+a}{2}\right) + C\right]}_{4p\left(\frac{a+b}{2}\right)} + \underbrace{(Ab^2 + Bb + C)}_{p(b)}$$

and you can write

$$\int_a^b p(x)\,dx = \left(\frac{b-a}{6}\right)\left[p(a) + 4p\left(\frac{a+b}{2}\right) + p(b)\right].$$

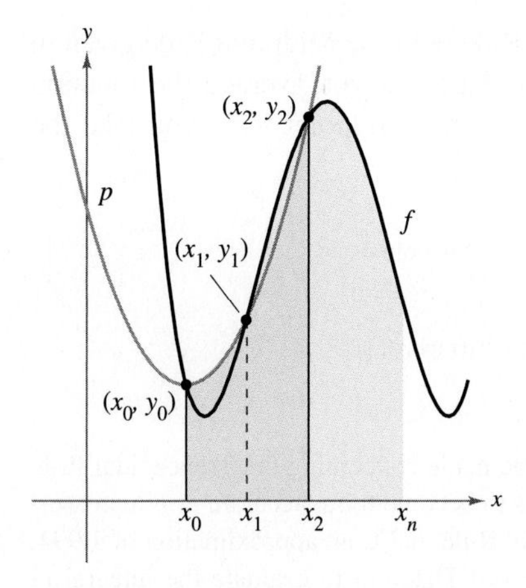

$$\int_{x_0}^{x_2} p(x)\,dx \approx \int_{x_0}^{x_2} f(x)\,dx$$

Figure 5.44

To develop Simpson's Rule for approximating a definite integral, you again partition the interval $[a, b]$ into n subintervals, each of width $\Delta x = (b - a)/n$. This time, however, n is required to be even, and the subintervals are grouped in pairs such that

$$a = \underbrace{x_0 < x_1 < x_2}_{[x_0,\, x_2]} < \underbrace{x_3 < x_4}_{[x_2,\, x_4]} < \cdot \cdot \cdot < \underbrace{x_{n-2} < x_{n-1} < x_n}_{[x_{n-2},\, x_n]} = b.$$

On each (double) subinterval $[x_{i-2}, x_i]$, you can approximate f by a polynomial p of degree less than or equal to 2. (See Exercise 56.) For example, on the subinterval $[x_0, x_2]$, choose the polynomial of least degree passing through the points (x_0, y_0), (x_1, y_1), and (x_2, y_2), as shown in Figure 5.44. Now, using p as an approximation of f on this subinterval, you have, by Theorem 5.17,

$$\begin{aligned}\int_{x_0}^{x_2} f(x)\,dx \approx \int_{x_0}^{x_2} p(x)\,dx &= \frac{x_2 - x_0}{6}\left[p(x_0) + 4p\left(\frac{x_0 + x_2}{2}\right) + p(x_2)\right]\\ &= \frac{2[(b - a)/n]}{6}\left[p(x_0) + 4p(x_1) + p(x_2)\right]\\ &= \frac{b - a}{3n}\left[f(x_0) + 4f(x_1) + f(x_2)\right].\end{aligned}$$

Repeating this procedure on the entire interval $[a, b]$ produces the following theorem.

NOTE Observe that the coefficients in Simpson's Rule have the following pattern.

1 4 2 4 2 4 . . . 4 2 4 1

THEOREM 5.18 Simpson's Rule (n is even)

Let f be continuous on $[a, b]$. Simpson's Rule for approximating $\int_a^b f(x)\,dx$ is

$$\int_a^b f(x)\,dx \approx \frac{b - a}{3n}\left[f(x_0) + 4f(x_1) + 2f(x_2) + 4f(x_3) + \cdot \cdot \cdot + 4f(x_{n-1}) + f(x_n)\right].$$

Moreover, as $n \to \infty$, the right-hand side approaches $\int_a^b f(x)\,dx$.

In Example 1, the Trapezoidal Rule was used to estimate $\int_0^{\pi} \sin x\,dx$. In the next example, Simpson's Rule is applied to the same integral.

EXAMPLE 2 Approximation with Simpson's Rule

NOTE In Example 1, the Trapezoidal Rule with $n = 8$ approximated $\int_0^{\pi} \sin x\,dx$ as 1.974. In Example 2, Simpson's Rule with $n = 8$ gave an approximation of 2.0003. The antiderivative would produce the true value of 2.

Use Simpson's Rule to approximate

$$\int_0^{\pi} \sin x\,dx.$$

Compare the results for $n = 4$ and $n = 8$.

Solution When $n = 4$, you have

$$\begin{aligned}\int_0^{\pi} \sin x\,dx &\approx \frac{\pi}{12}\left(\sin 0 + 4\sin\frac{\pi}{4} + 2\sin\frac{\pi}{2} + 4\sin\frac{3\pi}{4} + \sin\pi\right)\\ &\approx 2.005.\end{aligned}$$

When $n = 8$, you have $\int_0^{\pi} \sin x\,dx \approx 2.0003.$

Error Analysis

If you must use an approximation technique, it is important to know how accurate you can expect the approximation to be. The following theorem, which is listed without proof, gives the formulas for estimating the errors involved in the use of Simpson's Rule and the Trapezoidal Rule.

NOTE In Theorem 5.19, $\max|f''(x)|$ is the least upper bound of the absolute value of the second derivative on $[a, b]$, and $\max|f^{(4)}(x)|$ is the least upper bound of the absolute value of the fourth derivative on $[a, b]$.

THEOREM 5.19 Errors in the Trapezoidal Rule and Simpson's Rule

If f has a continuous second derivative on $[a, b]$, then the error E in approximating $\int_a^b f(x)\,dx$ by the Trapezoidal Rule is

$$E \leq \frac{(b-a)^3}{12n^2}\left[\max|f''(x)|\right], \quad a \leq x \leq b. \qquad \text{Trapezoidal Rule}$$

Moreover, if f has a continuous fourth derivative on $[a, b]$, then the error E in approximating $\int_a^b f(x)\,dx$ by Simpson's Rule is

$$E \leq \frac{(b-a)^5}{180n^4}\left[\max|f^{(4)}(x)|\right], \quad a \leq x \leq b. \qquad \text{Simpson's Rule}$$

TECHNOLOGY If you have access to a computer algebra system, use it to evaluate the definite integral in Example 3. You should obtain a value of

$$\int_0^1 \sqrt{1+x^2}\,dx = \tfrac{1}{2}\left[\sqrt{2} + \ln\left(1+\sqrt{2}\right)\right] \approx 1.14779.$$

Theorem 5.19 states that the errors generated by the Trapezoidal Rule and Simpson's Rule have upper bounds dependent on the extreme values of $f''(x)$ and $f^{(4)}(x)$ in the interval $[a, b]$. Furthermore, these errors can be made arbitrarily small by *increasing n*, provided that f'' and $f^{(4)}$ are continuous and therefore bounded in $[a, b]$.

EXAMPLE 3 The Approximate Error in the Trapezoidal Rule

Determine a value of n such that the Trapezoidal Rule will approximate the value of $\int_0^1 \sqrt{1+x^2}\,dx$ with an error that is less than 0.01.

Solution Begin by letting $f(x) = \sqrt{1+x^2}$ and finding the second derivative of f.

$$f'(x) = x(1+x^2)^{-1/2} \quad \text{and} \quad f''(x) = (1+x^2)^{-3/2}$$

The maximum value of $|f''(x)|$ on the interval $[0, 1]$ is $|f''(0)| = 1$. So, by Theorem 5.19, you can write

$$E \leq \frac{(b-a)^3}{12n^2}|f''(0)| = \frac{1}{12n^2}(1) = \frac{1}{12n^2}.$$

To obtain an error E that is less than 0.01, you must choose n such that $1/(12n^2) \leq 1/100$.

$$100 \leq 12n^2 \quad \Longrightarrow \quad n \geq \sqrt{\frac{100}{12}} \approx 2.89$$

So, you can choose $n = 3$ (because n must be greater than or equal to 2.89) and apply the Trapezoidal Rule, as shown in Figure 5.45, to obtain

$$\int_0^1 \sqrt{1+x^2}\,dx \approx \frac{1}{6}\left[\sqrt{1+0^2} + 2\sqrt{1+\left(\tfrac{1}{3}\right)^2} + 2\sqrt{1+\left(\tfrac{2}{3}\right)^2} + \sqrt{1+1^2}\right] \approx 1.154.$$

So, with an error no larger than 0.01, you know that

$$1.144 \leq \int_0^1 \sqrt{1+x^2}\,dx \leq 1.164.$$

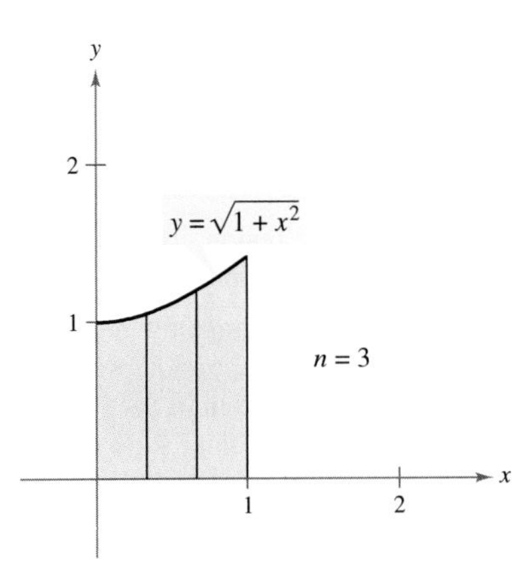

$1.144 \leq \int_0^1 \sqrt{1+x^2}\,dx \leq 1.164$

Figure 5.45

Exercises for Section 5.6

See www.CalcChat.com for worked-out solutions to odd-numbered exercises.

In Exercises 1–10, use the Trapezoidal Rule and Simpson's Rule to approximate the value of the definite integral for the given value of n. Round your answers to four decimal places and compare your results with the exact value of the definite integral.

1. $\int_0^2 x^2\,dx,\quad n = 4$
2. $\int_0^1 \left(\frac{x^2}{2} + 1\right) dx,\quad n = 4$
3. $\int_0^2 x^3\,dx,\quad n = 4$
4. $\int_1^2 \frac{2}{x^2}\,dx,\quad n = 4$
5. $\int_0^2 x^3\,dx,\quad n = 8$
6. $\int_0^8 \sqrt[3]{x}\,dx,\quad n = 8$
7. $\int_4^9 \sqrt{x}\,dx,\quad n = 8$
8. $\int_1^3 (4 - x^2)\,dx,\quad n = 4$
9. $\int_1^2 \frac{1}{(x + 1)^2}\,dx,\quad n = 4$
10. $\int_0^2 x\sqrt{x^2 + 1}\,dx,\quad n = 4$

In Exercises 11–24, approximate the definite integral using the Trapezoidal Rule and Simpson's Rule with $n = 4$. Compare these results with the approximation of the integral using a graphing utility.

11. $\int_0^2 \sqrt{1 + x^3}\,dx$
12. $\int_0^2 \frac{1}{\sqrt{1 + x^3}}\,dx$
13. $\int_0^1 \sqrt{x}\sqrt{1 - x}\,dx$
14. $\int_{\pi/2}^{\pi} \sqrt{x}\sin x\,dx$
15. $\int_0^{\sqrt{\pi/2}} \cos x^2\,dx$
16. $\int_0^{\sqrt{\pi/4}} \tan x^2\,dx$
17. $\int_1^{1.1} \sin x^2\,dx$
18. $\int_0^{\pi/2} \sqrt{1 + \cos^2 x}\,dx$
19. $\int_0^2 x\ln(x + 1)\,dx$
20. $\int_1^3 \ln x\,dx$
21. $\int_0^{\pi/4} x\tan x\,dx$
22. $\int_0^{\pi} f(x)\,dx,\quad f(x) = \begin{cases} \dfrac{\sin x}{x}, & x > 0 \\ 1, & x = 0 \end{cases}$
23. $\int_0^4 \sqrt{x}\,e^x\,dx$
24. $\int_0^2 xe^{-x}\,dx$

Writing About Concepts

25. If the function f is concave upward on the interval $[a, b]$, will the Trapezoidal Rule yield a result greater than or less than $\int_a^b f(x)\,dx$? Explain.
26. The Trapezoidal Rule and Simpson's Rule yield approximations of a definite integral $\int_a^b f(x)\,dx$ based on polynomial approximations of f. What degree polynomial is used for each?

In Exercises 27–30, use the error formulas in Theorem 5.19 to estimate the error in approximating the integral, with $n = 4$, using (a) the Trapezoidal Rule and (b) Simpson's Rule.

27. $\int_0^2 x^3\,dx$
28. $\int_1^3 (2x + 3)\,dx$
29. $\int_0^{\pi} \cos x\,dx$
30. $\int_0^1 \sin(\pi x)\,dx$

In Exercises 31–34, use the error formulas in Theorem 5.19 to find n such that the error in the approximation of the definite integral is less than 0.00001 using (a) the Trapezoidal Rule and (b) Simpson's Rule.

31. $\int_0^2 \sqrt{x + 2}\,dx$
32. $\int_1^3 \frac{1}{\sqrt{x}}\,dx$
33. $\int_0^1 \cos(\pi x)\,dx$
34. $\int_0^{\pi/2} \sin x\,dx$

In Exercises 35–38, use a computer algebra system and the error formulas to find n such that the error in the approximation of the definite integral is less than 0.00001 using (a) the Trapezoidal Rule and (b) Simpson's Rule.

35. $\int_0^2 \sqrt{1 + x}\,dx$
36. $\int_0^2 (x + 1)^{2/3}\,dx$
37. $\int_0^1 \tan x^2\,dx$
38. $\int_0^1 \sin x^2\,dx$

39. Approximate the area of the shaded region using (a) the Trapezoidal Rule and (b) Simpson's Rule with $n = 4$.

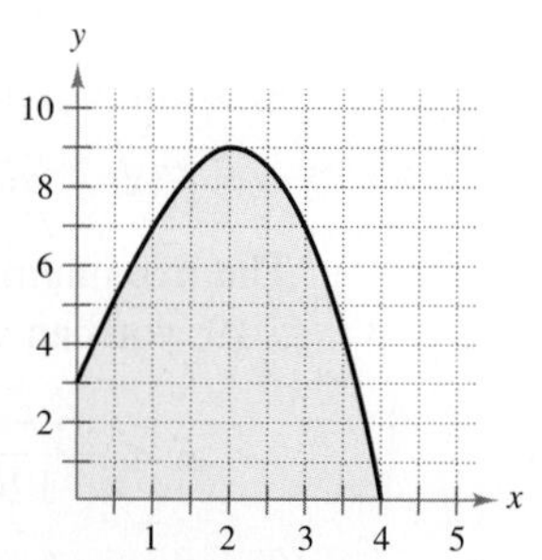

Figure for 39

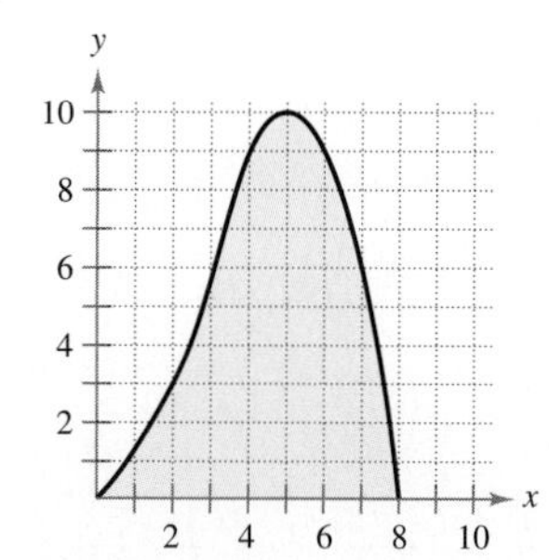

Figure for 40

40. Approximate the area of the shaded region using (a) the Trapezoidal Rule and (b) Simpson's Rule with $n = 8$.

41. ***Programming*** Write a program for a graphing utility to approximate a definite integral using the Trapezoidal Rule and Simpson's Rule. Start with the program written in Section 5.3, Exercises 59–64, and note that the Trapezoidal Rule can be written as $T(n) = \frac{1}{2}[L(n) + R(n)]$ and Simpson's Rule can be written as

$S(n) = \frac{1}{3}[T(n/2) + 2M(n/2)].$

[Recall that $L(n)$, $M(n)$, and $R(n)$ represent the Riemann sums using the left-hand endpoints, midpoints, and right-hand endpoints of subintervals of equal width.]

Programming **In Exercises 42–47, use the program in Exercise 41 to approximate the definite integral and complete the table.**

n	$L(n)$	$M(n)$	$R(n)$	$T(n)$	$S(n)$
4					
8					
10					
12					
16					
20					

42. $\displaystyle\int_0^4 \sqrt{2 + 3x^2}\, dx$

43. $\displaystyle\int_0^1 \sqrt{1 - x^2}\, dx$

44. $\displaystyle\int_0^4 \sin \sqrt{x}\, dx$

45. $\displaystyle\int_1^2 \frac{\sin x}{x}\, dx$

46. $\displaystyle\int_0^2 6e^{-x^2/2}\, dx$

47. $\displaystyle\int_0^3 \sqrt{x} \ln(x + 1)\, dx$

48. ***Work*** To determine the size of the motor required to operate a press, a company must know the amount of work done when the press moves an object linearly 5 feet. The variable force to move the object is

$$F(x) = 100x\sqrt{125 - x^3}$$

where F is given in pounds and x gives the position of the unit in feet. Use Simpson's Rule with $n = 12$ to approximate the work W (in foot-pounds) done through one cycle if

$$W = \int_0^5 F(x)\, dx.$$

49. The table lists several measurements gathered in an experiment to approximate an unknown continuous function $y = f(x)$.

(a) Approximate the integral $\int_0^2 f(x)\, dx$ using the Trapezoidal Rule and Simpson's Rule.

x	0.00	0.25	0.50	0.75	1.00
y	4.32	4.36	4.58	5.79	6.14

x	1.25	1.50	1.75	2.00
y	7.25	7.64	8.08	8.14

(b) Use a graphing utility to find a model of the form $y = ax^3 + bx^2 + cx + d$ for the data. Integrate the resulting polynomial over $[0, 2]$ and compare your result with your results in part (a).

Approximation of Pi **In Exercises 50 and 51, use Simpson's Rule with $n = 6$ to approximate π using the given equation. (In Section 5.8, you will be able to evaluate the integral using inverse trigonometric functions.)**

50. $\displaystyle\pi = \int_0^{1/2} \frac{6}{\sqrt{1 - x^2}}\, dx$

51. $\displaystyle\pi = \int_0^1 \frac{4}{1 + x^2}\, dx$

Area **In Exercises 52 and 53, use the Trapezoidal Rule to estimate the number of square meters of land in a lot where x and y are measured in meters, as shown in the figures. The land is bounded by a stream and two straight roads that meet at right angles.**

52.

x	y
0	125
100	125
200	120
300	112
400	90
500	90
600	95
700	88
800	75
900	35
1000	0

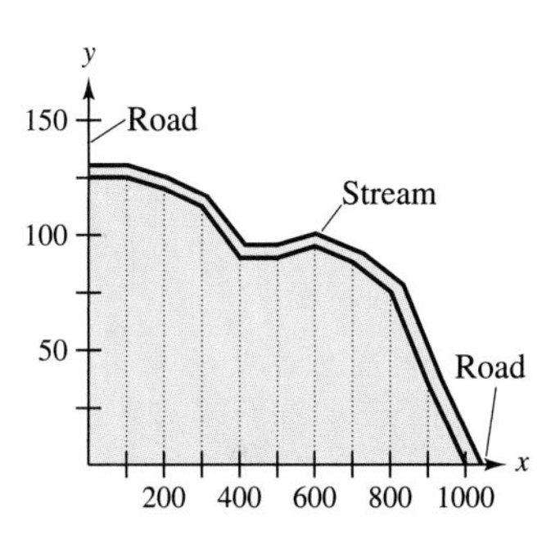

53.

x	y
0	75
10	81
20	84
30	76
40	67
50	68
60	69
70	72
80	68
90	56
100	42
110	23
120	0

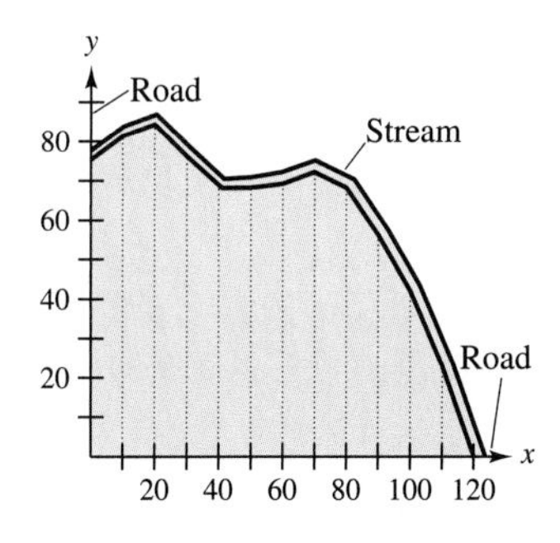

54. Prove that Simpson's Rule is exact when approximating the integral of a cubic polynomial function, and demonstrate the result for $\int_0^1 x^3\, dx$, $n = 2$.

55. Use Simpson's Rule with $n = 10$ and a computer algebra system to approximate t in the integral equation

$$\int_0^t \sin \sqrt{x}\, dx = 2.$$

56. Prove that you can find a polynomial $p(x) = Ax^2 + Bx + C$ that passes through any three points (x_1, y_1), (x_2, y_2), and (x_3, y_3), where the x_i's are distinct.

Section 5.7 The Natural Logarithmic Function: Integration

- **Use the Log Rule for Integration to integrate a rational function.**
- **Integrate trigonometric functions.**

EXPLORATION

Integrating Rational Functions Earlier in this chapter, you learned rules that allowed you to integrate *any* polynomial function. The Log Rule presented in this section goes a long way toward enabling you to integrate rational functions. For instance, each of the following functions can be integrated with the Log Rule.

$\dfrac{1}{2x}$	Example 1
$\dfrac{1}{4x-1}$	Example 2
$\dfrac{x}{x^2+1}$	Example 3
$\dfrac{3x^2+1}{x^3+x}$	Example 4(a)
$\dfrac{x+1}{x^2+2x}$	Example 4(c)
$\dfrac{1}{3x+2}$	Example 4(d)
$\dfrac{x^2+x+1}{x^2+1}$	Example 5
$\dfrac{2x}{(x+1)^2}$	Example 6

There are still some rational functions that cannot be integrated using the Log Rule. Give examples of these functions, and explain your reasoning.

Log Rule for Integration

In Chapter 3 you studied two differentiation rules for logarithms. The differentiation rule $d/dx[\ln x] = 1/x$ produces the Log Rule for Integration that you learned in Section 5.1. The differentiation rule $d/dx[\ln u] = u'/u$ produces the integration rule $\int 1/u = \ln|u| + C$. These rules are summarized below. (See Exercise 105.)

THEOREM 5.20 Log Rule for Integration

Let u be a differentiable function of x.

1. $\displaystyle\int \frac{1}{x}\,dx = \ln|x| + C$ **2.** $\displaystyle\int \frac{1}{u}\,du = \ln|u| + C$

Because $du = u'\,dx$, the second formula can also be written as

$$\int \frac{u'}{u}\,dx = \ln|u| + C. \qquad \text{Alternative form of Log Rule}$$

EXAMPLE 1 Using the Log Rule for Integration

To find $\int 1/(2x)\,dx$, let $u = 2x$. Then $du = 2\,dx$.

$$\int \frac{1}{2x}\,dx = \frac{1}{2}\int\left(\frac{1}{2x}\right)2\,dx \qquad \text{Multiply and divide by 2.}$$

$$= \frac{1}{2}\int \frac{1}{u}\,du \qquad \text{Substitute: } u = 2x.$$

$$= \frac{1}{2}\ln|u| + C \qquad \text{Apply Log Rule.}$$

$$= \frac{1}{2}\ln|2x| + C \qquad \text{Back-substitute.}$$

EXAMPLE 2 Using the Log Rule with a Change of Variables

To find $\int 1/(4x-1)\,dx$, let $u = 4x - 1$. Then $du = 4\,dx$.

$$\int \frac{1}{4x-1}\,dx = \frac{1}{4}\int\left(\frac{1}{4x-1}\right)4\,dx \qquad \text{Multiply and divide by 4.}$$

$$= \frac{1}{4}\int \frac{1}{u}\,du \qquad \text{Substitute: } u = 4x - 1.$$

$$= \frac{1}{4}\ln|u| + C \qquad \text{Apply Log Rule.}$$

$$= \frac{1}{4}\ln|4x-1| + C \qquad \text{Back-substitute.}$$

Example 3 uses the alternative form of the Log Rule. To apply this rule, look for quotients in which the numerator is the derivative of the denominator.

EXAMPLE 3 Finding Area with the Log Rule

Find the area of the region bounded by the graph of

$$y = \frac{x}{x^2 + 1}$$

the x-axis, and the line $x = 3$.

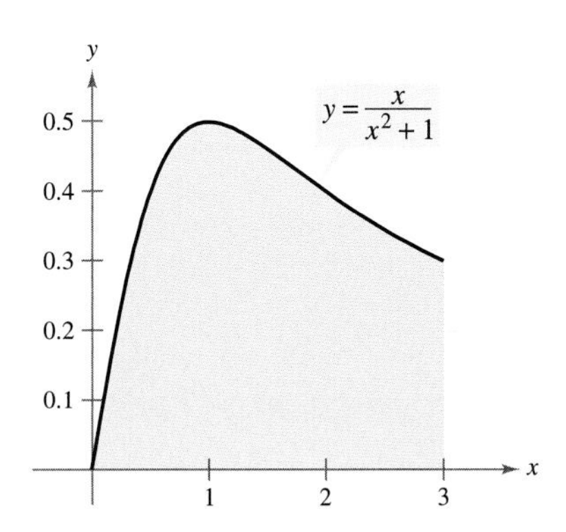

$\text{Area} = \int_0^3 \frac{x}{x^2 + 1}\, dx$

The area of the region bounded by the graph of y, the x-axis, and $x = 3$ is $\frac{1}{2} \ln 10$.

Figure 5.46

Solution From Figure 5.46, you can see that the area of the region is given by the definite integral

$$\int_0^3 \frac{x}{x^2 + 1}\, dx.$$

If you let $u = x^2 + 1$, then $u' = 2x$. To apply the Log Rule, multiply and divide by 2 as follows.

$$\begin{aligned}
\int_0^3 \frac{x}{x^2 + 1}\, dx &= \frac{1}{2} \int_0^3 \frac{2x}{x^2 + 1}\, dx && \text{Multiply and divide by 2.} \\
&= \frac{1}{2} \Big[\ln(x^2 + 1) \Big]_0^3 && \int \frac{u'}{u}\, dx = \ln|u| + C \\
&= \frac{1}{2} (\ln 10 - \ln 1) \\
&= \frac{1}{2} \ln 10 && \ln 1 = 0 \\
&\approx 1.151
\end{aligned}$$

EXAMPLE 4 Recognizing Quotient Forms of the Log Rule

a. $\displaystyle\int \frac{3x^2 + 1}{x^3 + x}\, dx = \ln|x^3 + x| + C$ $\qquad u = x^3 + x$

b. $\displaystyle\int \frac{\sec^2 x}{\tan x}\, dx = \ln|\tan x| + C$ $\qquad u = \tan x$

c. $\displaystyle\int \frac{x + 1}{x^2 + 2x}\, dx = \frac{1}{2} \int \frac{2x + 2}{x^2 + 2x}\, dx$ $\qquad u = x^2 + 2x$

$\displaystyle\qquad = \frac{1}{2} \ln|x^2 + 2x| + C$

d. $\displaystyle\int \frac{1}{3x + 2}\, dx = \frac{1}{3} \int \frac{3}{3x + 2}\, dx$ $\qquad u = 3x + 2$

$\displaystyle\qquad = \frac{1}{3} \ln|3x + 2| + C$

With antiderivatives involving logarithms, it is easy to obtain forms that look quite different but are still equivalent. For instance, which of the following are equivalent to the antiderivative listed in Example 4(d)?

$$\ln|(3x + 2)^{1/3}| + C, \quad \frac{1}{3} \ln\left|x + \tfrac{2}{3}\right| + C, \quad \ln|3x + 2|^{1/3} + C$$

Integrals to which the Log Rule can be applied often appear in disguised form. For instance, if a rational function has a *numerator of degree greater than or equal to that of the denominator*, division may reveal a form to which you can apply the Log Rule. This is illustrated in Example 5.

EXAMPLE 5 Using Long Division Before Integrating

Find $\int \frac{x^2+x+1}{x^2+1}\,dx.$

Solution Begin by using long division to rewrite the integrand.

$$\frac{x^2+x+1}{x^2+1} \quad \Longrightarrow \quad x^2+1\,\overline{\smash{)}\,x^2+x+1} \;\; (\text{quotient } 1;\; x^2 + 1;\; \text{remainder } x) \quad \Longrightarrow \quad 1+\frac{x}{x^2+1}$$

Now, you can integrate to obtain

$$\int \frac{x^2+x+1}{x^2+1}\,dx = \int\left(1+\frac{x}{x^2+1}\right)dx \qquad \text{Rewrite using long division.}$$

$$= \int dx + \frac{1}{2}\int\frac{2x}{x^2+1}\,dx \qquad \text{Rewrite as two integrals.}$$

$$= x + \frac{1}{2}\ln(x^2+1) + C. \qquad \text{Integrate.}$$

Check this result by differentiating to obtain the original integrand.

The next example gives another instance in which the use of the Log Rule is disguised. In this case, a change of variables helps you recognize the Log Rule.

EXAMPLE 6 Change of Variables with the Log Rule

Find $\int \frac{2x}{(x+1)^2}\,dx.$

Solution If you let $u = x + 1$, then $du = dx$ and $x = u - 1$.

$$\int\frac{2x}{(x+1)^2}\,dx = \int\frac{2(u-1)}{u^2}\,du \qquad \text{Substitute.}$$

$$= 2\int\left(\frac{u}{u^2}-\frac{1}{u^2}\right)du \qquad \text{Rewrite as two fractions.}$$

$$= 2\int\frac{du}{u} - 2\int u^{-2}\,du \qquad \text{Rewrite as two integrals.}$$

$$= 2\ln|u| - 2\left(\frac{u^{-1}}{-1}\right) + C \qquad \text{Integrate.}$$

$$= 2\ln|u| + \frac{2}{u} + C \qquad \text{Simplify.}$$

$$= 2\ln|x+1| + \frac{2}{x+1} + C \qquad \text{Back-substitute.}$$

Check this result by differentiating to obtain the original integrand.

TECHNOLOGY If you have access to a computer algebra system, try using it to find the indefinite integrals in Examples 5 and 6. How do the forms of the antiderivatives that it gives you compare with those given in Examples 5 and 6?

As you study the methods shown in Examples 5 and 6, be aware that both methods involve rewriting a disguised integrand so that it fits one or more of the basic integration formulas. Throughout the remaining sections of Chapter 5 and in Chapter 8, much time will be devoted to integration techniques. To master these techniques, you must recognize the "form-fitting" nature of integration. In this sense, integration is not nearly as straightforward as differentiation. Differentiation takes the form

"Here is the question; what is the answer?"

Integration is more like

"Here is the answer; what is the question?"

The following are guidelines you can use for integration.

Guidelines for Integration

1. Learn a basic list of integration formulas. (By the end of Section 5.8, this list will have expanded to 20 basic rules.)
2. Find an integration formula that resembles all or part of the integrand, and, by trial and error, find a choice of u that will make the integrand conform to the formula.
3. If you cannot find a u-substitution that works, try altering the integrand. You might try a trigonometric identity, multiplication and division by the same quantity, or addition and subtraction of the same quantity. Be creative.
4. If you have access to computer software that will find antiderivatives symbolically, use it.

EXAMPLE 7 *u*-Substitution and the Log Rule

Solve the differential equation

$$\frac{dy}{dx} = \frac{1}{x \ln x}.$$

Solution The solution can be written as an indefinite integral.

$$y = \int \frac{1}{x \ln x}\, dx$$

Because the integrand is a quotient whose denominator is raised to the first power, you should try the Log Rule. There are three basic choices for u. The choices $u = x$ and $u = x \ln x$ fail to fit the u'/u form of the Log Rule. However, the third choice does fit. Letting $u = \ln x$ produces $u' = 1/x$, and you obtain the following.

$$\int \frac{1}{x \ln x}\, dx = \int \frac{1/x}{\ln x}\, dx \qquad \text{Divide numerator and denominator by } x.$$

$$= \int \frac{u'}{u}\, dx \qquad \text{Substitute: } u = \ln x.$$

$$= \ln|u| + C \qquad \text{Apply Log Rule.}$$

$$= \ln|\ln x| + C \qquad \text{Back-substitute.}$$

So, the solution is $y = \ln|\ln x| + C$.

STUDY TIP Keep in mind that you can check your answer to an integration problem by differentiating the answer. For instance, in Example 7, the derivative of $y = \ln|\ln x| + C$ is $y' = 1/(x \ln x)$.

Integrals of Trigonometric Functions

In Section 5.1, you looked at six trigonometric integration rules—the six that correspond directly to differentiation rules. With the Log Rule, you can now complete the set of basic trigonometric integration formulas.

EXAMPLE 8 Using a Trigonometric Identity

Find $\int \tan x \, dx$.

Solution This integral does not seem to fit any formulas on our basic list. However, by using a trigonometric identity, you obtain the following.

$$\int \tan x \, dx = \int \frac{\sin x}{\cos x} \, dx$$

Knowing that $D_x[\cos x] = -\sin x$, you can let $u = \cos x$ and write

$$\int \tan x \, dx = -\int \frac{-\sin x}{\cos x} \, dx \qquad \text{Trigonometric identity}$$

$$= -\int \frac{u'}{u} \, dx \qquad \text{Substitute: } u = \cos x.$$

$$= -\ln|u| + C \qquad \text{Apply Log Rule.}$$

$$= -\ln|\cos x| + C. \qquad \text{Back-substitute.}$$

Example 8 uses a trigonometric identity to derive an integration rule for the tangent function. The next example takes a rather unusual step (multiplying and dividing by the same quantity) to derive an integration rule for the secant function.

EXAMPLE 9 Derivation of the Secant Formula

Find $\int \sec x \, dx$.

Solution Consider the following procedure.

$$\int \sec x \, dx = \int \sec x \left(\frac{\sec x + \tan x}{\sec x + \tan x} \right) dx$$

$$= \int \frac{\sec^2 x + \sec x \tan x}{\sec x + \tan x} \, dx$$

Letting u be the denominator of this quotient produces

$$u = \sec x + \tan x \quad \Rightarrow \quad u' = \sec x \tan x + \sec^2 x.$$

Therefore, you can conclude that

$$\int \sec x \, dx = \int \frac{\sec^2 x + \sec x \tan x}{\sec x + \tan x} \, dx \qquad \text{Rewrite integrand.}$$

$$= \int \frac{u'}{u} \, dx \qquad \text{Substitute: } u = \sec x + \tan x.$$

$$= \ln|u| + C \qquad \text{Apply Log Rule.}$$

$$= \ln|\sec x + \tan x| + C. \qquad \text{Back-substitute.}$$

With the results of Examples 8 and 9, you now have integration formulas for $\sin x$, $\cos x$, $\tan x$, and $\sec x$. All six trigonometric rules are summarized below.

NOTE Using trigonometric identities and properties of logarithms, you could rewrite these six integration rules in other forms. For instance, you could write

$$\int \csc u\,du = \ln|\csc u - \cot u| + C.$$

(See Exercises 85–88.)

Integrals of the Six Basic Trigonometric Functions

$$\int \sin u\,du = -\cos u + C \qquad \int \cos u\,du = \sin u + C$$

$$\int \tan u\,du = -\ln|\cos u| + C \qquad \int \cot u\,du = \ln|\sin u| + C$$

$$\int \sec u\,du = \ln|\sec u + \tan u| + C \qquad \int \csc u\,du = -\ln|\csc u + \cot u| + C$$

EXAMPLE 10 Integrating Trigonometric Functions

Evaluate $\displaystyle\int_0^{\pi/4} \sqrt{1 + \tan^2 x}\,dx$.

Solution Using $1 + \tan^2 x = \sec^2 x$, you can write

$$\begin{aligned}
\int_0^{\pi/4} \sqrt{1 + \tan^2 x}\,dx &= \int_0^{\pi/4} \sqrt{\sec^2 x}\,dx \\
&= \int_0^{\pi/4} \sec x\,dx && \sec x \ge 0 \text{ for } 0 \le x \le \frac{\pi}{4} \\
&= \ln|\sec x + \tan x|\Big]_0^{\pi/4} \\
&= \ln(\sqrt{2} + 1) - \ln 1 \\
&\approx 0.8814.
\end{aligned}$$

EXAMPLE 11 Finding an Average Value

Find the average value of $f(x) = \tan x$ on the interval $[0, \pi/4]$.

Solution

$$\begin{aligned}
\text{Average value} &= \frac{1}{(\pi/4) - 0}\int_0^{\pi/4} \tan x\,dx && \text{Average value} = \frac{1}{b-a}\int_a^b f(x)\,dx \\
&= \frac{4}{\pi}\int_0^{\pi/4} \tan x\,dx && \text{Simplify.} \\
&= \frac{4}{\pi}\Big[-\ln|\cos x|\Big]_0^{\pi/4} && \text{Integrate.} \\
&= -\frac{4}{\pi}\left[\ln\left(\frac{\sqrt{2}}{2}\right) - \ln(1)\right] \\
&= -\frac{4}{\pi}\ln\left(\frac{\sqrt{2}}{2}\right) \\
&\approx 0.441
\end{aligned}$$

The average value is about 0.441, as shown in Figure 5.47.

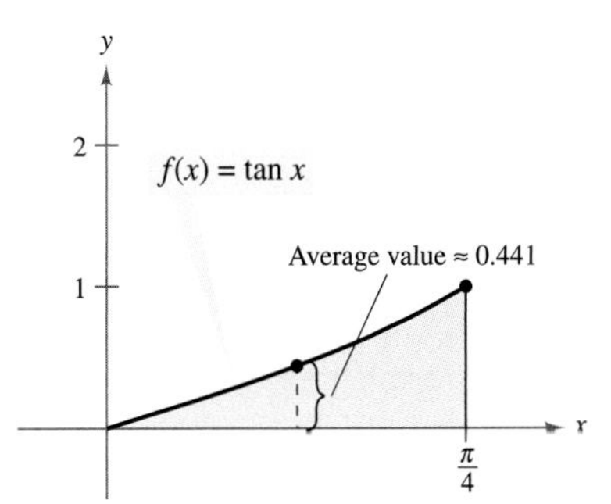

Figure 5.47

Exercises for Section 5.7

See www.CalcChat.com for worked-out solutions to odd-numbered exercises.

In Exercises 1–24, find the indefinite integral.

1. $\displaystyle\int \frac{5}{x}\,dx$
2. $\displaystyle\int \frac{10}{x}\,dx$
3. $\displaystyle\int \frac{1}{x+1}\,dx$
4. $\displaystyle\int \frac{1}{x-5}\,dx$
5. $\displaystyle\int \frac{1}{3-2x}\,dx$
6. $\displaystyle\int \frac{1}{3x+2}\,dx$
7. $\displaystyle\int \frac{x}{x^2+1}\,dx$
8. $\displaystyle\int \frac{x^2}{3-x^3}\,dx$
9. $\displaystyle\int \frac{x^2-4}{x}\,dx$
10. $\displaystyle\int \frac{x}{\sqrt{9-x^2}}\,dx$
11. $\displaystyle\int \frac{x^2+2x+3}{x^3+3x^2+9x}\,dx$
12. $\displaystyle\int \frac{x(x+2)}{x^3+3x^2-4}\,dx$
13. $\displaystyle\int \frac{x^2-3x+2}{x+1}\,dx$
14. $\displaystyle\int \frac{2x^2+7x-3}{x-2}\,dx$
15. $\displaystyle\int \frac{x^3-3x^2+5}{x-3}\,dx$
16. $\displaystyle\int \frac{x^3-6x-20}{x+5}\,dx$
17. $\displaystyle\int \frac{x^4+x-4}{x^2+2}\,dx$
18. $\displaystyle\int \frac{x^3-3x^2+4x-9}{x^2+3}\,dx$
19. $\displaystyle\int \frac{(\ln x)^2}{x}\,dx$
20. $\displaystyle\int \frac{1}{x\ln(x^3)}\,dx$
21. $\displaystyle\int \frac{1}{\sqrt{x+1}}\,dx$
22. $\displaystyle\int \frac{1}{x^{2/3}(1+x^{1/3})}\,dx$
23. $\displaystyle\int \frac{2x}{(x-1)^2}\,dx$
24. $\displaystyle\int \frac{x(x-2)}{(x-1)^3}\,dx$

In Exercises 25–28, find the indefinite integral by *u*-substitution. (*Hint:* Let *u* be the denominator of the integrand.)

25. $\displaystyle\int \frac{1}{1+\sqrt{2x}}\,dx$
26. $\displaystyle\int \frac{1}{1+\sqrt{3x}}\,dx$
27. $\displaystyle\int \frac{\sqrt{x}}{\sqrt{x}-3}\,dx$
28. $\displaystyle\int \frac{\sqrt[3]{x}}{\sqrt[3]{x}-1}\,dx$

In Exercises 29–38, find the indefinite integral.

29. $\displaystyle\int \frac{\cos\theta}{\sin\theta}\,d\theta$
30. $\displaystyle\int \tan 5\theta\,d\theta$
31. $\displaystyle\int \csc 2x\,dx$
32. $\displaystyle\int \sec\frac{x}{2}\,dx$
33. $\displaystyle\int \frac{\cos t}{1+\sin t}\,dt$
34. $\displaystyle\int \frac{\csc^2 t}{\cot t}\,dt$
35. $\displaystyle\int \frac{\sec x\tan x}{\sec x-1}\,dx$
36. $\displaystyle\int (\sec t+\tan t)\,dt$
37. $\displaystyle\int e^{-x}\tan(e^{-x})\,dx$
38. $\displaystyle\int \sec t(\sec t+\tan t)\,dt$

In Exercises 39–42, solve the differential equation. Use a graphing utility to graph three solutions, one of which passes through the given point.

39. $\dfrac{dy}{dx} = \dfrac{3}{2-x}$, $(1, 0)$
40. $\dfrac{dy}{dx} = \dfrac{2x}{x^2-9}$, $(0, 4)$
41. $\dfrac{ds}{d\theta} = \tan 2\theta$, $(0, 2)$
42. $\dfrac{dr}{dt} = \dfrac{\sec^2 t}{\tan t+1}$, $(\pi, 4)$

43. Determine the function f if $f''(x) = \dfrac{2}{x^2}$, $f(1) = 1$, and $f'(1) = 1$, $x > 0$.

44. Determine the function f if $f''(x) = -\dfrac{4}{(x-1)^2} - 2$, $f(2) = 3$, and $f'(2) = 0$, $x > 1$.

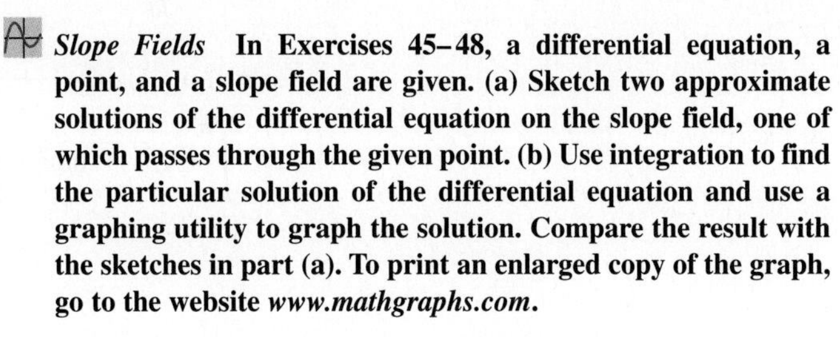
Slope Fields **In Exercises 45–48, a differential equation, a point, and a slope field are given. (a) Sketch two approximate solutions of the differential equation on the slope field, one of which passes through the given point. (b) Use integration to find the particular solution of the differential equation and use a graphing utility to graph the solution. Compare the result with the sketches in part (a). To print an enlarged copy of the graph, go to the website *www.mathgraphs.com*.**

45. $\dfrac{dy}{dx} = \dfrac{1}{x+2}$, $(0, 1)$

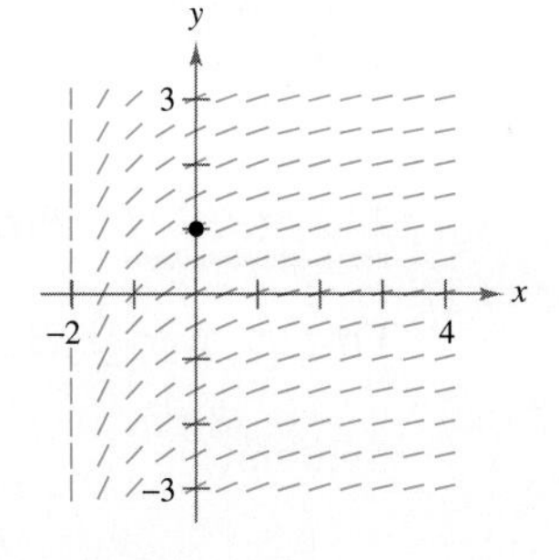

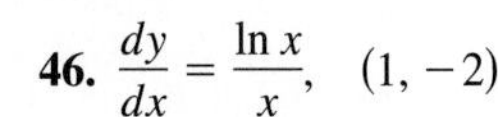
46. $\dfrac{dy}{dx} = \dfrac{\ln x}{x}$, $(1, -2)$

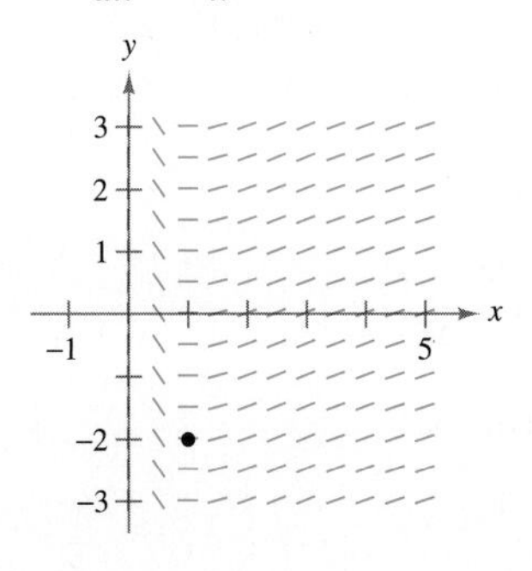

47. $\dfrac{dy}{dx} = 1 + \dfrac{1}{x}$, $(1, 4)$

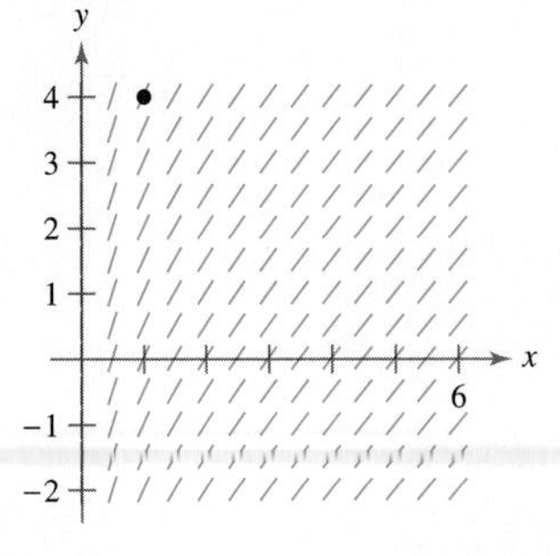

48. $\dfrac{dy}{dx} = \sec x$, $(0, 1)$

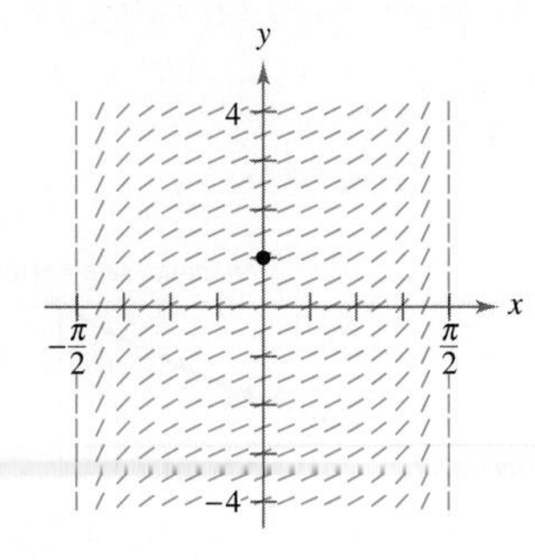

In Exercises 49–56, evaluate the definite integral. Use a graphing utility to verify your result.

49. $\displaystyle\int_0^4 \frac{5}{3x+1}\,dx$

50. $\displaystyle\int_{-1}^1 \frac{1}{x+2}\,dx$

51. $\displaystyle\int_1^e \frac{(1+\ln x)^2}{x}\,dx$

52. $\displaystyle\int_e^{e^2} \frac{1}{x\ln x}\,dx$

53. $\displaystyle\int_0^2 \frac{x^2-2}{x+1}\,dx$

54. $\displaystyle\int_0^1 \frac{x-1}{x+1}\,dx$

55. $\displaystyle\int_1^2 \frac{1-\cos\theta}{\theta-\sin\theta}\,d\theta$

56. $\displaystyle\int_{0.1}^{0.2} (\csc 2\theta - \cot 2\theta)^2\,d\theta$

In Exercises 57–62, use a computer algebra system to find or evaluate the integral.

57. $\displaystyle\int \frac{1}{1+\sqrt{x}}\,dx$

58. $\displaystyle\int \frac{1-\sqrt{x}}{1+\sqrt{x}}\,dx$

59. $\displaystyle\int \frac{\sqrt{x}}{x-1}\,dx$

60. $\displaystyle\int \frac{x^2}{x-1}\,dx$

61. $\displaystyle\int_{\pi/4}^{\pi/2} (\csc x - \sin x)\,dx$

62. $\displaystyle\int_{-\pi/4}^{\pi/4} \frac{\sin^2 x - \cos^2 x}{\cos x}\,dx$

In Exercises 63–66, find $F'(x)$.

63. $F(x) = \displaystyle\int_1^x \frac{1}{t}\,dt$

64. $F(x) = \displaystyle\int_0^x \tan t\,dt$

65. $F(x) = \displaystyle\int_1^{3x} \frac{1}{t}\,dt$

66. $F(x) = \displaystyle\int_1^{x^2} \frac{1}{t}\,dt$

Approximation **In Exercises 67 and 68, determine which value best approximates the area of the region between the x-axis and the graph of the function over the given interval. (Make your selection on the basis of a sketch of the region and not by performing any calculations.)**

67. $f(x) = \sec x,\quad [0, 1]$

(a) 6 (b) -6 (c) $\frac{1}{2}$ (d) 1.25 (e) 3

68. $f(x) = \dfrac{2x}{x^2+1},\quad [0, 4]$

(a) 3 (b) 7 (c) -2 (d) 5 (e) 1

Area **In Exercises 69–72, find the area of the given region. Use a graphing utility to verify your result.**

69. $y = \dfrac{4}{x}$

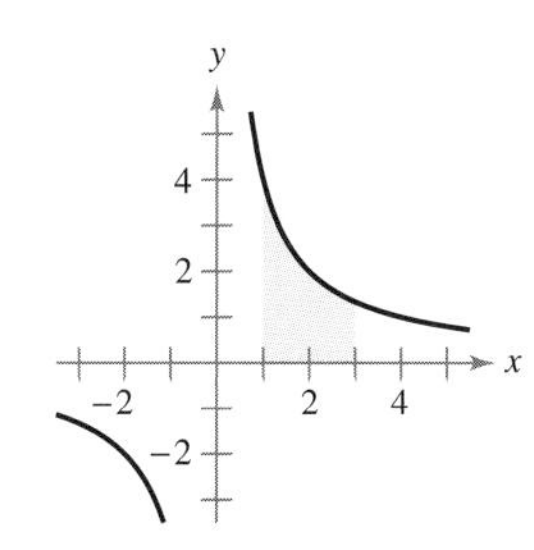

70. $y = \dfrac{2}{x\ln x}$

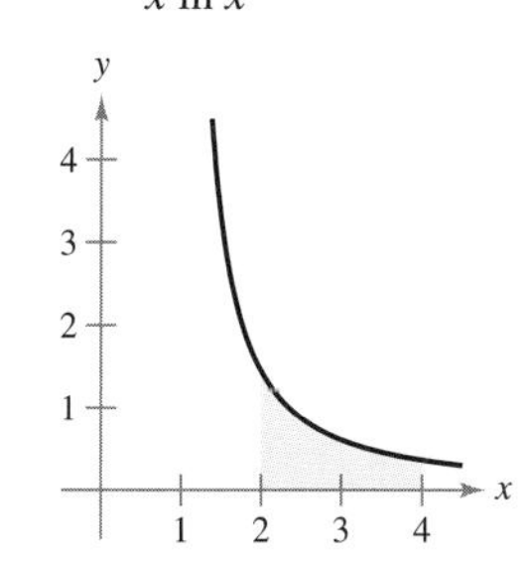

71. $y = \tan x$

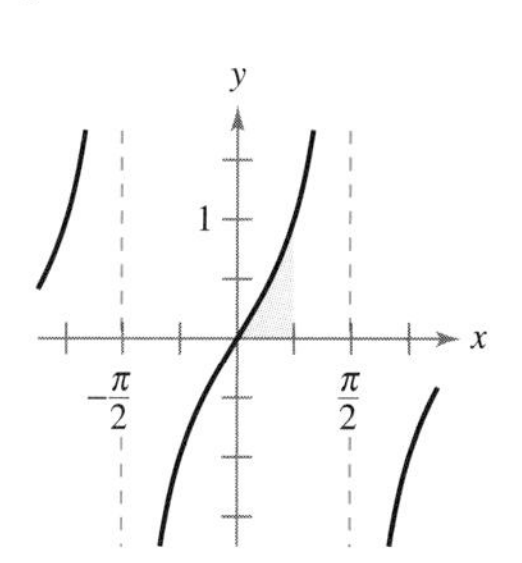

72. $y = \dfrac{\sin x}{1+\cos x}$

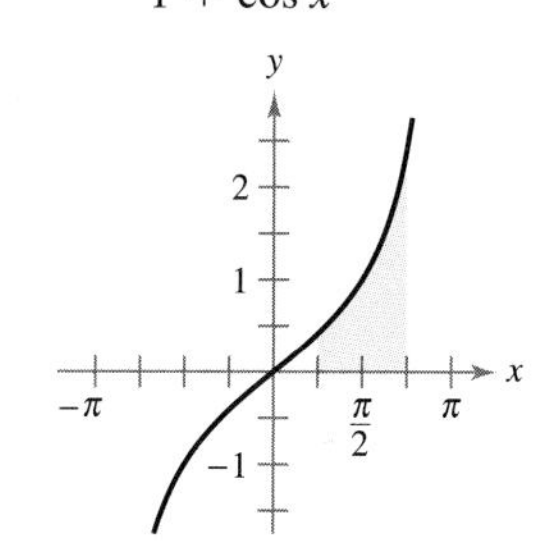

Area **In Exercises 73–76, find the area of the region bounded by the graphs of the equations. Use a graphing utility to verify your result.**

73. $y = \dfrac{x^2+4}{x},\ x = 1,\ x = 4,\ y = 0$

74. $y = \dfrac{x+4}{x},\ x = 1,\ x = 4,\ y = 0$

75. $y = 2\sec\dfrac{\pi x}{6},\ x = 0,\ x = 2,\ y = 0$

76. $y = 2x - \tan(0.3x),\ x = 1,\ x = 4,\ y = 0$

Numerical Integration **In Exercises 77–80, use the Trapezoidal Rule and Simpson's Rule to approximate the value of the definite integral. Let $n = 4$ and round your answers to four decimal places. Use a graphing utility to verify your result.**

77. $\displaystyle\int_1^5 \frac{12}{x}\,dx$

78. $\displaystyle\int_0^4 \frac{8x}{x^2+4}\,dx$

79. $\displaystyle\int_2^6 \ln x\,dx$

80. $\displaystyle\int_{-\pi/3}^{\pi/3} \sec x\,dx$

Writing About Concepts

In Exercises 81–84, state the integration formula you would use to perform the integration. Do not integrate.

81. $\displaystyle\int \sqrt[3]{x}\,dx$

82. $\displaystyle\int \frac{x}{(x^2+4)^3}\,dx$

83. $\displaystyle\int \frac{x}{x^2+4}\,dx$

84. $\displaystyle\int \frac{\sec^2 x}{\tan x}\,dx$

In Exercises 85–88, show that the two formulas are equivalent.

85. $\int \tan x\, dx = -\ln|\cos x| + C$

$\int \tan x\, dx = \ln|\sec x| + C$

86. $\int \cot x\, dx = \ln|\sin x| + C$

$\int \cot x\, dx = -\ln|\csc x| + C$

87. $\int \sec x\, dx = \ln|\sec x + \tan x| + C$

$\int \sec x\, dx = -\ln|\sec x - \tan x| + C$

88. $\int \csc x\, dx = -\ln|\csc x + \cot x| + C$

$\int \csc x\, dx = \ln|\csc x - \cot x| + C$

In Exercises 89–92, find the average value of the function over the given interval.

89. $f(x) = \dfrac{8}{x^2}$, $[2, 4]$

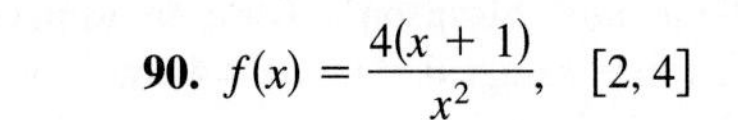

90. $f(x) = \dfrac{4(x+1)}{x^2}$, $[2, 4]$

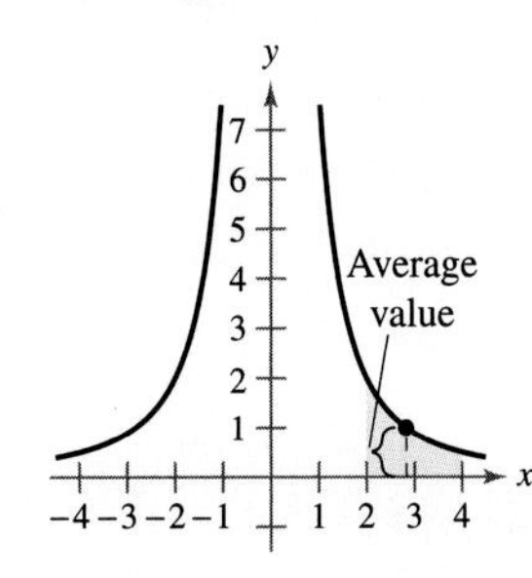

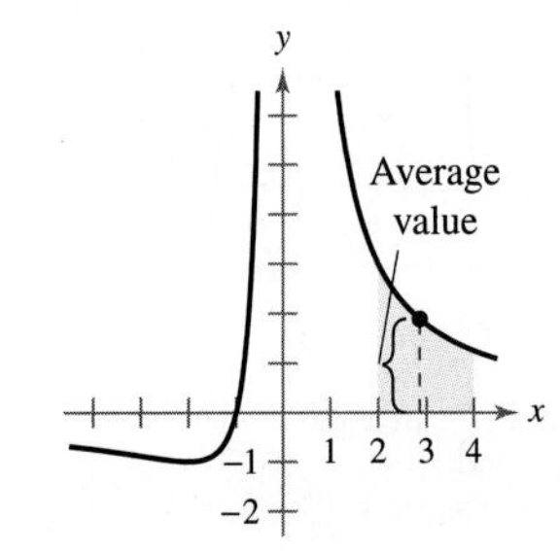

91. $f(x) = \dfrac{\ln x}{x}$, $[1, e]$

92. $f(x) = \sec \dfrac{\pi x}{6}$, $[0, 2]$

93. ***Population Growth*** A population of bacteria is changing at a rate of

$$\frac{dP}{dt} = \frac{3000}{1 + 0.25t}$$

where t is the time in days. The initial population (when $t = 0$) is 1000. Write an equation that gives the population at any time t, and find the population when $t = 3$ days.

94. ***Heat Transfer*** Find the time required for an object to cool from 300°F to 250°F by evaluating

$$t = \frac{10}{\ln 2}\int_{250}^{300} \frac{1}{T - 100}\, dT$$

where t is time in minutes.

95. ***Average Price*** The demand equation for a product is

$$p = \frac{90{,}000}{400 + 3x}.$$

Find the *average* price p on the interval $40 \le x \le 50$.

96. ***Sales*** The rate of change in sales S is inversely proportional to time t $(t > 1)$ measured in weeks. Find S as a function of t if sales after 2 and 4 weeks are 200 units and 300 units, respectively.

97. ***Orthogonal Trajectory***

(a) Use a graphing utility to graph the equation $2x^2 - y^2 = 8$.

(b) Evaluate the integral to find y^2 in terms of x.

$$y^2 = e^{-\int(1/x)\,dx}$$

For a particular value of the constant of integration, graph the result in the same viewing window used in part (a).

(c) Verify that the tangents to the graphs of parts (a) and (b) are perpendicular at the points of intersection.

98. Graph the function

$$f_k(x) = \frac{x^k - 1}{k}$$

for $k = 1, 0.5$, and 0.1 on $[0, 10]$. Find $\lim_{k\to 0^+} f_k(x)$.

True or False? **In Exercises 99–102, determine whether the statement is true or false. If it is false, explain why or give an example that shows it is false.**

99. $(\ln x)^{1/2} = \frac{1}{2}(\ln x)$

100. $\int \ln x\, dx = (1/x) + C$

101. $\int \frac{1}{x}\, dx = \ln|cx|, \quad c \ne 0$

102. $\int_{-1}^{2} \frac{1}{x}\, dx = \Big[\ln|x|\Big]_{-1}^{2} = \ln 2 - \ln 1 = \ln 2$

103. Graph the function

$$f(x) = \frac{x}{1 + x^2}$$

on the interval $[0, \infty)$.

(a) Find the area bounded by the graph of f and the line $y = \frac{1}{2}x$.

(b) Determine the values of the slope m such that the line $y = mx$ and the graph of f enclose a finite region.

(c) Calculate the area of this region as a function of m.

104. Prove that the function

$$F(x) = \int_{x}^{2x} \frac{1}{t}\, dt$$

is constant on the interval $(0, \infty)$.

105. Prove Theorem 5.20.

Section 5.8 Inverse Trigonometric Functions: Integration

- **Integrate functions whose antiderivatives involve inverse trigonometric functions.**
- **Use the method of completing the square to integrate a function.**
- **Review the basic integration rules involving elementary functions.**

Integrals Involving Inverse Trigonometric Functions

The derivatives of the six inverse trigonometric functions fall into three pairs. In each pair, the derivative of one function is the negative of the other. For example,

$$\frac{d}{dx}[\arcsin x] = \frac{1}{\sqrt{1 - x^2}}$$

and

$$\frac{d}{dx}[\arccos x] = -\frac{1}{\sqrt{1 - x^2}}.$$

When listing the *antiderivative* that corresponds to each of the inverse trigonometric functions, you need to use only one member from each pair. It is conventional to use $\arcsin x$ as the antiderivative of $1/\sqrt{1 - x^2}$, rather than $-\arccos x$. The next theorem gives one antiderivative formula for each of the three pairs. The proofs of these integration rules are left to you (see Exercises 79–81).

NOTE For a proof of part 2 of Theorem 5.21, see the article "A Direct Proof of the Integral Formula for Arctangent" by Arnold J. Insel in *The College Mathematics Journal*. To view this article, go to the website *www.matharticles.com.*

THEOREM 5.21 Integrals Involving Inverse Trigonometric Functions

Let u be a differentiable function of x, and let $a > 0$.

1. $\displaystyle\int \frac{du}{\sqrt{a^2 - u^2}} = \arcsin\frac{u}{a} + C$
2. $\displaystyle\int \frac{du}{a^2 + u^2} = \frac{1}{a}\arctan\frac{u}{a} + C$
3. $\displaystyle\int \frac{du}{u\sqrt{u^2 - a^2}} = \frac{1}{a}\,\text{arcsec}\,\frac{|u|}{a} + C$

EXAMPLE 1 Integration with Inverse Trigonometric Functions

a. $\displaystyle\int \frac{dx}{\sqrt{4 - x^2}} = \arcsin\frac{x}{2} + C$

b. $\displaystyle\int \frac{dx}{2 + 9x^2} = \frac{1}{3}\int \frac{3\,dx}{(\sqrt{2})^2 + (3x)^2}$ $\quad u = 3x,\ a = \sqrt{2}$

$$= \frac{1}{3\sqrt{2}}\arctan\frac{3x}{\sqrt{2}} + C$$

c. $\displaystyle\int \frac{dx}{x\sqrt{4x^2 - 9}} = \int \frac{2\,dx}{2x\sqrt{(2x)^2 - 3^2}}$ $\quad u = 2x,\ a = 3$

$$= \frac{1}{3}\,\text{arcsec}\,\frac{|2x|}{3} + C$$

The integrals in Example 1 are fairly straightforward applications of integration formulas. Unfortunately, this is not typical. The integration formulas for inverse trigonometric functions can be disguised in many ways.

TECHNOLOGY PITFALL
Computer software that can perform symbolic integration is useful for integrating functions such as the one in Example 2. When using such software, however, you must remember that it can fail to find an antiderivative for two reasons. First, some elementary functions simply do not have antiderivatives that are elementary functions. Second, every symbolic integration utility has limitations—you might have entered a function that the software was not programmed to handle. You should also remember that antiderivatives involving trigonometric functions or logarithmic functions can be written in many different forms. For instance, one symbolic integration utility found the integral in Example 2 to be

$$\int \frac{dx}{\sqrt{e^{2x}-1}} = \arctan \sqrt{e^{2x}-1} + C.$$

Try showing that this antiderivative is equivalent to that obtained in Example 2.

EXAMPLE 2 Integration by Substitution

Find $\displaystyle\int \frac{dx}{\sqrt{e^{2x}-1}}$.

Solution As it stands, this integral doesn't fit any of the three inverse trigonometric formulas. Using the substitution $u = e^x$, however, produces

$$u = e^x \implies du = e^x\,dx \implies dx = \frac{du}{e^x} = \frac{du}{u}.$$

With this substitution, you can integrate as shown.

$$\begin{aligned}
\int \frac{dx}{\sqrt{e^{2x}-1}} &= \int \frac{dx}{\sqrt{(e^x)^2-1}} && \text{Write } e^{2x} \text{ as } (e^x)^2.\\
&= \int \frac{du/u}{\sqrt{u^2-1}} && \text{Substitute.}\\
&= \int \frac{du}{u\sqrt{u^2-1}} && \text{Rewrite to fit Arcsecant Rule.}\\
&= \operatorname{arcsec} \frac{|u|}{1} + C && \text{Apply Arcsecant Rule.}\\
&= \operatorname{arcsec} e^x + C && \text{Back-substitute.}
\end{aligned}$$

EXAMPLE 3 Rewriting as the Sum of Two Quotients

Find $\displaystyle\int \frac{x+2}{\sqrt{4-x^2}}\,dx$.

Solution This integral does not appear to fit any of the basic integration formulas. By splitting the integrand into two parts, however, you can see that the first part can be found with the Power Rule, and the second part yields an inverse sine function.

$$\begin{aligned}
\int \frac{x+2}{\sqrt{4-x^2}}\,dx &= \int \frac{x}{\sqrt{4-x^2}}\,dx + \int \frac{2}{\sqrt{4-x^2}}\,dx\\
&= -\frac{1}{2}\int (4-x^2)^{-1/2}(-2x)\,dx + 2\int \frac{1}{\sqrt{4-x^2}}\,dx\\
&= -\frac{1}{2}\left[\frac{(4-x^2)^{1/2}}{1/2}\right] + 2\arcsin\frac{x}{2} + C\\
&= -\sqrt{4-x^2} + 2\arcsin\frac{x}{2} + C
\end{aligned}$$

Completing the Square

Completing the square helps when quadratic functions are involved in the integrand. For example, the quadratic $x^2 + bx + c$ can be written as the difference of two squares by adding and subtracting $(b/2)^2$.

$$\begin{aligned}
x^2 + bx + c &= x^2 + bx + \left(\frac{b}{2}\right)^2 - \left(\frac{b}{2}\right)^2 + c\\
&= \left(x + \frac{b}{2}\right)^2 - \left(\frac{b}{2}\right)^2 + c
\end{aligned}$$

EXAMPLE 4 Completing the Square

Find $\displaystyle\int \frac{dx}{x^2 - 4x + 7}$.

Solution You can write the denominator as the sum of two squares as shown.

$$\begin{aligned} x^2 - 4x + 7 &= (x^2 - 4x + 4) - 4 + 7 \\ &= (x - 2)^2 + 3 = u^2 + a^2 \end{aligned}$$

Now, in this completed square form, let $u = x - 2$ and $a = \sqrt{3}$.

$$\int \frac{dx}{x^2 - 4x + 7} = \int \frac{dx}{(x - 2)^2 + 3} = \frac{1}{\sqrt{3}} \arctan \frac{x - 2}{\sqrt{3}} + C$$

If the leading coefficient is not 1, it helps to factor before completing the square. For instance, you can complete the square for $2x^2 - 8x + 10$ by factoring first.

$$\begin{aligned} 2x^2 - 8x + 10 &= 2(x^2 - 4x + 5) \\ &= 2(x^2 - 4x + 4 - 4 + 5) \\ &= 2[(x - 2)^2 + 1] \end{aligned}$$

To complete the square when the coefficient of x^2 is negative, use the same factoring process shown above. For instance, you can complete the square for $3x - x^2$ as shown.

$$\begin{aligned} 3x - x^2 &= -(x^2 - 3x) \\ &= -\left[x^2 - 3x + \left(\tfrac{3}{2}\right)^2 - \left(\tfrac{3}{2}\right)^2\right] \\ &= \left(\tfrac{3}{2}\right)^2 - \left(x - \tfrac{3}{2}\right)^2 \end{aligned}$$

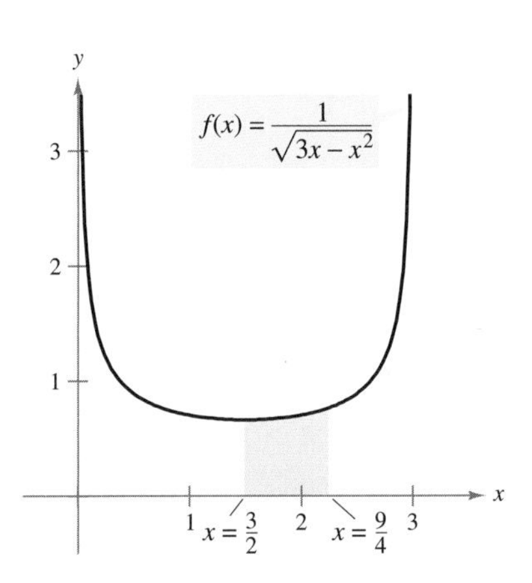

The area of the region bounded by the graph of f, the x-axis, $x = \frac{3}{2}$, and $x = \frac{9}{4}$ is $\pi/6$.
Figure 5.48

EXAMPLE 5 Completing the Square (Negative Leading Coefficient)

Find the area of the region bounded by the graph of

$$f(x) = \frac{1}{\sqrt{3x - x^2}}$$

the x-axis, and the lines $x = \frac{3}{2}$ and $x = \frac{9}{4}$.

Solution From Figure 5.48, you can see that the area is given by

$$\text{Area} = \int_{3/2}^{9/4} \frac{1}{\sqrt{3x - x^2}}\, dx.$$

Using the completed square form derived above, you can integrate as shown.

$$\begin{aligned} \int_{3/2}^{9/4} \frac{dx}{\sqrt{3x - x^2}} &= \int_{3/2}^{9/4} \frac{dx}{\sqrt{(3/2)^2 - [x - (3/2)]^2}} \\ &= \arcsin \frac{x - (3/2)}{3/2} \Bigg]_{3/2}^{9/4} \\ &= \arcsin \frac{1}{2} - \arcsin 0 \\ &= \frac{\pi}{6} \\ &\approx 0.524 \end{aligned}$$

TECHNOLOGY With definite integrals such as the one given in Example 5, remember that you can resort to a numerical solution. For instance, applying Simpson's Rule (with $n = 12$) to the integral in the example, you obtain

$$\int_{3/2}^{9/4} \frac{1}{\sqrt{3x - x^2}}\, dx \approx 0.523599.$$

This differs from the exact value of the integral ($\pi/6 \approx 0.5235988$) by less than one millionth.

Review of Basic Integration Rules

You have now completed the introduction of the **basic integration rules.** To be efficient at applying these rules, you should have practiced enough so that each rule is committed to memory.

Basic Integration Rules ($a > 0$)

1. $\displaystyle\int kf(u)\,du = k\int f(u)\,du$
2. $\displaystyle\int [f(u) \pm g(u)]\,du = \int f(u)\,du \pm \int g(u)\,du$
3. $\displaystyle\int du = u + C$
4. $\displaystyle\int u^n\,du = \frac{u^{n+1}}{n+1} + C, \quad n \neq -1$
5. $\displaystyle\int \frac{du}{u} = \ln|u| + C$
6. $\displaystyle\int e^u\,du = e^u + C$
7. $\displaystyle\int a^u\,du = \left(\frac{1}{\ln a}\right)a^u + C$
8. $\displaystyle\int \sin u\,du = -\cos u + C$
9. $\displaystyle\int \cos u\,du = \sin u + C$
10. $\displaystyle\int \tan u\,du = -\ln|\cos u| + C$
11. $\displaystyle\int \cot u\,du = \ln|\sin u| + C$
12. $\displaystyle\int \sec u\,du = \ln|\sec u + \tan u| + C$
13. $\displaystyle\int \csc u\,du = -\ln|\csc u + \cot u| + C$
14. $\displaystyle\int \sec^2 u\,du = \tan u + C$
15. $\displaystyle\int \csc^2 u\,du = -\cot u + C$
16. $\displaystyle\int \sec u \tan u\,du = \sec u + C$
17. $\displaystyle\int \csc u \cot u\,du = -\csc u + C$
18. $\displaystyle\int \frac{du}{\sqrt{a^2 - u^2}} = \arcsin\frac{u}{a} + C$
19. $\displaystyle\int \frac{du}{a^2 + u^2} = \frac{1}{a}\arctan\frac{u}{a} + C$
20. $\displaystyle\int \frac{du}{u\sqrt{u^2 - a^2}} = \frac{1}{a}\,\text{arcsec}\frac{|u|}{a} + C$

You can learn a lot about the nature of integration by comparing this list with the summary of differentiation rules given in Section 3.6. For differentiation, you now have rules that allow you to differentiate *any* elementary function. For integration, this is far from true.

The integration rules listed above are primarily those that were happened on when developing differentiation rules. So far, you have not learned any rules or techniques for finding the antiderivative of a general product or quotient, the natural logarithmic function, or the inverse trigonometric functions. More importantly, you cannot apply any of the rules in this list unless you can create the proper du corresponding to the u in the formula. The point is that you need to work more on integration techniques, which you will do in Chapter 8. The next two examples should give you a better feeling for the integration problems that you *can* and *cannot* do with the techniques and rules you now know.

EXAMPLE 6 Comparing Integration Problems

Find as many of the following integrals as you can using the formulas and techniques you have studied so far in the text.

a. $\displaystyle\int \frac{dx}{x\sqrt{x^2-1}}$ **b.** $\displaystyle\int \frac{x\,dx}{\sqrt{x^2-1}}$ **c.** $\displaystyle\int \frac{dx}{\sqrt{x^2-1}}$

Solution

a. You *can* find this integral (it fits the Arcsecant Rule).

$$\int \frac{dx}{x\sqrt{x^2-1}} = \operatorname{arcsec}|x| + C$$

b. You *can* find this integral (it fits the Power Rule).

$$\begin{aligned}\int \frac{x\,dx}{\sqrt{x^2-1}} &= \frac{1}{2}\int (x^2-1)^{-1/2}(2x)\,dx \\ &= \frac{1}{2}\left[\frac{(x^2-1)^{1/2}}{1/2}\right] + C \\ &= \sqrt{x^2-1} + C\end{aligned}$$

c. You *cannot* find this integral using present techniques. (You should scan the list of basic integration rules to verify this conclusion.)

EXAMPLE 7 Comparing Integration Problems

Find as many of the following integrals as you can using the formulas and techniques you have studied so far in the text.

a. $\displaystyle\int \frac{dx}{x\ln x}$ **b.** $\displaystyle\int \frac{\ln x\,dx}{x}$ **c.** $\displaystyle\int \ln x\,dx$

Solution

a. You *can* find this integral (it fits the Log Rule).

$$\begin{aligned}\int \frac{dx}{x\ln x} &= \int \frac{1/x}{\ln x}\,dx \\ &= \ln|\ln x| + C\end{aligned}$$

b. You *can* find this integral (it fits the Power Rule).

$$\begin{aligned}\int \frac{\ln x\,dx}{x} &= \int \left(\frac{1}{x}\right)(\ln x)^1\,dx \\ &= \frac{(\ln x)^2}{2} + C\end{aligned}$$

c. You *cannot* find this integral using present techniques.

NOTE Note in Examples 6 and 7 that the *simplest* functions are the ones that you cannot yet integrate.

Exercises for Section 5.8

See www.CalcChat.com for worked-out solutions to odd-numbered exercises.

In Exercises 1–20, find the integral.

1. $\displaystyle\int \frac{5}{\sqrt{9 - x^2}}\,dx$
2. $\displaystyle\int \frac{3}{\sqrt{1 - 4x^2}}\,dx$
3. $\displaystyle\int \frac{7}{16 + x^2}\,dx$
4. $\displaystyle\int \frac{4}{1 + 9x^2}\,dx$
5. $\displaystyle\int \frac{1}{x\sqrt{4x^2 - 1}}\,dx$
6. $\displaystyle\int \frac{1}{4 + (x - 1)^2}\,dx$
7. $\displaystyle\int \frac{x^3}{x^2 + 1}\,dx$
8. $\displaystyle\int \frac{x^4 - 1}{x^2 + 1}\,dx$
9. $\displaystyle\int \frac{1}{\sqrt{1 - (x + 1)^2}}\,dx$
10. $\displaystyle\int \frac{t}{t^4 + 16}\,dt$
11. $\displaystyle\int \frac{t}{\sqrt{1 - t^4}}\,dt$
12. $\displaystyle\int \frac{1}{x\sqrt{x^4 - 4}}\,dx$
13. $\displaystyle\int \frac{e^{2x}}{4 + e^{4x}}\,dx$
14. $\displaystyle\int \frac{1}{3 + (x - 2)^2}\,dx$
15. $\displaystyle\int \frac{1}{\sqrt{x}\sqrt{1 - x}}\,dx$
16. $\displaystyle\int \frac{3}{2\sqrt{x}(1 + x)}\,dx$
17. $\displaystyle\int \frac{x - 3}{x^2 + 1}\,dx$
18. $\displaystyle\int \frac{4x + 3}{\sqrt{1 - x^2}}\,dx$
19. $\displaystyle\int \frac{x + 5}{\sqrt{9 - (x - 3)^2}}\,dx$
20. $\displaystyle\int \frac{x - 2}{(x + 1)^2 + 4}\,dx$

In Exercises 21–30, evaluate the integral.

21. $\displaystyle\int_0^{1/6} \frac{1}{\sqrt{1 - 9x^2}}\,dx$
22. $\displaystyle\int_0^{1} \frac{dx}{\sqrt{4 - x^2}}$
23. $\displaystyle\int_0^{\sqrt{3}/2} \frac{1}{1 + 4x^2}\,dx$
24. $\displaystyle\int_{\sqrt{3}}^{3} \frac{1}{9 + x^2}\,dx$
25. $\displaystyle\int_0^{1/\sqrt{2}} \frac{\arcsin x}{\sqrt{1 - x^2}}\,dx$
26. $\displaystyle\int_0^{1/\sqrt{2}} \frac{\arccos x}{\sqrt{1 - x^2}}\,dx$
27. $\displaystyle\int_{-1/2}^{0} \frac{x}{\sqrt{1 - x^2}}\,dx$
28. $\displaystyle\int_{-\sqrt{3}}^{0} \frac{x}{1 + x^2}\,dx$
29. $\displaystyle\int_{\pi/2}^{\pi} \frac{\sin x}{1 + \cos^2 x}\,dx$
30. $\displaystyle\int_0^{\pi/2} \frac{\cos x}{1 + \sin^2 x}\,dx$

In Exercises 31–42, find or evaluate the integral. (Complete the square, if necessary.)

31. $\displaystyle\int_0^{2} \frac{dx}{x^2 - 2x + 2}$
32. $\displaystyle\int_{-2}^{2} \frac{dx}{x^2 + 4x + 13}$
33. $\displaystyle\int \frac{2x}{x^2 + 6x + 13}\,dx$
34. $\displaystyle\int \frac{2x - 5}{x^2 + 2x + 2}\,dx$
35. $\displaystyle\int \frac{1}{\sqrt{-x^2 - 4x}}\,dx$
36. $\displaystyle\int \frac{2}{\sqrt{-x^2 + 4x}}\,dx$
37. $\displaystyle\int \frac{x + 2}{\sqrt{-x^2 - 4x}}\,dx$
38. $\displaystyle\int \frac{x - 1}{\sqrt{x^2 - 2x}}\,dx$
39. $\displaystyle\int_2^{3} \frac{2x - 3}{\sqrt{4x - x^2}}\,dx$
40. $\displaystyle\int \frac{1}{(x - 1)\sqrt{x^2 - 2x}}\,dx$
41. $\displaystyle\int \frac{x}{x^4 + 2x^2 + 2}\,dx$
42. $\displaystyle\int \frac{x}{\sqrt{9 + 8x^2 - x^4}}\,dx$

In Exercises 43–46, use the specified substitution to find or evaluate the integral.

43. $\displaystyle\int \sqrt{e^t - 3}\,dt$

    $u = \sqrt{e^t - 3}$
44. $\displaystyle\int \frac{\sqrt{x - 2}}{x + 1}\,dx$

    $u = \sqrt{x - 2}$
45. $\displaystyle\int_1^{3} \frac{dx}{\sqrt{x}(1 + x)}$

    $u = \sqrt{x}$
46. $\displaystyle\int_0^{1} \frac{dx}{2\sqrt{3 - x}\sqrt{x + 1}}$

    $u = \sqrt{x + 1}$

Writing About Concepts

In Exercises 47–50, determine which of the integrals can be found using the basic integration formulas you have studied so far in the text.

47. (a) $\displaystyle\int \frac{1}{\sqrt{1 - x^2}}\,dx$ (b) $\displaystyle\int \frac{x}{\sqrt{1 - x^2}}\,dx$ (c) $\displaystyle\int \frac{1}{x\sqrt{1 - x^2}}\,dx$

48. (a) $\displaystyle\int e^{x^2}\,dx$ (b) $\displaystyle\int xe^{x^2}\,dx$ (c) $\displaystyle\int \frac{1}{x^2}e^{1/x}\,dx$

49. (a) $\displaystyle\int \sqrt{x - 1}\,dx$ (b) $\displaystyle\int x\sqrt{x - 1}\,dx$ (c) $\displaystyle\int \frac{x}{\sqrt{x - 1}}\,dx$

50. (a) $\displaystyle\int \frac{1}{1 + x^4}\,dx$ (b) $\displaystyle\int \frac{x}{1 + x^4}\,dx$ (c) $\displaystyle\int \frac{x^3}{1 + x^4}\,dx$

51. Determine which value best approximates the area of the region between the x-axis and the function

$$f(x) = \frac{1}{\sqrt{1 - x^2}}$$

over the interval $[-0.5, 0.5]$. (Make your selection on the basis of a sketch of the region and *not* by performing any calculations.)

(a) 4 (b) -3 (c) 1 (d) 2 (e) 3

52. Decide whether you can find the integral

$$\int \frac{2\,dx}{\sqrt{x^2 + 4}}$$

using the formulas and techniques you have studied so far. Explain your reasoning.

Differential Equations **In Exercises 53 and 54, use the differential equation and the specified initial condition to find y.**

53. $\displaystyle\frac{dy}{dx} = \frac{1}{\sqrt{4 - x^2}}$

    $y(0) = \pi$
54. $\displaystyle\frac{dy}{dx} = \frac{1}{4 + x^2}$

    $y(2) = \pi$

Slope Fields **In Exercises 55–58, a differential equation, a point, and a slope field are given. (a) Sketch two approximate solutions of the differential equation on the slope field, one of which passes through the given point. (b) Use integration to find the particular solution of the differential equation and use a graphing utility to graph the solution. Compare the result with the sketches in part (a). To print an enlarged copy of the graph, go to the website *www.mathgraphs.com*.**

55. $\dfrac{dy}{dx} = \dfrac{3}{1 + x^2}, \quad (0, 0)$

56. $\dfrac{dy}{dx} = \dfrac{2}{9 + x^2}, \quad (0, 2)$

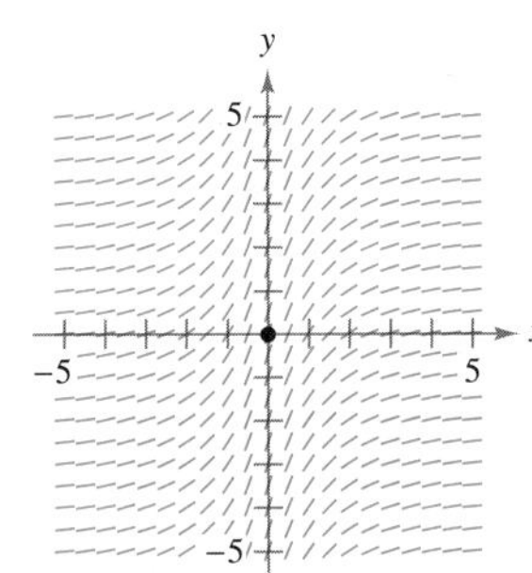

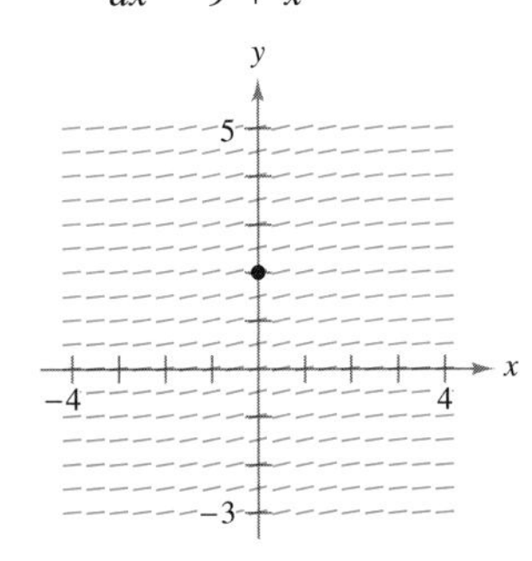

57. $\dfrac{dy}{dx} = \dfrac{1}{x\sqrt{x^2 - 4}}, \quad (2, 1)$

58. $\dfrac{dy}{dx} = \dfrac{2}{\sqrt{25 - x^2}}, \quad (5, \pi)$

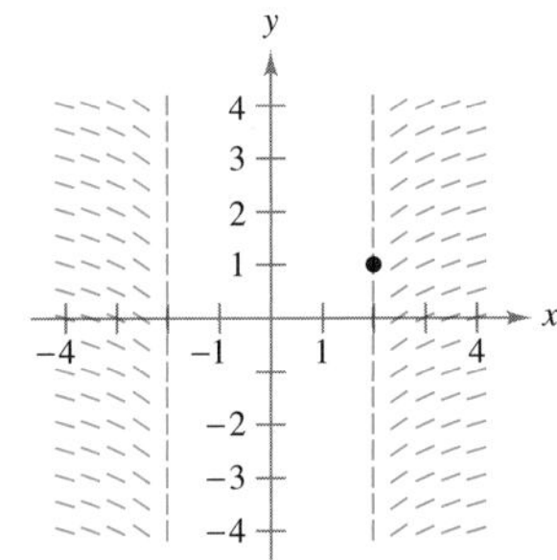

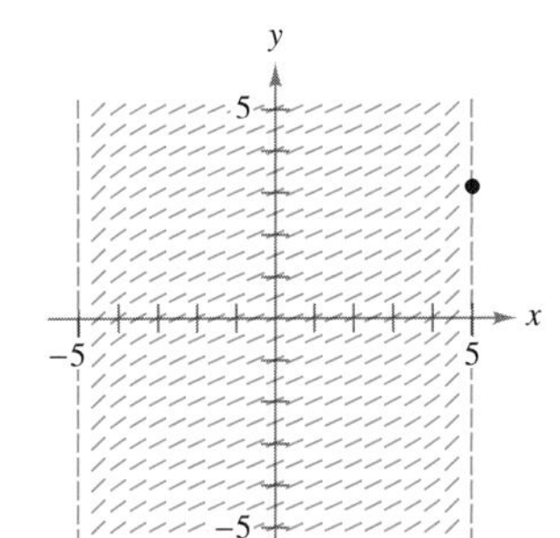

Slope Fields **In Exercises 59–62, use a computer algebra system to graph the slope field for the differential equation and graph the solution satisfying the specified initial condition.**

59. $\dfrac{dy}{dx} = \dfrac{10}{x\sqrt{x^2 - 1}}$

$y(3) = 0$

60. $\dfrac{dy}{dx} = \dfrac{1}{12 + x^2}$

$y(4) = 2$

61. $\dfrac{dy}{dx} = \dfrac{2y}{\sqrt{16 - x^2}}$

$y(0) = 2$

62. $\dfrac{dy}{dx} = \dfrac{\sqrt{y}}{1 + x^2}$

$y(0) = 4$

Area **In Exercises 63–68, find the area of the region.**

63. $y = \dfrac{1}{x^2 - 2x + 5}$

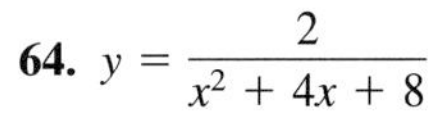

64. $y = \dfrac{2}{x^2 + 4x + 8}$

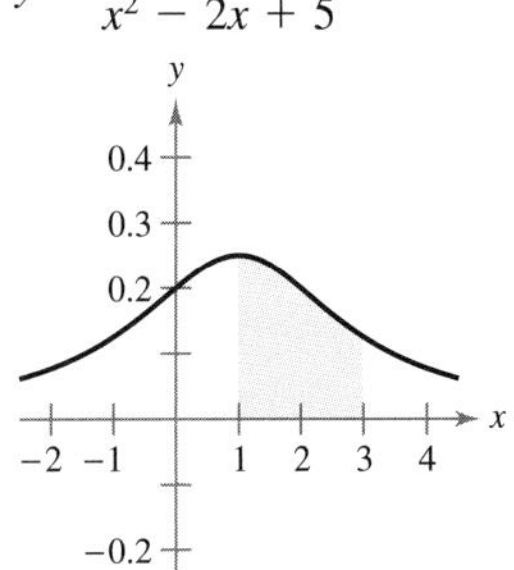

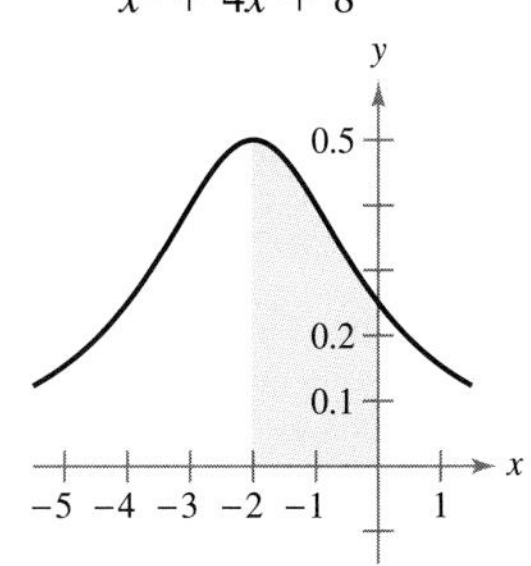

65. $y = \dfrac{1}{\sqrt{4 - x^2}}$

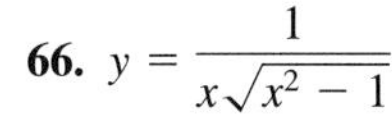

66. $y = \dfrac{1}{x\sqrt{x^2 - 1}}$

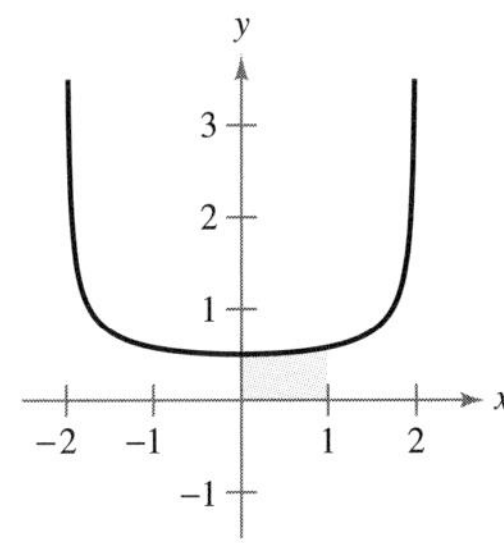

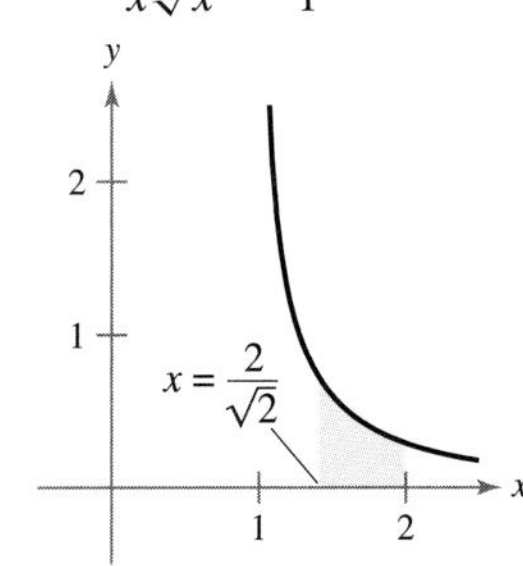

67. $y = \dfrac{3\cos x}{1 + \sin^2 x}$

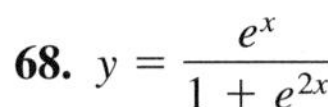

68. $y = \dfrac{e^x}{1 + e^{2x}}$

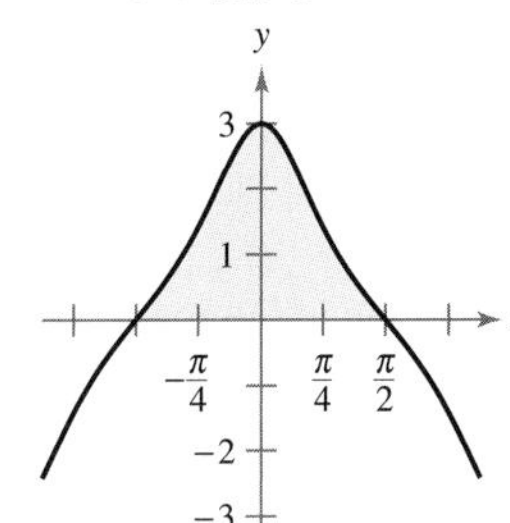

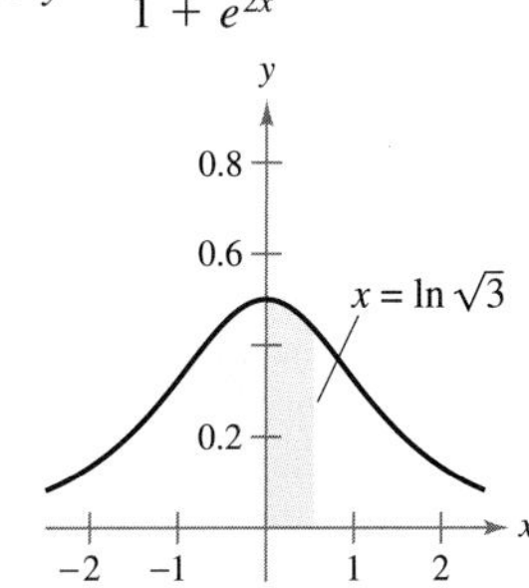

In Exercises 69 and 70, (a) verify the integration formula, and (b) use it to find the area of the region.

69. $\displaystyle\int \frac{\arctan x}{x^2}\,dx = \ln x - \frac{1}{2}\ln(1 + x^2) - \frac{\arctan x}{x} + C$

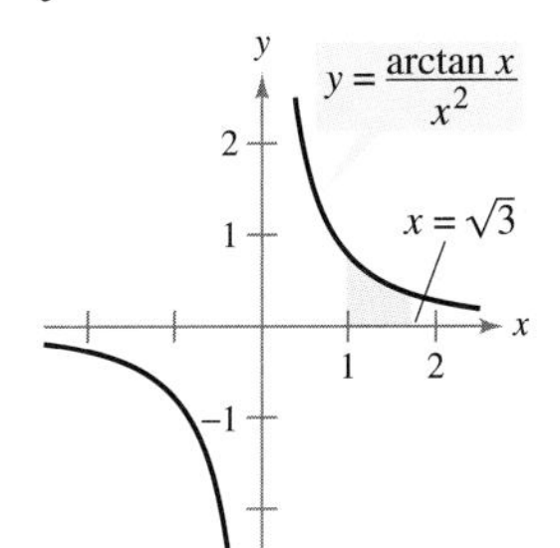

Figure for 69

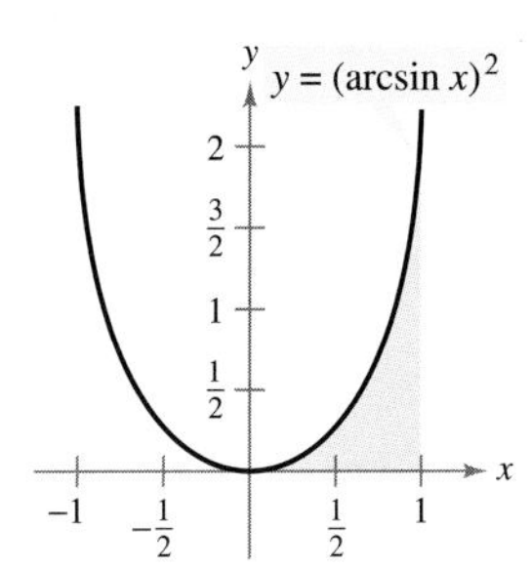

Figure for 70

70. $\displaystyle\int (\arcsin x)^2\,dx$

$= x(\arcsin x)^2 - 2x + 2\sqrt{1 - x^2}\arcsin x + C$

71. (a) Sketch the region whose area is represented by

$$\int_0^1 \arcsin x\, dx.$$

(b) Use the integration capabilities of a graphing utility to approximate the area.

(c) Find the exact area analytically.

72. (a) Show that $\displaystyle\int_0^1 \frac{4}{1+x^2}\, dx = \pi.$

(b) Approximate the number π using Simpson's Rule (with $n = 6$) and the integral in part (a).

(c) Approximate the number π by using the integration capabilities of a graphing utility.

73. ***Investigation*** Consider the function $F(x) = \dfrac{1}{2}\displaystyle\int_x^{x+2} \frac{2}{t^2+1}\, dt.$

(a) Write a short paragraph giving a geometric interpretation of the function $F(x)$ relative to the function $f(x) = \dfrac{2}{x^2+1}.$ Use what you have written to guess the value of x that will make F maximum.

(b) Perform the specified integration to find an alternative form of $F(x)$. Use calculus to locate the value of x that will make F maximum and compare the result with your guess in part (a).

74. Consider the integral $\displaystyle\int \frac{1}{\sqrt{6x - x^2}}\, dx.$

(a) Find the integral by completing the square of the radicand.

(b) Find the integral by making the substitution $u = \sqrt{x}$.

(c) The antiderivatives in parts (a) and (b) appear to be significantly different. Use a graphing utility to graph each antiderivative in the same viewing window and determine the relationship between them. Find the domain of each.

True or False? **In Exercises 75–78, determine whether the statement is true or false. If it is false, explain why or give an example that shows it is false.**

75. $\displaystyle\int \frac{dx}{3x\sqrt{9x^2-16}} = \frac{1}{4}\operatorname{arcsec}\frac{3x}{4} + C$

76. $\displaystyle\int \frac{dx}{25+x^2} = \frac{1}{25}\arctan\frac{x}{25} + C$

77. $\displaystyle\int \frac{dx}{\sqrt{4-x^2}} = -\arccos\frac{x}{2} + C$

78. One way to find $\displaystyle\int \frac{2e^{2x}}{\sqrt{9-e^{2x}}}\, dx$ is to use the Arcsine Rule.

Verifying Integration Rules **In Exercises 79–81, verify each rule by differentiating. Let $a > 0$.**

79. $\displaystyle\int \frac{du}{\sqrt{a^2-u^2}} = \arcsin\frac{u}{a} + C$

80. $\displaystyle\int \frac{du}{a^2+u^2} = \frac{1}{a}\arctan\frac{u}{a} + C$

81. $\displaystyle\int \frac{du}{u\sqrt{u^2-a^2}} = \frac{1}{a}\operatorname{arcsec}\frac{|u|}{a} + C$

82. ***Numerical Integration*** (a) Write an integral that represents the area of the region. (b) Then use the Trapezoidal Rule with $n = 8$ to estimate the area of the region. (c) Explain how you can use the results of parts (a) and (b) to estimate π.

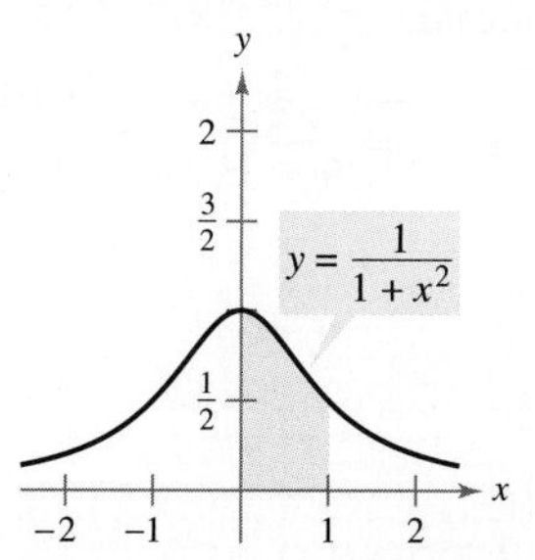

83. ***Vertical Motion*** An object is projected upward from ground level with an initial velocity of 500 feet per second. In this exercise, the goal is to analyze the motion of the object during its upward flight.

(a) If air resistance is neglected, find the velocity of the object as a function of time. Use a graphing utility to graph this function.

(b) Use the result in part (a) to find the position function and determine the maximum height attained by the object.

(c) If the air resistance is proportional to the square of the velocity, you obtain the equation

$$\frac{dv}{dt} = -(32 + kv^2)$$

where -32 feet per second per second is the acceleration due to gravity and k is a constant. Find the velocity as a function of time by solving the equation

$$\int \frac{dv}{32 + kv^2} = -\int dt.$$

(d) Use a graphing utility to graph the velocity function $v(t)$ in part (c) if $k = 0.001$. Use the graph to approximate the time t_0 at which the object reaches its maximum height.

(e) Use the integration capabilities of a graphing utility to approximate the integral

$$\int_0^{t_0} v(t)\, dt$$

where $v(t)$ and t_0 are those found in part (d). This is the approximation of the maximum height of the object.

(f) Explain the difference between the results in parts (b) and (e).

FOR FURTHER INFORMATION For more information on this topic, see "What Goes Up Must Come Down; Will Air Resistance Make It Return Sooner, or Later?" by John Lekner in *Mathematics Magazine*. To view this article, go to the website *www.matharticles.com*.

84. Graph $y_1 = \dfrac{x}{1+x^2}$, $y_2 = \arctan x$, and $y_3 = x$ on $[0, 10]$.

Prove that $\dfrac{x}{1+x^2} < \arctan x < x$ for $x > 0$.

Section 5.9 Hyperbolic Functions

- **Develop properties of hyperbolic functions.**
- **Differentiate and integrate hyperbolic functions.**
- **Develop properties of inverse hyperbolic functions.**
- **Differentiate and integrate functions involving inverse hyperbolic functions.**

American Institute of Physics/Emilio Segre Visual Archives, Physics Today Collection

JOHANN HEINRICH LAMBERT (1728–1777)

The first person to publish a comprehensive study on hyperbolic functions was Johann Heinrich Lambert, a Swiss-German mathematician and colleague of Euler.

Hyperbolic Functions

In this section you will look briefly at a special class of exponential functions called **hyperbolic functions.** The name *hyperbolic function* arose from comparison of the area of a semicircular region, as shown in Figure 5.49, with the area of a region under a hyperbola, as shown in Figure 5.50. The integral for the semicircular region involves an inverse trigonometric (circular) function:

$$\int_{-1}^{1} \sqrt{1 - x^2}\, dx = \frac{1}{2}\Big[x\sqrt{1 - x^2} + \arcsin x\Big]_{-1}^{1} = \frac{\pi}{2} \approx 1.571.$$

The integral for the hyperbolic region involves an inverse hyperbolic function:

$$\int_{-1}^{1} \sqrt{1 + x^2}\, dx = \frac{1}{2}\Big[x\sqrt{1 + x^2} + \sinh^{-1} x\Big]_{-1}^{1} \approx 2.296.$$

This is only one of many ways in which the hyperbolic functions are similar to the trigonometric functions.

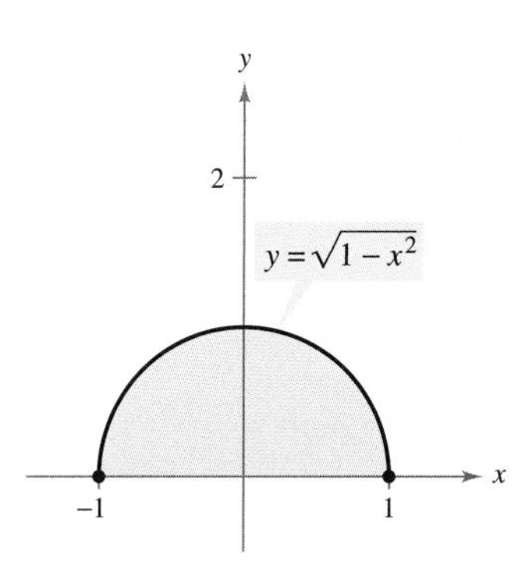

Circle: $x^2 + y^2 = 1$
Figure 5.49

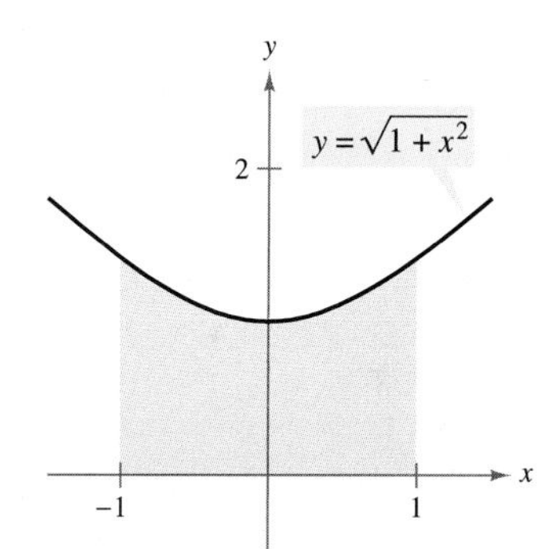

Hyperbola: $-x^2 + y^2 = 1$
Figure 5.50

FOR FURTHER INFORMATION For more information on the development of hyperbolic functions, see the article "An Introduction to Hyperbolic Functions in Elementary Calculus" by Jerome Rosenthal in *Mathematics Teacher*. To view this article, go to the website *www.matharticles.com*.

Definitions of the Hyperbolic Functions

$$\sinh x = \frac{e^x - e^{-x}}{2} \qquad \operatorname{csch} x = \frac{1}{\sinh x},\quad x \neq 0$$

$$\cosh x = \frac{e^x + e^{-x}}{2} \qquad \operatorname{sech} x = \frac{1}{\cosh x}$$

$$\tanh x = \frac{\sinh x}{\cosh x} \qquad \coth x = \frac{1}{\tanh x},\quad x \neq 0$$

NOTE $\sinh x$ is read as "the hyperbolic sine of x," $\cosh x$ as "the hyperbolic cosine of x," and so on.

The graphs of the six hyperbolic functions and their domains and ranges are shown in Figure 5.51. Note that the graph of $\sinh x$ can be obtained by *addition of ordinates* using the exponential functions $f(x) = \frac{1}{2}e^x$ and $g(x) = -\frac{1}{2}e^{-x}$. Likewise, the graph of $\cosh x$ can be obtained by *addition of ordinates* using the exponential functions $f(x) = \frac{1}{2}e^x$ and $h(x) = \frac{1}{2}e^{-x}$.

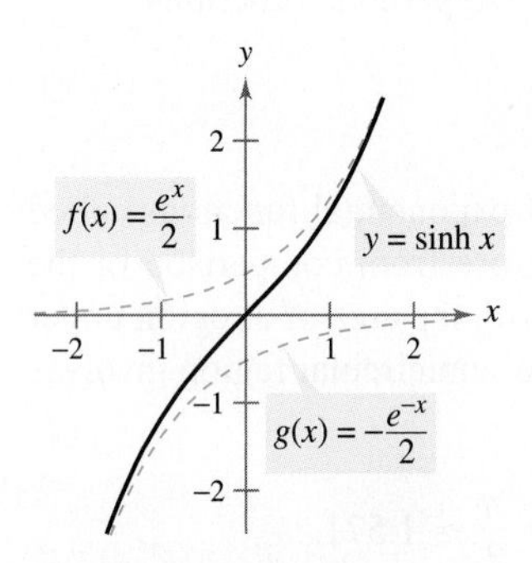

Domain: $(-\infty, \infty)$
Range: $(-\infty, \infty)$

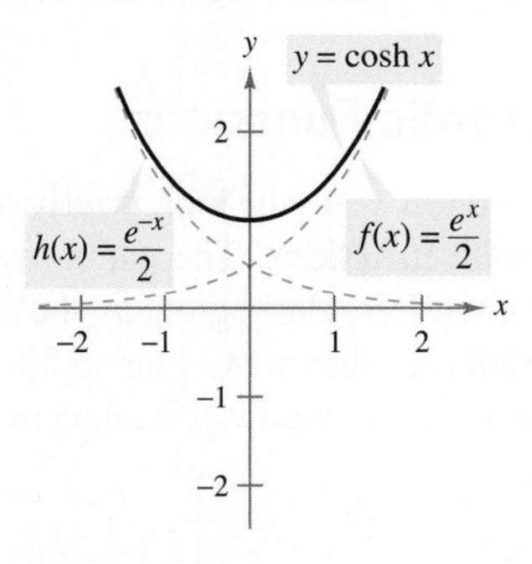

Domain: $(-\infty, \infty)$
Range: $[1, \infty)$

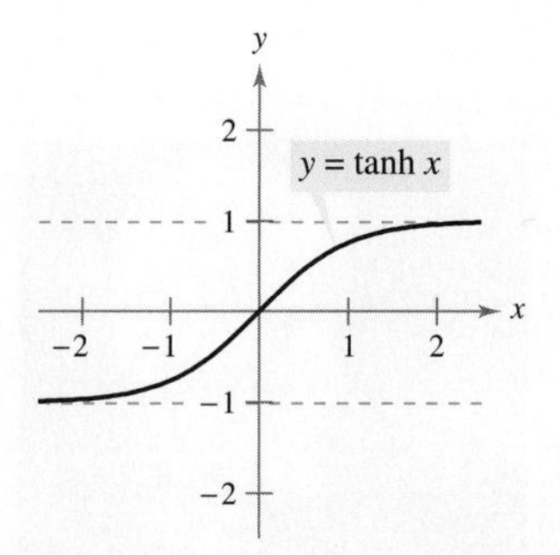

Domain: $(-\infty, \infty)$
Range: $(-1, 1)$

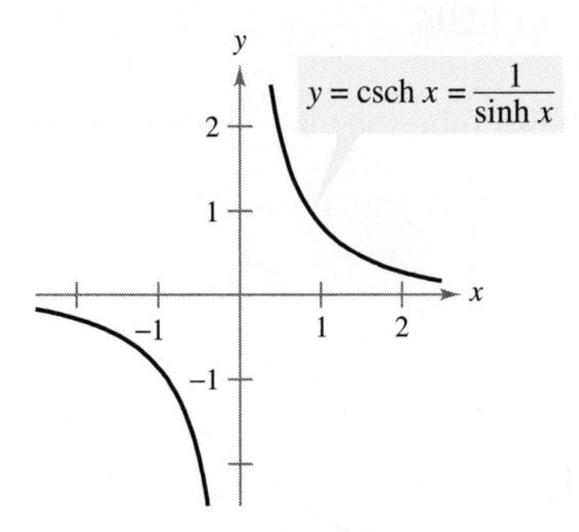

Domain: $(-\infty, 0) \cup (0, \infty)$
Range: $(-\infty, 0) \cup (0, \infty)$

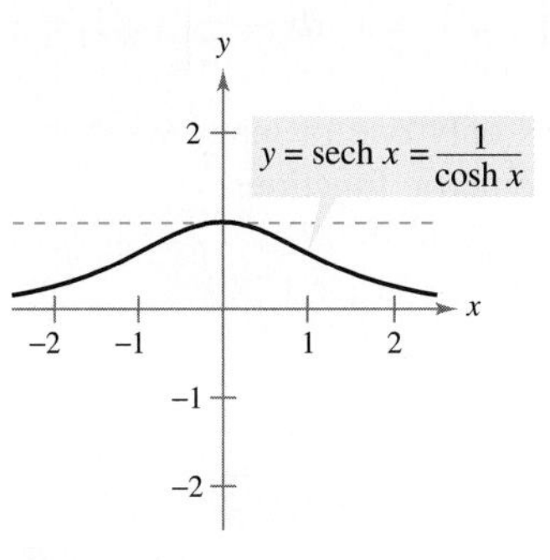

Domain: $(-\infty, \infty)$
Range: $(0, 1]$

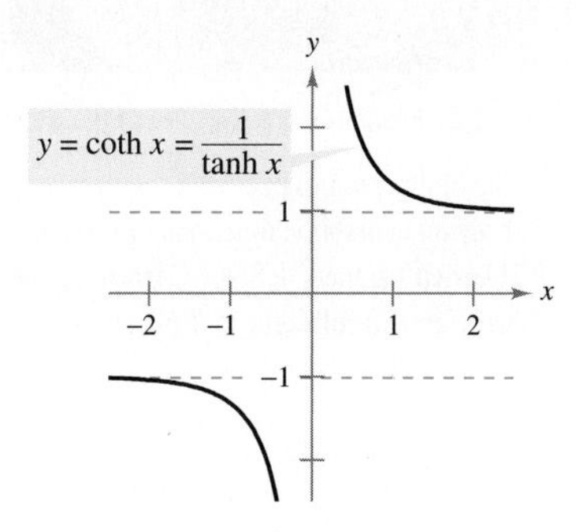

Domain: $(-\infty, 0) \cup (0, \infty)$
Range: $(-\infty, -1) \cup (1, \infty)$

Figure 5.51

Many of the trigonometric identities have corresponding *hyperbolic identities*. For instance,

$$\begin{aligned}\cosh^2 x - \sinh^2 x &= \left(\frac{e^x + e^{-x}}{2}\right)^2 - \left(\frac{e^x - e^{-x}}{2}\right)^2 \\ &= \frac{e^{2x} + 2 + e^{-2x}}{4} - \frac{e^{2x} - 2 + e^{-2x}}{4} \\ &= \frac{4}{4} \\ &= 1\end{aligned}$$

and

$$\begin{aligned}2 \sinh x \cosh x &= 2\left(\frac{e^x - e^{-x}}{2}\right)\left(\frac{e^x + e^{-x}}{2}\right) \\ &= \frac{e^{2x} - e^{-2x}}{2} \\ &= \sinh 2x.\end{aligned}$$

FOR FURTHER INFORMATION To understand geometrically the relationship between the hyperbolic and exponential functions, see the article "A Short Proof Linking the Hyperbolic and Exponential Functions" by Michael J. Seery in *The AMATYC Review*.

Hyperbolic Identities

$\cosh^2 x - \sinh^2 x = 1$

$\tanh^2 x + \text{sech}^2 x = 1$

$\coth^2 x - \text{csch}^2 x = 1$

$\sinh(x + y) = \sinh x \cosh y + \cosh x \sinh y$

$\sinh(x - y) = \sinh x \cosh y - \cosh x \sinh y$

$\cosh(x + y) = \cosh x \cosh y + \sinh x \sinh y$

$\cosh(x - y) = \cosh x \cosh y - \sinh x \sinh y$

$\sinh^2 x = \dfrac{-1 + \cosh 2x}{2}$

$\cosh^2 x = \dfrac{1 + \cosh 2x}{2}$

$\sinh 2x = 2 \sinh x \cosh x$

$\cosh 2x = \cosh^2 x + \sinh^2 x$

Differentiation and Integration of Hyperbolic Functions

Because the hyperbolic functions are written in terms of e^x and e^{-x}, you can easily derive rules for their derivatives. The following theorem lists these derivatives with the corresponding integration rules.

THEOREM 5.22 Derivatives and Integrals of Hyperbolic Functions

Let u be a differentiable function of x.

$\dfrac{d}{dx}[\sinh u] = (\cosh u)u'$ $\qquad$ $\displaystyle\int \cosh u\, du = \sinh u + C$

$\dfrac{d}{dx}[\cosh u] = (\sinh u)u'$ $\qquad$ $\displaystyle\int \sinh u\, du = \cosh u + C$

$\dfrac{d}{dx}[\tanh u] = (\text{sech}^2 u)u'$ $\qquad$ $\displaystyle\int \text{sech}^2 u\, du = \tanh u + C$

$\dfrac{d}{dx}[\coth u] = -(\text{csch}^2 u)u'$ $\qquad$ $\displaystyle\int \text{csch}^2 u\, du = -\coth u + C$

$\dfrac{d}{dx}[\text{sech}\, u] = -(\text{sech}\, u \tanh u)u'$ $\qquad$ $\displaystyle\int \text{sech}\, u \tanh u\, du = -\text{sech}\, u + C$

$\dfrac{d}{dx}[\text{csch}\, u] = -(\text{csch}\, u \coth u)u'$ $\qquad$ $\displaystyle\int \text{csch}\, u \coth u\, du = -\text{csch}\, u + C$

Proof

$$\begin{aligned}
\frac{d}{dx}[\sinh x] &= \frac{d}{dx}\left[\frac{e^x - e^{-x}}{2}\right] \\
&= \frac{e^x + e^{-x}}{2} = \cosh x \\
\frac{d}{dx}[\tanh x] &= \frac{d}{dx}\left[\frac{\sinh x}{\cosh x}\right] \\
&= \frac{\cosh x(\cosh x) - \sinh x(\sinh x)}{\cosh^2 x} \\
&= \frac{1}{\cosh^2 x} \\
&= \text{sech}^2 x
\end{aligned}$$

In Exercises 98 and 102, you are asked to prove some of the other differentiation rules.

EXAMPLE 1 Differentiation of Hyperbolic Functions

a. $\dfrac{d}{dx}[\sinh(x^2 - 3)] = 2x\cosh(x^2 - 3)$ **b.** $\dfrac{d}{dx}[\ln(\cosh x)] = \dfrac{\sinh x}{\cosh x} = \tanh x$

c. $\dfrac{d}{dx}[x \sinh x - \cosh x] = x\cosh x + \sinh x - \sinh x = x\cosh x$

EXAMPLE 2 Finding Relative Extrema

Find the relative extrema of $f(x) = (x - 1)\cosh x - \sinh x$.

Solution Begin by setting the first derivative of f equal to 0.

$$f'(x) = (x - 1)\sinh x + \cosh x - \cosh x = 0$$
$$(x - 1)\sinh x = 0$$

So, the critical numbers are $x = 1$ and $x = 0$. Using the Second Derivative Test, you can verify that the point $(0, -1)$ yields a relative maximum and the point $(1, -\sinh 1)$ yields a relative minimum, as shown in Figure 5.52. Try using a graphing utility to confirm this result. If your graphing utility does not have hyperbolic functions, you can use exponential functions as follows.

$$\begin{aligned} f(x) &= (x - 1)\left(\tfrac{1}{2}\right)(e^x + e^{-x}) - \tfrac{1}{2}(e^x - e^{-x}) \\ &= \tfrac{1}{2}(xe^x + xe^{-x} - e^x - e^{-x} - e^x + e^{-x}) \\ &= \tfrac{1}{2}(xe^x + xe^{-x} - 2e^x) \end{aligned}$$

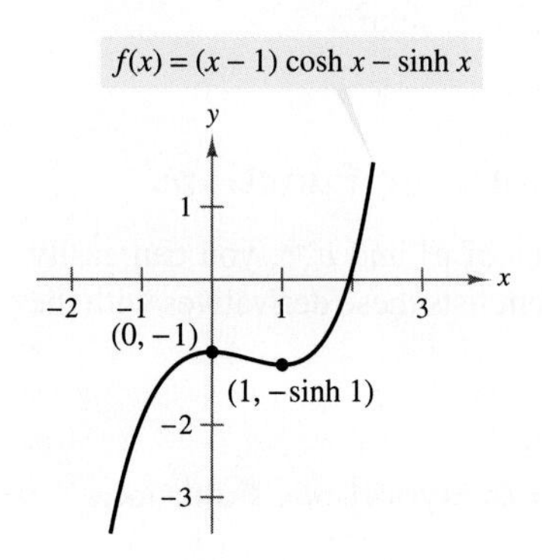

$f''(0) < 0$, so $(0, -1)$ is a relative maximum. $f''(1) > 0$, so $(1, -\sinh 1)$ is a relative minimum.
Figure 5.52

When a uniform flexible cable, such as a telephone wire, is suspended from two points, it takes the shape of a *catenary*, as discussed in Example 3.

EXAMPLE 3 Hanging Power Cables

Power cables are suspended between two towers, forming the catenary shown in Figure 5.53. The equation for this catenary is

$$y = a\cosh\frac{x}{a}.$$

The distance between the two towers is $2b$. Find the slope of the catenary at the point where the cable meets the right-hand tower.

Solution Differentiating produces

$$y' = a\left(\frac{1}{a}\right)\sinh\frac{x}{a} = \sinh\frac{x}{a}.$$

At the point $(b, a\cosh(b/a))$, the slope (from the left) is given by $m = \sinh\dfrac{b}{a}$.

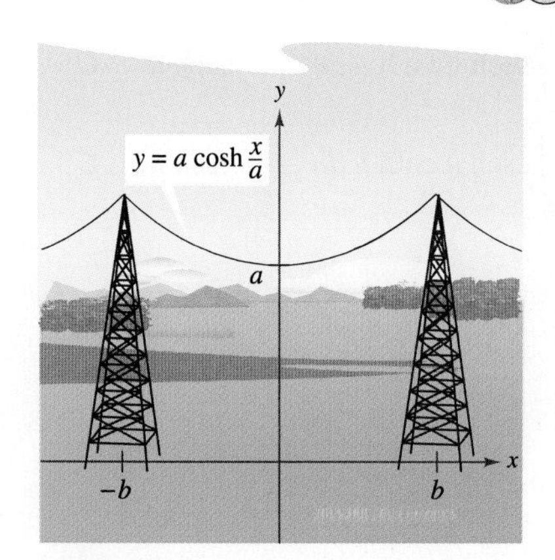

Catenary
Figure 5.53

FOR FURTHER INFORMATION In Example 3, the cable is a catenary between two supports at the same height. To learn about the shape of a cable hanging between supports of different heights, see the article "Reexamining the Catenary" by Paul Cella in *The College Mathematics Journal*. To view this article, go to the website *www.matharticles.com*.

EXAMPLE 4 Integrating a Hyperbolic Function

Find $\int \cosh 2x \sinh^2 2x\, dx$.

Solution

$$\begin{aligned}\int \cosh 2x \sinh^2 2x\, dx &= \frac{1}{2}\int (\sinh 2x)^2(2\cosh 2x)\, dx && u = \sinh 2x\\ &= \frac{1}{2}\left[\frac{(\sinh 2x)^3}{3}\right] + C\\ &= \frac{\sinh^3 2x}{6} + C\end{aligned}$$

Inverse Hyperbolic Functions

Unlike trigonometric functions, hyperbolic functions are not periodic. In fact, by looking back at Figure 5.51, you can see that four of the six hyperbolic functions are actually one-to-one (the hyperbolic sine, tangent, cosecant, and cotangent). So, you can conclude that these four functions have inverse functions. The other two (the hyperbolic cosine and secant) are one-to-one if their domains are restricted to the positive real numbers, and for this restricted domain they also have inverse functions. Because the hyperbolic functions are defined in terms of exponential functions, it is not surprising to find that the inverse hyperbolic functions can be written in terms of logarithmic functions, as shown in Theorem 5.23.

THEOREM 5.23 Inverse Hyperbolic Functions

Function	*Domain*
$\sinh^{-1} x = \ln\left(x + \sqrt{x^2+1}\right)$	$(-\infty, \infty)$
$\cosh^{-1} x = \ln\left(x + \sqrt{x^2-1}\right)$	$[1, \infty)$
$\tanh^{-1} x = \dfrac{1}{2}\ln\dfrac{1+x}{1-x}$	$(-1, 1)$
$\coth^{-1} x = \dfrac{1}{2}\ln\dfrac{x+1}{x-1}$	$(-\infty, -1) \cup (1, \infty)$
$\operatorname{sech}^{-1} x = \ln\dfrac{1+\sqrt{1-x^2}}{x}$	$(0, 1]$
$\operatorname{csch}^{-1} x = \ln\left(\dfrac{1}{x} + \dfrac{\sqrt{1+x^2}}{\lvert x\rvert}\right)$	$(-\infty, 0) \cup (0, \infty)$

Proof The proof of this theorem is a straightforward application of the properties of the exponential and logarithmic functions. For example, if

$$f(x) = \sinh x = \frac{e^x - e^{-x}}{2}$$

and

$$g(x) = \ln\left(x + \sqrt{x^2+1}\right)$$

you can show that $f(g(x)) = x$ and $g(f(x)) = x$, which implies that g is the inverse function of f.

2
$y_3 = y_4$
−3
3
$y_1 = y_2$
−2

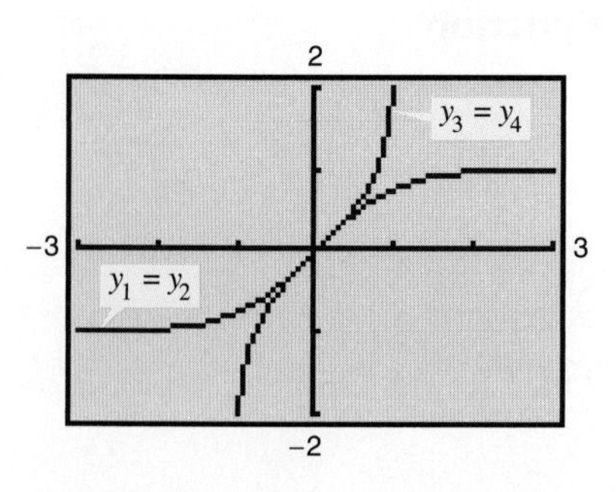

Graphs of the hyperbolic tangent function and the inverse hyperbolic tangent function
Figure 5.54

TECHNOLOGY You can use a graphing utility to confirm graphically the results of Theorem 5.23. For instance, graph the following functions.

$y_1 = \tanh x$ — Hyperbolic tangent

$y_2 = \dfrac{e^x - e^{-x}}{e^x + e^{-x}}$ — Definition of hyperbolic tangent

$y_3 = \tanh^{-1} x$ — Inverse hyperbolic tangent

$y_4 = \dfrac{1}{2} \ln \dfrac{1 + x}{1 - x}$ — Definition of inverse hyperbolic tangent

The resulting display is shown in Figure 5.54. As you watch the graphs being traced out, notice that $y_1 = y_2$ and $y_3 = y_4$. Also notice that the graph of y_1 is the reflection of the graph of y_3 in the line $y = x$.

The graphs of the inverse hyperbolic functions are shown in Figure 5.55.

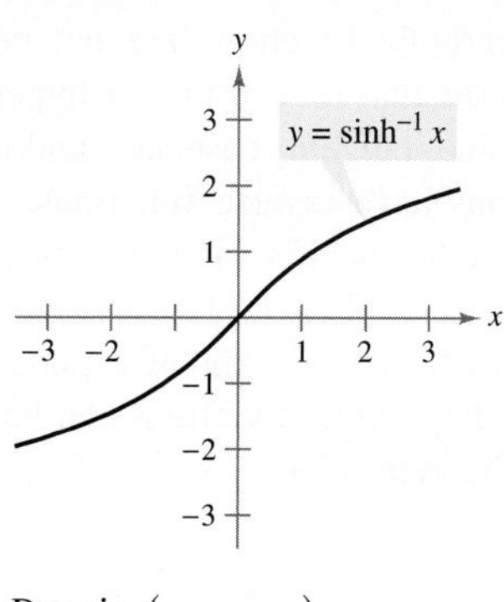

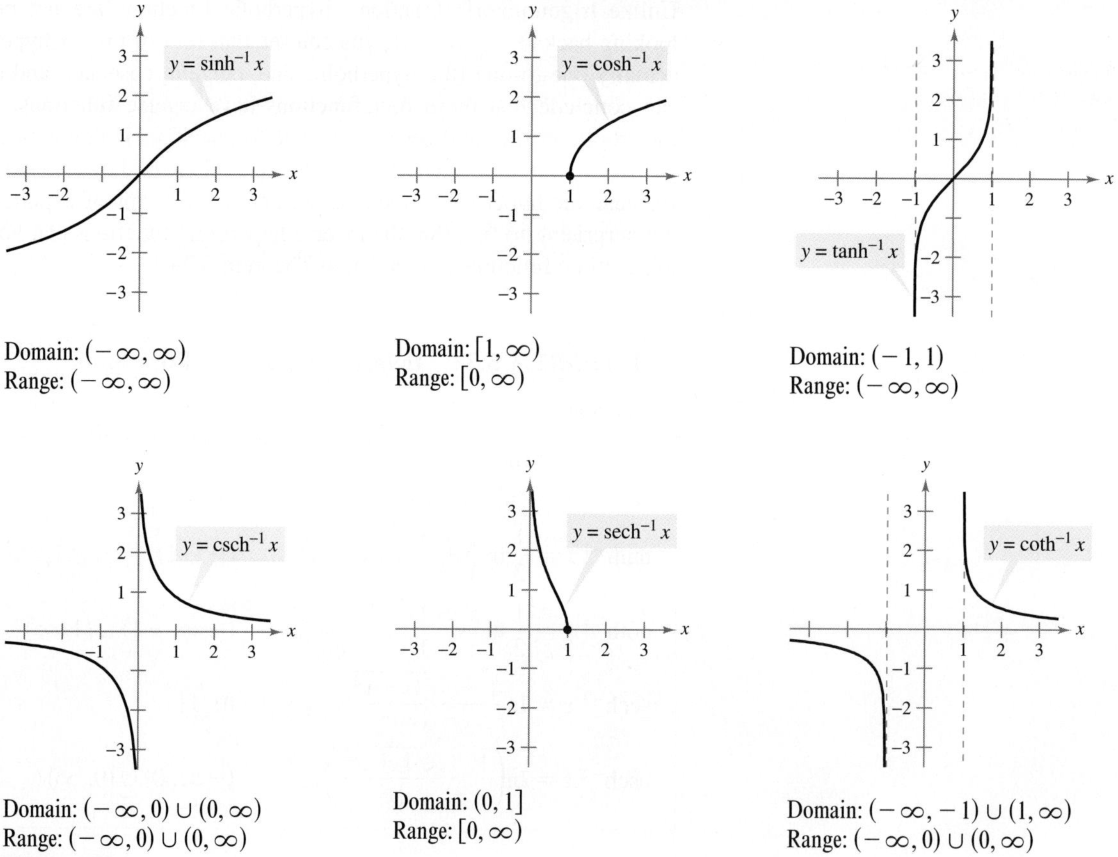

$y = \sinh^{-1} x$
Domain: $(-\infty, \infty)$
Range: $(-\infty, \infty)$
$y = \cosh^{-1} x$
Domain: $[1, \infty)$
Range: $[0, \infty)$
$y = \tanh^{-1} x$
Domain: $(-1, 1)$
Range: $(-\infty, \infty)$
$y = \operatorname{csch}^{-1} x$
Domain: $(-\infty, 0) \cup (0, \infty)$
Range: $(-\infty, 0) \cup (0, \infty)$
$y = \operatorname{sech}^{-1} x$
Domain: $(0, 1]$
Range: $[0, \infty)$
$y = \coth^{-1} x$
Domain: $(-\infty, -1) \cup (1, \infty)$
Range: $(-\infty, 0) \cup (0, \infty)$

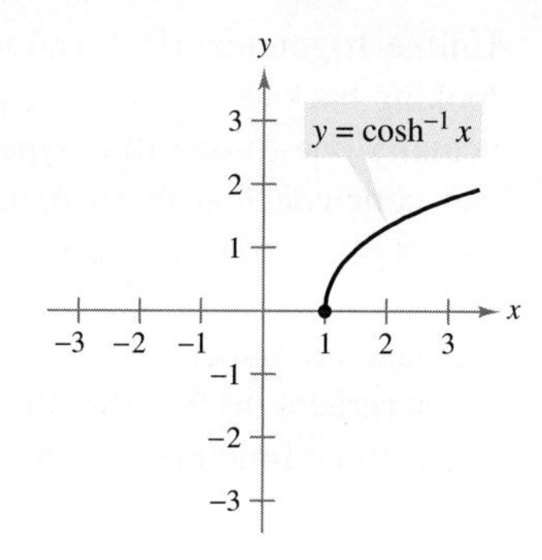

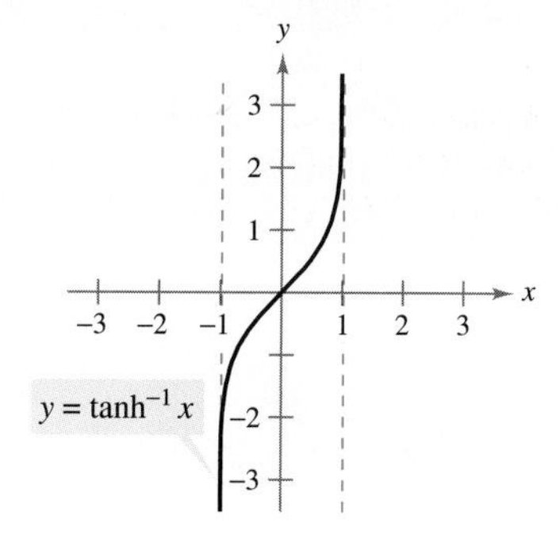

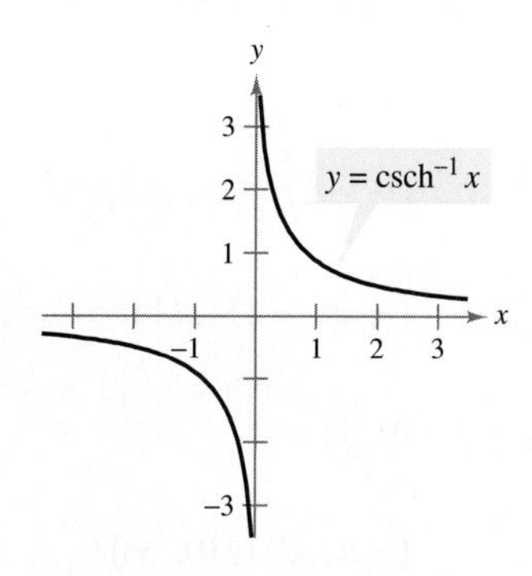

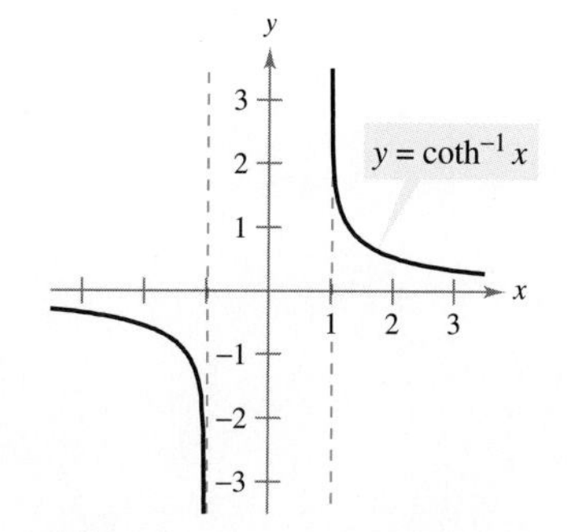

Figure 5.55

The inverse hyperbolic secant can be used to define a curve called a *tractrix* or *pursuit curve*, as discussed in Example 5.

EXAMPLE 5 A Tractrix

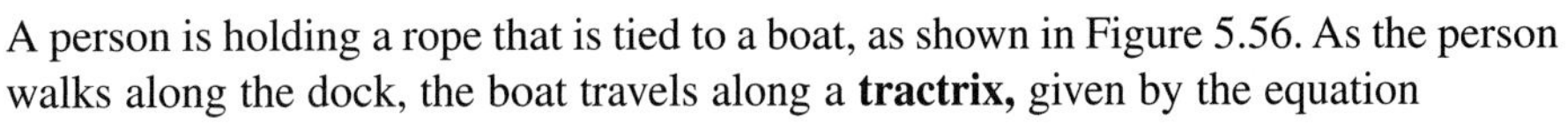

A person is holding a rope that is tied to a boat, as shown in Figure 5.56. As the person walks along the dock, the boat travels along a **tractrix,** given by the equation

$$y = a \operatorname{sech}^{-1}\frac{x}{a} - \sqrt{a^2 - x^2}$$

where a is the length of the rope. If $a = 20$ feet, find the distance the person must walk to bring the boat 5 feet from the dock.

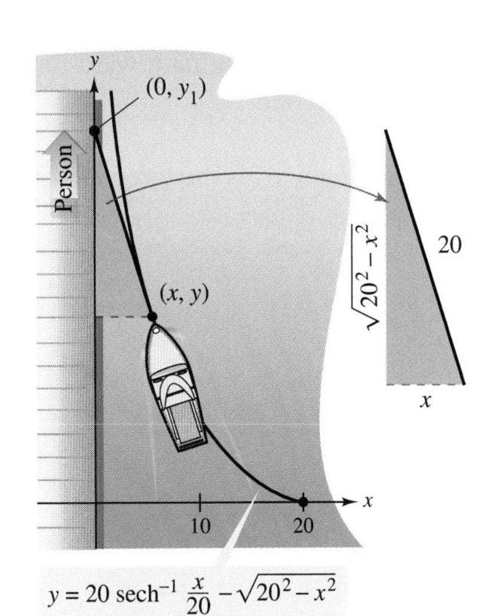

A person must walk 41.27 feet to bring the boat 5 feet from the dock.
Figure 5.56

Solution In Figure 5.56, notice that the distance the person has walked is given by

$$\begin{aligned} y_1 = y + \sqrt{20^2 - x^2} &= \left(20 \operatorname{sech}^{-1}\frac{x}{20} - \sqrt{20^2 - x^2}\right) + \sqrt{20^2 - x^2} \\ &= 20 \operatorname{sech}^{-1}\frac{x}{20}. \end{aligned}$$

When $x = 5$, this distance is

$$\begin{aligned} y_1 = 20 \operatorname{sech}^{-1}\frac{5}{20} &= 20 \ln \frac{1 + \sqrt{1 - (1/4)^2}}{1/4} \\ &= 20 \ln\left(4 + \sqrt{15}\right) \\ &\approx 41.27 \text{ feet.} \end{aligned}$$

Differentiation and Integration of Inverse Hyperbolic Functions

The derivatives of the inverse hyperbolic functions, which resemble the derivatives of the inverse trigonometric functions, are listed in Theorem 5.24 with the corresponding integration formulas (in logarithmic form). You can verify each of these formulas by applying the logarithmic definitions of the inverse hyperbolic functions. (See Exercises 99–101.)

THEOREM 5.24 Differentiation and Integration Involving Inverse Hyperbolic Functions

Let u be a differentiable function of x.

$$\frac{d}{dx}[\sinh^{-1} u] = \frac{u'}{\sqrt{u^2 + 1}} \qquad \frac{d}{dx}[\cosh^{-1} u] = \frac{u'}{\sqrt{u^2 - 1}}$$

$$\frac{d}{dx}[\tanh^{-1} u] = \frac{u'}{1 - u^2} \qquad \frac{d}{dx}[\coth^{-1} u] = \frac{u'}{1 - u^2}$$

$$\frac{d}{dx}[\operatorname{sech}^{-1} u] = \frac{-u'}{u\sqrt{1 - u^2}} \qquad \frac{d}{dx}[\operatorname{csch}^{-1} u] = \frac{-u'}{|u|\sqrt{1 + u^2}}$$

$$\int \frac{du}{\sqrt{u^2 \pm a^2}} = \ln\left(u + \sqrt{u^2 \pm a^2}\right) + C$$

$$\int \frac{du}{a^2 - u^2} = \frac{1}{2a} \ln\left|\frac{a + u}{a - u}\right| + C$$

$$\int \frac{du}{u\sqrt{a^2 \pm u^2}} = -\frac{1}{a} \ln \frac{a + \sqrt{a^2 \pm u^2}}{|u|} + C$$

EXAMPLE 6 More About a Tractrix

For the tractrix given in Example 5, show that the boat is always pointing toward the person.

Solution For a point (x, y) on a tractrix, the slope of the graph gives the direction of the boat, as shown in Figure 5.56.

$$
\begin{aligned}
y' &= \frac{d}{dx}\left[20 \operatorname{sech}^{-1}\frac{x}{20} - \sqrt{20^2 - x^2}\right] \\
&= -20\left(\frac{1}{20}\right)\left[\frac{1}{(x/20)\sqrt{1-(x/20)^2}}\right] - \left(\frac{1}{2}\right)\left(\frac{-2x}{\sqrt{20^2 - x^2}}\right) \\
&= \frac{-20^2}{x\sqrt{20^2 - x^2}} + \frac{x}{\sqrt{20^2 - x^2}} \\
&= -\frac{\sqrt{20^2 - x^2}}{x}
\end{aligned}
$$

However, from Figure 5.56, you can see that the slope of the line segment connecting the point $(0, y_1)$ with the point (x, y) is also

$$m = -\frac{\sqrt{20^2 - x^2}}{x}.$$

So, the boat is always pointing toward the person. (It is because of this property that a tractrix is called a *pursuit curve*.)

EXAMPLE 7 Integration Using Inverse Hyperbolic Functions

Find $\displaystyle\int \frac{dx}{x\sqrt{4 - 9x^2}}$.

Solution Let $a = 2$ and $u = 3x$.

$$
\begin{aligned}
\int \frac{dx}{x\sqrt{4 - 9x^2}} &= \int \frac{3\,dx}{(3x)\sqrt{4 - 9x^2}} \qquad \int \frac{du}{u\sqrt{a^2 - u^2}} \\
&= -\frac{1}{2}\ln\frac{2 + \sqrt{4 - 9x^2}}{|3x|} + C \qquad -\frac{1}{a}\ln\frac{a + \sqrt{a^2 - u^2}}{|u|} + C
\end{aligned}
$$

EXAMPLE 8 Integration Using Inverse Hyperbolic Functions

Find $\displaystyle\int \frac{dx}{5 - 4x^2}$.

Solution Let $a = \sqrt{5}$ and $u = 2x$.

$$
\begin{aligned}
\int \frac{dx}{5 - 4x^2} &= \frac{1}{2}\int \frac{2\,dx}{(\sqrt{5})^2 - (2x)^2} \qquad \int \frac{du}{a^2 - u^2} \\
&= \frac{1}{2}\left(\frac{1}{2\sqrt{5}}\ln\left|\frac{\sqrt{5} + 2x}{\sqrt{5} - 2x}\right|\right) + C \qquad \frac{1}{2a}\ln\left|\frac{a + u}{a - u}\right| + C \\
&= \frac{1}{4\sqrt{5}}\ln\left|\frac{\sqrt{5} + 2x}{\sqrt{5} - 2x}\right| + C
\end{aligned}
$$

Exercises for Section 5.9

See www.CalcChat.com for worked-out solutions to odd-numbered exercises.

In Exercises 1–6, evaluate the function. If the value is not a rational number, round your answer to three decimal places.

1. (a) $\sinh 3$ (b) $\tanh(-2)$

2. (a) $\cosh 0$ (b) $\text{sech}\, 1$

3. (a) $\text{csch}(\ln 2)$ (b) $\coth(\ln 5)$

4. (a) $\sinh^{-1} 0$ (b) $\tanh^{-1} 0$

5. (a) $\cosh^{-1} 2$ (b) $\text{sech}^{-1} \frac{2}{3}$

6. (a) $\text{csch}^{-1} 2$ (b) $\coth^{-1} 3$

In Exercises 7–12, verify the identity.

7. $\tanh^2 x + \text{sech}^2 x = 1$

8. $\cosh^2 x = \dfrac{1 + \cosh 2x}{2}$

9. $\sinh(x + y) = \sinh x \cosh y + \cosh x \sinh y$

10. $\sinh 2x = 2 \sinh x \cosh x$

11. $\sinh 3x = 3 \sinh x + 4 \sinh^3 x$

12. $\cosh x + \cosh y = 2 \cosh \dfrac{x + y}{2} \cosh \dfrac{x - y}{2}$

In Exercises 13 and 14, use the value of the given hyperbolic function to find the values of the other hyperbolic functions at x.

13. $\sinh x = \dfrac{3}{2}$

14. $\tanh x = \dfrac{1}{2}$

In Exercises 15–24, find the derivative of the function.

15. $y = \text{sech}(x + 1)$

16. $y = \coth 3x$

17. $f(x) = \ln(\sinh x)$

18. $g(x) = \ln(\cosh x)$

19. $y = \ln\left(\tanh \dfrac{x}{2}\right)$

20. $y = x \cosh x - \sinh x$

21. $h(x) = \dfrac{1}{4} \sinh 2x - \dfrac{x}{2}$

22. $h(t) = t - \coth t$

23. $f(t) = \arctan(\sinh t)$

24. $g(x) = \text{sech}^2\, 3x$

In Exercises 25–28, find an equation of the tangent line to the graph of the function at the given point.

25. $y = \sinh(1 - x^2)$

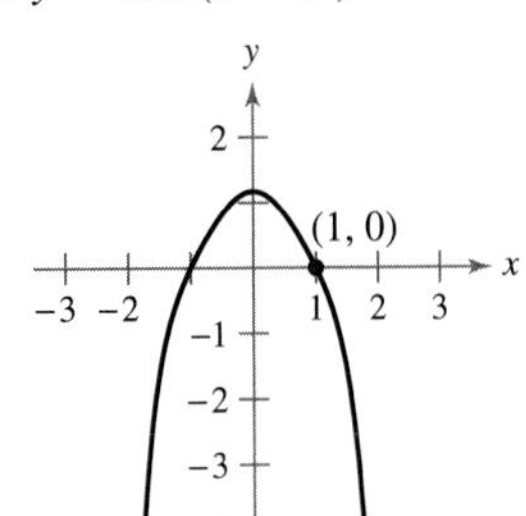

26. $y = x^{\cosh x}$

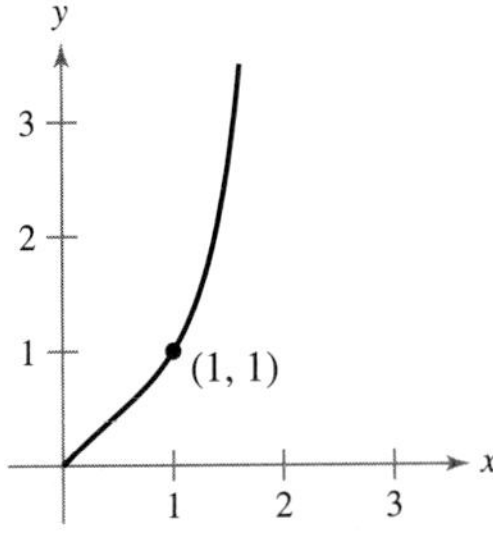

27. $y = (\cosh x - \sinh x)^2$

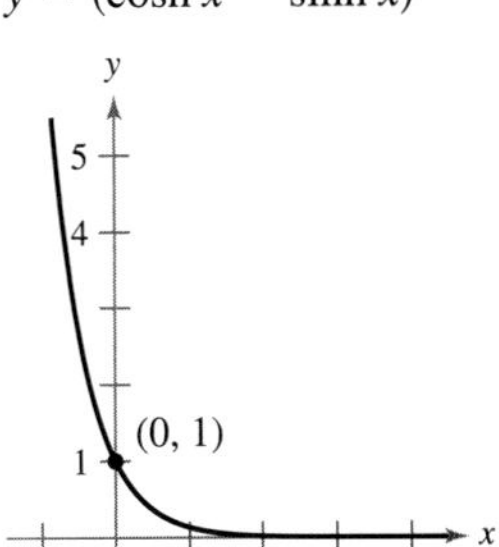

28. $y = e^{\sinh x}$

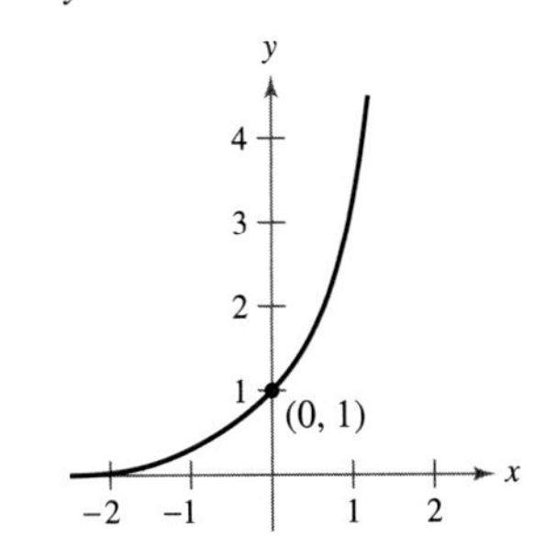

In Exercises 29–32, find any relative extrema of the function. Use a graphing utility to confirm your result.

29. $f(x) = \sin x \sinh x - \cos x \cosh x, \quad -4 \le x \le 4$

30. $f(x) = x \sinh(x - 1) - \cosh(x - 1)$

31. $g(x) = x\, \text{sech}\, x$

32. $h(x) = 2 \tanh x - x$

In Exercises 33 and 34, show that the function satisfies the differential equation.

Function	*Differential Equation*
33. $y = a \sinh x$	$y''' - y' = 0$
34. $y = a \cosh x$	$y'' - y = 0$

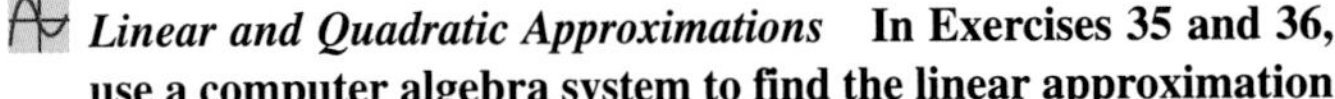

Linear and Quadratic Approximations **In Exercises 35 and 36, use a computer algebra system to find the linear approximation**

$$P_1(x) = f(a) + f'(a)(x - a)$$

and the quadratic approximation

$$P_2(x) = f(a) + f'(a)(x - a) + \tfrac{1}{2} f''(a)(x - a)^2$$

of the function f at $x = a$. Use a graphing utility to graph the function and its linear and quadratic approximations.

35. $f(x) = \tanh x, \quad a = 0$

36. $f(x) = \cosh x, \quad a = 0$

Catenary **In Exercises 37 and 38, a model for a power cable suspended between two towers is given. (a) Graph the model, (b) find the heights of the cable at the towers and at the midpoint between the towers, and (c) find the slope of the model at the point where the cable meets the right-hand tower.**

37. $y = 10 + 15 \cosh \dfrac{x}{15}, \quad -15 \le x \le 15$

38. $y = 18 + 25 \cosh \dfrac{x}{25}, \quad -25 \le x \le 25$

In Exercises 39–50, find the integral.

39. $\displaystyle\int \sinh(1 - 2x)\, dx$

40. $\displaystyle\int \frac{\cosh \sqrt{x}}{\sqrt{x}}\, dx$

41. $\displaystyle\int \cosh^2(x - 1) \sinh(x - 1)\, dx$

42. $\displaystyle\int \frac{\sinh x}{1 + \sinh^2 x}\, dx$

43. $\displaystyle\int \frac{\cosh x}{\sinh x}\,dx$

44. $\displaystyle\int \operatorname{sech}^2(2x-1)\,dx$

45. $\displaystyle\int x \operatorname{csch}^2 \frac{x^2}{2}\,dx$

46. $\displaystyle\int \operatorname{sech}^3 x \tanh x\,dx$

47. $\displaystyle\int \frac{\operatorname{csch}(1/x)\coth(1/x)}{x^2}\,dx$

48. $\displaystyle\int \frac{\cosh x}{\sqrt{9-\sinh^2 x}}\,dx$

49. $\displaystyle\int \frac{x}{x^4+1}\,dx$

50. $\displaystyle\int \frac{2}{x\sqrt{1+4x^2}}\,dx$

In Exercises 51–56, evaluate the integral.

51. $\displaystyle\int_0^{\ln 2} \tanh x\,dx$

52. $\displaystyle\int_0^1 \cosh^2 x\,dx$

53. $\displaystyle\int_0^4 \frac{1}{25-x^2}\,dx$

54. $\displaystyle\int_0^4 \frac{1}{\sqrt{25-x^2}}\,dx$

55. $\displaystyle\int_0^{\sqrt{2}/4} \frac{2}{\sqrt{1-4x^2}}\,dx$

56. $\displaystyle\int_0^{\ln 2} 2e^{-x}\cosh x\,dx$

In Exercises 57–64, find the derivative of the function.

57. $y = \cosh^{-1}(3x)$

58. $y = \tanh^{-1}\dfrac{x}{2}$

59. $y = \sinh^{-1}(\tan x)$

60. $y = \operatorname{sech}^{-1}(\cos 2x), \quad 0 < x < \pi/4$

61. $y = \tanh^{-1}(\sin 2x)$

62. $y = (\operatorname{csch}^{-1} x)^2$

63. $y = 2x\sinh^{-1}(2x) - \sqrt{1+4x^2}$

64. $y = x\tanh^{-1} x + \ln\sqrt{1-x^2}$

Writing About Concepts

65. Discuss several ways in which the hyperbolic functions are similar to the trigonometric functions.

66. Sketch the graph of each hyperbolic function. Then identify the domain and range of each function.

Limits **In Exercises 67–72, find the limit.**

67. $\displaystyle\lim_{x\to\infty} \sinh x$

68. $\displaystyle\lim_{x\to\infty} \tanh x$

69. $\displaystyle\lim_{x\to\infty} \operatorname{sech} x$

70. $\displaystyle\lim_{x\to-\infty} \operatorname{csch} x$

71. $\displaystyle\lim_{x\to 0} \frac{\sinh x}{x}$

72. $\displaystyle\lim_{x\to 0^-} \coth x$

In Exercises 73–80, find the indefinite integral using the formulas of Theorem 5.24.

73. $\displaystyle\int \frac{1}{\sqrt{1+e^{2x}}}\,dx$

74. $\displaystyle\int \frac{x}{9-x^4}\,dx$

75. $\displaystyle\int \frac{1}{\sqrt{x}\sqrt{1+x}}\,dx$

76. $\displaystyle\int \frac{\sqrt{x}}{\sqrt{1+x^3}}\,dx$

77. $\displaystyle\int \frac{-1}{4x-x^2}\,dx$

78. $\displaystyle\int \frac{dx}{(x+2)\sqrt{x^2+4x+8}}$

79. $\displaystyle\int \frac{1}{1-4x-2x^2}\,dx$

80. $\displaystyle\int \frac{dx}{(x+1)\sqrt{2x^2+4x+8}}$

In Exercises 81–84, solve the differential equation.

81. $\dfrac{dy}{dx} = \dfrac{1}{\sqrt{80+8x-16x^2}}$

82. $\dfrac{dy}{dx} = \dfrac{1}{(x-1)\sqrt{-4x^2+8x-1}}$

83. $\dfrac{dy}{dx} = \dfrac{x^3-21x}{5+4x-x^2}$

84. $\dfrac{dy}{dx} = \dfrac{1-2x}{4x-x^2}$

Area **In Exercises 85–88, find the area of the region.**

85. $y = \operatorname{sech}\dfrac{x}{2}$

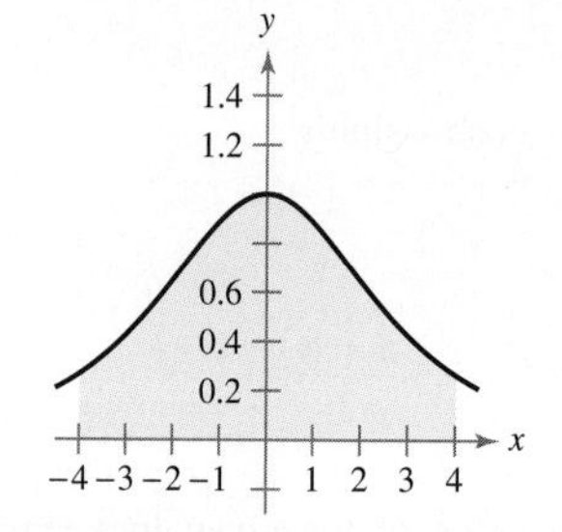

86. $y = \tanh 2x$

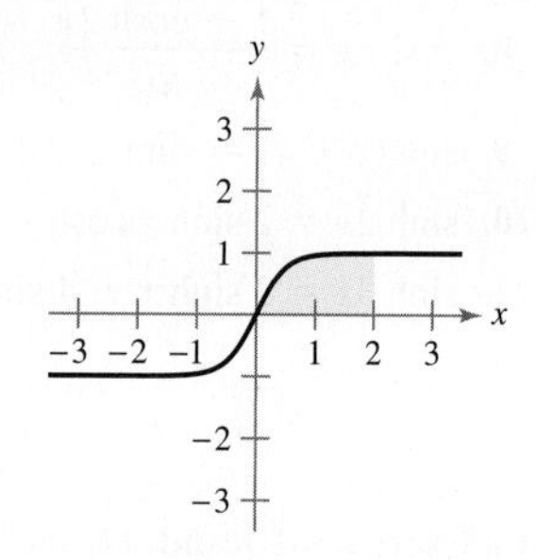

87. $y = \dfrac{5x}{\sqrt{x^4+1}}$

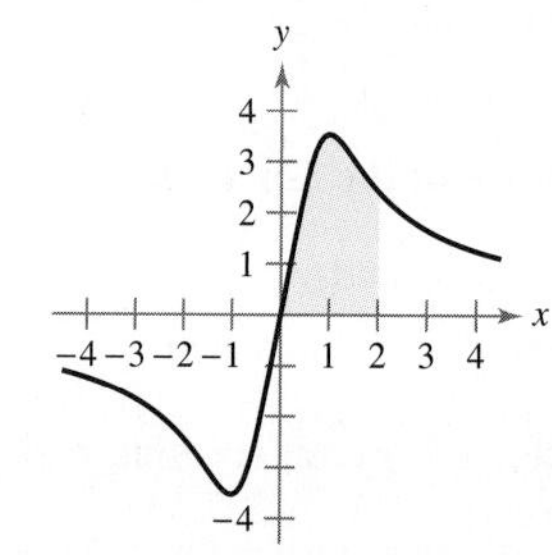

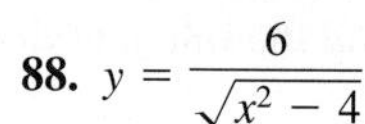

88. $y = \dfrac{6}{\sqrt{x^2-4}}$

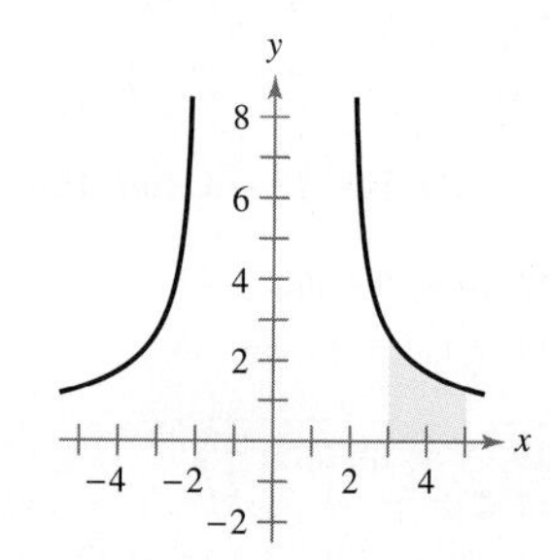

In Exercises 89 and 90, evaluate the integral in terms of (a) natural logarithms and (b) inverse hyperbolic functions.

89. $\displaystyle\int_0^{\sqrt{3}} \frac{dx}{\sqrt{x^2+1}}$

90. $\displaystyle\int_{-1/2}^{1/2} \frac{dx}{1-x^2}$

91. ***Chemical Reactions*** Chemicals A and B combine in a 3-to-1 ratio to form a compound. The amount of compound x being produced at any time t is proportional to the unchanged amounts of A and B remaining in the solution. So, if 3 kilograms of A is mixed with 2 kilograms of B, you have

$$\frac{dx}{dt} = k\left(3 - \frac{3x}{4}\right)\left(2 - \frac{x}{4}\right) = \frac{3k}{16}(x^2 - 12x + 32).$$

One kilogram of the compound is formed after 10 minutes. Find the amount formed after 20 minutes by solving the equation

$$\int \frac{3k}{16}\,dt = \int \frac{dx}{x^2-12x+32}.$$

92. ***Vertical Motion*** An object is dropped from a height of 400 feet.

(a) Find the velocity of the object as a function of time (neglect air resistance on the object).

(b) Use the result in part (a) to find the position function.

(c) If the air resistance is proportional to the square of the velocity, then $dv/dt = -32 + kv^2$, where -32 feet per second per second is the acceleration due to gravity and k is a constant. Show that the velocity v as a function of time t is

$$v(t) = -\sqrt{\frac{32}{k}}\tanh\left(\sqrt{32k}\,t\right)$$

by performing the following integration and simplifying the result.

$$\int \frac{dv}{32 - kv^2} = -\int dt$$

(d) Use the result in part (c) to find $\lim_{t\to\infty} v(t)$ and give its interpretation.

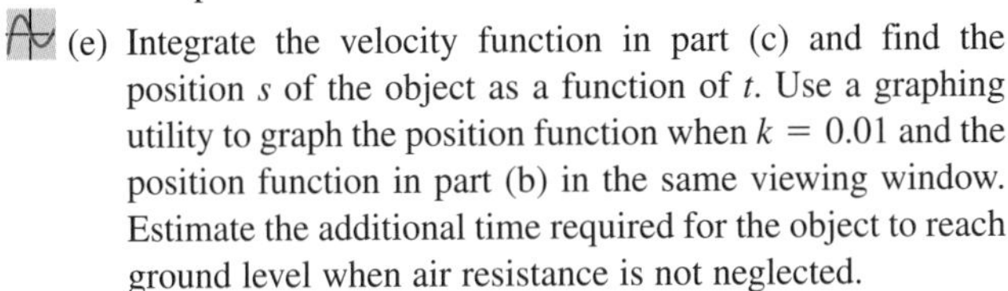

(e) Integrate the velocity function in part (c) and find the position s of the object as a function of t. Use a graphing utility to graph the position function when $k = 0.01$ and the position function in part (b) in the same viewing window. Estimate the additional time required for the object to reach ground level when air resistance is not neglected.

(f) Give a written description of what you believe would happen if k were increased. Then test your assertion with a particular value of k.

Tractrix **In Exercises 93 and 94, use the equation of the tractrix** $y = a\,\text{sech}^{-1}\dfrac{x}{a} - \sqrt{a^2 - x^2},\quad a > 0.$

93. Find dy/dx.

94. Let L be the tangent line to the tractrix at the point P. If L intersects the y-axis at the point Q, show that the distance between P and Q is a.

95. Prove that $\tanh^{-1} x = \dfrac{1}{2}\ln\left(\dfrac{1+x}{1-x}\right), \quad -1 < x < 1.$

96. Show that $\arctan(\sinh x) = \arcsin(\tanh x)$.

97. Let $x > 0$ and $b > 0$. Show that $\displaystyle\int_{-b}^{b} e^{xt}\,dt = \frac{2\sinh bx}{x}$.

In Exercises 98–102, verify the differentiation formula.

98. $\dfrac{d}{dx}[\cosh x] = \sinh x$

99. $\dfrac{d}{dx}[\text{sech}^{-1} x] = \dfrac{-1}{x\sqrt{1-x^2}}$

100. $\dfrac{d}{dx}[\cosh^{-1} x] = \dfrac{1}{\sqrt{x^2-1}}$

101. $\dfrac{d}{dx}[\sinh^{-1} x] = \dfrac{1}{\sqrt{x^2+1}}$

102. $\dfrac{d}{dx}[\text{sech}\, x] = -\text{sech}\, x \tanh x$

Putnam Exam Challenge

103. From the vertex $(0, c)$ of the catenary $y = c\cosh(x/c)$ a line L is drawn perpendicular to the tangent to the catenary at a point P. Prove that the length of L intercepted by the axes is equal to the ordinate y of the point P.

104. Prove or disprove that there is at least one straight line normal to the graph of $y = \cosh x$ at a point $(a, \cosh a)$ and also normal to the graph of $y = \sinh x$ at a point $(c, \sinh c)$.

[At a point on a graph, the normal line is the perpendicular to the tangent at that point. Also, $\cosh x = (e^x + e^{-x})/2$ and $\sinh x = (e^x - e^{-x})/2$.]

These problems were composed by the Committee on the Putnam Prize Competition.

Section Project: St. Louis Arch

Roy Ooms/Masterfile

The Gateway Arch in St. Louis, Missouri was constructed using the hyperbolic cosine function. The equation used for construction was

$y = 693.8597 - 68.7672\cosh 0.0100333x,$

$-299.2239 \le x \le 299.2239$

where x and y are measured in feet. Cross sections of the arch are equilateral triangles, and (x, y) traces the path of the centers of mass of the cross-sectional triangles. For each value of x, the area of the cross-sectional triangle is $A = 125.1406\cosh 0.0100333x$. (*Source:* Owner's Manual for the Gateway Arch, Saint Louis, MO, by William Thayer)

(a) How high above the ground is the center of the highest triangle? (At ground level, $y = 0$.)

(b) What is the height of the arch? (*Hint:* For an equilateral triangle, $A = \sqrt{3}c^2$, where c is one-half the base of the triangle, and the center of mass of the triangle is located at two-thirds the height of the triangle.)

(c) How wide is the arch at ground level?

Review Exercises for Chapter 5

See www.CalcChat.com for worked-out solutions to odd-numbered exercises.

In Exercises 1 and 2, use the graph of f' to sketch a graph of f. To print an enlarged copy of the graph, go to the website *www.mathgraphs.com*.

1.

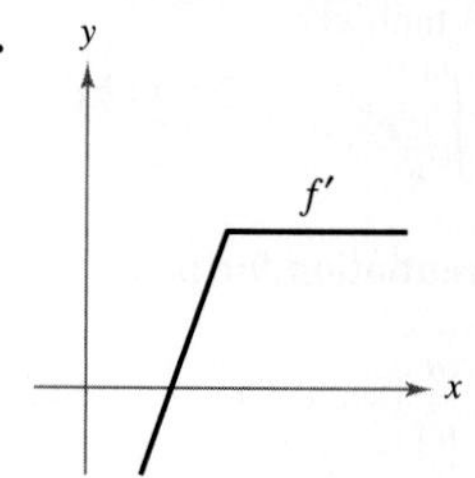

2.

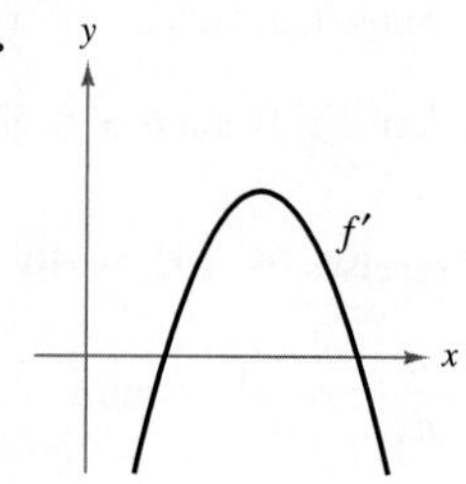

In Exercises 3–12, find the indefinite integral.

3. $\int (2x^2 + x - 1)\, dx$

4. $\int \frac{2}{\sqrt[3]{3x}}\, dx$

5. $\int \frac{x^3 + 1}{x^2}\, dx$

6. $\int \frac{x^3 - 2x^2 + 1}{x^2}\, dx$

7. $\int (4x - 3\sin x)\, dx$

8. $\int (5\cos x - 2\sec^2 x)\, dx$

9. $\int (5 - e^x)\, dx$

10. $\int (t + e^t)\, dt$

11. $\int \frac{5}{x}\, dx$

12. $\int \frac{10}{x}\, dx$

13. Find the particular solution of the differential equation $f'(x) = -2x$ whose graph passes through the point $(-1, 1)$.

14. Find the particular solution of the differential equation $f''(x) = 2e^x$ whose graph passes through the point $(0, 1)$ and is tangent to the line $3x - y - 5 = 0$ at that point.

15. ***Velocity and Acceleration*** An airplane taking off from a runway travels 3600 feet before lifting off. If it starts from rest, moves with constant acceleration, and makes the run in 30 seconds, with what speed does it lift off?

16. ***Velocity and Acceleration*** The speed of a car traveling in a straight line is reduced from 45 to 30 miles per hour in a distance of 264 feet. Find the distance in which the car can be brought to rest from 30 miles per hour, assuming the same constant deceleration.

17. ***Velocity and Acceleration*** A ball is thrown vertically upward from ground level with an initial velocity of 96 feet per second.

(a) How long will it take the ball to rise to its maximum height?

(b) What is the maximum height?

(c) When is the velocity of the ball one-half the initial velocity?

(d) What is the height of the ball when its velocity is one-half the initial velocity?

18. ***Velocity and Acceleration*** Repeat Exercise 17 for an initial velocity of 40 meters per second.

19. Write in sigma notation (a) the sum of the first 10 positive odd integers, (b) the sum of the cubes of the first n positive integers, and (c) $6 + 10 + 14 + 18 + \cdots + 42$.

20. Evaluate each sum for $x_1 = 2, x_2 = -1, x_3 = 5, x_4 = 3,$ and $x_5 = 7$.

(a) $\frac{1}{5}\sum_{i=1}^{5} x_i$ (b) $\sum_{i=1}^{5} \frac{1}{x_i}$

(c) $\sum_{i=1}^{5} (2x_i - x_i^2)$ (d) $\sum_{i=2}^{5} (x_i - x_{i-1})$

In Exercises 21 and 22, use upper and lower sums to approximate the area of the region using the indicated number of subintervals of equal width.

21. $y = \frac{10}{x^2 + 1}$

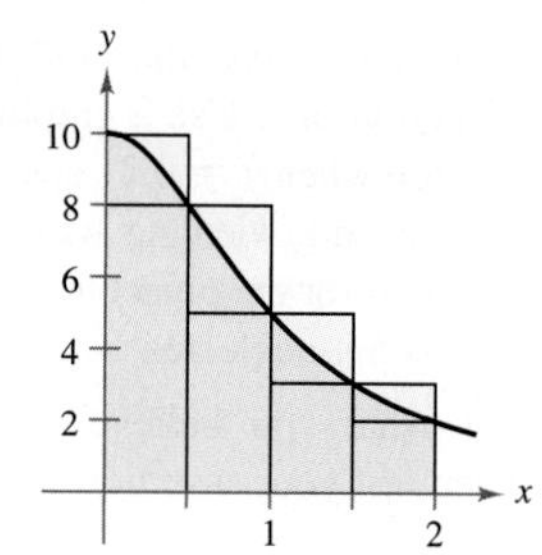

22. $y = 2^x$

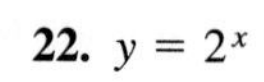

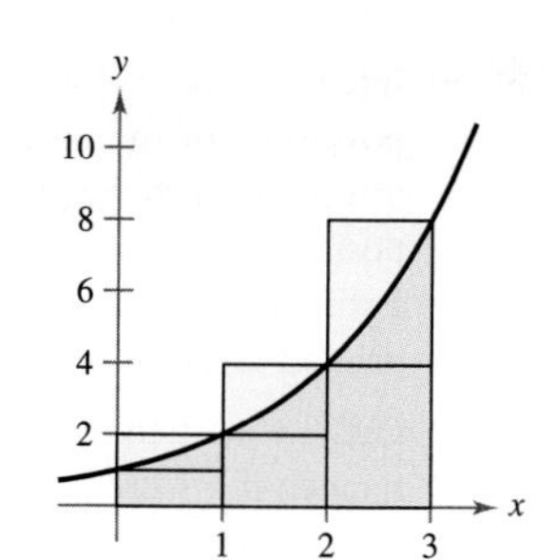

In Exercises 23–26, use the limit process to find the area of the region between the graph of the function and the x-axis over the indicated interval. Sketch the region.

	Function	*Interval*
23.	$y = 6 - x$	$[0, 4]$
24.	$y = x^2 + 3$	$[0, 2]$
25.	$y = 5 - x^2$	$[-2, 1]$
26.	$y = \frac{1}{4}x^3$	$[2, 4]$

27. Use the limit process to find the area of the region bounded by $x = 5y - y^2$, $x = 0$, $y = 2$, and $y = 5$.

28. Consider the region bounded by $y = mx$, $y = 0$, $x = 0$, and $x = b$.

(a) Find the upper and lower sums to approximate the area of the region when $\Delta x = b/4$.

(b) Find the upper and lower sums to approximate the area of the region when $\Delta x = b/n$.

(c) Find the area of the region by letting n approach infinity in both sums in part (b). Show that in each case you obtain the formula for the area of a triangle.

In Exercises 29 and 30, express the limit as a definite integral on the interval $[a, b]$, where c_i is any point in the ith subinterval.

Limit	*Interval*
29. $\lim_{\|\Delta\| \to 0} \sum_{i=1}^{n} (2c_i - 3)\Delta x_i$	$[4, 6]$
30. $\lim_{\|\Delta\| \to 0} \sum_{i=1}^{n} 3c_i(9 - c_i^2)\Delta x_i$	$[1, 3]$

In Exercises 31 and 32, sketch the region whose area is given by the definite integral. Then use a geometric formula to evaluate the integral.

31. $\int_0^5 (5 - |x - 5|)\, dx$

32. $\int_{-4}^4 \sqrt{16 - x^2}\, dx$

In Exercises 33 and 34, use the given values to evaluate each definite integral.

33. If $\int_2^6 f(x)\, dx = 10$ and $\int_2^6 g(x)\, dx = 3$, find

(a) $\int_2^6 [f(x) + g(x)]\, dx.$ (b) $\int_2^6 [f(x) - g(x)]\, dx.$

(c) $\int_2^6 [2f(x) - 3g(x)]\, dx.$ (d) $\int_2^6 5f(x)\, dx.$

34. If $\int_0^3 f(x)\, dx = 4$ and $\int_3^6 f(x)\, dx = -1$, find

(a) $\int_0^6 f(x)\, dx.$ (b) $\int_6^3 f(x)\, dx.$

(c) $\int_4^4 f(x)\, dx.$ (d) $\int_3^6 -10f(x)\, dx.$

In Exercises 35 and 36, select the correct value of the definite integral.

35. $\int_1^8 \left(\sqrt[3]{x} + 1\right) dx$

(a) $\frac{81}{4}$ (b) $\frac{331}{12}$

(c) $\frac{73}{4}$ (d) $\frac{355}{12}$

36. $\int_1^3 \frac{12}{x^3}\, dx$

(a) $\frac{320}{9}$ (b) $-\frac{16}{3}$

(c) $-\frac{5}{9}$ (d) $\frac{16}{3}$

In Exercises 37–46, use the Fundamental Theorem of Calculus to evaluate the definite integral.

37. $\int_0^4 (2 + x)\, dx$

38. $\int_{-1}^1 (t^2 + 2)\, dt$

39. $\int_{-1}^1 (4t^3 - 2t)\, dt$

40. $\int_{-2}^2 (x^4 + 2x^2 - 5)\, dx$

41. $\int_4^9 x\sqrt{x}\, dx$

42. $\int_1^2 \left(\frac{1}{x^2} - \frac{1}{x^3}\right) dx$

43. $\int_0^{3\pi/4} \sin\theta\, d\theta$

44. $\int_{-\pi/4}^{\pi/4} \sec^2 t\, dt$

45. $\int_0^2 (x + e^x)\, dx$

46. $\int_1^6 \frac{3}{x}\, dx$

In Exercises 47–52, sketch the graph of the region whose area is given by the integral, and find the area.

47. $\int_1^3 (2x - 1)\, dx$

48. $\int_0^2 (x + 4)\, dx$

49. $\int_3^4 (x^2 - 9)\, dx$

50. $\int_{-1}^2 (-x^2 + x + 2)\, dx$

51. $\int_0^1 (x - x^3)\, dx$

52. $\int_0^1 \sqrt{x}(1 - x)\, dx$

In Exercises 53–56, sketch the region bounded by the graphs of the equations, and determine its area.

53. $y = \frac{4}{\sqrt{x}},\quad y = 0,\quad x = 1,\quad x = 9$

54. $y = \sec^2 x,\quad y = 0,\quad x = 0,\quad x = \frac{\pi}{3}$

55. $y = \frac{2}{x},\quad y = 0,\quad x = 1,\quad x = 3$

56. $y = 1 + e^x,\quad y = 0,\quad x = 0,\quad x = 2$

In Exercises 57 and 58, find the average value of the function over the interval. Find the values of x at which the function assumes its average value, and graph the function.

57. $f(x) = \frac{1}{\sqrt{x}},\quad [4, 9]$

58. $f(x) = x^3,\quad [0, 2]$

In Exercises 59–62, use the Second Fundamental Theorem of Calculus to find $F'(x)$.

59. $F(x) = \int_0^x t^2\sqrt{1 + t^3}\, dt$

60. $F(x) = \int_1^x \frac{1}{t^2}\, dt$

61. $F(x) = \int_{-3}^x (t^2 + 3t + 2)\, dt$

62. $F(x) = \int_0^x \csc^2 t\, dt$

In Exercises 63–80, find the indefinite integral.

63. $\int (x^2 + 1)^3\, dx$

64. $\int \left(x + \frac{1}{x}\right)^2 dx$

65. $\int \frac{x^2}{\sqrt{x^3 + 3}}\, dx$

66. $\int x^2\sqrt{x^3 + 3}\, dx$

67. $\int x(1 - 3x^2)^4\, dx$

68. $\int \frac{x + 3}{(x^2 + 6x - 5)^2}\, dx$

69. $\int \sin^3 x \cos x\, dx$

70. $\int x \sin 3x^2\, dx$

71. $\int \frac{\sin\theta}{\sqrt{1 - \cos\theta}}\, d\theta$

72. $\int \frac{\cos x}{\sqrt{\sin x}}\, dx$

73. $\int \tan^n x \sec^2 x\, dx,\quad n \neq -1$

74. $\int \sec 2x \tan 2x\, dx$

75. $\int (1 + \sec \pi x)^2 \sec \pi x \tan \pi x\, dx$

76. $\int \cot^4 \alpha \csc^2 \alpha\, d\alpha$

77. $\int x e^{-3x^2}\, dx$

78. $\int \frac{e^{1/x}}{x^2}\, dx$

79. $\int (x + 1)5^{(x+1)^2}\, dx$

80. $\int \frac{1}{t^2}(2^{-1/t})\, dt$

In Exercises 81–88, evaluate the definite integral. Use a graphing utility to verify your result.

81. $\int_{-1}^{2} x(x^2 - 4)\, dx$

82. $\int_{0}^{1} x^2(x^3 + 1)^3\, dx$

83. $\int_{0}^{3} \frac{1}{\sqrt{1 + x}}\, dx$

84. $\int_{3}^{6} \frac{x}{3\sqrt{x^2 - 8}}\, dx$

85. $2\pi \int_{0}^{1} (y + 1)\sqrt{1 - y}\, dy$

86. $2\pi \int_{-1}^{0} x^2\sqrt{x + 1}\, dx$

87. $\int_{0}^{\pi} \cos\frac{x}{2}\, dx$

88. $\int_{-\pi/4}^{\pi/4} \sin 2x\, dx$

89. ***Fuel Cost*** Suppose that gasoline is increasing in price according to the equation $p = 1.20 + 0.04t$, where p is the dollar price per gallon and t is the time in years, with $t = 0$ representing 1990. If an automobile is driven 15,000 miles a year and gets M miles per gallon, the annual fuel cost is

$$C = \frac{15{,}000}{M} \int_{t}^{t+1} p\, ds = \frac{15{,}000}{M} \int_{t}^{t+1} (1.20 + 0.04s)ds.$$

Estimate the annual fuel cost for the year (a) 2005 and (b) 2007.

90. ***Respiratory Cycle*** After exercising for a few minutes, a person has a respiratory cycle for which the rate of air intake is $v = 1.75 \sin(\pi t/2)$. Find the volume, in liters, of air inhaled during one cycle by integrating the function over the interval $[0, 2]$.

In Exercises 91–95, use the Trapezoidal Rule and Simpson's Rule with $n = 4$, and use the integration capabilities of a graphing utility, to approximate the definite integral.

91. $\int_{1}^{2} \frac{1}{1 + x^3}\, dx$

92. $\int_{0}^{1} \frac{x^{3/2}}{3 - x^2}\, dx$

93. $\int_{0}^{\pi/2} \sqrt{x}\cos x\, dx$

94. $\int_{0}^{\pi} \sqrt{1 + \sin^2 x}\, dx$

95. $\int_{-1}^{1} e^{-x^2}\, dx$

96. Let $I = \int_0^4 f(x)\, dx$, where f is shown in the figure. Let $L(n)$ and $R(n)$ represent the Riemann sums using the left-hand endpoints and right-hand endpoints of n subintervals of equal width. (Assume n is even.) Let $T(n)$ and $S(n)$ be the corresponding values of the Trapezoidal Rule and Simpson's Rule.

(a) For any n, list $L(n)$, $R(n)$, $T(n)$, and I in increasing order.

(b) Approximate $S(4)$.

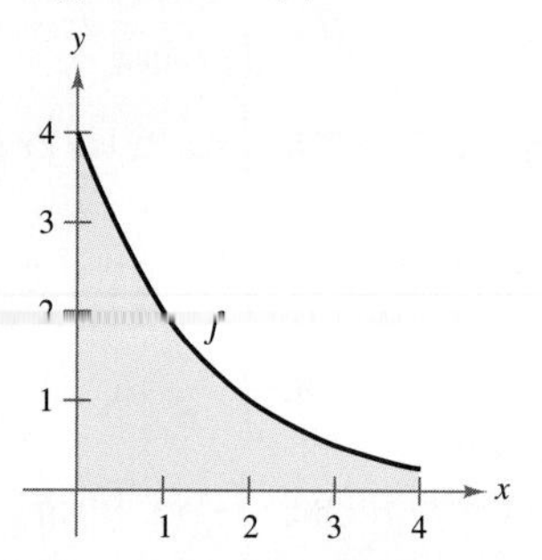

In Exercises 97–106, find or evaluate the integral.

97. $\int \frac{1}{7x - 2}\, dx$

98. $\int \frac{x}{x^2 - 1}\, dx$

99. $\int \frac{\sin x}{1 + \cos x}\, dx$

100. $\int \frac{\ln\sqrt{x}}{x}\, dx$

101. $\int_{1}^{4} \frac{x + 1}{x}\, dx$

102. $\int_{1}^{e} \frac{\ln x}{x}\, dx$

103. $\int_{0}^{\pi/3} \sec\theta\, d\theta$

104. $\int_{0}^{\pi/4} \tan\left(\frac{\pi}{4} - x\right) dx$

105. $\int \frac{e^{2x} - e^{-2x}}{e^{2x} + e^{-2x}}\, dx$

106. $\int \frac{e^{2x}}{e^{2x} + 1}\, dx$

In Exercises 107–114, find the indefinite integral.

107. $\int \frac{1}{e^{2x} + e^{-2x}}\, dx$

108. $\int \frac{1}{3 + 25x^2}\, dx$

109. $\int \frac{x}{\sqrt{1 - x^4}}\, dx$

110. $\int \frac{1}{16 + x^2}\, dx$

111. $\int \frac{x}{16 + x^2}\, dx$

112. $\int \frac{4 - x}{\sqrt{4 - x^2}}\, dx$

113. $\int \frac{\arctan(x/2)}{4 + x^2}\, dx$

114. $\int \frac{\arcsin x}{\sqrt{1 - x^2}}\, dx$

115. ***Harmonic Motion*** A weight of mass m is attached to a spring and oscillates with simple harmonic motion. By Hooke's Law, you can determine that

$$\int \frac{dy}{\sqrt{A^2 - y^2}} = \int \sqrt{\frac{k}{m}}\, dt$$

where A is the maximum displacement, t is the time, and k is a constant. Find y as a function of t, given that $y = 0$ when $t = 0$.

116. ***Think About It*** Sketch the region whose area is given by $\int_0^1 \arcsin x\, dx$. Then find the area of the region. Explain how you arrived at your answer.

In Exercises 117 and 118, find the derivative of the function.

117. $y = 2x - \cosh\sqrt{x}$

118. $y = x \tanh^{-1} 2x$

In Exercises 119 and 120, find the indefinite integral.

119. $\int \frac{x}{\sqrt{x^4 - 1}}\, dx$

120. $\int x^2 \operatorname{sech}^2 x^3\, dx$

P.S. Problem Solving

See www.CalcChat.com for worked-out solutions to odd-numbered exercises.

1. Let $L(x) = \int_1^x \frac{1}{t}\,dt,\ x > 0.$

(a) Find $L(1)$.

(b) Find $L'(x)$ and $L'(1)$.

(c) Use a graphing utility to approximate the value of x (to three decimal places) for which $L(x) = 1$.

(d) Prove that $L(x_1x_2) = L(x_1) + L(x_2)$ for all positive values of x_1 and x_2.

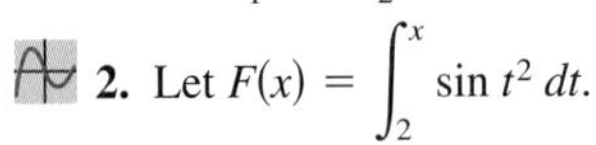

2. Let $F(x) = \int_2^x \sin t^2\,dt.$

(a) Use a graphing utility to complete the table.

x	0	1.0	1.5	1.9	2.0
$F(x)$					
x	2.1	2.5	3.0	4.0	5.0
$F(x)$					

(b) Let $G(x) = \frac{1}{x-2}F(x) = \frac{1}{x-2}\int_2^x \sin t^2\,dt$. Use a graphing utility to complete the table and estimate $\lim_{x\to 2} G(x)$.

x	1.9	1.95	1.99	2.01	2.1
$G(x)$					

(c) Use the definition of the derivative to find the exact value of $\lim_{x\to 2} G(x)$.

3. The **Fresnel function** S is defined by the integral

$$S(x) = \int_0^x \sin\left(\frac{\pi t^2}{2}\right) dt.$$

(a) Graph the function $y = \sin\left(\frac{\pi x^2}{2}\right)$ on the interval $[0, 3]$.

(b) Use the graph in part (a) to sketch the graph of S on the interval $[0, 3]$.

(c) Locate all relative extrema of S on the interval $(0, 3)$.

(d) Locate all points of inflection of S on the interval $(0, 3)$.

4. Galileo Galilei (1564–1642) stated the following proposition concerning falling objects:

The time in which any space is traversed by a uniformly accelerating body is equal to the time in which that same space would be traversed by the same body moving at a uniform speed whose value is the mean of the highest speed of the accelerating body and the speed just before acceleration began.

Use the techniques of this section to verify this proposition.

5. The graph of the function f consists of the three line segments joining the points $(0, 0)$, $(2, -2)$, $(6, 2)$, and $(8, 3)$. The function F is defined by the integral

$$F(x) = \int_0^x f(t)\,dt.$$

(a) Sketch the graph of f.

(b) Complete the table of values.

x	0	1	2	3	4	5	6	7	8
$F(x)$									

(c) Find the extrema of F on the interval $[0, 8]$.

(d) Determine all points of inflection of F on the interval $(0, 8)$.

6. A car is traveling in a straight line for 1 hour. Its velocity v in miles per hour at six-minute intervals is shown in the table.

t (hours)	0	0.1	0.2	0.3	0.4	0.5
v (mi/hr)	0	10	20	40	60	50

t (hours)	0.6	0.7	0.8	0.9	1.0
v (mi/hr)	40	35	40	50	65

(a) Produce a reasonable graph of the velocity function v by graphing these points and connecting them with a smooth curve.

(b) Find the open intervals over which the acceleration a is positive.

(c) Find the average acceleration of the car (in miles per hour squared) over the interval $[0, 0.4]$.

(d) What does the integral $\int_0^1 v(t)\,dt$ signify? Approximate this integral using the Trapezoidal Rule with five subintervals.

(e) Approximate the acceleration at $t = 0.8$.

7. The **Two-Point Gaussian Quadrature Approximation** for f is

$$\int_{-1}^1 f(x)\,dx \approx f\left(-\frac{1}{\sqrt{3}}\right) + f\left(\frac{1}{\sqrt{3}}\right).$$

(a) Use this formula to approximate

$$\int_{-1}^1 \cos x\,dx.$$

Find the error of the approximation.

(b) Use this formula to approximate

$$\int_{-1}^1 \frac{1}{1+x^2}\,dx.$$

(c) Prove that the Two-Point Gaussian Quadrature Approximation is exact for all polynomials of degree 3 or less.

8. Prove that $\int_0^x f(t)(x-t)\,dt = \int_0^x\left(\int_0^t f(v)\,dv\right)dt.$

9. Prove that $\int_a^b f(x)f'(x)\,dx = \frac{1}{2}([f(b)]^2 - [f(a)]^2).$

10. Use an appropriate Riemann sum to evaluate the limit

$$\lim_{n\to\infty} \frac{\sqrt{1}+\sqrt{2}+\sqrt{3}+\cdots+\sqrt{n}}{n^{3/2}}.$$

11. Use an appropriate Riemann sum to evaluate the limit

$$\lim_{n\to\infty} \frac{1^5+2^5+3^5+\cdots+n^5}{n^6}.$$

12. Archimedes showed that the area of a parabolic arch is equal to $\frac{2}{3}$ the product of the base and the height, as indicated in the figure.

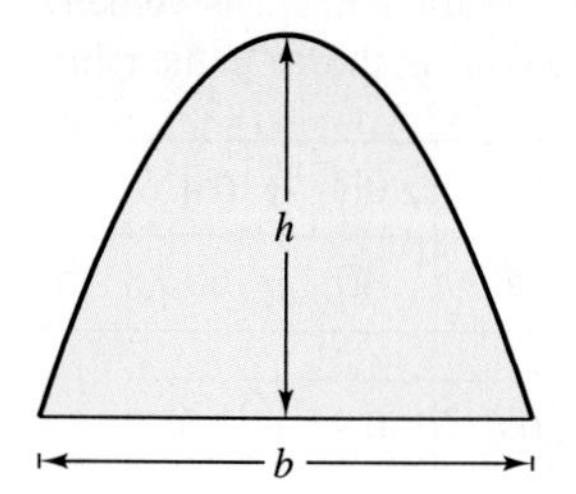

(a) Graph the parabolic arch bounded by $y = 9 - x^2$ and the x-axis. Use an appropriate integral to find the area A.

(b) Find the base and height of the arch and verify Archimedes' formula.

(c) Prove Archimedes' formula for a general parabola.

13. Suppose that f is integrable on $[a, b]$ and $0 < m \le f(x) \le M$ for all x in the interval $[a, b]$. Prove that

$$m(a-b) \le \int_a^b f(x)\,dx \le M(b-a).$$

Use this result to estimate $\int_0^1 \sqrt{1+x^4}\,dx.$

14. Verify that

$$\sum_{i=1}^n i^2 = \frac{n(n+1)(2n+1)}{6}$$

by showing the following.

(a) $(1+i)^3 - i^3 = 3i^2 + 3i + 1$

(b) $(n+1)^3 = \sum_{i=1}^n (3i^2 + 3i + 1) + 1$

(c) $\sum_{i=1}^n i^2 = \frac{n(n+1)(2n+1)}{6}$

15. Prove that if f is a continuous function on a closed interval $[a, b]$, then

$$\left|\int_a^b f(x)\,dx\right| \le \int_a^b |f(x)|\,dx.$$

16. Consider the three regions A, B, and C determined by the graph of $f(x) = \arcsin x$, as indicated in the figure.

(a) Calculate the areas of regions A and B.

(b) Use your answer in part (a) to evaluate the integral

$$\int_{1/2}^{\sqrt{2}/2} \arcsin x\,dx.$$

(c) Use your answer in part (a) to evaluate the integral

$$\int_1^3 \ln x\,dx.$$

(d) Use your answer in part (a) to evaluate the integral

$$\int_1^{\sqrt{3}} \arctan x\,dx.$$

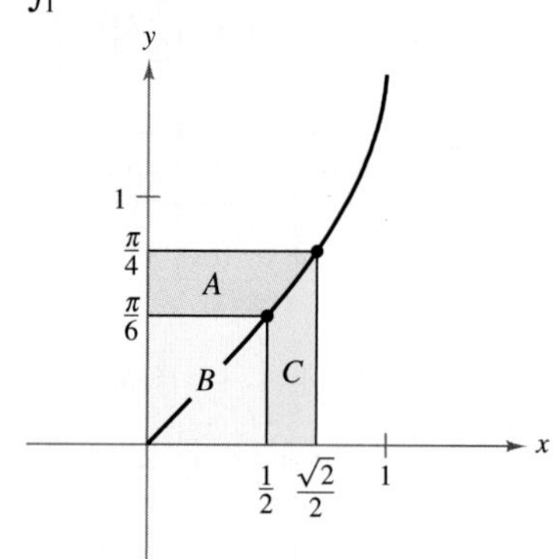

17. Use integration by substitution to find the area under the curve $y = \dfrac{1}{\sqrt{x}+x}$ between $x = 1$ and $x = 4$.

18. Use integration by substitution to find the area under the curve $y = \dfrac{1}{\sin^2 x + 4\cos^2 x}$ between $x = 0$ and $x = \pi/4$.

19. (a) Use a graphing utility to compare the graph of the function $y = e^x$ with the graphs of each of the given functions.

(i) $y_1 = 1 + \dfrac{x}{1!}$

(ii) $y_2 = 1 + \dfrac{x}{1!} + \dfrac{x^2}{2!}$

(iii) $y_3 = 1 + \dfrac{x}{1!} + \dfrac{x^2}{2!} + \dfrac{x^3}{3!}$

(b) Identify the pattern of successive polynomials in part (a) and extend the pattern one more term. Compare the graph of the resulting polynomial function with the graph of $y = e^x$.

(c) What do you think this pattern implies?

20. Let f be continuous on the interval $[0, b]$ such that $f(x) + f(b-x) \ne 0$. Show that

$$\int_0^b \frac{f(x)}{f(x)+f(b-x)}\,dx = \frac{b}{2}.$$

Use this result to evaluate $\int_0^1 \dfrac{\sin x}{\sin(1-x)+\sin x}\,dx.$

6 Differential Equations

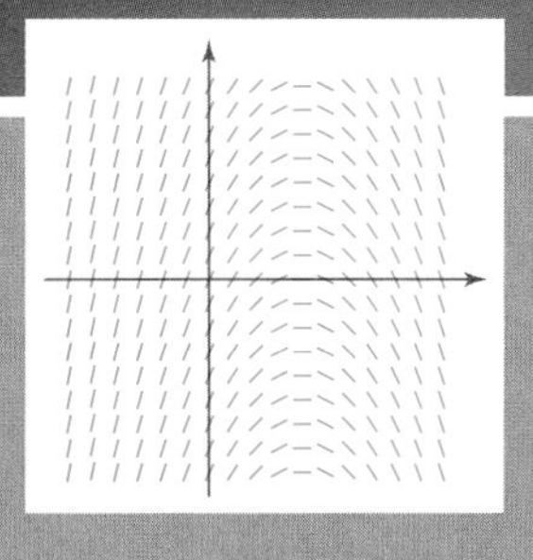

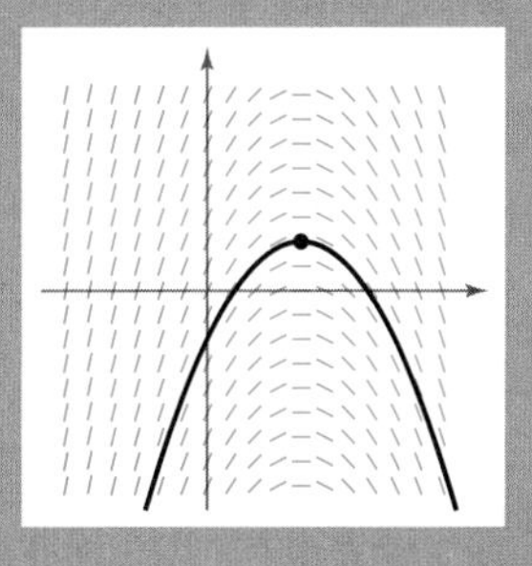

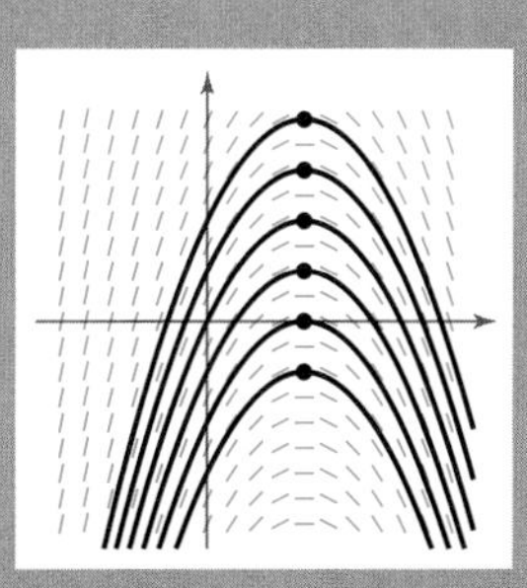

A function $y = f(x)$ is a solution of a differential equation if the equation is satisfied when y and its derivatives are replaced by $f(x)$ and its derivatives. One way to analyze a differential equation is to use slope fields, which show the general shape of all solutions of a differential equation. In Chapter 6, you will learn how to sketch slope fields and solve differential equations.

The SkyDome in Toronto is an entertainment center that has a retractable roof. How do you think the noise levels in the stadium compare when the roof is open and when the roof is closed? Explain your reasoning.

Image Gap/Alamy Images

Section 6.1 Slope Fields and Euler's Method

- **Use initial conditions to find particular solutions of differential equations.**
- **Use slope fields to approximate solutions of differential equations.**
- **Use Euler's Method to approximate solutions of differential equations.**

General and Particular Solutions

In this text, you will learn that physical phenomena can be described by differential equations. In Section 6.2, you will see that problems involving radioactive decay, population growth, and Newton's Law of Cooling can be formulated in terms of differential equations.

A function $y = f(x)$ is called a **solution** of a differential equation if the equation is satisfied when y and its derivatives are replaced by $f(x)$ and its derivatives. For example, differentiation and substitution would show that $y = e^{-2x}$ is a solution of the differential equation $y' + 2y = 0$. It can be shown that every solution of this differential equation is of the form

$$y = Ce^{-2x} \qquad \text{General solution of } y' + 2y = 0$$

where C is any real number. This solution is called the **general solution.** Some differential equations have **singular solutions** that cannot be written as special cases of the general solution. However, such solutions are not considered in this text. The **order** of a differential equation is determined by the highest-order derivative in the equation. For instance, $y' = 4y$ is a first-order differential equation.

NOTE First-order linear differential equations are discussed in Section 6.5.

In Section 5.1, Example 8, you saw that the second-order differential equation $s''(t) = -32$ has the general solution

$$s(t) = -16t^2 + C_1t + C_2 \qquad \text{General solution of } s''(t) = -32$$

which contains two arbitrary constants. It can be shown that a differential equation of order n has a general solution with n arbitrary constants.

EXAMPLE 1 Verifying Solutions

Determine whether the function is a solution of the differential equation $y'' - y = 0$.

a. $y = \sin x$ **b.** $y = 4e^{-x}$ **c.** $y = Ce^x$

Solution

a. Because $y = \sin x$, $y' = \cos x$, and $y'' = -\sin x$, it follows that

$$y'' - y = -\sin x - \sin x = -2 \sin x \neq 0.$$

So, $y = \sin x$ is *not* a solution.

b. Because $y = 4e^{-x}$, $y' = -4e^{-x}$, and $y'' = 4e^{-x}$, it follows that

$$y'' - y = 4e^{-x} - 4e^{-x} = 0.$$

So, $y = 4e^{-x}$ is a solution.

c. Because $y = Ce^x$, $y' = Ce^x$, and $y'' = Ce^x$, it follows that

$$y'' - y = Ce^x - Ce^x = 0.$$

So, $y = Ce^x$ is a solution for any value of C.

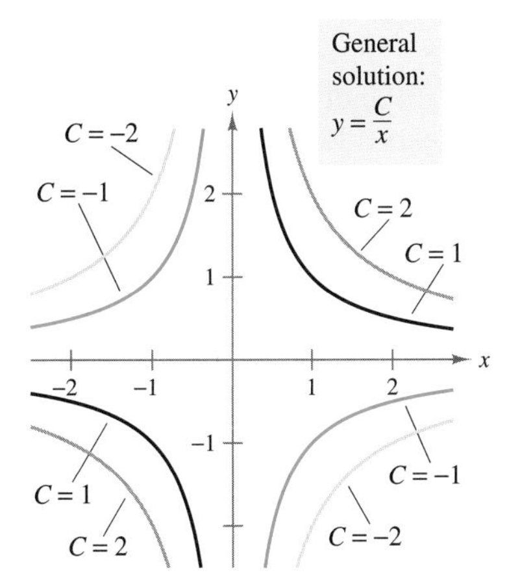

Solution curves for $xy' + y = 0$
Figure 6.1

Geometrically, the general solution of a first-order differential equation represents a family of curves known as **solution curves,** one for each value assigned to the arbitrary constant. For instance, you can verify that every function of the form

$$y = \frac{C}{x} \qquad \text{General solution of } xy' + y = 0$$

is a solution of the differential equation $xy' + y = 0$. Figure 6.1 shows four of the solution curves corresponding to different values of C.

As discussed in Section 5.1, **particular solutions** of a differential equation are obtained from **initial conditions** that give the value of the dependent variable or one of its derivatives for a particular value of the independent variable. The term "initial condition" stems from the fact that, often in problems involving time, the value of the dependent variable or one of its derivatives is known at the *initial* time $t = 0$. For instance, the second-order differential equation $s''(t) = -32$ having the general solution

$$s(t) = -16t^2 + C_1t + C_2 \qquad \text{General solution of } s''(t) = -32$$

might have the following initial conditions.

$$s(0) = 80, \qquad s'(0) = 64 \qquad \text{Initial conditions}$$

In this case, the initial conditions yield the particular solution

$$s(t) = -16t^2 + 64t + 80. \qquad \text{Particular solution}$$

EXAMPLE 2 Finding a Particular Solution

For the differential equation $xy' - 3y = 0$, verify that $y = Cx^3$ is a solution, and find the particular solution determined by the initial condition $y = 2$ when $x = -3$.

Solution You know that $y = Cx^3$ is a solution because $y' = 3Cx^2$ and

$$\begin{aligned} xy' - 3y &= x(3Cx^2) - 3(Cx^3) \\ &= 0. \end{aligned}$$

Furthermore, the initial condition $y = 2$ when $x = -3$ yields

$$\begin{aligned} y &= Cx^3 && \text{General solution} \\ 2 &= C(-3)^3 && \text{Substitute initial condition.} \\ -\frac{2}{27} &= C && \text{Solve for } C. \end{aligned}$$

and you can conclude that the particular solution is

$$y = -\frac{2x^3}{27}. \qquad \text{Particular solution}$$

Try checking this solution by substituting for y and y' in the original differential equation.

NOTE To determine a particular solution, the number of initial conditions must match the number of constants in the general solution.

---

indicates that in the HM mathSpace® *CD-ROM and the online* Eduspace® *system for this text, you will find an Open Exploration, which further explores this example using the computer algebra systems* Maple, Mathcad, Mathematica, *and* Derive.

Slope Fields

Solving a differential equation analytically can be difficult or even impossible. However, there is a graphical approach you can use to learn a lot about the solution of a differential equation. Consider a differential equation of the form

$$y' = F(x, y).$$ Differential equation

At each point (x, y) in the xy-plane where F is defined, the differential equation determines the slope $y' = F(x, y)$ of the solution at that point. If you draw a short line segment with slope $F(x, y)$ at selected points (x, y) in the domain of F, then these line segments form a **slope field,** or a *direction field* for the differential equation $y' = F(x, y)$. Each line segment has the same slope as the solution curve through that point. A slope field shows the general shape of all the solutions.

EXAMPLE 3 Sketching a Slope Field

Sketch a slope field for the differential equation $y' = x - y$ for the points $(-1, 1)$, $(0, 1)$, and $(1, 1)$.

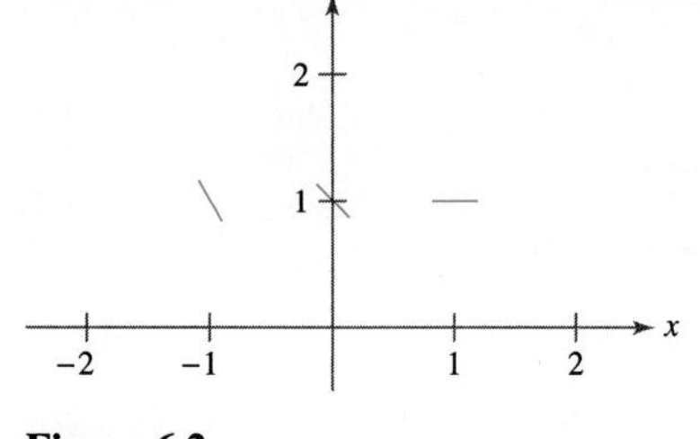

Figure 6.2

Solution

The slope of the solution curve at any point (x, y) is $F(x, y) = x - y$. So, the slope at $(-1, 1)$ is $y' = -1 - 1 = -2$, the slope at $(0, 1)$ is $y' = 0 - 1 = -1$, and the slope at $(1, 1)$ is $y' = 1 - 1 = 0$. Draw short line segments at the three points with their respective slopes, as shown in Figure 6.2.

EXAMPLE 4 Identifying Slope Fields for Differential Equations

Match each slope field with its differential equation.

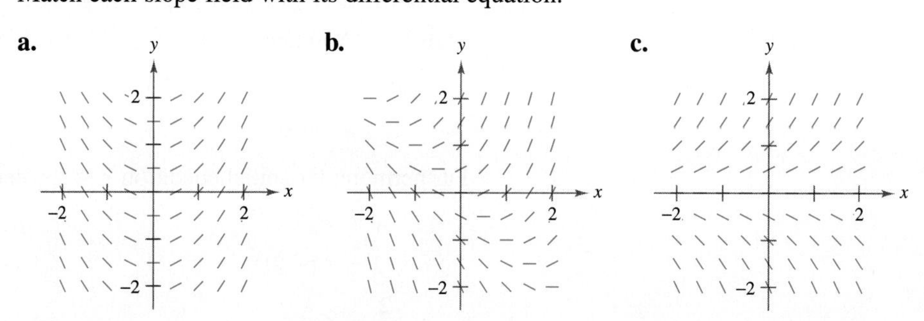

Figure 6.3

i. $y' = x + y$ **ii.** $y' = x$ **iii.** $y' = y$

Solution

a. From Figure 6.3(a), you can see that the slope at any point along the y-axis is 0. The only equation that satisfies this condition is $y' = x$. So, the graph matches (ii).

b. From Figure 6.3(b), you can see that the slope at the point $(1, -1)$ is 0. The only equation that satisfies this condition is $y' = x + y$. So, the graph matches (i).

c. From Figure 6.3(c), you can see that the slope at any point along the x-axis is 0. The only equation that satisfies this condition is $y' = y$. So, the graph matches (iii).

A solution curve of a differential equation $y' = F(x, y)$ is simply a curve in the xy-plane whose tangent line at each point (x, y) has slope equal to $F(x, y)$. This is illustrated in Example 5.

EXAMPLE 5 Sketching a Solution Using a Slope Field

Sketch a slope field for the differential equation

$$y' = 2x + y.$$

Use the slope field to sketch the solution that passes through the point $(1, 1)$.

Solution

Make a table showing the slopes at several points. The table shown is a small sample. The slopes at many other points should be calculated to get a representative slope field.

x	-2	-2	-1	-1	0	0	1	1	2	2
y	-1	1	-1	1	-1	1	-1	1	-1	1
$y' = 2x + y$	-5	-3	-3	-1	-1	1	1	3	3	5

Next draw line segments at the points with their respective slopes, as shown in Figure 6.4.

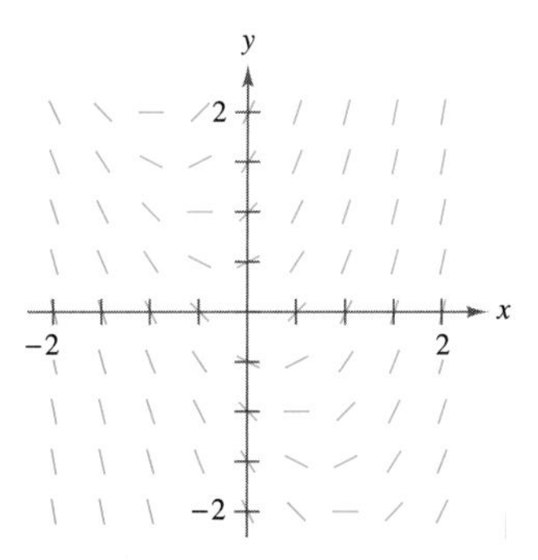

Slope field for $y' = 2x + y$
Figure 6.4

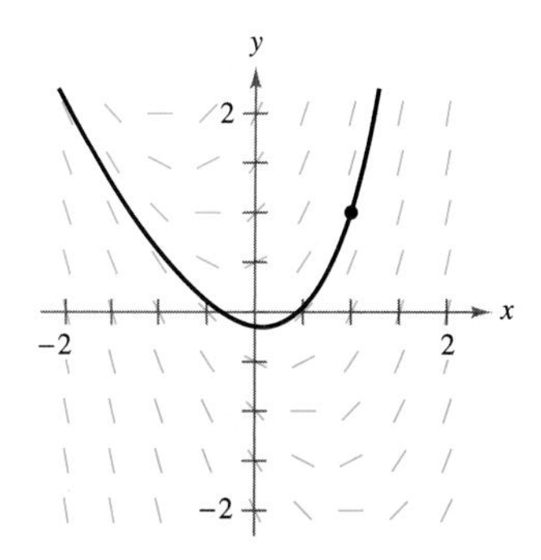

Particular solution for $y' = 2x + y$ passing through $(1, 1)$
Figure 6.5

After the slope field is drawn, start at the initial point $(1, 1)$ and move to the right in the direction of the line segment. Continue to draw the solution curve so that it moves parallel to the nearby line segments. Do the same to the left of $(1, 1)$. The resulting solution is shown in Figure 6.5.

From Example 5, note that the slope field shows that y' increases to infinity as x increases.

NOTE Drawing a slope field by hand is tedious. In practice, slope fields are usually drawn using a graphing utility.

Euler's Method

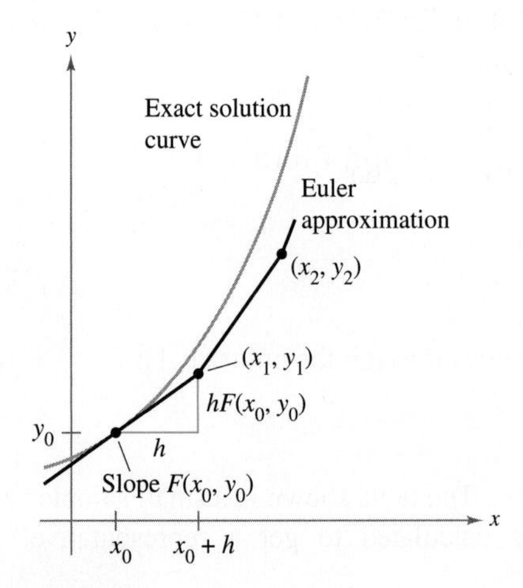

Figure 6.6

Euler's Method is a numerical approach to approximating the particular solution of the differential equation

$$y' = F(x, y)$$

that passes through the point (x_0, y_0). From the given information, you know that the graph of the solution passes through the point (x_0, y_0) and has a slope of $F(x_0, y_0)$ at this point. This gives you a "starting point" for approximating the solution.

From this starting point, you can proceed in the direction indicated by the slope. Using a small step h, move along the tangent line until you arrive at the point (x_1, y_1), where

$$x_1 = x_0 + h \quad \text{and} \quad y_1 = y_0 + hF(x_0, y_0)$$

as shown in Figure 6.6. If you think of (x_1, y_1) as a new starting point, you can repeat the process to obtain a second point (x_2, y_2). The values of x_i and y_i are as follows.

$$\begin{aligned} x_1 &= x_0 + h & y_1 &= y_0 + hF(x_0, y_0) \\ x_2 &= x_1 + h & y_2 &= y_1 + hF(x_1, y_1) \\ &\vdots & &\vdots \\ x_n &= x_{n-1} + h & y_n &= y_{n-1} + hF(x_{n-1}, y_{n-1}) \end{aligned}$$

NOTE You can obtain better approximations of the exact solution by choosing smaller and smaller step sizes.

EXAMPLE 6 Approximating a Solution Using Euler's Method

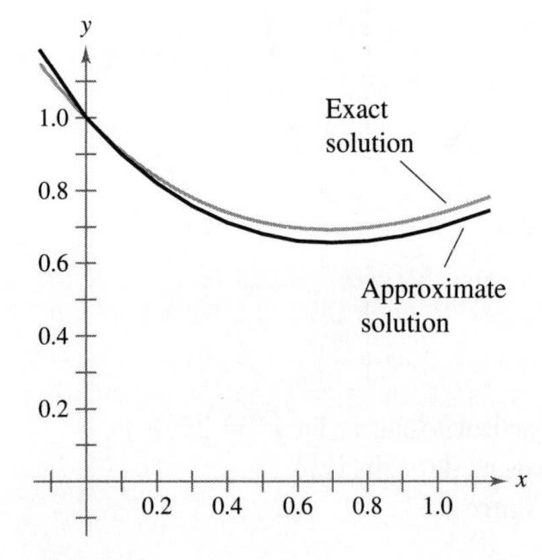

Figure 6.7

Use Euler's Method to approximate the particular solution of the differential equation

$$y' = x - y$$

passing through the point (0, 1). Use a step of $h = 0.1$.

Solution Using $h = 0.1$, $x_0 = 0$, $y_0 = 1$, and $F(x, y) = x - y$, you have $x_0 = 0$, $x_1 = 0.1$, $x_2 = 0.2$, $x_3 = 0.3, \ldots,$ and

$$\begin{aligned} y_1 &= y_0 + hF(x_0, y_0) = 1 + (0.1)(0 - 1) = 0.9 \\ y_2 &= y_1 + hF(x_1, y_1) = 0.9 + (0.1)(0.1 - 0.9) = 0.82 \\ y_3 &= y_2 + hF(x_2, y_2) = 0.82 + (0.1)(0.2 - 0.82) = 0.758. \end{aligned}$$

The first 10 approximations are shown in the table. You can plot these values to see a graph of the approximate solution, as shown in Figure 6.7.

n	0	1	2	3	4	5	6	7	8	9	10
x_n	0	0.1	0.2	0.3	0.4	0.5	0.6	0.7	0.8	0.9	1.0
y_n	1	0.900	0.820	0.758	0.712	0.681	0.663	0.657	0.661	0.675	0.697

NOTE For the differential equation in Example 6, you can verify the exact solution to be $y = x - 1 + 2e^{-x}$. Figure 6.7 compares this exact solution with the approximate solution obtained in Example 6.

Exercises for Section 6.1

See www.CalcChat.com for worked-out solutions to odd-numbered exercises.

In Exercises 1–8, verify the solution of the differential equation.

	Solution	*Differential Equation*
1.	$y = Ce^{4x}$	$y' = 4y$
2.	$y = e^{-x}$	$3y' + 4y = e^{-x}$
3.	$x^2 + y^2 = Cy$	$y' = 2xy/(x^2 - y^2)$
4.	$y^2 - 2 \ln y = x^2$	$\dfrac{dy}{dx} = \dfrac{xy}{y^2 - 1}$
5.	$y = C_1 \cos x + C_2 \sin x$	$y'' + y = 0$
6.	$y = C_1 e^{-x} \cos x + C_2 e^{-x} \sin x$	$y'' + 2y' + 2y = 0$
7.	$y = -\cos x \ln\lvert\sec x + \tan x\rvert$	$y'' + y = \tan x$
8.	$y = \frac{2}{5}(e^{-2x} + e^x)$	$y'' + 2y' = 2e^x$

In Exercises 9–12, verify the particular solution of the differential equation.

	Solution	*Differential Equation and Initial Condition*
9.	$y = \sin x \cos x - \cos^2 x$	$2y + y' = 2\sin(2x) - 1$ $y\left(\dfrac{\pi}{4}\right) = 0$
10.	$y = \frac{1}{2}x^2 - 4 \cos x + 2$	$y' = x + 4 \sin x$ $y(0) = -2$
11.	$y = 6e^{-2x^2}$	$y' = -4xy$ $y(0) = 6$
12.	$y = e^{-\cos x}$	$y' = y \sin x$ $y\left(\dfrac{\pi}{2}\right) = 1$

In Exercises 13–18, determine whether the function is a solution of the differential equation $y^{(4)} - 16y = 0$.

13. $y = 3 \cos x$

14. $y = 3 \cos 2x$

15. $y = e^{-2x}$

16. $y = 5 \ln x$

17. $y = C_1 e^{2x} + C_2 e^{-2x} + C_3 \sin 2x + C_4 \cos 2x$

18. $y = 3e^{2x} - 4 \sin 2x$

In Exercises 19–24, determine whether the function is a solution of the differential equation $xy' - 2y = x^3e^x$.

19. $y = x^2$

20. $y = x^2e^x$

21. $y = x^2(2 + e^x)$

22. $y = \sin x$

23. $y = \ln x$

24. $y = x^2e^x - 5x^2$

In Exercises 25–28, some of the curves corresponding to different values of C in the general solution of the differential equation are given. Find the particular solution that passes through the point shown on the graph.

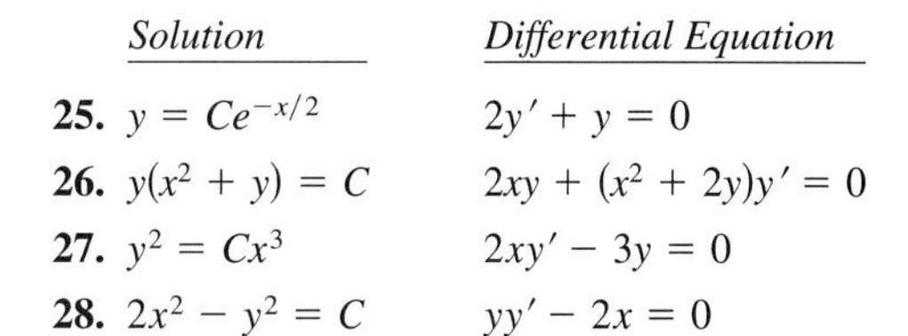

	Solution	*Differential Equation*
25.	$y = Ce^{-x/2}$	$2y' + y = 0$
26.	$y(x^2 + y) = C$	$2xy + (x^2 + 2y)y' = 0$
27.	$y^2 = Cx^3$	$2xy' - 3y = 0$
28.	$2x^2 - y^2 = C$	$yy' - 2x = 0$

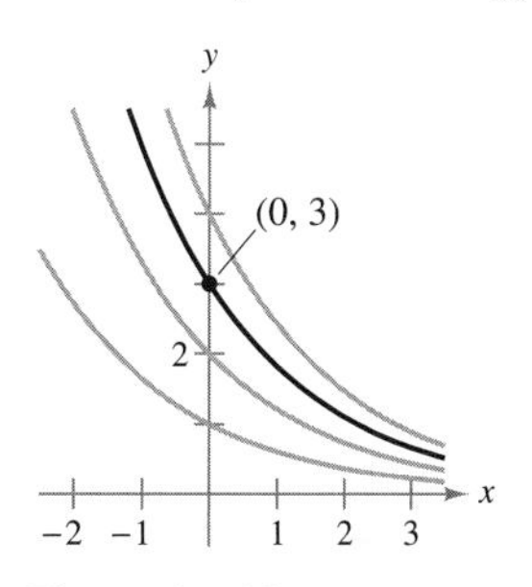

Figure for 25

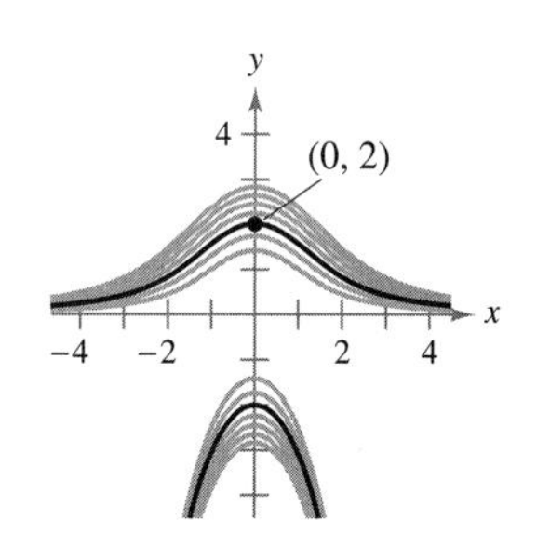

Figure for 26

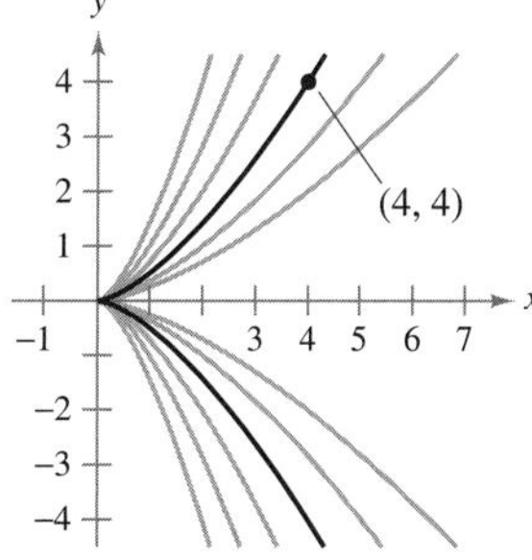

Figure for 27

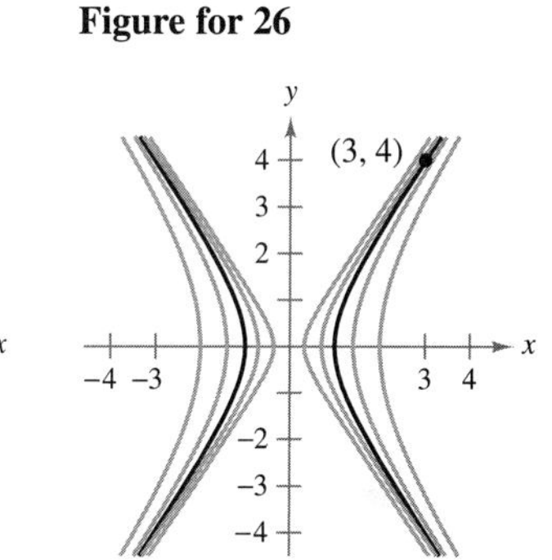

Figure for 28

In Exercises 29 and 30, the general solution of the differential equation is given. Use a graphing utility to graph the particular solutions for the given values of C.

29. $4yy' - x = 0$
$4y^2 - x^2 = C$
$C = 0, C = \pm 1, C = \pm 4$

30. $yy' + x = 0$
$x^2 + y^2 = C$
$C = 0, C = 1, C = 4$

In Exercises 31–36, verify that the general solution satisfies the differential equation. Then find the particular solution that satisfies the initial condition.

31. $y = Ce^{-2x}$
$y' + 2y = 0$
$y = 3$ when $x = 0$

32. $3x^2 + 2y^2 = C$
$3x + 2yy' = 0$
$y = 3$ when $x = 1$

33. $y = C_1 \sin 3x + C_2 \cos 3x$
$y'' + 9y = 0$
$y = 2$ when $x = \pi/6$
$y' = 1$ when $x = \pi/6$

34. $y = C_1 + C_2 \ln x$
$xy'' + y' = 0$
$y = 0$ when $x = 2$
$y' = \frac{1}{2}$ when $x = 2$

35. $y = C_1x + C_2x^3$
$x^2y'' - 3xy' + 3y = 0$
$y = 0$ when $x = 2$
$y' = 4$ when $x = 2$

36. $y = e^{2x/3}(C_1 + C_2x)$
$9y'' - 12y' + 4y = 0$
$y = 4$ when $x = 0$
$y = 0$ when $x = 3$

In Exercises 37–48, use integration to find a general solution of the differential equation.

37. $\dfrac{dy}{dx} = 3x^2$

38. $\dfrac{dy}{dx} = x^3 - 4x$

39. $\dfrac{dy}{dx} = \dfrac{x}{1 + x^2}$

40. $\dfrac{dy}{dx} = \dfrac{e^x}{1 + e^x}$

41. $\dfrac{dy}{dx} = \dfrac{x - 2}{x}$

42. $\dfrac{dy}{dx} = x \cos x^2$

43. $\dfrac{dy}{dx} = \sin 2x$

44. $\dfrac{dy}{dx} = \tan^2 x$

45. $\dfrac{dy}{dx} = x\sqrt{x - 3}$

46. $\dfrac{dy}{dx} = x\sqrt{5 - x}$

47. $\dfrac{dy}{dx} = xe^{x^2}$

48. $\dfrac{dy}{dx} = 5e^{-x/2}$

Slope Fields **In Exercises 49–52, a differential equation and its slope field are given. Determine the slopes (if possible) in the slope field at the points given in the table.**

x	-4	-2	0	2	4	8
y	2	0	4	4	6	8
dy/dx						

49. $\dfrac{dy}{dx} = \dfrac{x}{y}$

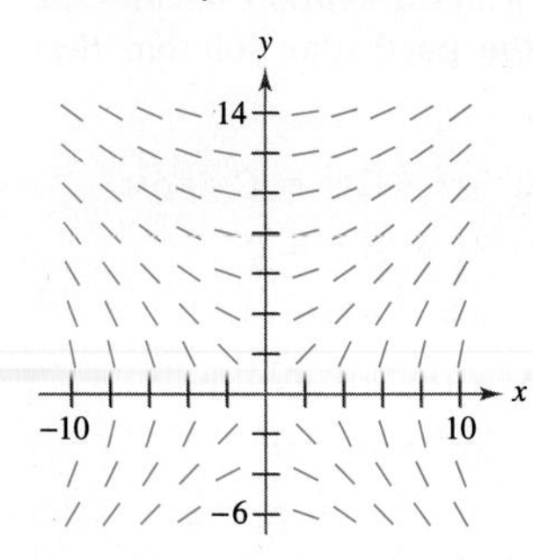

50. $\dfrac{dy}{dx} = x - y$

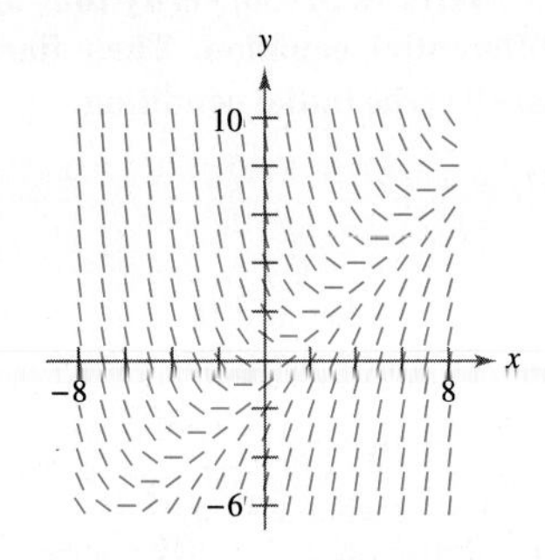

51. $\dfrac{dy}{dx} = x \cos \dfrac{\pi y}{8}$

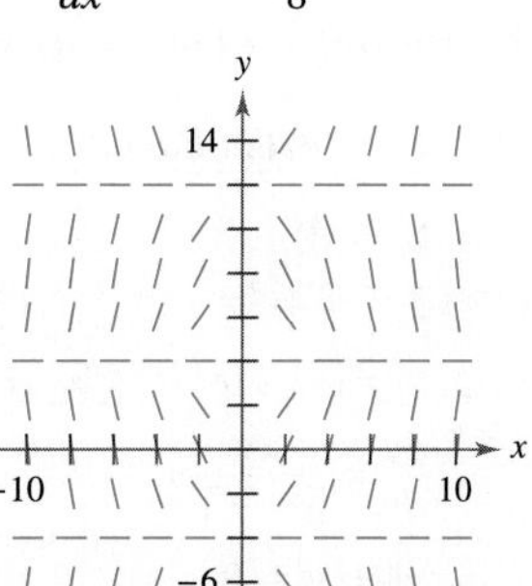

52. $\dfrac{dy}{dx} = \tan\left(\dfrac{\pi y}{6}\right)$

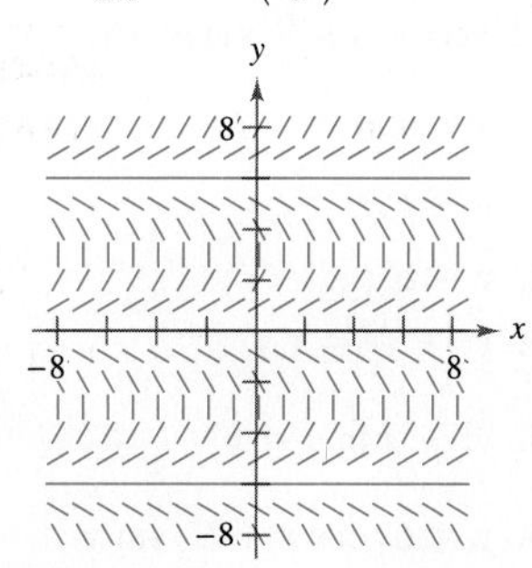

In Exercises 53–56, match the differential equation with its slope field. [The slope fields are labeled (a), (b), (c), and (d).]

(a)

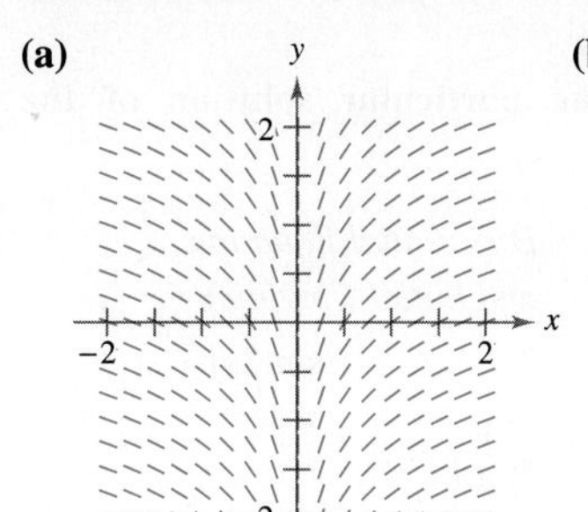

(b)

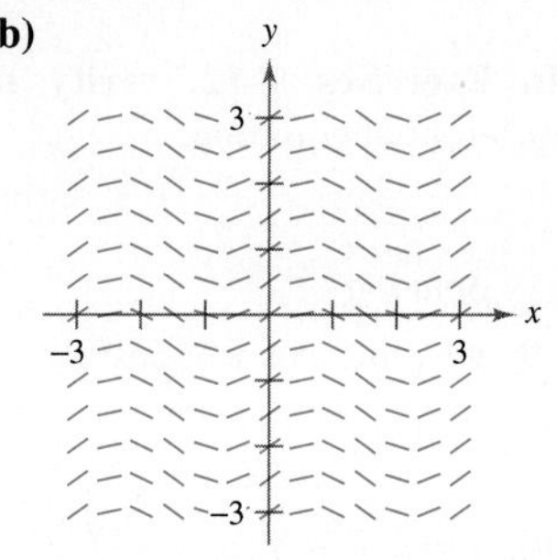

(c)

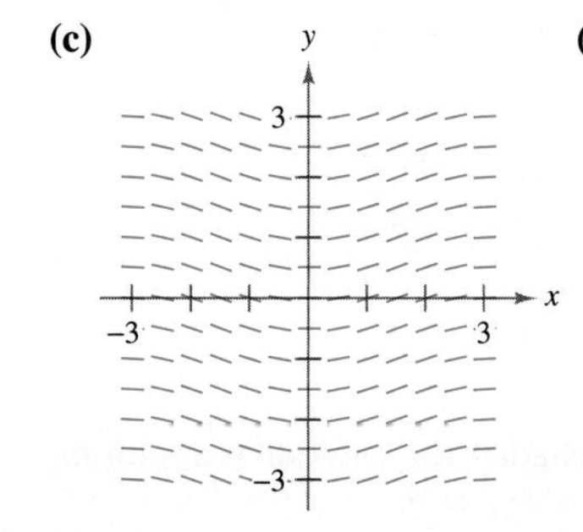

(d)

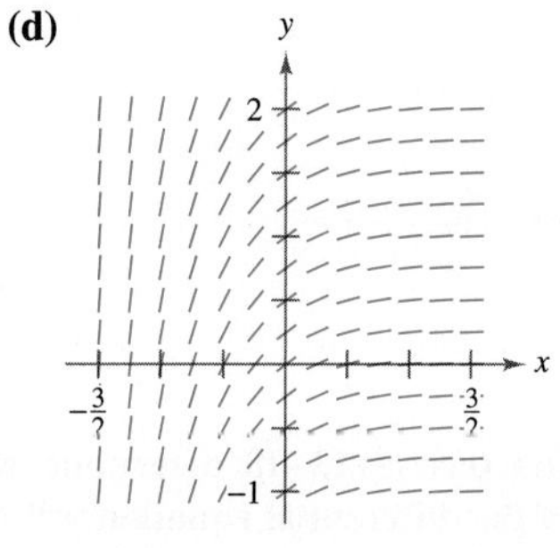

53. $\dfrac{dy}{dx} = \cos(2x)$

54. $\dfrac{dy}{dx} = \dfrac{1}{2} \sin x$

55. $\dfrac{dy}{dx} = e^{-2x}$

56. $\dfrac{dy}{dx} = \dfrac{1}{x}$

Slope Fields **In Exercises 57–60, (a) sketch the slope field for the differential equation, (b) use the slope field to sketch the solution that passes through the given point, and (c) discuss the graph of the solution as $x \to \infty$ and $x \to -\infty$.**

	Differential Equation	*Point*
57.	$y' = -x + 1$	(2, 4)
58.	$y' = \frac{1}{3}x^2 - \frac{1}{2}x$	(1, 1)
59.	$y' = y - 2x$	(1, 1)
60.	$y' = y + xy$	(0, 4)

61. ***Slope Field*** Use the slope field for the differential equation $y' = 1/x$, where $x > 0$, to sketch the graph of the solution that satisfies each given initial condition. Then make a conjecture about the behavior of a particular solution of $y' = 1/x$ as $x \to \infty$. To print an enlarged copy of the graph, go to the website *www.mathgraphs.com.*

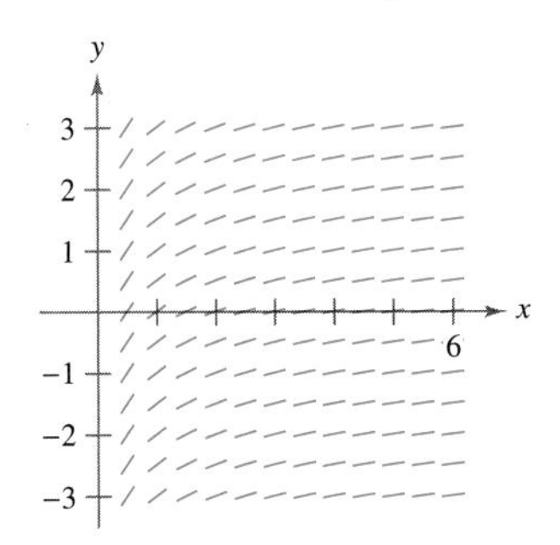

(a) $(1, 0)$ (b) $(2, -1)$

62. ***Slope Field*** Use the slope field for the differential equation $y' = 1/y$, where $y > 0$, to sketch the graph of the solution that satisfies each initial condition. Then make a conjecture about the behavior of a particular solution of $y' = 1/y$ as $x \to \infty$. To print an enlarged copy of the graph, go to the website *www.mathgraphs.com.*

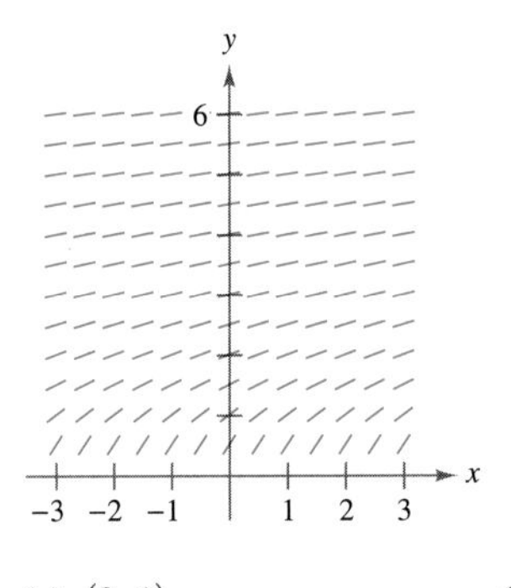

(a) $(0, 1)$ (b) $(1, 1)$

Slope Fields **In Exercises 63–68, use a computer algebra system to (a) graph the slope field for the differential equation and (b) graph the solution satisfying the specified initial condition.**

63. $\dfrac{dy}{dx} = 0.5y, \quad y(0) = 6$

64. $\dfrac{dy}{dx} = 2 - y, \quad y(0) = 4$

65. $\dfrac{dy}{dx} = 0.02y(10 - y), \quad y(0) = 2$

66. $\dfrac{dy}{dx} = 0.2x(2 - y), \quad y(0) = 9$

67. $\dfrac{dy}{dx} = 0.4y(3 - x), \quad y(0) = 1$

68. $\dfrac{dy}{dx} = \dfrac{1}{2}e^{-x/8} \sin \dfrac{\pi y}{4}, \quad y(0) = 2$

Euler's Method **In Exercises 69–74, use Euler's Method to make a table of values for the approximate solution of the differential equation with the specified initial value. Use n steps of size h.**

69. $y' = x + y, \quad y(0) = 2, \quad n = 10, \quad h = 0.1$

70. $y' = x + y, \quad y(0) = 2, \quad n = 20, \quad h = 0.05$

71. $y' = 3x - 2y, \quad y(0) = 3, \quad n = 10, \quad h = 0.05$

72. $y' = 0.5x(3 - y), \quad y(0) = 1, \quad n = 5, \quad h = 0.4$

73. $y' = e^{xy}, \quad y(0) = 1, \quad n = 10, \quad h = 0.1$

74. $y' = \cos x + \sin y, \quad y(0) = 5, \quad n = 10, \quad h = 0.1$

In Exercises 75–77, complete the table using the exact solution of the differential equation and two approximations obtained using Euler's Method to approximate the particular solution of the differential equation. Use $h = 0.2$ and 0.1 and compute each approximation to four decimal places.

x	0	0.2	0.4	0.6	0.8	1.0
$y(x)$ (exact)						
$y(x)$ ($h = 0.2$)						
$y(x)$ ($h = 0.1$)						

	Differential Equation	*Initial Condition*	*Exact Solution*
75.	$\dfrac{dy}{dx} = y$	$(0, 3)$	$y = 3e^x$
76.	$\dfrac{dy}{dx} = \dfrac{2x}{y}$	$(0, 2)$	$y = \sqrt{2x^2 + 4}$
77.	$\dfrac{dy}{dx} = y + \cos(x)$	$(0, 0)$	$y = \frac{1}{2}(\sin x - \cos x + e^x)$

78. Compare the values of the approximations in Exercises 75–77 with the values given by the exact solution. How does the error change as h increases?

79. ***Temperature*** At time $t = 0$ minutes, the temperature of an object is 140°F. The temperature of the object is changing at the rate given by the differential equation

$$\frac{dy}{dt} = -\frac{1}{2}(y - 72).$$

(a) Use a graphing utility and Euler's Method to approximate the particular solutions of this differential equation at $t = 1$, 2, and 3. Use a step size of $h = 0.1$. (A graphing utility program for Euler's Method is available on the website *college.hmco.com.*)

(b) Compare your results with the exact solution

$$y = 72 + 68e^{-t/2}.$$

80. ***Temperature*** Repeat Exercise 79 using a step size of $h = 0.05$. Compare the results.

Writing About Concepts

81. In your own words, describe the difference between a general solution of a differential equation and a particular solution.

82. Explain how to interpret a slope field.

83. Describe how to use Euler's Method to approximate the particular solution of a differential equation.

84. It is known that $y = Ce^{kx}$ is a solution of the differential equation $y' = 0.07y$. Is it possible to determine C or k from the information given? If so, find its value.

True or False? **In Exercises 85–88, determine whether the statement is true or false. If it is false, explain why or give an example that shows it is false.**

85. If $y = f(x)$ is a solution of a first-order differential equation, then $y = f(x) + C$ is also a solution.

86. The general solution of a differential equation is $y = -4.9x^2 + C_1x + C_2$. To find a particular solution, you must be given two initial conditions.

87. Slope fields represent the general solutions of differential equations.

88. A slope field shows that the slope at the point $(1, 1)$ is 6. This slope field represents the family of solutions for the differential equation $y' = 4x + 2y$.

89. ***Error and Euler's Method*** The exact solution of the differential equation

$$\frac{dy}{dx} = -2y$$

where $y(0) = 4$, is $y = 4e^{-2x}$.

(a) Use a graphing utility to complete the table, where y is the exact value of the solution, y_1 is the approximate solution using Euler's Method with $h = 0.1$, y_2 is the approximate solution using Euler's Method with $h = 0.2$, e_1 is the absolute error $|y - y_1|$, e_2 is the absolute error $|y - y_2|$, and r is the ratio e_1/e_2.

x	0	0.2	0.4	0.6	0.8	1
y						
y_1						
y_2						
e_1						
e_2						
r						

(b) What can you conclude about the ratio r as h changes?

(c) Predict the absolute error when $h = 0.05$.

90. ***Error and Euler's Method*** Repeat Exercise 89 where the exact solution of the differential equation

$$\frac{dy}{dx} = x - y$$

where $y(0) = 1$, is $y = x - 1 + 2e^{-x}$.

91. ***Electric Circuits*** The diagram shows a simple electric circuit consisting of a power source, a resistor, and an inductor.

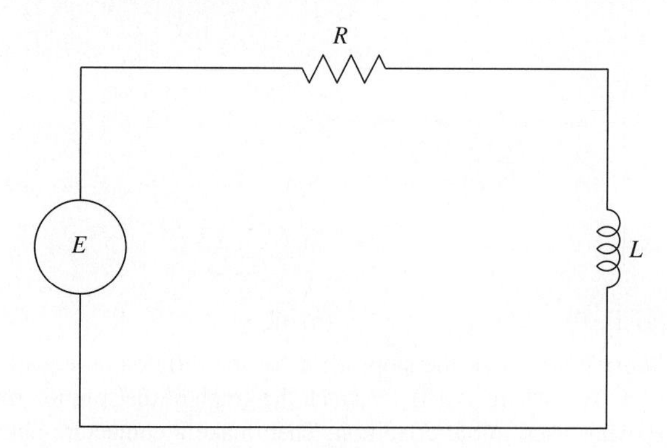

A model of the current I, in amperes (A), at time t is given by the first-order differential equation

$$L\frac{dI}{dt} + RI = E(t)$$

where $E(t)$ is the voltage (V) produced by the power source, R is the resistance in ohms (Ω), and L is the inductance in henrys (H). Suppose the electric circuit consists of a 24-V power source, a 12-Ω resistor, and a 4-H inductor.

(a) Sketch a slope field for the differential equation.

(b) What is the limiting value of the current? Explain.

92. ***Think About It*** It is known that $y = e^{kt}$ is a solution of the differential equation $y'' - 16y = 0$. Find the values of k.

93. ***Think About It*** It is known that $y = A \sin \omega t$ is a solution of the differential equation $y'' + 16y = 0$. Find the values of ω.

Putnam Exam Challenge

94. Let f be a twice-differentiable real-valued function satisfying

$$f(x) + f''(x) = -xg(x)f'(x)$$

where $g(x) \geq 0$ for all real x. Prove that $|f(x)|$ is bounded.

95. Prove that if the family of integral curves of the differential equation

$$\frac{dy}{dx} + p(x)y = q(x), \qquad p(x) \cdot q(x) \neq 0$$

is cut by the line $x = k$, the tangents at the points of intersection are concurrent.

These problems were composed by the Committee on the Putnam Prize Competition.

Section 6.2 Differential Equations: Growth and Decay

- **Use separation of variables to solve a simple differential equation.**
- **Use exponential functions to model growth and decay in applied problems.**

Differential Equations

In the preceding section, you learned to analyze visually the solutions of differential equations using slope fields and to approximate solutions numerically using Euler's Method. Analytically, you have learned to solve only two types of differential equations—those of the forms

$$y' = f(x) \quad \text{and} \quad y'' = f(x).$$

In this section, you will learn how to solve a more general type of differential equation. The strategy is to rewrite the equation so that each variable occurs on only one side of the equation. This strategy is called *separation of variables*. (You will study this strategy in detail in Section 6.3.)

NOTE When you integrate both sides of the equation in Example 1, you don't need to add a constant of integration to both sides of the equation. If you did, you would obtain the same result as in Example 1.

$$\int y\,dy = \int 2x\,dx$$
$$\tfrac{1}{2}y^2 + C_2 = x^2 + C_3$$
$$\tfrac{1}{2}y^2 = x^2 + (C_3 - C_2)$$
$$\tfrac{1}{2}y^2 = x^2 + C_1$$

EXAMPLE 1 Solving a Differential Equation

Solve the differential equation $y' = 2x/y$.

Solution

$$y' = \frac{2x}{y} \qquad \text{Write original equation.}$$
$$yy' = 2x \qquad \text{Multiply both sides by } y.$$
$$\int yy'\,dx = \int 2x\,dx \qquad \text{Integrate with respect to } x.$$
$$\int y\,dy = \int 2x\,dx \qquad dy = y'\,dx$$
$$\frac{1}{2}y^2 = x^2 + C_1 \qquad \text{Apply Power Rule.}$$
$$y^2 - 2x^2 = C \qquad \text{Rewrite, letting } C = 2C_1.$$

So, the general solution is given by

$$y^2 - 2x^2 = C.$$

You can use implicit differentiation to check this result.

EXPLORATION

In Example 1, the general solution of the differential equation is

$$y^2 - 2x^2 = C.$$

Use a graphing utility to sketch several particular solutions—those given by $C = \pm 2$, $C = \pm 1$, and $C = 0$. Describe the solutions graphically. Is the following statement true of each solution?

The slope of the graph at the point (x, y) is equal to twice the ratio of x and y.

Explain your reasoning. Are all curves for which this statement is true represented by the general solution?

In practice, most people prefer to use Leibniz notation and differentials when applying separation of variables. The solution of Example 1 is shown below using this notation.

$$\frac{dy}{dx} = \frac{2x}{y}$$
$$y\,dy = 2x\,dx$$
$$\int y\,dy = \int 2x\,dx$$
$$\frac{1}{2}y^2 = x^2 + C_1$$
$$y^2 - 2x^2 = C$$

Growth and Decay Models

In many applications, the rate of change of a variable y is proportional to the value of y. If y is a function of time t, the proportion can be written as shown.

Rate of change of y is proportional to y.

$$\frac{dy}{dt} = ky$$

The general solution of this differential equation is given in the following theorem.

THEOREM 6.1 Exponential Growth and Decay Model

If y is a differentiable function of t such that $y > 0$ and $y' = ky$, for some constant k, then

$$y = Ce^{kt}.$$

C is the **initial value** of y, and k is the **proportionality constant. Exponential growth** occurs when $k > 0$, and **exponential decay** occurs when $k < 0$.

NOTE Differentiate the function $y = Ce^{kt}$ with respect to t, and verify that $y' = ky$.

Proof

$$y' = ky \qquad \text{Write original equation.}$$

$$\frac{y'}{y} = k \qquad \text{Separate variables.}$$

$$\int \frac{y'}{y}\,dt = \int k\,dt \qquad \text{Integrate with respect to } t.$$

$$\int \frac{1}{y}\,dy = \int k\,dt \qquad dy = y'\,dt$$

$$\ln y = kt + C_1 \qquad \text{Find antiderivative of each side.}$$

$$y = e^{kt}e^{C_1} \qquad \text{Solve for } y.$$

$$y = Ce^{kt} \qquad \text{Let } C = e^{C_1}.$$

So, all solutions of $y' = ky$ are of the form $y = Ce^{kt}$.

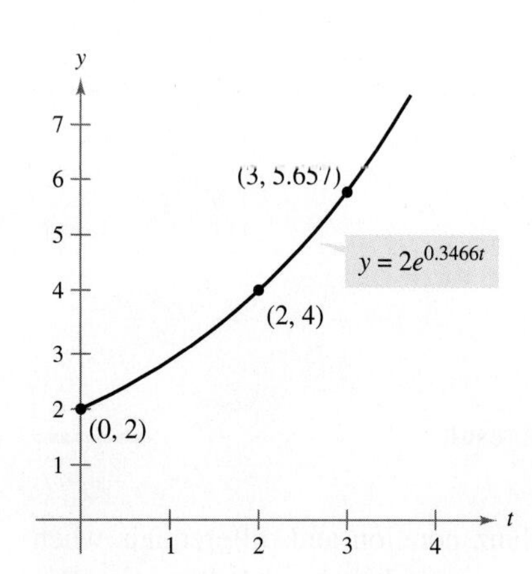

If the rate of change of y is proportional to y, then y follows an exponential model.
Figure 6.8

EXAMPLE 2 Using an Exponential Growth Model

The rate of change of y is proportional to y. When $t = 0$, $y = 2$. When $t = 2$, $y = 4$. What is the value of y when $t = 3$?

Solution Because $y' = ky$, you know that y and t are related by the equation $y = Ce^{kt}$. You can find the values of the constants C and k by applying the initial conditions.

$$2 = Ce^0 \quad \Rightarrow \quad C = 2 \qquad \text{When } t = 0, y = 2.$$

$$4 = 2e^{2k} \quad \Rightarrow \quad k = \frac{1}{2}\ln 2 \approx 0.3466 \qquad \text{When } t = 2, y = 4.$$

So, the model is $y \approx 2e^{0.3466t}$. When $t = 3$, the value of y is $2e^{0.3466(3)} \approx 5.657$ (see Figure 6.8).

STUDY TIP Using logarithmic properties, note that the value of k in Example 2 can also be written as $\ln(\sqrt{2})$. So, the model becomes $y = 2e^{(\ln\sqrt{2})t}$, which can then be rewritten as $y = 2(\sqrt{2})^t$.

TECHNOLOGY Most graphing utilities have curve-fitting capabilities that can be used to find models that represent data. Use the *exponential regression* feature of a graphing utility and the information in Example 2 to find a model for the data. How does your model compare with the given model?

Radioactive decay is measured in terms of *half-life*—the number of years required for half of the atoms in a sample of radioactive material to decay. The half-lives of some common radioactive isotopes are shown below.

Uranium (^{238}U)	4,470,000,000 years
Plutonium (^{239}Pu)	24,100 years
Carbon (^{14}C)	5715 years
Radium (^{226}Ra)	1599 years
Einsteinium (^{254}Es)	276 days
Nobelium (^{257}No)	25 seconds

EXAMPLE 3 Radioactive Decay

Sergei Supinsky/AFP/Getty Images

Suppose that 10 grams of the plutonium isotope Pu-239 was released in the Chernobyl nuclear accident. How long will it take for the 10 grams to decay to 1 gram?

Solution Let y represent the mass (in grams) of the plutonium. Because the rate of decay is proportional to y, you know that

$$y = Ce^{kt}$$

where t is the time in years. To find the values of the constants C and k, apply the initial conditions. Using the fact that $y = 10$ when $t = 0$, you can write

$$10 = Ce^{k(0)} = Ce^0$$

which implies that $C = 10$. Next, using the fact that $y = 5$ when $t = 24{,}100$, you can write

$$5 = 10e^{k(24{,}100)}$$

$$\frac{1}{2} = e^{24{,}100k}$$

$$\frac{1}{24{,}100}\ln\frac{1}{2} = k$$

$$-0.000028761 \approx k.$$

So, the model is

$$y = 10e^{-0.000028761t}. \qquad \text{Half-life model}$$

To find the time it would take for 10 grams to decay to 1 gram, you can solve for t in the equation

$$1 = 10e^{-0.000028761t}.$$

The solution is approximately 80,059 years.

NOTE The exponential decay model in Example 3 could also be written as $y = 10\left(\frac{1}{2}\right)^{t/24{,}100}$. This model is much easier to derive, but for some applications it is not as convenient to use.

From Example 3, notice that in an exponential growth or decay problem, it is easy to solve for C when you are given the value of y at $t = 0$. The next example demonstrates a procedure for solving for C and k when you do not know the value of y at $t = 0$.

EXAMPLE 4 Population Growth

Suppose an experimental population of fruit flies increases according to the law of exponential growth. There were 100 flies after the second day of the experiment and 300 flies after the fourth day. Approximately how many flies were in the original population?

Solution Let $y = Ce^{kt}$ be the number of flies at time t, where t is measured in days. Because $y = 100$ when $t = 2$ and $y = 300$ when $t = 4$, you can write

$$100 = Ce^{2k} \quad \text{and} \quad 300 = Ce^{4k}.$$

From the first equation, you know that $C = 100e^{-2k}$. Substituting this value into the second equation produces the following.

$$300 = 100e^{-2k}e^{4k}$$
$$300 = 100e^{2k}$$
$$\ln 3 = 2k$$
$$\frac{1}{2}\ln 3 = k$$
$$0.5493 \approx k$$

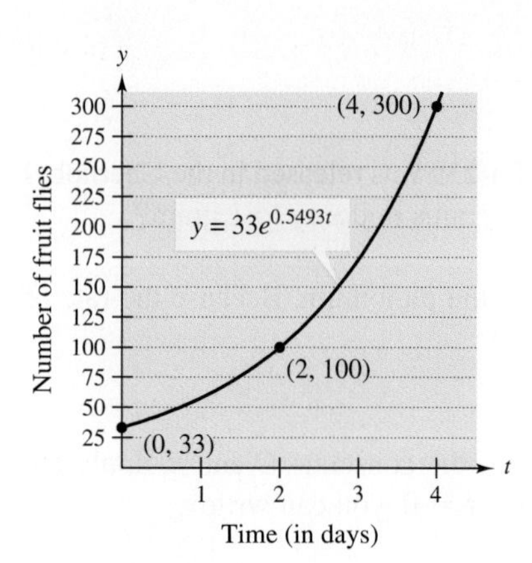

Figure 6.9

So, the exponential growth model is

$$y = Ce^{0.5493t}.$$

To solve for C, reapply the condition $y = 100$ when $t = 2$ and obtain

$$100 = Ce^{0.5493(2)}$$
$$C = 100e^{-1.0986} \approx 33.$$

So, the original population (when $t = 0$) consisted of approximately $y = C = 33$ flies, as shown in Figure 6.9.

EXAMPLE 5 Declining Sales

Four months after it stops advertising, a manufacturing company notices that its sales have dropped from 100,000 units per month to 80,000 units per month. If the sales follow an exponential pattern of decline, what will they be after another 2 months?

Solution Use the exponential decay model $y = Ce^{kt}$, where t is measured in months. From the initial condition ($t = 0$), you know that $C = 100{,}000$. Moreover, because $y = 80{,}000$ when $t = 4$, you have

$$80{,}000 = 100{,}000e^{4k}$$
$$0.8 = e^{4k}$$
$$\ln(0.8) = 4k$$
$$-0.0558 \approx k.$$

So, after 2 more months ($t = 6$), you can expect the monthly sales rate to be

$$y \approx 100{,}000e^{-0.0558(6)}$$
$$\approx 71{,}500 \text{ units.}$$

See Figure 6.10.

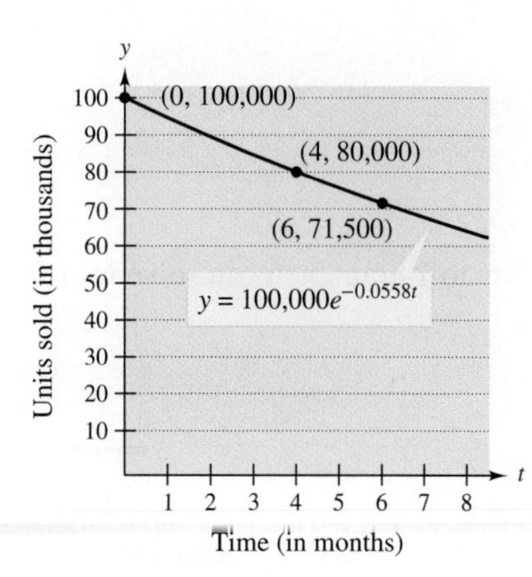

Figure 6.10

In Examples 2 through 5, you did not actually have to solve the differential equation

$$y' = ky.$$

(This was done once in the proof of Theorem 6.1.) The next example demonstrates a problem whose solution involves the separation of variables technique. The example concerns **Newton's Law of Cooling,** which states that the rate of change in the temperature of an object is proportional to the difference between the object's temperature and the temperature of the surrounding medium.

EXAMPLE 6 Newton's Law of Cooling

Let y represent the temperature (in °F) of an object in a room whose temperature is kept at a constant 60°. If the object cools from 100° to 90° in 10 minutes, how much longer will it take for its temperature to decrease to 80°?

Solution From Newton's Law of Cooling, you know that the rate of change in y is proportional to the difference between y and 60. This can be written as

$$y' = k(y - 60), \qquad 80 \le y \le 100.$$

To solve this differential equation, use separation of variables, as shown.

$$\frac{dy}{dt} = k(y - 60) \qquad \text{Differential equation}$$

$$\left(\frac{1}{y - 60}\right) dy = k\,dt \qquad \text{Separate variables.}$$

$$\int \frac{1}{y - 60}\,dy = \int k\,dt \qquad \text{Integrate each side.}$$

$$\ln|y - 60| = kt + C_1 \qquad \text{Find antiderivative of each side.}$$

Because $y > 60$, $|y - 60| = y - 60$, and you can omit the absolute value signs. Using exponential notation, you have

$$y - 60 = e^{kt + C_1} \implies y = 60 + Ce^{kt}. \qquad C = e^{C_1}$$

Using $y = 100$ when $t = 0$, you obtain $100 = 60 + Ce^{k(0)} = 60 + C$, which implies that $C = 40$. Because $y = 90$ when $t = 10$,

$$90 = 60 + 40e^{k(10)}$$
$$30 = 40e^{10k}$$
$$k = \tfrac{1}{10} \ln \tfrac{3}{4} \approx -0.02877.$$

So, the model is

$$y = 60 + 40e^{-0.02877t} \qquad \text{Cooling model}$$

and finally, when $y = 80$, you obtain

$$80 = 60 + 40e^{-0.02877t}$$
$$20 = 40e^{-0.02877t}$$
$$\tfrac{1}{2} = e^{-0.02877t}$$
$$\ln \tfrac{1}{2} = -0.02877t$$
$$t \approx 24.09 \text{ minutes.}$$

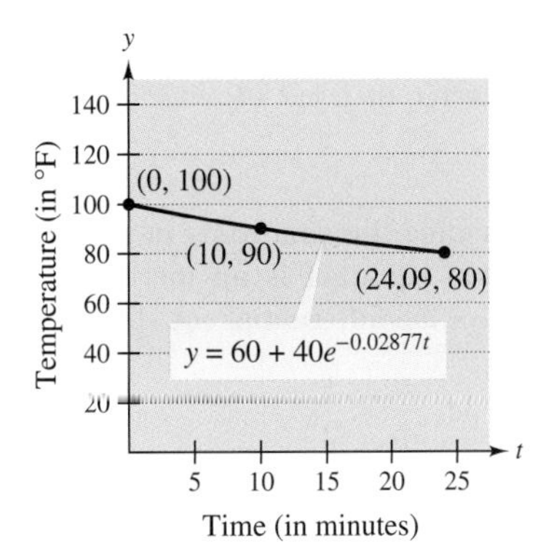

Figure 6.11

So, it will require about 14.09 *more* minutes for the object to cool to a temperature of 80° (see Figure 6.11).

Exercises for Section 6.2

See www.CalcChat.com for worked-out solutions to odd-numbered exercises.

In Exercises 1–10, solve the differential equation.

1. $\dfrac{dy}{dx} = x + 2$
2. $\dfrac{dy}{dx} = 4 - x$
3. $\dfrac{dy}{dx} = y + 2$
4. $\dfrac{dy}{dx} = 4 - y$
5. $y' = \dfrac{5x}{y}$
6. $y' = \dfrac{\sqrt{x}}{3y}$
7. $y' = \sqrt{x}\,y$
8. $y' = x(1 + y)$
9. $(1 + x^2)y' - 2xy = 0$
10. $xy + y' = 100x$

In Exercises 11–14, write and solve the differential equation that models the verbal statement.

11. The rate of change of Q with respect to t is inversely proportional to the square of t.
12. The rate of change of P with respect to t is proportional to $10 - t$.
13. The rate of change of N with respect to s is proportional to $250 - s$.
14. The rate of change of y with respect to x varies jointly as x and $L - y$.

Slope Fields **In Exercises 15 and 16, a differential equation, a point, and a slope field are given. (a) Sketch two approximate solutions of the differential equation on the slope field, one of which passes through the given point. (b) Use integration to find the particular solution of the differential equation and use a graphing utility to graph the solution. Compare the result with the sketch in part (a). To print an enlarged copy of the graph, go to the website *www.mathgraphs.com*.**

15. $\dfrac{dy}{dx} = x(6 - y)$, $(0, 0)$
16. $\dfrac{dy}{dx} = xy$, $\left(0, \frac{1}{2}\right)$

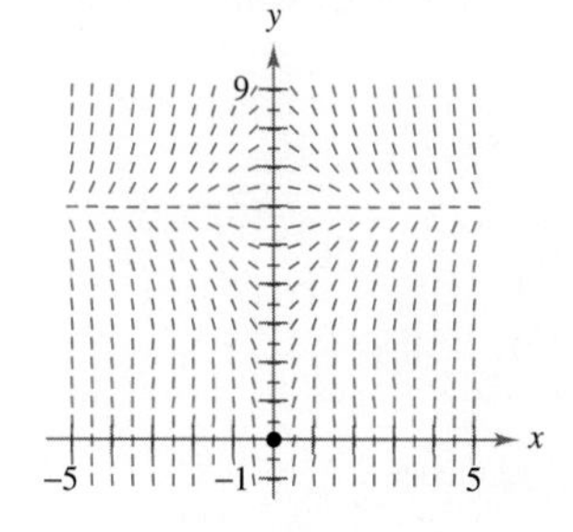

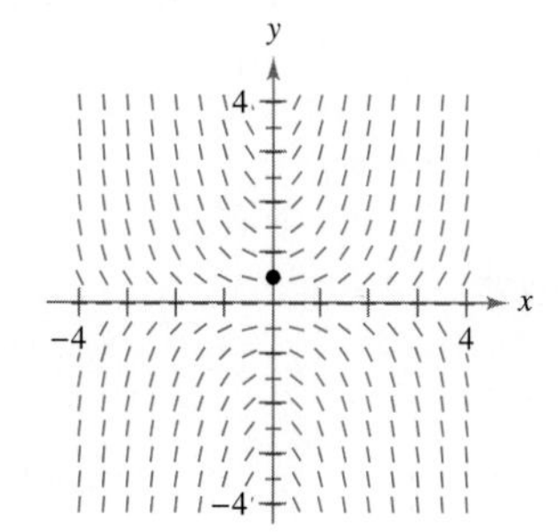

In Exercises 17–20, find the function $y = f(t)$ passing through the point $(0, 10)$ with the given first derivative. Use a graphing utility to graph the solution.

17. $\dfrac{dy}{dt} = \dfrac{1}{2}t$
18. $\dfrac{dy}{dt} = -\dfrac{3}{4}\sqrt{t}$
19. $\dfrac{dy}{dt} = -\dfrac{1}{2}y$
20. $\dfrac{dy}{dt} = \dfrac{3}{4}y$

In Exercises 21–24, write and solve the differential equation that models the verbal statement. Evaluate the solution at the specified value of the independent variable.

21. The rate of change of y is proportional to y. When $x = 0$, $y = 4$ and when $x = 3$, $y = 10$. What is the value of y when $x = 6$?
22. The rate of change of N is proportional to N. When $t = 0$, $N = 250$ and when $t = 1$, $N = 400$. What is the value of N when $t = 4$?
23. The rate of change of V is proportional to V. When $t = 0$, $V = 20{,}000$ and when $t = 4$, $V = 12{,}500$. What is the value of V when $t = 6$?
24. The rate of change of P is proportional to P. When $t = 0$, $P = 5000$ and when $t = 1$, $P = 4750$. What is the value of P when $t = 5$?

In Exercises 25–28, find the exponential function $y = Ce^{kt}$ that passes through the two given points.

25.

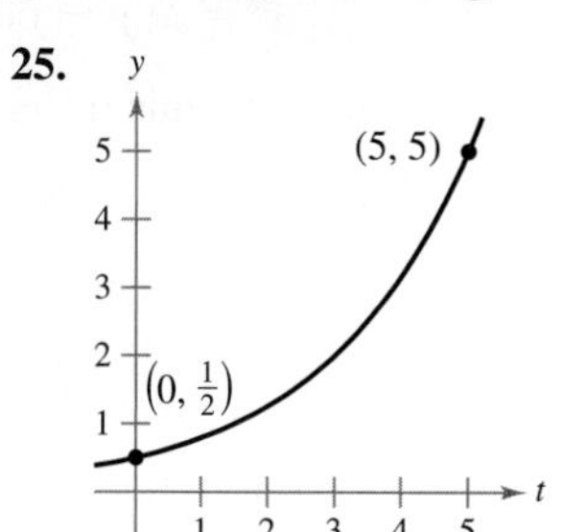

26.

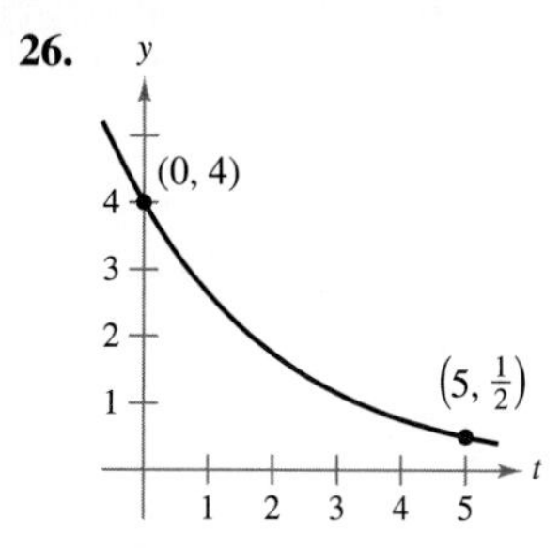

27.

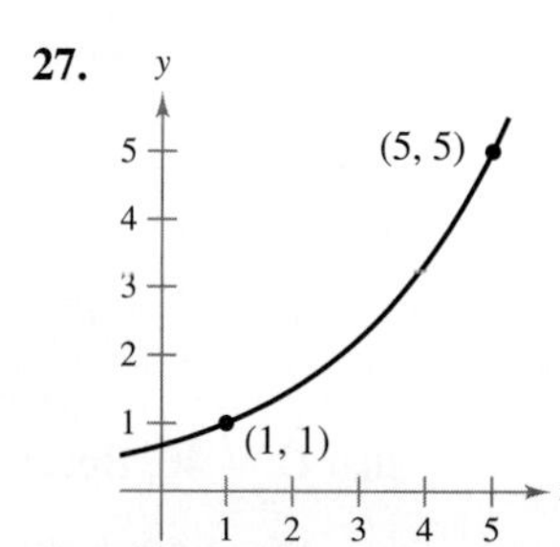

28.

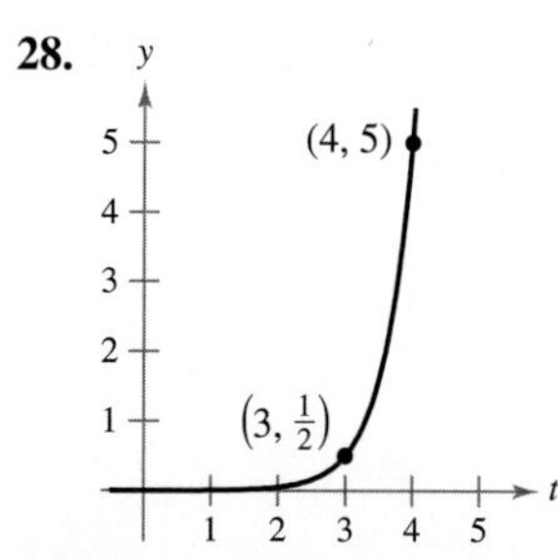

Writing About Concepts

29. Describe what the values of C and k represent in the exponential growth and decay model, $y = Ce^{kt}$.
30. Give the differential equation that models exponential growth and decay.

In Exercises 31 and 32, determine the quadrants in which the solution of the differential equation is an increasing function. Explain. (Do not solve the differential equation.)

31. $\dfrac{dy}{dx} = \dfrac{1}{2}xy$
32. $\dfrac{dy}{dx} = \dfrac{1}{2}x^2y$

Radioactive Decay **In Exercises 33–40, complete the table for the radioactive isotope.**

	Isotope	*Half-Life (in years)*	*Initial Quantity*	*Amount After 1000 Years*	*Amount After 10,000 Years*
33.	^{226}Ra	1599	10 g		
34.	^{226}Ra	1599		1.5 g	
35.	^{226}Ra	1599			0.5 g
36.	^{14}C	5715			2 g
37.	^{14}C	5715	5 g		
38.	^{14}C	5715		3.2 g	
39.	^{239}Pu	24,100		2.1 g	
40.	^{239}Pu	24,100			0.4 g

41. ***Radioactive Decay*** Radioactive radium has a half-life of approximately 1599 years. What percent of a given amount remains after 100 years?

42. ***Carbon Dating*** Carbon-14 dating assumes that the carbon dioxide on Earth today has the same radioactive content as it did centuries ago. If this is true, the amount of ^{14}C absorbed by a tree that grew several centuries ago should be the same as the amount of ^{14}C absorbed by a tree growing today. A piece of ancient charcoal contains only 15% as much of the radioactive carbon as a piece of modern charcoal. How long ago was the tree burned to make the ancient charcoal? (The half-life of ^{14}C is 5715 years.)

Compound Interest **In Exercises 43–48, complete the table for a savings account in which interest is compounded continuously.**

	Initial Investment	*Annual Rate*	*Time to Double*	*Amount After 10 Years*
43.	\$1000	6%		
44.	\$20,000	$5\frac{1}{2}\%$		
45.	\$750		$7\frac{3}{4}$ yr	
46.	\$10,000		5 yr	
47.	\$500			\$1292.85
48.	\$2000			\$5436.56

Compound Interest **In Exercises 49–52, find the principal P that must be invested at rate r, compounded monthly, so that \$500,000 will be available for retirement in t years.**

49. $r = 7\frac{1}{2}\%, \quad t = 20$

50. $r = 6\%, \quad t = 40$

51. $r = 8\%, \quad t = 35$

52. $r = 9\%, \quad t = 25$

Compound Interest **In Exercises 53–56, find the time necessary for \$1000 to double if it is invested at a rate r compounded (a) annually, (b) monthly, (c) daily, and (d) continuously.**

53. $r = 7\%$

54. $r = 6\%$

55. $r = 8.5\%$

56. $r = 5.5\%$

Population **In Exercises 57–60, the population (in millions) of a country in 2001 and the expected continuous annual rate of change k of the population for the years 2000 through 2010 are given.** ***(Source: U.S. Census Bureau, International Data Base)***

(a) Find the exponential growth model $P = Ce^{kt}$ for the population by letting $t = 0$ correspond to 2000.

(b) Use the model to predict the population of the country in 2015.

(c) Discuss the relationship between the sign of k and the change in population for the country.

	Country	*2001 Population*	k
57.	Bulgaria	7.7	−0.009
58.	Cambodia	12.7	0.018
59.	Jordan	5.2	0.026
60.	Lithuania	3.6	−0.002

61. ***Modeling Data*** One hundred bacteria are started in a culture and the number N of bacteria is counted each hour for 5 hours. The results are shown in the table, where t is the time in hours.

t	0	1	2	3	4	5
N	100	126	151	198	243	297

(a) Use the regression capabilities of a graphing utility to find an exponential model for the data.

(b) Use the model to estimate the time required for the population to quadruple in size.

62. ***Bacteria Growth*** The number of bacteria in a culture is increasing according to the law of exponential growth. There are 125 bacteria in the culture after 2 hours and 350 bacteria after 4 hours.

(a) Find the initial population.

(b) Write an exponential growth model for the bacteria population. Let t represent time in hours.

(c) Use the model to determine the number of bacteria after 8 hours.

(d) After how many hours will the bacteria count be 25,000?

63. ***Learning Curve*** The management at a certain factory has found that a worker can produce at most 30 units in a day. The learning curve for the number of units N produced per day after a new employee has worked t days is $N = 30(1 - e^{kt})$. After 20 days on the job, a particular worker produces 19 units.

(a) Find the learning curve for this worker.

(b) How many days should pass before this worker is producing 25 units per day?

64. ***Learning Curve*** If in Exercise 63 management requires a new employee to produce at least 20 units per day after 30 days on the job, find (a) the learning curve that describes this minimum requirement and (b) the number of days before a minimal achiever is producing 25 units per day.

65. ***Modeling Data*** The table shows the population P (in millions) of the United States from 1960 to 2000. *(Source: U.S. Census Bureau)*

Year	1960	1970	1980	1990	2000
Population, P	181	205	228	250	282

(a) Use the 1960 and 1970 data to find an exponential model P_1 for the data. Let $t = 0$ represent 1960.

(b) Use a graphing utility to find an exponential model P_2 for the data. Let $t = 0$ represent 1960.

(c) Use a graphing utility to plot the data and graph both models in the same viewing window. Compare the actual data with the predictions. Which model better fits the data?

(d) Estimate when the population will be 320 million.

66. ***Modeling Data*** The table shows the net receipts and the amounts required to service the national debt (interest on Treasury debt securities) of the United States from 1992 through 2001. The monetary amounts are given in billions of dollars. *(Source: U.S. Office of Management and Budget)*

Year	1992	1993	1994	1995	1996
Receipts	1091.3	1154.4	1258.6	1351.8	1453.1
Interest	292.3	292.5	296.3	332.4	343.9

Year	1997	1998	1999	2000	2001
Receipts	1579.3	1721.8	1827.5	2025.2	1991.2
Interest	355.8	363.8	353.5	361.9	359.5

(a) Use the regression capabilities of a graphing utility to find an exponential model R for the receipts and a quartic model I for the amount required to service the debt. Let t represent the time in years, with $t = 2$ corresponding to 1992.

(b) Use a graphing utility to plot the points corresponding to the receipts, and graph the corresponding model. Based on the model, what is the continuous rate of growth of the receipts?

(c) Use a graphing utility to plot the points corresponding to the amount required to service the debt, and graph the quartic model.

(d) Find a function $P(t)$ that approximates the percent of the receipts that is required to service the national debt. Use a graphing utility to graph this function.

67. ***Sound Intensity*** The level of sound β (in decibels), with an intensity of I is

$$\beta(I) = 10 \log_{10} \frac{I}{I_0}$$

where I_0 is an intensity of 10^{-16} watt per square centimeter, corresponding roughly to the faintest sound that can be heard. Determine $\beta(I)$ for the following.

(a) $I = 10^{-14}$ watt per square centimeter (whisper)

(b) $I = 10^{-9}$ watt per square centimeter (busy street corner)

(c) $I = 10^{-6.5}$ watt per square centimeter (air hammer)

(d) $I = 10^{-4}$ watt per square centimeter (threshold of pain)

68. ***Noise Level*** With the installation of noise suppression materials, the noise level in an auditorium was reduced from 93 to 80 decibels. Use the function in Exercise 67 to find the percent decrease in the intensity level of the noise as a result of the installation of these materials.

69. ***Forestry*** The value of a tract of timber is

$$V(t) = 100{,}000e^{0.8\sqrt{t}}$$

where t is the time in years, with $t = 0$ corresponding to 1998. If money earns interest continuously at 10%, the present value of the timber at any time t is $A(t) = V(t)e^{-0.10t}$. Find the year in which the timber should be harvested to maximize the present value function.

70. ***Earthquake Intensity*** On the Richter scale, the magnitude R of an earthquake of intensity I is

$$R = \frac{\ln I - \ln I_0}{\ln 10}$$

where I_0 is the minimum intensity used for comparison. Assume that $I_0 = 1$.

(a) Find the intensity of the 1906 San Francisco earthquake ($R = 8.3$).

(b) Find the factor by which the intensity is increased if the Richter scale measurement is doubled.

(c) Find dR/dI.

71. ***Newton's Law of Cooling*** When an object is removed from a furnace and placed in an environment with a constant temperature of 80°F, its core temperature is 1500°F. One hour after it is removed, the core temperature is 1120°F. Find the core temperature 5 hours after the object is removed from the furnace.

72. ***Newton's Law of Cooling*** A container of hot liquid is placed in a freezer that is kept at a constant temperature of 20°F. The initial temperature of the liquid is 160°F. After 5 minutes, the liquid's temperature is 60°F. How much longer will it take for the liquid's temperature to decrease to 30°F?

True or False? **In Exercises 73–76, determine whether the statement is true or false. If it is false, explain why or give an example that shows it is false.**

73. In exponential growth, the rate of growth is constant.

74. In linear growth, the rate of growth is constant.

75. If prices are rising at a rate of 0.5% per month, then they are rising at a rate of 6% per year.

76. The differential equation modeling exponential growth is $dy/dx = ky$, where k is a constant.

Section 6.3 Differential Equations: Separation of Variables

- **Recognize and solve differential equations that can be solved by separation of variables.**
- **Recognize and solve homogeneous differential equations.**
- **Use differential equations to model and solve applied problems.**

Separation of Variables

Consider a differential equation that can be written in the form

$$M(x) + N(y)\frac{dy}{dx} = 0$$

where M is a continuous function of x alone and N is a continuous function of y alone. As you saw in the preceding section, for this type of equation, all x terms can be collected with dx and all y terms with dy, and a solution can be obtained by integration. Such equations are said to be **separable,** and the solution procedure is called *separation of variables*. Below are some examples of differential equations that are separable.

Original Differential Equation	*Rewritten with Variables Separated*
$x^2 + 3y\dfrac{dy}{dx} = 0$	$3y\,dy = -x^2\,dx$
$(\sin x)y' = \cos x$	$dy = \cot x\,dx$
$\dfrac{xy'}{e^y + 1} = 2$	$\dfrac{1}{e^y + 1}\,dy = \dfrac{2}{x}\,dx$

EXAMPLE 1 Separation of Variables

Find the general solution of $(x^2 + 4)\dfrac{dy}{dx} = xy$.

Solution To begin, note that $y = 0$ is a solution. To find other solutions, assume that $y \neq 0$ and separate variables as shown.

$$(x^2 + 4)\,dy = xy\,dx \qquad \text{Differential form}$$

$$\frac{dy}{y} = \frac{x}{x^2 + 4}\,dx \qquad \text{Separate variables.}$$

Now, integrate to obtain

$$\int \frac{dy}{y} = \int \frac{x}{x^2 + 4}\,dx \qquad \text{Integrate.}$$

$$\ln|y| = \frac{1}{2}\ln(x^2 + 4) + C_1$$

$$\ln|y| = \ln\sqrt{x^2 + 4} + C_1$$

$$|y| = e^{C_1}\sqrt{x^2 + 4}$$

$$y = \pm e^{C_1}\sqrt{x^2 + 4}.$$

Because $y = 0$ is also a solution, you can write the general solution as

$$y = C\sqrt{x^2 + 4}. \qquad \text{General solution}$$

NOTE Be sure to check your solutions throughout this chapter. In Example 1, you can check the solution $y = C\sqrt{x^2 + 4}$ by differentiating and substituting into the original equation.

$$(x^2 + 4)\frac{dy}{dx} = xy$$

$$(x^2 + 4)\frac{Cx}{\sqrt{x^2 + 4}} \stackrel{?}{=} x\left(C\sqrt{x^2 + 4}\right)$$

$$Cx\sqrt{x^2 + 4} = Cx\sqrt{x^2 + 4}$$

So, the solution checks.

In some cases it is not feasible to write the general solution in the explicit form $y = f(x)$. The next example illustrates such a solution. Implicit differentiation can be used to verify this solution.

FOR FURTHER INFORMATION For an example (from engineering) of a differential equation that is separable, see the article "Designing a Rose Cutter" by J. S. Hartzler in *The College Mathematics Journal*. To view this article, go to the website *www.matharticles.com*.

EXAMPLE 2 Finding a Particular Solution

Given the initial condition $y(0) = 1$, find the particular solution of the equation

$$xy\,dx + e^{-x^2}(y^2 - 1)\,dy = 0.$$

Solution Note that $y = 0$ is a solution of the differential equation—but this solution does not satisfy the initial condition. So, you can assume that $y \neq 0$. To separate variables, you must rid the first term of y and the second term of e^{-x^2}. So, you should multiply by e^{x^2}/y and obtain the following.

$$xy\,dx + e^{-x^2}(y^2 - 1)\,dy = 0$$

$$e^{-x^2}(y^2 - 1)\,dy = -xy\,dx$$

$$\int\left(y - \frac{1}{y}\right)dy = \int -xe^{x^2}\,dx$$

$$\frac{y^2}{2} - \ln|y| = -\frac{1}{2}e^{x^2} + C$$

From the initial condition $y(0) = 1$, you have $\frac{1}{2} - 0 = -\frac{1}{2} + C$, which implies that $C = 1$. So, the particular solution has the implicit form

$$\frac{y^2}{2} - \ln|y| = -\frac{1}{2}e^{x^2} + 1$$

$$y^2 - \ln y^2 + e^{x^2} = 2.$$

You can check this by differentiating and rewriting to get the original equation.

EXAMPLE 3 Finding a Particular Solution Curve

Find the equation of the curve that passes through the point $(1, 3)$ and has a slope of y/x^2 at any point (x, y).

Solution Because the slope of the curve is given by y/x^2, you have

$$\frac{dy}{dx} = \frac{y}{x^2}$$

with the initial condition $y(1) = 3$. Separating variables and integrating produces

$$\int \frac{dy}{y} = \int \frac{dx}{x^2}, \quad y \neq 0$$

$$\ln|y| = -\frac{1}{x} + C_1$$

$$y = e^{-(1/x)+C_1} = Ce^{-1/x}.$$

Because $y = 3$ when $x = 1$, it follows that $3 = Ce^{-1}$ and $C = 3e$. So, the equation of the specified curve is

$$y = (3e)e^{-1/x} = 3e^{(x-1)/x}, \quad x > 0.$$

See Figure 6.12.

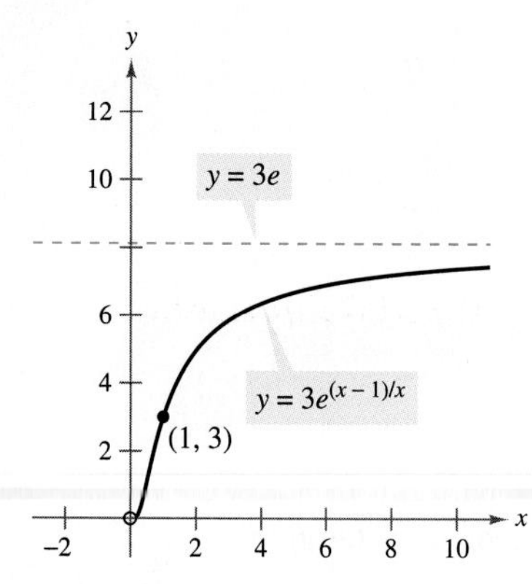

Figure 6.12

Homogeneous Differential Equations

Some differential equations that are not separable in x and y can be made separable by a change of variables. This is true for differential equations of the form $y' = f(x, y)$, where f is a **homogeneous function.** The function given by $f(x, y)$ is **homogeneous of degree n** if

NOTE The notation $f(x, y)$ is used to denote a function of two variables in much the same way that $f(x)$ denotes a function of one variable. You will study functions of two variables in detail in Chapter 13.

$$f(tx, ty) = t^n f(x, y) \qquad \text{Homogeneous function of degree } n$$

where n is a real number.

EXAMPLE 4 Verifying Homogeneous Functions

a. $f(x, y) = x^2y - 4x^3 + 3xy^2$ is a homogeneous function of degree 3 because

$$\begin{aligned} f(tx, ty) &= (tx)^2(ty) - 4(tx)^3 + 3(tx)(ty)^2 \\ &= t^3(x^2y) - t^3(4x^3) + t^3(3xy^2) \\ &= t^3(x^2y - 4x^3 + 3xy^2) \\ &= t^3 f(x, y). \end{aligned}$$

b. $f(x, y) = xe^{x/y} + y \sin(y/x)$ is a homogeneous function of degree 1 because

$$\begin{aligned} f(tx, ty) &= txe^{tx/ty} + ty \sin \frac{ty}{tx} \\ &= t\left(xe^{x/y} + y \sin \frac{y}{x}\right) \\ &= tf(x, y). \end{aligned}$$

c. $f(x, y) = x + y^2$ is *not* a homogeneous function because

$$f(tx, ty) = tx + t^2y^2 = t(x + ty^2) \neq t^n(x + y^2).$$

d. $f(x, y) = x/y$ is a homogeneous function of degree 0 because

$$f(tx, ty) = \frac{tx}{ty} = t^0 \frac{x}{y}.$$

Definition of Homogeneous Differential Equation

A **homogeneous differential equation** is an equation of the form

$$M(x, y)\, dx + N(x, y)\, dy = 0$$

where M and N are homogeneous functions of the same degree.

EXAMPLE 5 Testing for Homogeneous Differential Equations

a. $(x^2 + xy)\, dx + y^2\, dy = 0$ is homogeneous of degree 2.

b. $x^3\, dx = y^3\, dy$ is homogeneous of degree 3.

c. $(x^2 + 1)\, dx + y^2\, dy = 0$ is *not* a homogeneous differential equation.

To solve a homogeneous differential equation by the method of separation of variables, use the following change of variables theorem.

THEOREM 6.2 Change of Variables for Homogeneous Equations

If $M(x, y)\,dx + N(x, y)\,dy = 0$ is homogeneous, then it can be transformed into a differential equation whose variables are separable by the substitution

$$y = vx$$

where v is a differentiable function of x.

EXAMPLE 6 Solving a Homogeneous Differential Equation

Find the general solution of

$$(x^2 - y^2)\,dx + 3xy\,dy = 0.$$

STUDY TIP The substitution $y = vx$ will yield a differential equation that is separable with respect to the variables x and v. You must write your final solution, however, in terms of x and y.

Solution Because $(x^2 - y^2)$ and $3xy$ are both homogeneous of degree 2, let $y = vx$ to obtain $dy = x\,dv + v\,dx$. Then, by substitution, you have

$$(x^2 - v^2x^2)\,dx + 3x(vx)\overbrace{(x\,dv + v\,dx)}^{dy} = 0$$
$$(x^2 + 2v^2x^2)\,dx + 3x^3v\,dv = 0$$
$$x^2(1 + 2v^2)\,dx + x^2(3vx)\,dv = 0.$$

Dividing by x^2 and separating variables produces

$$(1 + 2v^2)\,dx = -3vx\,dv$$
$$\int \frac{dx}{x} = \int \frac{-3v}{1 + 2v^2}\,dv$$
$$\ln|x| = -\frac{3}{4}\ln(1 + 2v^2) + C_1$$
$$4\ln|x| = -3\ln(1 + 2v^2) + \ln|C|$$
$$\ln x^4 = \ln|C(1 + 2v^2)^{-3}|$$
$$x^4 = C(1 + 2v^2)^{-3}.$$

Substituting for v produces the following general solution.

$$x^4 = C\left[1 + 2\left(\frac{y}{x}\right)^2\right]^{-3}$$
$$\left(1 + \frac{2y^2}{x^2}\right)^3 x^4 = C$$
$$(x^2 + 2y^2)^3 = Cx^2 \qquad \text{General solution}$$

You can check this by differentiating and rewriting to get the original equation.

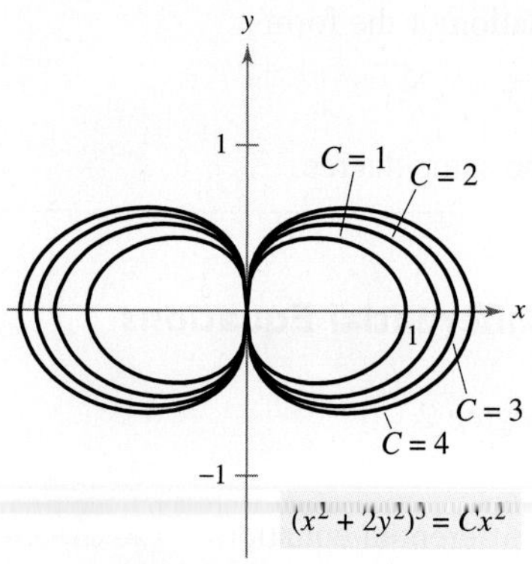

y
1
-1
1
x
C = 1
C = 2
C = 3
C = 4
$(x^2 + 2y^2)^3 = Cx^2$

General solutions of
$(x^2 - y^2)\,dx + 3xy\,dy = 0$
Figure 6.13

TECHNOLOGY If you have access to a graphing utility, try using it to graph several of the solutions in Example 6. For instance, Figure 6.13 shows the graphs of

$$(x^2 + 2y^2)^3 = Cx^2$$

for $C = 1, 2, 3,$ and 4.

Applications

EXAMPLE 7 Wildlife Population

The rate of change of the number of coyotes $N(t)$ in a population is directly proportional to $650 - N(t)$, where t is the time in years. When $t = 0$, the population is 300, and when $t = 2$, the population has increased to 500. Find the population when $t = 3$.

Solution Because the rate of change of the population is proportional to $650 - N(t)$, you can write the following differential equation.

$$\frac{dN}{dt} = k(650 - N)$$

You can solve this equation using separation of variables.

$$dN = k(650 - N)\,dt \qquad \text{Differential form}$$

$$\frac{dN}{650 - N} = k\,dt \qquad \text{Separate variables.}$$

$$-\ln|650 - N| = kt + C_1 \qquad \text{Integrate.}$$

$$\ln|650 - N| = -kt - C_1 \qquad \text{Multiply each side by } -1.$$

$$650 - N = e^{-kt - C_1} \qquad \text{Assume } N < 650.$$

$$N = 650 - Ce^{-kt} \qquad \text{General solution}$$

Using $N = 300$ when $t = 0$, you can conclude that $C = 350$, which produces

$$N = 650 - 350e^{-kt}.$$

Then, using $N = 500$ when $t = 2$, it follows that

$$500 = 650 - 350e^{-2k} \implies e^{-2k} = \frac{3}{7} \implies k \approx 0.4236.$$

So, the model for the coyote population is

$$N = 650 - 350e^{-0.4236t}. \qquad \text{Model for population}$$

When $t = 3$, you can approximate the population to be

$$N = 650 - 350e^{-0.4236(3)} \approx 552 \text{ coyotes.}$$

The model for the population is shown in Figure 6.14.

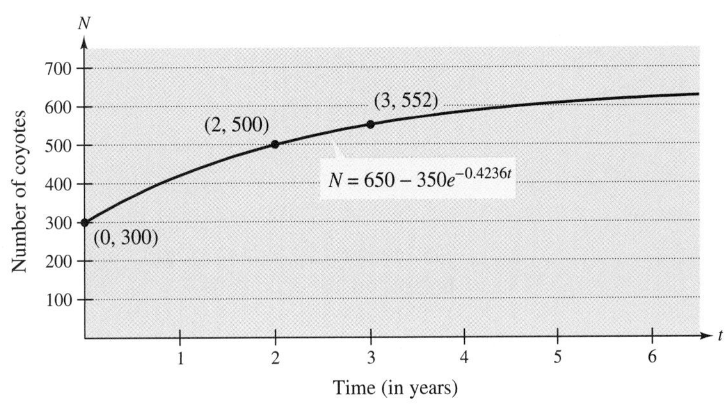

Figure 6.14

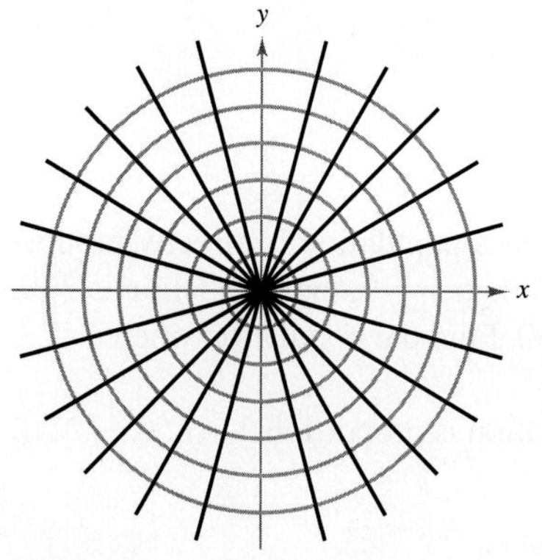

Each line $y = Kx$ is an orthogonal trajectory to the family of circles.
Figure 6.15

A common problem in electrostatics, thermodynamics, and hydrodynamics involves finding a family of curves, each of which is orthogonal to all members of a given family of curves. For example, Figure 6.15 shows a family of circles

$$x^2 + y^2 = C \qquad \text{Family of circles}$$

each of which intersects the lines in the family

$$y = Kx \qquad \text{Family of lines}$$

at right angles. Two such families of curves are said to be **mutually orthogonal,** and each curve in one of the families is called an **orthogonal trajectory** of the other family. In electrostatics, lines of force are orthogonal to the *equipotential curves.* In thermodynamics, the flow of heat across a plane surface is orthogonal to the *isothermal curves.* In hydrodynamics, the flow (stream) lines are orthogonal trajectories of the *velocity potential curves.*

EXAMPLE 8 Finding Orthogonal Trajectories

Describe the orthogonal trajectories for the family of curves given by

$$y = \frac{C}{x}$$

for $C \neq 0$. Sketch several members of each family.

Solution First, solve the given equation for C and write $xy = C$. Then, by differentiating implicitly with respect to x, you obtain the differential equation

$$xy' + y = 0 \qquad \text{Differential equation}$$

$$x\frac{dy}{dx} = -y$$

$$\frac{dy}{dx} = -\frac{y}{x}. \qquad \text{Slope of given family}$$

Because y' represents the slope of the given family of curves at (x, y), it follows that the orthogonal family has the negative reciprocal slope x/y. So,

$$\frac{dy}{dx} = \frac{x}{y}. \qquad \text{Slope of orthogonal family}$$

Now you can find the orthogonal family by separating variables and integrating.

$$\int y\,dy = \int x\,dx$$

$$\frac{y^2}{2} = \frac{x^2}{2} + C_1$$

$$y^2 - x^2 = K$$

The centers are at the origin, and the transverse axes are vertical for $K > 0$ and horizontal for $K < 0$. If $k = 0$, the orthogonal trajectories are the lines $y = \pm x$. If $K \neq 0$, the orthogonal trajectories are hyperbolas. Several trajectories are shown in Figure 6.16.

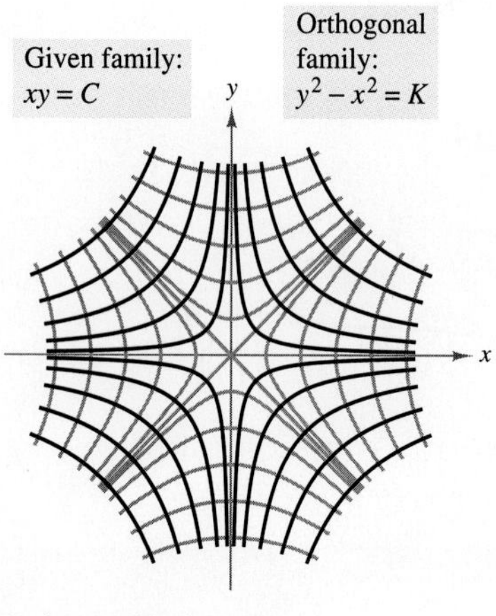

Given family: $xy = C$

Orthogonal family: $y^2 - x^2 = K$

Orthogonal trajectories
Figure 6.16

EXAMPLE 9 Modeling Advertising Awareness

A new cereal product is introduced through an advertising campaign to a population of 1 million potential customers. The rate at which the population hears about the product is assumed to be proportional to the number of people who are not yet aware of the product. By the end of 1 year, half of the population has heard of the product. How many will have heard of it by the end of 2 years?

Solution Let y be the number (in millions) of people at time t who have heard of the product. This means that $(1 - y)$ is the number of people who have not heard of it, and dy/dt is the rate at which the population hears about the product. From the given assumption, you can write the differential equation as shown.

$$\frac{dy}{dt} = k(1 - y)$$

Rate of change of y — is proportional to — the difference between 1 and y.

You can solve this equation using separation of variables.

$$dy = k(1 - y)\,dt \qquad \text{Differential form}$$

$$\frac{dy}{1 - y} = k\,dt \qquad \text{Separate variables.}$$

$$-\ln|1 - y| = kt + C_1 \qquad \text{Integrate.}$$

$$\ln|1 - y| = -kt - C_1 \qquad \text{Multiply each side by } -1.$$

$$1 - y = e^{-kt - C_1} \qquad \text{Assume } y < 1.$$

$$y = 1 - Ce^{-kt} \qquad \text{General solution}$$

To solve for the constants C and k, use the initial conditions. That is, because $y = 0$ when $t = 0$, you can determine that $C = 1$. Similarly, because $y = 0.5$ when $t = 1$, it follows that $0.5 = 1 - e^{-k}$, which implies that

$$k = \ln 2 \approx 0.693.$$

So, the particular solution is

$$y = 1 - e^{-0.693t}. \qquad \text{Particular solution}$$

This model is shown in Figure 6.17. Using the model, you can determine that the number of people who have heard of the product after 2 years is

$$y = 1 - e^{-0.693(2)}$$
$$\approx 0.75 \text{ or } 750{,}000 \text{ people.}$$

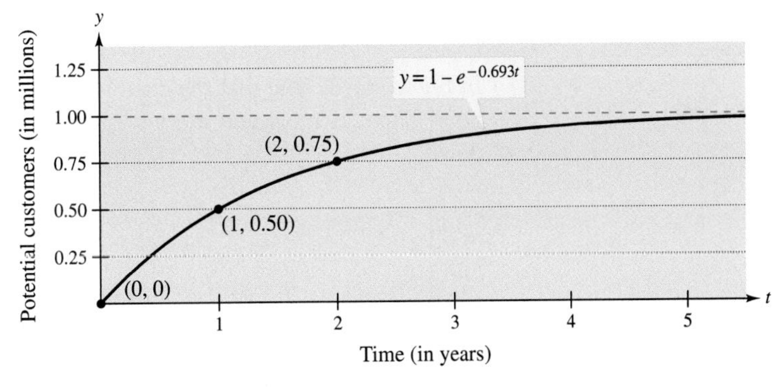

Figure 6.17

EXAMPLE 10 Modeling a Chemical Reaction

During a chemical reaction, substance A is converted into substance B at a rate that is proportional to the square of the amount of A. When $t = 0$, 60 grams of A is present, and after 1 hour ($t = 1$), only 10 grams of A remains unconverted. How much of A is present after 2 hours?

Solution Let y be the unconverted amount of substance A at any time t. From the given assumption about the conversion rate, you can write the differential equation as shown.

$$\frac{dy}{dt} = ky^2$$

Rate of change of y — is proportional to — the square of y.

You can solve this equation using separation of variables.

$$dy = ky^2\,dt \qquad \text{Differential form}$$

$$\frac{dy}{y^2} = k\,dt \qquad \text{Separate variables.}$$

$$-\frac{1}{y} = kt + C \qquad \text{Integrate.}$$

$$y = \frac{-1}{kt + C} \qquad \text{General solution}$$

To solve for the constants C and k, use the initial conditions. That is, because $y = 60$ when $t = 0$, you can determine that $C = -\frac{1}{60}$. Similarly, because $y = 10$ when $t = 1$, it follows that

$$10 = \frac{-1}{k - (1/60)}$$

which implies that $k = -\frac{1}{12}$. So, the particular solution is

$$y = \frac{-1}{(-1/12)t - (1/60)} \qquad \text{Substitute for } k \text{ and } C.$$

$$= \frac{60}{5t + 1}. \qquad \text{Particular solution}$$

Using the model, you can determine that the unconverted amount of substance A after 2 hours is

$$y = \frac{60}{5(2) + 1}$$

$$\approx 5.45 \text{ grams.}$$

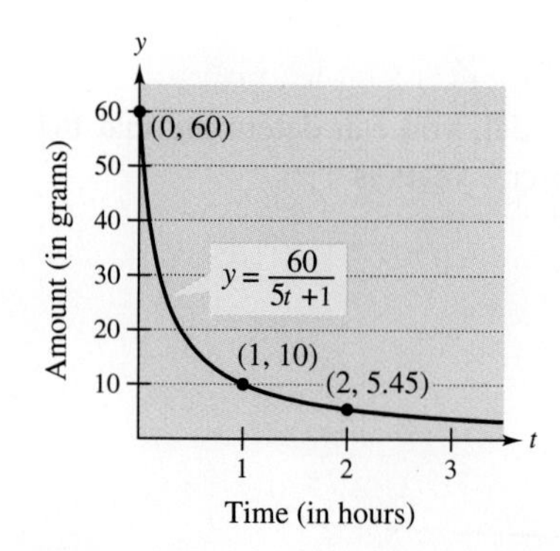

Figure 6.18

In Figure 6.18, note that the chemical conversion is occurring rapidly during the first hour. Then, as more and more of substance A is converted, the conversion rate slows down.

EXPLORATION

In Example 10, the rate of conversion was assumed to be proportional to the *square* of the unconverted amount. How would the result change if the rate of conversion were assumed to be proportional to the unconverted amount?

The next example describes a growth model called a **Gompertz growth model.** This model assumes that the rate of change of y is proportional to y and the natural log of L/y, where L is the population limit.

EXAMPLE 11 Modeling Population Growth

A population of 20 wolves has been introduced into a national park. The forest service estimates that the maximum population the park can sustain is 200 wolves. After 3 years, the population is estimated to be 40 wolves. If the population follows a Gompertz growth model, how many wolves will there be 10 years after their introduction?

Solution Let y be the number of wolves at any time t. From the given assumption about the rate of growth of the population, you can write the differential equation as shown.

$$\frac{dy}{dt} = ky \ln \frac{200}{y}$$

Rate of change of y — is proportional to — the product of y and — the log of the ratio of 200 and y.

Using separation of variables *or* a computer algebra system, you can find the general solution to be

$$y = 200e^{-Ce^{-kt}}. \qquad \text{General solution}$$

To solve for the constants C and k, use the initial conditions. That is, because $y = 20$ when $t = 0$, you can determine that

$$C = \ln 10 \approx 2.3026.$$

Similarly, because $y = 40$ when $t = 3$, it follows that

$$40 = 200e^{-2.3026e^{-3k}}$$

which implies that $k \approx 0.1194$. So, the particular solution is

$$y = 200e^{-2.3026e^{-0.1194t}}. \qquad \text{Particular solution}$$

Using the model, you can estimate the wolf population after 10 years to be

$$y = 200e^{-2.3026e^{-0.1194(10)}} \approx 100 \text{ wolves.}$$

In Figure 6.19, note that after 10 years the population has reached about half of the estimated maximum population. Try checking the growth model to see that it yields $y = 20$ when $t = 0$ and $y = 40$ when $t = 3$.

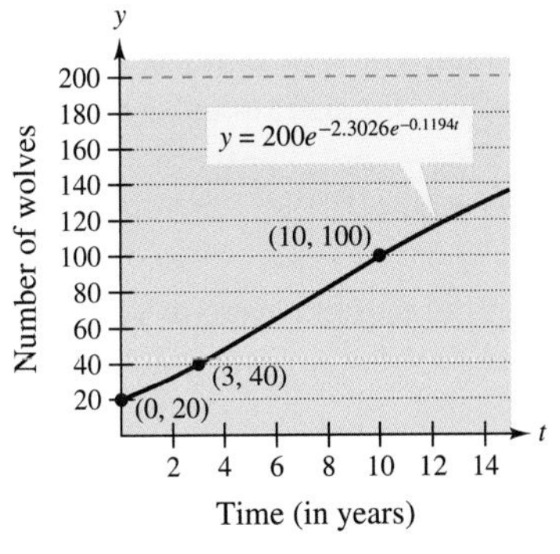

Figure 6.19

In genetics, a commonly used hybrid selection model is based on the differential equation

$$\frac{dy}{dt} = ky(1 - y)(a - by).$$

In this model, y represents the portion of the population that has a certain characteristic and t represents the time (measured in generations). The numbers a, b, and k are constants that depend on the genetic characteristic that is being studied.

EXAMPLE 12 Modeling Hybrid Selection

You are studying a population of beetles to determine how quickly characteristic D will pass from one generation to the next. At the beginning of your study ($t = 0$), you find that half the population has characteristic D. After four generations ($t = 4$), you find that 80% of the population has characteristic D. Use the hybrid selection model above with $a = 2$ and $b = 1$ to find the percent of the population that will have characteristic D after 10 generations.

Solution Using $a = 2$ and $b = 1$, the differential equation for the hybrid selection model is

$$\frac{dy}{dt} = ky(1 - y)(2 - y).$$

Using separation of variables *or* a computer algebra system, you can find the general solution to be

$$\frac{y(2 - y)}{(1 - y)^2} = Ce^{2kt}. \qquad \text{General solution}$$

To solve for the constants C and k, use the initial conditions. That is, because $y = 0.5$ when $t = 0$, you can determine that $C = 3$. Similarly, because $y = 0.8$ when $t = 4$, it follows that

$$\frac{0.8(1.2)}{(0.2)^2} = 3e^{8k}$$

which implies that

$$k = \frac{1}{8}\ln 8 \approx 0.2599.$$

So, the particular solution is

$$\frac{y(2 - y)}{(1 - y)^2} = 3e^{0.5199t}. \qquad \text{Particular solution}$$

Using the model, you can estimate the percent of the population that will have characteristic D after 10 generations to be given by

$$\frac{y(2 - y)}{(1 - y)^2} = 3e^{0.5199(10)}.$$

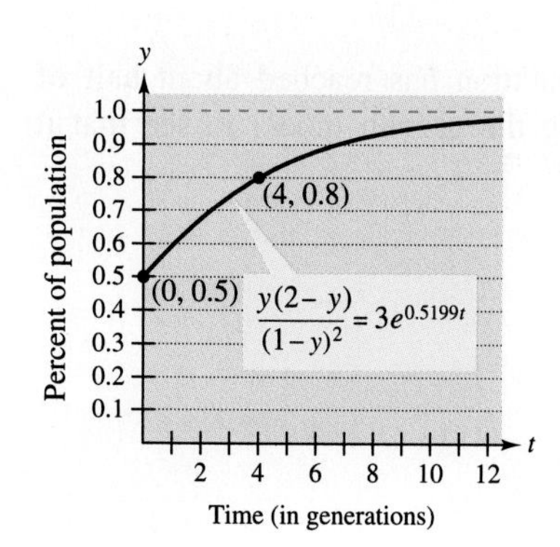

Figure 6.20

Using a computer algebra system, you can solve this equation for y to obtain $y \approx 0.96$ or 96% of the population. The graph of the model is shown in Figure 6.20.

Exercises for Section 6.3

See www.CalcChat.com for worked-out solutions to odd-numbered exercises.

In Exercises 1–12, find the general solution of the differential equation.

1. $\dfrac{dy}{dx} = \dfrac{x}{y}$
2. $\dfrac{dy}{dx} = \dfrac{x^2 + 2}{3y^2}$
3. $\dfrac{dr}{ds} = 0.05r$
4. $\dfrac{dr}{ds} = 0.05s$
5. $(2 + x)y' = 3y$
6. $xy' = y$
7. $yy' = \sin x$
8. $yy' = 6\cos(\pi x)$
9. $\sqrt{1 - 4x^2}\, y' = x$
10. $\sqrt{x^2 - 9}\, y' = 5x$
11. $y \ln x - xy' = 0$
12. $4yy' - 3e^x = 0$

In Exercises 13–22, find the particular solution that satisfies the initial condition.

	Differential Equation	*Initial Condition*
13.	$yy' - e^x = 0$	$y(0) = 4$
14.	$\sqrt{x} + \sqrt{y}\, y' = 0$	$y(1) = 4$
15.	$y(x + 1) + y' = 0$	$y(-2) = 1$
16.	$2xy' - \ln x^2 = 0$	$y(1) = 2$
17.	$y(1 + x^2)y' - x(1 + y^2) = 0$	$y(0) = \sqrt{3}$
18.	$y\sqrt{1 - x^2}\, y' - x\sqrt{1 - y^2} = 0$	$y(0) = 1$
19.	$\dfrac{du}{dv} = uv \sin v^2$	$u(0) = 1$
20.	$\dfrac{dr}{ds} = e^{r-2s}$	$r(0) = 0$
21.	$dP - kP\, dt = 0$	$P(0) = P_0$
22.	$dT + k(T - 70)\, dt = 0$	$T(0) = 140$

In Exercises 23 and 24, find an equation of the graph that passes through the point and has the given slope.

23. $(1, 1),\quad y' = -\dfrac{9x}{16y}$
24. $(8, 2),\quad y' = \dfrac{2y}{3x}$

In Exercises 25 and 26, find all functions *f* having the indicated property.

25. The tangent to the graph of f at the point (x, y) intersects the x-axis at $(x + 2, 0)$.
26. All tangents to the graph of f pass through the origin.

In Exercises 27–34, determine whether the function is homogeneous, and if it is, determine its degree.

27. $f(x, y) = x^3 - 4xy^2 + y^3$
28. $f(x, y) = x^3 + 3x^2y^2 - 2y^2$
29. $f(x, y) = \dfrac{x^2y^2}{\sqrt{x^2 + y^2}}$
30. $f(x, y) = \dfrac{xy}{\sqrt{x^2 + y^2}}$
31. $f(x, y) = 2 \ln xy$
32. $f(x, y) = \tan(x + y)$
33. $f(x, y) = 2 \ln \dfrac{x}{y}$
34. $f(x, y) = \tan \dfrac{y}{x}$

In Exercises 35–40, solve the homogeneous differential equation.

35. $y' = \dfrac{x + y}{2x}$
36. $y' = \dfrac{x^3 + y^3}{xy^2}$
37. $y' = \dfrac{x - y}{x + y}$
38. $y' = \dfrac{x^2 + y^2}{2xy}$
39. $y' = \dfrac{xy}{x^2 - y^2}$
40. $y' = \dfrac{2x + 3y}{x}$

In Exercises 41–44, find the particular solution that satisfies the initial condition.

	Differential Equation	*Initial Condition*
41.	$x\, dy - (2xe^{-y/x} + y)\, dx = 0$	$y(1) = 0$
42.	$-y^2\, dx + x(x + y)\, dy = 0$	$y(1) = 1$
43.	$\left(x \sec \dfrac{y}{x} + y\right) dx - x\, dy = 0$	$y(1) = 0$
44.	$(2x^2 + y^2)\, dx + xy\, dy = 0$	$y(1) = 0$

Slope Fields **In Exercises 45–48, sketch a few solutions of the differential equation on the slope field and then find the general solution analytically. To print an enlarged copy of the graph, go to the website *www.mathgraphs.com*.**

45. $\dfrac{dy}{dx} = x$

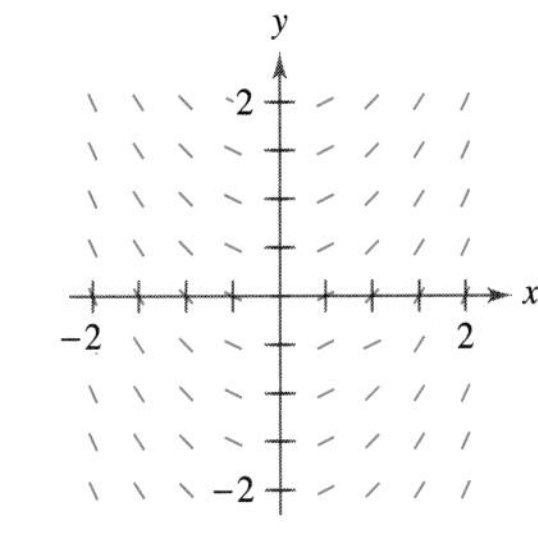

46. 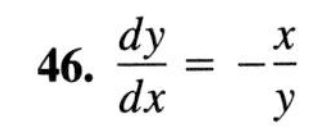 $\dfrac{dy}{dx} = -\dfrac{x}{y}$

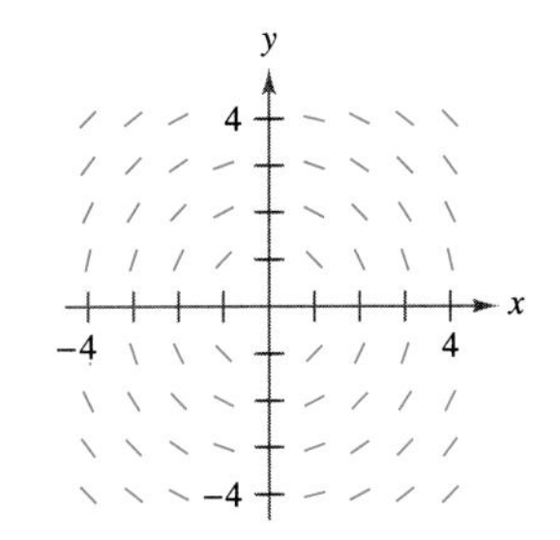

47. $\dfrac{dy}{dx} = 4 - y$

48. $\dfrac{dy}{dx} = 0.25x(4 - y)$

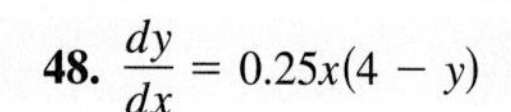

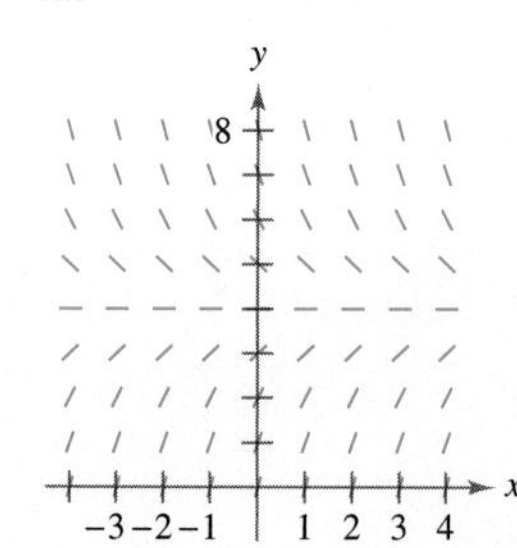

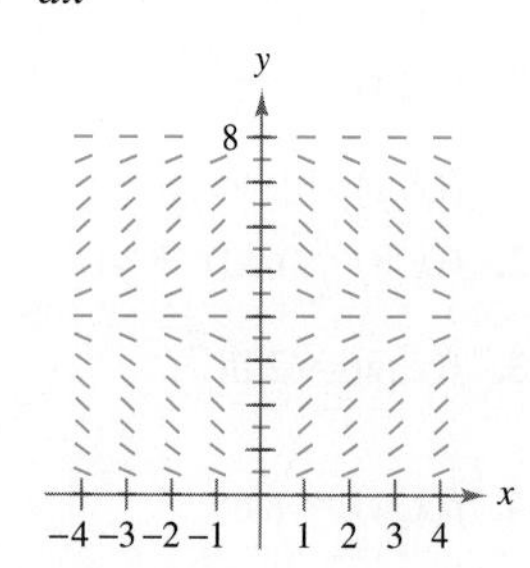

Euler's Method **In Exercises 49–52, (a) use Euler's Method with a step size of $h = 0.1$ to approximate the particular solution of the initial value problem at the given x-value, (b) find the exact solution of the differential equation analytically, and (c) compare the solutions at the given x-value.**

	Differential Equation	*Initial Condition*	*x-value*
49.	$\dfrac{dy}{dx} = -6xy$	$(0, 5)$	$x = 1$
50.	$\dfrac{dy}{dx} + 6xy^2 = 0$	$(0, 3)$	$x = 1$
51.	$\dfrac{dy}{dx} = \dfrac{2x + 12}{3y^2 - 4}$	$(1, 2)$	$x = 2$
52.	$\dfrac{dy}{dx} = 2x(1 + y^2)$	$(1, 0)$	$x = 1.5$

53. ***Radioactive Decay*** The rate of decomposition of radioactive radium is proportional to the amount present at any time. The half-life of radioactive radium is 1599 years. What percent of a present amount will remain after 25 years?

54. ***Chemical Reaction*** In a chemical reaction, a certain compound changes into another compound at a rate proportional to the unchanged amount. If initially there is 20 grams of the original compound, and there is 16 grams after 1 hour, when will 75 percent of the compound be changed?

Slope Fields **In Exercises 55–58, (a) write a differential equation for the statement, (b) match the differential equation with a possible slope field, and (c) verify your result by using a graphing utility to graph a slope field for the differential equation. [The slope fields are labeled (a), (b), (c), and (d).] To print an enlarged copy of the graph, go to the website *www.mathgraphs.com*.**

(a)

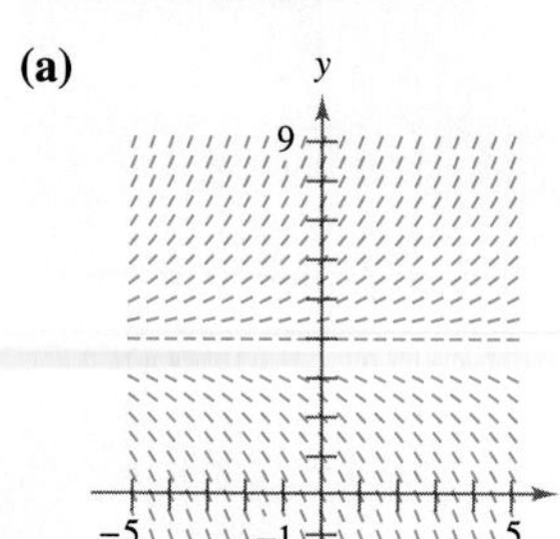

(b)

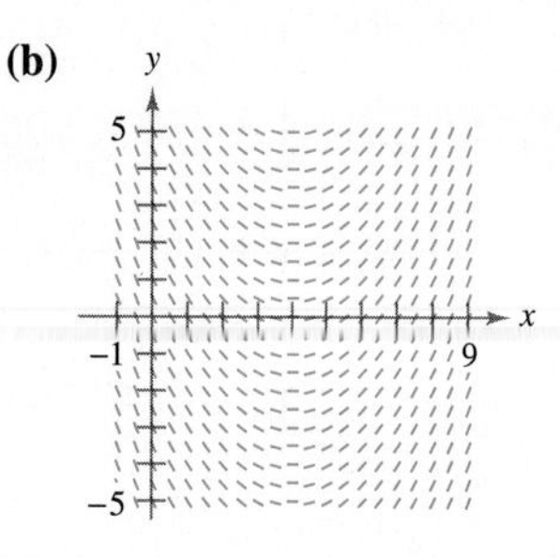

(c)

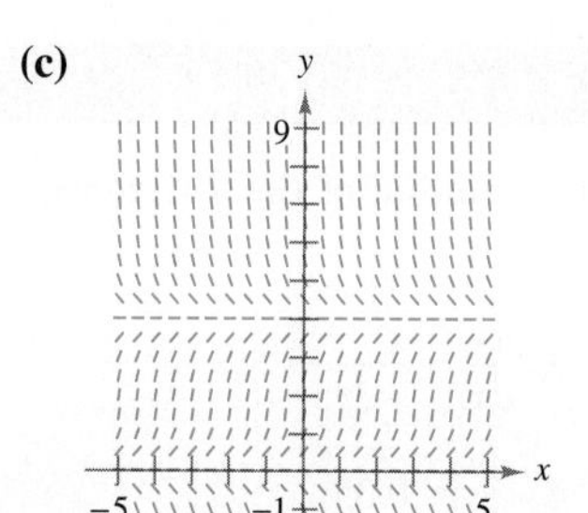

(d)

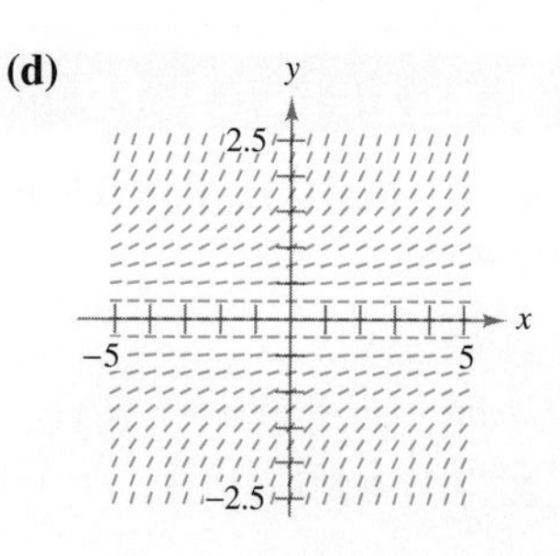

55. The rate of change of y with respect to x is proportional to the difference between y and 4.

56. The rate of change of y with respect to x is proportional to the difference between x and 4.

57. The rate of change of y with respect to x is proportional to the product of y and the difference between y and 4.

58. The rate of change of y with respect to x is proportional to y^2.

59. ***Weight Gain*** A calf that weighs 60 pounds at birth gains weight at the rate

$$\frac{dw}{dt} = k(1200 - w)$$

where w is weight in pounds and t is time in years. Solve the differential equation.

(a) Use a computer algebra system to solve the differential equation for $k = 0.8$, 0.9, and 1. Graph the three solutions.

(b) If the animal is sold when its weight reaches 800 pounds, find the time of sale for each of the models in part (a).

(c) What is the maximum weight of the animal for each of the models?

60. ***Weight Gain*** A calf that weighs w_0 pounds at birth gains weight at the rate

$$\frac{dw}{dt} = 1200 - w$$

where w is weight in pounds and t is time in years. Solve the differential equation.

In Exercises 61–66, find the orthogonal trajectories of the family. Use a graphing utility to graph several members of each family.

61. $x^2 + y^2 = C$

62. $x^2 - 2y^2 = C$

63. $x^2 = Cy$

64. $y^2 = 2Cx$

65. $y^2 = Cx^3$

66. $y = Ce^x$

67. ***Biology*** At any time t, the rate of growth of the population N of deer in a state park is proportional to the product of N and $L - N$, where $L = 500$ is the maximum number of deer the park can sustain. When $t = 0$, $N = 100$, and when $t = 4$, $N = 200$. Write N as a function of t.

68. ***Sales Growth*** The rate of change in sales S (in thousands of units) of a new product is proportional to the product of S and $L - S$. L (in thousands of units) is the estimated maximum level of sales, and $S = 10$ when $t = 0$. Write and solve the differential equation for this sales model.

Chemical Reaction **In Exercises 69 and 70, use the chemical reaction model given in Example 10 to find the amount y as a function of t, and use a graphing utility to graph the function.**

69. $y = 45$ grams when $t = 0$; $y = 4$ grams when $t = 2$

70. $y = 75$ grams when $t = 0$; $y = 12$ grams when $t = 1$

In Exercises 71 and 72, use the Gompertz growth model described in Example 11 to find the growth function, and sketch its graph.

71. $L = 500$; $y = 100$ when $t = 0$; $y = 150$ when $t = 2$

72. $L = 5000$; $y = 500$ when $t = 0$; $y = 625$ when $t = 1$

73. ***Biology*** A population of eight beavers has been introduced into a new wetlands area. Biologists estimate that the maximum population the wetlands can sustain is 60 beavers. After 3 years, the population is 15 beavers. If the population follows a Gompertz growth model, how many beavers will be present in the wetlands after 10 years?

74. ***Biology*** A population of 30 rabbits has been introduced into a new region. It is estimated that the maximum population the region can sustain is 400 rabbits. After 1 year, the population is estimated to be 90 rabbits. If the population follows a Gompertz growth model, how many rabbits will be present after 3 years?

Biology **In Exercises 75 and 76, use the hybrid selection model described in Example 12 to find the percent of the population that has the indicated characteristic.**

75. You are studying a population of mayflies to determine how quickly characteristic A will pass from one generation to the next. At the start of the study, half the population has characteristic A. After four generations, 75% of the population has characteristic A. Find the percent of the population that will have characteristic A after 10 generations. (Assume $a = 2$ and $b = 1$.)

76. A research team is studying a population of snails to determine how quickly characteristic B will pass from one generation to the next. At the start of the study, 40% of the snails have characteristic B. After five generations, 80% of the population has characteristic B. Find the percent of the population that will have characteristic B after eight generations. (Assume $a = 2$ and $b = 1$.)

77. ***Chemical Mixture*** A 100-gallon tank is full of a solution containing 25 pounds of a concentrate. Starting at time $t = 0$, distilled water is admitted to the tank at the rate of 5 gallons per minute, and the well-stirred solution is withdrawn at the same rate, as shown in the figure.

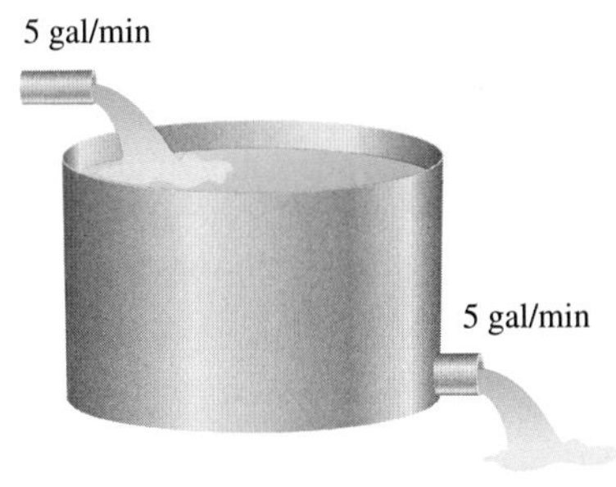

(a) Find the amount Q of the concentrate in the solution as a function of t. (*Hint:* $Q' + Q/20 = 0$)

(b) Find the time when the amount of concentrate in the tank reaches 15 pounds.

78. ***Chemical Mixture*** A 200-gallon tank is half full of distilled water. At time $t = 0$, a solution containing 0.5 pound of concentrate per gallon enters the tank at the rate of 5 gallons per minute, and the well-stirred mixture is withdrawn at the same rate. Find the amount Q of concentrate in the tank after 30 minutes. $\left(Hint\text{: } Q' + Q/20 = \frac{5}{2}\right)$

79. ***Chemical Reaction*** In a chemical reaction, a compound changes into another compound at a rate proportional to the unchanged amount, according to the model

$$\frac{dy}{dt} = ky.$$

(a) Solve the differential equation.

(b) If the initial amount of the original compound is 20 grams, and the amount remaining after 1 hour is 16 grams, when will 75% of the compound have been changed?

80. ***Safety*** Assume that the rate of change in the number of miles s of road cleared per hour by a snowplow is inversely proportional to the depth h of snow. That is,

$$\frac{ds}{dh} = \frac{k}{h}.$$

Find s as a function of h if $s = 25$ miles when $h = 2$ inches and $s = 12$ miles when $h = 6$ inches $(2 \le h \le 15)$.

81. ***Chemistry*** A wet towel hung from a clothesline to dry loses moisture through evaporation at a rate proportional to its moisture content. If after 1 hour the towel has lost 40% of its original moisture content, after how long will it have lost 80%?

82. ***Biology*** Let x and y be the sizes of two internal organs of a particular mammal at time t. Empirical data indicate that the relative growth rates of these two organs are equal, and can be modeled by

$$\frac{1}{x}\frac{dx}{dt} = \frac{1}{y}\frac{dy}{dt}.$$

Use this differential equation to write y as a function of x.

83. ***Population Growth*** When predicting population growth, demographers must consider birth and death rates as well as the net change caused by the difference between the rates of immigration and emigration. Let P be the population at time t and let N be the net increase per unit time due to the difference between immigration and emigration. So, the rate of growth of the population is given by

$$\frac{dP}{dt} = kP + N, \quad N \text{ is constant.}$$

Solve this differential equation to find P as a function of time.

84. ***Meteorology*** The barometric pressure y (in inches of mercury) at an altitude of x miles above sea level decreases at a rate proportional to the current pressure according to the model

$$\frac{dy}{dx} = -0.2y$$

where $y = 29.92$ inches when $x = 0$. Find the barometric pressure (a) at the top of Mt. St. Helens (8364 feet) and (b) at the top of Mt. McKinley (20,320 feet).

85. ***Investment*** A large corporation starts at time $t = 0$ to invest part of its receipts at a rate of P dollars per year in a fund for future corporate expansion. Assume that the fund earns r percent interest per year compounded continuously. So, the rate of growth of the amount A in the fund is given by

$$\frac{dA}{dt} = rA + P$$

where $A = 0$ when $t = 0$. Solve this differential equation for A as a function of t.

Investment **In Exercises 86–88, use the result of Exercise 85.**

86. Find A for each situation.

(a) $P = \$100{,}000$, $r = 6\%$, and $t = 5$ years

(b) $P = \$250{,}000$, $r = 5\%$, and $t = 10$ years

87. Find P if the corporation needs \$120,000,000 in 8 years and the fund earns $7\frac{1}{4}\%$ interest compounded continuously.

88. Find t if the corporation needs \$800,000 and it can invest \$75,000 per year in a fund earning 8% interest compounded continuously.

In Exercises 89 and 90, use the Gompertz growth model described in Example 11.

89. (a) Use a graphing utility to graph the slope field for the growth model when $k = 0.02$ and $L = 5000$.

(b) Describe the behavior of the graph as $t \to \infty$.

(c) Solve the growth model for $L = 5000$, $y_0 = 500$, and $k = 0.02$.

(d) Graph the equation you found in part (c). Determine the concavity of the graph.

90. (a) Use a graphing utility to graph the slope field for the growth model when $k = 0.05$ and $L = 1000$.

(b) Describe the behavior of the graph as $t \to \infty$.

(c) Solve the growth model for $L = 1000$, $y_0 = 100$, and $k = 0.05$.

(d) Graph the equation you found in part (c). Determine the concavity of the graph.

Writing About Concepts

91. In your own words, describe how to recognize and solve differential equations that can be solved by separation of variables.

92. State the test for determining if a differential equation is homogeneous. Give an example.

93. In your own words, describe the relationship between two families of curves that are mutually orthogonal.

94. ***Sailing*** Ignoring resistance, a sailboat starting from rest accelerates (dv/dt) at a rate proportional to the difference between the velocities of the wind and the boat.

(a) The wind is blowing at 20 knots, and after 1 minute the boat is moving at 5 knots. Write the velocity v as a function of time t.

(b) Use the result of part (a) to write the distance traveled by the boat as a function of time.

True or False? **In Exercises 95–98, determine whether the statement is true or false. If it is false, explain why or give an example that shows it is false.**

95. The function $y = 0$ is always a solution of a differential equation that can be solved by separation of variables.

96. The differential equation

$$y' = xy - 2y + x - 2$$

can be written in separated variables form.

97. The function

$$f(x, y) = x^2 + xy + 2$$

is homogeneous.

98. The families

$$x^2 + y^2 = 2Cy \text{ and } x^2 + y^2 = 2Kx$$

are mutually orthogonal.

Putnam Exam Challenge

99. A not uncommon calculus mistake is to believe that the product rule for derivatives says that $(fg)' = f'g'$. If $f(x) = e^{x^2}$, determine, with proof, whether there exists an open interval (a, b) and a nonzero function g defined on (a, b) such that this wrong product rule is true for x in (a, b).

This problem was composed by the Committee on the Putnam Prize Competition.

Section 6.4 The Logistic Equation

- **Solve and analyze logistic differential equations.**
- **Use logistic differential equations to model and solve applied problems.**

Logistic Differential Equation

In Section 6.2, the exponential growth model was derived from the fact that the rate of change of a variable y is proportional to the value of y. You observed that the differential equation $dy/dt = ky$ has the general solution $y = Ce^{kt}$. Exponential growth is unlimited, but when describing a population, there often exists some upper limit L past which growth cannot occur. This upper limit L is called the **carrying capacity,** which is the maximum population $y(t)$ that can be sustained or supported as time t increases. A model that is often used for this type of growth is the **logistic differential equation**

$$\frac{dy}{dt} = ky\left(1 - \frac{y}{L}\right)$$

Logistic differential equation

where k and L are positive constants. A population that satisfies this equation does not grow without bound, but approaches the carrying capacity L as t increases.

From the equation, you can see that if y is between 0 and the carrying capacity L, then $dy/dt > 0$, and the population increases. If y is greater than L, then $dy/dt < 0$, and the population decreases. The general solution of the logistic differential equation is derived in the next example.

EXAMPLE 1 Deriving the General Solution

Solve the logistic differential equation $\dfrac{dy}{dt} = ky\left(1 - \dfrac{y}{L}\right)$.

Solution Begin by separating variables.

$$\frac{dy}{dt} = ky\left(1 - \frac{y}{L}\right)$$

Write differential equation.

$$\frac{1}{y(1 - y/L)}\,dy = k\,dt$$

Separate variables.

$$\int \frac{1}{y(1 - y/L)}\,dy = \int k\,dt$$

Integrate each side.

$$\int \left(\frac{1}{y} + \frac{1}{L - y}\right) dy = \int k\,dt$$

Rewrite left side using partial fractions.

$$\ln|y| - \ln|L - y| = kt + C$$

Find antiderivative of each side.

$$\ln\left|\frac{L - y}{y}\right| = -kt - C$$

Multiply each side by -1 and simplify.

$$\left|\frac{L - y}{y}\right| = e^{-kt-C} = e^{-C}e^{-kt}$$

Exponentiate each side.

$$\frac{L - y}{y} = be^{-kt}$$

Let $\pm e^{-C} = b$.

Solving this equation for y produces $y = \dfrac{L}{1 + be^{-kt}}$.

From Example 1, you can conclude that all solutions of the logistic differential equation are of the general form

$$y = \frac{L}{1 + be^{-kt}}.$$

The graph of the function y is called the *logistic curve*, as shown in Figure 6.21. In the next example, you will verify a particular solution of a logistic differential equation and find the initial condition.

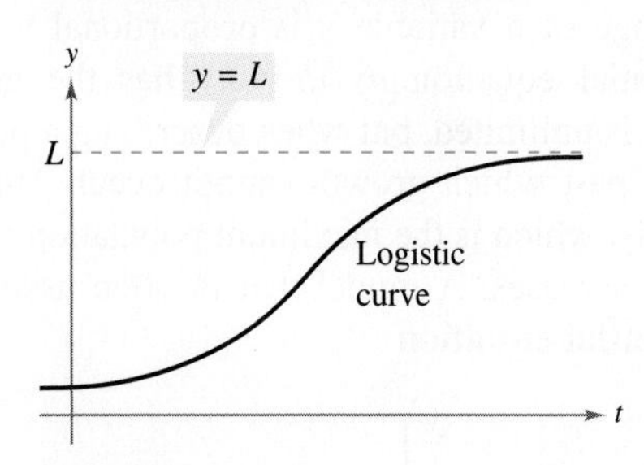

Note that as $t \to \infty$, $y \to L$.
Figure 6.21

EXAMPLE 2 Verifying a Particular Solution

Verify that the equation

$$y = \frac{4}{1 + 2e^{-3t}}$$

satisfies the logistic differential equation, and find the initial condition.

EXPLORATION

Use a graphing utility to investigate the effects of the values of L, b, and k on the graph of

$$y = \frac{L}{1 + be^{-kt}}.$$

Include some examples to support your results.

Solution Comparing the given equation with the general form derived in Example 1, you know that $L = 4$, $b = 2$, and $k = 3$. You can verify that y satisfies the logistic differential equation as follows.

$$y = 4(1 + 2e^{-3t})^{-1} \qquad \text{Rewrite using negative exponent.}$$

$$y' = 4(-1)(1 + 2e^{-3t})^{-2}(-6e^{-3t}) \qquad \text{Apply Power Rule.}$$

$$= 3\left(\frac{4}{1 + 2e^{-3t}}\right)\left(\frac{2e^{-3t}}{1 + 2e^{-3t}}\right) \qquad \text{Rewrite.}$$

$$= 3y\left(\frac{2e^{-3t}}{1 + 2e^{-3t}}\right) \qquad \text{Rewrite using } y = \frac{4}{1 + 2e^{-3t}}.$$

$$= 3y\left(1 - \frac{1}{1 + 2e^{-3t}}\right) \qquad \text{Rewrite fraction using long division.}$$

$$= 3y\left(1 - \frac{4}{4(1 + 2e^{-3t})}\right) \qquad \text{Multiply fraction by } \frac{4}{4}.$$

$$= 3y\left(1 - \frac{y}{4}\right) \qquad \text{Rewrite using } y = \frac{4}{1 + 2e^{-3t}}.$$

So, y satisfies the logistic differential equation $y' = 3y\left(1 - \frac{y}{4}\right)$. The initial condition can be found by letting $t = 0$ in the given equation.

$$y = \frac{4}{1 + 2e^{-3(0)}} = \frac{4}{3} \qquad \text{Let } t = 0 \text{ and simplify.}$$

So, the initial condition is $y(0) = \frac{4}{3}$.

EXAMPLE 3 Verifying the Upper Limit

Verify that the upper limit of $y = \dfrac{4}{1 + 2e^{-3t}}$ is 4.

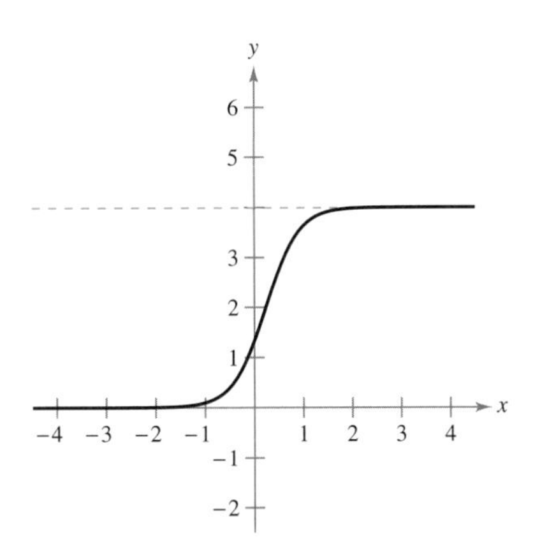

Figure 6.22

Solution In Figure 6.22, you can see that the values of y appear to approach 4 as t increases without bound. You can come to this conclusion numerically, as shown in the table.

t	0	1	2	5	10	100
y	1.3333	3.6378	3.9803	4.0000	4.0000	4.0000

Finally, you can obtain the same results analytically, as follows.

$$\lim_{t\to\infty} y = \lim_{t\to\infty} \frac{4}{1 + 2e^{-3t}} = \frac{\lim_{t\to\infty} 4}{\lim_{t\to\infty}(1 + 2e^{-3t})} = \frac{4}{1 + 0} = 4$$

The upper limit of y is 4, which is also the carrying capacity $L = 4$.

EXAMPLE 4 Determining the Point of Inflection

Sketch a graph of $y = \dfrac{4}{1 + 2^{-3t}}$. Calculate y'' in terms of y and y'. Then determine the point of inflection.

Solution From Example 2, you know that

$$y' = 3y\left(1 - \frac{y}{4}\right).$$

Now calculate y'' in terms of y and y'.

$$y'' = 3y\left(-\frac{y'}{4}\right) + \left(1 - \frac{y}{4}\right)3y' \qquad \text{Differentiate using Product Rule.}$$

$$y'' = 3y'\left(1 - \frac{y}{2}\right) \qquad \text{Factor and simplify.}$$

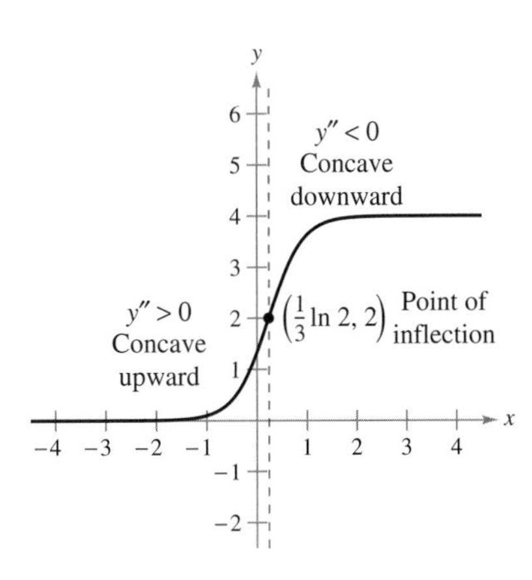

Figure 6.23

When $2 < y < 4$, $y'' < 0$ and the graph of y is concave downward. When $0 < y < 2$, $y'' > 0$ and the graph of y is concave upward. So, a point of inflection must occur at $y = 2$. The corresponding t-value is

$$2 = \frac{4}{1 + 2e^{-3t}} \implies 1 + 2e^{-3t} = 2 \implies e^{-3t} = \frac{1}{2} \implies t = \frac{1}{3}\ln 2.$$

The point of inflection is $\left(\frac{1}{3}\ln 2, 2\right)$, as shown in Figure 6.23.

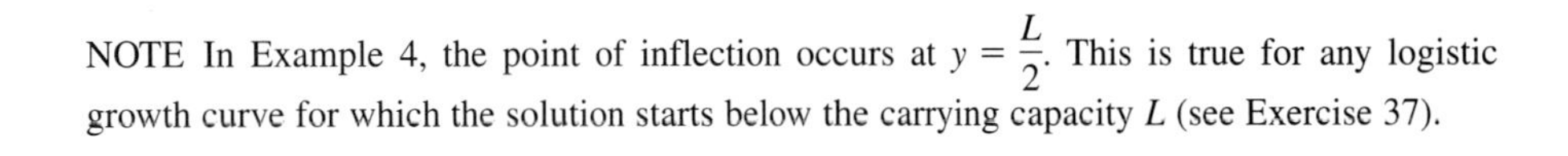

NOTE In Example 4, the point of inflection occurs at $y = \dfrac{L}{2}$. This is true for any logistic growth curve for which the solution starts below the carrying capacity L (see Exercise 37).

EXAMPLE 5 Graphing a Slope Field and Solution Curves

Graph a slope field for the logistic differential equation $y' = 0.05y\left(1 - \dfrac{y}{800}\right)$. Then graph solution curves for the initial conditions $y(0) = 200$, $y(0) = 1200$, and $y(0) = 800$.

Solution You can use a graphing utility to graph the slope field shown in Figure 6.24. The solution curves for the initial conditions $y(0) = 200$, $y(0) = 1200$, and $y(0) = 800$ are shown in Figures 6.25–6.27.

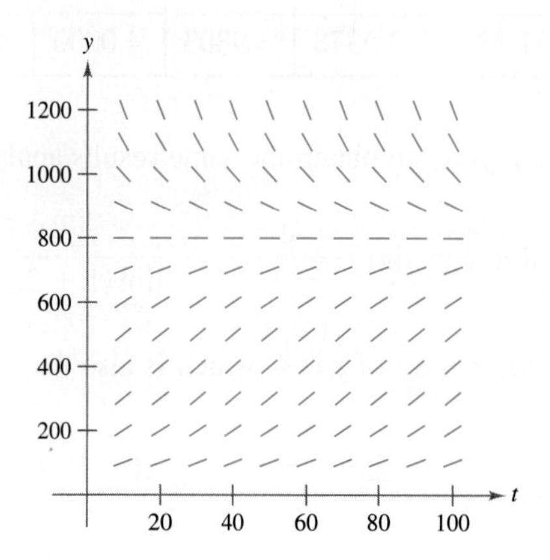

Slope field for

$y' = 0.05y\left(1 - \dfrac{y}{800}\right)$

Figure 6.24

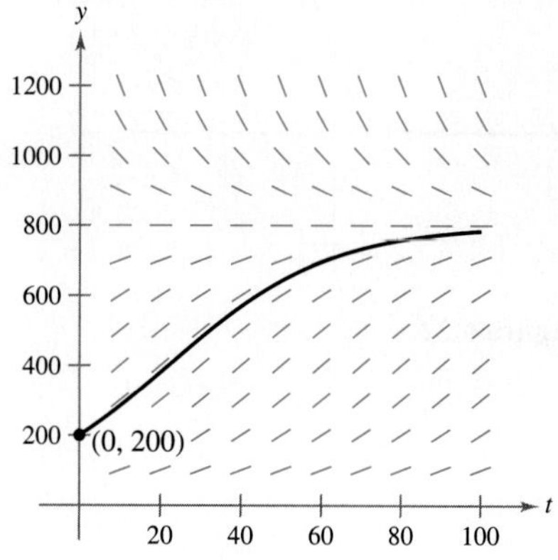

Particular solution for

$y' = 0.05y\left(1 - \dfrac{y}{800}\right)$

and initial condition $y(0) = 200$

Figure 6.25

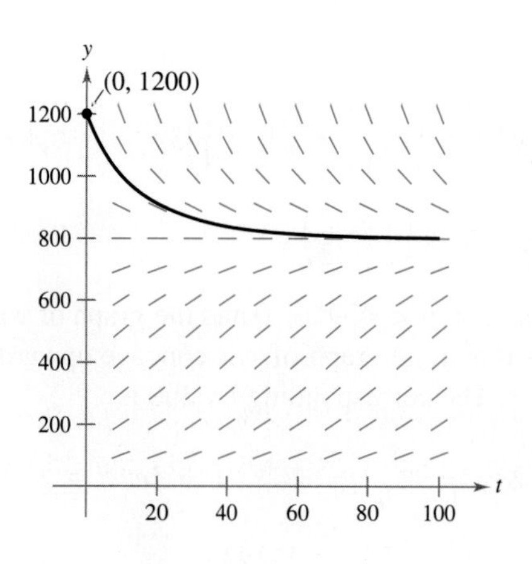

Particular solution for

$y' = 0.05y\left(1 - \dfrac{y}{800}\right)$

and initial condition $y(0) = 1200$

Figure 6.26

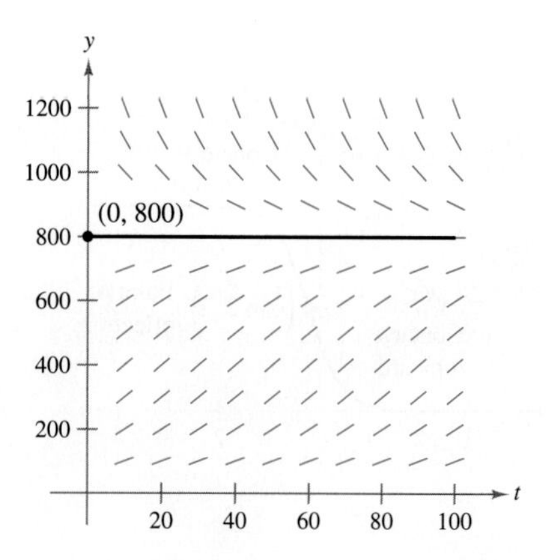

Particular solution for

$y' = 0.05y\left(1 - \dfrac{y}{800}\right)$

and initial condition $y(0) = 800$

Figure 6.27

Note that as t increases without bound, the solution curves in Figures 6.25–6.27 all tend to the same limit, which is the carrying capacity of 800.

EXAMPLE 6 Solving a Logistic Differential Equation

John Edwards/Getty Images

A state game commission releases 40 elk into a game refuge. After 5 years, the elk population is 104. The commission believes that the environment can support no more than 4000 elk. The growth rate of the elk population p is

$$\frac{dp}{dt} = kp\left(1 - \frac{p}{4000}\right), \quad 40 \le p \le 4000$$

where t is the number of years.

a. Write a model for the elk population in terms of t.

b. Graph the slope field of the differential equation and the solution that passes through the point $(0, 40)$.

c. Use the model to estimate the elk population after 15 years.

d. Find the limit of the model as $t \to \infty$.

EXPLORATION

Explain what happens if $p(0) = L$.

Solution

a. You know that $L = 4000$. So, the solution of the equation is of the form

$$p = \frac{4000}{1 + be^{-kt}}.$$

Because $p(0) = 40$, you can solve for b as shown.

$$40 = \frac{4000}{1 + be^{-k(0)}}$$

$$40 = \frac{4000}{1 + b} \quad \Rightarrow \quad b = 99$$

Then, because $p = 104$ when $t = 5$, you can solve for k.

$$104 = \frac{4000}{1 + 99e^{-k(5)}} \quad \Rightarrow \quad k \approx 0.194$$

So, a model for the elk population is given by $p = \dfrac{4000}{1 + 99e^{-0.194t}}$.

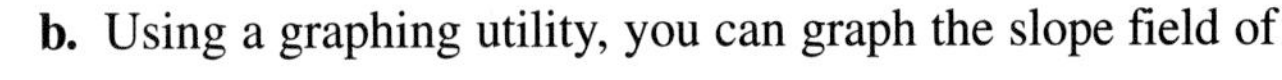

b. Using a graphing utility, you can graph the slope field of

$$\frac{dp}{dt} = 0.194p\left(1 - \frac{p}{4000}\right)$$

and the solution that passes through $(0, 40)$, as shown in Figure 6.28.

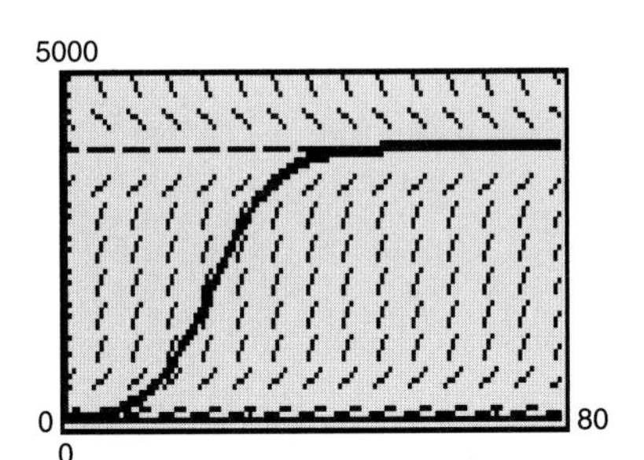

Figure 6.28

c. To estimate the elk population after 15 years, substitute 15 for t in the model.

$$p = \frac{4000}{1 + 99e^{-0.194(15)}} \qquad \text{Substitute 15 for } t.$$

$$= \frac{4000}{1 + 99e^{-2.91}} \approx 626 \qquad \text{Simplify.}$$

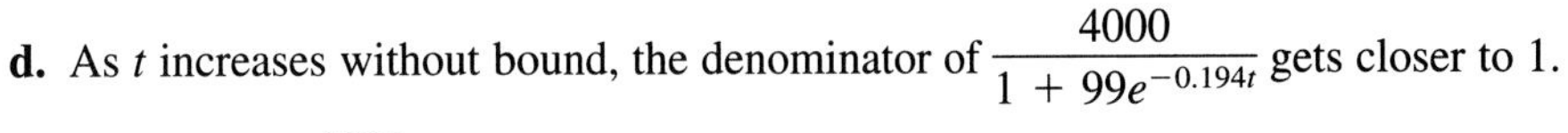

d. As t increases without bound, the denominator of $\dfrac{4000}{1 + 99e^{-0.194t}}$ gets closer to 1.

So, $\displaystyle\lim_{t\to\infty} \frac{4000}{1 + 99e^{-0.194t}} = 4000.$

Exercises for Section 6.4

See www.CalcChat.com for worked-out solutions to odd-numbered exercises.

In Exercises 1–4, match the logistic equation with its graph. [The graphs are labeled (a), (b), (c), and (d).]

(a)

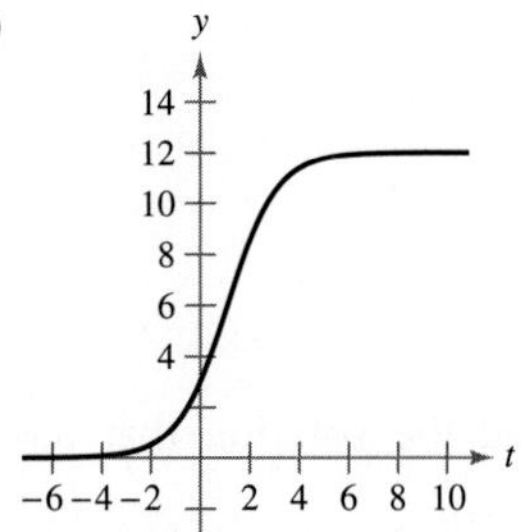

(b)

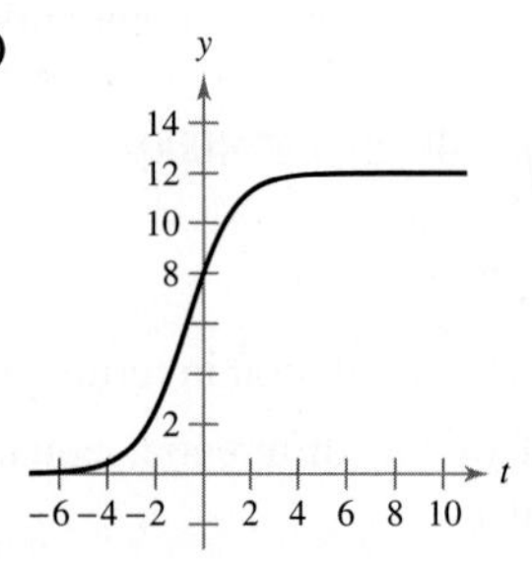

(c)

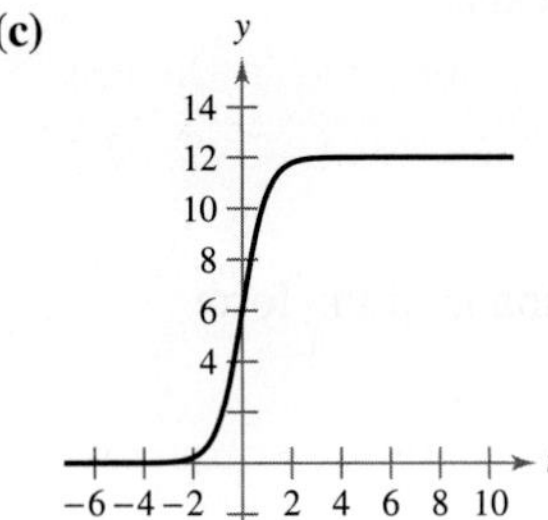

(d)

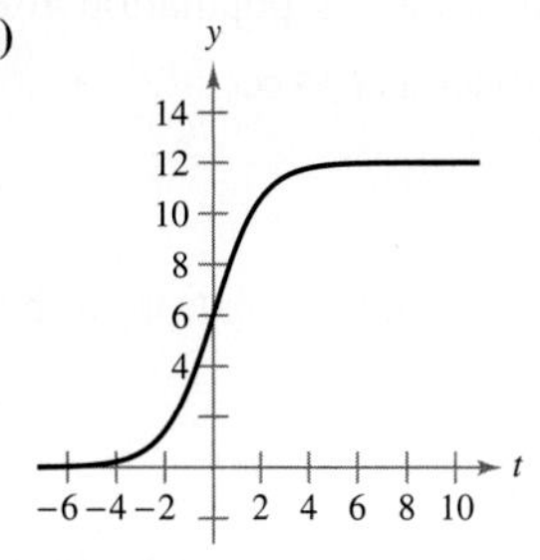

1. $y = \dfrac{12}{1 + e^{-t}}$

2. $y = \dfrac{12}{1 + 3e^{-t}}$

3. $y = \dfrac{12}{1 + \frac{1}{2}e^{-t}}$

4. $y = \dfrac{12}{1 + e^{-2t}}$

In Exercises 5–8, verify that the equation satisfies the logistic differential equation

$$\frac{dy}{dt} = ky\left(1 - \frac{y}{L}\right).$$

Then find the initial condition.

5. $y = \dfrac{4}{1 + e^{-2t}}$

6. $y = \dfrac{5}{1 + 3e^{-4t}}$

7. $y = \dfrac{12}{1 + 6e^{-t}}$

8. $y = \dfrac{14}{1 + 5e^{-3t}}$

In Exercises 9–12, the logistic equation models the growth of a population. Use the equation to (a) find the value of k, (b) find the carrying capacity, (c) find the initial population, (d) determine when the population will reach 50% of its carrying capacity, and (e) write a logistic differential equation that has the solution $P(t)$.

9. $P(t) = \dfrac{1500}{1 + 24e^{-0.75t}}$

10. $P(t) = \dfrac{5000}{1 + 39e^{-0.2t}}$

11. $P(t) = \dfrac{6000}{1 + 4999e^{-0.8t}}$

12. $P(t) = \dfrac{1000}{1 + 8e^{-0.2t}}$

In Exercises 13–16, the logistic differential equation models the growth rate of a population. Use the equation to (a) find the value of k, (b) find the carrying capacity, (c) use a computer algebra system to graph a slope field, and (d) determine the value of P at which the population growth rate is the greatest.

13. $\dfrac{dP}{dt} = 3P\left(1 - \dfrac{P}{100}\right)$

14. $\dfrac{dP}{dt} = 0.5P\left(1 - \dfrac{P}{250}\right)$

15. $\dfrac{dP}{dt} = 0.1P - 0.0004P^2$

16. $\dfrac{dP}{dt} = 0.4P - 0.00025P^2$

In Exercises 17–20, find the logistic equation that satisfies the initial condition. Then use the logistic equation to find y when $t = 5$ and $t = 100$.

	Logistic Differential Equation	*Initial Condition*
17.	$\dfrac{dy}{dt} = y\left(1 - \dfrac{y}{40}\right)$	$(0, 8)$
18.	$\dfrac{dy}{dt} = 1.2y\left(1 - \dfrac{y}{8}\right)$	$(0, 5)$
19.	$\dfrac{dy}{dt} = \dfrac{4y}{5} - \dfrac{y^2}{150}$	$(0, 8)$
20.	$\dfrac{dy}{dt} = \dfrac{3y}{20} - \dfrac{y^2}{1600}$	$(0, 15)$

In Exercises 21–24, match the logistic differential equation and initial condition with the graph of its solution. [The graphs are labeled (a), (b), (c), and (d).]

(a)

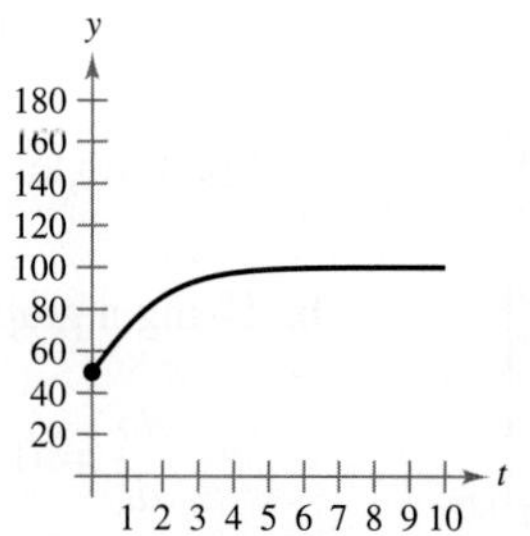

(b)

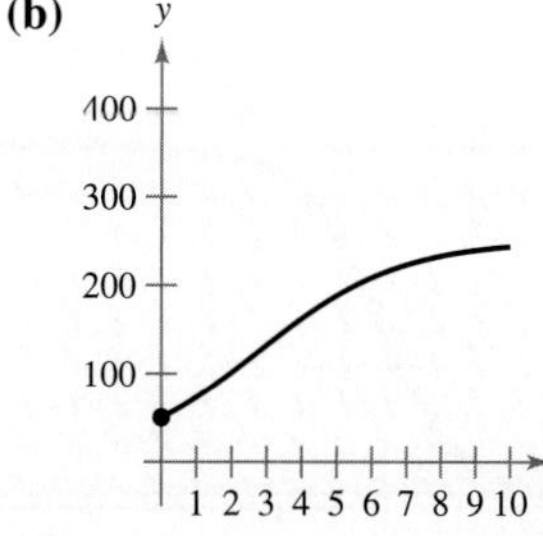

(c)

(d)

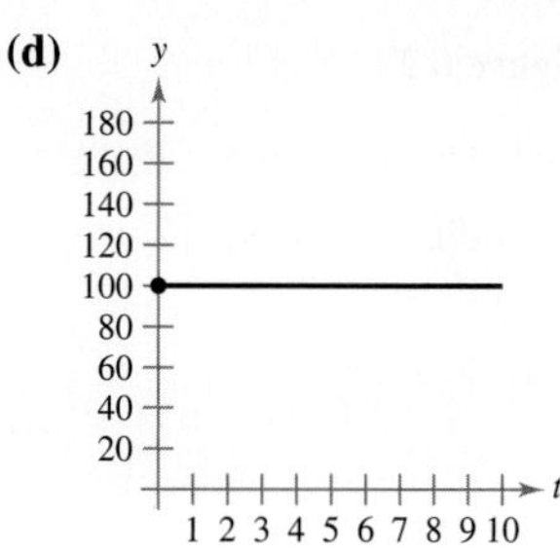

21. $\dfrac{dy}{dt} = 0.5y\left(1 - \dfrac{y}{250}\right)$, $(0, 350)$

22. $\dfrac{dy}{dt} = 0.9y\left(1 - \dfrac{y}{100}\right)$, $(0, 100)$

23. $\dfrac{dy}{dt} = 0.5y\left(1 - \dfrac{y}{250}\right)$, $(0, 50)$

24. $\dfrac{dy}{dt} = 0.9y\left(1 - \dfrac{y}{100}\right)$, $(0, 50)$

Slope Fields **In Exercises 25–28, a logistic differential equation, a point, and a slope field are given. (a) Sketch two approximate solutions of the differential equation on the slope field, one of which passes through the given point. (b) Find the particular solution of the differential equation and use a graphing utility to graph the solution. Compare the result with the sketch in part (a). To print an enlarged copy of the graph, go to the website *www.mathgraphs.com*.**

25. $\dfrac{dy}{dt} = 0.2y\left(1 - \dfrac{y}{1000}\right)$,
$(0, 105)$

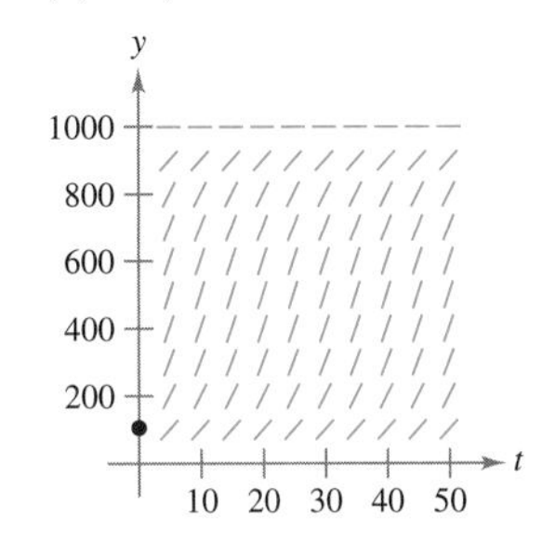

26. $\dfrac{dy}{dt} = 0.9y\left(1 - \dfrac{y}{200}\right)$,
$(0, 240)$

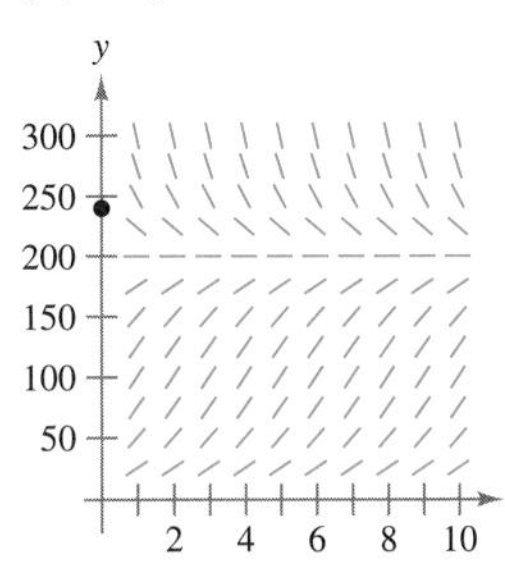

27. $\dfrac{dy}{dt} = 0.6y\left(1 - \dfrac{y}{700}\right)$,
$(0, 1000)$

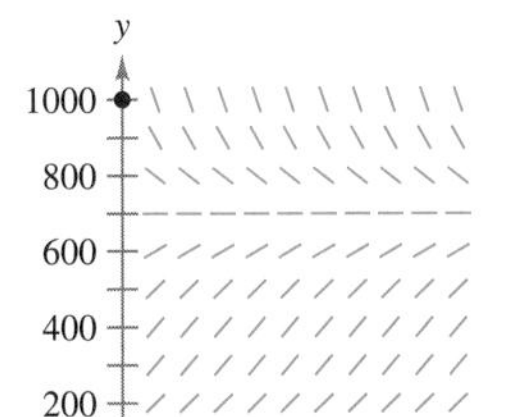

28. $\dfrac{dy}{dt} = 0.4y\left(1 - \dfrac{y}{500}\right)$,
$(0, 375)$

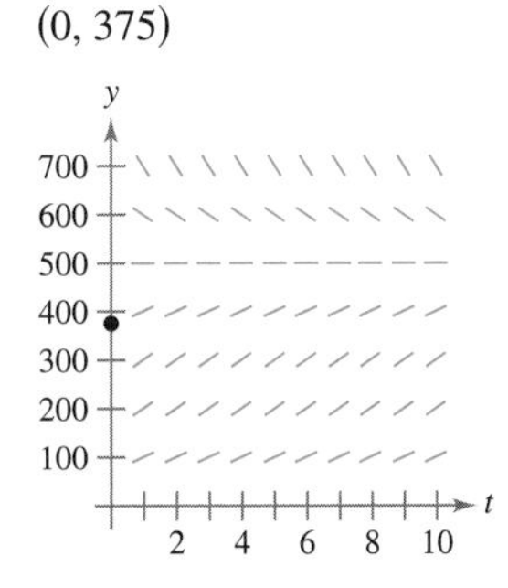

Writing About Concepts

29. Describe what the value of L represents in the logistic differential equation $\dfrac{dy}{dt} = ky\left(1 - \dfrac{y}{L}\right)$.

30. It is known that $y = \dfrac{L}{1 + be^{-kt}}$ is a solution of the logistic differential equation $\dfrac{dy}{dt} = 0.75y\left(1 - \dfrac{y}{2500}\right)$. Is it possible to determine L, k, and b from the information given? If so, find their values. If not, which value(s) cannot be determined and what information do you need to determine the value(s)?

31. ***Endangered Species*** A conservation organization releases 25 Florida panthers into a game preserve. After 2 years, there are 39 panthers in the preserve. The Florida preserve has a carrying capacity of 200 panthers.

(a) Write a logistic equation that models the population of panthers in the preserve.

(b) Find the population after 5 years.

(c) When will the population reach 100?

(d) Write a logistic differential equation that models the growth rate of the panther population. Then repeat part (b) using Euler's Method with a step size of $h = 1$. Compare the approximation with the exact answer.

(e) At what time is the panther population growing most rapidly? Explain.

32. Repeat Exercise 31, assuming that the organization releases 27 panthers into the game preserve and after 2 years there are 43 panthers.

33. ***Bacteria Growth*** At time $t = 0$, a bacterial culture weighs 1 gram. Two hours later, the culture weighs 2 grams. The maximum weight of the culture is 10 grams.

(a) Write a logistic equation that models the weight of the bacterial culture.

(b) Find the culture's weight after 5 hours.

(c) When will the culture's weight reach 8 grams?

(d) Write a logistic differential equation that models the growth rate of the culture's weight. Then repeat part (b) using Euler's Method with a step size of $h = 1$. Compare the approximation with the exact answer.

(e) At what time is the culture's weight increasing most rapidly? Explain.

34. Repeat Exercise 33 for another bacterial culture that weighs 1 gram at $t = 0$ and 1.2 grams after 10 hours. The maximum weight of the culture is 1.25 grams.

True or False? **In Exercises 35 and 36, determine whether the statement is true or false. If it is false, explain why or give an example that shows it is false.**

35. For the logistic differential equation $\dfrac{dy}{dt} = ky\left(1 - \dfrac{y}{L}\right)$, if $y > L$, then $dy/dt > 0$ and the population increases.

36. For the logistic differential equation $\dfrac{dy}{dt} = ky\left(1 - \dfrac{y}{L}\right)$, if $0 < y < L$, then $dy/dt > 0$ and the population increases.

37. For any logistic growth curve, show that the point of inflection occurs at $y = \dfrac{L}{2}$ when the solution starts below the carrying capacity L.

38. Show that if $y = \dfrac{1}{1 + be^{-kt}}$, then $\dfrac{dy}{dt} = ky(1 - y)$.

Section 6.5 First-Order Linear Differential Equations

- **Solve a first-order linear differential equation.**
- **Solve a Bernoulli differential equation.**
- **Use linear differential equations to solve applied problems.**

First-Order Linear Differential Equations

In this section, you will see how to solve a very important class of first-order differential equations—first-order linear differential equations.

Definition of First-Order Linear Differential Equation

A first-order linear differential equation is an equation of the form

$$\frac{dy}{dx} + P(x)y = Q(x)$$

where P and Q are continuous functions of x. This first-order linear differential equation is said to be in **standard form.**

NOTE It is instructive to see why the integrating factor helps solve a linear differential equation of the form $y' + P(x)y = Q(x)$. When both sides of the equation are multiplied by the integrating factor $u(x) = e^{\int P(x)\,dx}$, the left-hand side becomes the derivative of a product.

$$y'e^{\int P(x)\,dx} + P(x)ye^{\int P(x)\,dx} = Q(x)e^{\int P(x)\,dx}$$
$$\left[ye^{\int P(x)\,dx}\right]' = Q(x)e^{\int P(x)\,dx}$$

Integrating both sides of this second equation and dividing by $u(x)$ produces the general solution.

To solve a linear differential equation, write it in standard form to identify the functions $P(x)$ and $Q(x)$. Then integrate $P(x)$ and form the expression

$$u(x) = e^{\int P(x)\,dx} \qquad \text{Integrating factor}$$

which is called an **integrating factor.** The general solution of the equation is

$$y = \frac{1}{u(x)} \int Q(x)u(x)\,dx. \qquad \text{General solution}$$

EXAMPLE 1 Solving a Linear Differential Equation

Find the general solution of

$$y' + y = e^x.$$

Solution

For this equation, $P(x) = 1$ and $Q(x) = e^x$. So, the integrating factor is

$$\begin{aligned} u(x) &= e^{\int P(x)\,dx} && \text{Integrating factor} \\ &= e^{\int dx} \\ &= e^x. \end{aligned}$$

This implies that the general solution is

$$\begin{aligned} y &= \frac{1}{u(x)} \int Q(x)u(x)\,dx \\ &= \frac{1}{e^x} \int e^x(e^x)\,dx \\ &= e^{-x}\left(\frac{1}{2}e^{2x} + C\right) \\ &= \frac{1}{2}e^x + Ce^{-x}. && \text{General solution} \end{aligned}$$

THEOREM 6.3 Solution of a First-Order Linear Differential Equation

An integrating factor for the first-order linear differential equation

$$y' + P(x)y = Q(x)$$

is $u(x) = e^{\int P(x)\,dx}$. The solution of the differential equation is

$$ye^{\int P(x)\,dx} = \int Q(x)e^{\int P(x)\,dx}\,dx + C.$$

ANNA JOHNSON PELL WHEELER (1883–1966)

Anna Johnson Pell Wheeler was awarded a master's degree from the University of Iowa for her thesis *The Extension of Galois Theory to Linear Differential Equations* in 1904. Influenced by David Hilbert, she worked on integral equations while studying infinite linear spaces.

STUDY TIP Rather than memorizing the formula in Theorem 6.3, just remember that multiplication by the integrating factor $e^{\int P(x)\,dx}$ converts the left side of the differential equation into the derivative of the product $ye^{\int P(x)\,dx}$.

EXAMPLE 2 Solving a First-Order Linear Differential Equation

Find the general solution of

$$xy' - 2y = x^2.$$

Solution The standard form of the given equation is

$$y' + P(x)y = Q(x)$$

$$y' - \left(\frac{2}{x}\right)y = x. \qquad \text{Standard form}$$

So, $P(x) = -2/x$, and you have

$$\int P(x)\,dx = -\int \frac{2}{x}\,dx$$

$$= -\ln x^2$$

$$e^{\int P(x)\,dx} = e^{-\ln x^2}$$

$$= \frac{1}{e^{\ln x^2}}$$

$$= \frac{1}{x^2}. \qquad \text{Integrating factor}$$

So, multiplying each side of the standard form by $1/x^2$ yields

$$\frac{y'}{x^2} - \frac{2y}{x^3} = \frac{1}{x}$$

$$\frac{d}{dx}\left[\frac{y}{x^2}\right] = \frac{1}{x}$$

$$\frac{y}{x^2} = \int \frac{1}{x}\,dx$$

$$\frac{y}{x^2} = \ln|x| + C$$

$$y = x^2(\ln|x| + C). \qquad \text{General solution}$$

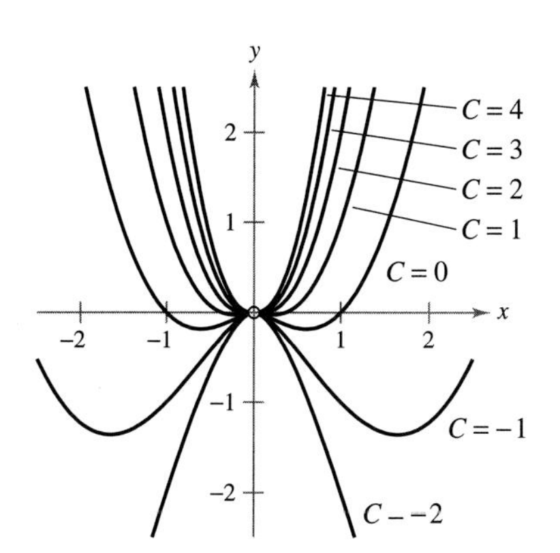

Figure 6.29

Several solution curves (for $C = -2, -1, 0, 1, 2, 3,$ and 4) are shown in Figure 6.29.

EXAMPLE 3 Solving a First-Order Linear Differential Equation

Find the general solution of

$$y' - y \tan t = 1, \quad -\frac{\pi}{2} < t < \frac{\pi}{2}.$$

Solution The equation is already in the standard form $y' + P(t)y = Q(t)$. So, $P(t) = -\tan t$, and

$$\int P(t)\,dt = -\int \tan t\,dt = \ln|\cos t|$$

which implies that the integrating factor is

$$\begin{aligned} e^{\int P(t)\,dt} &= e^{\ln|\cos t|} \\ &= |\cos t|. \end{aligned} \qquad \text{Integrating factor}$$

A quick check shows that $\cos t$ is also an integrating factor. So, multiplying $y' - y \tan t = 1$ by $\cos t$ produces

$$\frac{d}{dt}[y \cos t] = \cos t$$

$$y \cos t = \int \cos t\,dt$$

$$y \cos t = \sin t + C$$

$$y = \tan t + C \sec t. \qquad \text{General solution}$$

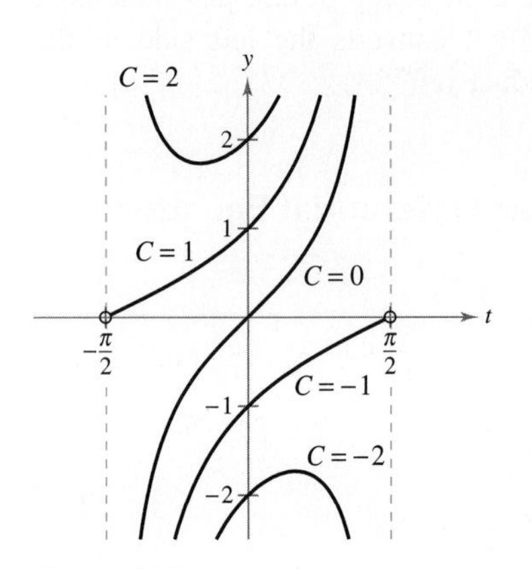

Figure 6.30

Several solution curves are shown in Figure 6.30.

Bernoulli Equation

A well-known nonlinear equation that reduces to a linear one with an appropriate substitution is the **Bernoulli equation,** named after James Bernoulli (1654–1705).

$$y' + P(x)y = Q(x)y^n \qquad \text{Bernoulli equation}$$

This equation is linear if $n = 0$, and has separable variables if $n = 1$. So, in the following development, assume that $n \neq 0$ and $n \neq 1$. Begin by multiplying by y^{-n} and $(1 - n)$ to obtain

$$y^{-n}y' + P(x)y^{1-n} = Q(x)$$

$$(1 - n)y^{-n}y' + (1 - n)P(x)y^{1-n} = (1 - n)Q(x)$$

$$\frac{d}{dx}[y^{1-n}] + (1 - n)P(x)y^{1-n} = (1 - n)Q(x)$$

which is a linear equation in the variable y^{1-n}. Letting $z = y^{1-n}$ produces the linear equation

$$\frac{dz}{dx} + (1 - n)P(x)z = (1 - n)Q(x).$$

Finally, by Theorem 6.3, the *general solution of the Bernoulli equation* is

$$y^{1-n}e^{\int(1-n)P(x)\,dx} = \int (1 - n)Q(x)e^{\int(1-n)P(x)\,dx}\,dx + C.$$

EXAMPLE 4 Solving a Bernoulli Equation

Find the general solution of

$$y' + xy = xe^{-x^2}y^{-3}.$$

Solution For this Bernoulli equation, let $n = -3$, and use the substitution

$$z = y^4 \qquad \text{Let } z = y^{1-n} = y^{1-(-3)}.$$

$$z' = 4y^3y'. \qquad \text{Differentiate.}$$

Multiplying the original equation by $4y^3$ produces

$$y' + xy = xe^{-x^2}y^{-3} \qquad \text{Write original equation.}$$

$$4y^3y' + 4xy^4 = 4xe^{-x^2} \qquad \text{Multiply each side by } 4y^3.$$

$$z' + 4xz = 4xe^{-x^2}. \qquad \text{Linear equation: } z' + P(x)z = Q(x)$$

This equation is linear in z. Using $P(x) = 4x$ produces

$$\int P(x)\,dx = \int 4x\,dx$$

$$= 2x^2$$

which implies that e^{2x^2} is an integrating factor. Multiplying the linear equation by this factor produces

$$z' + 4xz = 4xe^{-x^2} \qquad \text{Linear equation}$$

$$z'e^{2x^2} + 4xze^{2x^2} = 4xe^{x^2} \qquad \text{Multiply by integrating factor.}$$

$$\frac{d}{dx}[ze^{2x^2}] = 4xe^{x^2} \qquad \text{Write left side as derivative.}$$

$$ze^{2x^2} = \int 4xe^{x^2}\,dx \qquad \text{Integrate each side.}$$

$$ze^{2x^2} = 2e^{x^2} + C$$

$$z = 2e^{-x^2} + Ce^{-2x^2}. \qquad \text{Divide each side by } e^{2x^2}.$$

Finally, substituting $z = y^4$, the general solution is

$$y^4 = 2e^{-x^2} + Ce^{-2x^2}. \qquad \text{General solution}$$

So far you have studied several types of first-order differential equations. Of these, the separable variables case is usually the simplest, and solution by an integrating factor is ordinarily used only as a last resort.

Summary of First-Order Differential Equations

Method	*Form of Equation*
1. Separable variables:	$M(x)\,dx + N(y)\,dy = 0$
2. Homogeneous:	$M(x, y)\,dx + N(x, y)\,dy = 0$, where M and N are nth-degree homogeneous
3. Linear:	$y' + P(x)y = Q(x)$
4. Bernoulli equation:	$y' + P(x)y = Q(x)y^n$

Applications

One type of problem that can be described in terms of a differential equation involves chemical mixtures, as illustrated in the next example.

EXAMPLE 5 A Mixture Problem

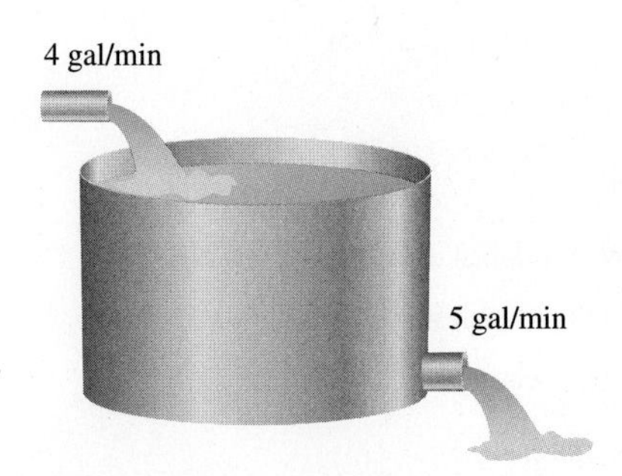

Figure 6.31

A tank contains 50 gallons of a solution composed of 90% water and 10% alcohol. A second solution containing 50% water and 50% alcohol is added to the tank at the rate of 4 gallons per minute. As the second solution is being added, the tank is being drained at a rate of 5 gallons per minute, as shown in Figure 6.31. Assuming the solution in the tank is stirred constantly, how much alcohol is in the tank after 10 minutes?

Solution Let y be the number of gallons of alcohol in the tank at any time t. You know that $y = 5$ when $t = 0$. Because the number of gallons of solution in the tank at any time is $50 - t$, and the tank loses 5 gallons of solution per minute, it must lose

$$\left(\frac{5}{50 - t}\right)y$$

gallons of alcohol per minute. Furthermore, because the tank is gaining 2 gallons of alcohol per minute, the rate of change of alcohol in the tank is given by

$$\frac{dy}{dt} = 2 - \left(\frac{5}{50 - t}\right)y \quad \Longrightarrow \quad \frac{dy}{dt} + \left(\frac{5}{50 - t}\right)y = 2.$$

To solve this linear equation, let $P(t) = 5/(50 - t)$ and obtain

$$\int P(t)\, dt = \int \frac{5}{50 - t}\, dt = -5 \ln|50 - t|.$$

Because $t < 50$, you can drop the absolute value signs and conclude that

$$e^{\int P(t)\, dt} = e^{-5 \ln(50 - t)} = \frac{1}{(50 - t)^5}.$$

So, the general solution is

$$\frac{y}{(50 - t)^5} = \int \frac{2}{(50 - t)^5}\, dt = \frac{1}{2(50 - t)^4} + C$$

$$y = \frac{50 - t}{2} + C(50 - t)^5.$$

Because $y = 5$ when $t = 0$, you have

$$5 = \frac{50}{2} + C(50)^5 \quad \Longrightarrow \quad -\frac{20}{50^5} = C$$

which means that the particular solution is

$$y = \frac{50 - t}{2} - 20\left(\frac{50 - t}{50}\right)^5.$$

Finally, when $t = 10$, the amount of alcohol in the tank is

$$y = \frac{50 - 10}{2} - 20\left(\frac{50 - 10}{50}\right)^5 \approx 13.45 \text{ gal}$$

which represents a solution containing 33.6% alcohol.

In most falling-body problems discussed so far in the text, air resistance has been neglected. The next example includes this factor. In the example, the air resistance on the falling object is assumed to be proportional to its velocity v. If g is the gravitational constant, the downward force F on a falling object of mass m is given by the difference $mg - kv$. But by Newton's Second Law of Motion, you know that

$$\begin{aligned} F &= ma \\ &= m\frac{dv}{dt} \end{aligned}$$

which yields the following differential equation.

$$m\frac{dv}{dt} = mg - kv \quad \Longrightarrow \quad \frac{dv}{dt} + \frac{k}{m}v = g$$

EXAMPLE 6 A Falling Object with Air Resistance

An object of mass m is dropped from a hovering helicopter. Find its velocity as a function of time t, assuming that the air resistance is proportional to the velocity of the object.

Solution The velocity v satisfies the equation

$$\frac{dv}{dt} + \frac{kv}{m} = g$$

where g is the gravitational constant and k is the constant of proportionality. Letting $b = k/m$, you can *separate variables* to obtain

$$\begin{aligned} dv &= (g - bv)\,dt \\ \int \frac{dv}{g - bv} &= \int dt \\ -\frac{1}{b}\ln|g - bv| &= t + C_1 \\ \ln|g - bv| &= -bt - bC_1 \\ g - bv &= Ce^{-bt}. \end{aligned}$$

Because the object was dropped, $v = 0$ when $t = 0$; so $g = C$, and it follows that

$$-bv = -g + ge^{-bt} \quad \Longrightarrow \quad v = \frac{g - ge^{-bt}}{b} = \frac{mg}{k}(1 - e^{-kt/m}).$$

NOTE Notice in Example 6 that the velocity approaches a limit of mg/k as a result of the air resistance. For falling-body problems in which air resistance is neglected, the velocity increases without bound.

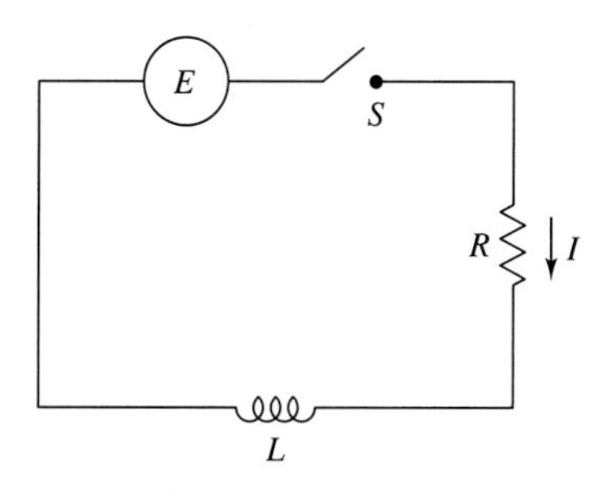

Figure 6.32

A simple electric circuit consists of electric current I (in amperes), a resistance R (in ohms), an inductance L (in henrys), and a constant electromotive force E (in volts), as shown in Figure 6.32. According to Kirchhoff's Second Law, if the switch S is closed when $t = 0$, the applied electromotive force (voltage) is equal to the sum of the voltage drops in the rest of the circuit. This in turn means that the current I satisfies the differential equation

$$L\frac{dI}{dt} + RI = E.$$

EXAMPLE 7 An Electric Circuit Problem

Find the current I as a function of time t (in seconds), given that I satisfies the differential equation $L(dI/dt) + RI = \sin 2t$, where R and L are nonzero constants.

TECHNOLOGY The integral in Example 7 was found using a computer algebra system. If you have access to *Derive*, *Maple*, *Mathcad*, *Mathematica*, or the *TI-89*, try using it to integrate

$$\frac{1}{L}\int e^{(R/L)t} \sin 2t\, dt.$$

In Chapter 8 you will learn how to integrate functions of this type using integration by parts.

Solution In standard form, the given linear equation is

$$\frac{dI}{dt} + \frac{R}{L}I = \frac{1}{L}\sin 2t.$$

Let $P(t) = R/L$, so that $e^{\int P(t)\,dt} = e^{(R/L)t}$, and, by Theorem 6.3,

$$Ie^{(R/L)t} = \frac{1}{L}\int e^{(R/L)t} \sin 2t\, dt$$

$$= \frac{1}{4L^2 + R^2} e^{(R/L)t}(R \sin 2t - 2L \cos 2t) + C.$$

So the general solution is

$$I = e^{-(R/L)t}\left[\frac{1}{4L^2 + R^2} e^{(R/L)t}(R \sin 2t - 2L \cos 2t) + C\right]$$

$$I = \frac{1}{4L^2 + R^2}(R \sin 2t - 2L \cos 2t) + Ce^{-(R/L)t}.$$

Exercises for Section 6.5

See www.CalcChat.com for worked-out solutions to odd-numbered exercises.

In Exercises 1–4, determine whether the differential equation is linear. Explain your reasoning.

1. $x^3y' + xy = e^x + 1$
2. $2xy - y' \ln x = y$
3. $y' + y \cos x = xy^2$
4. $\dfrac{1 - y'}{y} = 3x$

In Exercises 5–14, solve the first-order linear differential equation.

5. $\dfrac{dy}{dx} + \left(\dfrac{1}{x}\right)y = 3x + 4$
6. $\dfrac{dy}{dx} + \left(\dfrac{2}{x}\right)y = 3x + 2$
7. $y' - y = 10$
8. $y' + 2xy = 4x$
9. $(y + 1) \cos x\, dx - dy = 0$
10. $(y - 1) \sin x\, dx - dy = 0$
11. $(x - 1)y' + y = x^2 - 1$
12. $y' + 3y = e^{3x}$
13. $y' - 3x^2y = e^{x^3}$
14. $y' - y = \cos x$

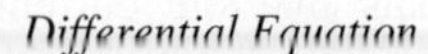

Slope Fields **In Exercises 15 and 16, (a) sketch an approximate solution of the differential equation satisfying the initial condition by hand on the slope field, (b) find the particular solution that satisfies the initial condition, and (c) use a graphing utility to graph the particular solution. Compare the graph with the hand-drawn graph of part (a). To print an enlarged copy of the graph, go to the website *www.mathgraphs.com*.**

Differential Equation	*Initial Condition*
15. $\dfrac{dy}{dx} = e^x - y$	$(0, 1)$
16. $y' + \left(\dfrac{1}{x}\right)y = \sin x^2$	$(\sqrt{\pi}, 0)$

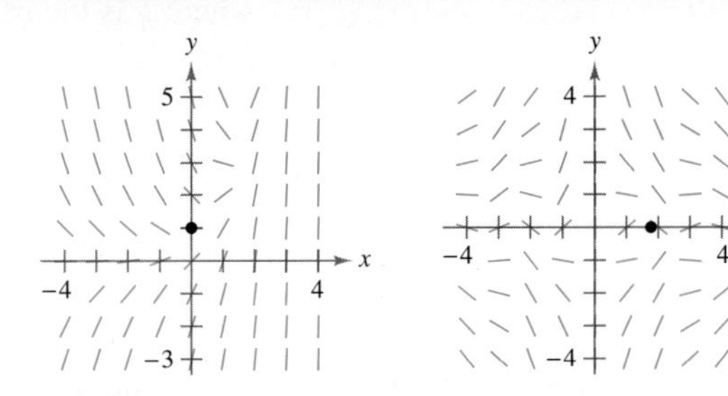

Figure for 15 **Figure for 16**

In Exercises 17–24, find the particular solution of the differential equation that satisfies the boundary condition.

Differential Equation	*Boundary Condition*
17. $y' \cos^2 x + y - 1 = 0$	$y(0) = 5$
18. $x^3y' + 2y = e^{1/x^2}$	$y(1) = e$
19. $y' + y \tan x = \sec x + \cos x$	$y(0) = 1$
20. $y' + y \sec x = \sec x$	$y(0) = 4$
21. $y' + \left(\dfrac{1}{x}\right)y = 0$	$y(2) = 2$
22. $y' + (2x - 1)y = 0$	$y(1) = 2$
23. $x\, dy = (x + y + 2)\, dx$	$y(1) = 10$
24. $2xy' - y = x^3 - x$	$y(4) = 2$

In Exercises 25–30, solve the Bernoulli differential equation.

25. $y' + 3x^2y = x^2y^3$
26. $y' + xy = xy^{-1}$

27. $y' + \left(\frac{1}{x}\right)y = xy^2$

28. $y' + \left(\frac{1}{x}\right)y = x\sqrt{y}$

29. $y' - y = e^x\sqrt[3]{y}$

30. $yy' - 2y^2 = e^x$

Slope Fields **In Exercises 31–34, (a) use a graphing utility to graph the slope field for the differential equation, (b) find the particular solutions of the differential equation passing through the given points, and (c) use a graphing utility to graph the particular solutions on the slope field.**

	Differential Equation	*Points*
31.	$\frac{dy}{dx} - \frac{1}{x}y = x^2$	$(-2, 4),\ (2, 8)$
32.	$\frac{dy}{dx} + 4x^3y = x^3$	$\left(0, \frac{7}{2}\right),\ \left(0, -\frac{1}{2}\right)$
33.	$\frac{dy}{dx} + (\cot x)y = 2$	$(1, 1),\ (3, -1)$
34.	$\frac{dy}{dx} + 2xy = xy^2$	$(0, 3),\ (0, 1)$

35. ***Intravenous Feeding*** Glucose is added intravenously to the bloodstream at the rate of q units per minute, and the body removes glucose from the bloodstream at a rate proportional to the amount present. Assume that $Q(t)$ is the amount of glucose in the bloodstream at time t.

(a) Determine the differential equation describing the rate of change of glucose in the bloodstream with respect to time.

(b) Solve the differential equation from part (a), letting $Q = Q_0$ when $t = 0$.

(c) Find the limit of $Q(t)$ as $t \to \infty$.

36. ***Learning Curve*** The management at a certain factory has found that the maximum number of units a worker can produce in a day is 40. The rate of increase in the number of units N produced with respect to time t in days by a new employee is proportional to $40 - N$.

(a) Determine the differential equation describing the rate of change of performance with respect to time.

(b) Solve the differential equation from part (a).

(c) Find the particular solution for a new employee who produced 10 units on the first day at the factory and 19 units on the twentieth day.

Mixture **In Exercises 37–42, consider a tank that at time $t = 0$ contains v_0 gallons of a solution that, by weight, contains q_0 pounds of soluble concentrate. Another solution containing q_1 pounds of the concentrate per gallon is running into the tank at the rate of r_1 gallons per minute. The solution in the tank is kept well stirred and is withdrawn at the rate of r_2 gallons per minute.**

37. If Q is the amount of concentrate in the solution at any time t, show that

$$\frac{dQ}{dt} + \frac{r_2Q}{v_0 + (r_1 - r_2)t} = q_1r_1.$$

38. If Q is the amount of concentrate in the solution at any time t, write the differential equation for the rate of change of Q with respect to t if $r_1 = r_2 = r$.

39. A 200-gallon tank is full of a solution containing 25 pounds of concentrate. Starting at time $t = 0$, distilled water is admitted to the tank at a rate of 10 gallons per minute, and the well-stirred solution is withdrawn at the same rate as shown in the figure.

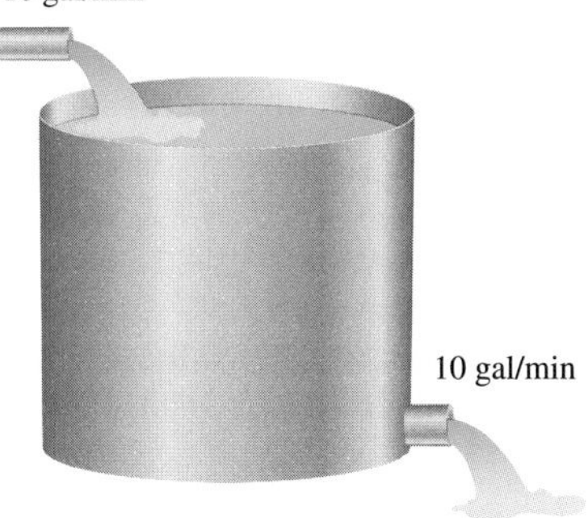

(a) Find the amount of concentrate Q in the solution as a function of t.

(b) Find the time at which the amount of concentrate in the tank reaches 15 pounds.

(c) Find the quantity of the concentrate in the solution as $t \to \infty$.

40. Repeat Exercise 39, assuming that the solution entering the tank contains 0.04 pound of concentrate per gallon.

41. A 200-gallon tank is half full of distilled water. At time $t = 0$, a solution containing 0.5 pound of concentrate per gallon enters the tank at the rate of 5 gallons per minute, and the well-stirred mixture is withdrawn at the rate of 3 gallons per minute as shown in the figure.

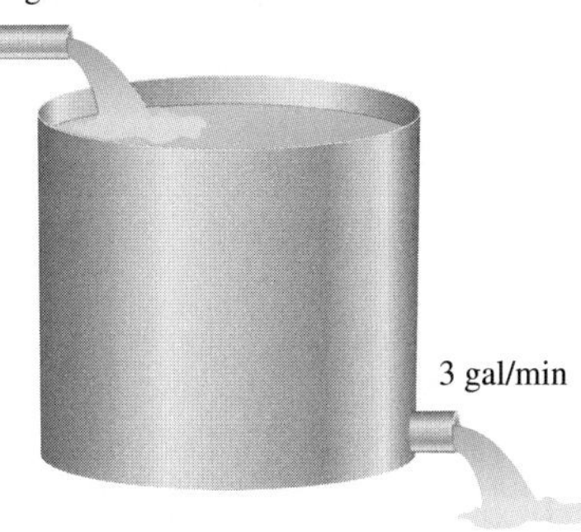

(a) At what time will the tank be full?

(b) At the time the tank is full, how many pounds of concentrate will it contain?

42. Repeat Exercise 41, assuming that the solution entering the tank contains 1 pound of concentrate per gallon.

Falling Object **In Exercises 43 and 44, consider an eight-pound object dropped from a height of 5000 feet, where the air resistance is proportional to the velocity.**

43. Write the velocity as a function of time if the object's velocity after 5 seconds is approximately -101 feet per second. What is the limiting value of the velocity function?

44. Use the result of Exercise 43 to write the position of the object as a function of time. Approximate the velocity of the object when it reaches ground level.

Electric Circuits **In Exercises 45 and 46, use the differential equation for electric circuits given by**

$$L\frac{dI}{dt} + RI = E.$$

In this equation, I is the current, R is the resistance, L is the inductance, and E is the electromotive force (voltage).

45. Solve the differential equation given a constant voltage E_0.

46. Use the result of Exercise 45 to find the equation for the current if $I(0) = 0$, $E_0 = 120$ volts, $R = 600$ ohms, and $L = 4$ henrys. When does the current reach 90% of its limiting value?

Writing About Concepts

47. Give the standard form of a first-order linear differential equation. What is its integrating factor?

48. Give the standard form of the Bernoulli equation. Describe how one reduces it to a linear equation.

In Exercises 49–52, match the differential equation with its solution.

Differential Equation	*Solution*
49. $y' - 2x = 0$	(a) $y = Ce^{x^2}$
50. $y' - 2y = 0$	(b) $y = -\frac{1}{2} + Ce^{x^2}$
51. $y' - 2xy = 0$	(c) $y = x^2 + C$
52. $y' - 2xy = x$	(d) $y = Ce^{2x}$

In Exercises 53–64, solve the first-order differential equation by any appropriate method.

53. $\dfrac{dy}{dx} = \dfrac{e^{2x+y}}{e^{x-y}}$

54. $\dfrac{dy}{dx} = \dfrac{x+1}{y(y+2)}$

55. $y \cos x - \cos x + \dfrac{dy}{dx} = 0$

56. $y' = 2x\sqrt{1 - y^2}$

57. $(3y^2 + 4xy)\,dx + (2xy + x^2)\,dy = 0$

58. $(x + y)\,dx - x\,dy = 0$

59. $(2y - e^x)\,dx + x\,dy = 0$

60. $(y^2 + xy)\,dx - x^2\,dy = 0$

61. $(x^2y^4 - 1)\,dx + x^3y^3\,dy = 0$

62. $y\,dx + (3x + 4y)\,dy = 0$

63. $3(y - 4x^2)\,dx + x\,dy = 0$

64. $x\,dx + (y + e^y)(x^2 + 1)\,dy = 0$

True or False? **In Exercises 65 and 66, determine whether the statement is true or false. If it is false, explain why or give an example that shows it is false.**

65. $y' + x\sqrt{y} = x^2$ is a first-order linear differential equation.

66. $y' + xy = e^x y$ is a first-order linear differential equation.

Section Project: Weight Loss

A person's weight depends on both the number of calories consumed and the energy used. Moreover, the amount of energy used depends on a person's weight—the average amount of energy used by a person is 17.5 calories per pound per day. So, the more weight a person loses, the less energy a person uses (assuming that the person maintains a constant level of activity). An equation that can be used to model weight loss is

$$\frac{dw}{dt} = \frac{C}{3500} - \frac{17.5}{3500}w$$

where w is the person's weight (in pounds), t is the time in days, and C is the constant daily calorie consumption.

(a) Find the general solution of the differential equation.

(b) Consider a person who weighs 180 pounds and begins a diet of 2500 calories per day. How long will it take the person to lose 10 pounds? How long will it take the person to lose 35 pounds?

(c) Use a graphing utility to graph the solution. What is the "limiting" weight of the person?

(d) Repeat parts (b) and (c) for a person who weighs 200 pounds when the diet is started.

FOR FURTHER INFORMATION For more information on modeling weight loss, see the article "A Linear Diet Model" by Arthur C. Segal in *The College Mathematics Journal*.

Section 6.6 Predator-Prey Differential Equations

- **Analyze predator-prey differential equations.**
- **Analyze competing-species differential equations.**

Predator–Prey Differential Equations

Although Alfred Lotka (1880-1949) and Vito Volterra (1860-1940) both worked on other problems, they are most known for their work on predator-prey equations. Lotka was also a statistician, and Volterra did work in the development of integral equations and functional analysis.

In the 1920s, mathematicians Alfred Lotka (1880–1949) and Vito Volterra (1860–1940) independently developed mathematical models to represent many of the different ways that two species can interact with each other. Two common ways that species interact with each other are as predator and prey, and as competing species.

Consider a predator-prey relationship involving foxes (predators) and rabbits (prey). Assume that the rabbits are the primary food source for the foxes, the rabbits have an unlimited food supply, and there is no threat to the rabbits other than from the foxes. Let x represent the number of rabbits, let y represent the number of foxes, and let t represent time. If there are no foxes, then the rabbit population will grow according to the exponential growth model $dx/dt = ax$, $a > 0$.

If there are foxes but no rabbits, the foxes have no food and their population will decay according to the exponential decay model $dy/dt = -my$, $m > 0$.

If both foxes and rabbits are present, there is an interaction rate of *decline* for the rabbit population given by $-bxy$, and an interaction rate of *increase* in the fox population given by nxy, where $b, n > 0$. So, the rates of change of each population can be modeled by the following predator-prey system of differential equations.

$$\frac{dx}{dt} = ax - bxy \qquad \text{Rate of change of prey}$$

$$\frac{dy}{dt} = -my + nxy \qquad \text{Rate of change of predators}$$

These equations are called **predator-prey equations** or **Lotka-Volterra equations.** The equations are **autonomous** because the rates of change do not depend explicitly on time t.

In general, it is not possible to solve the predator-prey equations explicitly for x and y. However, you can use techniques such as Euler's Method to approximate solutions. Also, you can discover properties of the solutions by analyzing the differential equations.

EXAMPLE 1 Analyzing Predator-Prey Equations

Write the predator-prey equations for $a = 0.04$, $b = 0.002$, $m = 0.08$, and $n = 0.0004$. Then find the values of x and y for which $dx/dt = dy/dt = 0$.

Solution For $a = 0.04$, $b = 0.002$, $m = 0.08$, and $n = 0.0004$, the predator-prey equations are:

$$\frac{dx}{dt} = 0.04x - 0.002xy \qquad \text{Rate of change of prey}$$

$$\frac{dy}{dt} = -0.08y + 0.0004xy \qquad \text{Rate of change of predators}$$

Solving $dx/dt = x(0.04 - 0.002y) = 0$ and $dy/dt = y(-0.08 + 0.0004x) = 0$, you can see that $dx/dt = dy/dt = 0$ when $(x, y) = (0, 0)$ and when $(x, y) = (200, 20)$.

In general, for the predator-prey equations

$$\frac{dx}{dt} = ax - bxy \qquad \text{and} \qquad \frac{dy}{dt} = -my + nxy$$

$\frac{dx}{dt} = 0$ when $x = 0$ or $y = \frac{a}{b}$ and $\frac{dy}{dt} = 0$ when $y = 0$ or $x = \frac{m}{n}$. So, at the points $(0, 0)$ and $\left(\frac{m}{n}, \frac{a}{b}\right)$, the prey and predator populations are constant. These points are called **critical points** or **equilibrium points** of the predator-prey equations.

EXAMPLE 2 Analyzing Predator-Prey Equations Graphically

Assume the predator-prey equations from Example 1

$$\frac{dx}{dt} = 0.04x - 0.002xy \qquad \text{and} \qquad \frac{dy}{dt} = -0.08y + 0.0004xy$$

model a predator-prey relationship involving foxes and rabbits where x is the number of rabbits and y is the number of foxes after t months. Use a graphing utility to graph the functions x and y when $0 \leq t \leq 240$ and the initial conditions are 200 rabbits and 10 foxes. What do you observe?

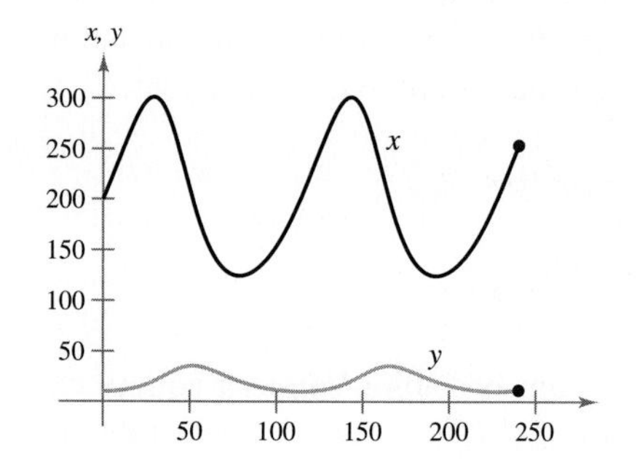

Figure 6.33

Solution The graphs of x and y are shown in Figure 6.33. Some observations are:

- The rabbit and fox populations oscillate periodically between their respective minimum and maximum values.
- The rabbit population oscillates from a minimum of about 125 rabbits to a maximum of about 300 rabbits.
- The fox population oscillates from about 10 foxes to about 35 foxes.
- About 20 months after the rabbit population peaks, the fox population peaks.
- The period of each population appears to be about 115 months.

You have used slope fields (or direction fields) to analyze solutions of differential equations. In Example 2, the graph shows the curves plotted together with time t along the horizontal axis. You can also use the predator-prey equations dy/dt and dx/dt to graph a slope field. The slope field is graphed using the x-axis to represent the prey and the y-axis to represent the predators.

NOTE If you are using a graphing utility, you may need to rewrite the equations as a function of x:

$$\frac{dy}{dx} = \frac{dy/dt}{dx/dt} = \frac{-my + nxy}{ax - bxy}.$$

EXAMPLE 3 Predator-Prey Equations and Slope Fields

Use a graphing utility to graph the slope field of the predator-prey equations given in Example 2.

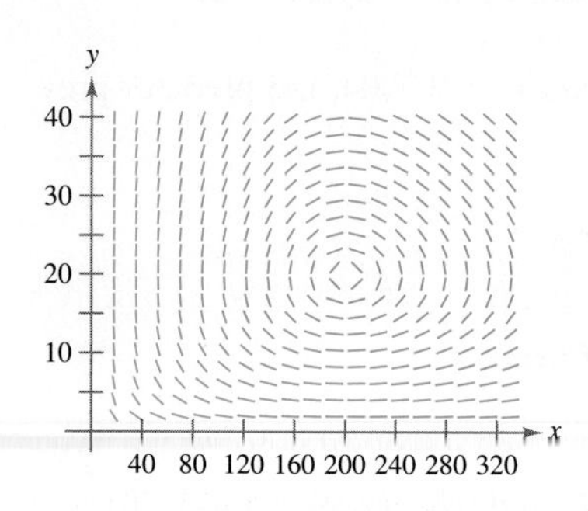

Figure 6.34

Solution The slope field is shown in Figure 6.34. The x-axis represents the rabbit population, and the y-axis represents the fox population.

EXAMPLE 4 Graphing a Solution Curve

Use the predator-prey equations

$$\frac{dx}{dt} = 0.04x - 0.002xy \qquad \text{and} \qquad \frac{dy}{dt} = -0.08y + 0.0004xy$$

and the slope field from Example 3 to graph the solution curve using the initial conditions of 200 rabbits and 10 foxes. Describe the changes in the populations as you trace the solution curve.

Solution The graph of the solution is a closed curve, as shown in Figures 6.35 and 6.36.

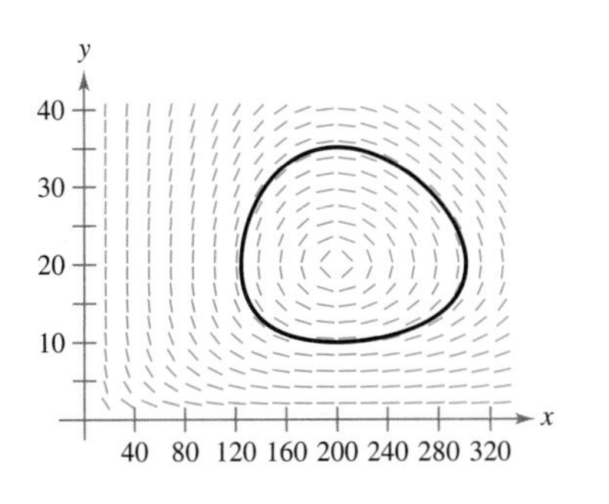

Figure 6.35

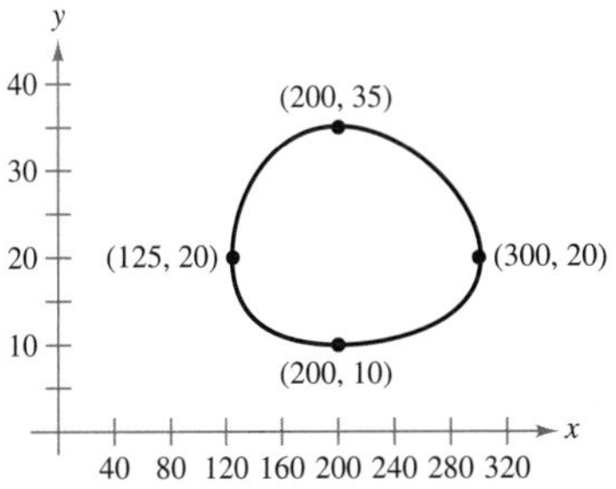

Figure 6.36

At (200, 10), $dy/dt = 0$ and $dx/dt = 4$. So, the rabbit population is increasing at (200, 10). This means that you should trace the curve counterclockwise as t increases. As you trace the curve, note the changes listed in Figure 6.37.

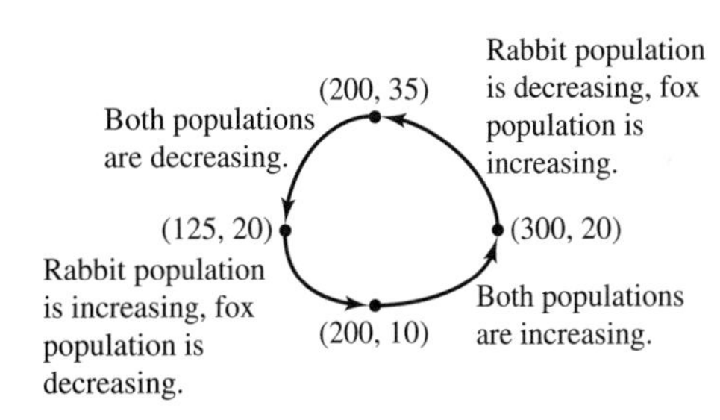

Figure 6.37

Although it is generally not possible to solve predator-prey equations explicitly for x and y, you can separate variables to derive an implicit solution. Begin by writing the equations dy/dt and dx/dt as a function of x.

$$\frac{dy}{dx} = \frac{y(-m + nx)}{x(a - by)} \qquad \text{Factor numerator and denominator.}$$

$$x(a - by)\,dy = y(-m + nx)\,dx \qquad \text{Differential form}$$

$$\frac{a - by}{y}\,dy = \frac{-m + nx}{x}\,dx \qquad \text{Separate variables.}$$

$$\int \frac{a - by}{y}\,dy = \int \frac{-m + nx}{x}\,dx \qquad \text{Integrate.}$$

$$a \ln y - by = -m \ln x + nx + C \qquad \text{Assume } x \text{ and } y \text{ are positive.}$$

$$a \ln y + m \ln x - by - nx = C \qquad \text{General solution}$$

The constant C is determined by the initial conditions.

NOTE The general solution $a \ln y + m \ln x - by - nx = C$ can be rewritten as $\ln(y^a x^m) = C + by + nx$ or as $y^a x^m = C_1 e^{by + nx}$.

Competing Species

Consider two species that compete with each other for the food available in their common environment. Assume that their populations are given by x and y at time t. If there were no interaction or competition between the species, then the populations x and y would each have logistic growth and would satisfy the following differential equations.

$$\frac{dx}{dt} = ax - bx^2$$ Rate of change of first species without interaction

$$\frac{dy}{dt} = my - ny^2$$ Rate of change of second species without interaction

If the species interact, then their competition for resources causes a rate of decline in each population proportional to the product xy. Using a negative interaction factor leads to the following **competing-species equations** (where a, b, c, m, n, and p are positive constants).

$$\frac{dx}{dt} = ax - bx^2 - cxy$$ Rate of change of first species with interaction

$$\frac{dy}{dt} = my - ny^2 - pxy$$ Rate of change of second species with interaction

In this text it is assumed that competing-species equations have four critical points, as shown in Example 5.

EXAMPLE 5 Deriving the Critical Points

Show that the critical points of the competing-species equations

$$\frac{dx}{dt} = ax - bx^2 - cxy \quad \text{and} \quad \frac{dy}{dt} = my - ny^2 - pxy$$

are $(0, 0)$, $(0, m/n)$, $(a/b, 0)$, and $\left(\dfrac{an - mc}{bn - cp}, \dfrac{bm - ap}{bn - cp}\right)$.

Solution Set dx/dt and dy/dt equal to 0 and then factor to obtain the following system.

$$x(a - bx - cy) = 0$$ Set dx/dt equal to 0 and factor out x.

$$y(m - ny - px) = 0$$ Set dy/dt equal to 0 and factor out y.

If $x = 0$, then $y = 0$ or $y = m/n$. If $y = 0$, then $x = 0$ or $x = a/b$. So three of the critical points are $(0, 0)$, $(0, m/n)$, and $(a/b, 0)$.

At each of these critical points, one of the populations is 0. These points represent the possibility that both species cannot coexist. The fourth critical point is obtained by solving the system

$$a - bx - cy = 0$$
$$m - ny - px = 0.$$

The solution of this system is

$$(x, y) = \left(\frac{an - mc}{bn - cp}, \frac{bm - ap}{bn - cp}\right).$$

Assuming this point exists and lies in Quadrant I of the xy-plane, the point represents the possibility that both species can coexist.

EXAMPLE 6 Competing Species: One Species Survives

Consider the competing-species equations given by

$$\frac{dx}{dt} = 10x - x^2 - 2xy \qquad \text{and} \qquad \frac{dy}{dt} = 10y - y^2 - 2xy.$$

a. Find the critical points.

b. Use a graphing utility to graph the solution of the equations when $0 \le t \le 3$ and the initial conditions are $x(0) = 10$ and $y(0) = 15$. What do you observe?

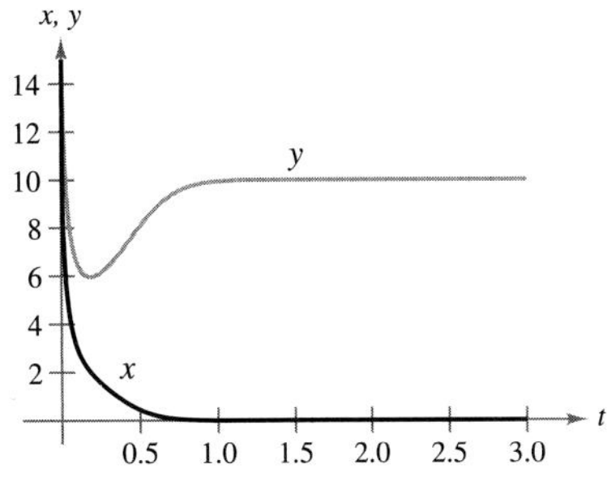

Solution

a. Note that $a = 10$, $b = 1$, $c = 2$, $m = 10$, $n = 1$, and $p = 2$. So, the critical points are (0,0), (0, 10), (10, 0), and $\left(\frac{10-20}{1-4}, \frac{10-20}{1-4}\right) = \left(\frac{10}{3}, \frac{10}{3}\right)$.

b. The solution of the competing-species equations is shown in Figure 6.38. From the graph, it appears that one species survives. The population of the surviving species, represented by the graph of y, appears to remain constant at 10.

Figure 6.38

EXAMPLE 7 Competing Species: Both Species Survive

Consider the competing-species equations given by

$$\frac{dx}{dt} = 10x - 3x^2 - xy \qquad \text{and} \qquad \frac{dy}{dt} = 14y - 3y^2 - xy.$$

a. Find the critical points.

b. Use a graphing utility to graph the solution of the equations when $0 \le t \le 15$ and the initial conditions are $x(0) = 10$ and $y(0) = 15$. What do you observe?

Solution

a. Note that $a = 10$, $b = 3$, $c = 1$, $m = 14$, $n = 3$, and $p = 1$. So, the critical points are $(0, 0)$, $\left(0, \frac{14}{3}\right)$, $\left(\frac{10}{3}, 0\right)$, and $\left(\frac{30-14}{9-1}, \frac{42-10}{9-1}\right) = (2, 4)$.

b. The solution of the competing-species equations is shown in Figure 6.39. From the graph, it appears that both species survive. The population represented by y appears to remain constant at 4. The population represented by x appears to remain constant at 2.

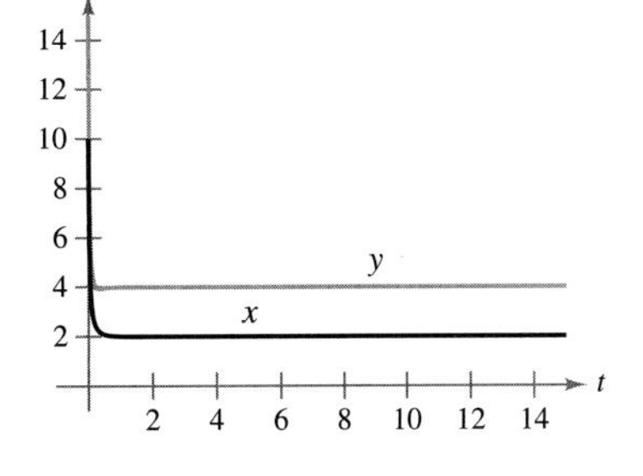

Figure 6.39

Examples 6 and 7 imply a general conclusion about competing-species equations that have precisely four critical points. In general, it can be shown that if $bn > cp$, both species survive. If $bn < cp$, then one species will survive and the other will not.

You can also use slope fields to analyze solutions of competing-species equations, as shown in Figures 6.40 (Example 6) and 6.41 (Example 7).

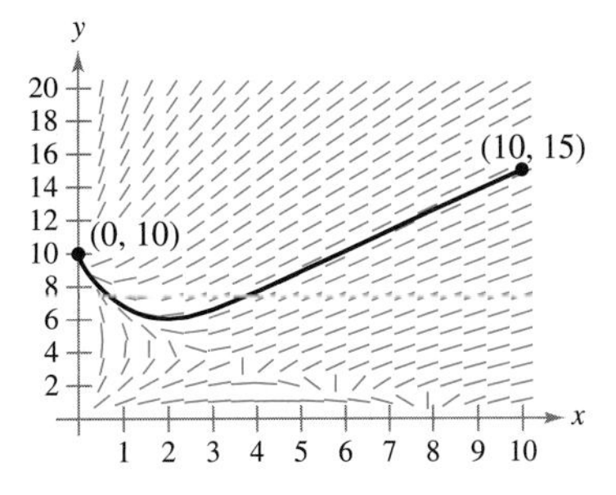

Figure 6.40

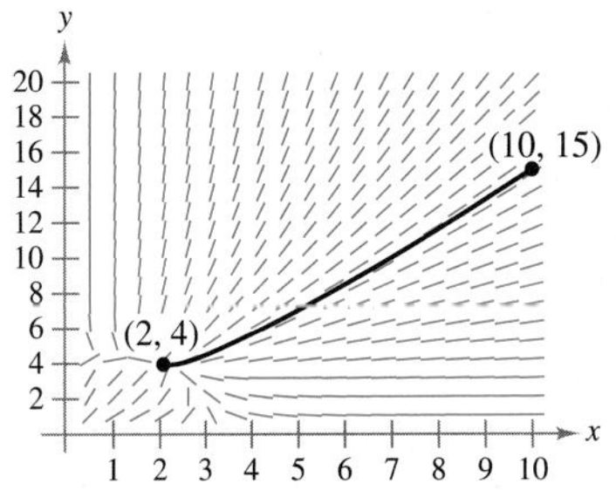

Figure 6.41

Exercises for Section 6.6

See www.CalcChat.com for worked-out solutions to odd-numbered exercises.

In Exercises 1–4, use the given values to write the predator-prey equations $dx/dt = ax - bxy$ and $dy/dt = -my + nxy$. Then find the values of x and y for which $dx/dt = dy/dt = 0$.

1. $a = 0.7, b = 0.05, m = 0.4, n = 0.007$
2. $a = 0.5, b = 0.004, m = 0.3\ n = 0.01$
3. $a = 0.3, b = 0.006, m = 0.5, n = 0.009$
4. $a = 0.6, b = 0.02, m = 0.6, n = 0.01$

Slope Fields **In Exercises 5 and 6, predator-prey equations, a point, and a slope field are given. (a) Sketch a solution of the predator-prey equations on the slope field that passes through the given point. (b) Use a graphing utility to graph the solution. Compare the result with the sketch in part (a). To print an enlarged copy of the graph, go to the website *www.mathgraphs.com*.**

5. $\dfrac{dx}{dt} = 0.04x - 0.002xy$

   $\dfrac{dy}{dt} = -0.08y + 0.0004xy$

   $(150, 30)$

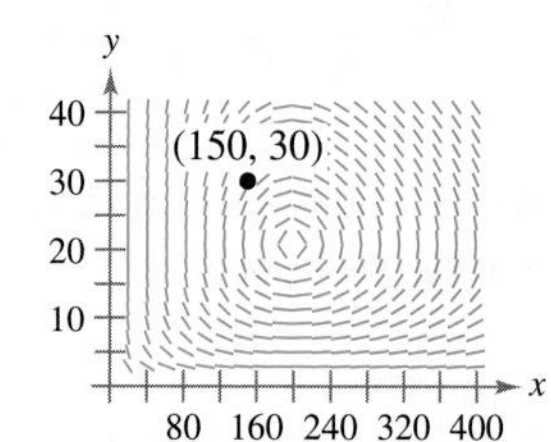

6. $\dfrac{dx}{dt} = 0.03x - 0.006xy$

   $\dfrac{dy}{dt} = -0.04y + 0.004xy$

   $(15, 3)$

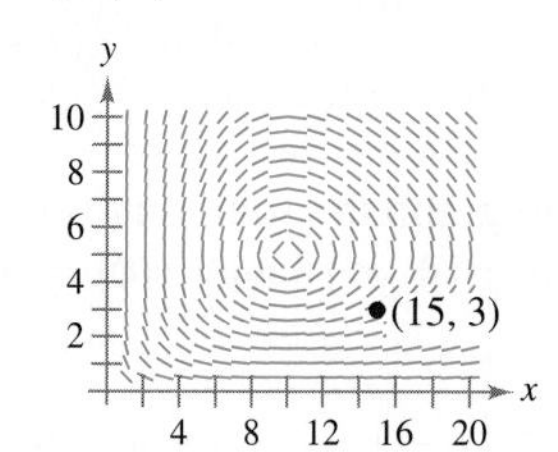

In Exercises 7 and 8, two graphs are given. The first is a graph of the functions x and y of a set of predator-prey equations where x is the number of prey and y is the number of predators at time t. The second graph is the corresponding slope field of the predator-prey equations. (a) Identify the initial conditions. (b) Sketch a solution of the predator-prey equations on the slope field that passes through the initial conditions. To print an enlarged copy of the graph, go to the website *www.mathgraphs.com*.

7.

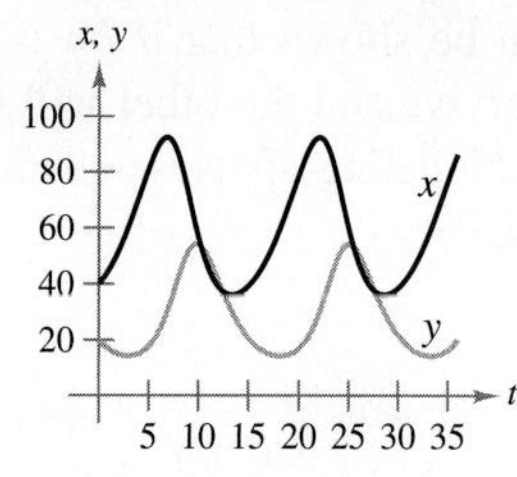

8.

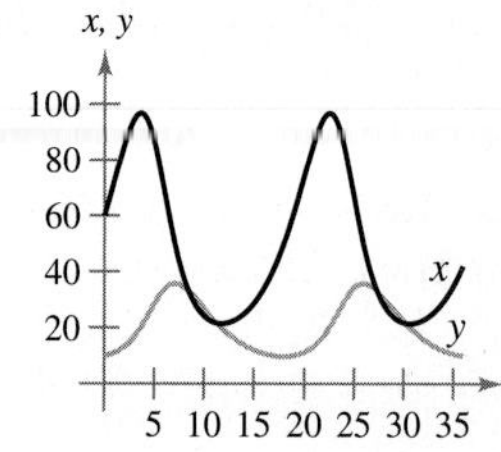

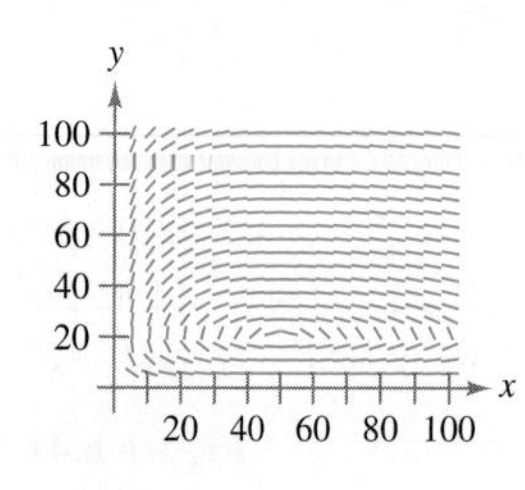

Rabbits and Foxes **In Exercises 9–12, consider a predator-prey relationship involving foxes (predators) and rabbits (prey). Let x represent the number of rabbits, let y represent the number of foxes, and let t represent the time in months. Assume that the following predator-prey equations model the rates of change of each population.**

$$\frac{dx}{dt} = 0.8x - 0.04xy \qquad \text{Rate of change of prey population}$$

$$\frac{dy}{dt} = -0.3y + 0.006xy \qquad \text{Rate of change of predator population}$$

When $t = 0, x = 55$ rabbits and $y = 10$ foxes.

9. Find the critical points of the predator-prey equations.

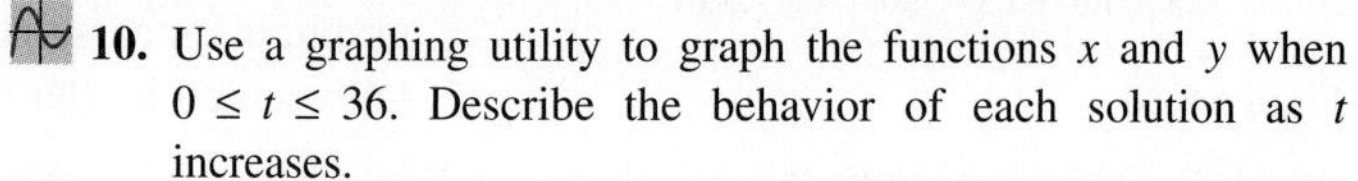

10. Use a graphing utility to graph the functions x and y when $0 \le t \le 36$. Describe the behavior of each solution as t increases.
11. Use a graphing utility to graph a slope field of the predator-prey equations when $0 \le x \le 150$ and $0 \le y \le 50$.
12. Use the predator-prey equations and the slope field in Exercise 11 to graph the solution curve using the initial conditions. Describe the changes in the rabbit and fox populations as you trace the solution curve.

Prairie Dogs and Black-Footed Ferrets **In Exercises 13–16, consider a predator-prey relationship involving black-footed ferrets (predators) and prairie dogs (prey). Let x represent the number of prairie dogs, let y represent the number of black-footed ferrets, and let t represent the time in months. Assume that the following predator-prey equations model the rates of change of each population.**

$$\frac{dx}{dt} = 0.1x - 0.00008xy \qquad \text{Rate of change of prey population}$$

$$\frac{dy}{dt} = -0.4y + 0.00004xy \qquad \text{Rate of change of predator population}$$

When $t = 0, x = 4000$ prairie dogs and $y = 1000$ black-footed ferrets.

13. Find the critical points of the predator-prey equations.

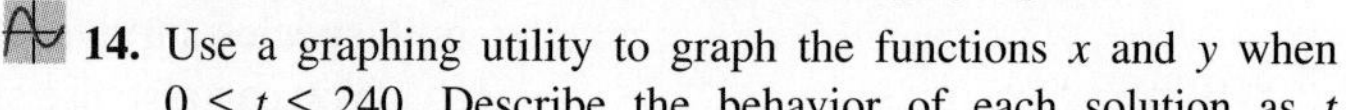

14. Use a graphing utility to graph the functions x and y when $0 \le t \le 240$. Describe the behavior of each solution as t increases.
15. Use a graphing utility to graph a slope field of the predator-prey equations when $0 \le x \le 25{,}000$ and $0 \le y \le 4000$.
16. Use the predator-prey equations and the slope field in Exercise 15 to graph the solution curve using the initial conditions. Describe the changes in the prairie dog and black-footed ferret populations as you trace the solution curve.
17. ***Critical Point as the Initial Condition*** In Exercise 9, you found the critical points of the predator-prey system. Assume that the critical point given by $(m/n, a/b)$ is the initial condition and repeat Exercises 10–12. Compare the results.

18. ***Critical Point as the Initial Condition*** In Exercise 13, you found the critical points of the predator-prey system. Assume that the critical point given by $(m/n, a/b)$ is the initial condition and repeat Exercises 14–16. Compare the results.

In Exercises 19–22, use the given values to write the competing-species equations $dx/dt = ax - bx^2 - cxy$ and $dy/dt = my - ny^2 - pxy$. Then find the values of x and y for which $dx/dt = dy/dt = 0$.

19. $a = 1, b = 2, c = 1, m = 1, n = 2, p = 1$

20. $a = 2, b = 1, c = 1, m = 5, n = 4, p = 1$

21. $a = 0.1, b = 0.4, c = 0.5, m = 0.1, n = 0.8, p = 0.3$

22. $a = 0.05, b = 0.2, c = 0.4, m = 0.06, n = 0.9, p = 0.2$

Bass and Trout **In Exercises 23 and 24, consider a competing-species relationship involving bass and trout. Assume the bass and trout compete for the same resources. Let x represent the number of bass (in thousands), let y represent the number of trout (in thousands), and let t represent the time in months. Assume that the following competing-species equations model the rates of change of each population.**

$$\frac{dx}{dt} = 0.8x - 0.4x^2 - 0.1xy \quad \text{Rate of change of bass population}$$

$$\frac{dy}{dt} = 0.3y - 0.6y^2 - 0.1xy \quad \text{Rate of change of trout population}$$

When $t = 0$, $x = 9$ and $y = 5$.

23. Find the critical points of the competing-species equations.

24. Use a graphing utility to graph the functions x and y when $0 \le t \le 36$. Describe the behavior of each solution as t increases.

Bass and Trout **In Exercises 25 and 26, consider a competing-species relationship involving bass and trout. Assume the bass and trout compete for the same resources. Let x represent the number of bass (in thousands), let y represent the number of trout (in thousands), and let t represent the time in months. Assume that the following competing-species equations model the rates of change of each population.**

$$\frac{dx}{dt} = 0.8x - 0.4x^2 - xy \quad \text{Rate of change of bass population}$$

$$\frac{dy}{dt} = 0.3y - 0.6y^2 - xy \quad \text{Rate of change of trout population}$$

When $t = 0$, $x = 7$ and $y = 6$.

25. Find the critical points of the competing-species equations.

26. Use a graphing utility to graph the functions x and y when $0 \le t \le 36$. Describe the behavior of each solution as t increases.

27. ***Critical Point as the Initial Condition*** In Exercise 23, you found the critical points of the competing-species system. Assume the critical point given by $\left(\frac{an - mc}{bn - cp}, \frac{bm - ap}{bn - cp}\right)$ is the initial condition and repeat Exercise 24. Compare the results.

28. ***Critical Point as the Initial Condition*** In Exercise 23, you found the critical points of the competing-species system. Assume the critical point given by $(0, m/n)$ is the initial condition and repeat Exercise 24. Compare the results.

Writing About Concepts

29. Given a set of predator-prey equations, describe how to determine initial values so that both populations remain constant for all $t \ge 0$.

30. Given a set of competing-species equations, describe how to determine initial values so that both populations remain constant for all $t > 0$.

True or False? **In Exercises 31–34, determine whether the statement is true or false. If it is false, explain why or give an example that shows it is false.**

31. The predator-prey equations are separable differential equations.

32. The predator-prey equations are linear differential equations.

33. The competing-species equations are a special case of the predator-prey equations.

34. The predator-prey equations can always be solved explicitly for x and y.

35. ***Revising the Predator-Prey Equations*** Consider a predator-prey relationship with x prey and y predators at time t. Assume both predator and prey are present. Then the rates of change of each population can be modeled by the following revised predator-prey system of differential equations.

$$\frac{dx}{dt} = ax\left(1 - \frac{x}{L}\right) - bxy \quad \text{Rate of change of prey population}$$

$$\frac{dy}{dt} = -my + nxy \quad \text{Rate of change of predator population}$$

(a) If there are no predators, the prey population will grow according to what model?

(b) Write the revised predator-prey equations for $a = 0.4$, $L = 100$, $b = 0.01$, $m = 0.3$ and $n = 0.005$. Find the critical numbers.

(c) Use a graphing utility to graph the functions x and y of the revised predator-prey equations when $0 \le t \le 72$ and the initial conditions are $x(0) = 40$ and $y(0) = 80$. Describe the behavior of each solution as t increases.

(d) Use a graphing utility to graph a slope field of the revised predator-prey equations when $0 \le x \le 100$ and $0 \le y \le 80$.

(e) Use the predator-prey equations and the slope field in part (d) to graph the solution curve using the initial conditions in part (c). Describe the changes in the prey and predator populations as you trace the solution curve.

36. ***Comparing Results*** Repeat Exercise 35, parts (b)–(e), using the original predator-prey equations. Use $0 \le t \le 72$, $0 \le x \le 140$, and $0 \le y \le 100$ to graph the solutions and the slope field. Compare the results.

Review Exercises for Chapter 6

See www.CalcChat.com for worked-out solutions to odd-numbered exercises.

1. Determine whether the function $y = x^3$ is a solution of the differential equation $x^2y' + 3y = 6x^3$.

2. Determine whether the function $y = 2 \sin 2x$ is a solution of the differential equation $y''' - 8y = 0$.

In Exercises 3–8, use integration to find a general solution of the differential equation.

3. $\dfrac{dy}{dx} = 2x^2 + 5$

4. $\dfrac{dy}{dx} = x^3 - 2x$

5. $\dfrac{dy}{dx} = \cos 2x$

6. $\dfrac{dy}{dx} = 2 \sin x$

7. $\dfrac{dy}{dx} = 2x\sqrt{x - 7}$

8. $\dfrac{dy}{dx} = 3e^{-x/3}$

Slope Fields **In Exercises 9 and 10, a differential equation and its slope field are given. Determine the slopes (if possible) in the slope field at the points given in the table.**

x	-4	-2	0	2	4	8
y	2	0	4	4	6	8
dy/dx						

9. $\dfrac{dy}{dx} = \dfrac{2x}{y}$

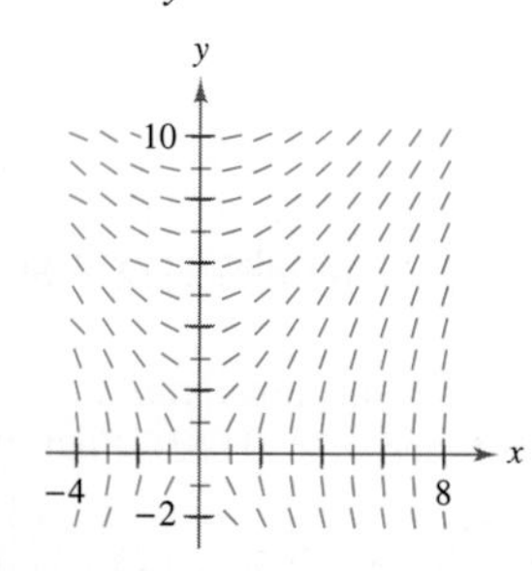

10. 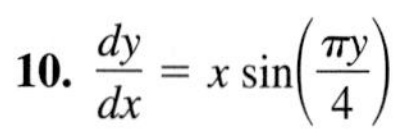 $\dfrac{dy}{dx} = x \sin\left(\dfrac{\pi y}{4}\right)$

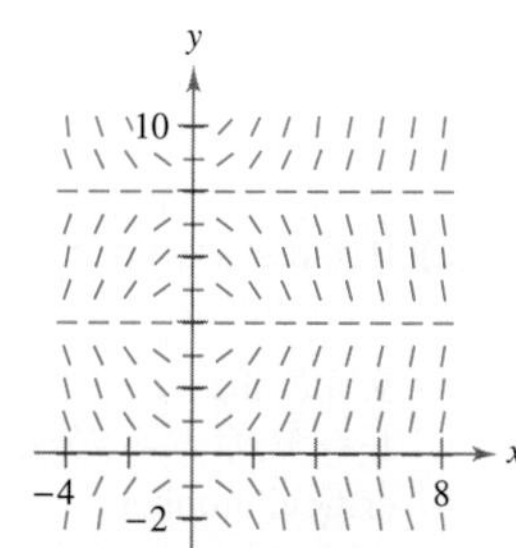

Slope Fields **In Exercises 11–16, (a) sketch the slope field for the differential equation, and (b) use the slope field to sketch the solution that passes through the given point.**

Differential Equation	*Point*
11. $y' = -x - 2$	$(-1, 1)$
12. $y' = 2x^2 - x$	$(0, 2)$
13. $y' = \frac{1}{4}x^2 - \frac{1}{3}x$	$(0, 3)$
14. $y' = y + 3x$	$(2, 1)$
15. $y' = \dfrac{xy}{x^2 + 4}$	$(0, 1)$
16. $y' = \dfrac{y}{x^2 + 1}$	$(0, -2)$

In Exercises 17–22, solve the differential equation.

17. $\dfrac{dy}{dx} = 6 - x$

18. $\dfrac{dy}{dx} = y + 6$

19. $\dfrac{dy}{dx} = (3 + y)^2$

20. $\dfrac{dy}{dx} = 4\sqrt{y}$

21. $(2 + x)y' - xy = 0$

22. $xy' - (x + 1)y = 0$

In Exercises 23–26, find the exponential function $y = Ce^{kt}$ that passes through the two points.

23.

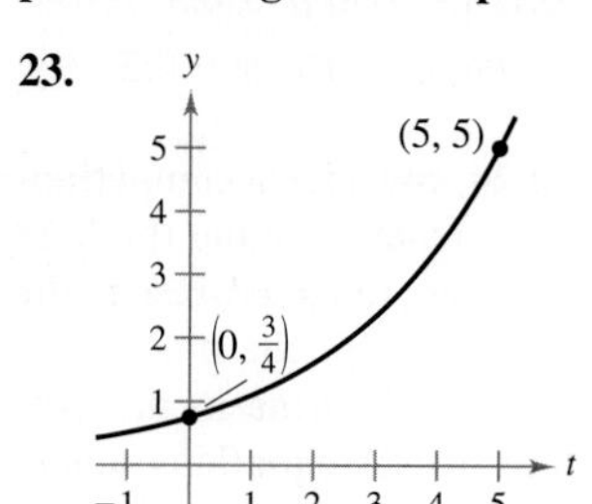

24.

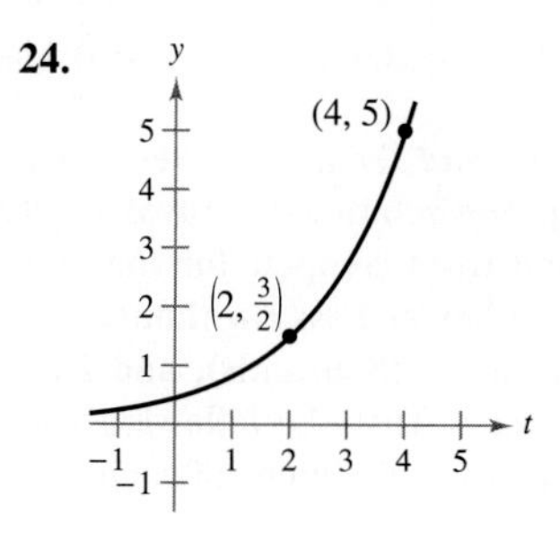

25. $(0, 5), \left(5, \dfrac{1}{6}\right)$

26. $(1, 9), (6, 2)$

27. ***Air Pressure*** Under ideal conditions, air pressure decreases continuously with the height above sea level at a rate proportional to the pressure at that height. The barometer reads 30 inches at sea level and 15 inches at 18,000 feet. Find the barometric pressure at 35,000 feet.

28. ***Radioactive Decay*** Radioactive radium has a half-life of approximately 1599 years. The initial quantity is 5 grams. How much remains after 600 years?

29. ***Sales*** The sales S (in thousands of units) of a new product after it has been on the market for t years is given by

$S = Ce^{k/t}$.

(a) Find S as a function of t if 5000 units have been sold after 1 year and the saturation point for the market is 30,000 units (that is, $\lim_{t\to\infty} S = 30$).

(b) How many units will have been sold after 5 years?

(c) Use a graphing utility to graph this sales function.

30. ***Sales*** The sales S (in thousands of units) of a new product after it has been on the market for t years is given by

$S = 25(1 - e^{kt})$.

(a) Find S as a function of t if 4000 units have been sold after 1 year.

(b) How many units will saturate this market?

(c) How many units will have been sold after 5 years?

(d) Use a graphing utility to graph this sales function.

31. ***Population Growth*** A population grows continuously at the rate of 1.5%. How long will it take the population to double?

32. ***Fuel Economy*** An automobile gets 28 miles per gallon of gasoline for speeds up to 50 miles per hour. Over 50 miles per hour, the number of miles per gallon drops at the rate of 12 percent for each 10 miles per hour.

(a) s is the speed and y is the number of miles per gallon. Find y as a function of s by solving the differential equation

$$\frac{dy}{ds} = -0.012y, \quad s > 50.$$

(b) Use the function in part (a) to complete the table.

Speed	50	55	60	65	70
Miles per Gallon					

In Exercises 33–38, solve the differential equation.

33. $\dfrac{dy}{dx} = \dfrac{x^2 + 3}{x}$

34. $\dfrac{dy}{dx} = \dfrac{e^{-2x}}{1 + e^{-2x}}$

35. $y' - 2xy = 0$

36. $y' - e^y \sin x = 0$

37. $\dfrac{dy}{dx} = \dfrac{x^2 + y^2}{2xy}$

38. $\dfrac{dy}{dx} = \dfrac{3(x + y)}{x}$

39. Verify that the general solution $y = C_1x + C_2x^3$ satisfies the differential equation $x^2y'' - 3xy' + 3y = 0$. Then find the particular solution that satisfies the initial conditions $y = 0$ and $y' = 4$ when $x = 2$.

40. ***Vertical Motion*** A falling object encounters air resistance that is proportional to its velocity. The acceleration due to gravity is -9.8 meters per second per second. The net change in velocity is $dv/dt = kv - 9.8$.

(a) Find the velocity of the object as a function of time if the initial velocity is v_0.

(b) Use the result of part (a) to find the limit of the velocity as t approaches infinity.

(c) Integrate the velocity function found in part (a) to find the position function s.

Slope Fields **In Exercises 41 and 42, sketch a few solutions of the differential equation on the slope field and then find the general solution analytically. To print an enlarged copy of the graph, go to the website *www.mathgraphs.com*.**

41. $\dfrac{dy}{dx} = -\dfrac{4x}{y}$

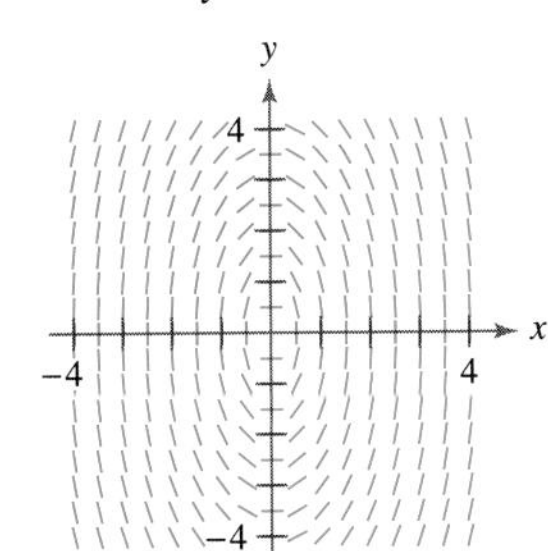

42. $\dfrac{dy}{dx} = 3 - 2y$

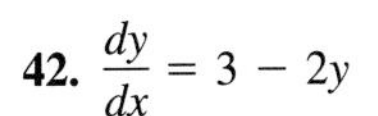

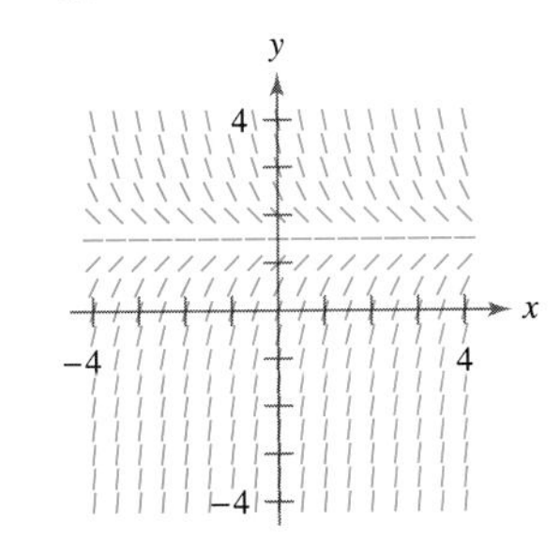

In Exercises 43 and 44, the logistic equation models the growth of a population. Use the equation to (a) find the value of k, (b) find the carrying capacity, (c) find the initial population, (d) determine when the population will reach 50% of its carrying capacity, and (e) write a logistic differential equation that has the solution $P(t)$.

43. $P(t) = \dfrac{7200}{1 + 44e^{-0.55t}}$

44. $P(t) = \dfrac{4800}{1 + 14e^{-0.15t}}$

45. ***Environment*** A conservation department releases 1200 brook trout into a lake. It is estimated that the carrying capacity of the lake for the species is 20,400. After the first year, there are 2000 brook trout in the lake.

(a) Write a logistic equation that models the number of brook trout in the lake.

(b) Find the number of brook trout in the lake after 8 years.

(c) When will the number of brook trout reach 10,000?

46. ***Environment*** Write a logistic differential equation that models the growth rate of the brook trout population in Exercise 45. Then repeat part (b) using Euler's Method with a step size of $h = 1$. Compare the approximation with the exact answers.

47. ***Sales Growth*** The rate of change in sales S (in thousands of units) of a new product is proportional to the difference between L and S (in thousands of units) at any time t. When $t = 0$, $S = 0$. Write and solve the differential equation for this sales model.

48. ***Sales Growth*** Use the result of Exercise 47 to write S as a function of t if (a) $L = 100$, $S = 25$ when $t = 2$, and (b) $L = 500$, $S = 50$ when $t = 1$.

Learning Theory **In Exercises 49 and 50, assume that the rate of change in the proportion P of correct responses after n trials is proportional to the product of P and $L - P$, where L is the limiting proportion of correct responses.**

49. Write and solve the differential equation for this learning theory model.

50. Use the solution of Exercise 49 to write P as a function of n, and then use a graphing utility to graph the solution.

(a) $L = 1.00$

$P = 0.50$ when $n = 0$

$P = 0.85$ when $n = 4$

(b) $L = 0.80$

$P = 0.25$ when $n = 0$

$P = 0.60$ when $n = 10$

In Exercises 51–54, (a) sketch an approximate solution of the differential equation satisfying the initial condition by hand on the slope field, (b) find the particular solution that satisfies the initial condition, and (c) use a graphing utility to graph the particular solution. Compare the graph with the hand-drawn graph of part (a). To print an enlarged copy of the graph, go to the website *www.mathgraphs.com*.

Differential Equation	*Initial Condition*
51. $\dfrac{dy}{dx} = e^{x/2} - y$	$(0, -1)$
52. $y' + 2y = \sin x$	$(0, 4)$
53. $y' = \csc x + y \cot x$	$(1, 1)$
54. $y' = \csc x - y \cot x$	$(1, 2)$

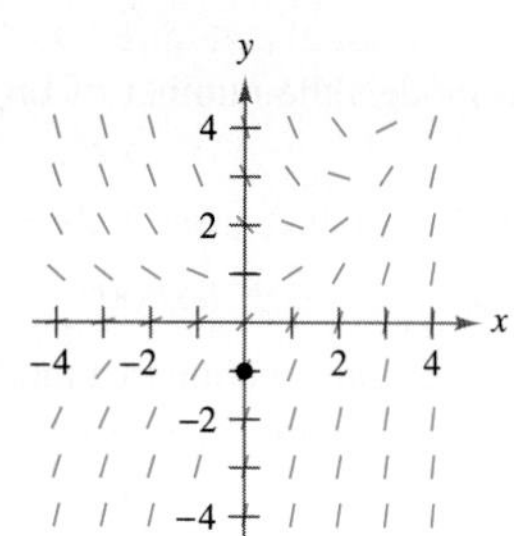

Figure for 51

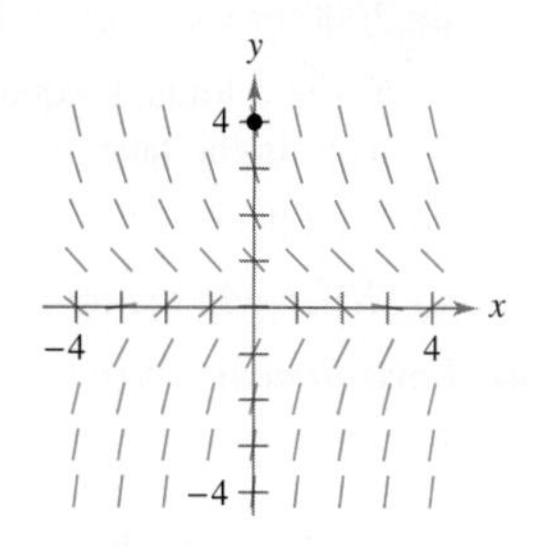

Figure for 52

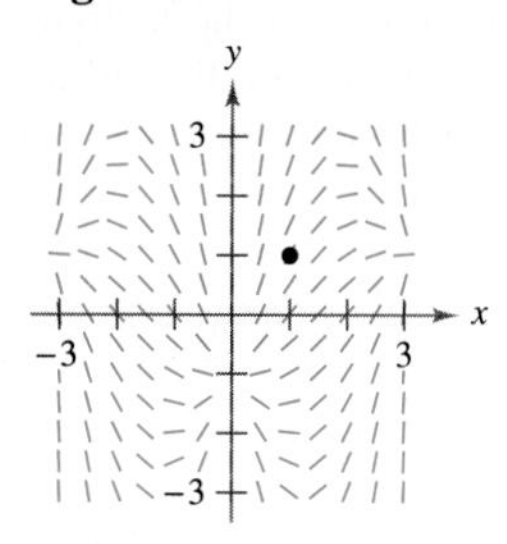

Figure for 53

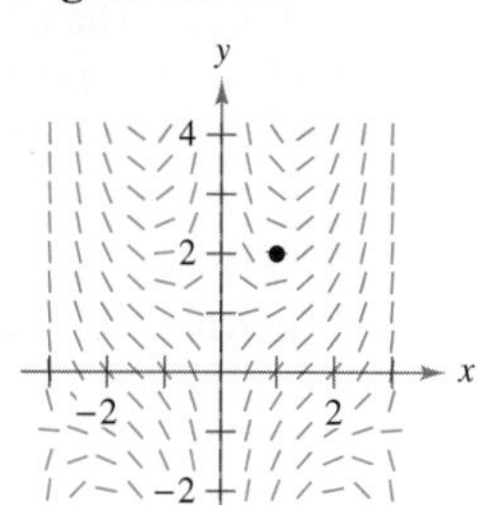

Figure for 54

In Exercises 55–64, solve the first-order linear differential equation.

55. $y' - y = 8$

56. $e^x y' + 4e^x y = 1$

57. $4y' = e^{x/4} + y$

58. $\dfrac{dy}{dx} - \dfrac{5y}{x^2} = \dfrac{1}{x^2}$

59. $(x - 2)y' + y = 1$

60. $(x + 3)y' + 2y = 2(x + 3)^2$

61. $(3y + \sin 2x)\, dx - dy = 0$

62. $dy = (y \tan x + 2e^x)\, dx$

63. $y' + 5y = e^{5x}$

64. $xy' - ay = bx^4$

In Exercises 65–68, solve the Bernoulli differential equation.

65. $y' + y = xy^2$ $\left[\textit{Hint: } \int xe^{-x}\, dx = (-x - 1)e^{-x}\right]$

66. $y' + 2xy = xy^2$

67. $y' + \left(\dfrac{1}{x}\right)y = \dfrac{y^3}{x^2}$

68. $xy' + y = xy^2$

In Exercises 69–72, write an example of the given differential equation. Then solve your equation.

69. Homogeneous differential equation

70. Logistic differential equation

71. First-order linear differential equation

72. Bernoulli differential equation

73. ***Investment*** Let $A(t)$ be the amount in a fund earning interest at an annual rate r compounded continuously. If a continuous cash flow of P dollars per year is withdrawn from the fund, then the rate of change of A is given by the differential equation

$$\frac{dA}{dt} = rA - P$$

where $A = A_0$ when $t = 0$. Solve this differential equation for A as a function of t.

74. ***Investment*** A retired couple plans to withdraw P dollars per year from a retirement account of \$500,000 earning 10% interest compounded continuously. Use the result of Exercise 73 and a graphing utility to graph the function A for each of the following continuous annual cash flows. Use the graphs to describe what happens to the balance in the fund for each case.

(a) $P = \$40{,}000$

(b) $P = \$50{,}000$

(c) $P = \$60{,}000$

75. ***Investment*** Use the result of Exercise 73 to find the time necessary to deplete a fund earning 14% interest compounded continuously if $A_0 = \$1{,}000{,}000$ and $P = \$200{,}000$.

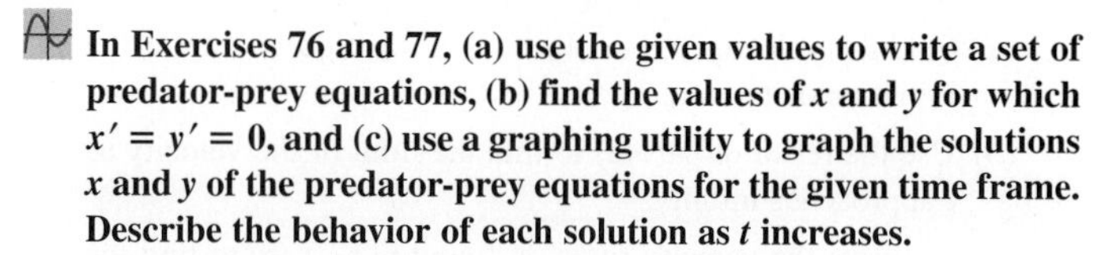

In Exercises 76 and 77, (a) use the given values to write a set of predator-prey equations, (b) find the values of x and y for which $x' = y' = 0$, and (c) use a graphing utility to graph the solutions x and y of the predator-prey equations for the given time frame. Describe the behavior of each solution as t increases.

76. Constants: $a = 0.3$, $b = 0.02$, $m = 0.4$, $n = 0.01$
Initial condition: $(20, 20)$
Time frame: $0 \le t \le 36$

77. Constants: $a = 0.4$, $b = 0.04$, $m = 0.6$, $n = 0.02$
Initial condition: $(30, 15)$
Time frame: $0 \le t \le 24$

In Exercises 78 and 79, (a) use the given values to write a set of competing-species equations, (b) find the values of x and y for which $x' = y' = 0$, and (c) use a graphing utility to graph the solutions x and y of the competing-species equations for the given time frame. Describe the behavior of each solution as t increases.

78. Constants: $a = 3$, $b = 1$, $c = 1$, $m = 2$, $n = 1$, $p = 0.5$
Initial condition: $(3, 2)$
Time frame: $0 \le t \le 6$

79. Constants: $a = 15$, $b = 2$, $c = 4$, $m = 17$, $n = 2$ $p = 4$
Initial condition: $(9, 10)$
Time frame: $0 \le t \le 4$

P.S. Problem Solving

See www.CalcChat.com for worked-out solutions to odd-numbered exercises.

1. The differential equation

$$\frac{dy}{dt} = ky^{1+\varepsilon}$$

where k and ε are positive constants, is called the **doomsday equation.**

(a) Solve the doomsday equation

$$\frac{dy}{dt} = y^{1.01}$$

given that $y(0) = 1$. Find the time T at which

$$\lim_{t \to T^-} y(t) = \infty.$$

(b) Solve the doomsday equation

$$\frac{dy}{dt} = ky^{1+\varepsilon}$$

given that $y(0) = y_0$. Explain why this equation is called the doomsday equation.

2. A thermometer is taken from a room at 72°F to the outdoors, where the temperature is 20°F. The reading drops to 48°F after 1 minute. Determine the reading on the thermometer after 5 minutes.

3. Let S represent sales of a new product (in thousands of units), let L represent the maximum level of sales (in thousands of units), and let t represent time (in months). The rate of change of S with respect to t varies jointly as the product of S and $L - S$.

(a) Write the differential equation for the sales model if $L = 100$, $S = 10$ when $t = 0$, and $S = 20$ when $t = 1$. Verify that

$$S = \frac{L}{1 + Ce^{-kt}}.$$

(b) At what time is the growth in sales increasing most rapidly?

(c) Use a graphing utility to graph the sales function.

(d) Sketch the solution from part (a) on the slope field shown in the figure below. To print an enlarged copy of the graph, go to the website *www.mathgraphs.com*.

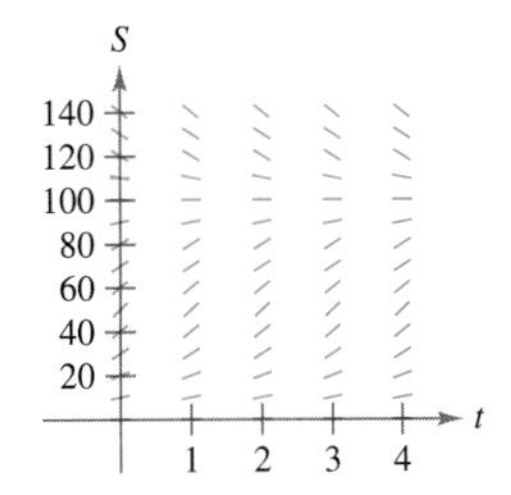

(e) If the estimated maximum level of sales is correct, use the slope field to describe the shape of the solution curves for sales if, at some period of time, sales exceed L.

4. Another numerical approach to approximating the particular solution of the differential equation $y' = F(x, y)$ is shown below.

$$x_n = x_{n-1} + h$$

$$y_n = y_{n-1} + hf\left(x_{n-1} + \frac{h}{2}, y_{n-1} + \frac{h}{2}f(x_{n-1}, y_{n-1})\right)$$

This approach is called **modified Euler's Method.**

(a) Use this method to approximate the solution of the differential equation $y' = x - y$ passing through the point $(0, 1)$. Use a step size of $h = 0.1$.

(b) Use a graphing utility to graph the exact solution and the approximations found using Euler's Method and modified Euler's Method (see Example 6, page 390). Compare the first 10 approximations found using modified Euler's Method to those found using Euler's Method and to the exact solution $y = x - 1 + 2e^{-x}$. Which approximation appears to be more accurate?

5. Show that the logistic equation

$$y = \frac{L}{1 + be^{-kt}}$$

can be written as

$$y = \frac{1}{2}L\left[1 + \tanh\left(\frac{1}{2}k\left(t - \frac{\ln b}{k}\right)\right)\right].$$

What can you conclude about the graph of the logistic equation?

6. **Torricelli's Law** states that water will flow from an opening at the bottom of a tank with the same speed that it would attain falling from the surface of the water to the opening. One of the forms of Torricelli's Law is

$$A(h)\frac{dh}{dt} = -k\sqrt{2gh}$$

where h is the height of the water in the tank, k is the area of the opening at the bottom of the tank, $A(h)$ is the horizontal cross-sectional area at height h, and g is the acceleration due to gravity ($g \approx 32$ feet per second per second). A hemispherical water tank has a radius of 6 feet. When the tank is full, a circular valve with a radius of 1 inch is opened at the bottom, as shown in the figure. How long will it take for the tank to drain completely?

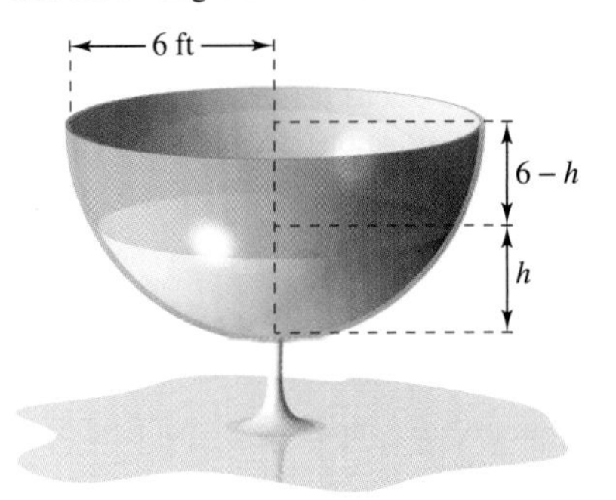

7. The cylindrical water tank shown in the figure has a height of 18 feet. When the tank is full, a circular valve is opened at the bottom of the tank. After 30 minutes, the depth of the water is 12 feet.

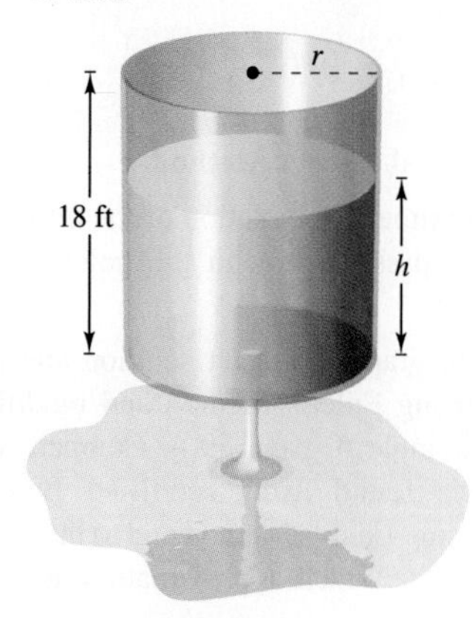

(a) How long will it take for the tank to drain completely?

(b) What is the depth of the water in the tank after 1 hour?

8. Suppose the tank in Exercise 7 has a height of 20 feet, a radius of 8 feet, and the valve is circular with a radius of 2 inches. The tank is full when the valve is opened. How long will it take for the tank to drain completely?

9. In hilly areas, radio reception may be poor. Consider a situation where an FM transmitter is located at the point $(-1, 1)$ behind a hill modeled by the graph of

$$y = x - x^2$$

and a radio receiver is located on the opposite side of the hill. (Assume that the x-axis represents ground level at the base of the hill.)

(a) What is the closest position $(x, 0)$ the radio can be to the hill so that reception is unobstructed?

(b) Write the closest position $(x, 0)$ of the radio, with x represented as a function of h, if the transmitter is located at $(-1, h)$.

(c) Use a graphing utility to graph the function for x in part (b). Determine the vertical asymptote of the function and interpret the result.

10. Biomass is a measure of the amount of living matter in an ecosystem. Suppose the biomass $s(t)$ in a given ecosystem increases at a rate of about 3.5 tons per year, and decreases by about 1.9% per year. This situation can be modeled by the differential equation

$$\frac{ds}{dt} = 3.5 - 0.019s.$$

(a) Solve the differential equation.

(b) Use a graphing utility to graph the slope field for the differential equation. What do you notice?

(c) Explain what happens as $t \to \infty$.

In Exercises 11–13, a medical researcher wants to determine the concentration C (in moles per liter) of a tracer drug injected into a moving fluid. Solve this problem by considering a single-compartment dilution model (see figure). Assume that the fluid is continuously mixed and that the volume of the fluid in the compartment is constant.

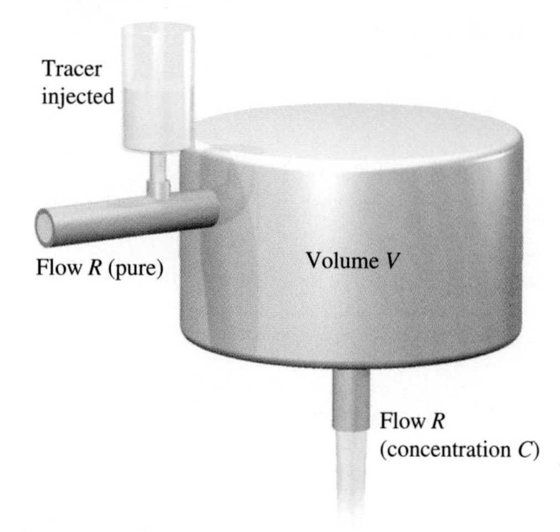

Figure for 11–13

11. If the tracer is injected instantaneously at time $t = 0$, then the concentration of the fluid in the compartment begins diluting according to the differential equation

$$\frac{dC}{dt} = \left(-\frac{R}{V}\right)C, \quad C = C_0 \text{ when } t = 0.$$

(a) Solve this differential equation to find the concentration C as a function of time t.

(b) Find the limit of C as $t \to \infty$.

12. Use the solution of the differential equation in Exercise 11 to find the concentration C as a function of time t, and use a graphing utility to graph the function.

(a) $V = 2$ liters, $R = 0.5$ liter per minute, and $C_0 = 0.6$ mole per liter

(b) $V = 2$ liters, $R = 1.5$ liters per minute, and $C_0 = 0.6$ mole per liter

13. In Exercises 11 and 12, it was assumed that there was a single initial injection of the tracer drug into the compartment. Now consider the case in which the tracer is continuously injected (beginning at $t = 0$) at the rate of Q moles per minute. Considering Q to be negligible compared with R, use the differential equation

$$\frac{dC}{dt} = \frac{Q}{V} - \left(\frac{R}{V}\right)C, \quad C = 0 \text{ when } t = 0.$$

(a) Solve this differential equation to find the concentration C as a function of time t.

(b) Find the limit of C as $t \to \infty$.

Appendices

The remaining appendices are located on the website that accompanies this text at *college.hmco.com*.

A Proofs of Selected Theorems

THEOREM 2.2 Properties of Limits (Properties 2, 3, 4, and 5) (page 79)

Let b and c be real numbers, let n be a positive integer, and let f and g be functions with the following limits.

$$\lim_{x \to c} f(x) = L \quad \text{and} \quad \lim_{x \to c} g(x) = K$$

2. Sum or difference: $\lim_{x \to c} [f(x) \pm g(x)] = L \pm K$

3. Product: $\lim_{x \to c} [f(x)g(x)] = LK$

4. Quotient: $\lim_{x \to c} \dfrac{f(x)}{g(x)} = \dfrac{L}{K}$, provided $K \neq 0$

5. Power: $\lim_{x \to c} [f(x)]^n = L^n$

Proof To prove Property 2, choose $\varepsilon > 0$. Because $\varepsilon/2 > 0$, you know that there exists $\delta_1 > 0$ such that $0 < |x - c| < \delta_1$ implies $|f(x) - L| < \varepsilon/2$. You also know that there exists $\delta_2 > 0$ such that $0 < |x - c| < \delta_2$ implies $|g(x) - K| < \varepsilon/2$. Let δ be the smaller of δ_1 and δ_2; then $0 < |x - c| < \delta$ implies that

$$|f(x) - L| < \frac{\varepsilon}{2} \quad \text{and} \quad |g(x) - K| < \frac{\varepsilon}{2}.$$

So, you can apply the triangle inequality to conclude that

$$|[f(x) + g(x)] - (L + K)| \leq |f(x) - L| + |g(x) - K| < \frac{\varepsilon}{2} + \frac{\varepsilon}{2} = \varepsilon$$

which implies that

$$\lim_{x \to c} [f(x) + g(x)] = L + K = \lim_{x \to c} f(x) + \lim_{x \to c} g(x).$$

The proof that

$$\lim_{x \to c} [f(x) - g(x)] = L - K$$

is similar.

To prove Property 3, given that

$$\lim_{x \to c} f(x) = L \quad \text{and} \quad \lim_{x \to c} g(x) = K$$

you can write

$$f(x)g(x) = [f(x) - L][g(x) - K] + [Lg(x) + Kf(x)] - LK.$$

Because the limit of $f(x)$ is L, and the limit of $g(x)$ is K, you have

$$\lim_{x \to c} [f(x) - L] = 0 \quad \text{and} \quad \lim_{x \to c} [g(x) - K] = 0.$$

Let $0 < \varepsilon < 1$. Then there exists $\delta > 0$ such that if $0 < |x - c| < \delta$, then

$$|f(x) - L - 0| < \varepsilon \quad \text{and} \quad |g(x) - K - 0| < \varepsilon$$

which implies that

$$|[f(x) - L][g(x) - K] - 0| = |f(x) - L|\,|g(x) - K| < \varepsilon\varepsilon < \varepsilon.$$

So,

$$\lim_{x \to c} [f(x) - L][g(x) - K] = 0.$$

Furthermore, by Property 1, you have

$$\lim_{x \to c} Lg(x) = LK \quad \text{and} \quad \lim_{x \to c} Kf(x) = KL.$$

Finally, by Property 2, you obtain

$$\begin{aligned}\lim_{x \to c} f(x)g(x) &= \lim_{x \to c} [f(x) - L][g(x) - K] + \lim_{x \to c} Lg(x) + \lim_{x \to c} Kf(x) - \lim_{x \to c} LK \\ &= 0 + LK + KL - LK \\ &= LK.\end{aligned}$$

To prove Property 4, note that it is sufficient to prove that

$$\lim_{x \to c} \frac{1}{g(x)} = \frac{1}{K}.$$

Then you can use Property 3 to write

$$\lim_{x \to c} \frac{f(x)}{g(x)} = \lim_{x \to c} f(x) \frac{1}{g(x)} = \lim_{x \to c} f(x) \cdot \lim_{x \to c} \frac{1}{g(x)} = \frac{L}{K}.$$

Let $\varepsilon > 0$. Because $\lim_{x \to c} g(x) = K$, there exists $\delta_1 > 0$ such that if

$$0 < |x - c| < \delta_1, \text{ then } |g(x) - K| < \frac{|K|}{2}$$

which implies that

$$|K| = |g(x) + [|K| - g(x)]| \le |g(x)| + ||K| - g(x)| < |g(x)| + \frac{|K|}{2}.$$

That is, for $0 < |x - c| < \delta_1$,

$$\frac{|K|}{2} < |g(x)| \quad \text{or} \quad \frac{1}{|g(x)|} < \frac{2}{|K|}.$$

Similarly, there exists a $\delta_2 > 0$ such that if $0 < |x - c| < \delta_2$, then

$$|g(x) - K| < \frac{|K|^2}{2}\varepsilon.$$

Let δ be the smaller of δ_1 and δ_2. For $0 < |x - c| < \delta$, you have

$$\left|\frac{1}{g(x)} - \frac{1}{K}\right| = \left|\frac{K - g(x)}{g(x)K}\right| = \frac{1}{|K|} \cdot \frac{1}{|g(x)|}|K - g(x)| < \frac{1}{|K|} \cdot \frac{2}{|K|} \frac{|K|^2}{2}\varepsilon = \varepsilon.$$

So, $\lim_{x \to c} \dfrac{1}{g(x)} = \dfrac{1}{K}$.

Finally, the proof of Property 5 can be obtained by a straightforward application of mathematical induction coupled with Property 3.

THEOREM 2.4 The Limit of a Function Involving a Radical (page 80)

Let n be a positive integer. The following limit is valid for all c if n is odd, and is valid for $c > 0$ if n is even.

$$\lim_{x \to c} \sqrt[n]{x} = \sqrt[n]{c}.$$

Proof Consider the case for which $c > 0$ and n is any positive integer. For a given $\varepsilon > 0$, you need to find $\delta > 0$ such that

$$\left|\sqrt[n]{x} - \sqrt[n]{c}\right| < \varepsilon \quad \text{whenever} \quad 0 < |x - c| < \delta$$

which is the same as saying

$$-\varepsilon < \sqrt[n]{x} - \sqrt[n]{c} < \varepsilon \quad \text{whenever} \quad -\delta < x - c < \delta.$$

Assume $\varepsilon < \sqrt[n]{c}$, which implies that $0 < \sqrt[n]{c} - \varepsilon < \sqrt[n]{c}$. Now, let δ be the smaller of the two numbers

$$c - \left(\sqrt[n]{c} - \varepsilon\right)^n \quad \text{and} \quad \left(\sqrt[n]{c} + \varepsilon\right)^n - c.$$

Then you have

$$\begin{aligned}
-\delta < x - c &< \delta \\
-\left[c - \left(\sqrt[n]{c} - \varepsilon\right)^n\right] < x - c &< \left(\sqrt[n]{c} + \varepsilon\right)^n - c \\
\left(\sqrt[n]{c} - \varepsilon\right)^n - c < x - c &< \left(\sqrt[n]{c} + \varepsilon\right)^n - c \\
\left(\sqrt[n]{c} - \varepsilon\right)^n < x &< \left(\sqrt[n]{c} + \varepsilon\right)^n \\
\sqrt[n]{c} - \varepsilon < \sqrt[n]{x} &< \sqrt[n]{c} + \varepsilon \\
-\varepsilon < \sqrt[n]{x} - \sqrt[n]{c} &< \varepsilon.
\end{aligned}$$

THEOREM 2.5 The Limit of a Composite Function (page 81)

If f and g are functions such that $\lim_{x \to c} g(x) = L$ and $\lim_{x \to L} f(x) = f(L)$, then

$$\lim_{x \to c} f(g(x)) = f\left(\lim_{x \to c} g(x)\right) = f(L).$$

Proof For a given $\varepsilon > 0$, you must find $\delta > 0$ such that

$$\left|f(g(x)) - f(L)\right| < \varepsilon \quad \text{whenever} \quad 0 < |x - c| < \delta.$$

Because the limit of $f(x)$ as $x \to L$ is $f(L)$, you know there exists $\delta_1 > 0$ such that

$$\left|f(u) - f(L)\right| < \varepsilon \quad \text{whenever} \quad |u - L| < \delta_1.$$

Moreover, because the limit of $g(x)$ as $x \to c$ is L, you know there exists $\delta > 0$ such that

$$\left|g(x) - L\right| < \delta_1 \quad \text{whenever} \quad 0 < |x - c| < \delta.$$

Finally, letting $u = g(x)$, you have

$$\left|f(g(x)) - f(L)\right| < \varepsilon \quad \text{whenever} \quad 0 < |x - c| < \delta.$$

THEOREM 2.7 Functions That Agree at All But One Point (page 82)

Let c be a real number and let $f(x) = g(x)$ for all $x \neq c$ in an open interval containing c. If the limit of $g(x)$ as x approaches c exists, then the limit of $f(x)$ also exists and

$$\lim_{x \to c} f(x) = \lim_{x \to c} g(x).$$

Proof Let L be the limit of $g(x)$ as $x \to c$. Then, for each $\varepsilon > 0$ there exists a $\delta > 0$ such that $f(x) = g(x)$ in the open intervals $(c - \delta, c)$ and $(c, c + \delta)$, and

$$|g(x) - L| < \varepsilon \quad \text{whenever} \quad 0 < |x - c| < \delta.$$

Because $f(x) = g(x)$ for all x in the open interval other than $x = c$, it follows that

$$|f(x) - L| < \varepsilon \quad \text{whenever} \quad 0 < |x - c| < \delta.$$

So, the limit of $f(x)$ as $x \to c$ is also L.

THEOREM 2.8 The Squeeze Theorem (page 85)

If $h(x) \leq f(x) \leq g(x)$ for all x in an open interval containing c, except possibly at c itself, and if

$$\lim_{x \to c} h(x) = L = \lim_{x \to c} g(x)$$

then $\lim_{x \to c} f(x)$ exists and is equal to L.

Proof For $\varepsilon > 0$ there exist $\delta_1 > 0$ and $\delta_2 > 0$ such that

$$|h(x) - L| < \varepsilon \quad \text{whenever} \quad 0 < |x - c| < \delta_1$$

and

$$|g(x) - L| < \varepsilon \quad \text{whenever} \quad 0 < |x - c| < \delta_2.$$

Because $h(x) \leq f(x) \leq g(x)$ for all x in an open interval containing c, except possibly at c itself, there exists $\delta_3 > 0$ such that $h(x) \leq f(x) \leq g(x)$ for $0 < |x - c| < \delta_3$. Let δ be the smallest of δ_1, δ_2, and δ_3. Then, if $0 < |x - c| < \delta$, it follows that $|h(x) - L| < \varepsilon$ and $|g(x) - L| < \varepsilon$, which implies that

$$-\varepsilon < h(x) - L < \varepsilon \quad \text{and} \quad -\varepsilon < g(x) - L < \varepsilon$$
$$L - \varepsilon < h(x) \quad \text{and} \quad g(x) < L + \varepsilon.$$

Now, because $h(x) \leq f(x) \leq g(x)$, it follows that $L - \varepsilon < f(x) < L + \varepsilon$, which implies that $|f(x) - L| < \varepsilon$. Therefore,

$$\lim_{x \to c} f(x) = L.$$

THEOREM 2.14 Vertical Asymptotes (page 105)

Let f and g be continuous on an open interval containing c. If $f(c) \neq 0$, $g(c) = 0$, and there exists an open interval containing c such that $g(x) \neq 0$ for all $x \neq c$ in the interval, then the graph of the function given by

$$h(x) = \frac{f(x)}{g(x)}$$

has a vertical asymptote at $x = c$.

Proof Consider the case for which $f(c) > 0$, and there exists $b > c$ such that $c < x < b$ implies $g(x) > 0$. Then for $M > 0$, choose δ_1 such that

$$0 < x - c < \delta_1 \quad \text{implies that} \quad \frac{f(c)}{2} < f(x) < \frac{3f(c)}{2}$$

and δ_2 such that

$$0 < x - c < \delta_2 \quad \text{implies that} \quad 0 < g(x) < \frac{f(c)}{2M}.$$

Now let δ be the smaller of δ_1 and δ_2. Then it follows that

$$0 < x - c < \delta \quad \text{implies that} \quad \frac{f(x)}{g(x)} > \frac{f(c)}{2}\left[\frac{2M}{f(c)}\right] = M.$$

So, it follows that

$$\lim_{x \to c^+} \frac{f(x)}{g(x)} = \infty$$

and the line $x = c$ is a vertical asymptote of the graph of h.

Alternative Form of the Derivative (page 121)

The derivative of f at c is given by

$$f'(c) = \lim_{x \to c} \frac{f(x) - f(c)}{x - c}$$

provided this limit exists.

Proof The derivative of f at c is given by

$$f'(c) = \lim_{\Delta x \to 0} \frac{f(c + \Delta x) - f(c)}{\Delta x}.$$

Let $x = c + \Delta x$. Then $x \to c$ as $\Delta x \to 0$. So, replacing $c + \Delta x$ by x, you have

$$f'(c) = \lim_{\Delta x \to 0} \frac{f(c + \Delta x) - f(c)}{\Delta x} = \lim_{x \to c} \frac{f(x) - f(c)}{x - c}.$$

THEOREM 3.11 The Chain Rule (page 152)

If $y = f(u)$ is a differentiable function of u, and $u = g(x)$ is a differentiable function of x, then $y = f(g(x))$ is a differentiable function of x and

$$\frac{dy}{dx} = \frac{dy}{du} \cdot \frac{du}{dx}$$

or, equivalently,

$$\frac{d}{dx}[f(g(x))] = f'(g(x))g'(x).$$

Proof In Section 3.4, you let $h(x) = f(g(x))$ and used the alternative form of the derivative to show that $h'(c) = f'(g(c))g'(c)$, provided $g(x) \neq g(c)$ for values of x other than c. Now consider a more general proof. Begin by considering the derivative of f.

$$f'(x) = \lim_{\Delta x \to 0} \frac{f(x + \Delta x) - f(x)}{\Delta x} = \lim_{\Delta x \to 0} \frac{\Delta y}{\Delta x}$$

For a fixed value of x, define a function η such that

$$\eta(\Delta x) = \begin{cases} 0, & \Delta x = 0 \\ \dfrac{\Delta y}{\Delta x} - f'(x), & \Delta x \neq 0 \end{cases}.$$

Because the limit of $\eta(\Delta x)$ as $\Delta x \to 0$ doesn't depend on the value of $\eta(0)$, you have

$$\lim_{\Delta x \to 0} \eta(\Delta x) = \lim_{\Delta x \to 0}\left[\frac{\Delta y}{\Delta x} - f'(x)\right] = 0$$

and you can conclude that η is continuous at 0. Moreover, because $\Delta y = 0$ when $\Delta x = 0$, the equation

$$\Delta y = \Delta x \eta(\Delta x) + \Delta x f'(x)$$

is valid whether Δx is zero or not. Now, by letting $\Delta u = g(x + \Delta x) - g(x)$, you can use the continuity of g to conclude that

$$\lim_{\Delta x \to 0} \Delta u = \lim_{\Delta x \to 0} [g(x + \Delta x) - g(x)] = 0$$

which implies that

$$\lim_{\Delta x \to 0} \eta(\Delta u) = 0.$$

Finally,

$$\Delta y = \Delta u \eta(\Delta u) + \Delta u f'(u) \to \frac{\Delta y}{\Delta x} = \frac{\Delta u}{\Delta x}\eta(\Delta u) + \frac{\Delta u}{\Delta x} f'(u), \quad \Delta x \neq 0$$

and taking the limit as $\Delta x \to 0$, you have

$$\begin{aligned} \frac{dy}{dx} = \frac{du}{dx}\left[\lim_{\Delta x \to 0} \eta(\Delta u)\right] + \frac{du}{dx} f'(u) &= \frac{dy}{dx}(0) + \frac{du}{dx} f'(u) \\ &= \frac{du}{dx} f'(u) \\ &= \frac{du}{dx} \cdot \frac{dy}{du}. \end{aligned}$$

THEOREM 3.16 Continuity and Differentiability of Inverse Functions (page 175)

Let f be a function whose domain is an interval I. If f has an inverse, then the following statements are true.

1. If f is continuous on its domain, then f^{-1} is continuous on its domain.
2. If f is differentiable on an interval containing c and $f'(c) \neq 0$, then f^{-1} is differentiable at $f(c)$.

Proof To prove Property 1, you first need to define what is meant by a *strictly increasing* function or a *strictly decreasing* function. A function f is **strictly increasing** on an entire interval I if for any two numbers x_1 and x_2 in the interval, $x_1 < x_2$ implies $f(x_1) < f(x_2)$. The function f is **strictly decreasing** on the entire interval I if $x_1 < x_2$ implies $f(x_1) > f(x_2)$. The function f is **strictly monotonic** on the interval I if it is either strictly increasing or strictly decreasing. Now show that if f is continuous on I, and has an inverse, then f is strictly monotonic on I. Suppose that f were not strictly monotonic. Then there would exist numbers x_1, x_2, x_3 in I such that $x_1 < x_2 < x_3$, but $f(x_2)$ is not between $f(x_1)$ and $f(x_3)$. Without loss of generality, assume $f(x_1) < f(x_3) < f(x_2)$. By the Intermediate Value Theorem, there exists a number x_0 between x_1 and x_2 such that $f(x_0) = f(x_3)$. So, f is not one-to-one and cannot have an inverse. So, f must be strictly monotonic.

Because f is continuous, the Intermediate Value Theorem implies that the set of values of f, $\{f(x): x \in I\}$, forms an interval J. Assume that a is an interior point of J. From the previous argument, $f^{-1}(a)$ is an interior point of I. Let $\varepsilon > 0$. There exists $0 < \varepsilon_1 < \varepsilon$ such that

$$I_1 = (f^{-1}(a) - \varepsilon_1, f^{-1}(a) + \varepsilon_1) \subseteq I.$$

Because f is strictly monotonic on I_1, the set of values $\{f(x): x \in I_1\}$ forms an interval $J_1 \subseteq J$. Let $\delta > 0$ such that $(a - \delta, a + \delta) \subseteq J_1$. Finally, if $|y - a| < \delta$, then $|f^{-1}(y) - f^{-1}(a)| < \varepsilon_1 < \varepsilon$. So, f^{-1} is continuous at a. A similar proof can be given if a is an endpoint.

To prove Property 2, consider the limit

$$(f^{-1})'(a) = \lim_{y \to a} \frac{f^{-1}(y) - f^{-1}(a)}{y - a}$$

where a is in the domain of f^{-1} and $f^{-1}(a) = c$. Because f is differentiable on an interval containing c, f is continuous on that interval, and so is f^{-1} at a. So, $y \to a$ implies that $x \to c$, and you have

$$(f^{-1})'(a) = \lim_{x \to c} \frac{x - c}{f(x) - f(c)} = \lim_{x \to c} \frac{1}{\left(\dfrac{f(x) - f(c)}{x - c}\right)} = \frac{1}{\displaystyle\lim_{x \to c} \frac{f(x) - f(c)}{x - c}} = \frac{1}{f'(c)}.$$

So, $(f^{-1})'(a)$ exists, and f^{-1} is differentiable at $f(c)$.

THEOREM 3.17 The Derivative of an Inverse Function (page 175)

Let f be a function that is differentiable on an interval I. If f has an inverse function g, then g is differentiable at any x for which $f'(g(x)) \neq 0$. Moreover,

$$g'(x) = \frac{1}{f'(g(x))}, \quad f'(g(x)) \neq 0.$$

Proof From the proof of Theorem 3.16, letting $a = x$, you know that g is differentiable. Using the Chain Rule, differentiate both sides of the equation $x = f(g(x))$ to obtain

$$1 = f'(g(x))\frac{d}{dx}[g(x)].$$

Because $f'(g(x)) \neq 0$, you can divide by this quantity to obtain

$$\frac{d}{dx}[g(x)] = \frac{1}{f'(g(x))}.$$

Concavity Interpretation (page 230)

1. Let f be differentiable on an open interval I. If the graph of f is concave *upward* on I, then the graph of f lies *above* all of its tangent lines on I.
2. Let f be differentiable on an open interval I. If the graph of f is concave *downward* on I, then the graph of f lies *below* all of its tangent lines on I.

Proof Assume that f is concave upward on $I = (a, b)$. Then, f' is increasing on (a, b). Let c be a point in the interval $I = (a, b)$. The equation of the tangent line to the graph of f at c is given by

$$g(x) = f(c) + f'(c)(x - c).$$

If x is in the open interval (c, b), then the directed distance from the point $(x, f(x))$ (on the graph of f) to the point $(x, g(x))$ (on the tangent line) is given by

$$\begin{aligned} d &= f(x) - [f(c) + f'(c)(x - c)] \\ &= f(x) - f(c) - f'(c)(x - c). \end{aligned}$$

Moreover, by the Mean Value Theorem, there exists a number z in (c, x) such that

$$f'(z) = \frac{f(x) - f(c)}{x - c}.$$

So, you have

$$\begin{aligned} d &= f(x) - f(c) - f'(c)(x - c) \\ &= f'(z)(x - c) - f'(c)(x - c) \\ &= [f'(z) - f'(c)](x - c). \end{aligned}$$

The second factor $(x - c)$ is positive because $c < x$. Moreover, because f' is increasing, it follows that the first factor $[f'(z) - f'(c)]$ is also positive. Therefore, $d > 0$ and you can conclude that the graph of f lies above the tangent line at x. If x is in the open interval (a, c), a similar argument can be given. This proves the first statement. The proof of the second statement is similar.

THEOREM 4.10 Limits at Infinity (page 239)

If r is a positive rational number and c is any real number, then

$$\lim_{x\to\infty} \frac{c}{x^r} = 0.$$

Furthermore, if x^r is defined when $x < 0$, then $\lim_{x\to-\infty} \frac{c}{x^r} = 0.$

Proof Begin by proving that

$$\lim_{x\to\infty} \frac{1}{x} = 0.$$

For $\varepsilon > 0$, let $M = 1/\varepsilon$. Then, for $x > M$, you have

$$x > M = \frac{1}{\varepsilon} \quad \Longrightarrow \quad \frac{1}{x} < \varepsilon \quad \Longrightarrow \quad \left|\frac{1}{x} - 0\right| < \varepsilon.$$

So, by the definition of a limit at infinity, you can conclude that the limit of $1/x$ as $x \to \infty$ is 0. Now, using this result and letting $r = m/n$, you can write the following.

$$\begin{aligned}
\lim_{x\to\infty} \frac{c}{x^r} &= \lim_{x\to\infty} \frac{c}{x^{m/n}} \\
&= c\left[\lim_{x\to\infty}\left(\frac{1}{\sqrt[n]{x}}\right)^m\right] \\
&= c\left(\lim_{x\to\infty}\sqrt[n]{\frac{1}{x}}\right)^m \\
&= c\left(\sqrt[n]{\lim_{x\to\infty}\frac{1}{x}}\right)^m \\
&= c\left(\sqrt[n]{0}\right)^m \\
&= 0
\end{aligned}$$

The proof of the second part of the theorem is similar.

THEOREM 5.2 Summation Formulas (page 296)

1. $\displaystyle\sum_{i=1}^{n} c = cn$
2. $\displaystyle\sum_{i=1}^{n} i = \frac{n(n+1)}{2}$
3. $\displaystyle\sum_{i=1}^{n} i^2 = \frac{n(n+1)(2n+1)}{6}$
4. $\displaystyle\sum_{i=1}^{n} i^3 = \frac{n^2(n+1)^2}{4}$

Proof The proof of Property 1 is straightforward. By adding c to itself n times, you obtain a sum of cn.

To prove Property 2, write the sum in increasing and decreasing order and add corresponding terms, as follows.

$$\begin{aligned}
\sum_{i=1}^{n} i &= 1 + 2 + 3 + \cdots + (n-1) + n \\
\sum_{i=1}^{n} i &= n + (n-1) + (n-2) + \cdots + 2 + 1 \\
2\sum_{i=1}^{n} i &= \underbrace{(n+1) + (n+1) + (n+1) + \cdots + (n+1) + (n+1)}_{n \text{ terms}}
\end{aligned}$$

So,

$$\sum_{i=1}^{n} i = \frac{n(n+1)}{2}.$$

To prove Property 3, use mathematical induction. First, if $n = 1$, the result is true because

$$\sum_{i=1}^{1} i^2 = 1^2 = 1 = \frac{1(1+1)(2+1)}{6}.$$

Now, assuming the result is true for $n = k$, you can show that it is true for $n = k + 1$, as follows.

$$\begin{aligned}
\sum_{i=1}^{k+1} i^2 &= \sum_{i=1}^{k} i^2 + (k+1)^2 \\
&= \frac{k(k+1)(2k+1)}{6} + (k+1)^2 \\
&= \frac{k+1}{6}(2k^2 + k + 6k + 6) \\
&= \frac{k+1}{6}[(2k+3)(k+2)] \\
&= \frac{(k+1)(k+2)[2(k+1)+1]}{6}
\end{aligned}$$

Property 4 can be proved using a similar argument with mathematical induction.

THEOREM 5.8 Preservation of Inequality (page 314)

1. If f is integrable and nonnegative on the closed interval $[a, b]$, then

$$0 \le \int_a^b f(x)\,dx.$$

2. If f and g are integrable on the closed interval $[a, b]$ and $f(x) \le g(x)$ for every x in $[a, b]$, then

$$\int_a^b f(x)\,dx \le \int_a^b g(x)\,dx.$$

Proof To prove Property 1, suppose, on the contrary, that

$$\int_a^b f(x)\,dx = I < 0.$$

Then, let $a = x_0 < x_1 < x_2 < \cdots < x_n = b$ be a partition of $[a, b]$, and let

$$R = \sum_{i=1}^{n} f(c_i)\,\Delta x_i$$

be a Riemann sum. Because $f(x) \ge 0$, it follows that $R \ge 0$. Now, for $\|\Delta\|$ sufficiently small, you have $|R - I| < -I/2$, which implies that

$$\sum_{i=1}^{n} f(c_i)\,\Delta x_i = R < I - \frac{I}{2} < 0$$

which is not possible. From this contradiction, you can conclude that

$$0 \le \int_a^b f(x)\,dx.$$

To prove Property 2 of the theorem, note that $f(x) \le g(x)$ implies that $g(x) - f(x) \ge 0$. So, you can apply the result of Property 1 to conclude that

$$0 \le \int_a^b [g(x) - f(x)]\,dx$$

$$0 \le \int_a^b g(x)\,dx - \int_a^b f(x)\,dx$$

$$\int_a^b f(x)\,dx \le \int_a^b g(x)\,dx.$$

THEOREM 8.3 The Extended Mean Value Theorem (page 568)

If f and g are differentiable on an open interval (a, b) and continuous on $[a, b]$ such that $g'(x) \neq 0$ for any x in (a, b), then there exists a point c in (a, b) such that

$$\frac{f'(c)}{g'(c)} = \frac{f(b) - f(a)}{g(b) - g(a)}.$$

Proof You can assume that $g(a) \neq g(b)$, because otherwise, by Rolle's Theorem, it would follow that $g'(x) = 0$ for some x in (a, b). Now, define $h(x)$ to be

$$h(x) = f(x) - \left[\frac{f(b) - f(a)}{g(b) - g(a)}\right] g(x).$$

Then

$$h(a) = f(a) - \left[\frac{f(b) - f(a)}{g(b) - g(a)}\right] g(a) = \frac{f(a)g(b) - f(b)g(a)}{g(b) - g(a)}$$

and

$$h(b) = f(b) - \left[\frac{f(b) - f(a)}{g(b) - g(a)}\right] g(b) = \frac{f(a)g(b) - f(b)g(a)}{g(b) - g(a)}$$

and, by Rolle's Theorem, there exists a point c in (a, b) such that

$$h'(c) = f'(c) - \frac{f(b) - f(a)}{g(b) - g(a)} g'(c) = 0$$

which implies that

$$\frac{f'(c)}{g'(c)} = \frac{f(b) - f(a)}{g(b) - g(a)}.$$

THEOREM 8.4 L'Hôpital's Rule (page 568)

Let f and g be functions that are differentiable on an open interval (a, b) containing c, except possibly at c itself. Assume that $g'(x) \neq 0$ for all x in (a, b), except possibly at c itself. If the limit of $f(x)/g(x)$ as x approaches c produces the indeterminate form $0/0$, then

$$\lim_{x\to c} \frac{f(x)}{g(x)} = \lim_{x\to c} \frac{f'(x)}{g'(x)}$$

provided the limit on the right exists (or is infinite). This result also applies if the limit of $f(x)/g(x)$ as x approaches c produces any one of the indeterminate forms ∞/∞, $(-\infty)/\infty$, $\infty/(-\infty)$, or $(-\infty)/(-\infty)$.

You can use the Extended Mean Value Theorem to prove L'Hôpital's Rule. Of the several different cases of this rule, the proof of only one case is illustrated. The remaining cases where $x \to c^-$ and $x \to c$ are left for you to prove.

Proof Consider the case for which

$$\lim_{x\to c^+} f(x) = 0 \quad \text{and} \quad \lim_{x\to c^+} g(x) = 0.$$

Define the following new functions:

$$F(x) = \begin{cases} f(x), & x \neq c \\ 0, & x = c \end{cases} \quad \text{and} \quad G(x) = \begin{cases} g(x), & x \neq c \\ 0, & x = c \end{cases}.$$

For any x, $c < x < b$, F and G are differentiable on $(c, x]$ and continuous on $[c, x]$. You can apply the Extended Mean Value Theorem to conclude that there exists a number z in (c, x) such that

$$\begin{aligned} \frac{F'(z)}{G'(z)} &= \frac{F(x) - F(c)}{G(x) - G(c)} \\ &= \frac{F(x)}{G(x)} \\ &= \frac{f'(z)}{g'(z)} \\ &= \frac{f(x)}{g(x)}. \end{aligned}$$

Finally, by letting x approach c from the right, $x \to c^+$, you have $z \to c^+$ because $c < z < x$, and

$$\begin{aligned} \lim_{x\to c^+} \frac{f(x)}{g(x)} &= \lim_{x\to c^+} \frac{f'(z)}{g'(z)} \\ &= \lim_{z\to c^+} \frac{f'(z)}{g'(z)} \\ &= \lim_{x\to c^+} \frac{f'(x)}{g'(x)}. \end{aligned}$$

THEOREM 9.19 Taylor's Theorem (page 654)

If a function f is differentiable through order $n + 1$ in an interval I containing c, then, for each x in I, there exists z between x and c such that

$$f(x) = f(c) + f'(c)(x - c) + \frac{f''(c)}{2!}(x - c)^2 + \cdot \cdot \cdot + \frac{f^{(n)}(c)}{n!}(x - c)^n + R_n(x)$$

where

$$R_n(x) = \frac{f^{(n+1)}(z)}{(n + 1)!}(x - c)^{n+1}.$$

Proof To find $R_n(x)$, fix x in I $(x \neq c)$ and write

$$R_n(x) = f(x) - P_n(x)$$

where $P_n(x)$ is the nth Taylor polynomial for $f(x)$. Then let g be a function of t defined by

$$g(t) = f(x) - f(t) - f'(t)(x - t) - \cdot \cdot \cdot - \frac{f^{(n)}(t)}{n!}(x - t)^n - R_n(x)\frac{(x - t)^{n+1}}{(x - c)^{n+1}}.$$

The reason for defining g in this way is that differentiation with respect to t has a telescoping effect. For example, you have

$$\begin{aligned}\frac{d}{dt}[-f(t) - f'(t)(x - t)] &= -f'(t) + f'(t) - f''(t)(x - t)\\ &= -f''(t)(x - t).\end{aligned}$$

The result is that the derivative $g'(t)$ simplifies to

$$g'(t) = -\frac{f^{(n+1)}(t)}{n!}(x - t)^n + (n + 1)R_n(x)\frac{(x - t)^n}{(x - c)^{n+1}}$$

for all t between c and x. Moreover, for a fixed x,

$$g(c) = f(x) - [P_n(x) + R_n(x)] = f(x) - f(x) = 0$$

and

$$g(x) = f(x) - f(x) - 0 - \cdot \cdot \cdot - 0 = f(x) - f(x) = 0.$$

Therefore, g satisfies the conditions of Rolle's Theorem, and it follows that there is a number z between c and x such that $g'(z) = 0$. Substituting z for t in the equation for $g'(t)$ and then solving for $R_n(x)$, you obtain

$$g'(z) = -\frac{f^{(n+1)}(z)}{n!}(x - z)^n + (n + 1)R_n(x)\frac{(x - z)^n}{(x - c)^{n+1}} = 0$$

$$R_n(x) = \frac{f^{(n+1)}(z)}{(n + 1)!}(x - c)^{n+1}.$$

Finally, because $g(c) = 0$, you have

$$0 = f(x) - f(c) - f'(c)(x - c) - \cdot \cdot \cdot - \frac{f^{(n)}(c)}{n!}(x - c)^n - R_n(x)$$

$$f(x) = f(c) + f'(c)(x - c) + \cdot \cdot \cdot + \frac{f^{(n)}(c)}{n!}(x - c)^n + R_n(x).$$

THEOREM 9.20 Convergence of a Power Series (page 660)

For a power series centered at c, precisely one of the following is true.

1. The series converges only at c.
2. There exists a real number $R > 0$ such that the series converges absolutely for $|x - c| < R$, and diverges for $|x - c| > R$.
3. The series converges absolutely for all x.

The number R is the **radius of convergence** of the power series. If the series converges only at c, the radius of convergence is $R = 0$, and if the series converges for all x, the radius of convergence is $R = \infty$. The set of all values of x for which the power series converges is the **interval of convergence** of the power series.

Proof In order to simplify the notation, the theorem for the power series $\Sigma\, a_n x^n$ centered at $x = 0$ will be proved. The proof for a power series centered at $x = c$ follows easily. A key step in this proof uses the completeness property of the set of real numbers: If a nonempty set S of real numbers has an upper bound, then it must have a least upper bound (see page 601).

It must be shown that if a power series $\Sigma\, a_n x^n$ converges at $x = d$, $d \neq 0$, then it converges for all b satisfying $|b| < |d|$. Because $\Sigma\, a_n x^n$ converges, $\lim_{x\to\infty} a_n d^n = 0$. So, there exists $N > 0$ such that $a_n d^n < 1$ for all $n \geq N$. Then for $n \geq N$,

$$|a_n b^n| = \left|a_n b^n \frac{d^n}{d^n}\right| = |a_n d^n|\left|\frac{b^n}{d^n}\right| < \left|\frac{b^n}{d^n}\right|.$$

So, for $|b| < |d|$, $\left|\dfrac{b}{d}\right| < 1$, which implies that

$$\Sigma \left|\frac{b^n}{d^n}\right|$$

is a convergent geometric series. By the Comparison Test, the series $\Sigma\, a_n b^n$ converges.

Similarly, if the power series $\Sigma\, a_n x^n$ diverges at $x = b$, where $b \neq 0$, then it diverges for all d satisfying $|d| > |b|$. If $\Sigma\, a_n d^n$ converged, then the argument above would imply that $\Sigma\, a_n b^n$ converged as well.

Finally, to prove the theorem, suppose that neither case 1 nor case 3 is true. Then there exist points b and d such that $\Sigma\, a_n x^n$ converges at b and diverges at d. Let $S = \{x\colon \Sigma\, a_n x^n \text{ converges}\}$. S is nonempty because $b \in S$. If $x \in S$, then $|x| \leq |d|$, which shows that $|d|$ is an upper bound for the nonempty set S. By the completeness property, S has a least upper bound, R.
Now, if $|x| > R$, then $x \notin S$ so $\Sigma\, a_n x^n$ diverges. And if $|x| < R$, then $|x|$ is not an upper bound for S, so there exists b in S satisfying $|b| > |x|$. Since $b \in S$, $\Sigma\, a_n b^n$ converges, which implies that $\Sigma\, a_n x^n$ converges.

THEOREM 10.16 Classification of Conics by Eccentricity (page 748)

Let F be a fixed point (*focus*) and let D be a fixed line (*directrix*) in the plane. Let P be another point in the plane and let e (*eccentricity*) be the ratio of the distance between P and F to the distance between P and D. The collection of all points P with a given eccentricity is a conic.

1. The conic is an ellipse if $0 < e < 1$.
2. The conic is a parabola if $e = 1$.
3. The conic is a hyperbola if $e > 1$.

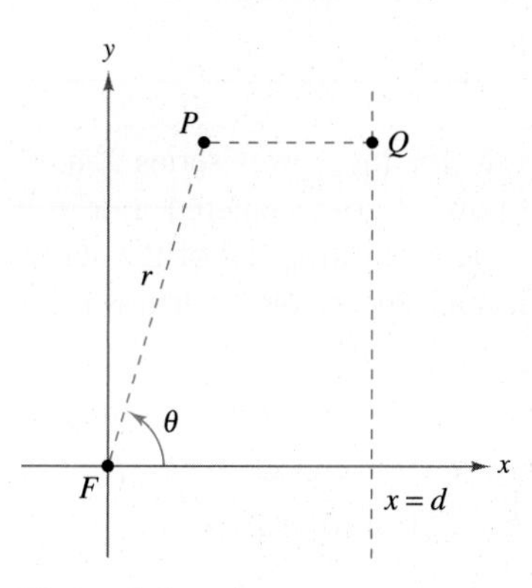

Figure A.1

Proof If $e = 1$, then, by definition, the conic must be a parabola. If $e \neq 1$, then you can consider the focus F to lie at the origin and the directrix $x = d$ to lie to the right of the origin, as shown in Figure A.1. For the point $P = (r, \theta) = (x, y)$, you have $|PF| = r$ and $|PQ| = d - r\cos\theta$. Given that $e = |PF|/|PQ|$, it follows that

$$|PF| = |PQ|e \quad \Longrightarrow \quad r = e(d - r\cos\theta).$$

By converting to rectangular coordinates and squaring each side, you obtain

$$x^2 + y^2 = e^2(d - x)^2 = e^2(d^2 - 2dx + x^2).$$

Completing the square produces

$$\left(x + \frac{e^2 d}{1 - e^2}\right)^2 + \frac{y^2}{1 - e^2} = \frac{e^2 d^2}{(1 - e^2)^2}.$$

If $e < 1$, this equation represents an ellipse. If $e > 1$, then $1 - e^2 < 0$, and the equation represents a hyperbola.

THEOREM 13.4 Sufficient Condition for Differentiability (page 917)

If f is a function of x and y, where f_x and f_y are continuous in an open region R, then f is differentiable on R.

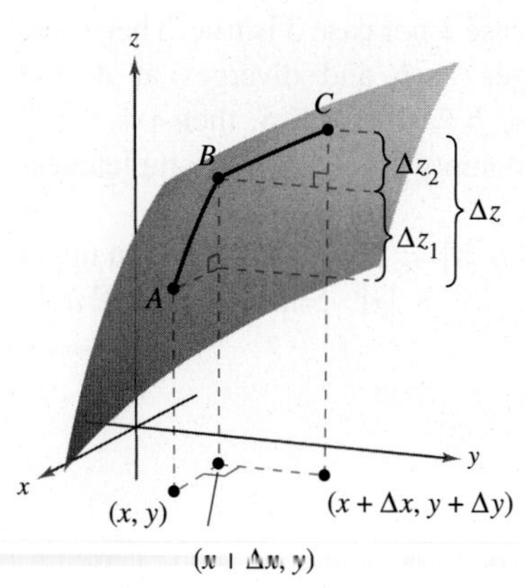

$\Delta z = f(x + \Delta x, y + \Delta y) - f(x, y)$
Figure A.2

Proof Let S be the surface defined by $z = f(x, y)$, where f, f_x, and f_y are continuous at (x, y). Let A, B, and C be points on surface S, as shown in Figure A.2. From this figure, you can see that the change in f from point A to point C is given by

$$\begin{aligned}\Delta z &= f(x + \Delta x, y + \Delta y) - f(x, y)\\ &= [f(x + \Delta x, y) - f(x, y)] + [f(x + \Delta x, y + \Delta y) - f(x + \Delta x, y)]\\ &= \Delta z_1 + \Delta z_2.\end{aligned}$$

Between A and B, y is fixed and x changes. So, by the Mean Value Theorem, there is a value x_1 between x and $x + \Delta x$ such that

$$\Delta z_1 = f(x + \Delta x, y) - f(x, y) = f_x(x_1, y)\,\Delta x.$$

Similarly, between B and C, x is fixed and y changes, and there is a value y_1 between y and $y + \Delta y$ such that

$$\Delta z_2 = f(x + \Delta x, y + \Delta y) - f(x + \Delta x, y) = f_y(x + \Delta x, y_1)\,\Delta y.$$

By combining these two results, you can write

$$\Delta z = \Delta z_1 + \Delta z_2 = f_x(x_1, y)\Delta x + f_y(x + \Delta x, y_1)\,\Delta y.$$

If you define ε_1 and ε_2 as

$$\varepsilon_1 = f_x(x_1, y) - f_x(x, y) \quad \text{and} \quad \varepsilon_2 = f_y(x + \Delta x, y_1) - f_y(x, y)$$

it follows that

$$\begin{aligned}\Delta z = \Delta z_1 + \Delta z_2 &= [\varepsilon_1 + f_x(x, y)]\,\Delta x + [\varepsilon_2 + f_y(x, y)]\,\Delta y\\ &= [f_x(x, y)\,\Delta x + f_y(x, y)\,\Delta y] + \varepsilon_1\Delta x + \varepsilon_2\Delta y.\end{aligned}$$

By the continuity of f_x and f_y and the fact that $x \le x_1 \le x + \Delta x$ and $y \le y_1 \le y + \Delta y$, it follows that $\varepsilon_1 \to 0$ and $\varepsilon_2 \to 0$ as $\Delta x \to 0$ and $\Delta y \to 0$. Therefore, by definition, f is differentiable.

THEOREM 13.6 Chain Rule: One Independent Variable (page 923)

Let $w = f(x, y)$, where f is a differentiable function of x and y. If $x = g(t)$ and $y = h(t)$, where g and h are differentiable functions of t, then w is a differentiable function of t, and

$$\frac{dw}{dt} = \frac{\partial w}{\partial x}\frac{dx}{dt} + \frac{\partial w}{\partial y}\frac{dy}{dt}.$$

Proof Because g and h are differentiable functions of t, you know that both Δx and Δy approach zero as Δt approaches zero. Moreover, because f is a differentiable function of x and y, you know that

$$\Delta w = \frac{\partial w}{\partial x}\Delta x + \frac{\partial w}{\partial y}\Delta y + \varepsilon_1\Delta x + \varepsilon_2\Delta y$$

where both ε_1 and $\varepsilon_2 \to 0$ as $(\Delta x, \Delta y) \to (0, 0)$. So, for $\Delta t \ne 0$,

$$\frac{\Delta w}{\Delta t} = \frac{\partial w}{\partial x}\frac{\Delta x}{\Delta t} + \frac{\partial w}{\partial y}\frac{\Delta y}{\Delta t} + \varepsilon_1\frac{\Delta x}{\Delta t} + \varepsilon_2\frac{\Delta y}{\Delta t}$$

from which it follows that

$$\begin{aligned}\frac{dw}{dt} = \lim_{\Delta t \to 0}\frac{\Delta w}{\Delta t} &= \frac{\partial w}{\partial x}\frac{dx}{dt} + \frac{\partial w}{\partial y}\frac{dy}{dt} + 0\left(\frac{dx}{dt}\right) + 0\left(\frac{dy}{dt}\right)\\ &= \frac{\partial w}{\partial x}\frac{dx}{dt} + \frac{\partial w}{\partial y}\frac{dy}{dt}.\end{aligned}$$

B Integration Tables

Forms Involving u^n

1. $\displaystyle\int u^n\,du = \frac{u^{n+1}}{n+1} + C,\ n \neq -1$

2. $\displaystyle\int \frac{1}{u}\,du = \ln|u| + C$

Forms Involving $a + bu$

3. $\displaystyle\int \frac{u}{a+bu}\,du = \frac{1}{b^2}\left(bu - a\ln|a+bu|\right) + C$

4. $\displaystyle\int \frac{u}{(a+bu)^2}\,du = \frac{1}{b^2}\left(\frac{a}{a+bu} + \ln|a+bu|\right) + C$

5. $\displaystyle\int \frac{u}{(a+bu)^n}\,du = \frac{1}{b^2}\left[\frac{-1}{(n-2)(a+bu)^{n-2}} + \frac{a}{(n-1)(a+bu)^{n-1}}\right] + C,\quad n \neq 1, 2$

6. $\displaystyle\int \frac{u^2}{a+bu}\,du = \frac{1}{b^3}\left[-\frac{bu}{2}(2a - bu) + a^2\ln|a+bu|\right] + C$

7. $\displaystyle\int \frac{u^2}{(a+bu)^2}\,du = \frac{1}{b^3}\left(bu - \frac{a^2}{a+bu} - 2a\ln|a+bu|\right) + C$

8. $\displaystyle\int \frac{u^2}{(a+bu)^3}\,du = \frac{1}{b^3}\left[\frac{2a}{a+bu} - \frac{a^2}{2(a+bu)^2} + \ln|a+bu|\right] + C$

9. $\displaystyle\int \frac{u^2}{(a+bu)^n}\,du = \frac{1}{b^3}\left[\frac{-1}{(n-3)(a+bu)^{n-3}} + \frac{2a}{(n-2)(a+bu)^{n-2}} - \frac{a^2}{(n-1)(a+bu)^{n-1}}\right] + C,\quad n \neq 1, 2, 3$

10. $\displaystyle\int \frac{1}{u(a+bu)}\,du = \frac{1}{a}\ln\left|\frac{u}{a+bu}\right| + C$

11. $\displaystyle\int \frac{1}{u(a+bu)^2}\,du = \frac{1}{a}\left(\frac{1}{a+bu} + \frac{1}{a}\ln\left|\frac{u}{a+bu}\right|\right) + C$

12. $\displaystyle\int \frac{1}{u^2(a+bu)}\,du = -\frac{1}{a}\left(\frac{1}{u} + \frac{b}{a}\ln\left|\frac{u}{a+bu}\right|\right) + C$

13. $\displaystyle\int \frac{1}{u^2(a+bu)^2}\,du = -\frac{1}{a^2}\left[\frac{a+2bu}{u(a+bu)} + \frac{2b}{a}\ln\left|\frac{u}{a+bu}\right|\right] + C$

Forms Involving $a + bu + cu^2,\ b^2 \neq 4ac$

14. $$\int \frac{1}{a + bu + cu^2}\,du = \begin{cases} \dfrac{2}{\sqrt{4ac - b^2}} \arctan \dfrac{2cu + b}{\sqrt{4ac - b^2}} + C, & b^2 < 4ac \\[2ex] \dfrac{1}{\sqrt{b^2 - 4ac}} \ln \left| \dfrac{2cu + b - \sqrt{b^2 - 4ac}}{2cu + b + \sqrt{b^2 - 4ac}} \right| + C, & b^2 > 4ac \end{cases}$$

15. $$\int \frac{u}{a + bu + cu^2}\,du = \frac{1}{2c}\left(\ln\left|a + bu + cu^2\right| - b\int \frac{1}{a + bu + cu^2}\,du\right)$$

Forms Involving $\sqrt{a + bu}$

16. $$\int u^n\sqrt{a + bu}\,du = \frac{2}{b(2n + 3)}\left[u^n(a + bu)^{3/2} - na\int u^{n-1}\sqrt{a + bu}\,du\right]$$

17. $$\int \frac{1}{u\sqrt{a + bu}}\,du = \begin{cases} \dfrac{1}{\sqrt{a}} \ln \left| \dfrac{\sqrt{a + bu} - \sqrt{a}}{\sqrt{a + bu} + \sqrt{a}} \right| + C, & a > 0 \\[2ex] \dfrac{2}{\sqrt{-a}} \arctan \sqrt{\dfrac{a + bu}{-a}} + C, & a < 0 \end{cases}$$

18. $$\int \frac{1}{u^n\sqrt{a + bu}}\,du = \frac{-1}{a(n - 1)}\left[\frac{\sqrt{a + bu}}{u^{n-1}} + \frac{(2n - 3)b}{2}\int \frac{1}{u^{n-1}\sqrt{a + bu}}\,du\right],\ n \neq 1$$

19. $$\int \frac{\sqrt{a + bu}}{u}\,du = 2\sqrt{a + bu} + a\int \frac{1}{u\sqrt{a + bu}}\,du$$

20. $$\int \frac{\sqrt{a + bu}}{u^n}\,du = \frac{-1}{a(n - 1)}\left[\frac{(a + bu)^{3/2}}{u^{n-1}} + \frac{(2n - 5)b}{2}\int \frac{\sqrt{a + bu}}{u^{n-1}}\,du\right],\ n \neq 1$$

21. $$\int \frac{u}{\sqrt{a + bu}}\,du = \frac{-2(2a - bu)}{3b^2}\sqrt{a + bu} + C$$

22. $$\int \frac{u^n}{\sqrt{a + bu}}\,du = \frac{2}{(2n + 1)b}\left(u^n\sqrt{a + bu} - na\int \frac{u^{n-1}}{\sqrt{a + bu}}\,du\right)$$

Forms Involving $a^2 \pm u^2,\ a > 0$

23. $$\int \frac{1}{a^2 + u^2}\,du = \frac{1}{a}\arctan\frac{u}{a} + C$$

24. $$\int \frac{1}{u^2 - a^2}\,du = -\int \frac{1}{a^2 - u^2}\,du = \frac{1}{2a}\ln\left|\frac{u - a}{u + a}\right| + C$$

25. $$\int \frac{1}{(a^2 \pm u^2)^n}\,du = \frac{1}{2a^2(n - 1)}\left[\frac{u}{(a^2 \pm u^2)^{n-1}} + (2n - 3)\int \frac{1}{(a^2 \pm u^2)^{n-1}}\,du\right],\ n \neq 1$$

Forms Involving $\sqrt{u^2 \pm a^2},\ a > 0$

26. $$\int \sqrt{u^2 \pm a^2}\,du = \frac{1}{2}\left(u\sqrt{u^2 \pm a^2} \pm a^2\ln\left|u + \sqrt{u^2 \pm a^2}\right|\right) + C$$

27. $$\int u^2\sqrt{u^2 \pm a^2}\,du = \frac{1}{8}\left[u(2u^2 \pm a^2)\sqrt{u^2 \pm a^2} - a^4\ln\left|u + \sqrt{u^2 \pm a^2}\right|\right] + C$$

28. $$\int \frac{\sqrt{u^2 + a^2}}{u}\,du = \sqrt{u^2 + a^2} - a\ln\left|\frac{a + \sqrt{u^2 + a^2}}{u}\right| + C$$

29. $\displaystyle\int \frac{\sqrt{u^2 - a^2}}{u}\,du = \sqrt{u^2 - a^2} - a \operatorname{arcsec}\frac{|u|}{a} + C$

30. $\displaystyle\int \frac{\sqrt{u^2 \pm a^2}}{u^2}\,du = \frac{-\sqrt{u^2 \pm a^2}}{u} + \ln\left|u + \sqrt{u^2 \pm a^2}\right| + C$

31. $\displaystyle\int \frac{1}{\sqrt{u^2 \pm a^2}}\,du = \ln\left|u + \sqrt{u^2 \pm a^2}\right| + C$

32. $\displaystyle\int \frac{1}{u\sqrt{u^2 + a^2}}\,du = \frac{-1}{a}\ln\left|\frac{a + \sqrt{u^2 + a^2}}{u}\right| + C$

33. $\displaystyle\int \frac{1}{u\sqrt{u^2 - a^2}}\,du = \frac{1}{a}\operatorname{arcsec}\frac{|u|}{a} + C$

34. $\displaystyle\int \frac{u^2}{\sqrt{u^2 \pm a^2}}\,du = \frac{1}{2}\left(u\sqrt{u^2 \pm a^2} \mp a^2 \ln\left|u + \sqrt{u^2 \pm a^2}\right|\right) + C$

35. $\displaystyle\int \frac{1}{u^2\sqrt{u^2 \pm a^2}}\,du = \mp\frac{\sqrt{u^2 \pm a^2}}{a^2 u} + C$

36. $\displaystyle\int \frac{1}{(u^2 \pm a^2)^{3/2}}\,du = \frac{\pm u}{a^2\sqrt{u^2 \pm a^2}} + C$

Forms Involving $\sqrt{a^2 - u^2},\ a > 0$

37. $\displaystyle\int \sqrt{a^2 - u^2}\,du = \frac{1}{2}\left(u\sqrt{a^2 - u^2} + a^2 \arcsin\frac{u}{a}\right) + C$

38. $\displaystyle\int u^2\sqrt{a^2 - u^2}\,du = \frac{1}{8}\left[u(2u^2 - a^2)\sqrt{a^2 - u^2} + a^4 \arcsin\frac{u}{a}\right] + C$

39. $\displaystyle\int \frac{\sqrt{a^2 - u^2}}{u}\,du = \sqrt{a^2 - u^2} - a\ln\left|\frac{a + \sqrt{a^2 - u^2}}{u}\right| + C$

40. $\displaystyle\int \frac{\sqrt{a^2 - u^2}}{u^2}\,du = \frac{-\sqrt{a^2 - u^2}}{u} - \arcsin\frac{u}{a} + C$

41. $\displaystyle\int \frac{1}{\sqrt{a^2 - u^2}}\,du = \arcsin\frac{u}{a} + C$

42. $\displaystyle\int \frac{1}{u\sqrt{a^2 - u^2}}\,du = \frac{-1}{a}\ln\left|\frac{a + \sqrt{a^2 - u^2}}{u}\right| + C$

43. $\displaystyle\int \frac{u^2}{\sqrt{a^2 - u^2}}\,du = \frac{1}{2}\left(-u\sqrt{a^2 - u^2} + a^2 \arcsin\frac{u}{a}\right) + C$

44. $\displaystyle\int \frac{1}{u^2\sqrt{a^2 - u^2}}\,du = \frac{-\sqrt{a^2 - u^2}}{a^2 u} + C$

45. $\displaystyle\int \frac{1}{(a^2 - u^2)^{3/2}}\,du = \frac{u}{a^2\sqrt{a^2 - u^2}} + C$

Forms Involving $\sin u$ or $\cos u$

46. $\displaystyle\int \sin u\,du = -\cos u + C$

47. $\displaystyle\int \cos u\,du = \sin u + C$

48. $\displaystyle\int \sin^2 u\,du = \frac{1}{2}(u - \sin u \cos u) + C$

49. $\displaystyle\int \cos^2 u\,du = \frac{1}{2}(u + \sin u \cos u) + C$

50. $\displaystyle\int \sin^n u\,du = -\frac{\sin^{n-1} u \cos u}{n} + \frac{n-1}{n}\int \sin^{n-2} u\,du$

51. $\displaystyle\int \cos^n u\,du = \frac{\cos^{n-1} u \sin u}{n} + \frac{n-1}{n}\int \cos^{n-2} u\,du$

52. $\displaystyle\int u \sin u\,du = \sin u - u\cos u + C$

53. $\displaystyle\int u \cos u\,du = \cos u + u\sin u + C$

54. $\displaystyle\int u^n \sin u\,du = -u^n \cos u + n\int u^{n-1}\cos u\,du$

55. $\displaystyle\int u^n \cos u\,du = u^n \sin u - n\int u^{n-1}\sin u\,du$

56. $\displaystyle\int \frac{1}{1 \pm \sin u}\,du = \tan u \mp \sec u + C$

57. $\displaystyle\int \frac{1}{1 \pm \cos u}\,du = -\cot u \pm \csc u + C$

58. $\displaystyle\int \frac{1}{\sin u \cos u}\,du = \ln|\tan u| + C$

Forms Involving $\tan u$, $\cot u$, $\sec u$, $\csc u$

59. $\displaystyle\int \tan u\,du = -\ln|\cos u| + C$

60. $\displaystyle\int \cot u\,du = \ln|\sin u| + C$

61. $\displaystyle\int \sec u\,du = \ln|\sec u + \tan u| + C$

62. $\displaystyle\int \csc u\,du = \ln|\csc u - \cot u| + C \quad \text{or} \quad \int \csc u\,du = -\ln|\csc u + \cot u| + C$

63. $\displaystyle\int \tan^2 u\,du = -u + \tan u + C$

64. $\displaystyle\int \cot^2 u\,du = -u - \cot u + C$

65. $\displaystyle\int \sec^2 u\,du = \tan u + C$

66. $\displaystyle\int \csc^2 u\,du = -\cot u + C$

67. $\displaystyle\int \tan^n u\,du = \frac{\tan^{n-1} u}{n-1} - \int \tan^{n-2} u\,du,\ n \neq 1$

68. $\displaystyle\int \cot^n u\,du = -\frac{\cot^{n-1} u}{n-1} - \int (\cot^{n-2} u)\,du,\ n \neq 1$

69. $\displaystyle\int \sec^n u\,du = \frac{\sec^{n-2} u \tan u}{n-1} + \frac{n-2}{n-1}\int \sec^{n-2} u\,du,\ n \neq 1$

70. $\displaystyle\int \csc^n u\,du = -\frac{\csc^{n-2} u \cot u}{n-1} + \frac{n-2}{n-1}\int \csc^{n-2} u\,du,\ n \neq 1$

71. $\displaystyle\int \frac{1}{1 \pm \tan u}\,du = \frac{1}{2}\left(u \pm \ln|\cos u \pm \sin u|\right) + C$

72. $\displaystyle\int \frac{1}{1 \pm \cot u}\,du = \frac{1}{2}\left(u \mp \ln|\sin u \pm \cos u|\right) + C$

73. $\displaystyle\int \frac{1}{1 \pm \sec u}\,du = u + \cot u \mp \csc u + C$

74. $\displaystyle\int \frac{1}{1 \pm \csc u}\,du = u - \tan u \pm \sec u + C$

Forms Involving Inverse Trigonometric Functions

75. $\int \arcsin u\, du = u \arcsin u + \sqrt{1 - u^2} + C$

76. $\int \arccos u\, du = u \arccos u - \sqrt{1 - u^2} + C$

77. $\int \arctan u\, du = u \arctan u - \ln\sqrt{1 + u^2} + C$

78. $\int \operatorname{arccot} u\, du = u \operatorname{arccot} u + \ln\sqrt{1 + u^2} + C$

79. $\int \operatorname{arcsec} u\, du = u \operatorname{arcsec} u - \ln\left|u + \sqrt{u^2 - 1}\right| + C$

80. $\int \operatorname{arccsc} u\, du = u \operatorname{arccsc} u + \ln\left|u + \sqrt{u^2 - 1}\right| + C$

Forms Involving e^u

81. $\int e^u\, du = e^u + C$

82. $\int ue^u\, du = (u - 1)e^u + C$

83. $\int u^n e^u\, du = u^n e^u - n\int u^{n-1}e^u\, du$

84. $\int \frac{1}{1 + e^u}\, du = u - \ln(1 + e^u) + C$

85. $\int e^{au} \sin bu\, du = \frac{e^{au}}{a^2 + b^2}(a \sin bu - b \cos bu) + C$

86. $\int e^{au} \cos bu\, du = \frac{e^{au}}{a^2 + b^2}(a \cos bu + b \sin bu) + C$

Forms Involving $\ln u$

87. $\int \ln u\, du = u(-1 + \ln u) + C$

88. $\int u \ln u\, du = \frac{u^2}{4}(-1 + 2 \ln u) + C$

89. $\int u^n \ln u\, du = \frac{u^{n+1}}{(n + 1)^2}[-1 + (n + 1) \ln u] + C,\ n \neq -1$

90. $\int (\ln u)^2\, du = u[2 - 2 \ln u + (\ln u)^2] + C$

91. $\int (\ln u)^n\, du = u(\ln u)^n - n\int (\ln u)^{n-1}\, du$

Forms Involving Hyperbolic Functions

92. $\int \cosh u\, du = \sinh u + C$

93. $\int \sinh u\, du = \cosh u + C$

94. $\int \operatorname{sech}^2 u\, du = \tanh u + C$

95. $\int \operatorname{csch}^2 u\, du = -\coth u + C$

96. $\int \operatorname{sech} u \tanh u\, du = -\operatorname{sech} u + C$

97. $\int \operatorname{csch} u \coth u\, du = -\operatorname{csch} u + C$

Forms Involving Inverse Hyperbolic Functions (in logarithmic form)

98. $\int \frac{du}{\sqrt{u^2 \pm a^2}} = \ln\left(u + \sqrt{u^2 \pm a^2}\right) + C$

99. $\int \frac{du}{a^2 - u^2} = \frac{1}{2a} \ln\left|\frac{a + u}{a - u}\right| + C$

100. $\int \frac{du}{u\sqrt{a^2 \pm u^2}} = -\frac{1}{a} \ln \frac{a + \sqrt{a^2 \pm u^2}}{|u|} + C$

C Business and Economic Applications

Previously, you learned that one of the most common ways to measure change is with respect to time. In this section, you will study some important rates of change in economics that are not measured with respect to time. For example, economists refer to **marginal profit, marginal revenue,** and **marginal cost** as the rates of change of the profit, revenue, and cost with respect to the number of units produced or sold.

Summary of Business Terms and Formulas

Basic Terms	*Basic Formulas*
x is the number of units produced (or sold).	
p is the price per unit.	
R is the total revenue from selling x units.	$R = xp$
C is the total cost of producing x units.	
$\overline{C}$ is the average cost per unit.	$\overline{C} = \dfrac{C}{x}$
P is the total profit from selling x units.	$P = R - C$

The **break-even point** is the number of units for which $R = C$.

Marginals

$\dfrac{dR}{dx}$ = Marginal revenue $\approx$ *extra* revenue from selling one additional unit

$\dfrac{dC}{dx}$ = Marginal cost $\approx$ *extra* cost of producing one additional unit

$\dfrac{dP}{dx}$ = Marginal profit $\approx$ *extra* profit from selling one additional unit

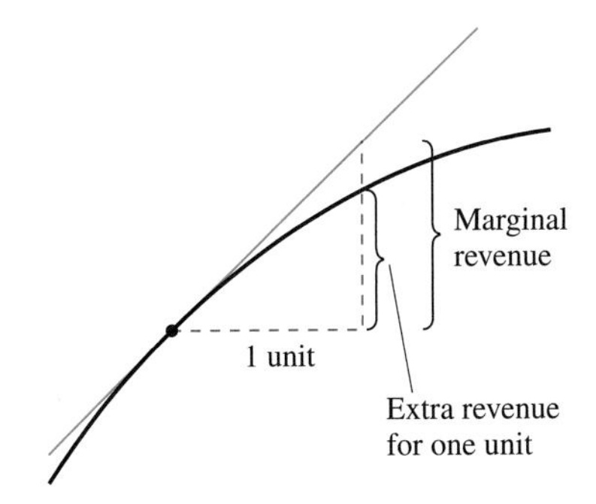

A revenue function
Figure C.1

In this summary, note that marginals can be used to approximate the *extra* revenue, cost, or profit associated with selling or producing one additional unit. This is illustrated graphically for marginal revenue in Figure C.1.

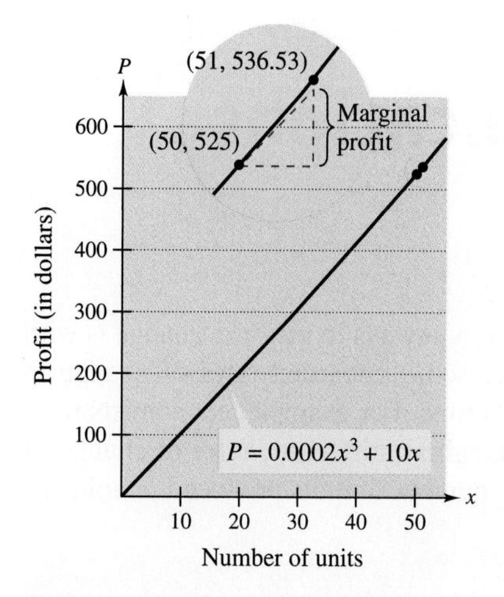

Marginal profit is the extra profit from selling one additional unit.
Figure C.2

EXAMPLE 1 Using Marginals as Approximations

A manufacturer determines that the profit P (in dollars) derived from selling x units of an item is given by

$$P = 0.0002x^3 + 10x.$$

a. Find the marginal profit for a production level of 50 units.

b. Compare this with the actual gain in profit obtained by increasing production from 50 to 51 units. (See Figure C.2.)

Solution

a. Because the profit is $P = 0.0002x^3 + 10x$, the marginal profit is given by the derivative

$$\frac{dP}{dx} = 0.0006x^2 + 10.$$

When $x = 50$, the marginal profit is

$$\begin{aligned}\frac{dP}{dx} &= (0.0006)(50)^2 + 10 \\ &= \$11.50.\end{aligned}$$ Marginal profit for $x = 50$

b. For $x = 50$ and 51, the actual profits are

$$\begin{aligned}P &= (0.0002)(50)^3 + 10(50) \\ &= 25 + 50 \\ &= \$525.00 \\ P &= (0.0002)(51)^3 + 10(51) \\ &= 26.53 + 510 \\ &= \$536.53.\end{aligned}$$

So, the additional profit obtained by increasing the production level from 50 to 51 units is

$$\$536.53 - \$525.00 = \$11.53.$$ Extra profit for one unit

The profit function in Example 1 is unusual in that the profit continues to increase as long as the number of units sold increases. In practice, it is more common to encounter situations in which sales can be increased only by lowering the price per item. Such reductions in price ultimately cause the profit to decline.

The number of units x that consumers are willing to purchase at a given price p per unit is defined as the **demand function**

$$p = f(x).$$ Demand function

EXAMPLE 2 Finding a Demand Function

A business sells 2000 items per month at a price of \$10 each. It is estimated that monthly sales will increase by 250 items for each \$0.25 reduction in price. Find the demand function corresponding to this estimate.

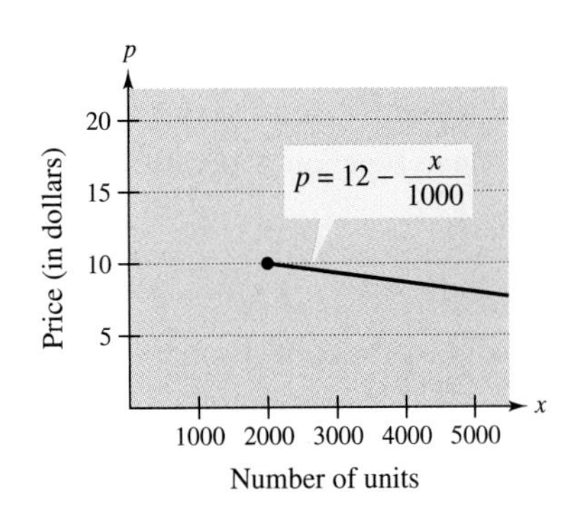

A demand function p
Figure C.3

Solution From the given estimate, x increases 250 units each time p drops \$0.25 from the original cost of \$10. This is described by the equation

$$\begin{aligned} x &= 2000 + 250\left(\frac{10 - p}{0.25}\right) \\ &= 12{,}000 - 1000p \end{aligned}$$

or

$$p = 12 - \frac{x}{1000}, \quad x \geq 2000. \qquad \text{Demand function}$$

The graph of the demand function is shown in Figure C.3.

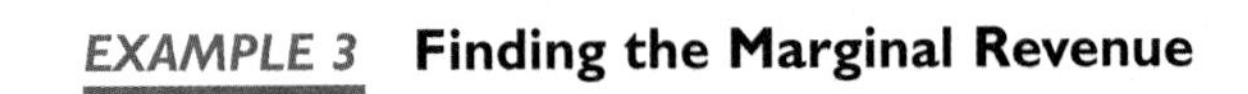

EXAMPLE 3 Finding the Marginal Revenue

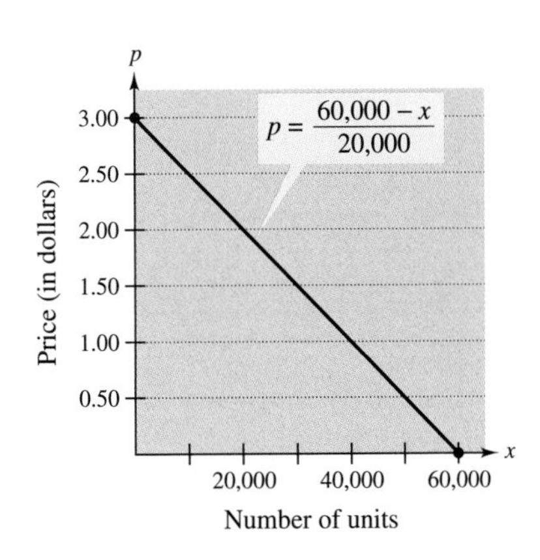

As the price decreases, more hamburgers are sold.
Figure C.4

A fast-food restaurant has determined that the monthly demand for its hamburgers is

$$p = \frac{60{,}000 - x}{20{,}000}.$$

Find the increase in revenue per hamburger (marginal revenue) for monthly sales of 20,000 hamburgers. (See Figure C.4.)

Solution Because the total revenue is given by $R = xp$, you have

$$R = xp = x\left(\frac{60{,}000 - x}{20{,}000}\right) = \frac{1}{20{,}000}(60{,}000x - x^2).$$

By differentiating, you can find the marginal revenue to be

$$\frac{dR}{dx} = \frac{1}{20{,}000}(60{,}000 - 2x).$$

When $x = 20{,}000$, the marginal revenue is

$$\begin{aligned} \frac{dR}{dx} &= \frac{1}{20{,}000}[60{,}000 - 2(20{,}000)] \\ &= \frac{20{,}000}{20{,}000} \\ &= \$1 \text{ per unit.} \end{aligned}$$

NOTE The demand function in Example 3 is typical in that a high demand corresponds to a low price, as shown in Figure C.4.

EXAMPLE 4 Finding the Marginal Profit

Suppose that in Example 3 the cost C (in dollars) of producing x hamburgers is

$$C = 5000 + 0.56x, \quad 0 \le x \le 50{,}000.$$

Find the total profit and the marginal profit for 20,000, 24,400, and 30,000 units.

Solution Because $P = R - C$, you can use the revenue function in Example 3 to obtain

$$P = \frac{1}{20{,}000}(60{,}000x - x^2) - 5000 - 0.56x$$

$$= 2.44x - \frac{x^2}{20{,}000} - 5000.$$

So, the marginal profit is

$$\frac{dP}{dx} = 2.44 - \frac{x}{10{,}000}.$$

The table shows the total profit and the marginal profit for each of the three indicated demands. Figure C.5 shows the graph of the profit function.

Demand	20,000	24,400	30,000
Profit	\$23,800	\$24,768	\$23,200
Marginal profit	\$0.44	\$0.00	−\$0.56

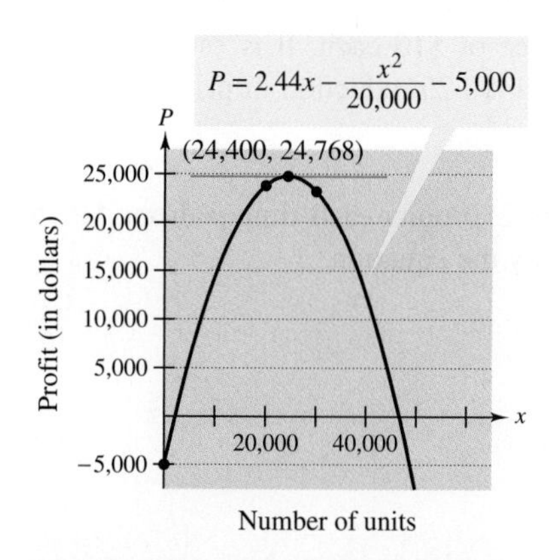

The maximum profit corresponds to the point where the marginal profit is 0. When more than 24,400 hamburgers are sold, the marginal profit is negative—increasing production beyond this point will *reduce* rather than increase profit.
Figure C.5

EXAMPLE 5 Finding the Maximum Profit

In marketing an item, a business has discovered that the demand for the item is represented by

$$p = \frac{50}{\sqrt{x}}. \qquad \text{Demand function}$$

The cost C (in dollars) of producing x items is given by $C = 0.5x + 500$. Find the price per unit that yields a maximum profit (see Figure C.6).

Solution From the given cost function, you obtain

$$P = R - C = xp - (0.5x + 500). \qquad \text{Primary equation}$$

Substituting for p (from the demand function) produces

$$P = x\left(\frac{50}{\sqrt{x}}\right) - (0.5x + 500) = 50\sqrt{x} - 0.5x - 500.$$

Setting the marginal profit equal to 0

$$\frac{dP}{dx} = \frac{25}{\sqrt{x}} - 0.5 = 0$$

yields $x = 2500$. From this, you can conclude that the maximum profit occurs when the price is

$$p = \frac{50}{\sqrt{2500}} = \frac{50}{50} = \$1.00.$$

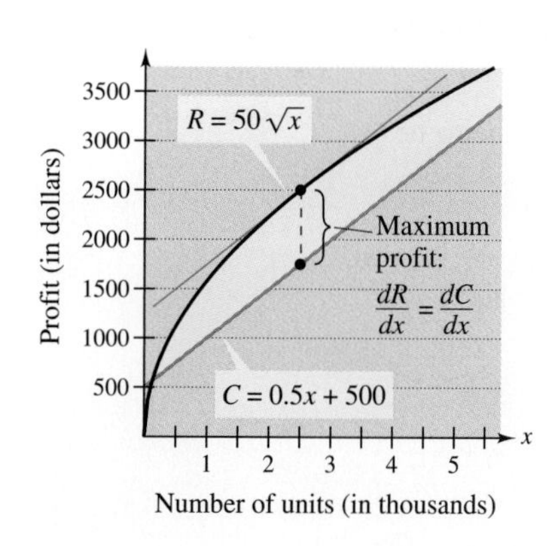

Maximum profit occurs when $\frac{dR}{dx} = \frac{dC}{dx}$.
Figure C.6

NOTE To find the maximum profit in Example 5, the profit function, $P = R - C$, was differentiated and set equal to 0. From the equation

$$\frac{dP}{dx} = \frac{dR}{dx} - \frac{dC}{dx} = 0$$

it follows that the maximum profit occurs when the marginal revenue is equal to the marginal cost, as shown in Figure C.6.

EXAMPLE 6 Minimizing the Average Cost

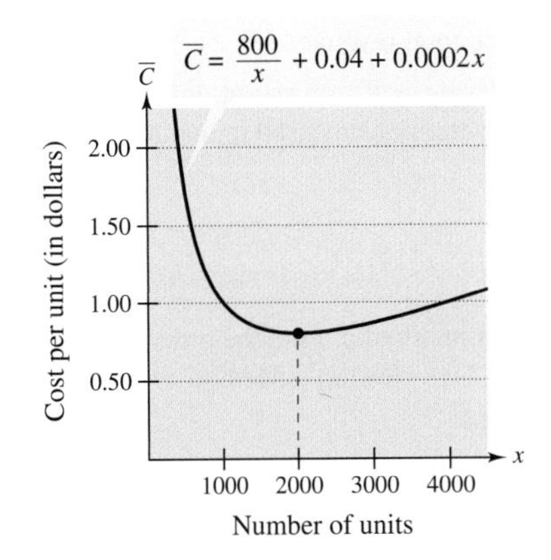

Minimum average cost occurs when $\frac{d\overline{C}}{dx} = 0$.

Figure C.7

A company estimates that the cost C (in dollars) of producing x units of a product is given by $C = 800 + 0.04x + 0.0002x^2$. Find the production level that minimizes the average cost per unit.

Solution Substituting from the given equation for C produces

$$\overline{C} = \frac{C}{x} = \frac{800 + 0.04x + 0.0002x^2}{x} = \frac{800}{x} + 0.04 + 0.0002x.$$

Setting the derivative $d\overline{C}/dx$ equal to 0 yields

$$\frac{d\overline{C}}{dx} = -\frac{800}{x^2} + 0.0002 = 0$$

$$x^2 = \frac{800}{0.0002} = 4{,}000{,}000 \Longrightarrow x = 2000 \text{ units.}$$

See Figure C.7.

Exercises for Appendix C

See www.CalcChat.com for worked-out solutions to odd-numbered exercises.

1. ***Think About It*** The figure shows the cost C of producing x units of a product.
   (a) What is $C(0)$ called?
   (b) Sketch a graph of the marginal cost function.
   (c) Does the marginal cost function have an extremum? If so, describe what it means in economic terms.

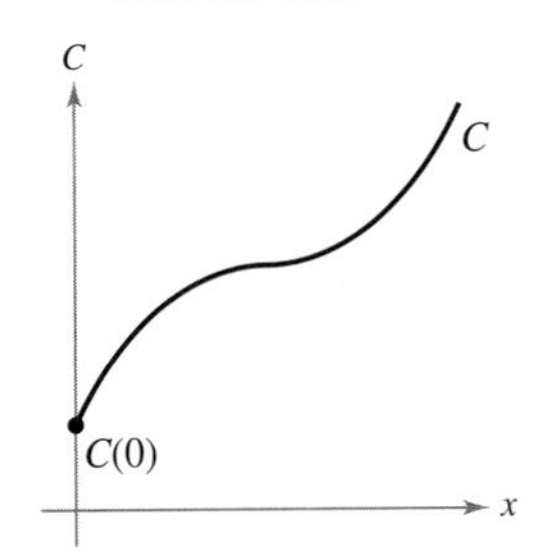

Figure for 1

Figure for 2

2. ***Think About It*** The figure shows the cost C and revenue R for producing and selling x units of a product.
   (a) Sketch a graph of the marginal revenue function.
   (b) Sketch a graph of the profit function. Approximate the position of the value of x for which profit is maximum.

In Exercises 3–6, find the number of units x that produces a maximum revenue R.

3. $R = 900x - 0.1x^2$
4. $R = 600x^2 - 0.02x^3$
5. $R = \frac{1{,}000{,}000x}{0.02x^2 + 1800}$
6. $R = 30x^{2/3} - 2x$

In Exercises 7–10, find the number of units x that produces the minimum average cost per unit $\overline{C}$.

7. $C = 0.125x^2 + 20x + 5000$
8. $C = 0.001x^3 - 5x + 250$
9. $C = 3000x - x^2\sqrt{300 - x}$
10. $C = \frac{2x^3 - x^2 + 5000x}{x^2 + 2500}$

In Exercises 11–14, find the price per unit p (in dollars) that produces the maximum profit P.

	Cost Function	*Demand Function*
11.	$C = 100 + 30x$	$p = 90 - x$
12.	$C = 2400x + 5200$	$p = 6000 - 0.4x^2$
13.	$C = 4000 - 40x + 0.02x^2$	$p = 50 - \dfrac{x}{100}$
14.	$C = 35x + 2\sqrt{x-1}$	$p = 40 - \sqrt{x-1}$

Average Cost **In Exercises 15 and 16, use the cost function to find the value of x at which the average cost is a minimum. For that value of x, show that the marginal cost and average cost are equal.**

15. $C = 2x^2 + 5x + 18$ **16.** $C = x^3 - 6x^2 + 13x$

17. Prove that the average cost is a minimum at the value of x where the average cost equals the marginal cost.

18. ***Maximum Profit*** The profit P for a company is

$$P = 230 + 20s - \tfrac{1}{2}s^2$$

where s is the amount (in hundreds of dollars) spent on advertising. What amount of advertising produces a maximum profit?

19. ***Numerical, Graphical, and Analytic Analysis*** The cost per unit for the production of a radio is \$60. The manufacturer charges \$90 per unit for orders of 100 or less. To encourage large orders, the manufacturer reduces the charge by \$0.15 per radio for each unit ordered in excess of 100 (for example, there would be a charge of \$87 per radio for an order size of 120).

(a) Analytically complete six rows of a table such as the one below. (The first two rows are shown.)

x	*Price*	*Profit*
102	$90 - 2(0.15)$	$102[90 - 2(0.15)] - 102(60) = 3029.40$
104	$90 - 4(0.15)$	$104[90 - 4(0.15)] - 104(60) = 3057.60$

(b) Use a graphing utility to generate additional rows of the table. Use the table to estimate the maximum profit. (*Hint:* Use the *table* feature of the graphing utility.)

(c) Write the profit P as a function of x.

(d) Use calculus to find the critical number of the function in part (c) and find the required order size.

(e) Use a graphing utility to graph the function in part (c) and verify the maximum profit from the graph.

20. ***Maximum Profit*** A real estate office handles 50 apartment units. When the rent is \$720 per month, all units are occupied. However, on the average, for each \$40 increase in rent, one unit becomes vacant. Each occupied unit requires an average of \$48 per month for service and repairs. What rent should be charged to obtain a maximum profit?

21. ***Minimum Cost*** A power station is on one side of a river that is $\frac{1}{2}$-mile wide, and a factory is 6 miles downstream on the other side. It costs \$12 per foot to run power lines over land and \$16 per foot to run them underwater. Find the most economical path for the transmission line from the power station to the factory.

22. ***Maximum Revenue*** When a wholesaler sold a product at \$25 per unit, sales were 800 units per week. After a price increase of \$5, the average number of units sold dropped to 775 per week. Assume that the demand function is linear, and find the price that will maximize the total revenue.

23. ***Minimum Cost*** The ordering and transportation cost C (in thousands of dollars) of the components used in manufacturing a product is

$$C = 100\left(\frac{200}{x^2} + \frac{x}{x+30}\right), \quad 1 \le x$$

where x is the order size (in hundreds). Find the order size that minimizes the cost. (*Hint:* Use Newton's Method or the *zero* feature of a graphing utility.)

24. ***Average Cost*** A company estimates that the cost C (in dollars) of producing x units of a product is

$$C = 800 + 0.4x + 0.02x^2 + 0.0001x^3.$$

Find the production level that minimizes the average cost per unit. (*Hint:* Use Newton's Method or the *zero* feature of a graphing utility.)

25. ***Revenue*** The revenue R for a company selling x units is

$$R = 900x - 0.1x^2.$$

Use differentials to approximate the change in revenue if sales increase from $x = 3000$ to $x = 3100$ units.

26. ***Analytic and Graphical Analysis*** A manufacturer of fertilizer finds that the national sales of fertilizer roughly follow the seasonal pattern

$$F = 100{,}000\left\{1 + \sin\left[\frac{2\pi(t-60)}{365}\right]\right\}$$

where F is measured in pounds. Time t is measured in days, with $t = 1$ corresponding to January 1.

(a) Use calculus to determine the day of the year when the maximum amount of fertilizer is sold.

(b) Use a graphing utility to graph the function and approximate the day of the year when sales are minimum.

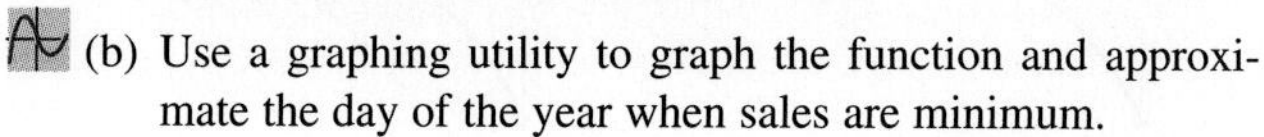

27. ***Modeling Data*** The table shows the monthly sales G (in thousands of gallons) of gasoline at a gas station in 2004. The time in months is represented by t, with $t = 1$ corresponding to January.

t	1	2	3	4	5	6
G	8.91	9.18	9.79	9.83	10.37	10.16

t	7	8	9	10	11	12
G	10.37	10.81	10.03	9.97	9.85	9.51

A model for these data is

$$G = 9.90 - 0.64 \cos\left(\frac{\pi t}{6} - 0.62\right).$$

(a) Use a graphing utility to plot the data and graph the model.

(b) Use the model to approximate the month when gasoline sales were greatest.

(c) What factor in the model causes the seasonal variation in sales of gasoline? What part of the model gives the average monthly sales of gasoline?

(d) Suppose the gas station added the term $0.02t$ to the model. What does the inclusion of this term mean? Use this model to estimate the maximum monthly sales in the year 2008.

28. ***Airline Revenues*** The annual revenue R (in millions of dollars) for an airline for the years 1995–2004 can be modeled by

$$R = 4.7t^4 - 193.5t^3 + 2941.7t^2 - 19{,}294.7t + 52{,}012$$

where $t = 5$ corresponds to 1995.

(a) During which year (between 1995 and 2004) was the airline's revenue the least?

(b) During which year was the revenue the greatest?

(c) Find the revenues for the years in which the revenue was the least and greatest.

(d) Use a graphing utility to confirm the results in parts (a) and (b).

29. ***Modeling Data*** The manager of a department store recorded the quarterly sales S (in thousands of dollars) of a new seasonal product over a period of 2 years, as shown in the table, where t is the time in quarters, with $t = 1$ corresponding to the winter quarter of 2002.

t	1	2	3	4	5	6	7	8
S	7.5	6.2	5.3	7.0	9.1	7.8	6.9	8.6

(a) Use a graphing utility to plot the data.

(b) Find a model of the form $S = a + bt + c \sin \beta t$ for the data. (*Hint:* Start by finding β. Next, use a graphing utility to find $a + bt$. Finally, approximate c.)

(c) Use a graphing utility to graph the model with the data and make any adjustments necessary to obtain a better fit.

(d) Use the model to predict the maximum quarterly sales in the year 2006.

30. ***Think About It*** Match each graph with the function it best represents—a demand function, a revenue function, a cost function, or a profit function. Explain your reasoning. [The graphs are labeled (a), (b), (c), and (d).]

(a)

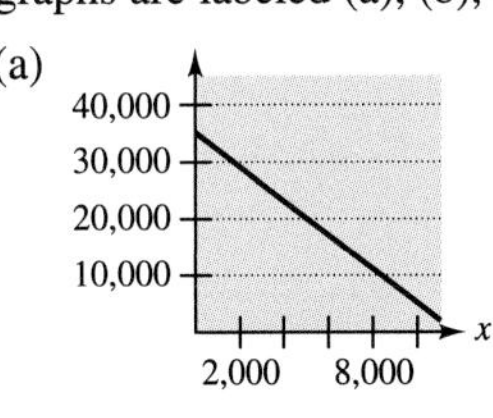

(b)

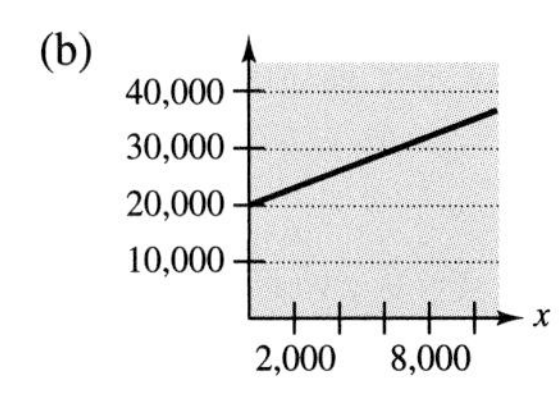

(c)

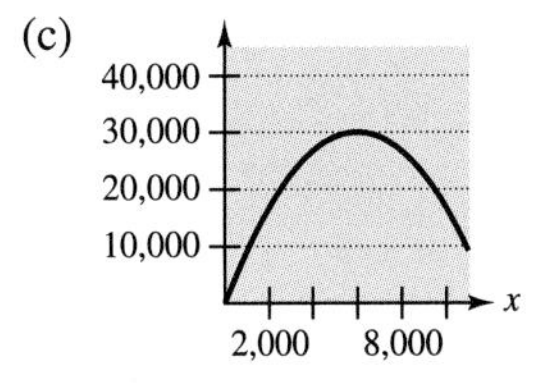

(d)

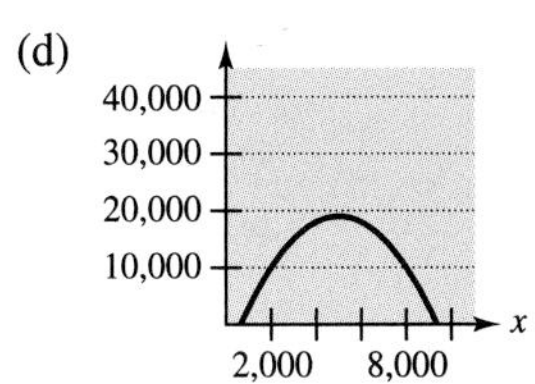

Elasticity **The relative responsiveness of consumers to a change in the price of an item is called the *price elasticity of demand.* If $p = f(x)$ is a differentiable demand function, the price elasticity of demand is**

$$\eta = \frac{p/x}{dp/dx}.$$

For a given price, if $|\eta| < 1$, the demand is *inelastic*, and if $|\eta| > 1$, the demand is *elastic*. In Exercises 31–34, find η for the demand function at the indicated x-value. Is the demand elastic, inelastic, or neither at the indicated x-value?

31. $p = 400 - 3x$
$x = 20$

32. $p = 5 - 0.03x$
$x = 100$

33. $p = 400 - 0.5x^2$
$x = 20$

34. $p = \dfrac{500}{x + 2}$
$x = 23$

Answers to Odd-Numbered Exercises

Chapter 1

Section 1.1 (page 8)

1. b **2.** d **3.** a **4.** c

5. Answers will vary.

x	-4	-2	0	2	4
y	-5	-2	1	4	7

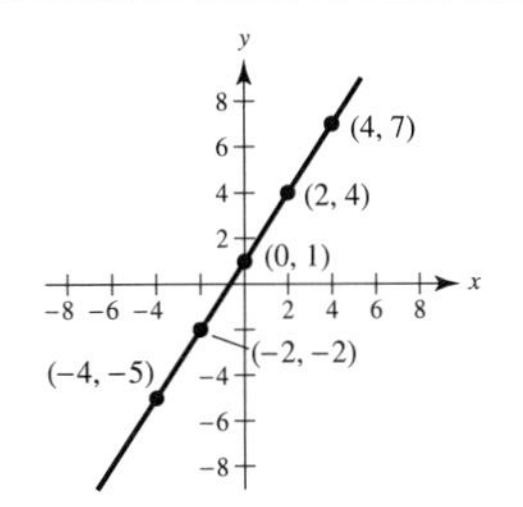

7. Answers will vary.

x	-3	-2	0	2	3
y	-5	0	4	0	-5

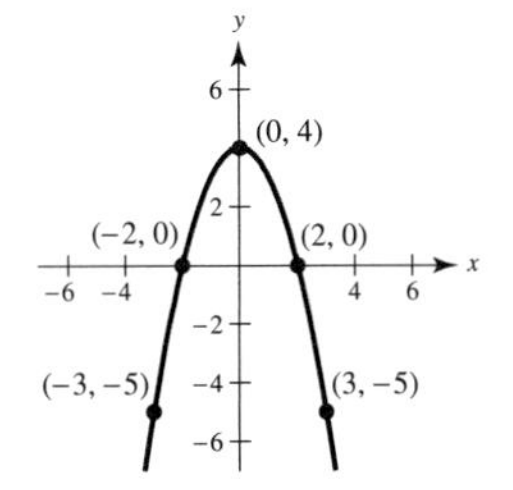

9. Answers will vary.

x	-5	-4	-3	-2	-1	0	1
y	3	2	1	0	1	2	3

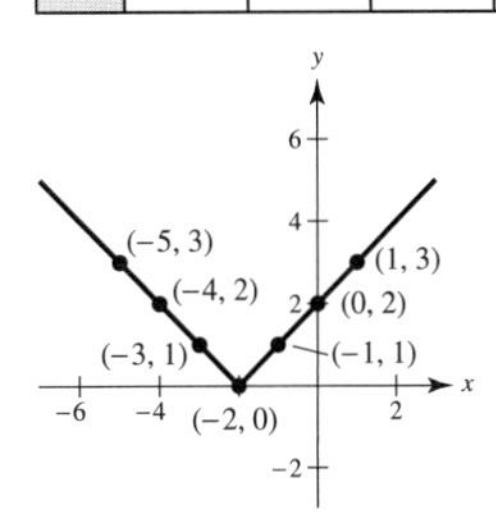

11. Answers will vary.

x	0	1	4	9	16
y	-4	-3	-2	-1	0

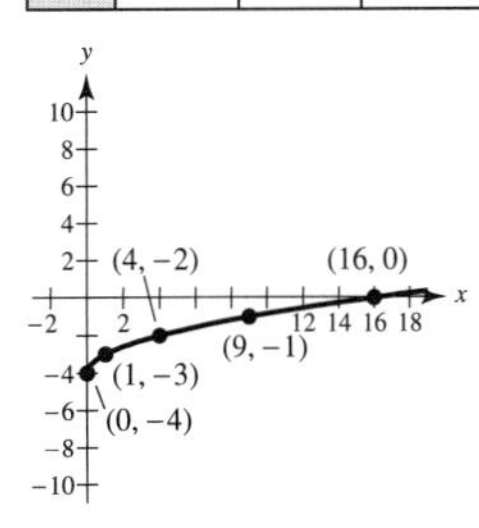

13. Answers will vary.

x	-3	-2	-1	0	1	2	3
y	$-\frac{2}{3}$	-1	-2	Undef.	2	1	$\frac{2}{3}$

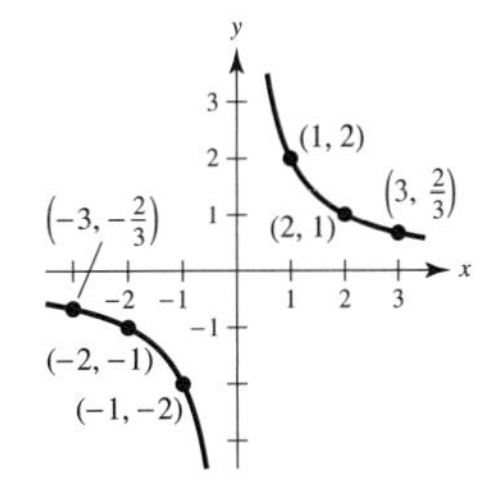

15.

Xmin = -3
Xmax = 5
Xscl = 1
Ymin = -3
Ymax = 5
Yscl = 1

17. $y = \sqrt{5 - x}$

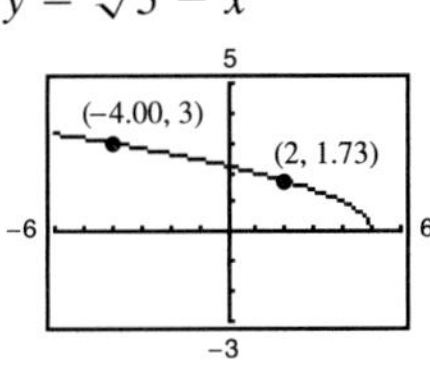

(a) $y \approx 1.73$ (b) $x = -4.00$

19. $(0, -2), (-2, 0), (1, 0)$ **21.** $(0, 0), (5, 0), (-5, 0)$

23. $(4, 0)$ **25.** $(0, 0)$ **27.** Symmetric with respect to the y-axis

29. Symmetric with respect to the x-axis

31. Symmetric with respect to the origin **33.** No symmetry

35. Symmetric with respect to the origin

37. Symmetric with respect to the y-axis

39. $y = -3x + 2$
Symmetry: none

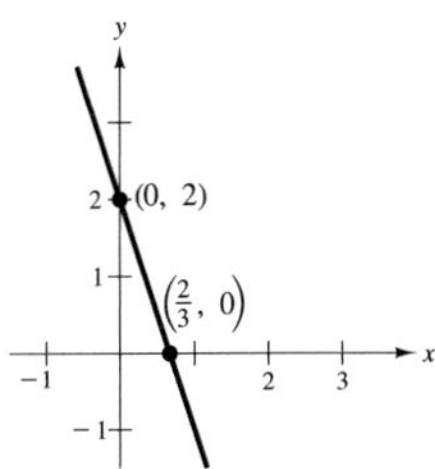

41. 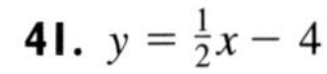 $y = \frac{1}{2}x - 4$
Symmetry: none

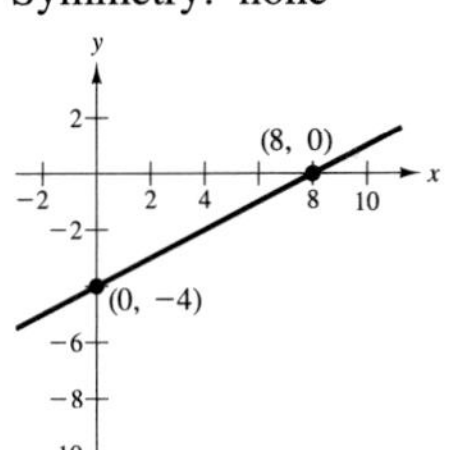

43. $y = 1 - x^2$
Symmetry: y-axis

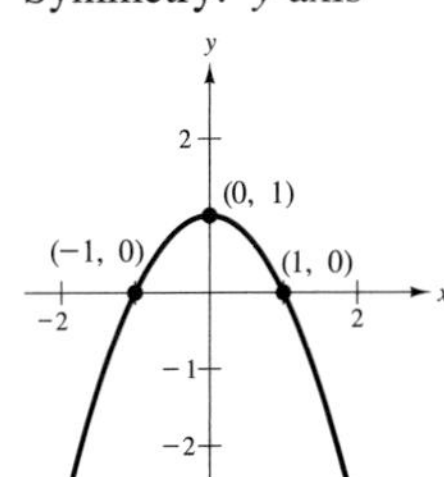

45. $y = (x + 3)^2$
Symmetry: none

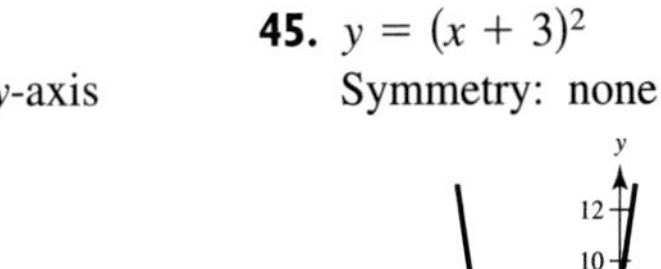

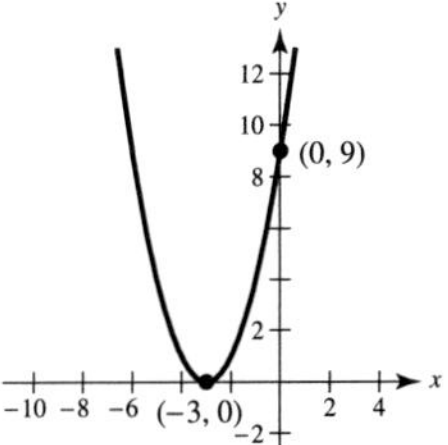

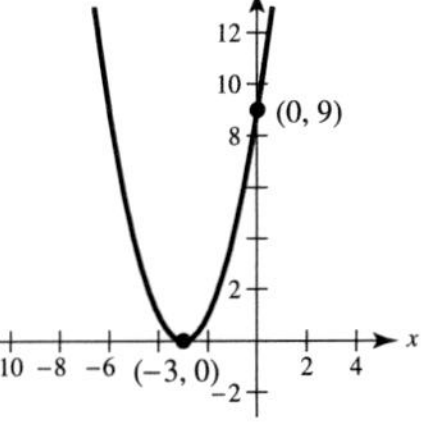

47. $y = x^3 + 2$
Symmetry: none

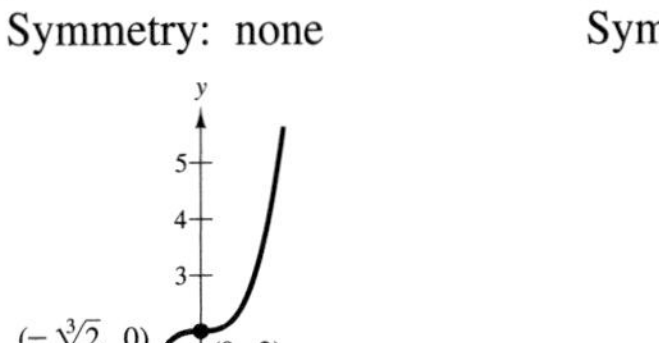

49. $y = x\sqrt{x + 2}$
Symmetry: none

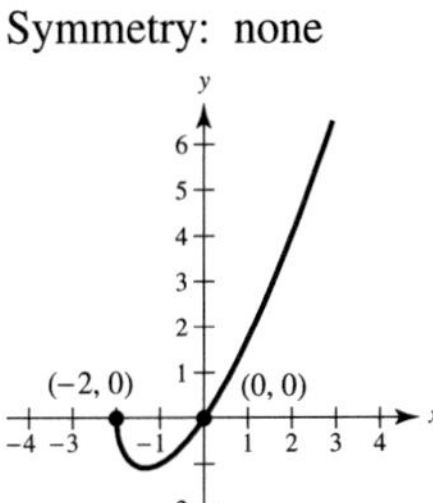

51. $x = y^3$
Symmetry: origin

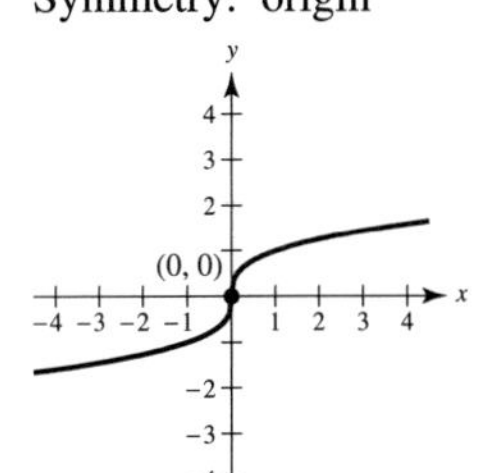

53. $y = 1/x$
Symmetry: origin

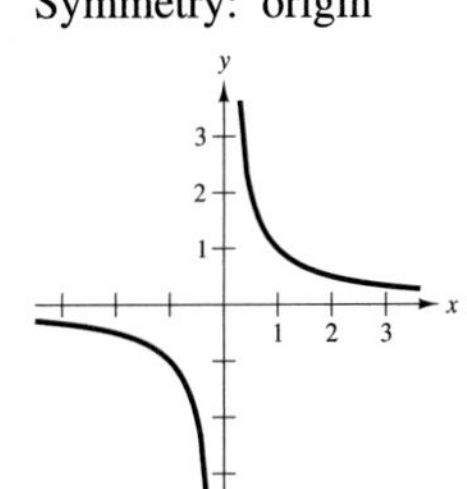

55. $y = 6 - |x|$
Symmetry: y-axis

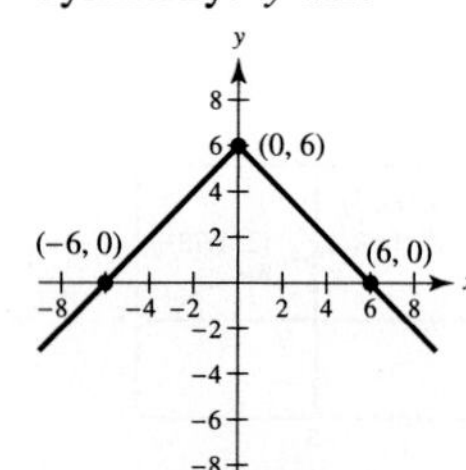

57. $y_1 = \sqrt{x+9}$
$y_2 = -\sqrt{x+9}$
Symmetry: x-axis

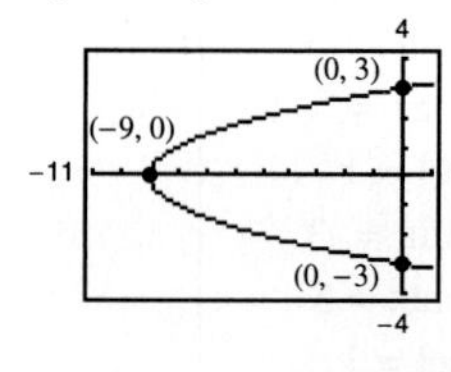

59. $y_1 = \sqrt{\dfrac{6-x}{3}}$

$y_2 = -\sqrt{\dfrac{6-x}{3}}$

Symmetry: x-axis

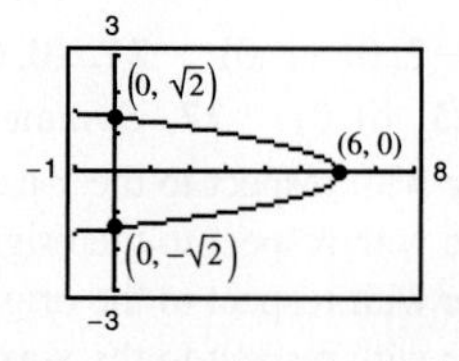

61. (1, 1) **63.** (−1, 5), (2, 2) **65.** (−1, −2), (2, 1)
67. (−1, −1), (0, 0), (1, 1)
69. (−1, −5), (0, −1), (2, 1) **71.** (−2, 2), $(-3, \sqrt{3})$
73. (a) $y = -0.007t^2 + 4.82t + 35.4$

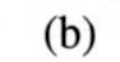
(b)

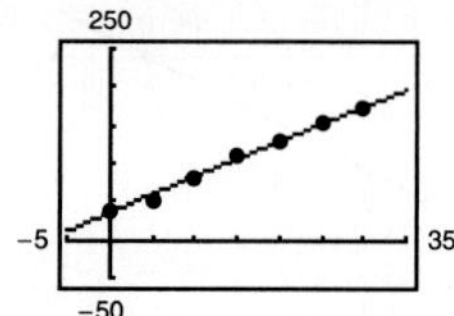

(c) 217.0

75. $x \approx 3133$ units **77.** $y = (x+2)(x-4)(x-6)$
79. (i) b; $k = 2$ (ii) d; $k = -10$ (iii) a; $k = 3$ (iv) c; $k = 36$
81. False. (−1, −2) is not a point on the graph of $x = \frac{1}{4}y^2$.
83. True **85.** $x^2 + (y-4)^2 = 4$

Section 1.2 (page 16)

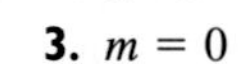
1. $m = 1$ **3.** $m = 0$ **5.** $m = -12$

7.

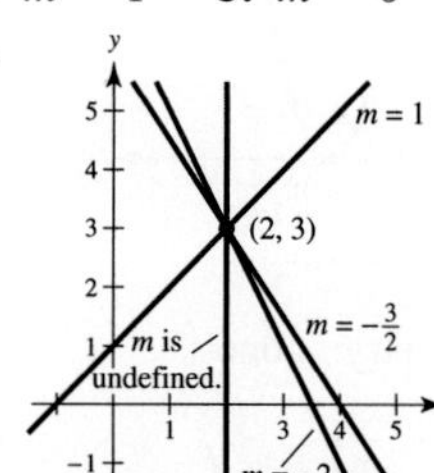

9. $m = 3$

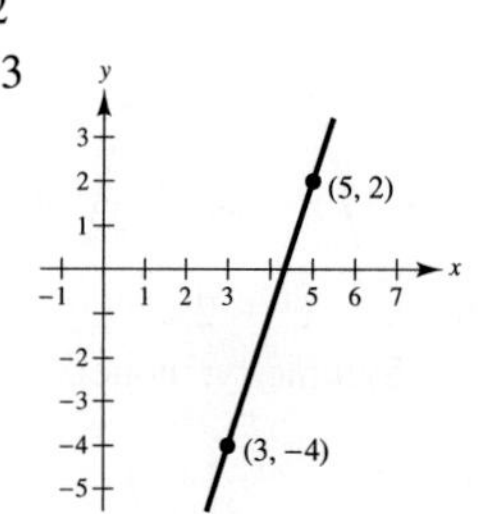

11. m is undefined.

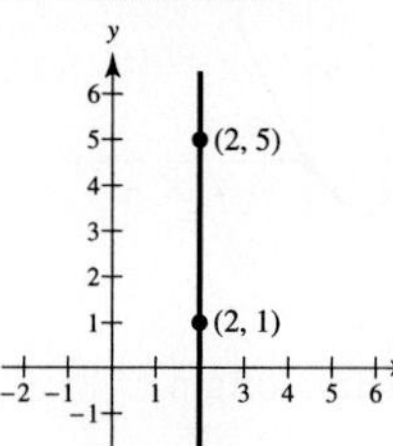

13. $m = 2$

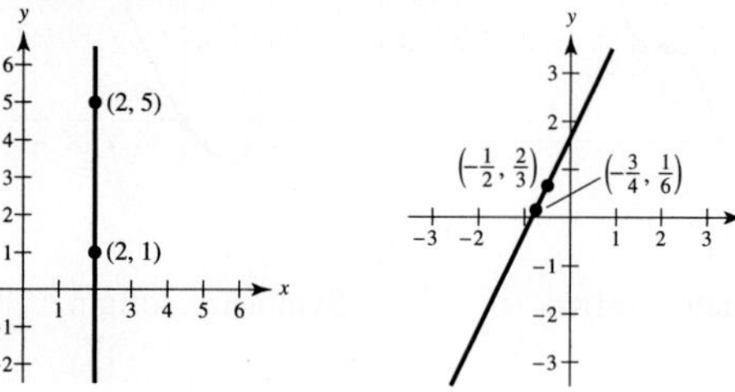

15. (0, 1), (1, 1), (3, 1) **17.** (0, 10), (2, 4), (3, 1)
19. (a) $\frac{1}{3}$ (b) $10\sqrt{10}$ ft

21. (a)

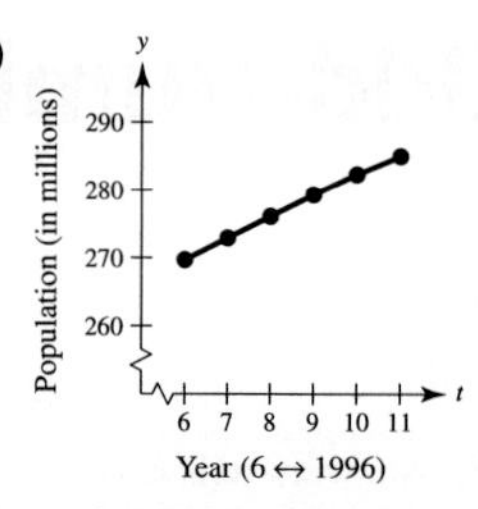

(b) Population increased least rapidly: 2000–2001

23. $m = -\frac{1}{5}$, (0, 4) **25.** m is undefined, no y-intercept
27. $3x - 4y + 12 = 0$ **29.** $2x - 3y = 0$

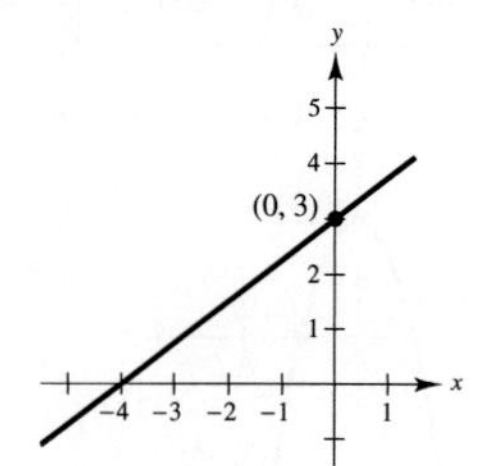

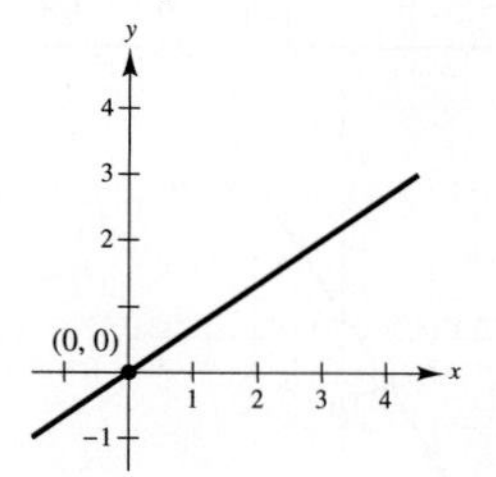

31. $3x - y - 11 = 0$

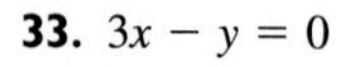
33. $3x - y = 0$

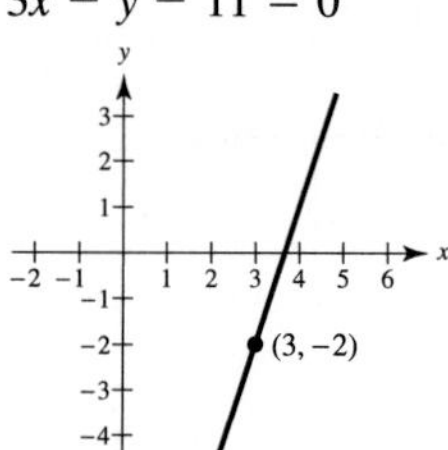

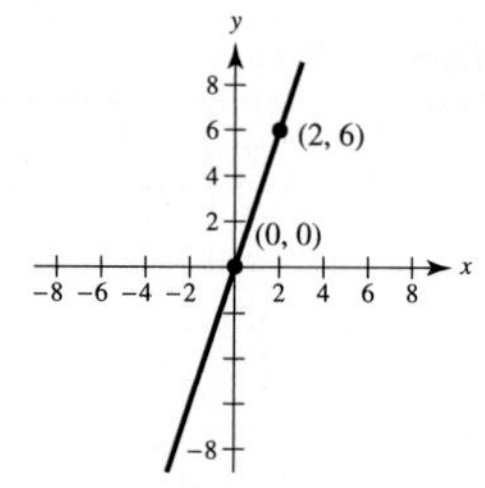

35. $2x - y - 3 = 0$

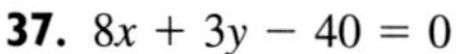
37. $8x + 3y - 40 = 0$

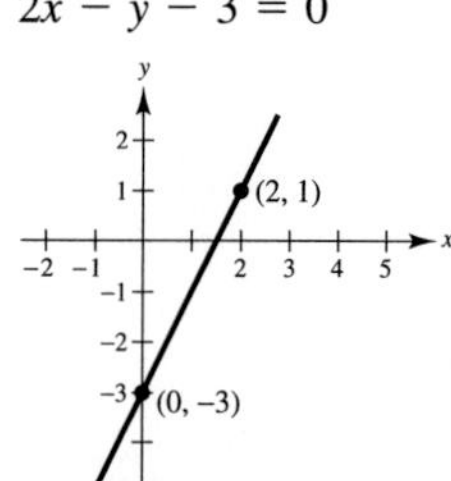

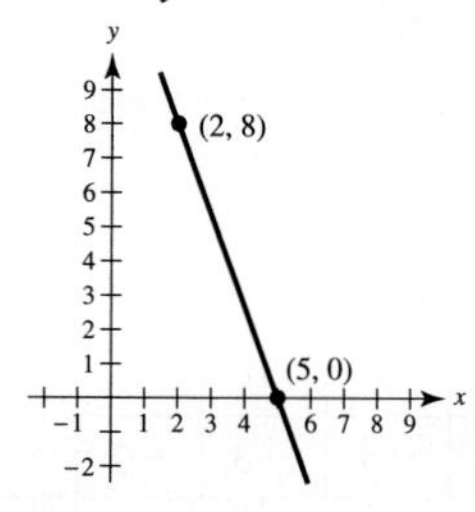

39. $x - 5 = 0$ **41.** $22x - 4y + 3 = 0$

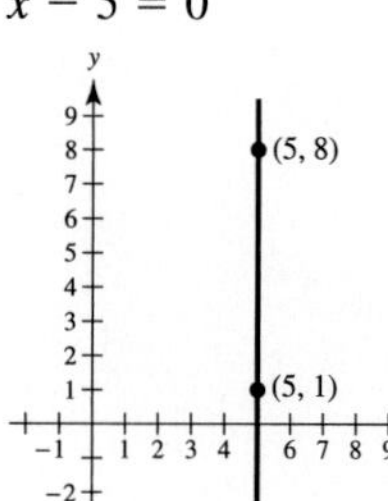

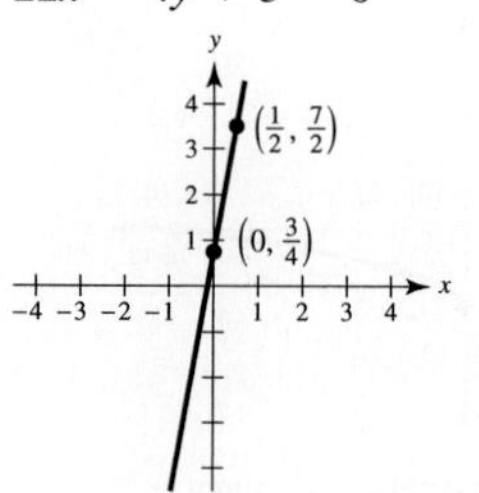

43. $x - 3 = 0$

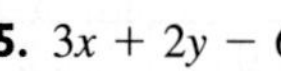
45. $3x + 2y - 6 = 0$ **47.** $x + y - 3 = 0$

49.

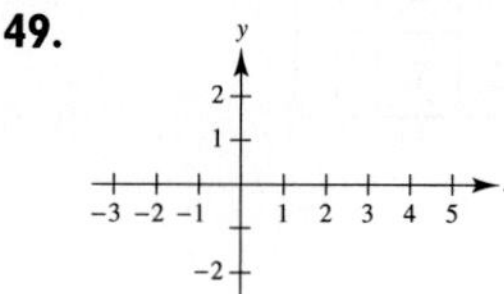
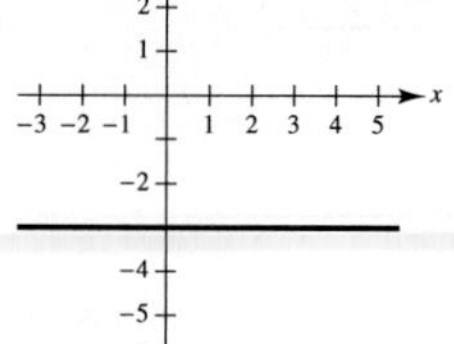

51.

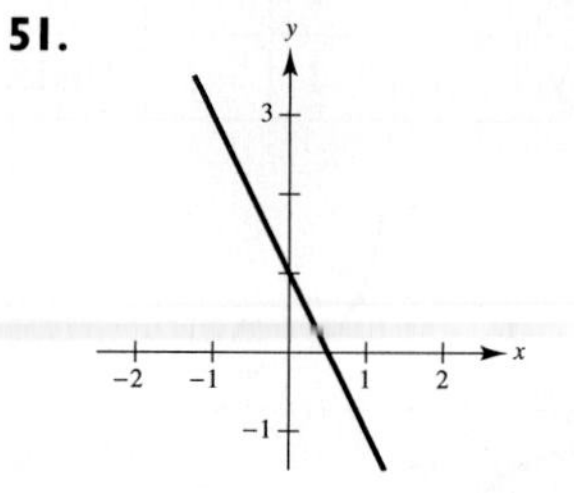

53.

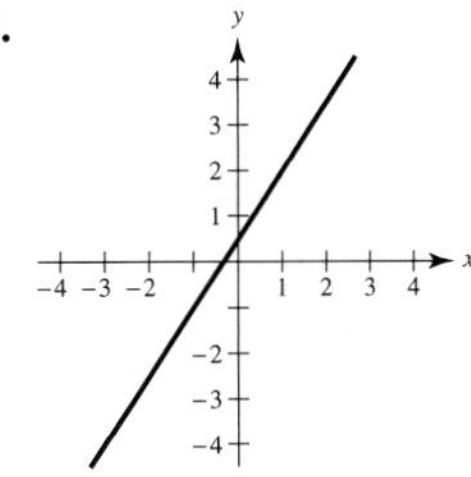

55.

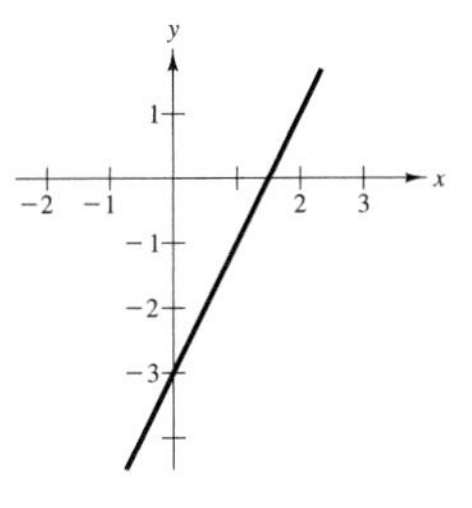

57. (a)

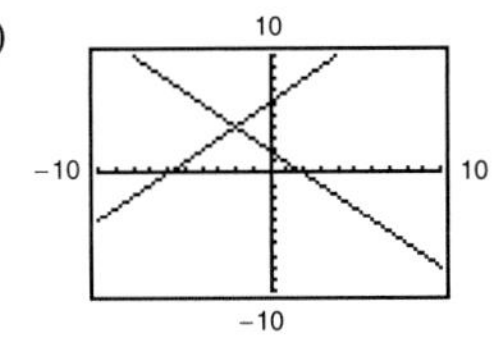

(b)

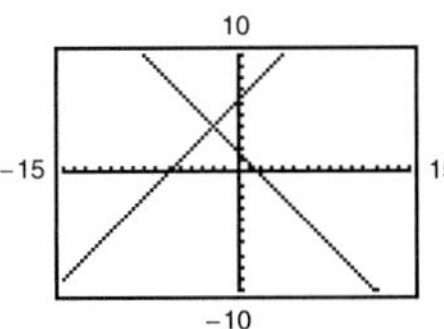

The lines in (a) do not appear perpendicular, but they do in (b) because a square setting is used. The lines are perpendicular.

59. (a) $2x - y - 3 = 0$ (b) $x + 2y - 4 = 0$

61. (a) $40x - 24y - 9 = 0$ (b) $24x + 40y - 53 = 0$

63. (a) $x - 2 = 0$ (b) $y - 5 = 0$

65. $V = 125t + 2040$ **67.** $V = -2000t + 28{,}400$

69. $y = 2x$ **71.** Not collinear, because $m_1 \neq m_2$

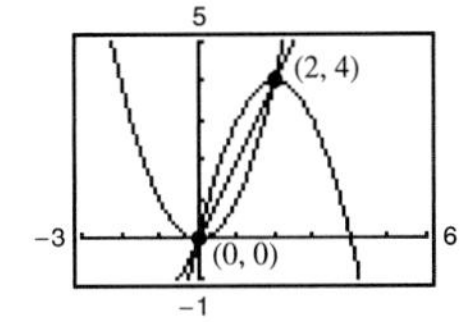

73. $\left(0, \dfrac{-a^2 + b^2 + c^2}{2c}\right)$ **75.** $\left(b, \dfrac{a^2 - b^2}{c}\right)$

77. $5F - 9C - 160 = 0$; $72°\text{F} \approx 22.2°\text{C}$

79. (a) $W_1 = 12.50 + 0.75x$; $W_2 = 9.20 + 1.30x$

(b)

(c) When six units are produced, wages are \$17.00 per hour with either option. Choose position 1 when less than six units are produced and position 2 otherwise.

81. (a) $x = (1330 - p)/15$

(b)

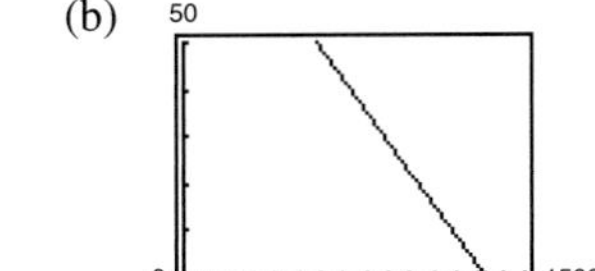

$x(655) = 45$ units

(c) 49 units

83. $12y + 5x - 169 = 0$ **85.** 2 **87.** $(5\sqrt{2})/2$ **89.** $2\sqrt{2}$

91. Proof **93.** Proof **95.** Proof **97.** True

Section 1.3 (page 27)

1. (a) Domain of f: $[-4, 4]$; Range of f: $[-3, 5]$
Domain of g: $[-3, 3]$; Range of g: $[-4, 4]$

(b) $f(-2) = -1$; $g(3) = -4$

(c) $x = -1$ (d) $x \approx 1$ (e) $x \approx -1, x \approx 1$, and $x \approx 2$

3. (a) -3 (b) -9 (c) $2b - 3$ (d) $2x - 5$

5. (a) 3 (b) 0 (c) -1 (d) $2 + 2t - t^2$

7. (a) 1 (b) 0 (c) $-\frac{1}{2}$ **9.** $3x^2 + 3x\,\Delta x + (\Delta x)^2$, $\Delta x \neq 0$

11. $\left(\sqrt{x-1} - x + 1\right)/[(x-2)(x-1)]$
$= -1/\left[\sqrt{x-1}\left(1 + \sqrt{x-1}\right)\right], x \neq 2$

13. Domain: $[-3, \infty)$; Range: $(-\infty, 0]$

15. Domain: All real numbers t such that $t \neq 4n + 2$, where n is an integer; Range: $(-\infty, -1] \cup [1, \infty)$

17. Domain: $(-\infty, 0) \cup (0, \infty)$; Range: $(-\infty, 0) \cup (0, \infty)$

19. Domain: $[0, 1]$

21. Domain: All real numbers x such that $x \neq 2n\pi$, where n is an integer

23. Domain: $(-\infty, -3) \cup (-3, \infty)$

25. (a) -1 (b) 2 (c) 6 (d) $2t^2 + 4$
Domain: $(-\infty, \infty)$; Range: $(-\infty, 1) \cup [2, \infty)$

27. (a) 4 (b) 0 (c) -2 (d) $-b^2$
Domain: $(-\infty, \infty)$; Range: $(-\infty, 0] \cup [1, \infty)$

29. $f(x) = \begin{cases} x + 2, & -2 \leq x \leq -1 \\ -\frac{1}{2}x + \frac{1}{2}, & -1 < x \leq 3 \end{cases}$

31. $f(x) = 4 - x$
Domain: $(-\infty, \infty)$
Range: $(-\infty, \infty)$

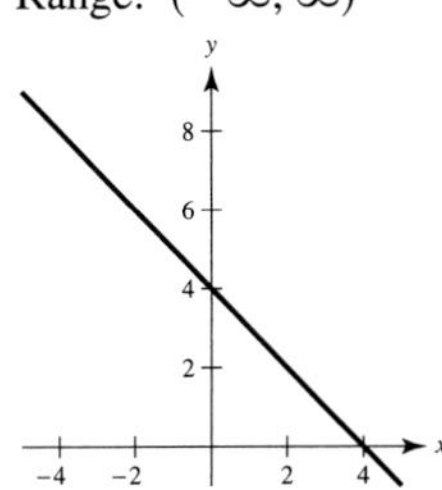

33. $h(x) = \sqrt{x - 1}$
Domain: $[1, \infty)$
Range: $[0, \infty)$

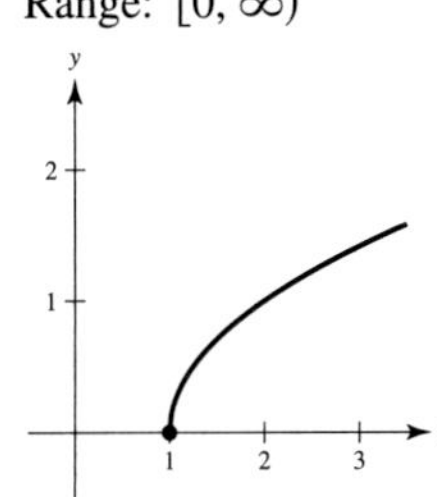

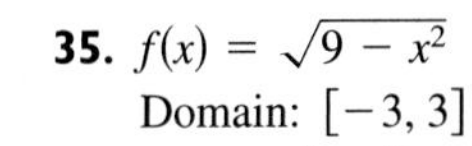

35. $f(x) = \sqrt{9 - x^2}$
Domain: $[-3, 3]$
Range: $[0, 3]$

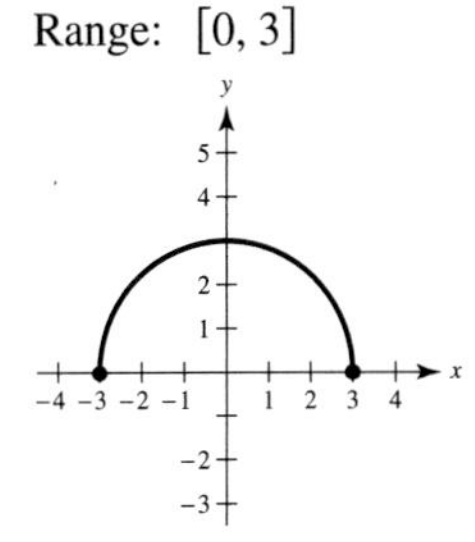

37. $g(t) = 2 \sin \pi t$
Domain: $(-\infty, \infty)$
Range: $[-2, 2]$

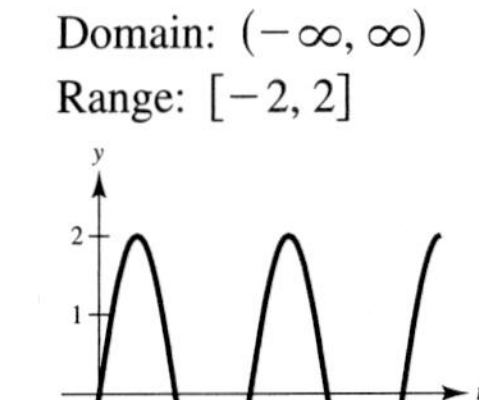

39. The student travels $\frac{1}{2}$ mi/min during the first 4 min, is stationary for the next 2 min, and travels 1 mi/min during the final 4 min.

41. y is not a function of x. **43.** y is a function of x.

45. y is not a function of x. **47.** y is not a function of x.

49. d **50.** b **51.** c **52.** a **53.** e **54.** g

55. (a)

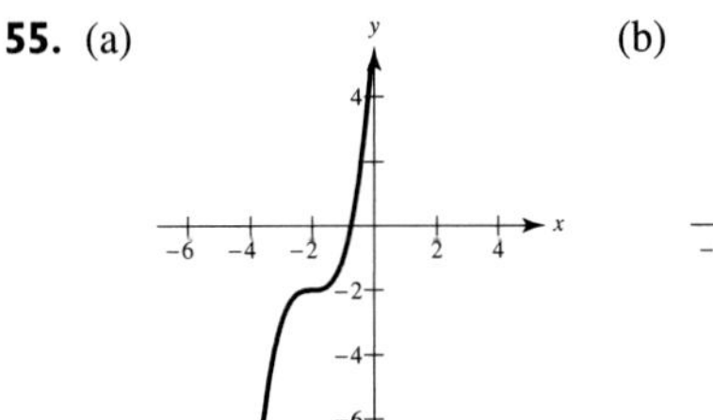

(b)

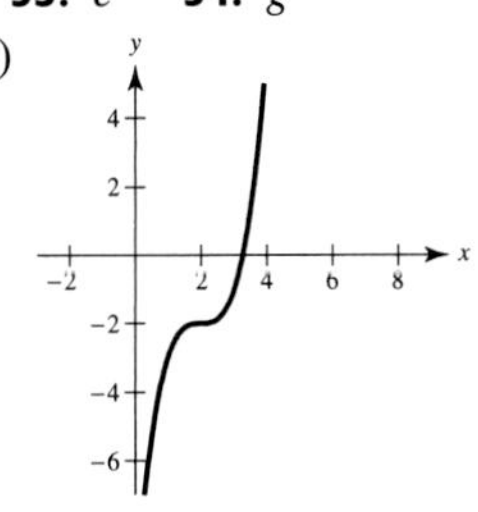

(c)

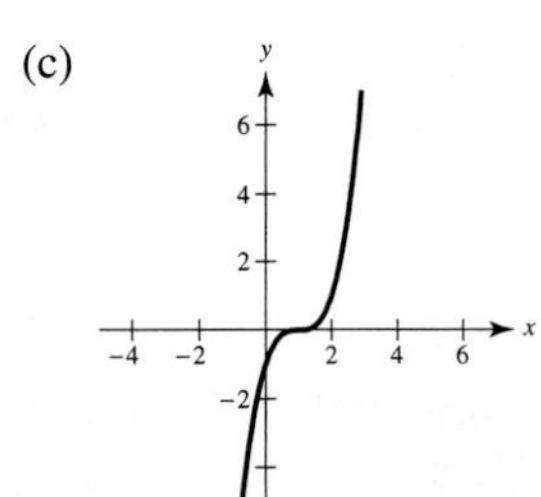

(d)

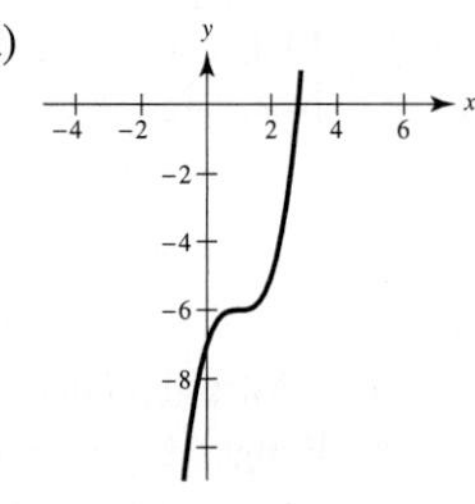

(e)

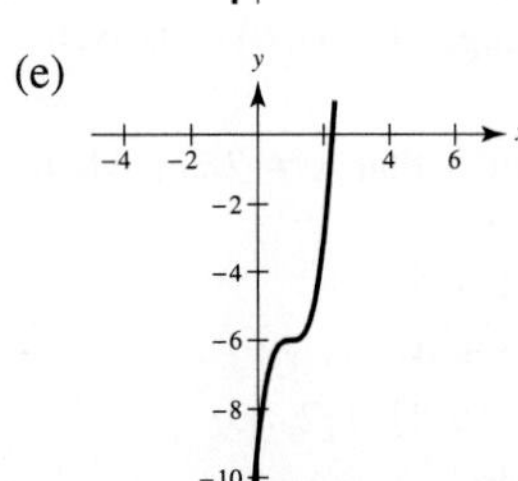

(f)

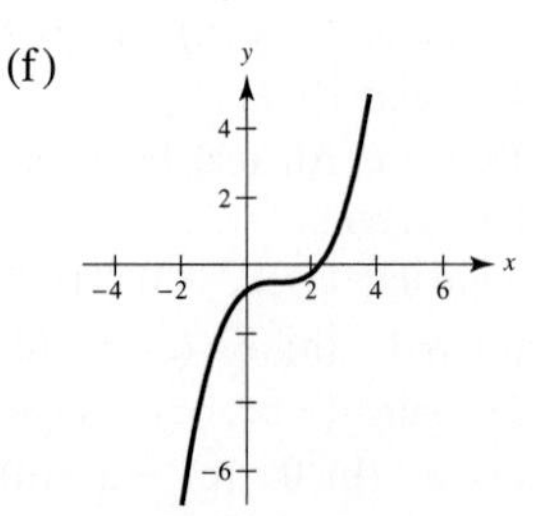

57. (a) Vertical translation

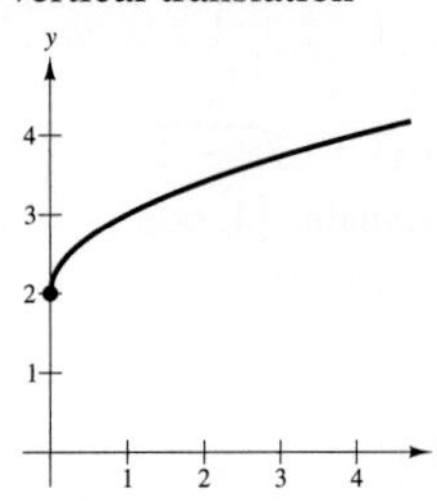

(b) Reflection about the x-axis

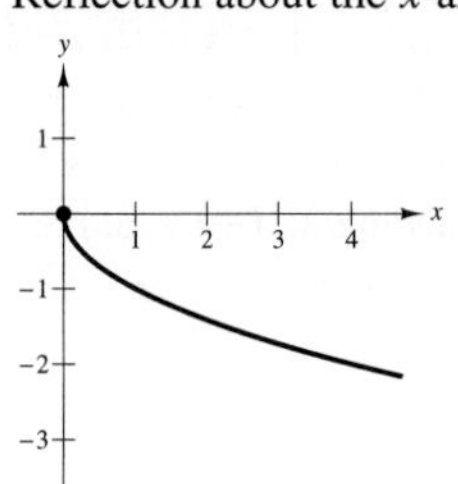

(c) Horizontal translation

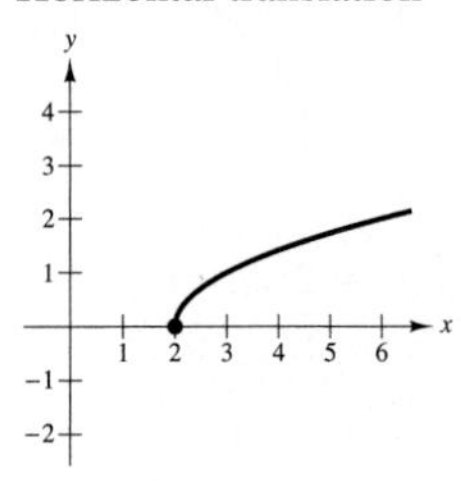

59. (a) 0 (b) 0 (c) -1 (d) $\sqrt{15}$
(e) $\sqrt{x^2 - 1}$ (f) $x - 1\ (x \geq 0)$

61. $(f \circ g)(x) = x$; Domain: $[0, \infty)$
$(g \circ f)(x) = |x|$; Domain: $(-\infty, \infty)$
No, their domains are different.

63. $(f \circ g)(x) = 3/(x^2 - 1)$; Domain: $(-\infty, -1) \cup (-1, 1) \cup (1, \infty)$
$(g \circ f)(x) = 9/x^2 - 1$; Domain: $(-\infty, 0) \cup (0, \infty)$
No

65. (a) 4 (b) -2
(c) Undefined. The graph of g does not exist at $x = -5$.
(d) 3 (e) 2
(f) Undefined. The graph of f does not exist at $x = -4$.

67. Answers will vary.
Example: $f(x) = \sqrt{x}$; $g(x) = x - 2$; $h(x) = 2x$

69. Even **71.** Odd **73.** (a) $\left(\frac{3}{2}, 4\right)$ (b) $\left(\frac{3}{2}, -4\right)$

75. f is even. g is neither even nor odd. h is odd.

77. $f(x) = 2x - 5$ **79.** $y = \sqrt{-x}$

81. ii, $c = -2$ **82.** i, $c = \frac{1}{4}$ **83.** iv, $c = 32$ **84.** iii, $c = 3$

85. (a) $T(4) = 16°$, $T(15) = 24°$
(b) The changes in temperature will occur 1 hr later.
(c) The temperatures are 1° lower.

87. (a) (b) $A(15) \approx 345$ acres/farm

89. $f(x) = |x| + |x - 2| = \begin{cases} 2x - 2, & \text{if } x \geq 2 \\ 2, & \text{if } 0 < x < 2 \\ -2x + 2, & \text{if } x \leq 0 \end{cases}$

91. Proof **93.** Proof

95. (a) $V(x) = x(24 - 2x)^2, x > 0$ (b) 4 cm × 16 cm × 16 cm

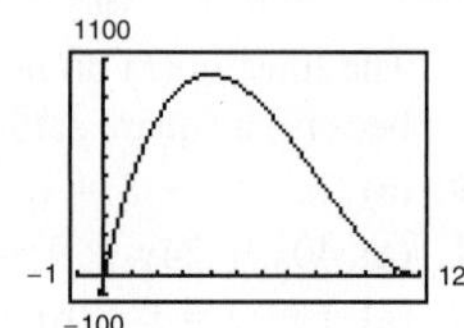

(c)

Height, x	Length and Width	Volume, V
1	$24 - 2(1)$	$1[24 - 2(1)]^2 = 484$
2	$24 - 2(2)$	$2[24 - 2(2)]^2 = 800$
3	$24 - 2(3)$	$3[24 - 2(3)]^2 = 972$
4	$24 - 2(4)$	$4[24 - 2(4)]^2 = 1024$
5	$24 - 2(5)$	$5[24 - 2(5)]^2 = 980$
6	$24 - 2(6)$	$6[24 - 2(6)]^2 = 864$

The dimensions of the box that yield a maximum volume are 4 cm × 16 cm × 16 cm.

97. False. For example, if $f(x) = x^2$, then $f(-1) = f(1)$.

99. True **101.** Putnam Problem A1, 1988

Section 1.4 (page 34)

1. Quadratic **3.** Linear

5. (a) and (b)

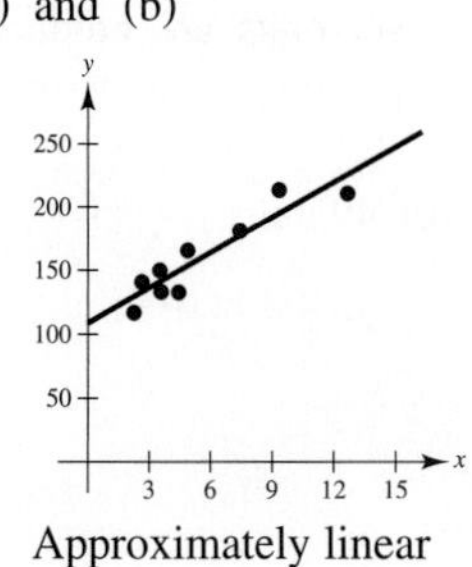

Approximately linear
(c) 136

7. (a) $d = 0.066F$
(b)

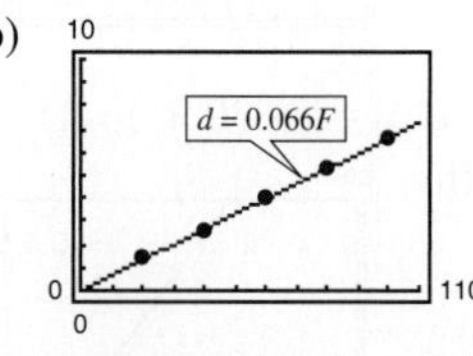

The model is a "good fit".
(c) 3.63 cm

9. (a) $y = 0.124x + 0.82$
$r \approx 0.838$
(b)

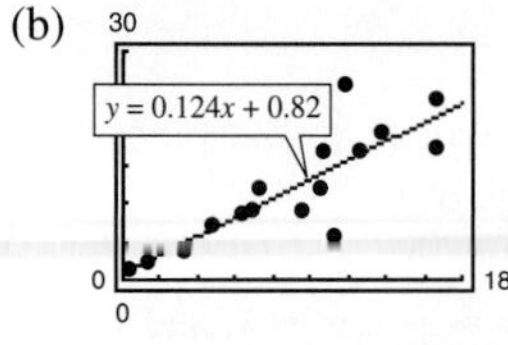

(c) Greater per capita electricity consumption by a country tends to relate to greater per capita gross national product of the country. Hong Kong, Venezuela, South Korea

(d) $y = 0.134x + 0.28$
$r \approx 0.968$

11. (a) $y_1 = 0.03434t^3 - 0.3451t^2 + 0.884t + 5.61$
$y_2 = 0.110t + 2.07$
$y_3 = 0.092t + 0.79$
(b) $y_1 + y_2 + y_3 = 0.03434t^3 - 0.3451t^2 + 1.086t + 8.47$

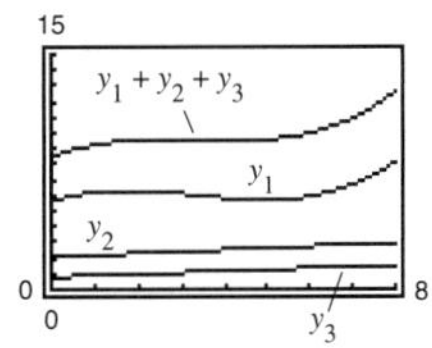

31.1 cents/mi

13. (a) Linear: $y_1 = 4.83t + 28.6$
Cubic: $y_2 = -0.1289t^3 + 2.235t^2 - 4.86t + 35.2$
(b)

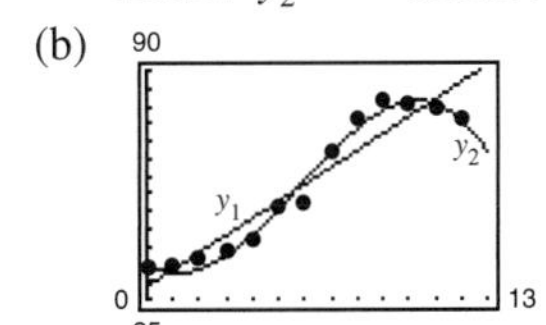

(c) Cubic

(d) $y = -0.084t^2 + 5.84t + 26.7$

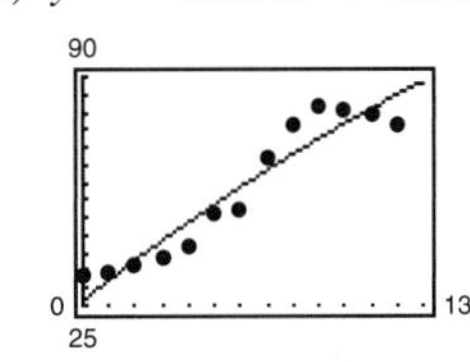

(e) Linear:
$N(14) \approx 96.2$ million people
Cubic:
$N(14) \approx 51.5$ million people
(f) Answers will vary.

15. (a) $y = -1.806x^3 + 14.58x^2 + 16.4x + 10$
(b)

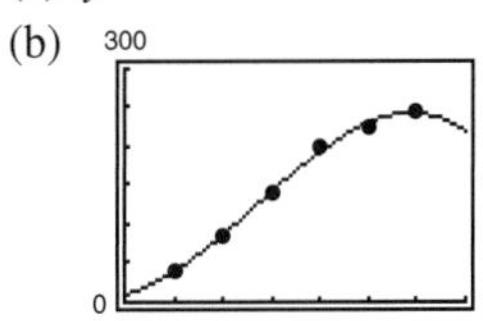

(c) 214

17. (a) Yes. At time t there is one and only one displacement y.
(b) Amplitude: 0.35; Period: 0.5 (c) $y = 0.35 \sin(4\pi t) + 2$
(d) The model appears to fit the data.

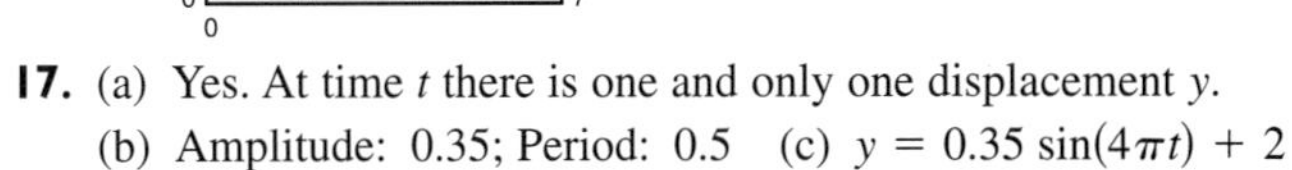

19. Answers will vary.

Section 1.5 (page 44)

1. (a) $f(g(x)) = 5[(x - 1)/5] + 1 = x$
$g(f(x)) = [(5x + 1) - 1]/5 = x$
(b)

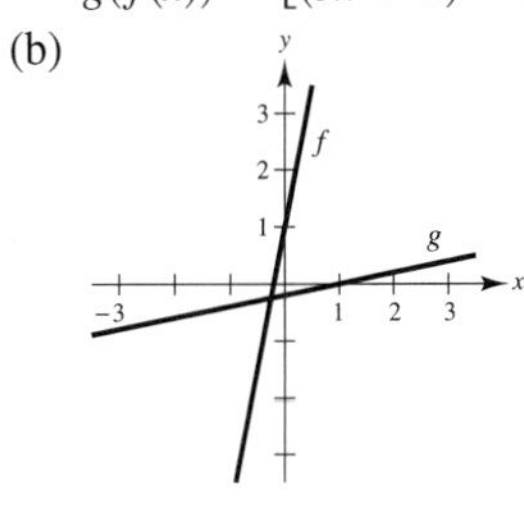

3. (a) $f(g(x)) = (\sqrt[3]{x})^3 = x$; $g(f(x)) = \sqrt[3]{x^3} = x$
(b)

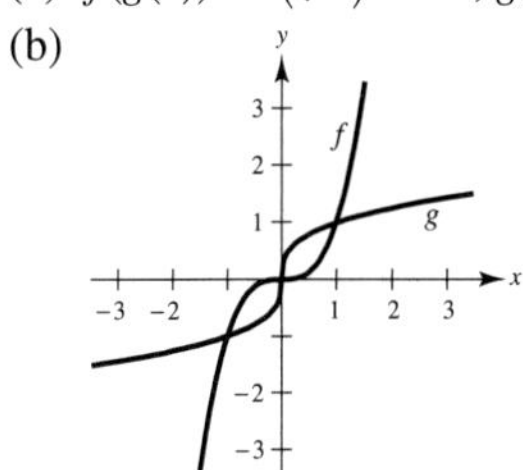

5. (a) $f(g(x)) = \sqrt{x^2 + 4 - 4} = x$;
$g(f(x)) = (\sqrt{x - 4})^2 + 4 = x$
(b)

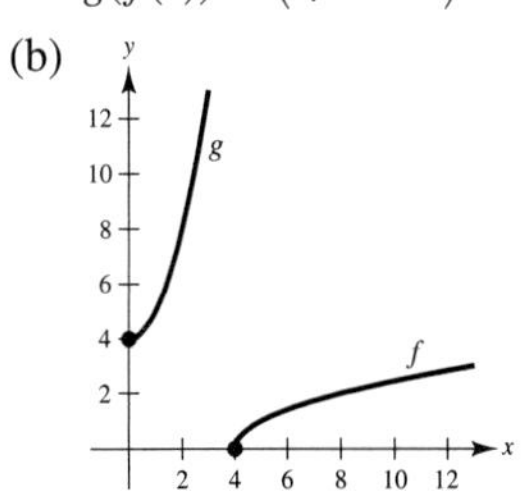

7. (a) $f(g(x)) = \dfrac{1}{1/x} = x$; $g(f(x)) = \dfrac{1}{1/x} = x$
(b)

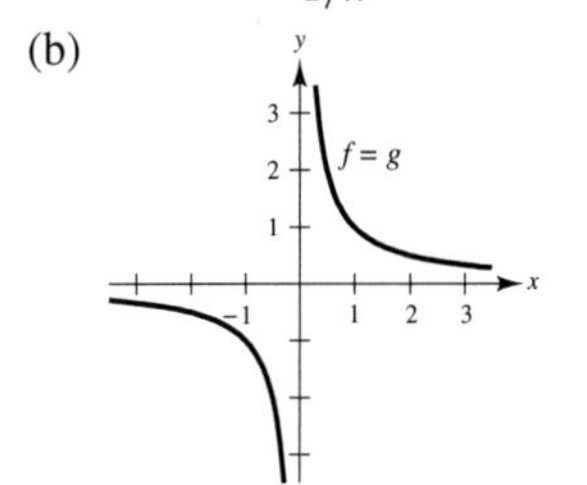

9. c **10.** b **11.** a **12.** d

13. Inverse exists. **15.** Inverse does not exist.

17. One-to-one

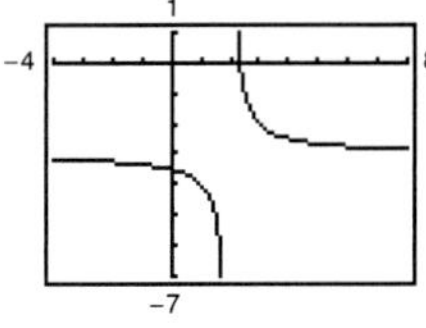

19. Not one-to-one; does not have an inverse

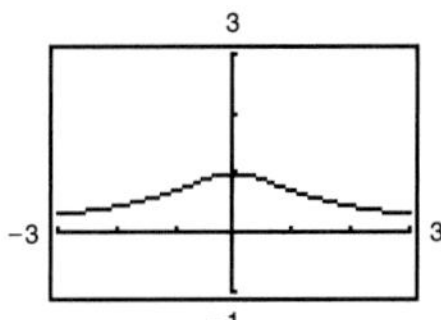

21. One-to-one

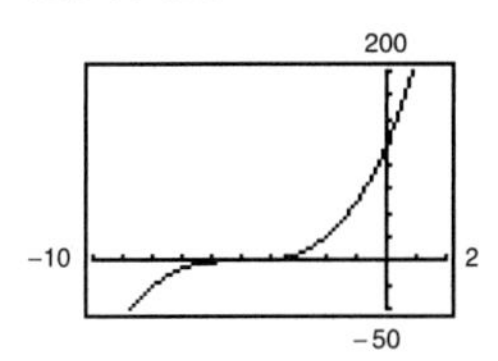

23. One-to-one **25.** Not one-to-one **27.** One-to-one

29. $f^{-1}(x) = (x + 3)/2$

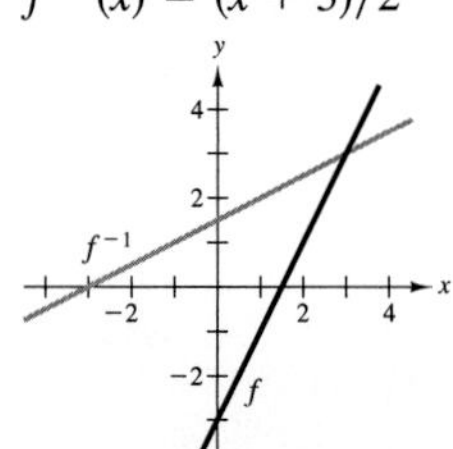

f and f^{-1} are symmetric about $y = x$.

31. $f^{-1}(x) = x^{1/5}$

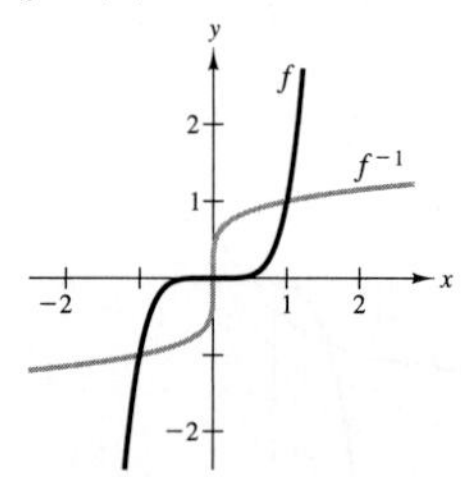

f and f^{-1} are symmetric about $y = x$.

33. $f^{-1}(x) = x^2,\ x \geq 0$

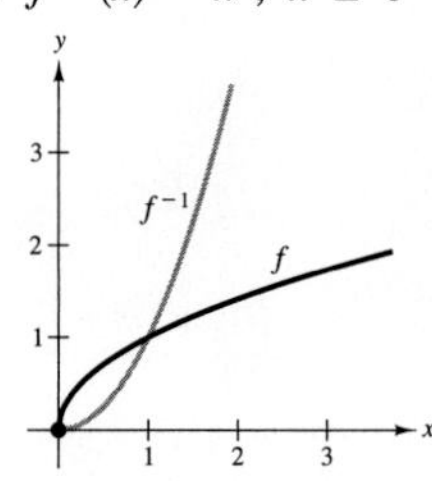

f and f^{-1} are symmetric about $y = x$.

35. $f^{-1}(x) = \sqrt{4 - x^2},\ 0 \leq x \leq 2$

f and f^{-1} are symmetric about $y = x$.

37. $f^{-1}(x) = x^3 + 1$

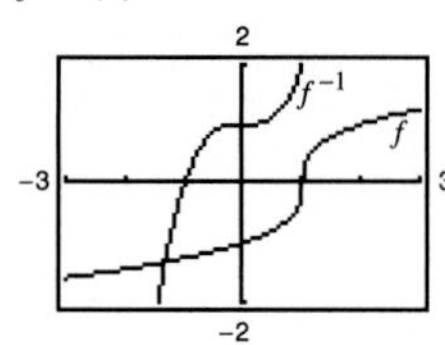

f and f^{-1} are symmetric about $y = x$.

39. $f^{-1}(x) = x^{3/2},\ x \geq 0$

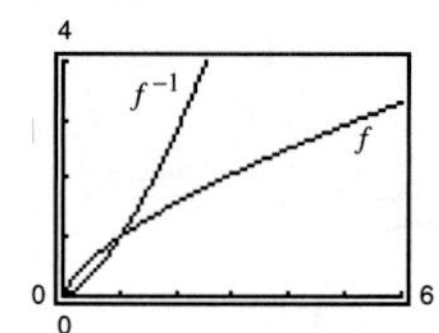

f and f^{-1} are symmetric about $y = x$.

41. $f^{-1}(x) = \sqrt{7}x/\sqrt{1 - x^2},\ -1 < x < 1$

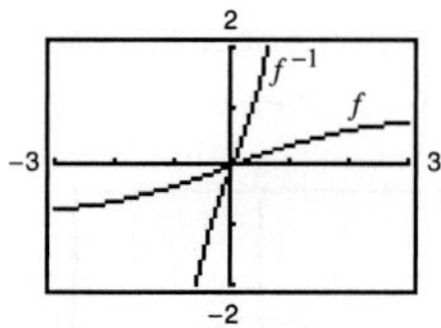

f and f^{-1} are symmetric about $y = x$.

43.

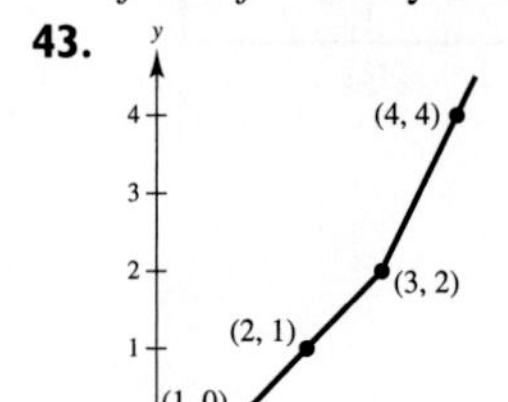

x	1	2	3	4
$f^{-1}(x)$	0	1	2	4

45. (a) Proof

(b) $y = \frac{20}{7}(80 - x)$

x: total cost

y: number of pounds of the less expensive commodity

(c) $[62.5, 80]$ (d) 20 pounds

47. $f^{-1}(x) = \begin{cases} \dfrac{1 - \sqrt{1 + 16x^2}}{2x}, & \text{if } x \neq 0 \\ 0, & \text{if } x = 0 \end{cases}$

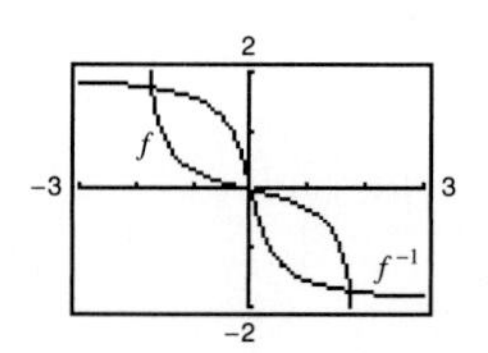

The graph of f^{-1} is a reflection of the graph of f in the line $y = x$.

49. (a) and (b)

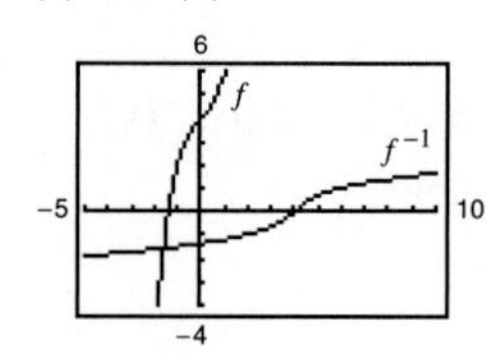

(c) Yes

51. (a) and (b)

(c) No, it is not an inverse function. It does not pass the Vertical Line Test.

53. The function f passes the Horizontal Line Test on $[4, \infty)$, so it is one-to-one on $[4, \infty)$.

55. The function f passes the Horizontal Line Test on $(0, \infty)$, so it is one-to-one on $(0, \infty)$.

57. The function f passes the Horizontal Line Test on $[0, \pi]$, so it is one-to-one on $[0, \pi]$.

59. One-to-one
$f^{-1}(x) = x^2 + 2,\ x \geq 0$

61. One-to-one
$f^{-1}(x) = 2 - x,\ x \geq 0$

63. $f^{-1}(x) = \sqrt{x} + 3,\ x \geq 0$
(Answer is not unique.)

65. $f^{-1}(x) = x - 3,\ x \geq 0$
(Answer is not unique.)

67. 1 **69.** $\frac{\pi}{6}$ **71.** 2 **73.** 32 **75.** 600

77. $(g^{-1} \circ f^{-1})(x) = \dfrac{x + 1}{2}$ **79.** $(f \circ g)^{-1}(x) = \dfrac{x + 1}{2}$

81. (a) f is one-to-one because it passes the Horizontal Line Test.

(b) $[-2, 2]$

(c) -4

83.

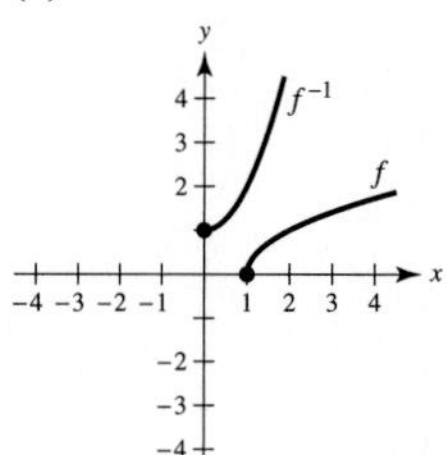

85. (a)

x	-1	-0.8	-0.6	-0.4	-0.2
y	-1.57	-0.93	-0.64	-0.41	-0.20

x	0	0.2	0.4	0.6	0.8	1
y	0	0.20	0.41	0.64	0.93	1.57

(b)

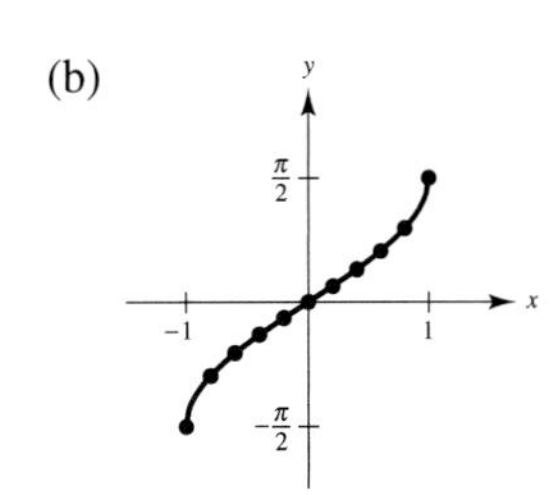

(c)

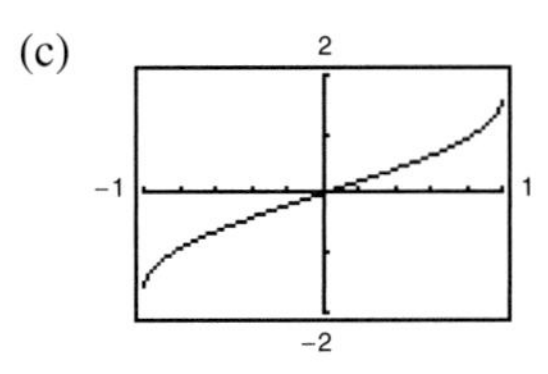

(d) Intercept: (0, 0);
Symmetry: origin

87. $\left(-\sqrt{2}/2, 3\pi/4\right), (1/2, \pi/3), \left(\sqrt{3}/2, \pi/6\right)$

89. $\pi/6$ **91.** $\pi/3$ **93.** $\pi/6$ **95.** $-\pi/4$ **97.** 2.50

99. $\arccos(1/1.269) \approx 0.66$

101. Let $y = f(x)$ be one-to-one. Solve for x as a function of y. Interchange x and y to get $y = f^{-1}(x)$. Let the domain of f^{-1} be the range of f. Verify that $f(f^{-1}(x)) = x$ and $f^{-1}(f(x)) = x$.
Example: $f(x) = x^3$
$y = x^3$
$x = \sqrt[3]{y}$
$y = \sqrt[3]{x}$
$f^{-1}(x) = \sqrt[3]{x}$

103. Answers will vary. Example: $y = x^4 - 2x^3$

105. If the domains were not restricted, then the trigonometric functions would not be one-to-one and hence would have no inverses.

107.

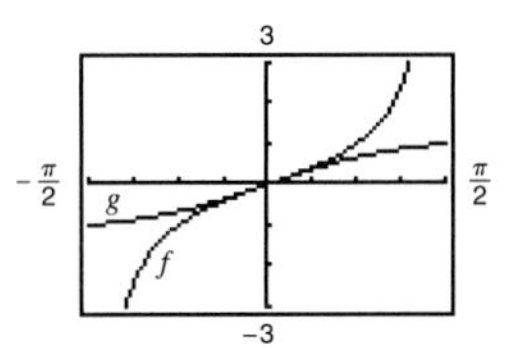

109. -0.1

111. (a) $\dfrac{1}{2}$

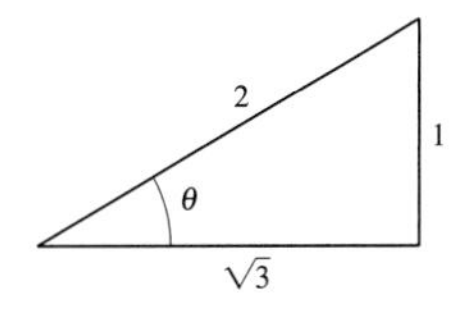

(b) $\dfrac{\sqrt{3}}{2}$

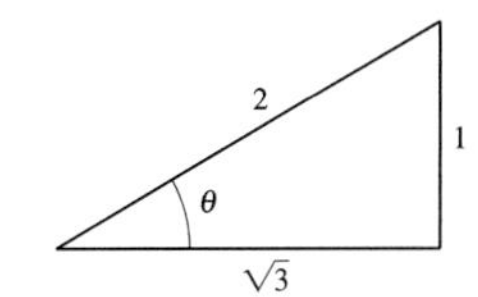

113. (a) $\frac{3}{5}$

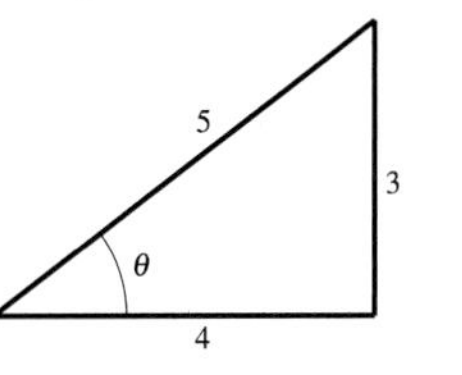
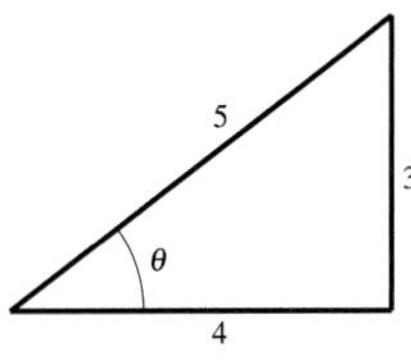

(b) $\frac{5}{3}$

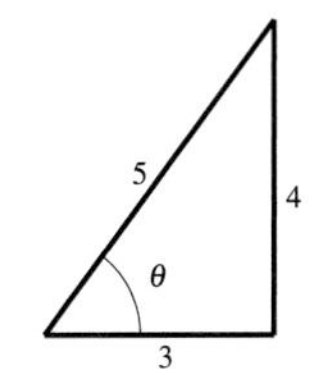

115. (a) $-\sqrt{3}$

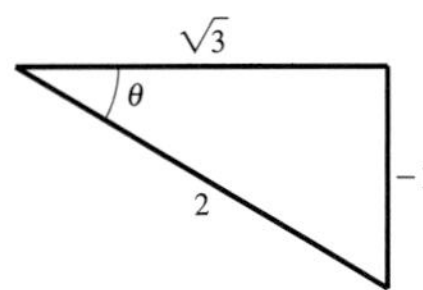

(b) $-\frac{13}{5}$

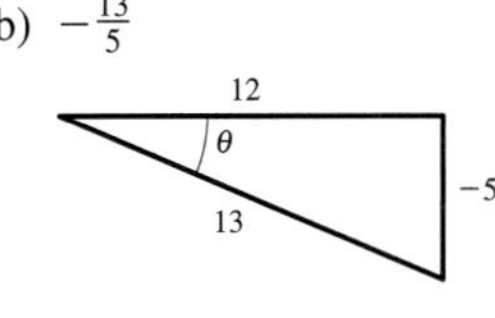

117. $x = \frac{1}{3}\left[\sin\left(\frac{1}{2}\right) + \pi\right] \approx 1.207$ **119.** $x = \frac{1}{3}$

121. (0.7862, 0.6662)

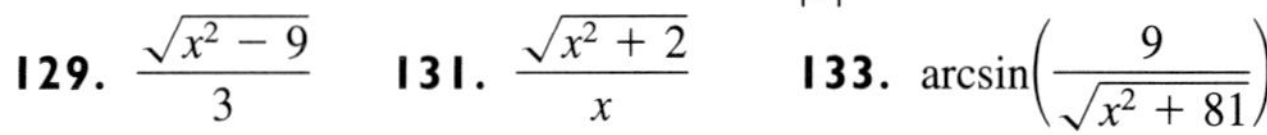

123. x **125.** $\sqrt{1 - 4x^2}$ **127.** $\dfrac{\sqrt{x^2 - 1}}{|x|}$

129. $\dfrac{\sqrt{x^2 - 9}}{3}$ **131.** $\dfrac{\sqrt{x^2 + 2}}{x}$ **133.** $\arcsin\left(\dfrac{9}{\sqrt{x^2 + 81}}\right)$

135. Proof

137.

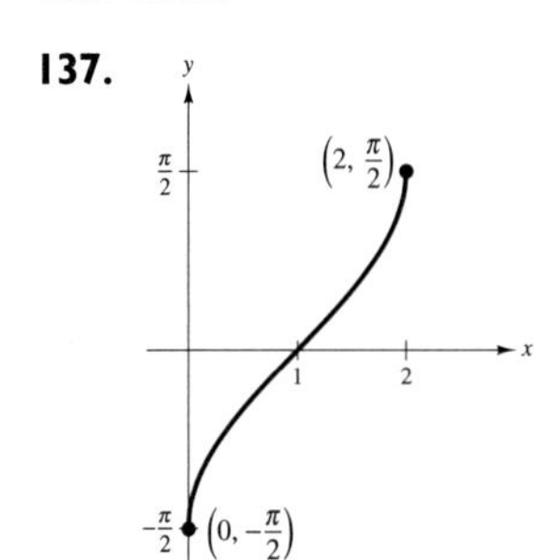

139.

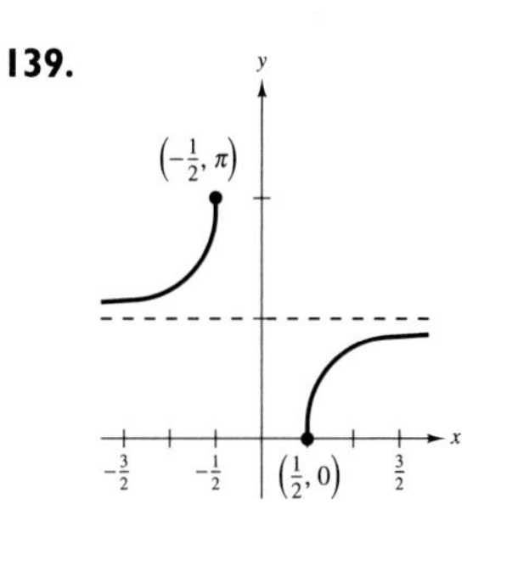

141. Proof **143.** Proof **145.** False: Let $f(x) = x^2$.

147. False: $\arcsin^2 0 + \arccos^2 0 = \left(\dfrac{\pi}{2}\right)^2 \neq 1$

149. True **151.** Proof

153. $f^{-1}(x) = \dfrac{-b - \sqrt{b^2 - 4ac + 4ax}}{2a}$

155. $ad - bc \neq 0$: $f^{-1}(x) = \dfrac{b - dx}{cx - a}$

Section 1.6 (page 54)

1. (a) 125 (b) 9 (c) $\frac{1}{9}$ (d) $\frac{1}{3}$

3. (a) 5^5 (b) $\frac{1}{5}$ (c) $\frac{1}{5}$ (d) 2^2

5. (a) e^6 (b) e^{12} (c) $\dfrac{1}{e^6}$ (d) e^2

7. $x = 4$ **9.** $x = -2$ **11.** $x = 2$

13. $x = 2^4 = 16$ **15.** $x = -\frac{5}{2}$ **17.** $2.7182805 < e$

19.

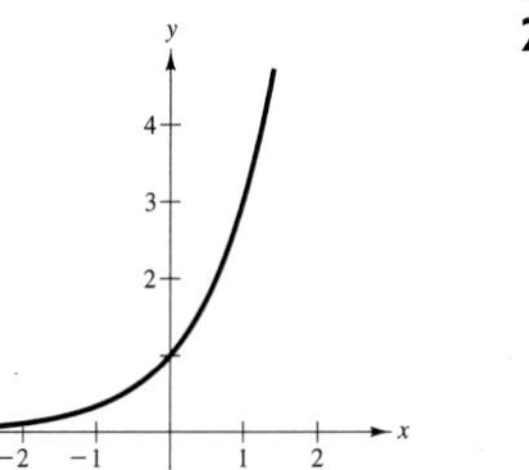

21.

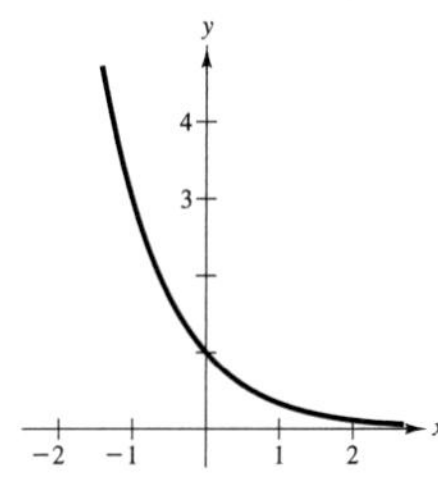

23.

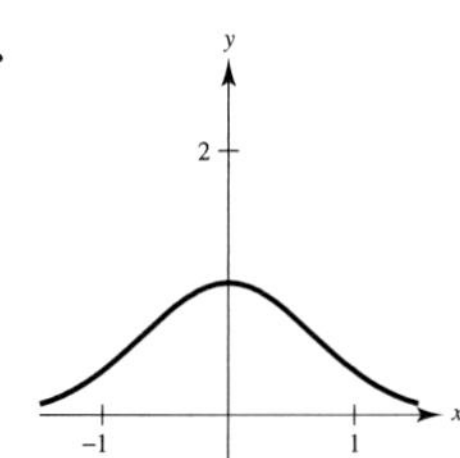

25.

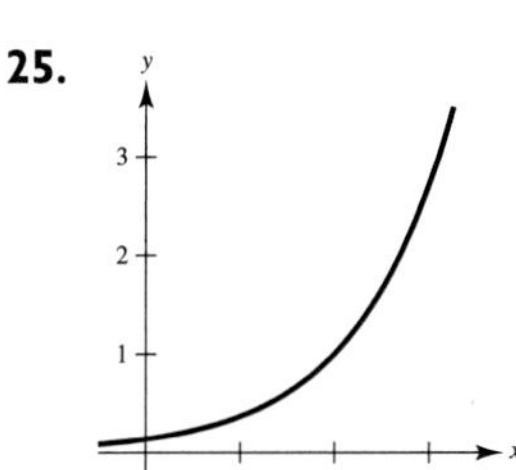

27.

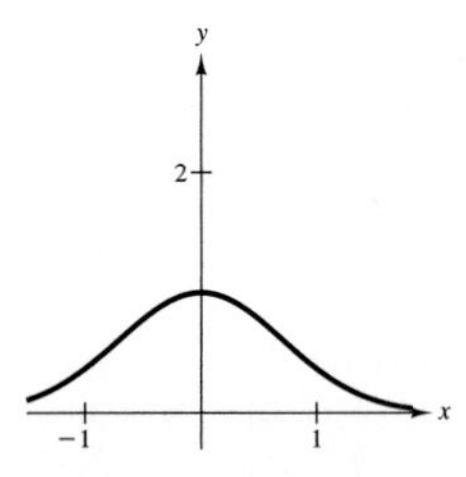

29. (a)

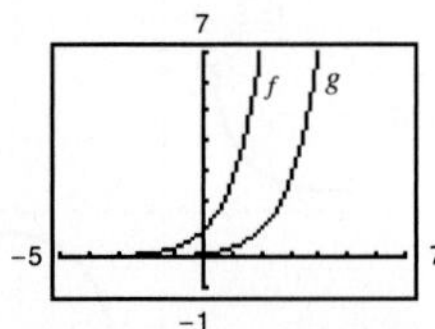

Translation two units to the right

(b)

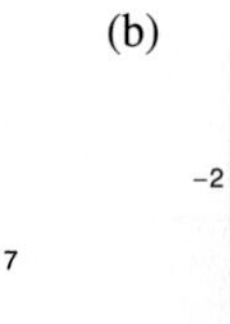

Reflection in the x-axis and vertical shrink

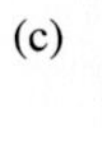

(c)

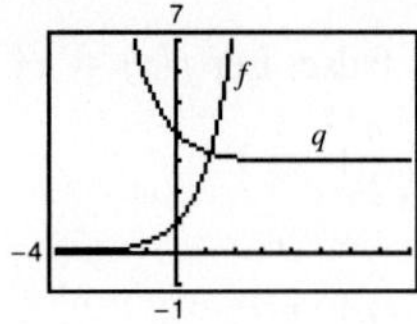

Reflection in the y-axis and translation three units upward

31. c **32.** d **33.** a **34.** b **35.** b **36.** d

37. a **38.** c **39.** $y = 2(3^x)$

41. $\ln 1 = 0$ **43.** $e^{0.6931\ldots} = 2$

45. Domain: $x > 0$

47. Domain: $x > 0$

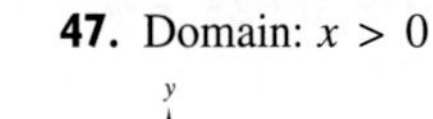

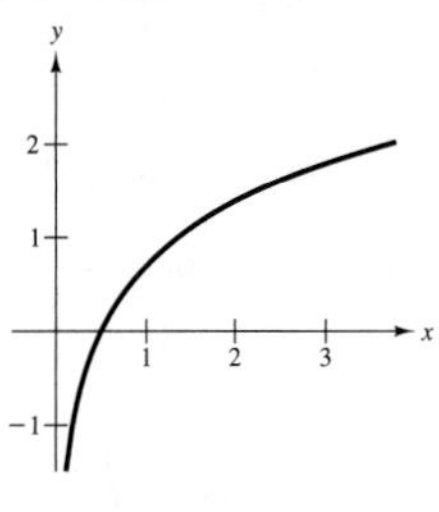

49. Domain: $x > 1$

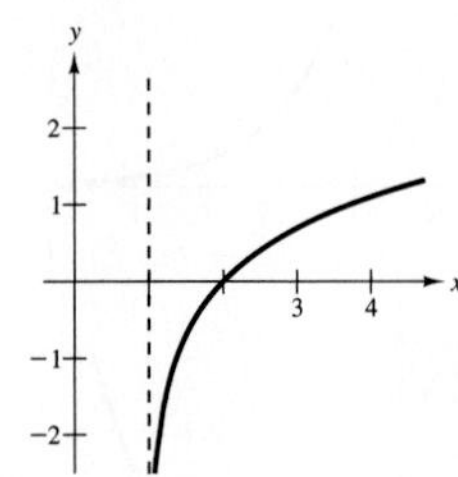

51.

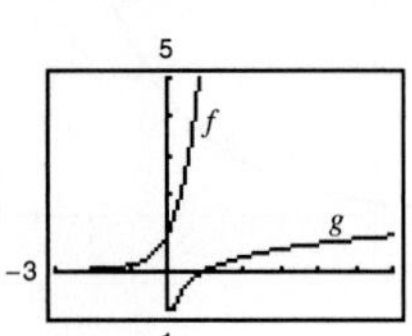

53.

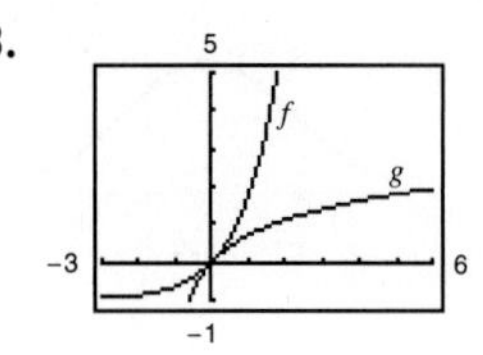

55. (a) $f^{-1}(x) = \dfrac{\ln x + 1}{4}$

(b)

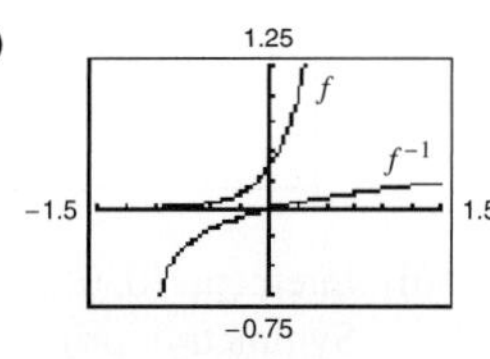

(c) Proof

57. (a) $f^{-1}(x) = e^{x/2} + 1$

(b)

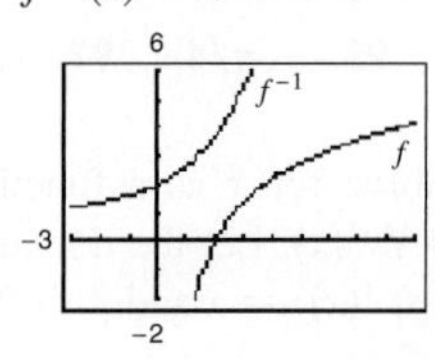

(c) Proof

59. x^2 **61.** $5x + 2$ **63.** $\sqrt{x}$

65. (a) 1.7917 (b) -0.4055 (c) 4.3944 (d) 0.5493

67. Answers will vary. **69.** Answers will vary.

71. $\ln 2 - \ln 3$ **73.** $\ln x + \ln y - \ln z$ **75.** $-\ln 5$

77. $3[\ln(x + 1) + \ln(x - 1) - 3 \ln x]$ **79.** $2 + \ln 3$

81. $\ln \dfrac{x - 2}{x + 2}$ **83.** $\ln \sqrt[3]{\dfrac{x(x + 3)^2}{x^2 - 1}}$ **85.** $\ln \dfrac{9}{\sqrt{x^2 + 1}}$

87. (a) $x = 4$ (b) $x = \frac{3}{2}$

89. (a) $x = e^2 \approx 7.389$ (b) $x = \ln 4 \approx 1.386$

91. $x > \ln 5$ **93.** $e^{-2} < x < 1$

95.

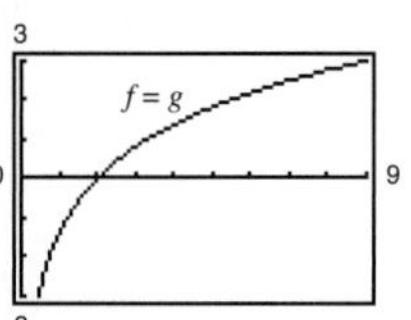

97. Proof

99.

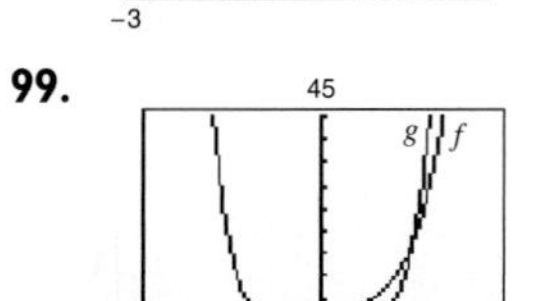

$(-0.7899, 0.2429)$, $(1.6242, 18.3615)$, $(6, 46656)$

$f(x) = 6^x$

101. (a)

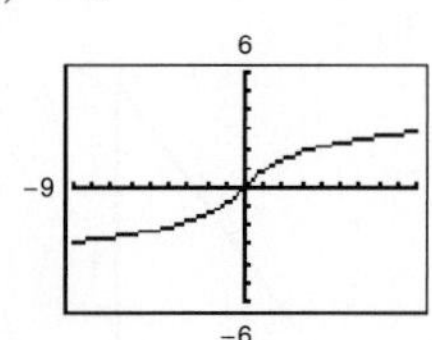

Domain: $(-\infty, \infty)$

(b) Proof

(c) $f^{-1}(x) = \dfrac{e^{2x} - 1}{2e^x}$

Review Exercises for Chapter 1 (page 57)

1. $\left(\frac{3}{2}, 0\right), (0, -3)$ **3.** $(1, 0), \left(0, \frac{1}{2}\right)$ **5.** y-axis symmetry

7.

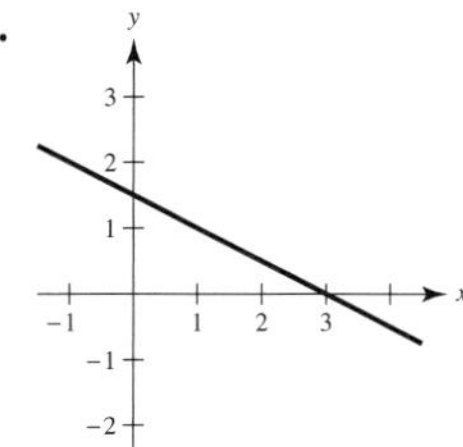

9.

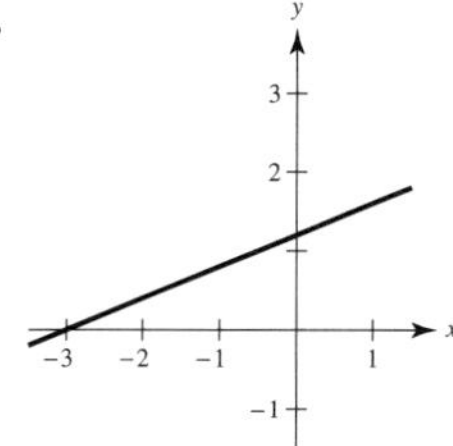

11.

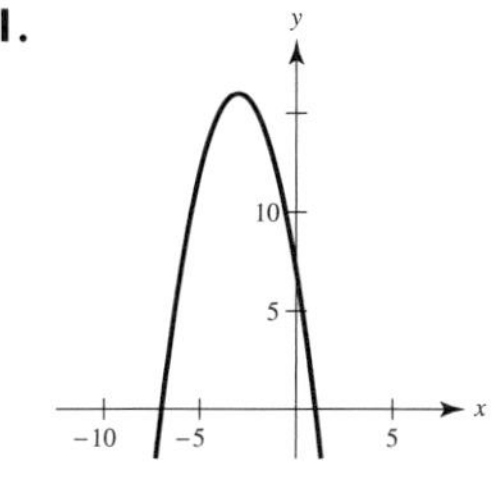

13.

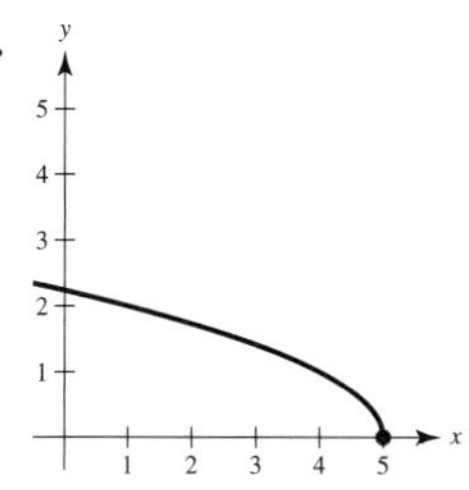

15. (4, 1)

17. $m = \frac{3}{7}$

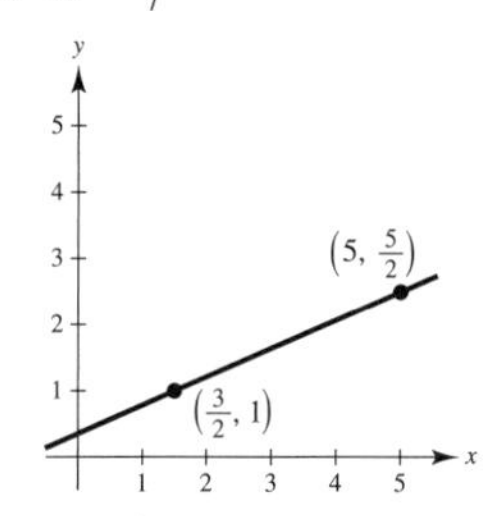

19. $t = \frac{7}{3}$

21. $y = \frac{3}{2}x - 5$ or
$3x - 2y - 10 = 0$

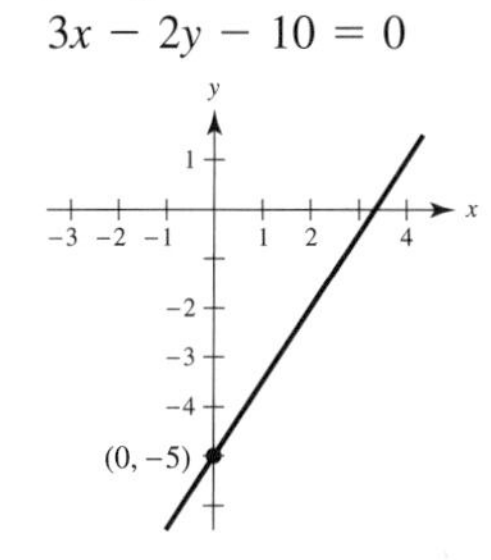

23. $y = -\frac{2}{3}x - 2$ or
$2x + 3y + 6 = 0$

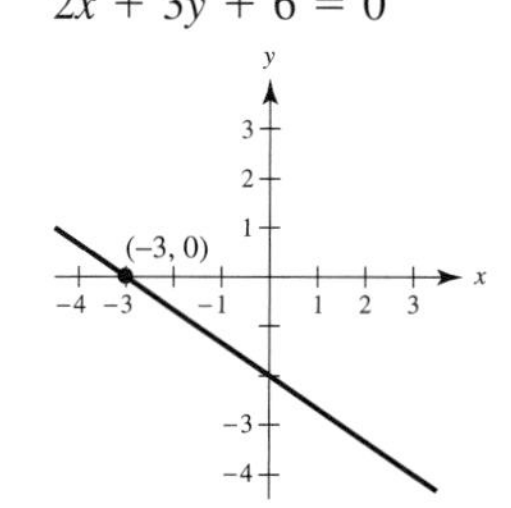

25. (a) $7x - 16y + 78 = 0$ (b) $5x - 3y + 22 = 0$
(c) $2x + y = 0$ (d) $x + 2 = 0$

27. $V = 12{,}500 - 850t$; \$9950

29. Not a function

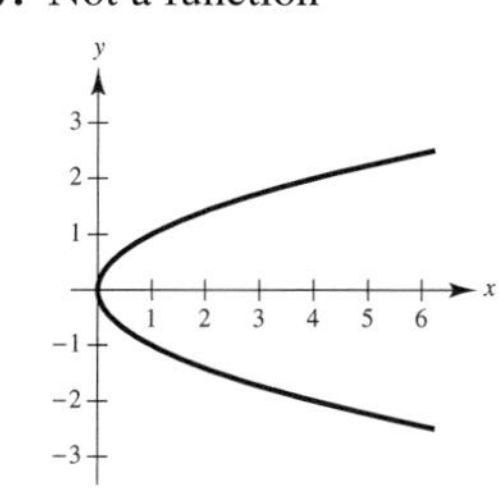

31. Function

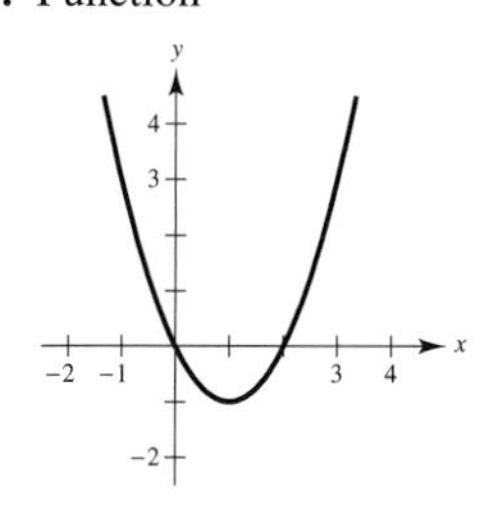

33. (a) Undefined (b) $-1/(1 + \Delta x), \Delta x \neq 0, -1$

35. (a) D: $[-6, 6]$; R: $[0, 6]$
(b) D: $(-\infty, 5) \cup (5, \infty)$; R: $(-\infty, 0) \cup (0, \infty)$
(c) D: $(-\infty, \infty)$; R: $(-\infty, \infty)$

37. (a)

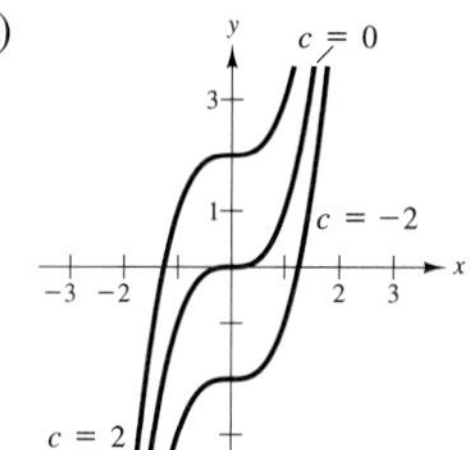

(b)

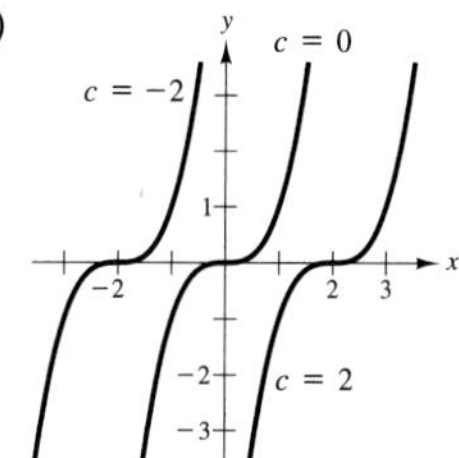

(c)

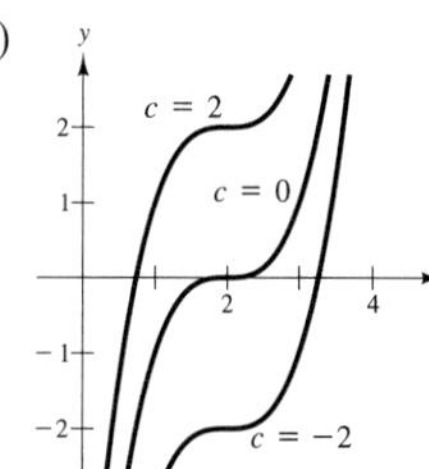

(d)

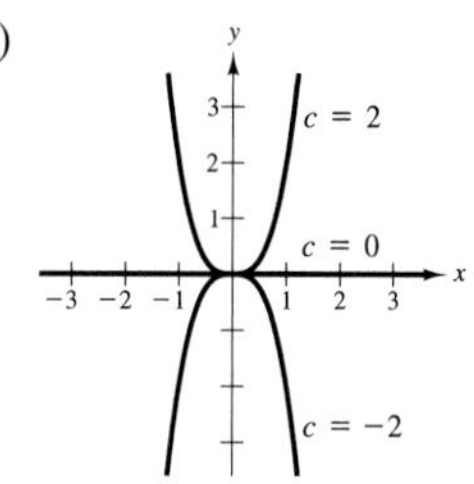

39. (a) Minimum degree: 3; Leading coefficient: negative
(b) Minimum degree: 4; Leading coefficient: positive
(c) Minimum degree: 2; Leading coefficient: negative
(d) Minimum degree: 5; Leading coefficient: positive

41. (a) Yes. For each time t there corresponds one and only one displacement y.
(b) Amplitude: 0.25; Period: 1.1 (c) $y = \frac{1}{4}\cos(5.7t)$
(d)

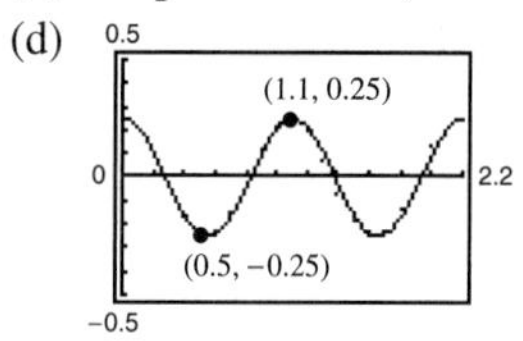

43. (a) $f^{-1}(x) = 2x + 6$
(b)

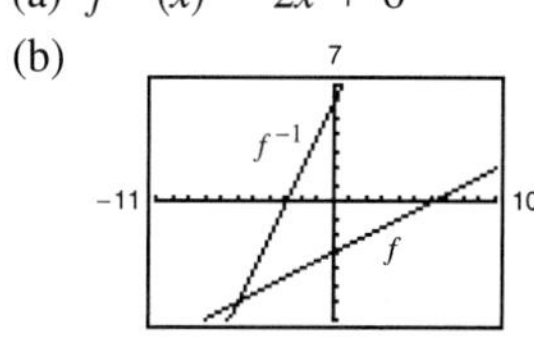

(c) Proof

45. (a) $f^{-1}(x) = x^2 - 1, \quad x \geq 0$
(b)

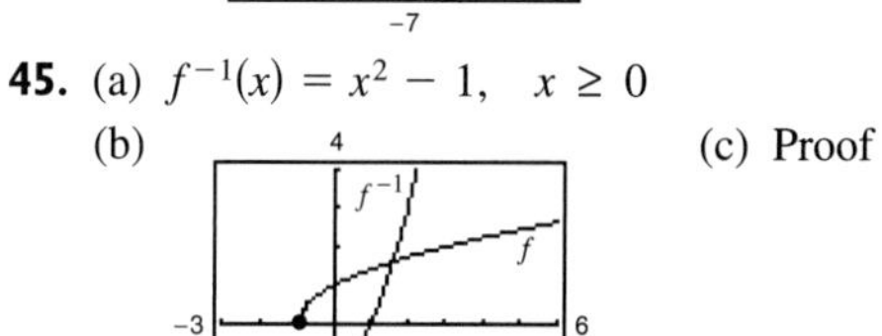

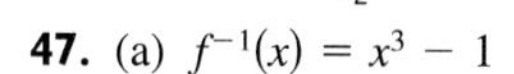

(c) Proof

47. (a) $f^{-1}(x) = x^3 - 1$
(b)

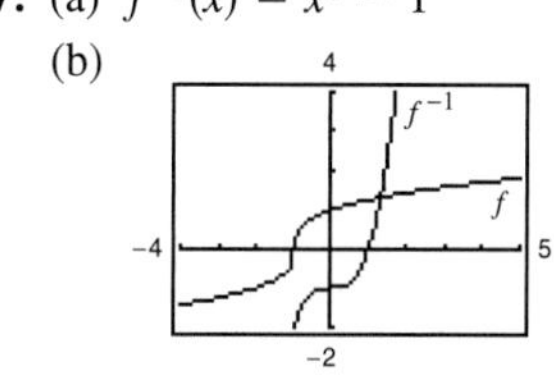

(c) Proof

49.

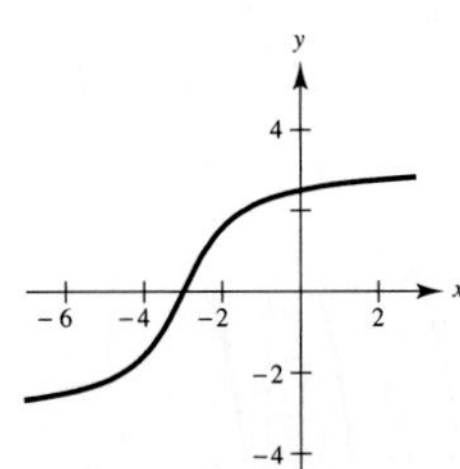

51. $\dfrac{1}{2}$

53.

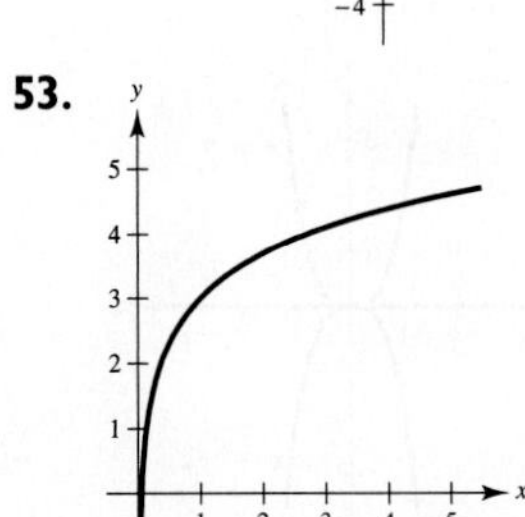

55. $\frac{1}{5}[\ln(2x+1)+\ln(2x-1)-\ln(4x^2+1)]$

57. $\ln\left(\dfrac{3\sqrt[3]{4-x^2}}{x}\right)$ **59.** $e^4-1\approx 53.598$

61. (a) $f^{-1}(x)=e^{2x}$

(b)

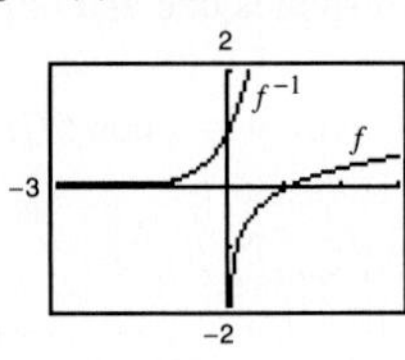

(c) Proof

63.

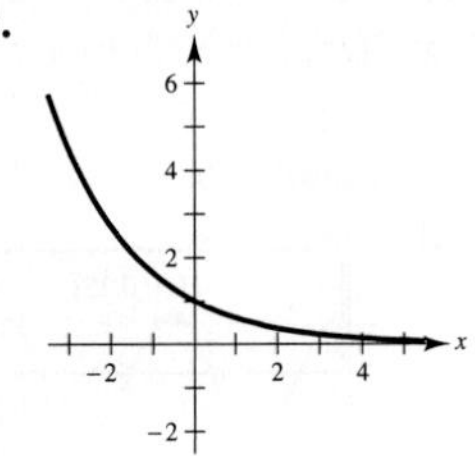

P.S. Problem Solving (page 59)

1. (a) Center: $(3, 4)$; Radius: 5

(b) $y=-\frac{3}{4}x$ (c) $y=\frac{3}{4}x-\frac{9}{2}$ (d) $\left(3,-\frac{9}{4}\right)$

3.

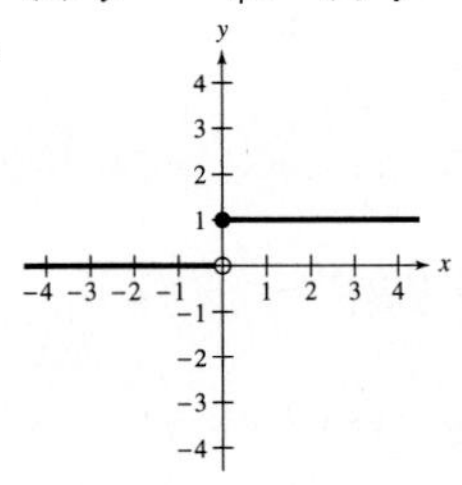

(a) $H(x)-2=\begin{cases}-1, & x\ge 0\\ -2, & x<0\end{cases}$ (b) $H(x-2)=\begin{cases}1, & x\ge 2\\ 0, & x<2\end{cases}$

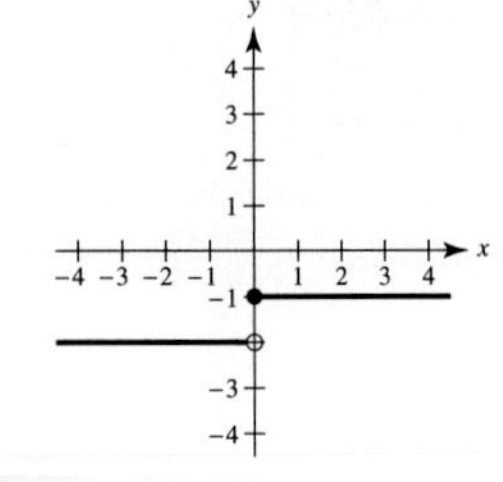

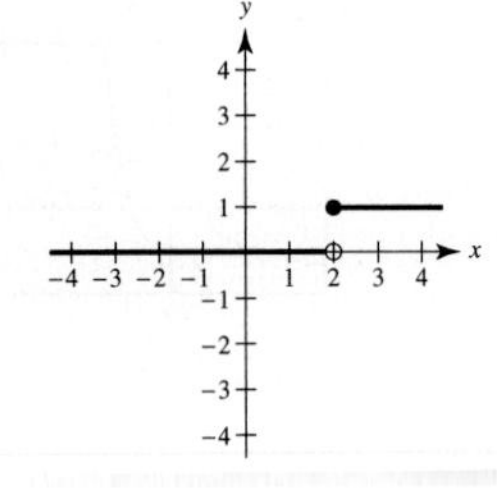

(c) $-H(x)=\begin{cases}-1, & x\ge 0\\ 0, & x<0\end{cases}$ (d) $H(-x)=\begin{cases}1, & x\le 0\\ 0, & x>0\end{cases}$

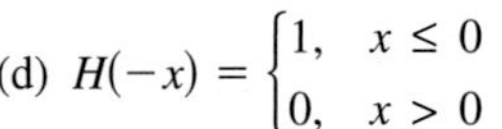

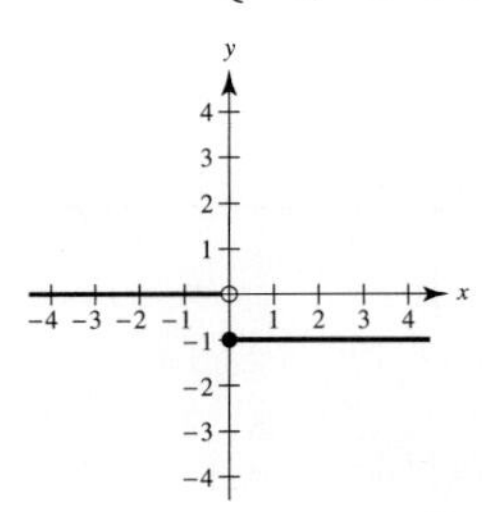

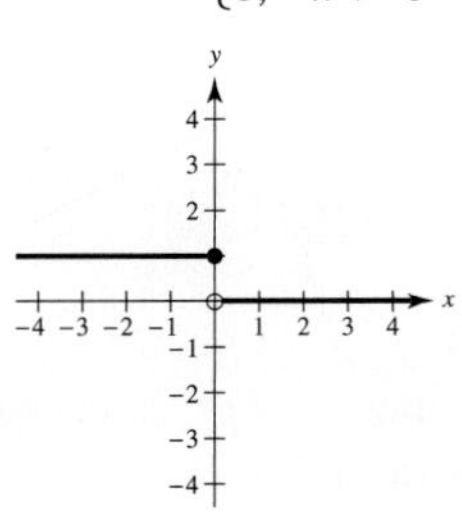

(e) $\frac{1}{2}H(x)=\begin{cases}\frac{1}{2}, & x\ge 0\\ 0, & x<0\end{cases}$ (f) $-H(x-2)+2=\begin{cases}1, & x\ge 2\\ 2, & x<2\end{cases}$

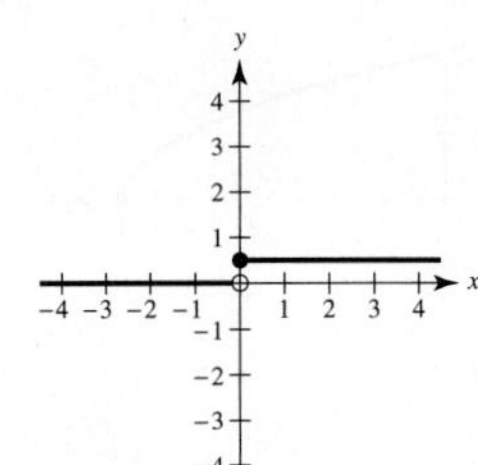

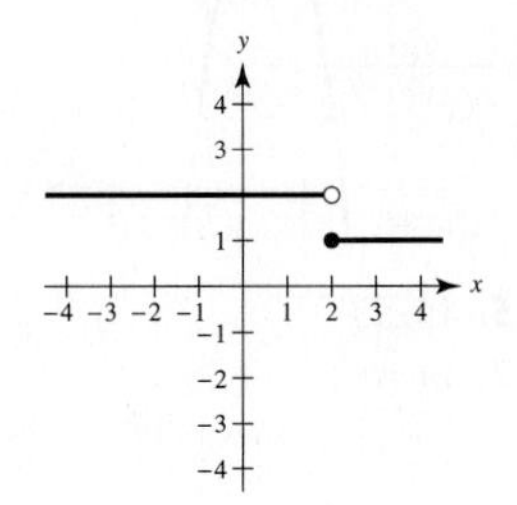

5. (a) $A(x)=x[(100-x)/2]$; Domain: $(0, 100)$

(b) Dimensions 50 m × 25 m yield maximum area of 1250 square meters.

(c) 50 m × 25 m; Area = 1250 square meters

7. $T(x)=\dfrac{2\sqrt{4+x^2}+\sqrt{(3-x)^2+1}}{4}$

9. (a) 5, less (b) 3, greater (c) 4.1, less (d) $4+h$

(e) 4; Answers will vary.

11. (a) $x=-3+\sqrt{18}\approx 1.2426$

$x=-3-\sqrt{18}\approx -7.2426$

(b) $(x+3)^2+y^2=18$

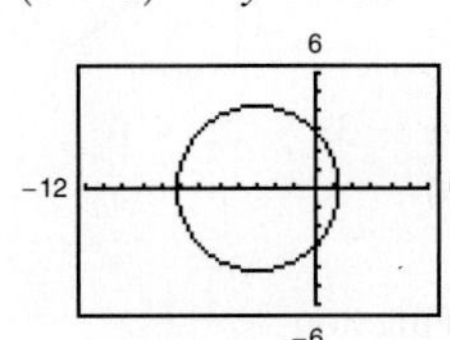

13. Answers will vary.

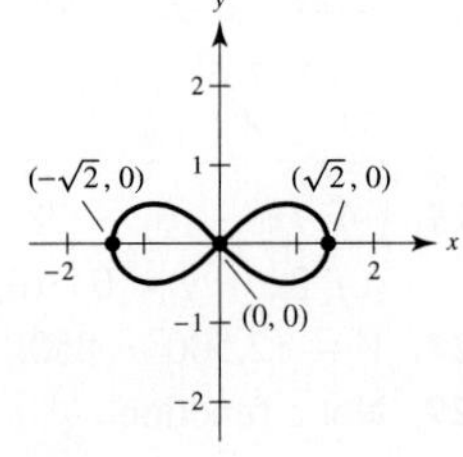

Chapter 2

Section 2.1 (page 67)

1. Precalculus: 300 ft

3. Calculus: Slope of the tangent line at $x=2$ is 0.16.

5. Precalculus: $\frac{15}{2}$ square units

7. (a)

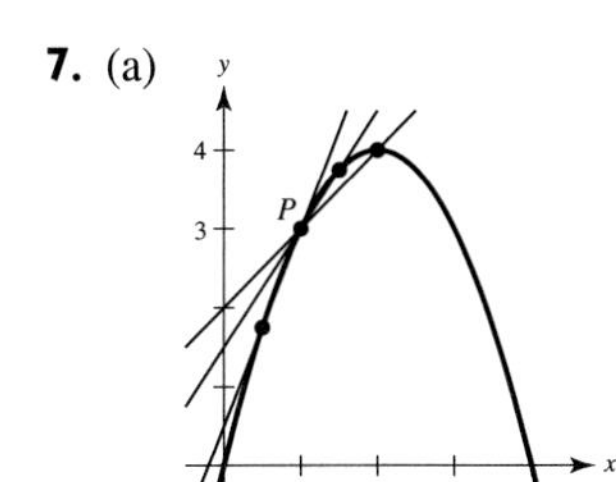

(b) $1; \frac{3}{2}; \frac{5}{2}$
(c) 2. Use points closer to P.

9. (a) Area $\approx$ 10.417; Area $\approx$ 9.145 (b) Use more rectangles.

11. (a) 5.66 (b) 6.11 (c) Increase the number of line segments.

Section 2.2 (page 74)

1.

x	1.9	1.99	1.999	2.001	2.01	2.1
$f(x)$	0.3448	0.3344	0.3334	0.3332	0.3322	0.3226

$\lim_{x\to 2} \frac{x-2}{x^2-x-2} \approx 0.3333$ $\left(\text{Actual limit is } \frac{1}{3}.\right)$

3.

x	2.9	2.99	2.999
$f(x)$	-0.0641	-0.0627	-0.0625

x	3.001	3.01	3.1
$f(x)$	-0.0625	-0.0623	-0.0610

$\lim_{x\to 3} \frac{[1/(x+1)] - (1/4)}{x-3} \approx -0.0625$ $\left(\text{Actual limit is } -\frac{1}{16}.\right)$

5.

x	-0.1	-0.01	-0.001	0.001	0.01	0.1
$f(x)$	0.9983	0.99998	1.0000	1.0000	0.99998	0.9983

$\lim_{x\to 0} \frac{\sin x}{x} \approx 1.0000$ (Actual limit is 1.)

7.

x	-0.1	-0.01	-0.001	0.001	0.01	0.1
$f(x)$	0.9516	0.9950	0.9995	1.0005	1.0050	1.0517

$\lim_{x\to 0} \frac{e^x - 1}{x} \approx 1$ (Actual limit is 1.)

9.

x	-0.1	-0.01	-0.001	0.001	0.01	0.1
$f(x)$	1.0536	1.0050	1.0005	0.9995	0.9950	0.9531

$\lim_{x\to 0} \frac{\ln(x+1)}{x} \approx 1$ (Actual limit is 1.)

11. 1

13. Limit does not exist. The function approaches 1 from the right side of 3 but it approaches -1 from the left side of 3.

15. 0

17. Limit does not exist. The function increases without bound as x approaches $\pi/2$ from the left and decreases without bound as x approaches $\pi/2$ from the right.

19. 1

21. (a) 2
(b) Limit does not exist. The function approaches 1 from the right side of 1 but it approaches 3.5 from the left side of 1.
(c) Value does not exist. The function is undefined at $x = 4$.
(d) 2

23. $\lim_{x\to c} f(x)$ exists at all points on the graph except where $c = -3$.

25.

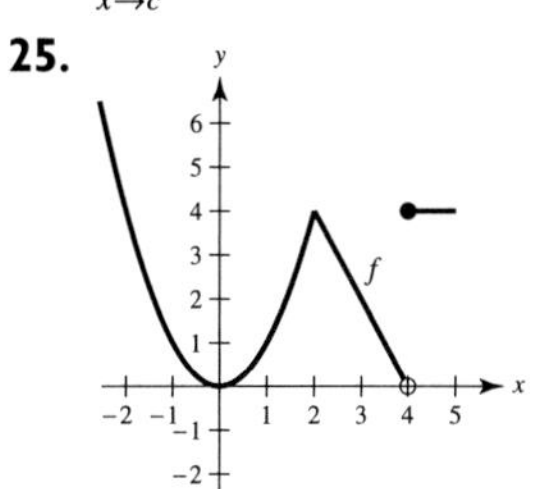

$\lim_{x\to c} f(x)$ exists at all points on the graph except where $c = 4$.

27.

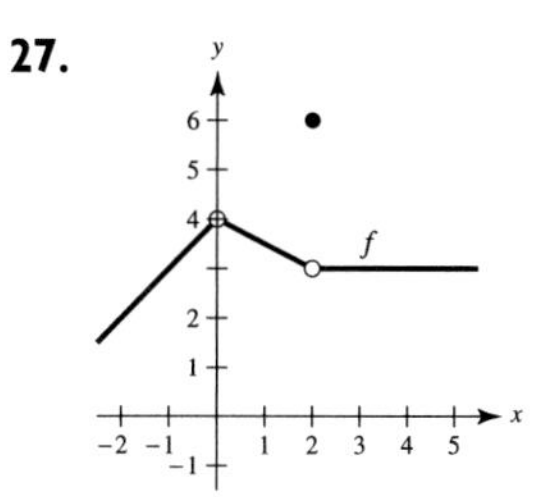

29. (a)

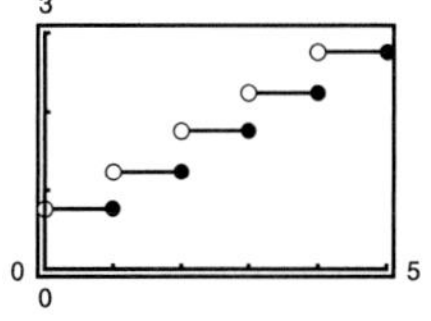

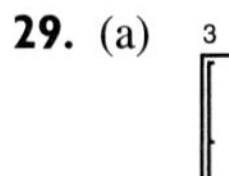

(b)

t	3	3.3	3.4	3.5	3.6	3.7	4
C	1.75	2.25	2.25	2.25	2.25	2.25	2.25

$\lim_{t\to 3.5} C(t) = 2.25$

(c)

t	2	2.5	2.9	3	3.1	3.5	4
C	1.25	1.75	1.75	1.75	2.25	2.25	2.25

The limit does not exist, because the limits from the right and left are not equal.

31. $\delta = 0.4$ **33.** $\delta = \frac{1}{11} \approx 0.091$

35. $L = 8$. Let $\delta = 0.01/3 \approx 0.0033$.

37. $L = 1$. Let $\delta = 0.01/5 = 0.002$.

39. 5 **41.** -3 **43.** 3 **45.** 0 **47.** 4 **49.** 2

51. Answers will vary.

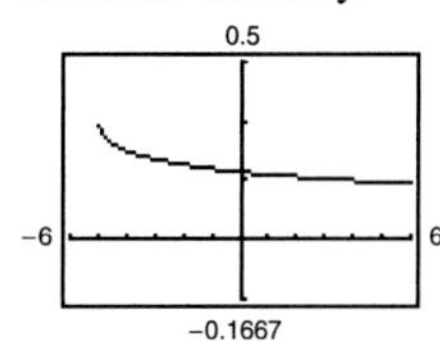

$\lim_{x\to 4} f(x) = \frac{1}{6}$

Domain: $[-5, 4) \cup (4, \infty)$

The graph has a hole at $x = 4$.

53. Answers will vary.

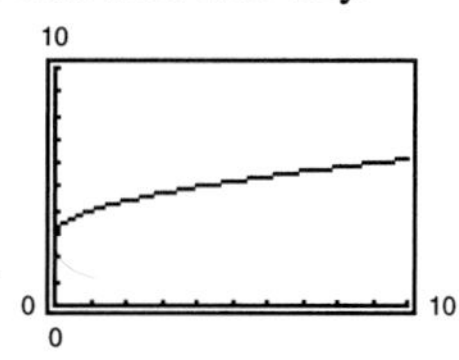

$\lim_{x\to 9} f(x) = 6$

Domain: $[0, 9) \cup (9, \infty)$

The graph has a hole at $x = 9$.

55. Answers will vary. Sample answer: As x approaches 8 from either side, $f(x)$ becomes arbitrarily close to 25.

57. No. The fact that $\lim_{x\to 2} f(x) = 4$ has no bearing on the value of f at 2.

59. (a) $r = \frac{3}{\pi} \approx 0.9549$ cm
(b) $\frac{5.5}{2\pi} \le r \le \frac{6.5}{2\pi}$, or approximately $0.8754 < r < 1.0345$
(c) $\lim_{r\to 3/\pi} 2\pi r = 6$; $\varepsilon = 0.5$; $\delta \approx 0.0796$

61.

x	-0.001	-0.0001	-0.00001
$f(x)$	2.7196	2.7184	2.7183

x	0.00001	0.0001	0.001
$f(x)$	2.7183	2.7181	2.7169

$\lim_{x\to 0} f(x) \approx 2.7183$

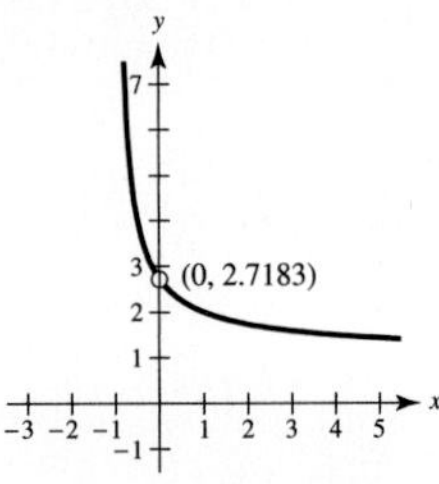

63. $\delta \approx 0.001$
(1.999, 2.001)

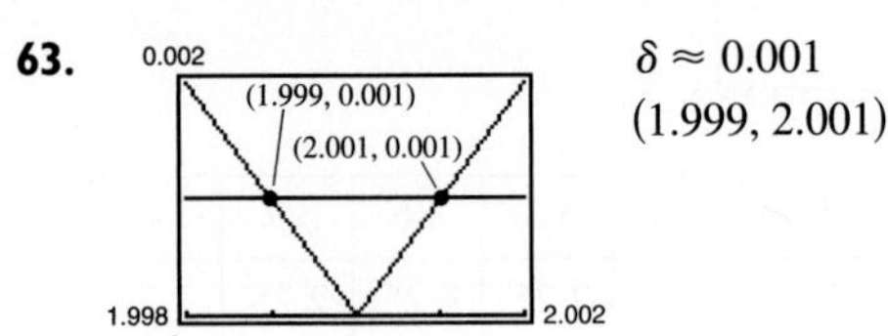

65. False. The existence or nonexistence of $f(x)$ at $x = c$ has no bearing on the existence of the limit of $f(x)$ as $x \to c$.

67. False. See Exercise 23.

69. (a) Yes. As x approaches 0.25 from either side, $\sqrt{x}$ becomes arbitrarily close to 0.5.
(b) No. $\lim_{x\to 0} \sqrt{x}$ does not exist because for $x < 0$, $\sqrt{x}$ does not exist.

71–73. Proofs **75.** Answers will vary.

77. Putnam Problem B1, 1986

Section 2.3 (page 87)

1.

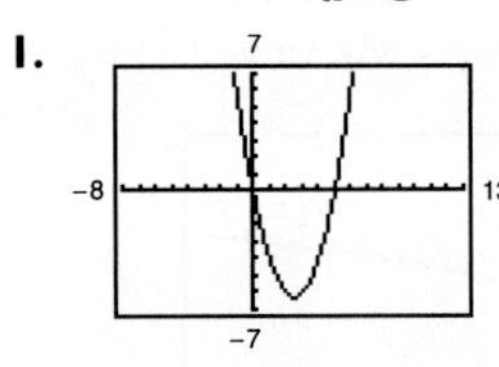

(a) 0 (b) 6

3.

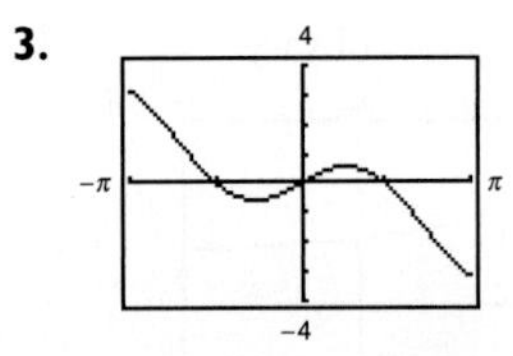

(a) 0 (b) ≈ 0.52 or $\pi/6$

5. 16 **7.** -1 **9.** 7 **11.** 1/2 **13.** $-2/5$
15. 35/3 **17.** 2 **19.** 1 **21.** $-1/2$ **23.** 1
25. 1/2 **27.** -1 **29.** 1 **31.** $\ln 3 + e$
33. (a) 4 (b) 64 (c) 64 **35.** (a) 3 (b) 2 (c) 2
37. (a) 15 (b) 5 (c) 6 (d) 2/3
39. (a) 64 (b) 2 (c) 12 (d) 8
41. (a) 1 (b) 3

$g(x) = \dfrac{-2x^2 + x}{x}$ and $f(x) = -2x + 1$ agree except at $x = 0$.

43. (a) 2 (b) 0

$g(x) = \dfrac{x^3 - x}{x - 1}$ and $f(x) = x^2 + x$ agree except at $x = 1$.

45. -2

$f(x) = \dfrac{x^2 - 1}{x + 1}$ and $g(x) = x - 1$ agree except at $x = -1$.

47. 12

$f(x) = \dfrac{x^3 - 8}{x - 2}$ and $g(x) = x^2 + 2x + 4$ agree except at $x = 2$.

49. $-\dfrac{\ln 2}{8} \approx -0.0866$

$f(x) = \dfrac{(x + 4)\ln(x + 6)}{x^2 - 16}$ and $g(x) = \dfrac{\ln(x + 6)}{x - 4}$ agree except at $x = -4$.

51. 1/10 **53.** 5/6 **55.** $\sqrt{5}/10$ **57.** 1/6 **59.** $-1/9$
61. 2 **63.** $2x - 2$

65. The graph has a hole at $x = 0$.

Answers will vary. Example:

x	-0.1	-0.01	-0.001	0.001	0.01	0.1
$f(x)$	0.358	0.354	0.354	0.354	0.353	0.349

$\lim_{x\to 0} \dfrac{\sqrt{x + 2} - \sqrt{2}}{x} \approx 0.354 \left(\text{Actual limit is } \dfrac{1}{2\sqrt{2}} = \dfrac{\sqrt{2}}{4}.\right)$

67. The graph has a hole at $x = 0$.

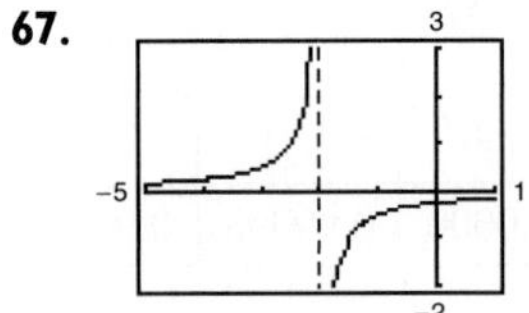

Answers will vary. Example:

x	-0.1	-0.01	-0.001
$f(x)$	-0.263	-0.251	-0.250

x	0.001	0.01	0.1
$f(x)$	-0.250	-0.249	-0.238

$\lim_{x\to 0} \dfrac{[1/(2 + x)] - (1/2)}{x} \approx -0.250 \left(\text{Actual limit is } -\dfrac{1}{4}.\right)$

69. 1/5 **71.** 0 **73.** 0 **75.** 0 **77.** 1 **79.** 1 **81.** 3/2

83. The graph has a hole at $t = 0$.

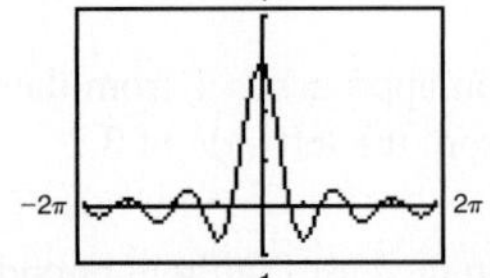

Answers will vary. Example:

t	-0.1	-0.01	0	0.01	0.1
$f(t)$	2.96	2.9996	?	2.9996	2.96

$\lim_{t\to 0} \dfrac{\sin 3t}{t} = 3$

85. 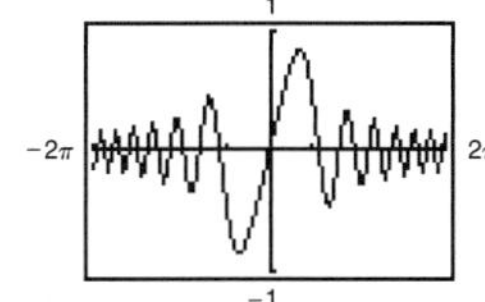

The graph has a hole at $x = 0$.

Answers will vary. Example:

x	-0.1	-0.01	-0.001	0	0.001	0.01	0.1
$f(x)$	-0.1	-0.01	-0.001	?	0.001	0.01	0.1

$\lim_{x\to 0} \frac{\sin x^2}{x} = 0$

87. 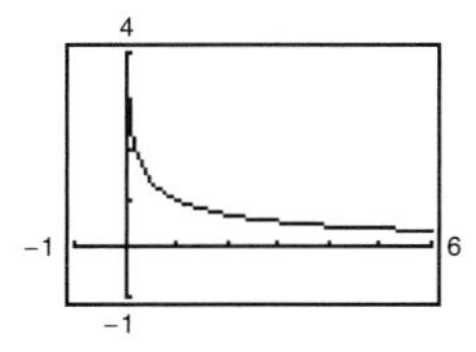

The graph has a hole at $x = 1$.

Answers will vary. Example:

x	0.5	0.9	0.99	1.01	1.1	1.5
$f(x)$	1.3863	1.0536	1.0050	0.9950	0.9531	0.8109

$\lim_{x\to 1} \frac{\ln x}{x - 1} = 1$

89. 2 **91.** $-4/x^2$ **93.** 4

95. 0

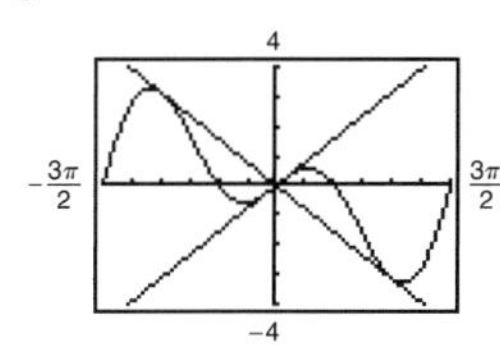

97. 0

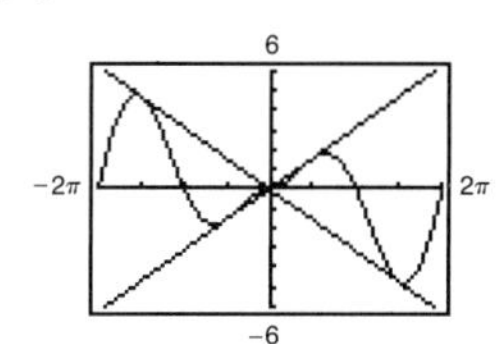

99. 0

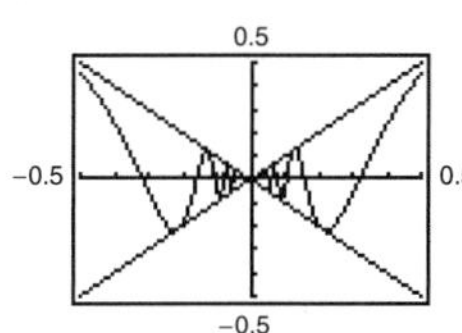

The graph has a hole at $x = 0$.

101. f and g agree at all but one point if c is a real number such that $f(x) = g(x)$ for all $x \neq c$.

103. An indeterminate form is obtained when the evaluation of a limit using direct substitution produces a meaningless fractional form, such as $\frac{0}{0}$.

105. 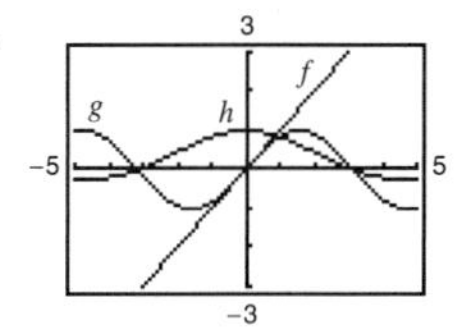

The magnitudes of $f(x)$ and $g(x)$ are approximately equal when x is "close to" 0. Therefore, their ratio is approximately 1.

107. 160 ft/sec **109.** -29.4 m/sec

111. Let $f(x) = 1/x$ and $g(x) = -1/x$.
$\lim_{x\to 0} f(x)$ and $\lim_{x\to 0} g(x)$ do not exist. However,

$$\lim_{x\to 0} [f(x) + g(x)] = \lim_{x\to 0} \left[\frac{1}{x} + \left(-\frac{1}{x}\right)\right] = \lim_{x\to 0} 0 = 0$$

and therefore does exist.

113. Proof **115.** Proof **117.** Proof

119. False. The limit does not exist because the function approaches 1 from the right side of 0 and approaches -1 from the left side of 0. (See graph below.)

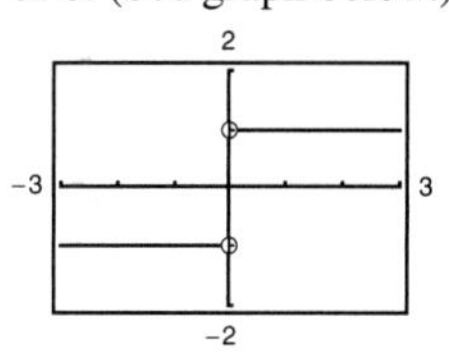

121. True. Theorem 2.7

123. False. The limit does not exist because $f(x)$ approaches 3 from the left side of 2 and approaches 0 from the right side of 2. (See graph below.)

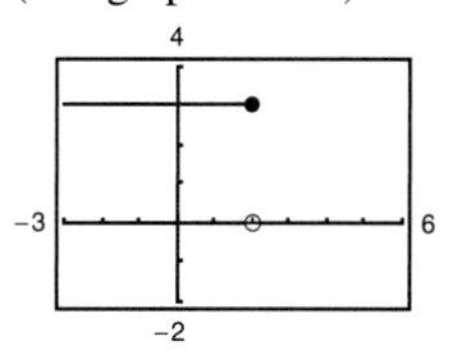

125. Answers will vary. Example:

Let $f(x) = \begin{cases} 4, & \text{if } x \geq 0 \\ -4, & \text{if } x < 0 \end{cases}$

$\lim_{x\to 0} |f(x)| = \lim_{x\to 0} 4 = 4$

$\lim_{x\to 0} f(x)$ does not exist because for $x < 0$, $f(x) = -4$ and for $x \geq 0$, $f(x) = 4$.

127. $\lim_{x\to 0} f(x)$ does not exist because $f(x)$ oscillates between two fixed values as x approaches 0.

$\lim_{x\to 0} g(x) = 0$ because, as x gets increasingly close to 0, the values of $g(x)$ become increasingly close to 0.

129. (a) 1/2

(b) Because $\frac{1 - \cos x}{x^2} \approx \frac{1}{2}$, it follows that $1 - \cos x \approx \frac{1}{2}x^2$

$\cos x \approx 1 - \frac{1}{2}x^2$ when $x \approx 0$.

(c) 0.995 (d) Calculator: $\cos(0.1) \approx 0.9950$

Section 2.4 (page 98)

1. (a) 1 (b) 1 (c) 1; $f(x)$ is continuous on $(-\infty, \infty)$.

3. (a) 0 (b) 0 (c) 0; Discontinuity at $x = 3$

5. (a) 2 (b) -2 (c) Limit does not exist.
Discontinuity at $x = 4$

7. $\frac{1}{10}$

9. Limit does not exist. The function decreases without bound as x approaches -3 from the left.

11. -1 **13.** $-1/x^2$ **15.** $5/2$ **17.** 2

19. Limit does not exist. The function decreases without bound as x approaches π from the left and increases without bound as x approaches π from the right.

21. 4

23. Limit does not exist. The function approaches 5 from the left side of 3 but approaches 6 from the right side of 3.

25. Does not exist. **27.** $\ln 4$

29. Discontinuous at $x = -2$ and $x = 2$

31. Discontinuous at every integer

33. Continuous on $[-5, 5]$ **35.** Continuous on $[-1, 4]$

37. Continuous for all real x **39.** Continuous for all real x

41. Nonremovable discontinuity at $x = 1$
Removable discontinuity at $x = 0$

43. Continuous for all real x

45. Removable discontinuity at $x = -2$
Nonremovable discontinuity at $x = 5$

47. Nonremovable discontinuity at $x = -2$

49. Continuous for all real x

51. Nonremovable discontinuity at $x = 2$

53. Continuous for all real x

55. Nonremovable discontinuity at $x = 0$

57. Nonremovable discontinuities at integer multiples of $\pi/2$

59. Nonremovable discontinuity at each integer

61.

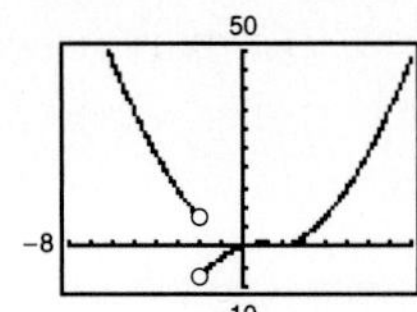

$\lim_{x\to0^+} f(x) = 0$

$\lim_{x\to0^-} f(x) = 0$

Discontinuity at $x = -2$

63. $a = 2$ **65.** $a = -1$, $b = 1$ **67.** Continuous for all real x

69. Nonremovable discontinuities at $x = 1$ and $x = -1$

71.

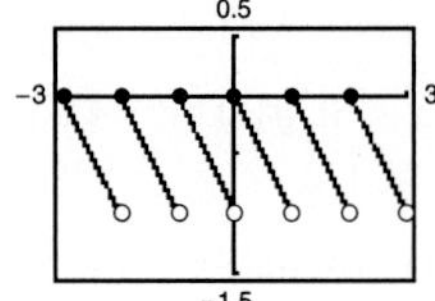

Nonremovable discontinuity at each integer

73.

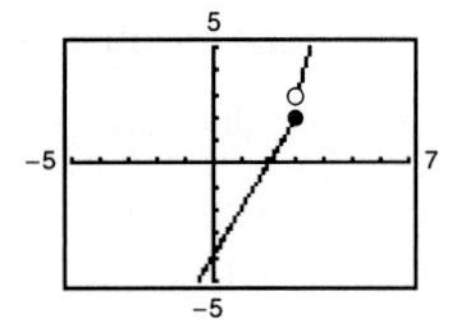

Discontinuous at $x = 3$

75. Continuous on $(-\infty, \infty)$

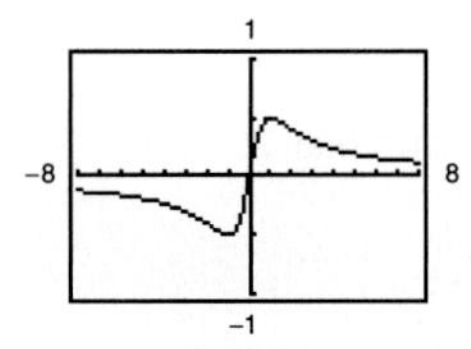

77. Continuous on . . . , $(-6, -2)$, $(-2, 2)$, $(2, 6)$, . . .

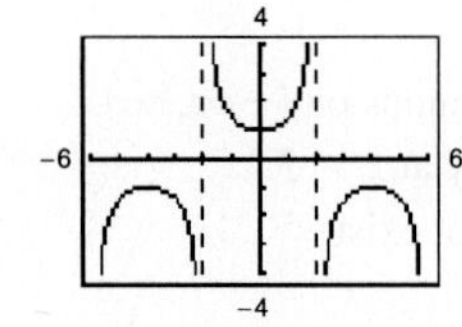

79.

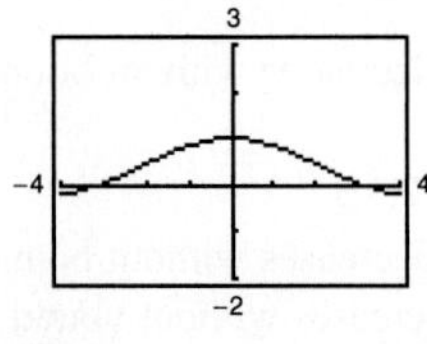

The graph has a hole at $x = 0$. The graph appears continuous but the function is not continuous on $[-4, 4]$. It is not obvious from the graph that the function has a discontinuity at $x = 0$.

81.

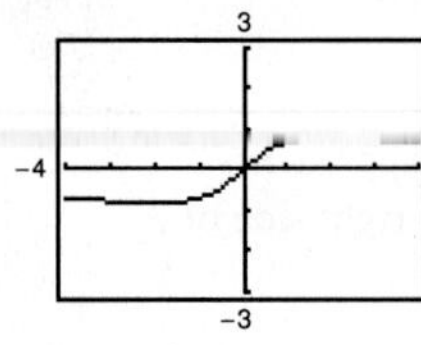

The graph has a hole at $x = 0$. The graph appears continuous but the function is not continuous on $[-4, 4]$. It is not obvious from the graph that the function has a discontinuity at $x = 0$.

83. $f(x)$ is continuous on $[2, 4]$.
$f(2) = -1$ and $f(4) = 3$
By the Intermediate Value Theorem, $f(c) = 0$ for at least one value of c between 2 and 4.

85. $h(x)$ is continuous on $[0, \pi/2]$.
$h(0) = -2 < 0$ and $h(\pi/2) \approx 0.9170 > 0$
By the Intermediate Value Theorem, $f(c) = 0$ for at least one value of c between 0 and $\pi/2$.

87. 0.68, 0.6823 **89.** 0.56, 0.5636

91. $f(3) = 11$ **93.** $f(2) = 4$

95. (a) The limit does not exist at $x = c$.
(b) The function is not defined at $x = c$.
(c) The limit exists, but it is not equal to the value of the function at $x = c$.
(d) The limit does not exist at $x = c$.

97.

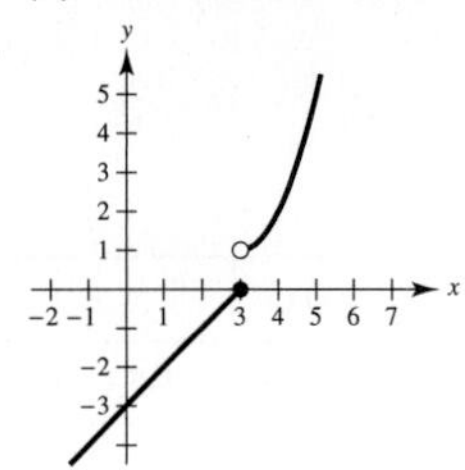

Not continuous because $\lim_{x\to3} f(x)$ does not exist.

99. True

101. False. A rational function can be written as $P(x)/Q(x)$, where P and Q are polynomials of degree m and n, respectively. It can have at most n discontinuities.

103. $\lim_{t\to4^-} f(t) \approx 28$; $\lim_{t\to4^+} f(t) \approx 56$
At the end of day 3, the amount of chlorine in the pool is about 28 oz. At the beginning of day 4, the amount of chlorine in the pool is about 56 oz.

105. $C = \begin{cases} 1.04, & 0 < t \le 2 \\ 1.04 + 0.36[\![t - 1]\!], & t > 2, t \text{ is not an integer} \\ 1.04 + 0.36(t - 2), & t > 2, t \text{ is an integer} \end{cases}$

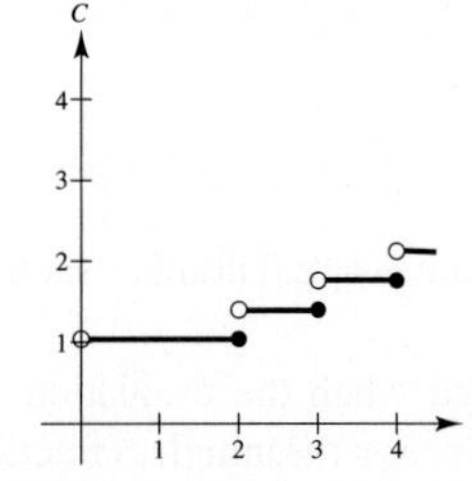

Nonremovable discontinuity at each integer greater than or equal to 2

107–109. Proofs **111.** Answers will vary.

113. (a)

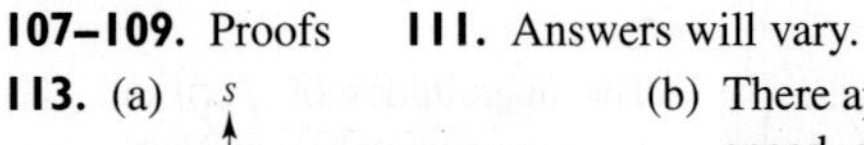

(b) There appears to be a limiting speed, and a possible cause is air resistance.

115. $c = \left(-1 \pm \sqrt{5}\right)/2$

117. Domain: $[-c^2, 0) \cup (0, \infty)$; Let $f(0) = 1/(2c)$.

119. $h(x)$ has a nonremovable discontinuity at every integer except 0.

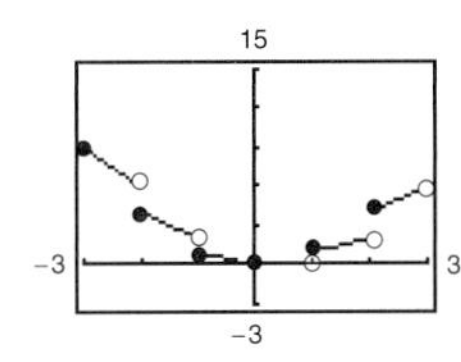

121. (a) Domain: $(-\infty, 0) \cup (0, \infty)$

(b)

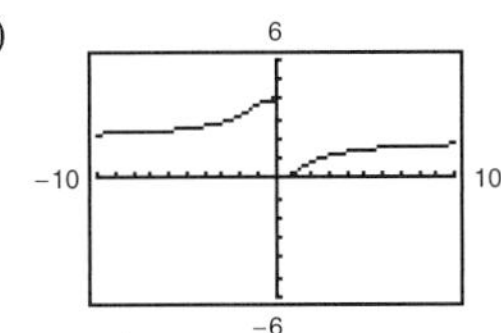

(c) $\lim_{x\to 0^-} f(x) = 4$; $\lim_{x\to 0^+} f(x) = 0$ (d) Answers will vary.

123. Putnam Problem A2, 1971

Section 2.5 (page 108)

1. $\lim_{x\to -2^+} 2\left|\frac{x}{x^2-4}\right| = \infty$ $\lim_{x\to -2^-} 2\left|\frac{x}{x^2-4}\right| = \infty$

3. $\lim_{x\to -2^+} \tan((\pi x)/4) = -\infty$ $\lim_{x\to -2^-} \tan((\pi x)/4) = \infty$

5.

x	-3.5	-3.1	-3.01	-3.001
$f(x)$	0.31	1.64	16.6	167

x	-2.999	-2.99	-2.9	-2.5
$f(x)$	-167	-16.7	-1.69	-0.36

$\lim_{x\to -3^+} f(x) = -\infty$ $\lim_{x\to -3^-} f(x) = \infty$

7.

x	-3.5	-3.1	-3.01	-3.001
$f(x)$	3.8	16	151	1501

x	-2.999	-2.99	-2.9	-2.5
$f(x)$	-1499	-149	-14	-2.3

$\lim_{x\to -3^+} f(x) = -\infty$ $\lim_{x\to -3^-} f(x) = \infty$

9. $x = 0$ **11.** $x = 2,\ x = -1$ **13.** $x = \pm 2$

15. No vertical asymptote **17.** $x = \pi/4 + (n\pi)/2$, n is an integer.

19. $t = 0$ **21.** $x = -2, x = 1$ **23.** No vertical asymptote

25. $x = 1$ **27.** $t = -2$ **29.** $x = 0$

31. $t = n\pi$, n is a nonzero integer.

33. Removable discontinuity at $x = -1$

35. Vertical asymptote at $x = -1$

37. Removable discontinuity at $x = -1$

39. $-\infty$ **41.** ∞ **43.** $\frac{4}{5}$

45. $\frac{1}{2}$ **47.** $-\infty$ **49.** ∞ **51.** $-\infty$ **53.** Does not exist

55. $\lim_{x\to 1^+} f(x) = \infty$

57. $\lim_{x\to 5^-} f(x) = -\infty$

59. Answers will vary.

61. Answers will vary. Example: $f(x) = \frac{x-3}{x^2-4x-12}$

63.

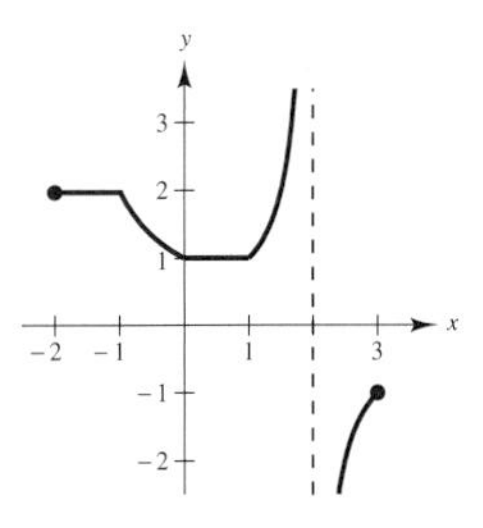

65. (a) $r = 200\pi/3$ (b) $r = 200\pi$ (c) ∞ **67.** ∞

69. (a) Proof; Domain: $x > 25$

(b)

x	30	40	50	60
y	150	66.667	50	42.857

Answers will vary.

(c) $\lim_{x\to 25^+} \frac{25x}{x-25} = \infty$

As x gets close to 25 mph, y becomes larger and larger.

71. (a) $A = 50\tan\theta - 50\theta$; Domain: $(0, \pi/2)$

(b)

θ	0.3	0.6	0.9	1.2	1.5
$f(\theta)$	0.47	4.21	18.01	68.61	630.07

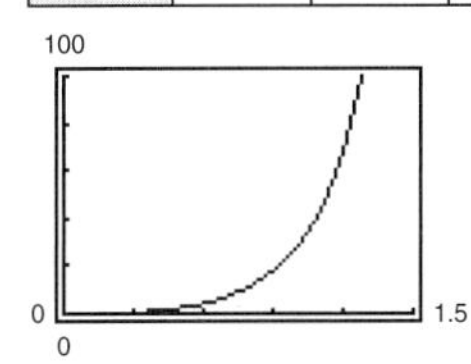

(c) $\lim_{\theta\to(\pi/2)^-} A = \infty$

73. False; let $f(x) = (x^2 - 1)/(x - 1)$. **75.** True

77. Let $f(x) = \frac{1}{x^2}$, $g(x) = \frac{1}{x^4}$, and $c = 0$. $\lim_{x\to 0}\frac{1}{x^2} = \infty$ and $\lim_{x\to 0}\frac{1}{x^4} = \infty$, but $\lim_{x\to 0}\left(\frac{1}{x^2} - \frac{1}{x^4}\right) = \lim_{x\to 0}\left(\frac{x^2-1}{x^4}\right) = -\infty \neq 0$.

79. Given $\lim_{x\to c} f(x) = \infty$, let $g(x) = 1$. Then $\lim_{x\to c}\frac{g(x)}{f(x)} = 0$ by Theorem 2.15.

81. Answers will vary.

Review Exercises for Chapter 2 (page 111)

1. Calculus Estimate: 8.268

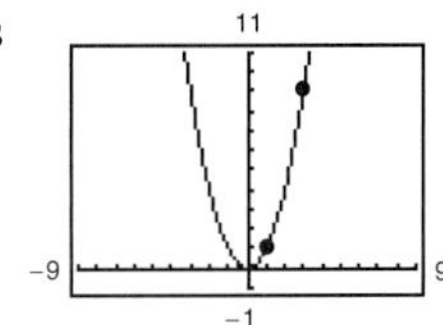

3.

x	-0.1	-0.01	-0.001
$f(x)$	-1.0526	-1.0050	-1.0005

x	0.001	0.01	0.1
$f(x)$	-0.9995	-0.9950	-0.9524

$\lim_{x\to 0} f(x) = -1$

5.

x	-0.1	-0.01	-0.001
$f(x)$	0.8867	0.0988	0.0100

x	0.001	0.01	0.1
$f(x)$	-0.0100	-0.1013	-1.1394

$\lim_{x\to 0} f(x) = 0$

7. (a) -2 (b) -3 **9.** (a) Limit does not exist. (b) 0
11. 2; Proof **13.** 1; Proof **15.** $\sqrt{6} \approx 2.45$ **17.** $-\frac{1}{4}$
19. $\frac{1}{4}$ **21.** -1 **23.** 75 **25.** 0 **27.** $\sqrt{3}/2$ **29.** 1
31. $-\frac{1}{2}$
33. (a)

x	1.1	1.01	1.001	1.0001
$f(x)$	0.5680	0.5764	0.5773	0.5773

$\lim_{x\to 1^+} f(x) \approx 0.5773$

(b)

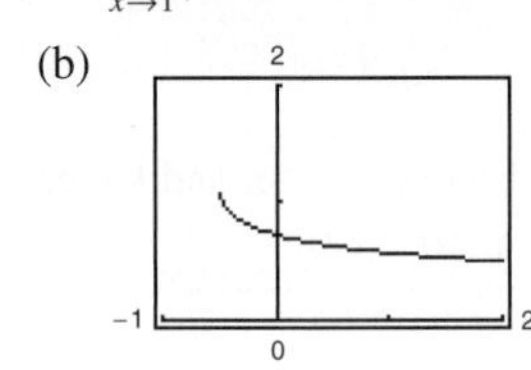

The graph has a hole at $x = 1$.
$\lim_{x\to 1^+} f(x) \approx 0.5774$

(c) $\sqrt{3}/3$

35. -39.2 m/sec **37.** -1 **39.** 0
41. Limit does not exist. The limit as t approaches 1 from the left is 2, whereas the limit as t approaches 1 from the right is 1.
43. Nonremovable discontinuity at each integer
Continuous on $(k, k + 1)$ for all integers k
45. Removable discontinuity at $x = 1$
Continuous on $(-\infty, 1) \cup (1, \infty)$
47. Nonremovable discontinuity at $x = 2$
Continuous on $(-\infty, 2) \cup (2, \infty)$
49. Nonremovable discontinuity at $x = -1$
Continuous on $(-\infty, -1) \cup (-1, \infty)$
51. Nonremovable discontinuity at each even integer
Continuous on $(2k, 2k + 2)$ for all integers k
53. Nonremovable discontinuity at each integer
Continuous on $(k, k + 1)$ for all integers k
55. $c = -\frac{1}{2}$ **57.** Proof
59. Nonremovable discontinuity every 6 months

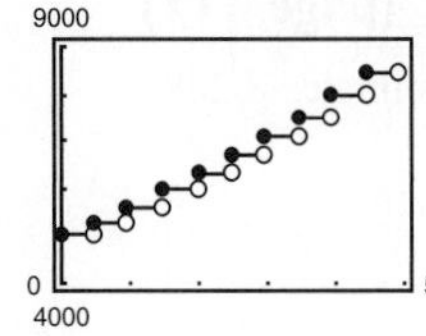

61. $x = 0$ **63.** $x = 10$ **65.** $x = -3, x = 3$ **67.** $-\infty$
69. $\frac{1}{3}$ **71.** $-\infty$ **73.** $\frac{4}{5}$ **75.** ∞ **77.** $-\infty$
79. (a) 2
(b) Yes, define as

$$f(x) = \begin{cases} \dfrac{\tan 2x}{x}, & x \neq 0 \\ 2, & x = 0 \end{cases}.$$

P.S. Problem Solving (page 113)

1. (a) Perimeter $\triangle PAO = 1 + \sqrt{(x^2 - 1)^2 + x^2} + \sqrt{x^4 + x^2}$
Perimeter $\triangle PBO = 1 + \sqrt{x^4 + (x - 1)^2} + \sqrt{x^4 + x^2}$

(b)

x	4	2	1
Perimeter $\triangle PAO$	33.0166	9.0777	3.4142
Perimeter $\triangle PBO$	33.7712	9.5952	3.4142
$r(x)$	0.9777	0.9461	1.0000

x	0.1	0.01
Perimeter $\triangle PAO$	2.0955	2.0100
Perimeter $\triangle PBO$	2.0006	2.0000
$r(x)$	1.0475	1.0050

(c) 1

3. (a) Area (hexagon) $= (3\sqrt{3})/2 \approx 2.5981$
Area (circle) $= \pi \approx 3.1416$
Area (circle) $-$ Area (hexagon) ≈ 0.5435
(b) $A_n = (n/2)\sin(2\pi/n)$
(c)

n	6	12	24	48	96
A_n	2.5981	3.0000	3.1058	3.1326	3.1394

(d) 3.1416 or π

5. (a) $m = -\frac{12}{5}$ (b) $y = \frac{5}{12}x - \frac{169}{12}$
(c) $m_x = \dfrac{-\sqrt{169 - x^2} + 12}{x - 5}$
(d) $\frac{5}{12}$; It is the same as the slope of the tangent line found in (b).

7. (a) Domain: $[-27, 1) \cup (1, \infty)$
(b)

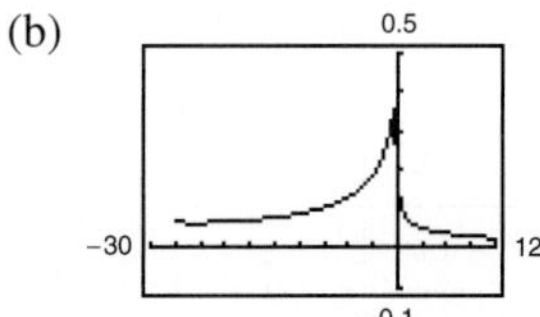

(c) $\frac{1}{14}$ (d) $\frac{1}{12}$

The graph has a hole at $x = 1$.

9. (a) g_1, g_4 (b) g_1 (c) g_1, g_3, g_4

11. The graph jumps at every integer.

(a) $f(1) = 0,\ f(0) = 0,\ f(\frac{1}{2}) = -1,\ f(-2.7) = -1$
(b) $\lim_{x\to 1^-} f(x) = -1,\ \lim_{x\to 1^+} f(x) = -1,\ \lim_{x\to 1/2} f(x) = -1$
(c) There is a discontinuity at each integer.

13. (a)

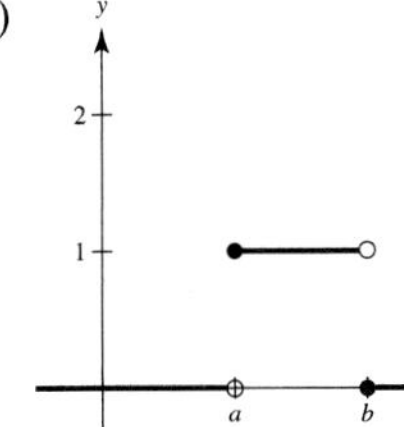

(b) (i) $\lim_{x \to a^+} P_{a,b}(x) = 1$
(ii) $\lim_{x \to a^-} P_{a,b}(x) = 0$
(iii) $\lim_{x \to b^+} P_{a,b}(x) = 0$
(iv) $\lim_{x \to b^-} P_{a,b}(x) = 1$

(c) Continuous for all positive real numbers except a and b
(d) The area under the curve gives a value of 1.

Chapter 3

Section 3.1 (page 123)

1. (a) $m_1 = 0, m_2 = 5/2$ (b) $m_1 = -5/2, m_2 = 2$

3.

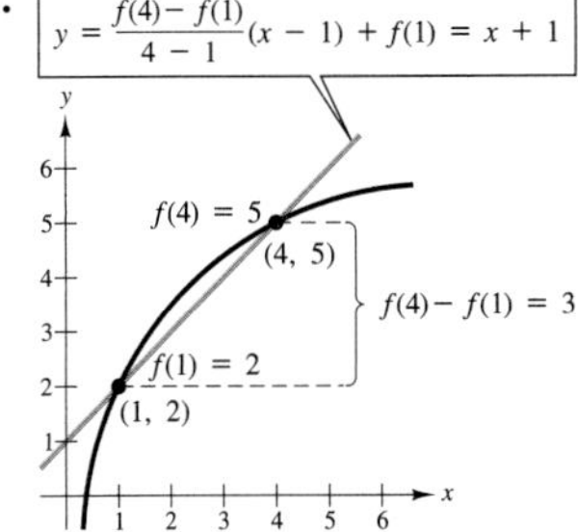

5. $m = -2$ **7.** $m = 2$

9. $m = 3$ **11.** $f'(x) = 0$ **13.** $f'(x) = -5$ **15.** $h'(s) = \frac{2}{3}$

17. $f'(x) = 4x + 1$ **19.** $f'(x) = 3x^2 - 12$

21. $f'(x) = -1/(x - 1)^2$ **23.** $f'(x) = 1/2\sqrt{x + 1}$

25. (a) Tangent line: $y = 4x - 3$
(b)

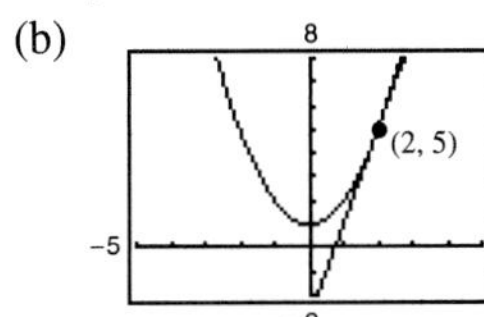

27. (a) Tangent line: $y = 12x - 16$
(b)

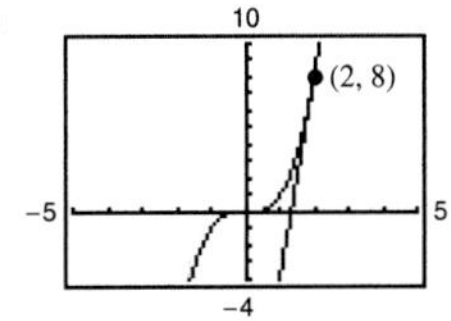

29. (a) Tangent line: $y = \frac{1}{2}x + \frac{1}{2}$
(b)

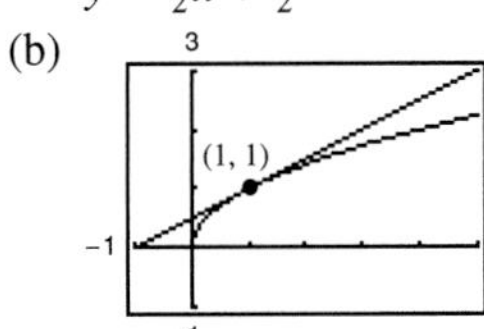

31. (a) Tangent line: $y = \frac{3}{4}x + 2$
(b)

33. $y = 3x - 2; y = 3x + 2$ **35.** $y = -\frac{1}{2}x + \frac{3}{2}$

37. b **38.** d **39.** a **40.** c **41.** $g(5) = 2; g'(5) = -\frac{1}{2}$

43. $f'(x) = 1$

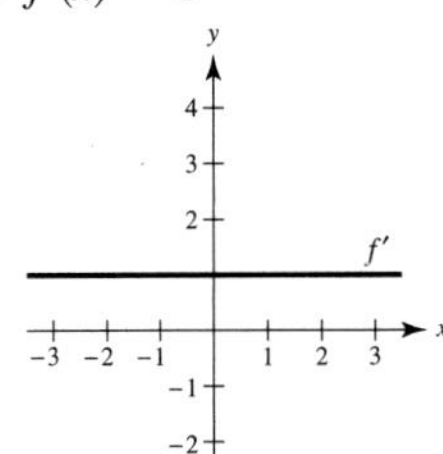

45. $f'(x) = 2x - 8$

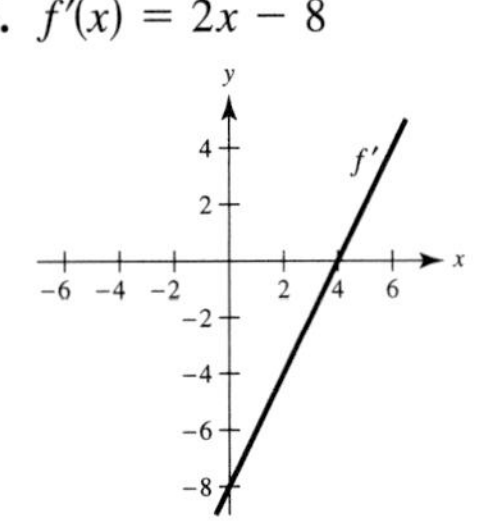

47. Answers will vary.
Sample answer: $y = -x$

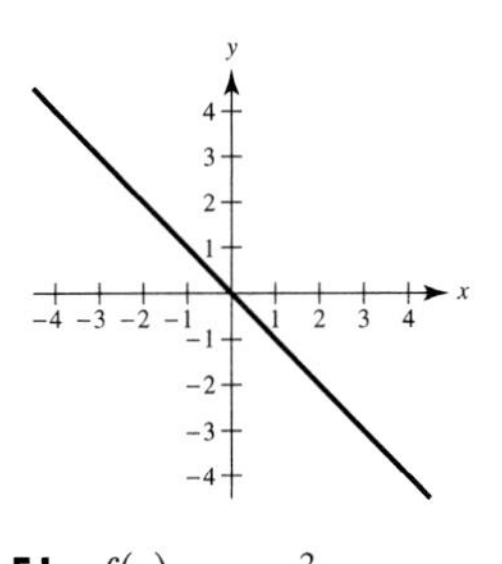

49. $f(x) = 5 - 3x$
$c = 1$

51. $f(x) = -x^2$
$c = 6$

53. $f(x) = -3x + 2$

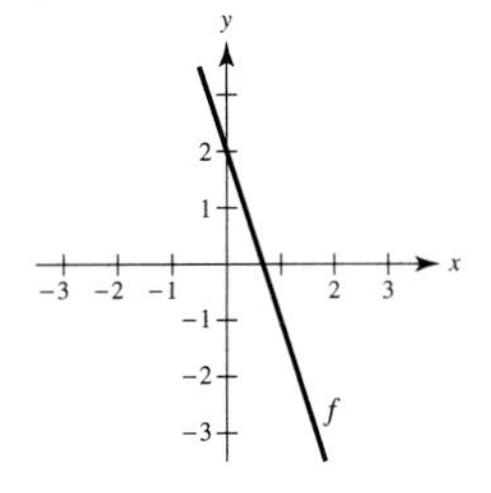

55. Answers will vary.
Sample answer: $f(x) = x^3$

57. $y = 2x + 1; y = -2x + 9$

59. (a) -3 (b) 0
(c) The graph is moving downward to the right when $x = 1$.
(d) The graph is moving upward to the right when $x = -4$.
(e) Positive. Because $g'(x) > 0$ on $[3, 6]$, the graph of g is moving upward to the right.
(f) No. Knowing only $g'(2)$ is not sufficient information. $g'(2)$ remains the same for any vertical translation of g.

61.

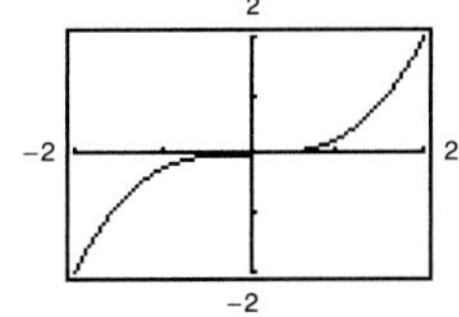

x	-2	-1.5	-1	-0.5	0	0.5	1	1.5	2
$f(x)$	-2	$-\frac{27}{32}$	$-\frac{1}{4}$	$-\frac{1}{32}$	0	$\frac{1}{32}$	$\frac{1}{4}$	$\frac{27}{32}$	2
$f'(x)$	3	$\frac{27}{16}$	$\frac{3}{4}$	$\frac{3}{16}$	0	$\frac{3}{16}$	$\frac{3}{4}$	$\frac{27}{16}$	3

63.

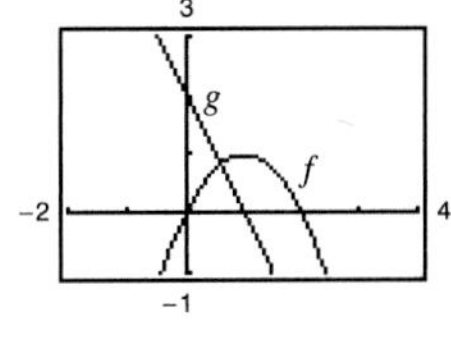

$g(x) \approx f'(x)$

65. $f(2) = 4; f(2.1) = 3.99; f'(2) \approx -0.1$

67.

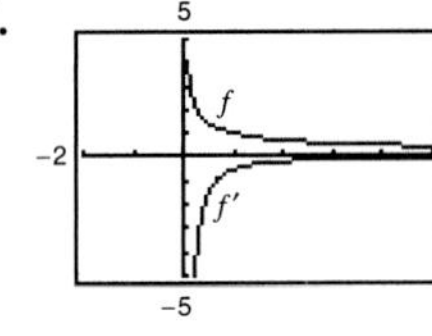

As x approaches infinity, the graph of f approaches a line of slope 0. Thus $f'(x)$ approaches 0.

69. (a)

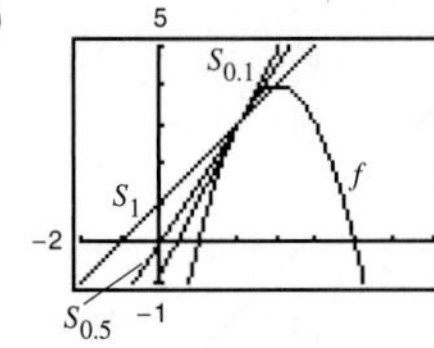

(b) The graphs of S for decreasing values of Δx are secant lines approaching the tangent line to the graph of f at the point $(2, f(2))$.

71. 4 **73.** 4 **75.** $g(x)$ is not differentiable at $x = 0$.
77. $f(x)$ is not differentiable at $x = 6$.
79. $h(x)$ is not differentiable at $x = -5$.
81. $(-\infty, -1) \cup (-1, \infty)$ **83.** $(-\infty, 3) \cup (3, \infty)$
85. $(1, \infty)$
87. $(-\infty, -3) \cup (-3, \infty)$ **89.** $(-\infty, 0) \cup (0, \infty)$

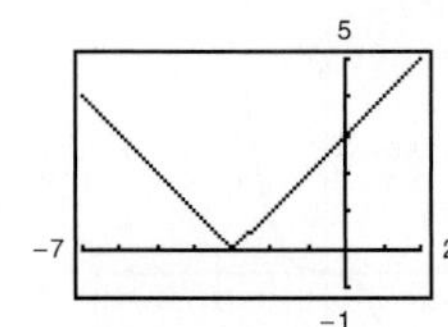

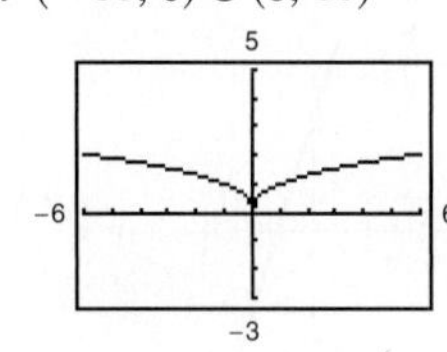

91. The derivative from the left is -1 and the derivative from the right is 1, so f is not differentiable at $x = 1$.
93. The derivatives from both the right and the left are 0, so $f'(1) = 0$.
95. f is differentiable at $x = 2$.
97. (a) $d = \left(3|m + 1|\right)/\sqrt{m^2 + 1}$
(b)

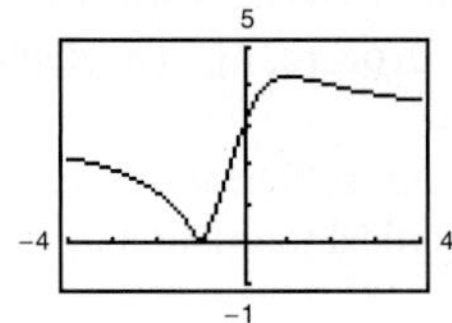

Not differentiable at $m = -1$

99. False. The slope is $\lim_{\Delta x \to 0} \dfrac{f(2 + \Delta x) - f(2)}{\Delta x}$.
101. False. For example: $f(x) = |x|$. The derivative from the left and the derivative from the right both exist but are not equal.
103. Proof

Section 3.2 (page 136)

1. (a) $\frac{1}{2}$ (b) 3 **3.** 0 **5.** $6x^5$ **7.** $1/(5x^{4/5})$ **9.** 1
11. $-4t + 3$ **13.** $2x + 12x^2$ **15.** $3t^2 - 2$ **17.** $6 - 5e^x$
19. $\dfrac{\pi}{2}\cos\theta + \sin\theta$ **21.** $2x + \dfrac{1}{2}\sin x$ **23.** $\dfrac{1}{2}e^x - 3\cos x$

	Function	*Rewrite*	*Differentiate*	*Simplify*
25.	$y = \dfrac{5}{2x^2}$	$y = \dfrac{5}{2}x^{-2}$	$y' = -5x^{-3}$	$y' = -\dfrac{5}{x^3}$
27.	$y = \dfrac{3}{(2x)^3}$	$y = \dfrac{3}{8}x^{-3}$	$y' = -\dfrac{9}{8}x^{-4}$	$y' = -\dfrac{9}{8x^4}$
29.	$y = \dfrac{\sqrt{x}}{x}$	$y = x^{-1/2}$	$y' = -\dfrac{1}{2}x^{-3/2}$	$y' = -\dfrac{1}{2x^{3/2}}$

31. -6 **33.** 0 **35.** 3 **37.** $\frac{3}{4}$ **39.** $2t + 12/t^4$
41. $(x^3 - 8)/x^3$ **43.** $3x^2 + 1$ **45.** $1/(2\sqrt{x}) - 2/x^{2/3}$
47. $4/5s^{1/5} - 2/3s^{1/3}$ **49.** $3/\sqrt{x} - 5\sin x$
51. $-2/x^3 - 2e^x$

53. (a) $5x + y + 3 = 0$
(b)

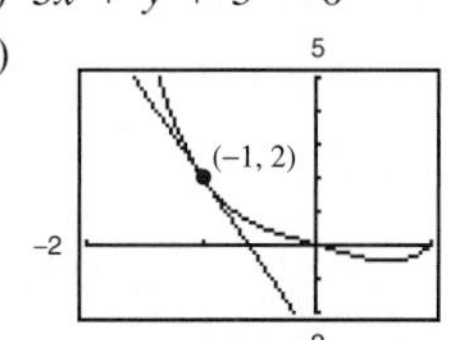

55. (a) $2x - y + 1 = 0$
(b)

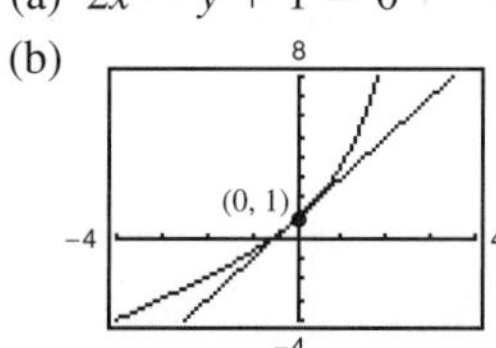

57. $(0, 2), (-2, -14), (2, -14)$ **59.** (π, π)
61. $(\ln 4, 4 - 4\ln 4)$ **63.** $k = 2, k = -10$ **65.** $k = 3$
67. (a) A and B (b) Greater than
(c)

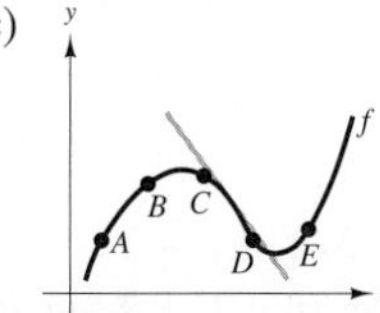

69. $g'(x) = f'(x)$
71.

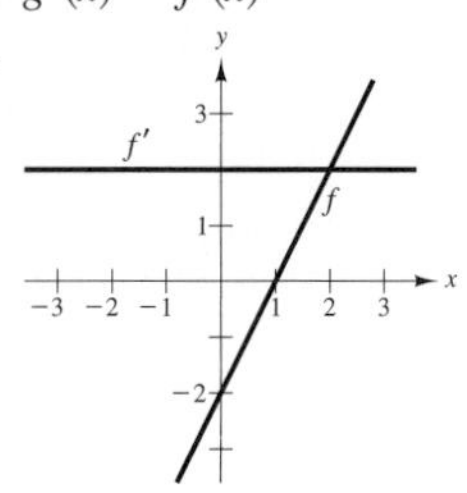

The rate of change of f is constant and therefore f' is a constant function.

73. $y = 2x - 1$ $y = 4x - 4$

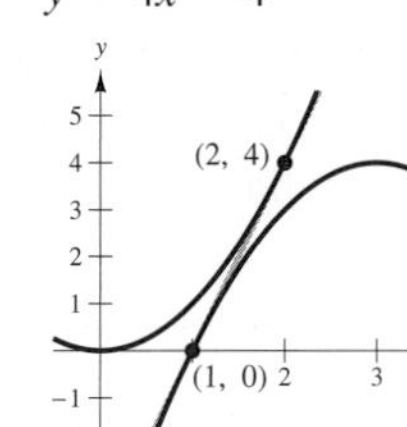

75. $f'(x) = 3 + \cos x \neq 0$ for all x.
77. $x - 4y + 4 = 0$
79.

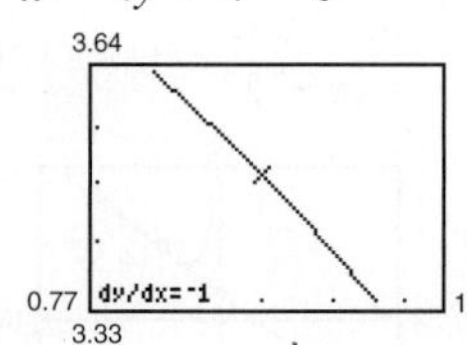

$f'(1)$ appears to be close to -1.
$f'(1) = -1$

81. (a)

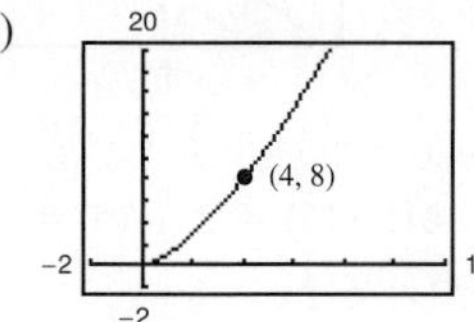

$(3.9, 7.7019)$; $S(x) = 2.981x - 3.924$
(b) $T(x) = 3(x - 4) + 8 = 3x - 4$
The slope (and equation) of the secant line approaches that of the tangent line at $(4, 8)$ as you choose points closer to $(4, 8)$.

(c) The approximation becomes less accurate.

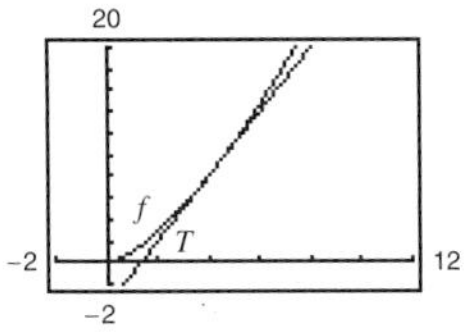

(d)

Δx	-3	-2	-1	-0.5	-0.1	0
$f(4 + \Delta x)$	1	2.828	5.196	6.458	7.702	8
$T(4 + \Delta x)$	-1	2	5	6.5	7.7	8

Δx	0.1	0.5	1	2	3
$f(4 + \Delta x)$	8.302	9.546	11.180	14.697	18.520
$T(4 + \Delta x)$	8.3	9.5	11	14	17

83. False: let $f(x) = x$ and $g(x) = x + 1$.

85. False: $dy/dx = 0$. **87.** True

89. Average rate: $\frac{1}{2}$
Instantaneous rates:
$f'(1) = 1$
$f'(2) = \frac{1}{4}$

91. Average rate: $e \approx 2.718$
Instantaneous rates:
$g'(0) = 1$
$g'(1) = 2 + e \approx 4.718$

93. (a) $s(t) = -16t^2 + 1362$; $v(t) = -32t$ (b) -48 ft/sec
(c) $s'(1) = -32$ ft/sec; $s'(2) = -64$ ft/sec
(d) $t = \sqrt{1362}/4 \approx 9.226$ sec (e) -295.242 ft/sec

95. $v(5) = 71$ m/sec; $v(10) = 22$ m/sec

97.

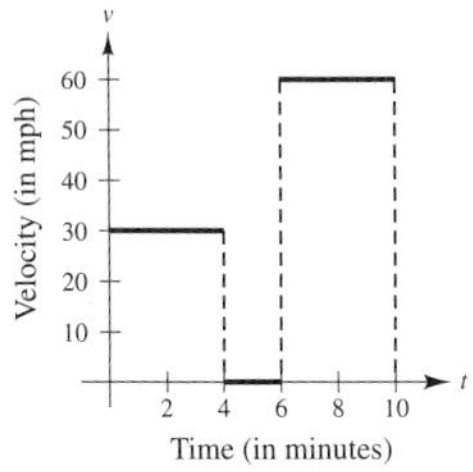

99.

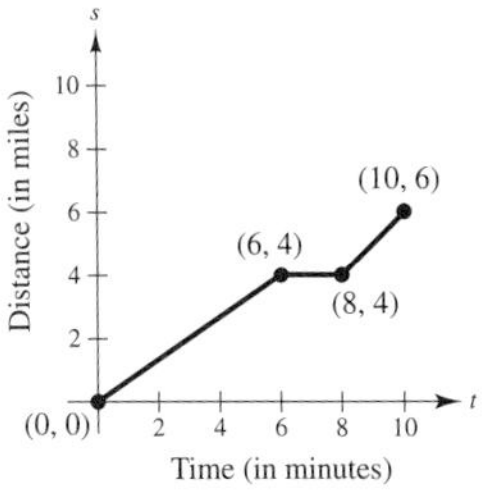

101. (a) $R(v) = 0.417v - 0.02$
(b) $B(v) = 0.0056v^2 + 0.001v + 0.04$
(c) $T(v) = 0.0056v^2 + 0.418v + 0.02$
(d)

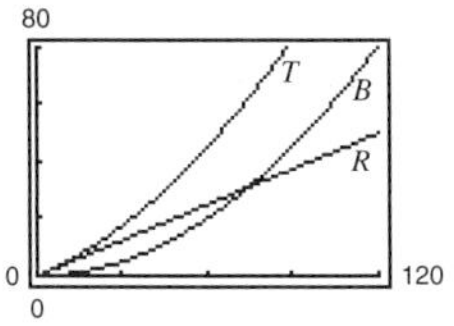

(e) $T'(v) = 0.0112v + 0.418$
$T'(40) = 0.866$
$T'(80) = 1.314$
$T'(100) = 1.538$

(f) As speed increases, the total stopping distance increases.

103. $V'(4) = 48 \text{ cm}^2$ **105.** Proof

107. (a) The rate of change of the number of gallons of gasoline sold when the price is \$1.479
(b) In general, the rate of change when $p = 1.479$ should be negative. As prices go up, sales go down.

109. $y = 2x^2 - 3x + 1$ **111.** $y = -9x$, $y = -\frac{9}{4}x - \frac{27}{4}$

113. $a = \frac{1}{3}$, $b = -\frac{4}{3}$

115. $f_1(x) = |\sin x|$ is differentiable for all $x \neq n\pi$, n an integer.
$f_2(x) = \sin|x|$ is differentiable for all $x \neq 0$.

Section 3.3 (page 147)

1. $2(2x^3 - 3x^2 + x - 1)$ **3.** $(7t^2 + 4)/(3t^{2/3})$

5. $x^2(3\cos x - x\sin x)$ **7.** $(1 - x^2)/(x^2 + 1)^2$

9. $(1 - 8x^3)/[3x^{2/3}(x^3 + 1)^2]$ **11.** $(x\cos x - 2\sin x)/x^3$

13. $f'(x) = (x^3 - 3x)(4x + 3) + (2x^2 + 3x + 5)(3x^2 - 3)$
$= 10x^4 + 12x^3 - 3x^2 - 18x - 15$
$f'(0) = -15$

15. $f'(x) = \dfrac{x^2 - 6x + 4}{(x - 3)^2}$
$f'(1) = -\dfrac{1}{4}$

17. $f'(x) = \cos x - x\sin x$
$f'\left(\dfrac{\pi}{4}\right) = \dfrac{\sqrt{2}}{8}(4 - \pi)$

19. $f'(x) = e^x(\cos x + \sin x)$
$f'(0) = 1$

	Function	*Rewrite*	*Differentiate*	*Simplify*
21.	$y = \dfrac{x^2 + 2x}{3}$	$y = \dfrac{1}{3}(x^2 + 2x)$	$y' = \dfrac{1}{3}(2x + 2)$	$y' = \dfrac{2(x + 1)}{3}$
23.	$y = \dfrac{7}{3x^3}$	$y = \dfrac{7}{3}x^{-3}$	$y' = -7x^{-4}$	$y' = -\dfrac{7}{x^4}$
25.	$y = \dfrac{4x^{3/2}}{x}$	$y = 4x^{1/2}$, $x > 0$	$y' = 2x^{-1/2}$	$y' = \dfrac{2}{\sqrt{x}}$, $x > 0$

27. $\dfrac{(x^2 - 1)(-2 - 2x) - (3 - 2x - x^2)(2x)}{(x^2 - 1)^2} = \dfrac{2}{(x + 1)^2}$, $x \neq 1$

29. $1 - 12/(x + 3)^2 = (x^2 + 6x - 3)/(x + 3)^2$

31. $\left[2\sqrt{x} - (2x + 5)/2\sqrt{x}\right]/x = (2x - 5)/2x^{3/2}$

33. $6s^2(s^3 - 2)$ **35.** $-(2x^2 - 2x + 3)/[x^2(x - 3)^2]$

37. $(3x^3 + 4x)[(x - 5) \cdot 1 + (x + 1) \cdot 1]$
$+ [(x - 5)(x + 1)](9x^2 + 4)$
$= 15x^4 - 48x^3 - 33x^2 - 32x - 20$

39. $\dfrac{(x^2 - c^2)(2x) - (x^2 + c^2)(2x)}{(x^2 - c^2)^2} = -\dfrac{4xc^2}{(x^2 - c^2)^2}$

41. $t(t\cos t + 2\sin t)$ **43.** $-(t\sin t + \cos t)/t^2$

45. $-e^x + \sec^2 x$ **47.** $\dfrac{1}{4t^{3/4}} + 8\sec t\tan t$

49. $\dfrac{-6\cos^2 x + 6\sin x - 6\sin^2 x}{4\cos^2 x} = \dfrac{3}{2}(-1 + \tan x\sec x - \tan^2 x)$
$= \dfrac{3}{2}\sec x(\tan x - \sec x)$

51. $\csc x\cot x - \cos x = \cos x\cot^2 x$ **53.** $x(x\sec^2 x + 2\tan x)$

55. $2x\cos x + 2\sin x + x^2e^x + 2xe^x$

57. $e^x/8x^{3/2}(2x - 1)$

59. $\left(\dfrac{x + 1}{x + 2}\right)(2) + (2x - 5)\left[\dfrac{(x + 2)(1) - (x + 1)(1)}{(x + 2)^2}\right]$
$= (2x^2 + 8x - 1)/(x + 2)^2$

61. $\dfrac{1 - \sin\theta + \theta\cos\theta}{(1 - \sin\theta)^2}$ **63.** $y' = \dfrac{-2\csc x\cot x}{(1 - \csc x)^2}$, $-4\sqrt{3}$

65. $h'(t) = \sec t(t\tan t - 1)/t^2$, $1/\pi^2$

67. (a) $y = -x - 2$
(b)

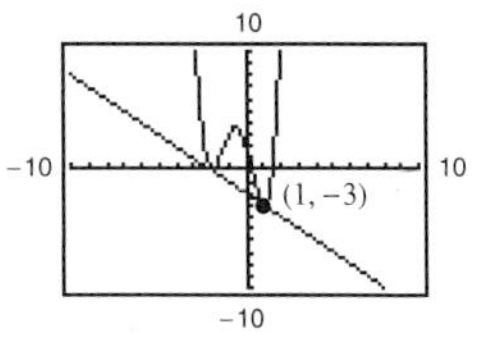

69. (a) $4x - 2y - \pi + 2 = 0$

(b)

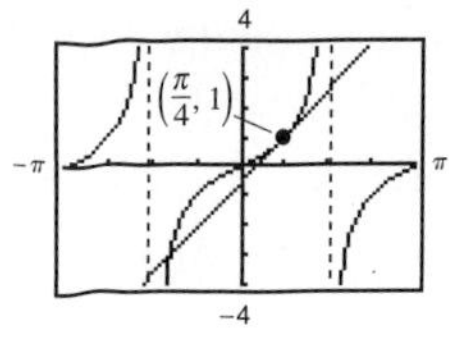

71. (a) $y = e(x - 1)$

(b)

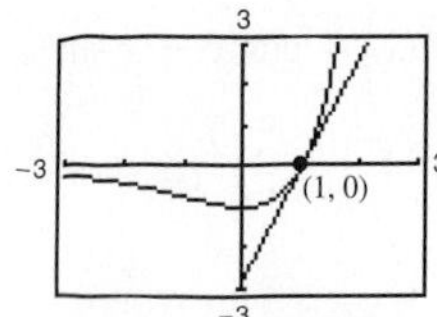

73. $2y + x - 4 = 0$

75. $25y - 12x + 16 = 0$ **77.** $(0, 0), (2, 4)$ **79.** $(3, 8e^{-3})$

81. Tangent lines: $2y + x = 7$; $2y + x = -1$

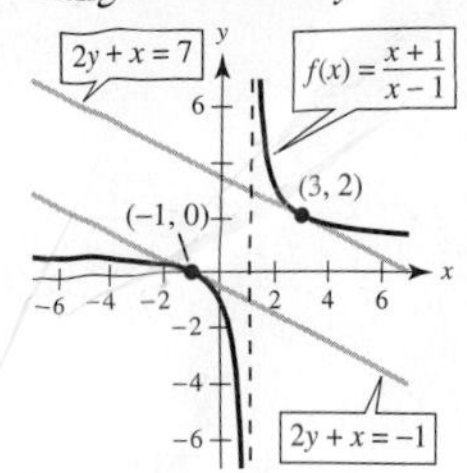

83. $f(x) + 2 = g(x)$ **85.** (a) $p'(1) = 1$ (b) $q'(4) = -1/3$

87. $(6t + 1)/(2\sqrt{t})$ cm²/sec

89. (a) About $-\$3.38$ per unit. The rate of change of the ordering and transportation cost C is decreasing at a rate of about \$3.38 per unit when the order size is 200 units.

(b) \$0.00 per unit. The rate of change of the ordering and transportation cost C is not changing when the order size is 250 units.

(c) About \$1.83 per unit. The rate of change of the ordering and transportation cost C is increasing at a rate of about \$1.83 per unit when the order size is 300 units.

91. 31.55 bacteria/hr **93.** Proof

95. (a) $n(t) = -3.5806t^3 + 82.577t^2 - 603.60t + 1667.5$

$v(t) = -0.1361t^3 + 3.165t^2 - 23.02t + 59.8$

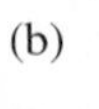

(b)

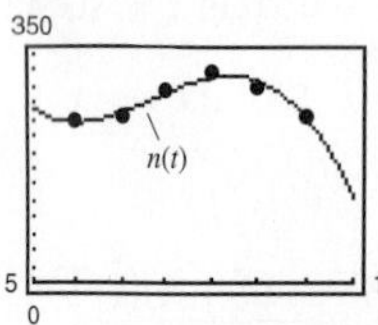

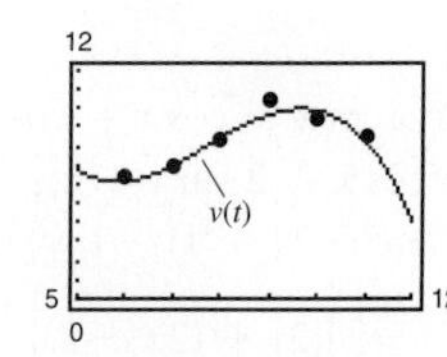

(c) $A = \dfrac{-0.1361t^3 + 3.165t^2 - 23.02t + 59.8}{-3.5806t^3 + 82.577t^2 - 603.60t + 1667.5}$

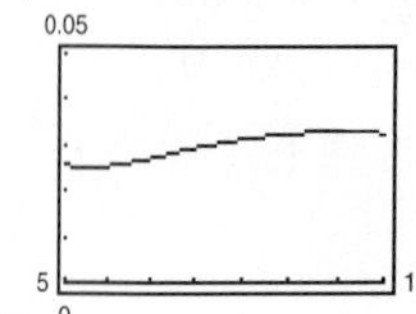

A represents the average retail value (in millions of dollars) per 1000 motor homes.

(d) $A'(t)$ represents the rate of change of the average retail value per 1000 motor homes for the given year.

97. $3/\sqrt{x}$ **99.** $2/(x - 1)^3$ **101.** $-3 \sin x$

103. $e^x/x^3(x^2 - 2x + 2)$ **105.** $2x$ **107.** $1/\sqrt{x}$

109. Answers will vary. For example: $(x - 2)^2$

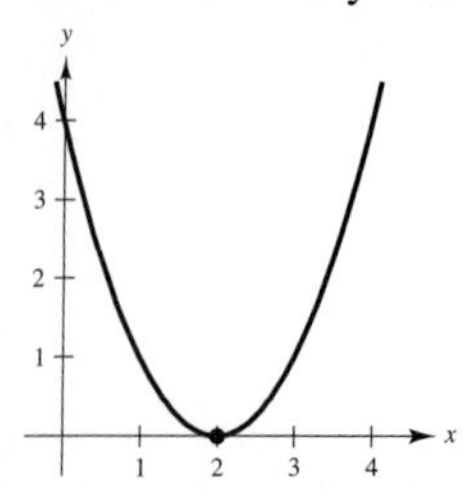

111. 0 **113.** -10

115.

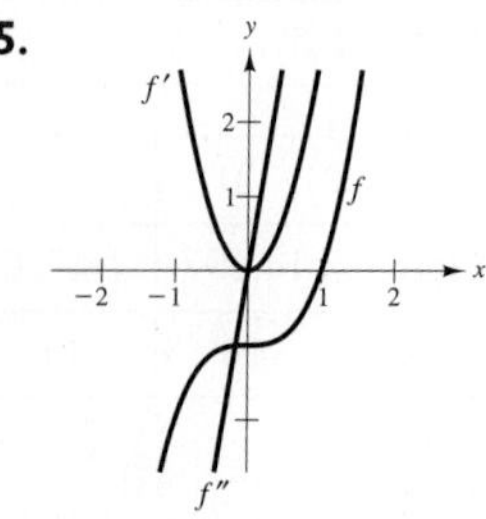

117.

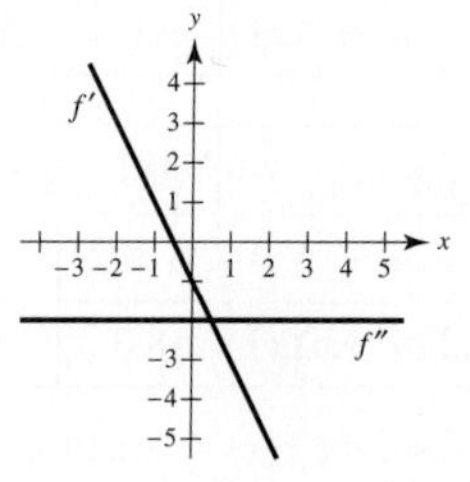

119.

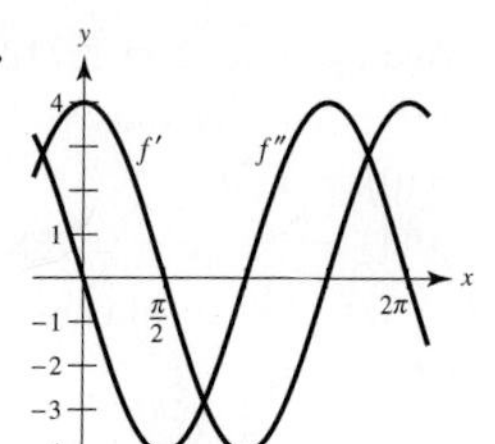

121. $v(3) = 27$ m/sec

$a(3) = -6$ m/sec²

The speed of the object is decreasing, but the rate of that decrease is increasing.

123. $f^{(n)}(x) = n(n - 1)(n - 2) \cdots (2)(1) = n!$

125. (a) $f''(x) = g(x)h''(x) + 2g'(x)h'(x) + g''(x)h(x)$

$f'''(x) = g(x)h'''(x) + 3g'(x)h''(x) + 3g''(x)h'(x) + g'''(x)h(x)$

$f^{(4)}(x) = g(x)h^{(4)}(x) + 4g'(x)h'''(x) + 6g''(x)h''(x) + 4g'''(x)h'(x) + g^{(4)}(x)h(x)$

(b) $f^{(n)}(x) = g(x)h^{(n)}(x) + \dfrac{n!}{1!(n-1)!}g'(x)h^{(n-1)}(x) + \dfrac{n!}{2!(n-2)!}g''(x)h^{(n-2)}(x) + \cdots + \dfrac{n!}{(n-1)!1!}g^{(n-1)}(x)h'(x) + g^{(n)}(x)h(x)$

127. $y' = -1/x^2$, $y'' = 2/x^3$,

$x^3y'' + 2x^2y' = x^3(2/x^3) + 2x^2(-1/x^2) = 2 - 2 = 0$

129. $y' = 2 \cos x$, $y'' = -2 \sin x$,

$y'' + y = -2 \sin x + 2 \sin x + 3 = 3$

131. (a) $P_1(x) = x - 1$

$P_2(x) = x - 1 - \frac{1}{2}(x - 1)^2$

(b)

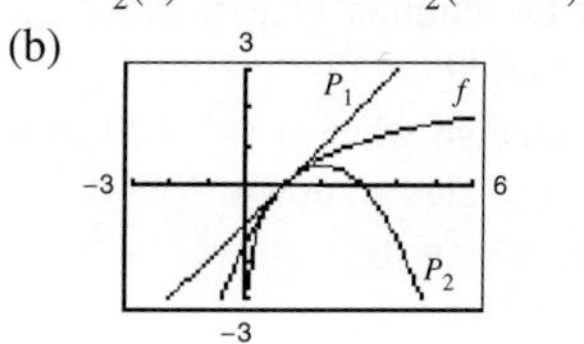

(c) P_1

(d) P_1 and P_2 become less accurate as you move farther from $x = a$.

133. False. $dy/dx = f(x)g'(x) + g(x)f'(x)$ **135.** True

137. True **139.** $f(x) = 3x^2 - 2x - 1$

141. $f'(x) = 2|x|$; $f''(0)$ does not exist.

Section 3.4 (page 161)

	$y = f(g(x))$	$u = g(x)$	$y = f(u)$
1.	$y = (6x - 5)^4$	$u = 6x - 5$	$y = u^4$
3.	$y = \sqrt{x^2 - 1}$	$u = x^2 - 1$	$y = \sqrt{u}$
5.	$y = \csc^3 x$	$u = \csc x$	$y = u^3$
7.	$y = e^{-2x}$	$u = -2x$	$y = e^u$

9. $6(2x - 7)^2$ **11.** $-108(4 - 9x)^3$

13. $\frac{2}{3}(9 - x^2)^{-1/3}(-2x) = -4x/3(9 - x^2)^{1/3}$

15. $\frac{1}{2}(1 - t)^{-1/2}(-1) = -1/(2\sqrt{1 - t})$

17. $\frac{1}{3}(9x^2 + 4)^{-2/3}(18x) = 6x/(9x^2 + 4)^{2/3}$

19. $\frac{1}{2}(4 - x^2)^{-3/4}(-2x) = -x/\sqrt[4]{(4 - x^2)^3}$ **21.** $-1/(x - 2)^2$

23. $-2(t - 3)^{-3}(1) = -2/(t - 3)^3$ **25.** $-1/[2(x + 2)^{3/2}]$

27. $x^2[4(x - 2)^3(1)] + (x - 2)^4(2x) = 2x(x - 2)^3(3x - 2)$

29. $x\left(\frac{1}{2}\right)(1 - x^2)^{-1/2}(-2x) + (1 - x^2)^{1/2}(1) = \dfrac{1 - 2x^2}{\sqrt{1 - x^2}}$

31. $\dfrac{(x^2 + 1)^{1/2}(1) - x(1/2)(x^2 + 1)^{-1/2}(2x)}{x^2 + 1} = \dfrac{1}{(x^2 + 1)^{3/2}}$

33. $\dfrac{-2(x + 5)(x^2 + 10x - 2)}{(x^2 + 2)^3}$ **35.** $\dfrac{-9(2v - 1)^2}{(v + 1)^4}$

37. $(1 - 3x^2 - 4x^{3/2})/[2\sqrt{x}(x^2 + 1)^2]$

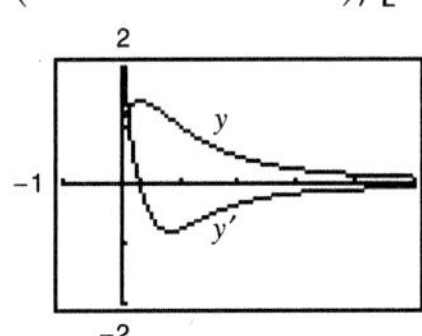

The zero of y' corresponds to the point on the graph of the function where the tangent line is horizontal.

39. $3t(t^2 + 3t - 2)/(t^2 + 2t - 1)^{3/2}$

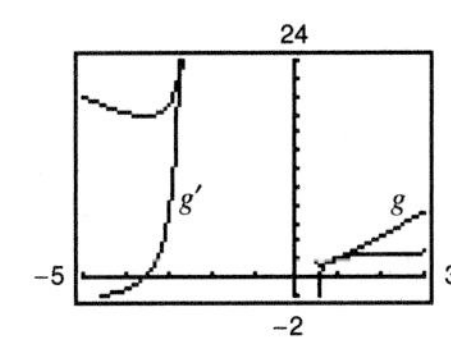

The zeros of $g'(t)$ correspond to the points on the graph of the function where the tangent line is horizontal.

41. $-\dfrac{\sqrt{\dfrac{x + 1}{x}}}{2x(x + 1)}$

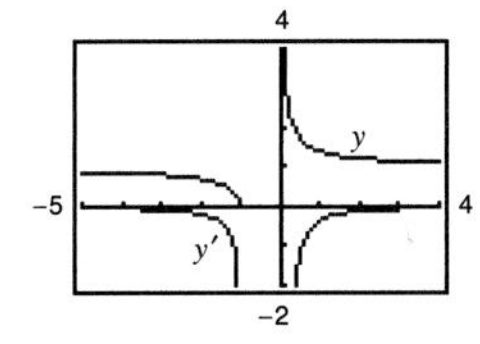

y' has no zeros.

43. $t/\sqrt{1 + t}$

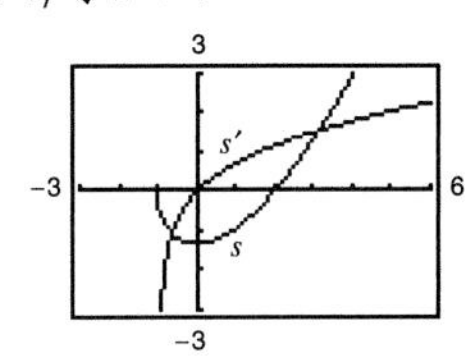

The zero of $s'(t)$ corresponds to the point on the graph of the function where the tangent line is horizontal.

45. $-[\pi x \sin(\pi x) + \cos(\pi x) + 1]/x^2$

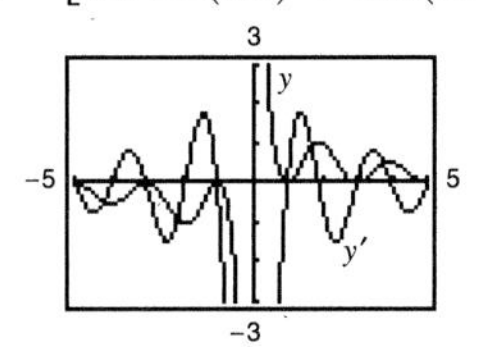

The zeros of y' correspond to the points on the graph of the function where the tangent lines are horizontal.

47. (a) 1 (b) 2; The slope of $\sin ax$ at the origin is a.

49. (a) 3 (b) -3 **51.** 3 **53.** 2 **55.** $-3 \sin 3x$

57. $12 \sec^2 4x$ **59.** $\sin 2\theta \cos 2\theta = \frac{1}{2} \sin 4\theta$

61. $1/2\sqrt{x} + 2x \cos(2x)^2$ **63.** $-\sin x \cos(\cos x)$ **65.** $2e^{2x}$

67. $e^{\sqrt{x}}/2\sqrt{x}$ **69.** $3(e^{-t} + e^t)^2(e^t - e^{-t})$ **71.** $2x$

73. $\dfrac{-2(e^x - e^{-x})}{(e^x + e^{-x})^2}$ **75.** x^2e^x **77.** $e^{-x}\left(\dfrac{1}{x} - \ln x\right)$

79. $2e^x \cos x$ **81.** $\dfrac{2}{x}$ **83.** $\dfrac{4(\ln x)^3}{x}$ **85.** $\dfrac{2x^2 - 1}{x(x^2 - 1)}$

87. $\dfrac{1 - x^2}{x(x^2 + 1)}$ **89.** $\dfrac{1 - 2 \ln t}{t^3}$ **91.** $\dfrac{1}{1 - x^2}$

93. $\dfrac{\sqrt{x^2 + 1}}{x^2}$ **95.** $\cot x$ **97.** $-\tan x + \dfrac{\sin x}{\cos x - 1}$

99. $\dfrac{3 \cos x}{(\sin x - 1)(\sin x + 2)}$ **101.** $12(5x^2 - 1)(x^2 - 1)$

103. $2(\cos x^2 - 2x^2 \sin x^2)$ **105.** $3(6x + 5)e^{-3x}$

107. $s'(t) = (t + 1)/\sqrt{t^2 + 2t + 8}$, $\frac{3}{4}$

109. $f'(x) = \dfrac{-9x^2}{(x^3 - 4)^2}, -\dfrac{9}{25}$ **111.** $f'(t) = \dfrac{-5}{(t - 1)^2}, -5$

113. $y' = -6 \sec^3(2x) \tan(2x), 0$

115. (a) $9x - 5y - 2 = 0$
(b)

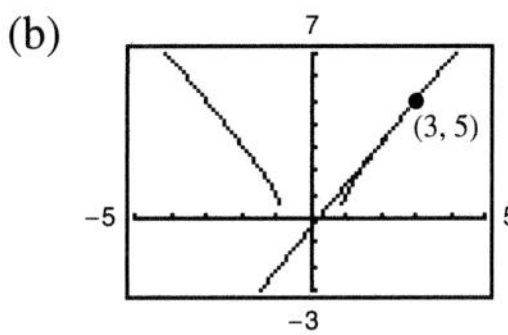

117. (a) $2x - y - 2\pi = 0$
(b)

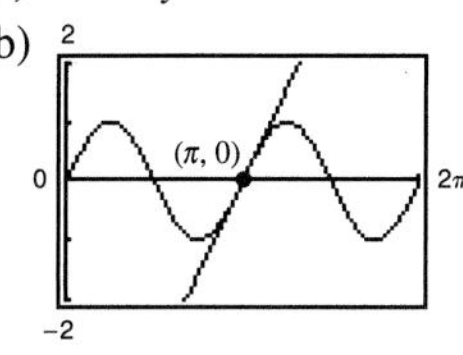

119. (a) $12x - y + 2 - 3\pi = 0$
(b)

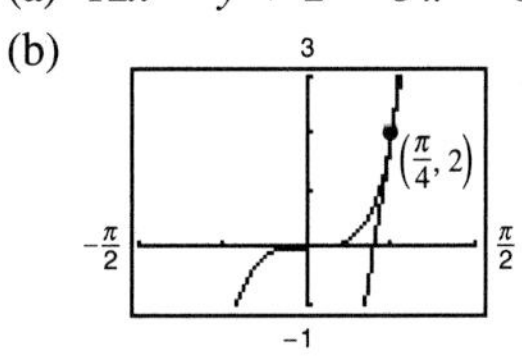

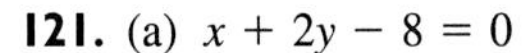

121. (a) $x + 2y - 8 = 0$
(b)

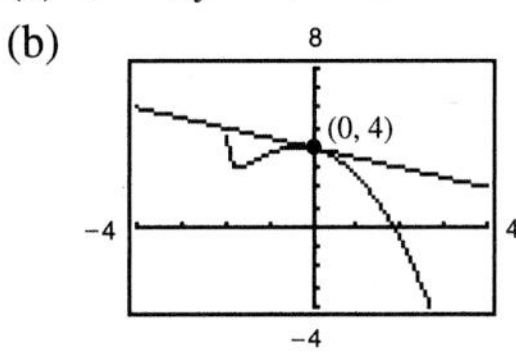

123. $(\ln 4)4^x$ **125.** $(\ln 5)5^{x-2}$ **127.** $t2^t(t \ln 2 + 2)$

129. $-2^{-\theta}[(\ln 2)\cos \pi\theta + \pi \sin \pi\theta]$ **131.** $1/x(\ln 3)$

133. $\dfrac{x - 2}{(\ln 2)x(x - 1)}$ **135.** $\dfrac{x}{(\ln 5)(x^2 - 1)}$

137. $\dfrac{5}{(\ln 2)t^2}(1 - \ln t)$

139.

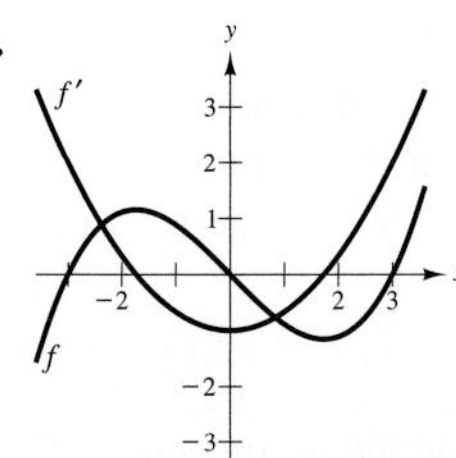

The zeros of f' correspond to the points where the graph of f has horizontal tangents.

141.

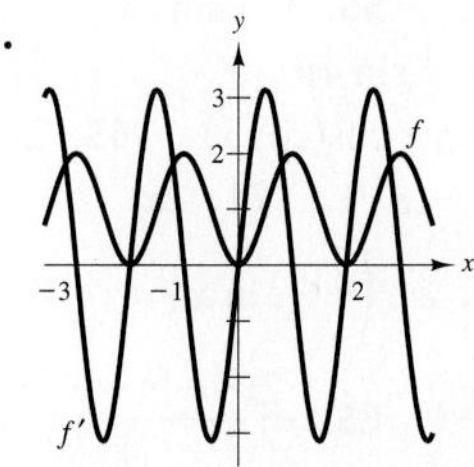

The zeros of f' correspond to the points where the graph of f has horizontal tangents.

143. $g'(x) = 3f'(3x)$

145. (a) 24 (b) Not possible because $g'(h(5))$ is not known. (c) $\frac{4}{3}$ (d) 162

147. (a) $g'(1/2) = -3$ (b) $3x + y - 3 = 0$ (c)

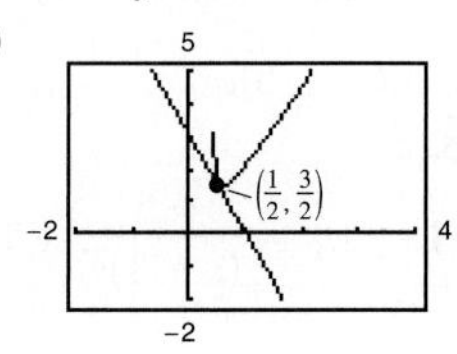

149. (a) $s'(0) = 0$ (b) $y = \frac{4}{3}$ (c)

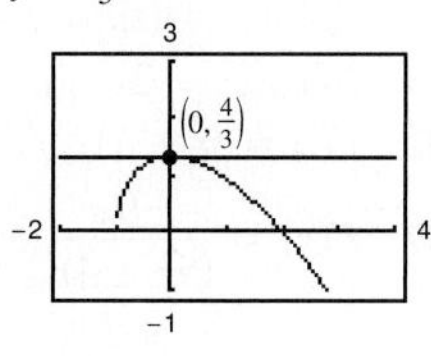

151. $3x + 4y - 25 = 0$

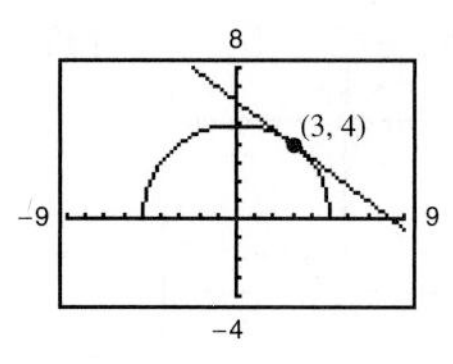

153. $\left(\frac{\pi}{6}, \frac{3\sqrt{3}}{2}\right), \left(\frac{5\pi}{6}, -\frac{3\sqrt{3}}{2}\right), \left(\frac{3\pi}{2}, 0\right)$

155. $h''(x) = 18x + 6$, 24

157. $f''(x) = -4x^2 \cos(x^2) - 2 \sin(x^2)$, 0

159. (a) 1.461 (b) -1.016

161. 0.2 rad, 1.45 rad/sec **163.** 0.04224

165. 768π in^3/sec

167. (a)

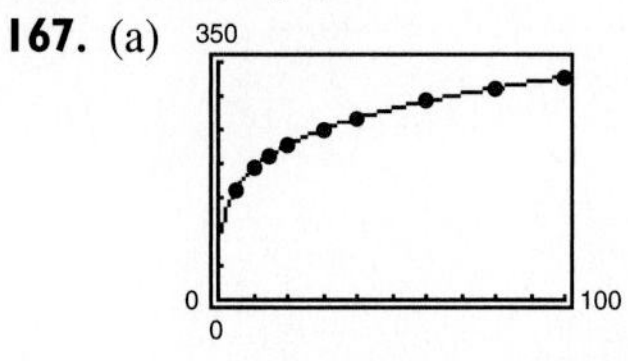

(b) $T'(10) \approx 4.75$ deg/lb/in^2
$T'(70) \approx 0.97$ deg/lb/in^2

169. (a) \$40.64 (b) $C'(1) \approx 0.051P$, $C'(8) \approx 0.072P$ (c) ln 1.05

171. (a) Yes; Proof (b) Yes; Proof **173.** Proof

175. $g'(x) = \left(\frac{2x - 3}{|2x - 3|}\right), \quad x \neq \frac{3}{2}$

177. $h'(x) = \frac{x}{|x|} \cos x - |x| \sin x, \quad x \neq 0$

179. (a) $P_1(x) = \frac{\pi}{2}(x - 1) + 1$

$P_2(x) = \frac{\pi^2}{8}(x - 1)^2 + \frac{\pi}{2}(x - 1) + 1$

(b)

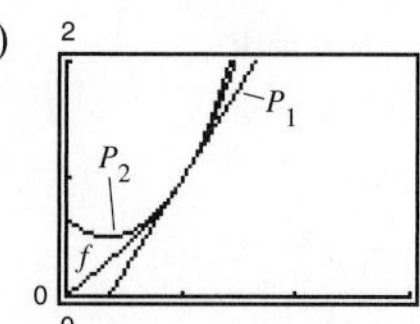

(c) P_2
(d) P_1 and P_2 become less accurate as you move farther from $x = 1$.

181. (a) $P_1 = 1$, $P_2 = 1 - \frac{x^2}{2}$

(b)

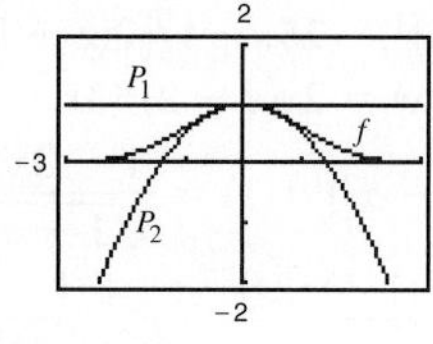

(c) P_2
(d) P_1 and P_2 become less accurate as you move farther from $x = 0$.

183. False, $y' = \frac{1}{2}(1 - x)^{-1/2}(-1)$ **185.** True

187. Putnam Problem A1, 2002

Section 3.5 (page 171)

1. $-x/y$ **3.** $-\sqrt{y/x}$ **5.** $(y - 3x^2)/(2y - x)$

7. $\frac{10 - e^y}{xe^y + 3}$ **9.** $\frac{1 - 3x^2y^3}{3x^3y^2 - 1}$ **11.** $\frac{4xy - 3x^2 - 3y^2}{6xy - 2x^2}$

13. $\cos x/4 \sin 2y$ **15.** $(\cos x - \tan y - 1)/x \sec^2 y$

17. $y \cos(xy)/[1 - x \cos(xy)]$

19. $2xy/(3 - 2y^2)$

21. (a) $y_1 = \sqrt{16 - x^2}$
$y_2 = -\sqrt{16 - x^2}$

(b)

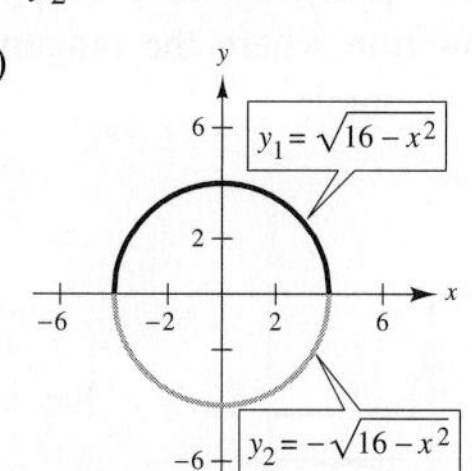

(c) $y' = \mp \frac{x}{\sqrt{16 - x^2}} = -\frac{x}{y}$
(d) $y' = -\frac{x}{y}$

23. (a)

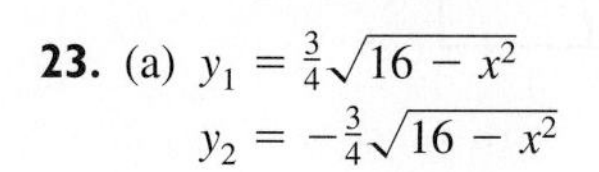

$y_1 = \frac{3}{4}\sqrt{16 - x^2}$
$y_2 = -\frac{3}{4}\sqrt{16 - x^2}$

(b)

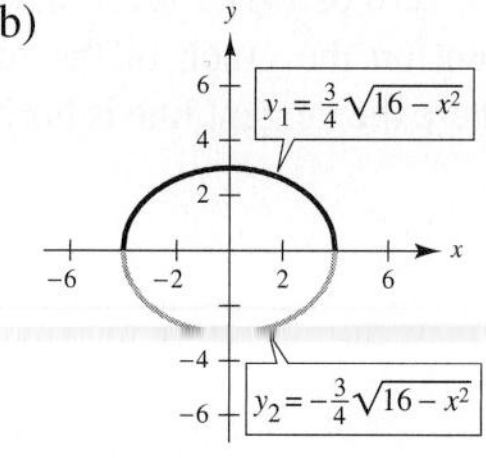

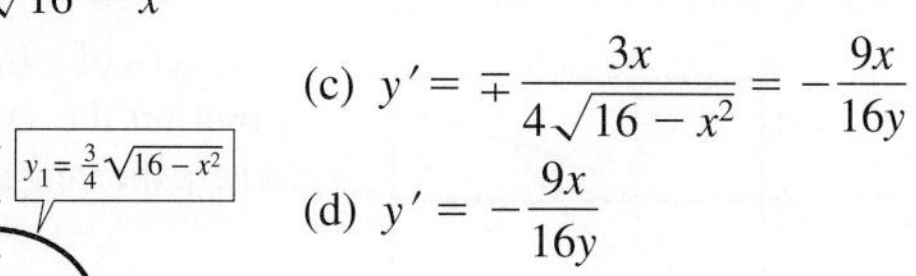

(c) $y' = \mp \frac{3x}{4\sqrt{16 - x^2}} = -\frac{9x}{16y}$
(d) $y' = -\frac{9x}{16y}$

25. $-\frac{y}{x}, -\frac{1}{4}$ **27.** $\frac{18x}{y(x^2+9)^2}$, Undefined **29.** $-\sqrt[3]{\frac{y}{x}}, -\frac{1}{2}$

31. $-\sin^2(x+y)$ or $-\frac{x^2}{x^2+1}, 0$ **33.** $-\frac{1-3ye^{xy}}{3xe^{xy}}, \frac{1}{9}$

35. $-\frac{1}{2}$ **37.** 0 **39.** $y=-x+4$ **41.** $y=-x+2$

43. $y=\frac{\sqrt{3}x}{6}+\frac{8\sqrt{3}}{3}$ **45.** $y=-\frac{2}{11}x+\frac{30}{11}$

47. (a) $y=-2x+4$ (b) Answers will vary.

49. $\cos^2 y$, $-\pi/2 < y < \pi/2$, $1/(1+x^2)$ **51.** $-36/y^3$

53. $-16/y^3$ **55.** $(3x)/(4y)$

57. $x+3y-12=0$

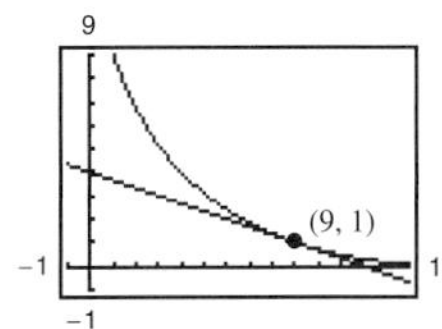

59. At $(4, 3)$:
Tangent line: $4x+3y-25=0$
Normal line: $3x-4y=0$

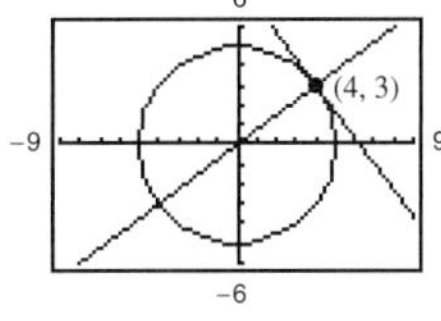

At $(-3, 4)$:
Tangent line: $3x-4y+25=0$
Normal line: $4x+3y=0$

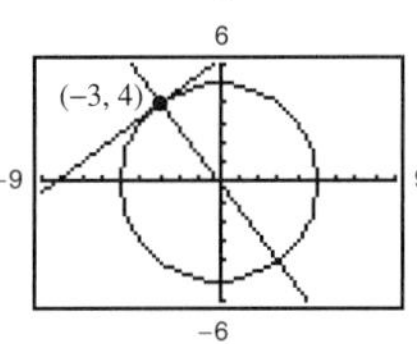

61. $x^2+y^2=r^2 \Rightarrow y'=-x/y \Rightarrow y/x=$ slope of normal line. Then, for (x_0, y_0) on the circle, $x_0 \neq 0$, an equation of the normal line is $y=(y_0/x_0)x$, which passes through the origin. If $x_0=0$, the normal line is vertical and passes through the origin.

63. Horizontal tangents: $(-4, 0), (-4, 10)$
Vertical tangents: $(0, 5), (-8, 5)$

65. $\frac{2x^2-1}{\sqrt{x^2-1}}$ **67.** $\frac{3x^3-15x^2+8x}{2(x-1)^3\sqrt{3x-2}}$

69. $(2x^2+2x-1)\sqrt{x-1}/(x+1)^{3/2}$

71. $2(1-\ln x)x^{(2/x)-2}$

73. $(x-2)^{x+1}\left[\frac{x+1}{x-2}+\ln(x-2)\right]$

75.

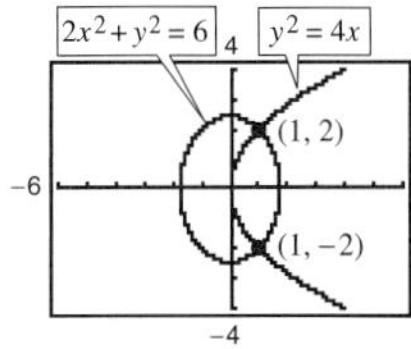

At $(1, 2)$:
Slope of ellipse: -1
Slope of parabola: 1
At $(1, -2)$:
Slope of ellipse: 1
Slope of parabola: -1

77.

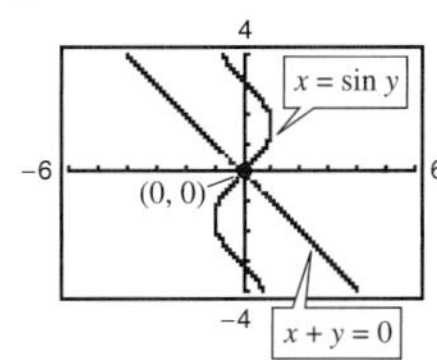

At $(0, 0)$:
Slope of line: -1
Slope of sine curve: 1

79. Derivatives: $\frac{dy}{dx}=-\frac{y}{x}, \frac{dy}{dx}=\frac{x}{y}$

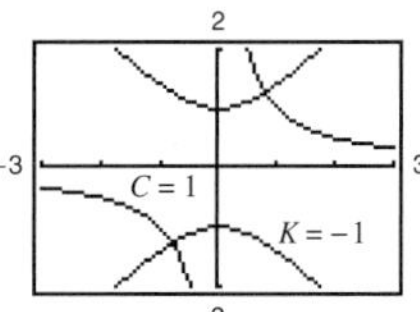

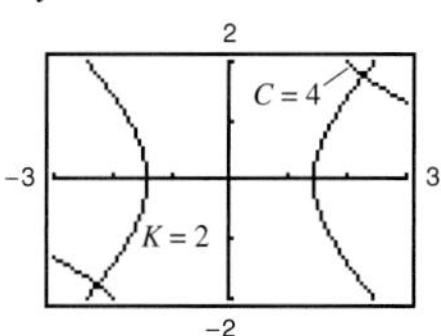

81. (a) $y\frac{dy}{dx}-3x^3=0$ (b) $y\frac{dy}{dt}-3x^3\frac{dx}{dt}=0$

83. (a) $-\pi\sin(\pi y)\left(\frac{dy}{dx}\right)-3\pi\cos(\pi x)=0$

(b) $-\pi\sin(\pi y)\left(\frac{dy}{dt}\right)-3\pi\cos(\pi x)\left(\frac{dx}{dt}\right)=0$

85. Answers will vary. In the explicit form of a function, the variable is explicitly written as a function of x. In an implicit equation, the function is only implied by an equation. An example of an implicit function is $x^2+xy=5$. In explicit form, this equation would be $y=(5-x^2)/x$.

87.

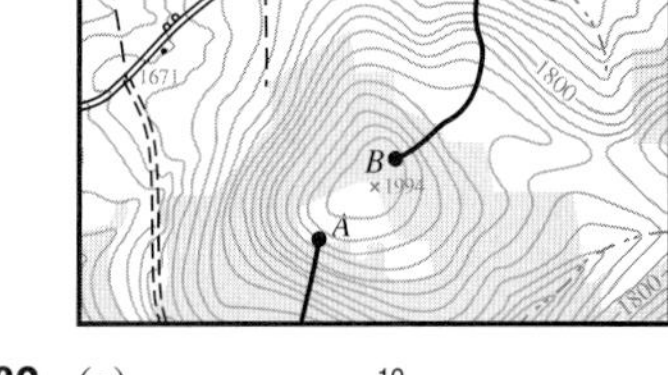

Use starting point B.

89. (a)

(b)

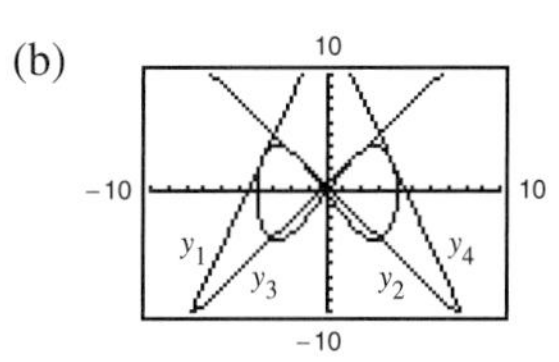

$y_1=\frac{1}{3}\left[\left(\sqrt{7}+7\right)x+\left(8\sqrt{7}+23\right)\right]$
$y_2=-\frac{1}{3}\left[\left(-\sqrt{7}+7\right)x-\left(23-8\sqrt{7}\right)\right]$
$y_3=-\frac{1}{3}\left[\left(\sqrt{7}-7\right)x-\left(23-8\sqrt{7}\right)\right]$
$y_4=-\frac{1}{3}\left[\left(\sqrt{7}+7\right)x-\left(8\sqrt{7}+23\right)\right]$

(c) $\left(8\sqrt{7}/7, 5\right)$

91. Proof **93.** $(0, \pm 1)$

95. (a) 1 (b) 1 (c) 3
$x_0=3/4$

Section 3.6 (page 179)

1. $\frac{1}{5}$ **3.** $\frac{2\sqrt{3}}{3}$ **5.** $\frac{1}{13}$ **7.** $f'\left(\frac{1}{2}\right)=\frac{3}{4}, (f^{-1})'\left(\frac{1}{8}\right)=\frac{4}{3}$

9. $f'(5)=\frac{1}{2}, (f^{-1})'(1)=2$

11. (a) $y = 2\sqrt{2}x + \frac{\pi}{4} - 1$

(b)

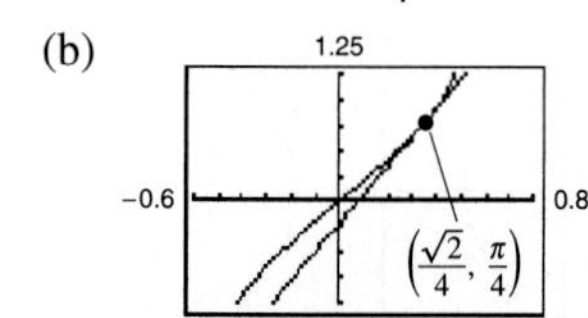

13. (a) $y = \frac{\pi}{2}$

(b)

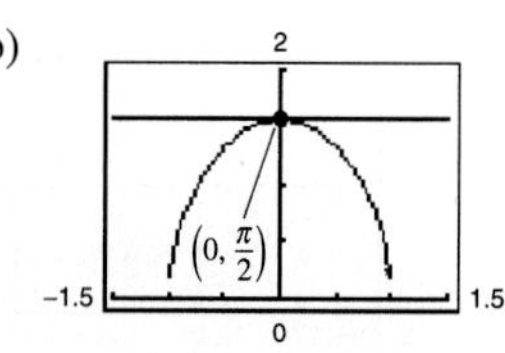

15. $-\frac{1}{11}$ **17.** $\frac{\pi + 2}{\pi}$ **19.** $\frac{2}{\sqrt{2x - x^2}}$ **21.** $-\frac{3}{\sqrt{4 - x^2}}$

23. $\frac{a}{a^2 + x^2}$ **25.** $\frac{3x - \sqrt{1 - 9x^2} \arcsin 3x}{x^2\sqrt{1 - 9x^2}}$

27. $-\frac{1}{(x + 1)\sqrt{1 - x^2}} - \frac{\arccos x}{(x + 1)^2}$ **29.** $-\frac{6}{1 + 36x^2}$

31. $-\frac{t}{\sqrt{1 - t^2}}$ **33.** $\arccos x$ **35.** $\frac{1}{1 - x^4}$

37. $\frac{1}{(1 - t^2)^{3/2}}$ **39.** $\arcsin x$ **41.** $\frac{x^2}{\sqrt{16 - x^2}}$ **43.** $\frac{2}{(1 + x^2)^2}$

45. $y = \frac{1}{3}(4\sqrt{3}x - 2\sqrt{3} + \pi)$

47. $y = \frac{1}{4}x + (\pi - 2)/4$ **49.** $y = (2\pi - 4)x + 4$

51. $y = -2x + \left(\frac{\pi}{6} + \sqrt{3}\right)$

$y = -2x + \left(\frac{5\pi}{6} - \sqrt{3}\right)$

53. $P_1(x) = \frac{\pi}{6} + \frac{2\sqrt{3}}{3}\left(x - \frac{1}{2}\right)$

$P_2(x) = \frac{\pi}{6} + \frac{2\sqrt{3}}{3}\left(x - \frac{1}{2}\right) + \frac{2\sqrt{3}}{9}\left(x - \frac{1}{2}\right)^2$

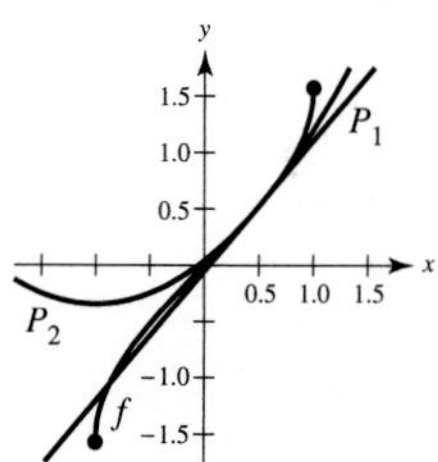

55. $P_1(x) = x$; $P_2(x) = x$

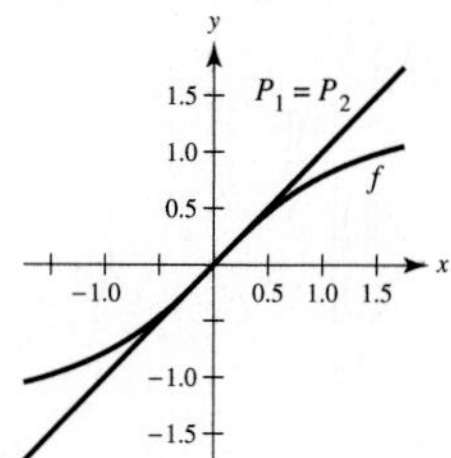

57. $y = [-2\pi x/(\pi + 8)] + 1 - [\pi^2/(2\pi + 16)]$

59. $y = -x + \sqrt{2}$

61. Many x-values yield the same y-value. For example, $f(\pi) = 0 = f(0)$. The graph is not continuous at $x = (2n - 1)\pi/2$, where n is an integer.

63. Theorem 3.17: Let f be differentiable on an interval I. If f has an inverse g, then g is differentiable at any x for which $f'(g(x)) \neq 0$. Moreover,

$g'(x) = 1/f'(g(x)), f'(g(x)) \neq 0.$

65. (a) $\theta = \operatorname{arccot}(x/5)$

(b) $x = 10$: 16 rad/hr; $x = 3$: 58.824 rad/hr

67. (a) $h(t) = -16t^2 + 256$; $t = 4$ sec

(b) $t = 1$: -0.0520 rad/sec; $t = 2$: -0.1116 rad/sec

69. 0.015 rad/sec **71.** Proof **73.** True **75.** Proof

77. (a)

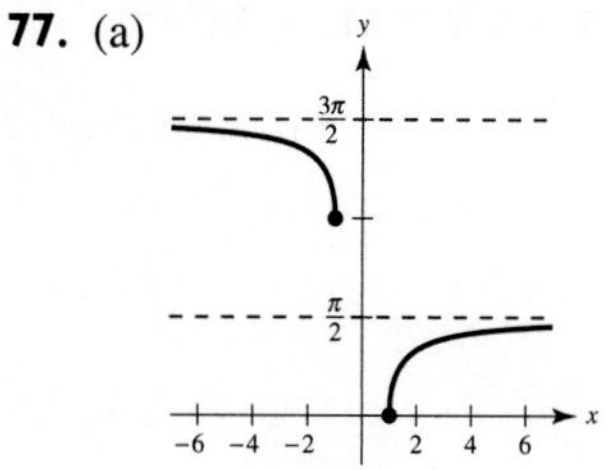

(b) Proof

79. $f'(x) = 0$ implies $f(x)$ is constant.

Section 3.7 (page 187)

1. (a) $\frac{3}{4}$ (b) 20 **3.** (a) $-\frac{5}{8}$ (b) $\frac{3}{2}$

5. (a) -4 cm/sec (b) 0 cm/sec (c) 4 cm/sec

7. (a) 8 cm/sec (b) 4 cm/sec (c) 2 cm/sec

9. (a) Positive (b) Negative

11. In a linear function, if x changes at a constant rate, so does y. However, unless $a = 1$, y does not change at the same rate as x.

13. $2(2x^3 + 3x)/\sqrt{x^4 + 3x^2 + 1}$

15. (a) 36π cm²/min

(b) 144π cm²/min

17. (a) Proof

(b) When $\theta = \frac{\pi}{6}, \frac{dA}{dt} = \frac{\sqrt{3}}{8}s^2.$

When $\theta = \frac{\pi}{3}, \frac{dA}{dt} = \frac{1}{8}s^2.$

(c) If s and $d\theta/dt$ are constant, dA/dt is proportional to $\cos \theta$.

19. $\frac{3}{32\pi}$ m/min **21.** (a) 36 cm²/sec (b) 360 cm²/sec

23. $\frac{8}{405\pi}$ ft/min **25.** (a) 12.5% (b) $\frac{1}{144}$ m/min

27. (a) $-\frac{7}{12}$ ft/sec; $-\frac{3}{2}$ ft/sec; $-\frac{48}{7}$ ft/sec

(b) $\frac{527}{24}$ ft²/sec (c) $\frac{1}{12}$ rad/sec

29. Rate of vertical change: $\frac{1}{5}$ m/sec

Rate of horizontal change: $-\frac{\sqrt{3}}{15}$ m/sec

31. (a) -750 mi/hr (b) 20 min

33. $-28/\sqrt{10} \approx -8.85$ ft/sec

35. (a) $\frac{25}{3}$ ft/sec (b) $\frac{10}{3}$ ft/sec

37. (a) 12 sec

(b) $\sqrt{3}/2$ m

(c) $\sqrt{5}\pi/120$ m/sec

39. Evaporation rate proportional to

$S \Longrightarrow \dfrac{dV}{dt} = k(4\pi r^2)$

$V = \left(\dfrac{4}{3}\right)\pi r^3 \Longrightarrow \dfrac{dV}{dt} = 4\pi r^2 \dfrac{dr}{dt}$

So, $k = \dfrac{dr}{dt}$.

41. $V^{0.3}\left(1.3p\dfrac{dV}{dt} + V\dfrac{dp}{dt}\right) = 0$ **43.** $\dfrac{1}{20}$ rad/sec

45. (a) $t = 65°$: $H \approx 99.8\%$ (b) $-4.7\%/\text{hr}$
$t = 80°$: $H \approx 60.2\%$

47. (a) $\dfrac{dx}{dt} = -400\pi \sin\theta$

(b)

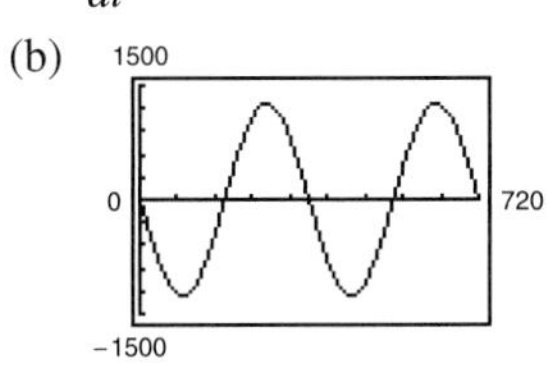

(c) $\theta = \dfrac{\pi}{2} + n\pi$ (or $90° + n \cdot 180°$); $\theta = n\pi$ (or $n \cdot 180°$)

(d) -200π cm/sec; $-200\pi\sqrt{3}$ cm/sec

49. $\dfrac{d\theta}{dt} = \dfrac{1}{25}\cos^2\theta, \quad -\dfrac{\pi}{4} \le \theta \le \dfrac{\pi}{4}$

51. (a) $\frac{1}{2}$ rad/min (b) $\frac{3}{2}$ rad/min
(c) 1.87 rad/min

53. -0.1808 ft/sec^2

55. (a) $m(s) = 0.3754s^3 - 18.780s^2 + 313.23s - 1707.8$

(b) $\dfrac{dm}{dt} = (1.1262s^2 - 37.560s + 313.23)\dfrac{ds}{dt}$; 1.12 million

Section 3.8 (page 195)

1.

n	x_n	$f(x_n)$	$f'(x_n)$	$\dfrac{f(x_n)}{f'(x_n)}$	$x_n - \dfrac{f(x_n)}{f'(x_n)}$
1	1.7000	−0.1100	3.4000	−0.0324	1.7324
2	1.7324	0.0012	3.4648	0.0003	1.7321

3.

n	x_n	$f(x_n)$	$f'(x_n)$	$\dfrac{f(x_n)}{f'(x_n)}$	$x_n - \dfrac{f(x_n)}{f'(x_n)}$
1	3	0.1411	−0.9900	−0.1425	3.1425
2	3.1425	−0.0009	−1.0000	0.0009	3.1416

5. 0.682 **7.** 1.146, 7.854 **9.** 0.567 **11.** −1.442
13. 0.900, 1.100, 1.900 **15.** −0.489 **17.** 0.569
19. 4.493 **21.** 0.567 **23.** 0.786
25. (a) Proof (b) $\sqrt{5} \approx 2.236$; $\sqrt{7} \approx 2.646$
27. $f'(x_1) = 0$
29. $2 = x_1 = x_3 = \ldots; 1 = x_2 = x_4 = \ldots$

31. Answers will vary. Sample answer: If f is a function that is continuous on $[a, b]$ and differentiable on (a, b), where $c \in [a, b]$ and $f(c) = 0$, Newton's Method uses tangent lines to approximate c. First, estimate an initial x_1 close to c. (See graph.) Then, determine x_2 by $x_2 = x_1 - f(x_1)/f'(x_1)$. Calculate a third estimate by $x_3 = x_2 - f(x_2)/f'(x_2)$. Continue this process until $|x_n - x_{n+1}|$ is within the desired accuracy, and let x_{n+1} be the final approximation of c.

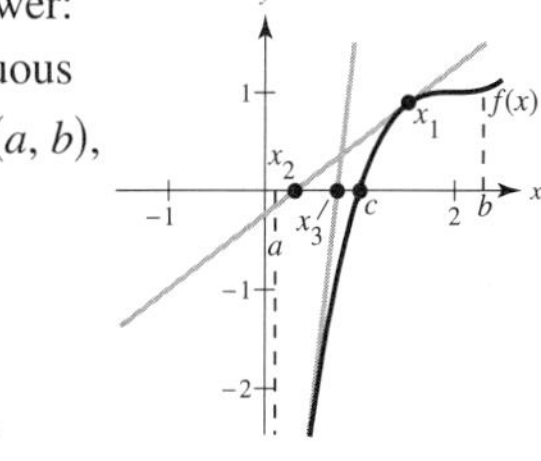

33. 0.74 **35.** 1.12

37. (a)

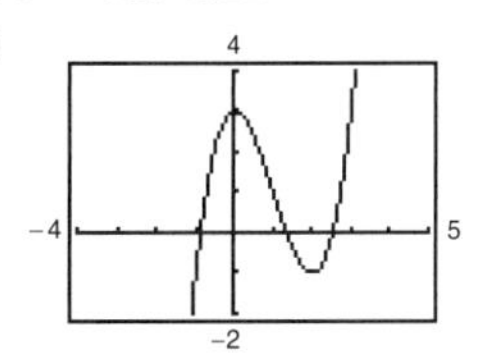

(b) 1.347 (c) 2.532

(d)

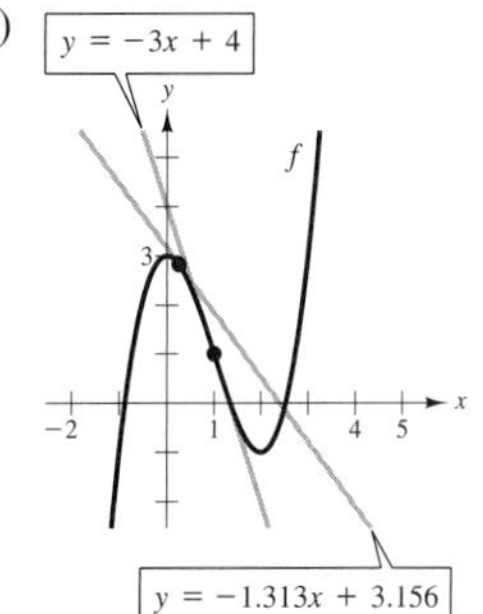

x-intercept of $y = -3x + 4$ is $\frac{4}{3}$.
x-intercept of
$y = -1.313x + 3.156$
is approximately 2.404.

(e) If the initial estimate $x = x_1$ is not sufficiently close to the desired zero of a function, the x-intercept of the corresponding tangent line to the function may approximate a second zero of the function.

39. Proof **41.** \$384,356
43. False. Let $f(x) = (x^2 - 1)/(x - 1)$.
45. True **47.** $x \approx 11.803$ **49.** 0.217

Review Exercises for Chapter 3 (page 197)

1. $f'(x) = 2x - 2$ **3.** $f'(x) = 1/2\sqrt{x} = \sqrt{x}/2x$
5. f is differentiable at all $x \neq -1$.

7.

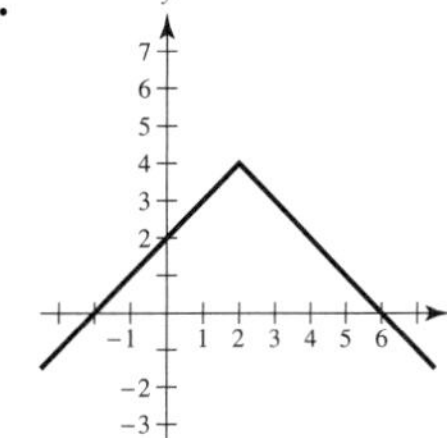

(a) Yes
(b) No, because the derivatives from the left and right are not equal.

9. $-\frac{3}{2}$

11. (a) $y = 3x + 1$ **13.** 8

(b)

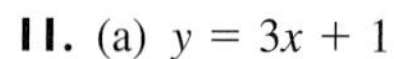
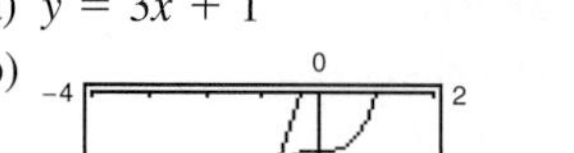
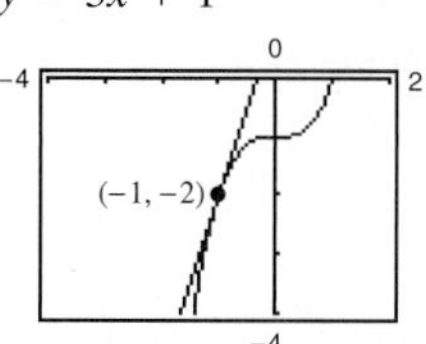

15.

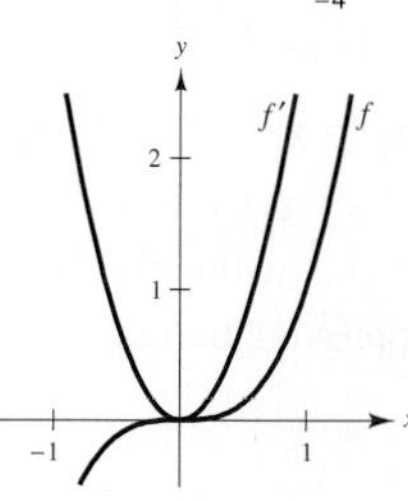

$f' > 0$ where the slopes of the tangent lines to the graph of f are positive.

17. 0 **19.** $8x^7$ **21.** $12t^3$ **23.** $3x(x - 2)$

25. $3/\sqrt{x} + 1/x^{2/3}$ **27.** $-4/3t^3$ **29.** $2 - 3\cos\theta$

31. $-3\sin t - 4e^t$

33. (a) 50 vibrations/sec/lb

(b) 33.33 vibrations/sec/lb

35. 414.74 m or 1354 ft

37. (a)

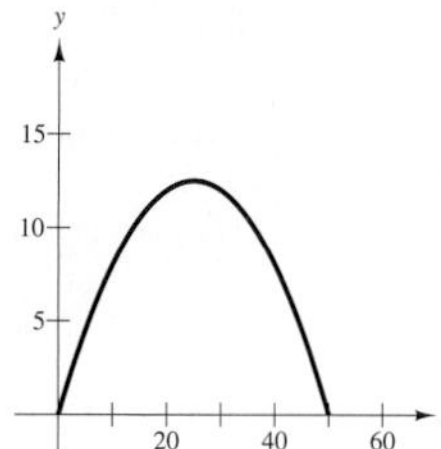

(b) 50 (c) $x = 25$

(d) $y' = 1 - 0.04x$

x	0	10	25	30	50
y'	1	0.6	0	−0.2	−1

(e) $y'(25) = 0$

39. (a) $x'(t) = 2t - 3$ (b) $(-\infty, 1.5)$ (c) $x = -\frac{1}{4}$ (d) 1

41. $2(6x^3 - 9x^2 + 16x - 7)$ **43.** $\sqrt{x}\cos x + \sin x/2\sqrt{x}$

45. $2 + \dfrac{2}{x^3}$ **47.** $-\dfrac{x^2 + 1}{(x^2 - 1)^2}$ **49.** $\dfrac{6x}{(4 - 3x^2)^2}$

51. $\dfrac{2x\cos x + x^2\sin x}{\cos^2 x}$ **53.** $3x^2\sec x\tan x + 6x\sec x$

55. $-x\sec^2 x - \tan x$ **57.** $4e^x(x + 1)$ **59.** $6t$

61. $6\sec^2\theta\tan\theta$

63. $y'' + y = -(2\sin x + 3\cos x) + (2\sin x + 3\cos x) = 0$

65. $x = \dfrac{3\pi}{4}, \dfrac{7\pi}{4}$ **67.** $\dfrac{-3x^2}{2\sqrt{1 - x^3}}$

69. $\dfrac{2(x - 3)(-x^2 + 6x + 1)}{(x^2 + 1)^3}$

71. $s(s^2 - 1)^{3/2}(8s^3 - 3s + 25)$ **73.** $-9\sin(3x + 1)$

75. $-\csc 2x\cot 2x$ **77.** $\frac{1}{2}(1 - \cos 2x) = \sin^2 x$

79. $\sin^{1/2} x\cos x - \sin^{5/2} x\cos x = \cos^3 x\sqrt{\sin x}$

81. $\dfrac{(x + 2)[\pi\cos(\pi x)] - \sin(\pi x)}{(x + 2)^2}$

83. $\frac{1}{4}te^{t/4}(t + 8)$ **85.** $\dfrac{e^{2x} - e^{-2x}}{\sqrt{e^{2x} + e^{-2x}}}$ **87.** $\dfrac{x(2 - x)}{e^x}$

89. $\dfrac{1}{2x}$ **91.** $\dfrac{1 + 2\ln x}{2\sqrt{\ln x}}$ **93.** $\dfrac{x}{(a + bx)^2}$ **95.** $\dfrac{1}{x(a + bx)}$

97. $t(t - 1)^4(7t - 2)$

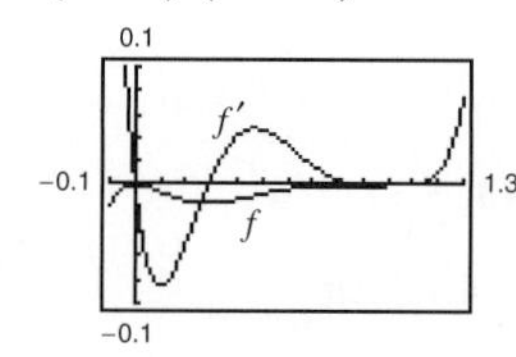

The zeros of f' correspond to the points on the graph of the function where the tangent line is horizontal.

99. $(x + 2)/(x + 1)^{3/2}$

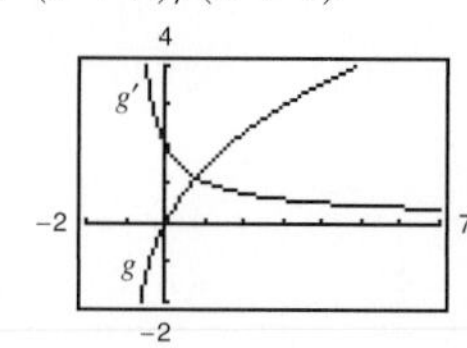

g' is not equal to zero for any x.

101. $5/6(t + 1)^{1/6}$

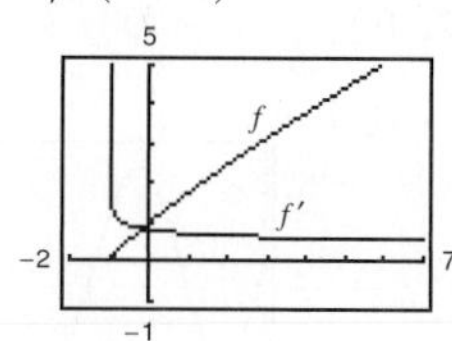

f' has no zeros.

103. $-\sec^2\sqrt{1 - x}/2\sqrt{1 - x}$

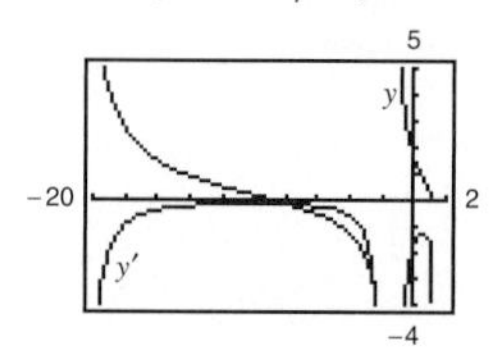

y' has no zeros.

105. $4 - 4\sin 2x$ **107.** $2\csc^2 x\cot x$

109. $\dfrac{2(t + 2)}{(1 - t)^4}$ **111.** $18\sec^2(3\theta)\tan(3\theta) + \sin(\theta - 1)$

113. $x(6\ln x + 5)$

115. (a) −18.667 degrees/hr

(b) −7.284 degrees/hr

(c) −3.240 degrees/hr

(d) −0.747 degree/hr

117. (a) $h = 0$ is not in the domain of the function.

(b) $h = 0.86 - 6.45\ln p$

(c)

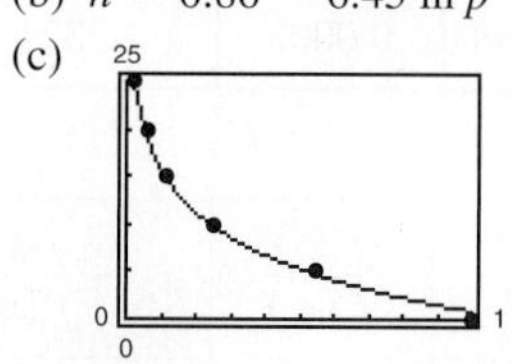

(d) 2.7 km

(e) 0.15 atm

(f) $h = 5: \dfrac{dp}{dh} = -0.085$

$h = 20: \dfrac{dp}{dh} = -0.009$

As the altitude increases, the pressure decreases at a lower rate.

119. $-\dfrac{2x + 3y}{3(x + y^2)}$ **121.** $-\dfrac{2x\sin x^2 + e^y}{xe^y}$

123. $\dfrac{2y\sqrt{x} - y\sqrt{y}}{2x\sqrt{y} - x\sqrt{x}}$ **125.** $\dfrac{y\sin x + \sin y}{\cos x - x\cos y}$

127. Tangent line: $x + 2y - 10 = 0$
Normal line: $2x - y = 0$

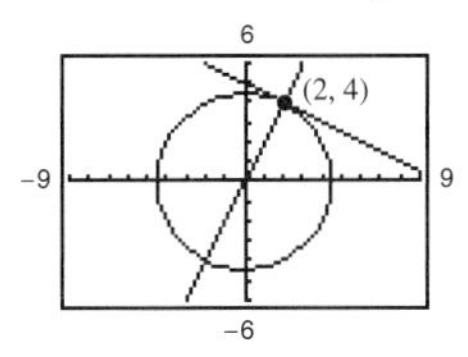

129. Tangent line: $xe^{-1} + y = 0$
Normal line: $xe - y - (e^2 + 1) = 0$

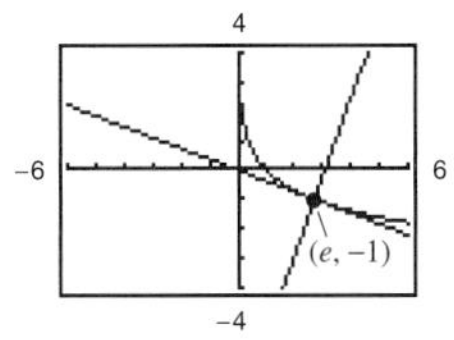

131. $\dfrac{x^3 + 8x^2 + 4}{(x + 4)^2\sqrt{x^2 + 1}}$ **133.** $\dfrac{1}{3(\sqrt[3]{-3})^2} \approx 0.160$ **135.** $\dfrac{3}{4}$

137. $(1 - x^2)^{-3/2}$ **139.** $\dfrac{x}{|x|\sqrt{x^2 - 1}} + \text{arcsec}\, x$

141. $(\arcsin x)^2$

143. (a) $2\sqrt{2}$ units/sec
(b) 4 units/sec
(c) 8 units/sec

145. $\frac{2}{25}$ m/min **147.** -38.34 m/sec

149. $-0.347, -1.532, 1.879$ **151.** 1.202

153. $-1.164, 1.453$

P.S. Problem Solving (page 201)

1. (a) $r = \frac{1}{2}$ (b) Center: $\left(0, \frac{5}{4}\right)$

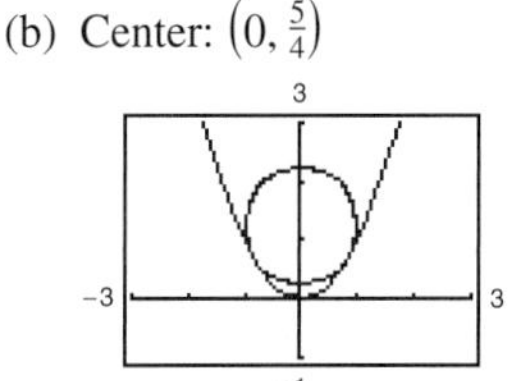

3. (a) $P_1(x) = 1$
(b) $P_2(x) = 1 - \frac{1}{2}x^2$
(c)

x	-1.0	-0.1	-0.001	0	0.001
$\cos x$	0.5403	0.9950	1.000	1	1
$P_2(x)$	0.5000	0.9950	1.000	1	1

x	0.1	1.0
$\cos x$	0.9950	0.5403
$P_2(x)$	0.9950	0.5000

$P_2(x)$ is a good approximation of $f(x) = \cos x$ when x is very close to 0.
(d) $P_3(x) = x - \frac{1}{6}x^3$

5. $p(x) = 2x^3 + 4x^2 - 5$

7. (a) Graph $\begin{cases} y_1 = 1/a\sqrt{x^2(a^2 - x^2)} \\ y_2 = -1/a\sqrt{x^2(a^2 - x^2)} \end{cases}$ as separate equations.
(b) Answers will vary.

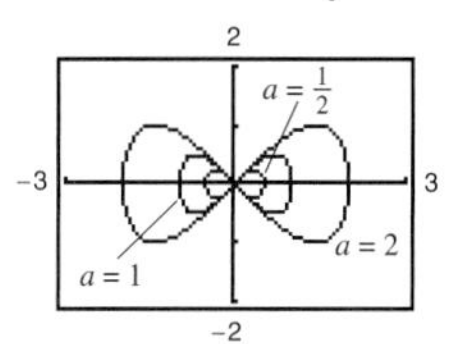

The intercepts will always be $(0, 0)$, $(a, 0)$, and $(-a, 0)$, and the maximum and minimum y-values appear to be $\pm\frac{1}{2}a$.
(c) $\left(\frac{a\sqrt{2}}{2}, \frac{a}{2}\right), \left(\frac{a\sqrt{2}}{2}, -\frac{a}{2}\right), \left(-\frac{a\sqrt{2}}{2}, \frac{a}{2}\right), \left(-\frac{a\sqrt{2}}{2}, -\frac{a}{2}\right)$

9. (a) When the man is 90 ft from the light, the tip of his shadow is $112\frac{1}{2}$ ft from the light. The tip of the child's shadow is $111\frac{1}{9}$ ft from the light, so the man's shadow extends $1\frac{7}{18}$ ft beyond the child's shadow.
(b) When the man is 60 ft from the light, the tip of his shadow is 75 ft from the light. The tip of the child's shadow is $77\frac{7}{9}$ ft from the light, so the child's shadow extends $2\frac{7}{9}$ ft beyond the man's shadow.
(c) $d = 80$ ft
(d) Let x be the distance from the light to the man and s be the distance from the light to the tip of the shadow.
If $0 < x < 80$, $ds/dt = -50/9$.
If $x > 80$, $ds/dt = -25/4$.
There is a discontinuity at $x = 80$.

11. Tangent line: $y = 1/ax + (b - 1)$
Passes through $(0, c)$, therefore, $c = b - 1$.
Distance between b and c is $b - c = 1$.

13. (a)

$z°$	0.1	0.01	0.0001
$\dfrac{\sin z}{z}$	0.0174532837	0.0174532924	0.0174532925

(b) $\pi/180$ (c) $\pi/180 \cos z$ (d) $\pi/180\, C(z)$
(e) Answers will vary.

15. (a) j is the rate of change of acceleration.
(b) a: position function; b: acceleration function;
c: jerk function; d: velocity function

Chapter 4

Section 4.1 (page 209)

1. A: none, B: absolute maximum (and relative maximum), C: none, D: none, E: relative maximum, F: relative minimum, G: none

3. $f'(0) = 0$ **5.** $f'(3) = 0$ **7.** $f'(-2)$ is undefined.

9. 2, absolute maximum (and relative maximum)

11. 1, absolute maximum (and relative maximum);
2, absolute minimum (and relative mimimum);
3, absolute maximum (and relative maximum)

13. $x = 0, x = 2$ **15.** $t = 8/3$ **17.** $x = \pi/3, \pi, 5\pi/3$
19. $x = 0$
21. Minimum: $(2, 2)$
Maximum: $(-1, 8)$
23. Minima: $(0, 0)$ and $(3, 0)$
Maximum: $\left(\frac{3}{2}, \frac{9}{4}\right)$
25. Minimum: $\left(-1, -\frac{5}{2}\right)$
Maximum: $(2, 2)$
27. Minimum: $(0, 0)$
Maximum: $(-1, 5)$
29. Minimum: $(0, 0)$
Maxima: $\left(-1, \frac{1}{4}\right)$ and $\left(1, \frac{1}{4}\right)$
31. Minimum: $(1, -1)$
Maximum: $\left(0, -\frac{1}{2}\right)$
33. Minima: $(0, 0)$ and $(\pi, 0)$
Maximum: $\left(3\pi/4, (\sqrt{2}/2)e^{3\pi/4}\right)$
35. Minimum: $\left(1/6, \sqrt{3}/2\right)$
Maximum: $(0, 1)$
37. Minimum: $(2, 3)$
Maximum: $\left(1, \sqrt{2} + 3\right)$
39. (a) Minimum: $(0, -3)$;
Maximum: $(2, 1)$
(b) Minimum: $(0, -3)$
(c) Maximum: $(2, 1)$
(d) No extrema
41.

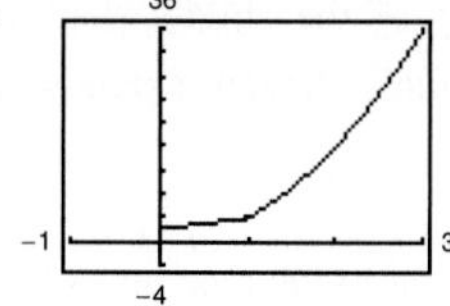

Minimum: $(0, 2)$
Maximum: $(3, 36)$

43.

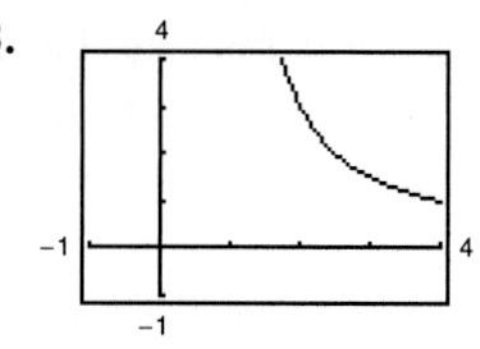

Minimum: $(4, 1)$

45.

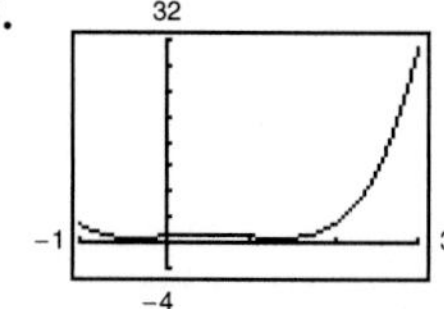

Minima: $\left((-\sqrt{3} + 1)/2, 3/4\right)$ and $\left((\sqrt{3} + 1)/2, 3/4\right)$
Maximum: $(3, 31)$

47. (a)

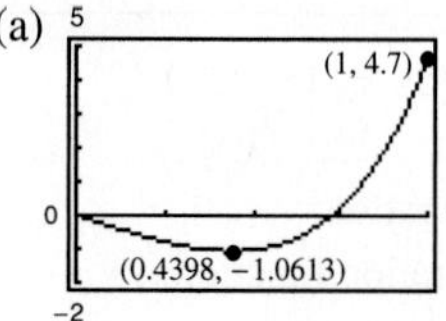

(b) Minimum: $(0.4398, -1.0613)$

49. (a) (b) Minimum: $(1.0863, -1.3972)$

51. (a) (b) Minimum: $(0.5327, -0.4657)$

53. Maximum: $\left|f''\left(\sqrt[3]{-10 + \sqrt{108}}\right)\right| = f''\left(\sqrt{3} - 1\right) \approx 1.47$

55. Maximum: $|f''(0)| = 1$ **57.** Maximum: $|f^{(4)}(0)| = \frac{56}{81}$
59. Because f is continuous on $[0, \pi/4]$, but not continuous on $[0, \pi]$.
61. Answers will vary. Example:

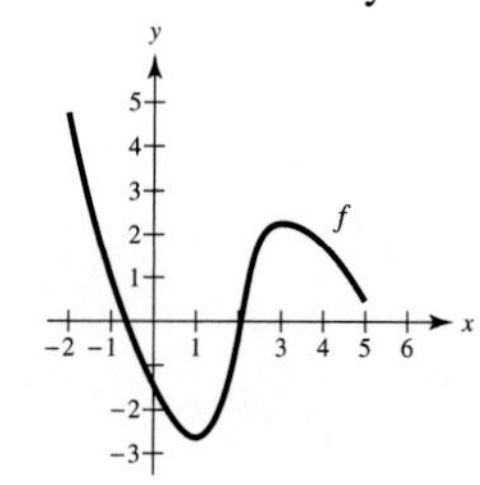

63. (a) Yes (b) No **65.** (a) No (b) Yes
67. $dx/dt = (v^2 \cos 2\theta/16)\,d\theta/dt$
In the interval $[\pi/4, 3\pi/4]$, $\theta = \pi/4, 3\pi/4$ indicate minima for $|dx/dt|$ and $\theta = \pi/2$ indicates a maximum for $|dx/dt|$. This implies that the sprinkler waters longest when $\theta = \pi/4$ and $3\pi/4$. So, the lawn farthest from the sprinkler gets the most water.
69. True **71.** True **73.** Proof
75. (a) $y = (3/40{,}000)x^2 - (3/200)x + 75/4$
(b)

x	-500	-400	-300	-200	-100	0
d	0	0.75	3	6.75	12	18.75

x	100	200	300	400	500
d	12	6.75	3	0.75	0

(c) Lowest point $\approx (100, 18)$; No

Section 4.2 (page 216)

1. $f(0) = f(2) = 0$; f is not differentiable on $(0, 2)$.
3. $f(-1) = f(1) = 1$; f is not continuous on $(-1, 1)$.
5. $(2, 0), (-1, 0); f'\left(\frac{1}{2}\right) = 0$ **7.** $(0, 0), (-4, 0); f'\left(-\frac{8}{3}\right) = 0$
9. $f'\left(-\frac{3}{2}\right) = 0$ **11.** $f'(1) = 0$
13. $f'\left((6 - \sqrt{3})/3\right) = 0$; $f'\left((6 + \sqrt{3})/3\right) = 0$
15. Not differentiable at $x = 0$ **17.** $f'\left(-2 + \sqrt{5}\right) = 0$
19. $f'\left(\sqrt{2}\right) = 0$ **21.** $f'(\pi/2) = 0$; $f'(3\pi/2) = 0$
23. $f'(0.249) \approx 0$ **25.** Not continuous on $[0, \pi]$
27.

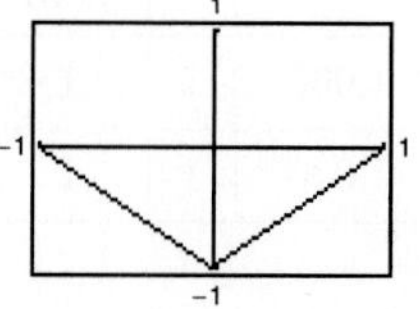

Rolle's Theorem does not apply.

29. $f'(\pm 0.1533) = 0$

31. Rolle's Theorem does not apply.

33. (a) $f(1) = f(2) = 64$
(b) Velocity $= 0$ for some t in $(1, 2)$; $t = \frac{3}{2}$ sec

35.

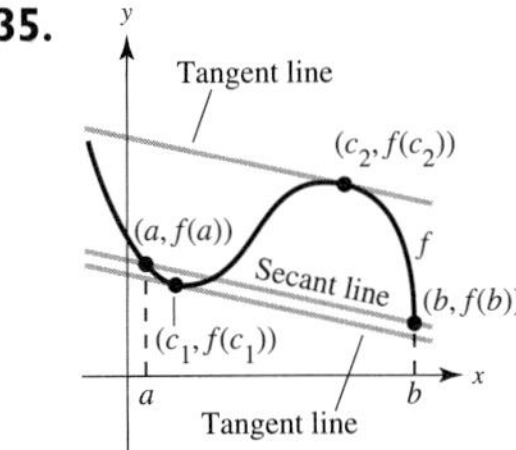

37. The function is not continuous on $[0, 6]$.

39. The function is not continuous on $[0, 6]$.

41. (a) Secant line: $x - y + 3 = 0$ (b) $c = \frac{1}{2}$
(c) Tangent line: $4x - 4y + 3 = 0$
(d)

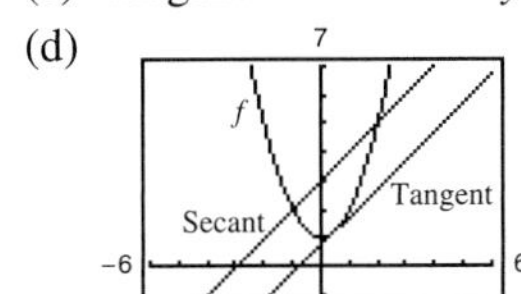

43. $f'\left(-\frac{1}{2}\right) = -1$ **45.** $f'\left(\frac{8}{27}\right) = 1$ **47.** $f'\left(-\frac{1}{4}\right) = -\frac{1}{3}$

49. $f'(\pi/2) = 0$ **51.** $f'(4e^{-1}) = 2$

53. Secant line: $2x - 3y - 2 = 0$
Tangent line: $c = \left(-2 + \sqrt{6}\right)/2,\ 2x - 3y + 5 - 2\sqrt{6} = 0$

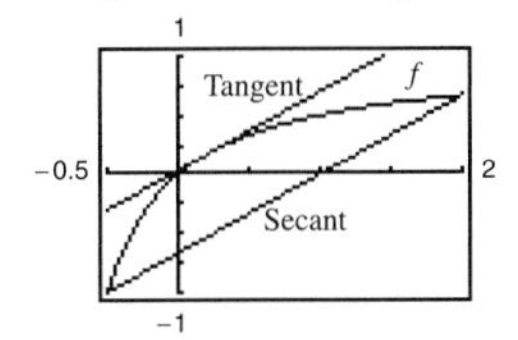

55. Secant line: $x - 4y + 3 = 0$
Tangent line: $c = 4,\quad x - 4y + 4 = 0$

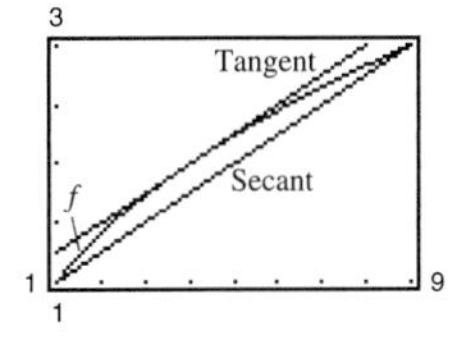

57. Secant line: $x + y - 2 = 0$
Tangent line: $c \approx 1.0161,\ x + y - 2.8161 = 0$

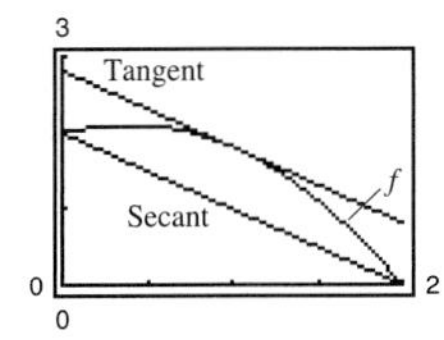

59. No. Let $f(x) = x^2$ on $[-1, 2]$.

61. No. $f(x)$ is not continuous on $[0, 1]$. So it does not satisfy the hypothesis of Rolle's Theorem.

63. By the Mean Value Theorem, there is a time when the speed of the plane must equal the average speed of 454.5 mph. The speed was 400 mph when the plane was accelerating to 454.5 mph and decelerating from 454.5 mph.

65. Proof

67. (a) f is continuous and changes sign in $[-10, 4]$ (Intermediate Value Theorem).
(b) There exist real numbers a and b such that $-10 < a < b < 4$ and $f(a) = f(b) = 2$. Therefore, f' has a zero in the interval by Rolle's Theorem.
(c) (d)

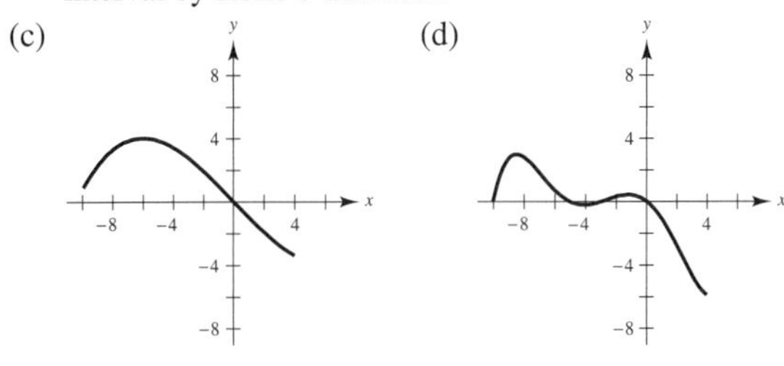

(e) No, by Theorem 3.1.

69.

71. Proof

73. $a = 6, b = 1, c = 2$ **75.** $f(x) = 5$ **77.** $f(x) = x^2 - 1$

79. False. f is not continuous on $[-1, 1]$. **81.** True

83–91. Proofs

Section 4.3 (page 226)

1. (a) $(0, 6)$ (b) $(6, 8)$

3. Increasing on $(3, \infty)$; Decreasing on $(-\infty, 3)$

5. Increasing on $(-\infty, -2)$ and $(2, \infty)$; Decreasing on $(-2, 2)$

7. Increasing on $(0, \pi/2)$ and $(3\pi/2, 2\pi)$;
Decreasing on $(\pi/2, 3\pi/2)$

9. Increasing on $(-\infty, 0)$; Decreasing on $(0, \infty)$

11. Increasing on $(1, \infty)$; Decreasing on $(-\infty, 1)$

13. Increasing on $\left(-2\sqrt{2}, 2\sqrt{2}\right)$;
Decreasing on $\left(-4, -2\sqrt{2}\right), \left(2\sqrt{2}, 4\right)$

15. Increasing on $(0, 7\pi/6)$ and $(11\pi/6, 2\pi)$;
Decreasing on $(7\pi/6, 11\pi/6)$

17. Critical number: $x = 3$
Increasing on $(3, \infty)$
Decreasing on $(-\infty, 3)$
Relative minimum: $(3, -9)$

19. Critical number: $x = 1$
Increasing on $(-\infty, 1)$
Decreasing on $(1, \infty)$
Relative maximum: $(1, 5)$

21. Critical numbers: $x = -2, 1$
Increasing on $(-\infty, -2)$ and $(1, \infty)$
Decreasing on $(-2, 1)$
Relative maximum: $(-2, 20)$
Relative minimum: $(1, -7)$

23. Critical numbers: $x = 0, 2$
Increasing on $(0, 2)$
Decreasing on $(-\infty, 0), (2, \infty)$
Relative maximum: $(2, 4)$
Relative minimum: $(0, 0)$

25. Critical numbers: $x = -1, 1$
Increasing on $(-\infty, -1)$ and $(1, \infty)$
Decreasing on $(-1, 1)$
Relative maximum: $\left(-1, \frac{4}{5}\right)$
Relative minimum: $\left(1, -\frac{4}{5}\right)$

27. Critical number: $x = 0$
Increasing on $(-\infty, \infty)$
No relative extrema

29. Critical number: $x = 1$
Increasing on $(1, \infty)$
Decreasing on $(-\infty, 1)$
Relative minimum: $(1, 0)$

31. Critical number: $x = 5$
Increasing on $(-\infty, 5)$
Decreasing on $(5, \infty)$
Relative maximum: $(5, 5)$

33. Critical numbers: $x = -1, 1$
Discontinuity: $x = 0$
Increasing on $(-\infty, -1)$ and $(1, \infty)$
Decreasing on $(-1, 0)$ and $(0, 1)$
Relative maximum: $(-1, -2)$
Relative minimum: $(1, 2)$

35. Critical number: $x = 0$
Discontinuities: $x = -3, 3$
Increasing on $(-\infty, -3)$ and $(-3, 0)$
Decreasing on $(0, 3)$ and $(3, \infty)$
Relative maximum: $(0, 0)$

37. Critical numbers: $x = -3, 1$
Discontinuity: $x = -1$
Increasing on $(-\infty, -3)$ and $(1, \infty)$
Decreasing on $(-3, -1)$ and $(-1, 1)$
Relative maximum: $(-3, -8)$
Relative minimum: $(1, 0)$

39. Critical number: $x = 2$
Increasing on $(-\infty, 2)$
Decreasing on $(2, \infty)$
Relative maximum: $(2, e^{-1})$

41. Critical number: $x = 0$
Decreasing on $[-1, 1]$

43. Critical numbers: $x = 1/\ln 3$
Increasing on $(-\infty, 1/\ln 3)$
Decreasing on $(1/\ln 3, \infty)$
Relative maximum: $(1/\ln 3, (3^{-1/\ln 3})/\ln 3)$ or $(1/\ln 3, 1/(e \ln 3))$

45. Critical number: $x = 1/\ln 4$
Increasing on $(1/\ln 4, \infty)$
Decreasing on $(0, 1/\ln 4)$
Relative minimum: $(1/\ln 4, (\ln(\ln 4) + 1)/\ln 4)$

47. (a) Critical numbers: $x = \pi/6, 5\pi/6$
Increasing on $(0, \pi/6), (5\pi/6, 2\pi)$
Decreasing on $(\pi/6, 5\pi/6)$
(b) Relative maximum: $\left(\pi/6, \left(\pi + 6\sqrt{3}\right)/12\right)$
Relative minimum: $\left(5\pi/6, \left(5\pi - 6\sqrt{3}\right)/12\right)$

49. (a) Critical numbers: $x = \pi/4, 5\pi/4$
Increasing on $(0, \pi/4), (5\pi/4, 2\pi)$
Decreasing on $(\pi/4, 5\pi/4)$
(b) Relative maximum: $\left(\pi/4, \sqrt{2}\right)$
Relative minimum: $\left(5\pi/4, -\sqrt{2}\right)$

51. (a) Critical numbers:
$x = \pi/4, \pi/2, 3\pi/4, \pi, 5\pi/4, 3\pi/2, 7\pi/4$
Increasing on $(\pi/4, \pi/2), (3\pi/4, \pi), (5\pi/4, 3\pi/2), (7\pi/4, 2\pi)$
Decreasing on $(0, \pi/4), (\pi/2, 3\pi/4), (\pi, 5\pi/4), (3\pi/2, 7\pi/4)$
(b) Relative maxima: $(\pi/2, 1), (\pi, 1), (3\pi/2, 1)$
Relative minima: $(\pi/4, 0), (3\pi/4, 0), (5\pi/4, 0), (7\pi/4, 0)$

53. (a) Critical numbers: $\pi/2, 7\pi/6, 3\pi/2, 11\pi/6$
Increasing on $(0, \pi/2), (7\pi/6, 3\pi/2), (11\pi/6, 2\pi)$
Decreasing on $(\pi/2, 7\pi/6), (3\pi/2, 11\pi/6)$
(b) Relative maxima: $(\pi/2, 2), (3\pi/2, 0)$
Relative minima: $(7\pi/6, -1/4), (11\pi/6, -1/4)$

55. (a) $f'(x) = (2(9 - 2x^2))/\sqrt{9 - x^2}$
(b)

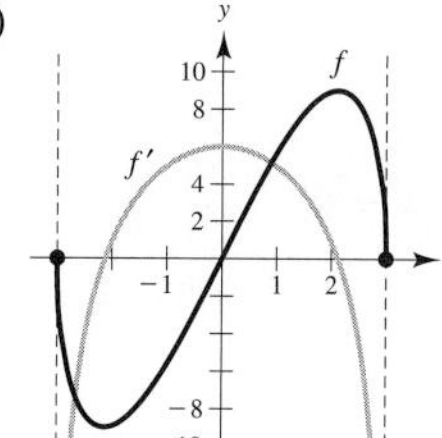

(c) $x = \pm 3\sqrt{2}/2$
(d) $f' > 0$ on $\left(-3\sqrt{2}/2, 3\sqrt{2}/2\right)$
$f' < 0$ on $\left(-3, -3\sqrt{2}/2\right), \left(3\sqrt{2}/2, 3\right)$

57. (a) $f'(t) = t(t \cos t + 2 \sin t)$
(b)

(c) Critical numbers: $t = 2.2889, 5.0870$
(d) $f' > 0$ on $(0, 2.2889), (5.0870, 2\pi)$
$f' < 0$ on $(2.2889, 5.0870)$

59. (a) $f'(x) = (2x^2 - 1)/2x$ (c) Critical numbers: $x = \sqrt{2}/2$
(b) (d) $f' > 0$ on $(\sqrt{2}/2, 3)$
$f' < 0$ on $(0, \sqrt{2}/2)$

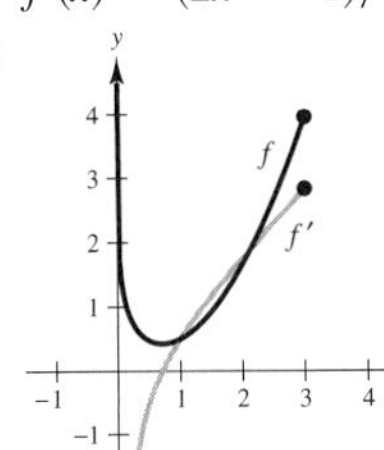

61. $f(x)$ is symmetric with respect to the origin.
Zeros: $(0, 0), (\pm\sqrt{3}, 0)$

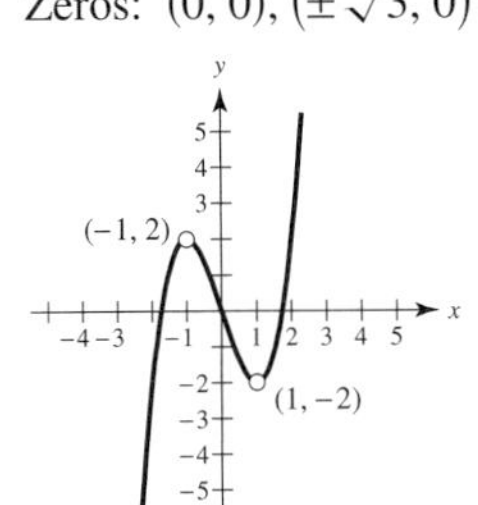

$g(x)$ is continuous on $(-\infty, \infty)$ and $f(x)$ has holes at $x = 1$ and $x = -1$.

63.

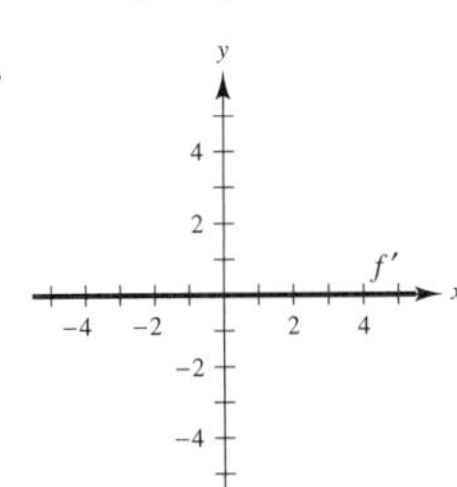

65.

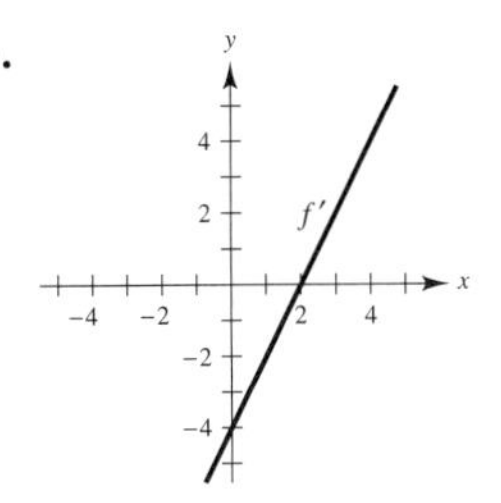

67.

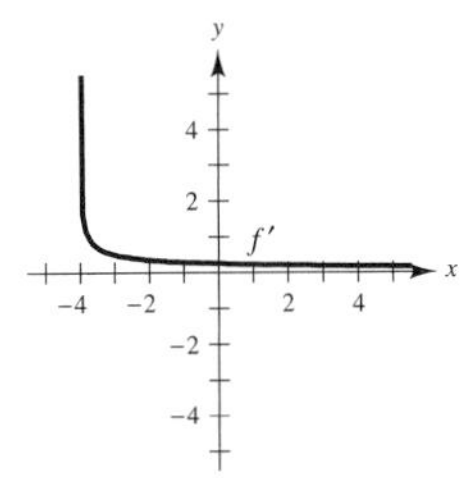

69. (a) Increasing on $(2, \infty)$; Decreasing on $(-\infty, 2)$
(b) Relative minimum: $x = 2$

71. (a) Increasing on $(-\infty, 0)$ and $(1, \infty)$; Decreasing on $(0, 1)$
(b) Relative maximum: $x = 0$; Relative minimum: $x = 1$

73. $g'(0) < 0$ **75.** $g'(-6) < 0$ **77.** $g'(0) > 0$

79.

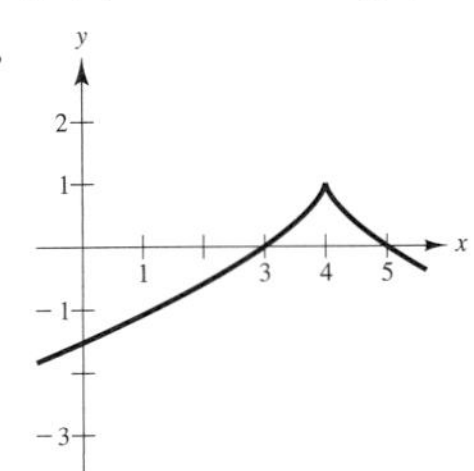

81.

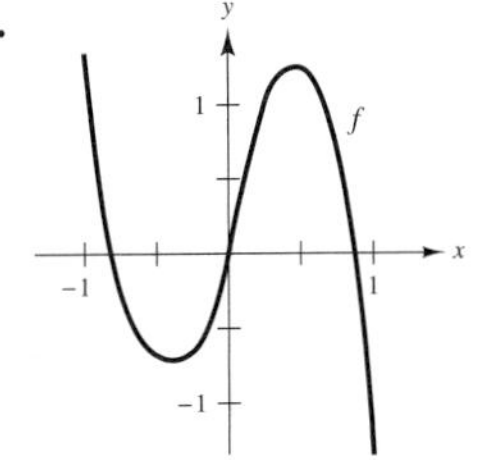

Relative minimum at the approximate critical number $x = -0.40$; Relative maximum at the approximate critical number $x = 0.48$

83. (a) $s'(t) = 9.8(\sin\theta)t$; speed $= |9.8(\sin\theta)t|$
(b)

θ	0	$\pi/4$	$\pi/3$	$\pi/2$	$2\pi/3$	$3\pi/4$	π
$s'(t)$	0	$4.9\sqrt{2}t$	$4.9\sqrt{3}t$	$9.8t$	$4.9\sqrt{3}t$	$4.9\sqrt{2}t$	0

The speed is maximum at $\theta = \pi/2$.

85. (a)

x	0.5	1	1.5	2	2.5	3
$f(x)$	0.5	1	1.5	2	2.5	3
$g(x)$	0.48	0.84	1.00	0.91	0.60	0.14

$f(x) > g(x)$

(b) (c) Proof

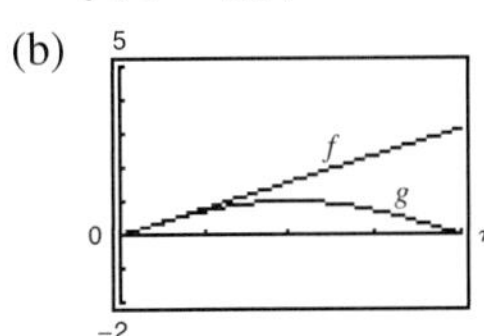

$f(x) > g(x)$

87. $r = 2R/3$

89. (a) $M = 5.267t^2 - 71.19t + 356.9$
(b) (c) $(6.8, 116.3)$

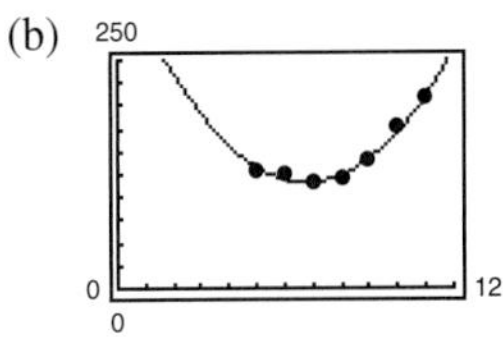

91. (a) $v(t) = 6 - 2t$ (b) $[0, 3)$ (c) $(3, \infty)$ (d) $t = 3$

93. (a) $v(t) = 3t^2 - 10t + 4$
(b) $[0, (5 - \sqrt{13})/3]$ and $[(5 + \sqrt{13})/3, \infty]$
(c) $[(5 - \sqrt{13})/3, (5 + \sqrt{13})/3]$
(d) $t = (5 \pm \sqrt{13})/3$

95. Answers will vary.

97. (a) 3
(b) $a_3(0)^3 + a_2(0)^2 + a_1(0) + a_0 = 0$
$a_3(2)^3 + a_2(2)^2 + a_1(2) + a_0 = 2$
$3a_3(0)^2 + 2a_2(0) + a_1 = 0$
$3a_3(2)^2 + 2a_2(2) + a_1 = 0$
(c) $f(x) = -\frac{1}{2}x^3 + \frac{3}{2}x^2$

99. (a) 4
(b) $a_4(0)^4 + a_3(0)^3 + a_2(0)^2 + a_1(0) + a_0 = 0$
$a_4(2)^4 + a_3(2)^3 + a_2(2)^2 + a_1(2) + a_0 = 4$
$a_4(4)^4 + a_3(4)^3 + a_2(4)^2 + a_1(4) + a_0 = 0$
$4a_4(0)^3 + 3a_3(0)^2 + 2a_2(0) + a_1 = 0$
$4a_4(2)^3 + 3a_3(2)^2 + 2a_2(2) + a_1 = 0$
(c) $f(x) = \frac{1}{4}x^4 - 2x^3 + 4x^2$

101. True **103.** False. Let $f(x) = x^3$.

105. False. Let $f(x) = x^3$. There is a critical number at $x = 0$, but not a relative extremum.

107–111. Proofs

Section 4.4 (page 235)

1. Concave upward: $(-\infty, \infty)$

3. Concave upward: $(-\infty, -2), (2, \infty)$
Concave downward: $(-2, 2)$

5. Concave upward: $(-\infty, -1), (1, \infty)$
Concave downward: $(-1, 1)$

7. Concave upward: $(-\infty, 1)$
Concave downward: $(1, \infty)$

9. Concave upward: $(-\pi/2, 0)$
Concave downward: $(0, \pi/2)$

11. Point of inflection: $(2, 8)$
Concave downward: $(-\infty, 2)$
Concave upward: $(2, \infty)$

13. Points of inflection: $\left(\pm 2/\sqrt{3}, -20/9\right)$
Concave upward: $\left(-\infty, -2/\sqrt{3}\right), \left(2/\sqrt{3}, \infty\right)$
Concave downward: $\left(-2/\sqrt{3}, 2/\sqrt{3}\right)$

15. Points of inflection: $(2, -16), (4, 0)$
Concave upward: $(-\infty, 2), (4, \infty)$
Concave downward: $(2, 4)$

17. Concave upward: $(-3, \infty)$

19. Points of inflection: $\left(-\sqrt{3}, -\sqrt{3}/4\right), (0, 0), \left(\sqrt{3}, \sqrt{3}/4\right)$
Concave upward: $\left(-\sqrt{3}, 0\right), \left(\sqrt{3}, \infty\right)$
Concave downward: $\left(-\infty, -\sqrt{3}\right), \left(0, \sqrt{3}\right)$

21. Point of inflection: $(2\pi, 0)$
Concave upward: $(2\pi, 4\pi)$
Concave downward: $(0, 2\pi)$

23. Concave upward: $(0, \pi), (2\pi, 3\pi)$
Concave downward: $(\pi, 2\pi), (3\pi, 4\pi)$

25. Points of inflection: $(\pi, 0), (1.823, 1.452), (4.46, -1.452)$
Concave upward: $(1.823, \pi), (4.46, 2\pi)$
Concave downward: $(0, 1.823), (\pi, 4.46)$

27. Concave upward: $(0, \infty)$

29. Relative minimum: $(3, -25)$

31. Relative minimum: $(5, 0)$

33. Relative maximum: $(0, 3)$
Relative minimum: $(2, -1)$

35. Relative maximum: $(2.4, 268.74)$
Relative minimum: $(0, 0)$

37. Relative minimum: $(0, -3)$

39. Relative maximum: $(-2, -4)$
Relative minimum: $(2, 4)$

41. No relative extrema, because f is nonincreasing.

43. Relative minimum: $\left(1, \frac{1}{2}\right)$

45. Relative minimum: (e, e) **47.** Relative minimum: $(0, 1)$

49. Relative minimum: $(0, 0)$
Relative maximum: $(2, 4e^{-2})$

51. Relative maximum: $(1/\ln 4, 4e^{-1}/\ln 2)$

53. Relative minimum: $(-1.272, 3.747)$
Relative maximum: $(1.272, -0.606)$

55. (a) $f'(x) = 0.2x(x - 3)^2(5x - 6)$
$f''(x) = 0.4(x - 3)(10x^2 - 24x + 9)$
(b) Relative maximum: $(0, 0)$
Relative minimum: $(1.2, -1.6796)$
Points of inflection: $(0.4652, -0.7048)$, $(1.9348, -0.9048)$, $(3, 0)$
(c)

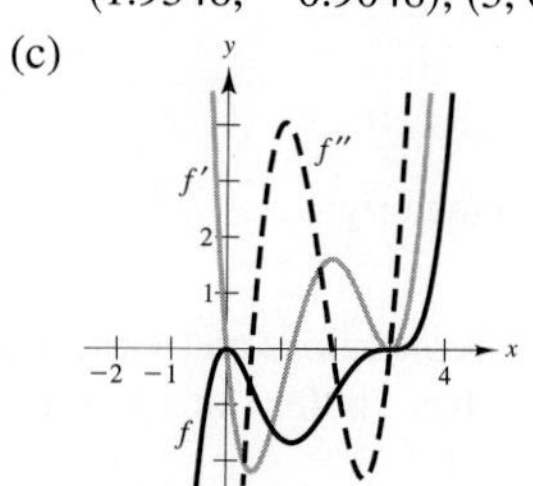

f is increasing when f' is positive, and decreasing when f' is negative. f is concave upward when f'' is positive, and concave downward when f'' is negative.

57. (a) $f'(x) = \cos x - \cos 3x + \cos 5x$
$f''(x) = -\sin x + 3 \sin 3x - 5 \sin 5x$
(b) Relative maximum: $(\pi/2, 1.53333)$
Points of inflection: $(0.5236, 0.2667)$, $(1.1731, 0.9637)$, $(1.9685, 0.9637)$, $(2.6180, 0.2667)$
(c)

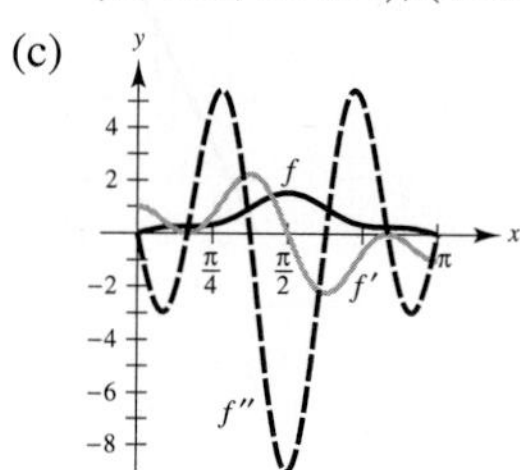

f is increasing when f' is positive, and decreasing when f' is negative. f is concave upward when f'' is positive, and concave downward when f'' is negative.

59. (a)

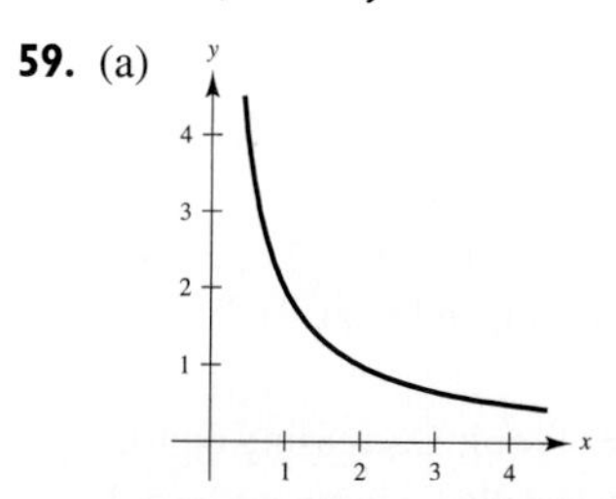

(b)

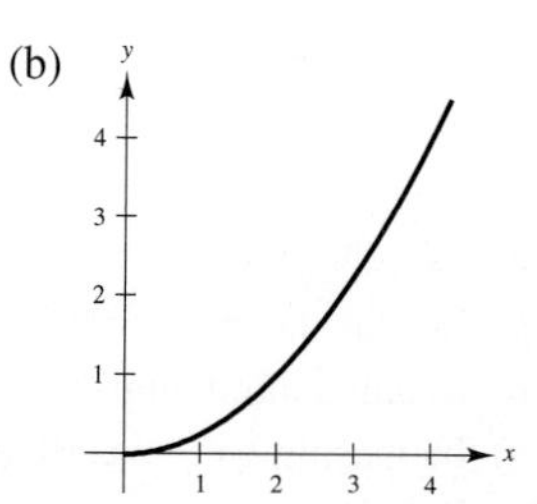

61. Answers will vary. Example: $f(x) = x^4$; $f''(0) = 0$, but $(0, 0)$ is not a point of inflection.

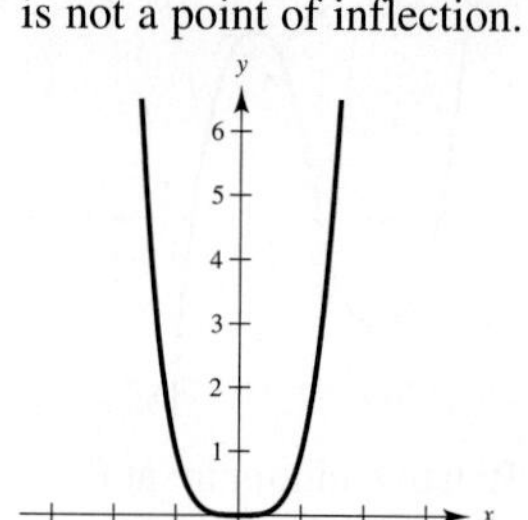

63.

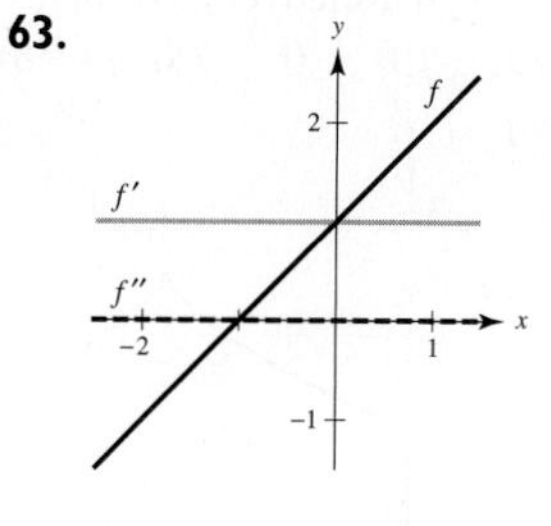

65.

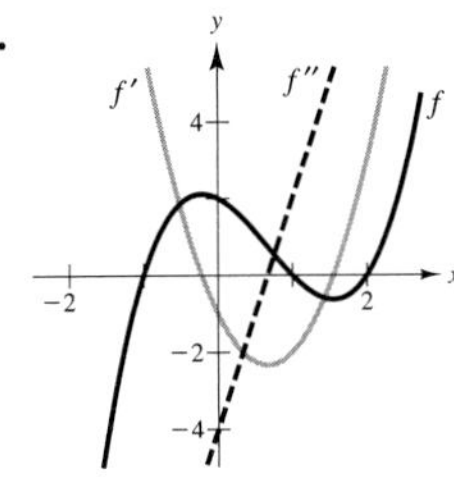

67.

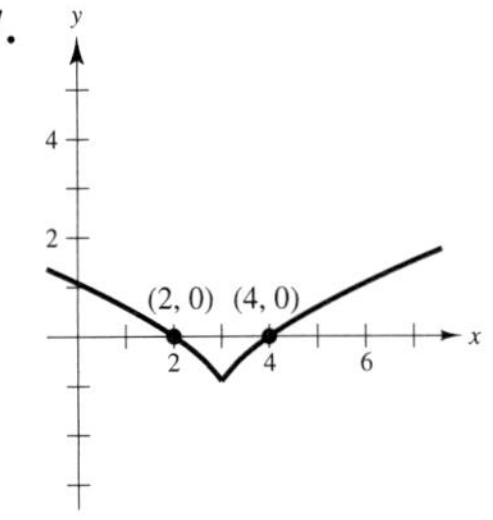

69.

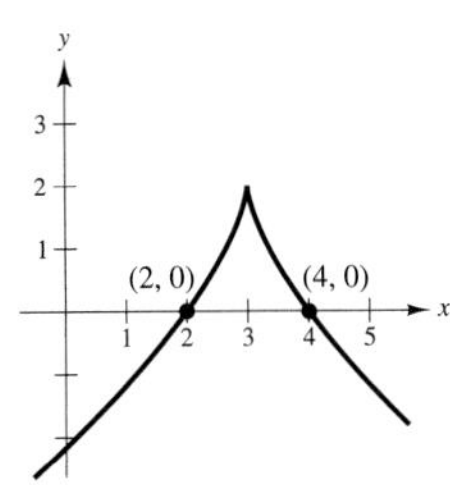

71. (a) $f(x) = (x - 2)^n$ has a point of inflection at $(2, 0)$ if n is odd and $n \geq 3$.

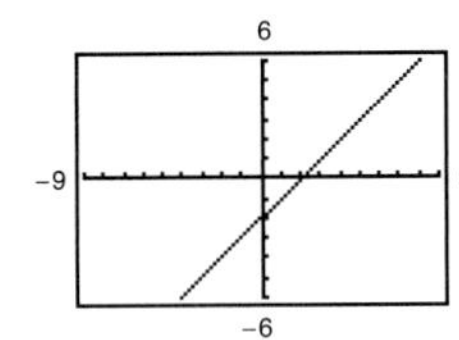

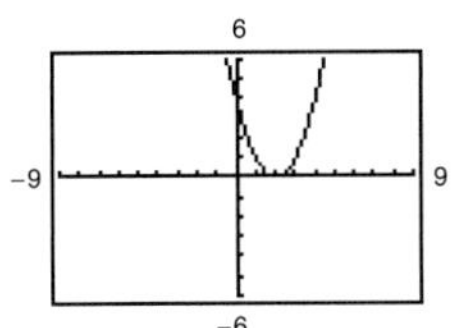

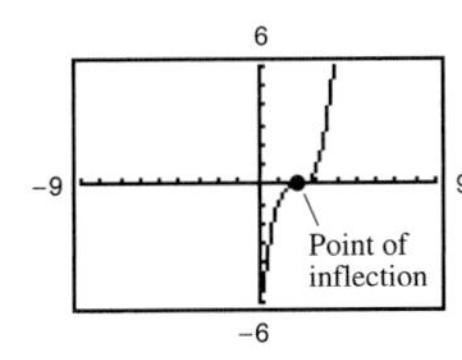

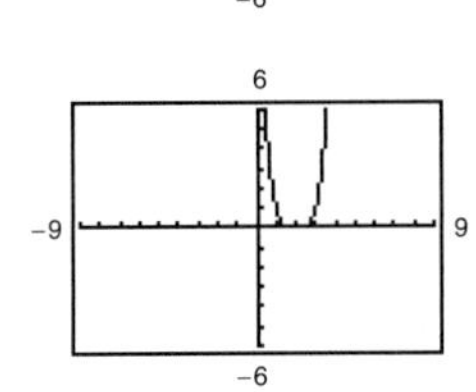

(b) Proof

73. $f(x) = \frac{1}{2}x^3 - 6x^2 + \frac{45}{2}x - 24$

75. (a) $f(x) = \frac{1}{32}x^3 + \frac{3}{16}x^2$ (b) 2 miles from touchdown

77. $x = \left[(15 - \sqrt{33})/16\right]L \approx 0.578L$ **79.** $x = 100$ units

81. $P_1(x) = 2\sqrt{2}$

$P_2(x) = 2\sqrt{2} - \sqrt{2}[x - (\pi/4)]^2$

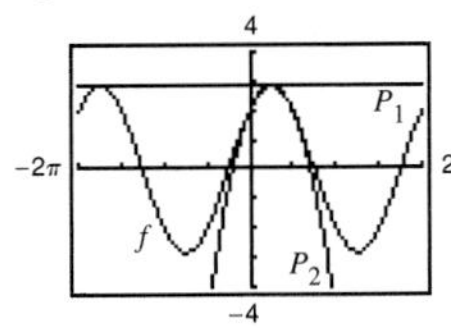

The values of f, P_1, and P_2 and their first derivatives are equal when $x = \pi/4$. The approximations worsen as you move away from $x = \pi/4$.

83. $P_1(x) = -\pi/4 + 1/2(x + 1)$

$P_2(x) = -\pi/4 + 1/2(x + 1) + 1/4(x + 1)^2$

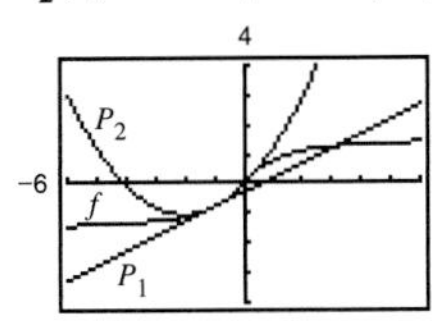

The values of f, P_1, and P_2 and their first derivatives are equal when $x = -1$. The approximations worsen as you move away from $x = -1$.

85.

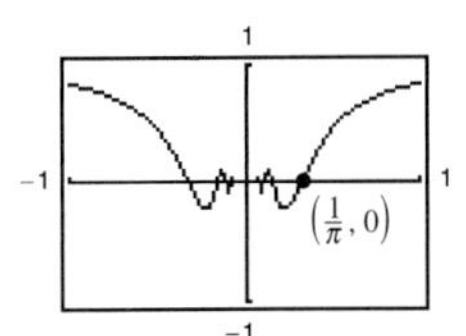

87. Proof **89.** True

91. False. The maximum value is $\sqrt{13} \approx 3.60555$.

93. False. f is concave upward at $x = c$ if $f''(c) > 0$. **95.** Proof

Section 4.5 (page 245)

1. As x becomes large, $f(x)$ approaches 4.

3. f **4.** c **5.** d **6.** a **7.** b **8.** e

9.

x	10^0	10^1	10^2	10^3
$f(x)$	7	2.2632	2.0251	2.0025

x	10^4	10^5	10^6
$f(x)$	2.0003	2.0000	2.0000

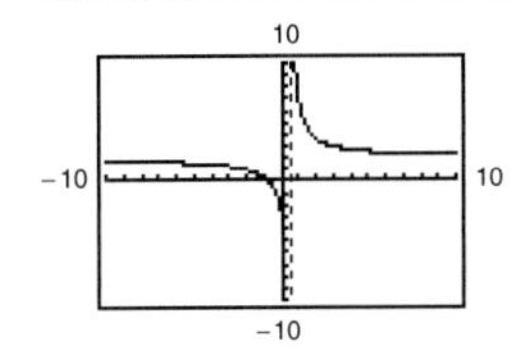

$$\lim_{x\to\infty} \frac{4x + 3}{2x - 1} = 2$$

11.

x	10^0	10^1	10^2	10^3
$f(x)$	−2	−2.9814	−2.9998	−3.0000

x	10^4	10^5	10^6
$f(x)$	−3.0000	−3.0000	−3.0000

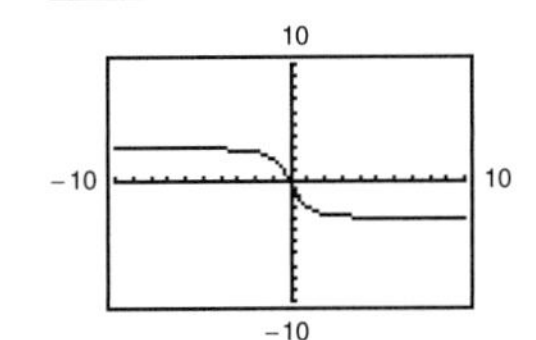

$$\lim_{x\to\infty} \frac{-6x}{\sqrt{4x^2 + 5}} = -3$$

13.

x	10^0	10^1	10^2	10^3
$f(x)$	4.5000	4.9901	4.9999	5.0000

x	10^4	10^5	10^6
$f(x)$	5.0000	5.0000	5.0000

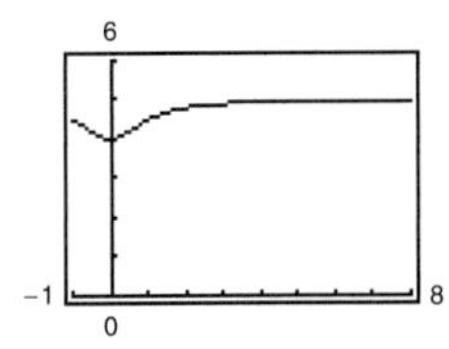

$$\lim_{x\to\infty} \left(5 - \frac{1}{x^2 + 1}\right) = 5$$

15. (a) ∞ (b) 5 (c) 0 **17.** (a) 0 (b) 1 (c) ∞

19. (a) 0 (b) $-\frac{2}{3}$ (c) $-\infty$ **21.** $\frac{2}{3}$ **23.** 0

25. $-\infty$ **27.** -1 **29.** -2 **31.** 0 **33.** 0

35. 2 **37.** 3 **39.** 0 **41.** $-\pi/2$

43. 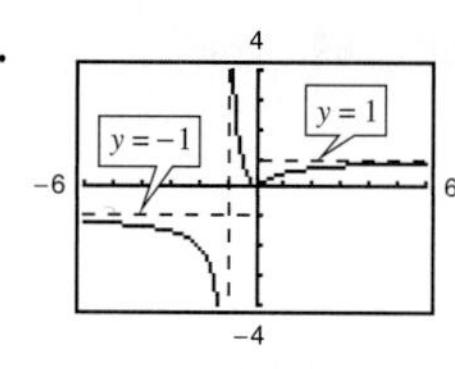

45. 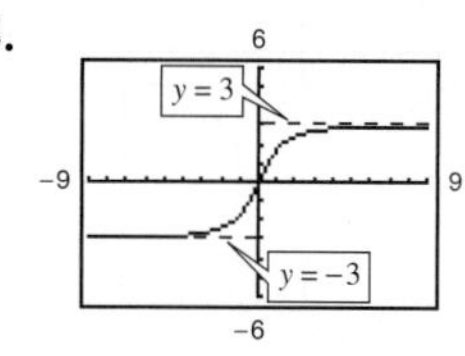

47. 1 **49.** 0 **51.** $-\frac{1}{2}$ **53.** $\frac{1}{8}$

55.

x	10^0	10^1	10^2	10^3	10^4	10^5	10^6
$f(x)$	1.000	0.513	0.501	0.500	0.500	0.500	0.500

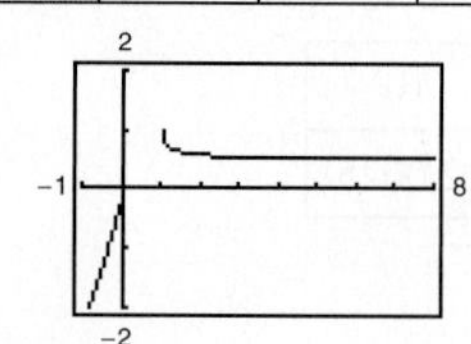

$\lim_{x\to\infty}\left[x - \sqrt{x(x-1)}\right] = \frac{1}{2}$

57.

x	10^0	10^1	10^2	10^3	10^4	10^5	10^6
$f(x)$	0.479	0.500	0.500	0.500	0.500	0.500	0.500

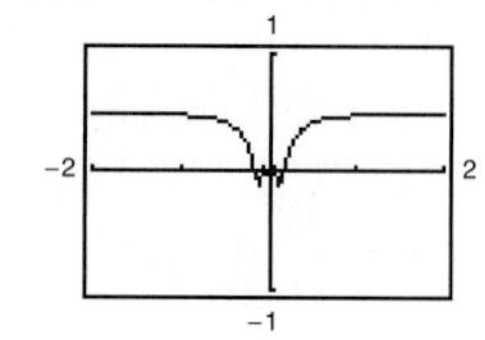

The graph has a hole at $x = 0$.

$\lim_{x\to\infty} x \sin \frac{1}{2x} = \frac{1}{2}$

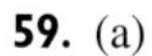

59. (a) 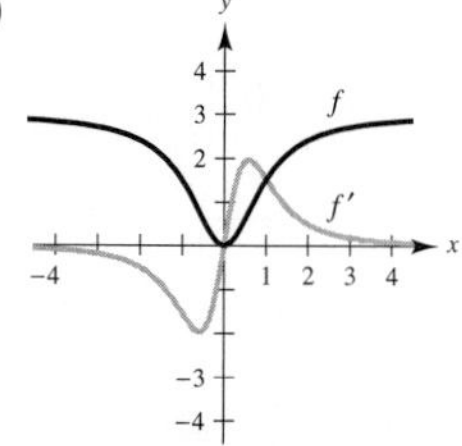

(b) $\lim_{x\to\infty} f(x) = 3,\ \lim_{x\to\infty} f'(x) = 0$

(c) $y = 3$ is a horizontal asymptote. The rate of increase of the function approaches 0 as the graph approaches $y = 3$.

61. Yes. For example, let $f(x) = \dfrac{6|x-2|}{\sqrt{(x-2)^2+1}}$.

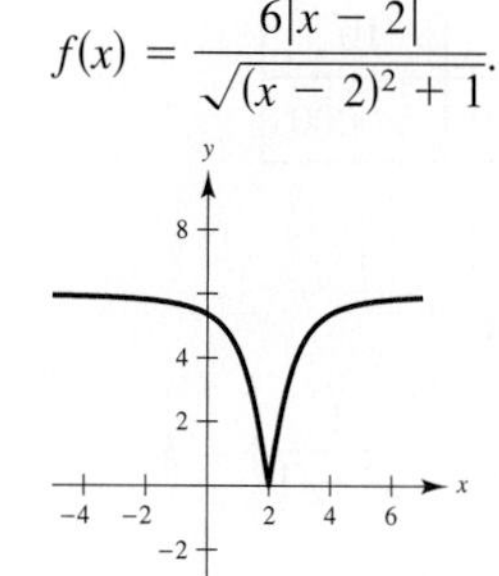

63. 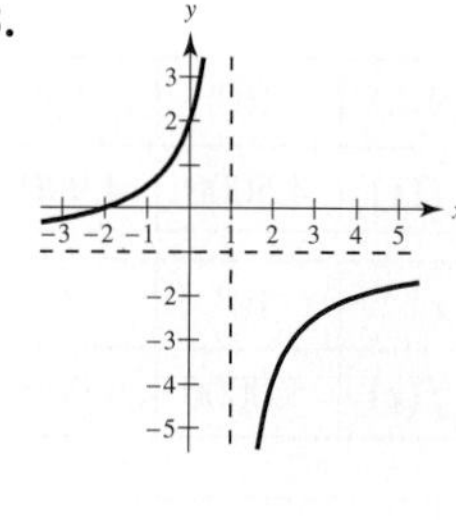

65. 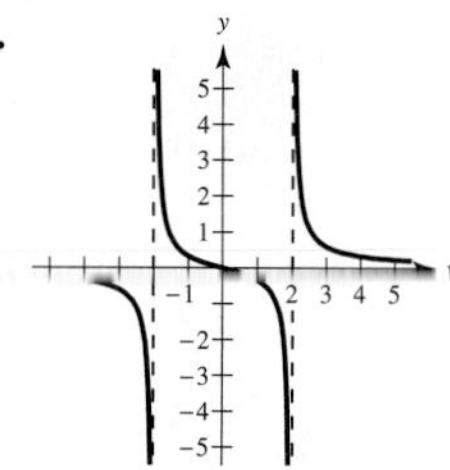

67. 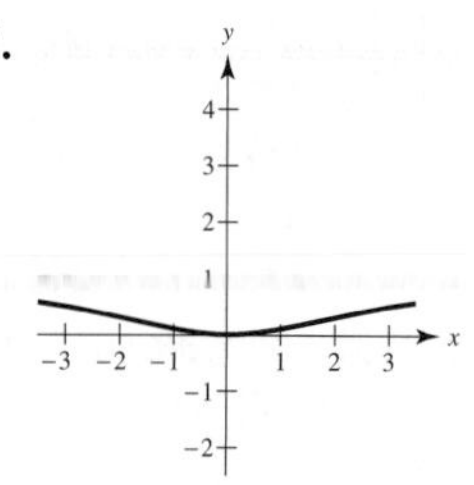

69. 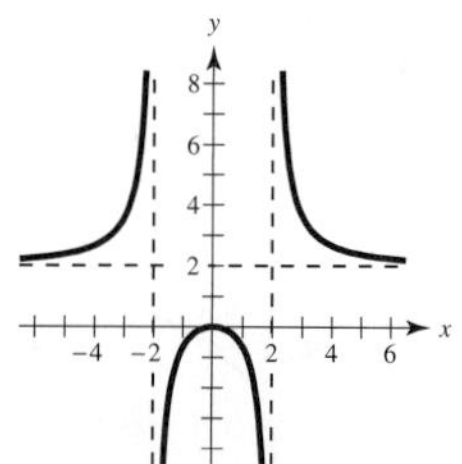

71. 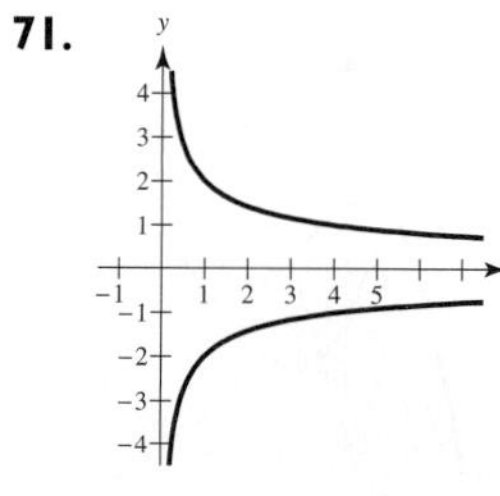

73. 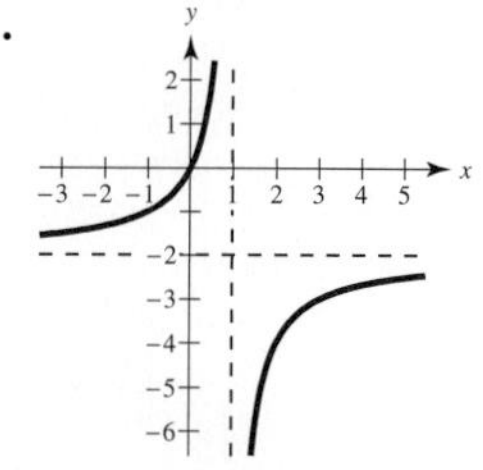

75. 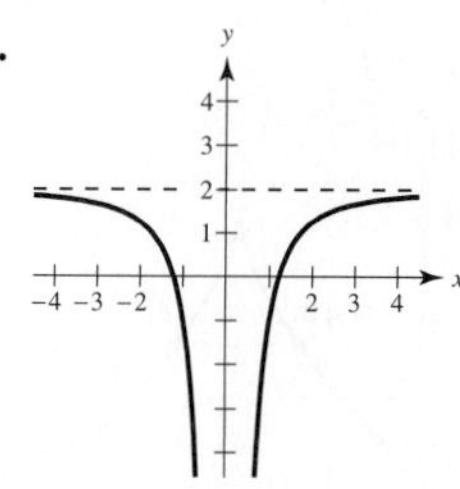

77. 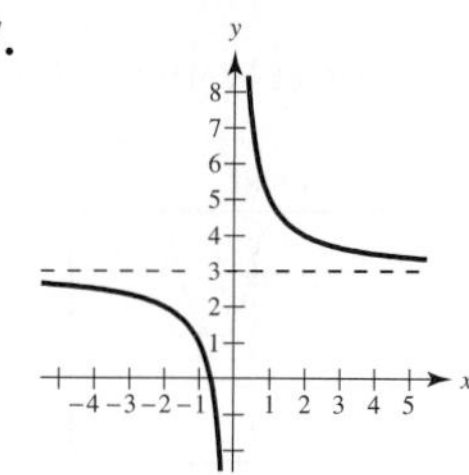

79. 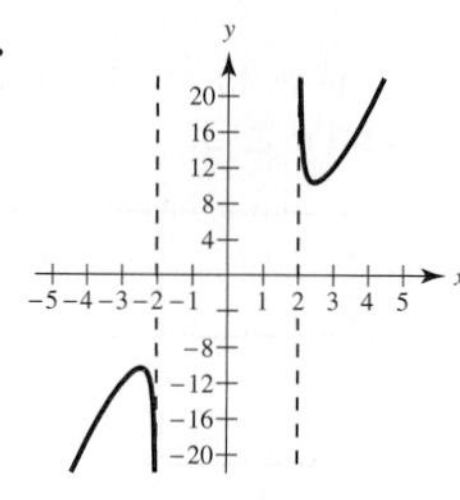

81. 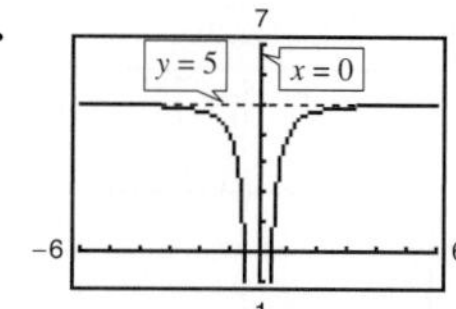

83. 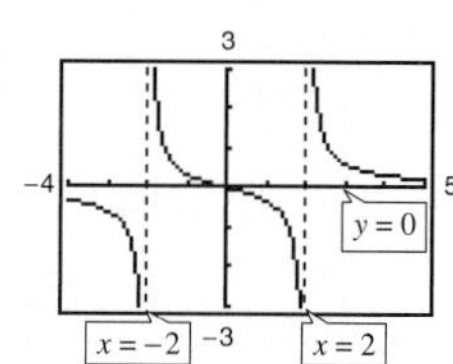

85. 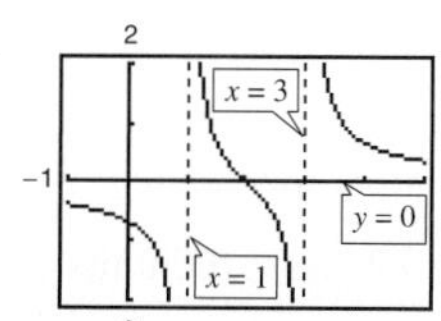

87. 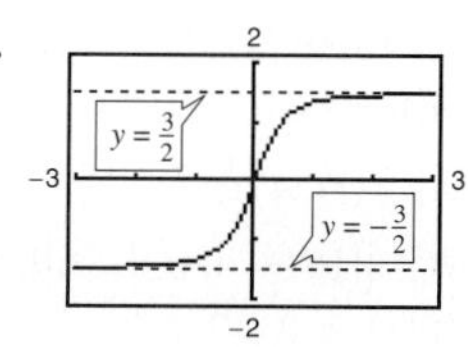

89. 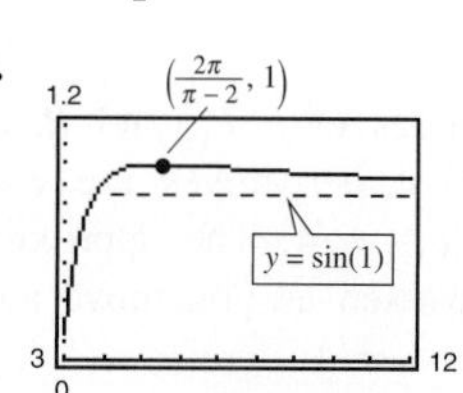

91. 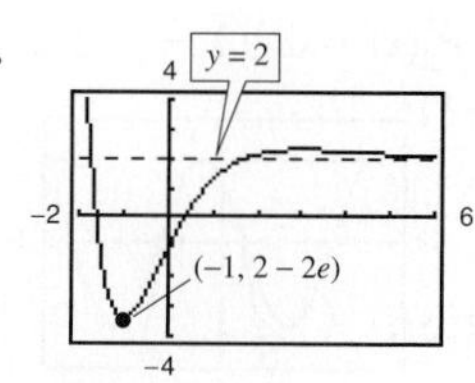

93. (a) 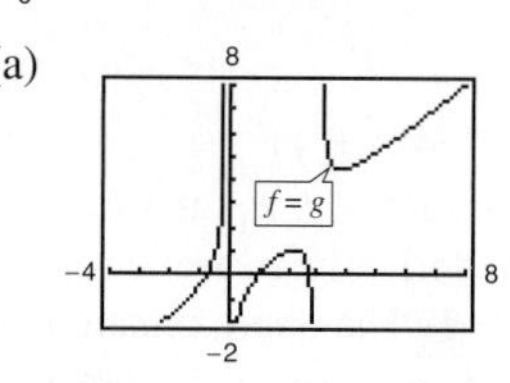

(b) Proof

(c) 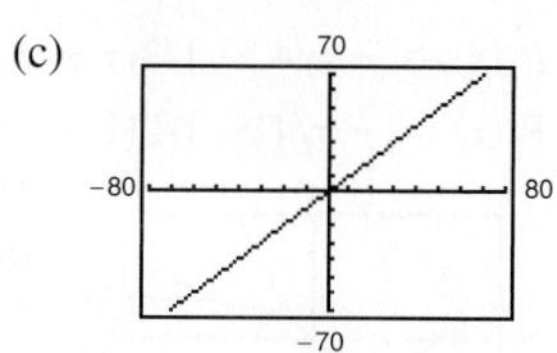

The slant asymptote $y = x$

95. $\frac{1}{2}$ **97.** $\lim_{t\to\infty} N(t) = +\infty$; $\lim_{t\to\infty} E(t) = c$

99. (a) $T_1 = -0.003t^2 + 0.68t + 26.6$

(b) (c)

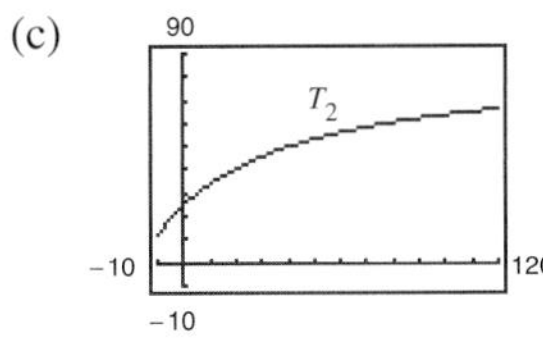

(d) $T_1(0) \approx 26.6°$, $T_2(0) \approx 25.0°$ (e) 86

(f) The limiting temperature is 86°.
No. T_1 has no horizontal asymptote.

101. (a) 7.1 million ft^3/acre

(b) $V'(20) \approx 0.077$
$V'(60) \approx 0.043$

103. (a)

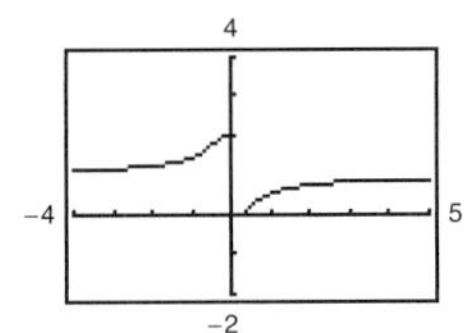

(b) Answers will vary.

105. (a) $d(m) = \dfrac{|3m + 3|}{\sqrt{m^2 + 1}}$

(b)

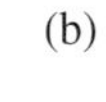

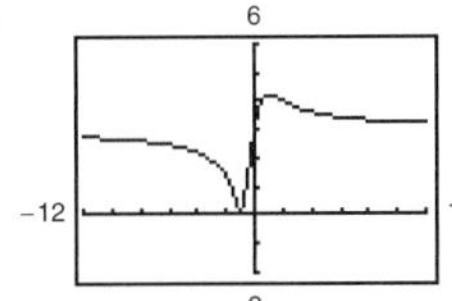

(c) $\lim_{m\to\infty} d(m) = 3$

$\lim_{m\to-\infty} d(m) = 3$

As m approaches $\pm\infty$, distance approaches 3.

107. (a) $\lim_{x\to\infty} f(x) = 2$ (b) $x_1 = \sqrt{\dfrac{4 - 2\varepsilon}{\varepsilon}}$, $x_2 = -\sqrt{\dfrac{4 - 2\varepsilon}{\varepsilon}}$

(c) $\sqrt{\dfrac{4 - 2\varepsilon}{\varepsilon}}$ (d) $-\sqrt{\dfrac{4 - 2\varepsilon}{\varepsilon}}$

109. (a) Answers will vary. $M = \dfrac{5\sqrt{33}}{11}$

(b) Answers will vary. $M = \dfrac{29\sqrt{177}}{59}$

111–115. Proofs

117. False. Let $f(x) = \dfrac{2x}{\sqrt{x^2 + 2}}$. $f'(x) > 0$ for all real numbers.

Section 4.6 (page 255)

1. d **2.** c **3.** a **4.** b

5. (a) $f'(x) = 0$ for $x = \pm 2$; $f'(x) > 0$ for $(-\infty, -2), (2, \infty)$
$f'(x) < 0$ for $(-2, 2)$

(b) $f''(x) = 0$ for $x = 0$; $f''(x) > 0$ for $(0, \infty)$
$f''(x) < 0$ for $(-\infty, 0)$

(c) $(0, \infty)$

(d) f' is minimum for $x = 0$.
f is decreasing at the fastest rate.

7.

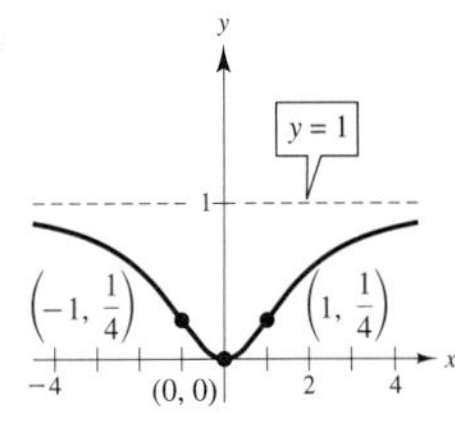

9.

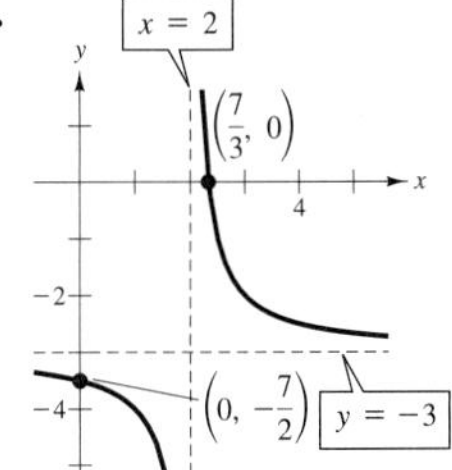

11.

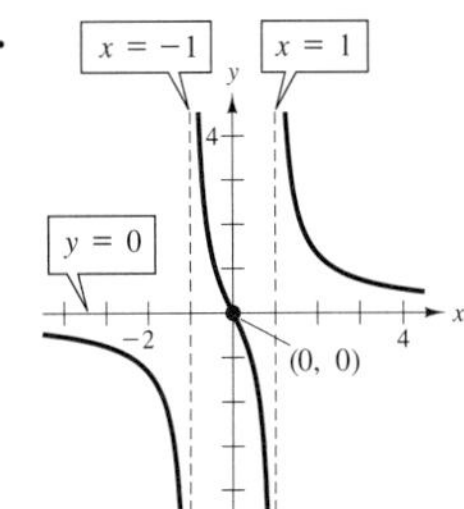

13.

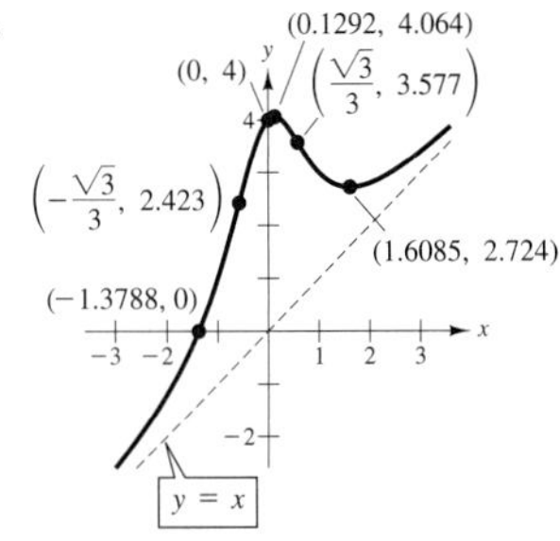

15.

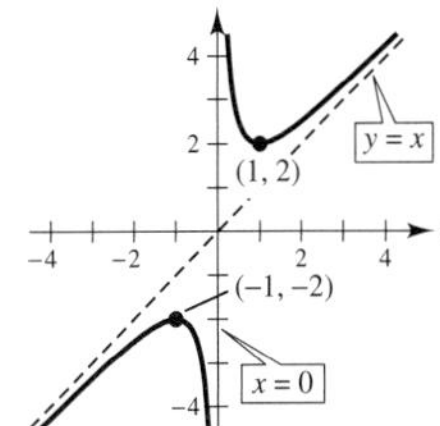

17.

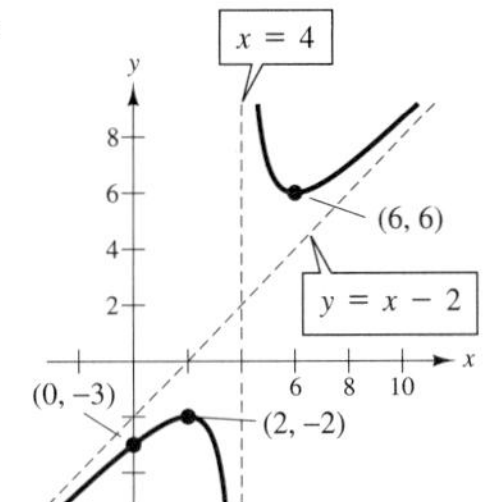

19.

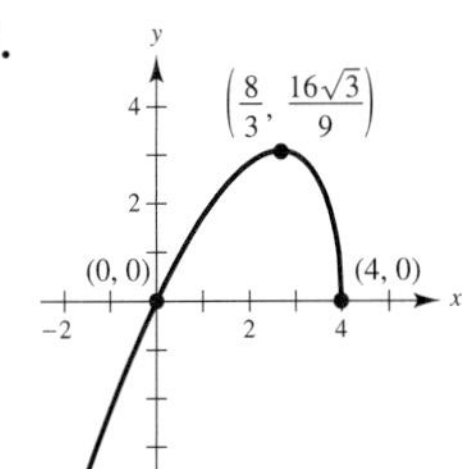

21.

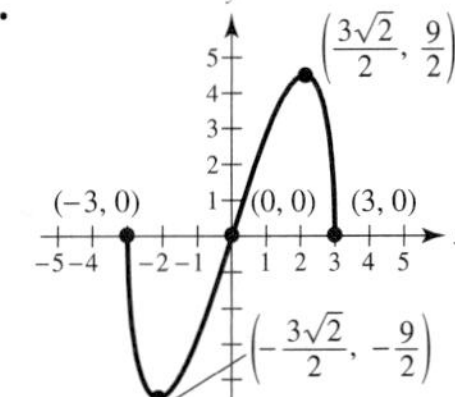

23.

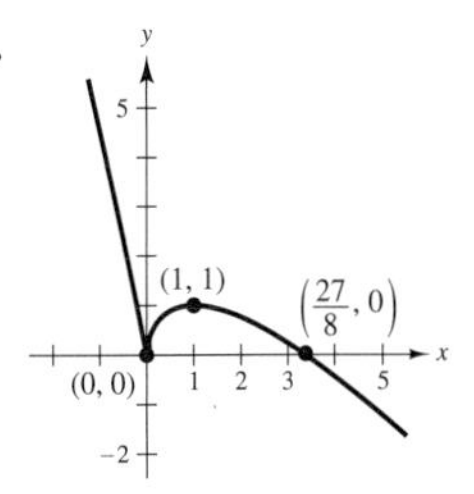

25.

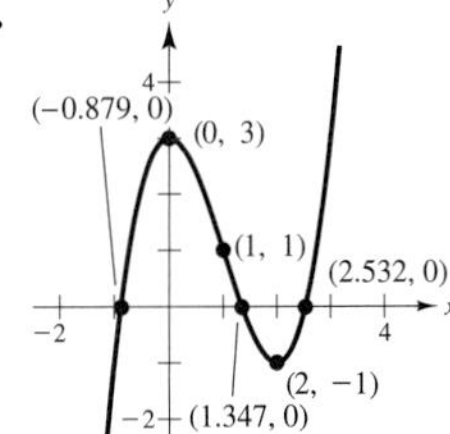

27.

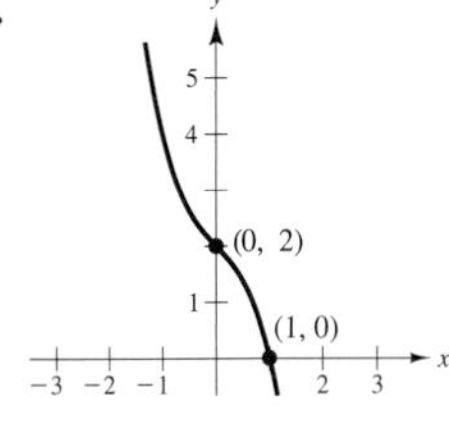

29.

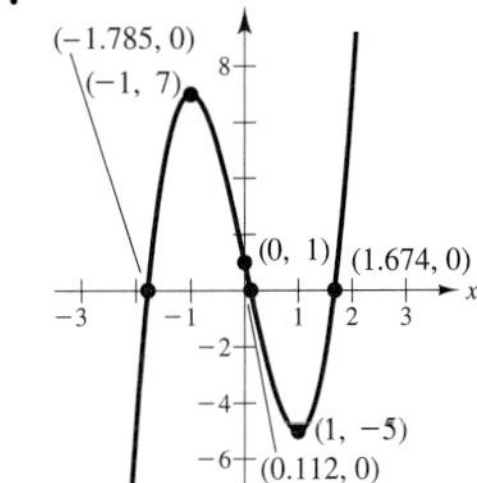

31.
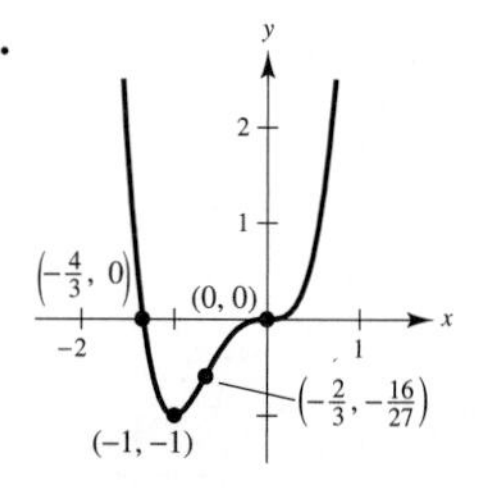

33.
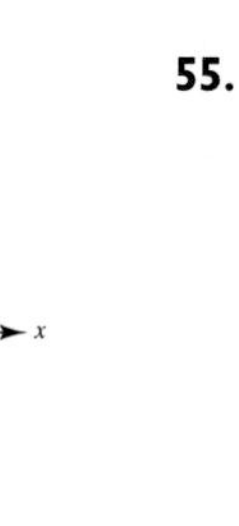

35.
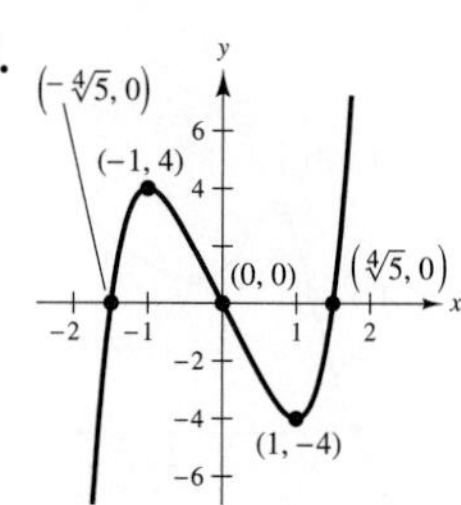

37.
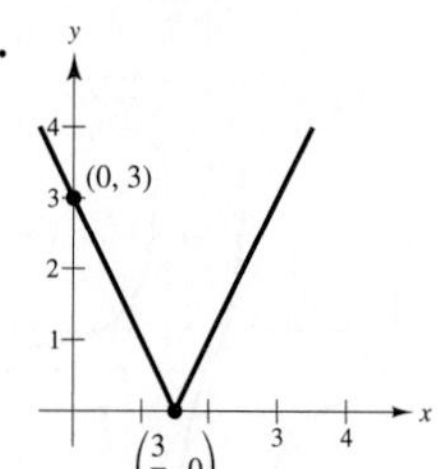

39.
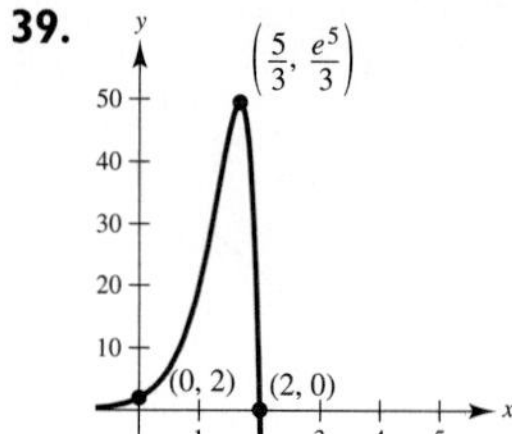

41.
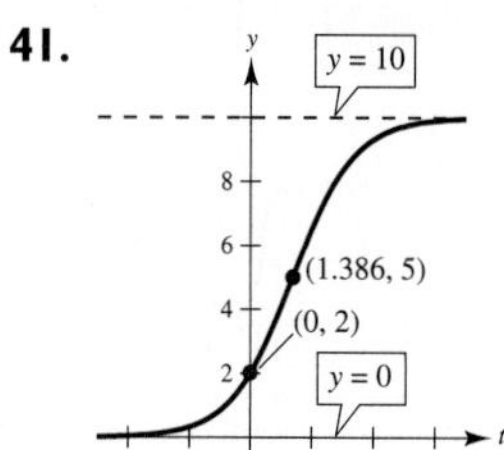

43.
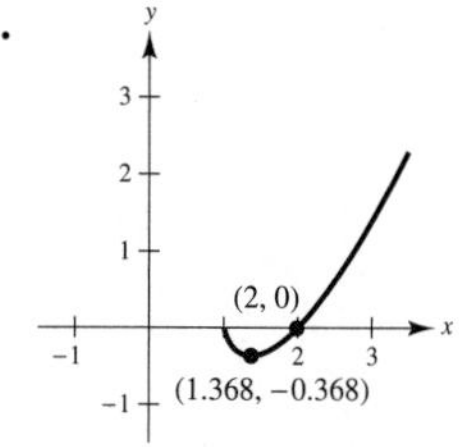

45.
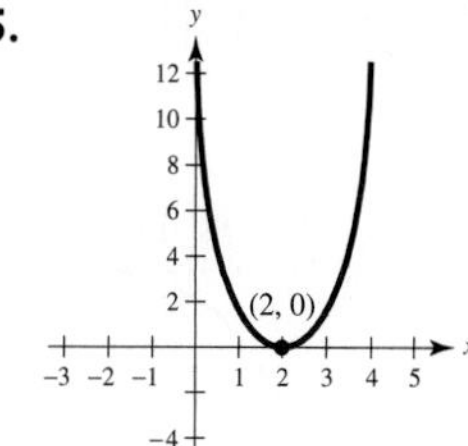

47.
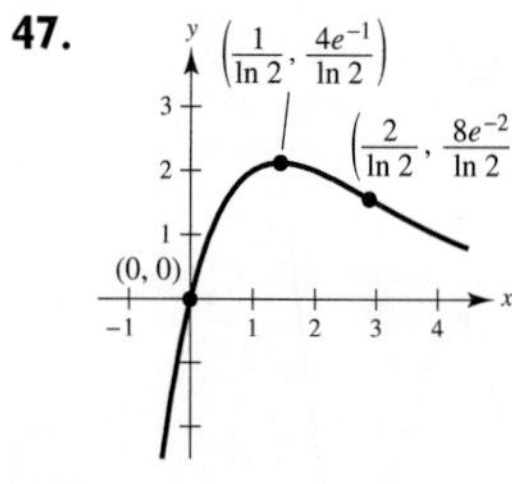

49.

51.
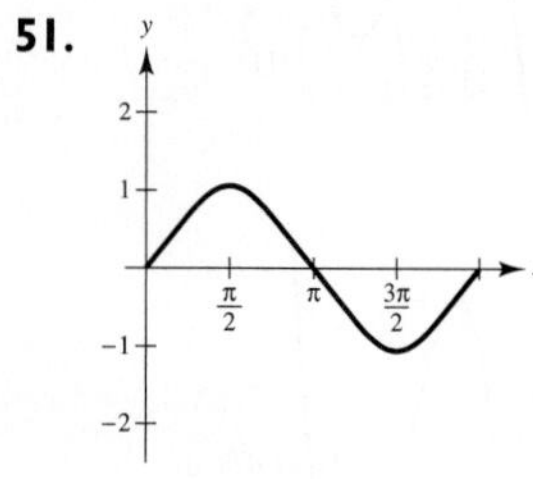

53.
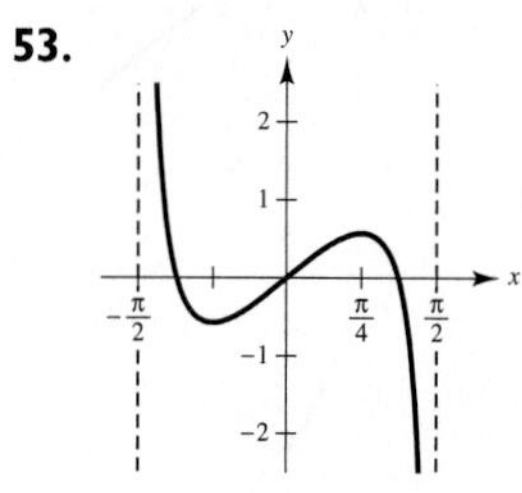

55.

57.
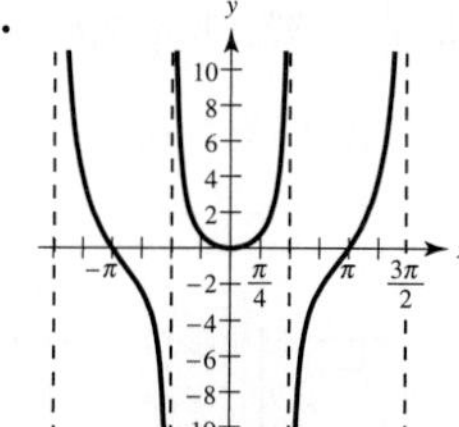

59.
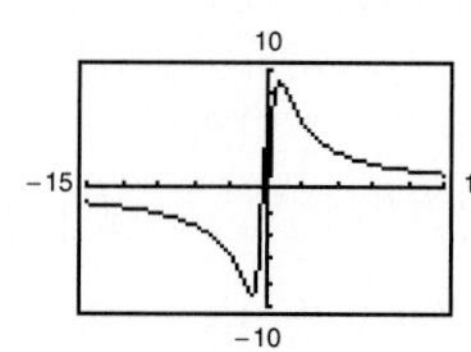

Minimum: $(-1.10, -9.05)$
Maximum: $(1.10, 9.05)$
Points of inflection: $(-1.84, -7.86), (1.84, 7.86)$
Vertical asymptote: $x = 0$
Horizontal asymptote: $y = 0$

61.
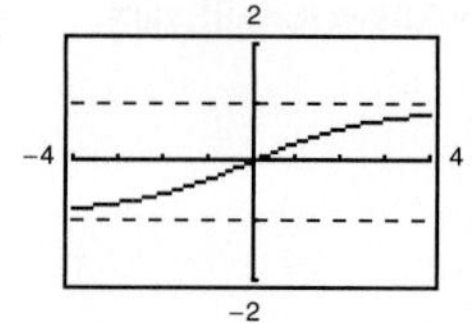

Point of inflection: $(0, 0)$
Horizontal asymptotes: $y = \pm 1$

63.
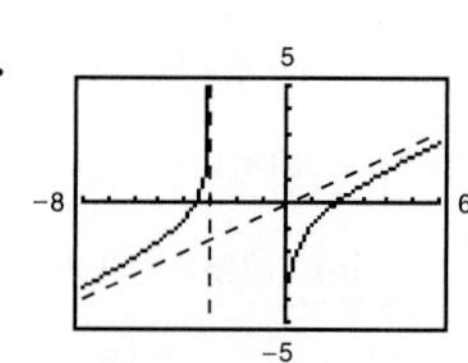

Vertical asymptotes: $x = -3$, $x = 0$
Slant asymptote: $y = x/2$

65. Answers will vary. Example: $y = 1/(x - 5)$

67. Answers will vary. Example: $y = (3x^2 - 13x - 9)/(x - 5)$

69.
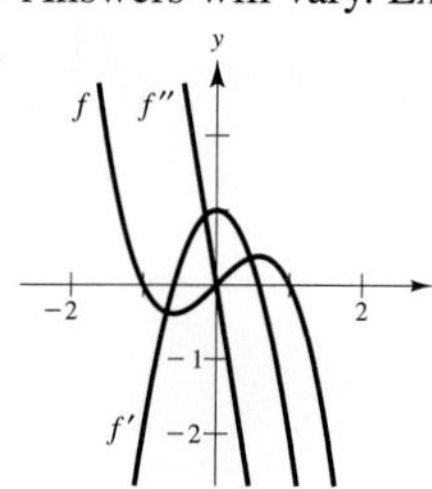
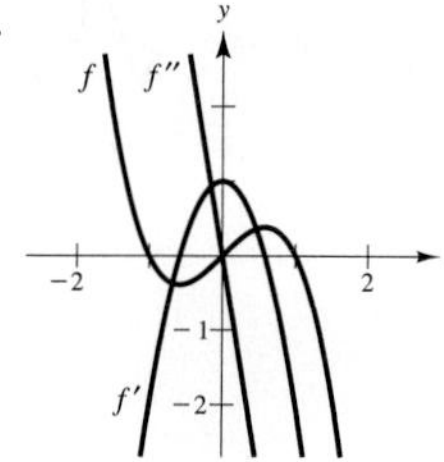

The zeros of f' correspond to the points where the graph of f has horizontal tangents. The zero of f'' corresponds to the point where the graph of f' has a horizontal tangent.

71.
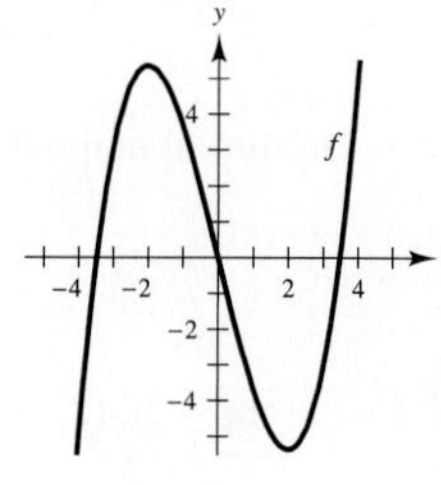

73.
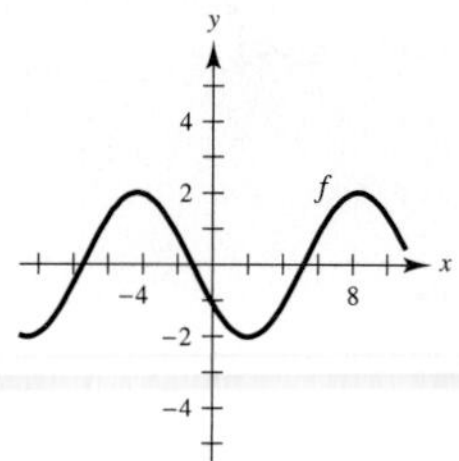

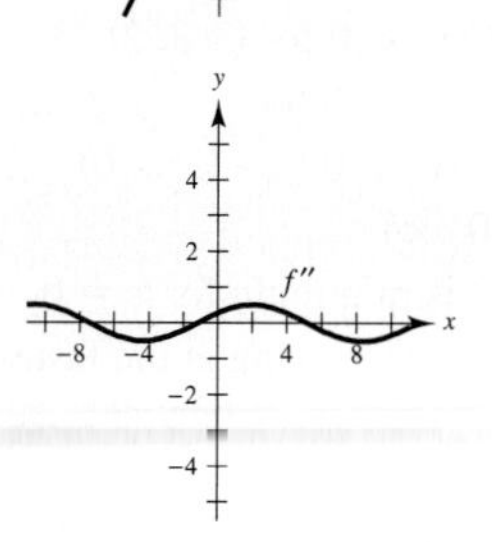

75. f is decreasing on $(2, 8)$ and therefore $f(3) > f(5)$.

77. 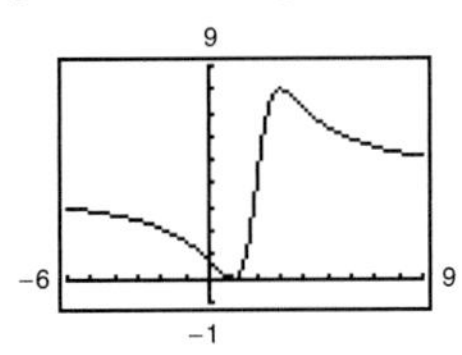

The graph crosses its horizontal asymptote $y = 4$. The graph of f does not cross its vertical asymptote $x = c$ because $f(c)$ does not exist.

79. 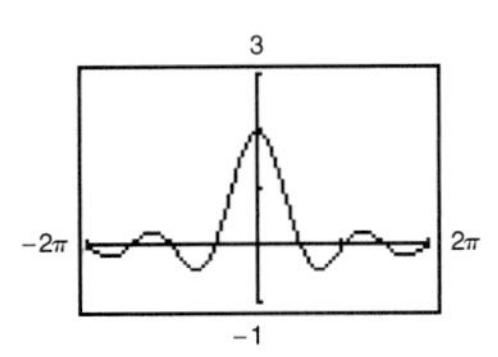

The graph has a hole at $x = 0$. The graph crosses its horizontal asymptote $y = 0$. The graph of a function f does not cross its vertical asymptote $x = c$ because $f(c)$ does not exist.

81. 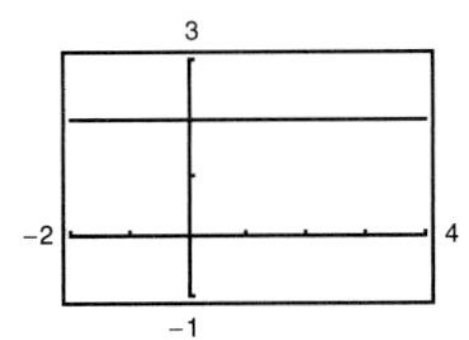

The graph has a hole at $x = 3$. The rational function is not reduced to lowest terms.

83. 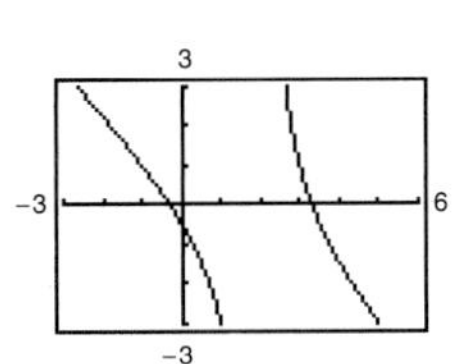

The graph appears to approach the line $y = -x + 1$, which is the slant asymptote.

85. 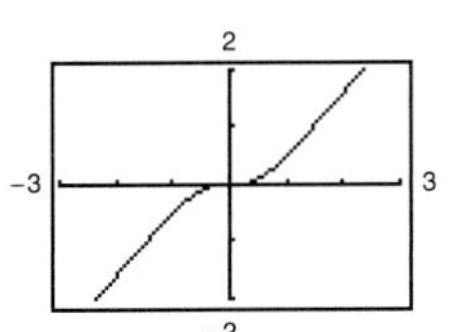

The graph appears to approach the line $y = x$, which is the slant asymptote.

87. (a) 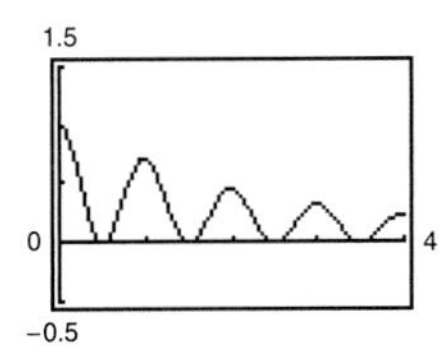

The graph has holes at $x = 0$ and at $x = 4$.
Visual approximation of critical numbers: $\frac{1}{2}, 1, \frac{3}{2}, 2, \frac{5}{2}, 3, \frac{7}{2}$

(b) $f'(x) = \dfrac{-x\cos^2(\pi x)}{(x^2+1)^{3/2}} - \dfrac{2\pi\sin(\pi x)\cos(\pi x)}{\sqrt{x^2+1}}$

Approximate critical numbers: $\frac{1}{2}, 0.97, \frac{3}{2}, 1.98, \frac{5}{2}, 2.98, \frac{7}{2}$
The critical numbers where maxima occur appear to be integers in part (a), but by approximating them using f', you see that they are not integers.

89. (a) The rate of change of f changes as a varies. If the sign of a is changed, the graph is reflected through the x-axis.
(b) The locations of the vertical asymptote and the minimum (if $a > 0$) or maximum (if $a < 0$) are changed.

91. (a) If n is even, f is symmetric with respect to the y-axis. If n is odd, f is symmetric with respect to the origin.
(b) $n = 0, 1, 2, 3$ (c) $n = 4$
(d) When $n = 5$, the slant asymptote is $y = 3x$.
(e) 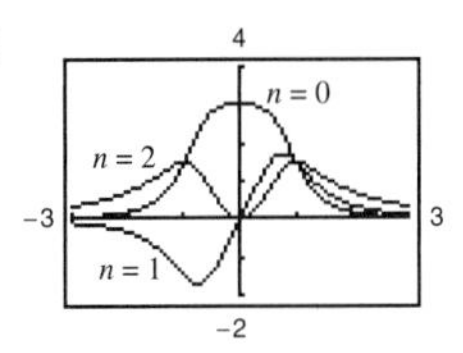

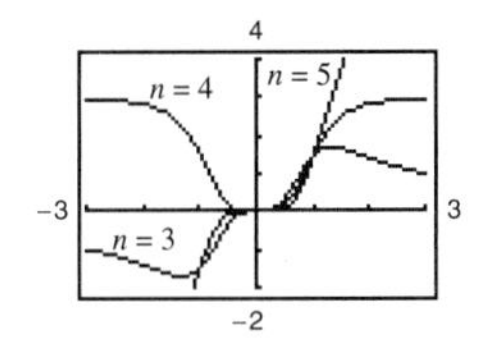

n	0	1	2	3	4	5
M	1	2	3	2	1	0
N	2	3	4	5	2	3

93. (a) 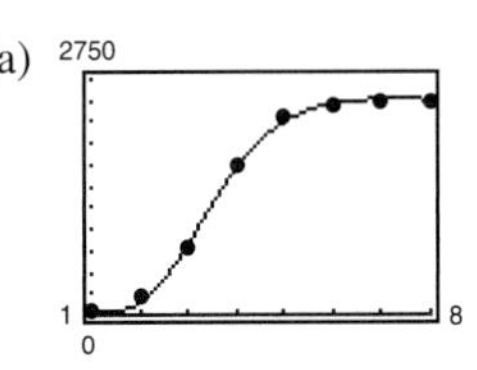

(b) 2434 bacteria
(c) The number of bacteria reaches its maximum early on the seventh day.
(d) The rate of increase in the number of bacteria is greatest approximately in the early part of the third day.
(e) 13,250/7

95. (a) Intercepts: $x = 0$ does not change and $|x| \to 0$ as $|a| \to \infty$; no change in extrema; no change in concavity
(b) 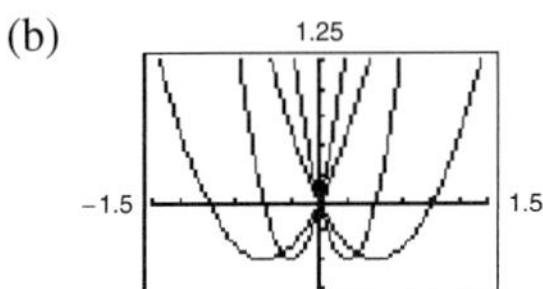

97. (a) 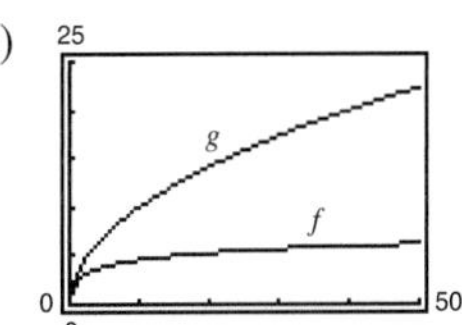

$\sqrt{x}$ increases at the faster rate.

(b)

$\sqrt[4]{x}$ increases at the faster rate.
$\ln x$ increases very slowly for "large" values of x.

99. $y = x + 3,\ y = -x - 3$

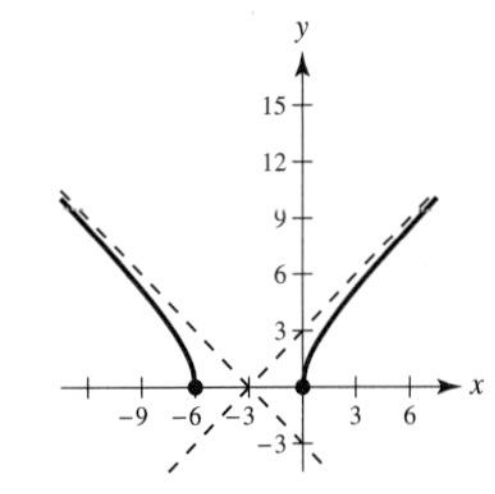

Section 4.7 (page 265)

1. (a) and (b)

First Number x	*Second Number*	*Product P*
10	110 − 10	10(110 − 10) = 1000
20	110 − 20	20(110 − 20) = 1800
30	110 − 30	30(110 − 30) = 2400
40	110 − 40	40(110 − 40) = 2800
50	110 − 50	50(110 − 50) = 3000
60	110 − 60	60(110 − 60) = 3000
70	110 − 70	70(110 − 70) = 2800
80	110 − 80	80(110 − 80) = 2400
90	110 − 90	90(110 − 90) = 1800
100	110 − 100	100(110 − 100) = 1000

(c) $P = x(110 - x)$

(d) 3500, (55, 3025), 0, 120, 0

(e) 55 and 55

3. $S/2$ and $S/2$ **5.** 24 and 8 **7.** 50 and 25

9. $l = w = 25$ m **11.** $l = w = 8$ ft **13.** $\left(\frac{7}{2}, \sqrt{\frac{7}{2}}\right)$

15. (1, 1) **17.** $x = Q_0/2$ **19.** 600 m × 300 m

21. (a) Proof (b) $V_1 = 99$ in.³, $V_2 = 125$ in.³, $V_3 = 117$ in.³
(c) 5 × 5 × 5 in.

23. Rectangular portion: $16/(\pi + 4)$ ft × $32/(\pi + 4)$ ft

25. (a) $L = \sqrt{x^2 + 4 + \dfrac{8}{x - 1} + \dfrac{4}{(x - 1)^2}}, \quad x > 1$

(b) Minimum when $x \approx 2.587$

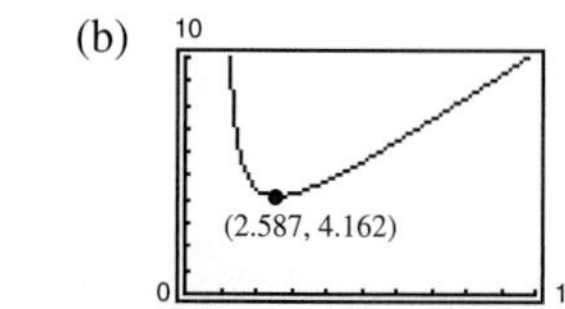

(c) (0, 0), (2, 0), (0, 4)

27. Width: $5\sqrt{2}/2$; Length: $5\sqrt{2}$

29. Dimensions of page: $(2 + \sqrt{30})$ in. × $(2 + \sqrt{30})$ in.

31. (a) $\frac{y}{2}$, y, x

(b)

Length x	Width y	Area xy
10	$2/\pi(100 - 10)$	$(10)(2/\pi)(100 - 10) \approx 573$
20	$2/\pi(100 - 20)$	$(20)(2/\pi)(100 - 20) \approx 1019$
30	$2/\pi(100 - 30)$	$(30)(2/\pi)(100 - 30) \approx 1337$
40	$2/\pi(100 - 40)$	$(40)(2/\pi)(100 - 40) \approx 1528$
50	$2/\pi(100 - 50)$	$(50)(2/\pi)(100 - 50) \approx 1592$
60	$2/\pi(100 - 60)$	$(60)(2/\pi)(100 - 60) \approx 1528$

The maximum area of the rectangle is approximately 1592 m².

(c) $A = 2/\pi(100x - x^2), \quad 0 < x < 100$

(d) $\dfrac{dA}{dx} = \dfrac{2}{\pi}(100 - 2x)$
$= 0$ when $x = 50$
The maximum value is approximately 1592 when $x = 50$.

(e)

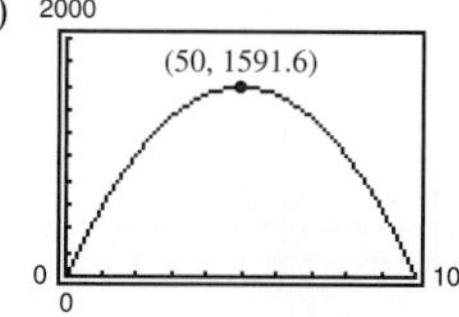

33. 18 × 18 × 36 in. **35.** $32\pi r^3/81$

37. Answers will vary. If area is expressed as a function of either length or width, the feasible domain is the interval (0, 10). No dimensions will yield a minimum area because the second derivative on this open interval is always negative.

39. $r = \sqrt[3]{9/\pi} \approx 1.42$ cm

41. Side of square: $\dfrac{10\sqrt{3}}{9 + 4\sqrt{3}}$; Side of triangle: $\dfrac{30}{9 + 4\sqrt{3}}$

43. $w = 8\sqrt{3}$ in., $h = 8\sqrt{6}$ in. **45.** $\theta = \pi/4$ **47.** $h = \sqrt{2}$ ft

49. One mile from the nearest point on the coast **51.** Proof

53. (a) Origin to y-intercept: 2
Origin to x-intercept: $\pi/2$

(b) $d = \sqrt{x^2 + (2 - 2\sin x)^2}$

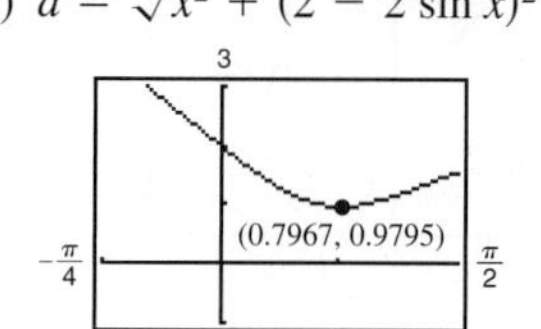

(c) Minimum distance is 0.9795 when $x \approx 0.7967$.

55. $\theta = \pi/3(6 - 2\sqrt{6}) \approx 1.15°$

57. 4045 units **59.** $A = \sqrt{2}e^{-1/2}$ **61.** Proof

63. $y = \frac{64}{141}x$; $S_1 \approx 6.1$ mi **65.** $y = \frac{3}{10}x$; $S_3 \approx 4.50$ mi

67. Putnam Problem A1, 1986

Section 4.8 (page 276)

1. $T(x) = 4x - 4$

x	1.9	1.99	2	2.01	2.1
$f(x)$	3.610	3.960	4	4.040	4.410
$T(x)$	3.600	3.960	4	4.040	4.400

3. $T(x) = 80x - 128$

x	1.9	1.99	2	2.01	2.1
$f(x)$	24.761	31.208	32	32.808	40.841
$T(x)$	24.000	31.200	32	32.800	40.000

5. $T(x) = (\cos 2)(x - 2) + \sin 2$

x	1.9	1.99	2	2.01	2.1
$f(x)$	0.946	0.913	0.909	0.905	0.863
$T(x)$	0.951	0.913	0.909	0.905	0.868

7. $\Delta y = 0.6305$; $dy = 0.6000$ **9.** $\Delta y = -0.039$; $dy = -0.040$

11. $6x\,dx$ **13.** $-\dfrac{3}{(2x - 1)^2}\,dx$ **15.** $\dfrac{1 - 2x^2}{\sqrt{1 - x^2}}\,dx$

17. $x/(x^2 - 4)\,dx$ **19.** $(2 + 2\cot x + 2\cot^3 x)\,dx$

21. $-\pi \sin\left(\dfrac{6\pi x - 1}{2}\right) dx$ **23.** $\left(\arcsin x + \dfrac{x}{\sqrt{1 - x^2}}\right) dx$

25. (a) 0.9 (b) 1.04 **27.** (a) 1.05 (b) 0.98

29. (a) 8.035 (b) 7.95 **31.** $\pm\frac{3}{8}$ in.^{2} **33.** $\pm 7\pi$ in.^{2}

35. (a) $\frac{2}{3}$% (b) 1.25%

37. (a) $\pm 2.88\pi$ in.^{3} (b) $\pm 0.96\pi$ in.^{2} (c) 1%, $\frac{2}{3}$%

39. 267.24, 3.1% **41.** 80π cm^3

43. (a) 0.87% (b) 2.16% **45.** 4961 ft

47. $f(x) = \sqrt{x},\ dy = \dfrac{1}{2\sqrt{x}}\,dx$

$f(99.4) \approx \sqrt{100} + \dfrac{1}{2\sqrt{100}}(-0.6) = 9.97$

Calculator: 9.97

49. $f(x) = \sqrt[4]{x},\ dy = \dfrac{1}{4x^{3/4}}\,dx$

$f(624) \approx \sqrt[4]{625} + \dfrac{1}{4(625)^{3/4}}(-1) = 4.998$

Calculator: 4.998

51. $f(x) = \sqrt{x};\ dy = \dfrac{1}{2\sqrt{x}}\,dx$

$f(4.02) \approx \sqrt{4} + \dfrac{1}{2\sqrt{4}}(0.02) = 2 + \dfrac{1}{4}(0.02)$

53. $y - f(0) = f'(0)(x - 0)$

$y - 2 = \frac{1}{4}x$

$y = 2 + x/4$

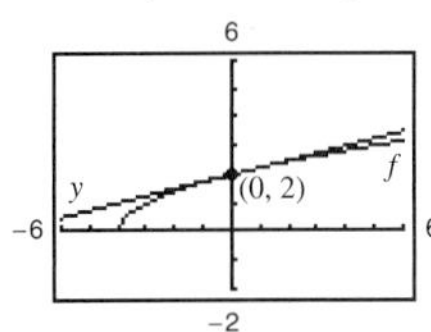

55. $y - f(0) = f'(0)(x - 0)$

$y - 0 = 1(x)$

$y = x$

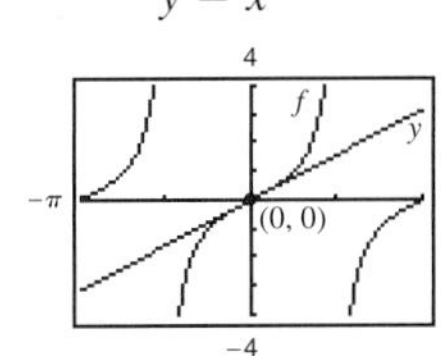

57. The value of dy becomes closer to the value of Δy as Δx decreases.

59. True **61.** True

Review Exercises for Chapter 4 (page 278)

1. Let f be defined at c. If $f'(c) = 0$ or if f' is undefined at c, then c is a critical number of f.

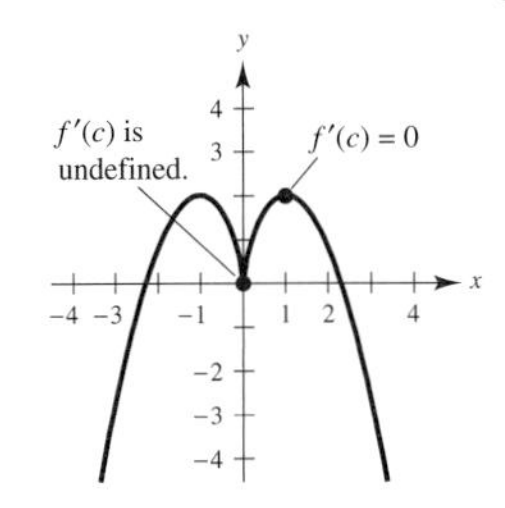

3. Maximum: $(2\pi, 17.57)$ **5.** $f'\left(\frac{1}{3}\right) = 0$

Minimum: $(2.73, 0.88)$

7. (a)

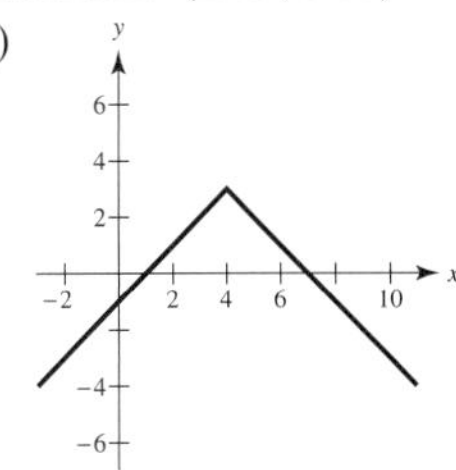

(b) f is not differentiable at $x = 4$.

9. $f'\left(\dfrac{2744}{729}\right) = \dfrac{3}{7}$ **11.** $f'(0) = 1$ **13.** $c = \dfrac{x_1 + x_2}{2}$

15. Critical numbers: $x = 1, \frac{7}{3}$

Increasing on $(-\infty, 1), \left(\frac{7}{3}, \infty\right)$; Decreasing on $\left(1, \frac{7}{3}\right)$

17. Critical number: $x = 1$

Increasing on $(1, \infty)$; Decreasing on $(0, 1)$

19. Critical number: $t = 2 - 1/\ln 2$

Increasing on $(-\infty, 2 - 1/\ln 2)$

Decreasing on $(2 - 1/\ln 2, \infty)$

21. Minimum: $(2, -12)$

23. (a) $y = \frac{1}{4}$ in.; $v = 4$ in./sec (b) Proof

(c) Period: $\pi/6$; Frequency: $6/\pi$

25. $(\pi/2, \pi/2), (3\pi/2, 3\pi/2)$

27. Relative maxima: $\left(\sqrt{2}/2, 1/2\right), \left(-\sqrt{2}/2, 1/2\right)$

Relative minimum: $(0, 0)$

29.

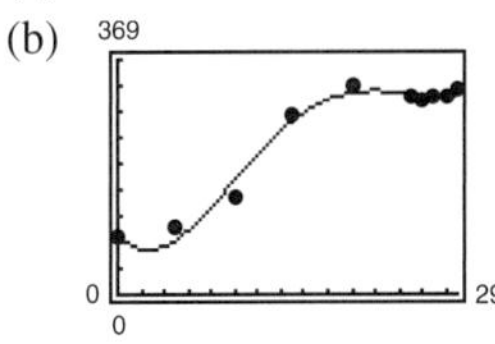

31. Increasing and concave down

33. (a) $D = 0.00340t^4 - 0.2352t^3 + 4.942t^2 - 20.86t + 94.4$

(b)

(c) Maximum occurs in 1991; Minimum occurs in 1972.

(d) 1979

35. $\frac{2}{3}$ **37.** $-\infty$ **39.** 0 **41.** 6

43. Vertical asymptote: $x = 4$; Horizontal asymptote: $y = 2$

45. Vertical asymptote: $x = 0$; Horizontal asymptote: $y = -2$

47. Horizontal asymptotes: $y = 0$; $y = \frac{5}{3}$

49. Horizontal asymptote: $y = 0$

51.

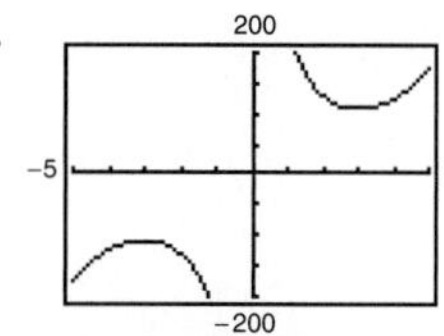

Vertical asymptote: $x = 0$
Relative minimum: $(3, 108)$
Relative maximum: $(-3, -108)$

53.

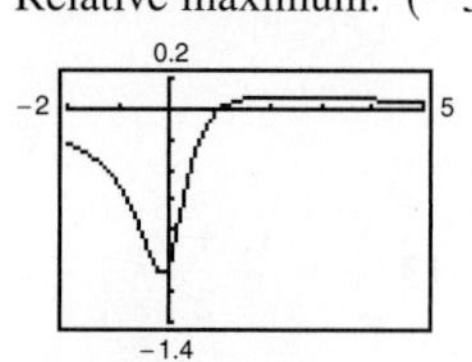

Horizontal asymptote: $y = 0$
Relative minimum: $(-0.155, -1.077)$
Relative maximum: $(2.155, 0.077)$

55.

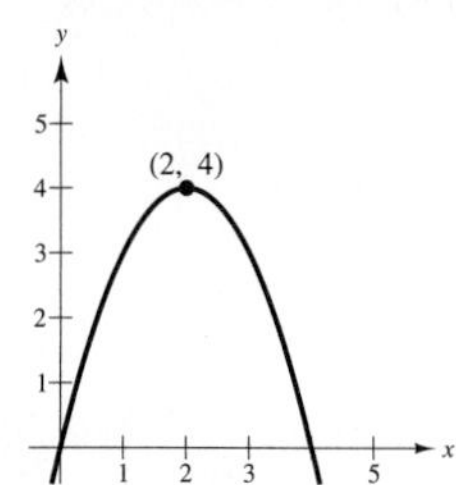

57.

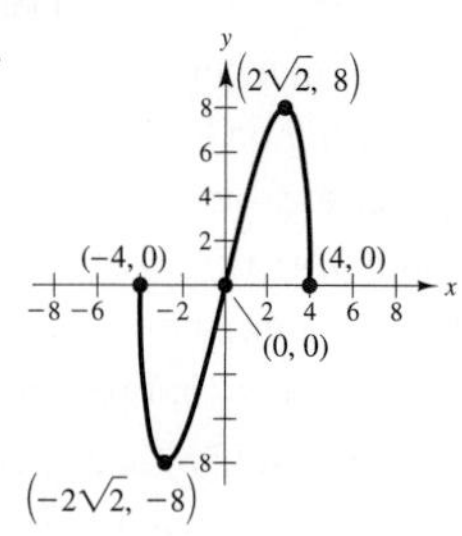

59.

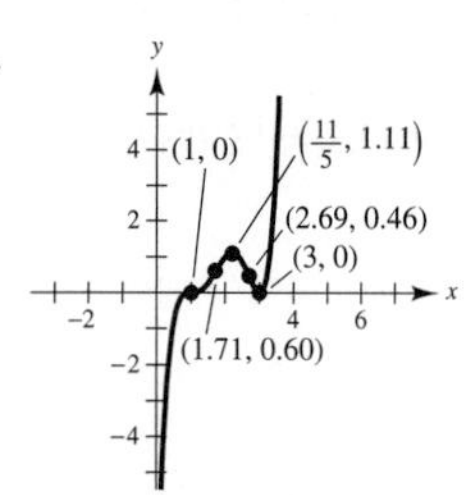

61.

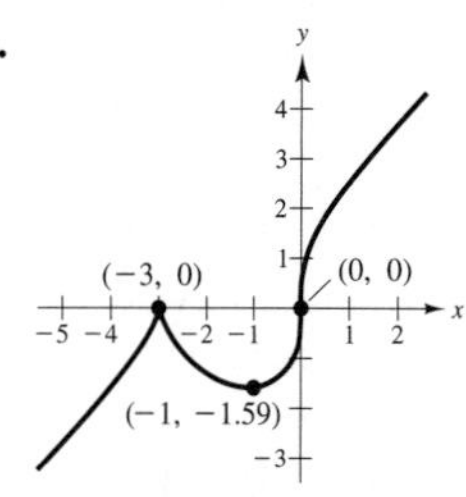

63.

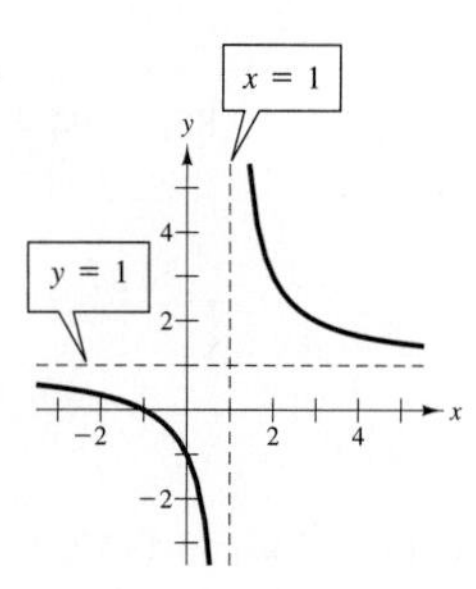

65.

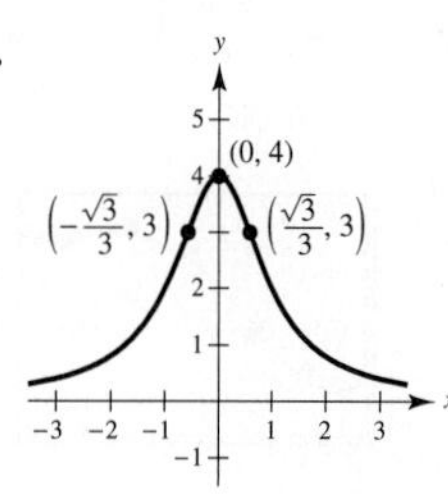

67.

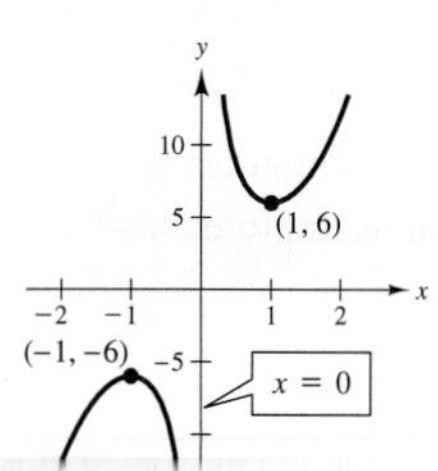

69.

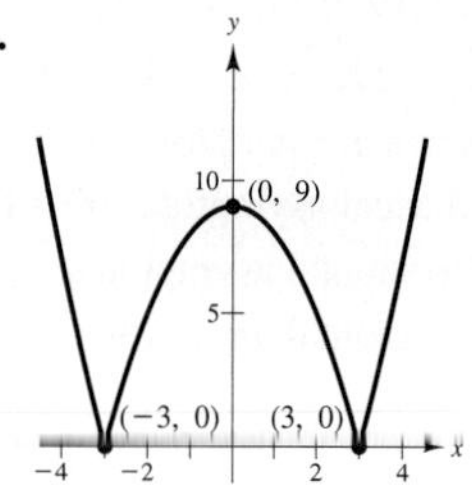

71.

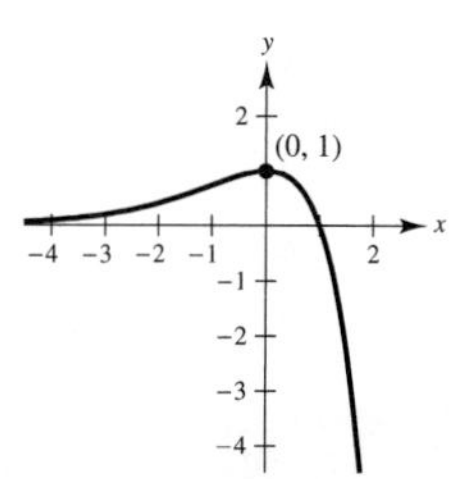

73.

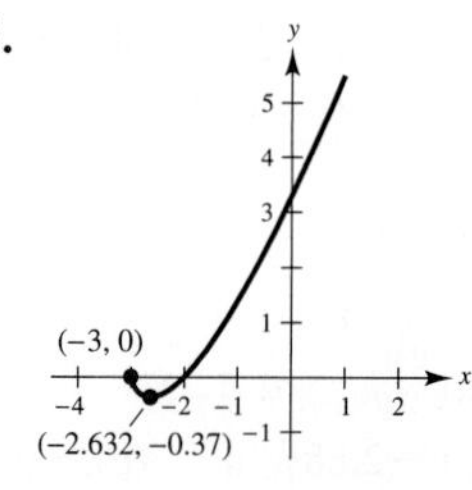

75.

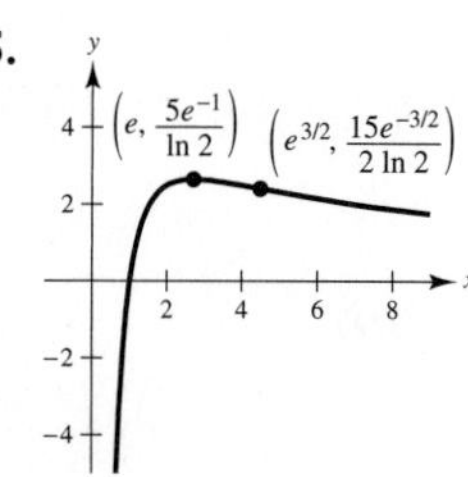

77.

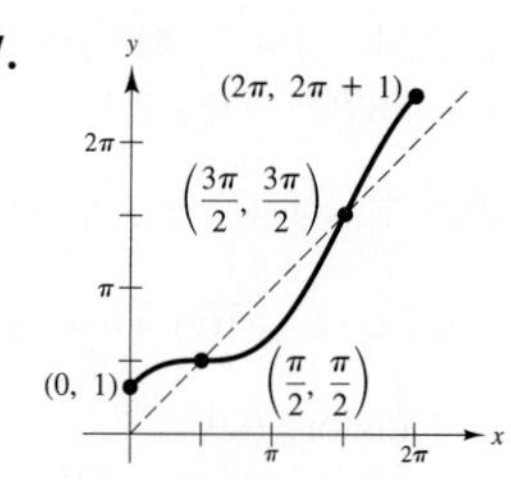

79.

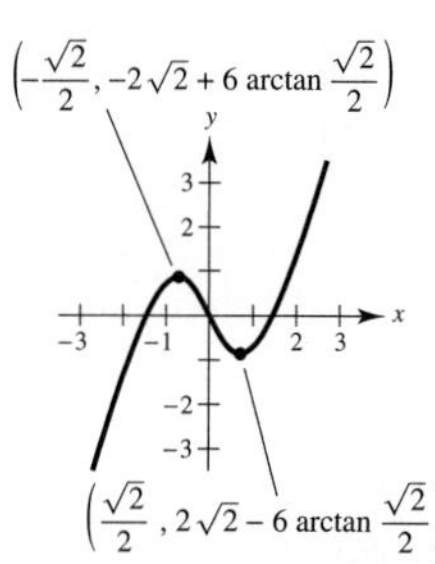

81. Maximum: $(1, 3)$
Minimum: $(1, 1)$

83. $t \approx 4.92 \approx$ 4:55 P.M.; $d \approx 64$ km

85. $(0, 0), (5, 0), (0, 10)$ **87.** Proof **89.** 14.05 ft

91. $3(3^{2/3} + 2^{2/3})^{3/2} \approx 21.07$ ft **93.** $v \approx 54.77$ mph

95. $dy = (1 - \cos x + x \sin x)\, dx$

97. $dS = \pm 1.8\pi \text{ cm}^2, \dfrac{dS}{S} \times 100 \approx \pm 0.56\%$

$dV = \pm 8.1\pi \text{ cm}^3, \dfrac{dV}{V} \times 100 \approx \pm 0.83\%$

P.S. Problem Solving (page 281)

1. Choices of a may vary.

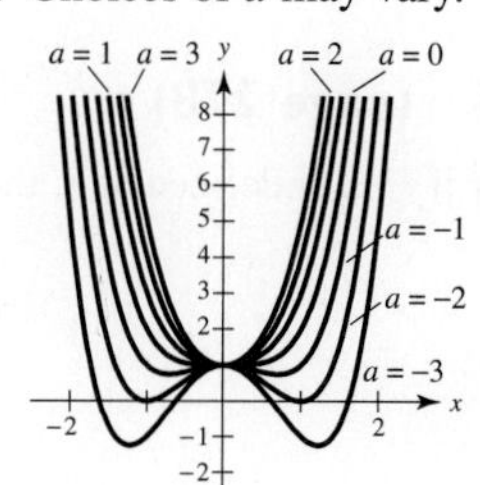

(a) One relative minimum at $(0, 1)$ for $a \geq 0$
(b) One relative maximum at $(0, 1)$ for $a < 0$
(c) Two relative minima for $a < 0$ when $x = \pm\sqrt{-a/2}$
(d) If $a < 0$, there are three critical points; if $a \geq 0$, there is only one critical point.

3. All c, where c is a real number **5–7.** Proofs

9. About 9.19 ft

11. Minimum: $(\sqrt{2} - 1)d$; There is no maximum.

13. (a)–(c) Proofs

15. (a)

x	0	0.5	1	2
$\sqrt{1 + x}$	1	1.2247	1.4142	1.7321
$\frac{1}{2}x + 1$	1	1.25	1.5	2

(b) Proof

17. (a)

v	20	40	60	80	100
s	5.56	11.11	16.67	22.22	27.78
d	5.1	13.7	27.2	44.2	66.4

$d(s) = 0.071s^2 + 0.389s + 0.727$

(b) The distance between the back of the first vehicle and the front of the second vehicle is $d(s)$, the safe stopping distance. The first vehicle passes the given point in $5.5/s$ seconds, and the second vehicle takes $d(s)/s$ more seconds. So, $T = d(s)/s + 5.5/s$.

(c)

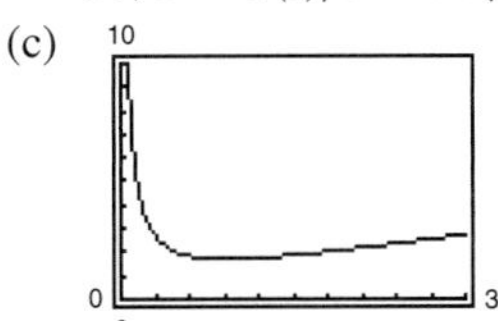

$s \approx 9.365$ m/sec

(d) $s \approx 9.365$ m/sec; 1.719 sec; 33.714 km/h (e) 10.597 m

19. (a) $(0, \infty)$

(b) Answers will vary. Sample answer: $x = e^{\pi/2}, x = e^{(\pi/2)+2\pi}$

(c) Answers will vary. Sample answer: $x = e^{-\pi/2}, x = e^{3\pi/2}$

(d) $[-1, 1]$ (e) $f'(x) = \dfrac{\cos(\ln x)}{x}$; Maximum $= e^{\pi/2}$

(f)

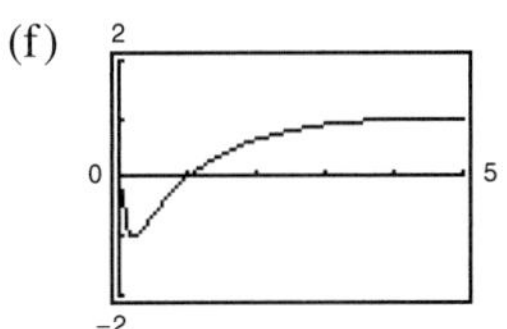

$\lim_{x \to 0^+} f(x)$ seems to be $-\frac{1}{2}$. (This is incorrect.)

(g) Limit does not exist.

Chapter 5

Section 5.1 (page 291)

1. Proof **3.** Proof **5.** $y = t^3 + C$ **7.** $y = \frac{2}{5}x^{5/2} + C$

	Original Integral	*Rewrite*	*Integrate*	*Simplify*
9.	$\int \sqrt[3]{x}\, dx$	$\int x^{1/3}\, dx$	$\dfrac{x^{4/3}}{4/3} + C$	$\dfrac{3}{4}x^{4/3} + C$
11.	$\int \dfrac{1}{x\sqrt{x}}\, dx$	$\int x^{-3/2}\, dx$	$\dfrac{x^{-1/2}}{-1/2} + C$	$-\dfrac{2}{\sqrt{x}} + C$
13.	$\int \dfrac{1}{2x^3}\, dx$	$\dfrac{1}{2}\int x^{-3}\, dx$	$\dfrac{1}{2}\left(\dfrac{x^{-2}}{-2}\right) + C$	$-\dfrac{1}{4x^2} + C$

15. $\frac{1}{2}x^2 + 3x + C$ **17.** $\frac{1}{4}x^4 + 5x + C$

19. $\frac{2}{5}x^{5/2} + x^2 + x + C$ **21.** $-1/(2x^2) + C$

23. $\frac{2}{15}x^{1/2}(3x^2 + 5x + 15) + C$ **25.** $x^3 + \frac{1}{2}x^2 - 2x + C$

27. $\frac{2}{7}y^{7/2} + C$ **29.** $x + C$ **31.** $-2\cos x + 3\sin x + C$

33. $t + \csc t + C$ **35.** $-2\cos x - 5e^x + C$

37. $\tan\theta + \cos\theta + C$ **39.** $\tan y + C$

41. $x^2 - 4^x/\ln 4 + C$ **43.** $\frac{1}{2}x^2 - 5\ln|x| + C$

45.

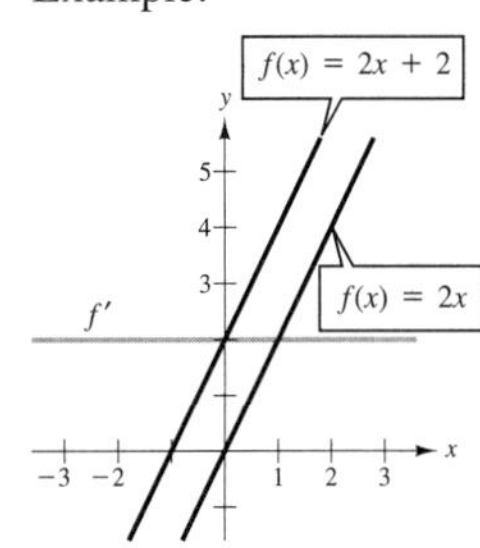

47.

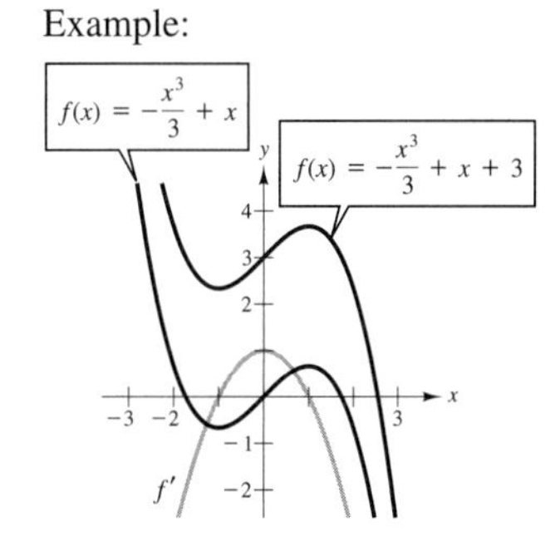

49. Answers will vary.
Example:

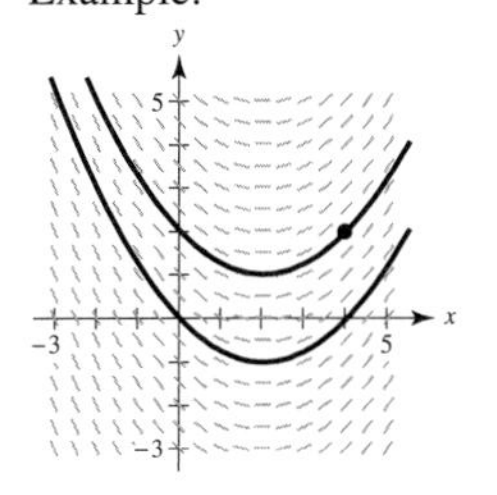

51. Answers will vary.
Example:

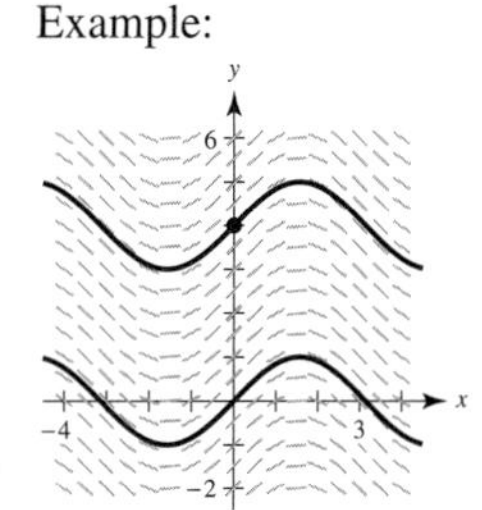

53. $y = x^2 - x + 1$ **55.** $y = \sin x + 4$

57. (a) Answers will vary.
Example:

(b) $y = \frac{1}{4}x^2 - x + 2$

59. (a) Answers will vary.
Example:

(b) $y = \sin x + 4$

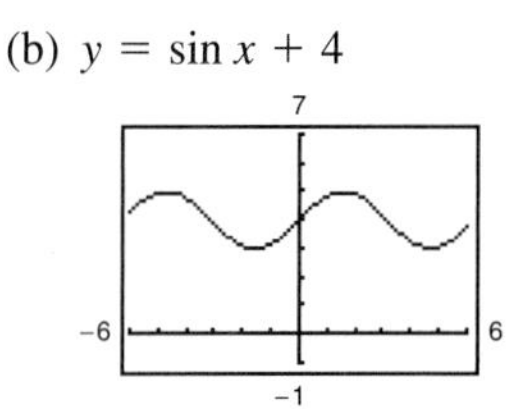

61. (a)

(b) $y = x^2 - 6$

(c)

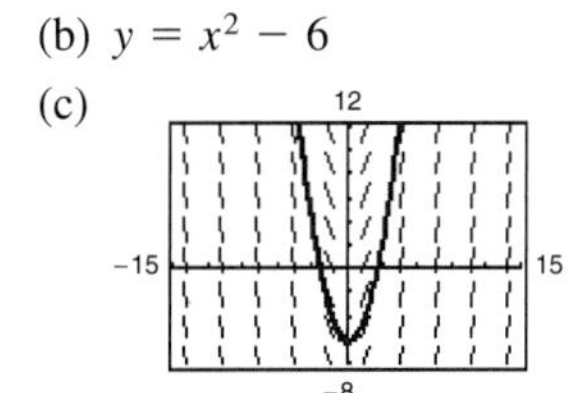

63. $f(x) = 2x^2 + 6$ **65.** $h(t) = 2t^4 + 5t - 11$

67. $f(x) = x^2 + x + 4$ **69.** $f(x) = -4\sqrt{x} + 3x$

71. $f(x) = e^x + x + 4$

73. (a) $h(t) = \frac{3}{4}t^2 + 5t + 12$ (b) 69 cm

75. (a) -1; $f'(4)$ represents the slope of f at $x = 4$.

(b) No. The slope of the tangent lines are greater than 2 on $[0, 2]$. Therefore, f must increase more than four units on $[0, 2]$.

(c) No. The function is decreasing on $[4, 5]$.

(d) 3.5; $f'(3.5) \approx 0$

(e) Concave upward: $(-\infty, 1), (5, \infty)$
Concave downward: $(1, 5)$
Points of inflection at $x \approx 1$ and $x \approx 5$

(f) 3 (g)

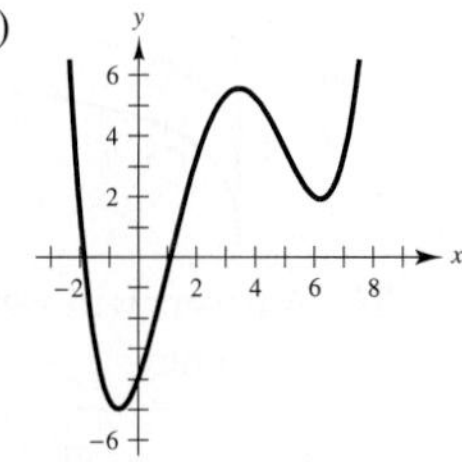

77. 62.25 ft **79.** $v_0 \approx 187.617$ ft/sec

81. $v(t) = -9.8t + C_1 = -9.8t + v_0$
$f(t) = -4.9t^2 + v_0 t + C_2 = -4.9t^2 + v_0 t + s_0$

83. 7.1 m **85.** 320 m; -32 m/sec

87. (a) $v(t) = 3t^2 - 12t + 9$; $a(t) = 6t - 12$
(b) $(0, 1), (3, 5)$ (c) -3

89. $a(t) = -1/(2t^{3/2})$; $x(t) = 2\sqrt{t} + 2$

91. (a) 1.18 m/sec² (b) 190 m

93. (a) 300 ft (b) 60 ft/sec $\approx$ 41 mph

95. True **97.** True

99. False. Let $f(x) = x$ and $g(x) = x + 1$.

101. $f(x) = \frac{1}{3}x^3 - 4x + \frac{16}{3}$

103. $f(x) = \begin{cases} x + 2, & 0 \le x < 2 \\ \frac{3}{2}x^2 - 2, & 2 \le x \le 5 \end{cases}$

f is not differentiable at $x = 2$ because the left- and right-hand derivatives at $x = 2$ do not agree.

105. Proof **107.** Putnam Problem B2, 1991

Section 5.2 (page 303)

1. 35 **3.** $\frac{158}{85}$ **5.** $4c$ **7.** $\sum_{i=1}^{9} \frac{1}{3i}$ **9.** $\sum_{j=1}^{8} \left[5\left(\frac{j}{8}\right) + 3\right]$

11. $\frac{2}{n}\sum_{i=1}^{n}\left[\left(\frac{2i}{n}\right)^3 - \left(\frac{2i}{n}\right)\right]$ **13.** $\frac{3}{n}\sum_{i=1}^{n}\left[2\left(1 + \frac{3i}{n}\right)^2\right]$

15. 420 **17.** 2470 **19.** 12,040 **21.** 2930

23. The area of the shaded region falls between 12.5 square units and 16.5 square units.

25. The area of the shaded region falls between 7 square units and 11 square units.

27. $A \approx S \approx 0.768$
$A \approx s \approx 0.518$

29. $A \approx S \approx 0.746$
$A \approx s \approx 0.646$

31. $\frac{81}{4}$ **33.** 9

35. $(n + 2)/n$
$n = 10; S = 1.2$
$n = 100; S = 1.02$
$n = 1000; S = 1.002$
$n = 10{,}000; S = 1.0002$

37. $[2(n + 1)(n - 1)]/n^2$
$n = 10; S = 1.98$
$n = 100; S = 1.9998$
$n = 1000; S = 1.999998$
$n = 10{,}000; S = 1.99999998$

39. $\lim_{n\to\infty}\left[8\left(\frac{n^2 + n}{n^2}\right)\right] = 8$ **41.** $\lim_{n\to\infty}\frac{1}{6}\left(\frac{2n^3 - 3n^2 + n}{n^3}\right) = \frac{1}{3}$

43. $\lim_{n\to\infty}[(3n + 1)/n] = 3$

45. (a)

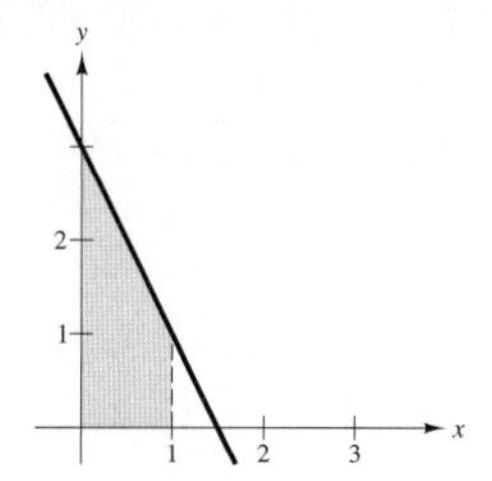

(b) $\Delta x = (2 - 0)/n = 2/n$

(c) $s(n) = \sum_{i=1}^{n} f(x_{i-1})\,\Delta x$
$= \sum_{i=1}^{n}\left[(i - 1)\left(\frac{2}{n}\right)\right]\left(\frac{2}{n}\right)$

(d) $S(n) = \sum_{i=1}^{n} f(x_i)\,\Delta x$
$= \sum_{i=1}^{n}\left[i\left(\frac{2}{n}\right)\right]\left(\frac{2}{n}\right)$

(e)

n	5	10	50	100
$s(n)$	1.6	1.8	1.96	1.98
$S(n)$	2.4	2.2	2.04	2.02

(f) $\lim_{n\to\infty}\sum_{i=1}^{n}\left[(i - 1)\left(\frac{2}{n}\right)\right]\left(\frac{2}{n}\right) = 2$; $\lim_{n\to\infty}\sum_{i=1}^{n}\left[i\left(\frac{2}{n}\right)\right]\left(\frac{2}{n}\right) = 2$

47. $A = 2$

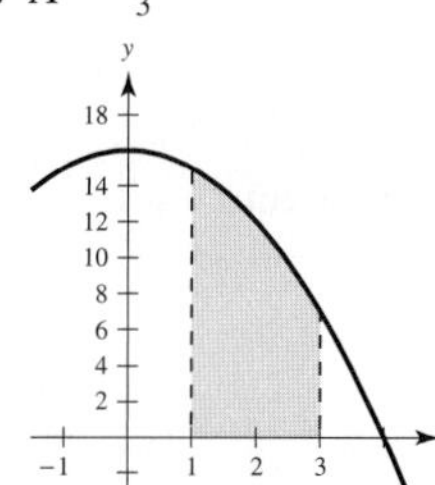

49. $A = \frac{7}{3}$

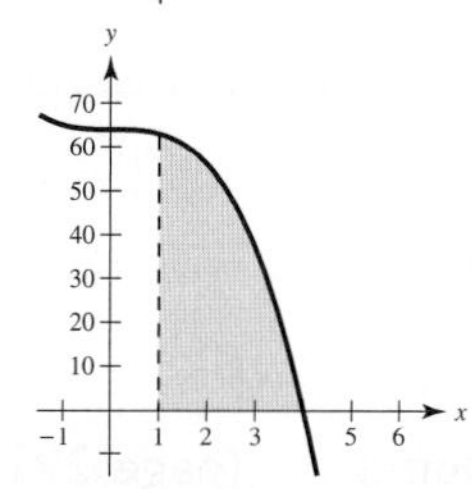

51. $A = \frac{70}{3}$

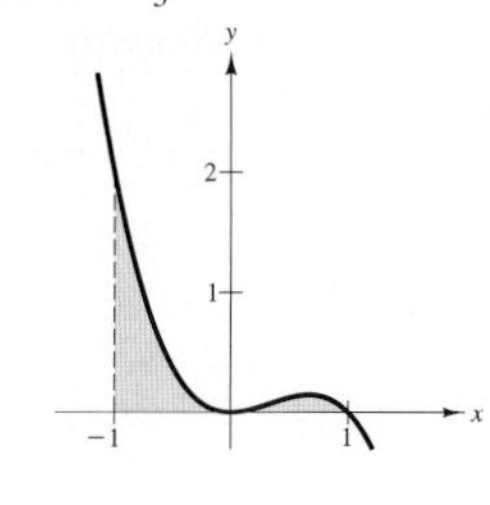

53. $A = \frac{513}{4}$

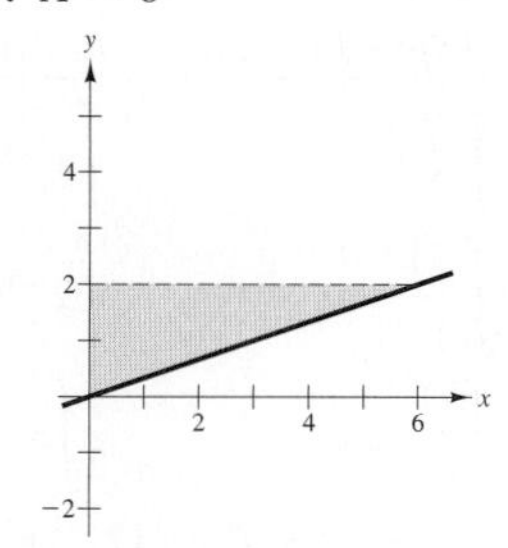

55. $A = \frac{2}{3}$

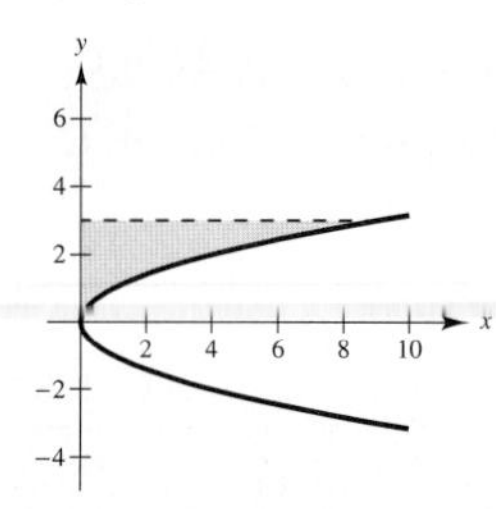

57. $A = 6$

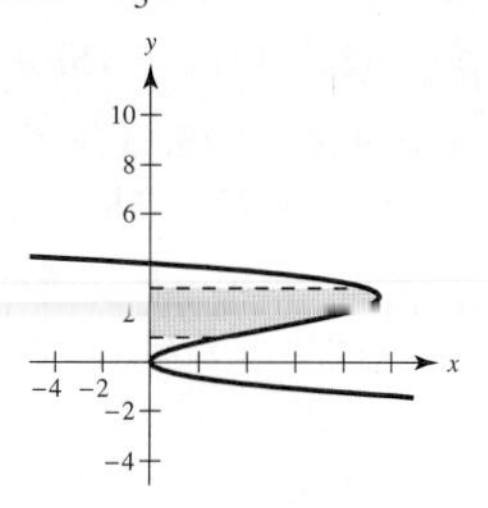

59. $A = 9$

61. $A = \frac{44}{3}$

63. $\frac{69}{8}$ **65.** 0.345

67.

n	4	8	12	16	20
Approximate Area	5.3838	5.3523	5.3439	5.3403	5.3384

69.

n	4	8	12	16	20
Approximate Area	2.2223	2.2387	2.2418	2.2430	2.2435

71.

n	4	8	12	16	20
Approximate Area	4.0786	4.0554	4.0509	4.0493	4.0485

73. b

75. You can use the line $y = x$ bounded by $x = a$ and $x = b$. The sum of the areas of the inscribed rectangles in the figure below is the lower sum.

The sum of the areas of the circumscribed rectangles in the figure below is the upper sum.

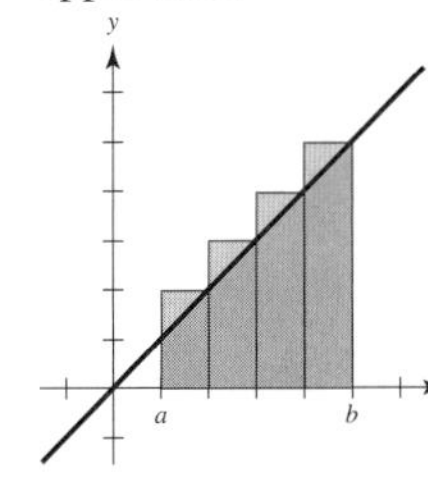

The rectangles in the first graph do not contain all of the area of the region, and the rectangles in the second graph cover more than the area of the region. The exact value of the area lies between these two sums.

77. (a)

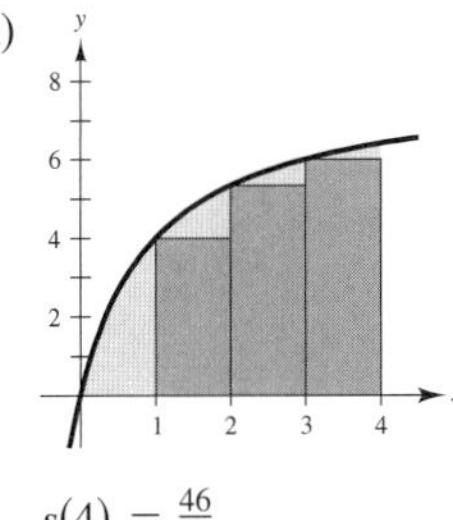

$s(4) = \frac{46}{3}$

(b)

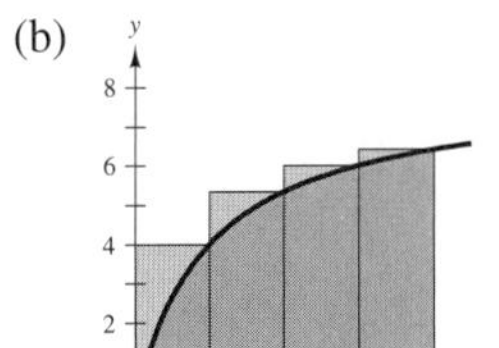

$S(4) = \frac{326}{15}$

(c)

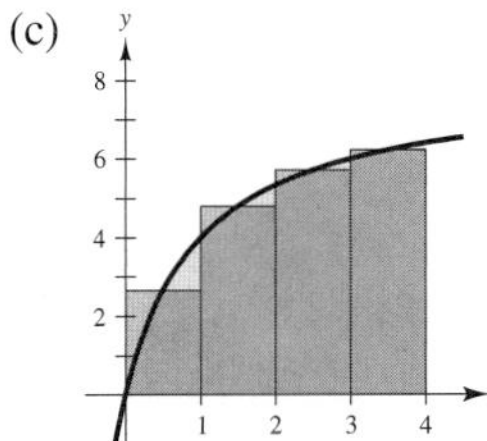

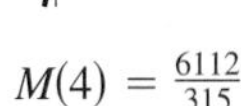

$M(4) = \frac{6112}{315}$

(d) Proof

(e)

n	4	8	20	100	200
$s(n)$	15.333	17.368	18.459	18.995	19.060
$S(n)$	21.733	20.568	19.739	19.251	19.188
$M(n)$	19.403	19.201	19.137	19.125	19.125

(f) Because f is an increasing function, $s(n)$ is always increasing and $S(n)$ is always decreasing.

79. True

81. Suppose there are n rows in the figure. The stars on the left total $1 + 2 + \cdots + n$, as do the stars on the right. There are $n(n + 1)$ stars in total. This means that $2[1 + 2 + \cdots + n] = n(n + 1)$. So $1 + 2 + \cdots + n = [n(n + 1)]/2$.

83. (a) $y = (-4.09 \times 10^{-5})x^3 + 0.016x^2 - 2.67x + 452.9$

(b)

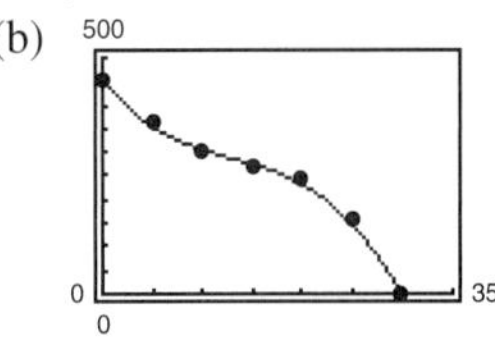

(c) $76,897.5 \text{ ft}^2$

85. Proof

Section 5.3 (page 314)

1. $2\sqrt{3} \approx 3.464$ **3.** 36 **5.** 0 **7.** $\frac{10}{3}$ **9.** $\int_{-1}^{5} (3x + 10)\, dx$

11. $\int_0^3 \sqrt{x^2 + 4}\, dx$ **13.** $\int_1^5 \left(1 + \frac{3}{x}\right) dx$ **15.** $\int_0^5 3\, dx$

17. $\int_1^4 \frac{2}{x} dx$ **19.** $\int_0^{\pi} \sin x\, dx$ **21.** $\int_0^2 y^3\, dy$

23.

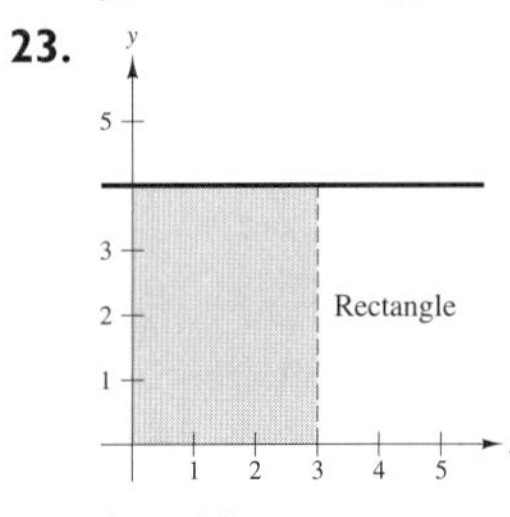

$A = 12$

25.

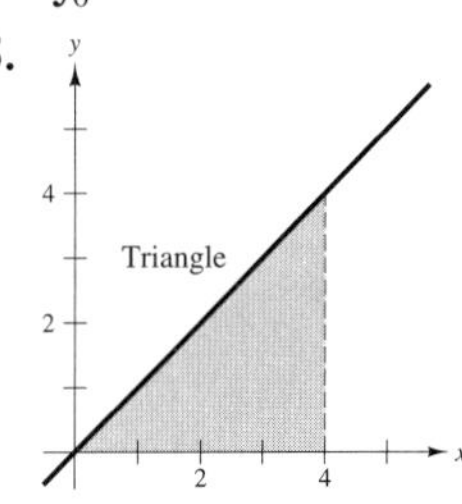

$A = 8$

27.

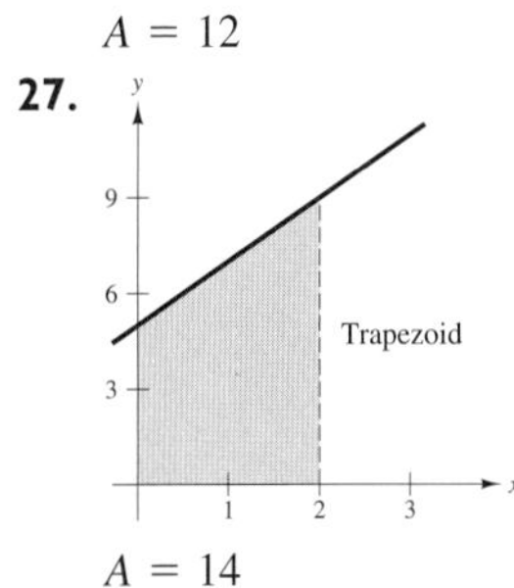

$A = 14$

29.

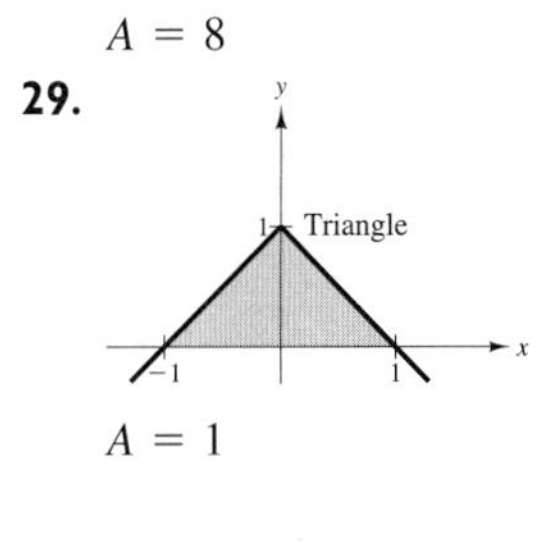

$A = 1$

31.

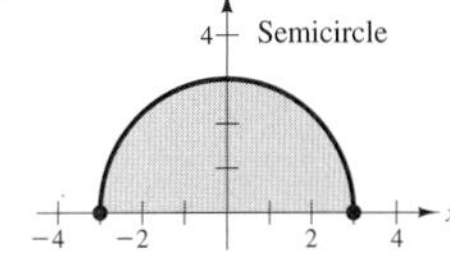

$A = 9\pi/2$

33. -6 **35.** 24 **37.** -10

39. 16 **41.** (a) 13 (b) -10 (c) 0 (d) 30

43. (a) 8 (b) -12 (c) -4 (d) 30 **45.** $-48, 88$

47. (a) $-\pi$ (b) 4 (c) $-(1 + 2\pi)$ (d) $3 - 2\pi$ (e) $5 + 2\pi$ (f) $23 - 2\pi$

49. (a) 14 (b) 4 (c) 8 (d) 0

51. $\sum_{i=1}^{n} f(x_i)\, \Delta x > \int_1^5 f(x)\, dx$

53. No. There is a discontinuity at $x = 4$. **55.** a **57.** c

59.

n	4	8	12	16	20
$L(n)$	3.6830	3.9956	4.0707	4.1016	4.1177
$M(n)$	4.3082	4.2076	4.1838	4.1740	4.1690
$R(n)$	3.6830	3.9956	4.0707	4.1016	4.1177

61.

n	4	8	12	16	20
$L(n)$	1.2833	1.1865	1.1562	1.1414	1.1327
$M(n)$	1.0898	1.0963	1.0976	1.0980	1.0982
$R(n)$	0.9500	1.0199	1.0451	1.0581	1.0660

63.

n	4	8	12	16	20
$L(n)$	0.5890	0.6872	0.7199	0.7363	0.7461
$M(n)$	0.7854	0.7854	0.7854	0.7854	0.7854
$R(n)$	0.9817	0.8836	0.8508	0.8345	0.8247

65. True **67.** True

69. False: $\int_0^2 (-x)\,dx = -2$ **71.** 272 **73.** Proof

75. No. No matter how small the subintervals, the number of both rational and irrational numbers within each subinterval is infinite and $f(c_i) = 0$ or $f(c_i) = 1$.

77. $a = -1$ and $b = 1$ maximize the integral. **79.** $\frac{1}{3}$

Section 5.4 (page 327)

1. Positive **3.** Zero

5. 1 **7.** $-\frac{5}{2}$ **9.** $-\frac{10}{3}$ **11.** $\frac{1}{3}$ **13.** $\frac{1}{2}$ **15.** $\frac{2}{3}$ **17.** -4

19. $-\frac{1}{18}$ **21.** $-\frac{27}{20}$ **23.** $\frac{9}{2}$ **25.** $\frac{23}{3}$ **27.** $\pi + 2$

29. $2\sqrt{3}/3$ **31.** $e^2 - 2$ **33.** 0 **35.** $(3/\ln 2) + 12$

37. $e - e^{-1}$ **39.** $\frac{1}{6}$ **41.** $12\sqrt{3}/5$ **43.** 1

45. 10 **47.** 6 **49.** 4 **51.** 0.4380, 1.7908

53. $\pm\arccos\sqrt{\pi}/2 \approx \pm 0.4817$ **55.** $3/\ln 4 \approx 2.1640$

57. Average value $= \frac{8}{3}$
$x = \pm 2\sqrt{3}/3 \approx \pm 1.155$

59. Average value $= e - e^{-1} \approx 2.3504$
$x = \ln((e - e^{-1})/2) \approx 0.1614$

61. Average value $= 2/\pi$
$x \approx 0.690, x \approx 2.451$

63. About 540 ft

65. The Fundamental Theorem of Calculus states that if a function f is continuous on $[a, b]$ and F is an antiderivative of f on $[a, b]$, then $\int_a^b f(x)\,dx = F(b) - F(a)$.

67. -1.5 **69.** 6.5 **71.** 15.5

73. (a) $F(x) = 500 \sec^2 x$ (b) $1500\sqrt{3}/\pi \approx 827$ N

75. ≈ 0.5318 L

77. (a) $v = -0.00086t^3 + 0.0782t^2 - 0.208t + 0.10$
(b) (c) 2475.6 m

79. $F(x) = \frac{1}{2}x^2 - 5x$
$F(2) = -8$
$F(5) = -\frac{25}{2}$
$F(8) = -8$

81. $F(x) = -10/x + 10$
$F(2) = 5$
$F(5) = 8$
$F(8) = \frac{35}{4}$

83. $F(x) = \sin x - \sin 1$
$F(2) = \sin 2 - \sin 1 \approx 0.0678$
$F(5) = \sin 5 - \sin 1 \approx -1.8004$
$F(8) = \sin 8 - \sin 1 \approx 0.1479$

85. (a) $g(0) = 0, g(2) \approx 7, g(4) \approx 9, g(6) \approx 8, g(8) \approx 5$
(b) Increasing: $(0, 4)$; Decreasing: $(4, 8)$
(c) A maximum occurs at $x = 4$.
(d)

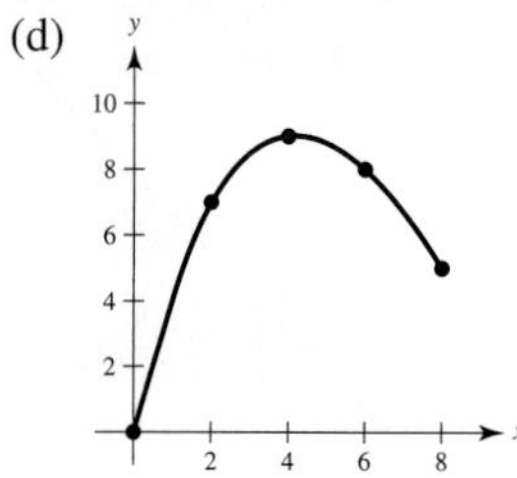

87. $\frac{1}{2}x^2 + 2x$ **89.** $\frac{3}{4}x^{4/3} - 12$ **91.** $\tan x - 1$

93. $e^x - e^{-1}$ **95.** $x^2 - 2x$ **97.** $\sqrt{x^4 + 1}$ **99.** $x \cos x$

101. 8 **103.** $\cos x\sqrt{\sin x}$ **105.** $3x^2 \sin x^6$

107.

An extremum of g occurs at $x = 2$.

109. 28 units **111.** 2 units **113.** True

115. $f(x) = x^{-2}$ has a nonremovable discontinuity at $x = 0$.

117. $f'(x) = \dfrac{1}{(1/x)^2 + 1}\left(-\dfrac{1}{x^2}\right) + \dfrac{1}{x^2 + 1} = 0$
Because $f'(x) = 0$, $f(x)$ is constant.

Section 5.5 (page 340)

	$\int f(g(x))g'(x)\,dx$	$u = g(x)$	$du = g'(x)\,dx$
1.	$\int (5x^2 + 1)^2(10x)\,dx$	$5x^2 + 1$	$10x\,dx$
3.	$\int \dfrac{x}{\sqrt{x^2 + 1}}\,dx$	$x^2 + 1$	$2x\,dx$
5.	$\int \tan^2 x \sec^2 x\,dx$	$\tan x$	$\sec^2 x\,dx$

7. $[(1 + 2x)^5]/5 + C$ **9.** $\frac{2}{3}(9 - x^2)^{3/2} + C$
11. $[(x^4 + 3)^3]/12 + C$ **13.** $[(x^3 - 1)^5]/15 + C$
15. $[(t^2 + 2)^{3/2}]/3 + C$ **17.** $-15/8(1 - x^2)^{4/3} + C$
19. $1/[4(1 - x^2)^2] + C$ **21.** $-1/[3(1 + x^3)] + C$
23. $-\sqrt{1 - x^2} + C$ **25.** $-\frac{1}{4}(1 + 1/t)^4 + C$ **27.** $\sqrt{2x} + C$
29. $\frac{2}{5}x^{5/2} + 2x^{3/2} + 14x^{1/2} + C = \frac{2}{5}\sqrt{x}(x^2 + 5x + 35) + C$
31. $\frac{1}{4}t^4 - t^2 + C$ **33.** $6y^{3/2} - \frac{2}{5}y^{5/2} + C = \frac{2}{5}y^{3/2}(15 - y) + C$
35. $2x^2 - 4\sqrt{16 - x^2} + C$ **37.** $-1/[2(x^2 + 2x - 3)] + C$

39. (a) Answers will vary
Example:

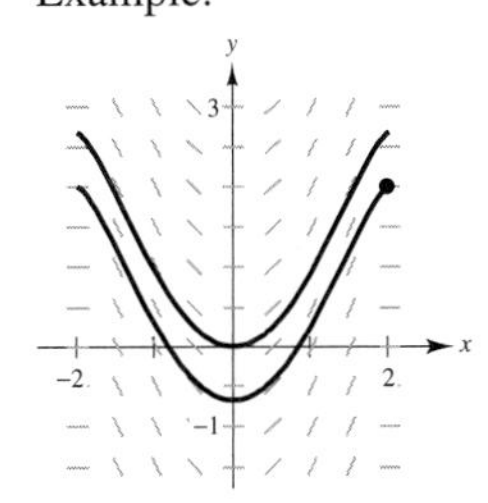

(b) $y = -\frac{1}{3}(4 - x^2)^{3/2} + 2$

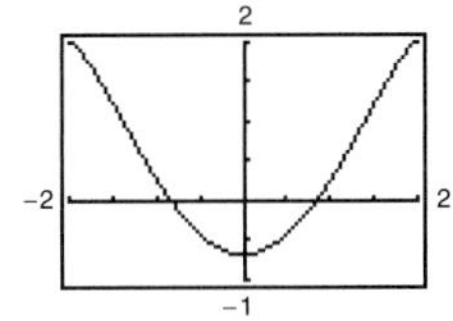

41. (a) Answers will vary.
Example:

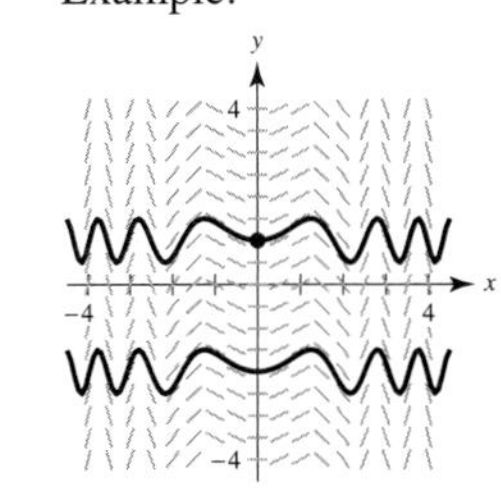

(b) $y = \frac{1}{2}\sin x^2 + 1$

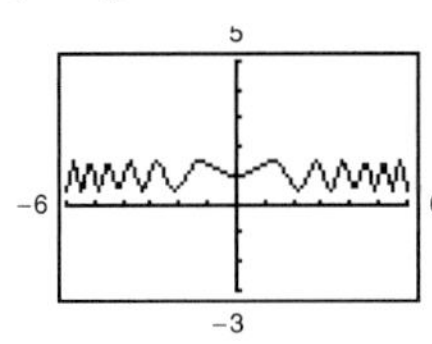

43. (a) Answers will vary.
Example:

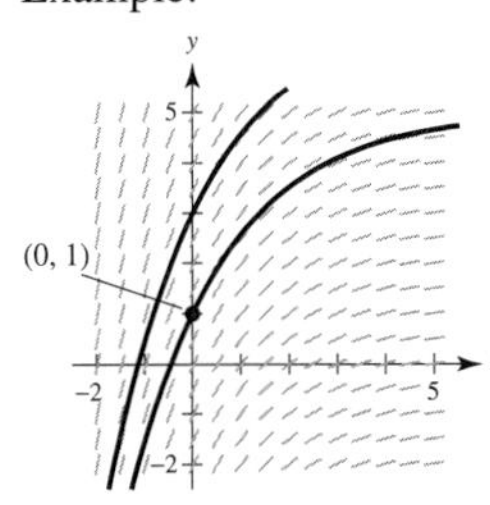

(b) $y = -4e^{-x/2} + 5$

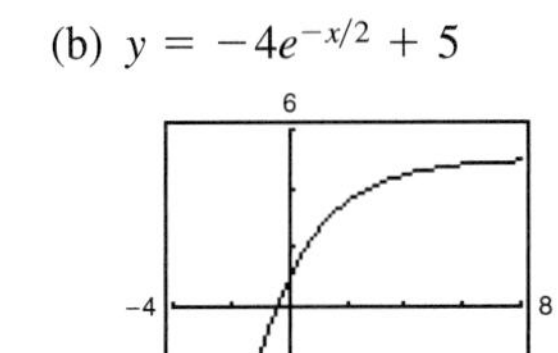

45. (a) Answers will vary.
Example:

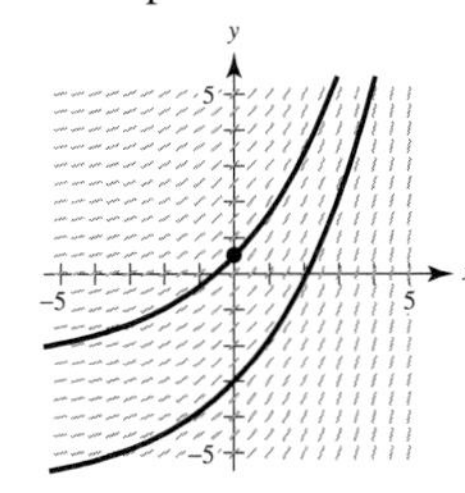

(b) $y = 3e^{x/3} - \frac{5}{2}$

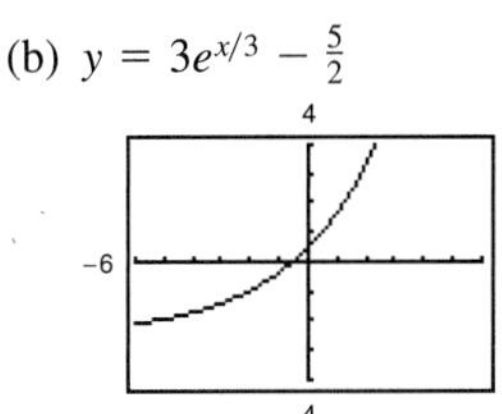

47. $-\cos(\pi x) + C$ **49.** $-\frac{1}{2}\cos 2x + C$ **51.** $-\sin(1/\theta) + C$
53. $e^{5x} + C$ **55.** $-\frac{1}{3}e^{-x^3} + C$
57. $\frac{1}{4}\sin^2 2x + C_1$ or $-\frac{1}{4}\cos^2 2x + C_2$ or $-\frac{1}{8}\cos 4x + C_3$
59. $\frac{1}{5}\tan^5 x + C$ **61.** $\frac{1}{2}\tan^2 x + C$ or $\frac{1}{2}\sec^2 x + C_2$
63. $-\cot x - x + C$ **65.** $\frac{1}{3}(e^x + 1)^3 + C$
67. $-\frac{2}{3}(1 - e^x)^{3/2} + C$ **69.** $-\frac{5}{2}e^{-2x} + e^{-x} + C$
71. $1/\pi\, e^{\sin \pi x} + C$ **73.** $-\tan(e^{-x}) + C$
75. $2/\ln 3(3^{x/2}) + C$ **77.** $-1/(2\ln 5)(5^{-x^2}) + C$
79. $f(x) = -\frac{1}{3}(4 - x^2)^{3/2} + 2$ **81.** $f(x) = 2\sin(x/2) + 3$
83. $f(x) = -8e^{-x/4} + 9$ **85.** $f(x) = \frac{1}{2}(e^x + e^{-x})$

87. $\frac{2}{15}(x + 2)^{3/2}(3x - 4) + C$
89. $-\frac{2}{105}(1 - x)^{3/2}(15x^2 + 12x + 8) + C$
91. $\left(\sqrt{2x - 1}/15\right)(3x^2 + 2x - 13) + C$
93. $-x - 1 - 2\sqrt{x + 1} + C$ or $-\left(x + 2\sqrt{x + 1}\right) + C_1$
95. 0 **97.** $12 - \left(8\sqrt{2}/9\right)$ **99.** 2 **101.** $(e^2 - 1)/2e^2$
103. $e/3(e^2 - 1)$ **105.** $\frac{1}{2}$ **107.** $\frac{4}{15}$ **109.** $3\sqrt{3}/4$
111. $7/\ln 4$ **113.** $f(x) = (2x^3 + 1)^3 + 3$
115. $f(x) = \sqrt{2x^2 - 1} - 3$
117. 1209/28 **119.** 4 **121.** $2\left(\sqrt{3} - 1\right)$

123. $e^5 - 1 \approx 147.413$

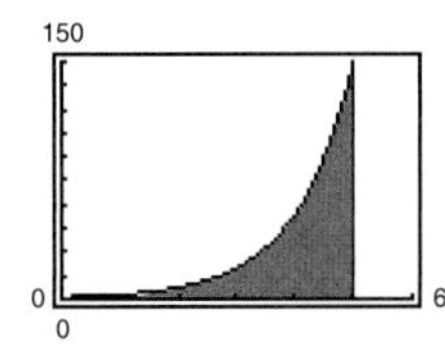

125. $2(1 - e^{-3/2}) \approx 1.554$

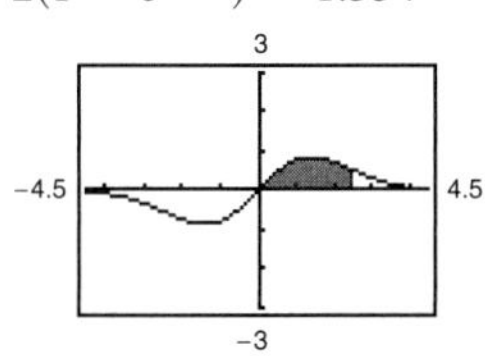

127. $\frac{10}{3}$

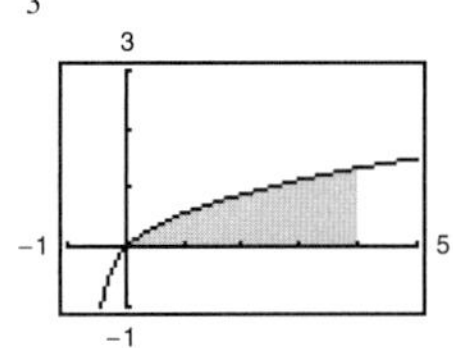

129. $\frac{144}{5}$

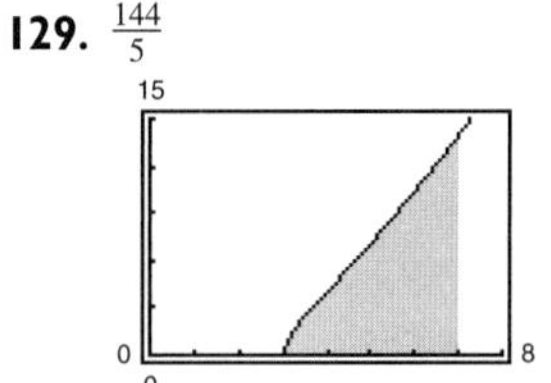

131. 7.38

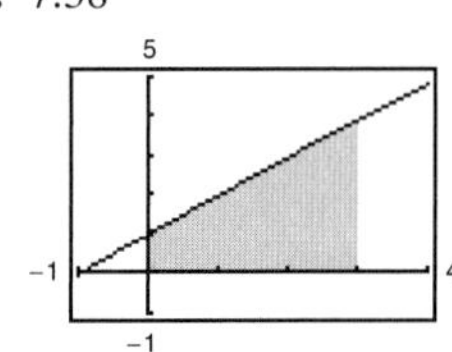

133. $1 - e^{-1} \approx 0.632$

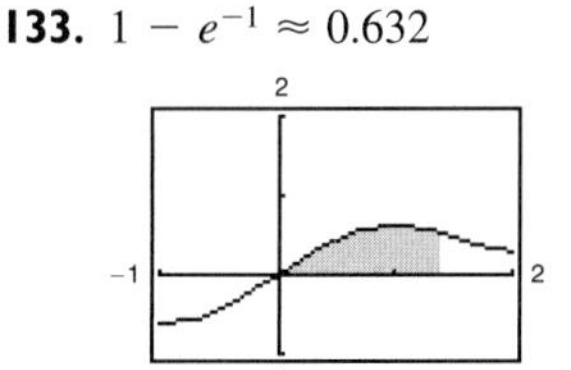

135. $\frac{1}{6}(2x - 1)^3 + C_1 = \frac{4}{3}x^3 - 2x^2 + x - \frac{1}{6} + C_1$
or $\frac{4}{3}x^3 - 2x^2 + x + C_2$
Answers differ by a constant: $C_2 = C_1 - \frac{1}{6}$
137. $\frac{272}{15}$ **139.** 0 **141.** (a) $\frac{8}{3}$ (b) $\frac{16}{3}$ (c) $-\frac{8}{3}$ (d) 8
143. $2\int_0^4 (6x^2 - 3)\, dx = 232$
145. If $u = 5 - x^2$, then $du = -2x\, dx$ and
$\int x(5 - x^2)^3\, dx = -\frac{1}{2}\int(5 - x^2)^3(-2x)\, dx = -\frac{1}{2}\int u^3\, du.$
147. $250,000
149. (a) Relative minimum: (6.4, 0.7) or July
Relative maximum: (0.4, 5.5) or January
(b) 37.47 in. (c) 4.33 in.
151. (a)

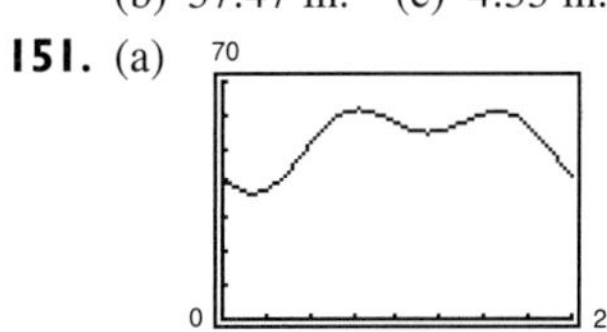

Maximum flow:
$R \approx 61.713$ at $t = 9.36$

(b) 1272 thousand gallons
153. (a) $P_{50,\,75} \approx 35.3\%$ (b) $b \approx 58.6\%$
155. 0.4772

157. (a)

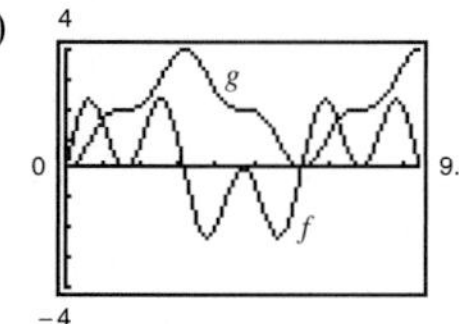

(b) g is nonnegative because the graph of f is positive at the beginning, and generally has more positive sections than negative ones.

(c) The points on g that correspond to the extrema of f are points of inflection of g.

(d) No, some zeros of f, such as $x = \pi/2$, do not correspond to extrema of g. The graph of g continues to increase after $x = \pi/2$ because f remains above the x-axis.

(e)

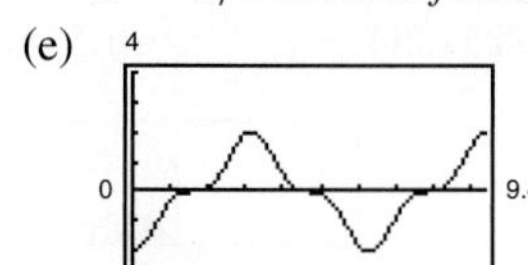

The graph of h is that of g shifted 2 units downward.

159. (a) Proof (b) Proof

161. False. $\int (2x + 1)^2\,dx = \frac{1}{6}(2x + 1)^3 + C$

163. True **165.** True **167–169.** Proofs

171. Putnam Problem A1, 1958

Section 5.6 (page 350)

	Trapezoidal	*Simpson's*	*Exact*
1.	2.7500	2.6667	2.6667
3.	4.2500	4.0000	4.0000
5.	4.0625	4.0000	4.0000
7.	12.6640	12.6667	12.6667
9.	0.1676	0.1667	0.1667

	Trapezoidal	*Simpson's*	*Graphing Utility*
11.	3.2833	3.2396	3.2413
13.	0.3415	0.3720	0.3927
15.	0.9567	0.9782	0.9775
17.	0.0891	0.0891	0.0891
19.	1.6845	1.6487	1.6479
21.	0.1940	0.1860	0.1858
23.	102.5553	93.3752	92.7437

25. The Trapezoidal Rule will yield a result greater than $\int_a^b f(x)\,dx$ if f is concave upward on $[a, b]$ because the graph of f will lie within the trapezoids.

27. (a) 0.500 (b) 0.000 **29.** (a) 0.1615 (b) 0.0066

31. (a) $n = 77$ (b) $n = 8$ **33.** (a) $n = 287$ (b) $n = 16$

35. (a) $n = 130$ (b) $n = 12$ **37.** (a) $n = 643$ (b) $n = 48$

39. (a) 24.5 (b) 25.67 **41.** Answers will vary.

43.

n	$L(n)$	$M(n)$	$R(n)$	$T(n)$	$S(n)$
4	0.8739	0.7960	0.6239	0.7489	0.7709
8	0.8350	0.7892	0.7100	0.7725	0.7803
10	0.8261	0.7881	0.7261	0.7761	0.7818
12	0.8200	0.7875	0.7367	0.7783	0.7826
16	0.8121	0.7867	0.7496	0.7808	0.7836
20	0.8071	0.7864	0.7571	0.7821	0.7841

45.

n	$L(n)$	$M(n)$	$R(n)$	$T(n)$	$S(n)$
4	0.7070	0.6597	0.6103	0.6586	0.6593
8	0.6833	0.6594	0.6350	0.6592	0.6593
10	0.6786	0.6594	0.6399	0.6592	0.6593
12	0.6754	0.6594	0.6431	0.6593	0.6593
16	0.6714	0.6594	0.6472	0.6593	0.6593
20	0.6690	0.6593	0.6496	0.6593	0.6593

47.

n	$L(n)$	$M(n)$	$R(n)$	$T(n)$	$S(n)$
4	2.5311	3.3953	4.3320	3.4316	3.4140
8	2.9632	3.4026	3.8637	3.4135	3.4074
10	3.0508	3.4037	3.7711	3.4109	3.4068
12	3.1094	3.4044	3.7100	3.4095	3.4065
16	3.1829	3.4050	3.6331	3.4080	3.4062
20	3.2273	3.4054	3.5874	3.4073	3.4061

49. (a) Trapezoidal Rule: 12.518
Simpson's Rule: 12.592

(b) $y = -1.3727x^3 + 4.0092x^2 - 0.6202x + 4.2844$; 12.53

51. 3.1416 **53.** 7435 m² **55.** $t \approx 2.477$

Section 5.7 (page 358)

1. $5 \ln|x| + C$ **3.** $\ln|x + 1| + C$ **5.** $-\frac{1}{2}\ln|3 - 2x| + C$

7. $\ln\sqrt{x^2 + 1} + C$ **9.** $x^2/2 - \ln(x^4) + C$

11. $\frac{1}{3}\ln|x^3 + 3x^2 + 9x| + C$

13. $\frac{1}{2}x^2 - 4x + 6\ln|x + 1| + C$ **15.** $\frac{1}{3}x^3 + 5\ln|x - 3| + C$

17. $\frac{1}{3}x^3 - 2x + \ln\sqrt{x^2 + 2} + C$ **19.** $\frac{1}{3}(\ln x)^3 + C$

21. $2\sqrt{x + 1} + C$ **23.** $2\ln|x - 1| - 2/(x - 1) + C$

25. $\sqrt{2x} - \ln\left|1 + \sqrt{2x}\right| + C$

27. $x + 6\sqrt{x} + 18\ln\left|\sqrt{x} - 3\right| + C$ **29.** $\ln|\sin\theta| + C$

31. $-\frac{1}{2}\ln|\csc 2x + \cot 2x| + C$ **33.** $\ln|1 + \sin t| + C$

35. $\ln|\sec x - 1| + C$ **37.** $\ln|\cos(e^{-x})| + C$

39. $y = -3\ln|2 - x| + C$ **41.** $s = -\frac{1}{2}\ln|\cos 2\theta| + C$

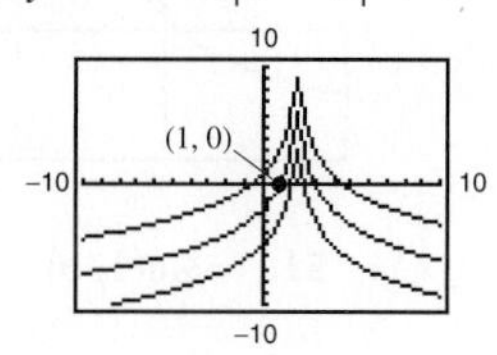

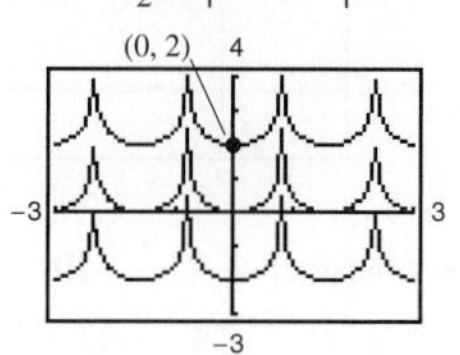

The graph has a hole at $x = 2$.

43. $f(x) = -2\ln x + 3x - 2$

45. (a) (b) $y = \ln|(x + 2)/2| + 1$

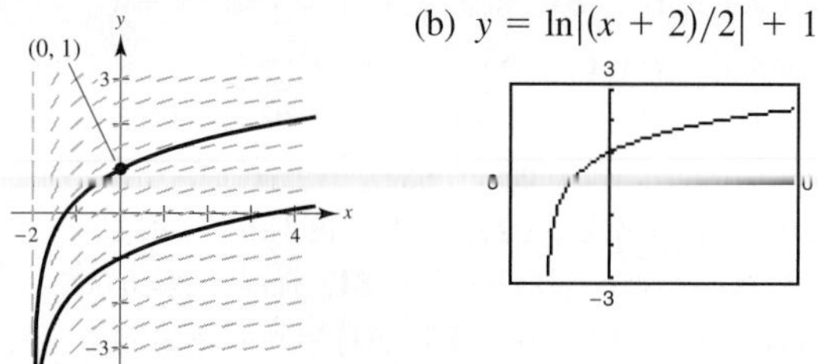

47. (a)

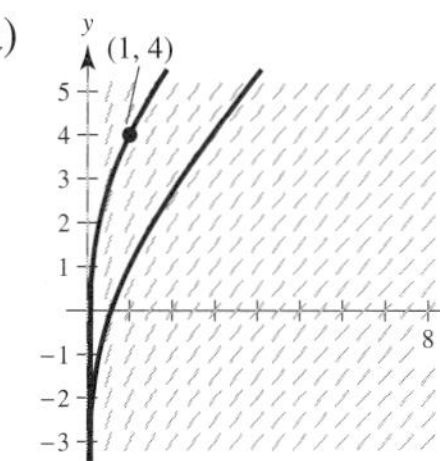

(b) $y = \ln x + x + 3$

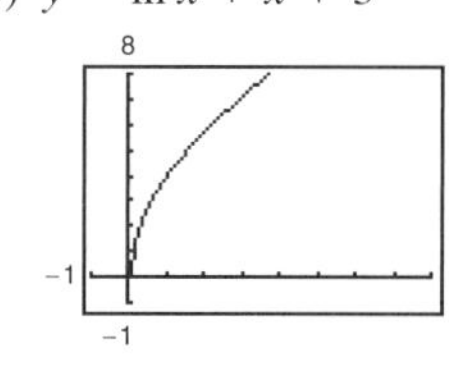

49. $\frac{5}{3}\ln 13 \approx 4.275$ **51.** $\frac{7}{3}$ **53.** $-\ln 3 \approx -1.099$

55. $\ln|(2 - \sin 2)/(1 - \sin 1)| \approx 1.929$

57. $2\left[\sqrt{x} - \ln\left(1 + \sqrt{x}\right)\right] + C$

59. $\ln\left[\left(\sqrt{x} - 1\right)/\left(\sqrt{x} + 1\right)\right] + 2\sqrt{x} + C$

61. $\ln\left(\sqrt{2} + 1\right) - \sqrt{2}/2 \approx 0.174$

63. $1/x$ **65.** $1/x$ **67.** d **69.** $4 \ln 3$ **71.** $\frac{1}{2}\ln 2$

73. $\frac{15}{2} + 8 \ln 2 \approx 13.045$

75. $(12/\pi)\left[2\ln\left(\sqrt{3} + 1\right) - \ln 2\right] \approx 5.03$

77. Trapezoidal Rule: 20.2
Simpson's Rule: 19.4667

79. Trapezoidal Rule: 5.3368
Simpson's Rule: 5.3632

81. Power Rule **83.** Log Rule

85. $-\ln|\cos x| + C = \ln|1/\cos x| + C = \ln|\sec x| + C$

87. $\ln|\sec x + \tan x| + C = \ln\left|\dfrac{\sec^2 x - \tan^2 x}{\sec x - \tan x}\right| + C = -\ln|\sec x - \tan x| + C$

89. 1 **91.** $1/[2(e - 1)] \approx 0.291$

93. $P(t) = 1000(12\ln|1 + 0.25t| + 1)$; $P(3) \approx 7715$

95. $168.27

97. (a)

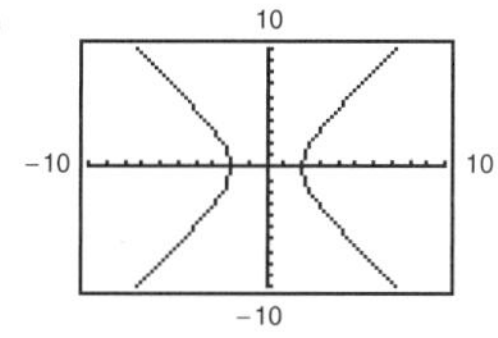

(b) Answers will vary. Example: $y^2 = e^{-\ln x + \ln 4} = 4/x$

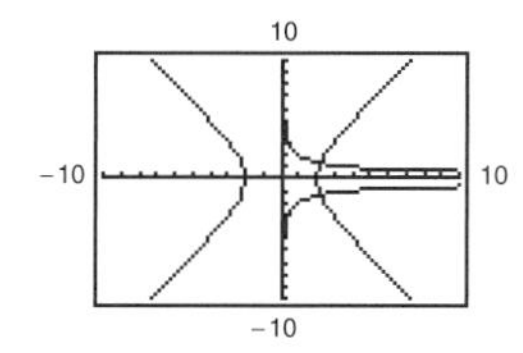

(c) Answers will vary.

99. False. $\frac{1}{2}(\ln x) = \ln x^{1/2}$

101. True

103.

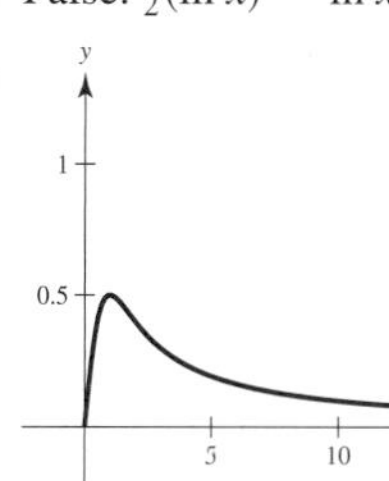

(a) $\frac{1}{2}\ln 2 - \frac{1}{4} \approx 0.0966$
(b) $0 < m < 1$
(c) $\frac{1}{2}(m - \ln m - 1)$

105. Proof

Section 5.8 (page 366)

1. $5 \arcsin \dfrac{x}{3} + C$ **3.** $\dfrac{7}{4}\arctan\dfrac{x}{4} + C$ **5.** $\operatorname{arcsec}|2x| + C$

7. $\frac{1}{2}x^2 - \frac{1}{2}\ln(x^2 + 1) + C$ **9.** $\arcsin(x + 1) + C$

11. $\frac{1}{2}\arcsin t^2 + C$ **13.** $\frac{1}{4}\arctan(e^{2x}/2) + C$

15. $2\arcsin\sqrt{x} + C$ **17.** $\frac{1}{2}\ln(x^2 + 1) - 3\arctan x + C$

19. $8\arcsin[(x - 3)/3] - \sqrt{6x - x^2} + C$ **21.** $\pi/18$

23. $\pi/6$ **25.** $\frac{1}{32}\pi^2 \approx 0.308$ **27.** $\frac{1}{2}\left(\sqrt{3} - 2\right) \approx -0.134$

29. $\pi/4$ **31.** $\pi/2$

33. $\ln|x^2 + 6x + 13| - 3\arctan[(x + 3)/2] + C$

35. $\arcsin[(x + 2)/2] + C$ **37.** $-\sqrt{-x^2 - 4x} + C$

39. $4 - 2\sqrt{3} + \frac{1}{6}\pi \approx 1.059$ **41.** $\frac{1}{2}\arctan(x^2 + 1) + C$

43. $2\sqrt{e^t - 3} - 2\sqrt{3}\arctan\left(\sqrt{e^t - 3}/\sqrt{3}\right) + C$

45. $\pi/6$ **47.** a and b **49.** a, b, and c **51.** c

53. $y = \arcsin(x/2) + \pi$

55. (a)

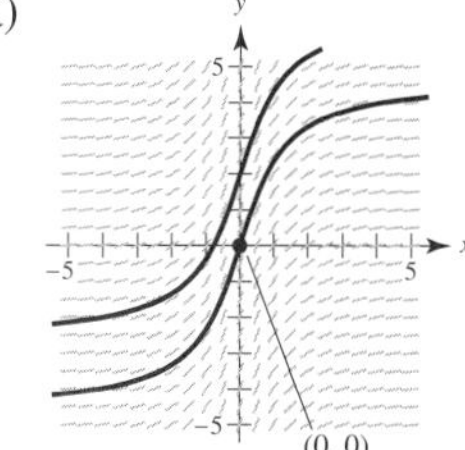

(b) $y = 3\arctan x$

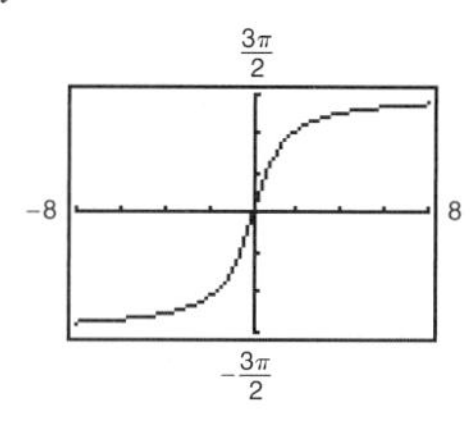

57. (a)

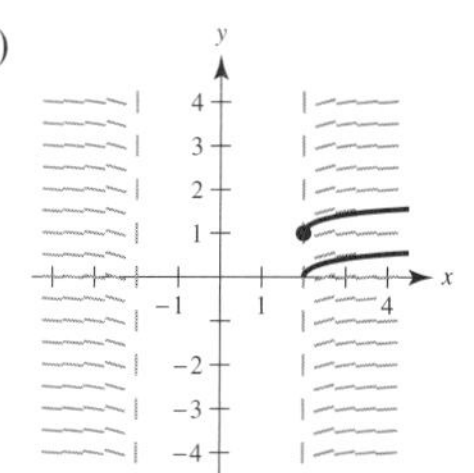

(b) $y = \frac{1}{2}\operatorname{arcsec}(x/2) + 1$, $x \geq 2$

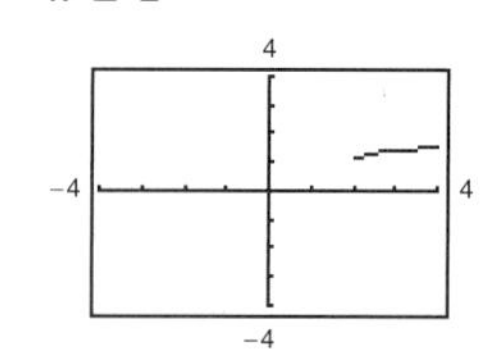

59.

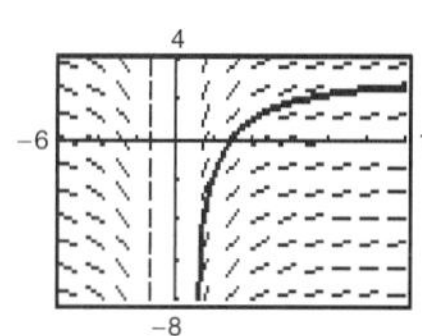

61.

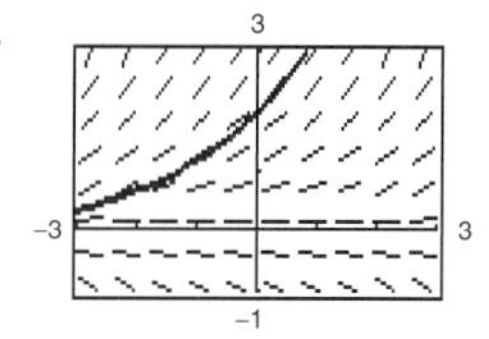

63. $\pi/8$ **65.** $\pi/6$ **67.** $3\pi/2$

69. (a) Proof (b) $\ln\left(\sqrt{6}/2\right) + \left(9\pi - 4\pi\sqrt{3}\right)/36$

71. (a)

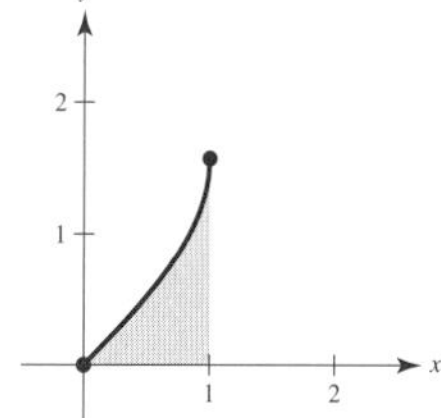

(b) 0.5708
(c) $(\pi - 2)/2$

73. (a) $F(x)$ represents the average value of $f(x)$ over the interval $[x, x + 2]$. Maximum at $x = -1$.
(b) $x = -1$

75. False. $\displaystyle\int \frac{dx}{3x\sqrt{9x^2 - 16}} = \frac{1}{12}\operatorname{arcsec}\frac{|3x|}{4} + C$

77. True **79–81.** Proofs

83. (a) $v(t) = -32t + 500$

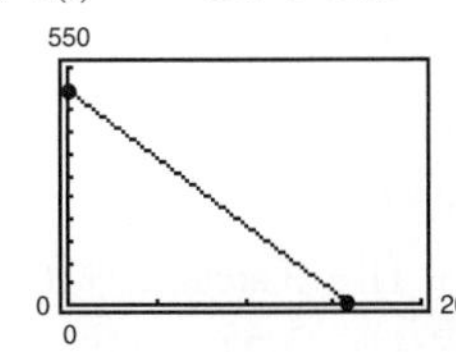

(b) $s(t) = -16t^2 + 500t$; 3906.25 ft

(c) $v(t) = \sqrt{\dfrac{32}{k}}\tan\left[\arctan\left(500\sqrt{\dfrac{k}{32}}\right) - \sqrt{32k}\,t\right]$

(d)

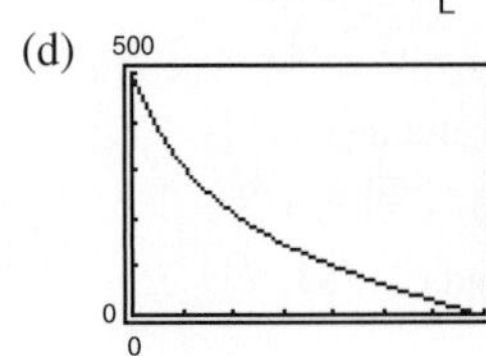

$t_0 = 6.86$ sec

(e) 1088 ft

(f) When air resistance is taken into account, the maximum height of the object is not as great.

Section 5.9 (page 377)

1. (a) 10.018 (b) -0.964 **3.** (a) $\frac{4}{3}$ (b) $\frac{13}{12}$

5. (a) 1.317 (b) 0.962 **7–11.** Proofs

13. $\cosh x = \sqrt{13}/2$; $\tanh x = 3\sqrt{13}/13$; $\operatorname{csch} x = 2/3$; $\operatorname{sech} x = 2\sqrt{13}/13$; $\coth x = \sqrt{13}/3$

15. $-\operatorname{sech}(x+1)\tanh(x+1)$ **17.** $\coth x$ **19.** $\operatorname{csch} x$

21. $\sinh^2 x$ **23.** $\operatorname{sech} t$ **25.** $y = -2x + 2$ **27.** $y = 1 - 2x$

29. Relative maxima: $(\pm\pi, \cosh \pi)$; Relative minimum: $(0, -1)$

31. Relative maximum: $(1.20, 0.66)$
Relative minimum: $(-1.20, -0.66)$

33. $y = a \sinh x$; $y' = a \cosh x$; $y'' = a \sinh x$; $y''' = a \cosh x$;
So, $y''' - y' = 0$.

35. $P_1(x) = x$; $P_2(x) = x$

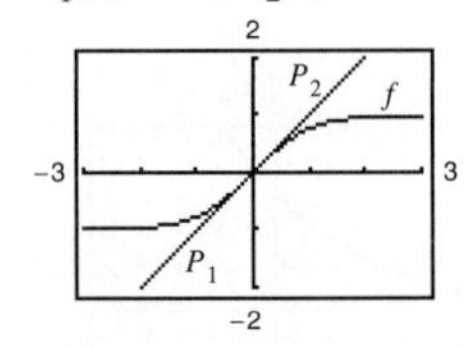

37. (a)

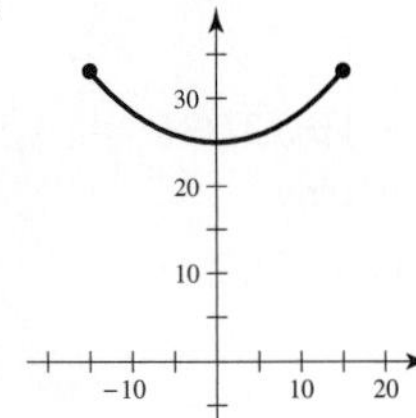

(b) 33.146 units; 25 units
(c) $m = \sinh(1) \approx 1.175$

39. $-\frac{1}{2}\cosh(1 - 2x) + C$ **41.** $\frac{1}{3}\cosh^3(x - 1) + C$

43. $\ln|\sinh x| + C$ **45.** $-\coth(x^2/2) + C$

47. $\operatorname{csch}(1/x) + C$ **49.** $\frac{1}{2}\arctan x^2 + C$ **51.** $\ln 5 - 2\ln 2$

53. $\frac{1}{6}\ln 3$ **55.** $\pi/4$ **57.** $3/\sqrt{9x^2 - 1}$ **59.** $|\sec x|$

61. $2\sec 2x$ **63.** $2\sinh^{-1}(2x)$ **65.** Answers will vary.

67. ∞ **69.** 0 **71.** 1 **73.** $\ln\left(\sqrt{e^{2x} + 1} - 1\right) - x + C$

75. $2\sinh^{-1}\sqrt{x} + C = 2\ln\left(\sqrt{x} + \sqrt{1 + x}\right) + C$

77. $\dfrac{1}{4}\ln\left|\dfrac{x - 4}{x}\right| + C$ **79.** $\dfrac{1}{2\sqrt{6}}\ln\left|\dfrac{\sqrt{2}(x + 1) + \sqrt{3}}{\sqrt{2}(x + 1) - \sqrt{3}}\right| + C$

81. $\dfrac{1}{4}\arcsin\left(\dfrac{4x - 1}{9}\right) + C$ **83.** $-\dfrac{x^2}{2} - 4x - \dfrac{10}{3}\ln\left|\dfrac{x - 5}{x + 1}\right| + C$

85. $8\arctan(e^2) - 2\pi \approx 5.207$ **87.** $\frac{5}{2}\ln\left(\sqrt{17} + 4\right) \approx 5.237$

89. (a) $\ln\left(\sqrt{3} + 2\right)$ (b) $\sinh^{-1}\sqrt{3}$

91. $\frac{52}{31}$ kg **93.** $-\sqrt{a^2 - x^2}/x$ **95–101.** Proofs

103. Putnam Problem 8, 1939

Review Exercises for Chapter 5 (page 380)

1.

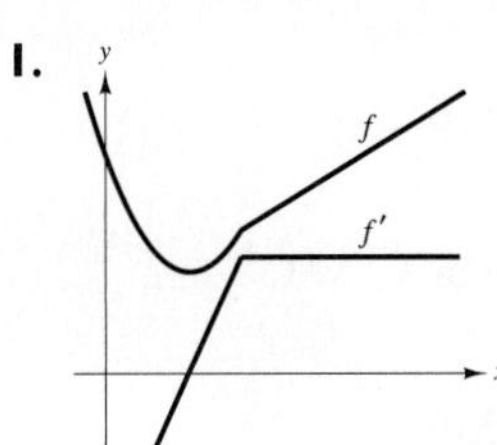
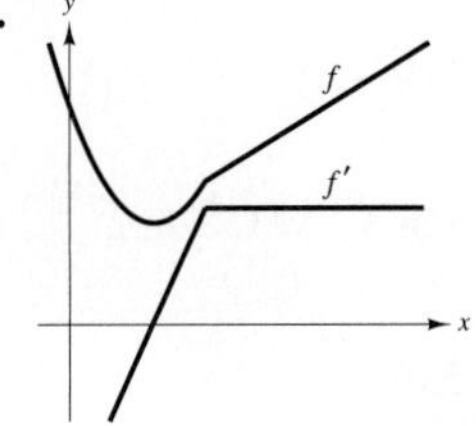

3. $\frac{2}{3}x^3 + \frac{1}{2}x^2 - x + C$

5. $x^2/2 - 1/x + C$ **7.** $2x^2 + 3\cos x + C$

9. $5x - e^x + C$ **11.** $5\ln|x| + C$ **13.** $y = 2 - x^2$

15. 240 ft/sec **17.** (a) 3 sec (b) 144 ft (c) $\frac{3}{2}$ sec (d) 108 ft

19. (a) $\displaystyle\sum_{i=1}^{10}(2i - 1)$ (b) $\displaystyle\sum_{i=1}^{n} i^3$ (c) $\displaystyle\sum_{i=1}^{10}(4i + 2)$

21. $9.038 < (\text{area of region}) < 13.038$

23. $A = 16$

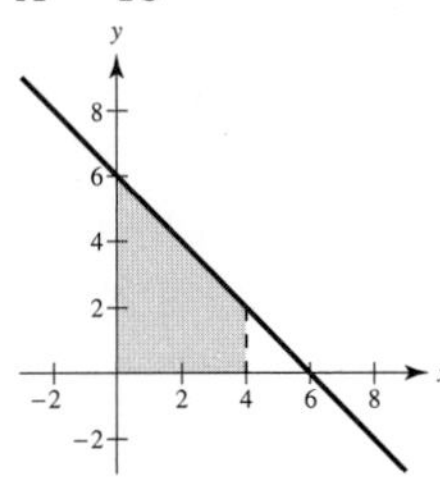

25. $A = 12$

27. $\frac{27}{2}$ **29.** $\displaystyle\int_4^6 (2x - 3)\,dx$

31. $A = \frac{25}{2}$

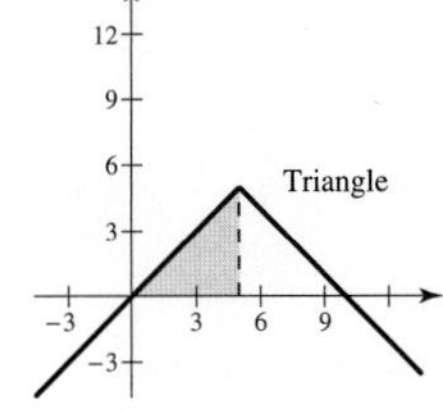

33. (a) 13 (b) 7 (c) 11 (d) 50 **35.** c **37.** 16 **39.** 0

41. $\frac{422}{5}$ **43.** $\left(\sqrt{2} + 2\right)/2$ **45.** $e^2 + 1$

47. $A = 6$

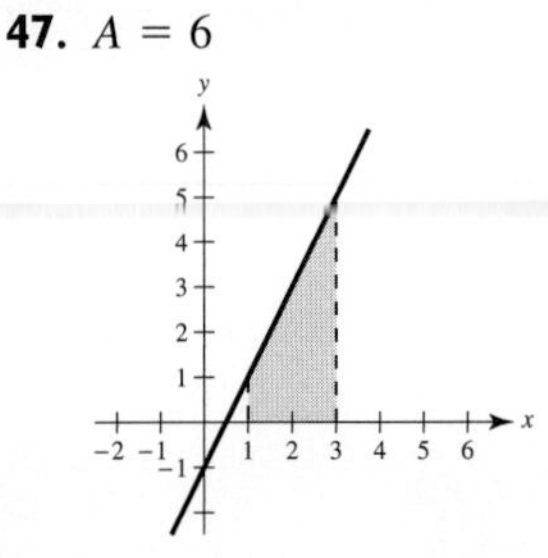

49. $A = \frac{10}{3}$

51. $A = \frac{1}{4}$

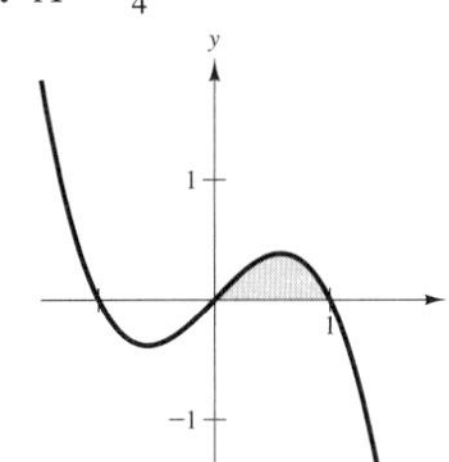

53. $A = 16$

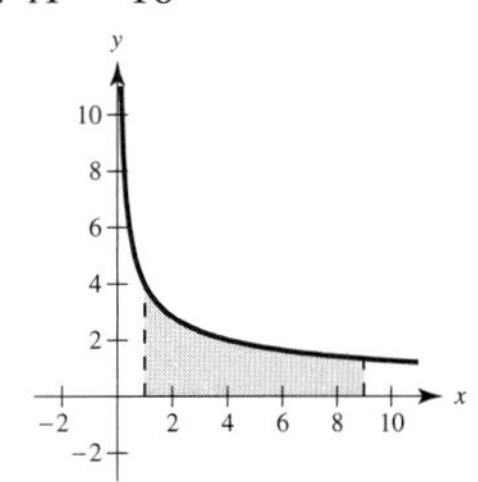

55. $A = 2 \ln 3 \approx 2.1972$

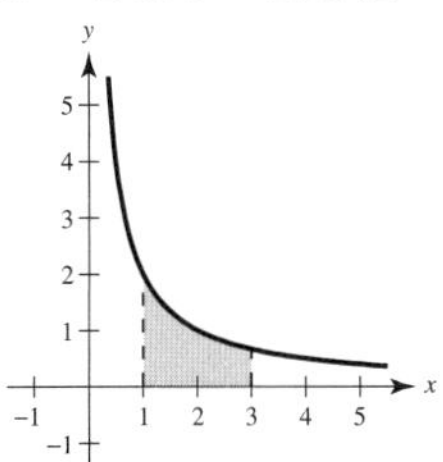

57. Average value $= \frac{2}{5}, x = \frac{25}{4}$

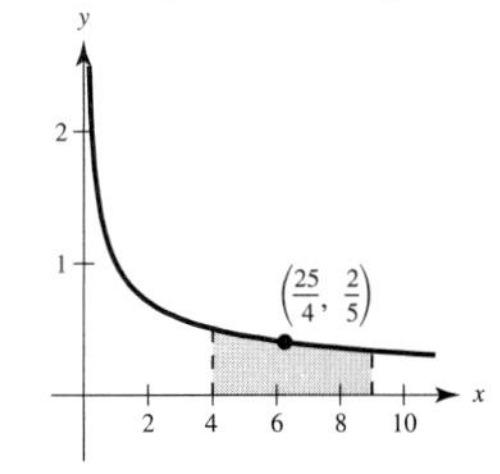

59. $x^2\sqrt{1 + x^3}$ **61.** $x^2 + 3x + 2$

63. $\frac{1}{7}x^7 + \frac{3}{5}x^5 + x^3 + x + C$ **65.** $\frac{2}{3}\sqrt{x^3 + 3} + C$

67. $-\frac{1}{30}(1 - 3x^2)^5 + C = (3x^2 - 1)^5/30 + C$

69. $\frac{1}{4}\sin^4 x + C$ **71.** $2\sqrt{1 - \cos \theta} + C$

73. $\dfrac{\tan^{n+1} x}{n + 1} + C, n \neq -1$ **75.** $\dfrac{1}{3\pi}(1 + \sec \pi x)^3 + C$

77. $-\frac{1}{6}e^{-3x^2} + C$ **79.** $\dfrac{1}{2 \ln 5}(5^{(x+1)^2}) + C$

81. $-9/4$ **83.** 2 **85.** $28\pi/15$ **87.** 2

89. (a) $27{,}300/M$ (b) $28{,}500/M$

91. Trapezoidal Rule: 0.257
Simpson's Rule: 0.254
Graphing utility: 0.254

93. Trapezoidal Rule: 0.637
Simpson's Rule: 0.685
Graphing utility: 0.704

95. Trapezoidal Rule: 1.463
Simpson's Rule: 1.494
Graphing utility: 1.494

97. $\frac{1}{7}\ln|7x - 2| + C$ **99.** $-\ln|1 + \cos x| + C$

101. $3 + \ln 4$ **103.** $\ln(2 + \sqrt{3})$ **105.** $\frac{1}{2}\ln(e^{2x} + e^{-2x}) + C$

107. $\frac{1}{2}\arctan(e^{2x}) + C$ **109.** $\frac{1}{2}\arcsin x^2 + C$

111. $\ln\sqrt{16 + x^2} + C$ **113.** $\frac{1}{4}(\arctan x/2)^2 + C$

115. $y = A \sin\left(\sqrt{k/m}\,t\right)$ **117.** $2 - \left[\left(\sinh\sqrt{x}\right)/\left(2\sqrt{x}\right)\right]$

119. $\frac{1}{2}\ln\left(\sqrt{x^4 - 1} + x^2\right) + C$

P.S. Problem Solving (page 383)

1. (a) $L(1) = 0$ (b) $L'(x) = 1/x, L'(1) = 1$
(c) $x \approx 2.718$ (d) Proof

3. (a)

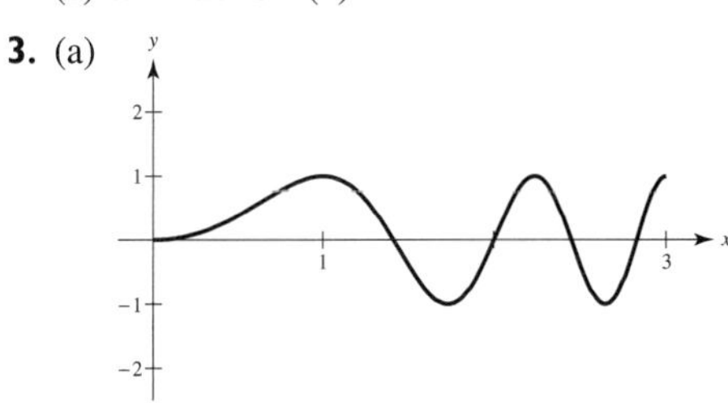

(b)

(c) Relative maxima at $x = \sqrt{2}, \sqrt{6}$
Relative minima at $x = 2, 2\sqrt{2}$
(d) Points of inflection at $x = 1, \sqrt{3}, \sqrt{5}, \sqrt{7}$

5. (a)

(b)

x	0	1	2	3	4	5	6	7	8
$F(x)$	0	$-\frac{1}{2}$	-2	$-\frac{7}{2}$	-4	$-\frac{7}{2}$	-2	$\frac{1}{4}$	3

(c) $x = 4, 8$ (d) $x = 2$

7. (a) 1.6758; Error of approximation ≈ 0.0071
(b) $\frac{3}{2}$ (c) Proof

9. Proof

11. $\displaystyle \lim_{n\to\infty} \sum_{t=1}^{n} \left(\frac{t}{n}\right)^5\left(\frac{1}{n}\right) = \frac{1}{6}$ **13.** $\displaystyle 1 \le \int_0^1 \sqrt{1 + x^4}\, dx \le \sqrt{2}$

15. Proof **17.** $2\ln\left(\frac{3}{2}\right) \approx 0.8109$

19. (a) (i)

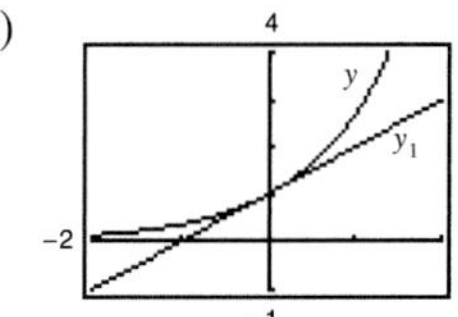

(ii)

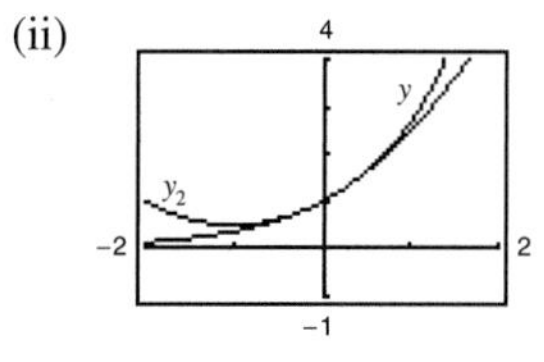

(iii)

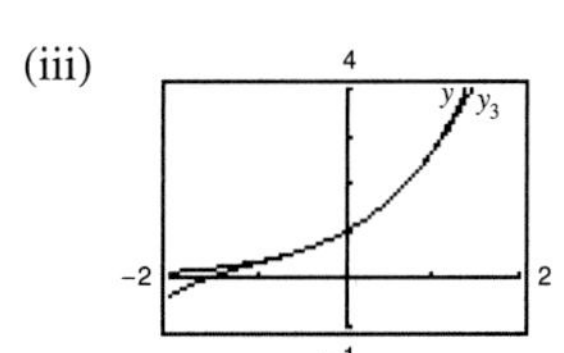

(b) $\displaystyle \sum_{n=0}^{\infty} \frac{x^n}{n!}$

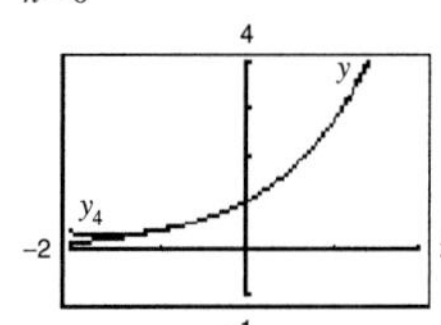

(c) $\displaystyle e^x = \sum_{n=0}^{\infty} \frac{x^n}{n!}$

Chapter 6

Section 6.1 (page 391)

1–11. Proofs **13.** Not a solution **15.** Solution
17. Solution **19.** Not a solution **21.** Solution
23. Not a solution **25.** $y = 3e^{-x/2}$ **27.** $4y^2 = x^3$

29.

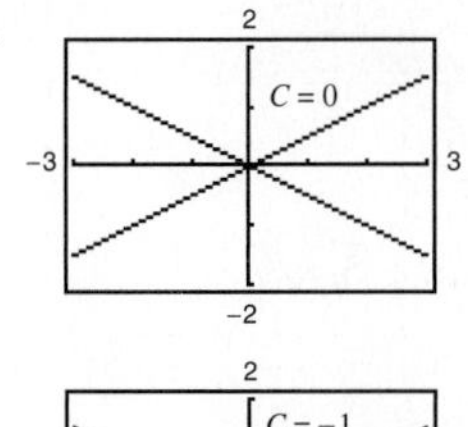

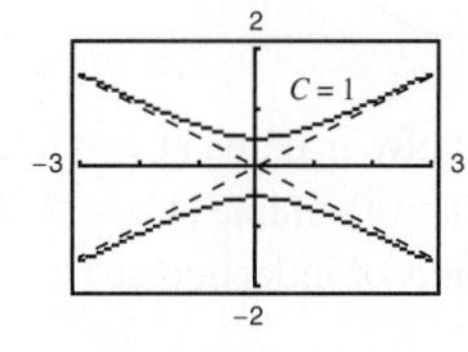

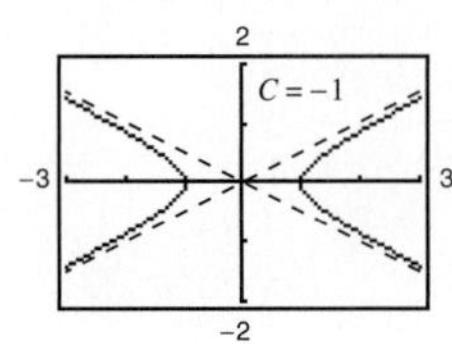

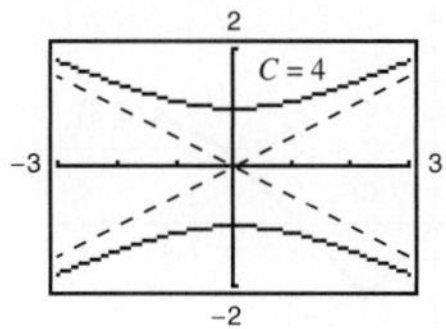

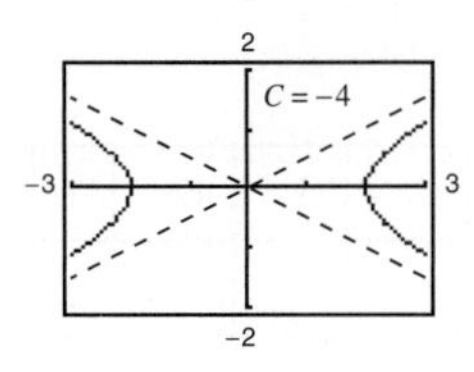

31. $y = 3e^{-2x}$ **33.** $y = 2\sin 3x - \frac{1}{3}\cos 3x$
35. $y = -2x + \frac{1}{2}x^3$ **37.** $y = x^3 + C$
39. $y = \frac{1}{2}\ln(1 + x^2) + C$ **41.** $y = x - \ln x^2 + C$
43. $y = -\frac{1}{2}\cos 2x + C$
45. $y = \frac{2}{5}(x - 3)^{5/2} + 2(x - 3)^{3/2} + C$
47. $y = \frac{1}{2}e^{x^2} + C$

49.

x	-4	-2	0	2	4	8
y	2	0	4	4	6	8
dy/dx	-2	Undef.	0	$\frac{1}{2}$	$\frac{2}{3}$	1

51.

x	-4	-2	0	2	4	8
y	2	0	4	4	6	8
dy/dx	$-2\sqrt{2}$	-2	0	0	$-2\sqrt{2}$	-8

53. b **54.** c **55.** d **56.** a

57. (a) and (b)

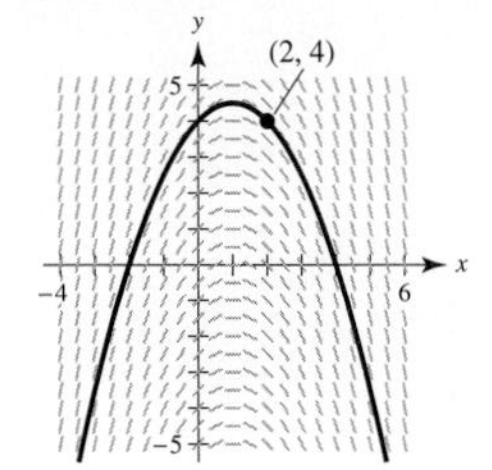

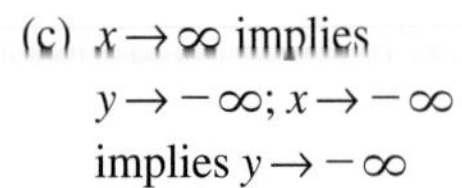
(c) $x \to \infty$ implies $y \to -\infty$; $x \to -\infty$ implies $y \to -\infty$

59. (a) and (b)

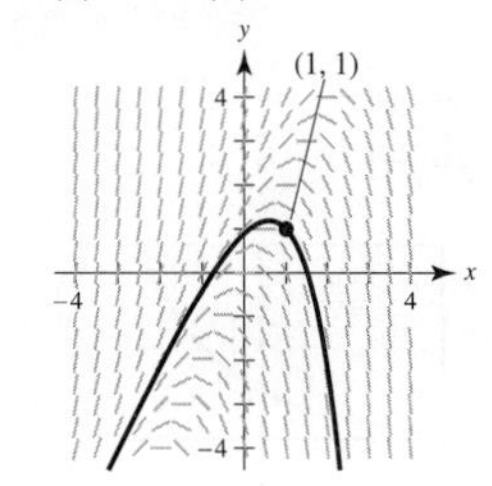

(c) $x \to \infty$ implies $y \to -\infty$; $x \to -\infty$ implies $y \to -\infty$

61. (a)

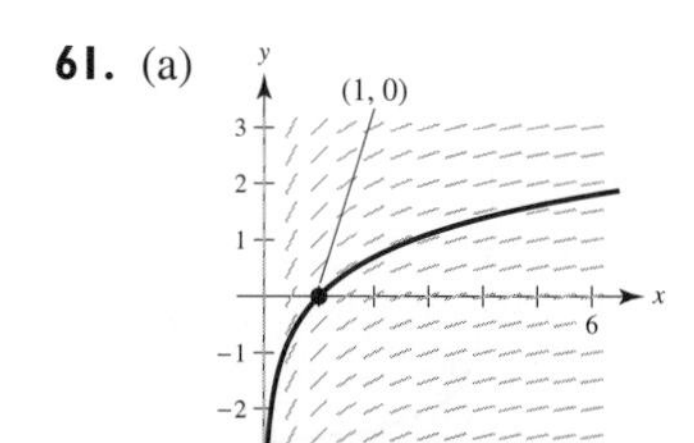

$x \to \infty, y \to \infty$

(b)

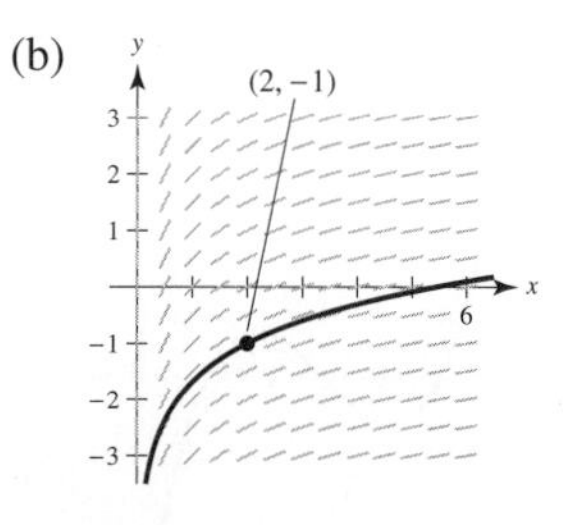

$x \to \infty, y \to \infty$

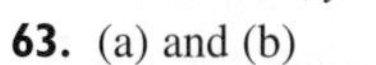
63. (a) and (b)

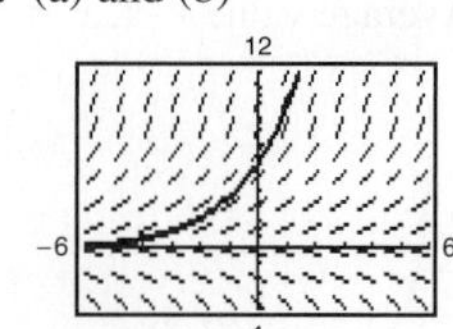

65. (a) and (b)

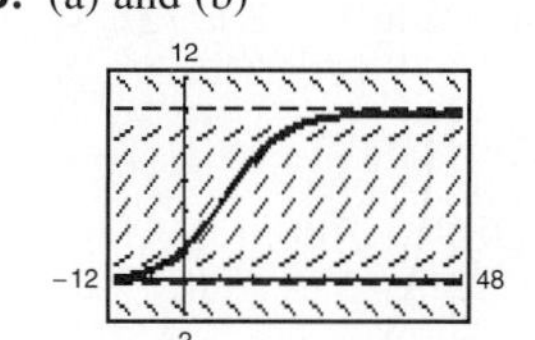

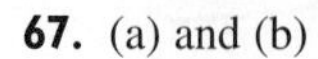
67. (a) and (b)

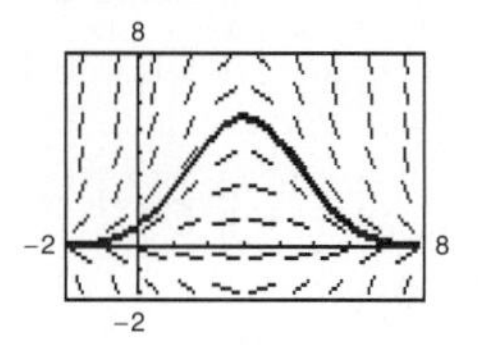

69.

n	0	1	2	3	4	5	6
x_n	0	0.1	0.2	0.3	0.4	0.5	0.6
y_n	2	2.2	2.43	2.693	2.992	3.332	3.715

n	7	8	9	10
x_n	0.7	0.8	0.9	1.0
y_n	4.146	4.631	5.174	5.781

71.

n	0	1	2	3	4	5	6
x_n	0	0.05	0.1	0.15	0.2	0.25	0.3
y_n	3	2.7	2.438	2.209	2.010	1.839	1.693

n	7	8	9	10
x_n	0.35	0.4	0.45	0.5
y_n	1.569	1.464	1.378	1.308

73.

n	0	1	2	3	4	5	6
x_n	0	0.1	0.2	0.3	0.4	0.5	0.6
y_n	1	1.1	1.212	1.339	1.488	1.670	1.900

n	7	8	9	10
x_n	0.7	0.8	0.9	1.0
y_n	2.213	2.681	3.510	5.958

75.

x	0	0.2	0.4	0.6	0.8	1
$y(x)$ (exact)	3.0000	3.6642	4.4755	5.4664	6.6766	8.1548
$y(x)$ $(h = 0.2)$	3.0000	3.6000	4.3200	5.1840	6.2208	7.4650
$y(x)$ $(h = 0.1)$	3.0000	3.6300	4.3923	5.3147	6.4308	7.7812

77.

x	0	0.2	0.4	0.6	0.8	1
$y(x)$ (exact)	0.0000	0.2200	0.4801	0.7807	1.1231	1.5097
$y(x)$ $(h = 0.2)$	0.0000	0.2000	0.4360	0.7074	1.0140	1.3561
$y(x)$ $(h = 0.1)$	0.0000	0.2095	0.4568	0.7418	1.0649	1.4273

79. (a) $y(1) = 112.7141°$; $y(2) = 96.3770°$; $y(3) = 86.5954°$
(b) $y(1) = 113.2441°$; $y(2) = 97.0158°$; $y(3) = 87.1729°$

81. The general solution is a family of curves that satisfies the differential equation. A particular solution is one member of the family that satisfies given conditions.

83. Begin with a point (x_0, y_0) that satisfies the initial condition $y(x_0) = y_0$. Then, using a small step size h, calculate the point $(x_1, y_1) = (x_0 + h, y_0 + hF(x_0, y_0))$. Continue generating the sequence of points $(x_n + h, y_n + hF(x_n, y_n))$ or (x_{n+1}, y_{n+1}).

85. False: $y = x^3$ is a solution of $xy' - 3y = 0$, but $y = x^3 + 1$ is not a solution.

87. True

89. (a)

x	0	0.2	0.4	0.6	0.8	1
y	4	2.6813	1.7973	1.2048	0.8076	0.5413
y_1	4	2.56	1.6384	1.0486	0.6711	0.4295
y_2	4	2.4	1.44	0.864	0.5184	0.3110
e_1	0	0.1213	0.1589	0.1562	0.1365	0.1118
e_2	0	0.2813	0.3573	0.3408	0.2892	0.2303
r	0	0.4312	0.4447	0.4583	0.4720	0.4855

(b) If h is halved, then the error is approximately halved because r is approximately 0.5.
(c) The error will again be halved.

91. (a)

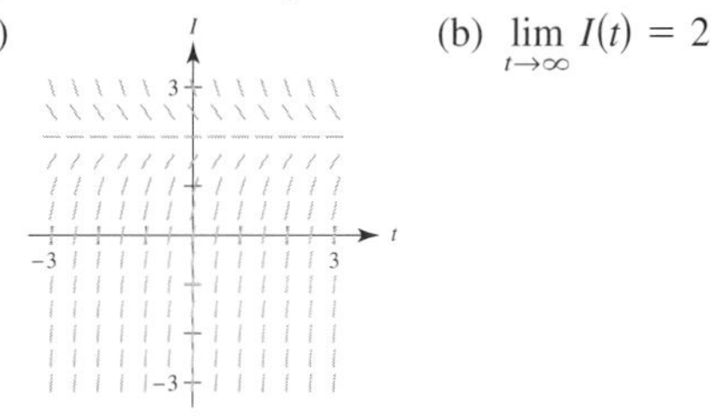

(b) $\lim_{t\to\infty} I(t) = 2$

93. $\omega = \pm 4$ rad/sec

95. Putnam Problem 3, Morning Session, 1954

Section 6.2 (page 400)

1. $y = \frac{1}{2}x^2 + 2x + C$ **3.** $y = Ce^x - 2$ **5.** $y^2 - 5x^2 = C$

7. $y = Ce^{(2x^{3/2})/3}$ **9.** $y = C(1 + x^2)$

11. $dQ/dt = k/t^2$
$Q = -k/t + C$

13. $dN/ds = k(250 - s)$
$N = -(k/2)(250 - s)^2 + C$

15. (a)

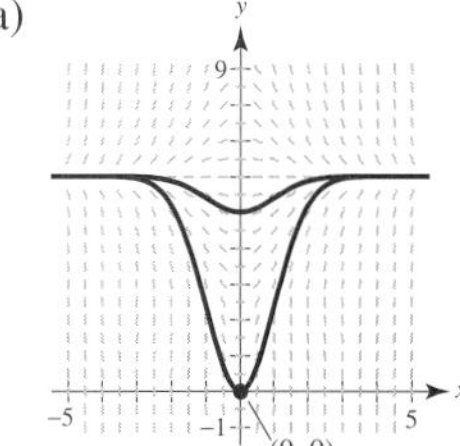

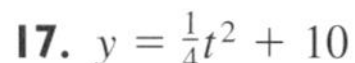

(b) $y = 6 - 6e^{-x^2/2}$

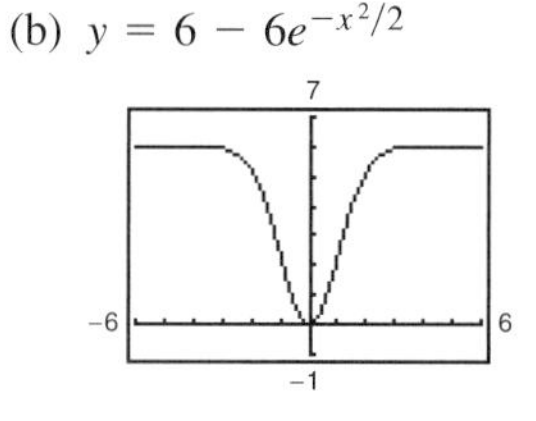

17. $y = \frac{1}{4}t^2 + 10$

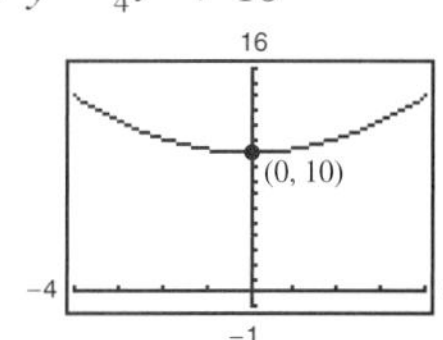

19. $y = 10e^{-t/2}$

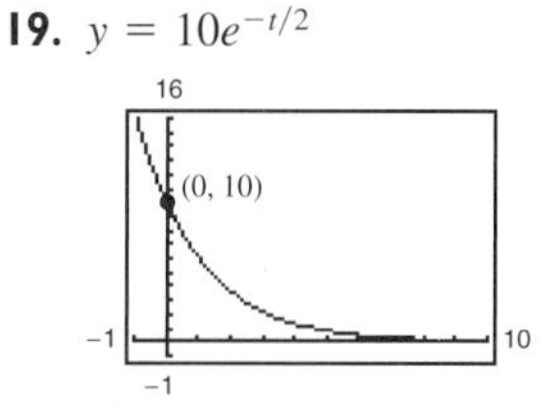

21. $dy/dx = ky$
$y = 4e^{0.3054x}$
$y(6) \approx 25$

23. $dV/dt = kV$
$V = 20{,}000e^{-0.1175t}$
$V(6) \approx 9882$

25. $y = \frac{1}{2}e^{0.4605t}$ **27.** $y = 0.6687e^{0.4024t}$

29. C is the initial value of y, and k is the proportionality constant.

31. Quadrants I and III; dy/dx is positive when both x and y are positive (Quadrant I) or when both x and y are negative (Quadrant III).

33. Amount after 1000 yrs: 6.48 g;
Amount after 10,000 yrs: 0.13 g

35. Initial quantity: 38.16 g;
Amount after 1000 yrs: 24.74 g

37. Amount after 1000 yrs: 4.43 g;
Amount after 10,000 yrs: 1.49 g

39. Initial quantity: 2.16 g;
Amount after 10,000 yrs: 1.62 g

41. 95.76%

43. Time to double: 11.55 yrs; Amount after 10 yrs: \$1822.12

45. Annual rate: 8.94%; Amount after 10 yrs: \$1833.67

47. Annual rate: 9.50%; Time to double: 7.30 yrs

49. \$112,087.09 **51.** \$30,688.87

53. (a) 10.24 yrs (b) 9.93 yrs (c) 9.90 yrs (d) 9.90 yrs

55. (a) 8.50 yrs (b) 8.18 yrs (c) 8.16 yrs (d) 8.15 yrs

57. (a) $P = 7.77e^{-0.009t}$ (b) 6.79 million
(c) Since $k < 0$, the population is decreasing.

59. (a) $P = 5.07e^{0.026t}$ (b) 7.48 million
(c) Since $k > 0$, the population is increasing.

61. (a) $N(t) = 100.1596(1.2455)^t$ (b) 6.3 hrs

63. (a) $N \approx 30(1 - e^{-0.0502t})$ (b) 36 days

65. (a) $P_1 = 181e^{0.01245t}$ or $P_1 = 181(1.01253)^t$
(b) $P_2 = 182.3248(1.01091)^t$

(c) 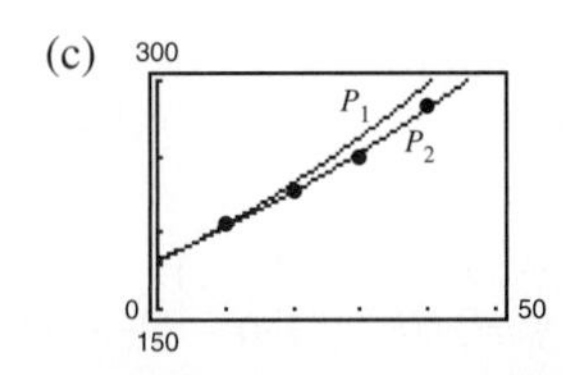

P_2 is a better approximation.

(d) 2011

67. (a) 20 decibels (b) 70 decibels (c) 95 decibels (d) 120 decibels

69. 2014 ($t = 16$) **71.** 379.2° F

73. False. The rate of growth dy/dx is proportional to y.

75. True

Section 6.3 (page 413)

1. $y^2 - x^2 = C$ **3.** $r = Ce^{0.05s}$ **5.** $y = C(x + 2)^3$

7. $y^2 = C - 2\cos x$ **9.** $y = -\frac{1}{4}\sqrt{1 - 4x^2} + C$

11. $y = Ce^{(\ln x)^2/2}$ **13.** $y^2 = 2e^x + 14$ **15.** $y = e^{-(x^2+2x)/2}$

17. $y^2 = 4x^2 + 3$ **19.** $u = e^{(1-\cos v^2)/2}$ **21.** $P = P_0 e^{kt}$

23. $9x^2 + 16y^2 = 25$ **25.** $f(x) = Ce^{-x/2}$

27. Homogeneous of degree 3 **29.** Homogeneous of degree 3

31. Not homogeneous **33.** Homogeneous of degree 0

35. $|x| = C(x - y)^2$ **37.** $|y^2 + 2xy - x^2| = C$

39. $y = Ce^{-x^2/2y^2}$ **41.** $e^{y/x} = 1 + \ln x^2$ **43.** $x = e^{\sin(y/x)}$

45. $y = \frac{1}{2}x^2 + C$

47. 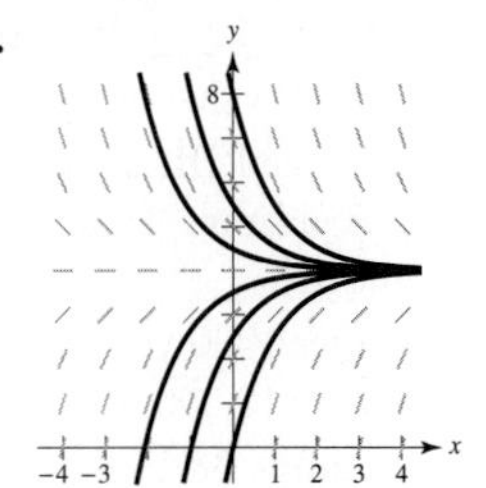

$y = 4 + Ce^{-x}$

49. (a) $y = 0.1602$ (b) $y = 5e^{-3x^2}$ (c) $y = 0.2489$

51. (a) $y = 3.0318$ (b) $y^3 - 4y = x^2 + 12x - 13$ (c) $y = 3$

53. 98.9% of the original amount

55. (a) $dy/dx = k(y - 4)$ (b) a (c) Proof

56. (a) $dy/dx = k(x - 4)$ (b) b (c) Proof

57. (a) $dy/dx = ky(y - 4)$ (b) c (c) Proof

58. (a) $dy/dx = ky^2$ (b) d (c) Proof

59. $w = 1200 - 1140e^{-kt}$

(a) $w = 1200 - 1140e^{-0.8t}$

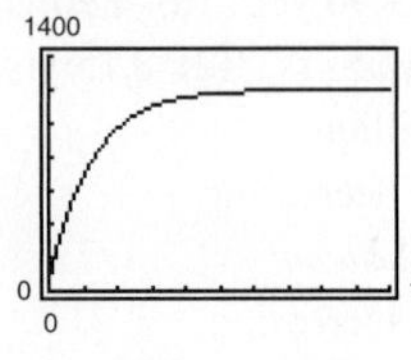

$w = 1200 - 1140e^{-0.9t}$

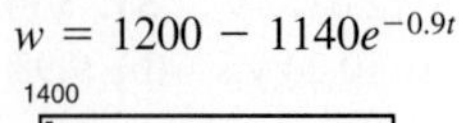

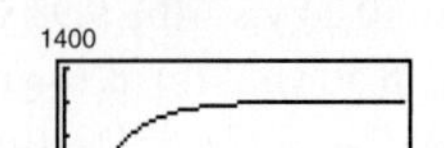

$w = 1200 - 1140e^{-t}$

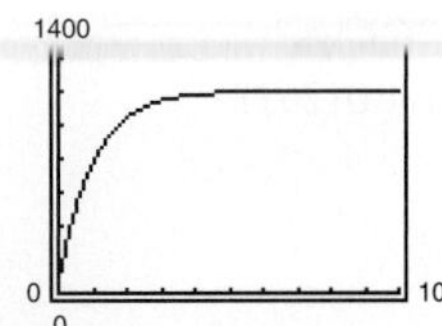

(b) 1.31 yrs; 1.16 yrs; 1.05 yrs (c) 1200 lb

61. Circles: $x^2 + y^2 = C$
Lines: $y = Kx$
Graphs will vary.

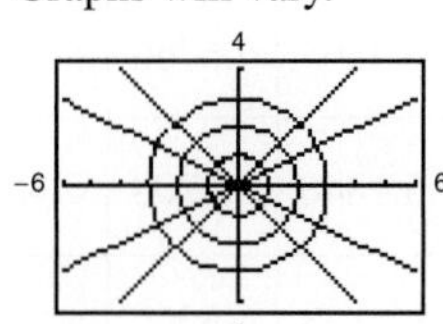

63. Parabolas: $x^2 = Cy$
Ellipses: $x^2 + 2y^2 = K$
Graphs will vary.

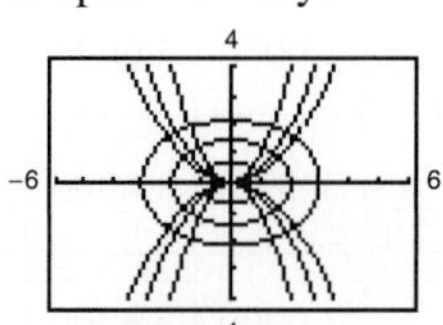

65. Curves: $y^2 = Cx^3$
Ellipses: $2x^2 + 3y^2 = K$
Graphs will vary.

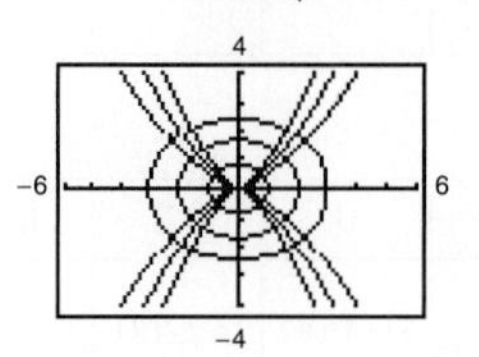

67. $N = 500/(1 + 4e^{-0.2452t})$

69. $y = 360/(8 + 41t)$ **71.** $y = 500e^{-1.6094e^{-0.1451t}}$

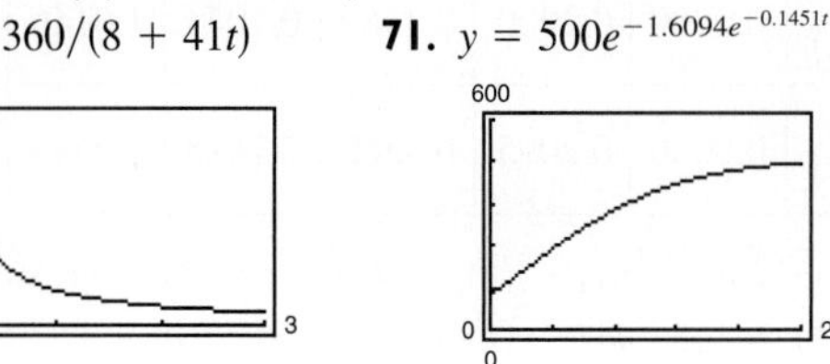

73. 34 beavers **75.** 92%

77. (a) $Q = 25e^{-(1/20)t}$ (b) $t \approx 10.2$ min

79. (a) $y = Ce^{kt}$ (b) ≈ 6.2 hrs **81.** ≈ 3.15 hrs

83. $P = Ce^{kt} - N/k$ **85.** $A = P/r(e^{rt} - 1)$

87. \$11,068,161.12

89. (a) 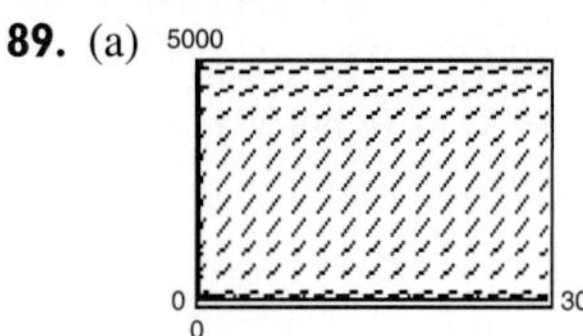

(b) As $t \to \infty$, $y \to L$.

(c) $y = 5000e^{-2.303e^{-0.02t}}$

(d) 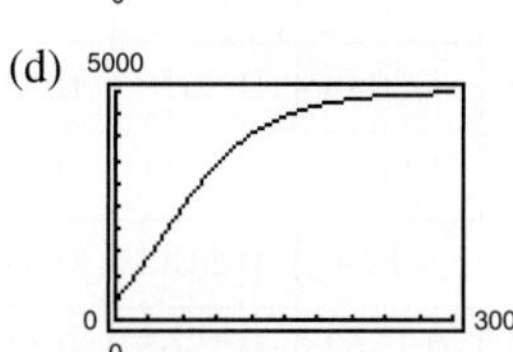

The graph is concave upward on $(0, 41.7)$ and downward on $(41.7, \infty)$.

91. Answers will vary.

93. Two families of curves are mutually orthogonal if each curve in the first family intersects each curve in the second family at right angles.

95. False. $y' = x/y$ is separable, but $y = 0$ is not a solution.

97. False. $f(tx, ty) \neq t^n f(x, y)$

99. Putnam Problem A2, 1988

Section 6.4 (page 422)

1. d **2.** a **3.** b **4.** c **5.** $y(0) = 2$ **7.** $y(0) = \frac{12}{7}$

9. (a) 0.75 (b) 1500 units (c) 60 units (d) 4.24 yrs (e) $dP/dt = 0.75P(1 - (P/1500))$

11. (a) 0.8 (b) 6000 units (c) 1.2 units (d) 10.65 yrs (e) $dP/dt = 0.8P(1 - (P/6000))$

13. (a) 3 (b) 100 units

(c)

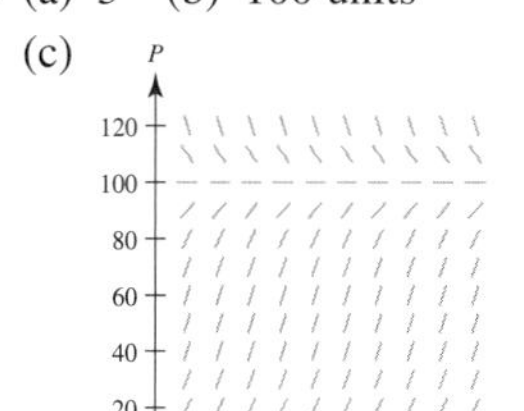

(d) 50 units

15. (a) 0.1 (b) 250 units

(c)

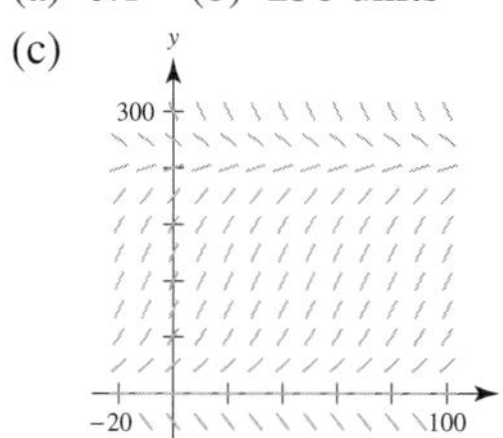

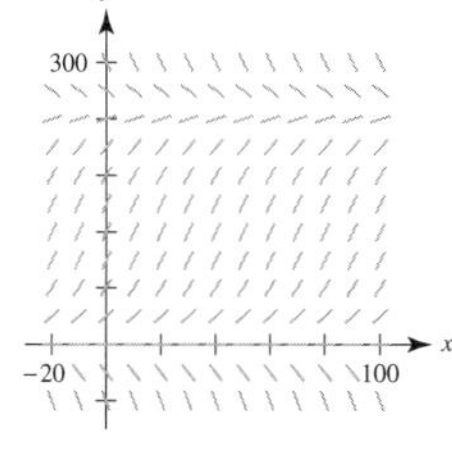

(d) 125 units

17. $y = 40/(1 + 4e^{-t})$; 38.95; 40

19. $y = 120/(1 + 14e^{-0.8t})$; 95.51; 120

21. c **22.** d **23.** b **24.** a

25. (a)

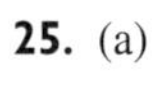

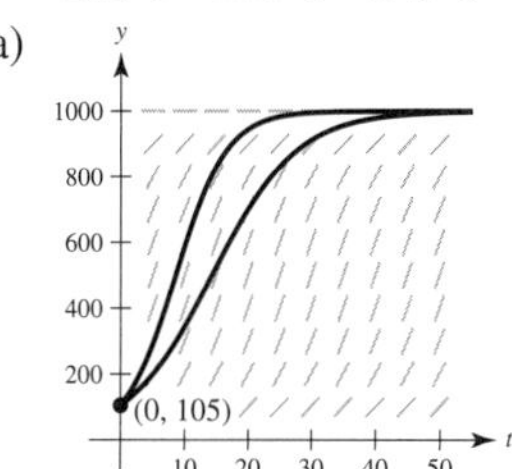

(b) $y = \dfrac{1000}{1 + (179/21)e^{-0.2t}}$

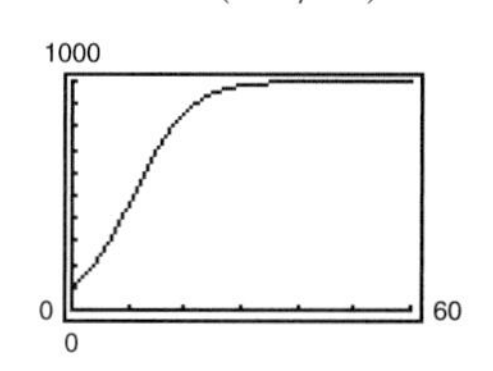

27. (a)

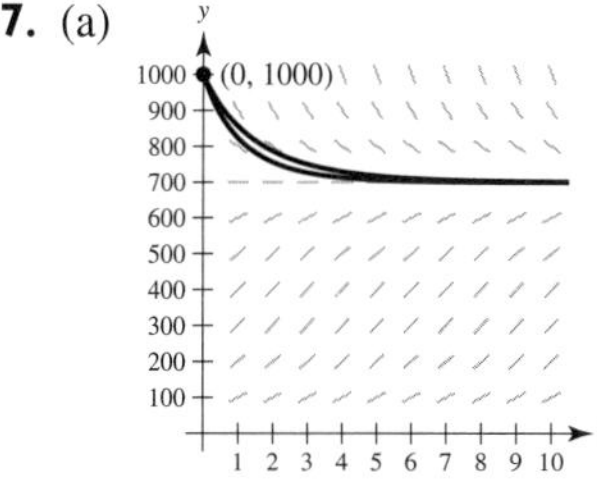

(b) $y = \dfrac{700}{1 - 0.3e^{-0.6t}}$

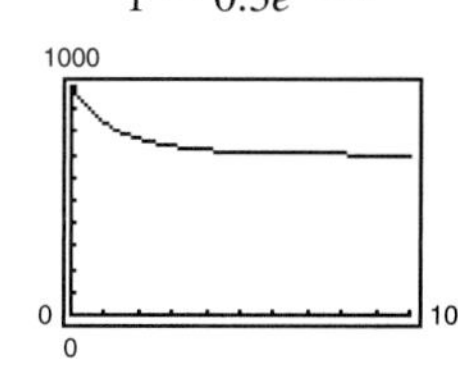

29. L represents the value that y approaches as t approaches infinity. L is the carrying capacity.

31. (a) $P = \dfrac{200}{1 + 7e^{-0.2640t}}$ (b) 70 panthers (c) 7.37 yrs

(d) $dP/dt = 0.2640P(1 - P/200)$; 69.25 panthers (e) 100 yrs

33. (a) $P = \dfrac{10}{1 + 9e^{-0.4055t}}$ (b) 4.58 g (c) 8.84 hrs

(d) $dP/dt = 0.4055P(1 - (P/10))$; 4.09 g (e) 5.42 hrs

35. False. $dy/dt < 0$ and the population decreases to approach L.

37. Proof

Section 6.5 (page 430)

1. Linear; can be written in the form $dy/dx + P(x)y = Q(x)$

3. Not linear; can't be written in the form $dy/dx + P(x)y = Q(x)$

5. $y = x^2 + 2x + C/x$ **7.** $y = -10 + Ce^x$

9. $y = -1 + Ce^{\sin x}$ **11.** $y = (x^3 - 3x + C)/[3(x - 1)]$

13. $y = e^{x^3}(x + C)$

15. (a) Answers will vary.

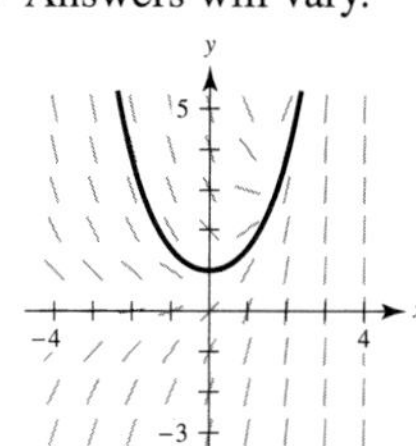

(c)

(b) $y = \frac{1}{2}(e^x + e^{-x})$

17. $y = 1 + 4/e^{\tan x}$ **19.** $y = \sin x + (x + 1)\cos x$

21. $xy = 4$ **23.** $y = -2 + x\ln|x| + 12x$

25. $1/y^2 = Ce^{2x^3} + \frac{1}{3}$ **27.** $y = 1/(Cx - x^2)$

29. $y^{2/3} = 2e^x + Ce^{2x/3}$

31. (a)

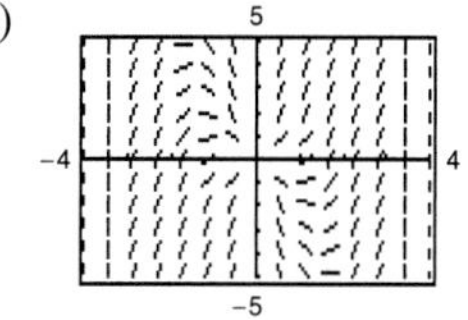

(c)

(b) $(-2, 4)$: $y = \frac{1}{2}x(x^2 - 8)$

$(2, 8)$: $y = \frac{1}{2}x(x^2 + 4)$

33. (a)

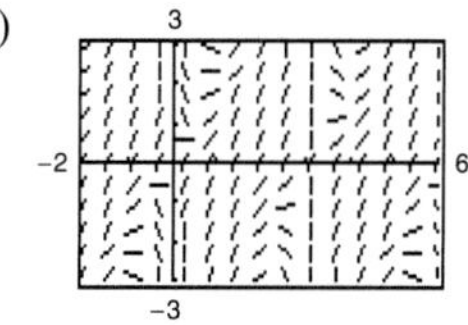

(b) $(1, 1)$: $y = (2\cos 1 + \sin 1)\csc x - 2\cot x$

$(3, -1)$: $y = (2\cos 3 - \sin 3)\csc x - 2\cot x$

(c)

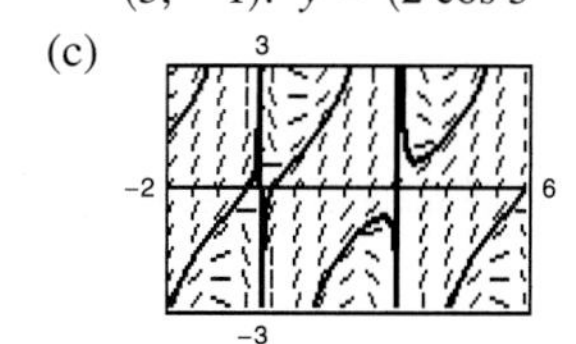

35. (a) $dQ/dt = q - kQ$ (b) $Q = q/k + (Q_0 - q/k)e^{-kt}$

(c) q/k

37. Proof

39. (a) $Q = 25e^{-t/20}$ (b) $-20\ln\left(\frac{3}{5}\right) \approx 10.2$ min (c) 0

41. (a) $t = 50$ min (b) $100 - 25/\sqrt{2} \approx 82.32$ lb

43. $V(t) = -159.47(1 - e^{-0.2007t})$; -159.47 ft/sec

45. $I = E_0/R + Ce^{-Rt/L}$

47. $dy/dx + P(x)y = Q(x)$; $u(x) = e^{\int P(x)\,dx}$

49. c **50.** d **51.** a **52.** b **53.** $2e^x + e^{-2y} = C$

55. $y = Ce^{-\sin x} + 1$ **57.** $x^3y^2 + x^4y = C$

59. $y = [e^x(x - 1) + C]/x^2$ **61.** $x^4y^4 - 2x^2 = C$

63. $y = \frac{12}{5}x^2 + C/x^3$ **65.** False. The equation contains $\sqrt{y}$.

Section 6.6 (page 438)

1. $dx/dt = 0.7x - 0.05xy$
$dy/dt = -0.4y + 0.007xy$; $(0, 0)$ and $(400/7, 14)$

3. $dx/dt = 0.3x - 0.006xy$
$dy/dt = -0.5y + 0.009xy$; $(0, 0)$ and $(500/9, 50)$

5. (a)

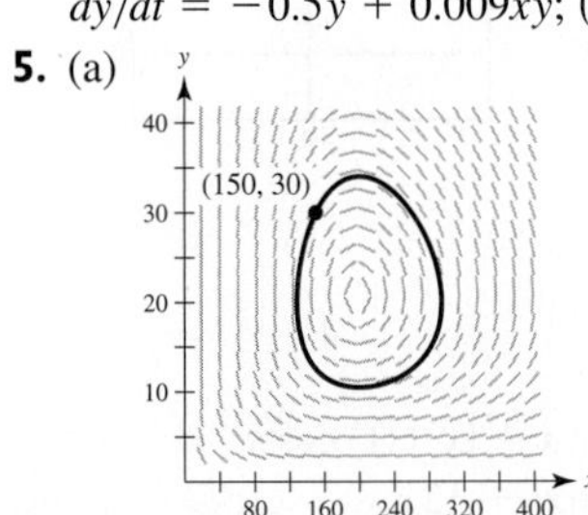

(b)

7. (a) $(40, 20)$

(b)

9. $(0, 0)$, $(50, 20)$

11.

13. $(0, 0)$, $(10{,}000, 1250)$

15.

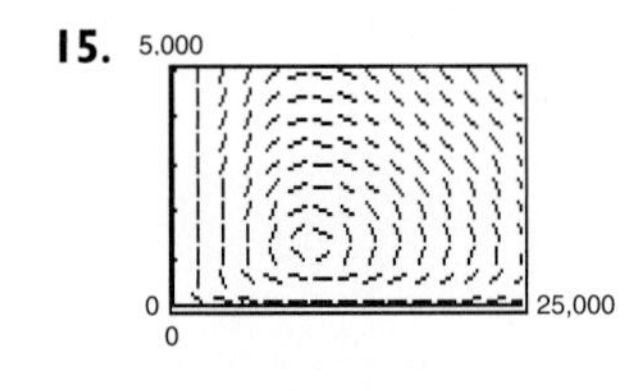

17.

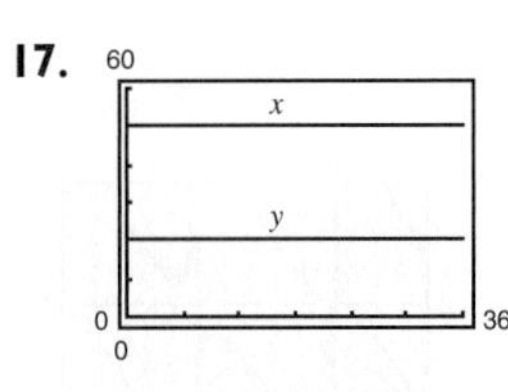

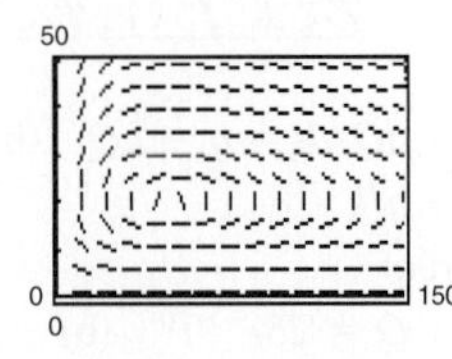

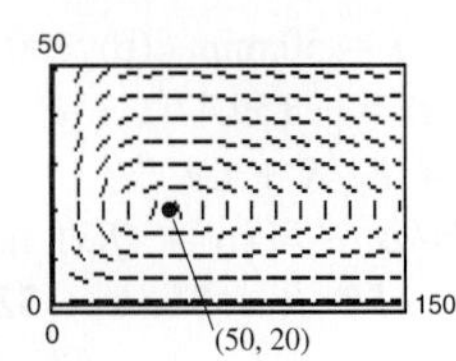

19. $dx/dt = x - 2x^2 - xy$, $dy/dt = y - 2y^2 - xy$;
$(x, y) = (0, 0)$, $(1/3, 1/3)$, $(0, 1/2)$, and $(1/2, 0)$

21. $dx/dt = 0.1x - 0.4x^2 - 0.5xy$, $dy/dt = 0.1y - 0.8y^2 - 0.3xy$;
$(x, y) = (0, 0)$, $(0, 1/8)$ $(1/4, 0)$, and $(3/17, 1/17)$

23. $(0, 0)$, $(0, 0.5)$, $(2, 0)$ and $(45/23, 4/23)$

25. $(0, 0)$, $(0, 0.5)$, $(2, 0)$ and $(-9/38, 17/19)$

27.

29. Solve $dx/dt = ax - bxy = 0$ and $dy/dt = -my + nxy = 0$. This will yield the points $(0, 0)$ and $(m/n, a/b)$ at which the prey and predator populations are constant.

31. True

33. False. The predator-prey equations are a special case of the competing-species equations.

35. (a) $dx/dt = ax(1 - x/L)$. The equation is logistic.

(b) $dx/dt = 0.4x(1 - (x/100)) - 0.01xy$,
$dy/dt = -0.3y = 0.005xy$
Critical points: $(0, 0)$, $(60, 16)$, $(100, 0)$

(c)

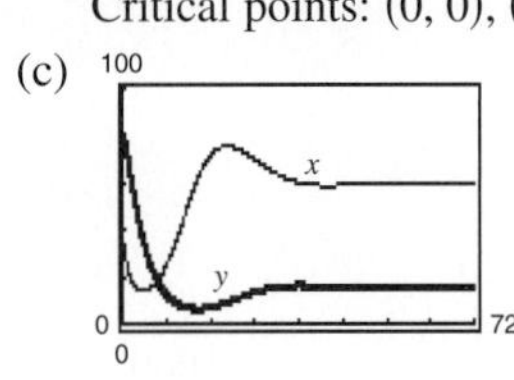

(d)

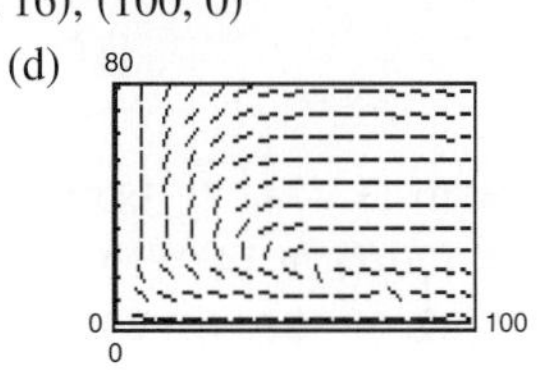

(e)

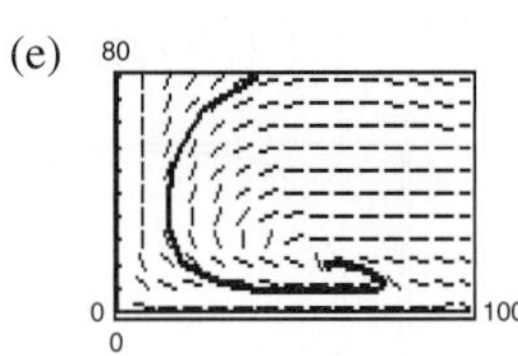

Review Exercises for Chapter 6 (page 440)

1. No **3.** $y = \frac{2}{3}x^3 + 5x + C$ **5.** $y = \frac{1}{2}\sin 2x + C$

7. $y = 4(x - 7)^{3/2}(3x + 14)/15 + C$

9.

x	-4	-2	0	2	4	8
y	2	0	4	4	6	8
dy/dx	-4	Undef.	0	1	$\frac{4}{3}$	2

11. (a) and (b)

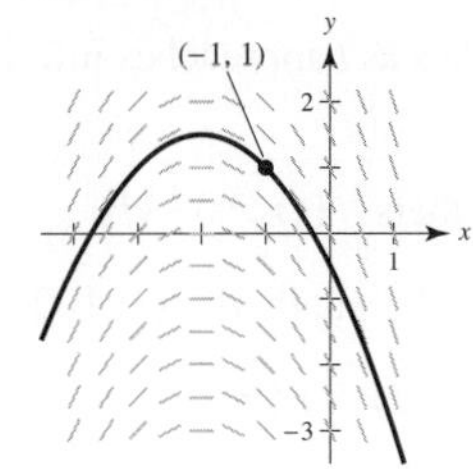

13. (a) and (b)

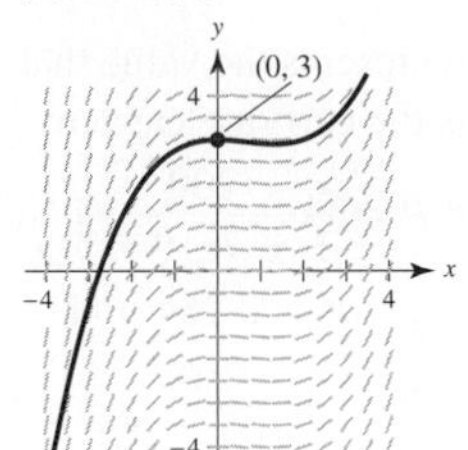

15. (a) and (b)

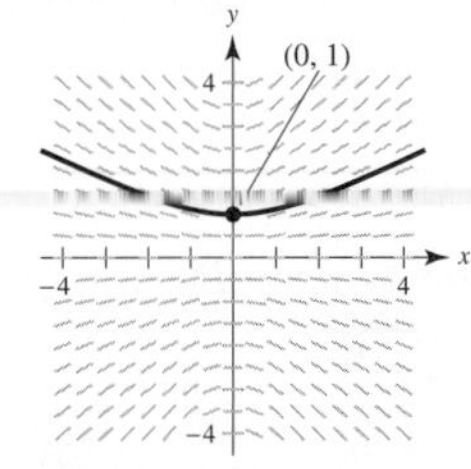

17. $y = 6x - \frac{1}{2}x^2 + C$

19. $y = -3 - 1/(x + C)$ **21.** $y = Ce^x/(2 + x)^2$

23. $y = \frac{3}{4}e^{0.379t}$ **25.** $y = 5e^{-0.680t}$ **27.** ≈ 7.79 in.

29. (a) $S \approx 30e^{-1.7918/t}$ (b) 20,965 units

(c)

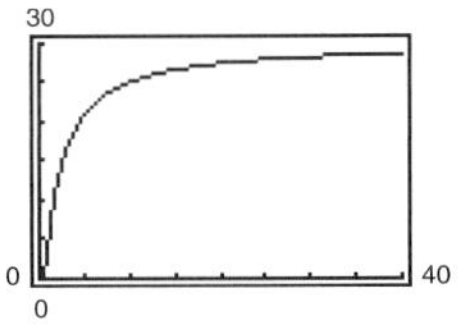

31. About 46.2 yrs **33.** $y = \frac{1}{2}x^2 + 3 \ln|x| + C$

35. $y = Ce^{x^2}$ **37.** $x/(x^2 - y^2) = C$

39. Proof; $y = -2x + \frac{1}{2}x^3$

41.

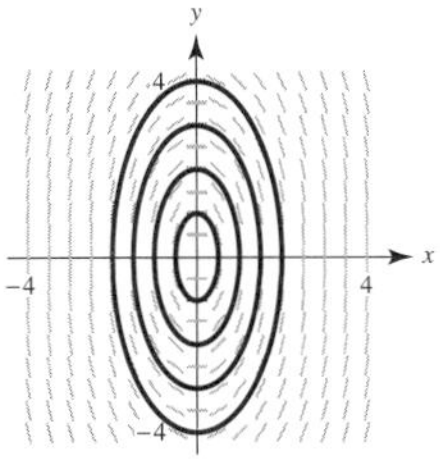

Graphs will vary.

$4x^2 + y^2 = C$

43. (a) 0.55 (b) 7200 (c) 160 (d) 6.88

(e) $dP/dt = 0.55P(1 - (P/7200))$

45. (a) $P(t) = \dfrac{20{,}400}{1 + 16e^{-0.553t}}$ (b) 17,118 trout (c) 4.94 yrs

47. $dS/dt = k(L - S)$; $S = L(1 - e^{-kt})$

49. $dP/dn = kP(L - P)$, $P = CL/(e^{-Lkn} + C)$

51. (a) Answers will vary. (b) $y = \frac{1}{3}(2e^{x/2} - 5e^{-x})$

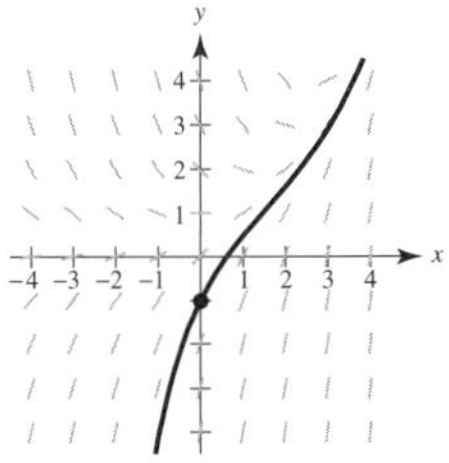

(c)

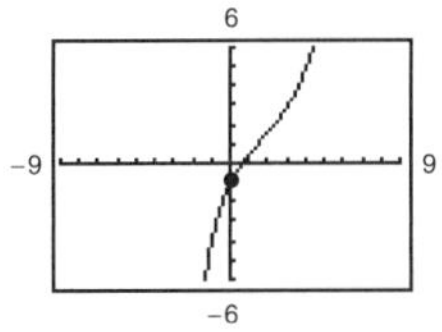

53. (a) Answers will vary. (b) $y = -\cos x + 1.8305 \sin x$

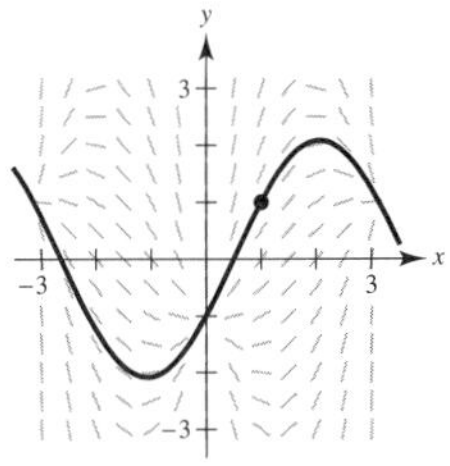

(c)

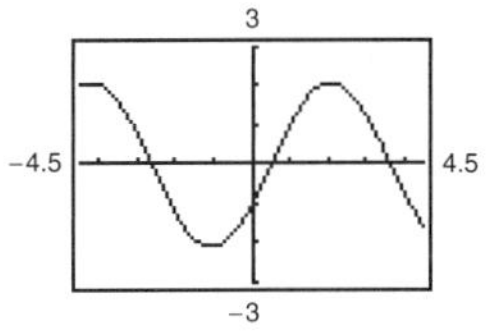

55. $y = -8 + Ce^x$ **57.** $y = e^{x/4}\left[\frac{1}{4}(x + C)\right]$

59. $y = (x + C)/(x - 2)$

61. $y = Ce^{3x} - \frac{1}{13}(2 \cos 2x + 3 \sin 2x)$

63. $y = e^{5x}/10 + Ce^{-5x}$ **65.** $y = 1/(1 + x + Ce^x)$

67. $y^{-2} = Cx^2 + 2/(3x)$

69. Answers will vary. Sample answer:

$(x^2 + 3y^2)\,dx - 2xy\,dy = 0$; $x^3 = C(x^2 + y^2)$

71. Answers will vary. Sample answer:

$x^3y' + 2x^2y = 1$; $x^2y = \ln|x| + C$

73. $A = P/r + (A_0 - (P/r))e^{rt}$

75. $t \approx 8.6$ yrs

77. (a) Predator: $dx/dt = 0.4x - 0.04xy$;

Prey: $dy/dt = -0.6y + 0.02xy$

(b) (0, 0) and (30, 10)

(c)

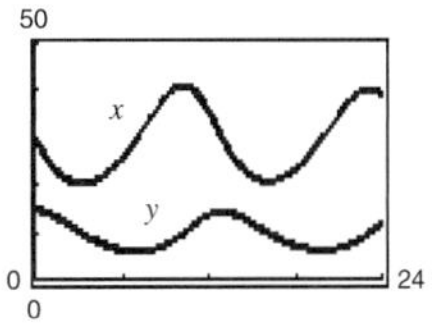

As t increases, both x and y oscillate.

79. (a) Species 1: $dx/dt = 15x - 2x^2 - 4xy$;

Species 2: $dy/dt = 17y - 2y^2 - 4xy$

(b) (0, 0), (19/6, 13/6), (0, 17/2), and (15/2, 0)

(c)

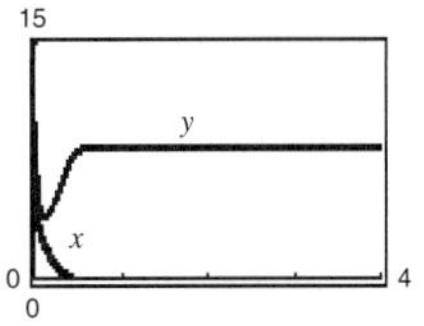

As t increases, x becomes extinct and y remains constant at approximately 8.5.

P.S. Problem Solving (page 443)

1. (a) $y = 1/(1 - 0.01t)^{100}$; $T = 100$

(b) $y = 1\Big/\left[\left(\dfrac{1}{y_0}\right)^{\varepsilon} - k\varepsilon t\right]^{1/\varepsilon}$; Answers will vary.

3. (a) $dS/dt = kS(L - S)$; $S = 100/(1 + 9e^{-0.8109t})$

(b) 2.7 months

(c)

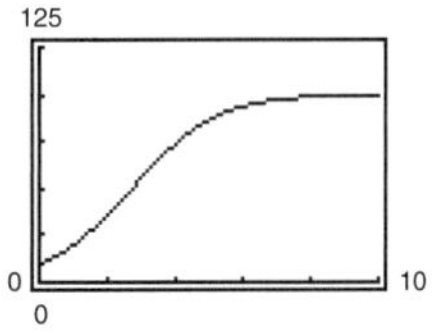

(d)

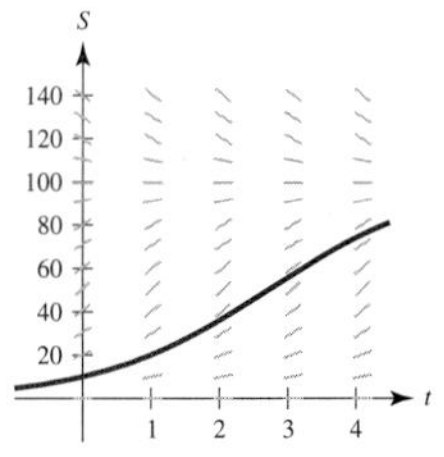

(e) Sales will decrease toward the line $S = L$.

5. Proof **7.** (a) 9809.1 sec (b) 7.21 ft

9. (a) (1.155, 0) (b) $\left(\dfrac{h\sqrt{2 + h}}{2h + 4 - 3\sqrt{2 + h}} - 1, 0\right)$

(c)

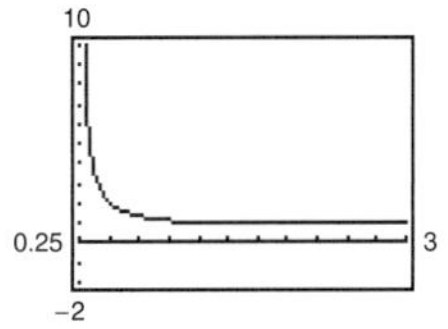

There is a vertical asymptote at $h = \frac{1}{4}$, which is the height of the hill.

11. (a) $C = C_0e^{-Rt/V}$ (b) $\lim_{t\to\infty} C(t) = 0$

13. (a) $C = (Q/R)(1 - e^{-Rt/V})$ (b) $\lim_{t\to\infty} C(t) = Q/R$

APPENDIX C (page A27)

1. (a) Fixed cost

(b)

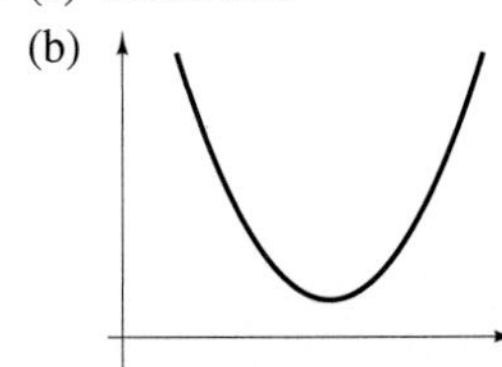

(c) Yes; the extremum occurs when production costs are increasing at their slowest rate.

3. 4500 **5.** 300 **7.** 200 **9.** 200

11. \$60 **13.** \$35 **15.** $x = 3$ **17.** Proof

19. (a)

Order size, x	*Price*	*Profit, P*
102	$90 - 2(0.15)$	$102[90 - 2(0.15)] - 102(60) = 3029.40$
104	$90 - 4(0.15)$	$104[90 - 4(0.15)] - 104(60) = 3057.60$
106	$90 - 6(0.15)$	$106[90 - 6(0.15)] - 106(60) = 3084.60$
108	$90 - 8(0.15)$	$108[90 - 8(0.15)] - 108(60) = 3110.40$
110	$90 - 10(0.15)$	$110[90 - 10(0.15)] - 110(60) = 3135.00$
112	$90 - 12(0.15)$	$112[90 - 12(0.15)] - 112(60) = 3158.40$

(b)

Order size, x	*Price*	*Profit, P*
⋮	⋮	⋮
146	$90 - 46(0.15)$	$146[90 - 46(0.15)] - 146(60) = 3372.60$
148	$90 - 48(0.15)$	$148[90 - 48(0.15)] - 148(60) = 3374.40$
150	$90 - 50(0.15)$	$150[90 - 50(0.15)] - 150(60) = 3375.00$
152	$90 - 52(0.15)$	$152[90 - 52(0.15)] - 152(60) = 3374.40$
154	$90 - 54(0.15)$	$154[90 - 54(0.15)] - 154(60) = 3372.60$
⋮	⋮	⋮

Maximum profit: \$3375.00

(c) $P = x[90 - (x - 100)(0.15)] - x(60) = 45x - 0.15x^2,$ $x \geq 100$

(d) 150 units

(e)

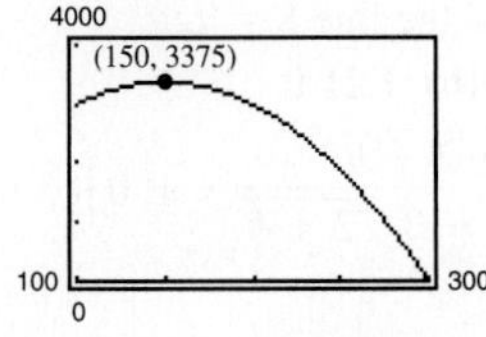

21. The line should run from the power station to a point across the river $3/(2\sqrt{7})$ mile downstream.

23. $x \approx 40$ units **25.** \$30,000

27. (a)

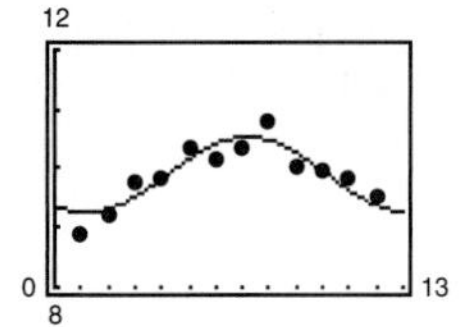

(b) July

(c) The cosine factor; 9.90

(d) The term $0.02t$ would mean a steady growth of sales over time. In this case, the maximum sales in 2008 (that is, on $49 \leq t \leq 60$) would be about 11.6 thousand gallons.

29. (a)

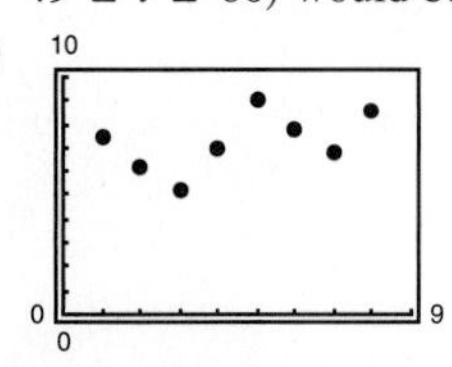

(b) $S = 6.2 + 0.25t + 1.5 \sin\left(\frac{\pi}{2}t\right)$

(c)

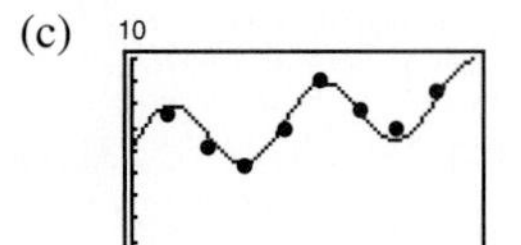

(d) \$12,000

31. $\eta = -\frac{17}{3}$; elastic **33.** $\eta = -\frac{1}{2}$; inelastic

Index of Applications

Engineering and Physical Sciences

Business and Economics

Social and Behavioral Sciences

Life Sciences

General

Index

P

INDEX

U

V

W

X

Y

Z

ALGEBRA

Factors and Zeros of Polynomials

Let $p(x) = a_n x^n + a_{n-1}x^{n-1} + \cdots + a_1 x + a_0$ be a polynomial. If $p(a) = 0$, then a is a *zero* of the polynomial and a solution of the equation $p(x) = 0$. Furthermore, $(x - a)$ is a *factor* of the polynomial.

Fundamental Theorem of Algebra

An nth degree polynomial has n (not necessarily distinct) zeros. Although all of these zeros may be imaginary, a real polynomial of odd degree must have at least one real zero.

Quadratic Formula

If $p(x) = ax^2 + bx + c$, and $0 \le b^2 - 4ac$, then the real zeros of p are $x = \left(-b \pm \sqrt{b^2 - 4ac}\right)/2a$.

Special Factors

$x^2 - a^2 = (x - a)(x + a)$

$x^3 - a^3 = (x - a)(x^2 + ax + a^2)$

$x^3 + a^3 = (x + a)(x^2 - ax + a^2)$

$x^4 - a^4 = (x^2 - a^2)(x^2 + a^2)$

Binomial Theorem

$(x + y)^2 = x^2 + 2xy + y^2$

$(x - y)^2 = x^2 - 2xy + y^2$

$(x + y)^3 = x^3 + 3x^2y + 3xy^2 + y^3$

$(x - y)^3 = x^3 - 3x^2y + 3xy^2 - y^3$

$(x + y)^4 = x^4 + 4x^3y + 6x^2y^2 + 4xy^3 + y^4$

$(x - y)^4 = x^4 - 4x^3y + 6x^2y^2 - 4xy^3 + y^4$

$$(x + y)^n = x^n + nx^{n-1}y + \frac{n(n-1)}{2!}x^{n-2}y^2 + \cdots + nxy^{n-1} + y^n$$

$$(x - y)^n = x^n - nx^{n-1}y + \frac{n(n-1)}{2!}x^{n-2}y^2 - \cdots \pm nxy^{n-1} \mp y^n$$

Rational Zero Theorem

If $p(x) = a_n x^n + a_{n-1}x^{n-1} + \cdots + a_1 x + a_0$ has integer coefficients, then every *rational zero* of p is of the form $x = r/s$, where r is a factor of a_0 and s is a factor of a_n.

Factoring by Grouping

$acx^3 + adx^2 + bcx + bd = ax^2(cx + d) + b(cx + d) = (ax^2 + b)(cx + d)$

Arithmetic Operations

$ab + ac = a(b + c)$

$\dfrac{a}{b} + \dfrac{c}{d} = \dfrac{ad + bc}{bd}$

$\dfrac{a + b}{c} = \dfrac{a}{c} + \dfrac{b}{c}$

$\dfrac{\left(\frac{a}{b}\right)}{\left(\frac{c}{d}\right)} = \left(\dfrac{a}{b}\right)\left(\dfrac{d}{c}\right) = \dfrac{ad}{bc}$

$\dfrac{\left(\frac{a}{b}\right)}{c} = \dfrac{a}{bc}$

$\dfrac{a}{\left(\frac{b}{c}\right)} = \dfrac{ac}{b}$

$a\left(\dfrac{b}{c}\right) = \dfrac{ab}{c}$

$\dfrac{a - b}{c - d} = \dfrac{b - a}{d - c}$

$\dfrac{ab + ac}{a} = b + c$

Exponents and Radicals

$a^0 = 1, \quad a \ne 0$

$(ab)^x = a^x b^x$

$a^x a^y = a^{x+y}$

$\sqrt{a} = a^{1/2}$

$\dfrac{a^x}{a^y} = a^{x-y}$

$\sqrt[n]{a} = a^{1/n}$

$\left(\dfrac{a}{b}\right)^x = \dfrac{a^x}{b^x}$

$\sqrt[n]{a^m} = a^{m/n}$

$a^{-x} = \dfrac{1}{a^x}$

$\sqrt[n]{ab} = \sqrt[n]{a}\,\sqrt[n]{b}$

$(a^x)^y = a^{xy}$

$\sqrt[n]{\dfrac{a}{b}} = \dfrac{\sqrt[n]{a}}{\sqrt[n]{b}}$